S0-BQI-273

Stanford Studies
in Mathematics and Statistics, I

TABLES OF THE NON-CENTRAL t-DISTRIBUTION

TABLES OF THE NON-CENTRAL t-DISTRIBUTION

DENSITY FUNCTION, CUMULATIVE DISTRIBUTION FUNCTION AND PERCENTAGE POINTS

George J. Resnikoff
and
Gerald J. Lieberman

STANFORD UNIVERSITY PRESS
Stanford, California

Stanford University Press, Stanford, California

Printed in the United States of America
ISBN 0-8047-0492-9
Original edition 1957
Last figure below indicates year of this printing:
87 86 85 84 83 82 81 80 79 78

Preface

The computation of the tables contained herein was undertaken to facilitate the construction of variables sampling plans in industrial acceptance sampling under Office of Naval Research contract N6onr-25126.

Because it is believed that the tables have numerous other applications to both practical and theoretical statistics, it was decided to make them generally available by publishing them in book form.

The authors wish to acknowledge their indebtedness to the following persons: Albert H. Bowker, for originally suggesting the computation of the tables and for encouragement and advice during the course of the work; Joseph Carter, for assistance in much of the programming; Gladys R. Garabedian, for invaluable assistance in preparing the text material; and last but not least, Herbert Solomon, without whose inspiration the work could never have been initiated.

The authors imply no responsibility on the part of these people for any inacccuracies that may exist in the tables.

January 25, 1957

GEORGE J. RESNIKOFF
Research Associate, Applied Mathematics and Statistics Laboratory, Stanford University

GERALD J. LIEBERMAN
Associate Professor of Statistics and Industrial Engineering, Stanford University

Contents

Introduction 1

Description of the Tables 2

Examples of the Uses of the Probability Density Function of the Non-Central t-Statistic 4

Examples of the Use of the Probability Integral and Percentage Points of the Non-Central t-Statistic 10

Computational Methods 25

Table of the Probability Density Function of the Non-Central t-Statistic 33

Table of the Probability Integral of the Non-Central t-Statistic 179

Table of the Percentage Points of the Non-Central t-Statistic 383

TABLES OF THE NON-CENTRAL t-DISTRIBUTION

TABLES OF THE NON-CENTRAL t-DISTRIBUTION: DENSITY FUNCTION, CUMULATIVE DISTRIBUTION FUNCTION AND PERCENTAGE POINTS

1. Introduction

Let z be a random variable distributed normally about zero with unit standard deviation, and let w be a random variable distributed independently of z as χ^2/f with f degrees of freedom. If t is defined by

$$t = \frac{z+\delta}{\sqrt{w}}$$

where δ is some constant, then t is said to have the non-central t-distribution with f degrees of freedom and non-centrality parameter δ.

The probability density of t is given by

$$h(f,\delta,t) = \frac{f!}{2^{\frac{f-1}{2}}\,\Gamma(\frac{f}{2})\,\sqrt{\pi f}}\, e^{-\frac{1}{2}\frac{f\delta^2}{f+t^2}} \left(\frac{f}{f+t^2}\right)^{\frac{f+1}{2}} Hh_f\left(\frac{-\delta t}{\sqrt{f+t^2}}\right)$$

where

$$Hh_f(y) = \int_0^\infty \frac{v^f}{f!}\, e^{-\frac{1}{2}(v+y)^2}\, dv.$$

The tables contained herein give values of the probability integral, of the probability density function and of the percentage points of the non-central t-statistic for selected values of the parameters f and δ.

Important tables related to the non-central t-statistic have been published previously by Johnson and Welch [4]. Their tables do not deal directly with the probability integral, nor is it possible to obtain from them values of the density function. The Johnson and Welch tables facilitate the computation of $\delta(f,t_o,\epsilon)$, that is, that value of the parameter δ for which $\Pr[t \leq t_o | f,\delta] = 1-\epsilon$, for seventeen values of ϵ. By a more extended computation it is possible to obtain the percentage points corresponding to these seventeen values of ϵ. The percentage points corresponding to the special values $\epsilon = .05$ and $\epsilon = .95$ may be obtained with somewhat less labor. The tables may also be used to compute values of the probability integral as a function of δ, for fixed f and t.

Other tables of the probability integral are those of Neyman [9] and Neyman and Tokarska [10], which were computed for the purpose of obtaining the power curve of the Student t-test.

2. Description of the Tables

Three tables of the non-central t-statistic are given: the probability integral, the probability density function, and the percentage points of t.

Tabulation of the non-central t-distribution requires a table of triple entry since the distribution depends on the two parameters f and δ, where f is the number of degrees of freedom and δ is the non-centrality parameter.

The argument used throughout the tables is $x = t/\sqrt{f}$. The reason for using $t/\sqrt{f}$ as the argument instead of t itself is that the range for this argument is roughly the same whatever the values of the parameters f and δ. This enables a somewhat more compact form of tabulation than would be possible if the argument were t itself.

The ranges of the parameters f and δ are as follows: f ranges from 2 to 24 by steps of 1 and from 24 to 49 by steps of 5; and $\delta = \sqrt{f+1}\,K_p$ where K_p is the standardized normal random variable exceeded with probability p; that is,

$$\int_{K_p}^{\infty} \frac{e^{-\frac{y^2}{2}}}{\sqrt{2\pi}}\, dy = p.$$

The values of p which were used are $p = .2500, .1500, .1000, .0650, .0400, .0250, .0100, .0040, .0025$ and $.0010$. A table of $\sqrt{f+1}\,K_p$ for these values of f and p is given on page 3.

These values of the parameters were chosen so as to cover the useful range adequately. In particular, the choice of $f = 2(1)24$ and $24(5)49$ appears to fill the needs for most practical applications. It is felt that for f greater than 49, a normal approximation [12] is adequate. The values for f of $24(5)49$ were chosen rather than f of $25(5)50$ because most practical applications require that $f = n-1$ where n is the sample size. Sample sizes which are multiples of 5 are prevalent in industrial statistics.

The indexing of $\delta = \sqrt{f+1}\,K_p$ as a function of K_p requires a word of explanation. General examples of the usefulness of these tables appear in the next section. However, the development of these tables was motivated by the necessity of using the non-central t-statistic for problems involving the fraction of normally distributed random variables falling above and/or below specified values. The

Table of $\sqrt{f+1}\,K_p$

K_p	0.674490	1.036433	1.281552	1.514102	1.750686	1.959964	2.326348	2.652070	2.807034	3.090232
f \ p	.2500	.1500	.1000	.0650	.0400	.0250	.0100	.0040	.0025	.0010
2	1.168251	1.795155	2.219712	2.622501	3.032277	3.394757	4.029353	4.593520	4.861925	5.352439
3	1.348980	2.072867	2.563103	3.028204	3.501372	3.919928	4.652696	5.304140	5.614068	6.180465
4	1.508205	2.317536	2.865636	3.385635	3.914653	4.382613	5.201872	5.930208	6.276718	6.909970
5	1.652156	2.538733	3.139147	3.708777	4.288287	4.800912	5.698365	6.496218	6.875800	7.569492
6	1.784532	2.742145	3.390667	4.005937	4.631880	5.185577	6.154938	7.016717	7.426713	8.175986
7	1.907745	2.931476	3.624775	4.282527	4.951688	5.543615	6.579905	7.501186	7.939490	8.740497
8	2.023469	3.109300	3.844655	4.542306	5.252058	5.879892	6.979044	7.956209	8.421101	9.270697
9	2.132924	3.277490	4.052622	4.788011	5.536156	6.197950	7.356558	8.386581	8.876620	9.772173
10	2.237029	3.437461	4.250426	5.021708	5.806369	6.500465	7.715623	8.795921	9.309878	10.249141
11	2.336501	3.590311	4.439425	5.245003	6.064554	6.789514	8.058705	9.187039	9.723850	10.704879
12	2.431907	3.736914	4.620700	5.459172	6.312188	7.066751	8.387767	9.562174	10.120904	11.141991
13	2.523710	3.877979	4.795127	5.665251	6.550467	7.333514	8.704397	9.923137	10.502959	11.562591
14	2.612288	4.014089	4.963428	5.864091	6.780378	7.590908	9.009906	10.271422	10.871595	11.968418
15	2.697959	4.145734	5.126206	6.056408	7.002744	7.839856	9.305391	10.608279	11.228135	12.360929
16	2.780992	4.273324	5.283972	6.242802	7.218264	8.081138	9.591778	10.934764	11.573697	12.741354
17	2.861618	4.397214	5.437163	6.423790	7.427532	8.315423	9.869858	11.251779	11.909236	13.110745
18	2.940033	4.517708	5.586154	6.599817	7.631064	8.543285	10.140315	11.560104	12.235576	13.470010
19	3.016410	4.635071	5.731273	6.771270	7.829306	8.765225	10.403744	11.860417	12.553437	13.819939
20	3.090900	4.749534	5.872807	6.938487	8.022651	8.981683	10.660665	12.153311	12.863445	14.161223
21	3.163637	4.861304	6.011010	7.101767	8.211446	9.193046	10.911539	12.439310	13.166156	14.494474
22	3.234739	4.970560	6.146105	7.261378	8.395995	9.399657	11.156772	12.718880	13.462061	14.820233
23	3.304311	5.077466	6.278295	7.417554	8.576575	9.601823	11.396731	12.992436	13.751601	15.138985
24	3.372449	5.182167	6.407758	7.570509	8.753430	9.799820	11.631739	13.260349	14.035169	15.451162
29	3.694333	5.676779	7.019347	8.293078	9.588903	10.735165	12.741932	14.525985	15.374757	16.925900
34	3.990335	6.131623	7.581761	8.957548	10.357198	11.595303	13.762860	15.689857	16.606636	18.282061
39	4.265848	6.554980	8.105244	9.576021	11.072311	12.395901	14.713116	16.773162	17.753240	19.544345
44	4.524615	6.952607	8.596909	10.156904	11.743959	13.147838	15.605616	17.790625	18.830155	20.729908
49	4.769363	7.328691	9.061938	10.706317	12.379220	13.859038	16.449764	18.752965	19.848726	21.851242

solution of such problems involves K_p. The particular values of p chosen coincide with the Acceptable Quality Levels (AQL's) of Military Standard 105A, Sampling Procedures and Tables for Inspection by Attributes.[1] It should be noted that this choice does not restrict the usefulness of these tables, but rather gives a method of adequately representing the useful range of δ for the applications.

In the subsequent sections the probability integral of t will be denoted by $P(f,\delta,x) = \Pr[t/\sqrt{f} \leq x \mid f,\delta]$. The probability density of t, tabled with argument $t/\sqrt{f}$, will be denoted by $P'(f,\delta,t/\sqrt{f})$. The percentage points of t will be denoted by $x(f,\delta,\epsilon)$, where x is the value such that $\Pr[t/\sqrt{f} > x \mid f,\delta] = \epsilon$.

For negative values of δ the following relationships are useful:

$$P'(f, -\delta,x) = P'(f,\delta, -x),$$

$$P(f, -\delta,x) = 1-P(f,\delta, -x)$$

and

$$x(f, -\delta,\epsilon) = -x(f,\delta,1-\epsilon).$$

3. Examples of the Uses of the Probability Density Function of the Non-Central t-Statistic

a. The WAGR Sequential Test

The WAGR test is a sequential procedure for testing the null hypothesis H_o, that the proportion of a normal population exceeding a given constant U is p_o (given), against the alternative hypothesis H_1, that the proportion is p_1 (given). The name WAGR stems from the initials of several of the individuals who proposed this test: Wald, Arnold, Goldberg and Rushton. A description of the WAGR test follows.

Let $Y_1, Y_2, \ldots$ be a sequence of independent observations on a normally distributed random variable with mean μ and variance σ^2.

Let $\bar{Y}$ be the arithmetic mean of the first n of the observations. Define u_n by

$$u_n = \frac{1}{\sqrt{n}} \left(\frac{\sum_{i=1}^{n} (U-Y_i)}{\sqrt{\sum_{i=1}^{n} (Y_i-\bar{Y})^2}} \right) \qquad n=2,3,\ldots$$

---

[1] These same values of p will be the AQL's of a new military standard for sampling inspection by variables.

Then $t_n = \sqrt{n-1}\, u_n$ is a non-central t-statistic with parameters $f = n-1$ and $\delta = \sqrt{n}\, \frac{U-\mu}{\sigma} = \sqrt{n}\, K_p$. The ratio of the probability density function of t_n under H_1 to the probability density function of t_n under H_o is, in the present notation, given by

$$\lambda_n = \frac{P'(n-1, \sqrt{n}\, K_{p_1}, u_n)}{P'(n-1, \sqrt{n}\, K_{p_o}, u_n)} .$$

The test procedure is to observe λ_n sequentially for $n = 2,3,\ldots$. The first time that the inequality

$$\frac{\beta}{1-\alpha} \leq \lambda_n \leq \frac{1-\beta}{\alpha} , \quad 0 < \alpha + \beta < 1, \quad 0 < \alpha, \quad 0 < \beta$$

is violated, the null hypothesis is either accepted or rejected. If $\lambda_n > \frac{1-\beta}{\alpha}$, reject H_o; if $\lambda_n < \frac{\beta}{1-\alpha}$, accept H_o.

It has been shown [8] that the WAGR is a true sequential test as defined by Wald and hence that the probability of accepting H_o is approximately $1-\alpha$ when p_o is the true population proportion, and is approximately β when p_1 is the true proportion. A proof that this test reaches a decision with probability one has been given [2].

In actually carrying out the test, one initially observes Y_1 and Y_2 and computes u_2. The ratio λ_2 is not available from the present tables since it corresponds to $f = n-1 = 1$. However λ_2 may be expressed as follows:

$$\lambda_2 = \left[\frac{\varphi(z_1/u_2)}{\varphi(z_o/u_2)}\right] \left[\frac{z_1\, \Phi(z_1) + \varphi(z_1)}{z_o\, \Phi(z_o) + \varphi(z_o)}\right]$$

where

$$z_j = \frac{\sqrt{2}\, K_{p_j}\, u_2}{\sqrt{u_2^2 + 1}} , \quad j = 0,1 ,$$

$$u_2 = \frac{(2U - Y_1 - Y_2)}{|Y_1 - Y_2|} ,$$

$$\varphi(z) = \frac{e^{-\frac{z^2}{2}}}{\sqrt{2\pi}} ,$$

and

$$\Phi(z) = \int_{-\infty}^{z} \varphi(v)dv.$$

The functions $\varphi(z)$ and $\Phi(z)$ are conveniently available in a single volume of tables of the normal probability function [7].

After this first step, if the inequality is not violated, one observes Y_n (and hence u_n) for $n = 3,4,\ldots$ until the test is terminated. Each time u_n is computed, the ratio λ_n is obtained directly from the appropriate table of the non-central t-density function, if p_1 and p_o are among the ten tabulated values of p; otherwise, interpolation on the non-centrality parameter is required.

A numerical example of the application of the tables of the probability density to the WAGR test follows. Let $U = 10$, $p_o = .01$, $p_1 = .065$, $\alpha = .05$ and $\beta = .10$. Then $\frac{\beta}{1-\alpha} = .1053$ and $\frac{1-\beta}{\alpha} = 18.0000$. Suppose the first observations are $Y_1 = 8.10$ and $Y_2 = 8.70$; then

$$u_2 = \frac{[2(10) - 8.10 - 8.70]}{|8.10 - 8.70|} = 5.3333 ,$$

$$z_o = \frac{(3.2900)(5.3333)}{\sqrt{29.4441}} = 3.2337$$

and

$$z_1 = \frac{(2.1413)(5.3333)}{\sqrt{29.4441}} = 2.1046.$$

From the tables of the normal probability function, we obtain: $\varphi(z_o) = .002139$, $\varphi(z_1) = .04356$, $\Phi(z_o) = .9994$, $\Phi(z_1) = .9823$, $\varphi(z_o/u_2) = \varphi(.6063) = .3320$ and $\varphi(z_1/u_2) = \varphi(.3946) = .3691$. Hence

$$\lambda_2 = [\frac{.3691}{.3320}] [\frac{(2.1046)(.9823) + (.04356)}{(3.2337)(.9994) + (.002139)}] = .7257 .$$

Thus, the inequality is not violated at the $n=2$ level and we take another observation. Let $Y_3 = 7.20$; then

$$u_3 = \frac{1}{\sqrt{3}} (\frac{6.00}{\sqrt{1.14}}) = 3.2444.$$

From the tables of the density function of t, we obtain

$$P'(2, \sqrt{3}\, K_{.01}, 3.2444) = .1414$$

and

$$P'(2, \sqrt{3}\, K_{.065}, 3.2444) = .0976.$$

Hence

$$\lambda_3 = \frac{.0976}{.1414} = .6902$$

and the inequality still holds at the $n=3$ level. One would then take another observation and proceed as for $n=3$ above, obtaining the values of the density function from the appropriate table. This operation is continued until the inequality is violated.

b. An Application of the Probability Density Function to the Computation of Moments of a Function of the Non-Central t - Statistic

The first two moments of the non-central t - statistic are given by the expressions

$$E(t) = \frac{\sqrt{\frac{f}{2}}\, \Gamma(\frac{f-1}{2})\delta}{\Gamma(\frac{f}{2})}$$

and

$$E(t^2) = \frac{f(1+\delta^2)}{f-2} .$$

The moments of several useful functions of t are known only approximately, for large values of the parameter f. One application of the tables of the probability density to the problem of computing the moments of an arbitrary function $g(t)$ of the non-central t - statistic is by straightforward numerical integration.

The kth moment of $g(t)$ can be written as

$$E[(g(t))^k] = \int_{-\infty}^{\infty} (g(t))^k\, h(f,\delta,t)dt$$

where $h(f,\delta,t)$ is the analytical expression for the non-central t - density given in the introductory section.

This may be approximated by

$$E[(g(t))^k] \sim \sum_{t_i} (g(t_i))^k \, P'(f,\delta, \frac{t_i}{\sqrt{f}})$$

where $\sum$ refers to summation of a grid over the t_i, such as the well-known trapezoidal rule, Simpson's rule, or other more elaborate integration formulas.

Particular examples of useful functions of the non-central t-statistic are two estimates of the proportion p of a normal population which lies above a given limit U. Let the population mean be μ and the population variance be σ^2; then p is expressed as

$$p = \int_{\frac{U-\mu}{\sigma}}^{\infty} \frac{e^{-\frac{v^2}{2}}}{\sqrt{2\pi}} \, dv .$$

An estimate of p which is often given is the biased estimate $\tilde{p}(\overline{Y},s)$ which is obtained by replacing μ and σ by their sample estimates in the expression for p. Let $Y_1, Y_2, \ldots, Y_n$ be a sample of n observations and let

$$\overline{Y} = \frac{\sum Y_i}{n}$$

and

$$s = \sqrt{\frac{\sum (Y_i - \overline{Y})^2}{n-1}} .$$

Then

$$\tilde{p} = \int_{\frac{U-\overline{Y}}{s}}^{\infty} \frac{e^{-\frac{v^2}{2}}}{\sqrt{2\pi}} \, dv .$$

Another estimate of p which has appeared recently in the literature [6] is the uniformly minimum variance unbiased estimate defined by

$$\hat{p} = \begin{cases} I_z(\frac{n-2}{2}, \frac{n-2}{2}) & \text{for } 0 < z < 1 \\ 0 & \text{for } z \leq 0 \\ 1 & \text{for } z \geq 1 \end{cases}$$

where $z = \frac{1}{2} - \frac{1}{2} \frac{U-\bar{Y}}{s} \frac{\sqrt{n}}{n-1}$ and $I_z(a,b)$ is the Incomplete Beta Function ratio with parameters a and b.

The quantity

$$\frac{\sqrt{n}\,(U-\bar{Y})}{s}$$

has the non-central t-distribution with degrees of freedom $f = n-1$ and $\delta = \sqrt{n}\, K_p$, so that both estimates are functions of the non-central t-statistic.

The two estimates $\tilde{p}$ and $\hat{p}$ are asymptotically equivalent and asymptotically efficient. The latter estimate, $\hat{p}$, is unbiased and has the smallest variance among all unbiased estimates of p. The former estimate, $\tilde{p}$, is biased so that its mean square error $E(\tilde{p}-p)^2$ is more relevant than its variance. No small sample comparison of the mean square errors of these estimates is available. It has sometimes been assumed that $\tilde{p}$ is a "best" estimate in the sense of least mean square error. Numerical integrations performed by the writers using the present tables of the probability density show that this is not the case, and in fact their relative merit depends on the value of the parameter p at which the mean square error is computed.

Suppose we wish to obtain the first moment of the estimate $\tilde{p}$ of the proportion defective of a normal population for $p = .0250$. Let $n = 17$. Then we use the expression

$$E[g(t)] \sim \sum_{t_i} g(t_i)\, P'(16, \sqrt{17}\, K_{.0250}, \frac{t_i}{\sqrt{16}})$$

where

$$g(t) = \tilde{p} = \int_{\frac{U-\bar{Y}}{s} = \frac{t}{\sqrt{n}}}^{\infty} \frac{e^{-\frac{v^2}{2}}}{\sqrt{2\pi}}\, dv.$$

Note that we use $t/\sqrt{f}$ as argument for computation purposes to avoid interpolation in the tables of the non-central t-distribution. But the increment to be used in applying the numerical integration formula must be the increment in t itself. In the example below, the increment for argument $t/\sqrt{f}$ is .25, but the increment in t is $.25\sqrt{f} = 1.00$. We choose arbitrarily to use every fifth point given in the probability density tables, but the integration may be done with as fine an interval as desired. The values for $g(t_i)$ were obtained from tables of the normal probability integral. Applying the trapezoid rule, we obtain

$$\frac{.25\sqrt{16}}{2} \sum a_i g(t_i)\, P'(16, \sqrt{17}\, K_{.0250}, \frac{t_i}{\sqrt{16}}) = .0294$$

where the a_i are the appropriate coefficients for the trapezoid rule.

$\frac{t_i}{\sqrt{16}}$	$P'(16, \sqrt{17}\, K_{.0250}, \frac{t_i}{\sqrt{16}})$	$\frac{t_i}{\sqrt{17}}$	$g(t_i)$
.75	.0000	.7276	.2334
1.00	.0016	.9701	.1660
1.25	.0240	1.2127	.1126
1.50	.1044	1.4552	.0728
1.75	.2009	1.6977	.0448
2.00	.2277	1.9403	.0262
2.25	.1837	2.1828	.0145
2.50	.1195	2.4254	.0076
2.75	.0679	2.6679	.0038
3.00	.0356	2.9104	.0018
3.25	.0178	3.1530	.0008
3.50	.0087	3.3955	.0003
3.75	.0042	3.6380	.0001
4.00	.0020	3.8806	.0001
4.25	.0010	4.1231	.0000
4.50	.0005	4.3656	.0000
4.75	.0002	4.6082	.0000
5.00	.0001	4.8507	.0000
5.25	.0001	5.0932	.0000
5.50	.0000	5.3358	.0000

4. Examples of the Use of the Probability Integral and Percentage Points of the Non-Central t-Statistic

a. Sampling Inspection by Variables for Fraction Defective, with One Standard Given

A class of problems in which the non-central t-statistic plays an important role is in the computation of operating characteristic (OC) curves for sampling inspection by variables procedures. Suppose an object is classified as defective or non-defective according to whether the value of a characteristic exceeds or falls short of a fixed standard U. The percent defective, p, in a lot is defined as the percentage of the objects falling above U. The general sampling inspection problem is the formulation of a procedure which will accept the lot if p is sufficiently small. In particular, sampling inspection by variables can be used if the measured characteristic, Y, is a normally distributed random variable. (Note that this variable Y is distinct from that random variable which depends solely on whether the object is defective or non-defective.) The acceptance procedure for the case in which the mean, μ, and the standard deviation, σ, are unknown is to accept the lot if

$$\bar{Y} + ks \leq U$$

where k is a constant, and $\bar{Y}$ and s, the sample mean and sample standard deviation computed from a sample of size n, are given by:

$$\bar{Y} = \frac{\sum Y_i}{n}$$

and

$$s = \sqrt{\frac{\sum (Y_i - \bar{Y})^2}{n-1}} .$$

The acceptance criterion can be rewritten as

$$\frac{\sqrt{n}\,(U-\bar{Y})}{s} \geq \sqrt{n}\,k$$

and further as

$$\left\{ \frac{\sqrt{n}\,(U-\mu)}{\sigma} - \frac{\sqrt{n}\,(\bar{Y}-\mu)}{\sigma} \right\} \Big/ \frac{s}{\sigma} \geq \sqrt{n}\,k .$$

But the expression on the left has a non-central t-distribution with $f = n-1$ and $\delta = \frac{\sqrt{n}\,(U-\mu)}{\sigma} = \sqrt{n}\,K_p$. Hence the acceptance criterion can be written as

$$t \geq \sqrt{n}\,k.$$

The OC curve for this procedure, that is, the probability of accepting the lot, can be written as

$$\Pr\left\{ t \geq \sqrt{n}\, k \right\} = 1-\Pr\left\{ t \leq \sqrt{n}\, k \right\} = 1-\Pr\left\{ \frac{t}{\sqrt{n-1}} \leq \sqrt{\frac{n}{n-1}}\, k \right\}$$

$$= 1-P(n-1, \sqrt{n}\, K_p, \sqrt{\frac{n}{n-1}}\, k) .$$

Ten points on the OC curve corresponding to the ten values of p found in the table can be obtained immediately.

For example, for $n = 10$ and $k = 1.72$, $\Pr\left\{ t \geq \sqrt{n}\, k \right\} = 1-P(9, \sqrt{10}\, K_p, 1.813)$. By interpolation in the Table of the Probability Integral for $f = 9$, the following points are obtained:

$p = .2500 \qquad 1-P(9, \sqrt{10}\, K_{.2500}, 1.813) = .0215$

$p = .1500 \qquad 1-P(9, \sqrt{10}\, K_{.1500}, 1.813) = .1038$

$p = .1000 \qquad 1-P(9, \sqrt{10}\, K_{.1000}, 1.813) = .2242$

$p = .0650 \qquad 1-P(9, \sqrt{10}\, K_{.0650}, 1.813) = .3851$

$p = .0400 \qquad 1-P(9, \sqrt{10}\, K_{.0400}, 1.813) = .5685$

$p = .0250 \qquad 1-P(9, \sqrt{10}\, K_{.0250}, 1.813) = .7177$

$p = .0100 \qquad 1-P(9, \sqrt{10}\, K_{.0100}, 1.813) = .8976$

$p = .0040 \qquad 1-P(9, \sqrt{10}\, K_{.0040}, 1.813) = .9692$

$p = .0025 \qquad 1-P(9, \sqrt{10}\, K_{.0025}, 1.813) = .9843$

$p = .0010 \qquad 1-P(9, \sqrt{10}\, K_{.0010}, 1.813) = .9961$

These tables can also be used to design sampling inspection plans. If two points on the OC curve, $(p_1, 1-\alpha)$ and (p_2, β), are given, the required values of n and k can be found. These values are obtained from the relationships:

$$P(n-1, \sqrt{n}\, K_{p_1}, \sqrt{\frac{n}{n-1}}\, k) = \alpha$$

$$P(n-1, \sqrt{n}\, K_{p_2}, \sqrt{\frac{n}{n-1}}\, k) = 1-\beta.$$

The values of n and k are found by trial and error. Theoretically, the correct value of f has the property that there exists a value of x for which the entry in the p_1 column equals α and simultaneously the entry in the p_2 column equals $1-\beta$; that is,

$$P(n-1, \sqrt{n}\, K_{p_1}, x) = \alpha$$

$$P(n-1, \sqrt{n}\, K_{p_2}, x) = 1-\beta.$$

However, because of the necessary discreteness in the sample size, an integral value of f possessing the above property is not usually obtainable. Therefore, the following procedure for $f \leq 24$ is established. Two consecutive values of f, say f^* and $f^* + 1$, are found such that

(1) for f^* there exists an x' for which

$$P(n-1, \sqrt{n}\, K_{p_1}, x') \geq \alpha$$

and simultaneously

$$P(n-1, \sqrt{n}\, K_{p_2}, x') \leq 1-\beta$$

and

(2) for $f^* + 1$ there exists an x'' for which

$$P(n-1, \sqrt{n}\, K_{p_1}, x'') \leq \alpha$$

and simultaneously

$$P(n-1, \sqrt{n}\, K_{p_2}, x'') \geq 1-\beta.$$

The value $f^* + 1$ is the desired f, and the appropriate value of x is found by interpolating in the p_1 column with degrees of freedom $f^* + 1$ for the given value α. The value of k is obtained by substituting this value of x in $k = \sqrt{\frac{n-1}{n}}\, x$. Note that this procedure always yields a value of β less than the desired one. This has the effect of establishing a sampling plan that is a little stricter than the one desired in the sense that lots of incoming quality p_2 will be accepted less frequently than prescribed.

For example, suppose an OC curve is to be constructed passing through the points $(p_1, 1-\alpha) = (.01, .99)$ and $(p_2, \beta) = (.15, .10)$. Thus, $\alpha = .01$ and $1-\beta = .90$. $f^* = 16$ and $f^* + 1 = 17$ have the properties described above, that is,

$$P(16, \sqrt{17}\, K_{.01}, 1.55) = .0104 \geq .01$$

$$P(16, \sqrt{17}\, K_{.15}, 1.55) = .8936 \leq .90$$

and

$$P(17, \sqrt{18}\, K_{.01}, 1.55) = .0090 \leq .01$$

$$P(17, \sqrt{18}\, K_{.15}, 1.55) = .9022 \geq .90 \; .$$

Thus, $f = 17$ is chosen. It should be noted that x' and x'' need not have the same numerical value. Because of the relative coarseness of the interval in the argument x, the value of x, corresponding to a specified value of α for f, will often be identical to the value of x corresponding to α for $f+1$. With this f and by linear interpolation in the p_1 column, the value of $x = 1.560$ provides an $\alpha = .01$. The value of k is found from $k = \sqrt{\frac{17}{18}}\, 1.560 = 1.516$. Thus, for the values of $n = 18$ and $k = 1.516$, the OC curve will pass through $(p_1, 1-\alpha) = (.01, .99)$ but will not pass through $(p_2, \beta) = (.15, .10)$ exactly. Instead, for these values of n and k, the curve will pass through $(.15, .0941)$.

For $24 \leq f \leq 49$, it is suggested that a first approximation for $n = f+1$ be obtained by the normal approximation which is given by

$$n = \frac{2(K_\alpha + K_\beta)^2 + (K_\alpha K_{p_2} + K_\beta K_{p_1})^2}{2(K_{p_1} - K_{p_2})^2}$$

where the numerical value obtained is always rounded to the next higher integer and where K_ϵ is a normal deviate exceeded with probability ϵ; that is, K_ϵ is defined by

$$\frac{1}{\sqrt{2\pi}} \int_{K_\epsilon}^{\infty} e^{-\frac{y^2}{2}}\, dy = \epsilon \; .$$

Call this value $\tilde{f}$. Then interpolate in the Tables of the Probability Integral of t for the x corresponding to p_1 and α for the two tabled values of f adjacent to $\tilde{f}$; then obtain the $(1-\beta)$ value corresponding to these x's in the p_2 column. If $\tilde{f}$ is already one of the tabulated values of f, the desired values of x and $1-\beta$ are obtained by these interpolations. Finally interpolate on f for the x corresponding to p_1 and α and for $(1-\beta)$.

For example, suppose a sampling plan is to be constructed which will have an OC curve passing through (.065, .99) and (.25, .04). From the normal approximation to f, the value of $\tilde{f} = 36$ is obtained. The two tabled values of f adjacent to 36 are 34 and 39. For $f = 34$, by interpolation it is found that $x = 1.043$ corresponds to $p_1 = .065$ and $\alpha = .01$; and for this value of x, $1-\beta = .9507$. Similarly for $f = 39$, $x = 1.066$ and $1-\beta = .9698$ are found. Then, by interpolation for $\tilde{f} = 36$ on these pairs of x and $(1-\beta)$ values, $x = 1.052$ and $1-\beta = .9583$ are obtained. Then $k = \sqrt{\frac{36}{37}}\, 1.052 = 1.038$. Thus for the values $n = 37$ and $k = 1.038$, the OC curve will pass through (.065, .99) and (.25, .0417) which is close enough for most practical applications. It must be noted that these results are only as accurate as the method of interpolation used.

If the value of f required exceeds the range of the table the normal approximation for both n and k [12] can be used.

The solution to the problem of finding values of n and k corresponding to two given points on the OC curve can also be obtained by using the Table of the Percentage Points of t. This table can be used (and should be used) whenever $1-\alpha$ and β are values of ϵ given in this table. Theoretically, the solution is obtained by finding values of n and x such that

$$x(n-1, \sqrt{n}\, K_{p_1}, 1-\alpha) = x(n-1, \sqrt{n}\, K_{p_2}, \beta) .$$

However, again the discreteness of f does not allow an exact solution and it is necessary to proceed as above, finding two consecutive values of f, f^* and $f^* + 1$, such that for f^*

$$x(n-1, \sqrt{n}\, K_{p_1}, 1-\alpha) < x(n-1, \sqrt{n}\, K_{p_2}, \beta)$$

and for $f^* + 1$

$$x(n-1, \sqrt{n}\, K_{p_1}, 1-\alpha) > x(n-1, \sqrt{n}\, K_{p_2}, \beta).$$

The proper value of f is $f^* + 1$ and x is the value $x(n-1, \sqrt{n}\, K_{p_1}, 1-\alpha)$ for this f. This procedure yields a slightly stricter OC curve than the one specified.

For example, if once again a sampling plan is to be constructed passing through the points $(.01, .99)$ and $(.15, .10)$. The Tables of the Percentage Points of t may be used since $\epsilon = .99$ and $\epsilon = .10$ are both tabled values. For $f^* = 16$

$$x(16, \sqrt{17}\, K_{.01}, .99) = 1.545 < x(16, \sqrt{17}\, K_{.15}, .10) = 1.566$$

whereas for $f^* + 1 = 17$

$$x(17, \sqrt{18}\, K_{.01}, .99) = 1.561 > x(17, \sqrt{18}\, K_{.15}, .10) = 1.545.$$

Thus $f = 17$ is chosen and $k = \sqrt{\frac{17}{18}}\, 1.561 = 1.517$. The difference between this solution and the one using the probability integral is due to the fact that the percentage point table yields more accurate results than does inverse linear interpolation in the Tables of the Probability Integral.

If the values of p_1 and p_2 chosen are not those found in the table, a solution can be obtained by interpolation on p or K_p.

Finally, these tables can also be used to design sampling plans when the value of n is given and a single point $(p_1, 1-\alpha)$ is specified. The solution yields the appropriate value of k and is obtained from the expression

$$P(n-1, \sqrt{n}\, K_{p_1}, \sqrt{\frac{n}{n-1}}\, k) = \alpha$$

when the Table of the Probability Integral is used. The value of x which satisfies the above expression is used to determine $k = \sqrt{\frac{n-1}{n}}\, x$.

If the Table of Percentage Points is used, the value of x corresponding to $x(n-1, \sqrt{n}\, K_{p_1}, 1-\alpha)$ yields the appropriate solution and $k = \sqrt{\frac{n-1}{n}}\, x$.

For example, if $p_1 = .01$, $\alpha = .05$, and $n = 15$, and the Table of Percentage Points is used, the value of x corresponding to

$$x(14, \sqrt{15}\, K_{.01}, .95) = 1.736$$

and

$$k = \sqrt{\frac{14}{15}}\, 1.736 = 1.677.$$

b. Sampling Inspection by Variables for Fraction Defective, with Two Standards Given

An unsolved problem is that of constructing tests of significance about the proportion defective p, with given significance level α, when p is characterized by two specifications, a lower standard L, and an upper standard U, so that

$$p = p_U + p_L = \int_{\frac{U-\mu}{\sigma}}^{\infty} \frac{e^{-\frac{v^2}{2}}}{\sqrt{2\pi}} dv + \int_{-\infty}^{\frac{L-\mu}{\sigma}} \frac{e^{-\frac{v^2}{2}}}{\sqrt{2\pi}} dv .$$

This is sometimes called the two-sided proportion defective. However it is possible to devise such tests of significance for p with significance level very near a given value α by the use of a procedure based on the uniformly minimum variance estimate $\hat{p}$, defined by

$$\hat{p} = \hat{p}_U + \hat{p}_L = I_{z_U}(\frac{n-2}{2}, \frac{n-2}{2}) + I_{z_L}(\frac{n-2}{2}, \frac{n-2}{2})$$

where

$$z_U = \frac{1}{2} - \frac{1}{2} \frac{U-\bar{Y}}{s} \frac{\sqrt{n}}{n-1} \quad ; \quad z_L = \frac{1}{2} - \frac{1}{2} \frac{\bar{Y}-L}{s} \frac{\sqrt{n}}{n-1} .$$

I_z is the Incomplete Beta Function Ratio.

If $z < 0$, then I_z is taken to be zero; if $z > 1$, I_z is taken to be one.

If $L = -\infty$ so that $p_L = 0$,

$$p = p_U = \int_{\frac{U-\mu}{\sigma}}^{\infty} \frac{e^{-\frac{v^2}{2}}}{\sqrt{2\pi}} dv.$$

Then $\hat{p}_L = 0$ and the uniformly minimum variance unbiased estimate of the one-sided proportion defective p_U is

$$p_U = \begin{cases} I_{z_U}(\frac{n-2}{2}, \frac{n-2}{2}) & \text{for } 0 < z_U < 1 \\ 0 & \text{for } z_U \leq 0 \\ 1 & \text{for } z_U \geq 1 \end{cases}$$

If $U = +\infty$ so that $p_U = 0$, then $\hat{p}_U = 0$ and the uniformly minimum variance unbiased estimate of p_L is

$$p_L = \begin{cases} I_{z_L}(\frac{n-2}{2}, \frac{n-2}{2}) & \text{for} \quad 0 < z_L < 1 \\ 0 & \text{for} \quad z_L \leq 0 \\ 1 & \text{for} \quad z_L \geq 1 \end{cases}$$

To test $p_U \leq p_o$ against $p_U > 0$ the procedure is as follows:

1. Choose a significance level α and sample size n.

2. Using the Tables of the Percentage Points, solve the equation

$$x(n-1, \sqrt{n}\, K_{p_o}, 1-\alpha) = \sqrt{n-1} \quad (1-2\, \beta_{p*})$$

for β_{p*}.

3. β_{p*} is defined by $I_{\beta_{p*}}(\frac{n-2}{2}, \frac{n-2}{2}) = p*$. By interpolation in the tables of the Incomplete Beta Function determine $p*$.

4. Draw a sample of n and compute

$$z_U = \frac{1}{2} - \frac{1}{2} \frac{U-\bar{Y}}{s} \frac{\sqrt{n}}{n-1} .$$

5. From the Table of the Incomplete Beta Function, using z_U as argument determine $\hat{p}$.

6. If $\hat{p} > p*$, the hypothesis $p_U \leq p_o$ is rejected with probability exactly α.

If in step 3 the critical region $\hat{p}_U > p*$ is such that $p* = 0$, then for this combination of p_o and n, the significance level α is unobtainable with the test statistic $\hat{p}_U$. This situation occurs only for the combination of very small p_o and very small n, so that $x(n-1, \sqrt{n}\, K_{p_o}, 1-\alpha) > \sqrt{n-1}$.

The extension of this procedure to the case of testing $p_L \leq p_o$ involves only replacing z_U by

$$z_L = \frac{1}{2} - \frac{1}{2} \frac{\bar{Y}-L}{s} \frac{\sqrt{n}}{n-1} .$$

The operating characteristic function (probability of acceptance) for these one-sided procedures is obtained from the Table of the Probability Integral and is given by

$$1-P[n-1, \sqrt{n}\, K_p, x(n-1, \sqrt{n}\, K_{p_o}, 1-\alpha)].$$

Thus it is seen to be equivalent to the one-sided acceptance procedure given in the preceding section where the criterion is $\bar{Y} + ks \leq U$.

The relation between k and p^* for given α is

$$k = \frac{n-1}{\sqrt{n}} (1-2\beta_{p^*}).$$

Unless one is interested in simultaneously testing an hypothesis about proportion defective p and obtaining an unbiased estimate of p, the foregoing procedure affords no advantages over the criterion

$$\bar{Y} + ks \leq U.$$

However in the two-sided case the uniformly minimum variance unbiased estimate provides a procedure for testing $p = p_U + p_L \leq p_o$. Computations in [11] show that the probability distribution of $\hat{p} = \hat{p}_U + \hat{p}_L$ is essentially dependent only on p and only very slightly on the partition of p into its components p_U and p_L. To this extent the following statement holds:

$$\Pr\left\{ \hat{p} \leq p^* | p \right\} \sim 1-P[n-1, \sqrt{n}\, K_p, \sqrt{n-1}\, (1-2\beta_{p^*})].$$

The two-sided test procedure is as follows:

1. Choose a significance level α and sample size n.

2. Using the Table of the Percentage Points, solve the equation

$$x(n-1, \sqrt{n}\, K_{p_o}, 1-\alpha) = \sqrt{n-1}\, (1-2\beta_{p^*})$$

for β_{p^*}.

3. By interpolation in the Table of the Incomplete Beta Function determine p^*.

4. Draw a sample of size n and compute both

$$z_U = \frac{1}{2} - \frac{1}{2} \frac{U-\bar{Y}}{s} \frac{\sqrt{n}}{n-1}$$

$$z_L = \frac{1}{2} - \frac{1}{2} \frac{\bar{Y}-L}{s} \frac{\sqrt{n}}{n-1}.$$

5. From the Table of the Incomplete Beta Function, using z_U and z_L as

arguments, determine $\hat{p}_U$ and $\hat{p}_L$ and hence $\hat{p} = \hat{p}_U + \hat{p}_L$.

6. If $\hat{p} > p^*$ the hypothesis $p \leq p_o$ is rejected with probability very nearly equal to α.

If the critical region $\hat{p} > p^*$ is such that $p^* = 0$ the significance level α is unachievable for this combination of p_o and n.

The operating characteristic function is obtained from the Table of the Probability Integral and is given approximately by

$$1-P[n-1, \sqrt{n}\, K_p, x(n-1, \sqrt{n}\, K_{p_o}, 1-\alpha)].$$

For the two-sided test procedure steps 1, 2 and 3 and the OC curve are identical with those of the one-sided case. Therefore any test criterion p^* and OC curve for a one-sided test may be used as approximations for the two-sided test.

An example of this procedure follows. It is assumed that the hypothesis to be tested is that the proportion defective p_o is less than or equal to .01. A sample size of 10 is to be used with significance level $\alpha = .10$. To find the test criterion the equation

$$x(9, \sqrt{10}\, K_{.01}, .90) = \sqrt{9}\,(1-2\beta_{p^*})$$

is solved for β_{p^*}. From the Tables of the Percentage Points $x(9, \sqrt{10}\, K_{.01}, .90) = 1.807$ so that $\beta_{p^*} = .1988$. By entering the Tables of the Incomplete Beta Function Ratio the value $p^* = .0327$ is found.

As a result of sampling 10 items the following statistics are observed

$$\frac{U-\bar{Y}}{s} = 2.034 \quad \text{and} \quad \frac{\bar{Y}-L}{s} = 1.925.$$

From these sample values

$$z_U = .14266 \quad \text{and} \quad z_L = .16181$$

are computed. The estimates $\hat{p}_U = .0101$ and $\hat{p}_L = .0159$ are found by entering the Tables of the Incomplete Beta Function Ratio. Thus $\hat{p} = .0101 + .0159 = .0260 < p^* = .0327$ so that the hypothesis is not rejected on the basis of this sample. The OC curve for this procedure is given by $1-P(9, \sqrt{10}\, K_p, 1.807)$. Ten points on this curve obtained from the Tables of the Probability Integral are given below.

p	OC function
.0010	.9963
.0025	.9849
.0040	.9702
.0100	.9000
.0250	.7217
.0400	.5728
.0650	.3890
.1000	.2270
.1500	.1054
.2500	.0219

c. Confidence Intervals for Proportions from a Normal Population

Denote by p the proportion of a normal population with unknown mean μ and unknown standard deviation σ which lies above a fixed standard U. In variables acceptance sampling p is termed the proportion defective.

The limits p_1 and p_2 of a confidence interval for p, with confidence coefficient γ, may be found by estimating $\sqrt{n}\, K_p = \sqrt{n}\, \frac{U-\mu}{\sigma}$ by the non-central t-statistic based on a sample of n observations, $Y_1, Y_2, \ldots, Y_n$. This non-central t-statistic is

$$\frac{\sqrt{n}\,(U-\bar{Y})}{s}$$

where

$$\bar{Y} = \frac{\sum Y_i}{n}$$

and

$$s = \sqrt{\frac{\sum (Y_i-\bar{Y})^2}{n-1}} .$$

Confidence limits for $K_p = \frac{U-\mu}{\sigma}$ are determined by solving the equations

$$x(n-1, \sqrt{n}\, K_{p_1}, 1-\gamma_1) = \sqrt{\frac{n}{n-1}}\ \frac{U-\bar{Y}}{s}$$

$$x(n-1, \sqrt{n}\, K_{p_2}, \gamma_2) = \sqrt{\frac{n}{n-1}}\ \frac{U-\bar{Y}}{s}$$

for K_{p_1} and K_{p_2} using the Tables of the Percentage Points of the non-central t-statistic. Here $\gamma_1 + \gamma_2 = 1-\gamma$. It is usually necessary to interpolate on K_p in these tables for this purpose. One then converts K_{p_2} and K_{p_1} into p_2 and p_1, the upper and lower limits, respectively, of the confidence interval for p with coefficient γ. The most convenient table for converting K_p to p is [5]. However, any table of the cumulative normal distribution will serve.

A numerical example of the use of the Tables of the Percentage Points to obtain confidence limits for proportion defective p follows: Suppose a sample of size n = 20 is drawn and on the basis of the sample mean $\overline{Y}$ and sample standard deviation s, the statistic $\frac{U-\overline{Y}}{s}$ is found to be 1.834. A confidence interval for p is to be constructed with confidence coefficient $\gamma = .90$. Let $\gamma_1 = .05$ and $\gamma_2 = .05$. Then

$$\sqrt{\frac{n}{n-1}}\ \frac{U-\overline{Y}}{s} = \sqrt{\frac{20}{19}}\ 1.834 = 1.882.$$

Solve the equations

$$x(19, \sqrt{20}\ K_{p_1}, .95) = 1.882$$

$$x(19, \sqrt{20}\ K_{p_2}, .05) = 1.882$$

for p_1 and p_2. Interpolation should be performed on K_p, not on p. Values of K_p can be found in the table on page 3 . For this example, the following results are obtained:

$$K_{p_1} = 2.429021; \quad p_1 = .0076$$

$$K_{p_2} = 1.209322; \quad p_2 = .1133.$$

Thus the confidence interval for p with confidence coefficient $\gamma = .90$ based on this sample of 20 items is

$$[.0076, .1133].$$

If a lower specification L is given so that

$$p = \int_{-\infty}^{\frac{L-\overline{Y}}{s}} \frac{e^{-\frac{v^2}{2}}}{\sqrt{2\pi}}\ dv$$

the equations which yield confidence limits for K_p and hence for p are

$$x(n-1, \sqrt{n}\, K_{p_1}, 1-\gamma_1) = \sqrt{\frac{n}{n-1}}\, \frac{(\bar{Y}-L)}{s}$$

$$x(n-1, \sqrt{n}\, K_{p_2}, \gamma_2) = \sqrt{\frac{n}{n-1}}\, \frac{(\bar{Y}-L)}{s} .$$

d. The Power of Student's t-Test

In testing the hypothesis H_o that the mean of a normal distribution is $\mu = \mu_o$ against the alternative that $\mu > \mu_o$, Student's t-test consists of calculating

$$t = \frac{\sqrt{n}\, (\bar{Y}-\mu_o)}{s}$$

where $\bar{Y} = \Sigma Y_i/n$ and $s = \sqrt{\frac{\Sigma (Y_i-\bar{Y})^2}{n-1}}$ and rejecting the hypotheses if

$$t > t_o .$$

If the level of significance is α, the following probability statement holds,

$$P(n-1, 0, t_o/\sqrt{n-1}) = 1-\alpha .$$

The value of t_o is obtained from a Table of the Percentage Points of Student's t-distribution.

Ten other points on the power curve (probability of rejection) can be obtained as follows:

When the true mean is μ, the quantity t has a non-central t-distribution with degrees of freedom $n-1$, and non-centrality parameter $\frac{\sqrt{n}\, (\mu-\mu_o)}{\sigma}$, that is,

$$t = \left\{ \frac{\sqrt{n}\, (\bar{Y}-\mu)}{\sigma} + \frac{\sqrt{n}\, (\mu-\mu_o)}{\sigma} \right\} \Big/ \frac{s}{\sigma} .$$

The power of the test is therefore given by

$$1-P(n-1, \frac{\sqrt{n}\, (\mu-\mu_o)}{\sigma}, t_o/\sqrt{n-1}) .$$

The values $\frac{\mu-\mu_o}{\sigma}$ (the abscissa of the power curve) which correspond to a given power are found by equating $\frac{\mu-\mu_o}{\sigma} = K_p$. Thus, for the ten values of p given in the probability integral table, ten values of $\frac{\mu-\mu_o}{\sigma}$ are obtained, and one minus the power associated with these points are read out of the table. These values appear in the entry corresponding to $x = t_o/\sqrt{n-1}$.

As an example, the power of the t-test at ten points will be obtained from the Table of the Probability Integral for $f = 4 (n = 5)$. A level of significance equal to five percent is chosen. For this case, $t_o = 2.132$ and $\frac{t_o}{\sqrt{n-1}} = 1.066$.

$\frac{\mu-\mu_o}{\sigma} = K_p$	1 - Power	Power
0.6745	.6512	.3488
1.0364	.3948	.6052
1.2816	.2409	.7591
1.5141	.1329	.8671
1.7507	.0636	.9364
1.9600	.0296	.9704
2.3263	.0060	.9940
2.6521	.0012	.9988
2.8070	.0005	.9995
3.0902	.0001	.9999

e. The Coefficient of Variation

This example of the use of the Tables of the Probability Integral of the non-central t-statistic deals with the coefficient of variation,

$$V = \sigma/\mu ,$$

where σ is the standard deviation of a normal distribution and μ is the mean of a normal distribution. An estimate of V is provided by the sample coefficient of variation $v = \frac{s}{\bar{Y}}$, where $\bar{Y} = \Sigma Y_i/n$ and $s = \sqrt{\frac{\Sigma(Y_i-\bar{Y})^2}{n-1}}$.

The following statistic

$$\frac{\sqrt{n}}{v} = \frac{\sqrt{n}\,\bar{Y}}{s} = \left\{ \frac{\sqrt{n}\,(\bar{Y}-\mu)}{\sigma} + \frac{\sqrt{n}\,\mu}{\sigma} \right\} \Big/ \frac{s}{\sigma}$$

has a non-central t-distribution with n-1 degrees of freedom and non-centrality parameter $\delta = \frac{\sqrt{n}\,\mu}{\sigma}$. Thus

$$\Pr\left\{ v > v_o \right\} = \Pr\left\{ \frac{\sqrt{n}}{v} \leq \frac{\sqrt{n}}{v_o} \right\}$$

$$= \Pr\left\{ \sqrt{\frac{n}{n-1}}\ \frac{1}{v} \leq \sqrt{\frac{n}{n-1}}\ \frac{1}{v_o} \right\} = P\left(n-1, \frac{\sqrt{n}\ \mu}{\sigma}, \sqrt{\frac{n}{n-1}}\ \frac{1}{v_o}\right).$$

Suppose a test of the hypothesis that $V = V_o$ at the ϵ level of significance is to be made against the alternative that $V > V_o$. The procedure used will be to reject the hypothesis if $v > v_o$. The critical value of the test v_o can be obtained from the identity

$$x\left(n-1, \frac{\sqrt{n}}{V_o}, 1-\epsilon\right) = \sqrt{\frac{n}{n-1}}\ \frac{1}{v_o}\ .$$

The solution will usually require an interpolation on the non-centrality parameter. Similarly, the power of the test can be obtained from the relationship Power $= P\left(n-1, \frac{\sqrt{n}\ \mu}{\sigma}, \sqrt{\frac{n}{n-1}}\ \frac{1}{v_o}\right)$. Ten points on the power curve can be obtained by choosing the abscissa values corresponding to $\frac{1}{V} = K_p$.

As an example let $V_o = 1$, $n = 9$, and $\epsilon = .05$. Using the Table of the Percentage Points the value $v_o = 0.495$ for $f = 8$ and $\epsilon = .05$ is obtained from the expression

$$x(8, 3, .95) = 2.141 = \sqrt{\frac{9}{8}}\ \frac{1}{v_o}\ .$$

The ten points on the power curve are as follows:

$V = \frac{1}{K_p}$	Power
1.4826	.9880
.9649	.9428
.7803	.8719
.6605	.7661
.5712	.6256
.5102	.4873
.4299	.2635
.3771	.1245
.3563	.0814
.3236	.0332

5. Computational Methods

The tables were computed on the IBM Card Programmed Computer, Model II. This machine does not readily lend itself to table look-up, so that it was decided to generate the probability density function on the machine directly, without using existing tables of the Hh_f function [1]. In any case the range of the latter tables would not have sufficed for the purpose at hand.

The Hh_f function in the non-central t-density obeys the recurrence relation

$$fHh_f(x) = Hh_{f-2}(x) - xHh_{f-1}(x).$$

In particular

$$\sqrt{2\pi}\ Hh_o(x) = \int_{-\infty}^{x} \frac{e^{-\frac{t^2}{2}}}{\sqrt{2\pi}}\,dt \quad \text{and} \quad \sqrt{2\pi}\ Hh_{-1}(x) = \frac{e^{-\frac{x^2}{2}}}{\sqrt{2\pi}}$$

Repeated application of the recurrence formula shows that Hh_f may be expressed as

$$Hh_f(x) = P_f(x)\ Hh_o(x) + Q_f(x)\ Hh_{-1}(x)$$

where P_f and Q_f are polynomials. It can easily be demonstrated that P_f and Q_f obey the same recurrence laws, that is, that

$$fP_f(x) = P_{f-2}(x) - xP_{f-1}(x)$$

with

$$P_o(x) = 1 \quad \text{and} \quad P_{-1}(x) = 0$$

and that

$$fQ_f(x) = Q_{f-2}(x) - xQ_{f-1}(x)$$

with

$$Q_o(x) = 0 \quad \text{and} \quad Q_{-1}(x) = 1.$$

A table of the polynomials P_f and Q_f was prepared for $f = 1$ to $f = 20$. The first few are given below:

f	$P_f(x)$	$Q_f(x)$
1	1	$-x$
2	$\frac{1}{2!}(-x)$	$\frac{1}{2!}(1+x^2)$
3	$\frac{1}{3!}(2+x^2)$	$\frac{1}{3!}(-3x-x^3)$

The polynomials and exponential functions in the non-central t-density were generated on the machine by standard programs. In order to compute the normal probability integral Hh_0, a rational function approximation due to Cecil Hastings [3] was used. This approximation yields an order of accuracy consistent with the number of significant figures available on the machine.

For f greater than 20, the polynomials P_f and Q_f became too cumbersome, causing the running time on the machine to be excessive. It was decided to use an asymptotic expansion for the Hh_f function. The following expression was derived using the method of Steepest Descent:

$$Hh_f(x) = \frac{1}{f!} t^f e^{-\frac{1}{2}(t+x)^2} \sqrt{\frac{2\pi t^2}{f+t^2}} \left[1 - \frac{3f}{4(f+t^2)^2} + \frac{5f^2}{6(f+t^2)^3}\right]$$

where

$$t = \frac{-x + \sqrt{x^2 + 4f}}{2} .$$

This approximation was tested in the following ways. The Hh_{20} functions were computed both by the method described previously and by the asymptotic expansion, and the two methods were compared. Both methods were then compared with Airey's tables of the Hh_{20} function.

The final results for the density function were adjudged to be accurate to at least five decimal places. (Seven places were retained in all functions during the computations.) In order to limit the size of the printed tables the final tabulation was rounded down to the four places shown.

The probability integral was computed by summing the probability density function by numerical integration. The results were spot-checked in scores of places against the tables of Johnson and Welch. The results usually agreed to four decimal places. In rare cases they differed by no more than one or two units in the third decimal place.

The discrepancies between the values obtained from the Johnson and Welch tables and the present Table of the Probability Integral may be due to interpolation in the former tables. Johnson and Welch indicate that results obtainable from their tables may be in error to this extent.

The Tables of the Percentage Points were obtained by inverse interpolation on the probability integral using six-point Lagrangian interpolation polynomials. The percentage points were checked in numerous instances by comparison with results obtained from the Johnson and Welch tables. The percentage points are believed to be correct in the second decimal place throughout, and to differ occasionally from the true values by no more than one or two units in the third decimal place.

Bibliography

[1] British Association for the Advancement of Science, Mathematical Tables, Vol. 1, Cambridge, England: University Press (1946).

[2] David, Herbert T., and Kruskal, William H., "The WAGR sequential t-test reaches a decision with probability one," Annals of Mathematical Statistics, 27(1956), 797-805.

[3] Hastings, Jr., Cecil, Approximations for Digital Computers, Princeton, New Jersey: Princeton University Press (1955).

[4] Johnson, N. L., and Welch, B. L., "Applications of the non-central t-distribution," Biometrika, 31(1940), 362- 89.

[5] Kelley, Truman Lee, The Kelley Statistical Tables, Cambridge, Mass.: Harvard University Press (1948).

[6] Lieberman, Gerald J., and Resnikoff, George J., "Sampling plans for inspection by variables," Journal of the American Statistical Association, 50(1955), 457-516.

[7] Mathematical Tables Project, Work Projects Administration for the City of New York, F.W.A., Tables of Probability Functions, Vol. II, Washington, D. C.: Government Printing Office (1942).

[8] National Bureau of Standards, Tables to Facilitate Sequential t-Tests, Applied Mathematics, Series 7, Washington, D. C.: Government Printing Office (1951).

[9] Neyman, J., "Statistical problems in agricultural experimentation," Supplement to the Journal of the Royal Statistical Society,2(1935), 107-180.

[10] Neyman, J., and Tokarska, B., "Errors of the second kind in testing 'Student's' hypothesis," Journal of the American Statistical Association, 31(1936), 318-326.

[11] Resnikoff, George J., A New Two-Sided Acceptance Region for Sampling by Variables, Technical Report No. 8, Office of Naval Research, Contract No. N6onr-25126, Applied Mathematics and Statistics Laboratory, Stanford University, (1952).

[12] Wallis, W. Allen, "Use of variables in acceptance inspection for percent defective," Chapter 1 in Statistical Research Group, Techniques of Statistical Analysis for Scientific and Industrial Research and Production and Management Engineering, New York: McGraw-Hill Book Company, Inc., (1947), 6-93.

PROBABILITY DENSITY OF t, THE NON-CENTRAL t-STATISTIC

PROBABILITY DENSITY OF t, THE NON-CENTRAL t-STATISTIC, TABLED AS A FUNCTION OF $t/\sqrt{f}$.

f is the number of degrees of freedom; the non-centrality parameter is $\sqrt{f+1}\,K_p$.
K_p is the standardized normal deviate exceeded with probability p.

DEGREES OF FREEDOM **2**

$t/\sqrt{f}$ \ p	.2500	.1500	.1000	.0650	.0400	.0250	.0100	.0040	.0025	.0010
-9.00	0000	0000	0000	0000	0000	0000	0000	0000	0000	0000
-8.95	0001	0000	0000	0000	0000	0000	0000	0000	0000	0000
90	0001	0000	0000	0000	0000	0000	0000	0000	0000	0000
85	0001	0000	0000	0000	0000	0000	0000	0000	0000	0000
80	0001	0000	0000	0000	0000	0000	0000	0000	0000	0000
75	0001	0000	0000	0000	0000	0000	0000	0000	0000	0000
70	0001	0000	0000	0000	0000	0000	0000	0000	0000	0000
65	0001	0000	0000	0000	0000	0000	0000	0000	0000	0000
60	0001	0000	0000	0000	0000	0000	0000	0000	0000	0000
55	0001	0000	0000	0000	0000	0000	0000	0000	0000	0000
50	0001	0000	0000	0000	0000	0000	0000	0000	0000	0000
45	0001	0000	0000	0000	0000	0000	0000	0000	0000	0000
40	0001	0000	0000	0000	0000	0000	0000	0000	0000	0000
35	0001	0000	0000	0000	0000	0000	0000	0000	0000	0000
30	0001	0000	0000	0000	0000	0000	0000	0000	0000	0000
25	0001	0000	0000	0000	0000	0000	0000	0000	0000	0000
20	0001	0000	0000	0000	0000	0000	0000	0000	0000	0000
15	0001	0000	0000	0000	0000	0000	0000	0000	0000	0000
10	0001	0000	0000	0000	0000	0000	0000	0000	0000	0000
05	0001	0000	0000	0000	0000	0000	0000	0000	0000	0000
-8.00	0001	0000	0000	0000	0000	0000	0000	0000	0000	0000
-7.95	0001	0000	0000	0000	0000	0000	0000	0000	0000	0000
90	0001	0000	0000	0000	0000	0000	0000	0000	0000	0000
85	0001	0000	0000	0000	0000	0000	0000	0000	0000	0000
80	0001	0000	0000	0000	0000	0000	0000	0000	0000	0000
75	0001	0000	0000	0000	0000	0000	0000	0000	0000	0000
70	0001	0000	0000	0000	0000	0000	0000	0000	0000	0000
65	0001	0000	0000	0000	0000	0000	0000	0000	0000	0000
60	0001	0000	0000	0000	0000	0000	0000	0000	0000	0000
55	0001	0000	0000	0000	0000	0000	0000	0000	0000	0000
-7.50	0001	0000	0000	0000	0000	0000	0000	0000	0000	0000
45	0001	0000	0000	0000	0000	0000	0000	0000	0000	0000
40	0001	0000	0000	0000	0000	0000	0000	0000	0000	0000
35	0001	0000	0000	0000	0000	0000	0000	0000	0000	0000
30	0001	0000	0000	0000	0000	0000	0000	0000	0000	0000
25	0001	0000	0000	0000	0000	0000	0000	0000	0000	0000
20	0001	0000	0000	0000	0000	0000	0000	0000	0000	0000
15	0001	0000	0000	0000	0000	0000	0000	0000	0000	0000
10	0001	0000	0000	0000	0000	0000	0000	0000	0000	0000
05	0001	0000	0000	0000	0000	0000	0000	0000	0000	0000
-7.00	0001	0000	0000	0000	0000	0000	0000	0000	0000	0000
-6.95	0001	0000	0000	0000	0000	0000	0000	0000	0000	0000
90	0001	0000	0000	0000	0000	0000	0000	0000	0000	0000
85	0001	0000	0000	0000	0000	0000	0000	0000	0000	0000
80	0001	0000	0000	0000	0000	0000	0000	0000	0000	0000
75	0001	0000	0000	0000	0000	0000	0000	0000	0000	0000
70	0001	0000	0000	0000	0000	0000	0000	0000	0000	0000
65	0001	0000	0000	0000	0000	0000	0000	0000	0000	0000
60	0001	0000	0000	0000	0000	0000	0000	0000	0000	0000
55	0001	0000	0000	0000	0000	0000	0000	0000	0000	0000
50	0001	0000	0000	0000	0000	0000	0000	0000	0000	0000
45	0001	0000	0000	0000	0000	0000	0000	0000	0000	0000
40	0001	0000	0000	0000	0000	0000	0000	0000	0000	0000
35	0001	0000	0000	0000	0000	0000	0000	0000	0000	0000
30	0001	0000	0000	0000	0000	0000	0000	0000	0000	0000
25	0001	0000	0000	0000	0000	0000	0000	0000	0000	0000
20	0001	0000	0000	0000	0000	0000	0000	0000	0000	0000
15	0002	0000	0000	0000	0000	0000	0000	0000	0000	0000
10	0002	0000	0000	0000	0000	0000	0000	0000	0000	0000
05	0002	0000	0000	0000	0000	0000	0000	0000	0000	0000

PROBABILITY DENSITY OF t, THE NON-CENTRAL t-STATISTIC, TABLED AS A FUNCTION OF $t/\sqrt{f}$.

f is the number of degrees of freedom; the non-centrality parameter is $\sqrt{f+1}\, K_p$.
K_p is the standardized normal deviate exceeded with probability p.

DEGREES OF FREEDOM **2**

$t/\sqrt{f}$ \ p	.2500	.1500	.1000	.0650	.0400	.0250	.0100	.0040	.0025	.0010
-6.00	0002	0000	0000	0000	0000	0000	0000	0000	0000	0000
-5.95	0002	0000	0000	0000	0000	0000	0000	0000	0000	0000
90	0002	0000	0000	0000	0000	0000	0000	0000	0000	0000
85	0002	0000	0000	0000	0000	0000	0000	0000	0000	0000
80	0002	0000	0000	0000	0000	0000	0000	0000	0000	0000
75	0002	0000	0000	0000	0000	0000	0000	0000	0000	0000
70	0002	0000	0000	0000	0000	0000	0000	0000	0000	0000
65	0002	0000	0000	0000	0000	0000	0000	0000	0000	0000
60	0002	0000	0000	0000	0000	0000	0000	0000	0000	0000
55	0002	0000	0000	0000	0000	0000	0000	0000	0000	0000
50	0002	0000	0000	0000	0000	0000	0000	0000	0000	0000
45	0002	0000	0000	0000	0000	0000	0000	0000	0000	0000
40	0002	0000	0000	0000	0000	0000	0000	0000	0000	0000
35	0002	0000	0000	0000	0000	0000	0000	0000	0000	0000
30	0002	0000	0000	0000	0000	0000	0000	0000	0000	0000
25	0002	0000	0000	0000	0000	0000	0000	0000	0000	0000
20	0003	0001	0000	0000	0000	0000	0000	0000	0000	0000
15	0003	0001	0000	0000	0000	0000	0000	0000	0000	0000
10	0003	0001	0000	0000	0000	0000	0000	0000	0000	0000
05	0003	0001	0000	0000	0000	0000	0000	0000	0000	0000
-5.00	0003	0001	0000	0000	0000	0000	0000	0000	0000	0000
-4.95	0003	0001	0000	0000	0000	0000	0000	0000	0000	0000
90	0003	0001	0000	0000	0000	0000	0000	0000	0000	0000
85	0003	0001	0000	0000	0000	0000	0000	0000	0000	0000
80	0003	0001	0000	0000	0000	0000	0000	0000	0000	0000
75	0003	0001	0000	0000	0000	0000	0000	0000	0000	0000
70	0003	0001	0000	0000	0000	0000	0000	0000	0000	0000
65	0003	0001	0000	0000	0000	0000	0000	0000	0000	0000
60	0004	0001	0000	0000	0000	0000	0000	0000	0000	0000
55	0004	0001	0000	0000	0000	0000	0000	0000	0000	0000
-4.50	0004	0001	0000	0000	0000	0000	0000	0000	0000	0000
45	0004	0001	0000	0000	0000	0000	0000	0000	0000	0000
40	0004	0001	0000	0000	0000	0000	0000	0000	0000	0000
35	0004	0001	0000	0000	0000	0000	0000	0000	0000	0000
30	0004	0001	0000	0000	0000	0000	0000	0000	0000	0000
25	0005	0001	0000	0000	0000	0000	0000	0000	0000	0000
20	0005	0001	0000	0000	0000	0000	0000	0000	0000	0000
15	0005	0001	0000	0000	0000	0000	0000	0000	0000	0000
10	0005	0001	0000	0000	0000	0000	0000	0000	0000	0000
05	0005	0001	0000	0000	0000	0000	0000	0000	0000	0000
-4.00	0005	0001	0000	0000	0000	0000	0000	0000	0000	0000
-3.95	0006	0001	0000	0000	0000	0000	0000	0000	0000	0000
90	0006	0001	0000	0000	0000	0000	0000	0000	0000	0000
85	0006	0001	0000	0000	0000	0000	0000	0000	0000	0000
80	0006	0001	0000	0000	0000	0000	0000	0000	0000	0000
75	0007	0001	0000	0000	0000	0000	0000	0000	0000	0000
70	0007	0001	0000	0000	0000	0000	0000	0000	0000	0000
65	0007	0001	0000	0000	0000	0000	0000	0000	0000	0000
60	0007	0002	0000	0000	0000	0000	0000	0000	0000	0000
55	0008	0002	0000	0000	0000	0000	0000	0000	0000	0000
50	0008	0002	0000	0000	0000	0000	0000	0000	0000	0000
45	0008	0002	0000	0000	0000	0000	0000	0000	0000	0000
40	0009	0002	0001	0000	0000	0000	0000	0000	0000	0000
35	0009	0002	0001	0000	0000	0000	0000	0000	0000	0000
30	0009	0002	0001	0000	0000	0000	0000	0000	0000	0000
25	0010	0002	0001	0000	0000	0000	0000	0000	0000	0000
20	0010	0002	0001	0000	0000	0000	0000	0000	0000	0000
15	0011	0002	0001	0000	0000	0000	0000	0000	0000	0000
10	0011	0002	0001	0000	0000	0000	0000	0000	0000	0000
05	0012	0002	0001	0000	0000	0000	0000	0000	0000	0000

PROBABILITY DENSITY OF t, THE NON-CENTRAL t-STATISTIC, TABLED AS A FUNCTION OF $t/\sqrt{f}$.

f is the number of degrees of freedom; the non-centrality parameter is $\sqrt{f+1}\,K_p$.
K_p is the standardized normal deviate exceeded with probability p.

DEGREES OF FREEDOM **2**

$t/\sqrt{f}$ \ p	.2500	.1500	.1000	.0650	.0400	.0250	.0100	.0040	.0025	.0010
-3.00	0012	0003	0001	0000	0000	0000	0000	0000	0000	0000
-2.95	0013	0003	0001	0000	0000	0000	0000	0000	0000	0000
90	0014	0003	0001	0000	0000	0000	0000	0000	0000	0000
85	0014	0003	0001	0000	0000	0000	0000	0000	0000	0000
80	0015	0003	0001	0000	0000	0000	0000	0000	0000	0000
75	0016	0003	0001	0000	0000	0000	0000	0000	0000	0000
70	0017	0003	0001	0000	0000	0000	0000	0000	0000	0000
65	0017	0004	0001	0000	0000	0000	0000	0000	0000	0000
60	0018	0004	0001	0000	0000	0000	0000	0000	0000	0000
55	0019	0004	0001	0000	0000	0000	0000	0000	0000	0000
50	0021	0004	0001	0000	0000	0000	0000	0000	0000	0000
45	0022	0005	0001	0000	0000	0000	0000	0000	0000	0000
40	0023	0005	0001	0000	0000	0000	0000	0000	0000	0000
35	0024	0005	0001	0000	0000	0000	0000	0000	0000	0000
30	0026	0005	0002	0000	0000	0000	0000	0000	0000	0000
25	0027	0006	0002	0000	0000	0000	0000	0000	0000	0000
20	0029	0006	0002	0000	0000	0000	0000	0000	0000	0000
15	0031	0007	0002	0001	0000	0000	0000	0000	0000	0000
10	0033	0007	0002	0001	0000	0000	0000	0000	0000	0000
05	0035	0007	0002	0001	0000	0000	0000	0000	0000	0000
-2.00	0038	0008	0002	0001	0000	0000	0000	0000	0000	0000
-1.95	0040	0009	0002	0001	0000	0000	0000	0000	0000	0000
90	0043	0009	0003	0001	0000	0000	0000	0000	0000	0000
85	0046	0010	0003	0001	0000	0000	0000	0000	0000	0000
80	0050	0011	0003	0001	0000	0000	0000	0000	0000	0000
75	0053	0011	0003	0001	0000	0000	0000	0000	0000	0000
70	0057	0012	0004	0001	0000	0000	0000	0000	0000	0000
65	0062	0013	0004	0001	0000	0000	0000	0000	0000	0000
60	0067	0014	0004	0001	0000	0000	0000	0000	0000	0000
55	0072	0016	0005	0001	0000	0000	0000	0000	0000	0000
-1.50	0078	0017	0005	0001	0000	0000	0000	0000	0000	0000
45	0085	0019	0006	0002	0000	0000	0000	0000	0000	0000
40	0093	0020	0006	0002	0000	0000	0000	0000	0000	0000
35	0101	0022	0007	0002	0000	0000	0000	0000	0000	0000
30	0110	0024	0007	0002	0000	0000	0000	0000	0000	0000
25	0121	0027	0008	0002	0001	0000	0000	0000	0000	0000
20	0132	0030	0009	0003	0001	0000	0000	0000	0000	0000
15	0145	0033	0010	0003	0001	0000	0000	0000	0000	0000
10	0160	0036	0011	0003	0001	0000	0000	0000	0000	0000
05	0176	0041	0012	0003	0001	0000	0000	0000	0000	0000
-1.00	0195	0045	0014	0004	0001	0000	0000	0000	0000	0000
-0.95	0216	0051	0016	0004	0001	0000	0000	0000	0000	0000
90	0239	0057	0018	0005	0001	0000	0000	0000	0000	0000
85	0266	0064	0020	0006	0001	0000	0000	0000	0000	0000
80	0297	0072	0023	0007	0002	0000	0000	0000	0000	0000
75	0331	0082	0026	0008	0002	0000	0000	0000	0000	0000
70	0371	0093	0030	0009	0002	0001	0000	0000	0000	0000
65	0415	0106	0034	0010	0003	0001	0000	0000	0000	0000
60	0466	0121	0040	0012	0003	0001	0000	0000	0000	0000
55	0524	0139	0046	0014	0004	0001	0000	0000	0000	0000
50	0589	0160	0054	0016	0004	0001	0000	0000	0000	0000
45	0663	0185	0063	0020	0005	0001	0000	0000	0000	0000
40	0745	0214	0074	0023	0006	0002	0000	0000	0000	0000
35	0838	0248	0088	0028	0007	0002	0000	0000	0000	0000
30	0942	0288	0104	0034	0009	0003	0000	0000	0000	0000
25	1057	0335	0124	0041	0011	0003	0000	0000	0000	0000
20	1182	0390	0148	0050	0014	0004	0000	0000	0000	0000
15	1319	0453	0176	0061	0018	0005	0000	0000	0000	0000
10	1467	0527	0211	0075	0022	0007	0001	0000	0000	0000
05	1623	0611	0252	0092	0028	0009	0001	0000	0000	0000

PROBABILITY DENSITY OF t, THE NON-CENTRAL t-STATISTIC, TABLED AS A FUNCTION OF $t/\sqrt{f}$.

f is the number of degrees of freedom; the non-centrality parameter is $\sqrt{f+1}\,K_p$.
K_p is the standardized normal deviate exceeded with probability p.

DEGREES OF FREEDOM **2**

$t/\sqrt{f}$ \ p	.2500	.1500	.1000	.0650	.0400	.0250	.0100	.0040	.0025	.0010
0.00	1787	0706	0301	0114	0036	0011	0001	0000	0000	0000
05	1955	0813	0359	0140	0045	0015	0001	0000	0000	0000
10	2126	0932	0427	0172	0058	0019	0002	0000	0000	0000
15	2295	1062	0505	0212	0074	0026	0003	0000	0000	0000
20	2458	1202	0595	0259	0095	0034	0004	0000	0000	0000
25	2613	1350	0696	0315	0120	0045	0006	0001	0000	0000
30	2755	1506	0808	0382	0152	0059	0008	0001	0000	0000
35	2881	1665	0931	0458	0191	0077	0012	0002	0001	0000
40	2989	1824	1062	0546	0238	0100	0017	0002	0001	0000
45	3076	1981	1201	0643	0294	0130	0023	0004	0001	0000
50	3142	2132	1346	0751	0359	0165	0032	0006	0002	0000
55	3186	2275	1493	0868	0434	0209	0045	0009	0004	0001
60	3208	2405	1639	0991	0518	0260	0060	0013	0005	0001
65	3210	2522	1782	1120	0611	0320	0081	0018	0008	0002
70	3192	2623	1920	1253	0713	0389	0106	0026	0012	0003
75	3157	2707	2049	1386	0821	0466	0137	0036	0018	0004
80	3107	2773	2168	1517	0935	0551	0175	0050	0026	0007
85	3043	2823	2275	1645	1052	0643	0219	0067	0036	0010
90	2969	2855	2368	1766	1170	0741	0271	0089	0049	0015
95	2886	2871	2448	1880	1288	0843	0329	0116	0066	0021
1.00	2795	2872	2513	1984	1404	0948	0394	0148	0087	0030
05	2700	2859	2563	2078	1515	1054	0464	0185	0112	0041
10	2602	2833	2599	2161	1621	1160	0540	0228	0143	0055
15	2501	2797	2622	2232	1720	1264	0621	0277	0178	0073
20	2400	2752	2632	2290	1811	1364	0705	0331	0218	0094
25	2299	2698	2631	2337	1893	1460	0791	0389	0264	0119
30	2200	2639	2619	2373	1966	1550	0878	0452	0314	0148
35	2102	2574	2598	2397	2029	1634	0965	0519	0368	0182
40	2007	2504	2569	2412	2082	1711	1051	0589	0426	0219
45	1914	2432	2532	2416	2126	1779	1135	0660	0488	0261
1.50	1824	2358	2490	2412	2160	1840	1215	0733	0552	0306
55	1738	2283	2442	2400	2185	1893	1292	0806	0618	0355
60	1655	2206	2390	2381	2202	1937	1364	0879	0685	0407
65	1576	2130	2334	2356	2212	1974	1432	0951	0752	0461
70	1500	2055	2276	2325	2213	2003	1494	1020	0820	0517
75	1428	1980	2217	2290	2209	2025	1550	1088	0886	0574
80	1359	1907	2155	2251	2198	2040	1601	1152	0951	0632
85	1294	1835	2094	2208	2182	2049	1645	1213	1013	0691
90	1231	1766	2031	2163	2161	2052	1684	1270	1073	0749
95	1173	1698	1969	2116	2136	2049	1717	1322	1131	0806
2.00	1117	1632	1907	2067	2107	2042	1745	1371	1184	0862
05	1064	1568	1846	2016	2075	2029	1767	1415	1235	0917
10	1013	1506	1786	1965	2041	2013	1784	1455	1282	0969
15	0966	1447	1727	1914	2004	1994	1796	1490	1324	1019
20	0921	1390	1669	1862	1966	1971	1804	1521	1363	1067
25	0878	1335	1612	1811	1926	1945	1807	1548	1398	1112
30	0838	1282	1557	1760	1885	1917	1807	1570	1430	1154
35	0800	1231	1503	1709	1843	1888	1803	1589	1457	1193
40	0764	1183	1451	1659	1800	1856	1796	1603	1480	1228
45	0729	1136	1401	1610	1758	1823	1786	1614	1500	1261
50	0697	1092	1352	1562	1715	1789	1773	1622	1517	1291
55	0666	1049	1305	1514	1672	1754	1758	1626	1530	1317
60	0637	1008	1259	1468	1630	1718	1740	1628	1540	1341
65	0610	0969	1215	1423	1587	1682	1721	1627	1547	1361
70	0584	0932	1173	1379	1546	1646	1700	1623	1551	1379
75	0559	0896	1132	1336	1505	1610	1677	1616	1553	1393
80	0536	0862	1092	1294	1464	1573	1654	1608	1552	1405
85	0513	0829	1054	1254	1424	1537	1629	1597	1549	1415
90	0492	0798	1018	1215	1385	1500	1603	1585	1544	1422
95	0472	0768	0983	1177	1347	1465	1577	1572	1536	1427

PROBABILITY DENSITY OF t, THE NON-CENTRAL t-STATISTIC, TABLED AS A FUNCTION OF $t/\sqrt{f}$.

f is the number of degrees of freedom; the non-centrality parameter is $\sqrt{f+1}\,K_p$.
K_p is the standardized normal deviate exceeded with probability p.

DEGREES OF FREEDOM **2**

$t/\sqrt{f}$ \ p	.2500	.1500	.1000	.0650	.0400	.0250	.0100	.0040	.0025	.0010
3.00	0453	0740	0949	1140	1310	1429	1550	1556	1528	1430
05	0435	0713	0917	1104	1273	1394	1523	1540	1517	1430
10	0418	0687	0885	1069	1237	1360	1495	1522	1505	1429
15	0401	0662	0855	1036	1203	1326	1467	1504	1492	1426
20	0386	0638	0827	1004	1169	1292	1439	1484	1477	1422
25	0371	0615	0799	0973	1136	1260	1411	1464	1462	1416
30	0357	0593	0772	0943	1104	1227	1383	1443	1446	1408
35	0344	0572	0747	0913	1073	1196	1355	1422	1429	1400
40	0331	0552	0722	0885	1042	1165	1327	1400	1411	1390
45	0319	0533	0699	0858	1013	1135	1299	1378	1392	1379
50	0307	0515	0676	0832	0985	1106	1272	1356	1373	1367
55	0296	0497	0654	0807	0957	1078	1245	1334	1354	1354
60	0285	0481	0633	0782	0930	1050	1218	1311	1334	1341
65	0275	0465	0613	0759	0904	1022	1192	1288	1314	1327
70	0266	0449	0593	0736	0879	0996	1166	1265	1294	1312
75	0256	0434	0575	0714	0854	0970	1140	1243	1273	1297
80	0248	0420	0557	0693	0831	0945	1115	1220	1253	1281
85	0239	0407	0540	0673	0808	0921	1090	1198	1232	1265
90	0231	0394	0523	0653	0785	0897	1066	1175	1211	1248
95	0223	0381	0507	0634	0764	0874	1042	1153	1191	1232
4.00	0216	0369	0492	0616	0743	0852	1019	1131	1170	1215
05	0209	0357	0477	0598	0723	0830	0996	1110	1150	1198
10	0202	0346	0463	0581	0703	0809	0974	1088	1129	1180
15	0196	0336	0449	0564	0684	0788	0952	1067	1109	1163
20	0189	0325	0436	0549	0666	0768	0930	1046	1089	1145
25	0183	0316	0423	0533	0648	0749	0909	1026	1069	1128
30	0178	0306	0411	0518	0631	0730	0889	1005	1050	1110
35	0172	0297	0399	0504	0614	0711	0869	0985	1030	1093
40	0167	0288	0387	0490	0598	0694	0849	0966	1011	1076
45	0162	0280	0376	0477	0583	0676	0830	0946	0992	1058
4.50	0157	0272	0366	0464	0568	0659	0811	0927	0974	1041
55	0152	0264	0356	0451	0553	0643	0793	0909	0955	1024
60	0148	0256	0346	0439	0539	0627	0776	0891	0937	1007
65	0143	0249	0336	0427	0525	0612	0758	0873	0920	0990
70	0139	0242	0327	0416	0512	0597	0741	0855	0902	0974
75	0135	0235	0318	0405	0499	0583	0725	0838	0885	0957
80	0131	0229	0310	0395	0486	0569	0709	0821	0868	0941
85	0128	0222	0301	0384	0474	0555	0693	0804	0851	0925
90	0124	0216	0293	0375	0462	0542	0678	0788	0835	0909
95	0121	0211	0286	0365	0451	0529	0663	0772	0819	0893
5.00	0117	0205	0278	0356	0440	0516	0648	0757	0803	0878
05	0114	0199	0271	0347	0429	0504	0634	0742	0788	0862
10	0111	0194	0264	0338	0419	0492	0620	0727	0773	0847
15	0108	0189	0257	0330	0409	0481	0607	0712	0758	0833
20	0105	0184	0251	0321	0399	0470	0594	0698	0743	0818
25	0102	0179	0244	0314	0389	0459	0581	0684	0729	0804
30	0100	0175	0238	0306	0380	0448	0569	0670	0715	0789
35	0097	0170	0232	0299	0371	0438	0556	0657	0702	0776
40	0095	0166	0227	0291	0363	0428	0545	0644	0688	0762
45	0092	0162	0221	0284	0354	0419	0533	0631	0675	0749
50	0090	0158	0216	0278	0346	0409	0522	0619	0662	0735
55	0088	0154	0210	0271	0338	0400	0511	0607	0650	0722
60	0085	0150	0205	0265	0330	0391	0500	0595	0637	0710
65	0083	0147	0201	0259	0323	0383	0490	0583	0625	0697
70	0081	0143	0196	0253	0316	0374	0480	0572	0614	0685
75	0079	0140	0191	0247	0308	0366	0470	0561	0602	0673
80	0077	0136	0187	0241	0302	0358	0460	0550	0591	0661
85	0075	0133	0182	0236	0295	0350	0451	0539	0580	0649
90	0074	0130	0178	0230	0288	0343	0441	0529	0569	0638
95	0072	0127	0174	0225	0282	0335	0432	0519	0558	0627

PROBABILITY DENSITY OF t, THE NON-CENTRAL t-STATISTIC, TABLED AS A FUNCTION OF $t/\sqrt{f}$.

f is the number of degrees of freedom; the non-centrality parameter is $\sqrt{f+1}\,K_p$.
K_p is the standardized normal deviate exceeded with probability p.

DEGREES OF FREEDOM **2**

$t/\sqrt{f}$ \ p	.2500	.1500	.1000	.0650	.0400	.0250	.0100	.0040	.0025	.0010
6.00	0070	0124	0170	0220	0276	0328	0424	0509	0548	0616
05	0069	0121	0166	0215	0270	0321	0415	0499	0538	0605
10	0067	0118	0163	0211	0264	0315	0407	0489	0528	0595
15	0065	0116	0159	0206	0259	0308	0399	0480	0518	0584
20	0064	0113	0156	0202	0253	0302	0391	0471	0509	0574
25	0062	0111	0152	0197	0248	0296	0383	0462	0499	0564
30	0061	0108	0149	0193	0243	0289	0376	0454	0490	0555
35	0060	0106	0146	0189	0238	0284	0368	0445	0481	0545
40	0058	0104	0142	0185	0233	0278	0361	0437	0473	0536
45	0057	0101	0139	0181	0228	0272	0354	0429	0464	0526
50	0056	0099	0136	0177	0223	0267	0347	0421	0456	0517
55	0055	0097	0134	0174	0219	0261	0341	0413	0448	0509
60	0053	0095	0131	0170	0214	0256	0334	0406	0440	0500
65	0052	0093	0128	0166	0210	0251	0328	0398	0432	0491
70	0051	0091	0125	0163	0206	0246	0322	0391	0424	0483
75	0050	0089	0123	0160	0202	0241	0316	0384	0417	0475
80	0049	0087	0120	0157	0198	0237	0310	0377	0409	0467
85	0048	0085	0118	0153	0194	0232	0304	0370	0402	0459
90	0047	0084	0115	0150	0190	0228	0298	0364	0395	0452
95	0046	0082	0113	0147	0186	0223	0293	0357	0388	0444
7.00	0045	0080	0111	0144	0183	0219	0288	0351	0382	0437
05	0044	0079	0109	0142	0179	0215	0282	0345	0375	0429
10	0043	0077	0107	0139	0176	0211	0277	0339	0369	0422
15	0042	0076	0104	0136	0172	0207	0272	0333	0362	0415
20	0042	0074	0102	0134	0169	0203	0267	0327	0356	0409
25	0041	0073	0100	0131	0166	0199	0262	0322	0350	0402
30	0040	0071	0099	0129	0163	0196	0258	0316	0344	0395
35	0039	0070	0097	0126	0160	0192	0253	0311	0338	0389
40	0038	0069	0095	0124	0157	0189	0249	0305	0333	0383
45	0038	0067	0093	0121	0154	0185	0244	0300	0327	0377
7.50	0037	0066	0091	0119	0151	0182	0240	0295	0322	0370
55	0036	0065	0090	0117	0148	0179	0236	0290	0316	0365
60	0036	0063	0088	0115	0146	0175	0232	0285	0311	0359
65	0035	0062	0086	0113	0143	0172	0228	0280	0306	0353
70	0034	0061	0085	0111	0141	0169	0224	0276	0301	0347
75	0034	0060	0083	0109	0138	0166	0220	0271	0296	0342
80	0033	0059	0082	0107	0136	0163	0216	0267	0291	0337
85	0032	0058	0080	0105	0133	0161	0213	0262	0287	0331
90	0032	0057	0079	0103	0131	0158	0209	0258	0282	0326
95	0031	0056	0077	0101	0129	0155	0206	0254	0278	0321
8.00	0031	0055	0076	0099	0126	0152	0202	0250	0273	0316
05	0030	0054	0075	0098	0124	0150	0199	0246	0269	0311
10	0030	0053	0073	0096	0122	0147	0196	0242	0264	0307
15	0029	0052	0072	0094	0120	0145	0192	0238	0260	0302
20	0028	0051	0071	0093	0118	0142	0189	0234	0256	0297
25	0028	0050	0070	0091	0116	0140	0186	0230	0252	0293
30	0027	0049	0068	0090	0114	0138	0183	0227	0248	0288
35	0027	0048	0067	0088	0112	0135	0180	0223	0244	0284
40	0027	0048	0066	0087	0110	0133	0177	0220	0241	0280
45	0026	0047	0065	0085	0108	0131	0175	0216	0237	0275
50	0026	0046	0064	0084	0107	0129	0172	0213	0233	0271
55	0025	0045	0063	0082	0105	0127	0169	0210	0230	0267
60	0025	0044	0062	0081	0103	0125	0166	0207	0226	0263
65	0024	0044	0061	0080	0102	0123	0164	0203	0223	0260
70	0024	0043	0060	0078	0100	0121	0161	0200	0220	0256
75	0024	0042	0059	0077	0098	0119	0159	0197	0216	0252
80	0023	0042	0058	0076	0097	0117	0156	0194	0213	0248
85	0023	0041	0057	0075	0095	0115	0154	0191	0210	0245
90	0022	0040	0056	0073	0094	0113	0152	0188	0207	0241
95	0022	0040	0055	0072	0092	0112	0149	0186	0204	0238

PROBABILITY DENSITY OF t, THE NON-CENTRAL t-STATISTIC, TABLED AS A FUNCTION OF $t/\sqrt{f}$.

f is the number of degrees of freedom; the non-centrality parameter is $\sqrt{f+1}\,K_p$.
K_p is the standardized normal deviate exceeded with probability p.

DEGREES OF FREEDOM **2**

$t/\sqrt{f}$ \ p	.2500	.1500	.1000	.0650	.0400	.0250	.0100	.0040	.0025	.0010
9.00	0022	0039	0054	0071	0091	0110	0147	0183	0201	0234
05	0021	0038	0053	0070	0089	0108	0145	0180	0198	0231
10	0021	0038	0052	0069	0088	0107	0143	0178	0195	0228
15	0021	0037	0052	0068	0087	0105	0140	0175	0192	0224
20	0020	0037	0051	0067	0085	0103	0138	0172	0189	0221
25	0020	0036	0050	0066	0084	0102	0136	0170	0187	0218
30	0020	0035	0049	0065	0083	0100	0134	0167	0184	0215
35	0019	0035	0049	0064	0081	0099	0132	0165	0181	0212
40	0019	0034	0048	0063	0080	0097	0130	0163	0179	0209
45	0019	0034	0047	0062	0079	0096	0128	0160	0176	0206
50	0019	0033	0046	0061	0078	0094	0127	0158	0174	0203
55	0018	0033	0046	0060	0077	0093	0125	0156	0171	0201
60	0018	0032	0045	0059	0076	0092	0123	0154	0169	0198
65	0018	0032	0044	0058	0074	0090	0121	0151	0167	0195
70	0017	0031	0044	0057	0073	0089	0119	0149	0164	0192
75	0017	0031	0043	0057	0072	0088	0118	0147	0162	0190
80	0017	0030	0042	0056	0071	0086	0116	0145	0160	0187
85	0017	0030	0042	0055	0070	0085	0114	0143	0158	0185
90	0016	0030	0041	0054	0069	0084	0113	0141	0155	0182
95	0016	0029	0041	0053	0068	0083	0111	0139	0153	0180
10.00	0016	0029	0040	0053	0067	0082	0110	0137	0151	0178
05	0016	0028	0039	0052	0066	0081	0108	0136	0149	0175
10	0015	0028	0039	0051	0065	0079	0107	0134	0147	0173
15	0015	0027	0038	0050	0064	0078	0105	0132	0145	0171
20	0015	0027	0038	0050	0064	0077	0104	0130	0143	0168
25	0015	0027	0037	0049	0063	0076	0102	0128	0141	0166
30	0015	0026	0037	0048	0062	0075	0101	0127	0140	0164
35	0014	0026	0036	0048	0061	0074	0100	0125	0138	0162
40	0014	0026	0036	0047	0060	0073	0098	0123	0136	0160
45	0014	0025	0035	0046	0059	0072	0097	0122	0134	0158
10.50	0014	0025	0035	0046	0058	0071	0096	0120	0132	0156
55	0014	0025	0034	0045	0058	0070	0094	0119	0131	0154
60	0013	0024	0034	0044	0057	0069	0093	0117	0129	0152
65	0013	0024	0033	0044	0056	0068	0092	0115	0127	0150
70	0013	0024	0033	0043	0055	0067	0091	0114	0126	0148
75	0013	0023	0032	0043	0055	0066	0090	0113	0124	0146
80	0013	0023	0032	0042	0054	0066	0088	0111	0122	0144
85	0013	0023	0031	0042	0053	0065	0087	0110	0121	0142
90	0012	0022	0031	0041	0053	0064	0086	0108	0119	0141
95	0012	0022	0031	0040	0052	0063	0085	0107	0118	0139
11.00	0012	0022	0030	0040	0051	0062	0084	0106	0116	0137
05	0012	0021	0030	0039	0050	0061	0083	0104	0115	0136
10	0012	0021	0029	0039	0050	0061	0082	0103	0114	0134
15	0012	0021	0029	0038	0049	0060	0081	0102	0112	0132
20	0011	0021	0029	0038	0049	0059	0080	0100	0111	0131
25	0011	0020	0028	0037	0048	0058	0079	0099	0109	0129
30	0011	0020	0028	0037	0047	0058	0078	0098	0108	0128
35	0011	0020	0028	0036	0047	0057	0077	0097	0107	0126
40	0011	0020	0027	0036	0046	0056	0076	0096	0105	0124
45	0011	0019	0027	0036	0046	0055	0075	0094	0104	0123
50	0011	0019	0027	0035	0045	0055	0074	0093	0103	0122
55	0010	0019	0026	0035	0044	0054	0073	0092	0102	0120
60	0010	0019	0026	0034	0044	0053	0072	0091	0100	0119
65	0010	0018	0026	0034	0043	0053	0071	0090	0099	0117
70	0010	0018	0025	0033	0043	0052	0070	0089	0098	0116
75	0010	0018	0025	0033	0042	0051	0070	0088	0097	0115
80	0010	0018	0025	0033	0042	0051	0069	0087	0096	0113
85	0010	0017	0024	0032	0041	0050	0068	0086	0095	0112
90	0010	0017	0024	0032	0041	0050	0067	0085	0094	0111
95	0009	0017	0024	0031	0040	0049	0066	0084	0092	0109

PROBABILITY DENSITY OF t, THE NON-CENTRAL t-STATISTIC, TABLED AS A FUNCTION OF $t/\sqrt{f}$.

f is the number of degrees of freedom; the non-centrality parameter is $\sqrt{f+1}\,K_p$.
K_p is the standardized normal deviate exceeded with probability p.

DEGREES OF FREEDOM **2**

$t/\sqrt{f}$ \ p	.2500	.1500	.1000	.0650	.0400	.0250	.0100	.0040	.0025	.0010
12.00	0009	0017	0023	0031	0040	0048	0066	0083	0091	0108
05	0009	0017	0023	0031	0039	0048	0065	0082	0090	0107
10	0009	0016	0023	0030	0039	0047	0064	0081	0089	0106
15	0009	0016	0023	0030	0038	0047	0063	0080	0088	0104
20	0009	0016	0022	0030	0038	0046	0063	0079	0087	0103
25	0009	0016	0022	0029	0037	0046	0062	0078	0086	0102
30	0009	0016	0022	0029	0037	0045	0061	0077	0085	0101
35	0009	0015	0022	0028	0037	0045	0060	0076	0084	0100
40	0008	0015	0021	0028	0036	0044	0060	0075	0083	0099
45	0008	0015	0021	0028	0036	0044	0059	0074	0082	0098
50	0008	0015	0021	0027	0035	0043	0058	0074	0081	0096
55	0008	0015	0021	0027	0035	0043	0058	0073	0081	0095
60	0008	0015	0020	0027	0034	0042	0057	0072	0080	0094
65	0008	0014	0020	0027	0034	0042	0056	0071	0079	0093
70	0008	0014	0020	0026	0034	0041	0056	0070	0078	0092
75	0008	0014	0020	0026	0033	0041	0055	0070	0077	0091
80	0008	0014	0019	0026	0033	0040	0055	0069	0076	0090
85	0008	0014	0019	0025	0033	0040	0054	0068	0075	0089
90	0007	0014	0019	0025	0032	0039	0053	0067	0075	0088
95	0007	0013	0019	0025	0032	0039	0053	0067	0074	0087
13.00	0007	0013	0019	0024	0031	0038	0052	0066	0073	0086
05	0007	0013	0018	0024	0031	0038	0052	0065	0072	0086
10	0007	0013	0018	0024	0031	0038	0051	0064	0071	0085
15	0007	0013	0018	0024	0030	0037	0050	0064	0071	0084
20	0007	0013	0018	0023	0030	0037	0050	0063	0070	0083
25	0007	0013	0018	0023	0030	0036	0049	0062	0069	0082
30	0007	0012	0017	0023	0029	0036	0049	0062	0068	0081
35	0007	0012	0017	0023	0029	0036	0048	0061	0068	0080
40	0007	0012	0017	0022	0029	0035	0048	0060	0067	0079
45	0007	0012	0017	0022	0029	0035	0047	0060	0066	0079
13.50	0007	0012	0017	0022	0028	0034	0047	0059	0065	0078
55	0006	0012	0016	0022	0028	0034	0046	0059	0065	0077
60	0006	0012	0016	0021	0028	0034	0046	0058	0064	0076
65	0006	0011	0016	0021	0027	0033	0045	0057	0063	0075
70	0006	0011	0016	0021	0027	0033	0045	0057	0063	0075
75	0006	0011	0016	0021	0027	0033	0044	0056	0062	0074
80	0006	0011	0016	0021	0026	0032	0044	0056	0062	0073
85	0006	0011	0015	0020	0026	0032	0043	0055	0061	0072
90	0006	0011	0015	0020	0026	0032	0043	0054	0060	0072
95	0006	0011	0015	0020	0026	0031	0043	0054	0060	0071
14.00	0006	0011	0015	0020	0025	0031	0042	0053	0059	0070
05	0006	0011	0015	0020	0025	0031	0042	0053	0058	0069
10	0006	0010	0015	0019	0025	0030	0041	0052	0058	0069
15	0006	0010	0014	0019	0025	0030	0041	0052	0057	0068
20	0006	0010	0014	0019	0024	0030	0040	0051	0057	0067
25	0006	0010	0014	0019	0024	0029	0040	0051	0056	0067
30	0006	0010	0014	0019	0024	0029	0040	0050	0056	0066
35	0005	0010	0014	0018	0024	0029	0039	0050	0055	0065
40	0005	0010	0014	0018	0023	0029	0039	0049	0054	0065
45	0005	0010	0014	0018	0023	0028	0038	0049	0054	0064
50	0005	0010	0013	0018	0023	0028	0038	0048	0053	0064
55	0005	0009	0013	0018	0023	0028	0038	0048	0053	0063
60	0005	0009	0013	0017	0022	0027	0037	0047	0052	0062
65	0005	0009	0013	0017	0022	0027	0037	0047	0052	0062
70	0005	0009	0013	0017	0022	0027	0037	0046	0051	0061
75	0005	0009	0013	0017	0022	0027	0036	0046	0051	0060
80	0005	0009	0013	0017	0022	0026	0036	0045	0050	0060
85	0005	0009	0013	0017	0021	0026	0035	0045	0050	0059
90	0005	0009	0012	0016	0021	0026	0035	0045	0049	0059
95	0005	0009	0012	0016	0021	0026	0035	0044	0049	0058

PROBABILITY DENSITY OF t, THE NON-CENTRAL t-STATISTIC, TABLED AS A FUNCTION OF $t/\sqrt{f}$.

f is the number of degrees of freedom; the non-centrality parameter is $\sqrt{f+1}\,K_p$.
K_p is the standardized normal deviate exceeded with probability p.

DEGREES OF FREEDOM **2**

$t/\sqrt{f}$ \ p	.2500	.1500	.1000	.0650	.0400	.0250	.0100	.0040	.0025	.0010
15.00	0005	0009	0012	0016	0021	0025	0034	0044	0048	0058
05	0005	0009	0012	0016	0021	0025	0034	0043	0048	0057
10	0005	0009	0012	0016	0020	0025	0034	0043	0048	0057
15	0005	0008	0012	0016	0020	0025	0033	0042	0047	0056
20	0005	0008	0012	0015	0020	0024	0033	0042	0047	0056
25	0005	0008	0012	0015	0020	0024	0033	0042	0046	0055
30	0005	0008	0011	0015	0020	0024	0033	0041	0046	0054
35	0004	0008	0011	0015	0019	0024	0032	0041	0045	0054
40	0004	0008	0011	0015	0019	0023	0032	0041	0045	0053
45	0004	0008	0011	0015	0019	0023	0032	0040	0045	0053
50	0004	0008	0011	0015	0019	0023	0031	0040	0044	0053
55	0004	0008	0011	0014	0019	0023	0031	0039	0044	0052
60	0004	0008	0011	0014	0018	0023	0031	0039	0043	0052
65	0004	0008	0011	0014	0018	0022	0030	0039	0043	0051
70	0004	0008	0011	0014	0018	0022	0030	0038	0042	0051
75	0004	0008	0011	0014	0018	0022	0030	0038	0042	0050
80	0004	0007	0010	0014	0018	0022	0030	0038	0042	0050
85	0004	0007	0010	0014	0018	0022	0029	0037	0041	0049
90	0004	0007	0010	0014	0017	0021	0029	0037	0041	0049
95	0004	0007	0010	0013	0017	0021	0029	0037	0041	0048
16.00	0004	0007	0010	0013	0017	0021	0029	0036	0040	0048
05	0004	0007	0010	0013	0017	0021	0028	0036	0040	0048
10	0004	0007	0010	0013	0017	0021	0028	0036	0040	0047
15	0004	0007	0010	0013	0017	0020	0028	0035	0039	0047
20	0004	0007	0010	0013	0017	0020	0028	0035	0039	0046
25	0004	0007	0010	0013	0016	0020	0027	0035	0038	0046
30	0004	0007	0010	0013	0016	0020	0027	0034	0038	0045
35	0004	0007	0009	0012	0016	0020	0027	0034	0038	0045
40	0004	0007	0009	0012	0016	0019	0027	0034	0037	0045
45	0004	0007	0009	0012	0016	0019	0026	0033	0037	0044
16.50	0004	0007	0009	0012	0016	0019	0026	0033	0037	0044
55	0004	0006	0009	0012	0016	0019	0026	0033	0036	0044
60	0004	0006	0009	0012	0015	0019	0026	0033	0036	0043
65	0004	0006	0009	0012	0015	0019	0025	0032	0036	0043
70	0003	0006	0009	0012	0015	0018	0025	0032	0036	0042
75	0003	0006	0009	0012	0015	0018	0025	0032	0035	0042
80	0003	0006	0009	0012	0015	0018	0025	0031	0035	0042
85	0003	0006	0009	0011	0015	0018	0025	0031	0035	0041
90	0003	0006	0009	0011	0015	0018	0024	0031	0034	0041
95	0003	0006	0008	0011	0014	0018	0024	0031	0034	0041
17.00	0003	0006	0008	0011	0014	0018	0024	0030	0034	0040
05	0003	0006	0008	0011	0014	0017	0024	0030	0033	0040
10	0003	0006	0008	0011	0014	0017	0024	0030	0033	0040
15	0003	0006	0008	0011	0014	0017	0023	0030	0033	0039
20	0003	0006	0008	0011	0014	0017	0023	0029	0033	0039
25	0003	0006	0008	0011	0014	0017	0023	0029	0032	0039
30	0003	0006	0008	0011	0014	0017	0023	0029	0032	0038
35	0003	0006	0008	0010	0013	0017	0023	0029	0032	0038
40	0003	0006	0008	0010	0013	0016	0022	0028	0032	0038
45	0003	0006	0008	0010	0013	0016	0022	0028	0031	0037
50	0003	0005	0008	0010	0013	0016	0022	0028	0031	0037
55	0003	0005	0008	0010	0013	0016	0022	0028	0031	0037
60	0003	0005	0008	0010	0013	0016	0022	0027	0031	0036
65	0003	0005	0008	0010	0013	0016	0021	0027	0030	0036
70	0003	0005	0007	0010	0013	0016	0021	0027	0030	0036
75	0003	0005	0007	0010	0013	0015	0021	0027	0030	0036
80	0003	0005	0007	0010	0013	0015	0021	0027	0030	0035
85	0003	0005	0007	0010	0012	0015	0021	0026	0029	0035
90	0003	0005	0007	0010	0012	0015	0021	0026	0029	0035
95	0003	0005	0007	0009	0012	0015	0020	0026	0029	0034

PROBABILITY DENSITY OF t, THE NON-CENTRAL t-STATISTIC, TABLED AS A FUNCTION OF $t/\sqrt{f}$.

f is the number of degrees of freedom; the non-centrality parameter is $\sqrt{f+1}\,K_p$.
K_p is the standardized normal deviate exceeded with probability p. DEGREES OF FREEDOM **2**

$t/\sqrt{f}$ \ p	.2500	.1500	.1000	.0650	.0400	.0250	.0100	.0040	.0025	.0010
18.00	0003	0005	0007	0009	0012	0015	0020	0026	0029	0034
05	0003	0005	0007	0009	0012	0015	0020	0026	0028	0034
10	0003	0005	0007	0009	0012	0015	0020	0025	0028	0034
15	0003	0005	0007	0009	0012	0014	0020	0025	0028	0033
20	0003	0005	0007	0009	0012	0014	0020	0025	0028	0033
25	0003	0005	0007	0009	0012	0014	0019	0025	0027	0033
30	0003	0005	0007	0009	0012	0014	0019	0025	0027	0033
35	0003	0005	0007	0009	0011	0014	0019	0024	0027	0032
40	0003	0005	0007	0009	0011	0014	0019	0024	0027	0032
45	0003	0005	0007	0009	0011	0014	0019	0024	0027	0032
50	0003	0005	0007	0009	0011	0014	0019	0024	0026	0032
55	0003	0005	0006	0009	0011	0014	0019	0024	0026	0031
60	0003	0005	0006	0009	0011	0013	0018	0023	0026	0031
65	0003	0005	0006	0008	0011	0013	0018	0023	0026	0031
70	0002	0005	0006	0008	0011	0013	0018	0023	0026	0031
75	0002	0004	0006	0008	0011	0013	0018	0023	0025	0030
80	0002	0004	0006	0008	0011	0013	0018	0023	0025	0030
85	0002	0004	0006	0008	0011	0013	0018	0022	0025	0030
90	0002	0004	0006	0008	0010	0013	0018	0022	0025	0030
95	0002	0004	0006	0008	0010	0013	0017	0022	0025	0029
19.00	0002	0004	0006	0008	0010	0013	0017	0022	0024	0029
05	0002	0004	0006	0008	0010	0013	0017	0022	0024	0029
10	0002	0004	0006	0008	0010	0012	0017	0022	0024	0029
15	0002	0004	0006	0008	0010	0012	0017	0021	0024	0029
20	0002	0004	0006	0008	0010	0012	0017	0021	0024	0028
25	0002	0004	0006	0008	0010	0012	0017	0021	0024	0028
30	0002	0004	0006	0008	0010	0012	0016	0021	0023	0028
35	0002	0004	0006	0008	0010	0012	0016	0021	0023	0028
40	0002	0004	0006	0008	0010	0012	0016	0021	0023	0027
45	0002	0004	0006	0007	0010	0012	0016	0021	0023	0027
19.50	0002	0004	0006	0007	0010	0012	0016	0020	0023	0027
55	0002	0004	0006	0007	0009	0012	0016	0020	0022	0027
60	0002	0004	0005	0007	0009	0012	0016	0020	0022	0027
65	0002	0004	0005	0007	0009	0011	0016	0020	0022	0026
70	0002	0004	0005	0007	0009	0011	0016	0020	0022	0026
75	0002	0004	0005	0007	0009	0011	0015	0020	0022	0026
80	0002	0004	0005	0007	0009	0011	0015	0019	0022	0026
85	0002	0004	0005	0007	0009	0011	0015	0019	0021	0026
90	0002	0004	0005	0007	0009	0011	0015	0019	0021	0026
95	0002	0004	0005	0007	0009	0011	0015	0019	0021	0025
20.00	0002	0004	0005	0007	0009	0011	0015	0019	0021	0025
05	0002	0004	0005	0007	0009	0011	0015	0019	0021	0025
10	0002	0004	0005	0007	0009	0011	0015	0019	0021	0025
15	0002	0004	0005	0007	0009	0011	0015	0019	0021	0025
20	0002	0004	0005	0007	0009	0011	0014	0018	0020	0024
25	0002	0004	0005	0007	0009	0010	0014	0018	0020	0024
30	0002	0004	0005	0007	0008	0010	0014	0018	0020	0024
35	0002	0004	0005	0007	0008	0010	0014	0018	0020	0024
40	0002	0003	0005	0006	0008	0010	0014	0018	0020	0024
45	0002	0003	0005	0006	0008	0010	0014	0018	0020	0024
50	0002	0003	0005	0006	0008	0010	0014	0018	0020	0023
55	0002	0003	0005	0006	0008	0010	0014	0017	0019	0023
60	0002	0003	0005	0006	0008	0010	0014	0017	0019	0023
65	0002	0003	0005	0006	0008	0010	0014	0017	0019	0023
70	0002	0003	0005	0006	0008	0010	0013	0017	0019	0023
75	0002	0003	0005	0006	0008	0010	0013	0017	0019	0023
80	0002	0003	0005	0006	0008	0010	0013	0017	0019	0022
85	0002	0003	0005	0006	0008	0010	0013	0017	0019	0022
90	0002	0003	0005	0006	0008	0010	0013	0017	0018	0022
95	0002	0003	0005	0006	0008	0009	0013	0017	0018	0022

PROBABILITY DENSITY OF t, THE NON-CENTRAL t-STATISTIC, TABLED AS A FUNCTION OF $t/\sqrt{f}$.

f is the number of degrees of freedom; the non-centrality parameter is $\sqrt{f+1}\,K_p$.
K_p is the standardized normal deviate exceeded with probability p.

DEGREES OF FREEDOM **2**

$t/\sqrt{f}$ \ p	.2500	.1500	.1000	.0650	.0400	.0250	.0100	.0040	.0025	.0010
21.00	0002	0003	0004	0006	0008	0009	0013	0016	0018	0022
05	0002	0003	0004	0006	0008	0009	0013	0016	0018	0022
10	0002	0003	0004	0006	0008	0009	0013	0016	0018	0021
15	0002	0003	0004	0006	0008	0009	0013	0016	0018	0021
20	0002	0003	0004	0006	0007	0009	0012	0016	0018	0021
25	0002	0003	0004	0006	0007	0009	0012	0016	0018	0021
30	0002	0003	0004	0006	0007	0009	0012	0016	0018	0021
35	0002	0003	0004	0006	0007	0009	0012	0016	0017	0021
40	0002	0003	0004	0006	0007	0009	0012	0016	0017	0021
45	0002	0003	0004	0006	0007	0009	0012	0015	0017	0020
50	0002	0003	0004	0006	0007	0009	0012	0015	0017	0020
55	0002	0003	0004	0005	0007	0009	0012	0015	0017	0020
60	0002	0003	0004	0005	0007	0009	0012	0015	0017	0020
65	0002	0003	0004	0005	0007	0009	0012	0015	0017	0020
70	0002	0003	0004	0005	0007	0009	0012	0015	0017	0020
75	0002	0003	0004	0005	0007	0008	0012	0015	0016	0020
80	0002	0003	0004	0005	0007	0008	0012	0015	0016	0020
85	0002	0003	0004	0005	0007	0008	0011	0015	0016	0019
90	0002	0003	0004	0005	0007	0008	0011	0014	0016	0019
95	0002	0003	0004	0005	0007	0008	0011	0014	0016	0019
22.00	0002	0003	0004	0005	0007	0008	0011	0014	0016	0019
05	0002	0003	0004	0005	0007	0008	0011	0014	0016	0019
10	0002	0003	0004	0005	0007	0008	0011	0014	0016	0019
15	0001	0003	0004	0005	0007	0008	0011	0014	0016	0019
20	0001	0003	0004	0005	0006	0008	0011	0014	0015	0019
25	0001	0003	0004	0005	0006	0008	0011	0014	0015	0018
30	0001	0003	0004	0005	0006	0008	0011	0014	0015	0018
35	0001	0003	0004	0005	0006	0008	0011	0014	0015	0018
40	0001	0003	0004	0005	0006	0008	0011	0014	0015	0018
45	0001	0003	0004	0005	0006	0008	0011	0013	0015	0018
22.50	0001	0003	0004	0005	0006	0008	0010	0013	0015	0018
55	0001	0003	0004	0005	0006	0008	0010	0013	0015	0018
60	0001	0003	0004	0005	0006	0008	0010	0013	0015	0018
65	0001	0003	0004	0005	0006	0008	0010	0013	0015	0017
70	0001	0003	0004	0005	0006	0007	0010	0013	0014	0017
75	0001	0003	0004	0005	0006	0007	0010	0013	0014	0017
80	0001	0002	0004	0005	0006	0007	0010	0013	0014	0017
85	0001	0002	0003	0005	0006	0007	0010	0013	0014	0017
90	0001	0002	0003	0005	0006	0007	0010	0013	0014	0017
95	0001	0002	0003	0005	0006	0007	0010	0013	0014	0017
23.00	0001	0002	0003	0005	0006	0007	0010	0013	0014	0017
05	0001	0002	0003	0004	0006	0007	0010	0012	0014	0017
10	0001	0002	0003	0004	0006	0007	0010	0012	0014	0016
15	0001	0002	0003	0004	0006	0007	0010	0012	0014	0016
20	0001	0002	0003	0004	0006	0007	0010	0012	0014	0016
25	0001	0002	0003	0004	0006	0007	0010	0012	0014	0016
30	0001	0002	0003	0004	0006	0007	0009	0012	0013	0016
35	0001	0002	0003	0004	0006	0007	0009	0012	0013	0016
40	0001	0002	0003	0004	0006	0007	0009	0012	0013	0016
45	0001	0002	0003	0004	0006	0007	0009	0012	0013	0016
50	0001	0002	0003	0004	0005	0007	0009	0012	0013	0016
55	0001	0002	0003	0004	0005	0007	0009	0012	0013	0016
60	0001	0002	0003	0004	0005	0007	0009	0012	0013	0015
65	0001	0002	0003	0004	0005	0007	0009	0012	0013	0015
70	0001	0002	0003	0004	0005	0007	0009	0011	0013	0015
75	0001	0002	0003	0004	0005	0007	0009	0011	0013	0015
80	0001	0002	0003	0004	0005	0006	0009	0011	0013	0015
85	0001	0002	0003	0004	0005	0006	0009	0011	0013	0015
90	0001	0002	0003	0004	0005	0006	0009	0011	0012	0015
95	0001	0002	0003	0004	0005	0006	0009	0011	0012	0015

PROBABILITY DENSITY OF t, THE NON-CENTRAL t-STATISTIC, TABLED AS A FUNCTION OF $t/\sqrt{f}$.

f is the number of degrees of freedom; the non-centrality parameter is $\sqrt{f+1}\,K_p$.
K_p is the standardized normal deviate exceeded with probability p.

DEGREES OF FREEDOM **2**

$t/\sqrt{f}$ \ p	.2500	.1500	.1000	.0650	.0400	.0250	.0100	.0040	.0025	.0010
24.00	0001	0002	0003	0004	0005	0006	0009	0011	0012	0015
05	0001	0002	0003	0004	0005	0006	0009	0011	0012	0015
10	0001	0002	0003	0004	0005	0006	0009	0011	0012	0015
15	0001	0002	0003	0004	0005	0006	0008	0011	0012	0014
20	0001	0002	0003	0004	0005	0006	0008	0011	0012	0014
25	0001	0002	0003	0004	0005	0006	0008	0011	0012	0014
30	0001	0002	0003	0004	0005	0006	0008	0011	0012	0014
35	0001	0002	0003	0004	0005	0006	0008	0011	0012	0014
40	0001	0002	0003	0004	0005	0006	0008	0011	0012	0014
45	0001	0002	0003	0004	0005	0006	0008	0010	0012	0014
50	0001	0002	0003	0004	0005	0006	0008	0010	0012	0014
55	0001	0002	0003	0004	0005	0006	0008	0010	0011	0014
60	0001	0002	0003	0004	0005	0006	0008	0010	0011	0014
65	0001	0002	0003	0004	0005	0006	0008	0010	0011	0014
70	0001	0002	0003	0004	0005	0006	0008	0010	0011	0014
75	0001	0002	0003	0004	0005	0006	0008	0010	0011	0013
80	0001	0002	0003	0004	0005	0006	0008	0010	0011	0013
85	0001	0002	0003	0004	0005	0006	0008	0010	0011	0013
90	0001	0002	0003	0004	0005	0006	0008	0010	0011	0013
95	0001	0002	0003	0004	0005	0006	0008	0010	0011	0013
25.00	0001	0002	0003	0004	0005	0006	0008	0010	0011	0013
05	0001	0002	0003	0004	0005	0006	0008	0010	0011	0013
10	0001	0002	0003	0003	0005	0006	0008	0010	0011	0013
15	0001	0002	0003	0003	0004	0005	0008	0010	0011	0013
20	0001	0002	0003	0003	0004	0005	0007	0010	0011	0013
25	0001	0002	0003	0003	0004	0005	0007	0010	0011	0013
30	0001	0002	0003	0003	0004	0005	0007	0009	0011	0013
35	0001	0002	0003	0003	0004	0005	0007	0009	0010	0013
40	0001	0002	0003	0003	0004	0005	0007	0009	0010	0012
45	0001	0002	0003	0003	0004	0005	0007	0009	0010	0012
25.50	0001	0002	0003	0003	0004	0005	0007	0009	0010	0012
55	0001	0002	0002	0003	0004	0005	0007	0009	0010	0012
60	0001	0002	0002	0003	0004	0005	0007	0009	0010	0012
65	0001	0002	0002	0003	0004	0005	0007	0009	0010	0012
70	0001	0002	0002	0003	0004	0005	0007	0009	0010	0012
75	0001	0002	0002	0003	0004	0005	0007	0009	0010	0012
80	0001	0002	0002	0003	0004	0005	0007	0009	0010	0012
85	0001	0002	0002	0003	0004	0005	0007	0009	0010	0012
90	0001	0002	0002	0003	0004	0005	0007	0009	0010	0012
95	0001	0002	0002	0003	0004	0005	0007	0009	0010	0012
00	0001	0002	0002	0003	0004	0005	0007	0009	0010	0012
05	0001	0002	0002	0003	0004	0005	0007	0009	0010	0012
10	0001	0002	0002	0003	0004	0005	0007	0009	0010	0012
15	0001	0002	0002	0003	0004	0005	0007	0009	0010	0011
20	0001	0002	0002	0003	0004	0005	0007	0009	0009	0011
25	0001	0002	0002	0003	0004	0005	0007	0008	0009	0011
30	0001	0002	0002	0003	0004	0005	0007	0008	0009	0011
35	0001	0002	0002	0003	0004	0005	0007	0008	0009	0011
40	0001	0002	0002	0003	0004	0005	0007	0008	0009	0011
45	0001	0002	0002	0003	0004	0005	0006	0008	0009	0011
26.50	0001	0002	0002	0003	0004	0005	0006	0008	0009	0011
55	0001	0002	0002	0003	0004	0005	0006	0008	0009	0011
60	0001	0002	0002	0003	0004	0005	0006	0008	0009	0011
65	0001	0002	0002	0003	0004	0005	0006	0008	0009	0011
70	0001	0002	0002	0003	0004	0005	0006	0008	0009	0011
75	0001	0002	0002	0003	0004	0005	0006	0008	0009	0011
80	0001	0002	0002	0003	0004	0005	0006	0008	0009	0011
85	0001	0002	0002	0003	0004	0005	0006	0008	0009	0011
90	0001	0002	0002	0003	0004	0004	0006	0008	0009	0011
95	0001	0002	0002	0003	0004	0004	0006	0008	0009	0010

PROBABILITY DENSITY OF t, THE NON-CENTRAL t-STATISTIC, TABLED AS A FUNCTION OF $t/\sqrt{f}$.

f is the number of degrees of freedom; the non-centrality parameter is $\sqrt{f+1}\,K_p$.
K_p is the standardized normal deviate exceeded with probability p.

DEGREES OF FREEDOM 2

$t/\sqrt{f}$ \ p	.2500	.1500	.1000	.0650	.0400	.0250	.0100	.0040	.0025	.0010
27.00	0001	0002	0002	0003	0004	0004	0006	0008	0009	0010
05	0001	0001	0002	0003	0004	0004	0006	0008	0009	0010
10	0001	0001	0002	0003	0004	0004	0006	0008	0009	0010
15	0001	0001	0002	0003	0004	0004	0006	0008	0009	0010
20	0001	0001	0002	0003	0004	0004	0006	0008	0008	0010
25	0001	0001	0002	0003	0004	0004	0006	0008	0008	0010
30	0001	0001	0002	0003	0004	0004	0006	0008	0008	0010
35	0001	0001	0002	0003	0003	0004	0006	0008	0008	0010
40	0001	0001	0002	0003	0003	0004	0006	0007	0008	0010
45	0001	0001	0002	0003	0003	0004	0006	0007	0008	0010
50	0001	0001	0002	0003	0003	0004	0006	0007	0008	0010
55	0001	0001	0002	0003	0003	0004	0006	0007	0008	0010
60	0001	0001	0002	0003	0003	0004	0006	0007	0008	0010
65	0001	0001	0002	0003	0003	0004	0006	0007	0008	0010
70	0001	0001	0002	0003	0003	0004	0006	0007	0008	0010
75	0001	0001	0002	0003	0003	0004	0006	0007	0008	0010
80	0001	0001	0002	0003	0003	0004	0006	0007	0008	0010
85	0001	0001	0002	0003	0003	0004	0006	0007	0008	0009
90	0001	0001	0002	0003	0003	0004	0006	0007	0008	0009
95	0001	0001	0002	0003	0003	0004	0006	0007	0008	0009
28.00	0001	0001	0002	0003	0003	0004	0005	0007	0008	0009
05	0001	0001	0002	0003	0003	0004	0005	0007	0008	0009
10	0001	0001	0002	0002	0003	0004	0005	0007	0008	0009
15	0001	0001	0002	0002	0003	0004	0005	0007	0008	0009
20	0001	0001	0002	0002	0003	0004	0005	0007	0008	0009
25	0001	0001	0002	0002	0003	0004	0005	0007	0008	0009
30	0001	0001	0002	0002	0003	0004	0005	0007	0008	0009
35	0001	0001	0002	0002	0003	0004	0005	0007	0008	0009
40	0001	0001	0002	0002	0003	0004	0005	0007	0007	0009
45	0001	0001	0002	0002	0003	0004	0005	0007	0007	0009
28.50	0001	0001	0002	0002	0003	0004	0005	0007	0007	0009
55	0001	0001	0002	0002	0003	0004	0005	0007	0007	0009
60	0001	0001	0002	0002	0003	0004	0005	0007	0007	0009
65	0001	0001	0002	0002	0003	0004	0005	0007	0007	0009
70	0001	0001	0002	0002	0003	0004	0005	0007	0007	0009
75	0001	0001	0002	0002	0003	0004	0005	0006	0007	0009
80	0001	0001	0002	0002	0003	0004	0005	0006	0007	0009
85	0001	0001	0002	0002	0003	0004	0005	0006	0007	0009
90	0001	0001	0002	0002	0003	0004	0005	0006	0007	0009
95	0001	0001	0002	0002	0003	0004	0005	0006	0007	0008
29.00	0001	0001	0002	0002	0003	0004	0005	0006	0007	0008
05	0001	0001	0002	0002	0003	0004	0005	0006	0007	0008
10	0001	0001	0002	0002	0003	0004	0005	0006	0007	0008
15	0001	0001	0002	0002	0003	0004	0005	0006	0007	0008
20	0001	0001	0002	0002	0003	0004	0005	0006	0007	0008
25	0001	0001	0002	0002	0003	0004	0005	0006	0007	0008
30	0001	0001	0002	0002	0003	0003	0005	0006	0007	0008
35	0001	0001	0002	0002	0003	0003	0005	0006	0007	0008
40	0001	0001	0002	0002	0003	0003	0005	0006	0007	0008
45	0001	0001	0002	0002	0003	0003	0005	0006	0007	0008
50	0001	0001	0002	0002	0003	0003	0005	0006	0007	0008
55	0001	0001	0002	0002	0003	0003	0005	0006	0007	0008
60	0001	0001	0002	0002	0003	0003	0005	0006	0007	0008
65	0001	0001	0002	0002	0003	0003	0005	0006	0007	0008
70	0001	0001	0002	0002	0003	0003	0005	0006	0007	0008
75	0001	0001	0002	0002	0003	0003	0005	0006	0007	0008
80	0001	0001	0002	0002	0003	0003	0005	0006	0006	0008
85	0001	0001	0002	0002	0003	0003	0005	0006	0006	0008
90	0001	0001	0002	0002	0003	0003	0005	0006	0006	0008
95	0001	0001	0002	0002	0003	0003	0004	0006	0006	0008

PROBABILITY DENSITY OF t, THE NON-CENTRAL t-STATISTIC, TABLED AS A FUNCTION OF $t/\sqrt{f}$.

f is the number of degrees of freedom; the non-centrality parameter is $\sqrt{f+1}\,K_p$.
K_p is the standardized normal deviate exceeded with probability p.

DEGREES OF FREEDOM **2**

$t/\sqrt{f}$ \ P	.2500	.1500	.1000	.0650	.0400	.0250	.0100	.0040	.0025	.0010
30.00	0001	0001	0002	0002	0003	0003	0004	0006	0006	0008
05	0001	0001	0002	0002	0003	0003	0004	0006	0006	0008
10	0001	0001	0002	0002	0003	0003	0004	0006	0006	0008
15	0001	0001	0002	0002	0003	0003	0004	0006	0006	0008
20	0001	0001	0002	0002	0003	0003	0004	0006	0006	0007
25	0001	0001	0002	0002	0003	0003	0004	0006	0006	0007
30	0001	0001	0001	0002	0003	0003	0004	0006	0006	0007
35	0001	0001	0001	0002	0003	0003	0004	0006	0006	0007
40	0001	0001	0001	0002	0003	0003	0004	0005	0006	0007
45	0001	0001	0001	0002	0003	0003	0004	0005	0006	0007
50	0001	0001	0001	0002	0003	0003	0004	0005	0006	0007
55	0001	0001	0001	0002	0003	0003	0004	0005	0006	0007
60	0001	0001	0001	0002	0002	0003	0004	0005	0006	0007
65	0001	0001	0001	0002	0002	0003	0004	0005	0006	0007
70	0001	0001	0001	0002	0002	0003	0004	0005	0006	0007
75	0001	0001	0001	0002	0002	0003	0004	0005	0006	0007
80	0001	0001	0001	0002	0002	0003	0004	0005	0006	0007
85	0001	0001	0001	0002	0002	0003	0004	0005	0006	0007
90	0001	0001	0001	0002	0002	0003	0004	0005	0006	0007
95	0001	0001	0001	0002	0002	0003	0004	0005	0006	0007
31.00	0001	0001	0001	0002	0002	0003	0004	0005	0006	0007
05	0001	0001	0001	0002	0002	0003	0004	0005	0006	0007
10	0001	0001	0001	0002	0002	0003	0004	0005	0006	0007
15	0001	0001	0001	0002	0002	0003	0004	0005	0006	0007
20	0001	0001	0001	0002	0002	0003	0004	0005	0006	0007
25	0001	0001	0001	0002	0002	0003	0004	0005	0006	0007
30	0001	0001	0001	0002	0002	0003	0004	0005	0006	0007
35	0001	0001	0001	0002	0002	0003	0004	0005	0006	0007
40	0001	0001	0001	0002	0002	0003	0004	0005	0006	0007
45	0001	0001	0001	0002	0002	0003	0004	0005	0006	0007
31.50	0001	0001	0001	0002	0002	0003	0004	0005	0005	0007
55	0001	0001	0001	0002	0002	0003	0004	0005	0005	0007
60	0001	0001	0001	0002	0002	0003	0004	0005	0005	0007
65	0001	0001	0001	0002	0002	0003	0004	0005	0005	0007
70	0001	0001	0001	0002	0002	0003	0004	0005	0005	0006
75	0001	0001	0001	0002	0002	0003	0004	0005	0005	0006
80	0001	0001	0001	0002	0002	0003	0004	0005	0005	0006
85	0001	0001	0001	0002	0002	0003	0004	0005	0005	0006
90	0001	0001	0001	0002	0002	0003	0004	0005	0005	0006
95	0001	0001	0001	0002	0002	0003	0004	0005	0005	0006
32.00	0000	0001	0001	0002	0002	0003	0004	0005	0005	0006
05	0000	0001	0001	0002	0002	0003	0004	0005	0005	0006
10	0000	0001	0001	0002	0002	0003	0004	0005	0005	0006
15	0000	0001	0001	0002	0002	0003	0004	0005	0005	0006
20	0000	0001	0001	0002	0002	0003	0004	0005	0005	0006
25	0000	0001	0001	0002	0002	0003	0004	0005	0005	0006
30	0000	0001	0001	0002	0002	0003	0004	0005	0005	0006
35	0000	0001	0001	0002	0002	0003	0004	0005	0005	0006
40	0000	0001	0001	0002	0002	0003	0004	0005	0005	0006
45	0000	0001	0001	0002	0002	0003	0004	0005	0005	0006
50	0000	0001	0001	0002	0002	0003	0004	0004	0005	0006
55	0000	0001	0001	0002	0002	0003	0003	0004	0005	0006
60	0000	0001	0001	0002	0002	0003	0003	0004	0005	0006
65	0000	0001	0001	0002	0002	0003	0003	0004	0005	0006
70	0000	0001	0001	0002	0002	0003	0003	0004	0005	0006
75	0000	0001	0001	0002	0002	0003	0003	0004	0005	0006
80	0000	0001	0001	0002	0002	0002	0003	0004	0005	0006
85	0000	0001	0001	0002	0002	0002	0003	0004	0005	0006
90	0000	0001	0001	0002	0002	0002	0003	0004	0005	0006
95	0000	0001	0001	0002	0002	0002	0003	0004	0005	0006

PROBABILITY DENSITY OF t, THE NON-CENTRAL t-STATISTIC, TABLED AS A FUNCTION OF $t/\sqrt{f}$.

f is the number of degrees of freedom; the non-centrality parameter is $\sqrt{f+1}\,K_p$.
K_p is the standardized normal deviate exceeded with probability p.

DEGREES OF FREEDOM **2**

$t/\sqrt{f}$ \ p	.2500	.1500	.1000	.0650	.0400	.0250	.0100	.0040	.0025	.0010
33.00	0000	0001	0001	0002	0002	0002	0003	0004	0005	0006
05	0000	0001	0001	0002	0002	0002	0003	0004	0005	0006
10	0000	0001	0001	0002	0002	0002	0003	0004	0005	0006
15	0000	0001	0001	0002	0002	0002	0003	0004	0005	0006
20	0000	0001	0001	0002	0002	0002	0003	0004	0005	0006
25	0000	0001	0001	0002	0002	0002	0003	0004	0005	0006
30	0000	0001	0001	0002	0002	0002	0003	0004	0005	0006
35	0000	0001	0001	0001	0002	0002	0003	0004	0005	0006
40	0000	0001	0001	0001	0002	0002	0003	0004	0005	0006
45	0000	0001	0001	0001	0002	0002	0003	0004	0005	0006
50	0000	0001	0001	0001	0002	0002	0003	0004	0005	0005
55	0000	0001	0001	0001	0002	0002	0003	0004	0005	0005
60	0000	0001	0001	0001	0002	0002	0003	0004	0005	0005
65	0000	0001	0001	0001	0002	0002	0003	0004	0005	0005
70	0000	0001	0001	0001	0002	0002	0003	0004	0004	0005
75	0000	0001	0001	0001	0002	0002	0003	0004	0004	0005
80	0000	0001	0001	0001	0002	0002	0003	0004	0004	0005
85	0000	0001	0001	0001	0002	0002	0003	0004	0004	0005
90	0000	0001	0001	0001	0002	0002	0003	0004	0004	0005
95	0000	0001	0001	0001	0002	0002	0003	0004	0004	0005
34.00	0000	0001	0001	0001	0002	0002	0003	0004	0004	0005
05	0000	0001	0001	0001	0002	0002	0003	0004	0004	0005
10	0000	0001	0001	0001	0002	0002	0003	0004	0004	0005
15	0000	0001	0001	0001	0002	0002	0003	0004	0004	0005
20	0000	0001	0001	0001	0002	0002	0003	0004	0004	0005
25	0000	0001	0001	0001	0002	0002	0003	0004	0004	0005
30	0000	0001	0001	0001	0002	0002	0003	0004	0004	0005
35	0000	0001	0001	0001	0002	0002	0003	0004	0004	0005
40	0000	0001	0001	0001	0002	0002	0003	0004	0004	0005
45	0000	0001	0001	0001	0002	0002	0003	0004	0004	0005
34.50	0000	0001	0001	0001	0002	0002	0003	0004	0004	0005
55	0000	0001	0001	0001	0002	0002	0003	0004	0004	0005
60	0000	0001	0001	0001	0002	0002	0003	0004	0004	0005
65	0000	0001	0001	0001	0002	0002	0003	0004	0004	0005
70	0000	0001	0001	0001	0002	0002	0003	0004	0004	0005
75	0000	0001	0001	0001	0002	0002	0003	0004	0004	0005
80	0000	0001	0001	0001	0002	0002	0003	0004	0004	0005
85	0000	0001	0001	0001	0002	0002	0003	0004	0004	0005
90	0000	0001	0001	0001	0002	0002	0003	0004	0004	0005
95	0000	0001	0001	0001	0002	0002	0003	0004	0004	0005
35.00	0000	0001	0001	0001	0002	0002	0003	0004	0004	0005
05	0000	0001	0001	0001	0002	0002	0003	0004	0004	0005
10	0000	0001	0001	0001	0002	0002	0003	0004	0004	0005
15	0000	0001	0001	0001	0002	0002	0003	0004	0004	0005
20	0000	0001	0001	0001	0002	0002	0003	0004	0004	0005
25	0000	0001	0001	0001	0002	0002	0003	0004	0004	0005
30	0000	0001	0001	0001	0002	0002	0003	0004	0004	0005
35	0000	0001	0001	0001	0002	0002	0003	0004	0004	0005
40	0000	0001	0001	0001	0002	0002	0003	0003	0004	0005
45	0000	0001	0001	0001	0002	0002	0003	0003	0004	0005
50	0000	0001	0001	0001	0002	0002	0003	0003	0004	0005
55	0000	0001	0001	0001	0002	0002	0003	0003	0004	0005
60	0000	0001	0001	0001	0002	0002	0003	0003	0004	0005
65	0000	0001	0001	0001	0002	0002	0003	0003	0004	0005
70	0000	0001	0001	0001	0002	0002	0003	0003	0004	0005
75	0000	0001	0001	0001	0002	0002	0003	0003	0004	0005
80	0000	0001	0001	0001	0002	0002	0003	0003	0004	0005
85	0000	0001	0001	0001	0002	0002	0003	0003	0004	0004
90	0000	0001	0001	0001	0002	0002	0003	0003	0004	0004
95	0000	0001	0001	0001	0002	0002	0003	0003	0004	0004

PROBABILITY DENSITY OF t, THE NON-CENTRAL t-STATISTIC, TABLED AS A FUNCTION OF $t/\sqrt{f}$.

f is the number of degrees of freedom; the non-centrality parameter is $\sqrt{f+1}\,K_p$.
K_p is the standardized normal deviate exceeded with probability p.

DEGREES OF FREEDOM **2**

$t/\sqrt{f}$ \ P	.2500	.1500	.1000	.0650	.0400	.0250	.0100	.0040	.0025	.0010
36.00	0000	0001	0001	0001	0002	0002	0003	0003	0004	0004
05	0000	0001	0001	0001	0002	0002	0003	0003	0004	0004
10	0000	0001	0001	0001	0002	0002	0003	0003	0004	0004
15	0000	0001	0001	0001	0002	0002	0003	0003	0004	0004
20	0000	0001	0001	0001	0002	0002	0003	0003	0004	0004
25	0000	0001	0001	0001	0002	0002	0003	0003	0004	0004
30	0000	0001	0001	0001	0001	0002	0003	0003	0004	0004
35	0000	0001	0001	0001	0001	0002	0003	0003	0004	0004
40	0000	0001	0001	0001	0001	0002	0003	0003	0004	0004
45	0000	0001	0001	0001	0001	0002	0002	0003	0004	0004
50	0000	0001	0001	0001	0001	0002	0002	0003	0004	0004
55	0000	0001	0001	0001	0001	0002	0002	0003	0004	0004
60	0000	0001	0001	0001	0001	0002	0002	0003	0004	0004
65	0000	0001	0001	0001	0001	0002	0002	0003	0004	0004
70	0000	0001	0001	0001	0001	0002	0002	0003	0003	0004
75	0000	0001	0001	0001	0001	0002	0002	0003	0003	0004
80	0000	0001	0001	0001	0001	0002	0002	0003	0003	0004
85	0000	0001	0001	0001	0001	0002	0002	0003	0003	0004
90	0000	0001	0001	0001	0001	0002	0002	0003	0003	0004
95	0000	0001	0001	0001	0001	0002	0002	0003	0003	0004
37.00	0000	0001	0001	0001	0001	0002	0002	0003	0003	0004
05	0000	0001	0001	0001	0001	0002	0002	0003	0003	0004
10	0000	0001	0001	0001	0001	0002	0002	0003	0003	0004
15	0000	0001	0001	0001	0001	0002	0002	0003	0003	0004
20	0000	0001	0001	0001	0001	0002	0002	0003	0003	0004
25	0000	0001	0001	0001	0001	0002	0002	0003	0003	0004
30	0000	0001	0001	0001	0001	0002	0002	0003	0003	0004
35	0000	0001	0001	0001	0001	0002	0002	0003	0003	0004
40	0000	0001	0001	0001	0001	0002	0002	0003	0003	0004
45	0000	0001	0001	0001	0001	0002	0002	0003	0003	0004
37.50	0000	0001	0001	0001	0001	0002	0002	0003	0003	0004
55	0000	0001	0001	0001	0001	0002	0002	0003	0003	0004
60	0000	0001	0001	0001	0001	0002	0002	0003	0003	0004
65	0000	0001	0001	0001	0001	0002	0002	0003	0003	0004
70	0000	0001	0001	0001	0001	0002	0002	0003	0003	0004
75	0000	0001	0001	0001	0001	0002	0002	0003	0003	0004
80	0000	0001	0001	0001	0001	0002	0002	0003	0003	0004
85	0000	0001	0001	0001	0001	0002	0002	0003	0003	0004
90	0000	0001	0001	0001	0001	0002	0002	0003	0003	0004
95	0000	0001	0001	0001	0001	0002	0002	0003	0003	0004
38.00	0000	0001	0001	0001	0001	0002	0002	0003	0003	0004
05	0000	0001	0001	0001	0001	0002	0002	0003	0003	0004
10	0000	0001	0001	0001	0001	0002	0002	0003	0003	0004
15	0000	0001	0001	0001	0001	0002	0002	0003	0003	0004
20	0000	0001	0001	0001	0001	0002	0002	0003	0003	0004
25	0000	0001	0001	0001	0001	0002	0002	0003	0003	0004
30	0000	0001	0001	0001	0001	0002	0002	0003	0003	0004
35	0000	0001	0001	0001	0001	0002	0002	0003	0003	0004
40	0000	0001	0001	0001	0001	0002	0002	0003	0003	0004
45	0000	0001	0001	0001	0001	0002	0002	0003	0003	0004
50	0000	0001	0001	0001	0001	0002	0002	0003	0003	0004
55	0000	0001	0001	0001	0001	0002	0002	0003	0003	0004
60	0000	0001	0001	0001	0001	0002	0002	0003	0003	0004
65	0000	0001	0001	0001	0001	0002	0002	0003	0003	0004
70	0000	0001	0001	0001	0001	0002	0002	0003	0003	0004
75	0000	0001	0001	0001	0001	0002	0002	0003	0003	0004
80	0000	0001	0001	0001	0001	0002	0002	0003	0003	0004
85	0000	0001	0001	0001	0001	0002	0002	0003	0003	0004
90	0000	0001	0001	0001	0001	0001	0002	0003	0003	0004
95	0000	0001	0001	0001	0001	0001	0002	0003	0003	0004

PROBABILITY DENSITY OF t, THE NON-CENTRAL t-STATISTIC, TABLED AS A FUNCTION OF $t/\sqrt{f}$.

f is the number of degrees of freedom; the non-centrality parameter is $\sqrt{f+1}\,K_p$.
K_p is the standardized normal deviate exceeded with probability p.

DEGREES OF FREEDOM **2**

$t/\sqrt{f}$ \ p	.2500	.1500	.1000	.0650	.0400	.0250	.0100	.0040	.0025	.0010
39.00	0000	0001	0001	0001	0001	0001	0002	0003	0003	0003
05	0000	0000	0001	0001	0001	0001	0002	0003	0003	0003
10	0000	0000	0001	0001	0001	0001	0002	0003	0003	0003
15	0000	0000	0001	0001	0001	0001	0002	0003	0003	0003
20	0000	0000	0001	0001	0001	0001	0002	0003	0003	0003
25	0000	0000	0001	0001	0001	0001	0002	0003	0003	0003
30	0000	0000	0001	0001	0001	0001	0002	0003	0003	0003
35	0000	0000	0001	0001	0001	0001	0002	0003	0003	0003
40	0000	0000	0001	0001	0001	0001	0002	0003	0003	0003
45	0000	0000	0001	0001	0001	0001	0002	0003	0003	0003
50	0000	0000	0001	0001	0001	0001	0002	0003	0003	0003
55	0000	0000	0001	0001	0001	0001	0002	0003	0003	0003
60	0000	0000	0001	0001	0001	0001	0002	0002	0003	0003
65	0000	0000	0001	0001	0001	0001	0002	0002	0003	0003
70	0000	0000	0001	0001	0001	0001	0002	0002	0003	0003
75	0000	0000	0001	0001	0001	0001	0002	0002	0003	0003
80	0000	0000	0001	0001	0001	0001	0002	0002	0003	0003
85	0000	0000	0001	0001	0001	0001	0002	0002	0003	0003
90	0000	0000	0001	0001	0001	0001	0002	0002	0003	0003
95	0000	0000	0001	0001	0001	0001	0002	0002	0003	0003
40.00	0000	0000	0001	0001	0001	0001	0002	0002	0003	0003
05	0000	0000	0001	0001	0001	0001	0002	0002	0003	0003
10	0000	0000	0001	0001	0001	0001	0002	0002	0003	0003
15	0000	0000	0001	0001	0001	0001	0002	0002	0003	0003
20	0000	0000	0001	0001	0001	0001	0002	0002	0003	0003
25	0000	0000	0001	0001	0001	0001	0002	0002	0003	0003
30	0000	0000	0001	0001	0001	0001	0002	0002	0003	0003
35	0000	0000	0001	0001	0001	0001	0002	0002	0003	0003
40	0000	0000	0001	0001	0001	0001	0002	0002	0003	0003
45	0000	0000	0001	0001	0001	0001	0002	0002	0003	0003
40.50	0000	0000	0001	0001	0001	0001	0002	0002	0003	0003
55	0000	0000	0001	0001	0001	0001	0002	0002	0003	0003
60	0000	0000	0001	0001	0001	0001	0002	0002	0003	0003
65	0000	0000	0001	0001	0001	0001	0002	0002	0003	0003
70	0000	0000	0001	0001	0001	0001	0002	0002	0003	0003
75	0000	0000	0001	0001	0001	0001	0002	0002	0003	0003
80	0000	0000	0001	0001	0001	0001	0002	0002	0003	0003
85	0000	0000	0001	0001	0001	0001	0002	0002	0003	0003
90	0000	0000	0001	0001	0001	0001	0002	0002	0003	0003
95	0000	0000	0001	0001	0001	0001	0002	0002	0003	0003
41.00	0000	0000	0001	0001	0001	0001	0002	0002	0003	0003
05	0000	0000	0001	0001	0001	0001	0002	0002	0002	0003
10	0000	0000	0001	0001	0001	0001	0002	0002	0002	0003
15	0000	0000	0001	0001	0001	0001	0002	0002	0002	0003
20	0000	0000	0001	0001	0001	0001	0002	0002	0002	0003
25	0000	0000	0001	0001	0001	0001	0002	0002	0002	0003
30	0000	0000	0001	0001	0001	0001	0002	0002	0002	0003
35	0000	0000	0001	0001	0001	0001	0002	0002	0002	0003
40	0000	0000	0001	0001	0001	0001	0002	0002	0002	0003
45	0000	0000	0001	0001	0001	0001	0002	0002	0002	0003
50	0000	0000	0001	0001	0001	0001	0002	0002	0002	0003
55	0000	0000	0001	0001	0001	0001	0002	0002	0002	0003
60	0000	0000	0001	0001	0001	0001	0002	0002	0002	0003
65	0000	0000	0001	0001	0001	0001	0002	0002	0002	0003
70	0000	0000	0001	0001	0001	0001	0002	0002	0002	0003
75	0000	0000	0001	0001	0001	0001	0002	0002	0002	0003
80	0000	0000	0001	0001	0001	0001	0002	0002	0002	0003
85	0000	0000	0001	0001	0001	0001	0002	0002	0002	0003
90	0000	0000	0001	0001	0001	0001	0002	0002	0002	0003
95	0000	0000	0001	0001	0001	0001	0002	0002	0002	0003

PROBABILITY DENSITY OF t, THE NON-CENTRAL t-STATISTIC, TABLED AS A FUNCTION OF $t/\sqrt{f}$.

f is the number of degrees of freedom; the non-centrality parameter is $\sqrt{f+1}\,K_p$.
K_p is the standardized normal deviate exceeded with probability p.

DEGREES OF FREEDOM **2**

$t/\sqrt{f}$ \ p	.2500	.1500	.1000	.0650	.0400	.0250	.0100	.0040	.0025	.0010
42.00	0000	0000	0001	0001	0001	0001	0002	0002	0002	0003
05	0000	0000	0001	0001	0001	0001	0002	0002	0002	0003
10	0000	0000	0001	0001	0001	0001	0002	0002	0002	0003
15	0000	0000	0001	0001	0001	0001	0002	0002	0002	0003
20	0000	0000	0001	0001	0001	0001	0002	0002	0002	0003
25	0000	0000	0001	0001	0001	0001	0002	0002	0002	0003
30	0000	0000	0001	0001	0001	0001	0002	0002	0002	0003
35	0000	0000	0001	0001	0001	0001	0002	0002	0002	0003
40	0000	0000	0001	0001	0001	0001	0002	0002	0002	0003
45	0000	0000	0001	0001	0001	0001	0002	0002	0002	0003
50	0000	0000	0001	0001	0001	0001	0002	0002	0002	0003
55	0000	0000	0001	0001	0001	0001	0002	0002	0002	0003
60	0000	0000	0001	0001	0001	0001	0002	0002	0002	0003
65	0000	0000	0001	0001	0001	0001	0002	0002	0002	0003
70	0000	0000	0001	0001	0001	0001	0002	0002	0002	0003
75	0000	0000	0001	0001	0001	0001	0002	0002	0002	0003
80	0000	0000	0001	0001	0001	0001	0002	0002	0002	0003
85	0000	0000	0001	0001	0001	0001	0002	0002	0002	0003
90	0000	0000	0001	0001	0001	0001	0002	0002	0002	0003
95	0000	0000	0001	0001	0001	0001	0002	0002	0002	0003
43.00	0000	0000	0001	0001	0001	0001	0002	0002	0002	0003
05	0000	0000	0001	0001	0001	0001	0002	0002	0002	0003
10	0000	0000	0001	0001	0001	0001	0002	0002	0002	0003
15	0000	0000	0001	0001	0001	0001	0002	0002	0002	0003
20	0000	0000	0001	0001	0001	0001	0002	0002	0002	0003
25	0000	0000	0001	0001	0001	0001	0001	0002	0002	0003
30	0000	0000	0001	0001	0001	0001	0001	0002	0002	0003
35	0000	0000	0001	0001	0001	0001	0001	0002	0002	0003
40	0000	0000	0001	0001	0001	0001	0001	0002	0002	0003
45	0000	0000	0001	0001	0001	0001	0001	0002	0002	0003
43.50	0000	0000	0001	0001	0001	0001	0001	0002	0002	0003
55	0000	0000	0001	0001	0001	0001	0001	0002	0002	0003
60	0000	0000	0001	0001	0001	0001	0001	0002	0002	0003
65	0000	0000	0001	0001	0001	0001	0001	0002	0002	0002
70	0000	0000	0001	0001	0001	0001	0001	0002	0002	0002
75	0000	0000	0000	0001	0001	0001	0001	0002	0002	0002
80	0000	0000	0000	0001	0001	0001	0001	0002	0002	0002
85	0000	0000	0000	0001	0001	0001	0001	0002	0002	0002
90	0000	0000	0000	0001	0001	0001	0001	0002	0002	0002
95	0000	0000	0000	0001	0001	0001	0001	0002	0002	0002
44.00	0000	0000	0000	0001	0001	0001	0001	0002	0002	0002
05	0000	0000	0000	0001	0001	0001	0001	0002	0002	0002
10	0000	0000	0000	0001	0001	0001	0001	0002	0002	0002
15	0000	0000	0000	0001	0001	0001	0001	0002	0002	0002
20	0000	0000	0000	0001	0001	0001	0001	0002	0002	0002
25	0000	0000	0000	0001	0001	0001	0001	0002	0002	0002
30	0000	0000	0000	0001	0001	0001	0001	0002	0002	0002
35	0000	0000	0000	0001	0001	0001	0001	0002	0002	0002
40	0000	0000	0000	0001	0001	0001	0001	0002	0002	0002
45	0000	0000	0000	0001	0001	0001	0001	0002	0002	0002
50	0000	0000	0000	0001	0001	0001	0001	0002	0002	0002
55	0000	0000	0000	0001	0001	0001	0001	0002	0002	0002
60	0000	0000	0000	0001	0001	0001	0001	0002	0002	0002
65	0000	0000	0000	0001	0001	0001	0001	0002	0002	0002
70	0000	0000	0000	0001	0001	0001	0001	0002	0002	0002
75	0000	0000	0000	0001	0001	0001	0001	0002	0002	0002
80	0000	0000	0000	0001	0001	0001	0001	0002	0002	0002
85	0000	0000	0000	0001	0001	0001	0001	0002	0002	0002
90	0000	0000	0000	0001	0001	0001	0001	0002	0002	0002
95	0000	0000	0000	0001	0001	0001	0001	0002	0002	0002

PROBABILITY DENSITY OF t, THE NON-CENTRAL t-STATISTIC, TABLED AS A FUNCTION OF $t/\sqrt{f}$.

f is the number of degrees of freedom; the non-centrality parameter is $\sqrt{f+1}\,K_p$.
K_p is the standardized normal deviate exceeded with probability p.

DEGREES OF FREEDOM **2**

$t/\sqrt{f}$ \ p	.2500	.1500	.1000	.0650	.0400	.0250	.0100	.0040	.0025	.0010
45.00	0000	0000	0000	0001	0001	0001	0001	0002	0002	0002
05	0000	0000	0000	0001	0001	0001	0001	0002	0002	0002
10	0000	0000	0000	0001	0001	0001	0001	0002	0002	0002
15	0000	0000	0000	0001	0001	0001	0001	0002	0002	0002
20	0000	0000	0000	0001	0001	0001	0001	0002	0002	0002
25	0000	0000	0000	0001	0001	0001	0001	0002	0002	0002
30	0000	0000	0000	0001	0001	0001	0001	0002	0002	0002
35	0000	0000	0000	0001	0001	0001	0001	0002	0002	0002
40	0000	0000	0000	0001	0001	0001	0001	0002	0002	0002
45	0000	0000	0000	0001	0001	0001	0001	0002	0002	0002
50	0000	0000	0000	0001	0001	0001	0001	0002	0002	0002
55	0000	0000	0000	0001	0001	0001	0001	0002	0002	0002
60	0000	0000	0000	0001	0001	0001	0001	0002	0002	0002
65	0000	0000	0000	0001	0001	0001	0001	0002	0002	0002
70	0000	0000	0000	0001	0001	0001	0001	0002	0002	0002
75	0000	0000	0000	0001	0001	0001	0001	0002	0002	0002
80	0000	0000	0000	0001	0001	0001	0001	0002	0002	0002
85	0000	0000	0000	0001	0001	0001	0001	0002	0002	0002
90	0000	0000	0000	0001	0001	0001	0001	0002	0002	0002
95	0000	0000	0000	0001	0001	0001	0001	0002	0002	0002
46.00	0000	0000	0000	0001	0001	0001	0001	0002	0002	0002
05	0000	0000	0000	0001	0001	0001	0001	0002	0002	0002
10	0000	0000	0000	0001	0001	0001	0001	0002	0002	0002
15	0000	0000	0000	0001	0001	0001	0001	0002	0002	0002
20	0000	0000	0000	0001	0001	0001	0001	0002	0002	0002
25	0000	0000	0000	0001	0001	0001	0001	0002	0002	0002
30	0000	0000	0000	0001	0001	0001	0001	0002	0002	0002
35	0000	0000	0000	0001	0001	0001	0001	0002	0002	0002
40	0000	0000	0000	0001	0001	0001	0001	0002	0002	0002
45	0000	0000	0000	0001	0001	0001	0001	0002	0002	0002
46.50	0000	0000	0000	0001	0001	0001	0001	0002	0002	0002
55	0000	0000	0000	0001	0001	0001	0001	0002	0002	0002
60	0000	0000	0000	0001	0001	0001	0001	0002	0002	0002
65	0000	0000	0000	0001	0001	0001	0001	0002	0002	0002
70	0000	0000	0000	0001	0001	0001	0001	0002	0002	0002
75	0000	0000	0000	0001	0001	0001	0001	0002	0002	0002
80	0000	0000	0000	0001	0001	0001	0001	0002	0002	0002
85	0000	0000	0000	0001	0001	0001	0001	0002	0002	0002
90	0000	0000	0000	0001	0001	0001	0001	0002	0002	0002
95	0000	0000	0000	0001	0001	0001	0001	0002	0002	0002
47.00	0000	0000	0000	0001	0001	0001	0001	0001	0002	0002
05	0000	0000	0000	0001	0001	0001	0001	0001	0002	0002
10	0000	0000	0000	0001	0001	0001	0001	0001	0002	0002
15	0000	0000	0000	0001	0001	0001	0001	0001	0002	0002
20	0000	0000	0000	0001	0001	0001	0001	0001	0002	0002
25	0000	0000	0000	0001	0001	0001	0001	0001	0002	0002
30	0000	0000	0000	0001	0001	0001	0001	0001	0002	0002
35	0000	0000	0000	0001	0001	0001	0001	0001	0002	0002
40	0000	0000	0000	0001	0001	0001	0001	0001	0002	0002
45	0000	0000	0000	0001	0001	0001	0001	0001	0002	0002
50	0000	0000	0000	0001	0001	0001	0001	0001	0002	0002
55	0000	0000	0000	0001	0001	0001	0001	0001	0002	0002
60	0000	0000	0000	0001	0001	0001	0001	0001	0002	0002
65	0000	0000	0000	0001	0001	0001	0001	0001	0002	0002
70	0000	0000	0000	0001	0001	0001	0001	0001	0002	0002
75	0000	0000	0000	0001	0001	0001	0001	0001	0002	0002
80	0000	0000	0000	0001	0001	0001	0001	0001	0002	0002
85	0000	0000	0000	0001	0001	0001	0001	0001	0002	0002
90	0000	0000	0000	0001	0001	0001	0001	0001	0002	0002
95	0000	0000	0000	0001	0001	0001	0001	0001	0002	0002

PROBABILITY DENSITY OF t, THE NON-CENTRAL t-STATISTIC, TABLED AS A FUNCTION OF $t/\sqrt{f}$.

f is the number of degrees of freedom; the non-centrality parameter is $\sqrt{f+1}\,K_p$.
K_p is the standardized normal deviate exceeded with probability p.

DEGREES OF FREEDOM **2**

$t/\sqrt{f}$ \ p	.2500	.1500	.1000	.0650	.0400	.0250	.0100	.0040	.0025	.0010
48.00	0000	0000	0000	0001	0001	0001	0001	0001	0002	0002
05	0000	0000	0000	0001	0001	0001	0001	0001	0002	0002
10	0000	0000	0000	0000	0001	0001	0001	0001	0002	0002
15	0000	0000	0000	0000	0001	0001	0001	0001	0002	0002
20	0000	0000	0000	0000	0001	0001	0001	0001	0002	0002
25	0000	0000	0000	0000	0001	0001	0001	0001	0002	0002
30	0000	0000	0000	0000	0001	0001	0001	0001	0002	0002
35	0000	0000	0000	0000	0001	0001	0001	0001	0002	0002
40	0000	0000	0000	0000	0001	0001	0001	0001	0002	0002
45	0000	0000	0000	0000	0001	0001	0001	0001	0002	0002
50	0000	0000	0000	0000	0001	0001	0001	0001	0002	0002
55	0000	0000	0000	0000	0001	0001	0001	0001	0002	0002
60	0000	0000	0000	0000	0001	0001	0001	0001	0002	0002
65	0000	0000	0000	0000	0001	0001	0001	0001	0002	0002
70	0000	0000	0000	0000	0001	0001	0001	0001	0001	0002
75	0000	0000	0000	0000	0001	0001	0001	0001	0001	0002
80	0000	0000	0000	0000	0001	0001	0001	0001	0001	0002
85	0000	0000	0000	0000	0001	0001	0001	0001	0001	0002
90	0000	0000	0000	0000	0001	0001	0001	0001	0001	0002
95	0000	0000	0000	0000	0001	0001	0001	0001	0001	0002
49.00	0000	0000	0000	0000	0001	0001	0001	0001	0001	0002
05	0000	0000	0000	0000	0001	0001	0001	0001	0001	0002
10	0000	0000	0000	0000	0001	0001	0001	0001	0001	0002
15	0000	0000	0000	0000	0001	0001	0001	0001	0001	0002
20	0000	0000	0000	0000	0001	0001	0001	0001	0001	0002
25	0000	0000	0000	0000	0001	0001	0001	0001	0001	0002
30	0000	0000	0000	0000	0001	0001	0001	0001	0001	0002
35	0000	0000	0000	0000	0001	0001	0001	0001	0001	0002
40	0000	0000	0000	0000	0001	0001	0001	0001	0001	0002
45	0000	0000	0000	0000	0001	0001	0001	0001	0001	0002
49.50	0000	0000	0000	0000	0001	0001	0001	0001	0001	0002
55	0000	0000	0000	0000	0001	0001	0001	0001	0001	0002
60	0000	0000	0000	0000	0001	0001	0001	0001	0001	0002
65	0000	0000	0000	0000	0001	0001	0001	0001	0001	0002
70	0000	0000	0000	0000	0001	0001	0001	0001	0001	0002
75	0000	0000	0000	0000	0001	0001	0001	0001	0001	0002
80	0000	0000	0000	0000	0001	0001	0001	0001	0001	0002
85	0000	0000	0000	0000	0001	0001	0001	0001	0001	0002
90	0000	0000	0000	0000	0001	0001	0001	0001	0001	0002
95	0000	0000	0000	0000	0001	0001	0001	0001	0001	0002
50.00	0000	0000	0000	0000	0001	0001	0001	0001	0001	0002
.05	0000	0000	0000	0000	0001	0001	0001	0001	0001	0002
10	0000	0000	0000	0000	0001	0001	0001	0001	0001	0002
15	0000	0000	0000	0000	0001	0001	0001	0001	0001	0002
20	0000	0000	0000	0000	0001	0001	0001	0001	0001	0002
25	0000	0000	0000	0000	0001	0001	0001	0001	0001	0002
30	0000	0000	0000	0000	0001	0001	0001	0001	0001	0002
35	0000	0000	0000	0000	0001	0001	0001	0001	0001	0002
40	0000	0000	0000	0000	0001	0001	0001	0001	0001	0002
45	0000	0000	0000	0000	0001	0001	0001	0001	0001	0002
50	0000	0000	0000	0000	0001	0001	0001	0001	0001	0002
55	0000	0000	0000	0000	0001	0001	0001	0001	0001	0002
60	0000	0000	0000	0000	0001	0001	0001	0001	0001	0002
65	0000	0000	0000	0000	0001	0001	0001	0001	0001	0002
70	0000	0000	0000	0000	0001	0001	0001	0001	0001	0002
75	0000	0000	0000	0000	0001	0001	0001	0001	0001	0002
80	0000	0000	0000	0000	0001	0001	0001	0001	0001	0002
85	0000	0000	0000	0000	0001	0001	0001	0001	0001	0002
90	0000	0000	0000	0000	0001	0001	0001	0001	0001	0002
95	0000	0000	0000	0000	0001	0001	0001	0001	0001	0002

PROBABILITY DENSITY OF t, THE NON-CENTRAL t-STATISTIC, TABLED AS A FUNCTION OF $t/\sqrt{f}$.

f is the number of degrees of freedom; the non-centrality parameter is $\sqrt{f+1}\,K_p$.
K_p is the standardized normal deviate exceeded with probability p.

DEGREES OF FREEDOM **2**

$t/\sqrt{f}$ \ p	.2500	.1500	.1000	.0650	.0400	.0250	.0100	.0040	.0025	.0010
51.00	0000	0000	0000	0000	0001	0001	0001	0001	0001	0002
05	0000	0000	0000	0000	0001	0001	0001	0001	0001	0002
10	0000	0000	0000	0000	0001	0001	0001	0001	0001	0002
15	0000	0000	0000	0000	0001	0001	0001	0001	0001	0002
20	0000	0000	0000	0000	0001	0001	0001	0001	0001	0002
25	0000	0000	0000	0000	0001	0001	0001	0001	0001	0002
30	0000	0000	0000	0000	0001	0001	0001	0001	0001	0002
35	0000	0000	0000	0000	0001	0001	0001	0001	0001	0002
40	0000	0000	0000	0000	0001	0001	0001	0001	0001	0002
45	0000	0000	0000	0000	0001	0001	0001	0001	0001	0002
50	0000	0000	0000	0000	0001	0001	0001	0001	0001	0002
55	0000	0000	0000	0000	0001	0001	0001	0001	0001	0002
60	0000	0000	0000	0000	0001	0001	0001	0001	0001	0002
65	0000	0000	0000	0000	0001	0001	0001	0001	0001	0002
70	0000	0000	0000	0000	0001	0001	0001	0001	0001	0002
75	0000	0000	0000	0000	0001	0001	0001	0001	0001	0002
80	0000	0000	0000	0000	0001	0001	0001	0001	0001	0001
85	0000	0000	0000	0000	0001	0001	0001	0001	0001	0001
90	0000	0000	0000	0000	0001	0001	0001	0001	0001	0001
95	0000	0000	0000	0000	0001	0001	0001	0001	0001	0001
52.00	0000	0000	0000	0000	0001	0001	0001	0001	0001	0001
05	0000	0000	0000	0000	0001	0001	0001	0001	0001	0001
10	0000	0000	0000	0000	0001	0001	0001	0001	0001	0001
15	0000	0000	0000	0000	0001	0001	0001	0001	0001	0001
20	0000	0000	0000	0000	0001	0001	0001	0001	0001	0001
25	0000	0000	0000	0000	0001	0001	0001	0001	0001	0001
30	0000	0000	0000	0000	0001	0001	0001	0001	0001	0001
35	0000	0000	0000	0000	0001	0001	0001	0001	0001	0001
40	0000	0000	0000	0000	0000	0001	0001	0001	0001	0001
45	0000	0000	0000	0000	0000	0001	0001	0001	0001	0001
52.50	0000	0000	0000	0000	0000	0001	0001	0001	0001	0001
55	0000	0000	0000	0000	0000	0001	0001	0001	0001	0001
60	0000	0000	0000	0000	0000	0001	0001	0001	0001	0001
65	0000	0000	0000	0000	0000	0001	0001	0001	0001	0001
70	0000	0000	0000	0000	0000	0001	0001	0001	0001	0001
75	0000	0000	0000	0000	0000	0001	0001	0001	0001	0001
80	0000	0000	0000	0000	0000	0001	0001	0001	0001	0001
85	0000	0000	0000	0000	0000	0001	0001	0001	0001	0001
90	0000	0000	0000	0000	0000	0001	0001	0001	0001	0001
95	0000	0000	0000	0000	0000	0001	0001	0001	0001	0001
53.00	0000	0000	0000	0000	0000	0001	0001	0001	0001	0001
05	0000	0000	0000	0000	0000	0001	0001	0001	0001	0001
10	0000	0000	0000	0000	0000	0001	0001	0001	0001	0001
15	0000	0000	0000	0000	0000	0001	0001	0001	0001	0001
20	0000	0000	0000	0000	0000	0001	0001	0001	0001	0001
25	0000	0000	0000	0000	0000	0001	0001	0001	0001	0001
30	0000	0000	0000	0000	0000	0001	0001	0001	0001	0001
35	0000	0000	0000	0000	0000	0001	0001	0001	0001	0001
40	0000	0000	0000	0000	0000	0001	0001	0001	0001	0001
45	0000	0000	0000	0000	0000	0001	0001	0001	0001	0001
50	0000	0000	0000	0000	0000	0001	0001	0001	0001	0001
55	0000	0000	0000	0000	0000	0001	0001	0001	0001	0001
60	0000	0000	0000	0000	0000	0001	0001	0001	0001	0001
65	0000	0000	0000	0000	0000	0001	0001	0001	0001	0001
70	0000	0000	0000	0000	0000	0001	0001	0001	0001	0001
75	0000	0000	0000	0000	0000	0001	0001	0001	0001	0001
80	0000	0000	0000	0000	0000	0001	0001	0001	0001	0001
85	0000	0000	0000	0000	0000	0001	0001	0001	0001	0001
90	0000	0000	0000	0000	0000	0001	0001	0001	0001	0001
95	0000	0000	0000	0000	0000	0001	0001	0001	0001	0001

PROBABILITY DENSITY OF t, THE NON-CENTRAL t-STATISTIC, TABLED AS A FUNCTION OF $t/\sqrt{f}$.

f is the number of degrees of freedom; the non-centrality parameter is $\sqrt{f+1}\,K_p$.
K_p is the standardized normal deviate exceeded with probability p.

DEGREES OF FREEDOM **2**

$t/\sqrt{f}$ \ p	.2500	.1500	.1000	.0650	.0400	.0250	.0100	.0040	.0025	.0010
54.00	0000	0000	0000	0000	0000	0001	0001	0001	0001	0001
05	0000	0000	0000	0000	0000	0001	0001	0001	0001	0001
10	0000	0000	0000	0000	0000	0001	0001	0001	0001	0001
15	0000	0000	0000	0000	0000	0001	0001	0001	0001	0001
20	0000	0000	0000	0000	0000	0001	0001	0001	0001	0001
25	0000	0000	0000	0000	0000	0001	0001	0001	0001	0001
30	0000	0000	0000	0000	0000	0001	0001	0001	0001	0001
35	0000	0000	0000	0000	0000	0001	0001	0001	0001	0001
40	0000	0000	0000	0000	0000	0001	0001	0001	0001	0001
45	0000	0000	0000	0000	0000	0001	0001	0001	0001	0001
50	0000	0000	0000	0000	0000	0001	0001	0001	0001	0001
55	0000	0000	0000	0000	0000	0001	0001	0001	0001	0001
60	0000	0000	0000	0000	0000	0001	0001	0001	0001	0001
65	0000	0000	0000	0000	0000	0001	0001	0001	0001	0001
70	0000	0000	0000	0000	0000	0001	0001	0001	0001	0001
75	0000	0000	0000	0000	0000	0001	0001	0001	0001	0001
80	0000	0000	0000	0000	0000	0001	0001	0001	0001	0001
85	0000	0000	0000	0000	0000	0001	0001	0001	0001	0001
90	0000	0000	0000	0000	0000	0001	0001	0001	0001	0001
95	0000	0000	0000	0000	0000	0001	0001	0001	0001	0001
55.00	0000	0000	0000	0000	0000	0001	0001	0001	0001	0001
05	0000	0000	0000	0000	0000	0001	0001	0001	0001	0001
10	0000	0000	0000	0000	0000	0001	0001	0001	0001	0001
15	0000	0000	0000	0000	0000	0001	0001	0001	0001	0001
20	0000	0000	0000	0000	0000	0001	0001	0001	0001	0001
25	0000	0000	0000	0000	0000	0001	0001	0001	0001	0001
30	0000	0000	0000	0000	0000	0001	0001	0001	0001	0001
35	0000	0000	0000	0000	0000	0001	0001	0001	0001	0001
40	0000	0000	0000	0000	0000	0001	0001	0001	0001	0001
45	0000	0000	0000	0000	0000	0001	0001	0001	0001	0001
55.50	0000	0000	0000	0000	0000	0001	0001	0001	0001	0001
55	0000	0000	0000	0000	0000	0001	0001	0001	0001	0001
60	0000	0000	0000	0000	0000	0001	0001	0001	0001	0001
65	0000	0000	0000	0000	0000	0001	0001	0001	0001	0001
70	0000	0000	0000	0000	0000	0001	0001	0001	0001	0001
75	0000	0000	0000	0000	0000	0001	0001	0001	0001	0001
80	0000	0000	0000	0000	0000	0001	0001	0001	0001	0001
85	0000	0000	0000	0000	0000	0001	0001	0001	0001	0001
90	0000	0000	0000	0000	0000	0001	0001	0001	0001	0001
95	0000	0000	0000	0000	0000	0001	0001	0001	0001	0001
56.00	0000	0000	0000	0000	0000	0001	0001	0001	0001	0001
05	0000	0000	0000	0000	0000	0001	0001	0001	0001	0001
10	0000	0000	0000	0000	0000	0001	0001	0001	0001	0001
15	0000	0000	0000	0000	0000	0000	0001	0001	0001	0001
20	0000	0000	0000	0000	0000	0000	0001	0001	0001	0001
25	0000	0000	0000	0000	0000	0000	0001	0001	0001	0001
30	0000	0000	0000	0000	0000	0000	0001	0001	0001	0001
35	0000	0000	0000	0000	0000	0000	0001	0001	0001	0001
40	0000	0000	0000	0000	0000	0000	0001	0001	0001	0001
45	0000	0000	0000	0000	0000	0000	0001	0001	0001	0001
50	0000	0000	0000	0000	0000	0000	0001	0001	0001	0001
55	0000	0000	0000	0000	0000	0000	0001	0001	0001	0001
60	0000	0000	0000	0000	0000	0000	0001	0001	0001	0001
65	0000	0000	0000	0000	0000	0000	0001	0001	0001	0001
70	0000	0000	0000	0000	0000	0000	0001	0001	0001	0001
75	0000	0000	0000	0000	0000	0000	0001	0001	0001	0001
80	0000	0000	0000	0000	0000	0000	0001	0001	0001	0001
85	0000	0000	0000	0000	0000	0000	0001	0001	0001	0001
90	0000	0000	0000	0000	0000	0000	0001	0001	0001	0001
95	0000	0000	0000	0000	0000	0000	0001	0001	0001	0001

PROBABILITY DENSITY OF t, THE NON-CENTRAL t-STATISTIC, TABLED AS A FUNCTION OF $t/\sqrt{f}$.

f is the number of degrees of freedom; the non-centrality parameter is $\sqrt{f+1}\,K_p$.
K_p is the standardized normal deviate exceeded with probability p.

DEGREES OF FREEDOM **2**

$t/\sqrt{f}$ \ p	.2500	.1500	.1000	.0650	.0400	.0250	.0100	.0040	.0025	.0010
57.00	0000	0000	0000	0000	0000	0000	0001	0001	0001	0001
05	0000	0000	0000	0000	0000	0000	0001	0001	0001	0001
10	0000	0000	0000	0000	0000	0000	0001	0001	0001	0001
15	0000	0000	0000	0000	0000	0000	0001	0001	0001	0001
20	0000	0000	0000	0000	0000	0000	0001	0001	0001	0001
25	0000	0000	0000	0000	0000	0000	0001	0001	0001	0001
30	0000	0000	0000	0000	0000	0000	0001	0001	0001	0001
35	0000	0000	0000	0000	0000	0000	0001	0001	0001	0001
40	0000	0000	0000	0000	0000	0000	0001	0001	0001	0001
45	0000	0000	0000	0000	0000	0000	0001	0001	0001	0001
50	0000	0000	0000	0000	0000	0000	0001	0001	0001	0001
55	0000	0000	0000	0000	0000	0000	0001	0001	0001	0001
60	0000	0000	0000	0000	0000	0000	0001	0001	0001	0001
65	0000	0000	0000	0000	0000	0000	0001	0001	0001	0001
70	0000	0000	0000	0000	0000	0000	0001	0001	0001	0001
75	0000	0000	0000	0000	0000	0000	0001	0001	0001	0001
80	0000	0000	0000	0000	0000	0000	0001	0001	0001	0001
85	0000	0000	0000	0000	0000	0000	0001	0001	0001	0001
90	0000	0000	0000	0000	0000	0000	0001	0001	0001	0001
95	0000	0000	0000	0000	0000	0000	0001	0001	0001	0001
58.00	0000	0000	0000	0000	0000	0000	0001	0001	0001	0001
05	0000	0000	0000	0000	0000	0000	0001	0001	0001	0001
10	0000	0000	0000	0000	0000	0000	0001	0001	0001	0001
15	0000	0000	0000	0000	0000	0000	0001	0001	0001	0001
20	0000	0000	0000	0000	0000	0000	0001	0001	0001	0001
25	0000	0000	0000	0000	0000	0000	0001	0001	0001	0001
30	0000	0000	0000	0000	0000	0000	0001	0001	0001	0001
35	0000	0000	0000	0000	0000	0000	0001	0001	0001	0001
40	0000	0000	0000	0000	0000	0000	0001	0001	0001	0001
45	0000	0000	0000	0000	0000	0000	0001	0001	0001	0001
58.50	0000	0000	0000	0000	0000	0000	0001	0001	0001	0001
55	0000	0000	0000	0000	0000	0000	0001	0001	0001	0001
60	0000	0000	0000	0000	0000	0000	0001	0001	0001	0001
65	0000	0000	0000	0000	0000	0000	0001	0001	0001	0001
70	0000	0000	0000	0000	0000	0000	0001	0001	0001	0001
75	0000	0000	0000	0000	0000	0000	0001	0001	0001	0001
80	0000	0000	0000	0000	0000	0000	0001	0001	0001	0001
85	0000	0000	0000	0000	0000	0000	0001	0001	0001	0001
90	0000	0000	0000	0000	0000	0000	0001	0001	0001	0001
95	0000	0000	0000	0000	0000	0000	0001	0001	0001	0001
59.00	0000	0000	0000	0000	0000	0000	0001	0001	0001	0001
05	0000	0000	0000	0000	0000	0000	0001	0001	0001	0001
10	0000	0000	0000	0000	0000	0000	0001	0001	0001	0001
15	0000	0000	0000	0000	0000	0000	0001	0001	0001	0001
20	0000	0000	0000	0000	0000	0000	0001	0001	0001	0001
25	0000	0000	0000	0000	0000	0000	0001	0001	0001	0001
30	0000	0000	0000	0000	0000	0000	0001	0001	0001	0001
35	0000	0000	0000	0000	0000	0000	0001	0001	0001	0001
40	0000	0000	0000	0000	0000	0000	0001	0001	0001	0001
45	0000	0000	0000	0000	0000	0000	0001	0001	0001	0001
50	0000	0000	0000	0000	0000	0000	0001	0001	0001	0001
55	0000	0000	0000	0000	0000	0000	0001	0001	0001	0001
60	0000	0000	0000	0000	0000	0000	0001	0001	0001	0001
65	0000	0000	0000	0000	0000	0000	0001	0001	0001	0001
70	0000	0000	0000	0000	0000	0000	0001	0001	0001	0001
75	0000	0000	0000	0000	0000	0000	0001	0001	0001	0001
80	0000	0000	0000	0000	0000	0000	0001	0001	0001	0001
85	0000	0000	0000	0000	0000	0000	0001	0001	0001	0001
90	0000	0000	0000	0000	0000	0000	0001	0001	0001	0001
95	0000	0000	0000	0000	0000	0000	0001	0001	0001	0001

PROBABILITY DENSITY OF t, THE NON-CENTRAL t-STATISTIC, TABLED AS A FUNCTION OF $t/\sqrt{f}$.

f is the number of degrees of freedom; the non-centrality parameter is $\sqrt{f+1}\,K_p$.
K_p is the standardized normal deviate exceeded with probability p.

DEGREES OF FREEDOM **2**

$t/\sqrt{f}$ \ p	.2500	.1500	.1000	.0650	.0400	.0250	.0100	.0040	.0025	.0010
60.00	0000	0000	0000	0000	0000	0000	0001	0001	0001	0001
05	0000	0000	0000	0000	0000	0000	0001	0001	0001	0001
10	0000	0000	0000	0000	0000	0000	0001	0001	0001	0001
15	0000	0000	0000	0000	0000	0000	0001	0001	0001	0001
20	0000	0000	0000	0000	0000	0000	0001	0001	0001	0001
25	0000	0000	0000	0000	0000	0000	0001	0001	0001	0001
30	0000	0000	0000	0000	0000	0000	0001	0001	0001	0001
35	0000	0000	0000	0000	0000	0000	0001	0001	0001	0001
40	0000	0000	0000	0000	0000	0000	0001	0001	0001	0001
45	0000	0000	0000	0000	0000	0000	0001	0001	0001	0001
50	0000	0000	0000	0000	0000	0000	0001	0001	0001	0001
55	0000	0000	0000	0000	0000	0000	0001	0001	0001	0001
60	0000	0000	0000	0000	0000	0000	0001	0001	0001	0001
65	0000	0000	0000	0000	0000	0000	0001	0001	0001	0001
70	0000	0000	0000	0000	0000	0000	0001	0001	0001	0001
75	0000	0000	0000	0000	0000	0000	0001	0001	0001	0001
80	0000	0000	0000	0000	0000	0000	0001	0001	0001	0001
85	0000	0000	0000	0000	0000	0000	0001	0001	0001	0001
90	0000	0000	0000	0000	0000	0000	0001	0001	0001	0001
95	0000	0000	0000	0000	0000	0000	0001	0001	0001	0001
61.00	0000	0000	0000	0000	0000	0000	0001	0001	0001	0001
05	0000	0000	0000	0000	0000	0000	0001	0001	0001	0001
10	0000	0000	0000	0000	0000	0000	0001	0001	0001	0001
15	0000	0000	0000	0000	0000	0000	0001	0001	0001	0001
20	0000	0000	0000	0000	0000	0000	0001	0001	0001	0001
25	0000	0000	0000	0000	0000	0000	0001	0001	0001	0001
30	0000	0000	0000	0000	0000	0000	0001	0001	0001	0001
35	0000	0000	0000	0000	0000	0000	0001	0001	0001	0001
40	0000	0000	0000	0000	0000	0000	0001	0001	0001	0001
45	0000	0000	0000	0000	0000	0000	0001	0001	0001	0001
61.50	0000	0000	0000	0000	0000	0000	0001	0001	0001	0001
55	0000	0000	0000	0000	0000	0000	0001	0001	0001	0001
60	0000	0000	0000	0000	0000	0000	0001	0001	0001	0001
65	0000	0000	0000	0000	0000	0000	0001	0001	0001	0001
70	0000	0000	0000	0000	0000	0000	0001	0001	0001	0001
75	0000	0000	0000	0000	0000	0000	0001	0001	0001	0001
80	0000	0000	0000	0000	0000	0000	0001	0001	0001	0001
85	0000	0000	0000	0000	0000	0000	0001	0001	0001	0001
90	0000	0000	0000	0000	0000	0000	0001	0001	0001	0001
95	0000	0000	0000	0000	0000	0000	0001	0001	0001	0001
62.00	0000	0000	0000	0000	0000	0000	0001	0001	0001	0001
05	0000	0000	0000	0000	0000	0000	0001	0001	0001	0001
10	0000	0000	0000	0000	0000	0000	0001	0001	0001	0001
15	0000	0000	0000	0000	0000	0000	0001	0001	0001	0001
20	0000	0000	0000	0000	0000	0000	0001	0001	0001	0001
25	0000	0000	0000	0000	0000	0000	0001	0001	0001	0001
30	0000	0000	0000	0000	0000	0000	0001	0001	0001	0001
35	0000	0000	0000	0000	0000	0000	0001	0001	0001	0001
40	0000	0000	0000	0000	0000	0000	0001	0001	0001	0001
45	0000	0000	0000	0000	0000	0000	0000	0001	0001	0001
50	0000	0000	0000	0000	0000	0000	0000	0001	0001	0001
55	0000	0000	0000	0000	0000	0000	0000	0001	0001	0001
60	0000	0000	0000	0000	0000	0000	0000	0001	0001	0001
65	0000	0000	0000	0000	0000	0000	0000	0001	0001	0001
70	0000	0000	0000	0000	0000	0000	0000	0001	0001	0001
75	0000	0000	0000	0000	0000	0000	0000	0001	0001	0001
80	0000	0000	0000	0000	0000	0000	0000	0001	0001	0001
85	0000	0000	0000	0000	0000	0000	0000	0001	0001	0001
90	0000	0000	0000	0000	0000	0000	0000	0001	0001	0001
95	0000	0000	0000	0000	0000	0000	0000	0001	0001	0001

PROBABILITY DENSITY OF t, THE NON-CENTRAL t-STATISTIC, TABLED AS A FUNCTION OF $t/\sqrt{f}$.

f is the number of degrees of freedom; the non-centrality parameter is $\sqrt{f+1}\,K_p$.
K_p is the standardized normal deviate exceeded with probability p.

DEGREES OF FREEDOM **2**

$t/\sqrt{f}$ \ p	.2500	.1500	.1000	.0650	.0400	.0250	.0100	.0040	.0025	.0010
6 3.00	0000	0000	0000	0000	0000	0000	0000	0001	0001	0001
05	0000	0000	0000	0000	0000	0000	0000	0001	0001	0001
10	0000	0000	0000	0000	0000	0000	0000	0001	0001	0001
15	0000	0000	0000	0000	0000	0000	0000	0001	0001	0001
20	0000	0000	0000	0000	0000	0000	0000	0001	0001	0001
25	0000	0000	0000	0000	0000	0000	0000	0001	0001	0001
30	0000	0000	0000	0000	0000	0000	0000	0001	0001	0001
35	0000	0000	0000	0000	0000	0000	0000	0001	0001	0001
40	0000	0000	0000	0000	0000	0000	0000	0001	0001	0001
45	0000	0000	0000	0000	0000	0000	0000	0001	0001	0001
50	0000	0000	0000	0000	0000	0000	0000	0001	0001	0001
55	0000	0000	0000	0000	0000	0000	0000	0001	0001	0001
60	0000	0000	0000	0000	0000	0000	0000	0001	0001	0001
65	0000	0000	0000	0000	0000	0000	0000	0001	0001	0001
70	0000	0000	0000	0000	0000	0000	0000	0001	0001	0001
75	0000	0000	0000	0000	0000	0000	0000	0001	0001	0001
80	0000	0000	0000	0000	0000	0000	0000	0001	0001	0001
85	0000	0000	0000	0000	0000	0000	0000	0001	0001	0001
90	0000	0000	0000	0000	0000	0000	0000	0001	0001	0001
95	0000	0000	0000	0000	0000	0000	0000	0001	0001	0001
6 4.00	0000	0000	0000	0000	0000	0000	0000	0001	0001	0001
05	0000	0000	0000	0000	0000	0000	0000	0001	0001	0001
10	0000	0000	0000	0000	0000	0000	0000	0001	0001	0001
15	0000	0000	0000	0000	0000	0000	0000	0001	0001	0001
20	0000	0000	0000	0000	0000	0000	0000	0001	0001	0001
25	0000	0000	0000	0000	0000	0000	0000	0001	0001	0001
30	0000	0000	0000	0000	0000	0000	0000	0001	0001	0001
35	0000	0000	0000	0000	0000	0000	0000	0001	0001	0001
40	0000	0000	0000	0000	0000	0000	0000	0001	0001	0001
45	0000	0000	0000	0000	0000	0000	0000	0001	0001	0001
6 4.50	0000	0000	0000	0000	0000	0000	0000	0001	0001	0001
55	0000	0000	0000	0000	0000	0000	0000	0001	0001	0001
60	0000	0000	0000	0000	0000	0000	0000	0001	0001	0001
65	0000	0000	0000	0000	0000	0000	0000	0001	0001	0001
70	0000	0000	0000	0000	0000	0000	0000	0001	0001	0001
75	0000	0000	0000	0000	0000	0000	0000	0001	0001	0001
80	0000	0000	0000	0000	0000	0000	0000	0001	0001	0001
85	0000	0000	0000	0000	0000	0000	0000	0001	0001	0001
90	0000	0000	0000	0000	0000	0000	0000	0001	0001	0001
95	0000	0000	0000	0000	0000	0000	0000	0001	0001	0001
6 5.00	0000	0000	0000	0000	0000	0000	0000	0001	0001	0001
05	0000	0000	0000	0000	0000	0000	0000	0001	0001	0001
10	0000	0000	0000	0000	0000	0000	0000	0001	0001	0001
15	0000	0000	0000	0000	0000	0000	0000	0001	0001	0001
20	0000	0000	0000	0000	0000	0000	0000	0001	0001	0001
25	0000	0000	0000	0000	0000	0000	0000	0001	0001	0001
30	0000	0000	0000	0000	0000	0000	0000	0001	0001	0001
35	0000	0000	0000	0000	0000	0000	0000	0001	0001	0001
40	0000	0000	0000	0000	0000	0000	0000	0001	0001	0001
45	0000	0000	0000	0000	0000	0000	0000	0001	0001	0001
50	0000	0000	0000	0000	0000	0000	0000	0001	0001	0001
55	0000	0000	0000	0000	0000	0000	0000	0001	0001	0001
60	0000	0000	0000	0000	0000	0000	0000	0001	0001	0001
65	0000	0000	0000	0000	0000	0000	0000	0001	0001	0001
70	0000	0000	0000	0000	0000	0000	0000	0001	0001	0001
75	0000	0000	0000	0000	0000	0000	0000	0001	0001	0001
80	0000	0000	0000	0000	0000	0000	0000	0001	0001	0001
85	0000	0000	0000	0000	0000	0000	0000	0001	0001	0001
90	0000	0000	0000	0000	0000	0000	0000	0001	0001	0001
95	0000	0000	0000	0000	0000	0000	0000	0001	0001	0001

PROBABILITY DENSITY OF t, THE NON-CENTRAL t-STATISTIC, TABLED AS A FUNCTION OF $t/\sqrt{f}$.

f is the number of degrees of freedom; the non-centrality parameter is $\sqrt{f+1}\,K_p$.
K_p is the standardized normal deviate exceeded with probability p.

DEGREES OF FREEDOM **2**

p / $t/\sqrt{f}$	.2500	.1500	.1000	.0650	.0400	.0250	.0100	.0040	.0025	.0010
66.00	0000	0000	0000	0000	0000	0000	0000	0001	0001	0001
05	0000	0000	0000	0000	0000	0000	0000	0001	0001	0001
10	0000	0000	0000	0000	0000	0000	0000	0001	0001	0001
15	0000	0000	0000	0000	0000	0000	0000	0001	0001	0001
20	0000	0000	0000	0000	0000	0000	0000	0001	0001	0001
25	0000	0000	0000	0000	0000	0000	0000	0001	0001	0001
30	0000	0000	0000	0000	0000	0000	0000	0001	0001	0001
35	0000	0000	0000	0000	0000	0000	0000	0001	0001	0001
40	0000	0000	0000	0000	0000	0000	0000	0001	0001	0001
45	0000	0000	0000	0000	0000	0000	0000	0001	0001	0001
50	0000	0000	0000	0000	0000	0000	0000	0001	0001	0001
55	0000	0000	0000	0000	0000	0000	0000	0001	0001	0001
60	0000	0000	0000	0000	0000	0000	0000	0001	0001	0001
65	0000	0000	0000	0000	0000	0000	0000	0001	0001	0001
70	0000	0000	0000	0000	0000	0000	0000	0001	0001	0001
75	0000	0000	0000	0000	0000	0000	0000	0001	0001	0001
80	0000	0000	0000	0000	0000	0000	0000	0001	0001	0001
85	0000	0000	0000	0000	0000	0000	0000	0001	0001	0001
90	0000	0000	0000	0000	0000	0000	0000	0001	0001	0001
95	0000	0000	0000	0000	0000	0000	0000	0001	0001	0001
67.00	0000	0000	0000	0000	0000	0000	0000	0001	0001	0001
05	0000	0000	0000	0000	0000	0000	0000	0001	0001	0001
10	0000	0000	0000	0000	0000	0000	0000	0001	0001	0001
15	0000	0000	0000	0000	0000	0000	0000	0001	0001	0001
20	0000	0000	0000	0000	0000	0000	0000	0001	0001	0001
25	0000	0000	0000	0000	0000	0000	0000	0001	0001	0001
30	0000	0000	0000	0000	0000	0000	0000	0001	0001	0001
35	0000	0000	0000	0000	0000	0000	0000	0001	0001	0001
40	0000	0000	0000	0000	0000	0000	0000	0001	0001	0001
45	0000	0000	0000	0000	0000	0000	0000	0001	0001	0001
67.50	0000	0000	0000	0000	0000	0000	0000	0001	0001	0001
55	0000	0000	0000	0000	0000	0000	0000	0001	0001	0001
60	0000	0000	0000	0000	0000	0000	0000	0001	0001	0001
65	0000	0000	0000	0000	0000	0000	0000	0001	0001	0001
70	0000	0000	0000	0000	0000	0000	0000	0001	0001	0001
75	0000	0000	0000	0000	0000	0000	0000	0001	0001	0001
80	0000	0000	0000	0000	0000	0000	0000	0000	0001	0001
85	0000	0000	0000	0000	0000	0000	0000	0000	0001	0001
90	0000	0000	0000	0000	0000	0000	0000	0000	0001	0001
95	0000	0000	0000	0000	0000	0000	0000	0000	0001	0001
68.00	0000	0000	0000	0000	0000	0000	0000	0000	0001	0001
05	0000	0000	0000	0000	0000	0000	0000	0000	0001	0001
10	0000	0000	0000	0000	0000	0000	0000	0000	0001	0001
15	0000	0000	0000	0000	0000	0000	0000	0000	0001	0001
2.0	0000	0000	0000	0000	0000	0000	0000	0000	0001	0001
25	0000	0000	0000	0000	0000	0000	0000	0000	0001	0001
30	0000	0000	0000	0000	0000	0000	0000	0000	0001	0001
35	0000	0000	0000	0000	0000	0000	0000	0000	0001	0001
40	0000	0000	0000	0000	0000	0000	0000	0000	0001	0001
45	0000	0000	0000	0000	0000	0000	0000	0000	0001	0001
50	0000	0000	0000	0000	0000	0000	0000	0000	0001	0001
55	0000	0000	0000	0000	0000	0000	0000	0000	0001	0001
60	0000	0000	0000	0000	0000	0000	0000	0000	0001	0001
65	0000	0000	0000	0000	0000	0000	0000	0000	0001	0001
70	0000	0000	0000	0000	0000	0000	0000	0000	0001	0001
75	0000	0000	0000	0000	0000	0000	0000	0000	0001	0001
80	0000	0000	0000	0000	0000	0000	0000	0000	0001	0001
85	0000	0000	0000	0000	0000	0000	0000	0000	0001	0001
90	0000	0000	0000	0000	0000	0000	0000	0000	0001	0001
95	0000	0000	0000	0000	0000	0000	0000	0000	0001	0001

PROBABILITY DENSITY OF t, THE NON-CENTRAL t-STATISTIC, TABLED AS A FUNCTION OF $t/\sqrt{f}$.

f is the number of degrees of freedom; the non-centrality parameter is $\sqrt{f+1}\,K_p$.
K_p is the standardized normal deviate exceeded with probability p.

DEGREES OF FREEDOM **2**

$t/\sqrt{f}$ \ P	.2500	.1500	.1000	.0650	.0400	.0250	.0100	.0040	.0025	.0010
69.00	0000	0000	0000	0000	0000	0000	0000	0000	0001	0001
05	0000	0000	0000	0000	0000	0000	0000	0000	0001	0001
10	0000	0000	0000	0000	0000	0000	0000	0000	0001	0001
15	0000	0000	0000	0000	0000	0000	0000	0000	0001	0001
20	0000	0000	0000	0000	0000	0000	0000	0000	0001	0001
25	0000	0000	0000	0000	0000	0000	0000	0000	0001	0001
30	0000	0000	0000	0000	0000	0000	0000	0000	0001	0001
35	0000	0000	0000	0000	0000	0000	0000	0000	0001	0001
40	0000	0000	0000	0000	0000	0000	0000	0000	0001	0001
45	0000	0000	0000	0000	0000	0000	0000	0000	0001	0001
50	0000	0000	0000	0000	0000	0000	0000	0000	0001	0001
55	0000	0000	0000	0000	0000	0000	0000	0000	0001	0001
60	0000	0000	0000	0000	0000	0000	0000	0000	0001	0001
65	0000	0000	0000	0000	0000	0000	0000	0000	0001	0001
70	0000	0000	0000	0000	0000	0000	0000	0000	0001	0001
75	0000	0000	0000	0000	0000	0000	0000	0000	0001	0001
80	0000	0000	0000	0000	0000	0000	0000	0000	0001	0001
85	0000	0000	0000	0000	0000	0000	0000	0000	0001	0001
90	0000	0000	0000	0000	0000	0000	0000	0000	0001	0001
95	0000	0000	0000	0000	0000	0000	0000	0000	0001	0001
70.00	0000	0000	0000	0000	0000	0000	0000	0000	0001	0001
05	0000	0000	0000	0000	0000	0000	0000	0000	0001	0001
10	0000	0000	0000	0000	0000	0000	0000	0000	0001	0001
15	0000	0000	0000	0000	0000	0000	0000	0000	0001	0001
20	0000	0000	0000	0000	0000	0000	0000	0000	0001	0001
25	0000	0000	0000	0000	0000	0000	0000	0000	0001	0001
30	0000	0000	0000	0000	0000	0000	0000	0000	0001	0001
35	0000	0000	0000	0000	0000	0000	0000	0000	0000	0001
40	0000	0000	0000	0000	0000	0000	0000	0000	0000	0001
45	0000	0000	0000	0000	0000	0000	0000	0000	0000	0001
70.50	0000	0000	0000	0000	0000	0000	0000	0000	0000	0001
55	0000	0000	0000	0000	0000	0000	0000	0000	0000	0001
60	0000	0000	0000	0000	0000	0000	0000	0000	0000	0001
65	0000	0000	0000	0000	0000	0000	0000	0000	0000	0001
70	0000	0000	0000	0000	0000	0000	0000	0000	0000	0001
75	0000	0000	0000	0000	0000	0000	0000	0000	0000	0001
80	0000	0000	0000	0000	0000	0000	0000	0000	0000	0001
85	0000	0000	0000	0000	0000	0000	0000	0000	0000	0001
90	0000	0000	0000	0000	0000	0000	0000	0000	0000	0001
95	0000	0000	0000	0000	0000	0000	0000	0000	0000	0001
71.00	0000	0000	0000	0000	0000	0000	0000	0000	0000	0001
05	0000	0000	0000	0000	0000	0000	0000	0000	0000	0001
10	0000	0000	0000	0000	0000	0000	0000	0000	0000	0001
15	0000	0000	0000	0000	0000	0000	0000	0000	0000	0001
20	0000	0000	0000	0000	0000	0000	0000	0000	0000	0001
25	0000	0000	0000	0000	0000	0000	0000	0000	0000	0001
30	0000	0000	0000	0000	0000	0000	0000	0000	0000	0001
35	0000	0000	0000	0000	0000	0000	0000	0000	0000	0001
40	0000	0000	0000	0000	0000	0000	0000	0000	0000	0001
45	0000	0000	0000	0000	0000	0000	0000	0000	0000	0001
50	0000	0000	0000	0000	0000	0000	0000	0000	0000	0001
55	0000	0000	0000	0000	0000	0000	0000	0000	0000	0001
60	0000	0000	0000	0000	0000	0000	0000	0000	0000	0001
65	0000	0000	0000	0000	0000	0000	0000	0000	0000	0001
70	0000	0000	0000	0000	0000	0000	0000	0000	0000	0001
75	0000	0000	0000	0000	0000	0000	0000	0000	0000	0001
80	0000	0000	0000	0000	0000	0000	0000	0000	0000	0001
85	0000	0000	0000	0000	0000	0000	0000	0000	0000	0001
90	0000	0000	0000	0000	0000	0000	0000	0000	0000	0001
95	0000	0000	0000	0000	0000	0000	0000	0000	0000	0001

PROBABILITY DENSITY OF t, THE NON-CENTRAL t-STATISTIC, TABLED AS A FUNCTION OF $t/\sqrt{f}$.

f is the number of degrees of freedom; the non-centrality parameter is $\sqrt{f+1}\,K_p$.
K_p is the standardized normal deviate exceeded with probability p.

DEGREES OF FREEDOM **2**

$t/\sqrt{f}$ \ p	.2500	.1500	.1000	.0650	.0400	.0250	.0100	.0040	.0025	.0010
72.00	0000	0000	0000	0000	0000	0000	0000	0000	0000	0001
05	0000	0000	0000	0000	0000	0000	0000	0000	0000	0001
10	0000	0000	0000	0000	0000	0000	0000	0000	0000	0001
15	0000	0000	0000	0000	0000	0000	0000	0000	0000	0001
20	0000	0000	0000	0000	0000	0000	0000	0000	0000	0001
25	0000	0000	0000	0000	0000	0000	0000	0000	0000	0001
30	0000	0000	0000	0000	0000	0000	0000	0000	0000	0001
35	0000	0000	0000	0000	0000	0000	0000	0000	0000	0001
40	0000	0000	0000	0000	0000	0000	0000	0000	0000	0001
45	0000	0000	0000	0000	0000	0000	0000	0000	0000	0001
50	0000	0000	0000	0000	0000	0000	0000	0000	0000	0001
55	0000	0000	0000	0000	0000	0000	0000	0000	0000	0001
60	0000	0000	0000	0000	0000	0000	0000	0000	0000	0001
65	0000	0000	0000	0000	0000	0000	0000	0000	0000	0001
70	0000	0000	0000	0000	0000	0000	0000	0000	0000	0001
75	0000	0000	0000	0000	0000	0000	0000	0000	0000	0001
80	0000	0000	0000	0000	0000	0000	0000	0000	0000	0001
85	0000	0000	0000	0000	0000	0000	0000	0000	0000	0001
90	0000	0000	0000	0000	0000	0000	0000	0000	0000	0001
95	0000	0000	0000	0000	0000	0000	0000	0000	0000	0001
73.00	0000	0000	0000	0000	0000	0000	0000	0000	0000	0001
05	0000	0000	0000	0000	0000	0000	0000	0000	0000	0001
10	0000	0000	0000	0000	0000	0000	0000	0000	0000	0001
15	0000	0000	0000	0000	0000	0000	0000	0000	0000	0001
20	0000	0000	0000	0000	0000	0000	0000	0000	0000	0001
25	0000	0000	0000	0000	0000	0000	0000	0000	0000	0001
30	0000	0000	0000	0000	0000	0000	0000	0000	0000	0001
35	0000	0000	0000	0000	0000	0000	0000	0000	0000	0001
40	0000	0000	0000	0000	0000	0000	0000	0000	0000	0001
45	0000	0000	0000	0000	0000	0000	0000	0000	0000	0001
73.50	0000	0000	0000	0000	0000	0000	0000	0000	0000	0001
55	0000	0000	0000	0000	0000	0000	0000	0000	0000	0001
60	0000	0000	0000	0000	0000	0000	0000	0000	0000	0001
65	0000	0000	0000	0000	0000	0000	0000	0000	0000	0001
70	0000	0000	0000	0000	0000	0000	0000	0000	0000	0001
75	0000	0000	0000	0000	0000	0000	0000	0000	0000	0001
80	0000	0000	0000	0000	0000	0000	0000	0000	0000	0001
85	0000	0000	0000	0000	0000	0000	0000	0000	0000	0001
90	0000	0000	0000	0000	0000	0000	0000	0000	0000	0001
95	0000	0000	0000	0000	0000	0000	0000	0000	0000	0001
74.00	0000	0000	0000	0000	0000	0000	0000	0000	0000	0001
05	0000	0000	0000	0000	0000	0000	0000	0000	0000	0001
10	0000	0000	0000	0000	0000	0000	0000	0000	0000	0001
15	0000	0000	0000	0000	0000	0000	0000	0000	0000	0001
20	0000	0000	0000	0000	0000	0000	0000	0000	0000	0001
25	0000	0000	0000	0000	0000	0000	0000	0000	0000	0001
30	0000	0000	0000	0000	0000	0000	0000	0000	0000	0001
35	0000	0000	0000	0000	0000	0000	0000	0000	0000	0001
40	0000	0000	0000	0000	0000	0000	0000	0000	0000	0001
45	0000	0000	0000	0000	0000	0000	0000	0000	0000	0001
50	0000	0000	0000	0000	0000	0000	0000	0000	0000	0001
55	0000	0000	0000	0000	0000	0000	0000	0000	0000	0001
60	0000	0000	0000	0000	0000	0000	0000	0000	0000	0001
65	0000	0000	0000	0000	0000	0000	0000	0000	0000	0001
70	0000	0000	0000	0000	0000	0000	0000	0000	0000	0001
75	0000	0000	0000	0000	0000	0000	0000	0000	0000	0001
80	0000	0000	0000	0000	0000	0000	0000	0000	0000	0000
85	0000	0000	0000	0000	0000	0000	0000	0000	0000	0000

PROBABILITY DENSITY OF t, THE NON-CENTRAL t-STATISTIC, TABLED AS A FUNCTION OF $t/\sqrt{f}$.

f is the number of degrees of freedom; the non-centrality parameter is $\sqrt{f+1}\,K_p$.
K_p is the standardized normal deviate exceeded with probability p.

DEGREES OF FREEDOM **3**

$t/\sqrt{f}$ \ p	.2500	.1500	.1000	.0650	.0400	.0250	.0100	.0040	.0025	.0010
-4.25	0000	0000	0000	0000	0000	0000	0000	0000	0000	0000
20	0001	0000	0000	0000	0000	0000	0000	0000	0000	0000
15	0001	0000	0000	0000	0000	0000	0000	0000	0000	0000
10	0001	0000	0000	0000	0000	0000	0000	0000	0000	0000
05	0001	0000	0000	0000	0000	0000	0000	0000	0000	0000
-4.00	0001	0000	0000	0000	0000	0000	0000	0000	0000	0000
-3.95	0001	0000	0000	0000	0000	0000	0000	0000	0000	0000
90	0001	0000	0000	0000	0000	0000	0000	0000	0000	0000
85	0001	0000	0000	0000	0000	0000	0000	0000	0000	0000
80	0001	0000	0000	0000	0000	0000	0000	0000	0000	0000
75	0001	0000	0000	0000	0000	0000	0000	0000	0000	0000
70	0001	0000	0000	0000	0000	0000	0000	0000	0000	0000
65	0001	0000	0000	0000	0000	0000	0000	0000	0000	0000
60	0001	0000	0000	0000	0000	0000	0000	0000	0000	0000
55	0001	0000	0000	0000	0000	0000	0000	0000	0000	0000
50	0001	0000	0000	0000	0000	0000	0000	0000	0000	0000
45	0001	0000	0000	0000	0000	0000	0000	0000	0000	0000
40	0001	0000	0000	0000	0000	0000	0000	0000	0000	0000
35	0001	0000	0000	0000	0000	0000	0000	0000	0000	0000
30	0001	0000	0000	0000	0000	0000	0000	0000	0000	0000
25	0001	0000	0000	0000	0000	0000	0000	0000	0000	0000
20	0001	0000	0000	0000	0000	0000	0000	0000	0000	0000
15	0002	0000	0000	0000	0000	0000	0000	0000	0000	0000
10	0002	0000	0000	0000	0000	0000	0000	0000	0000	0000
05	0002	0000	0000	0000	0000	0000	0000	0000	0000	0000
-3.00	0002	0000	0000	0000	0000	0000	0000	0000	0000	0000
-2.95	0002	0000	0000	0000	0000	0000	0000	0000	0000	0000
90	0002	0000	0000	0000	0000	0000	0000	0000	0000	0000
85	0002	0000	0000	0000	0000	0000	0000	0000	0000	0000
80	0002	0000	0000	0000	0000	0000	0000	0000	0000	0000
-2.75	0003	0000	0000	0000	0000	0000	0000	0000	0000	0000
70	0003	0000	0000	0000	0000	0000	0000	0000	0000	0000
65	0003	0000	0000	0000	0000	0000	0000	0000	0000	0000
60	0003	0000	0000	0000	0000	0000	0000	0000	0000	0000
55	0003	0000	0000	0000	0000	0000	0000	0000	0000	0000
50	0004	0000	0000	0000	0000	0000	0000	0000	0000	0000
45	0004	0000	0000	0000	0000	0000	0000	0000	0000	0000
40	0004	0001	0000	0000	0000	0000	0000	0000	0000	0000
35	0005	0001	0000	0000	0000	0000	0000	0000	0000	0000
30	0005	0001	0000	0000	0000	0000	0000	0000	0000	0000
25	0005	0001	0000	0000	0000	0000	0000	0000	0000	0000
20	0006	0001	0000	0000	0000	0000	0000	0000	0000	0000
15	0006	0001	0000	0000	0000	0000	0000	0000	0000	0000
10	0007	0001	0000	0000	0000	0000	0000	0000	0000	0000
05	0008	0001	0000	0000	0000	0000	0000	0000	0000	0000
-2.00	0008	0001	0000	0000	0000	0000	0000	0000	0000	0000
-1.95	0009	0001	0000	0000	0000	0000	0000	0000	0000	0000
90	0010	0001	0000	0000	0000	0000	0000	0000	0000	0000
85	0011	0001	0000	0000	0000	0000	0000	0000	0000	0000
80	0012	0002	0000	0000	0000	0000	0000	0000	0000	0000
75	0013	0002	0000	0000	0000	0000	0000	0000	0000	0000
70	0015	0002	0000	0000	0000	0000	0000	0000	0000	0000
65	0016	0002	0000	0000	0000	0000	0000	0000	0000	0000
60	0018	0002	0000	0000	0000	0000	0000	0000	0000	0000
55	0020	0003	0001	0000	0000	0000	0000	0000	0000	0000
50	0022	0003	0001	0000	0000	0000	0000	0000	0000	0000
45	0025	0003	0001	0000	0000	0000	0000	0000	0000	0000
40	0028	0004	0001	0000	0000	0000	0000	0000	0000	0000
35	0031	0004	0001	0000	0000	0000	0000	0000	0000	0000
30	0035	0005	0001	0000	0000	0000	0000	0000	0000	0000

PROBABILITY DENSITY OF t, THE NON-CENTRAL t-STATISTIC, TABLED AS A FUNCTION OF $t/\sqrt{f}$.

f is the number of degrees of freedom; the non-centrality parameter is $\sqrt{f+1}\,K_p$.

K_p is the standardized normal deviate exceeded with probability p.

DEGREES OF FREEDOM **3**

$t/\sqrt{f}$ \ p	.2500	.1500	.1000	.0650	.0400	.0250	.0100	.0040	.0025	.0010
-1.25	0039	0005	0001	0000	0000	0000	0000	0000	0000	0000
20	0044	0006	0001	0000	0000	0000	0000	0000	0000	0000
15	0050	0007	0001	0000	0000	0000	0000	0000	0000	0000
10	0057	0008	0002	0000	0000	0000	0000	0000	0000	0000
05	0065	0009	0002	0000	0000	0000	0000	0000	0000	0000
-1.00	0075	0011	0002	0000	0000	0000	0000	0000	0000	0000
-0.95	0086	0012	0003	0000	0000	0000	0000	0000	0000	0000
90	0099	0014	0003	0001	0000	0000	0000	0000	0000	0000
85	0114	0017	0004	0001	0000	0000	0000	0000	0000	0000
80	0131	0020	0004	0001	0000	0000	0000	0000	0000	0000
75	0152	0023	0005	0001	0000	0000	0000	0000	0000	0000
70	0177	0028	0006	0001	0000	0000	0000	0000	0000	0000
65	0206	0033	0007	0001	0000	0000	0000	0000	0000	0000
60	0241	0040	0009	0002	0000	0000	0000	0000	0000	0000
55	0282	0048	0011	0002	0000	0000	0000	0000	0000	0000
50	0330	0058	0013	0003	0000	0000	0000	0000	0000	0000
45	0387	0070	0017	0003	0001	0000	0000	0000	0000	0000
40	0453	0085	0021	0004	0001	0000	0000	0000	0000	0000
35	0531	0104	0026	0006	0001	0000	0000	0000	0000	0000
30	0622	0127	0033	0007	0001	0000	0000	0000	0000	0000
25	0726	0156	0041	0009	0002	0000	0000	0000	0000	0000
20	0845	0192	0052	0012	0002	0000	0000	0000	0000	0000
15	0980	0235	0067	0016	0003	0001	0000	0000	0000	0000
10	1132	0288	0085	0021	0004	0001	0000	0000	0000	0000
05	1299	0352	0108	0028	0006	0001	0000	0000	0000	0000
0.00	1480	0429	0138	0038	0008	0002	0000	0000	0000	0000
05	1673	0520	0175	0050	0011	0002	0000	0000	0000	0000
10	1875	0626	0221	0066	0016	0004	0000	0000	0000	0000
15	2081	0748	0279	0088	0022	0005	0000	0000	0000	0000
20	2287	0886	0348	0115	0030	0008	0000	0000	0000	0000
0.25	2488	1040	0432	0151	0042	0011	0001	0000	0000	0000
30	2676	1207	0530	0196	0058	0016	0001	0000	0000	0000
35	2848	1385	0643	0252	0079	0024	0002	0000	0000	0000
40	2998	1571	0771	0320	0107	0034	0003	0000	0000	0000
45	3123	1760	0913	0401	0142	0048	0005	0000	0000	0000
50	3219	1949	1067	0495	0187	0067	0008	0001	0000	0000
55	3286	2132	1230	0604	0243	0092	0012	0001	0000	0000
60	3324	2304	1400	0725	0309	0125	0018	0002	0001	0000
65	3332	2462	1572	0858	0388	0166	0027	0004	0001	0000
70	3314	2602	1742	1000	0479	0216	0039	0006	0002	0000
75	3271	2721	1907	1149	0581	0277	0056	0010	0004	0001
80	3206	2818	2063	1302	0694	0348	0078	0015	0006	0001
85	3124	2892	2207	1456	0816	0430	0105	0022	0010	0002
90	3026	2943	2336	1607	0945	0522	0140	0033	0015	0003
95	2917	2972	2447	1753	1079	0623	0183	0047	0022	0005
1.00	2800	2979	2541	1890	1215	0732	0234	0065	0032	0008
05	2677	2967	2615	2016	1350	0847	0293	0088	0046	0012
10	2550	2937	2671	2129	1482	0965	0360	0117	0064	0018
15	2422	2892	2708	2228	1608	1086	0435	0152	0086	0026
20	2294	2834	2728	2313	1727	1207	0517	0194	0113	0037
25	2169	2765	2731	2381	1837	1325	0605	0242	0146	0052
30	2046	2687	2719	2435	1936	1439	0697	0296	0184	0069
35	1927	2602	2694	2473	2024	1547	0793	0357	0229	0091
40	1812	2512	2658	2497	2100	1648	0891	0424	0279	0118
45	1703	2419	2611	2508	2163	1741	0989	0495	0335	0149
50	1598	2323	2556	2506	2214	1824	1087	0571	0396	0185
55	1499	2227	2494	2493	2252	1898	1182	0649	0462	0226
60	1405	2130	2426	2470	2279	1961	1273	0730	0531	0272
65	1317	2035	2354	2438	2295	2014	1360	0812	0603	0321
70	1234	1941	2278	2399	2301	2057	1442	0994	0677	0375

PROBABILITY DENSITY OF t, THE NON-CENTRAL t-STATISTIC, TABLED AS A FUNCTION OF $t/\sqrt{f}$.

f is the number of degrees of freedom; the non-centrality parameter is $\sqrt{f+1}\,K_p$.
K_p is the standardized normal deviate exceeded with probability p.

DEGREES OF FREEDOM 3

$t/\sqrt{f}$ \ p	.2500	.1500	.1000	.0650	.0400	.0250	.0100	.0040	.0025	.0010
1.75	1156	1849	2201	2353	2297	2090	1517	0975	0753	0433
80	1082	1759	2122	2302	2285	2114	1586	1055	0828	0493
85	1014	1673	2043	2246	2265	2129	1647	1131	0903	0555
90	0950	1589	1964	2187	2239	2135	1702	1204	0977	0619
95	0890	1509	1885	2126	2206	2134	1749	1273	1048	0684
2.00	0834	1432	1808	2062	2169	2125	1788	1338	1117	0750
05	0782	1359	1732	1997	2127	2111	1821	1397	1182	0815
10	0733	1289	1658	1931	2082	2090	1846	1452	1243	0878
15	0688	1222	1586	1866	2034	2065	1865	1500	1300	0941
20	0646	1158	1517	1800	1983	2035	1877	1544	1353	1001
25	0606	1098	1449	1735	1931	2001	1884	1581	1401	1058
30	0570	1041	1384	1671	1878	1965	1885	1613	1444	1113
35	0535	0987	1322	1608	1824	1925	1881	1640	1482	1164
40	0504	0936	1261	1546	1769	1884	1872	1662	1516	1212
45	0474	0887	1204	1486	1714	1841	1860	1678	1544	1256
50	0446	0841	1149	1428	1660	1796	1843	1690	1568	1297
55	0420	0798	1096	1371	1606	1750	1823	1697	1587	1333
60	0396	0757	1045	1316	1552	1704	1800	1700	1602	1366
65	0373	0718	0997	1263	1500	1657	1774	1699	1613	1395
70	0352	0682	0952	1211	1448	1610	1746	1695	1620	1420
75	0333	0647	0908	1162	1397	1564	1717	1687	1623	1441
80	0314	0615	0866	1114	1348	1517	1685	1676	1622	1458
85	0297	0584	0827	1068	1300	1471	1653	1663	1619	1472
90	0281	0555	0789	1024	1253	1426	1619	1647	1612	1483
95	0266	0528	0753	0982	1207	1381	1584	1628	1603	1491
3.00	0252	0502	0719	0942	1163	1337	1549	1608	1592	1495
05	0238	0478	0686	0903	1121	1294	1513	1586	1578	1497
10	0226	0455	0656	0865	1079	1252	1477	1563	1562	1496
15	0214	0433	0626	0830	1039	1211	1441	1538	1544	1492
20	0203	0412	0598	0796	1001	1171	1405	1512	1525	1487
3.25	0193	0393	0572	0763	0964	1132	1369	1485	1504	1479
30	0183	0374	0547	0732	0928	1094	1333	1458	1482	1469
35	0174	0357	0523	0702	0893	1057	1297	1429	1459	1457
40	0165	0341	0500	0674	0860	1021	1262	1401	1436	1444
45	0157	0325	0479	0646	0828	0986	1228	1372	1411	1429
50	0150	0310	0458	0620	0797	0953	1194	1343	1386	1413
55	0143	0296	0438	0596	0768	0920	1160	1314	1360	1396
60	0136	0283	0420	0572	0739	0889	1127	1284	1334	1378
65	0130	0270	0402	0549	0712	0859	1095	1255	1308	1359
70	0124	0259	0385	0527	0686	0829	1063	1226	1281	1340
75	0118	0247	0369	0507	0661	0801	1033	1197	1254	1319
80	0113	0237	0354	0487	0636	0774	1003	1168	1228	1298
85	0107	0227	0340	0468	0613	0747	0973	1140	1201	1277
90	0103	0217	0326	0450	0591	0722	0945	1112	1175	1255
95	0098	0208	0313	0432	0569	0697	0917	1084	1148	1233
4.00	0094	0199	0300	0416	0549	0674	0890	1057	1122	1210
05	0090	0191	0288	0400	0529	0651	0863	1030	1096	1188
10	0086	0183	0277	0385	0510	0629	0838	1004	1071	1165
15	0082	0176	0266	0370	0492	0608	0813	0978	1045	1143
20	0079	0168	0255	0357	0475	0587	0789	0953	1021	1120
25	0075	0162	0246	0343	0458	0568	0765	0928	0996	1098
30	0072	0155	0236	0331	0442	0549	0742	0904	0972	1075
35	0069	0149	0227	0319	0426	0530	0720	0880	0948	1053
40	0067	0143	0219	0307	0412	0513	0699	0857	0925	1031
45	0064	0138	0210	0296	0397	0496	0678	0834	0902	1009
50	0061	0132	0203	0285	0384	0480	0658	0812	0880	0987
55	0059	0127	0195	0275	0371	0464	0638	0791	0858	0966
60	0057	0123	0188	0265	0358	0449	0619	0770	0836	0945
65	0054	0118	0181	0256	0346	0434	0601	0749	0815	0924
70	0052	0114	0174	0247	0334	0420	0583	0729	0795	0903

PROBABILITY DENSITY OF t, THE NON-CENTRAL t-STATISTIC, TABLED AS A FUNCTION OF $t/\sqrt{f}$.

f is the number of degrees of freedom; the non-centrality parameter is $\sqrt{f+1}\,K_p$.
K_p is the standardized normal deviate exceeded with probability p.

DEGREES OF FREEDOM 3

$t/\sqrt{f}$ \ p	.2500	.1500	.1000	.0650	.0400	.0250	.0100	.0040	.0025	.0010
4.75	0050	0109	0168	0238	0323	0407	0566	0710	0775	0883
80	0048	0105	0162	0230	0312	0394	0550	0691	0755	0863
85	0047	0101	0156	0222	0302	0381	0534	0672	0736	0844
90	0045	0098	0151	0215	0292	0369	0518	0654	0717	0824
95	0043	0094	0146	0207	0283	0358	0503	0637	0699	0806
5.00	0042	0091	0141	0200	0273	0346	0488	0620	0682	0787
05	0040	0088	0136	0194	0264	0336	0474	0604	0664	0769
10	0039	0085	0131	0187	0256	0325	0460	0588	0647	0751
15	0037	0082	0127	0181	0248	0315	0447	0572	0631	0734
20	0036	0079	0122	0175	0240	0306	0434	0557	0615	0717
25	0035	0076	0118	0170	0232	0296	0422	0542	0599	0700
30	0033	0074	0114	0164	0225	0287	0410	0528	0584	0684
35	0032	0071	0111	0159	0218	0279	0398	0514	0569	0668
40	0031	0069	0107	0154	0211	0270	0387	0500	0555	0652
45	0030	0067	0104	0149	0205	0262	0376	0487	0541	0637
50	0029	0064	0100	0144	0199	0254	0366	0474	0527	0622
55	0028	0062	0097	0140	0193	0247	0356	0462	0514	0608
60	0027	0060	0094	0135	0187	0240	0346	0450	0501	0594
65	0026	0058	0091	0131	0181	0233	0336	0438	0489	0580
70	0025	0056	0088	0127	0176	0226	0327	0427	0476	0566
75	0025	0055	0085	0123	0170	0219	0318	0416	0464	0553
80	0024	0053	0083	0119	0165	0213	0309	0405	0453	0540
85	0023	0051	0080	0116	0161	0207	0301	0395	0442	0528
90	0022	0050	0078	0112	0156	0201	0293	0385	0431	0515
95	0022	0048	0075	0109	0151	0195	0285	0375	0420	0503
6.00	0021	0047	0073	0106	0147	0190	0277	0366	0410	0492
05	0020	0045	0071	0103	0143	0184	0270	0356	0400	0480
10	0020	0044	0069	0100	0139	0179	0263	0347	0390	0469
15	0019	0043	0067	0097	0135	0174	0256	0339	0380	0458
20	0019	0041	0065	0094	0131	0170	0249	0330	0371	0448
6.25	0018	0040	0063	0091	0127	0165	0243	0322	0362	0438
30	0017	0039	0061	0089	0124	0160	0236	0314	0353	0428
35	0017	0038	0059	0086	0120	0156	0230	0306	0345	0418
40	0016	0037	0058	0084	0117	0152	0224	0299	0337	0408
45	0016	0036	0056	0082	0114	0148	0218	0291	0329	0399
50	0015	0035	0054	0079	0111	0144	0213	0284	0321	0390
55	0015	0034	0053	0077	0108	0140	0208	0277	0313	0381
60	0015	0033	0051	0075	0105	0136	0202	0271	0306	0373
65	0014	0032	0050	0073	0102	0133	0197	0264	0298	0364
70	0014	0031	0049	0071	0099	0129	0192	0258	0291	0356
75	0013	0030	0047	0069	0097	0126	0187	0252	0285	0348
80	0013	0029	0046	0067	0094	0123	0183	0246	0278	0340
85	0013	0028	0045	0065	0092	0120	0178	0240	0272	0333
90	0012	0028	0044	0064	0089	0117	0174	0234	0265	0325
95	0012	0027	0042	0062	0087	0114	0170	0229	0259	0318
7.00	0012	0026	0041	0060	0085	0111	0166	0223	0253	0311
05	0011	0025	0040	0059	0083	0108	0162	0218	0247	0304
10	0011	0025	0039	0057	0081	0105	0158	0213	0242	0298
15	0011	0024	0038	0056	0079	0103	0154	0208	0236	0291
20	0010	0023	0037	0054	0077	0100	0150	0203	0231	0285
25	0010	0023	0036	0053	0075	0098	0147	0199	0226	0279
30	0010	0022	0035	0052	0073	0095	0143	0194	0221	0273
35	0010	0022	0034	0050	0071	0093	0140	0190	0216	0267
40	0009	0021	0034	0049	0069	0091	0136	0185	0211	0261
45	0009	0021	0033	0048	0068	0089	0133	0181	0206	0255
50	0009	0020	0032	0047	0066	0086	0130	0177	0202	0250
55	0009	0020	0031	0046	0064	0084	0127	0173	0197	0245
60	0008	0019	0030	0045	0063	0082	0124	0169	0193	0240
65	0008	0019	0030	0043	0061	0080	0121	0166	0189	0234
70	0008	0018	0029	0042	0060	0079	0119	0162	0185	0230

PROBABILITY DENSITY OF t, THE NON-CENTRAL t-STATISTIC, TABLED AS A FUNCTION OF $t/\sqrt{f}$.

f is the number of degrees of freedom; the non-centrality parameter is $\sqrt{f+1}\,K_p$.

K_p is the standardized normal deviate exceeded with probability p.

DEGREES OF FREEDOM **3**

$t/\sqrt{f}$ \ p	.2500	.1500	.1000	.0650	.0400	.0250	.0100	.0040	.0025	.0010
7.75	0008	0018	0028	0041	0058	0077	0116	0158	0181	0225
80	0008	0017	0027	0040	0057	0075	0113	0155	0177	0220
85	0007	0017	0027	0039	0056	0073	0111	0151	0173	0215
90	0007	0016	0026	0039	0054	0072	0108	0148	0169	0211
95	0007	0016	0026	0038	0053	0070	0106	0145	0166	0207
8.00	0007	0016	0025	0037	0052	0068	0104	0142	0162	0202
05	0007	0015	0024	0036	0051	0067	0101	0139	0159	0198
10	0007	0015	0024	0035	0050	0065	0099	0136	0155	0194
15	0006	0015	0023	0034	0048	0064	0097	0133	0152	0190
20	0006	0014	0023	0033	0047	0062	0095	0130	0149	0186
25	0006	0014	0022	0033	0046	0061	0093	0127	0146	0183
30	0006	0014	0022	0032	0045	0060	0091	0125	0143	0179
35	0006	0013	0021	0031	0044	0058	0089	0122	0140	0175
40	0006	0013	0021	0031	0043	0057	0087	0120	0137	0172
45	0006	0013	0020	0030	0042	0056	0085	0117	0134	0168
50	0005	0012	0020	0029	0041	0055	0083	0115	0132	0165
55	0005	0012	0019	0029	0041	0053	0082	0112	0129	0162
60	0005	0012	0019	0028	0040	0052	0080	0110	0126	0159
65	0005	0012	0019	0027	0039	0051	0078	0108	0124	0156
70	0005	0011	0018	0027	0038	0050	0077	0106	0121	0153
75	0005	0011	0018	0026	0037	0049	0075	0104	0119	0150
80	0005	0011	0017	0026	0036	0048	0073	0101	0117	0147
85	0005	0011	0017	0025	0036	0047	0072	0099	0114	0144
90	0005	0010	0017	0025	0035	0046	0071	0097	0112	0141
95	0004	0010	0016	0024	0034	0045	0069	0096	0110	0138
9.00	0004	0010	0016	0024	0033	0044	0068	0094	0108	0136
05	0004	0010	0016	0023	0033	0043	0066	0092	0106	0133
10	0004	0010	0015	0023	0032	0042	0065	0090	0104	0131
15	0004	0009	0015	0022	0031	0042	0064	0088	0102	0128
20	0004	0009	0015	0022	0031	0041	0062	0087	0100	0126
9.25	0004	0009	0014	0021	0030	0040	0061	0085	0098	0123
30	0004	0009	0014	0021	0030	0039	0060	0083	0096	0121
35	0004	0009	0014	0020	0029	0038	0059	0082	0094	0119
40	0004	0008	0013	0020	0028	0038	0058	0080	0092	0117
45	0004	0008	0013	0020	0028	0037	0057	0079	0091	0115
50	0004	0008	0013	0019	0027	0036	0055	0077	0089	0112
55	0003	0008	0013	0019	0027	0035	0054	0076	0087	0110
60	0003	0008	0012	0018	0026	0035	0053	0074	0086	0108
65	0003	0008	0012	0018	0026	0034	0052	0073	0084	0106
70	0003	0007	0012	0018	0025	0033	0051	0071	0082	0105
75	0003	0007	0012	0017	0025	0033	0050	0070	0081	0103
80	0003	0007	0011	0017	0024	0032	0049	0069	0079	0101
85	0003	0007	0011	0017	0024	0031	0049	0068	0078	0099
90	0003	0007	0011	0016	0023	0031	0048	0066	0077	0097
95	0003	0007	0011	0016	0023	0030	0047	0065	0075	0096
10.00	0003	0007	0011	0016	0022	0030	0046	0064	0074	0094
05	0003	0006	0010	0015	0022	0029	0045	0063	0072	0092
10	0003	0006	0010	0015	0022	0029	0044	0062	0071	0091
15	0003	0006	0010	0015	0021	0028	0043	0061	0070	0089
20	0003	0006	0010	0015	0021	0028	0043	0060	0069	0087
25	0003	0006	0010	0014	0020	0027	0042	0058	0067	0086
30	0003	0006	0009	0014	0020	0027	0041	0057	0066	0084
35	0003	0006	0009	0014	0020	0026	0040	0056	0065	0083
40	0002	0006	0009	0014	0019	0026	0040	0055	0064	0082
45	0002	0006	0009	0013	0019	0025	0039	0054	0063	0080
50	0002	0005	0009	0013	0019	0025	0038	0053	0062	0079
55	0002	0005	0009	0013	0018	0024	0038	0053	0061	0077
60	0002	0005	0008	0013	0018	0024	0037	0052	0060	0076
65	0002	0005	0008	0012	0018	0023	0036	0051	0059	0075
70	0002	0005	0008	0012	0017	0023	0036	0050	0058	0074

PROBABILITY DENSITY OF t, THE NON-CENTRAL t-STATISTIC, TABLED AS A FUNCTION OF $t/\sqrt{f}$.

f is the number of degrees of freedom; the non-centrality parameter is $\sqrt{f+1}\,K_p$.
K_p is the standardized normal deviate exceeded with probability p.

DEGREES OF FREEDOM **3**

$t/\sqrt{f}$ \ p	.2500	.1500	.1000	.0650	.0400	.0250	.0100	.0040	.0025	.0010
10.75	0002	0005	0008	0012	0017	0023	0035	0049	0057	0072
80	0002	0005	0008	0012	0017	0022	0034	0048	0056	0071
85	0002	0005	0008	0011	0016	0022	0034	0047	0055	0070
90	0002	0005	0008	0011	0016	0021	0033	0047	0054	0069
95	0002	0005	0007	0011	0016	0021	0033	0046	0053	0068
11.00	0002	0005	0007	0011	0016	0021	0032	0045	0052	0067
05	0002	0004	0007	0011	0015	0020	0032	0044	0051	0066
10	0002	0004	0007	0010	0015	0020	0031	0044	0050	0064
15	0002	0004	0007	0010	0015	0020	0031	0043	0050	0063
20	0002	0004	0007	0010	0015	0019	0030	0042	0049	0062
25	0002	0004	0007	0010	0014	0019	0030	0041	0048	0061
30	0002	0004	0007	0010	0014	0019	0029	0041	0047	0060
35	0002	0004	0006	0010	0014	0018	0029	0040	0046	0059
40	0002	0004	0006	0009	0014	0018	0028	0039	0046	0058
45	0002	0004	0006	0009	0013	0018	0028	0039	0045	0058
50	0002	0004	0006	0009	0013	0017	0027	0038	0044	0057
55	0002	0004	0006	0009	0013	0017	0027	0038	0043	0056
60	0002	0004	0006	0009	0013	0017	0026	0037	0043	0055
65	0002	0004	0006	0009	0012	0017	0026	0036	0042	0054
70	0002	0004	0006	0009	0012	0016	0025	0036	0041	0053
75	0002	0004	0006	0008	0012	0016	0025	0035	0041	0052
80	0002	0003	0006	0008	0012	0016	0025	0035	0040	0052
85	0001	0003	0005	0008	0012	0016	0024	0034	0040	0051
90	0001	0003	0005	0008	0011	0015	0024	0034	0039	0050
95	0001	0003	0005	0008	0011	0015	0023	0033	0038	0049
12.00	0001	0003	0005	0008	0011	0015	0023	0033	0038	0048
05	0001	0003	0005	0008	0011	0015	0023	0032	0037	0048
10	0001	0003	0005	0008	0011	0014	0022	0032	0037	0047
15	0001	0003	0005	0007	0011	0014	0022	0031	0036	0046
20	0001	0003	0005	0007	0010	0014	0022	0031	0035	0046
12.25	0001	0003	0005	0007	0010	0014	0021	0030	0035	0045
30	0001	0003	0005	0007	0010	0013	0021	0030	0034	0044
35	0001	0003	0005	0007	0010	0013	0021	0029	0034	0044
40	0001	0003	0005	0007	0010	0013	0020	0029	0033	0043
45	0001	0003	0005	0007	0010	0013	0020	0028	0033	0042
50	0001	0003	0004	0007	0009	0013	0020	0028	0032	0042
55	0001	0003	0004	0007	0009	0012	0019	0027	0032	0041
60	0001	0003	0004	0006	0009	0012	0019	0027	0031	0040
65	0001	0003	0004	0006	0009	0012	0019	0027	0031	0040
70	0001	0003	0004	0006	0009	0012	0019	0026	0030	0039
75	0001	0003	0004	0006	0009	0012	0018	0026	0030	0039
80	0001	0003	0004	0006	0009	0012	0018	0025	0030	0038
85	0001	0002	0004	0006	0009	0011	0018	0025	0029	0038
90	0001	0002	0004	0006	0008	0011	0018	0025	0029	0037
95	0001	0002	0004	0006	0008	0011	0017	0024	0028	0036
13.00	0001	0002	0004	0006	0008	0011	0017	0024	0028	0036
05	0001	0002	0004	0006	0008	0011	0017	0024	0028	0035
10	0001	0002	0004	0006	0008	0011	0017	0023	0027	0035
15	0001	0002	0004	0005	0008	0010	0016	0023	0027	0034
20	0001	0002	0004	0005	0008	0010	0016	0023	0026	0034
25	0001	0002	0004	0005	0008	0010	0016	0022	0026	0033
30	0001	0002	0003	0005	0007	0010	0016	0022	0026	0033
35	0001	0002	0003	0005	0007	0010	0015	0022	0025	0033
40	0001	0002	0003	0005	0007	0010	0015	0021	0025	0032
45	0001	0002	0003	0005	0007	0010	0015	0021	0025	0032
50	0001	0002	0003	0005	0007	0009	0015	0021	0024	0031
55	0001	0002	0003	0005	0007	0009	0015	0021	0024	0031
60	0001	0002	0003	0005	0007	0009	0014	0020	0024	0030
65	0001	0002	0003	0005	0007	0009	0014	0020	0023	0030
70	0001	0002	0003	0005	0007	0009	0014	0020	0023	0030

PROBABILITY DENSITY OF t, THE NON-CENTRAL t-STATISTIC, TABLED AS A FUNCTION OF $t/\sqrt{f}$.

f is the number of degrees of freedom; the non-centrality parameter is $\sqrt{f+1}\,K_p$.
K_p is the standardized normal deviate exceeded with probability p.

DEGREES OF FREEDOM **3**

$t/\sqrt{f}$ \ p	.2500	.1500	.1000	.0650	.0400	.0250	.0100	.0040	.0025	.0010
13.75	0001	0002	0003	0005	0007	0009	0014	0019	0023	0029
80	0001	0002	0003	0004	0006	0009	0014	0019	0022	0029
85	0001	0002	0003	0004	0006	0009	0013	0019	0022	0028
90	0001	0002	0003	0004	0006	0008	0013	0019	0022	0028
95	0001	0002	0003	0004	0006	0008	0013	0018	0021	0028
14.00	0001	0002	0003	0004	0006	0008	0013	0018	0021	0027
05	0001	0002	0003	0004	0006	0008	0013	0018	0021	0027
10	0001	0002	0003	0004	0006	0008	0012	0018	0021	0027
15	0001	0002	0003	0004	0006	0008	0012	0017	0020	0026
20	0001	0002	0003	0004	0006	0008	0012	0017	0020	0026
25	0001	0002	0003	0004	0006	0008	0012	0017	0020	0025
30	0001	0002	0003	0004	0006	0008	0012	0017	0019	0025
35	0001	0002	0003	0004	0006	0007	0012	0016	0019	0025
40	0001	0002	0003	0004	0005	0007	0011	0016	0019	0024
45	0001	0002	0003	0004	0005	0007	0011	0016	0019	0024
50	0001	0002	0002	0004	0005	0007	0011	0016	0018	0024
55	0001	0002	0002	0004	0005	0007	0011	0016	0018	0024
60	0001	0001	0002	0004	0005	0007	0011	0015	0018	0023
65	0001	0001	0002	0004	0005	0007	0011	0015	0018	0023
70	0001	0001	0002	0004	0005	0007	0011	0015	0018	0023
75	0001	0001	0002	0003	0005	0007	0010	0015	0017	0022
80	0001	0001	0002	0003	0005	0007	0010	0015	0017	0022
85	0001	0001	0002	0003	0005	0006	0010	0014	0017	0022
90	0001	0001	0002	0003	0005	0006	0010	0014	0017	0022
95	0001	0001	0002	0003	0005	0006	0010	0014	0016	0021
15.00	0001	0001	0002	0003	0005	0006	0010	0014	0016	0021
05	0001	0001	0002	0003	0005	0006	0010	0014	0016	0021
10	0001	0001	0002	0003	0005	0006	0010	0014	0016	0020
15	0001	0001	0002	0003	0004	0006	0009	0013	0016	0020
20	0001	0001	0002	0003	0004	0006	0009	0013	0015	0020
15.25	0001	0001	0002	0003	0004	0006	0009	0013	0015	0020
30	0001	0001	0002	0003	0004	0006	0009	0013	0015	0019
35	0001	0001	0002	0003	0004	0006	0009	0013	0015	0019
40	0001	0001	0002	0003	0004	0006	0009	0013	0015	0019
45	0001	0001	0002	0003	0004	0006	0009	0012	0014	0019
50	0001	0001	0002	0003	0004	0005	0009	0012	0014	0019
55	0001	0001	0002	0003	0004	0005	0009	0012	0014	0018
60	0001	0001	0002	0003	0004	0005	0008	0012	0014	0018
65	0000	0001	0002	0003	0004	0005	0008	0012	0014	0018
70	0000	0001	0002	0003	0004	0005	0008	0012	0014	0018
75	0000	0001	0002	0003	0004	0005	0008	0012	0013	0017
80	0000	0001	0002	0003	0004	0005	0008	0011	0013	0017
85	0000	0001	0002	0003	0004	0005	0008	0011	0013	0017
90	0000	0001	0002	0003	0004	0005	0008	0011	0013	0017
95	0000	0001	0002	0003	0004	0005	0008	0011	0013	0017
16.00	0000	0001	0002	0003	0004	0005	0008	0011	0013	0016
05	0000	0001	0002	0002	0004	0005	0008	0011	0012	0016
10	0000	0001	0002	0002	0004	0005	0007	0011	0012	0016
15	0000	0001	0002	0002	0003	0005	0007	0010	0012	0016
20	0000	0001	0002	0002	0003	0005	0007	0010	0012	0016
25	0000	0001	0002	0002	0003	0005	0007	0010	0012	0015
30	0000	0001	0002	0002	0003	0005	0007	0010	0012	0015
35	0000	0001	0002	0002	0003	0004	0007	0010	0012	0015
40	0000	0001	0002	0002	0003	0004	0007	0010	0011	0015
45	0000	0001	0002	0002	0003	0004	0007	0010	0011	0015
50	0000	0001	0001	0002	0003	0004	0007	0010	0011	0015
55	0000	0001	0001	0002	0003	0004	0007	0010	0011	0014
60	0000	0001	0001	0002	0003	0004	0007	0009	0011	0014
65	0000	0001	0001	0002	0003	0004	0007	0009	0011	0014
70	0000	0001	0001	0002	0003	0004	0006	0009	0011	0014

PROBABILITY DENSITY OF t, THE NON-CENTRAL t-STATISTIC, TABLED AS A FUNCTION OF $t/\sqrt{f}$.

f is the number of degrees of freedom; the non-centrality parameter is $\sqrt{f+1}\,K_p$.
K_p is the standardized normal deviate exceeded with probability p.

$t/\sqrt{f}$ \ p	.2500	.1500	.1000	.0650	.0400	.0250	.0100	.0040	.0025	.0010
16.75	0000	0001	0001	0002	0003	0004	0006	0009	0011	0014
80	0000	0001	0001	0002	0003	0004	0006	0009	0010	0014
85	0000	0001	0001	0002	0003	0004	0006	0009	0010	0013
90	0000	0001	0001	0002	0003	0004	0006	0009	0010	0013
95	0000	0001	0001	0002	0003	0004	0006	0009	0010	0013
17.00	0000	0001	0001	0002	0003	0004	0006	0009	0010	0013
05	0000	0001	0001	0002	0003	0004	0006	0008	0010	0013
10	0000	0001	0001	0002	0003	0004	0006	0008	0010	0013
15	0000	0001	0001	0002	0003	0004	0006	0008	0010	0013
20	0000	0001	0001	0002	0003	0004	0006	0008	0010	0012
25	0000	0001	0001	0002	0003	0004	0006	0008	0009	0012
30	0000	0001	0001	0002	0003	0004	0006	0008	0009	0012
35	0000	0001	0001	0002	0003	0004	0006	0008	0009	0012
40	0000	0001	0001	0002	0003	0003	0005	0008	0009	0012
45	0000	0001	0001	0002	0003	0003	0005	0008	0009	0012
50	0000	0001	0001	0002	0003	0003	0005	0008	0009	0012
55	0000	0001	0001	0002	0003	0003	0005	0008	0009	0011
60	0000	0001	0001	0002	0002	0003	0005	0007	0009	0011
65	0000	0001	0001	0002	0002	0003	0005	0007	0009	0011
70	0000	0001	0001	0002	0002	0003	0005	0007	0009	0011
75	0000	0001	0001	0002	0002	0003	0005	0007	0008	0011
80	0000	0001	0001	0002	0002	0003	0005	0007	0008	0011
85	0000	0001	0001	0002	0002	0003	0005	0007	0008	0011
90	0000	0001	0001	0002	0002	0003	0005	0007	0008	0011
95	0000	0001	0001	0002	0002	0003	0005	0007	0008	0011
18.00	0000	0001	0001	0002	0002	0003	0005	0007	0008	0010
05	0000	0001	0001	0002	0002	0003	0005	0007	0008	0010
10	0000	0001	0001	0002	0002	0003	0005	0007	0008	0010
15	0000	0001	0001	0002	0002	0003	0005	0007	0008	0010
20	0000	0001	0001	0002	0002	0003	0005	0007	0008	0010
18.25	0000	0001	0001	0001	0002	0003	0005	0007	0008	0010
30	0000	0001	0001	0001	0002	0003	0005	0006	0008	0010
35	0000	0001	0001	0001	0002	0003	0004	0006	0007	0010
40	0000	0001	0001	0001	0002	0003	0004	0006	0007	0010
45	0000	0001	0001	0001	0002	0003	0004	0006	0007	0009
50	0000	0001	0001	0001	0002	0003	0004	0006	0007	0009
55	0000	0001	0001	0001	0002	0003	0004	0006	0007	0009
60	0000	0001	0001	0001	0002	0003	0004	0006	0007	0009
65	0000	0001	0001	0001	0002	0003	0004	0006	0007	0009
70	0000	0001	0001	0001	0002	0003	0004	0006	0007	0009
75	0000	0001	0001	0001	0002	0003	0004	0006	0007	0009
80	0000	0001	0001	0001	0002	0003	0004	0006	0007	0009
85	0000	0001	0001	0001	0002	0003	0004	0006	0007	0009
90	0000	0001	0001	0001	0002	0003	0004	0006	0007	0009
95	0000	0001	0001	0001	0002	0002	0004	0006	0007	0009
19.00	0000	0001	0001	0001	0002	0002	0004	0006	0006	0008
05	0000	0001	0001	0001	0002	0002	0004	0006	0006	0008
10	0000	0001	0001	0001	0002	0002	0004	0005	0006	0008
15	0000	0001	0001	0001	0002	0002	0004	0005	0006	0008
20	0000	0001	0001	0001	0002	0002	0004	0005	0006	0008
25	0000	0001	0001	0001	0002	0002	0004	0005	0006	0008
30	0000	0000	0001	0001	0002	0002	0004	0005	0006	0008
35	0000	0000	0001	0001	0002	0002	0004	0005	0006	0008
40	0000	0000	0001	0001	0002	0002	0004	0005	0006	0008
45	0000	0000	0001	0001	0002	0002	0004	0005	0006	0008
50	0000	0000	0001	0001	0002	0002	0004	0005	0006	0008
55	0000	0000	0001	0001	0002	0002	0003	0005	0006	0008
60	0000	0000	0001	0001	0002	0002	0003	0005	0006	0007
65	0000	0000	0001	0001	0002	0002	0003	0005	0006	0007
70	0000	0000	0001	0001	0002	0002	0003	0005	0006	0007

PROBABILITY DENSITY OF t, THE NON-CENTRAL t-STATISTIC, TABLED AS A FUNCTION OF $t/\sqrt{f}$.

f is the number of degrees of freedom; the non-centrality parameter is $\sqrt{f+1}\,K_p$.
K_p is the standardized normal deviate exceeded with probability p.

DEGREES OF FREEDOM 3

$t/\sqrt{f}$ \ p	.2500	.1500	.1000	.0650	.0400	.0250	.0100	.0040	.0025	.0010
19.75	0000	0000	0001	0001	0002	0002	0003	0005	0006	0007
80	0000	0000	0001	0001	0002	0002	0003	0005	0006	0007
85	0000	0000	0001	0001	0002	0002	0003	0005	0005	0007
90	0000	0000	0001	0001	0002	0002	0003	0005	0005	0007
95	0000	0000	0001	0001	0002	0002	0003	0005	0005	0007
20.00	0000	0000	0001	0001	0002	0002	0003	0005	0005	0007
05	0000	0000	0001	0001	0001	0002	0003	0005	0005	0007
10	0000	0000	0001	0001	0001	0002	0003	0004	0005	0007
15	0000	0000	0001	0001	0001	0002	0003	0004	0005	0007
20	0000	0000	0001	0001	0001	0002	0003	0004	0005	0007
25	0000	0000	0001	0001	0001	0002	0003	0004	0005	0007
30	0000	0000	0001	0001	0001	0002	0003	0004	0005	0007
35	0000	0000	0001	0001	0001	0002	0003	0004	0005	0006
40	0000	0000	0001	0001	0001	0002	0003	0004	0005	0006
45	0000	0000	0001	0001	0001	0002	0003	0004	0005	0006
50	0000	0000	0001	0001	0001	0002	0003	0004	0005	0006
55	0000	0000	0001	0001	0001	0002	0003	0004	0005	0006
60	0000	0000	0001	0001	0001	0002	0003	0004	0005	0006
65	0000	0000	0001	0001	0001	0002	0003	0004	0005	0006
70	0000	0000	0001	0001	0001	0002	0003	0004	0005	0006
75	0000	0000	0001	0001	0001	0002	0003	0004	0005	0006
80	0000	0000	0001	0001	0001	0002	0003	0004	0005	0006
85	0000	0000	0001	0001	0001	0002	0003	0004	0005	0006
90	0000	0000	0001	0001	0001	0002	0003	0004	0004	0006
95	0000	0000	0001	0001	0001	0002	0003	0004	0004	0006
21.00	0000	0000	0001	0001	0001	0002	0003	0004	0004	0006
05	0000	0000	0001	0001	0001	0002	0003	0004	0004	0006
10	0000	0000	0001	0001	0001	0002	0003	0004	0004	0006
15	0000	0000	0001	0001	0001	0002	0003	0004	0004	0006
20	0000	0000	0001	0001	0001	0002	0003	0004	0004	0006
21.25	0000	0000	0001	0001	0001	0002	0003	0004	0004	0005
30	0000	0000	0001	0001	0001	0002	0002	0004	0004	0005
35	0000	0000	0001	0001	0001	0002	0002	0004	0004	0005
40	0000	0000	0001	0001	0001	0002	0002	0003	0004	0005
45	0000	0000	0001	0001	0001	0002	0002	0003	0004	0005
50	0000	0000	0001	0001	0001	0002	0002	0003	0004	0005
55	0000	0000	0001	0001	0001	0002	0002	0003	0004	0005
60	0000	0000	0001	0001	0001	0001	0002	0003	0004	0005
65	0000	0000	0001	0001	0001	0001	0002	0003	0004	0005
70	0000	0000	0001	0001	0001	0001	0002	0003	0004	0005
75	0000	0000	0000	0001	0001	0001	0002	0003	0004	0005
80	0000	0000	0000	0001	0001	0001	0002	0003	0004	0005
85	0000	0000	0000	0001	0001	0001	0002	0003	0004	0005
90	0000	0000	0000	0001	0001	0001	0002	0003	0004	0005
95	0000	0000	0000	0001	0001	0001	0002	0003	0004	0005
22.00	0000	0000	0000	0001	0001	0001	0002	0003	0004	0005
05	0000	0000	0000	0001	0001	0001	0002	0003	0004	0005
10	0000	0000	0000	0001	0001	0001	0002	0003	0004	0005
15	0000	0000	0000	0001	0001	0001	0002	0003	0004	0005
20	0000	0000	0000	0001	0001	0001	0002	0003	0004	0005
25	0000	0000	0000	0001	0001	0001	0002	0003	0004	0005
30	0000	0000	0000	0001	0001	0001	0002	0003	0003	0005
35	0000	0000	0000	0001	0001	0001	0002	0003	0003	0004
40	0000	0000	0000	0001	0001	0001	0002	0003	0003	0004
45	0000	0000	0000	0001	0001	0001	0002	0003	0003	0004
50	0000	0000	0000	0001	0001	0001	0002	0003	0003	0004
55	0000	0000	0000	0001	0001	0001	0002	0003	0003	0004
60	0000	0000	0000	0001	0001	0001	0002	0003	0003	0004
65	0000	0000	0000	0001	0001	0001	0002	0003	0003	0004
70	0000	0000	0000	0001	0001	0001	0002	0003	0003	0004

PROBABILITY DENSITY OF t, THE NON-CENTRAL t-STATISTIC, TABLED AS A FUNCTION OF $t/\sqrt{f}$.

f is the number of degrees of freedom; the non-centrality parameter is $\sqrt{f+1}\,K_p$.
K_p is the standardized normal deviate exceeded with probability p.

DEGREES OF FREEDOM **3**

$t/\sqrt{f}$ \ p	.2500	.1500	.1000	.0650	.0400	.0250	.0100	.0040	.0025	.0010
22.75	0000	0000	0000	0001	0001	0001	0002	0003	0003	0004
80	0000	0000	0000	0001	0001	0001	0002	0003	0003	0004
85	0000	0000	0000	0001	0001	0001	0002	0003	0003	0004
90	0000	0000	0000	0001	0001	0001	0002	0003	0003	0004
95	0000	0000	0000	0001	0001	0001	0002	0003	0003	0004
23.00	0000	0000	0000	0001	0001	0001	0002	0003	0003	0004
05	0000	0000	0000	0001	0001	0001	0002	0003	0003	0004
10	0000	0000	0000	0001	0001	0001	0002	0003	0003	0004
15	0000	0000	0000	0001	0001	0001	0002	0003	0003	0004
20	0000	0000	0000	0001	0001	0001	0002	0003	0003	0004
25	0000	0000	0000	0001	0001	0001	0002	0003	0003	0004
30	0000	0000	0000	0001	0001	0001	0002	0002	0003	0004
35	0000	0000	0000	0001	0001	0001	0002	0002	0003	0004
40	0000	0000	0000	0001	0001	0001	0002	0002	0003	0004
45	0000	0000	0000	0001	0001	0001	0002	0002	0003	0004
50	0000	0000	0000	0001	0001	0001	0002	0002	0003	0004
55	0000	0000	0000	0001	0001	0001	0002	0002	0003	0004
60	0000	0000	0000	0001	0001	0001	0002	0002	0003	0004
65	0000	0000	0000	0001	0001	0001	0002	0002	0003	0004
70	0000	0000	0000	0001	0001	0001	0002	0002	0003	0004
75	0000	0000	0000	0001	0001	0001	0002	0002	0003	0004
80	0000	0000	0000	0001	0001	0001	0002	0002	0003	0004
85	0000	0000	0000	0001	0001	0001	0002	0002	0003	0003
90	0000	0000	0000	0001	0001	0001	0002	0002	0003	0003
95	0000	0000	0000	0001	0001	0001	0002	0002	0003	0003
24.00	0000	0000	0000	0001	0001	0001	0002	0002	0003	0003
05	0000	0000	0000	0001	0001	0001	0002	0002	0003	0003
10	0000	0000	0000	0000	0001	0001	0002	0002	0003	0003
15	0000	0000	0000	0000	0001	0001	0002	0002	0003	0003
20	0000	0000	0000	0000	0001	0001	0002	0002	0003	0003
24.25	0000	0000	0000	0000	0001	0001	0001	0002	0002	0003
30	0000	0000	0000	0000	0001	0001	0001	0002	0002	0003
35	0000	0000	0000	0000	0001	0001	0001	0002	0002	0003
40	0000	0000	0000	0000	0001	0001	0001	0002	0002	0003
45	0000	0000	0000	0000	0001	0001	0001	0002	0002	0003
50	0000	0000	0000	0000	0001	0001	0001	0002	0002	0003
55	0000	0000	0000	0000	0001	0001	0001	0002	0002	0003
60	0000	0000	0000	0000	0001	0001	0001	0002	0002	0003
65	0000	0000	0000	0000	0001	0001	0001	0002	0002	0003
70	0000	0000	0000	0000	0001	0001	0001	0002	0002	0003
75	0000	0000	0000	0000	0001	0001	0001	0002	0002	0003
80	0000	0000	0000	0000	0001	0001	0001	0002	0002	0003
85	0000	0000	0000	0000	0001	0001	0001	0002	0002	0003
90	0000	0000	0000	0000	0001	0001	0001	0002	0002	0003
95	0000	0000	0000	0000	0001	0001	0001	0002	0002	0003
25.00	0000	0000	0000	0000	0001	0001	0001	0002	0002	0003
05	0000	0000	0000	0000	0001	0001	0001	0002	0002	0003
10	0000	0000	0000	0000	0001	0001	0001	0002	0002	0003
15	0000	0000	0000	0000	0001	0001	0001	0002	0002	0003
20	0000	0000	0000	0000	0001	0001	0001	0002	0002	0003
25	0000	0000	0000	0000	0001	0001	0001	0002	0002	0003
30	0000	0000	0000	0000	0001	0001	0001	0002	0002	0003
35	0000	0000	0000	0000	0001	0001	0001	0002	0002	0003
40	0000	0000	0000	0000	0001	0001	0001	0002	0002	0003
45	0000	0000	0000	0000	0001	0001	0001	0002	0002	0003
50	0000	0000	0000	0000	0001	0001	0001	0002	0002	0003
55	0000	0000	0000	0000	0001	0001	0001	0002	0002	0003
60	0000	0000	0000	0000	0001	0001	0001	0002	0002	0003
65	0000	0000	0000	0000	0001	0001	0001	0002	0002	0003
70	0000	0000	0000	0000	0001	0001	0001	0002	0002	0003

PROBABILITY DENSITY OF t, THE NON-CENTRAL t-STATISTIC, TABLED AS A FUNCTION OF $t/\sqrt{f}$.

f is the number of degrees of freedom; the non-centrality parameter is $\sqrt{f+1}\,K_{\bar{p}}$.
K_p is the standardized normal deviate exceeded with probability p.

DEGREES OF FREEDOM **3**

$t/\sqrt{f}$ \ p	.2500	.1500	.1000	.0650	.0400	.0250	.0100	.0040	.0025	.0010
25.75	0000	0000	0000	0000	0001	0001	0001	0002	0002	0003
80	0000	0000	0000	0000	0001	0001	0001	0002	0002	0003
85	0000	0000	0000	0000	0001	0001	0001	0002	0002	0003
90	0000	0000	0000	0000	0001	0001	0001	0002	0002	0003
95	0000	0000	0000	0000	0001	0001	0001	0002	0002	0003
26.00	0000	0000	0000	0000	0001	0001	0001	0002	0002	0002
05	0000	0000	0000	0000	0001	0001	0001	0002	0002	0002
10	0000	0000	0000	0000	0001	0001	0001	0002	0002	0002
15	0000	0000	0000	0000	0001	0001	0001	0002	0002	0002
20	0000	0000	0000	0000	0001	0001	0001	0002	0002	0002
25	0000	0000	0000	0000	0001	0001	0001	0002	0002	0002
30	0000	0000	0000	0000	0001	0001	0001	0002	0002	0002
35	0000	0000	0000	0000	0001	0001	0001	0002	0002	0002
40	0000	0000	0000	0000	0000	0001	0001	0002	0002	0002
45	0000	0000	0000	0000	0000	0001	0001	0002	0002	0002
50	0000	0000	0000	0000	0000	0001	0001	0002	0002	0002
55	0000	0000	0000	0000	0000	0001	0001	0001	0002	0002
60	0000	0000	0000	0000	0000	0001	0001	0001	0002	0002
65	0000	0000	0000	0000	0000	0001	0001	0001	0002	0002
70	0000	0000	0000	0000	0000	0001	0001	0001	0002	0002
75	0000	0000	0000	0000	0000	0001	0001	0001	0002	0002
80	0000	0000	0000	0000	0000	0001	0001	0001	0002	0002
85	0000	0000	0000	0000	0000	0001	0001	0001	0002	0002
90	0000	0000	0000	0000	0000	0001	0001	0001	0002	0002
95	0000	0000	0000	0000	0000	0001	0001	0001	0002	0002
27.00	0000	0000	0000	0000	0000	0001	0001	0001	0002	0002
05	0000	0000	0000	0000	0000	0001	0001	0001	0002	0002
10	0000	0000	0000	0000	0000	0001	0001	0001	0002	0002
15	0000	0000	0000	0000	0000	0001	0001	0001	0002	0002
20	0000	0000	0000	0000	0000	0001	0001	0001	0002	0002
27.25	0000	0000	0000	0000	0000	0001	0001	0001	0002	0002
30	0000	0000	0000	0000	0000	0001	0001	0001	0002	0002
35	0000	0000	0000	0000	0000	0001	0001	0001	0002	0002
40	0000	0000	0000	0000	0000	0001	0001	0001	0002	0002
45	0000	0000	0000	0000	0000	0001	0001	0001	0002	0002
50	0000	0000	0000	0000	0000	0001	0001	0001	0002	0002
55	0000	0000	0000	0000	0000	0001	0001	0001	0002	0002
60	0000	0000	0000	0000	0000	0001	0001	0001	0002	0002
65	0000	0000	0000	0000	0000	0001	0001	0001	0001	0002
70	0000	0000	0000	0000	0000	0001	0001	0001	0001	0002
75	0000	0000	0000	0000	0000	0001	0001	0001	0001	0002
80	0000	0000	0000	0000	0000	0001	0001	0001	0001	0002
85	0000	0000	0000	0000	0000	0001	0001	0001	0001	0002
90	0000	0000	0000	0000	0000	0001	0001	0001	0001	0002
95	0000	0000	0000	0000	0000	0001	0001	0001	0001	0002
28.00	0000	0000	0000	0000	0000	0001	0001	0001	0001	0002
05	0000	0000	0000	0000	0000	0001	0001	0001	0001	0002
10	0000	0000	0000	0000	0000	0001	0001	0001	0001	0002
15	0000	0000	0000	0000	0000	0001	0001	0001	0001	0002
20	0000	0000	0000	0000	0000	0001	0001	0001	0001	0002
25	0000	0000	0000	0000	0000	0001	0001	0001	0001	0002
30	0000	0000	0000	0000	0000	0001	0001	0001	0001	0002
35	0000	0000	0000	0000	0000	0001	0001	0001	0001	0002
40	0000	0000	0000	0000	0000	0001	0001	0001	0001	0002
45	0000	0000	0000	0000	0000	0000	0001	0001	0001	0002
50	0000	0000	0000	0000	0000	0000	0001	0001	0001	0002
55	0000	0000	0000	0000	0000	0000	0001	0001	0001	0002
60	0000	0000	0000	0000	0000	0000	0001	0001	0001	0002
65	0000	0000	0000	0000	0000	0000	0001	0001	0001	0002
70	0000	0000	0000	0000	0000	0000	0001	0001	0001	0002

PROBABILITY DENSITY OF t, THE NON-CENTRAL t-STATISTIC, TABLED AS A FUNCTION OF $t/\sqrt{f}$.

f is the number of degrees of freedom; the non-centrality parameter is $\sqrt{f+1}\,K_p$.
K_p is the standardized normal deviate exceeded with probability p.

DEGREES OF FREEDOM 3

$t/\sqrt{f}$ \ p	.2500	.1500	.1000	.0650	.0400	.0250	.0100	.0040	.0025	.0010
28.75	0000	0000	0000	0000	0000	0000	0001	0001	0001	0002
80	0000	0000	0000	0000	0000	0000	0001	0001	0001	0002
85	0000	0000	0000	0000	0000	0000	0001	0001	0001	0002
90	0000	0000	0000	0000	0000	0000	0001	0001	0001	0002
95	0000	0000	0000	0000	0000	0000	0001	0001	0001	0002
29.00	0000	0000	0000	0000	0000	0000	0001	0001	0001	0002
05	0000	0000	0000	0000	0000	0000	0001	0001	0001	0002
10	0000	0000	0000	0000	0000	0000	0001	0001	0001	0002
15	0000	0000	0000	0000	0000	0000	0001	0001	0001	0002
20	0000	0000	0000	0000	0000	0000	0001	0001	0001	0002
25	0000	0000	0000	0000	0000	0000	0001	0001	0001	0002
30	0000	0000	0000	0000	0000	0000	0001	0001	0001	0002
35	0000	0000	0000	0000	0000	0000	0001	0001	0001	0002
40	0000	0000	0000	0000	0000	0000	0001	0001	0001	0002
45	0000	0000	0000	0000	0000	0000	0001	0001	0001	0002
50	0000	0000	0000	0000	0000	0000	0001	0001	0001	0002
55	0000	0000	0000	0000	0000	0000	0001	0001	0001	0001
60	0000	0000	0000	0000	0000	0000	0001	0001	0001	0001
65	0000	0000	0000	0000	0000	0000	0001	0001	0001	0001
70	0000	0000	0000	0000	0000	0000	0001	0001	0001	0001
75	0000	0000	0000	0000	0000	0000	0001	0001	0001	0001
80	0000	0000	0000	0000	0000	0000	0001	0001	0001	0001
85	0000	0000	0000	0000	0000	0000	0001	0001	0001	0001
90	0000	0000	0000	0000	0000	0000	0001	0001	0001	0001
95	0000	0000	0000	0000	0000	0000	0001	0001	0001	0001
30.00	0000	0000	0000	0000	0000	0000	0001	0001	0001	0001
05	0000	0000	0000	0000	0000	0000	0001	0001	0001	0001
10	0000	0000	0000	0000	0000	0000	0001	0001	0001	0001
15	0000	0000	0000	0000	0000	0000	0001	0001	0001	0001
20	0000	0000	0000	0000	0000	0000	0001	0001	0001	0001
30.25	0000	0000	0000	0000	0000	0000	0001	0001	0001	0001
30	0000	0000	0000	0000	0000	0000	0001	0001	0001	0001
35	0000	0000	0000	0000	0000	0000	0001	0001	0001	0001
40	0000	0000	0000	0000	0000	0000	0001	0001	0001	0001
45	0000	0000	0000	0000	0000	0000	0001	0001	0001	0001
50	0000	0000	0000	0000	0000	0000	0001	0001	0001	0001
55	0000	0000	0000	0000	0000	0000	0001	0001	0001	0001
60	0000	0000	0000	0000	0000	0000	0001	0001	0001	0001
65	0000	0000	0000	0000	0000	0000	0001	0001	0001	0001
70	0000	0000	0000	0000	0000	0000	0001	0001	0001	0001
75	0000	0000	0000	0000	0000	0000	0001	0001	0001	0001
80	0000	0000	0000	0000	0000	0000	0001	0001	0001	0001
85	0000	0000	0000	0000	0000	0000	0001	0001	0001	0001
90	0000	0000	0000	0000	0000	0000	0001	0001	0001	0001
95	0000	0000	0000	0000	0000	0000	0001	0001	0001	0001
31.00	0000	0000	0000	0000	0000	0000	0001	0001	0001	0001
05	0000	0000	0000	0000	0000	0000	0001	0001	0001	0001
10	0000	0000	0000	0000	0000	0000	0001	0001	0001	0001
15	0000	0000	0000	0000	0000	0000	0001	0001	0001	0001
20	0000	0000	0000	0000	0000	0000	0001	0001	0001	0001
25	0000	0000	0000	0000	0000	0000	0001	0001	0001	0001
30	0000	0000	0000	0000	0000	0000	0001	0001	0001	0001
35	0000	0000	0000	0000	0000	0000	0001	0001	0001	0001
40	0000	0000	0000	0000	0000	0000	0001	0001	0001	0001
45	0000	0000	0000	0000	0000	0000	0001	0001	0001	0001
50	0000	0000	0000	0000	0000	0000	0001	0001	0001	0001
55	0000	0000	0000	0000	0000	0000	0001	0001	0001	0001
60	0000	0000	0000	0000	0000	0000	0001	0001	0001	0001
65	0000	0000	0000	0000	0000	0000	0001	0001	0001	0001
70	0000	0000	0000	0000	0000	0000	0001	0001	0001	0001

PROBABILITY DENSITY OF t, THE NON-CENTRAL t-STATISTIC, TABLED AS A FUNCTION OF $t/\sqrt{f}$.

f is the number of degrees of freedom; the non-centrality parameter is $\sqrt{f+1}\,K_p$.
K_p is the standardized normal deviate exceeded with probability p.

DEGREES OF FREEDOM **3**

$t/\sqrt{f}$ \ p	.2500	.1500	.1000	.0650	.0400	.0250	.0100	.0040	.0025	.0010
31.75	0000	0000	0000	0000	0000	0000	0001	0001	0001	0001
80	0000	0000	0000	0000	0000	0000	0001	0001	0001	0001
85	0000	0000	0000	0000	0000	0000	0001	0001	0001	0001
90	0000	0000	0000	0000	0000	0000	0001	0001	0001	0001
95	0000	0000	0000	0000	0000	0000	0000	0001	0001	0001
32.00	0000	0000	0000	0000	0000	0000	0000	0001	0001	0001
05	0000	0000	0000	0000	0000	0000	0000	0001	0001	0001
10	0000	0000	0000	0000	0000	0000	0000	0001	0001	0001
15	0000	0000	0000	0000	0000	0000	0000	0001	0001	0001
20	0000	0000	0000	0000	0000	0000	0000	0001	0001	0001
25	0000	0000	0000	0000	0000	0000	0000	0001	0001	0001
30	0000	0000	0000	0000	0000	0000	0000	0001	0001	0001
35	0000	0000	0000	0000	0000	0000	0000	0001	0001	0001
40	0000	0000	0000	0000	0000	0000	0000	0001	0001	0001
45	0000	0000	0000	0000	0000	0000	0000	0001	0001	0001
50	0000	0000	0000	0000	0000	0000	0000	0001	0001	0001
55	0000	0000	0000	0000	0000	0000	0000	0001	0001	0001
60	0000	0000	0000	0000	0000	0000	0000	0001	0001	0001
65	0000	0000	0000	0000	0000	0000	0000	0001	0001	0001
70	0000	0000	0000	0000	0000	0000	0000	0001	0001	0001
75	0000	0000	0000	0000	0000	0000	0000	0001	0001	0001
80	0000	0000	0000	0000	0000	0000	0000	0001	0001	0001
85	0000	0000	0000	0000	0000	0000	0000	0001	0001	0001
90	0000	0000	0000	0000	0000	0000	0000	0001	0001	0001
95	0000	0000	0000	0000	0000	0000	0000	0001	0001	0001
33.00	0000	0000	0000	0000	0000	0000	0000	0001	0001	0001
05	0000	0000	0000	0000	0000	0000	0000	0001	0001	0001
10	0000	0000	0000	0000	0000	0000	0000	0001	0001	0001
15	0000	0000	0000	0000	0000	0000	0000	0001	0001	0001
20	0000	0000	0000	0000	0000	0000	0000	0001	0001	0001
33.25	0000	0000	0000	0000	0000	0000	0000	0001	0001	0001
30	0000	0000	0000	0000	0000	0000	0000	0001	0001	0001
35	0000	0000	0000	0000	0000	0000	0000	0001	0001	0001
40	0000	0000	0000	0000	0000	0000	0000	0001	0001	0001
45	0000	0000	0000	0000	0000	0000	0000	0001	0001	0001
50	0000	0000	0000	0000	0000	0000	0000	0001	0001	0001
55	0000	0000	0000	0000	0000	0000	0000	0001	0001	0001
60	0000	0000	0000	0000	0000	0000	0000	0001	0001	0001
65	0000	0000	0000	0000	0000	0000	0000	0001	0001	0001
70	0000	0000	0000	0000	0000	0000	0000	0001	0001	0001
75	0000	0000	0000	0000	0000	0000	0000	0001	0001	0001
80	0000	0000	0000	0000	0000	0000	0000	0001	0001	0001
85	0000	0000	0000	0000	0000	0000	0000	0001	0001	0001
90	0000	0000	0000	0000	0000	0000	0000	0001	0001	0001
95	0000	0000	0000	0000	0000	0000	0000	0001	0001	0001
34.00	0000	0000	0000	0000	0000	0000	0000	0001	0001	0001
05	0000	0000	0000	0000	0000	0000	0000	0001	0001	0001
10	0000	0000	0000	0000	0000	0000	0000	0001	0001	0001
15	0000	0000	0000	0000	0000	0000	0000	0001	0001	0001
20	0000	0000	0000	0000	0000	0000	0000	0001	0001	0001
25	0000	0000	0000	0000	0000	0000	0000	0001	0001	0001
30	0000	0000	0000	0000	0000	0000	0000	0001	0001	0001
35	0000	0000	0000	0000	0000	0000	0000	0001	0001	0001
40	0000	0000	0000	0000	0000	0000	0000	0001	0001	0001
45	0000	0000	0000	0000	0000	0000	0000	0001	0001	0001
50	0000	0000	0000	0000	0000	0000	0000	0001	0001	0001
55	0000	0000	0000	0000	0000	0000	0000	0001	0001	0001
60	0000	0000	0000	0000	0000	0000	0000	0001	0001	0001
65	0000	0000	0000	0000	0000	0000	0000	0001	0001	0001
70	0000	0000	0000	0000	0000	0000	0000	0001	0001	0001

…BABILITY DENSITY OF t, THE NON-CENTRAL t-STATISTIC, TABLED AS A FUNCTION OF $t/\sqrt{f}$.

f is the number of degrees of freedom; the non-centrality parameter is $\sqrt{f+1}\,K_p$.
K_p is the standardized normal deviate exceeded with probability p.

DEGREES OF FREEDOM 3

$t/\sqrt{f}$ \ p	.2500	.1500	.1000	.0650	.0400	.0250	.0100	.0040	.0025	.0010
34.75	0000	0000	0000	0000	0000	0000	0000	0001	0001	0001
80	0000	0000	0000	0000	0000	0000	0000	0001	0001	0001
85	0000	0000	0000	0000	0000	0000	0000	0001	0001	0001
90	0000	0000	0000	0000	0000	0000	0000	0001	0001	0001
95	0000	0000	0000	0000	0000	0000	0000	0001	0001	0001
35.00	0000	0000	0000	0000	0000	0000	0000	0000	0001	0001
05	0000	0000	0000	0000	0000	0000	0000	0000	0001	0001
10	0000	0000	0000	0000	0000	0000	0000	0000	0001	0001
15	0000	0000	0000	0000	0000	0000	0000	0000	0001	0001
20	0000	0000	0000	0000	0000	0000	0000	0000	0001	0001
25	0000	0000	0000	0000	0000	0000	0000	0000	0001	0001
30	0000	0000	0000	0000	0000	0000	0000	0000	0001	0001
35	0000	0000	0000	0000	0000	0000	0000	0000	0001	0001
40	0000	0000	0000	0000	0000	0000	0000	0000	0001	0001
45	0000	0000	0000	0000	0000	0000	0000	0000	0001	0001
50	0000	0000	0000	0000	0000	0000	0000	0000	0001	0001
55	0000	0000	0000	0000	0000	0000	0000	0000	0001	0001
60	0000	0000	0000	0000	0000	0000	0000	0000	0001	0001
65	0000	0000	0000	0000	0000	0000	0000	0000	0001	0001
70	0000	0000	0000	0000	0000	0000	0000	0000	0001	0001
75	0000	0000	0000	0000	0000	0000	0000	0000	0001	0001
80	0000	0000	0000	0000	0000	0000	0000	0000	0001	0001
85	0000	0000	0000	0000	0000	0000	0000	0000	0001	0001
90	0000	0000	0000	0000	0000	0000	0000	0000	0001	0001
95	0000	0000	0000	0000	0000	0000	0000	0000	0001	0001
36.00	0000	0000	0000	0000	0000	0000	0000	0000	0001	0001
05	0000	0000	0000	0000	0000	0000	0000	0000	0001	0001
10	0000	0000	0000	0000	0000	0000	0000	0000	0001	0001
15	0000	0000	0000	0000	0000	0000	0000	0000	0001	0001
20	0000	0000	0000	0000	0000	0000	0000	0000	0001	0001
36.25	0000	0000	0000	0000	0000	0000	0000	0000	0001	0001
30	0000	0000	0000	0000	0000	0000	0000	0000	0001	0001
35	0000	0000	0000	0000	0000	0000	0000	0000	0001	0001
40	0000	0000	0000	0000	0000	0000	0000	0000	0001	0001
45	0000	0000	0000	0000	0000	0000	0000	0000	0000	0001
50	0000	0000	0000	0000	0000	0000	0000	0000	0000	0001
55	0000	0000	0000	0000	0000	0000	0000	0000	0000	0001
60	0000	0000	0000	0000	0000	0000	0000	0000	0000	0001
65	0000	0000	0000	0000	0000	0000	0000	0000	0000	0001
70	0000	0000	0000	0000	0000	0000	0000	0000	0000	0001
75	0000	0000	0000	0000	0000	0000	0000	0000	0000	0001
80	0000	0000	0000	0000	0000	0000	0000	0000	0000	0001
85	0000	0000	0000	0000	0000	0000	0000	0000	0000	0001
90	0000	0000	0000	0000	0000	0000	0000	0000	0000	0001
95	0000	0000	0000	0000	0000	0000	0000	0000	0000	0001
37.00	0000	0000	0000	0000	0000	0000	0000	0000	0000	0001
05	0000	0000	0000	0000	0000	0000	0000	0000	0000	0001
10	0000	0000	0000	0000	0000	0000	0000	0000	0000	0001
15	0000	0000	0000	0000	0000	0000	0000	0000	0000	0001
20	0000	0000	0000	0000	0000	0000	0000	0000	0000	0001
25	0000	0000	0000	0000	0000	0000	0000	0000	0000	0001
30	0000	0000	0000	0000	0000	0000	0000	0000	0000	0001
35	0000	0000	0000	0000	0000	0000	0000	0000	0000	0001
40	0000	0000	0000	0000	0000	0000	0000	0000	0000	0001
45	0000	0000	0000	0000	0000	0000	0000	0000	0000	0001
50	0000	0000	0000	0000	0000	0000	0000	0000	0000	0001
55	0000	0000	0000	0000	0000	0000	0000	0000	0000	0001
60	0000	0000	0000	0000	0000	0000	0000	0000	0000	0001
65	0000	0000	0000	0000	0000	0000	0000	0000	0000	0001
70	0000	0000	0000	0000	0000	0000	0000	0000	0000	0001

PROBABILITY DENSITY OF t, THE NON-CENTRAL t-STATISTIC, TABLED AS A FUNCTION OF $t/\sqrt{f}$.

f is the number of degrees of freedom; the non-centrality parameter is $\sqrt{f+1}\,K_p$.
K_p is the standardized normal deviate exceeded with probability p.

DEGREES OF FREEDOM 3

$t/\sqrt{f}$ \ p	.2500	.1500	.1000	.0650	.0400	.0250	.0100	.0040	.0025	.0010
37.75	0000	0000	0000	0000	0000	0000	0000	0000	0000	0001
80	0000	0000	0000	0000	0000	0000	0000	0000	0000	0001
85	0000	0000	0000	0000	0000	0000	0000	0000	0000	0001
90	0000	0000	0000	0000	0000	0000	0000	0000	0000	0001
95	0000	0000	0000	0000	0000	0000	0000	0000	0000	0001
38.00	0000	0000	0000	0000	0000	0000	0000	0000	0000	0001
05	0000	0000	0000	0000	0000	0000	0000	0000	0000	0001
10	0000	0000	0000	0000	0000	0000	0000	0000	0000	0001
15	0000	0000	0000	0000	0000	0000	0000	0000	0000	0001
20	0000	0000	0000	0000	0000	0000	0000	0000	0000	0001
25	0000	0000	0000	0000	0000	0000	0000	0000	0000	0001
30	0000	0000	0000	0000	0000	0000	0000	0000	0000	0001
35	0000	0000	0000	0000	0000	0000	0000	0000	0000	0001
40	0000	0000	0000	0000	0000	0000	0000	0000	0000	0001
45	0000	0000	0000	0000	0000	0000	0000	0000	0000	0001
50	0000	0000	0000	0000	0000	0000	0000	0000	0000	0001
55	0000	0000	0000	0000	0000	0000	0000	0000	0000	0001
60	0000	0000	0000	0000	0000	0000	0000	0000	0000	0001
65	0000	0000	0000	0000	0000	0000	0000	0000	0000	0001
70	0000	0000	0000	0000	0000	0000	0000	0000	0000	0001
75	0000	0000	0000	0000	0000	0000	0000	0000	0000	0001
80	0000	0000	0000	0000	0000	0000	0000	0000	0000	0001
85	0000	0000	0000	0000	0000	0000	0000	0000	0000	0001
90	0000	0000	0000	0000	0000	0000	0000	0000	0000	0001
95	0000	0000	0000	0000	0000	0000	0000	0000	0000	0001
39.00	0000	0000	0000	0000	0000	0000	0000	0000	0000	0000

PROBABILITY DENSITY OF t, THE NON-CENTRAL t-STATISTIC, TABLED AS A FUNCTION OF $t/\sqrt{f}$.

f is the number of degrees of freedom; the non-centrality parameter is $\sqrt{f+1}\,K_p$.
K_p is the standardized normal deviate exceeded with probability p.

DEGREES OF FREEDOM **4**

$t/\sqrt{f}$ \ p	.2500	.1500	.1000	.0650	.0400	.0250	.0100	.0040	.0025	.0010
-2.70	0000	0000	0000	0000	0000	0000	0000	0000	0000	0000
65	0000	0000	0000	0000	0000	0000	0000	0000	0000	0000
60	0001	0000	0000	0000	0000	0000	0000	0000	0000	0000
55	0001	0000	0000	0000	0000	0000	0000	0000	0000	0000
50	0001	0000	0000	0000	0000	0000	0000	0000	0000	0000
45	0001	0000	0000	0000	0000	0000	0000	0000	0000	0000
40	0001	0000	0000	0000	0000	0000	0000	0000	0000	0000
35	0001	0000	0000	0000	0000	0000	0000	0000	0000	0000
30	0001	0000	0000	0000	0000	0000	0000	0000	0000	0000
25	0001	0000	0000	0000	0000	0000	0000	0000	0000	0000
20	0001	0000	0000	0000	0000	0000	0000	0000	0000	0000
15	0001	0000	0000	0000	0000	0000	0000	0000	0000	0000
10	0001	0000	0000	0000	0000	0000	0000	0000	0000	0000
05	0002	0000	0000	0000	0000	0000	0000	0000	0000	0000
-2.00	0002	0000	0000	0000	0000	0000	0000	0000	0000	0000
-1.95	0002	0000	0000	0000	0000	0000	0000	0000	0000	0000
90	0002	0000	0000	0000	0000	0000	0000	0000	0000	0000
85	0003	0000	0000	0000	0000	0000	0000	0000	0000	0000
80	0003	0000	0000	0000	0000	0000	0000	0000	0000	0000
75	0003	0000	0000	0000	0000	0000	0000	0000	0000	0000
70	0004	0000	0000	0000	0000	0000	0000	0000	0000	0000
65	0004	0000	0000	0000	0000	0000	0000	0000	0000	0000
60	0005	0000	0000	0000	0000	0000	0000	0000	0000	0000
55	0005	0000	0000	0000	0000	0000	0000	0000	0000	0000
50	0006	0000	0000	0000	0000	0000	0000	0000	0000	0000
45	0007	0001	0000	0000	0000	0000	0000	0000	0000	0000
40	0008	0001	0000	0000	0000	0000	0000	0000	0000	0000
35	0009	0001	0000	0000	0000	0000	0000	0000	0000	0000
30	0011	0001	0000	0000	0000	0000	0000	0000	0000	0000
-1.25	0013	0001	0000	0000	0000	0000	0000	0000	0000	0000
20	0015	0001	0000	0000	0000	0000	0000	0000	0000	0000
15	0017	0001	0000	0000	0000	0000	0000	0000	0000	0000
10	0020	0002	0000	0000	0000	0000	0000	0000	0000	0000
05	0024	0002	0000	0000	0000	0000	0000	0000	0000	0000
-1.00	0028	0002	0000	0000	0000	0000	0000	0000	0000	0000
-0.95	0033	0003	0000	0000	0000	0000	0000	0000	0000	0000
90	0040	0004	0000	0000	0000	0000	0000	0000	0000	0000
85	0048	0004	0001	0000	0000	0000	0000	0000	0000	0000
80	0057	0005	0001	0000	0000	0000	0000	0000	0000	0000
75	0069	0007	0001	0000	0000	0000	0000	0000	0000	0000
70	0083	0008	0001	0000	0000	0000	0000	0000	0000	0000
65	0101	0010	0002	0000	0000	0000	0000	0000	0000	0000
60	0122	0013	0002	0000	0000	0000	0000	0000	0000	0000
55	0149	0016	0003	0000	0000	0000	0000	0000	0000	0000
50	0181	0020	0003	0000	0000	0000	0000	0000	0000	0000
45	0221	0026	0004	0001	0000	0000	0000	0000	0000	0000
40	0270	0033	0006	0001	0000	0000	0000	0000	0000	0000
35	0330	0043	0008	0001	0000	0000	0000	0000	0000	0000
30	0402	0055	0010	0002	0000	0000	0000	0000	0000	0000
25	0489	0071	0013	0002	0000	0000	0000	0000	0000	0000
20	0593	0092	0018	0003	0000	0000	0000	0000	0000	0000
15	0715	0120	0025	0004	0001	0000	0000	0000	0000	0000
10	0857	0155	0033	0006	0001	0000	0000	0000	0000	0000
05	1019	0199	0045	0008	0001	0000	0000	0000	0000	0000
0.00	1203	0256	0062	0012	0002	0000	0000	0000	0000	0000
05	1405	0326	0084	0017	0003	0000	0000	0000	0000	0000
10	1623	0413	0113	0025	0004	0001	0000	0000	0000	0000
15	1853	0517	0151	0036	0006	0001	0000	0000	0000	0000
20	2089	0641	0200	0051	0009	0002	0000	0000	0000	0000

PROBABILITY DENSITY OF t, THE NON-CENTRAL t-STATISTIC, TABLED AS A FUNCTION OF $t/\sqrt{f}$.

f is the number of degrees of freedom; the non-centrality parameter is $\sqrt{f+1}\,K_p$.
K_p is the standardized normal deviate exceeded with probability p.

DEGREES OF FREEDOM **4**

$t/\sqrt{f}$ \ p	.2500	.1500	.1000	.0650	.0400	.0250	.0100	.0040	.0025	.0010
0.25	2325	0786	0263	0071	0014	0003	0000	0000	0000	0000
30	2552	0950	0341	0099	0022	0004	0000	0000	0000	0000
35	2764	1132	0436	0136	0032	0007	0000	0000	0000	0000
40	2953	1329	0549	0184	0047	0011	0001	0000	0000	0000
45	3113	1536	0681	0245	0068	0018	0001	0000	0000	0000
50	3239	1749	0830	0321	0096	0027	0002	0000	0000	0000
55	3329	1962	0996	0413	0133	0040	0003	0000	0000	0000
60	3381	2168	1174	0521	0181	0059	0005	0000	0000	0000
65	3397	2361	1361	0645	0242	0085	0009	0000	0000	0000
70	3378	2535	1553	0784	0316	0118	0014	0001	0000	0000
75	3327	2687	1743	0936	0404	0162	0022	0003	0001	0000
80	3249	2813	1928	1098	0507	0216	0034	0004	0001	0000
85	3149	2911	2103	1266	0622	0283	0050	0007	0003	0000
90	3030	2981	2262	1437	0750	0362	0072	0012	0004	0001
95	2897	3022	2403	1605	0888	0452	0100	0018	0007	0001
1.00	2754	3036	2524	1768	1032	0555	0136	0028	0012	0002
05	2606	3024	2621	1920	1180	0668	0181	0041	0018	0004
10	2454	2990	2696	2061	1330	0789	0235	0059	0028	0006
15	2303	2937	2747	2186	1476	0916	0299	0082	0040	0009
20	2154	2866	2776	2293	1618	1047	0372	0111	0057	0015
25	2009	2782	2784	2383	1750	1180	0453	0147	0079	0022
30	1869	2687	2773	2453	1873	1311	0543	0190	0106	0032
35	1735	2584	2744	2506	1983	1438	0639	0241	0140	0045
40	1608	2475	2701	2539	2079	1559	0741	0299	0179	0062
45	1488	2363	2644	2556	2160	1672	0846	0364	0226	0083
50	1375	2248	2577	2557	2227	1775	0953	0435	0279	0110
55	1270	2134	2501	2543	2278	1868	1060	0513	0338	0141
60	1172	2020	2418	2516	2315	1948	1165	0595	0403	0178
65	1081	1909	2331	2478	2338	2017	1267	0680	0473	0220
70	0997	1800	2240	2430	2348	2073	1365	0768	0548	0267
1.75	0919	1695	2146	2375	2345	2117	1456	0857	0626	0319
80	0847	1594	2052	2312	2332	2149	1541	0946	0707	0376
85	0781	1497	1958	2244	2309	2170	1618	1034	0789	0437
90	0720	1405	1865	2172	2277	2180	1686	1120	0871	0502
95	0663	1318	1773	2097	2237	2180	1746	1202	0952	0569
2.00	0612	1235	1684	2020	2191	2171	1797	1279	1031	0638
05	0565	1156	1597	1942	2140	2153	1839	1352	1108	0709
10	0521	1083	1512	1863	2085	2129	1873	1419	1182	0780
15	0481	1013	1431	1785	2026	2098	1898	1480	1251	0850
20	0445	0948	1354	1708	1964	2061	1915	1535	1316	0919
25	0411	0887	1279	1632	1901	2020	1925	1583	1375	0986
30	0380	0830	1208	1558	1836	1975	1928	1624	1429	1051
35	0352	0777	1141	1485	1771	1927	1924	1659	1477	1113
40	0326	0727	1077	1415	1706	1876	1914	1687	1520	1171
45	0302	0680	1016	1347	1641	1823	1898	1709	1557	1225
50	0280	0637	0958	1281	1576	1768	1878	1725	1588	1276
55	0260	0596	0904	1218	1513	1713	1853	1735	1613	1321
60	0242	0558	0853	1158	1451	1657	1825	1740	1632	1363
65	0225	0523	0804	1100	1390	1601	1793	1739	1647	1399
70	0209	0490	0758	1045	1331	1546	1759	1734	1656	1431
75	0194	0460	0715	0992	1273	1490	1722	1725	1661	1459
80	0181	0431	0674	0942	1218	1436	1683	1712	1661	1482
85	0169	0404	0636	0894	1164	1382	1643	1695	1657	1500
90	0157	0379	0600	0848	1112	1329	1602	1675	1650	1514
95	0147	0356	0566	0805	1062	1277	1560	1653	1639	1524
3.00	0137	0335	0535	0764	1014	1227	1517	1628	1624	1531
05	0128	0314	0505	0725	0968	1178	1473	1600	1607	1533
10	0120	0296	0477	0688	0924	1131	1430	1571	1587	1532
15	0112	0278	0450	0652	0881	1085	1387	1541	1565	1528
20	0105	0262	0425	0619	0841	1040	1344	1509	1541	1522

PROBABILITY DENSITY OF t, THE NON-CENTRAL t-STATISTIC, TABLED AS A FUNCTION OF $t/\sqrt{f}$.

f is the number of degrees of freedom; the non-centrality parameter is $\sqrt{f+1}\,K_p$.
K_p is the standardized normal deviate exceeded with probability p.

DEGREES OF FREEDOM 4

$t/\sqrt{f}$ \ P	.2500	.1500	.1000	.0650	.0400	.0250	.0100	.0040	.0025	.0010
3.25	0098	0246	0402	0588	0802	0997	1301	1476	1516	1512
30	0092	0232	0380	0558	0765	0956	1259	1442	1489	1500
35	0087	0219	0360	0530	0730	0916	1217	1408	1460	1485
40	0081	0206	0340	0503	0696	0878	1177	1373	1431	1469
45	0076	0194	0322	0478	0664	0841	1137	1338	1400	1451
50	0072	0184	0305	0454	0634	0805	1098	1302	1370	1431
55	0068	0173	0289	0431	0604	0771	1059	1267	1338	1409
60	0064	0164	0273	0410	0577	0738	1022	1232	1306	1387
65	0060	0155	0259	0390	0550	0707	0986	1198	1274	1363
70	0056	0146	0246	0371	0525	0677	0951	1163	1242	1339
75	0053	0138	0233	0353	0501	0649	0917	1129	1210	1314
80	0050	0131	0221	0336	0478	0621	0884	1096	1179	1288
85	0047	0124	0210	0319	0457	0595	0852	1063	1147	1262
90	0045	0117	0199	0304	0436	0570	0821	1030	1116	1235
95	0042	0111	0189	0289	0416	0546	0790	0998	1085	1208
4.00	0040	0105	0180	0276	0398	0523	0761	0967	1054	1181
05	0038	0100	0171	0263	0380	0501	0733	0937	1024	1154
10	0036	0095	0163	0250	0363	0480	0706	0907	0994	1126
15	0034	0090	0155	0239	0347	0460	0680	0878	0965	1099
20	0032	0086	0147	0228	0332	0440	0655	0850	0937	1072
25	0030	0081	0140	0217	0317	0422	0631	0823	0909	1046
30	0029	0077	0133	0207	0303	0405	0608	0796	0881	1019
35	0027	0074	0127	0198	0290	0388	0585	0770	0855	0993
40	0026	0070	0121	0189	0278	0372	0564	0745	0829	0967
45	0025	0067	0115	0180	0266	0357	0543	0720	0803	0941
50	0024	0063	0110	0172	0254	0342	0523	0697	0778	0916
55	0022	0060	0105	0165	0244	0328	0503	0674	0754	0892
60	0021	0058	0100	0157	0233	0315	0485	0652	0731	0867
65	0020	0055	0096	0150	0224	0302	0467	0630	0708	0844
70	0019	0052	0091	0144	0214	0290	0450	0609	0686	0820
4.75	0018	0050	0087	0138	0205	0279	0434	0589	0664	0797
80	0018	0048	0083	0132	0197	0268	0418	0569	0644	0775
85	0017	0045	0080	0126	0189	0257	0403	0551	0623	0753
90	0016	0043	0076	0121	0181	0247	0388	0532	0604	0732
95	0015	0042	0073	0116	0174	0237	0374	0515	0585	0711
5.00	0014	0040	0070	0111	0167	0228	0360	0498	0566	0691
05	0014	0038	0067	0106	0160	0219	0347	0481	0548	0671
10	0013	0036	0064	0102	0154	0211	0335	0465	0531	0652
15	0013	0035	0061	0098	0147	0203	0323	0450	0514	0633
20	0012	0033	0059	0094	0142	0195	0312	0435	0498	0615
25	0012	0032	0056	0090	0136	0188	0300	0421	0482	0597
30	0011	0030	0054	0086	0131	0180	0290	0407	0467	0580
35	0011	0029	0052	0083	0126	0174	0280	0394	0452	0563
40	0010	0028	0050	0080	0121	0167	0270	0381	0438	0547
45	0010	0027	0048	0076	0116	0161	0260	0368	0424	0531
50	0009	0026	0046	0073	0112	0155	0251	0356	0411	0516
55	0009	0025	0044	0071	0108	0149	0243	0345	0398	0501
60	0009	0024	0042	0068	0104	0144	0234	0334	0386	0486
65	0008	0023	0041	0065	0100	0139	0226	0323	0374	0472
70	0008	0022	0039	0063	0096	0134	0218	0313	0362	0458
75	0008	0021	0037	0060	0092	0129	0211	0303	0351	0445
80	0007	0020	0036	0058	0089	0124	0204	0293	0340	0432
85	0007	0019	0035	0056	0086	0120	0197	0283	0330	0420
90	0007	0019	0033	0054	0083	0116	0190	0274	0319	0407
95	0006	0018	0032	0052	0080	0111	0184	0266	0310	0396
6.00	0006	0017	0031	0050	0077	0108	0178	0257	0300	0384
05	0006	0017	0030	0048	0074	0104	0172	0249	0291	0373
10	0006	0016	0029	0046	0071	0100	0166	0242	0282	0362
15	0005	0015	0028	0045	0069	0097	0161	0234	0274	0352
20	0005	0015	0027	0043	0067	0093	0156	0227	0265	0342

PROBABILITY DENSITY OF t, THE NON-CENTRAL t-STATISTIC, TABLED AS A FUNCTION OF $t/\sqrt{f}$.

f is the number of degrees of freedom; the non-centrality parameter is $\sqrt{f+1}\,K_p$.
K_p is the standardized normal deviate exceeded with probability p.

DEGREES OF FREEDOM **4**

$t/\sqrt{f}$ \ p	.2500	.1500	.1000	.0650	.0400	.0250	.0100	.0040	.0025	.0010
6.25	0005	0014	0026	0042	0064	0090	0151	0220	0257	0332
30	0005	0014	0025	0040	0062	0087	0146	0213	0249	0323
35	0005	0013	0024	0039	0060	0084	0141	0206	0242	0314
40	0005	0013	0023	0037	0058	0081	0136	0200	0235	0305
45	0004	0012	0022	0036	0056	0079	0132	0194	0228	0296
50	0004	0012	0021	0035	0054	0076	0128	0188	0221	0288
55	0004	0011	0021	0034	0052	0074	0124	0182	0215	0280
60	0004	0011	0020	0032	0050	0071	0120	0177	0208	0272
65	0004	0011	0019	0031	0049	0069	0116	0172	0202	0264
70	0004	0010	0019	0030	0047	0067	0113	0166	0196	0257
75	0004	0010	0018	0029	0046	0064	0109	0161	0190	0250
80	0003	0010	0017	0028	0044	0062	0106	0157	0185	0243
85	0003	0009	0017	0027	0043	0060	0102	0152	0180	0236
90	0003	0009	0016	0026	0041	0059	0099	0148	0174	0229
95	0003	0009	0016	0026	0040	0057	0096	0143	0169	0223
7.00	0003	0008	0015	0025	0039	0055	0093	0139	0165	0217
05	0003	0008	0015	0024	0037	0053	0091	0135	0160	0211
10	0003	0008	0014	0023	0036	0052	0088	0131	0155	0205
15	0003	0008	0014	0022	0035	0050	0085	0127	0151	0200
20	0003	0007	0013	0022	0034	0048	0083	0124	0147	0194
25	0002	0007	0013	0021	0033	0047	0080	0120	0143	0189
30	0002	0007	0012	0020	0032	0045	0078	0117	0139	0184
35	0002	0007	0012	0020	0031	0044	0076	0113	0135	0179
40	0002	0006	0012	0019	0030	0043	0073	0110	0131	0174
45	0002	0006	0011	0019	0029	0042	0071	0107	0127	0170
50	0002	0006	0011	0018	0028	0040	0069	0104	0124	0165
55	0002	0006	0011	0017	0027	0039	0067	0101	0120	0161
60	0002	0006	0010	0017	0027	0038	0065	0098	0117	0157
65	0002	0005	0010	0016	0026	0037	0063	0096	0114	0152
70	0002	0005	0010	0016	0025	0036	0062	0093	0111	0148
7.75	0002	0005	0009	0015	0024	0035	0060	0091	0108	0145
80	0002	0005	0009	0015	0024	0034	0058	0088	0105	0141
85	0002	0005	0009	0015	0023	0033	0057	0086	0102	0137
90	0002	0005	0009	0014	0022	0032	0055	0083	0100	0134
95	0002	0005	0008	0014	0022	0031	0053	0081	0097	0130
8.00	0002	0004	0008	0013	0021	0030	0052	0079	0094	0127
05	0001	0004	0008	0013	0020	0029	0051	0077	0092	0124
10	0001	0004	0008	0013	0020	0028	0049	0075	0089	0120
15	0001	0004	0007	0012	0019	0028	0048	0073	0087	0117
20	0001	0004	0007	0012	0019	0027	0047	0071	0085	0114
25	0001	0004	0007	0012	0018	0026	0045	0069	0083	0112
30	0001	0004	0007	0011	0018	0025	0044	0067	0081	0109
35	0001	0004	0007	0011	0017	0025	0043	0065	0078	0106
40	0001	0003	0006	0011	0017	0024	0042	0064	0076	0103
45	0001	0003	0006	0010	0016	0023	0041	0062	0075	0101
50	0001	0003	0006	0010	0016	0023	0040	0061	0073	0098
55	0001	0003	0006	0010	0015	0022	0039	0059	0071	0096
60	0001	0003	0006	0009	0015	0022	0038	0058	0069	0094
65	0001	0003	0006	0009	0015	0021	0037	0056	0067	0091
70	0001	0003	0005	0009	0014	0020	0036	0055	0066	0089
75	0001	0003	0005	0009	0014	0020	0035	0053	0064	0087
80	0001	0003	0005	0008	0013	0019	0034	0052	0062	0085
85	0001	0003	0005	0008	0013	0019	0033	0051	0061	0083
90	0001	0003	0005	0008	0013	0018	0032	0049	0059	0081
95	0001	0003	0005	0008	0012	0018	0031	0048	0058	0079
9.00	0001	0003	0005	0008	0012	0017	0031	0047	0057	0077
05	0001	0002	0004	0007	0012	0017	0030	0046	0055	0075
10	0001	0002	0004	0007	0011	0017	0029	0045	0054	0073
15	0001	0002	0004	0007	0011	0016	0028	0044	0053	0072
20	0001	0002	0004	0007	0011	0016	0028	0043	0051	0070

PROBABILITY DENSITY OF t, THE NON-CENTRAL t-STATISTIC, TABLED AS A FUNCTION OF $t/\sqrt{f}$

f is the number of degrees of freedom; the non-centrality parameter is $\sqrt{f+1}\,K_p$.
K_p is the standardized normal deviate exceeded with probability p.

DEGREES OF FREEDOM **4**

$t/\sqrt{f}$ \ P	.2500	.1500	.1000	.0650	.0400	.0250	.0100	.0040	.0025	.0010
9.25	0001	0002	0004	0007	0011	0015	0027	0042	0050	0068
30	0001	0002	0004	0007	0010	0015	0026	0041	0049	0067
35	0001	0002	0004	0006	0010	0015	0026	0040	0048	0065
40	0001	0002	0004	0006	0010	0014	0025	0039	0047	0064
45	0001	0002	0004	0006	0010	0014	0024	0038	0046	0062
50	0001	0002	0004	0006	0009	0014	0024	0037	0044	0061
55	0001	0002	0003	0006	0009	0013	0023	0036	0043	0060
60	0001	0002	0003	0006	0009	0013	0023	0035	0042	0058
65	0001	0002	0003	0005	0009	0013	0022	0034	0041	0057
70	0001	0002	0003	0005	0008	0012	0022	0034	0040	0056
75	0001	0002	0003	0005	0008	0012	0021	0033	0040	0054
80	0001	0002	0003	0005	0008	0012	0021	0032	0039	0053
85	0001	0002	0003	0005	0008	0011	0020	0031	0038	0052
90	0001	0002	0003	0005	0008	0011	0020	0031	0037	0051
95	0001	0002	0003	0005	0008	0011	0019	0030	0036	0050
10.00	0001	0002	0003	0005	0007	0011	0019	0029	0035	0049
05	0001	0001	0003	0004	0007	0010	0018	0029	0034	0047
10	0000	0001	0003	0004	0007	0010	0018	0028	0034	0046
15	0000	0001	0003	0004	0007	0010	0018	0027	0033	0045
20	0000	0001	0003	0004	0007	0010	0017	0027	0032	0044
25	0000	0001	0002	0004	0007	0009	0017	0026	0032	0043
30	0000	0001	0002	0004	0006	0009	0016	0025	0031	0043
35	0000	0001	0002	0004	0006	0009	0016	0025	0030	0042
40	0000	0001	0002	0004	0006	0009	0016	0024	0030	0041
45	0000	0001	0002	0004	0006	0009	0015	0024	0029	0040
50	0000	0001	0002	0004	0006	0008	0015	0023	0028	0039
55	0000	0001	0002	0004	0006	0008	0015	0023	0028	0038
60	0000	0001	0002	0003	0006	0008	0014	0022	0027	0037
65	0000	0001	0002	0003	0005	0008	0014	0022	0026	0037
70	0000	0001	0002	0003	0005	0008	0014	0021	0026	0036
10.75	0000	0001	0002	0003	0005	0008	0013	0021	0025	0035
80	0000	0001	0002	0003	0005	0007	0013	0020	0025	0034
85	0000	0001	0002	0003	0005	0007	0013	0020	0024	0034
90	0000	0001	0002	0003	0005	0007	0013	0020	0024	0033
95	0000	0001	0002	0003	0005	0007	0012	0019	0023	0032
11.00	0000	0001	0002	0003	0005	0007	0012	0019	0023	0032
05	0000	0001	0002	0003	0005	0007	0012	0018	0022	0031
10	0000	0001	0002	0003	0004	0006	0012	0018	0022	0030
15	0000	0001	0002	0003	0004	0006	0011	0018	0021	0030
20	0000	0001	0002	0003	0004	0006	0011	0017	0021	0029
25	0000	0001	0002	0003	0004	0006	0011	0017	0021	0029
30	0000	0001	0002	0003	0004	0006	0011	0017	0020	0028
35	0000	0001	0001	0002	0004	0006	0010	0016	0020	0027
40	0000	0001	0001	0002	0004	0006	0010	0016	0019	0027
45	0000	0001	0001	0002	0004	0006	0010	0016	0019	0026
50	0000	0001	0001	0002	0004	0005	0010	0015	0019	0026
55	0000	0001	0001	0002	0004	0005	0010	0015	0018	0025
60	0000	0001	0001	0002	0004	0005	0009	0015	0018	0025
65	0000	0001	0001	0002	0004	0005	0009	0014	0018	0024
70	0000	0001	0001	0002	0003	0005	0009	0014	0017	0024
75	0000	0001	0001	0002	0003	0005	0009	0014	0017	0023
80	0000	0001	0001	0002	0003	0005	0009	0014	0016	0023
85	0000	0001	0001	0002	0003	0005	0008	0013	0016	0023
90	0000	0001	0001	0002	0003	0005	0008	0013	0016	0022
95	0000	0001	0001	0002	0003	0005	0008	0013	0016	0022
12.00	0000	0001	0001	0002	0003	0004	0008	0013	0015	0021
05	0000	0001	0001	0002	0003	0004	0008	0012	0015	0021
10	0000	0001	0001	0002	0003	0004	0008	0012	0015	0020
15	0000	0001	0001	0002	0003	0004	0008	0012	0014	0020
20	0000	0001	0001	0002	0003	0004	0007	0012	0014	0020

PROBABILITY DENSITY OF t, THE NON-CENTRAL t-STATISTIC, TABLED AS A FUNCTION OF $t/\sqrt{f}$.

f is the number of degrees of freedom; the non-centrality parameter is $\sqrt{f+1}\,K_p$.
K_p is the standardized normal deviate exceeded with probability p.

DEGREES OF FREEDOM **4**

$t/\sqrt{f}$ \ p	.2500	.1500	.1000	.0650	.0400	.0250	.0100	.0040	.0025	.0010
12.25	0000	0001	0001	0002	0003	0004	0007	0011	0014	0019
30	0000	0001	0001	0002	0003	0004	0007	0011	0014	0019
35	0000	0001	0001	0002	0003	0004	0007	0011	0013	0019
40	0000	0001	0001	0002	0003	0004	0007	0011	0013	0018
45	0000	0001	0001	0002	0003	0004	0007	0011	0013	0018
50	0000	0001	0001	0002	0003	0004	0007	0010	0013	0018
55	0000	0000	0001	0002	0002	0004	0006	0010	0012	0017
60	0000	0000	0001	0002	0002	0004	0006	0010	0012	0017
65	0000	0000	0001	0001	0002	0003	0006	0010	0012	0017
70	0000	0000	0001	0001	0002	0003	0006	0010	0012	0016
75	0000	0000	0001	0001	0002	0003	0006	0009	0011	0016
80	0000	0000	0001	0001	0002	0003	0006	0009	0011	0016
85	0000	0000	0001	0001	0002	0003	0006	0009	0011	0015
90	0000	0000	0001	0001	0002	0003	0006	0009	0011	0015
95	0000	0000	0001	0001	0002	0003	0006	0009	0011	0015
13.00	0000	0000	0001	0001	0002	0003	0005	0009	0010	0015
05	0000	0000	0001	0001	0002	0003	0005	0008	0010	0014
10	0000	0000	0001	0001	0002	0003	0005	0008	0010	0014
15	0000	0000	0001	0001	0002	0003	0005	0008	0010	0014
20	0000	0000	0001	0001	0002	0003	0005	0008	0010	0014
25	0000	0000	0001	0001	0002	0003	0005	0008	0010	0013
30	0000	0000	0001	0001	0002	0003	0005	0008	0009	0013
35	0000	0000	0001	0001	0002	0003	0005	0008	0009	0013
40	0000	0000	0001	0001	0002	0003	0005	0007	0009	0013
45	0000	0000	0001	0001	0002	0003	0005	0007	0009	0012
50	0000	0000	0001	0001	0002	0003	0005	0007	0009	0012
55	0000	0000	0001	0001	0002	0002	0004	0007	0009	0012
60	0000	0000	0001	0001	0002	0002	0004	0007	0008	0012
65	0000	0000	0001	0001	0002	0002	0004	0007	0008	0012
70	0000	0000	0001	0001	0002	0002	0004	0007	0008	0011
13.75	0000	0000	0001	0001	0002	0002	0004	0007	0008	0011
80	0000	0000	0001	0001	0002	0002	0004	0006	0008	0011
85	0000	0000	0001	0001	0002	0002	0004	0006	0008	0011
90	0000	0000	0001	0001	0001	0002	0004	0006	0008	0011
95	0000	0000	0001	0001	0001	0002	0004	0006	0007	0011
14.00	0000	0000	0001	0001	0001	0002	0004	0006	0007	0010
05	0000	0000	0001	0001	0001	0002	0004	0006	0007	0010
10	0000	0000	0001	0001	0001	0002	0004	0006	0007	0010
15	0000	0000	0001	0001	0001	0002	0004	0006	0007	0010
20	0000	0000	0000	0001	0001	0002	0004	0006	0007	0010
25	0000	0000	0000	0001	0001	0002	0003	0006	0007	0010
30	0000	0000	0000	0001	0001	0002	0003	0005	0007	0009
35	0000	0000	0000	0001	0001	0002	0003	0005	0007	0009
40	0000	0000	0000	0001	0001	0002	0003	0005	0006	0009
45	0000	0000	0000	0001	0001	0002	0003	0005	0006	0009
50	0000	0000	0000	0001	0001	0002	0003	0005	0006	0009
55	0000	0000	0000	0001	0001	0002	0003	0005	0006	0009
60	0000	0000	0000	0001	0001	0002	0003	0005	0006	0008
65	0000	0000	0000	0001	0001	0002	0003	0005	0006	0008
70	0000	0000	0000	0001	0001	0002	0003	0005	0006	0008
75	0000	0000	0000	0001	0001	0002	0003	0005	0006	0008
80	0000	0000	0000	0001	0001	0002	0003	0005	0006	0008
85	0000	0000	0000	0001	0001	0002	0003	0005	0006	0008
90	0000	0000	0000	0001	0001	0002	0003	0004	0005	0008
95	0000	0000	0000	0001	0001	0002	0003	0004	0005	0008
15.00	0000	0000	0000	0001	0001	0002	0003	0004	0005	0007
05	0000	0000	0000	0001	0001	0001	0003	0004	0005	0007
10	0000	0000	0000	0001	0001	0001	0003	0004	0005	0007
15	0000	0000	0000	0001	0001	0001	0003	0004	0005	0007
20	0000	0000	0000	0001	0001	0001	0003	0004	0005	0007

PROBABILITY DENSITY OF t, THE NON-CENTRAL t-STATISTIC, TABLED AS A FUNCTION OF $t/\sqrt{f}$.

f is the number of degrees of freedom; the non-centrality parameter is $\sqrt{f+1}\,K_p$.
K_p is the standardized normal deviate exceeded with probability p.

DEGREES OF FREEDOM **4**

$t/\sqrt{f}$ \ p	.2500	.1500	.1000	.0650	.0400	.0250	.0100	.0040	.0025	.0010
15.25	0000	0000	0000	0001	0001	0001	0003	0004	0005	0007
30	0000	0000	0000	0001	0001	0001	0002	0004	0005	0007
35	0000	0000	0000	0001	0001	0001	0002	0004	0005	0007
40	0000	0000	0000	0001	0001	0001	0002	0004	0005	0007
45	0000	0000	0000	0001	0001	0001	0002	0004	0005	0006
50	0000	0000	0000	0001	0001	0001	0002	0004	0005	0006
55	0000	0000	0000	0001	0001	0001	0002	0004	0004	0006
60	0000	0000	0000	0001	0001	0001	0002	0004	0004	0006
65	0000	0000	0000	0001	0001	0001	0002	0004	0004	0006
70	0000	0000	0000	0001	0001	0001	0002	0003	0004	0006
75	0000	0000	0000	0001	0001	0001	0002	0003	0004	0006
80	0000	0000	0000	0000	0001	0001	0002	0003	0004	0006
85	0000	0000	0000	0000	0001	0001	0002	0003	0004	0006
90	0000	0000	0000	0000	0001	0001	0002	0003	0004	0006
95	0000	0000	0000	0000	0001	0001	0002	0003	0004	0006
16.00	0000	0000	0000	0000	0001	0001	0002	0003	0004	0005
05	0000	0000	0000	0000	0001	0001	0002	0003	0004	0005
10	0000	0000	0000	0000	0001	0001	0002	0003	0004	0005
15	0000	0000	0000	0000	0001	0001	0002	0003	0004	0005
20	0000	0000	0000	0000	0001	0001	0002	0003	0004	0005
25	0000	0000	0000	0000	0001	0001	0002	0003	0004	0005
30	0000	0000	0000	0000	0001	0001	0002	0003	0004	0005
35	0000	0000	0000	0000	0001	0001	0002	0003	0004	0005
40	0000	0000	0000	0000	0001	0001	0002	0003	0003	0005
45	0000	0000	0000	0000	0001	0001	0002	0003	0003	0005
50	0000	0000	0000	0000	0001	0001	0002	0003	0003	0005
55	0000	0000	0000	0000	0001	0001	0002	0003	0003	0005
60	0000	0000	0000	0000	0001	0001	0002	0003	0003	0005
65	0000	0000	0000	0000	0001	0001	0002	0003	0003	0005
70	0000	0000	0000	0000	0001	0001	0002	0003	0003	0004
16.75	0000	0000	0000	0000	0001	0001	0002	0003	0003	0004
80	0000	0000	0000	0000	0001	0001	0002	0003	0003	0004
85	0000	0000	0000	0000	0001	0001	0002	0002	0003	0004
90	0000	0000	0000	0000	0001	0001	0002	0002	0003	0004
95	0000	0000	0000	0000	0001	0001	0002	0002	0003	0004
17.00	0000	0000	0000	0000	0001	0001	0001	0002	0003	0004
05	0000	0000	0000	0000	0001	0001	0001	0002	0003	0004
10	0000	0000	0000	0000	0001	0001	0001	0002	0003	0004
15	0000	0000	0000	0000	0001	0001	0001	0002	0003	0004
20	0000	0000	0000	0000	0001	0001	0001	0002	0003	0004
25	0000	0000	0000	0000	0001	0001	0001	0002	0003	0004
30	0000	0000	0000	0000	0001	0001	0001	0002	0003	0004
35	0000	0000	0000	0000	0001	0001	0001	0002	0003	0004
40	0000	0000	0000	0000	0000	0001	0001	0002	0003	0004
45	0000	0000	0000	0000	0000	0001	0001	0002	0003	0004
50	0000	0000	0000	0000	0000	0001	0001	0002	0003	0004
55	0000	0000	0000	0000	0000	0001	0001	0002	0002	0004
60	0000	0000	0000	0000	0000	0001	0001	0002	0002	0003
65	0000	0000	0000	0000	0000	0001	0001	0002	0002	0003
70	0000	0000	0000	0000	0000	0001	0001	0002	0002	0003
75	0000	0000	0000	0000	0000	0001	0001	0002	0002	0003
80	0000	0000	0000	0000	0000	0001	0001	0002	0002	0003
85	0000	0000	0000	0000	0000	0001	0001	0002	0002	0003
90	0000	0000	0000	0000	0000	0001	0001	0002	0002	0003
95	0000	0000	0000	0000	0000	0001	0001	0002	0002	0003
18.00	0000	0000	0000	0000	0000	0001	0001	0002	0002	0003
05	0000	0000	0000	0000	0000	0001	0001	0002	0002	0003
10	0000	0000	0000	0000	0000	0001	0001	0002	0002	0003
15	0000	0000	0000	0000	0000	0001	0001	0002	0002	0003
20	0000	0000	0000	0000	0000	0001	0001	0002	0002	0003

PROBABILITY DENSITY OF t, THE NON-CENTRAL t-STATISTIC, TABLED AS A FUNCTION OF $t/\sqrt{f}$.

f is the number of degrees of freedom; the non-centrality parameter is $\sqrt{f+1}\,K_p$.
K_p is the standardized normal deviate exceeded with probability p.

DEGREES OF FREEDOM **4**

$t/\sqrt{f}$ \ p	.2500	.1500	.1000	.0650	.0400	.0250	.0100	.0040	.0025	.0010
18.25	0000	0000	0000	0000	0000	0001	0001	0002	0002	0003
30	0000	0000	0000	0000	0000	0001	0001	0002	0002	0003
35	0000	0000	0000	0000	0000	0001	0001	0002	0002	0003
40	0000	0000	0000	0000	0000	0001	0001	0002	0002	0003
45	0000	0000	0000	0000	0000	0001	0001	0002	0002	0003
50	0000	0000	0000	0000	0000	0001	0001	0002	0002	0003
55	0000	0000	0000	0000	0000	0001	0001	0002	0002	0003
60	0000	0000	0000	0000	0000	0001	0001	0002	0002	0003
65	0000	0000	0000	0000	0000	0001	0001	0002	0002	0003
70	0000	0000	0000	0000	0000	0001	0001	0001	0002	0003
75	0000	0000	0000	0000	0000	0001	0001	0001	0002	0003
80	0000	0000	0000	0000	0000	0000	0001	0001	0002	0003
85	0000	0000	0000	0000	0000	0000	0001	0001	0002	0002
90	0000	0000	0000	0000	0000	0000	0001	0001	0002	0002
95	0000	0000	0000	0000	0000	0000	0001	0001	0002	0002
19.00	0000	0000	0000	0000	0000	0000	0001	0001	0002	0002
05	0000	0000	0000	0000	0000	0000	0001	0001	0002	0002
10	0000	0000	0000	0000	0000	0000	0001	0001	0002	0002
15	0000	0000	0000	0000	0000	0000	0001	0001	0002	0002
20	0000	0000	0000	0000	0000	0000	0001	0001	0002	0002
25	0000	0000	0000	0000	0000	0000	0001	0001	0002	0002
30	0000	0000	0000	0000	0000	0000	0001	0001	0002	0002
35	0000	0000	0000	0000	0000	0000	0001	0001	0002	0002
40	0000	0000	0000	0000	0000	0000	0001	0001	0002	0002
45	0000	0000	0000	0000	0000	0000	0001	0001	0002	0002
50	0000	0000	0000	0000	0000	0000	0001	0001	0001	0002
55	0000	0000	0000	0000	0000	0000	0001	0001	0001	0002
60	0000	0000	0000	0000	0000	0000	0001	0001	0001	0002
65	0000	0000	0000	0000	0000	0000	0001	0001	0001	0002
70	0000	0000	0000	0000	0000	0000	0001	0001	0001	0002
19.75	0000	0000	0000	0000	0000	0000	0001	0001	0001	0002
80	0000	0000	0000	0000	0000	0000	0001	0001	0001	0002
85	0000	0000	0000	0000	0000	0000	0001	0001	0001	0002
90	0000	0000	0000	0000	0000	0000	0001	0001	0001	0002
95	0000	0000	0000	0000	0000	0000	0001	0001	0001	0002
20.00	0000	0000	0000	0000	0000	0000	0001	0001	0001	0002
05	0000	0000	0000	0000	0000	0000	0001	0001	0001	0002
10	0000	0000	0000	0000	0000	0000	0001	0001	0001	0002
15	0000	0000	0000	0000	0000	0000	0001	0001	0001	0002
20	0000	0000	0000	0000	0000	0000	0001	0001	0001	0002
25	0000	0000	0000	0000	0000	0000	0001	0001	0001	0002
30	0000	0000	0000	0000	0000	0000	0001	0001	0001	0002
35	0000	0000	0000	0000	0000	0000	0001	0001	0001	0002
40	0000	0000	0000	0000	0000	0000	0001	0001	0001	0002
45	0000	0000	0000	0000	0000	0000	0001	0001	0001	0002
50	0000	0000	0000	0000	0000	0000	0001	0001	0001	0002
55	0000	0000	0000	0000	0000	0000	0001	0001	0001	0002
60	0000	0000	0000	0000	0000	0000	0001	0001	0001	0002
65	0000	0000	0000	0000	0000	0000	0001	0001	0001	0002
70	0000	0000	0000	0000	0000	0000	0001	0001	0001	0002
75	0000	0000	0000	0000	0000	0000	0001	0001	0001	0002
80	0000	0000	0000	0000	0000	0000	0001	0001	0001	0002
85	0000	0000	0000	0000	0000	0000	0001	0001	0001	0002
90	0000	0000	0000	0000	0000	0000	0001	0001	0001	0002
95	0000	0000	0000	0000	0000	0000	0001	0001	0001	0001
21.00	0000	0000	0000	0000	0000	0000	0001	0001	0001	0001
05	0000	0000	0000	0000	0000	0000	0001	0001	0001	0001
10	0000	0000	0000	0000	0000	0000	0001	0001	0001	0001
15	0000	0000	0000	0000	0000	0000	0001	0001	0001	0001
20	0000	0000	0000	0000	0000	0000	0001	0001	0001	0001

PROBABILITY DENSITY OF t, THE NON-CENTRAL t-STATISTIC, TABLED AS A FUNCTION OF $t/\sqrt{f}$.

f is the number of degrees of freedom; the non-centrality parameter is $\sqrt{f+1}\,K_p$.
K_p is the standardized normal deviate exceeded with probability p.

DEGREES OF FREEDOM **4**

$t/\sqrt{f}$ \ P	.2500	.1500	.1000	.0650	.0400	.0250	.0100	.0040	.0025	.0010
2 1.25	0000	0000	0000	0000	0000	0000	0000	0001	0001	0001
30	0000	0000	0000	0000	0000	0000	0000	0001	0001	0001
35	0000	0000	0000	0000	0000	0000	0000	0001	0001	0001
40	0000	0000	0000	0000	0000	0000	0000	0001	0001	0001
45	0000	0000	0000	0000	0000	0000	0000	0001	0001	0001
50	0000	0000	0000	0000	0000	0000	0000	0001	0001	0001
55	0000	0000	0000	0000	0000	0000	0000	0001	0001	0001
60	0000	0000	0000	0000	0000	0000	0000	0001	0001	0001
65	0000	0000	0000	0000	0000	0000	0000	0001	0001	0001
70	0000	0000	0000	0000	0000	0000	0000	0001	0001	0001
75	0000	0000	0000	0000	0000	0000	0000	0001	0001	0001
80	0000	0000	0000	0000	0000	0000	0000	0001	0001	0001
85	0000	0000	0000	0000	0000	0000	0000	0001	0001	0001
90	0000	0000	0000	0000	0000	0000	0000	0001	0001	0001
95	0000	0000	0000	0000	0000	0000	0000	0001	0001	0001
2 2.00	0000	0000	0000	0000	0000	0000	0000	0001	0001	0001
05	0000	0000	0000	0000	0000	0000	0000	0001	0001	0001
10	0000	0000	0000	0000	0000	0000	0000	0001	0001	0001
15	0000	0000	0000	0000	0000	0000	0000	0001	0001	0001
20	0000	0000	0000	0000	0000	0000	0000	0001	0001	0001
25	0000	0000	0000	0000	0000	0000	0000	0001	0001	0001
30	0000	0000	0000	0000	0000	0000	0000	0001	0001	0001
35	0000	0000	0000	0000	0000	0000	0000	0001	0001	0001
40	0000	0000	0000	0000	0000	0000	0000	0001	0001	0001
45	0000	0000	0000	0000	0000	0000	0000	0001	0001	0001
50	0000	0000	0000	0000	0000	0000	0000	0001	0001	0001
55	0000	0000	0000	0000	0000	0000	0000	0001	0001	0001
60	0000	0000	0000	0000	0000	0000	0000	0001	0001	0001
65	0000	0000	0000	0000	0000	0000	0000	0001	0001	0001
70	0000	0000	0000	0000	0000	0000	0000	0001	0001	0001
2 2.75	0000	0000	0000	0000	0000	0000	0000	0001	0001	0001
80	0000	0000	0000	0000	0000	0000	0000	0001	0001	0001
85	0000	0000	0000	0000	0000	0000	0000	0001	0001	0001
90	0000	0000	0000	0000	0000	0000	0000	0001	0001	0001
95	0000	0000	0000	0000	0000	0000	0000	0001	0001	0001
2 3.00	0000	0000	0000	0000	0000	0000	0000	0001	0001	0001
05	0000	0000	0000	0000	0000	0000	0000	0001	0001	0001
10	0000	0000	0000	0000	0000	0000	0000	0001	0001	0001
15	0000	0000	0000	0000	0000	0000	0000	0001	0001	0001
20	0000	0000	0000	0000	0000	0000	0000	0001	0001	0001
25	0000	0000	0000	0000	0000	0000	0000	0001	0001	0001
30	0000	0000	0000	0000	0000	0000	0000	0001	0001	0001
35	0000	0000	0000	0000	0000	0000	0000	0001	0001	0001
40	0000	0000	0000	0000	0000	0000	0000	0000	0001	0001
45	0000	0000	0000	0000	0000	0000	0000	0000	0001	0001
50	0000	0000	0000	0000	0000	0000	0000	0000	0001	0001
55	0000	0000	0000	0000	0000	0000	0000	0000	0001	0001
60	0000	0000	0000	0000	0000	0000	0000	0000	0001	0001
65	0000	0000	0000	0000	0000	0000	0000	0000	0001	0001
70	0000	0000	0000	0000	0000	0000	0000	0000	0001	0001
75	0000	0000	0000	0000	0000	0000	0000	0000	0001	0001
80	0000	0000	0000	0000	0000	0000	0000	0000	0001	0001
85	0000	0000	0000	0000	0000	0000	0000	0000	0001	0001
90	0000	0000	0000	0000	0000	0000	0000	0000	0001	0001
95	0000	0000	0000	0000	0000	0000	0000	0000	0001	0001
2 4.00	0000	0000	0000	0000	0000	0000	0000	0000	0001	0001
05	0000	0000	0000	0000	0000	0000	0000	0000	0001	0001
10	0000	0000	0000	0000	0000	0000	0000	0000	0001	0001
15	0000	0000	0000	0000	0000	0000	0000	0000	0001	0001
20	0000	0000	0000	0000	0000	0000	0000	0000	0001	0001

PROBABILITY DENSITY OF t, THE NON-CENTRAL t-STATISTIC, TABLED AS A FUNCTION OF $t/\sqrt{f}$.

f is the number of degrees of freedom; the non-centrality parameter is $\sqrt{f+1}\,K_p$.
K_p is the standardized normal deviate exceeded with probability p.

DEGREES OF FREEDOM **4**

$t/\sqrt{f}$ \ p	.2500	.1500	.1000	.0650	.0400	.0250	.0100	.0040	.0025	.0010
24.25	0000	0000	0000	0000	0000	0000	0000	0000	0001	0001
30	0000	0000	0000	0000	0000	0000	0000	0000	0001	0001
35	0000	0000	0000	0000	0000	0000	0000	0000	0001	0001
40	0000	0000	0000	0000	0000	0000	0000	0000	0000	0001
45	0000	0000	0000	0000	0000	0000	0000	0000	0000	0001
50	0000	0000	0000	0000	0000	0000	0000	0000	0000	0001
55	0000	0000	0000	0000	0000	0000	0000	0000	0000	0001
60	0000	0000	0000	0000	0000	0000	0000	0000	0000	0001
65	0000	0000	0000	0000	0000	0000	0000	0000	0000	0001
70	0000	0000	0000	0000	0000	0000	0000	0000	0000	0001
75	0000	0000	0000	0000	0000	0000	0000	0000	0000	0001
80	0000	0000	0000	0000	0000	0000	0000	0000	0000	0001
85	0000	0000	0000	0000	0000	0000	0000	0000	0000	0001
90	0000	0000	0000	0000	0000	0000	0000	0000	0000	0001
95	0000	0000	0000	0000	0000	0000	0000	0000	0000	0001
25.00	0000	0000	0000	0000	0000	0000	0000	0000	0000	0001
05	0000	0000	0000	0000	0000	0000	0000	0000	0000	0001
10	0000	0000	0000	0000	0000	0000	0000	0000	0000	0001
15	0000	0000	0000	0000	0000	0000	0000	0000	0000	0001
20	0000	0000	0000	0000	0000	0000	0000	0000	0000	0001
25	0000	0000	0000	0000	0000	0000	0000	0000	0000	0001
30	0000	0000	0000	0000	0000	0000	0000	0000	0000	0001
35	0000	0000	0000	0000	0000	0000	0000	0000	0000	0001
40	0000	0000	0000	0000	0000	0000	0000	0000	0000	0001
45	0000	0000	0000	0000	0000	0000	0000	0000	0000	0001
50	0000	0000	0000	0000	0000	0000	0000	0000	0000	0001
55	0000	0000	0000	0000	0000	0000	0000	0000	0000	0001
60	0000	0000	0000	0000	0000	0000	0000	0000	0000	0001
65	0000	0000	0000	0000	0000	0000	0000	0000	0000	0001
70	0000	0000	0000	0000	0000	0000	0000	0000	0000	0001
25.75	0000	0000	0000	0000	0000	0000	0000	0000	0000	0001
80	0000	0000	0000	0000	0000	0000	0000	0000	0000	0001
85	0000	0000	0000	0000	0000	0000	0000	0000	0000	0001
90	0000	0000	0000	0000	0000	0000	0000	0000	0000	0001
95	0000	0000	0000	0000	0000	0000	0000	0000	0000	0001
26.00	0000	0000	0000	0000	0000	0000	0000	0000	0000	0001
05	0000	0000	0000	0000	0000	0000	0000	0000	0000	0001
10	0000	0000	0000	0000	0000	0000	0000	0000	0000	0001
15	0000	0000	0000	0000	0000	0000	0000	0000	0000	0001
20	0000	0000	0000	0000	0000	0000	0000	0000	0000	0000

PROBABILITY DENSITY OF t, THE NON-CENTRAL t-STATISTIC, TABLED AS A FUNCTION OF $t/\sqrt{f}$.

f is the number of degrees of freedom; the non-centrality parameter is $\sqrt{f+1}\,K_p$.
K_p is the standardized normal deviate exceeded with probability p.

DEGREES OF FREEDOM **5**

$t/\sqrt{f}$ \ P	.2500	.1500	.1000	.0650	.0400	.0250	.0100	.0040	.0025	.0010
-2.00	0000	0000	0000	0000	0000	0000	0000	0000	0000	0000
-1.95	0000	0000	0000	0000	0000	0000	0000	0000	0000	0000
90	0001	0000	0000	0000	0000	0000	0000	0000	0000	0000
85	0001	0000	0000	0000	0000	0000	0000	0000	0000	0000
80	0001	0000	0000	0000	0000	0000	0000	0000	0000	0000
75	0001	0000	0000	0000	0000	0000	0000	0000	0000	0000
70	0001	0000	0000	0000	0000	0000	0000	0000	0000	0000
65	0001	0000	0000	0000	0000	0000	0000	0000	0000	0000
60	0001	0000	0000	0000	0000	0000	0000	0000	0000	0000
55	0001	0000	0000	0000	0000	0000	0000	0000	0000	0000
50	0002	0000	0000	0000	0000	0000	0000	0000	0000	0000
45	0002	0000	0000	0000	0000	0000	0000	0000	0000	0000
40	0002	0000	0000	0000	0000	0000	0000	0000	0000	0000
35	0003	0000	0000	0000	0000	0000	0000	0000	0000	0000
30	0003	0000	0000	0000	0000	0000	0000	0000	0000	0000
25	0004	0000	0000	0000	0000	0000	0000	0000	0000	0000
20	0005	0000	0000	0000	0000	0000	0000	0000	0000	0000
15	0006	0000	0000	0000	0000	0000	0000	0000	0000	0000
10	0007	0000	0000	0000	0000	0000	0000	0000	0000	0000
05	0009	0000	0000	0000	0000	0000	0000	0000	0000	0000
-1.00	0010	0001	0000	0000	0000	0000	0000	0000	0000	0000
-0.95	0013	0001	0000	0000	0000	0000	0000	0000	0000	0000
90	0016	0001	0000	0000	0000	0000	0000	0000	0000	0000
85	0020	0001	0000	0000	0000	0000	0000	0000	0000	0000
80	0025	0001	0000	0000	0000	0000	0000	0000	0000	0000
75	0031	0002	0000	0000	0000	0000	0000	0000	0000	0000
70	0039	0002	0000	0000	0000	0000	0000	0000	0000	0000
65	0049	0003	0000	0000	0000	0000	0000	0000	0000	0000
60	0061	0004	0000	0000	0000	0000	0000	0000	0000	0000
55	0078	0005	0001	0000	0000	0000	0000	0000	0000	0000
-0.50	0099	0007	0001	0000	0000	0000	0000	0000	0000	0000
45	0126	0010	0001	0000	0000	0000	0000	0000	0000	0000
40	0160	0013	0002	0000	0000	0000	0000	0000	0000	0000
35	0204	0017	0002	0000	0000	0000	0000	0000	0000	0000
30	0258	0024	0003	0000	0000	0000	0000	0000	0000	0000
25	0327	0032	0004	0000	0000	0000	0000	0000	0000	0000
20	0413	0044	0006	0001	0000	0000	0000	0000	0000	0000
15	0517	0060	0009	0001	0000	0000	0000	0000	0000	0000
10	0644	0082	0013	0002	0000	0000	0000	0000	0000	0000
05	0794	0112	0019	0003	0000	0000	0000	0000	0000	0000
0.00	0970	0151	0028	0004	0000	0000	0000	0000	0000	0000
05	1170	0203	0040	0006	0001	0000	0000	0000	0000	0000
10	1394	0270	0057	0009	0001	0000	0000	0000	0000	0000
15	1637	0355	0081	0014	0002	0000	0000	0000	0000	0000
20	1894	0461	0114	0022	0003	0000	0000	0000	0000	0000
25	2156	0589	0159	0033	0005	0001	0000	0000	0000	0000
30	2416	0742	0218	0050	0008	0001	0000	0000	0000	0000
35	2662	0918	0293	0073	0013	0002	0000	0000	0000	0000
40	2886	1115	0388	0105	0021	0004	0000	0000	0000	0000
45	3079	1330	0504	0149	0032	0006	0000	0000	0000	0000
50	3233	1558	0641	0206	0049	0011	0000	0000	0000	0000
55	3346	1792	0800	0280	0073	0017	0001	0000	0000	0000
60	3413	2024	0977	0371	0106	0028	0002	0000	0000	0000
65	3436	2246	1170	0482	0150	0043	0003	0000	0000	0000
70	3417	2452	1373	0611	0207	0064	0005	0000	0000	0000
75	3359	2634	1582	0757	0279	0094	0009	0001	0000	0000
80	3268	2787	1789	0919	0367	0133	0015	0001	0000	0000
85	3150	2909	1989	1093	0471	0184	0023	0002	0001	0000
90	3010	2996	2175	1274	0591	0248	0036	0004	0001	0000
95	2855	3050	2343	1459	0724	0326	0054	0007	0002	0000

PROBABILITY DENSITY OF t, THE NON-CENTRAL t-STATISTIC, TABLED AS A FUNCTION OF $t/\sqrt{f}$

f is the number of degrees of freedom; the non-centrality parameter is $\sqrt{f+1}\,K_p$.
K_p is the standardized normal deviate exceeded with probability p.

DEGREES OF FREEDOM **5**

$t/\sqrt{f}$ \ p	.2500	.1500	.1000	.0650	.0400	.0250	.0100	.0040	.0025	.0010
1.00	2689	3070	2488	1641	0870	0417	0079	0012	0004	0001
05	2518	3060	2608	1816	1024	0522	0111	0019	0007	0001
10	2345	3022	2700	1979	1184	0639	0152	0029	0012	0002
15	2174	2960	2766	2127	1345	0767	0204	0044	0019	0003
20	2007	2878	2804	2257	1503	0902	0265	0063	0029	0006
25	1846	2779	2817	2366	1655	1042	0337	0089	0042	0009
30	1694	2668	2806	2453	1797	1184	0419	0121	0061	0015
35	1550	2547	2774	2519	1927	1325	0511	0161	0084	0022
40	1415	2421	2724	2562	2042	1462	0611	0209	0114	0032
45	1290	2291	2658	2585	2141	1592	0717	0265	0151	0046
50	1174	2160	2579	2588	2223	1713	0828	0329	0194	0064
55	1068	2029	2490	2574	2287	1823	0942	0401	0245	0087
60	0970	1901	2393	2543	2333	1920	1057	0480	0304	0115
65	0881	1777	2290	2499	2363	2003	1171	0565	0369	0149
70	0799	1657	2185	2443	2377	2072	1281	0654	0440	0188
75	0725	1542	2077	2377	2376	2127	1386	0747	0517	0233
80	0657	1433	1969	2304	2361	2167	1484	0842	0598	0285
85	0596	1330	1862	2224	2334	2194	1575	0937	0682	0341
90	0541	1233	1757	2140	2297	2207	1657	1032	0769	0403
95	0491	1142	1655	2052	2251	2209	1729	1124	0857	0469
2.00	0446	1056	1556	1963	2196	2199	1791	1213	0944	0539
05	0405	0977	1460	1873	2136	2179	1843	1297	1030	0611
10	0368	0903	1369	1784	2071	2150	1884	1375	1113	0686
15	0334	0834	1282	1695	2002	2114	1916	1447	1193	0761
20	0304	0771	1199	1608	1930	2071	1938	1512	1268	0836
25	0277	0712	1121	1523	1856	2022	1950	1570	1338	0911
30	0252	0657	1047	1441	1781	1969	1954	1620	1402	0984
35	0230	0607	0978	1361	1706	1912	1950	1663	1459	1054
40	0210	0561	0912	1285	1632	1852	1939	1698	1511	1121
45	0191	0518	0851	1211	1558	1790	1921	1725	1555	1185
2.50	0175	0478	0794	1141	1485	1727	1897	1745	1593	1244
55	0160	0442	0740	1075	1414	1663	1868	1758	1624	1298
60	0146	0409	0690	1011	1345	1599	1834	1765	1649	1347
65	0134	0378	0643	0951	1278	1535	1797	1765	1667	1391
70	0123	0350	0600	0894	1214	1471	1756	1759	1679	1430
75	0113	0324	0559	0840	1151	1408	1713	1749	1685	1463
80	0104	0300	0521	0789	1091	1347	1667	1733	1686	1492
85	0095	0278	0486	0742	1034	1287	1620	1713	1682	1514
90	0088	0257	0453	0697	0979	1229	1572	1690	1673	1532
95	0081	0239	0423	0654	0927	1172	1522	1663	1660	1545
3.00	0074	0221	0395	0614	0877	1117	1472	1633	1643	1553
05	0068	0205	0368	0577	0829	1064	1423	1600	1622	1556
10	0063	0191	0344	0542	0784	1013	1373	1566	1599	1556
15	0058	0177	0321	0509	0741	0964	1323	1530	1573	1551
20	0054	0165	0300	0478	0701	0917	1274	1492	1544	1543
25	0050	0153	0280	0449	0662	0872	1226	1453	1514	1532
30	0046	0143	0262	0422	0626	0829	1179	1414	1481	1517
35	0043	0133	0245	0397	0592	0787	1133	1374	1448	1500
40	0040	0124	0230	0373	0559	0748	1088	1333	1413	1480
45	0037	0116	0215	0351	0528	0710	1044	1293	1378	1459
50	0034	0108	0201	0330	0499	0675	1001	1252	1341	1435
55	0032	0101	0189	0310	0472	0641	0959	1212	1305	1410
60	0030	0094	0177	0292	0446	0608	0919	1172	1268	1383
65	0027	0088	0166	0275	0422	0578	0880	1133	1231	1355
70	0026	0082	0155	0259	0399	0548	0843	1094	1194	1326
75	0024	0077	0146	0244	0377	0521	0807	1056	1158	1296
80	0022	0072	0137	0229	0357	0494	0772	1018	1121	1266
85	0021	0067	0129	0216	0337	0470	0739	0982	1085	1235
90	0019	0063	0121	0204	0319	0446	0707	0946	1050	1204
95	0018	0059	0114	0192	0302	0423	0676	0911	1015	1173

PROBABILITY DENSITY OF t, THE NON-CENTRAL t-STATISTIC, TABLED AS A FUNCTION OF $t/\sqrt{f}$.

f is the number of degrees of freedom; the non-centrality parameter is $\sqrt{f+1}\,K_p$.
K_p is the standardized normal deviate exceeded with probability p.

DEGREES OF FREEDOM **5**

$t/\sqrt{f}$ \ p	.2500	.1500	.1000	.0650	.0400	.0250	.0100	.0040	.0025	.0010
4.00	0017	0055	0107	0181	0286	0402	0646	0877	0981	1141
05	0016	0052	0101	0171	0271	0382	0618	0845	0947	1110
10	0015	0049	0095	0162	0256	0363	0591	0813	0915	1079
15	0014	0046	0089	0153	0243	0345	0565	0782	0883	1048
20	0013	0043	0084	0144	0230	0328	0540	0752	0852	1017
25	0012	0041	0079	0136	0218	0311	0516	0723	0821	0987
30	0011	0038	0075	0129	0207	0296	0493	0695	0792	0957
35	0011	0036	0071	0122	0196	0281	0471	0668	0763	0928
40	0010	0034	0067	0115	0186	0268	0451	0642	0735	0899
45	0010	0032	0063	0109	0176	0255	0431	0617	0709	0870
50	0009	0030	0059	0103	0167	0242	0412	0592	0683	0843
55	0008	0028	0056	0098	0159	0230	0394	0569	0657	0816
60	0008	0027	0053	0093	0151	0219	0376	0547	0633	0789
65	0007	0025	0050	0088	0143	0209	0360	0525	0609	0763
70	0007	0024	0047	0083	0136	0199	0344	0504	0587	0738
75	0007	0023	0045	0079	0129	0189	0329	0484	0565	0714
80	0006	0021	0043	0075	0123	0180	0315	0465	0543	0690
85	0006	0020	0040	0071	0117	0172	0301	0447	0523	0667
90	0006	0019	0038	0067	0111	0164	0288	0429	0503	0644
95	0005	0018	0036	0064	0106	0156	0276	0412	0484	0622
5.00	0005	0017	0034	0061	0101	0149	0264	0396	0466	0601
05	0005	0016	0033	0058	0096	0142	0252	0380	0449	0580
10	0004	0015	0031	0055	0091	0135	0242	0365	0432	0561
15	0004	0015	0029	0052	0087	0129	0231	0351	0415	0541
20	0004	0014	0028	0050	0083	0123	0221	0337	0400	0523
25	0004	0013	0027	0047	0079	0118	0212	0324	0385	0505
30	0004	0013	0025	0045	0075	0112	0203	0311	0370	0487
35	0003	0012	0024	0043	0072	0107	0195	0299	0356	0470
40	0003	0011	0023	0041	0069	0103	0186	0287	0343	0454
45	0003	0011	0022	0039	0065	0098	0179	0276	0330	0438
5.50	0003	0010	0021	0037	0062	0094	0171	0265	0318	0423
55	0003	0010	0020	0035	0060	0090	0164	0255	0306	0408
60	0003	0009	0019	0034	0057	0086	0157	0245	0294	0394
65	0003	0009	0018	0032	0054	0082	0151	0236	0283	0381
70	0002	0008	0017	0031	0052	0078	0145	0227	0273	0367
75	0002	0008	0016	0029	0050	0075	0139	0218	0263	0355
80	0002	0008	0016	0028	0048	0072	0133	0210	0253	0342
85	0002	0007	0015	0027	0045	0069	0128	0202	0244	0331
90	0002	0007	0014	0026	0043	0066	0123	0194	0235	0319
95	0002	0007	0014	0024	0042	0063	0118	0187	0226	0308
6.00	0002	0006	0013	0023	0040	0060	0113	0180	0218	0297
05	0002	0006	0012	0022	0038	0058	0109	0173	0210	0287
10	0002	0006	0012	0021	0037	0056	0104	0166	0202	0277
15	0002	0006	0011	0020	0035	0053	0100	0160	0195	0268
20	0001	0005	0011	0020	0034	0051	0096	0154	0188	0259
25	0001	0005	0010	0019	0032	0049	0093	0149	0181	0250
30	0001	0005	0010	0018	0031	0047	0089	0143	0175	0241
35	0001	0005	0009	0017	0030	0045	0086	0138	0168	0233
40	0001	0004	0009	0017	0028	0043	0082	0133	0162	0225
45	0001	0004	0009	0016	0027	0042	0079	0128	0157	0218
50	0001	0004	0008	0015	0026	0040	0076	0123	0151	0210
55	0001	0004	0008	0015	0025	0038	0073	0119	0146	0203
60	0001	0004	0008	0014	0024	0037	0071	0115	0140	0196
65	0001	0004	0007	0013	0023	0035	0068	0110	0136	0190
70	0001	0003	0007	0013	0022	0034	0065	0106	0131	0183
75	0001	0003	0007	0012	0021	0033	0063	0103	0126	0177
80	0001	0003	0006	0012	0020	0031	0061	0099	0122	0171
85	0001	0003	0006	0011	0020	0030	0058	0096	0118	0166
90	0001	0003	0006	0011	0019	0029	0056	0092	0114	0160
95	0001	0003	0006	0011	0018	0028	0054	0089	0110	0155

PROBABILITY DENSITY OF t, THE NON-CENTRAL t-STATISTIC, TABLED AS A FUNCTION OF $t/\sqrt{f}$.

f is the number of degrees of freedom; the non-centrality parameter is $\sqrt{f+1}\,K_p$.
K_p is the standardized normal deviate exceeded with probability p.

$t/\sqrt{f}$ \ p	.2500	.1500	.1000	.0650	.0400	.0250	.0100	.0040	.0025	.0010
7.00	0001	0003	0005	0010	0017	0027	0052	0086	0106	0150
05	0001	0003	0005	0010	0017	0026	0050	0083	0102	0145
10	0001	0002	0005	0009	0016	0025	0049	0080	0099	0140
15	0001	0002	0005	0009	0016	0024	0047	0077	0096	0136
20	0001	0002	0005	0009	0015	0023	0045	0075	0092	0131
25	0001	0002	0005	0008	0014	0022	0044	0072	0089	0127
30	0001	0002	0004	0008	0014	0022	0042	0070	0086	0123
35	0001	0002	0004	0008	0013	0021	0041	0067	0083	0119
40	0001	0002	0004	0007	0013	0020	0039	0065	0081	0115
45	0001	0002	0004	0007	0012	0019	0038	0063	0078	0112
50	0001	0002	0004	0007	0012	0019	0036	0061	0075	0108
55	0000	0002	0004	0007	0012	0018	0035	0059	0073	0105
60	0000	0002	0003	0006	0011	0017	0034	0057	0071	0101
65	0000	0002	0003	0006	0011	0017	0033	0055	0068	0098
70	0000	0002	0003	0006	0010	0016	0032	0053	0066	0095
75	0000	0001	0003	0006	0010	0016	0031	0051	0064	0092
80	0000	0001	0003	0006	0010	0015	0030	0050	0062	0089
85	0000	0001	0003	0005	0009	0015	0029	0048	0060	0086
90	0000	0001	0003	0005	0009	0014	0028	0046	0058	0084
95	0000	0001	0003	0005	0009	0014	0027	0045	0056	0081
8.00	0000	0001	0003	0005	0008	0013	0026	0044	0054	0079
05	0000	0001	0002	0005	0008	0013	0025	0042	0053	0076
10	0000	0001	0002	0004	0008	0012	0024	0041	0051	0074
15	0000	0001	0002	0004	0008	0012	0023	0040	0049	0072
20	0000	0001	0002	0004	0007	0011	0023	0038	0048	0070
25	0000	0001	0002	0004	0007	0011	0022	0037	0046	0068
30	0000	0001	0002	0004	0007	0011	0021	0036	0045	0066
35	0000	0001	0002	0004	0007	0010	0021	0035	0044	0064
40	0000	0001	0002	0004	0006	0010	0020	0034	0042	0062
45	0000	0001	0002	0004	0006	0010	0019	0033	0041	0060
8.50	0000	0001	0002	0003	0006	0009	0019	0032	0040	0058
55	0000	0001	0002	0003	0006	0009	0018	0031	0039	0056
60	0000	0001	0002	0003	0006	0009	0018	0030	0037	0055
65	0000	0001	0002	0003	0005	0008	0017	0029	0036	0053
70	0000	0001	0002	0003	0005	0008	0016	0028	0035	0052
75	0000	0001	0002	0003	0005	0008	0016	0027	0034	0050
80	0000	0001	0001	0003	0005	0008	0015	0026	0033	0049
85	0000	0001	0001	0003	0005	0007	0015	0026	0032	0047
90	0000	0001	0001	0003	0005	0007	0015	0025	0031	0046
95	0000	0001	0001	0003	0004	0007	0014	0024	0030	0045
9.00	0000	0001	0001	0002	0004	0007	0014	0023	0029	0043
05	0000	0001	0001	0002	0004	0007	0013	0023	0029	0042
10	0000	0001	0001	0002	0004	0006	0013	0022	0028	0041
15	0000	0001	0001	0002	0004	0006	0012	0021	0027	0040
20	0000	0001	0001	0002	0004	0006	0012	0021	0026	0039
25	0000	0001	0001	0002	0004	0006	0012	0020	0025	0038
30	0000	0001	0001	0002	0004	0006	0011	0020	0025	0037
35	0000	0001	0001	0002	0003	0005	0011	0019	0024	0036
40	0000	0000	0001	0002	0003	0005	0011	0018	0023	0035
45	0000	0000	0001	0002	0003	0005	0010	0018	0023	0034
50	0000	0000	0001	0002	0003	0005	0010	0017	0022	0033
55	0000	0000	0001	0002	0003	0005	0010	0017	0021	0032
60	0000	0000	0001	0002	0003	0005	0010	0016	0021	0031
65	0000	0000	0001	0002	0003	0005	0009	0016	0020	0030
70	0000	0000	0001	0002	0003	0004	0009	0016	0020	0029
75	0000	0000	0001	0002	0003	0004	0009	0015	0019	0028
80	0000	0000	0001	0002	0003	0004	0009	0015	0019	0028
85	0000	0000	0001	0001	0003	0004	0008	0014	0018	0027
90	0000	0000	0001	0001	0003	0004	0008	0014	0018	0026
95	0000	0000	0001	0001	0002	0004	0008	0014	0017	0026

PROBABILITY DENSITY OF t, THE NON-CENTRAL t-STATISTIC, TABLED AS A FUNCTION OF $t/\sqrt{f}$.

f is the number of degrees of freedom; the non-centrality parameter is $\sqrt{f+1}\,K_p$.
K_p is the standardized normal deviate exceeded with probability p.

DEGREES OF FREEDOM **5**

$t/\sqrt{f}$ \ p	.2500	.1500	.1000	.0650	.0400	.0250	.0100	.0040	.0025	.0010
10.00	0000	0000	0001	0001	0002	0004	0008	0013	0017	0025
05	0000	0000	0001	0001	0002	0004	0007	0013	0016	0024
10	0000	0000	0001	0001	0002	0004	0007	0012	0016	0024
15	0000	0000	0001	0001	0002	0003	0007	0012	0015	0023
20	0000	0000	0001	0001	0002	0003	0007	0012	0015	0022
25	0000	0000	0001	0001	0002	0003	0007	0012	0015	0022
30	0000	0000	0001	0001	0002	0003	0006	0011	0014	0021
35	0000	0000	0001	0001	0002	0003	0006	0011	0014	0021
40	0000	0000	0001	0001	0002	0003	0006	0011	0014	0020
45	0000	0000	0001	0001	0002	0003	0006	0010	0013	0020
50	0000	0000	0001	0001	0002	0003	0006	0010	0013	0019
55	0000	0000	0001	0001	0002	0003	0006	0010	0012	0019
60	0000	0000	0001	0001	0002	0003	0006	0010	0012	0018
65	0000	0000	0000	0001	0002	0003	0005	0009	0012	0018
70	0000	0000	0000	0001	0002	0003	0005	0009	0012	0017
75	0000	0000	0000	0001	0002	0002	0005	0009	0011	0017
80	0000	0000	0000	0001	0002	0002	0005	0009	0011	0016
85	0000	0000	0000	0001	0001	0002	0005	0008	0011	0016
90	0000	0000	0000	0001	0001	0002	0005	0008	0010	0016
95	0000	0000	0000	0001	0001	0002	0005	0008	0010	0015
11.00	0000	0000	0000	0001	0001	0002	0004	0008	0010	0015
05	0000	0000	0000	0001	0001	0002	0004	0008	0010	0015
10	0000	0000	0000	0001	0001	0002	0004	0007	0009	0014
15	0000	0000	0000	0001	0001	0002	0004	0007	0009	0014
20	0000	0000	0000	0001	0001	0002	0004	0007	0009	0013
25	0000	0000	0000	0001	0001	0002	0004	0007	0009	0013
30	0000	0000	0000	0001	0001	0002	0004	0007	0009	0013
35	0000	0000	0000	0001	0001	0002	0004	0007	0008	0013
40	0000	0000	0000	0001	0001	0002	0004	0006	0008	0012
45	0000	0000	0000	0001	0001	0002	0004	0006	0008	0012
11.50	0000	0000	0000	0001	0001	0002	0003	0006	0008	0012
55	0000	0000	0000	0001	0001	0002	0003	0006	0008	0011
60	0000	0000	0000	0001	0001	0002	0003	0006	0007	0011
65	0000	0000	0000	0001	0001	0002	0003	0006	0007	0011
70	0000	0000	0000	0001	0001	0002	0003	0006	0007	0011
75	0000	0000	0000	0001	0001	0001	0003	0005	0007	0010
80	0000	0000	0000	0001	0001	0001	0003	0005	0007	0010
85	0000	0000	0000	0001	0001	0001	0003	0005	0007	0010
90	0000	0000	0000	0000	0001	0001	0003	0005	0006	0010
95	0000	0000	0000	0000	0001	0001	0003	0005	0006	0009
12.00	0000	0000	0000	0000	0001	0001	0003	0005	0006	0009
05	0000	0000	0000	0000	0001	0001	0003	0005	0006	0009
10	0000	0000	0000	0000	0001	0001	0003	0005	0006	0009
15	0000	0000	0000	0000	0001	0001	0003	0004	0006	0009
20	0000	0000	0000	0000	0001	0001	0002	0004	0006	0008
25	0000	0000	0000	0000	0001	0001	0002	0004	0005	0008
30	0000	0000	0000	0000	0001	0001	0002	0004	0005	0008
35	0000	0000	0000	0000	0001	0001	0002	0004	0005	0008
40	0000	0000	0000	0000	0001	0001	0002	0004	0005	0008
45	0000	0000	0000	0000	0001	0001	0002	0004	0005	0008
50	0000	0000	0000	0000	0001	0001	0002	0004	0005	0007
55	0000	0000	0000	0000	0001	0001	0002	0004	0005	0007
60	0000	0000	0000	0000	0001	0001	0002	0004	0005	0007
65	0000	0000	0000	0000	0001	0001	0002	0004	0005	0007
70	0000	0000	0000	0000	0001	0001	0002	0003	0004	0007
75	0000	0000	0000	0000	0001	0001	0002	0003	0004	0007
80	0000	0000	0000	0000	0001	0001	0002	0003	0004	0006
85	0000	0000	0000	0000	0001	0001	0002	0003	0004	0006
90	0000	0000	0000	0000	0001	0001	0002	0003	0004	0006
95	0000	0000	0000	0000	0001	0001	0002	0003	0004	0006

PROBABILITY DENSITY OF t, THE NON-CENTRAL t-STATISTIC, TABLED AS A FUNCTION OF $t/\sqrt{f}$.

f is the number of degrees of freedom; the non-centrality parameter is $\sqrt{f+1}\,K_p$.
K_p is the standardized normal deviate exceeded with probability p.

DEGREES OF FREEDOM **5**

$t/\sqrt{f}$ \ P	.2500	.1500	.1000	.0650	.0400	.0250	.0100	.0040	.0025	.0010
13.00	0000	0000	0000	0000	0001	0001	0002	0003	0004	0006
05	0000	0000	0000	0000	0001	0001	0002	0003	0004	0006
10	0000	0000	0000	0000	0001	0001	0002	0003	0004	0006
15	0000	0000	0000	0000	0000	0001	0002	0003	0004	0006
20	0000	0000	0000	0000	0000	0001	0002	0003	0004	0005
25	0000	0000	0000	0000	0000	0001	0002	0003	0003	0005
30	0000	0000	0000	0000	0000	0001	0002	0003	0003	0005
35	0000	0000	0000	0000	0000	0001	0001	0003	0003	0005
40	0000	0000	0000	0000	0000	0001	0001	0003	0003	0005
45	0000	0000	0000	0000	0000	0001	0001	0003	0003	0005
50	0000	0000	0000	0000	0000	0001	0001	0002	0003	0005
55	0000	0000	0000	0000	0000	0001	0001	0002	0003	0005
60	0000	0000	0000	0000	0000	0001	0001	0002	0003	0005
65	0000	0000	0000	0000	0000	0001	0001	0002	0003	0004
70	0000	0000	0000	0000	0000	0001	0001	0002	0003	0004
75	0000	0000	0000	0000	0000	0001	0001	0002	0003	0004
80	0000	0000	0000	0000	0000	0001	0001	0002	0003	0004
85	0000	0000	0000	0000	0000	0001	0001	0002	0003	0004
90	0000	0000	0000	0000	0000	0001	0001	0002	0003	0004
95	0000	0000	0000	0000	0000	0001	0001	0002	0003	0004
14.00	0000	0000	0000	0000	0000	0001	0001	0002	0003	0004
05	0000	0000	0000	0000	0000	0001	0001	0002	0003	0004
10	0000	0000	0000	0000	0000	0001	0001	0002	0002	0004
15	0000	0000	0000	0000	0000	0001	0001	0002	0002	0004
20	0000	0000	0000	0000	0000	0001	0001	0002	0002	0004
25	0000	0000	0000	0000	0000	0000	0001	0002	0002	0004
30	0000	0000	0000	0000	0000	0000	0001	0002	0002	0003
35	0000	0000	0000	0000	0000	0000	0001	0002	0002	0003
40	0000	0000	0000	0000	0000	0000	0001	0002	0002	0003
45	0000	0000	0000	0000	0000	0000	0001	0002	0002	0003
14.50	0000	0000	0000	0000	0000	0000	0001	0002	0002	0003
55	0000	0000	0000	0000	0000	0000	0001	0002	0002	0003
60	0000	0000	0000	0000	0000	0000	0001	0002	0002	0003
65	0000	0000	0000	0000	0000	0000	0001	0002	0002	0003
70	0000	0000	0000	0000	0000	0000	0001	0002	0002	0003
75	0000	0000	0000	0000	0000	0000	0001	0001	0002	0003
80	0000	0000	0000	0000	0000	0000	0001	0001	0002	0003
85	0000	0000	0000	0000	0000	0000	0001	0001	0002	0003
90	0000	0000	0000	0000	0000	0000	0001	0001	0002	0003
95	0000	0000	0000	0000	0000	0000	0001	0001	0002	0003
15.00	0000	0000	0000	0000	0000	0000	0001	0001	0002	0003
05	0000	0000	0000	0000	0000	0000	0001	0001	0002	0003
10	0000	0000	0000	0000	0000	0000	0001	0001	0002	0003
15	0000	0000	0000	0000	0000	0000	0001	0001	0002	0002
20	0000	0000	0000	0000	0000	0000	0001	0001	0002	0002
25	0000	0000	0000	0000	0000	0000	0001	0001	0002	0002
30	0000	0000	0000	0000	0000	0000	0001	0001	0002	0002
35	0000	0000	0000	0000	0000	0000	0001	0001	0002	0002
40	0000	0000	0000	0000	0000	0000	0001	0001	0001	0002
45	0000	0000	0000	0000	0000	0000	0001	0001	0001	0002
50	0000	0000	0000	0000	0000	0000	0001	0001	0001	0002
55	0000	0000	0000	0000	0000	0000	0001	0001	0001	0002
60	0000	0000	0000	0000	0000	0000	0001	0001	0001	0002
65	0000	0000	0000	0000	0000	0000	0001	0001	0001	0002
70	0000	0000	0000	0000	0000	0000	0001	0001	0001	0002
75	0000	0000	0000	0000	0000	0000	0001	0001	0001	0002
80	0000	0000	0000	0000	0000	0000	0001	0001	0001	0002
85	0000	0000	0000	0000	0000	0000	0001	0001	0001	0002
90	0000	0000	0000	0000	0000	0000	0001	0001	0001	0002
95	0000	0000	0000	0000	0000	0000	0001	0001	0001	0002

PROBABILITY DENSITY OF t, THE NON-CENTRAL t-STATISTIC, TABLED AS A FUNCTION OF $t/\sqrt{f}$.

f is the number of degrees of freedom; the non-centrality parameter is $\sqrt{f+1}\,K_p$.
K_p is the standardized normal deviate exceeded with probability p.

$t/\sqrt{f}$ \ p	.2500	.1500	.1000	.0650	.0400	.0250	.0100	.0040	.0025	.0010
16.00	0000	0000	0000	0000	0000	0000	0001	0001	0001	0002
05	0000	0000	0000	0000	0000	0000	0001	0001	0001	0002
10	0000	0000	0000	0000	0000	0000	0001	0001	0001	0002
15	0000	0000	0000	0000	0000	0000	0000	0001	0001	0002
20	0000	0000	0000	0000	0000	0000	0000	0001	0001	0002
25	0000	0000	0000	0000	0000	0000	0000	0001	0001	0002
30	0000	0000	0000	0000	0000	0000	0000	0001	0001	0002
35	0000	0000	0000	0000	0000	0000	0000	0001	0001	0002
40	0000	0000	0000	0000	0000	0000	0000	0001	0001	0002
45	0000	0000	0000	0000	0000	0000	0000	0001	0001	0002
50	0000	0000	0000	0000	0000	0000	0000	0001	0001	0002
55	0000	0000	0000	0000	0000	0000	0000	0001	0001	0002
60	0000	0000	0000	0000	0000	0000	0000	0001	0001	0001
65	0000	0000	0000	0000	0000	0000	0000	0001	0001	0001
70	0000	0000	0000	0000	0000	0000	0000	0001	0001	0001
75	0000	0000	0000	0000	0000	0000	0000	0001	0001	0001
80	0000	0000	0000	0000	0000	0000	0000	0001	0001	0001
85	0000	0000	0000	0000	0000	0000	0000	0001	0001	0001
90	0000	0000	0000	0000	0000	0000	0000	0001	0001	0001
95	0000	0000	0000	0000	0000	0000	0000	0001	0001	0001
17.00	0000	0000	0000	0000	0000	0000	0000	0001	0001	0001
05	0000	0000	0000	0000	0000	0000	0000	0001	0001	0001
10	0000	0000	0000	0000	0000	0000	0000	0001	0001	0001
15	0000	0000	0000	0000	0000	0000	0000	0001	0001	0001
20	0000	0000	0000	0000	0000	0000	0000	0001	0001	0001
25	0000	0000	0000	0000	0000	0000	0000	0001	0001	0001
30	0000	0000	0000	0000	0000	0000	0000	0001	0001	0001
35	0000	0000	0000	0000	0000	0000	0000	0001	0001	0001
40	0000	0000	0000	0000	0000	0000	0000	0001	0001	0001
45	0000	0000	0000	0000	0000	0000	0000	0001	0001	0001
17.50	0000	0000	0000	0000	0000	0000	0000	0001	0001	0001
55	0000	0000	0000	0000	0000	0000	0000	0001	0001	0001
60	0000	0000	0000	0000	0000	0000	0000	0001	0001	0001
65	0000	0000	0000	0000	0000	0000	0000	0001	0001	0001
70	0000	0000	0000	0000	0000	0000	0000	0001	0001	0001
75	0000	0000	0000	0000	0000	0000	0000	0001	0001	0001
80	0000	0000	0000	0000	0000	0000	0000	0001	0001	0001
85	0000	0000	0000	0000	0000	0000	0000	0000	0001	0001
90	0000	0000	0000	0000	0000	0000	0000	0000	0001	0001
95	0000	0000	0000	0000	0000	0000	0000	0000	0001	0001
18.00	0000	0000	0000	0000	0000	0000	0000	0000	0001	0001
05	0000	0000	0000	0000	0000	0000	0000	0000	0001	0001
10	0000	0000	0000	0000	0000	0000	0000	0000	0001	0001
15	0000	0000	0000	0000	0000	0000	0000	0000	0001	0001
20	0000	0000	0000	0000	0000	0000	0000	0000	0001	0001
25	0000	0000	0000	0000	0000	0000	0000	0000	0001	0001
30	0000	0000	0000	0000	0000	0000	0000	0000	0001	0001
35	0000	0000	0000	0000	0000	0000	0000	0000	0001	0001
40	0000	0000	0000	0000	0000	0000	0000	0000	0001	0001
45	0000	0000	0000	0000	0000	0000	0000	0000	0001	0001
50	0000	0000	0000	0000	0000	0000	0000	0000	0001	0001
55	0000	0000	0000	0000	0000	0000	0000	0000	0001	0001
60	0000	0000	0000	0000	0000	0000	0000	0000	0000	0001
65	0000	0000	0000	0000	0000	0000	0000	0000	0000	0001
70	0000	0000	0000	0000	0000	0000	0000	0000	0000	0001
75	0000	0000	0000	0000	0000	0000	0000	0000	0000	0001
80	0000	0000	0000	0000	0000	0000	0000	0000	0000	0001
85	0000	0000	0000	0000	0000	0000	0000	0000	0000	0001
90	0000	0000	0000	0000	0000	0000	0000	0000	0000	0001
95	0000	0000	0000	0000	0000	0000	0000	0000	0000	0001

PROBABILITY DENSITY OF t, THE NON-CENTRAL t-STATISTIC, TABLED AS A FUNCTION OF $t/\sqrt{f}$.

f is the number of degrees of freedom; the non-centrality parameter is $\sqrt{f+1}\,K_{p}$.
K_p is the standardized normal deviate exceeded with probability p.

DEGREES OF FREEDOM **5**

$t/\sqrt{f}$ \ P	.2500	.1500	.1000	.0650	.0400	.0250	.0100	.0040	.0025	.0010
19.00	0000	0000	0000	0000	0000	0000	0000	0000	0000	0001
05	0000	0000	0000	0000	0000	0000	0000	0000	0000	0001
10	0000	0000	0000	0000	0000	0000	0000	0000	0000	0001
15	0000	0000	0000	0000	0000	0000	0000	0000	0000	0001
20	0000	0000	0000	0000	0000	0000	0000	0000	0000	0001
25	0000	0000	0000	0000	0000	0000	0000	0000	0000	0001
30	0000	0000	0000	0000	0000	0000	0000	0000	0000	0001
35	0000	0000	0000	0000	0000	0000	0000	0000	0000	0001
40	0000	0000	0000	0000	0000	0000	0000	0000	0000	0001
45	0000	0000	0000	0000	0000	0000	0000	0000	0000	0001
50	0000	0000	0000	0000	0000	0000	0000	0000	0000	0001
55	0000	0000	0000	0000	0000	0000	0000	0000	0000	0001
60	0000	0000	0000	0000	0000	0000	0000	0000	0000	0001
65	0000	0000	0000	0000	0000	0000	0000	0000	0000	0001
70	0000	0000	0000	0000	0000	0000	0000	0000	0000	0001
75	0000	0000	0000	0000	0000	0000	0000	0000	0000	0001
80	0000	0000	0000	0000	0000	0000	0000	0000	0000	0001
85	0000	0000	0000	0000	0000	0000	0000	0000	0000	0001
90	0000	0000	0000	0000	0000	0000	0000	0000	0000	0001
95	0000	0000	0000	0000	0000	0000	0000	0000	0000	0001
20.00	0000	0000	0000	0000	0000	0000	0000	0000	0000	0001
05	0000	0000	0000	0000	0000	0000	0000	0000	0000	0000

PROBABILITY DENSITY OF t, THE NON-CENTRAL t-STATISTIC, TABLED AS A FUNCTION OF $t/\sqrt{f}$.

f is the number of degrees of freedom; the non-centrality parameter is $\sqrt{f+1}\,K_p$.
K_p is the standardized normal deviate exceeded with probability p.

DEGREES OF FREEDOM 6

$t/\sqrt{f}$ \ p	.2500	.1500	.1000	.0650	.0400	.0250	.0100	.0040	.0025	.0010
-1.50	0000	0000	0000	0000	0000	0000	0000	0000	0000	0000
45	0001	0000	0000	0000	0000	0000	0000	0000	0000	0000
40	0001	0000	0000	0000	0000	0000	0000	0000	0000	0000
35	0001	0000	0000	0000	0000	0000	0000	0000	0000	0000
30	0001	0000	0000	0000	0000	0000	0000	0000	0000	0000
25	0001	0000	0000	0000	0000	0000	0000	0000	0000	0000
20	0002	0000	0000	0000	0000	0000	0000	0000	0000	0000
15	0002	0000	0000	0000	0000	0000	0000	0000	0000	0000
10	0002	0000	0000	0000	0000	0000	0000	0000	0000	0000
05	0003	0000	0000	0000	0000	0000	0000	0000	0000	0000
-1.00	0004	0000	0000	0000	0000	0000	0000	0000	0000	0000
-0.95	0005	0000	0000	0000	0000	0000	0000	0000	0000	0000
90	0006	0000	0000	0000	0000	0000	0000	0000	0000	0000
85	0008	0000	0000	0000	0000	0000	0000	0000	0000	0000
80	0011	0000	0000	0000	0000	0000	0000	0000	0000	0000
75	0014	0001	0000	0000	0000	0000	0000	0000	0000	0000
70	0018	0001	0000	0000	0000	0000	0000	0000	0000	0000
65	0023	0001	0000	0000	0000	0000	0000	0000	0000	0000
60	0031	0001	0000	0000	0000	0000	0000	0000	0000	0000
55	0041	0002	0000	0000	0000	0000	0000	0000	0000	0000
50	0054	0002	0000	0000	0000	0000	0000	0000	0000	0000
45	0071	0004	0000	0000	0000	0000	0000	0000	0000	0000
40	0094	0005	0000	0000	0000	0000	0000	0000	0000	0000
35	0125	0007	0001	0000	0000	0000	0000	0000	0000	0000
30	0165	0010	0001	0000	0000	0000	0000	0000	0000	0000
25	0218	0015	0001	0000	0000	0000	0000	0000	0000	0000
20	0286	0021	0002	0000	0000	0000	0000	0000	0000	0000
15	0373	0030	0003	0000	0000	0000	0000	0000	0000	0000
10	0482	0044	0005	0000	0000	0000	0000	0000	0000	0000
05	0616	0063	0008	0001	0000	0000	0000	0000	0000	0000
0.00	0779	0089	0012	0001	0000	0000	0000	0000	0000	0000
05	0971	0126	0019	0002	0000	0000	0000	0000	0000	0000
10	1193	0176	0029	0003	0000	0000	0000	0000	0000	0000
15	1441	0243	0043	0006	0001	0000	0000	0000	0000	0000
20	1710	0330	0065	0009	0001	0000	0000	0000	0000	0000
25	1992	0440	0096	0015	0002	0000	0000	0000	0000	0000
30	2277	0577	0139	0025	0003	0000	0000	0000	0000	0000
35	2554	0741	0197	0039	0005	0001	0000	0000	0000	0000
40	2809	0932	0273	0060	0009	0001	0000	0000	0000	0000
45	3033	1148	0372	0090	0015	0002	0000	0000	0000	0000
50	3215	1383	0494	0132	0025	0004	0000	0000	0000	0000
55	3350	1631	0640	0189	0039	0007	0000	0000	0000	0000
60	3432	1883	0810	0264	0061	0013	0000	0000	0000	0000
65	3463	2130	1002	0358	0092	0021	0001	0000	0000	0000
70	3443	2362	1210	0473	0135	0035	0002	0000	0000	0000
75	3378	2571	1430	0610	0192	0054	0003	0000	0000	0000
80	3274	2751	1653	0766	0265	0082	0006	0000	0000	0000
85	3139	2895	1873	0940	0355	0120	0011	0001	0000	0000
90	2979	3000	2083	1126	0463	0170	0018	0002	0000	0000
95	2802	3066	2275	1321	0589	0234	0029	0003	0001	0000
1.00	2615	3093	2443	1517	0731	0313	0045	0005	0002	0000
05	2423	3084	2584	1710	0885	0407	0068	0009	0003	0000
10	2231	3042	2695	1893	1050	0516	0098	0015	0005	0001
15	2043	2972	2774	2062	1220	0639	0138	0023	0009	0001
20	1863	2878	2821	2212	1391	0774	0188	0036	0014	0002
25	1691	2765	2839	2340	1558	0917	0250	0053	0023	0004
30	1530	2638	2828	2444	1718	1066	0323	0077	0035	0007
35	1380	2501	2793	2522	1865	1217	0407	0107	0051	0011
40	1241	2358	2736	2575	1998	1366	0501	0146	0072	0017
45	1115	2212	2661	2604	2113	1511	0606	0192	0100	0026

PROBABILITY DENSITY OF t, THE NON-CENTRAL t-STATISTIC, TABLED AS A FUNCTION OF $t/\sqrt{f}$.

f is the number of degrees of freedom; the non-centrality parameter is $\sqrt{f+1}\,K_p$.
K_p is the standardized normal deviate exceeded with probability p.

DEGREES OF FREEDOM **6**

$t/\sqrt{f}$ \ p	.2500	.1500	.1000	.0650	.0400	.0250	.0100	.0040	.0025	.0010
1.50	0999	2066	2570	2610	2209	1647	0717	0248	0135	0038
55	0894	1922	2468	2595	2286	1772	0835	0313	0177	0054
60	0800	1783	2358	2561	2342	1884	0955	0386	0227	0074
65	0715	1648	2242	2510	2378	1981	1077	0467	0286	0100
70	0638	1520	2123	2446	2396	2062	1197	0554	0351	0132
75	0570	1398	2002	2371	2396	2127	1313	0648	0424	0170
80	0508	1284	1882	2287	2380	2176	1424	0746	0503	0214
85	0454	1177	1764	2196	2350	2208	1526	0846	0588	0265
90	0405	1078	1649	2100	2308	2226	1620	0947	0676	0322
95	0362	0986	1538	2001	2254	2228	1704	1047	0768	0385
2.00	0323	0900	1432	1901	2192	2218	1777	1144	0860	0452
05	0289	0822	1330	1800	2123	2196	1838	1238	0953	0525
10	0258	0750	1234	1701	2049	2163	1887	1327	1044	0600
15	0231	0684	1143	1603	1970	2121	1925	1409	1132	0678
20	0207	0624	1058	1508	1888	2072	1952	1484	1216	0758
25	0186	0569	0978	1416	1805	2016	1967	1551	1295	0837
30	0166	0518	0904	1327	1721	1955	1973	1610	1368	0916
35	0149	0472	0834	1242	1637	1890	1969	1660	1435	0994
40	0134	0431	0770	1162	1554	1821	1956	1701	1494	1068
45	0121	0393	0710	1085	1473	1751	1936	1733	1547	1139
50	0109	0358	0654	1012	1394	1680	1908	1757	1591	1206
55	0098	0327	0603	0944	1317	1608	1875	1773	1628	1269
60	0088	0298	0556	0879	1242	1536	1836	1781	1657	1325
65	0080	0272	0512	0819	1170	1464	1793	1782	1679	1377
70	0072	0249	0472	0762	1102	1394	1746	1776	1694	1422
75	0065	0227	0435	0709	1036	1326	1696	1764	1701	1461
80	0059	0208	0401	0659	0974	1259	1644	1746	1703	1494
85	0053	0190	0370	0613	0915	1194	1590	1723	1698	1521
90	0048	0174	0341	0570	0858	1131	1535	1696	1688	1542
95	0044	0159	0314	0530	0805	1070	1479	1665	1673	1558
3.00	0040	0146	0290	0492	0755	1012	1423	1630	1653	1568
05	0036	0134	0268	0458	0708	0957	1367	1593	1630	1572
10	0033	0123	0247	0425	0663	0903	1312	1553	1603	1572
15	0030	0113	0228	0395	0621	0853	1257	1512	1572	1567
20	0028	0103	0211	0368	0582	0804	1203	1469	1539	1557
25	0025	0095	0195	0342	0545	0759	1151	1425	1504	1544
30	0023	0087	0180	0318	0510	0715	1099	1380	1467	1528
35	0021	0081	0167	0296	0477	0674	1049	1334	1429	1508
40	0019	0074	0154	0275	0447	0635	1001	1289	1389	1485
45	0018	0068	0143	0256	0419	0598	0954	1243	1349	1460
50	0016	0063	0132	0238	0392	0563	0908	1198	1308	1432
55	0015	0058	0123	0222	0367	0530	0865	1154	1266	1403
60	0014	0054	0114	0207	0344	0499	0823	1110	1225	1372
65	0013	0050	0106	0193	0322	0470	0782	1066	1183	1340
70	0012	0046	0098	0180	0302	0442	0744	1024	1142	1307
75	0011	0042	0091	0168	0283	0416	0707	0983	1102	1273
80	0010	0039	0085	0156	0265	0392	0672	0942	1062	1238
85	0009	0036	0079	0146	0248	0369	0638	0903	1022	1203
90	0008	0034	0073	0136	0233	0347	0606	0865	0983	1168
95	0008	0031	0068	0127	0218	0327	0575	0828	0946	1133
4.00	0007	0029	0063	0119	0205	0308	0546	0792	0909	1098
05	0007	0027	0059	0111	0192	0290	0518	0758	0873	1063
10	0006	0025	0055	0104	0180	0273	0492	0725	0838	1028
15	0006	0023	0051	0097	0169	0258	0466	0693	0804	0994
20	0005	0022	0048	0091	0159	0243	0443	0662	0771	0960
25	0005	0020	0045	0085	0149	0229	0420	0632	0739	0927
30	0005	0019	0042	0080	0140	0216	0398	0604	0708	0894
35	0004	0018	0039	0075	0132	0203	0378	0576	0678	0862
40	0004	0016	0036	0070	0124	0192	0359	0550	0650	0831
45	0004	0015	0034	0066	0117	0181	0340	0525	0622	0801

PROBABILITY DENSITY OF t, THE NON-CENTRAL t-STATISTIC, TABLED AS A FUNCTION OF $t/\sqrt{f}$.

f is the number of degrees of freedom; the non-centrality parameter is $\sqrt{f+1}\,K_p$.
K_p is the standardized normal deviate exceeded with probability p.

DEGREES OF FREEDOM **6**

$t/\sqrt{f}$ \ p	.2500	.1500	.1000	.0650	.0400	.0250	.0100	.0040	.0025	.0010
4.50	0003	0014	0032	0062	0110	0171	0323	0501	0596	0771
55	0003	0013	0030	0058	0103	0161	0306	0478	0570	0742
60	0003	0012	0028	0054	0097	0152	0291	0457	0546	0714
65	0003	0012	0026	0051	0092	0144	0276	0436	0522	0687
70	0003	0011	0025	0048	0086	0136	0262	0416	0499	0661
75	0002	0010	0023	0045	0081	0128	0249	0397	0478	0635
80	0002	0010	0022	0042	0077	0121	0236	0378	0457	0611
85	0002	0009	0020	0040	0072	0114	0224	0361	0437	0587
90	0002	0008	0019	0038	0068	0108	0213	0344	0418	0564
95	0002	0008	0018	0035	0064	0102	0202	0328	0399	0542
5.00	0002	0007	0017	0033	0061	0097	0192	0313	0382	0520
05	0002	0007	0016	0031	0057	0092	0182	0299	0365	0500
10	0002	0007	0015	0030	0054	0087	0173	0285	0349	0480
15	0001	0006	0014	0028	0051	0082	0165	0272	0334	0461
20	0001	0006	0013	0026	0048	0078	0157	0260	0319	0442
25	0001	0005	0012	0025	0046	0074	0149	0248	0305	0424
30	0001	0005	0012	0023	0043	0070	0142	0237	0292	0407
35	0001	0005	0011	0022	0041	0066	0135	0226	0279	0391
40	0001	0005	0010	0021	0039	0063	0128	0216	0267	0375
45	0001	0004	0010	0020	0037	0059	0122	0206	0255	0360
50	0001	0004	0009	0019	0035	0056	0116	0197	0244	0346
55	0001	0004	0009	0018	0033	0054	0111	0188	0234	0332
60	0001	0004	0008	0017	0031	0051	0105	0179	0224	0318
65	0001	0003	0008	0016	0030	0048	0100	0171	0214	0305
70	0001	0003	0007	0015	0028	0046	0095	0164	0205	0293
75	0001	0003	0007	0014	0027	0044	0091	0156	0196	0281
80	0001	0003	0007	0013	0025	0041	0087	0149	0187	0270
85	0001	0003	0006	0013	0024	0039	0083	0143	0179	0259
90	0001	0003	0006	0012	0023	0037	0079	0136	0172	0249
95	0001	0002	0006	0011	0022	0036	0075	0130	0164	0239
6.00	0001	0002	0005	0011	0021	0034	0072	0125	0157	0229
05	0000	0002	0005	0010	0020	0032	0068	0119	0151	0220
10	0000	0002	0005	0010	0019	0031	0065	0114	0144	0211
15	0000	0002	0005	0009	0018	0029	0062	0109	0138	0203
20	0000	0002	0004	0009	0017	0028	0059	0104	0132	0195
25	0000	0002	0004	0008	0016	0026	0057	0100	0127	0187
30	0000	0002	0004	0008	0015	0025	0054	0096	0122	0180
35	0000	0002	0004	0008	0014	0024	0052	0092	0116	0173
40	0000	0002	0004	0007	0014	0023	0049	0088	0112	0166
45	0000	0001	0003	0007	0013	0022	0047	0084	0107	0159
50	0000	0001	0003	0007	0013	0021	0045	0080	0103	0153
55	0000	0001	0003	0006	0012	0020	0043	0077	0098	0147
60	0000	0001	0003	0006	0011	0019	0041	0074	0094	0141
65	0000	0001	0003	0006	0011	0018	0039	0071	0091	0136
70	0000	0001	0003	0005	0010	0017	0038	0068	0087	0130
75	0000	0001	0003	0005	0010	0017	0036	0065	0083	0125
80	0000	0001	0002	0005	0009	0016	0035	0062	0080	0121
85	0000	0001	0002	0005	0009	0015	0033	0060	0077	0116
90	0000	0001	0002	0004	0009	0014	0032	0057	0074	0111
95	0000	0001	0002	0004	0008	0014	0030	0055	0071	0107
7.00	0000	0001	0002	0004	0008	0013	0029	0053	0068	0103
05	0000	0001	0002	0004	0008	0013	0028	0051	0065	0099
10	0000	0001	0002	0004	0007	0012	0027	0049	0063	0095
15	0000	0001	0002	0004	0007	0012	0026	0047	0060	0092
20	0000	0001	0002	0003	0007	0011	0025	0045	0058	0088
25	0000	0001	0002	0003	0006	0011	0024	0043	0056	0085
30	0000	0001	0002	0003	0006	0010	0023	0041	0053	0082
35	0000	0001	0001	0003	0006	0010	0022	0040	0051	0079
40	0000	0001	0001	0003	0006	0009	0021	0038	0049	0076
45	0000	0001	0001	0003	0005	0009	0020	0037	0047	0073

PROBABILITY DENSITY OF t, THE NON-CENTRAL t-STATISTIC, TABLED AS A FUNCTION OF $t/\sqrt{f}$.

f is the number of degrees of freedom; the non-centrality parameter is $\sqrt{f+1}\,K_p$.
K_p is the standardized normal deviate exceeded with probability p.

DEGREES OF FREEDOM **6**

$t/\sqrt{f}$ \ p	.2500	.1500	.1000	.0650	.0400	.0250	.0100	.0040	.0025	.0010
7.50	0000	0001	0001	0003	0005	0009	0019	0035	0046	0070
55	0000	0001	0001	0003	0005	0008	0018	0034	0044	0068
60	0000	0000	0001	0002	0005	0008	0018	0033	0042	0065
65	0000	0000	0001	0002	0004	0008	0017	0031	0041	0063
70	0000	0000	0001	0002	0004	0007	0016	0030	0039	0061
75	0000	0000	0001	0002	0004	0007	0016	0029	0038	0058
80	0000	0000	0001	0002	0004	0007	0015	0028	0036	0056
85	0000	0000	0001	0002	0004	0006	0014	0027	0035	0054
90	0000	0000	0001	0002	0004	0006	0014	0026	0034	0052
95	0000	0000	0001	0002	0003	0006	0013	0025	0032	0050
8.00	0000	0000	0001	0002	0003	0006	0013	0024	0031	0049
05	0000	0000	0001	0002	0003	0005	0012	0023	0030	0047
10	0000	0000	0001	0002	0003	0005	0012	0022	0029	0045
15	0000	0000	0001	0002	0003	0005	0011	0021	0028	0044
20	0000	0000	0001	0001	0003	0005	0011	0021	0027	0042
25	0000	0000	0001	0001	0003	0005	0011	0020	0026	0041
30	0000	0000	0001	0001	0003	0004	0010	0019	0025	0039
35	0000	0000	0001	0001	0003	0004	0010	0018	0024	0038
40	0000	0000	0001	0001	0002	0004	0009	0018	0023	0037
45	0000	0000	0001	0001	0002	0004	0009	0017	0022	0035
50	0000	0000	0001	0001	0002	0004	0009	0016	0022	0034
55	0000	0000	0001	0001	0002	0004	0008	0016	0021	0033
60	0000	0000	0001	0001	0002	0004	0008	0015	0020	0032
65	0000	0000	0000	0001	0002	0003	0008	0015	0019	0031
70	0000	0000	0000	0001	0002	0003	0008	0014	0019	0030
75	0000	0000	0000	0001	0002	0003	0007	0014	0018	0029
80	0000	0000	0000	0001	0002	0003	0007	0013	0018	0028
85	0000	0000	0000	0001	0002	0003	0007	0013	0017	0027
90	0000	0000	0000	0001	0002	0003	0007	0012	0016	0026
95	0000	0000	0000	0001	0002	0003	0006	0012	0016	0025
9.00	0000	0000	0000	0001	0002	0003	0006	0012	0015	0024
05	0000	0000	0000	0001	0001	0003	0006	0011	0015	0023
10	0000	0000	0000	0001	0001	0002	0006	0011	0014	0023
15	0000	0000	0000	0001	0001	0002	0005	0010	0014	0022
20	0000	0000	0000	0001	0001	0002	0005	0010	0013	0021
25	0000	0000	0000	0001	0001	0002	0005	0010	0013	0021
30	0000	0000	0000	0001	0001	0002	0005	0009	0012	0020
35	0000	0000	0000	0001	0001	0002	0005	0009	0012	0019
40	0000	0000	0000	0001	0001	0002	0005	0009	0012	0019
45	0000	0000	0000	0001	0001	0002	0004	0009	0011	0018
50	0000	0000	0000	0001	0001	0002	0004	0008	0011	0017
55	0000	0000	0000	0001	0001	0002	0004	0008	0011	0017
60	0000	0000	0000	0001	0001	0002	0004	0008	0010	0016
65	0000	0000	0000	0000	0001	0002	0004	0007	0010	0016
70	0000	0000	0000	0000	0001	0002	0004	0007	0010	0015
75	0000	0000	0000	0000	0001	0002	0004	0007	0009	0015
80	0000	0000	0000	0000	0001	0002	0004	0007	0009	0014
85	0000	0000	0000	0000	0001	0001	0003	0007	0009	0014
90	0000	0000	0000	0000	0001	0001	0003	0006	0008	0014
95	0000	0000	0000	0000	0001	0001	0003	0006	0008	0013
10.00	0000	0000	0000	0000	0001	0001	0003	0006	0008	0013
05	0000	0000	0000	0000	0001	0001	0003	0006	0008	0012
10	0000	0000	0000	0000	0001	0001	0003	0006	0007	0012
15	0000	0000	0000	0000	0001	0001	0003	0005	0007	0012
20	0000	0000	0000	0000	0001	0001	0003	0005	0007	0011
25	0000	0000	0000	0000	0001	0001	0003	0005	0007	0011
30	0000	0000	0000	0000	0001	0001	0003	0005	0007	0011
35	0000	0000	0000	0000	0001	0001	0002	0005	0006	0010
40	0000	0000	0000	0000	0001	0001	0002	0005	0006	0010
45	0000	0000	0000	0000	0001	0001	0002	0004	0006	0010

PROBABILITY DENSITY OF t, THE NON-CENTRAL t-STATISTIC, TABLED AS A FUNCTION OF $t/\sqrt{f}$.

f is the number of degrees of freedom; the non-centrality parameter is $\sqrt{f+1}\,K_p$.
K_p is the standardized normal deviate exceeded with probability p.

DEGREES OF FREEDOM **6**

$t/\sqrt{f}$ \ p	.2500	.1500	.1000	.0650	.0400	.0250	.0100	.0040	.0025	.0010
10.50	0000	0000	0000	0000	0001	0001	0002	0004	0006	0009
55	0000	0000	0000	0000	0001	0001	0002	0004	0006	0009
60	0000	0000	0000	0000	0001	0001	0002	0004	0005	0009
65	0000	0000	0000	0000	0001	0001	0002	0004	0005	0009
70	0000	0000	0000	0000	0000	0001	0002	0004	0005	0008
75	0000	0000	0000	0000	0000	0001	0002	0004	0005	0008
80	0000	0000	0000	0000	0000	0001	0002	0004	0005	0008
85	0000	0000	0000	0000	0000	0001	0002	0004	0005	0008
90	0000	0000	0000	0000	0000	0001	0002	0003	0005	0007
95	0000	0000	0000	0000	0000	0001	0002	0003	0004	0007
11.00	0000	0000	0000	0000	0000	0001	0002	0003	0004	0007
05	0000	0000	0000	0000	0000	0001	0002	0003	0004	0007
10	0000	0000	0000	0000	0000	0001	0002	0003	0004	0007
15	0000	0000	0000	0000	0000	0001	0002	0003	0004	0006
20	0000	0000	0000	0000	0000	0001	0001	0003	0004	0006
25	0000	0000	0000	0000	0000	0001	0001	0003	0004	0006
30	0000	0000	0000	0000	0000	0001	0001	0003	0004	0006
35	0000	0000	0000	0000	0000	0001	0001	0003	0003	0006
40	0000	0000	0000	0000	0000	0001	0001	0003	0003	0006
45	0000	0000	0000	0000	0000	0001	0001	0002	0003	0005
50	0000	0000	0000	0000	0000	0001	0001	0002	0003	0005
55	0000	0000	0000	0000	0000	0001	0001	0002	0003	0005
60	0000	0000	0000	0000	0000	0000	0001	0002	0003	0005
65	0000	0000	0000	0000	0000	0000	0001	0002	0003	0005
70	0000	0000	0000	0000	0000	0000	0001	0002	0003	0005
75	0000	0000	0000	0000	0000	0000	0001	0002	0003	0005
80	0000	0000	0000	0000	0000	0000	0001	0002	0003	0004
85	0000	0000	0000	0000	0000	0000	0001	0002	0003	0004
90	0000	0000	0000	0000	0000	0000	0001	0002	0003	0004
95	0000	0000	0000	0000	0000	0000	0001	0002	0003	0004
12.00	0000	0000	0000	0000	0000	0000	0001	0002	0002	0004
05	0000	0000	0000	0000	0000	0000	0001	0002	0002	0004
10	0000	0000	0000	0000	0000	0000	0001	0002	0002	0004
15	0000	0000	0000	0000	0000	0000	0001	0002	0002	0004
20	0000	0000	0000	0000	0000	0000	0001	0002	0002	0004
25	0000	0000	0000	0000	0000	0000	0001	0002	0002	0003
30	0000	0000	0000	0000	0000	0000	0001	0002	0002	0003
35	0000	0000	0000	0000	0000	0000	0001	0002	0002	0003
40	0000	0000	0000	0000	0000	0000	0001	0001	0002	0003
45	0000	0000	0000	0000	0000	0000	0001	0001	0002	0003
50	0000	0000	0000	0000	0000	0000	0001	0001	0002	0003
55	0000	0000	0000	0000	0000	0000	0001	0001	0002	0003
60	0000	0000	0000	0000	0000	0000	0001	0001	0002	0003
65	0000	0000	0000	0000	0000	0000	0001	0001	0002	0003
70	0000	0000	0000	0000	0000	0000	0001	0001	0002	0003
75	0000	0000	0000	0000	0000	0000	0001	0001	0002	0003
80	0000	0000	0000	0000	0000	0000	0001	0001	0002	0003
85	0000	0000	0000	0000	0000	0000	0001	0001	0002	0003
90	0000	0000	0000	0000	0000	0000	0001	0001	0002	0002
95	0000	0000	0000	0000	0000	0000	0001	0001	0001	0002
13.00	0000	0000	0000	0000	0000	0000	0001	0001	0001	0002
05	0000	0000	0000	0000	0000	0000	0001	0001	0001	0002
10	0000	0000	0000	0000	0000	0000	0001	0001	0001	0002
15	0000	0000	0000	0000	0000	0000	0001	0001	0001	0002
20	0000	0000	0000	0000	0000	0000	0000	0001	0001	0002
25	0000	0000	0000	0000	0000	0000	0000	0001	0001	0002
30	0000	0000	0000	0000	0000	0000	0000	0001	0001	0002
35	0000	0000	0000	0000	0000	0000	0000	0001	0001	0002
40	0000	0000	0000	0000	0000	0000	0000	0001	0001	0002
45	0000	0000	0000	0000	0000	0000	0000	0001	0001	0002

PROBABILITY DENSITY OF t, THE NON-CENTRAL t-STATISTIC, TABLED AS A FUNCTION OF $t/\sqrt{f}$.

f is the number of degrees of freedom; the non-centrality parameter is $\sqrt{f+1}\,K_p$.
K_p is the standardized normal deviate exceeded with probability p.

DEGREES OF FREEDOM **6**

$t/\sqrt{f}$ \ p	.2500	.1500	.1000	.0650	.0400	.0250	.0100	.0040	.0025	.0010
13.50	0000	0000	0000	0000	0000	0000	0000	0001	0001	0002
55	0000	0000	0000	0000	0000	0000	0000	0001	0001	0002
60	0000	0000	0000	0000	0000	0000	0000	0001	0001	0002
65	0000	0000	0000	0000	0000	0000	0000	0001	0001	0002
70	0000	0000	0000	0000	0000	0000	0000	0001	0001	0002
75	0000	0000	0000	0000	0000	0000	0000	0001	0001	0002
80	0000	0000	0000	0000	0000	0000	0000	0001	0001	0002
85	0000	0000	0000	0000	0000	0000	0000	0001	0001	0002
90	0000	0000	0000	0000	0000	0000	0000	0001	0001	0002
95	0000	0000	0000	0000	0000	0000	0000	0001	0001	0001
14.00	0000	0000	0000	0000	0000	0000	0000	0001	0001	0001
05	0000	0000	0000	0000	0000	0000	0000	0001	0001	0001
10	0000	0000	0000	0000	0000	0000	0000	0001	0001	0001
15	0000	0000	0000	0000	0000	0000	0000	0001	0001	0001
20	0000	0000	0000	0000	0000	0000	0000	0001	0001	0001
25	0000	0000	0000	0000	0000	0000	0000	0001	0001	0001
30	0000	0000	0000	0000	0000	0000	0000	0001	0001	0001
35	0000	0000	0000	0000	0000	0000	0000	0001	0001	0001
40	0000	0000	0000	0000	0000	0000	0000	0001	0001	0001
45	0000	0000	0000	0000	0000	0000	0000	0001	0001	0001
50	0000	0000	0000	0000	0000	0000	0000	0001	0001	0001
55	0000	0000	0000	0000	0000	0000	0000	0001	0001	0001
60	0000	0000	0000	0000	0000	0000	0000	0000	0001	0001
65	0000	0000	0000	0000	0000	0000	0000	0000	0001	0001
70	0000	0000	0000	0000	0000	0000	0000	0000	0001	0001
75	0000	0000	0000	0000	0000	0000	0000	0000	0001	0001
80	0000	0000	0000	0000	0000	0000	0000	0000	0001	0001
85	0000	0000	0000	0000	0000	0000	0000	0000	0001	0001
90	0000	0000	0000	0000	0000	0000	0000	0000	0001	0001
95	0000	0000	0000	0000	0000	0000	0000	0000	0001	0001
15.00	0000	0000	0000	0000	0000	0000	0000	0000	0001	0001
05	0000	0000	0000	0000	0000	0000	0000	0000	0001	0001
10	0000	0000	0000	0000	0000	0000	0000	0000	0001	0001
15	0000	0000	0000	0000	0000	0000	0000	0000	0001	0001
20	0000	0000	0000	0000	0000	0000	0000	0000	0001	0001
25	0000	0000	0000	0000	0000	0000	0000	0000	0000	0001
30	0000	0000	0000	0000	0000	0000	0000	0000	0000	0001
35	0000	0000	0000	0000	0000	0000	0000	0000	0000	0001
40	0000	0000	0000	0000	0000	0000	0000	0000	0000	0001
45	0000	0000	0000	0000	0000	0000	0000	0000	0000	0001
50	0000	0000	0000	0000	0000	0000	0000	0000	0000	0001
55	0000	0000	0000	0000	0000	0000	0000	0000	0000	0001
60	0000	0000	0000	0000	0000	0000	0000	0000	0000	0001
65	0000	0000	0000	0000	0000	0000	0000	0000	0000	0001
70	0000	0000	0000	0000	0000	0000	0000	0000	0000	0001
75	0000	0000	0000	0000	0000	0000	0000	0000	0000	0001
80	0000	0000	0000	0000	0000	0000	0000	0000	0000	0001
85	0000	0000	0000	0000	0000	0000	0000	0000	0000	0001
90	0000	0000	0000	0000	0000	0000	0000	0000	0000	0001
95	0000	0000	0000	0000	0000	0000	0000	0000	0000	0001
16.00	0000	0000	0000	0000	0000	0000	0000	0000	0000	0001
05	0000	0000	0000	0000	0000	0000	0000	0000	0000	0001
10	0000	0000	0000	0000	0000	0000	0000	0000	0000	0001
15	0000	0000	0000	0000	0000	0000	0000	0000	0000	0001
20	0000	0000	0000	0000	0000	0000	0000	0000	0000	0001
25	0000	0000	0000	0000	0000	0000	0000	0000	0000	0001
30	0000	0000	0000	0000	0000	0000	0000	0000	0000	0001
35	0000	0000	0000	0000	0000	0000	0000	0000	0000	0001
40	0000	0000	0000	0000	0000	0000	0000	0000	0000	0001
45	0000	0000	0000	0000	0000	0000	0000	0000	0000	0000

PROBABILITY DENSITY OF t, THE NON-CENTRAL t-STATISTIC, TABLED AS A FUNCTION OF $t/\sqrt{f}$.

f is the number of degrees of freedom; the non-centrality parameter is $\sqrt{f+1}\,K_p$.
K_p is the standardized normal deviate exceeded with probability p.

DEGREES OF FREEDOM **7**

$t/\sqrt{f}$ \ p	.2500	.1500	.1000	.0650	.0400	.0250	.0100	.0040	.0025	.0010
-1.25	0000	0000	0000	0000	0000	0000	0000	0000	0000	0000
20	0001	0000	0000	0000	0000	0000	0000	0000	0000	0000
15	0001	0000	0000	0000	0000	0000	0000	0000	0000	0000
10	0001	0000	0000	0000	0000	0000	0000	0000	0000	0000
05	0001	0000	0000	0000	0000	0000	0000	0000	0000	0000
-1.00	0001	0000	0000	0000	0000	0000	0000	0000	0000	0000
-0.95	0002	0000	0000	0000	0000	0000	0000	0000	0000	0000
90	0003	0000	0000	0000	0000	0000	0000	0000	0000	0000
85	0003	0000	0000	0000	0000	0000	0000	0000	0000	0000
80	0005	0000	0000	0000	0000	0000	0000	0000	0000	0000
75	0006	0000	0000	0000	0000	0000	0000	0000	0000	0000
70	0008	0000	0000	0000	0000	0000	0000	0000	0000	0000
65	0011	0000	0000	0000	0000	0000	0000	0000	0000	0000
60	0015	0000	0000	0000	0000	0000	0000	0000	0000	0000
55	0021	0001	0000	0000	0000	0000	0000	0000	0000	0000
50	0029	0001	0000	0000	0000	0000	0000	0000	0000	0000
45	0040	0001	0000	0000	0000	0000	0000	0000	0000	0000
40	0055	0002	0000	0000	0000	0000	0000	0000	0000	0000
35	0077	0003	0000	0000	0000	0000	0000	0000	0000	0000
30	0105	0004	0000	0000	0000	0000	0000	0000	0000	0000
25	0145	0007	0000	0000	0000	0000	0000	0000	0000	0000
20	0198	0010	0001	0000	0000	0000	0000	0000	0000	0000
15	0268	0015	0001	0000	0000	0000	0000	0000	0000	0000
10	0359	0023	0002	0000	0000	0000	0000	0000	0000	0000
05	0477	0035	0003	0000	0000	0000	0000	0000	0000	0000
0.00	0624	0052	0005	0000	0000	0000	0000	0000	0000	0000
05	0804	0078	0009	0001	0000	0000	0000	0000	0000	0000
10	1018	0114	0014	0001	0000	0000	0000	0000	0000	0000
15	1265	0166	0023	0002	0000	0000	0000	0000	0000	0000
20	1540	0235	0037	0004	0000	0000	0000	0000	0000	0000
0.25	1836	0328	0057	0007	0001	0000	0000	0000	0000	0000
30	2142	0448	0088	0012	0001	0000	0000	0000	0000	0000
35	2444	0597	0132	0021	0002	0000	0000	0000	0000	0000
40	2729	0778	0192	0034	0004	0000	0000	0000	0000	0000
45	2982	0988	0274	0054	0007	0001	0000	0000	0000	0000
50	3190	1225	0379	0084	0012	0002	0000	0000	0000	0000
55	3346	1481	0511	0128	0021	0003	0000	0000	0000	0000
60	3444	1747	0670	0187	0035	0006	0000	0000	0000	0000
65	3481	2014	0856	0266	0057	0011	0000	0000	0000	0000
70	3462	2271	1064	0366	0088	0019	0001	0000	0000	0000
75	3390	2505	1289	0490	0132	0031	0001	0000	0000	0000
80	3273	2709	1524	0637	0190	0050	0003	0000	0000	0000
85	3121	2874	1761	0806	0267	0078	0005	0000	0000	0000
90	2942	2997	1990	0993	0362	0116	0009	0001	0000	0000
95	2745	3076	2204	1193	0478	0167	0016	0001	0000	0000
1.00	2537	3109	2394	1400	0612	0234	0026	0002	0001	0000
05	2327	3101	2555	1607	0764	0316	0041	0004	0001	0000
10	2119	3056	2683	1807	0929	0416	0063	0007	0002	0000
15	1917	2977	2775	1994	1104	0531	0094	0012	0004	0000
20	1725	2872	2832	2163	1284	0662	0133	0020	0007	0001
25	1545	2745	2855	2309	1464	0805	0184	0032	0012	0002
30	1378	2603	2845	2428	1638	0957	0247	0049	0020	0003
35	1225	2451	2806	2520	1801	1114	0323	0071	0030	0005
40	1086	2292	2742	2582	1950	1273	0411	0101	0046	0009
45	0961	2132	2658	2617	2081	1429	0510	0139	0066	0014
50	0848	1973	2556	2626	2191	1579	0619	0186	0093	0022
55	0747	1817	2442	2610	2279	1718	0737	0243	0128	0033
60	0658	1668	2319	2572	2345	1844	0861	0309	0170	0048
65	0579	1525	2190	2516	2388	1954	0988	0385	0221	0067
70	0508	1391	2058	2444	2410	2048	1115	0469	0280	0092

PROBABILITY DENSITY OF t, THE NON-CENTRAL t-STATISTIC, TABLED AS A FUNCTION OF $t/\sqrt{f}$.

f is the number of degrees of freedom; the non-centrality parameter is $\sqrt{f+1}\,K_p$.
K_p is the standardized normal deviate exceeded with probability p.

DEGREES OF FREEDOM **7**

$t/\sqrt{f}$ \ p	.2500	.1500	.1000	.0650	.0400	.0250	.0100	.0040	.0025	.0010
1.75	0447	1265	1926	2359	2411	2123	1241	0561	0347	0123
80	0392	1148	1795	2264	2394	2179	1362	0658	0423	0161
85	0345	1039	1668	2162	2361	2218	1476	0761	0505	0205
90	0303	0940	1544	2056	2313	2239	1581	0866	0593	0257
95	0266	0849	1427	1946	2253	2243	1675	0972	0686	0315
2.00	0234	0766	1315	1836	2183	2232	1758	1077	0782	0379
05	0206	0690	1209	1726	2105	2207	1828	1179	0879	0449
10	0181	0622	1110	1618	2022	2171	1885	1276	0976	0524
15	0160	0560	1017	1512	1934	2124	1929	1368	1071	0603
20	0141	0504	0931	1411	1843	2068	1960	1452	1163	0684
25	0124	0453	0852	1313	1751	2005	1979	1528	1250	0768
30	0110	0408	0778	1220	1659	1936	1986	1594	1332	0851
35	0097	0367	0710	1131	1567	1863	1982	1652	1407	0934
40	0086	0330	0648	1048	1477	1787	1968	1699	1474	1015
45	0076	0297	0591	0969	1389	1708	1945	1737	1534	1093
50	0067	0267	0539	0896	1304	1629	1914	1765	1585	1167
55	0060	0241	0491	0827	1223	1550	1876	1784	1627	1237
60	0053	0217	0447	0763	1144	1471	1832	1794	1661	1300
65	0047	0195	0407	0703	1069	1394	1784	1795	1687	1358
70	0042	0176	0371	0648	0998	1318	1731	1789	1704	1410
75	0038	0159	0338	0597	0931	1244	1675	1775	1713	1455
80	0034	0144	0308	0549	0867	1173	1616	1755	1715	1493
85	0030	0130	0281	0505	0807	1104	1556	1729	1710	1524
90	0027	0117	0256	0465	0751	1038	1495	1698	1699	1549
95	0024	0106	0233	0428	0698	0975	1434	1663	1682	1567
3.00	0022	0096	0213	0394	0648	0915	1372	1624	1660	1578
05	0019	0087	0194	0362	0602	0858	1311	1582	1633	1584
10	0017	0079	0177	0333	0559	0804	1250	1537	1602	1584
15	0016	0071	0162	0307	0519	0752	1191	1490	1568	1578
20	0014	0065	0148	0282	0482	0704	1133	1442	1531	1568
3.25	0013	0059	0135	0260	0447	0658	1077	1393	1491	1553
30	0011	0053	0124	0239	0414	0616	1022	1343	1449	1534
35	0010	0049	0113	0220	0384	0575	0969	1293	1406	1511
40	0009	0044	0103	0203	0357	0537	0918	1243	1362	1485
45	0008	0040	0095	0187	0331	0502	0869	1193	1317	1457
50	0008	0037	0087	0172	0307	0469	0822	1144	1271	1426
55	0007	0034	0080	0159	0285	0437	0777	1095	1226	1393
60	0006	0031	0073	0146	0264	0408	0735	1048	1180	1358
65	0006	0028	0067	0135	0245	0381	0694	1001	1135	1322
70	0005	0026	0062	0125	0228	0356	0655	0956	1090	1285
75	0005	0023	0057	0115	0211	0332	0618	0912	1046	1247
80	0004	0021	0052	0106	0196	0310	0583	0869	1002	1208
85	0004	0020	0048	0098	0182	0289	0549	0828	0960	1169
90	0004	0018	0044	0091	0169	0270	0518	0788	0919	1130
95	0003	0017	0041	0084	0157	0252	0488	0750	0878	1092
4.00	0003	0015	0037	0078	0146	0235	0460	0713	0839	1053
05	0003	0014	0035	0072	0136	0220	0433	0678	0802	1015
10	0003	0013	0032	0067	0126	0205	0408	0644	0765	0977
15	0002	0012	0029	0062	0118	0192	0384	0612	0730	0940
20	0002	0011	0027	0057	0109	0179	0362	0581	0696	0904
25	0002	0010	0025	0053	0102	0168	0341	0551	0663	0868
30	0002	0009	0023	0049	0095	0157	0321	0523	0632	0833
35	0002	0009	0021	0046	0088	0146	0302	0496	0601	0800
40	0002	0008	0020	0042	0082	0137	0285	0471	0573	0767
45	0001	0007	0018	0039	0077	0128	0268	0446	0545	0735
50	0001	0007	0017	0037	0072	0120	0252	0423	0518	0704
55	0001	0006	0016	0034	0067	0112	0238	0401	0493	0674
60	0001	0006	0015	0032	0062	0105	0224	0380	0469	0645
65	0001	0005	0014	0030	0058	0098	0211	0360	0446	0617
70	0001	0005	0013	0028	0054	0092	0199	0342	0424	0590

PROBABILITY DENSITY OF t, THE NON-CENTRAL t-STATISTIC, TABLED AS A FUNCTION OF $t/\sqrt{f}$.

f is the number of degrees of freedom; the non-centrality parameter is $\sqrt{f+1}\,K_p$.
K_p is the standardized normal deviate exceeded with probability p.

DEGREES OF FREEDOM **7**

$t/\sqrt{f}$ \ p	.2500	.1500	.1000	.0650	.0400	.0250	.0100	.0040	.0025	.0010
4.75	0001	0005	0012	0026	0051	0086	0187	0324	0403	0564
80	0001	0004	0011	0024	0048	0081	0177	0307	0383	0539
85	0001	0004	0010	0022	0044	0076	0166	0291	0364	0515
90	0001	0004	0010	0021	0042	0071	0157	0276	0346	0492
95	0001	0003	0009	0019	0039	0067	0148	0261	0329	0470
5.00	0001	0003	0008	0018	0036	0063	0140	0247	0312	0449
05	0001	0003	0008	0017	0034	0059	0132	0235	0297	0429
10	0001	0003	0007	0016	0032	0055	0124	0222	0282	0409
15	0000	0003	0007	0015	0030	0052	0117	0211	0268	0391
20	0000	0002	0006	0014	0028	0049	0111	0200	0254	0373
25	0000	0002	0006	0013	0026	0046	0104	0189	0242	0356
30	0000	0002	0005	0012	0025	0043	0099	0180	0230	0340
35	0000	0002	0005	0011	0023	0041	0093	0170	0218	0324
40	0000	0002	0005	0011	0022	0038	0088	0161	0207	0309
45	0000	0002	0004	0010	0021	0036	0083	0153	0197	0295
50	0000	0002	0004	0009	0019	0034	0079	0145	0187	0281
55	0000	0001	0004	0009	0018	0032	0074	0138	0178	0269
60	0000	0001	0004	0008	0017	0030	0070	0131	0169	0256
65	0000	0001	0003	0008	0016	0028	0066	0124	0161	0244
70	0000	0001	0003	0007	0015	0027	0063	0118	0153	0233
75	0000	0001	0003	0007	0014	0025	0059	0112	0146	0223
80	0000	0001	0003	0006	0013	0024	0056	0106	0138	0212
85	0000	0001	0003	0006	0013	0022	0053	0101	0132	0203
90	0000	0001	0003	0006	0012	0021	0050	0096	0125	0193
95	0000	0001	0002	0005	0011	0020	0048	0091	0119	0185
6.00	0000	0001	0002	0005	0011	0019	0045	0086	0113	0176
05	0000	0001	0002	0005	0010	0018	0043	0082	0108	0168
10	0000	0001	0002	0005	0009	0017	0041	0078	0103	0160
15	0000	0001	0002	0004	0009	0016	0038	0074	0098	0153
20	0000	0001	0002	0004	0008	0015	0036	0071	0093	0146
6.25	0000	0001	0002	0004	0008	0014	0035	0067	0089	0140
30	0000	0001	0002	0004	0008	0014	0033	0064	0084	0133
35	0000	0001	0001	0003	0007	0013	0031	0061	0080	0127
40	0000	0001	0001	0003	0007	0012	0030	0058	0077	0122
45	0000	0000	0001	0003	0006	0011	0028	0055	0073	0116
50	0000	0000	0001	0003	0006	0011	0027	0052	0070	0111
55	0000	0000	0001	0003	0006	0010	0025	0050	0066	0106
60	0000	0000	0001	0003	0005	0010	0024	0047	0063	0101
65	0000	0000	0001	0002	0005	0009	0023	0045	0060	0097
70	0000	0000	0001	0002	0005	0009	0022	0043	0057	0093
75	0000	0000	0001	0002	0005	0008	0021	0041	0055	0088
80	0000	0000	0001	0002	0004	0008	0020	0039	0052	0085
85	0000	0000	0001	0002	0004	0008	0019	0037	0050	0081
90	0000	0000	0001	0002	0004	0007	0018	0036	0048	0077
95	0000	0000	0001	0002	0004	0007	0017	0034	0045	0074
7.00	0000	0000	0001	0002	0004	0006	0016	0032	0043	0071
05	0000	0000	0001	0002	0003	0006	0015	0031	0041	0068
10	0000	0000	0001	0001	0003	0006	0015	0029	0040	0065
15	0000	0000	0001	0001	0003	0006	0014	0028	0038	0062
20	0000	0000	0001	0001	0003	0005	0013	0027	0036	0059
25	0000	0000	0001	0001	0003	0005	0013	0026	0035	0057
30	0000	0000	0001	0001	0003	0005	0012	0024	0033	0054
35	0000	0000	0000	0001	0002	0005	0012	0023	0032	0052
40	0000	0000	0000	0001	0002	0004	0011	0022	0030	0050
45	0000	0000	0000	0001	0002	0004	0010	0021	0029	0048
50	0000	0000	0000	0001	0002	0004	0010	0020	0028	0046
55	0000	0000	0000	0001	0002	0004	0010	0019	0026	0044
60	0000	0000	0000	0001	0002	0004	0009	0019	0025	0042
65	0000	0000	0000	0001	0002	0003	0009	0018	0024	0040
70	0000	0000	0000	0001	0002	0003	0008	0017	0023	0039

PROBABILITY DENSITY OF t, THE NON-CENTRAL t-STATISTIC, TABLED AS A FUNCTION OF $t/\sqrt{f}$.

f is the number of degrees of freedom; the non-centrality parameter is $\sqrt{f+1}\,K_p$.
K_p is the standardized normal deviate exceeded with probability p.

DEGREES OF FREEDOM **7**

$t/\sqrt{f}$ \ p	.2500	.1500	.1000	.0650	.0400	.0250	.0100	.0040	.0025	.0010
7.75	0000	0000	0000	0001	0002	0003	0008	0016	0022	0037
80	0000	0000	0000	0001	0002	0003	0008	0016	0021	0035
85	0000	0000	0000	0001	0002	0003	0007	0015	0020	0034
90	0000	0000	0000	0001	0001	0003	0007	0014	0019	0033
95	0000	0000	0000	0001	0001	0003	0007	0014	0019	0031
8.00	0000	0000	0000	0001	0001	0002	0006	0013	0018	0030
05	0000	0000	0000	0001	0001	0002	0006	0013	0017	0029
10	0000	0000	0000	0001	0001	0002	0006	0012	0016	0028
15	0000	0000	0000	0001	0001	0002	0006	0011	0016	0027
20	0000	0000	0000	0001	0001	0002	0005	0011	0015	0025
25	0000	0000	0000	0000	0001	0002	0005	0011	0014	0024
30	0000	0000	0000	0000	0001	0002	0005	0010	0014	0023
35	0000	0000	0000	0000	0001	0002	0005	0010	0013	0023
40	0000	0000	0000	0000	0001	0002	0004	0009	0013	0022
45	0000	0000	0000	0000	0001	0002	0004	0009	0012	0021
50	0000	0000	0000	0000	0001	0002	0004	0009	0012	0020
55	0000	0000	0000	0000	0001	0002	0004	0008	0011	0019
60	0000	0000	0000	0000	0001	0001	0004	0008	0011	0018
65	0000	0000	0000	0000	0001	0001	0004	0008	0010	0018
70	0000	0000	0000	0000	0001	0001	0003	0007	0010	0017
75	0000	0000	0000	0000	0001	0001	0003	0007	0010	0016
80	0000	0000	0000	0000	0001	0001	0003	0007	0009	0016
85	0000	0000	0000	0000	0001	0001	0003	0006	0009	0015
90	0000	0000	0000	0000	0001	0001	0003	0006	0009	0015
95	0000	0000	0000	0000	0001	0001	0003	0006	0008	0014
9.00	0000	0000	0000	0000	0001	0001	0003	0006	0008	0014
05	0000	0000	0000	0000	0001	0001	0003	0005	0008	0013
10	0000	0000	0000	0000	0001	0001	0002	0005	0007	0013
15	0000	0000	0000	0000	0000	0001	0002	0005	0007	0012
20	0000	0000	0000	0000	0000	0001	0002	0005	0007	0012
9.25	0000	0000	0000	0000	0000	0001	0002	0005	0006	0011
30	0000	0000	0000	0000	0000	0001	0002	0005	0006	0011
35	0000	0000	0000	0000	0000	0001	0002	0004	0006	0010
40	0000	0000	0000	0000	0000	0001	0002	0004	0006	0010
45	0000	0000	0000	0000	0000	0001	0002	0004	0006	0010
50	0000	0000	0000	0000	0000	0001	0002	0004	0005	0009
55	0000	0000	0000	0000	0000	0001	0002	0004	0005	0009
60	0000	0000	0000	0000	0000	0001	0002	0004	0005	0009
65	0000	0000	0000	0000	0000	0001	0002	0003	0005	0008
70	0000	0000	0000	0000	0000	0001	0002	0003	0005	0008
75	0000	0000	0000	0000	0000	0001	0002	0003	0004	0008
80	0000	0000	0000	0000	0000	0001	0001	0003	0004	0007
85	0000	0000	0000	0000	0000	0001	0001	0003	0004	0007
90	0000	0000	0000	0000	0000	0001	0001	0003	0004	0007
95	0000	0000	0000	0000	0000	0000	0001	0003	0004	0007
10.00	0000	0000	0000	0000	0000	0000	0001	0003	0004	0006
05	0000	0000	0000	0000	0000	0000	0001	0003	0004	0006
10	0000	0000	0000	0000	0000	0000	0001	0002	0003	0006
15	0000	0000	0000	0000	0000	0000	0001	0002	0003	0006
20	0000	0000	0000	0000	0000	0000	0001	0002	0003	0006
25	0000	0000	0000	0000	0000	0000	0001	0002	0003	0005
30	0000	0000	0000	0000	0000	0000	0001	0002	0003	0005
35	0000	0000	0000	0000	0000	0000	0001	0002	0003	0005
40	0000	0000	0000	0000	0000	0000	0001	0002	0003	0005
45	0000	0000	0000	0000	0000	0000	0001	0002	0003	0005
50	0000	0000	0000	0000	0000	0000	0001	0002	0003	0005
55	0000	0000	0000	0000	0000	0000	0001	0002	0003	0004
60	0000	0000	0000	0000	0000	0000	0001	0002	0002	0004
65	0000	0000	0000	0000	0000	0000	0001	0002	0002	0004
70	0000	0000	0000	0000	0000	0000	0001	0002	0002	0004

PROBABILITY DENSITY OF t, THE NON-CENTRAL t-STATISTIC, TABLED AS A FUNCTION OF $t/\sqrt{f}$.

f is the number of degrees of freedom; the non-centrality parameter is $\sqrt{f+1}\,K_p$.
K_p is the standardized normal deviate exceeded with probability p.

DEGREES OF FREEDOM **7**

$t/\sqrt{f}$ \ p	.2500	.1500	.1000	.0650	.0400	.0250	.0100	.0040	.0025	.0010
10.75	0000	0000	0000	0000	0000	0000	0001	0002	0002	0004
80	0000	0000	0000	0000	0000	0000	0001	0002	0002	0004
85	0000	0000	0000	0000	0000	0000	0001	0001	0002	0004
90	0000	0000	0000	0000	0000	0000	0001	0001	0002	0003
95	0000	0000	0000	0000	0000	0000	0001	0001	0002	0003
11.00	0000	0000	0000	0000	0000	0000	0001	0001	0002	0003
05	0000	0000	0000	0000	0000	0000	0001	0001	0002	0003
10	0000	0000	0000	0000	0000	0000	0001	0001	0002	0003
15	0000	0000	0000	0000	0000	0000	0001	0001	0002	0003
20	0000	0000	0000	0000	0000	0000	0001	0001	0002	0003
25	0000	0000	0000	0000	0000	0000	0001	0001	0002	0003
30	0000	0000	0000	0000	0000	0000	0000	0001	0002	0003
35	0000	0000	0000	0000	0000	0000	0000	0001	0001	0003
40	0000	0000	0000	0000	0000	0000	0000	0001	0001	0003
45	0000	0000	0000	0000	0000	0000	0000	0001	0001	0002
50	0000	0000	0000	0000	0000	0000	0000	0001	0001	0002
55	0000	0000	0000	0000	0000	0000	0000	0001	0001	0002
60	0000	0000	0000	0000	0000	0000	0000	0001	0001	0002
65	0000	0000	0000	0000	0000	0000	0000	0001	0001	0002
70	0000	0000	0000	0000	0000	0000	0000	0001	0001	0002
75	0000	0000	0000	0000	0000	0000	0000	0001	0001	0002
80	0000	0000	0000	0000	0000	0000	0000	0001	0001	0002
85	0000	0000	0000	0000	0000	0000	0000	0001	0001	0002
90	0000	0000	0000	0000	0000	0000	0000	0001	0001	0002
95	0000	0000	0000	0000	0000	0000	0000	0001	0001	0002
12.00	0000	0000	0000	0000	0000	0000	0000	0001	0001	0002
05	0000	0000	0000	0000	0000	0000	0000	0001	0001	0002
10	0000	0000	0000	0000	0000	0000	0000	0001	0001	0002
15	0000	0000	0000	0000	0000	0000	0000	0001	0001	0002
20	0000	0000	0000	0000	0000	0000	0000	0001	0001	0002
12.25	0000	0000	0000	0000	0000	0000	0000	0001	0001	0001
30	0000	0000	0000	0000	0000	0000	0000	0001	0001	0001
35	0000	0000	0000	0000	0000	0000	0000	0001	0001	0001
40	0000	0000	0000	0000	0000	0000	0000	0001	0001	0001
45	0000	0000	0000	0000	0000	0000	0000	0001	0001	0001
50	0000	0000	0000	0000	0000	0000	0000	0001	0001	0001
55	0000	0000	0000	0000	0000	0000	0000	0000	0001	0001
60	0000	0000	0000	0000	0000	0000	0000	0000	0001	0001
65	0000	0000	0000	0000	0000	0000	0000	0000	0001	0001
70	0000	0000	0000	0000	0000	0000	0000	0000	0001	0001
75	0000	0000	0000	0000	0000	0000	0000	0000	0001	0001
80	0000	0000	0000	0000	0000	0000	0000	0000	0001	0001
85	0000	0000	0000	0000	0000	0000	0000	0000	0001	0001
90	0000	0000	0000	0000	0000	0000	0000	0000	0001	0001
95	0000	0000	0000	0000	0000	0000	0000	0000	0001	0001
13.00	0000	0000	0000	0000	0000	0000	0000	0000	0001	0001
05	0000	0000	0000	0000	0000	0000	0000	0000	0001	0001
10	0000	0000	0000	0000	0000	0000	0000	0000	0001	0001
15	0000	0000	0000	0000	0000	0000	0000	0000	0000	0001
20	0000	0000	0000	0000	0000	0000	0000	0000	0000	0001
25	0000	0000	0000	0000	0000	0000	0000	0000	0000	0001
30	0000	0000	0000	0000	0000	0000	0000	0000	0000	0001
35	0000	0000	0000	0000	0000	0000	0000	0000	0000	0001
40	0000	0000	0000	0000	0000	0000	0000	0000	0000	0001
45	0000	0000	0000	0000	0000	0000	0000	0000	0000	0001
50	0000	0000	0000	0000	0000	0000	0000	0000	0000	0001
55	0000	0000	0000	0000	0000	0000	0000	0000	0000	0001
60	0000	0000	0000	0000	0000	0000	0000	0000	0000	0001
65	0000	0000	0000	0000	0000	0000	0000	0000	0000	0001
70	0000	0000	0000	0000	0000	0000	0000	0000	0000	0001

PROBABILITY DENSITY OF t, THE NON-CENTRAL t-STATISTIC, TABLED AS A FUNCTION OF $t/\sqrt{f}$

f is the number of degrees of freedom; the non-centrality parameter is $\sqrt{f+1}\,K_p$.
K_p is the standardized normal deviate exceeded with probability p.

DEGREES OF FREEDOM **7**

$t/\sqrt{f}$ \ p	.2500	.1500	.1000	.0650	.0400	.0250	.0100	.0040	.0025	.0010
13.75	0000	0000	0000	0000	0000	0000	0000	0000	0000	0001
80	0000	0000	0000	0000	0000	0000	0000	0000	0000	0001
85	0000	0000	0000	0000	0000	0000	0000	0000	0000	0001
90	0000	0000	0000	0000	0000	0000	0000	0000	0000	0001
95	0000	0000	0000	0000	0000	0000	0000	0000	0000	0001
14.00	0000	0000	0000	0000	0000	0000	0000	0000	0000	0001
05	0000	0000	0000	0000	0000	0000	0000	0000	0000	0001
10	0000	0000	0000	0000	0000	00-00	0000	0000	0000	0001
15	0000	0000	0000	0000	0000	0000	0000	0000	0000	0001
20	0000	000Q	0000	0000	0000	0000	0000	0000	0000	0000

PROBABILITY DENSITY OF t, THE NON-CENTRAL t-STATISTIC, TABLED AS A FUNCTION OF $t/\sqrt{f}$.

f is the number of degrees of freedom; the non-centrality parameter is $\sqrt{f+1}\,K_p$.
K_p is the standardized normal deviate exceeded with probability p.

DEGREES OF FREEDOM **8**

$t/\sqrt{f}$ \ p	.2500	.1500	.1000	.0650	.0400	.0250	.0100	.0040	.0025	.0010
-1.05	0000	0000	0000	0000	0000	0000	0000	0000	0000	0000
-1.00	0001	0000	0000	0000	0000	0000	0000	0000	0000	0000
-0.95	0001	0000	0000	0000	0000	0000	0000	0000	0000	0000
90	0001	0000	0000	0000	0000	0000	0000	0000	0000	0000
85	0001	0000	0000	0000	0000	0000	0000	0000	0000	0000
80	0002	0000	0000	0000	0000	0000	0000	0000	0000	0000
75	0003	0000	0000	0000	0000	0000	0000	0000	0000	0000
70	0004	0000	0000	0000	0000	0000	0000	0000	0000	0000
65	0005	0000	0000	0000	0000	0000	0000	0000	0000	0000
60	0008	0000	0000	0000	0000	0000	0000	0000	0000	0000
55	0011	0000	0000	0000	0000	0000	0000	0000	0000	0000
50	0016	0000	0000	0000	0000	0000	0000	0000	0000	0000
45	0023	0000	0000	0000	0000	0000	0000	0000	0000	0000
40	0033	0001	0000	0000	0000	0000	0000	0000	0000	0000
35	0047	0001	0000	0000	0000	0000	0000	0000	0000	0000
30	0067	0002	0000	0000	0000	0000	0000	0000	0000	0000
25	0096	0003	0000	0000	0000	0000	0000	0000	0000	0000
20	0136	0005	0000	0000	0000	0000	0000	0000	0000	0000
15	0192	0008	0000	0000	0000	0000	0000	0000	0000	0000
10	0268	0012	0001	0000	0000	0000	0000	0000	0000	0000
05	0368	0019	0001	0000	0000	0000	0000	0000	0000	0000
0.00	0499	0031	0002	0000	0000	0000	0000	0000	0000	0000
05	0665	0048	0004	0000	0000	0000	0000	0000	0000	0000
10	0868	0074	0007	0000	0000	0000	0000	0000	0000	0000
15	1109	0113	0012	0001	0000	0000	0000	0000	0000	0000
20	1385	0168	0021	0002	0000	0000	0000	0000	0000	0000
0.25	1690	0244	0034	0003	0000	0000	0000	0000	0000	0000
30	2012	0347	0056	0006	0000	0000	0000	0000	0000	0000
35	2336	0481	0088	0011	0001	0000	0000	0000	0000	0000
40	2647	0648	0135	0019	0002	0000	0000	0000	0000	0000
45	2927	0849	0201	0033	0003	0000	0000	0000	0000	0000
50	3161	1083	0291	0054	0006	0001	0000	0000	0000	0000
55	3338	1342	0408	0086	0012	0001	0000	0000	0000	0000
60	3450	1620	0554	0132	0020	0003	0000	0000	0000	0000
65	3496	1903	0730	0197	0035	0005	0000	0000	0000	0000
70	3476	2180	0934	0283	0057	0010	0000	0000	0000	0000
75	3397	2437	1161	0393	0090	0018	0001	0000	0000	0000
80	3268	2664	1404	0529	0137	0031	0001	0000	0000	0000
85	3099	2850	1653	0691	0200	0050	0002	0000	0000	0000
90	2901	2990	1899	0874	0283	0079	0005	0000	0000	0000
95	2685	3081	2132	1076	0387	0120	0008	0000	0000	0000
1.00	2459	3121	2342	1289	0512	0175	0015	0001	0000	0000
05	2231	3115	2522	1507	0658	0246	0025	0002	0000	0000
10	2009	3065	2667	1722	0821	0334	0041	0004	0001	0000
15	1796	2979	2773	1926	0998	0441	0063	0007	0002	0000
20	1595	2862	2839	2112	1184	0565	0094	0011	0004	0000
25	1410	2722	2866	2275	1373	0705	0136	0019	0006	0001
30	1240	2565	2857	2409	1559	0858	0190	0031	0011	0001
35	1087	2398	2815	2513	1737	1019	0256	0047	0018	0003
40	0949	2225	2745	2586	1900	1185	0336	0070	0029	0005
45	0827	2051	2651	2627	2046	1351	0429	0101	0044	0008
50	0719	1881	2538	2637	2169	1511	0534	0140	0065	0013
55	0624	1715	2412	2621	2269	1663	0650	0188	0092	0020
60	0540	1558	2277	2580	2344	1802	0775	0247	0127	0031
65	0468	1409	2136	2517	2394	1925	0905	0316	0170	0045
70	0405	1271	1992	2437	2420	2030	1038	0396	0223	0064

PROBABILITY DENSITY OF t, THE NON-CENTRAL t-STATISTIC, TABLED AS A FUNCTION OF $t/\sqrt{f}$.

f is the number of degrees of freedom; the non-centrality parameter is $\sqrt{f+1}\,K_p$.
K_p is the standardized normal deviate exceeded with probability p.

DEGREES OF FREEDOM **8**

$t/\sqrt{f}$ \ p	.2500	.1500	.1000	.0650	.0400	.0250	.0100	.0040	.0025	.0010
1.75	0350	1142	1849	2343	2423	2115	1171	0484	0284	0089
80	0302	1024	1709	2239	2404	2180	1301	0581	0354	0121
85	0261	0917	1574	2126	2368	2224	1424	0683	0433	0159
90	0226	0819	1444	2009	2315	2248	1540	0791	0520	0204
95	0195	0730	1321	1890	2248	2254	1644	0901	0612	0257
2.00	0169	0650	1205	1770	2171	2242	1737	1012	0709	0317
05	0146	0579	1097	1652	2084	2215	1815	1121	0809	0384
10	0127	0515	0997	1536	1992	2175	1880	1226	0911	0457
15	0110	0457	0904	1425	1895	2122	1930	1325	1012	0535
20	0096	0406	0819	1318	1796	2060	1966	1418	1111	0617
25	0083	0361	0740	1216	1696	1990	1988	1502	1205	0702
30	0072	0320	0669	1119	1596	1914	1996	1577	1295	0789
35	0063	0284	0604	1029	1498	1833	1992	1641	1377	0877
40	0055	0253	0545	0944	1401	1750	1977	1695	1452	0963
45	0048	0224	0491	0865	1308	1664	1952	1738	1519	1047
50	0042	0199	0442	0791	1219	1578	1917	1770	1576	1127
55	0037	0177	0399	0723	1133	1492	1875	1791	1624	1203
60	0032	0158	0359	0661	1052	1408	1826	1803	1662	1274
65	0028	0140	0323	0603	0975	1325	1772	1805	1691	1338
70	0025	0125	0291	0550	0903	1244	1713	1798	1711	1395
75	0022	0111	0262	0501	0834	1166	1651	1783	1722	1446
80	0019	0099	0236	0457	0771	1091	1587	1760	1725	1489
85	0017	0088	0213	0416	0711	1020	1521	1732	1719	1524
90	0015	0079	0192	0379	0656	0952	1454	1697	1707	1552
95	0013	0070	0173	0345	0604	0887	1387	1658	1688	1573
3.00	0012	0063	0156	0314	0556	0826	1320	1614	1663	1586
05	0010	0056	0141	0286	0512	0768	1254	1567	1633	1592
10	0009	0050	0127	0261	0471	0714	1190	1518	1599	1593
15	0008	0045	0115	0237	0433	0663	1127	1466	1561	1587
20	0007	0041	0103	0216	0398	0615	1065	1413	1519	1575
3.25	0006	0036	0094	0197	0366	0571	1006	1359	1475	1558
30	0006	0033	0085	0179	0336	0529	0949	1305	1429	1537
35	0005	0029	0077	0163	0309	0490	0894	1250	1382	1512
40	0005	0026	0069	0149	0284	0454	0841	1196	1333	1483
45	0004	0024	0063	0136	0261	0421	0791	1142	1283	1451
50	0004	0021	0057	0124	0240	0389	0743	1089	1234	1416
55	0003	0019	0052	0113	0221	0360	0698	1038	1184	1380
60	0003	0017	0047	0103	0203	0334	0655	0987	1135	1341
65	0003	0016	0043	0094	0186	0309	0614	0938	1086	1301
70	0002	0014	0039	0086	0171	0286	0575	0891	1038	1260
75	0002	0013	0035	0079	0158	0264	0539	0845	0991	1219
80	0002	0012	0032	0072	0145	0245	0505	0801	0945	1176
85	0002	0011	0029	0066	0133	0226	0472	0758	0900	1134
90	0002	0010	0027	0060	0123	0210	0442	0717	0857	1092
95	0001	0009	0024	0055	0113	0194	0413	0678	0815	1050
4.00	0001	0008	0022	0051	0104	0180	0387	0641	0774	1008
05	0001	0007	0020	0046	0096	0166	0362	0606	0735	0967
10	0001	0007	0018	0043	0089	0154	0338	0572	0697	0927
15	0001	0006	0017	0039	0082	0143	0316	0540	0661	0888
20	0001	0005	0015	0036	0075	0132	0295	0509	0627	0849
25	0001	0005	0014	0033	0069	0123	0276	0480	0594	0812
30	0001	0005	0013	0030	0064	0114	0258	0452	0562	0775
35	0001	0004	0012	0028	0059	0105	0241	0426	0532	0740
40	0001	0004	0011	0026	0055	0098	0226	0402	0504	0706
45	0001	0003	0010	0024	0051	0091	0211	0379	0476	0673
50	0000	0003	0009	0022	0047	0084	0197	0357	0450	0641
55	0000	0003	0008	0020	0043	0078	0184	0336	0426	0611
60	0000	0003	0008	0019	0040	0073	0172	0316	0402	0581
65	0000	0002	0007	0017	0037	0067	0161	0298	0380	0553
70	0000	0002	0007	0016	0034	0063	0151	0280	0359	0526

PROBABILITY DENSITY OF t, THE NON-CENTRAL t-STATISTIC, TABLED AS A FUNCTION OF $t/\sqrt{f}$.

f is the number of degrees of freedom; the non-centrality parameter is $\sqrt{f+1}\,K_p$.
K_p is the standardized normal deviate exceeded with probability p.

DEGREES OF FREEDOM **8**

$t/\sqrt{f}$ \ p	.2500	.1500	.1000	.0650	.0400	.0250	.0100	.0040	.0025	.0010
4.75	0000	0002	0006	0015	0032	0058	0141	0264	0339	0500
80	0000	0002	0006	0013	0029	0054	0132	0248	0320	0475
85	0000	0002	0005	0012	0027	0050	0123	0234	0302	0452
90	0000	0002	0005	0012	0025	0047	0115	0220	0286	0429
95	0000	0001	0004	0011	0024	0044	0108	0207	0270	0407
5.00	0000	0001	0004	0010	0022	0041	0101	0195	0255	0387
05	0000	0001	0004	0009	0020	0038	0095	0184	0240	0367
10	0000	0001	0003	0008	0019	0035	0089	0173	0227	0349
15	0000	0001	0003	0008	0018	0033	0083	0163	0214	0331
20	0000	0001	0003	0007	0016	0031	0078	0153	0202	0314
25	0000	0001	0003	0007	0015	0029	0073	0144	0191	0298
30	0000	0001	0003	0006	0014	0027	0068	0136	0180	0283
35	0000	0001	0002	0006	0013	0025	0064	0128	0170	0268
40	0000	0001	0002	0005	0012	0023	0060	0121	0161	0254
45	0000	0001	0002	0005	0011	0022	0056	0114	0152	0241
50	0000	0001	0002	0005	0011	0020	0053	0107	0143	0229
55	0000	0001	0002	0004	0010	0019	0050	0101	0135	0217
60	0000	0001	0002	0004	0009	0018	0047	0095	0128	0206
65	0000	0001	0002	0004	0009	0017	0044	0090	0121	0195
70	0000	0000	0001	0004	0008	0016	0041	0085	0114	0185
75	0000	0000	0001	0003	0008	0015	0039	0080	0108	0176
80	0000	0000	0001	0003	0007	0014	0036	0075	0102	0167
85	0000	0000	0001	0003	0007	0013	0034	0071	0096	0158
90	0000	0000	0001	0003	0006	0012	0032	0067	0091	0150
95	0000	0000	0001	0003	0006	0011	0030	0063	0086	0142
6.00	0000	0000	0001	0002	0005	0011	0028	0060	0082	0135
05	0000	0000	0001	0002	0005	0010	0027	0056	0077	0128
10	0000	0000	0001	0002	0005	0009	0025	0053	0073	0122
15	0000	0000	0001	0002	0004	0009	0024	0050	0069	0115
20	0000	0000	0001	0002	0004	0008	0022	0048	0065	0110
6.25	0000	0000	0001	0002	0004	0008	0021	0045	0062	0104
30	0000	0000	0001	0002	0004	0007	0020	0042	0058	0099
35	0000	0000	0001	0001	0003	0007	0019	0040	0055	0094
40	0000	0000	0001	0001	0003	0006	0018	0038	0052	0089
45	0000	0000	0001	0001	0003	0006	0017	0036	0050	0085
50	0000	0000	0000	0001	0003	0006	0016	0034	0047	0080
55	0000	0000	0000	0001	0003	0005	0015	0032	0045	0076
60	0000	0000	0000	0001	0003	0005	0014	0030	0042	0072
65	0000	0000	0000	0001	0002	0005	0013	0029	0040	0069
70	0000	0000	0000	0001	0002	0004	0013	0027	0038	0065
75	0000	0000	0000	0001	0002	0004	0012	0026	0036	0062
80	0000	0000	0000	0001	0002	0004	0011	0024	0034	0059
85	0000	0000	0000	0001	0002	0004	0011	0023	0032	0056
90	0000	0000	0000	0001	0002	0004	0010	0022	0031	0053
95	0000	0000	0000	0001	0002	0003	0009	0021	0029	0051
7.00	0000	0000	0000	0001	0002	0003	0009	0020	0028	0048
05	0000	0000	0000	0001	0001	0003	0008	0019	0026	0046
10	0000	0000	0000	0001	0001	0003	0008	0018	0025	0044
15	0000	0000	0000	0001	0001	0003	0008	0017	0024	0042
20	0000	0000	0000	0001	0001	0003	0007	0016	0023	0040
25	0000	0000	0000	0000	0001	0002	0007	0015	0021	0038
30	0000	0000	0000	0000	0001	0002	0006	0014	0020	0036
35	0000	0000	0000	0000	0001	0002	0006	0014	0019	0034
40	0000	0000	0000	0000	0001	0002	0006	0013	0018	0033
45	0000	0000	0000	0000	0001	0002	0006	0012	0017	0031
50	0000	0000	0000	0000	0001	0002	0005	0012	0017	0030
55	0000	0000	0000	0000	0001	0002	0005	0011	0016	0028
60	0000	0000	0000	0000	0001	0002	0005	0011	0015	0027
65	0000	0000	0000	0000	0001	0002	0004	0010	0014	0026
70	0000	0000	0000	0000	0001	0001	0004	0010	0014	0024

PROBABILITY DENSITY OF t, THE NON-CENTRAL t-STATISTIC, TABLED AS A FUNCTION OF $t/\sqrt{f}$.

f is the number of degrees of freedom; the non-centrality parameter is $\sqrt{f+1}\,K_p$.
K_p is the standardized normal deviate exceeded with probability p.

DEGREES OF FREEDOM **8**

$t/\sqrt{f}$ \ p	.2500	.1500	.1000	.0650	.0400	.0250	.0100	.0040	.0025	.0010
7.75	0000	0000	0000	0000	0001	0001	0004	0009	0013	0023
80	0000	0000	0000	0000	0001	0001	0004	0009	0012	0022
85	0000	0000	0000	0000	0001	0001	0004	0008	0012	0021
90	0000	0000	0000	0000	0001	0001	0003	0008	0011	0020
95	0000	0000	0000	0000	0001	0001	0003	0008	0011	0019
8.00	0000	0000	0000	0000	0001	0001	0003	0007	0010	0018
05	0000	0000	0000	0000	0000	0001	0003	0007	0010	0018
10	0000	0000	0000	0000	0000	0001	0003	0006	0009	0017
15	0000	0000	0000	0000	0000	0001	0003	0006	0009	0016
20	0000	0000	0000	0000	0000	0001	0003	0006	0008	0015
25	0000	0000	0000	0000	0000	0001	0002	0006	0008	0015
30	0000	0000	0000	0000	0000	0001	0002	0005	0008	0014
35	0000	0000	0000	0000	0000	0001	0002	0005	0007	0013
40	0000	0000	0000	0000	0000	0001	0002	0005	0007	0013
45	0000	0000	0000	0000	0000	0001	0002	0005	0007	0012
50	0000	0000	0000	0000	0000	0001	0002	0004	0006	0012
55	0000	0000	0000	0000	0000	0001	0002	0004	0006	0011
60	0000	0000	0000	0000	0000	0001	0002	0004	0006	0011
65	0000	0000	0000	0000	0000	0001	0002	0004	0006	0010
70	0000	0000	0000	0000	0000	0001	0002	0004	0005	0010
75	0000	0000	0000	0000	0000	0001	0002	0004	0005	0009
80	0000	0000	0000	0000	0000	0000	0001	0003	0005	0009
85	0000	0000	0000	0000	0000	0000	0001	0003	0005	0009
90	0000	0000	0000	0000	0000	0000	0001	0003	0004	0008
95	0000	0000	0000	0000	0000	0000	0001	0003	0004	0008
9.00	0000	0000	0000	0000	0000	0000	0001	0003	0004	0008
05	0000	0000	0000	0000	0000	0000	0001	0003	0004	0007
10	0000	0000	0000	0000	0000	0000	0001	0003	0004	0007
15	0000	0000	0000	0000	0000	0000	0001	0002	0004	0007
20	0000	0000	0000	0000	0000	0000	0001	0002	0003	0006
9.25	0000	0000	0000	0000	0000	0000	0001	0002	0003	0006
30	0000	0000	0000	0000	0000	0000	0001	0002	0003	0006
35	0000	0000	0000	0000	0000	0000	0001	0002	0003	0006
40	0000	0000	0000	0000	0000	0000	0001	0002	0003	0005
45	0000	0000	0000	0000	0000	0000	0001	0002	0003	0005
50	0000	0000	0000	0000	0000	0000	0001	0002	0003	0005
55	0000	0000	0000	0000	0000	0000	0001	0002	0003	0005
60	0000	0000	0000	0000	0000	0000	0001	0002	0002	0005
65	0000	0000	0000	0000	0000	0000	0001	0002	0002	0004
70	0000	0000	0000	0000	0000	0000	0001	0002	0002	0004
75	0000	0000	0000	0000	0000	0000	0001	0001	0002	0004
80	0000	0000	0000	0000	0000	0000	0001	0001	0002	0004
85	0000	0000	0000	0000	0000	0000	0001	0001	0002	0004
90	0000	0000	0000	0000	0000	0000	0001	0001	0002	0004
95	0000	0000	0000	0000	0000	0000	0001	0001	0002	0003
10.00	0000	0000	0000	0000	0000	0000	0001	0001	0002	0003
05	0000	0000	0000	0000	0000	0000	0000	0001	0002	0003
10	0000	0000	0000	0000	0000	0000	0000	0001	0002	0003
15	0000	0000	0000	0000	0000	0000	0000	0001	0002	0003
20	0000	0000	0000	0000	0000	0000	0000	0001	0001	0003
25	0000	0000	0000	0000	0000	0000	0000	0001	0001	0003
30	0000	0000	0000	0000	0000	0000	0000	0001	0001	0003
35	0000	0000	0000	0000	0000	0000	0000	0001	0001	0002
40	0000	0000	0000	0000	0000	0000	0000	0001	0001	0002
45	0000	0000	0000	0000	0000	0000	0000	0001	0001	0002
50	0000	0000	0000	0000	0000	0000	0000	0001	0001	0002
55	0000	0000	0000	0000	0000	0000	0000	0001	0001	0002
60	0000	0000	0000	0000	0000	0000	0000	0001	0001	0002
65	0000	0000	0000	0000	0000	0000	0000	0001	0001	0002
70	0000	0000	0000	0000	0000	0000	0000	0001	0001	0002

PROBABILITY DENSITY OF t, THE NON-CENTRAL t-STATISTIC, TABLED AS A FUNCTION OF $t/\sqrt{f}$.

f is the number of degrees of freedom; the non-centrality parameter is $\sqrt{f+1}\,K_p$.
K_p is the standardized normal deviate exceeded with probability p.

DEGREES OF FREEDOM **8**

$t/\sqrt{f}$ \ p	.2500	.1500	.1000	.0650	.0400	.0250	.0100	.0040	.0025	.0010
10.75	0000	0000	0000	0000	0000	0000	0000	0001	0001	0002
80	0000	0000	0000	0000	0000	0000	0000	0001	0001	0002
85	0000	0000	0000	0000	0000	0000	0000	0001	0001	0002
90	0000	0000	0000	0000	0000	0000	0000	0001	0001	0002
95	0000	0000	0000	0000	0000	0000	0000	0001	0001	0002
11.00	0000	0000	0000	0000	0000	0000	0000	0001	0001	0002
05	0000	0000	0000	0000	0000	0000	0000	0001	0001	0001
10	0000	0000	0000	0000	0000	0000	0000	0001	0001	0001
15	0000	0000	0000	0000	0000	0000	0000	0000	0001	0001
20	0000	0000	0000	0000	0000	0000	0000	0000	0001	0001
25	0000	0000	0000	0000	0000	0000	0000	0000	0001	0001
30	0000	0000	0000	0000	0000	0000	0000	0000	0001	0001
35	0000	0000	0000	0000	0000	0000	0000	0000	0001	0001
40	0000	0000	0000	0000	0000	0000	0000	0000	0001	0001
45	0000	0000	0000	0000	0000	0000	0000	0000	0001	0001
50	0000	0000	0000	0000	0000	0000	0000	0000	0001	0001
55	0000	0000	0000	0000	0000	0000	0000	0000	0001	0001
60	0000	0000	0000	0000	0000	0000	0000	0000	0001	0001
65	0000	0000	0000	0000	0000	0000	0000	0000	0000	0001
70	0000	0000	0000	0000	0000	0000	0000	0000	0000	0001
75	0000	0000	0000	0000	0000	0000	0000	0000	0000	0001
80	0000	0000	0000	0000	0000	0000	0000	0000	0000	0001
85	0000	0000	0000	0000	0000	0000	0000	0000	0000	0001
90	0000	0000	0000	0000	0000	0000	0000	0000	0000	0001
95	0000	0000	0000	0000	0000	0000	0000	0000	0000	0001
12.00	0000	0000	0000	0000	0000	0000	0000	0000	0000	0001
05	0000	0000	0000	0000	0000	0000	0000	0000	0000	0001
10	0000	0000	0000	0000	0000	0000	0000	0000	0000	0001
15	0000	0000	0000	0000	0000	0000	0000	0000	0000	0001
20	0000	0000	0000	0000	0000	0000	0000	0000	0000	0001
12.25	0000	0000	0000	0000	0000	0000	0000	0000	0000	0001
30	0000	0000	0000	0000	0000	0000	0000	0000	0000	0001
35	0000	0000	0000	0000	0000	0000	0000	0000	0000	0001
40	0000	0000	0000	0000	0000	0000	0000	0000	0000	0001
45	0000	0000	0000	0000	0000	0000	0000	0000	0000	0001
50	0000	0000	0000	0000	0000	0000	0000	0000	0000	0001
55	0000	0000	0000	0000	0000	0000	0000	0000	0000	0001
60	0000	0000	0000	0000	0000	0000	0000	0000	0000	0000

PROBABILITY DENSITY OF t, THE NON-CENTRAL t-STATISTIC, TABLED AS A FUNCTION OF $t/\sqrt{f}$.

f is the number of degrees of freedom; the non-centrality parameter is $\sqrt{f+1}\,K_p$.
K_p is the standardized normal deviate exceeded with probability p.

DEGREES OF FREEDOM **9**

$t/\sqrt{f}$ \ p	.2500	.1500	.1000	.0650	.0400	.0250	.0100	.0040	.0025	.0010
-0.90	0000	0000	0000	0000	0000	0000	0000	0000	0000	0000
85	0001	0000	0000	0000	0000	0000	0000	0000	0000	0000
80	0001	0000	0000	0000	0000	0000	0000	0000	0000	0000
75	0001	0000	0000	0000	0000	0000	0000	0000	0000	0000
70	0002	0000	0000	0000	0000	0000	0000	0000	0000	0000
65	0003	0000	0000	0000	0000	0000	0000	0000	0000	0000
60	0004	0000	0000	0000	0000	0000	0000	0000	0000	0000
55	0006	0000	0000	0000	0000	0000	0000	0000	0000	0000
50	0009	0000	0000	0000	0000	0000	0000	0000	0000	0000
45	0013	0000	0000	0000	0000	0000	0000	0000	0000	0000
40	0019	0000	0000	0000	0000	0000	0000	0000	0000	0000
35	0029	0000	0000	0000	0000	0000	0000	0000	0000	0000
30	0043	0001	0000	0000	0000	0000	0000	0000	0000	0000
25	0064	0001	0000	0000	0000	0000	0000	0000	0000	0000
20	0094	0002	0000	0000	0000	0000	0000	0000	0000	0000
15	0138	0004	0000	0000	0000	0000	0000	0000	0000	0000
10	0199	0006	0000	0000	0000	0000	0000	0000	0000	0000
05	0285	0011	0001	0000	0000	0000	0000	0000	0000	0000
0.00	0399	0018	0001	0000	0000	0000	0000	0000	0000	0000
05	0549	0030	0002	0000	0000	0000	0000	0000	0000	0000
10	0739	0048	0004	0000	0000	0000	0000	0000	0000	0000
15	0971	0077	0007	0000	0000	0000	0000	0000	0000	0000
20	1245	0119	0012	0001	0000	0000	0000	0000	0000	0000
25	1554	0182	0021	0002	0000	0000	0000	0000	0000	0000
30	1888	0269	0035	0003	0000	0000	0000	0000	0000	0000
35	2231	0386	0059	0006	0000	0000	0000	0000	0000	0000
40	2565	0539	0095	0011	0001	0000	0000	0000	0000	0000
45	2870	0729	0148	0020	0002	0000	0000	0000	0000	0000
0.50	3129	0956	0223	0034	0003	0000	0000	0000	0000	0000
55	3327	1216	0325	0058	0006	0001	0000	0000	0000	0000
60	3454	1500	0457	0094	0012	0001	0000	0000	0000	0000
65	3506	1796	0622	0146	0021	0003	0000	0000	0000	0000
70	3486	2090	0819	0218	0037	0005	0000	0000	0000	0000
75	3401	2369	1045	0316	0062	0010	0000	0000	0000	0000
80	3259	2617	1291	0439	0098	0019	0000	0000	0000	0000
85	3074	2824	1550	0591	0150	0032	0001	0000	0000	0000
90	2858	2981	1810	0769	0221	0054	0002	0000	0000	0000
95	2623	3083	2060	0969	0313	0085	0005	0000	0000	0000
1.00	2380	3131	2289	1187	0428	0130	0009	0000	0000	0000
05	2138	3125	2488	1413	0566	0190	0015	0001	0000	0000
10	1903	3071	2649	1640	0724	0269	0026	0002	0000	0000
15	1680	2977	2768	1859	0901	0366	0043	0003	0001	0000
20	1474	2849	2844	2061	1090	0483	0067	0006	0002	0000
25	1285	2696	2875	2239	1286	0617	0100	0011	0003	0000
30	1115	2525	2867	2388	1483	0768	0145	0019	0006	0001
35	0963	2344	2821	2505	1673	0931	0203	0031	0011	0001
40	0829	2158	2744	2586	1850	1101	0274	0049	0018	0002
45	0711	1972	2641	2633	2009	1275	0360	0073	0029	0004
50	0609	1791	2518	2647	2146	1445	0460	0105	0045	0007
55	0520	1618	2381	2629	2257	1609	0573	0146	0066	0012
60	0444	1454	2233	2585	2342	1759	0696	0198	0095	0020
65	0378	1301	2081	2516	2398	1894	0828	0260	0131	0030
70	0322	1160	1927	2429	2428	2010	0965	0334	0177	0045
75	0274	1031	1775	2326	2432	2105	1104	0418	0232	0065
80	0233	0913	1626	2211	2412	2177	1241	0511	0297	0090
85	0198	0807	1484	2089	2372	2227	1373	0613	0371	0123
90	0169	0712	1349	1962	2314	2255	1498	0722	0454	0162
95	0143	0627	1222	1834	2241	2262	1612	0834	0545	0209

PROBABILITY DENSITY OF t, THE NON-CENTRAL t-STATISTIC, TABLED AS A FUNCTION OF $t/\sqrt{f}$.

f is the number of degrees of freedom; the non-centrality parameter is $\sqrt{f+1}\,K_p$.
K_p is the standardized normal deviate exceeded with probability p.

DEGREES OF FREEDOM **9**

$t/\sqrt{f}$ \ p	.2500	.1500	.1000	.0650	.0400	.0250	.0100	.0040	.0025	.0010
2.00	0122	0552	1104	1706	2156	2250	1714	0949	0643	0265
05	0104	0485	0995	1580	2062	2221	1801	1064	0745	0327
10	0089	0426	0894	1458	1961	2176	1873	1176	0849	0397
15	0076	0373	0803	1341	1856	2119	1929	1283	0955	0474
20	0065	0327	0719	1230	1748	2050	1969	1383	1059	0556
25	0055	0287	0643	1125	1641	1974	1994	1475	1160	0642
30	0047	0251	0575	1026	1534	1890	2004	1557	1257	0731
35	0041	0220	0513	0934	1430	1802	2000	1629	1346	0822
40	0035	0193	0457	0849	1328	1712	1984	1688	1428	0912
45	0030	0169	0408	0771	1231	1619	1956	1736	1502	1001
50	0026	0149	0363	0699	1138	1527	1918	1772	1565	1087
55	0022	0130	0323	0632	1050	1435	1871	1797	1618	1169
60	0019	0114	0288	0572	0967	1345	1818	1810	1661	1246
65	0017	0100	0256	0517	0889	1258	1758	1812	1694	1316
70	0014	0088	0228	0466	0815	1173	1694	1805	1716	1380
75	0012	0078	0203	0421	0747	1092	1626	1789	1729	1435
80	0011	0068	0181	0380	0684	1015	1556	1764	1732	1483
85	0009	0060	0161	0342	0626	0941	1485	1732	1727	1522
90	0008	0053	0144	0309	0572	0872	1412	1694	1713	1554
95	0007	0047	0128	0278	0522	0806	1340	1651	1692	1577
3.00	0006	0041	0114	0251	0476	0745	1269	1603	1665	1592
05	0005	0036	0102	0226	0434	0687	1199	1551	1632	1599
10	0005	0032	0091	0203	0396	0633	1131	1497	1594	1600
15	0004	0029	0081	0183	0361	0584	1064	1441	1552	1593
20	0004	0025	0072	0165	0329	0537	1000	1383	1506	1580
25	0003	0022	0065	0149	0299	0494	0939	1325	1458	1562
30	0003	0020	0058	0134	0273	0454	0880	1266	1408	1539
35	0002	0018	0052	0121	0248	0417	0823	1208	1356	1511
40	0002	0016	0046	0109	0226	0383	0770	1150	1303	1479
45	0002	0014	0042	0099	0206	0352	0719	1093	1249	1444
3.50	0002	0012	0037	0089	0187	0323	0671	1037	1196	1406
55	0002	0011	0033	0081	0171	0297	0626	0982	1142	1365
60	0001	0010	0030	0073	0155	0272	0583	0929	1090	1323
65	0001	0009	0027	0066	0142	0250	0543	0878	1038	1279
70	0001	0008	0024	0060	0129	0229	0505	0829	0987	1235
75	0001	0007	0022	0054	0118	0210	0470	0782	0938	1190
80	0001	0006	0020	0049	0107	0193	0437	0737	0889	1144
85	0001	0006	0018	0044	0098	0177	0406	0693	0843	1099
90	0001	0005	0016	0040	0089	0163	0377	0652	0798	1053
95	0001	0005	0014	0036	0081	0149	0350	0613	0754	1008
4.00	0001	0004	0013	0033	0074	0137	0325	0576	0713	0964
05	0000	0004	0012	0030	0068	0126	0302	0540	0673	0921
10	0000	0003	0011	0027	0062	0115	0280	0507	0635	0878
15	0000	0003	0010	0025	0057	0106	0260	0475	0599	0837
20	0000	0003	0009	0023	0052	0097	0241	0446	0564	0797
25	0000	0002	0008	0021	0047	0090	0224	0417	0531	0758
30	0000	0002	0007	0019	0043	0082	0207	0391	0500	0720
35	0000	0002	0007	0017	0040	0076	0192	0366	0470	0684
40	0000	0002	0006	0016	0036	0070	0179	0343	0442	0649
45	0000	0002	0005	0014	0033	0064	0166	0321	0416	0616
50	0000	0001	0005	0013	0030	0059	0154	0300	0391	0583
55	0000	0001	0004	0012	0028	0054	0143	0281	0367	0553
60	0000	0001	0004	0011	0026	0050	0132	0263	0345	0523
65	0000	0001	0004	0010	0024	0046	0123	0246	0324	0495
70	0000	0001	0003	0009	0022	0042	0114	0230	0304	0468
75	0000	0001	0003	0008	0020	0039	0106	0215	0285	0443
80	0000	0001	0003	0008	0018	0036	0098	0201	0268	0418
85	0000	0001	0003	0007	0017	0033	0091	0188	0251	0395
90	0000	0001	0002	0006	0015	0031	0085	0176	0236	0373
95	0000	0001	0002	0006	0014	0028	0079	0164	0221	0353

PROBABILITY DENSITY OF t, THE NON-CENTRAL t-STATISTIC, TABLED AS A FUNCTION OF $t/\sqrt{f}$.

f is the number of degrees of freedom; the non-centrality parameter is $\sqrt{f+1}\,K_p$.
K_p is the standardized normal deviate exceeded with probability p.

DEGREES OF FREEDOM **9**

$t/\sqrt{f}$ \ p	.2500	.1500	.1000	.0650	.0400	.0250	.0100	.0040	.0025	.0010
5.00	0000	0001	0002	0005	0013	0026	0073	0154	0207	0333
05	0000	0001	0002	0005	0012	0024	0068	0144	0195	0314
10	0000	0000	0002	0005	0011	0022	0063	0134	0183	0297
15	0000	0000	0002	0004	0010	0021	0059	0126	0171	0280
20	0000	0000	0001	0004	0009	0019	0055	0118	0161	0264
25	0000	0000	0001	0004	0009	0018	0051	0110	0151	0249
30	0000	0000	0001	0003	0008	0016	0047	0103	0141	0235
35	0000	0000	0001	0003	0007	0015	0044	0096	0133	0222
40	0000	0000	0001	0003	0007	0014	0041	0090	0125	0209
45	0000	0000	0001	0003	0006	0013	0038	0084	0117	0197
50	0000	0000	0001	0002	0006	0012	0036	0079	0110	0186
55	0000	0000	0001	0002	0005	0011	0033	0074	0103	0175
60	0000	0000	0001	0002	0005	0010	0031	0069	0097	0165
65	0000	0000	0001	0002	0005	0010	0029	0065	0091	0156
70	0000	0000	0001	0002	0004	0009	0027	0061	0085	0147
75	0000	0000	0001	0002	0004	0008	0025	0057	0080	0139
80	0000	0000	0001	0001	0004	0008	0024	0053	0075	0131
85	0000	0000	0000	0001	0003	0007	0022	0050	0071	0123
90	0000	0000	0000	0001	0003	0007	0021	0047	0066	0116
95	0000	0000	0000	0001	0003	0006	0019	0044	0062	0110
6.00	0000	0000	0000	0001	0003	0006	0018	0041	0059	0103
05	0000	0000	0000	0001	0003	0005	0017	0039	0055	0098
10	0000	0000	0000	0001	0002	0005	0016	0036	0052	0092
15	0000	0000	0000	0001	0002	0005	0015	0034	0049	0087
20	0000	0000	0000	0001	0002	0004	0014	0032	0046	0082
25	0000	0000	0000	0001	0002	0004	0013	0030	0043	0077
30	0000	0000	0000	0001	0002	0004	0012	0028	0041	0073
35	0000	0000	0000	0001	0002	0004	0011	0027	0038	0069
40	0000	0000	0000	0001	0002	0003	0011	0025	0036	0065
45	0000	0000	0000	0001	0001	0003	0010	0023	0034	0061
6.50	0000	0000	0000	0001	0001	0003	0009	0022	0032	0058
55	0000	0000	0000	0000	0001	0003	0009	0021	0030	0055
60	0000	0000	0000	0000	0001	0003	0008	0020	0028	0052
65	0000	0000	0000	0000	0001	0002	0008	0018	0027	0049
70	0000	0000	0000	0000	0001	0002	0007	0017	0025	0046
75	0000	0000	0000	0000	0001	0002	0007	0016	0024	0044
80	0000	0000	0000	0000	0001	0002	0006	0015	0022	0041
85	0000	0000	0000	0000	0001	0002	0006	0014	0021	0039
90	0000	0000	0000	0000	0001	0002	0006	0014	0020	0037
95	0000	0000	0000	0000	0001	0002	0005	0013	0019	0035
7.00	0000	0000	0000	0000	0001	0002	0005	0012	0018	0033
05	0000	0000	0000	0000	0001	0001	0005	0011	0017	0031
10	0000	0000	0000	0000	0001	0001	0004	0011	0016	0030
15	0000	0000	0000	0000	0001	0001	0004	0010	0015	0028
20	0000	0000	0000	0000	0001	0001	0004	0010	0014	0027
25	0000	0000	0000	0000	0001	0001	0004	0009	0013	0025
30	0000	0000	0000	0000	0000	0001	0003	0009	0013	0024
35	0000	0000	0000	0000	0000	0001	0003	0008	0012	0023
40	0000	0000	0000	0000	0000	0001	0003	0008	0011	0021
45	0000	0000	0000	0000	0000	0001	0003	0007	0011	0020
50	0000	0000	0000	0000	0000	0001	0003	0007	0010	0019
55	0000	0000	0000	0000	0000	0001	0003	0006	0009	0018
60	0000	0000	0000	0000	0000	0001	0002	0006	0009	0017
65	0000	0000	0000	0000	0000	0001	0002	0006	0008	0016
70	0000	0000	0000	0000	0000	0001	0002	0005	0008	0016
75	0000	0000	0000	0000	0000	0001	0002	0005	0008	0015
80	0000	0000	0000	0000	0000	0001	0002	0005	0007	0014
85	0000	0000	0000	0000	0000	0001	0002	0005	0007	0013
90	0000	0000	0000	0000	0000	0001	0002	0004	0006	0013
95	0000	0000	0000	0000	0000	0000	0002	0004	0006	0012

PROBABILITY DENSITY OF t, THE NON-CENTRAL t-STATISTIC, TABLED AS A FUNCTION OF $t/\sqrt{f}$.

f is the number of degrees of freedom; the non-centrality parameter is $\sqrt{f+1}\,K_p$.
K_p is the standardized normal deviate exceeded with probability p.

DEGREES OF FREEDOM **9**

$t/\sqrt{f}$ \ p	.2500	.1500	.1000	.0650	.0400	.0250	.0100	.0040	.0025	.0010
8.00	0000	0000	0000	0000	0000	0000	0002	0004	0006	0011
05	0000	0000	0000	0000	0000	0000	0001	0004	0006	0011
10	0000	0000	0000	0000	0000	0000	0001	0004	0005	0010
15	0000	0000	0000	0000	0000	0000	0001	0003	0005	0010
20	0000	0000	0000	0000	0000	0000	0001	0003	0005	0009
25	0000	0000	0000	0000	0000	0000	0001	0003	0004	0009
30	0000	0000	0000	0000	0000	0000	0001	0003	0004	0008
35	0000	0000	0000	0000	0000	0000	0001	0003	0004	0008
40	0000	0000	0000	0000	0000	0000	0001	0003	0004	0008
45	0000	0000	0000	0000	0000	0000	0001	0002	0004	0007
50	0000	0000	0000	0000	0000	0000	0001	0002	0003	0007
55	0000	0000	0000	0000	0000	0000	0001	0002	0003	0006
60	0000	0000	0000	0000	0000	0000	0001	0002	0003	0006
65	0000	0000	0000	0000	0000	0000	0001	0002	0003	0006
70	0000	0000	0000	0000	0000	0000	0001	0002	0003	0006
75	0000	0000	0000	0000	0000	0000	0001	0002	0003	0005
80	0000	0000	0000	0000	0000	0000	0001	0002	0003	0005
85	0000	0000	0000	0000	0000	0000	0001	0002	0002	0005
90	0000	0000	0000	0000	0000	0000	0001	0002	0002	0005
95	0000	0000	0000	0000	0000	0000	0001	0001	0002	0004
9.00	0000	0000	0000	0000	0000	0000	0001	0001	0002	0004
05	0000	0000	0000	0000	0000	0000	0001	0001	0002	0004
10	0000	0000	0000	0000	0000	0000	0000	0001	0002	0004
15	0000	0000	0000	0000	0000	0000	0000	0001	0002	0004
20	0000	0000	0000	0000	0000	0000	0000	0001	0002	0003
25	0000	0000	0000	0000	0000	0000	0000	0001	0002	0003
30	0000	0000	0000	0000	0000	0000	0000	0001	0002	0003
35	0000	0000	0000	0000	0000	0000	0000	0001	0001	0003
40	0000	0000	0000	0000	0000	0000	0000	0001	0001	0003
45	0000	0000	0000	0000	0000	0000	0000	0001	0001	0003
9.50	0000	0000	0000	0000	0000	0000	0000	0001	0001	0003
55	0000	0000	0000	0000	0000	0000	0000	0001	0001	0002
60	0000	0000	0000	0000	0000	0000	0000	0001	0001	0002
65	0000	0000	0000	0000	0000	0000	0000	0001	0001	0002
70	0000	0000	0000	0000	0000	0000	0000	0001	0001	0002
75	0000	0000	0000	0000	0000	0000	0000	0001	0001	0002
80	0000	0000	0000	0000	0000	0000	0000	0001	0001	0002
85	0000	0000	0000	0000	0000	0000	0000	0001	0001	0002
90	0000	0000	0000	0000	0000	0000	0000	0001	0001	0002
95	0000	0000	0000	0000	0000	0000	0000	0001	0001	0002
10.00	0000	0000	0000	0000	0000	0000	0000	0001	0001	0002
05	0000	0000	0000	0000	0000	0000	0000	0001	0001	0002
10	0000	0000	0000	0000	0000	0000	0000	0000	0001	0002
15	0000	0000	0000	0000	0000	0000	0000	0000	0001	0001
20	0000	0000	0000	0000	0000	0000	0000	0000	0001	0001
25	0000	0000	0000	0000	0000	0000	0000	0000	0001	0001
30	0000	0000	0000	0000	0000	0000	0000	0000	0001	0001
35	0000	0000	0000	0000	0000	0000	0000	0000	0001	0001
40	0000	0000	0000	0000	0000	0000	0000	0000	0001	0001
45	0000	0000	0000	0000	0000	0000	0000	0000	0001	0001
50	0000	0000	0000	0000	0000	0000	0000	0000	0001	0001
55	0000	0000	0000	0000	0000	0000	0000	0000	0001	0001
60	0000	0000	0000	0000	0000	0000	0000	0000	0000	0001
65	0000	0000	0000	0000	0000	0000	0000	0000	0000	0001
70	0000	0000	0000	0000	0000	0000	0000	0000	0000	0001
75	0000	0000	0000	0000	0000	0000	0000	0000	0000	0001
80	0000	0000	0000	0000	0000	0000	0000	0000	0000	0001
85	0000	0000	0000	0000	0000	0000	0000	0000	0000	0001
90	0000	0000	0000	0000	0000	0000	0000	0000	0000	0001
95	0000	0000	0000	0000	0000	0000	0000	0000	0000	0001

PROBABILITY DENSITY OF t, THE NON-CENTRAL t-STATISTIC, TABLED AS A FUNCTION OF $t/\sqrt{f}$.

f is the number of degrees of freedom; the non-centrality parameter is $\sqrt{f+1}\,K_p$.
K_p is the standardized normal deviate exceeded with probability p.

DEGREES OF FREEDOM **9**

$t/\sqrt{f}$ \ p	.2500	.1500	.1000	.0650	.0400	.0250	.0100	.0040	.0025	.0010
11.00	0000	0000	0000	0000	0000	0000	0000	0000	0000	0001
05	0000	0000	0000	0000	0000	0000	0000	0000	0000	0001
10	0000	0000	0000	0000	0000	0000	0000	0000	0000	0001
15	0000	0000	0000	0000	0000	0000	0000	0000	0000	0001
20	0000	0000	0000	0000	0000	0000	0000	0000	0000	0001
25	0000	0000	0000	0000	0000	0000	0000	0000	0000	0001
30	0000	0000	0000	0000	0000	0000	0000	0000	0000	0001
35	0000	0000	0000	0000	0000	0000	0000	0000	0000	0001
40	0000	0000	0000	0000	0000	0000	0000	0000	0000	0001
45	0000	0000	0000	0000	0000	0000	0000	0000	0000	0000

PROBABILITY DENSITY OF t, THE NON-CENTRAL t-STATISTIC, TABLED AS A FUNCTION OF $t/\sqrt{f}$.

f is the number of degrees of freedom; the non-centrality parameter is $\sqrt{f+1}\,K_p$.
K_p is the standardized normal deviate exceeded with probability p.

DEGREES OF FREEDOM **10**

$t/\sqrt{f}$ \ p	.2500	.1500	.1000	.0650	.0400	.0250	.0100	.0040	.0025	.0010
-0.80	0000	0000	0000	0000	0000	0000	0000	0000	0000	0000
75	0001	0000	0000	0000	0000	0000	0000	0000	0000	0000
70	0001	0000	0000	0000	0000	0000	0000	0000	0000	0000
65	0001	0000	0000	0000	0000	0000	0000	0000	0000	0000
60	0002	0000	0000	0000	0000	0000	0000	0000	0000	0000
55	0003	0000	0000	0000	0000	0000	0000	0000	0000	0000
50	0005	0000	0000	0000	0000	0000	0000	0000	0000	0000
45	0007	0000	0000	0000	0000	0000	0000	0000	0000	0000
40	0011	0000	0000	0000	0000	0000	0000	0000	0000	0000
35	0017	0000	0000	0000	0000	0000	0000	0000	0000	0000
30	0027	0000	0000	0000	0000	0000	0000	0000	0000	0000
25	0042	0001	0000	0000	0000	0000	0000	0000	0000	0000
20	0065	0001	0000	0000	0000	0000	0000	0000	0000	0000
15	0099	0002	0000	0000	0000	0000	0000	0000	0000	0000
10	0148	0003	0000	0000	0000	0000	0000	0000	0000	0000
05	0220	0006	0000	0000	0000	0000	0000	0000	0000	0000
0.00	0319	0011	0000	0000	0000	0000	0000	0000	0000	0000
05	0453	0018	0001	0000	0000	0000	0000	0000	0000	0000
10	0629	0031	0002	0000	0000	0000	0000	0000	0000	0000
15	0850	0052	0004	0000	0000	0000	0000	0000	0000	0000
20	1118	0085	0007	0000	0000	0000	0000	0000	0000	0000
25	1428	0135	0012	0001	0000	0000	0000	0000	0000	0000
30	1770	0208	0022	0002	0000	0000	0000	0000	0000	0000
35	2129	0310	0039	0003	0000	0000	0000	0000	0000	0000
40	2484	0448	0066	0006	0000	0000	0000	0000	0000	0000
45	2813	0626	0108	0012	0001	0000	0000	0000	0000	0000
0.50	3096	0844	0170	0022	0002	0000	0000	0000	0000	0000
55	3314	1101	0258	0039	0003	0000	0000	0000	0000	0000
60	3455	1388	0377	0066	0007	0001	0000	0000	0000	0000
65	3515	1693	0530	0108	0013	0001	0000	0000	0000	0000
70	3495	2003	0718	0169	0024	0003	0000	0000	0000	0000
75	3402	2301	0939	0253	0042	0006	0000	0000	0000	0000
80	3248	2569	1187	0364	0071	0011	0000	0000	0000	0000
85	3048	2796	1453	0505	0113	0021	0001	0000	0000	0000
90	2814	2969	1725	0676	0172	0037	0001	0000	0000	0000
95	2562	3084	1990	0873	0253	0061	0002	0000	0000	0000
1.00	2303	3138	2236	1091	0357	0097	0005	0000	0000	0000
05	2047	3133	2453	1324	0486	0148	0009	0000	0000	0000
10	1801	3076	2630	1560	0639	0216	0017	0001	0000	0000
15	1572	2973	2762	1792	0813	0303	0029	0002	0000	0000
20	1361	2834	2846	2009	1003	0412	0047	0004	0001	0000
25	1171	2668	2883	2203	1205	0540	0074	0007	0002	0000
30	1002	2484	2874	2366	1409	0687	0111	0012	0004	0000
35	0853	2289	2826	2494	1610	0850	0160	0021	0007	0001
40	0723	2091	2742	2585	1800	1023	0224	0034	0011	0001
45	0611	1895	2630	2638	1972	1202	0302	0052	0019	0002
50	0515	1705	2497	2654	2121	1381	0396	0078	0031	0004
55	0433	1525	2348	2636	2244	1554	0504	0113	0047	0008
60	0364	1356	2189	2588	2337	1716	0625	0158	0070	0013
65	0305	1200	2026	2514	2400	1863	0757	0214	0101	0020
70	0256	1058	1862	2418	2434	1989	0896	0281	0140	0031
75	0214	0929	1701	2306	2439	2093	1039	0360	0189	0047
80	0179	0814	1546	2183	2418	2174	1183	0450	0248	0068
85	0150	0711	1398	2051	2375	2229	1323	0550	0318	0095
90	0126	0619	1260	1915	2312	2261	1456	0658	0397	0129
95	0105	0539	1130	1778	2232	2269	1579	0772	0486	0171

PROBABILITY DENSITY OF t, THE NON-CENTRAL t-STATISTIC, TABLED AS A FUNCTION OF $t/\sqrt{f}$.

f is the number of degrees of freedom; the non-centrality parameter is $\sqrt{f+1}\,K_p$.
K_p is the standardized normal deviate exceeded with probability p.

DEGREES OF FREEDOM **10**

$t/\sqrt{f}$ \ p	.2500	.1500	.1000	.0650	.0400	.0250	.0100	.0040	.0025	.0010
2.00	0088	0468	1011	1642	2140	2257	1690	0890	0582	0221
05	0074	0406	0902	1510	2037	2225	1785	1009	0684	0279
10	0062	0352	0802	1382	1929	2176	1864	1127	0791	0346
15	0052	0305	0712	1261	1816	2114	1927	1241	0900	0419
20	0044	0264	0631	1147	1701	2039	1971	1348	1009	0500
25	0037	0228	0558	1039	1586	1956	1999	1448	1116	0586
30	0031	0197	0493	0940	1473	1866	2011	1537	1219	0676
35	0026	0171	0435	0848	1364	1771	2007	1615	1315	0769
40	0022	0148	0384	0764	1258	1673	1989	1681	1404	0863
45	0019	0128	0338	0686	1157	1574	1958	1733	1484	0957
50	0016	0111	0298	0616	1062	1476	1917	1773	1553	1048
55	0014	0096	0262	0552	0972	1379	1866	1801	1612	1135
60	0012	0083	0231	0494	0887	1284	1808	1816	1659	1218
65	0010	0072	0203	0442	0809	1193	1743	1819	1695	1294
70	0008	0062	0179	0395	0736	1105	1673	1811	1720	1363
75	0007	0054	0157	0353	0669	1022	1600	1793	1734	1423
80	0006	0047	0138	0315	0607	0942	1525	1766	1738	1476
85	0005	0041	0122	0281	0550	0868	1448	1731	1732	1519
90	0004	0036	0107	0251	0498	0797	1371	1690	1718	1554
95	0004	0031	0095	0224	0451	0732	1294	1642	1695	1580
3.00	0003	0027	0083	0200	0408	0671	1219	1590	1665	1596
05	0003	0024	0074	0178	0369	0614	1145	1534	1629	1605
10	0002	0021	0065	0159	0333	0562	1074	1476	1587	1605
15	0002	0018	0057	0142	0301	0513	1005	1415	1541	1598
20	0002	0016	0051	0126	0271	0468	0939	1353	1492	1584
25	0002	0014	0045	0113	0245	0427	0875	1290	1440	1564
30	0001	0012	0040	0101	0221	0390	0815	1228	1385	1539
35	0001	0011	0035	0090	0199	0355	0758	1166	1329	1508
40	0001	0009	0031	0080	0180	0323	0704	1104	1272	1473
45	0001	0008	0027	0072	0162	0294	0653	1044	1215	1435
3.50	0001	0007	0024	0064	0146	0268	0605	0986	1158	1394
55	0001	0006	0022	0057	0132	0244	0560	0929	1101	1350
60	0001	0006	0019	0051	0119	0222	0518	0874	1046	1304
65	0001	0005	0017	0046	0107	0202	0479	0822	0991	1257
70	0000	0004	0015	0041	0097	0184	0443	0771	0938	1209
75	0000	0004	0014	0037	0088	0167	0409	0723	0886	1160
80	0000	0003	0012	0033	0079	0152	0377	0677	0837	1112
85	0000	0003	0011	0030	0071	0138	0348	0634	0789	1063
90	0000	0003	0010	0027	0065	0126	0321	0592	0742	1015
95	0000	0002	0009	0024	0058	0115	0296	0553	0698	0968
4.00	0000	0002	0008	0022	0053	0104	0273	0516	0656	0921
05	0000	0002	0007	0019	0048	0095	0251	0482	0616	0876
10	0000	0002	0006	0017	0043	0086	0231	0449	0578	0831
15	0000	0002	0006	0016	0039	0079	0213	0418	0541	0789
20	0000	0001	0005	0014	0036	0072	0196	0390	0507	0747
25	0000	0001	0004	0013	0032	0065	0181	0363	0475	0707
30	0000	0001	0004	0012	0029	0060	0167	0337	0444	0669
35	0000	0001	0004	0010	0026	0054	0153	0314	0415	0632
40	0000	0001	0003	0009	0024	0050	0141	0292	0388	0596
45	0000	0001	0003	0008	0022	0045	0130	0271	0363	0563
50	0000	0001	0003	0008	0020	0041	0120	0252	0339	0530
55	0000	0001	0002	0007	0018	0038	0110	0234	0316	0500
60	0000	0001	0002	0006	0016	0034	0102	0218	0295	0470
65	0000	0001	0002	0006	0015	0031	0094	0202	0275	0443
70	0000	0000	0002	0005	0014	0029	0086	0188	0257	0417
75	0000	0000	0002	0005	0012	0026	0080	0175	0240	0392
80	0000	0000	0001	0004	0011	0024	0073	0162	0224	0368
85	0000	0000	0001	0004	0010	0022	0068	0151	0208	0346
90	0000	0000	0001	0004	0009	0020	0062	0140	0194	0325
95	0000	0000	0001	0003	0009	0018	0058	0130	0181	0305

PROBABILITY DENSITY OF t, THE NON-CENTRAL t-STATISTIC, TABLED AS A FUNCTION OF $t/\sqrt{f}$.

f is the number of degrees of freedom; the non-centrality parameter is $\sqrt{f+1}\,K_p$.
K_p is the standardized normal deviate exceeded with probability p.

DEGREES OF FREEDOM **10**

$t/\sqrt{f}$ \ p	.2500	.1500	.1000	.0650	.0400	.0250	.0100	.0040	.0025	.0010
5.00	0000	0000	0001	0003	0008	0017	0053	0121	0169	0286
05	0000	0000	0001	0003	0007	0016	0049	0112	0157	0269
10	0000	0000	0001	0002	0007	0014	0045	0104	0147	0252
15	0000	0000	0001	0002	0006	0013	0042	0097	0137	0236
20	0000	0000	0001	0002	0005	0012	0039	0090	0128	0222
25	0000	0000	0001	0002	0005	0011	0036	0084	0119	0208
30	0000	0000	0001	0002	0005	0010	0033	0078	0111	0195
35	0000	0000	0000	0002	0004	0009	0030	0072	0103	0183
40	0000	0000	0000	0001	0004	0009	0028	0067	0096	0172
45	0000	0000	0000	0001	0004	0008	0026	0062	0090	0161
50	0000	0000	0000	0001	0003	0007	0024	0058	0084	0151
55	0000	0000	0000	0001	0003	0007	0022	0054	0078	0141
60	0000	0000	0000	0001	0003	0006	0021	0050	0073	0133
65	0000	0000	0000	0001	0003	0006	0019	0047	0068	0124
70	0000	0000	0000	0001	0002	0005	0018	0044	0063	0116
75	0000	0000	0000	0001	0002	0005	0016	0041	0059	0109
80	0000	0000	0000	0001	0002	0004	0015	0038	0055	0102
85	0000	0000	0000	0001	0002	0004	0014	0035	0052	0096
90	0000	0000	0000	0001	0002	0004	0013	0033	0048	0090
95	0000	0000	0000	0001	0002	0004	0012	0031	0045	0084
6.00	0000	0000	0000	0001	0001	0003	0011	0028	0042	0079
05	0000	0000	0000	0000	0001	0003	0010	0027	0039	0074
10	0000	0000	0000	0000	0001	0003	0010	0025	0037	0070
15	0000	0000	0000	0000	0001	0003	0009	0023	0034	0065
20	0000	0000	0000	0000	0001	0002	0008	0022	0032	0061
25	0000	0000	0000	0000	0001	0002	0008	0020	0030	0058
30	0000	0000	0000	0000	0001	0002	0007	0019	0028	0054
35	0000	0000	0000	0000	0001	0002	0007	0018	0026	0051
40	0000	0000	0000	0000	0001	0002	0006	0016	0025	0048
45	0000	0000	0000	0000	0001	0002	0006	0015	0023	0045
6.50	0000	0000	0000	0000	0001	0002	0005	0014	0021	0042
55	0000	0000	0000	0000	0001	0001	0005	0013	0020	0039
60	0000	0000	0000	0000	0001	0001	0005	0012	0019	0037
65	0000	0000	0000	0000	0001	0001	0004	0012	0018	0035
70	0000	0000	0000	0000	0000	0001	0004	0011	0017	0033
75	0000	0000	0000	0000	0000	0001	0004	0010	0015	0031
80	0000	0000	0000	0000	0000	0001	0004	0010	0015	0029
85	0000	0000	0000	0000	0000	0001	0003	0009	0014	0027
90	0000	0000	0000	0000	0000	0001	0003	0008	0013	0026
95	0000	0000	0000	0000	0000	0001	0003	0008	0012	0024
7.00	0000	0000	0000	0000	0000	0001	0003	0007	0011	0023
05	0000	0000	0000	0000	0000	0001	0003	0007	0011	0021
10	0000	0000	0000	0000	0000	0001	0002	0006	0010	0020
15	0000	0000	0000	0000	0000	0001	0002	0006	0009	0019
20	0000	0000	0000	0000	0000	0001	0002	0006	0009	0018
25	0000	0000	0000	0000	0000	0001	0002	0005	0008	0017
30	0000	0000	0000	0000	0000	0000	0002	0005	0008	0016
35	0000	0000	0000	0000	0000	0000	0002	0005	0007	0015
40	0000	0000	0000	0000	0000	0000	0002	0004	0007	0014
45	0000	0000	0000	0000	0000	0000	0002	0004	0006	0013
50	0000	0000	0000	0000	0000	0000	0001	0004	0006	0012
55	0000	0000	0000	0000	0000	0000	0001	0004	0006	0012
60	0000	0000	0000	0000	0000	0000	0001	0003	0005	0011
65	0000	0000	0000	0000	0000	0000	0001	0003	0005	0010
70	0000	0000	0000	0000	0000	0000	0001	0003	0005	0010
75	0000	0000	0000	0000	0000	0000	0001	0003	0004	0009
80	0000	0000	0000	0000	0000	0000	0001	0003	0004	0009
85	0000	0000	0000	0000	0000	0000	0001	0003	0004	0008
90	0000	0000	0000	0000	0000	0000	0001	0002	0004	0008
95	0000	0000	0000	0000	0000	0000	0001	0002	0004	0007

PROBABILITY DENSITY OF t, THE NON-CENTRAL t-STATISTIC, TABLED AS A FUNCTION OF $t/\sqrt{f}$.

f is the number of degrees of freedom; the non-centrality parameter is $\sqrt{f+1}\,K_p$.
K_p is the standardized normal deviate exceeded with probability p.

DEGREES OF FREEDOM **10**

$t/\sqrt{f}$ \ p	.2500	.1500	.1000	.0650	.0400	.0250	.0100	.0040	.0025	.0010
8.00	0000	0000	0000	0000	0000	0000	0001	0002	0003	0007
05	0000	0000	0000	0000	0000	0000	0001	0002	0003	0007
10	0000	0000	0000	0000	0000	0000	0001	0002	0003	0006
15	0000	0000	0000	0000	0000	0000	0001	0002	0003	0006
20	0000	0000	0000	0000	0000	0000	0001	0002	0003	0006
25	0000	0000	0000	0000	0000	0000	0001	0002	0002	0005
30	0000	0000	0000	0000	0000	0000	0001	0001	0002	0005
35	0000	0000	0000	0000	0000	0000	0000	0001	0002	0005
40	0000	0000	0000	0000	0000	0000	0000	0001	0002	0004
45	0000	0000	0000	0000	0000	0000	0000	0001	0002	0004
50	0000	0000	0000	0000	0000	0000	0000	0001	0002	0004
55	0000	0000	0000	0000	0000	0000	0000	0001	0002	0004
60	0000	0000	0000	0000	0000	0000	0000	0001	0002	0004
65	0000	0000	0000	0000	0000	0000	0000	0001	0002	0003
70	0000	0000	0000	0000	0000	0000	0000	0001	0001	0003
75	0000	0000	0000	0000	0000	0000	0000	0001	0001	0003
80	0000	0000	0000	0000	0000	0000	0000	0001	0001	0003
85	0000	0000	0000	0000	0000	0000	0000	0001	0001	0003
90	0000	0000	0000	0000	0000	0000	0000	0001	0001	0003
95	0000	0000	0000	0000	0000	0000	0000	0001	0001	0002
9.00	0000	0000	0000	0000	0000	0000	0000	0001	0001	0002
05	0000	0000	0000	0000	0000	0000	0000	0001	0001	0002
10	0000	0000	0000	0000	0000	0000	0000	0001	0001	0002
15	0000	0000	0000	0000	0000	0000	0000	0001	0001	0002
20	0000	0000	0000	0000	0000	0000	0000	0001	0001	0002
25	0000	0000	0000	0000	0000	0000	0000	0001	0001	0002
30	0000	0000	0000	0000	0000	0000	0000	0000	0001	0002
35	0000	0000	0000	0000	0000	0000	0000	0000	0001	0002
40	0000	0000	0000	0000	0000	0000	0000	0000	0001	0002
45	0000	0000	0000	0000	0000	0000	0000	0000	0001	0001
9.50	0000	0000	0000	0000	0000	0000	0000	0000	0001	0001
55	0000	0000	0000	0000	0000	0000	0000	0000	0001	0001
60	0000	0000	0000	0000	0000	0000	0000	0000	0001	0001
65	0000	0000	0000	0000	0000	0000	0000	0000	0001	0001
70	0000	0000	0000	0000	0000	0000	0000	0000	0001	0001
75	0000	0000	0000	0000	0000	0000	0000	0000	0000	0001
80	0000	0000	0000	0000	0000	0000	0000	0000	0000	0001
85	0000	0000	0000	0000	0000	0000	0000	0000	0000	0001
90	0000	0000	0000	0000	0000	0000	0000	0000	0000	0001
95	0000	0000	0000	0000	0000	0000	0000	0000	0000	0001
10.00	0000	0000	0000	0000	0000	0000	0000	0000	0000	0001
05	0000	0000	0000	0000	0000	0000	0000	0000	0000	0001
10	0000	0000	0000	0000	0000	0000	0000	0000	0000	0001
15	0000	0000	0000	0000	0000	0000	0000	0000	0000	0001
20	0000	0000	0000	0000	0000	0000	0000	0000	0000	0001
25	0000	0000	0000	0000	0000	0000	0000	0000	0000	0001
30	0000	0000	0000	0000	0000	0000	0000	0000	0000	0001
35	0000	0000	0000	0000	0000	0000	0000	0000	0000	0001
40	0000	0000	0000	0000	0000	0000	0000	0000	0000	0001
45	0000	0000	0000	0000	0000	0000	0000	0000	0000	0001
50	0000	0000	0000	0000	0000	0000	0000	0000	0000	0001
55	0000	0000	0000	0000	0000	0000	0000	0000	0000	0000

PROBABILITY DENSITY OF t, THE NON-CENTRAL t-STATISTIC, TABLED AS A FUNCTION OF $t/\sqrt{f}$.

f is the number of degrees of freedom; the non-centrality parameter is $\sqrt{f+1}\,K_p$.
K_p is the standardized normal deviate exceeded with probability p.

DEGREES OF FREEDOM **11**

$t/\sqrt{f}$ \ p	.2500	.1500	.1000	.0650	.0400	.0250	.0100	.0040	.0025	.0010
-0.70	0000	0000	0000	0000	0000	0000	0000	0000	0000	0000
65	0001	0000	0000	0000	0000	0000	0000	0000	0000	0000
60	0001	0000	0000	0000	0000	0000	0000	0000	0000	0000
55	0002	0000	0000	0000	0000	0000	0000	0000	0000	0000
50	0002	0000	0000	0000	0000	0000	0000	0000	0000	0000
45	0004	0000	0000	0000	0000	0000	0000	0000	0000	0000
40	0007	0000	0000	0000	0000	0000	0000	0000	0000	0000
35	0011	0000	0000	0000	0000	0000	0000	0000	0000	0000
30	0017	0000	0000	0000	0000	0000	0000	0000	0000	0000
25	0028	0000	0000	0000	0000	0000	0000	0000	0000	0000
20	0045	0001	0000	0000	0000	0000	0000	0000	0000	0000
15	0071	0001	0000	0000	0000	0000	0000	0000	0000	0000
10	0110	0002	0000	0000	0000	0000	0000	0000	0000	0000
05	0169	0003	0000	0000	0000	0000	0000	0000	0000	0000
0.00	0254	0006	0000	0000	0000	0000	0000	0000	0000	0000
05	0374	0011	0000	0000	0000	0000	0000	0000	0000	0000
10	0535	0020	0001	0000	0000	0000	0000	0000	0000	0000
15	0744	0035	0002	0000	0000	0000	0000	0000	0000	0000
20	1003	0061	0004	0000	0000	0000	0000	0000	0000	0000
25	1311	0100	0007	0000	0000	0000	0000	0000	0000	0000
30	1659	0161	0014	0001	0000	0000	0000	0000	0000	0000
35	2030	0249	0026	0002	0000	0000	0000	0000	0000	0000
40	2404	0373	0046	0003	0000	0000	0000	0000	0000	0000
45	2756	0537	0079	0007	0000	0000	0000	0000	0000	0000
50	3061	0745	0130	0014	0001	0000	0000	0000	0000	0000
55	3299	0996	0206	0026	0002	0000	0000	0000	0000	0000
60	3454	1284	0311	0047	0004	0000	0000	0000	0000	0000
65	3522	1596	0451	0080	0008	0001	0000	0000	0000	0000
70	3502	1919	0629	0130	0016	0002	0000	0000	0000	0000
0.75	3402	2234	0844	0202	0029	0003	0000	0000	0000	0000
80	3236	2521	1091	0302	0051	0007	0000	0000	0000	0000
85	3020	2767	1361	0432	0084	0014	0000	0000	0000	0000
90	2769	2956	1642	0594	0134	0025	0001	0000	0000	0000
95	2501	3083	1921	0786	0205	0043	0001	0000	0000	0000
1.00	2227	3143	2184	1003	0298	0072	0003	0000	0000	0000
05	1959	3140	2417	1239	0418	0114	0006	0000	0000	0000
10	1704	3079	2609	1484	0563	0173	0011	0000	0000	0000
15	1469	2968	2754	1727	0733	0251	0019	0001	0000	0000
20	1256	2819	2847	1958	0923	0351	0033	0002	0000	0000
25	1066	2640	2889	2165	1127	0472	0054	0004	0001	0000
30	0899	2442	2881	2342	1339	0615	0085	0008	0002	0000
35	0755	2235	2829	2483	1549	0775	0127	0014	0004	0000
40	0631	2025	2739	2583	1750	0950	0183	0023	0007	0001
45	0525	1819	2618	2642	1934	1133	0254	0038	0013	0001
50	0436	1622	2474	2660	2096	1319	0341	0059	0021	0002
55	0361	1436	2315	2642	2229	1501	0443	0087	0034	0005
60	0298	1264	2145	2590	2331	1674	0561	0126	0052	0008
65	0246	1107	1972	2510	2401	1830	0691	0175	0078	0014
70	0203	0965	1799	2407	2438	1967	0832	0237	0111	0022
75	0167	0838	1631	2286	2445	2081	0978	0310	0154	0034
80	0138	0725	1469	2153	2423	2169	1127	0396	0208	0050
85	0113	0625	1317	2012	2376	2230	1274	0493	0272	0073
90	0094	0538	1175	1868	2308	2265	1415	0599	0347	0102
95	0077	0462	1044	1723	2222	2275	1546	0714	0432	0139
2.00	0064	0396	0925	1580	2122	2262	1665	0834	0526	0184
05	0052	0340	0816	1442	2013	2228	1768	0957	0629	0238
10	0043	0290	0719	1310	1896	2175	1855	1080	0737	0300
15	0036	0248	0631	1185	1775	2107	1923	1199	0848	0371
20	0030	0212	0553	1069	1654	2027	1972	1314	0961	0450

PROBABILITY DENSITY OF t, THE NON-CENTRAL t-STATISTIC, TABLED AS A FUNCTION OF $t/\sqrt{f}$.

f is the number of degrees of freedom; the non-centrality parameter is $\sqrt{f+1}\,K_p$.
K_p is the standardized normal deviate exceeded with probability p.

DEGREES OF FREEDOM **11**

$t/\sqrt{f}$ \ p	.2500	.1500	.1000	.0650	.0400	.0250	.0100	.0040	.0025	.0010
2.25	0025	0181	0484	0960	1533	1937	2003	1420	1073	0535
30	0020	0155	0423	0861	1415	1840	2016	1516	1181	0626
35	0017	0132	0369	0769	1300	1739	2012	1600	1284	0720
40	0014	0113	0322	0686	1191	1635	1993	1672	1379	0817
45	0012	0096	0281	0611	1087	1530	1960	1730	1465	0914
50	0010	0082	0244	0543	0990	1426	1915	1774	1540	1009
55	0008	0070	0213	0482	0899	1324	1860	1804	1604	1102
60	0007	0060	0185	0427	0814	1226	1797	1820	1656	1189
65	0006	0051	0161	0378	0736	1131	1727	1824	1696	1271
70	0005	0044	0140	0335	0664	1041	1652	1815	1723	1345
75	0004	0038	0122	0296	0598	0955	1574	1796	1739	1411
80	0003	0032	0106	0262	0538	0875	1493	1767	1743	1468
85	0003	0028	0092	0231	0484	0799	1411	1729	1737	1515
90	0002	0024	0080	0204	0434	0729	1330	1684	1721	1553
95	0002	0021	0070	0180	0389	0664	1249	1633	1696	1581
3.00	0002	0018	0061	0159	0349	0604	1170	1577	1664	1600
05	0002	0015	0053	0140	0312	0549	1093	1517	1625	1609
10	0001	0013	0046	0124	0280	0498	1019	1454	1580	1610
15	0001	0011	0040	0109	0250	0451	0948	1389	1530	1603
20	0001	0010	0035	0097	0224	0408	0880	1323	1477	1588
25	0001	0009	0031	0085	0200	0370	0816	1256	1421	1566
30	0001	0007	0027	0075	0179	0334	0755	1190	1362	1538
35	0001	0006	0024	0067	0160	0302	0697	1124	1302	1505
40	0001	0006	0021	0059	0143	0273	0643	1060	1242	1467
45	0000	0005	0018	0052	0128	0246	0593	0997	1181	1426
50	0000	0004	0016	0046	0114	0222	0546	0937	1121	1381
55	0000	0004	0014	0041	0102	0201	0502	0878	1061	1334
60	0000	0003	0012	0036	0091	0181	0461	0822	1003	1285
65	0000	0003	0011	0032	0081	0163	0423	0768	0946	1234
70	0000	0002	0010	0028	0073	0147	0388	0717	0891	1183
3.75	0000	0002	0008	0025	0065	0133	0356	0668	0838	1131
80	0000	0002	0007	0022	0058	0120	0326	0622	0786	1080
85	0000	0002	0007	0020	0052	0108	0299	0578	0737	1028
90	0000	0001	0006	0018	0047	0097	0273	0538	0690	0978
95	0000	0001	0005	0016	0042	0088	0250	0499	0646	0928
4.00	0000	0001	0005	0014	0038	0079	0229	0463	0603	0880
05	0000	0001	0004	0012	0034	0072	0209	0429	0563	0832
10	0000	0001	0004	0011	0030	0065	0191	0397	0525	0787
15	0000	0001	0003	0010	0027	0058	0175	0368	0489	0743
20	0000	0001	0003	0009	0024	0053	0160	0340	0456	0700
25	0000	0001	0002	0008	0022	0048	0146	0315	0424	0659
30	0000	0001	0002	0007	0020	0043	0134	0291	0394	0620
35	0000	0000	0002	0006	0018	0039	0122	0269	0367	0583
40	0000	0000	0002	0006	0016	0035	0112	0249	0341	0548
45	0000	0000	0002	0005	0014	0032	0102	0230	0316	0514
50	0000	0000	0001	0005	0013	0029	0093	0212	0293	0482
55	0000	0000	0001	0004	0012	0026	0085	0196	0272	0452
60	0000	0000	0001	0004	0011	0024	0078	0181	0253	0423
65	0000	0000	0001	0003	0009	0021	0071	0167	0234	0396
70	0000	0000	0001	0003	0009	0019	0065	0154	0217	0370
75	0000	0000	0001	0003	0008	0018	0060	0142	0201	0346
80	0000	0000	0001	0002	0007	0016	0055	0131	0186	0324
85	0000	0000	0001	0002	0006	0015	0050	0121	0173	0302
90	0000	0000	0001	0002	0006	0013	0046	0112	0160	0282
95	0000	0000	0001	0002	0005	0012	0042	0103	0148	0264
5.00	0000	0000	0000	0002	0005	0011	0038	0095	0137	0246
05	0000	0000	0000	0001	0004	0010	0035	0088	0127	0230
10	0000	0000	0000	0001	0004	0009	0032	0081	0118	0214
15	0000	0000	0000	0001	0004	0008	0030	0075	0109	0200
20	0000	0000	0000	0001	0003	0008	0027	0069	0101	0186

PROBABILITY DENSITY OF t, THE NON-CENTRAL t-STATISTIC, TABLED AS A FUNCTION OF $t/\sqrt{f}$.

f is the number of degrees of freedom; the non-centrality parameter is $\sqrt{f+1}\,K_p$.
K_p is the standardized normal deviate exceeded with probability p.

DEGREES OF FREEDOM **11**

$t/\sqrt{f}$ \ P	.2500	.1500	.1000	.0650	.0400	.0250	.0100	.0040	.0025	.0010
5.25	0000	0000	0000	0001	0003	0007	0025	0064	0094	0174
30	0000	0000	0000	0001	0003	0006	0023	0059	0087	0162
35	0000	0000	0000	0001	0002	0006	0021	0054	0080	0151
40	0000	0000	0000	0001	0002	0005	0019	0050	0074	0141
45	0000	0000	0000	0001	0002	0005	0018	0046	0069	0131
50	0000	0000	0000	0001	0002	0004	0016	0043	0064	0122
55	0000	0000	0000	0001	0002	0004	0015	0040	0059	0114
60	0000	0000	0000	0000	0002	0004	0014	0037	0055	0106
65	0000	0000	0000	0000	0001	0003	0013	0034	0051	0099
70	0000	0000	0000	0000	0001	0003	0012	0031	0047	0092
75	0000	0000	0000	0000	0001	0003	0011	0029	0044	0086
80	0000	0000	0000	0000	0001	0003	0010	0027	0041	0080
85	0000	0000	0000	0000	0001	0002	0009	0025	0038	0075
90	0000	0000	0000	0000	0001	0002	0008	0023	0035	0070
95	0000	0000	0000	0000	0001	0002	0008	0021	0032	0065
6.00	0000	0000	0000	0000	0001	0002	0007	0020	0030	0061
05	0000	0000	0000	0000	0001	0002	0007	0018	0028	0056
10	0000	0000	0000	0000	0001	0002	0006	0017	0026	0053
15	0000	0000	0000	0000	0001	0001	0006	0016	0024	0049
20	0000	0000	0000	0000	0001	0001	0005	0014	0022	0046
25	0000	0000	0000	0000	0000	0001	0005	0013	0021	0043
30	0000	0000	0000	0000	0000	0001	0004	0012	0019	0040
35	0000	0000	0000	0000	0000	0001	0004	0012	0018	0037
40	0000	0000	0000	0000	0000	0001	0004	0011	0017	0035
45	0000	0000	0000	0000	0000	0001	0003	0010	0016	0032
50	0000	0000	0000	0000	0000	0001	0003	0009	0014	0030
55	0000	0000	0000	0000	0000	0001	0003	0009	0013	0028
60	0000	0000	0000	0000	0000	0001	0003	0008	0013	0026
65	0000	0000	0000	0000	0000	0001	0003	0007	0012	0025
70	0000	0000	0000	0000	0000	0001	0002	0007	0011	0023
6.75	0000	0000	0000	0000	0000	0001	0002	0006	0010	0022
80	0000	0000	0000	0000	0000	0000	0002	0006	0009	0020
85	0000	0000	0000	0000	0000	0000	0002	0006	0009	0019
90	0000	0000	0000	0000	0000	0000	0002	0005	0008	0018
95	0000	0000	0000	0000	0000	0000	0002	0005	0008	0016
7.00	0000	0000	0000	0000	0000	0000	0002	0004	0007	0015
05	0000	0000	0000	0000	0000	0000	0001	0004	0007	0014
10	0000	0000	0000	0000	0000	0000	0001	0004	0006	0014
15	0000	0000	0000	0000	0000	0000	0001	0004	0006	0013
20	0000	0000	0000	0000	0000	0000	0001	0003	0005	0012
25	0000	0000	0000	0000	0000	0000	0001	0003	0005	0011
30	0000	0000	0000	0000	0000	0000	0001	0003	0005	0010
35	0000	0000	0000	0000	0000	0000	0001	0003	0004	0010
40	0000	0000	0000	0000	0000	0000	0001	0003	0004	0009
45	0000	0000	0000	0000	0000	0000	0001	0002	0004	0009
50	0000	0000	0000	0000	0000	0000	0001	0002	0004	0008
55	0000	0000	0000	0000	0000	0000	0001	0002	0003	0008
60	0000	0000	0000	0000	0000	0000	0001	0002	0003	0007
65	0000	0000	0000	0000	0000	0000	0001	0002	0003	0007
70	0000	0000	0000	0000	0000	0000	0001	0002	0003	0006
75	0000	0000	0000	0000	0000	0000	0001	0002	0003	0006
80	0000	0000	0000	0000	0000	0000	0000	0002	0002	0005
85	0000	0000	0000	0000	0000	0000	0000	0001	0002	0005
90	0000	0000	0000	0000	0000	0000	0000	0001	0002	0005
95	0000	0000	0000	0000	0000	0000	0000	0001	0002	0005
8.00	0000	0000	0000	0000	0000	0000	0000	0001	0002	0004
05	0000	0000	0000	0000	0000	0000	0000	0001	0002	0004
10	0000	0000	0000	0000	0000	0000	0000	0001	0002	0004
15	0000	0000	0000	0000	0000	0000	0000	0001	0002	0004
20	0000	0000	0000	0000	0000	0000	0000	0001	0001	0003

PROBABILITY DENSITY OF t, THE NON-CENTRAL t-STATISTIC, TABLED AS A FUNCTION OF $t/\sqrt{f}$.

f is the number of degrees of freedom; the non-centrality parameter is $\sqrt{f+1}\,K_p$.
K_p is the standardized normal deviate exceeded with probability p.

DEGREES OF FREEDOM **11**

$t/\sqrt{f}$ \ p	.2500	.1500	.1000	.0650	.0400	.0250	.0100	.0040	.0025	.0010
8.25	0000	0000	0000	0000	0000	0000	0000	0001	0001	0003
30	0000	0000	0000	0000	0000	0000	0000	0001	0001	0003
35	0000	0000	0000	0000	0000	0000	0000	0001	0001	0003
40	0000	0000	0000	0000	0000	0000	0000	0001	0001	0003
45	0000	0000	0000	0000	0000	0000	0000	0001	0001	0002
50	0000	0000	0000	0000	0000	0000	0000	0001	0001	0002
55	0000	0000	0000	0000	0000	0000	0000	0001	0001	0002
60	0000	0000	0000	0000	0000	0000	0000	0001	0001	0002
65	0000	0000	0000	0000	0000	0000	0000	0001	0001	0002
70	0000	0000	0000	0000	0000	0000	0000	0000	0001	0002
75	0000	0000	0000	0000	0000	0000	0000	0000	0001	0002
80	0000	0000	0000	0000	0000	0000	0000	0000	0001	0002
85	0000	0000	0000	0000	0000	0000	0000	0000	0001	0002
90	0000	0000	0000	0000	0000	0000	0000	0000	0001	0001
95	0000	0000	0000	0000	0000	0000	0000	0000	0001	0001
9.00	0000	0000	0000	0000	0000	0000	0000	0000	0001	0001
05	0000	0000	0000	0000	0000	0000	0000	0000	0001	0001
10	0000	0000	0000	0000	0000	0000	0000	0000	0000	0001
15	0000	0000	0000	0000	0000	0000	0000	0000	0000	0001
20	0000	0000	0000	0000	0000	0000	0000	0000	0000	0001
25	0000	0000	0000	0000	0000	0000	0000	0000	0000	0001
30	0000	0000	0000	0000	0000	0000	0000	0000	0000	0001
35	0000	0000	0000	0000	0000	0000	0000	0000	0000	0001
40	0000	0000	0000	0000	0000	0000	0000	0000	0000	0001
45	0000	0000	0000	0000	0000	0000	0000	0000	0000	0001
50	0000	0000	0000	0000	0000	0000	0000	0000	0000	0001
55	0000	0000	0000	0000	0000	0000	0000	0000	0000	0001
60	0000	0000	0000	0000	0000	0000	0000	0000	0000	0001
65	0000	0000	0000	0000	0000	0000	0000	0000	0000	0001
70	0000	0000	0000	0000	0000	0000	0000	0000	0000	0001
9.75	0000	0000	0000	0000	0000	0000	0000	0000	0000	0001
80	0000	0000	0000	0000	0000	0000	0000	0000	0000	0001
85	0000	0000	0000	0000	0000	0000	0000	0000	0000	0001
90	0000	0000	0000	0000	0000	0000	0000	0000	0000	0000

PROBABILITY DENSITY OF t, THE NON-CENTRAL t-STATISTIC, TABLED AS A FUNCTION OF $t/\sqrt{f}$.

f is the number of degrees of freedom; the non-centrality parameter is $\sqrt{f+1}\,K_p$.
K_p is the standardized normal deviate exceeded with probability p.

DEGREES OF FREEDOM **12**

$t/\sqrt{f}$ \ p	.2500	.1500	.1000	.0650	.0400	.0250	.0100	.0040	.0025	.0010
-1.25	0000	0000	0000	0000	0000	0000	0000	0000	0000	0000
20	0000	0000	0000	0000	0000	0000	0000	0000	0000	0000
15	0000	0000	0000	0000	0000	0000	0000	0000	0000	0000
10	0000	0000	0000	0000	0000	0000	0000	0000	0000	0000
05	0000	0000	0000	0000	0000	0000	0000	0000	0000	0000
-1.00	0000	0000	0000	0000	0000	0000	0000	0000	0000	0000
-0.95	0000	0000	0000	0000	0000	0000	0000	0000	0000	0000
90	0000	0000	0000	0000	0000	0000	0000	0000	0000	0000
85	0000	0000	0000	0000	0000	0000	0000	0000	0000	0000
80	0000	0000	0000	0000	0000	0000	0000	0000	0000	0000
75	0000	0000	0000	0000	0000	0000	0000	0000	0000	0000
70	0000	0000	0000	0000	0000	0000	0000	0000	0000	0000
65	0000	0000	0000	0000	0000	0000	0000	0000	0000	0000
60	0000	0000	0000	0000	0000	0000	0000	0000	0000	0000
55	0001	0000	0000	0000	0000	0000	0000	0000	0000	0000
50	0001	0000	0000	0000	0000	0000	0000	0000	0000	0000
45	0002	0000	0000	0000	0000	0000	0000	0000	0000	0000
40	0004	0000	0000	0000	0000	0000	0000	0000	0000	0000
35	0007	0000	0000	0000	0000	0000	0000	0000	0000	0000
30	0011	0000	0000	0000	0000	0000	0000	0000	0000	0000
25	0018	0000	0000	0000	0000	0000	0000	0000	0000	0000
20	0031	0000	0000	0000	0000	0000	0000	0000	0000	0000
15	0051	0000	0000	0000	0000	0000	0000	0000	0000	0000
10	0082	0001	0000	0000	0000	0000	0000	0000	0000	0000
05	0130	0002	0000	0000	0000	0000	0000	0000	0000	0000
0.00	0203	0004	0000	0000	0000	0000	0000	0000	0000	0000
05	0308	0007	0000	0000	0000	0000	0000	0000	0000	0000
10	0455	0013	0000	0000	0000	0000	0000	0000	0000	0000
15	0650	0024	0001	0000	0000	0000	0000	0000	0000	0000
20	0900	0043	0002	0000	0000	0000	0000	0000	0000	0000
0.25	1204	0074	0004	0000	0000	0000	0000	0000	0000	0000
30	1554	0124	0009	0000	0000	0000	0000	0000	0000	0000
35	1936	0200	0017	0001	0000	0000	0000	0000	0000	0000
40	2326	0310	0032	0002	0000	0000	0000	0000	0000	0000
45	2698	0460	0058	0004	0000	0000	0000	0000	0000	0000
50	3025	0657	0100	0009	0000	0000	0000	0000	0000	0000
55	3282	0901	0164	0018	0001	0000	0000	0000	0000	0000
60	3453	1187	0256	0033	0002	0000	0000	0000	0000	0000
65	3527	1504	0384	0059	0005	0000	0000	0000	0000	0000
70	3507	1838	0551	0100	0010	0001	0000	0000	0000	0000
75	3401	2167	0758	0162	0020	0002	0000	0000	0000	0000
80	3223	2473	1002	0250	0036	0004	0000	0000	0000	0000
85	2991	2736	1274	0369	0063	0009	0000	0000	0000	0000
90	2724	2942	1563	0521	0105	0017	0000	0000	0000	0000
95	2440	3080	1854	0707	0165	0031	0001	0000	0000	0000
1.00	2153	3148	2131	0922	0249	0054	0002	0000	0000	0000
05	1874	3146	2380	1160	0359	0088	0003	0000	0000	0000
10	1612	3081	2587	1411	0496	0139	0007	0000	0000	0000
15	1373	2962	2744	1664	0661	0208	0013	0001	0000	0000
20	1159	2802	2847	1907	0849	0299	0023	0001	0000	0000
25	0970	2611	2893	2128	1055	0413	0040	0002	0001	0000
30	0807	2401	2886	2318	1271	0549	0065	0005	0001	0000
35	0668	2181	2830	2470	1490	0707	0100	0009	0002	0000
40	0550	1961	2734	2580	1701	0881	0149	0016	0005	0000
45	0451	1746	2605	2644	1897	1068	0213	0027	0008	0001
50	0368	1542	2451	2665	2070	1259	0293	0044	0015	0001
55	0300	1352	2281	2646	2213	1450	0390	0068	0024	0003
60	0244	1178	2101	2591	2324	1631	0503	0100	0039	0005
65	0199	1020	1918	2505	2401	1798	0632	0144	0060	0009
70	0161	0879	1737	2394	2442	1945	0772	0199	0088	0015

PROBABILITY DENSITY OF t, THE NON-CENTRAL t-STATISTIC, TABLED AS A FUNCTION OF $t/\sqrt{f}$.

f is the number of degrees of freedom; the non-centrality parameter is $\sqrt{f+1}\,K_p$.
K_p is the standardized normal deviate exceeded with probability p.

DEGREES OF FREEDOM **12**

$t/\sqrt{f}$ \ p	.2500	.1500	.1000	.0650	.0400	.0250	.0100	.0040	.0025	.0010
1.75	0131	0755	1562	2265	2450	2067	0920	0267	0126	0024
80	0106	0645	1396	2123	2427	2163	1073	0348	0174	0038
85	0086	0550	1240	1974	2377	2230	1226	0441	0232	0056
90	0070	0468	1096	1821	2303	2269	1374	0546	0303	0081
95	0056	0397	0965	1669	2211	2280	1513	0660	0384	0113
2.00	0046	0336	0846	1520	2104	2266	1640	0781	0476	0153
05	0037	0284	0739	1377	1987	2229	1751	0907	0577	0203
10	0030	0240	0644	1241	1863	2173	1844	1034	0686	0261
15	0025	0202	0560	1114	1735	2100	1918	1159	0799	0328
20	0020	0171	0485	0996	1607	2014	1972	1279	0915	0404
25	0016	0144	0420	0887	1481	1918	2006	1392	1031	0488
30	0013	0121	0363	0788	1357	1814	2020	1494	1145	0578
35	0011	0102	0313	0698	1239	1706	2016	1585	1253	0674
40	0009	0086	0270	0617	1127	1596	1996	1662	1354	0772
45	0007	0073	0233	0544	1021	1486	1961	1725	1446	0872
50	0006	0061	0200	0479	0923	1377	1912	1773	1527	0972
55	0005	0052	0172	0421	0831	1271	1853	1806	1596	1068
60	0004	0044	0148	0369	0747	1169	1785	1824	1652	1161
65	0003	0037	0127	0324	0670	1072	1710	1828	1695	1248
70	0003	0031	0110	0284	0599	0979	1630	1819	1725	1327
75	0002	0026	0094	0248	0535	0893	1547	1799	1742	1398
80	0002	0022	0081	0217	0477	0812	1461	1767	1747	1459
85	0002	0019	0070	0190	0425	0736	1375	1727	1741	1511
90	0001	0016	0060	0166	0378	0667	1290	1678	1724	1552
95	0001	0014	0052	0145	0336	0602	1205	1623	1697	1582
3.00	0001	0012	0045	0127	0298	0544	1123	1563	1662	1603
05	0001	0010	0038	0111	0265	0490	1043	1499	1620	1613
10	0001	0008	0033	0097	0235	0441	0967	1431	1572	1614
15	0001	0007	0029	0084	0208	0396	0894	1362	1519	1606
20	0000	0006	0025	0074	0184	0356	0825	1292	1461	1590
3.25	0000	0005	0021	0065	0163	0319	0760	1222	1401	1566
30	0000	0005	0018	0056	0145	0286	0698	1152	1339	1536
35	0000	0004	0016	0049	0128	0257	0641	1084	1275	1501
40	0000	0003	0014	0043	0113	0230	0588	1017	1211	1460
45	0000	0003	0012	0038	0100	0206	0538	0952	1148	1416
50	0000	0002	0010	0033	0089	0184	0492	0889	1084	1368
55	0000	0002	0009	0029	0079	0165	0449	0829	1022	1317
60	0000	0002	0008	0025	0070	0147	0410	0772	0961	1265
65	0000	0002	0007	0022	0062	0132	0373	0718	0903	1211
70	0000	0001	0006	0020	0055	0118	0340	0666	0846	1157
75	0000	0001	0005	0017	0049	0105	0309	0617	0791	1102
80	0000	0001	0005	0015	0043	0094	0281	0571	0739	1048
85	0000	0001	0004	0013	0038	0084	0256	0528	0689	0994
90	0000	0001	0003	0012	0034	0075	0233	0488	0642	0941
95	0000	0001	0003	0010	0030	0067	0211	0450	0597	0890
4.00	0000	0001	0003	0009	0027	0060	0192	0415	0555	0839
05	0000	0001	0002	0008	0024	0054	0174	0382	0515	0791
10	0000	0000	0002	0007	0021	0048	0158	0352	0477	0744
15	0000	0000	0002	0006	0019	0043	0143	0324	0442	0699
20	0000	0000	0002	0006	0017	0039	0130	0297	0409	0656
25	0000	0000	0001	0005	0015	0035	0118	0273	0379	0614
30	0000	0000	0001	0004	0013	0031	0107	0251	0350	0575
35	0000	0000	0001	0004	0012	0028	0097	0230	0323	0538
40	0000	0000	0001	0003	0011	0025	0088	0212	0299	0503
45	0000	0000	0001	0003	0009	0023	0080	0194	0276	0469
50	0000	0000	0001	0003	0008	0020	0073	0178	0254	0438
55	0000	0000	0001	0002	0008	0018	0066	0163	0234	0408
60	0000	0000	0001	0002	0007	0016	0060	0150	0216	0380
65	0000	0000	0001	0002	0006	0015	0054	0137	0199	0354
70	0000	0000	0000	0002	0005	0013	0049	0126	0183	0329

PROBABILITY DENSITY OF t, THE NON-CENTRAL t-STATISTIC, TABLED AS A FUNCTION OF $t/\sqrt{f}$.

f is the number of degrees of freedom; the non-centrality parameter is $\sqrt{f+1}\,K_p$.
K_p is the standardized normal deviate exceeded with probability p.

DEGREES OF FREEDOM **12**

$t/\sqrt{f}$ \ p	.2500	.1500	.1000	.0650	.0400	.0250	.0100	.0040	.0025	.0010
4.75	0000	0000	0000	0002	0005	0012	0045	0115	0169	0306
80	0000	0000	0000	0001	0004	0011	0041	0106	0156	0284
85	0000	0000	0000	0001	0004	0010	0037	0097	0143	0264
90	0000	0000	0000	0001	0003	0009	0034	0089	0132	0245
95	0000	0000	0000	0001	0003	0008	0031	0081	0121	0228
5.00	0000	0000	0000	0001	0003	0007	0028	0075	0112	0211
05	0000	0000	0000	0001	0003	0006	0025	0068	0103	0196
10	0000	0000	0000	0001	0002	0006	0023	0063	0095	0182
15	0000	0000	0000	0001	0002	0005	0021	0057	0087	0169
20	0000	0000	0000	0001	0002	0005	0019	0053	0080	0156
25	0000	0000	0000	0001	0002	0004	0017	0048	0074	0145
30	0000	0000	0000	0000	0001	0004	0016	0044	0068	0134
35	0000	0000	0000	0000	0001	0003	0014	0041	0063	0124
40	0000	0000	0000	0000	0001	0003	0013	0037	0058	0115
45	0000	0000	0000	0000	0001	0003	0012	0034	0053	0107
50	0000	0000	0000	0000	0001	0003	0011	0031	0049	0099
55	0000	0000	0000	0000	0001	0002	0010	0029	0045	0092
60	0000	0000	0000	0000	0001	0002	0009	0027	0041	0085
65	0000	0000	0000	0000	0001	0002	0008	0024	0038	0079
70	0000	0000	0000	0000	0001	0002	0008	0022	0035	0073
75	0000	0000	0000	0000	0001	0002	0007	0021	0032	0068
80	0000	0000	0000	0000	0001	0001	0006	0019	0030	0063
85	0000	0000	0000	0000	0001	0001	0006	0017	0028	0058
90	0000	0000	0000	0000	0000	0001	0005	0016	0025	0054
95	0000	0000	0000	0000	0000	0001	0005	0015	0023	0050
6.00	0000	0000	0000	0000	0000	0001	0004	0014	0022	0046
05	0000	0000	0000	0000	0000	0001	0004	0012	0020	0043
10	0000	0000	0000	0000	0000	0001	0004	0011	0018	0040
15	0000	0000	0000	0000	0000	0001	0003	0011	0017	0037
20	0000	0000	0000	0000	0000	0001	0003	0010	0016	0034
6.25	0000	0000	0000	0000	0000	0001	0003	0009	0014	0032
30	0000	0000	0000	0000	0000	0001	0003	0008	0013	0029
35	0000	0000	0000	0000	0000	0001	0002	0008	0012	0027
40	0000	0000	0000	0000	0000	0000	0002	0007	0011	0025
45	0000	0000	0000	0000	0000	0000	0002	0006	0011	0024
50	0000	0000	0000	0000	0000	0000	0002	0006	0010	0022
55	0000	0000	0000	0000	0000	0000	0002	0006	0009	0020
60	0000	0000	0000	0000	0000	0000	0002	0005	0008	0019
65	0000	0000	0000	0000	0000	0000	0001	0005	0008	0018
70	0000	0000	0000	0000	0000	0000	0001	0004	0007	0016
75	0000	0000	0000	0000	0000	0000	0001	0004	0007	0015
80	0000	0000	0000	0000	0000	0000	0001	0004	0006	0014
85	0000	0000	0000	0000	0000	0000	0001	0003	0006	0013
90	0000	0000	0000	0000	0000	0000	0001	0003	0005	0012
95	0000	0000	0000	0000	0000	0000	0001	0003	0005	0011
7.00	0000	0000	0000	0000	0000	0000	0001	0003	0005	0011
05	0000	0000	0000	0000	0000	0000	0001	0003	0004	0010
10	0000	0000	0000	0000	0000	0000	0001	0002	0004	0009
15	0000	0000	0000	0000	0000	0000	0001	0002	0004	0008
20	0000	0000	0000	0000	0000	0000	0001	0002	0003	0008
25	0000	0000	0000	0000	0000	0000	0001	0002	0003	0007
30	0000	0000	0000	0000	0000	0000	0001	0002	0003	0007
35	0000	0000	0000	0000	0000	0000	0000	0002	0003	0006
40	0000	0000	0000	0000	0000	0000	0000	0001	0003	0006
45	0000	0000	0000	0000	0000	0000	0000	0001	0002	0006
50	0000	0000	0000	0000	0000	0000	0000	0001	0002	0005
55	0000	0000	0000	0000	0000	0000	0000	0001	0002	0005
60	0000	0000	0000	0000	0000	0000	0000	0001	0002	0005
65	0000	0000	0000	0000	0000	0000	0000	0001	0002	0004
70	0000	0000	0000	0000	0000	0000	0000	0001	0002	0004

PROBABILITY DENSITY OF t, THE NON-CENTRAL t-STATISTIC, TABLED AS A FUNCTION OF $t/\sqrt{f}$.

f is the number of degrees of freedom; the non-centrality parameter is $\sqrt{f+1}\,K_{\bar{p}}$.
K_p is the standardized normal deviate exceeded with probability p.

DEGREES OF FREEDOM 12

$t/\sqrt{f}$ \ p	.2500	.1500	.1000	.0650	.0400	.0250	.0100	.0040	.0025	.0010
7.75	0000	0000	0000	0000	0000	0000	0000	0001	0002	0004
80	0000	0000	0000	0000	0000	0000	0000	0001	0001	0003
85	0000	0000	0000	0000	0000	0000	0000	0001	0001	0003
90	0000	0000	0000	0000	0000	0000	0000	0001	0001	0003
95	0000	0000	0000	0000	0000	0000	0000	0001	0001	0003
8.00	0000	0000	0000	0000	0000	0000	0000	0001	0001	0003
05	0000	0000	0000	0000	0000	0000	0000	0001	0001	0002
10	0000	0000	0000	0000	0000	0000	0000	0001	0001	0002
15	0000	0000	0000	0000	0000	0000	0000	0001	0001	0002
20	0000	0000	0000	0000	0000	0000	0000	0000	0001	0002
25	0000	0000	0000	0000	0000	0000	0000	0000	0001	0002
30	0000	0000	0000	0000	0000	0000	0000	0000	0001	0002
35	0000	0000	0000	0000	0000	0000	0000	0000	0001	0002
40	0000	0000	0000	0000	0000	0000	0000	0000	0001	0002
45	0000	0000	0000	0000	0000	0000	0000	0000	0001	0001
50	0000	0000	0000	0000	0000	0000	0000	0000	0001	0001
55	0000	0000	0000	0000	0000	0000	0000	0000	0001	0001
60	0000	0000	0000	0000	0000	0000	0000	0000	0000	0001
65	0000	0000	0000	0000	0000	0000	0000	0000	0000	0001
70	0000	0000	0000	0000	0000	0000	0000	0000	0000	0001
75	0000	0000	0000	0000	0000	0000	0000	0000	0000	0001
80	0000	0000	0000	0000	0000	0000	0000	0000	0000	0001
85	0000	0000	0000	0000	0000	0000	0000	0000	0000	0001
90	0000	0000	0000	0000	0000	0000	0000	0000	0000	0001
95	0000	0000	0000	0000	0000	0000	0000	0000	0000	0001
9.00	0000	0000	0000	0000	0000	0000	0000	0000	0000	0001
05	0000	0000	0000	0000	0000	0000	0000	0000	0000	0001
10	0000	0000	0000	0000	0000	0000	0000	0000	0000	0001
15	0000	0000	0000	0000	0000	0000	0000	0000	0000	0001
20	0000	0000	0000	0000	0000	0000	0000	0000	0000	0001
9.25	0000	0000	0000	0000	0000	0000	0000	0000	0000	0001
30	0000	0000	0000	0000	0000	0000	0000	0000	0000	0000

PROBABILITY DENSITY OF t, THE NON-CENTRAL t-STATISTIC, TABLED AS A FUNCTION OF $t/\sqrt{f}$

f is the number of degrees of freedom; the non-centrality parameter is $\sqrt{f+1}\,K_p$.
K_p is the standardized normal deviate exceeded with probability p.

DEGREES OF FREEDOM **13**

$t/\sqrt{f}$ \ p	.2500	.1500	.1000	.0650	.0400	.0250	.0100	.0040	.0025	.0010
-0.55	0000	0000	0000	0000	0000	0000	0000	0000	0000	0000
50	0001	0000	0000	0000	0000	0000	0000	0000	0000	0000
45	0001	0000	0000	0000	0000	0000	0000	0000	0000	0000
40	0002	0000	0000	0000	0000	0000	0000	0000	0000	0000
35	0004	0000	0000	0000	0000	0000	0000	0000	0000	0000
30	0007	0000	0000	0000	0000	0000	0000	0000	0000	0000
25	0012	0000	0000	0000	0000	0000	0000	0000	0000	0000
20	0021	0000	0000	0000	0000	0000	0000	0000	0000	0000
15	0036	0000	0000	0000	0000	0000	0000	0000	0000	0000
10	0061	0000	0000	0000	0000	0000	0000	0000	0000	0000
05	0101	0001	0000	0000	0000	0000	0000	0000	0000	0000
0.00	0162	0002	0000	0000	0000	0000	0000	0000	0000	0000
05	0254	0004	0000	0000	0000	0000	0000	0000	0000	0000
10	0386	0008	0000	0000	0000	0000	0000	0000	0000	0000
15	0568	0016	0001	0000	0000	0000	0000	0000	0000	0000
20	0807	0031	0001	0000	0000	0000	0000	0000	0000	0000
25	1105	0055	0003	0000	0000	0000	0000	0000	0000	0000
30	1456	0096	0006	0000	0000	0000	0000	0000	0000	0000
35	1845	0161	0012	0000	0000	0000	0000	0000	0000	0000
40	2250	0257	0023	0001	0000	0000	0000	0000	0000	0000
45	2641	0395	0043	0003	0000	0000	0000	0000	0000	0000
50	2989	0579	0076	0006	0000	0000	0000	0000	0000	0000
55	3266	0814	0130	0012	0001	0000	0000	0000	0000	0000
60	3450	1097	0211	0023	0001	0000	0000	0000	0000	0000
65	3532	1417	0327	0044	0003	0000	0000	0000	0000	0000
70	3512	1759	0483	0077	0007	0000	0000	0000	0000	0000
0.75	3398	2103	0681	0130	0013	0001	0000	0000	0000	0000
80	3208	2426	0920	0207	0026	0003	0000	0000	0000	0000
85	2962	2706	1193	0315	0047	0006	0000	0000	0000	0000
90	2679	2927	1487	0458	0082	0012	0000	0000	0000	0000
95	2380	3077	1789	0636	0134	0022	0000	0000	0000	0000
1.00	2080	3151	2079	0847	0208	0040	0001	0000	0000	0000
05	1792	3150	2343	1085	0308	0068	0002	0000	0000	0000
10	1525	3082	2565	1341	0437	0111	0004	0000	0000	0000
15	1283	2955	2735	1602	0595	0172	0009	0000	0000	0000
20	1069	2784	2846	1857	0780	0255	0017	0001	0000	0000
25	0883	2581	2897	2091	0986	0361	0029	0001	0000	0000
30	0724	2359	2890	2294	1207	0491	0049	0003	0001	0000
35	0591	2128	2831	2457	1432	0644	0079	0006	0001	0000
40	0479	1898	2729	2575	1653	0818	0121	0011	0003	0000
45	0387	1676	2591	2646	1859	1006	0178	0020	0006	0000
50	0311	1466	2427	2670	2043	1202	0252	0033	0010	0001
55	0250	1273	2246	2650	2197	1399	0343	0052	0017	0002
60	0200	1097	2057	2591	2317	1589	0451	0080	0029	0003
65	0160	0940	1865	2499	2400	1766	0577	0118	0046	0006
70	0128	0801	1677	2381	2445	1922	0716	0167	0070	0011
75	0102	0680	1496	2244	2454	2053	0866	0230	0102	0018
80	0081	0574	1325	2093	2430	2156	1022	0306	0145	0028
85	0065	0484	1167	1935	2377	2229	1180	0395	0199	0043
90	0052	0406	1022	1775	2298	2271	1334	0497	0264	0064
95	0041	0340	0891	1616	2200	2284	1481	0610	0342	0092
2.00	0033	0284	0773	1461	2086	2270	1615	0731	0430	0128
05	0026	0237	0669	1314	1961	2231	1733	0859	0530	0172
10	0021	0198	0577	1175	1830	2170	1833	0990	0638	0226
15	0017	0165	0496	1046	1696	2092	1913	1119	0752	0290
20	0014	0137	0425	0927	1561	2000	1971	1245	0871	0363

PROBABILITY DENSITY OF t, THE NON-CENTRAL t-STATISTIC, TABLED AS A FUNCTION OF $t/\sqrt{f}$.

f is the number of degrees of freedom; the non-centrality parameter is $\sqrt{f+1}\,K_p$.
K_p is the standardized normal deviate exceeded with probability p.

DEGREES OF FREEDOM **13**

$t/\sqrt{f}$ \ p	.2500	.1500	.1000	.0650	.0400	.0250	.0100	.0040	.0025	.0010
2.25	0011	0114	0364	0819	1430	1898	2008	1364	0991	0445
30	0009	0095	0311	0721	1302	1788	2024	1473	1109	0534
35	0007	0079	0265	0633	1181	1674	2020	1569	1222	0630
40	0006	0066	0226	0554	1066	1558	1998	1652	1329	0730
45	0005	0055	0193	0484	0959	1443	1961	1720	1426	0832
50	0004	0046	0164	0422	0859	1330	1909	1771	1513	0935
55	0003	0038	0139	0367	0768	1220	1846	1807	1587	1036
60	0002	0032	0119	0319	0685	1115	1773	1827	1647	1133
65	0002	0026	0101	0277	0609	1015	1693	1832	1694	1224
70	0002	0022	0086	0240	0540	0922	1608	1822	1727	1309
75	0001	0018	0073	0208	0478	0834	1520	1800	1745	1384
80	0001	0015	0062	0180	0423	0753	1430	1767	1751	1450
85	0001	0013	0053	0156	0373	0678	1340	1723	1744	1505
90	0001	0011	0045	0135	0329	0609	1250	1671	1726	1550
95	0001	0009	0038	0117	0290	0546	1162	1613	1697	1583
3.00	0001	0008	0033	0101	0255	0489	1077	1548	1660	1605
05	0000	0006	0028	0087	0224	0437	0995	1480	1615	1616
10	0000	0005	0024	0075	0197	0390	0917	1409	1563	1617
15	0000	0005	0020	0065	0173	0348	0843	1336	1506	1609
20	0000	0004	0017	0056	0152	0310	0773	1262	1445	1592
25	0000	0003	0015	0049	0133	0276	0708	1188	1382	1567
30	0000	0003	0013	0042	0117	0245	0646	1116	1316	1534
35	0000	0002	0011	0037	0103	0218	0589	1045	1249	1496
40	0000	0002	0009	0032	0090	0194	0536	0975	1181	1453
45	0000	0002	0008	0027	0079	0172	0488	0909	1115	1405
50	0000	0001	0007	0024	0069	0153	0443	0844	1049	1354
55	0000	0001	0006	0021	0061	0135	0402	0783	0984	1300
60	0000	0001	0005	0018	0053	0120	0364	0725	0921	1245
65	0000	0001	0004	0016	0047	0106	0329	0670	0861	1188
70	0000	0001	0004	0014	0041	0094	0298	0618	0803	1131
3.75	0000	0001	0003	0012	0036	0084	0269	0570	0747	1074
80	0000	0001	0003	0010	0032	0074	0243	0524	0694	1017
85	0000	0000	0002	0009	0028	0066	0219	0482	0644	0961
90	0000	0000	0002	0008	0025	0058	0198	0442	0596	0906
95	0000	0000	0002	0007	0022	0052	0178	0406	0552	0853
4.00	0000	0000	0002	0006	0019	0046	0161	0372	0510	0801
05	0000	0000	0001	0005	0017	0041	0145	0340	0470	0751
10	0000	0000	0001	0005	0015	0036	0131	0311	0434	0703
15	0000	0000	0001	0004	0013	0032	0118	0284	0399	0657
20	0000	0000	0001	0003	0011	0029	0106	0260	0368	0614
25	0000	0000	0001	0003	0010	0025	0095	0237	0338	0572
30	0000	0000	0001	0003	0009	0023	0086	0216	0311	0533
35	0000	0000	0001	0002	0008	0020	0077	0197	0285	0496
40	0000	0000	0001	0002	0007	0018	0070	0180	0262	0461
45	0000	0000	0000	0002	0006	0016	0063	0164	0240	0428
50	0000	0000	0000	0002	0005	0014	0057	0149	0220	0397
55	0000	0000	0000	0001	0005	0013	0051	0136	0202	0368
60	0000	0000	0000	0001	0004	0011	0046	0124	0185	0341
65	0000	0000	0000	0001	0004	0010	0041	0113	0169	0316
70	0000	0000	0000	0001	0003	0009	0037	0103	0155	0292
75	0000	0000	0000	0001	0003	0008	0034	0094	0142	0270
80	0000	0000	0000	0001	0003	0007	0030	0085	0130	0250
85	0000	0000	0000	0001	0002	0006	0027	0078	0119	0231
90	0000	0000	0000	0001	0002	0006	0025	0071	0109	0213
95	0000	0000	0000	0001	0002	0005	0022	0064	0099	0197
5.00	0000	0000	0000	0000	0002	0005	0020	0059	0091	0181
05	0000	0000	0000	0000	0001	0004	0018	0053	0083	0167
10	0000	0000	0000	0000	0001	0004	0016	0049	0076	0154
15	0000	0000	0000	0000	0001	0003	0015	0044	0069	0142
20	0000	0000	0000	0000	0001	0003	0013	0040	0064	0131

PROBABILITY DENSITY OF t, THE NON-CENTRAL t-STATISTIC, TABLED AS A FUNCTION OF $t/\sqrt{f}$.

f is the number of degrees of freedom; the non-centrality parameter is $\sqrt{f+1}\,K_p$.
K_p is the standardized normal deviate exceeded with probability p.

DEGREES OF FREEDOM **13**

$t/\sqrt{f}$ \ p	.2500	.1500	.1000	.0650	.0400	.0250	.0100	.0040	.0025	.0010
5.25	0000	0000	0000	0000	0001	0003	0012	0037	0058	0121
30	0000	0000	0000	0000	0001	0002	0011	0033	0053	0111
35	0000	0000	0000	0000	0001	0002	0010	0030	0049	0103
40	0000	0000	0000	0000	0001	0002	0009	0028	0044	0095
45	0000	0000	0000	0000	0001	0002	0008	0025	0041	0087
50	0000	0000	0000	0000	0001	0002	0007	0023	0037	0080
55	0000	0000	0000	0000	0000	0001	0007	0021	0034	0074
60	0000	0000	0000	0000	0000	0001	0006	0019	0031	0068
65	0000	0000	0000	0000	0000	0001	0005	0018	0029	0063
70	0000	0000	0000	0000	0000	0001	0005	0016	0026	0058
75	0000	0000	0000	0000	0000	0001	0005	0015	0024	0053
80	0000	0000	0000	0000	0000	0001	0004	0013	0022	0049
85	0000	0000	0000	0000	0000	0001	0004	0012	0020	0045
90	0000	0000	0000	0000	0000	0001	0003	0011	0018	0042
95	0000	0000	0000	0000	0000	0001	0003	0010	0017	0038
6.00	0000	0000	0000	0000	0000	0001	0003	0009	0015	0035
05	0000	0000	0000	0000	0000	0001	0003	0009	0014	0033
10	0000	0000	0000	0000	0000	0000	0002	0008	0013	0030
15	0000	0000	0000	0000	0000	0000	0002	0007	0012	0028
20	0000	0000	0000	0000	0000	0000	0002	0007	0011	0026
25	0000	0000	0000	0000	0000	0000	0002	0006	0010	0024
30	0000	0000	0000	0000	0000	0000	0002	0005	0009	0022
35	0000	0000	0000	0000	0000	0000	0001	0005	0008	0020
40	0000	0000	0000	0000	0000	0000	0001	0005	0008	0019
45	0000	0000	0000	0000	0000	0000	0001	0004	0007	0017
50	0000	0000	0000	0000	0000	0000	0001	0004	0007	0016
55	0000	0000	0000	0000	0000	0000	0001	0004	0006	0015
60	0000	0000	0000	0000	0000	0000	0001	0003	0006	0013
65	0000	0000	0000	0000	0000	0000	0001	0003	0005	0012
70	0000	0000	0000	0000	0000	0000	0001	0003	0005	0011
6.75	0000	0000	0000	0000	0000	0000	0001	0003	0004	0011
80	0000	0000	0000	0000	0000	0000	0001	0002	0004	0010
85	0000	0000	0000	0000	0000	0000	0001	0002	0004	0009
90	0000	0000	0000	0000	0000	0000	0001	0002	0003	0008
95	0000	0000	0000	0000	0000	0000	0001	0002	0003	0008
7.00	0000	0000	0000	0000	0000	0000	0000	0002	0003	0007
05	0000	0000	0000	0000	0000	0000	0000	0002	0003	0007
10	0000	0000	0000	0000	0000	0000	0000	0001	0002	0006
15	0000	0000	0000	0000	0000	0000	0000	0001	0002	0006
20	0000	0000	0000	0000	0000	0000	0000	0001	0002	0005
25	0000	0000	0000	0000	0000	0000	0000	0001	0002	0005
30	0000	0000	0000	0000	0000	0000	0000	0001	0002	0005
35	0000	0000	0000	0000	0000	0000	0000	0001	0002	0004
40	0000	0000	0000	0000	0000	0000	0000	0001	0002	0004
45	0000	0000	0000	0000	0000	0000	0000	0001	0001	0004
50	0000	0000	0000	0000	0000	0000	0000	0001	0001	0003
55	0000	0000	0000	0000	0000	0000	0000	0001	0001	0003
60	0000	0000	0000	0000	0000	0000	0000	0001	0001	0003
65	0000	0000	0000	0000	0000	0000	0000	0001	0001	0003
70	0000	0000	0000	0000	0000	0000	0000	0001	0001	0002
75	0000	0000	0000	0000	0000	0000	0000	0001	0001	0002
80	0000	0000	0000	0000	0000	0000	0000	0000	0001	0002
85	0000	0000	0000	0000	0000	0000	0000	0000	0001	0002
90	0000	0000	0000	0000	0000	0000	0000	0000	0001	0002
95	0000	0000	0000	0000	0000	0000	0000	0000	0001	0002
8.00	0000	0000	0000	0000	0000	0000	0000	0000	0001	0002
05	0000	0000	0000	0000	0000	0000	0000	0000	0001	0001
10	0000	0000	0000	0000	0000	0000	0000	0000	0001	0001
15	0000	0000	0000	0000	0000	0000	0000	0000	0000	0001
20	0000	0000	0000	0000	0000	0000	0000	0000	0000	0001

PROBABILITY DENSITY OF t, THE NON-CENTRAL t-STATISTIC, TABLED AS A FUNCTION OF $t/\sqrt{f}$.

f is the number of degrees of freedom; the non-centrality parameter is $\sqrt{f+1}\,K_p$.
K_p is the standardized normal deviate exceeded with probability p.

DEGREES OF FREEDOM **13**

$t/\sqrt{f}$ \ p	.2500	.1500	.1000	.0650	.0400	.0250	.0100	.0040	.0025	.0010
8.25	0000	0000	0000	0000	0000	0000	0000	0000	0000	0001
30	0000	0000	0000	0000	0000	0000	0000	0000	0000	0001
35	0000	0000	0000	0000	0000	0000	0000	0000	0000	0001
40	0000	0000	0000	0000	0000	0000	0000	0000	0000	0001
45	0000	0000	0000	0000	0000	0000	0000	0000	0000	0001
50	0000	0000	0000	0000	0000	0000	0000	0000	0000	0001
55	0000	0000	0000	0000	0000	0000	0000	0000	0000	0001
60	0000	0000	0000	0000	0000	0000	0000	0000	0000	0001
65	0000	0000	0000	0000	0000	0000	0000	0000	0000	0001
70	0000	0000	0000	0000	0000	0000	0000	0000	0000	0001
75	0000	0000	0000	0000	0000	0000	0000	0000	0000	0001
80	0000	0000	0000	0000	0000	0000	0000	0000	0000	0001
85	0000	0000	0000	0000	0000	0000	0000	0000	0000	0000

PROBABILITY DENSITY OF t, THE NON-CENTRAL t-STATISTIC, TABLED AS A FUNCTION OF $t/\sqrt{f}$.

f is the number of degrees of freedom; the non-centrality parameter is $\sqrt{f+1}\,K_p$.
K_p is the standardized normal deviate exceeded with probability p.

DEGREES OF FREEDOM **14**

$t/\sqrt{f}$ \ P	.2500	.1500	.1000	.0650	.0400	.0250	.0100	.0040	.0025	.0010
-0.50	0000	0000	0000	0000	0000	0000	0000	0000	0000	0000
45	0001	0000	0000	0000	0000	0000	0000	0000	0000	0000
40	0001	0000	0000	0000	0000	0000	0000	0000	0000	0000
35	0002	0000	0000	0000	0000	0000	0000	0000	0000	0000
30	0004	0000	0000	0000	0000	0000	0000	0000	0000	0000
25	0008	0000	0000	0000	0000	0000	0000	0000	0000	0000
20	0015	0000	0000	0000	0000	0000	0000	0000	0000	0000
15	0026	0000	0000	0000	0000	0000	0000	0000	0000	0000
10	0045	0000	0000	0000	0000	0000	0000	0000	0000	0000
05	0077	0001	0000	0000	0000	0000	0000	0000	0000	0000
0.00	0129	0001	0000	0000	0000	0000	0000	0000	0000	0000
05	0209	0003	0000	0000	0000	0000	0000	0000	0000	0000
10	0328	0005	0000	0000	0000	0000	0000	0000	0000	0000
15	0497	0011	0000	0000	0000	0000	0000	0000	0000	0000
20	0724	0022	0001	0000	0000	0000	0000	0000	0000	0000
25	1014	0041	0002	0000	0000	0000	0000	0000	0000	0000
30	1363	0074	0004	0000	0000	0000	0000	0000	0000	0000
35	1758	0129	0008	0000	0000	0000	0000	0000	0000	0000
40	2176	0214	0016	0001	0000	0000	0000	0000	0000	0000
45	2585	0338	0031	0002	0000	0000	0000	0000	0000	0000
50	2953	0511	0058	0004	0000	0000	0000	0000	0000	0000
55	3248	0736	0103	0008	0000	0000	0000	0000	0000	0000
60	3446	1014	0174	0016	0001	0000	0000	0000	0000	0000
65	3536	1334	0278	0032	0002	0000	0000	0000	0000	0000
70	3515	1684	0423	0059	0004	0000	0000	0000	0000	0000
75	3395	2040	0611	0104	0009	0001	0000	0000	0000	0000
80	3193	2378	0845	0172	0019	0002	0000	0000	0000	0000
85	2932	2676	1116	0269	0035	0004	0000	0000	0000	0000
90	2634	2912	1415	0402	0064	0008	0000	0000	0000	0000
95	2321	3073	1725	0572	0108	0016	0000	0000	0000	0000
1.00	2010	3154	2029	0778	0173	0030	0001	0000	0000	0000
05	1714	3154	2306	1015	0265	0053	0001	0000	0000	0000
10	1441	3082	2542	1274	0385	0089	0003	0000	0000	0000
15	1198	2948	2724	1543	0536	0143	0006	0000	0000	0000
20	0985	2766	2844	1807	0717	0217	0012	0000	0000	0000
25	0803	2551	2900	2054	0922	0315	0022	0001	0000	0000
30	0650	2317	2894	2269	1145	0439	0038	0002	0000	0000
35	0522	2076	2832	2444	1376	0587	0063	0004	0001	0000
40	0417	1837	2723	2571	1606	0758	0099	0008	0002	0000
45	0332	1608	2576	2647	1822	0947	0149	0014	0004	0000
50	0263	1394	2403	2673	2017	1147	0216	0024	0007	0000
55	0208	1198	2212	2653	2181	1350	0301	0040	0013	0001
60	0164	1022	2014	2590	2309	1548	0405	0064	0022	0002
65	0129	0866	1814	2493	2398	1734	0526	0097	0035	0004
70	0101	0730	1618	2368	2447	1899	0664	0141	0055	0007
75	0080	0612	1432	2222	2458	2039	0814	0198	0083	0013
80	0063	0511	1258	2063	2433	2149	0973	0269	0121	0021
85	0049	0425	1098	1897	2376	2227	1135	0353	0170	0033
90	0039	0353	0953	1729	2293	2273	1295	0452	0230	0051
95	0030	0292	0823	1564	2188	2288	1448	0563	0304	0075
2.00	0024	0241	0707	1405	2067	2273	1590	0685	0389	0106
05	0019	0198	0605	1254	1935	2231	1715	0814	0486	0147
10	0015	0163	0516	1112	1797	2167	1822	0947	0593	0197
15	0012	0134	0439	0982	1657	2084	1907	1081	0708	0256
20	0009	0110	0373	0863	1517	1986	1970	1212	0828	0326
25	0007	0091	0316	0756	1380	1878	2010	1336	0951	0406
30	0006	0074	0267	0659	1249	1762	2027	1451	1073	0494
35	0005	0061	0225	0573	1125	1642	2023	1554	1192	0589
40	0004	0050	0190	0497	1008	1521	2000	1642	1304	0690
45	0003	0041	0160	0430	0900	1401	1960	1714	1407	0794

PROBABILITY DENSITY OF t, THE NON-CENTRAL t-STATISTIC, TABLED AS A FUNCTION OF $t/\sqrt{f}$.

f is the number of degrees of freedom; the non-centrality parameter is $\sqrt{f+1}\,K_p$.
K_p is the standardized normal deviate exceeded with probability p.

$t/\sqrt{f}$ \ p	.2500	.1500	.1000	.0650	.0400	.0250	.0100	.0040	.0025	.0010
2.50	0002	0034	0134	0371	0801	1283	1905	1770	1499	0899
55	0002	0028	0113	0320	0710	1171	1838	1808	1578	1004
60	0001	0023	0095	0275	0628	1063	1761	1829	1642	1105
65	0001	0019	0080	0237	0553	0962	1676	1835	1692	1201
70	0001	0016	0067	0203	0487	0867	1586	1825	1727	1290
75	0001	0013	0056	0174	0428	0779	1493	1802	1748	1370
80	0001	0011	0047	0149	0375	0698	1399	1766	1754	1440
85	0001	0009	0040	0128	0328	0624	1304	1720	1747	1500
90	0000	0007	0033	0109	0287	0557	1211	1664	1727	1547
95	0000	0006	0028	0094	0250	0495	1121	1602	1697	1583
3.00	0000	0005	0024	0080	0218	0440	1033	1534	1657	1607
05	0000	0004	0020	0069	0190	0390	0949	1461	1609	1619
10	0000	0003	0017	0059	0165	0346	0870	1386	1554	1620
15	0000	0003	0014	0050	0144	0306	0795	1309	1494	1611
20	0000	0002	0012	0043	0125	0270	0724	1232	1429	1593
25	0000	0002	0010	0037	0109	0238	0659	1156	1362	1566
30	0000	0002	0009	0032	0095	0210	0598	1080	1292	1532
35	0000	0001	0007	0027	0082	0185	0541	1006	1222	1491
40	0000	0001	0006	0023	0071	0163	0490	0935	1152	1445
45	0000	0001	0005	0020	0062	0144	0442	0867	1082	1394
50	0000	0001	0004	0017	0054	0126	0399	0802	1014	1340
55	0000	0001	0004	0015	0047	0111	0359	0740	0947	1283
60	0000	0001	0003	0013	0041	0098	0323	0681	0883	1225
65	0000	0000	0003	0011	0035	0086	0290	0626	0821	1165
70	0000	0000	0002	0009	0031	0076	0261	0574	0761	1105
75	0000	0000	0002	0008	0027	0066	0234	0526	0705	1045
80	0000	0000	0002	0007	0023	0058	0210	0481	0652	0986
85	0000	0000	0001	0006	0020	0051	0188	0439	0601	0928
90	0000	0000	0001	0005	0018	0045	0168	0401	0554	0872
95	0000	0000	0001	0004	0015	0040	0151	0365	0510	0817
4.00	0000	0000	0001	0004	0013	0035	0135	0333	0468	0764
05	0000	0000	0001	0003	0012	0031	0121	0303	0430	0713
10	0000	0000	0001	0003	0010	0027	0108	0275	0394	0665
15	0000	0000	0001	0003	0009	0024	0096	0250	0361	0618
20	0000	0000	0001	0002	0008	0021	0086	0227	0330	0575
25	0000	0000	0000	0002	0007	0018	0077	0206	0302	0533
30	0000	0000	0000	0002	0006	0016	0069	0186	0275	0494
35	0000	0000	0000	0001	0005	0014	0062	0169	0251	0457
40	0000	0000	0000	0001	0005	0013	0055	0153	0229	0423
45	0000	0000	0000	0001	0004	0011	0049	0139	0209	0391
50	0000	0000	0000	0001	0004	0010	0044	0125	0190	0360
55	0000	0000	0000	0001	0003	0009	0039	0114	0173	0332
60	0000	0000	0000	0001	0003	0008	0035	0103	0158	0306
65	0000	0000	0000	0001	0002	0007	0031	0093	0144	0282
70	0000	0000	0000	0001	0002	0006	0028	0084	0131	0259
75	0000	0000	0000	0000	0002	0005	0025	0076	0119	0239
80	0000	0000	0000	0000	0002	0005	0023	0069	0108	0219
85	0000	0000	0000	0000	0001	0004	0020	0062	0098	0201
90	0000	0000	0000	0000	0001	0004	0018	0056	0089	0185
95	0000	0000	0000	0000	0001	0003	0016	0051	0081	0170
5.00	0000	0000	0000	0000	0001	0003	0014	0046	0074	0156
05	0000	0000	0000	0000	0001	0003	0013	0042	0067	0143
10	0000	0000	0000	0000	0001	0002	0012	0038	0061	0131
15	0000	0000	0000	0000	0001	0002	0010	0034	0055	0120
20	0000	0000	0000	0000	0001	0002	0009	0031	0050	0110
25	0000	0000	0000	0000	0001	0002	0008	0028	0046	0101
30	0000	0000	0000	0000	0000	0001	0008	0025	0042	0092
35	0000	0000	0000	0000	0000	0001	0007	0023	0038	0085
40	0000	0000	0000	0000	0000	0001	0006	0021	0034	0077
45	0000	0000	0000	0000	0000	0001	0005	0019	0031	0071

PROBABILITY DENSITY OF t, THE NON-CENTRAL t-STATISTIC, TABLED AS A FUNCTION OF $t/\sqrt{f}$.

f is the number of degrees of freedom; the non-centrality parameter is $\sqrt{f+1}\,K_p$.
K_p is the standardized normal deviate exceeded with probability p.

DEGREES OF FREEDOM **14**

$t/\sqrt{f}$ \ P	.2500	.1500	.1000	.0650	.0400	.0250	.0100	.0040	.0025	.0010
5.50	0000	0000	0000	0000	0000	0001	0005	0017	0028	0065
55	0000	0000	0000	0000	0000	0001	0004	0015	0026	0060
60	0000	0000	0000	0000	0000	0001	0004	0014	0023	0055
65	0000	0000	0000	0000	0000	0001	0004	0013	0021	0050
70	0000	0000	0000	0000	0000	0001	0003	0011	0019	0046
75	0000	0000	0000	0000	0000	0001	0003	0010	0018	0042
80	0000	0000	0000	0000	0000	0000	0003	0009	0016	0038
85	0000	0000	0000	0000	0000	0000	0002	0009	0015	0035
90	0000	0000	0000	0000	0000	0000	0002	0008	0013	0032
95	0000	0000	0000	0000	0000	0000	0002	0007	0012	0029
6.00	0000	0000	0000	0000	0000	0000	0002	0006	0011	0027
05	0000	0000	0000	0000	0000	0000	0002	0006	0010	0025
10	0000	0000	0000	0000	0000	0000	0001	0005	0009	0023
15	0000	0000	0000	0000	0000	0000	0001	0005	0008	0021
20	0000	0000	0000	0000	0000	0000	0001	0004	0008	0019
25	0000	0000	0000	0000	0000	0000	0001	0004	0007	0017
30	0000	0000	0000	0000	0000	0000	0001	0004	0006	0016
35	0000	0000	0000	0000	0000	0000	0001	0003	0006	0015
40	0000	0000	0000	0000	0000	0000	0001	0003	0005	0014
45	0000	0000	0000	0000	0000	0000	0001	0003	0005	0012
50	0000	0000	0000	0000	0000	0000	0001	0003	0004	0011
55	0000	0000	0000	0000	0000	0000	0001	0002	0004	0010
60	0000	0000	0000	0000	0000	0000	0001	0002	0004	0010
65	0000	0000	0000	0000	0000	0000	0000	0002	0003	0009
70	0000	0000	0000	0000	0000	0000	0000	0002	0003	0008
75	0000	0000	0000	0000	0000	0000	0000	0002	0003	0007
80	0000	0000	0000	0000	0000	0000	0000	0001	0003	0007
85	0000	0000	0000	0000	0000	0000	0000	0001	0002	0006
90	0000	0000	0000	0000	0000	0000	0000	0001	0002	0006
95	0000	0000	0000	0000	0000	0000	0000	0001	0002	0005
7.00	0000	0000	0000	0000	0000	0000	0000	0001	0002	0005
05	0000	0000	0000	0000	0000	0000	0000	0001	0002	0005
10	0000	0000	0000	0000	0000	0000	0000	0001	0002	0004
15	0000	0000	0000	0000	0000	0000	0000	0001	0001	0004
20	0000	0000	0000	0000	0000	0000	0000	0001	0001	0004
25	0000	0000	0000	0000	0000	0000	0000	0001	0001	0003
30	0000	0000	0000	0000	0000	0000	0000	0001	0001	0003
35	0000	0000	0000	0000	0000	0000	0000	0001	0001	0003
40	0000	0000	0000	0000	0000	0000	0000	0001	0001	0003
45	0000	0000	0000	0000	0000	0000	0000	0000	0001	0002
50	0000	0000	0000	0000	0000	0000	0000	0000	0001	0002
55	0000	0000	0000	0000	0000	0000	0000	0000	0001	0002
60	0000	0000	0000	0000	0000	0000	0000	0000	0001	0002
65	0000	0000	0000	0000	0000	0000	0000	0000	0001	0002
70	0000	0000	0000	0000	0000	0000	0000	0000	0001	0002
75	0000	0000	0000	0000	0000	0000	0000	0000	0001	0001
80	0000	0000	0000	0000	0000	0000	0000	0000	0000	0001
85	0000	0000	0000	0000	0000	0000	0000	0000	0000	0001
90	0000	0000	0000	0000	0000	0000	0000	0000	0000	0001
95	0000	0000	0000	0000	0000	0000	0000	0000	0000	0001
8.00	0000	0000	0000	0000	0000	0000	0000	0000	0000	0001
05	0000	0000	0000	0000	0000	0000	0000	0000	0000	0001
10	0000	0000	0000	0000	0000	0000	0000	0000	0000	0001
15	0000	0000	0000	0000	0000	0000	0000	0000	0000	0001
20	0000	0000	0000	0000	0000	0000	0000	0000	0000	0001
25	0000	0000	0000	0000	0000	0000	0000	0000	0000	0001
30	0000	0000	0000	0000	0000	0000	0000	0000	0000	0001
35	0000	0000	0000	0000	0000	0000	0000	0000	0000	0001
40	0000	0000	0000	0000	0000	0000	0000	0000	0000	0001
45	0000	0000	0000	0000	0000	0000	0000	0000	0000	0000

PROBABILITY DENSITY OF t, THE NON-CENTRAL t-STATISTIC, TABLED AS A FUNCTION OF $t/\sqrt{f}$.

f is the number of degrees of freedom; the non-centrality parameter is $\sqrt{f+1}\,K_p$.
K_p is the standardized normal deviate exceeded with probability p.

DEGREES OF FREEDOM **15**

$t/\sqrt{f}$ \ p	.2500	.1500	.1000	.0650	.0400	.0250	.0100	.0040	.0025	.0010
-0.45	0000	0000	0000	0000	0000	0000	0000	0000	0000	0000
40	0001	0000	0000	0000	0000	0000	0000	0000	0000	0000
35	0001	0000	0000	0000	0000	0000	0000	0000	0000	0000
30	0003	0000	0000	0000	0000	0000	0000	0000	0000	0000
25	0005	0000	0000	0000	0000	0000	0000	0000	0000	0000
20	0010	0000	0000	0000	0000	0000	0000	0000	0000	0000
15	0019	0000	0000	0000	0000	0000	0000	0000	0000	0000
10	0034	0000	0000	0000	0000	0000	0000	0000	0000	0000
05	0060	0000	0000	0000	0000	0000	0000	0000	0000	0000
0.00	0103	0001	0000	0000	0000	0000	0000	0000	0000	0000
05	0172	0002	0000	0000	0000	0000	0000	0000	0000	0000
10	0279	0004	0000	0000	0000	0000	0000	0000	0000	0000
15	0434	0008	0000	0000	0000	0000	0000	0000	0000	0000
20	0649	0015	0000	0000	0000	0000	0000	0000	0000	0000
25	0930	0030	0001	0000	0000	0000	0000	0000	0000	0000
30	1276	0057	0002	0000	0000	0000	0000	0000	0000	0000
35	1675	0103	0005	0000	0000	0000	0000	0000	0000	0000
40	2104	0177	0011	0000	0000	0000	0000	0000	0000	0000
45	2530	0290	0023	0001	0000	0000	0000	0000	0000	0000
50	2917	0450	0045	0002	0000	0000	0000	0000	0000	0000
55	3230	0665	0082	0005	0000	0000	0000	0000	0000	0000
60	3442	0936	0143	0012	0000	0000	0000	0000	0000	0000
65	3539	1257	0237	0024	0001	0000	0000	0000	0000	0000
70	3518	1611	0370	0046	0003	0000	0000	0000	0000	0000
75	3392	1978	0549	0083	0006	0000	0000	0000	0000	0000
80	3178	2331	0775	0142	0013	0001	0000	0000	0000	0000
85	2902	2645	1045	0230	0027	0002	0000	0000	0000	0000
90	2590	2896	1346	0353	0050	0005	0000	0000	0000	0000
95	2263	3069	1664	0514	0087	0011	0000	0000	0000	0000
1.00	1941	3156	1979	0714	0144	0022	0000	0000	0000	0000
05	1638	3158	2270	0949	0227	0041	0001	0000	0000	0000
10	1362	3081	2520	1210	0339	0071	0002	0000	0000	0000
15	1119	2940	2713	1485	0483	0118	0004	0000	0000	0000
20	0908	2748	2842	1759	0659	0185	0008	0000	0000	0000
25	0731	2522	2903	2017	0862	0275	0016	0001	0000	0000
30	0583	2276	2897	2244	1087	0392	0029	0001	0000	0000
35	0462	2025	2832	2430	1323	0535	0049	0003	0000	0000
40	0364	1777	2716	2565	1559	0703	0081	0005	0001	0000
45	0285	1542	2561	2648	1785	0892	0125	0010	0002	0000
50	0222	1325	2379	2677	1990	1094	0186	0018	0005	0000
55	0173	1127	2178	2655	2164	1302	0265	0031	0009	0001
60	0134	0952	1971	2589	2301	1508	0363	0051	0016	0001
65	0104	0798	1763	2486	2396	1702	0480	0079	0027	0003
70	0080	0665	1562	2354	2449	1876	0615	0118	0044	0005
75	0062	0551	1371	2200	2461	2024	0765	0170	0068	0009
80	0048	0455	1194	2033	2435	2142	0925	0236	0101	0016
85	0037	0374	1033	1860	2375	2225	1091	0316	0145	0026
90	0029	0306	0888	1685	2286	2275	1257	0411	0201	0040
95	0022	0250	0759	1514	2176	2291	1416	0520	0270	0061
2.00	0017	0204	0646	1350	2048	2275	1565	0641	0352	0089
05	0013	0166	0548	1196	1910	2231	1697	0770	0446	0125
10	0010	0135	0462	1053	1765	2163	1810	0906	0551	0171
15	0008	0109	0389	0922	1618	2075	1901	1043	0666	0227
20	0006	0089	0327	0804	1473	1972	1968	1179	0788	0293
25	0005	0072	0273	0697	1332	1858	2011	1309	0913	0370
30	0004	0058	0229	0603	1198	1736	2030	1429	1039	0456
35	0003	0047	0191	0520	1071	1610	2026	1538	1162	0550
40	0002	0038	0159	0446	0953	1484	2002	1631	1279	0651
45	0002	0031	0132	0382	0845	1360	1959	1708	1387	0757

PROBABILITY DENSITY OF t, THE NON-CENTRAL t-STATISTIC, TABLED AS A FUNCTION OF $t/\sqrt{f}$.

f is the number of degrees of freedom; the non-centrality parameter is $\sqrt{f+1}\,K_p$.
K_p is the standardized normal deviate exceeded with probability p.

DEGREES OF FREEDOM **15**

$t/\sqrt{f}$ \ p	.2500	.1500	.1000	.0650	.0400	.0250	.0100	.0040	.0025	.0010
2.50	0001	0025	0110	0327	0746	1239	1901	1767	1484	0865
55	0001	0020	0091	0279	0656	1123	1829	1808	1568	0973
60	0001	0017	0076	0238	0575	1013	1748	1832	1637	1078
65	0001	0013	0063	0202	0503	0911	1659	1838	1690	1179
70	0001	0011	0052	0172	0439	0815	1564	1827	1728	1272
75	0000	0009	0044	0146	0382	0728	1467	1802	1750	1356
80	0000	0007	0036	0124	0332	0647	1368	1765	1757	1431
85	0000	0006	0030	0105	0288	0574	1270	1715	1749	1494
90	0000	0005	0025	0089	0249	0508	1173	1657	1729	1544
95	0000	0004	0021	0075	0216	0449	1080	1591	1696	1582
3.00	0000	0003	0017	0064	0186	0396	0990	1519	1654	1608
05	0000	0003	0014	0054	0161	0348	0905	1442	1603	1621
10	0000	0002	0012	0046	0139	0306	0824	1364	1545	1623
15	0000	0002	0010	0039	0120	0268	0749	1283	1481	1613
20	0000	0001	0008	0033	0103	0235	0678	1203	1413	1594
25	0000	0001	0007	0028	0089	0206	0613	1123	1342	1565
30	0000	0001	0006	0024	0077	0180	0553	1045	1269	1529
35	0000	0001	0005	0020	0066	0157	0497	0970	1196	1486
40	0000	0001	0004	0017	0057	0137	0447	0897	1123	1437
45	0000	0001	0003	0014	0049	0120	0401	0827	1050	1383
50	0000	0000	0003	0012	0042	0105	0359	0761	0980	1326
55	0000	0000	0002	0010	0036	0091	0321	0698	0911	1266
60	0000	0000	0002	0009	0031	0080	0287	0639	0845	1205
65	0000	0000	0002	0008	0027	0069	0256	0584	0782	1142
70	0000	0000	0001	0006	0023	0060	0228	0533	0722	1080
75	0000	0000	0001	0005	0020	0053	0203	0485	0665	1018
80	0000	0000	0001	0005	0017	0046	0181	0441	0612	0956
85	0000	0000	0001	0004	0015	0040	0161	0401	0562	0896
90	0000	0000	0001	0003	0013	0035	0143	0363	0515	0838
95	0000	0000	0001	0003	0011	0030	0127	0329	0471	0782
4.00	0000	0000	0001	0003	0010	0027	0113	0298	0430	0728
05	0000	0000	0000	0002	0008	0023	0100	0269	0392	0677
10	0000	0000	0000	0002	0007	0020	0089	0243	0358	0628
15	0000	0000	0000	0002	0006	0018	0079	0219	0326	0581
20	0000	0000	0000	0001	0005	0015	0070	0198	0296	0538
25	0000	0000	0000	0001	0005	0013	0062	0178	0269	0496
30	0000	0000	0000	0001	0004	0012	0055	0161	0244	0458
35	0000	0000	0000	0001	0004	0010	0049	0145	0222	0421
40	0000	0000	0000	0001	0003	0009	0043	0130	0201	0388
45	0000	0000	0000	0001	0003	0008	0039	0117	0182	0356
50	0000	0000	0000	0001	0002	0007	0034	0105	0165	0327
55	0000	0000	0000	0000	0002	0006	0030	0095	0149	0300
60	0000	0000	0000	0000	0002	0005	0027	0085	0135	0275
65	0000	0000	0000	0000	0002	0005	0024	0076	0122	0252
70	0000	0000	0000	0000	0001	0004	0021	0069	0110	0230
75	0000	0000	0000	0000	0001	0004	0019	0062	0100	0211
80	0000	0000	0000	0000	0001	0003	0017	0055	0090	0192
85	0000	0000	0000	0000	0001	0003	0015	0050	0081	0176
90	0000	0000	0000	0000	0001	0002	0013	0045	0074	0160
95	0000	0000	0000	0000	0001	0002	0012	0040	0066	0146
5.00	0000	0000	0000	0000	0001	0002	0010	0036	0060	0134
05	0000	0000	0000	0000	0001	0002	0009	0032	0054	0122
10	0000	0000	0000	0000	0000	0001	0008	0029	0049	0111
15	0000	0000	0000	0000	0000	0001	0007	0026	0044	0101
20	0000	0000	0000	0000	0000	0001	0007	0024	0040	0092
25	0000	0000	0000	0000	0000	0001	0006	0021	0036	0084
30	0000	0000	0000	0000	0000	0001	0005	0019	0033	0077
35	0000	0000	0000	0000	0000	0001	0005	0017	0029	0070
40	0000	0000	0000	0000	0000	0001	0004	0015	0027	0063
45	0000	0000	0000	0000	0000	0001	0004	0014	0024	0058

PROBABILITY DENSITY OF t, THE NON-CENTRAL t-STATISTIC, TABLED AS A FUNCTION OF $t/\sqrt{f}$.

f is the number of degrees of freedom; the non-centrality parameter is $\sqrt{f+1}\, K_p$.
K_p is the standardized normal deviate exceeded with probability p.

DEGREES OF FREEDOM **15**

$t/\sqrt{f}$ \ p	.2500	.1500	.1000	.0650	.0400	.0250	.0100	.0040	.0025	.0010
5.50	0000	0000	0000	0000	0000	0001	0003	0012	0022	0053
55	0000	0000	0000	0000	0000	0000	0003	0011	0020	0048
60	0000	0000	0000	0000	0000	0000	0003	0010	0018	0044
65	0000	0000	0000	0000	0000	0000	0002	0009	0016	0040
70	0000	0000	0000	0000	0000	0000	0002	0008	0014	0036
75	0000	0000	0000	0000	0000	0000	0002	0007	0013	0033
80	0000	0000	0000	0000	0000	0000	0002	0007	0012	0030
85	0000	0000	0000	0000	0000	0000	0002	0006	0011	0027
90	0000	0000	0000	0000	0000	0000	0001	0005	0010	0025
95	0000	0000	0000	0000	0000	0000	0001	0005	0009	0023
6.00	0000	0000	0000	0000	0000	0000	0001	0004	0008	0021
05	0000	0000	0000	0000	0000	0000	0001	0004	0007	0019
10	0000	0000	0000	0000	0000	0000	0001	0004	0007	0017
15	0000	0000	0000	0000	0000	0000	0001	0003	0006	0016
20	0000	0000	0000	0000	0000	0000	0001	0003	0005	0014
25	0000	0000	0000	0000	0000	0000	0001	0003	0005	0013
30	0000	0000	0000	0000	0000	0000	0001	0002	0004	0012
35	0000	0000	0000	0000	0000	0000	0001	0002	0004	0011
40	0000	0000	0000	0000	0000	0000	0000	0002	0004	0010
45	0000	0000	0000	0000	0000	0000	0000	0002	0003	0009
50	0000	0000	0000	0000	0000	0000	0000	0002	0003	0008
55	0000	0000	0000	0000	0000	0000	0000	0001	0003	0007
60	0000	0000	0000	0000	0000	0000	0000	0001	0002	0007
65	0000	0000	0000	0000	0000	0000	0000	0001	0002	0006
70	0000	0000	0000	0000	0000	0000	0000	0001	0002	0006
75	0000	0000	0000	0000	0000	0000	0000	0001	0002	0005
80	0000	0000	0000	0000	0000	0000	0000	0001	0002	0005
85	0000	0000	0000	0000	0000	0000	0000	0001	0002	0004
90	0000	0000	0000	0000	0000	0000	0000	0001	0001	0004
95	0000	0000	0000	0000	0000	0000	0000	0001	0001	0004
7.00	0000	0000	0000	0000	0000	0000	0000	0001	0001	0003
05	0000	0000	0000	0000	0000	0000	0000	0001	0001	0003
10	0000	0000	0000	0000	0000	0000	0000	0001	0001	0003
15	0000	0000	0000	0000	0000	0000	0000	0000	0001	0003
20	0000	0000	0000	0000	0000	0000	0000	0000	0001	0002
25	0000	0000	0000	0000	0000	0000	0000	0000	0001	0002
30	0000	0000	0000	0000	0000	0000	0000	0000	0001	0002
35	0000	0000	0000	0000	0000	0000	0000	0000	0001	0002
40	0000	0000	0000	0000	0000	0000	0000	0000	0001	0002
45	0000	0000	0000	0000	0000	0000	0000	0000	0001	0002
50	0000	0000	0000	0000	0000	0000	0000	0000	0000	0001
55	0000	0000	0000	0000	0000	0000	0000	0000	0000	0001
60	0000	0000	0000	0000	0000	0000	0000	0000	0000	0001
65	0000	0000	0000	0000	0000	0000	0000	0000	0000	0001
70	0000	0000	0000	0000	0000	0000	0000	0000	0000	0001
75	0000	0000	0000	0000	0000	0000	0000	0000	0000	0001
80	0000	0000	0000	0000	0000	0000	0000	0000	0000	0001
85	0000	0000	0000	0000	0000	0000	0000	0000	0000	0001
90	0000	0000	0000	0000	0000	0000	0000	0000	0000	0001
95	0000	0000	0000	0000	0000	0000	0000	0000	0000	0001
8.00	0000	0000	0000	0000	0000	0000	0000	0000	0000	0001
05	0000	0000	0000	0000	0000	0000	0000	0000	0000	0001
10	0000	0000	0000	0000	0000	0000	0000	0000	0000	0001
15	0000	0000	0000	0000	0000	0000	0000	0000	0000	0000

PROBABILITY DENSITY OF t, THE NON-CENTRAL t-STATISTIC, TABLED AS A FUNCTION OF $t/\sqrt{f}$.

f is the number of degrees of freedom; the non-centrality parameter is $\sqrt{f+1}\,K_p$.
K_p is the standardized normal deviate exceeded with probability p.

DEGREES OF FREEDOM **16**

$t/\sqrt{f}$ \ p	.2500	.1500	.1000	.0650	.0400	.0250	.0100	.0040	.0025	.0010
-0.40	0000	0000	0000	0000	0000	0000	0000	0000	0000	0000
35	0001	0000	0000	0000	0000	0000	0000	0000	0000	0000
30	0002	0000	0000	0000	0000	0000	0000	0000	0000	0000
25	0004	0000	0000	0000	0000	0000	0000	0000	0000	0000
20	0007	0000	0000	0000	0000	0000	0000	0000	0000	0000
15	0013	0000	0000	0000	0000	0000	0000	0000	0000	0000
10	0025	0000	0000	0000	0000	0000	0000	0000	0000	0000
05	0046	0000	0000	0000	0000	0000	0000	0000	0000	0000
0.00	0082	0000	0000	0000	0000	0000	0000	0000	0000	0000
05	0142	0001	0000	0000	0000	0000	0000	0000	0000	0000
10	0237	0002	0000	0000	0000	0000	0000	0000	0000	0000
15	0379	0005	0000	0000	0000	0000	0000	0000	0000	0000
20	0582	0011	0000	0000	0000	0000	0000	0000	0000	0000
25	0853	0023	0001	0000	0000	0000	0000	0000	0000	0000
30	1195	0044	0001	0000	0000	0000	0000	0000	0000	0000
35	1596	0083	0003	0000	0000	0000	0000	0000	0000	0000
40	2034	0147	0008	0000	0000	0000	0000	0000	0000	0000
45	2475	0248	0017	0001	0000	0000	0000	0000	0000	0000
50	2881	0397	0034	0001	0000	0000	0000	0000	0000	0000
55	3212	0601	0065	0004	0000	0000	0000	0000	0000	0000
60	3438	0865	0118	0008	0000	0000	0000	0000	0000	0000
65	3542	1183	0201	0018	0001	0000	0000	0000	0000	0000
70	3521	1541	0324	0035	0002	0000	0000	0000	0000	0000
75	3387	1918	0493	0066	0004	0000	0000	0000	0000	0000
80	3162	2285	0712	0118	0010	0001	0000	0000	0000	0000
85	2873	2614	0977	0196	0020	0002	0000	0000	0000	0000
90	2546	2880	1280	0309	0039	0004	0000	0000	0000	0000
95	2206	3064	1604	0462	0070	0008	0000	0000	0000	0000
1.00	1875	3158	1929	0656	0120	0016	0000	0000	0000	0000
05	1566	3161	2234	0888	0195	0032	0000	0000	0000	0000
10	1288	3081	2497	1150	0299	0057	0001	0000	0000	0000
15	1044	2931	2702	1430	0435	0098	0003	0000	0000	0000
20	0837	2729	2840	1712	0605	0157	0006	0000	0000	0000
25	0664	2492	2905	1980	0806	0240	0012	0000	0000	0000
30	0523	2235	2900	2219	1031	0350	0022	0001	0000	0000
35	0408	1974	2831	2415	1271	0488	0039	0002	0000	0000
40	0317	1720	2709	2560	1514	0652	0066	0004	0001	0000
45	0245	1479	2546	2648	1749	0839	0105	0007	0002	0000
50	0188	1259	2354	2679	1963	1044	0160	0014	0003	0000
55	0144	1061	2145	2657	2147	1256	0232	0024	0006	0000
60	0110	0886	1928	2588	2292	1468	0325	0041	0012	0001
65	0084	0735	1714	2479	2394	1670	0438	0065	0021	0002
70	0064	0605	1507	2339	2451	1853	0570	0100	0035	0004
75	0049	0496	1312	2178	2464	2009	0719	0146	0055	0007
80	0037	0404	1133	2003	2436	2134	0880	0207	0084	0012
85	0028	0328	0972	1822	2373	2223	1049	0283	0124	0020
90	0021	0266	0828	1641	2280	2276	1219	0374	0175	0032
95	0016	0214	0701	1465	2163	2293	1385	0480	0240	0049
2.00	0012	0173	0590	1298	2029	2277	1540	0600	0318	0074
05	0009	0139	0495	1141	1884	2231	1679	0729	0409	0106
10	0007	0111	0414	0997	1733	2159	1798	0867	0513	0148
15	0006	0089	0345	0866	1580	2066	1895	1007	0627	0200
20	0004	0071	0286	0748	1430	1958	1966	1147	0749	0263
25	0003	0057	0237	0644	1285	1837	2012	1282	0877	0337
30	0002	0046	0196	0551	1148	1710	2032	1408	1006	0421
35	0002	0036	0162	0471	1020	1579	2028	1521	1133	0514
40	0001	0029	0133	0401	0901	1448	2003	1620	1255	0615
45	0001	0023	0110	0340	0793	1319	1958	1702	1368	0722

PROBABILITY DENSITY OF t, THE NON-CENTRAL t-STATISTIC, TABLED AS A FUNCTION OF $t/\sqrt{f}$.

f is the number of degrees of freedom; the non-centrality parameter is $\sqrt{f+1}\,K_p$.
K_p is the standardized normal deviate exceeded with probability p.

DEGREES OF FREEDOM **16**

$t/\sqrt{f}$ \ p	.2500	.1500	.1000	.0650	.0400	.0250	.0100	.0040	.0025	.0010
2.50	0001	0019	0090	0288	0694	1195	1896	1765	1470	0832
55	0001	0015	0074	0243	0606	1077	1821	1809	1558	0943
60	0001	0012	0061	0205	0527	0966	1735	1833	1631	1051
65	0000	0010	0050	0173	0457	0862	1641	1840	1688	1156
70	0000	0008	0041	0145	0395	0767	1542	1829	1728	1254
75	0000	0006	0034	0122	0341	0679	1440	1803	1752	1342
80	0000	0005	0028	0103	0294	0600	1338	1763	1759	1421
85	0000	0004	0023	0086	0253	0528	1236	1711	1751	1487
90	0000	0003	0019	0072	0217	0464	1137	1649	1729	1541
95	0000	0003	0015	0061	0186	0407	1041	1579	1695	1582
3.00	0000	0002	0013	0051	0159	0356	0949	1504	1650	1609
05	0000	0002	0010	0043	0136	0311	0863	1424	1597	1623
10	0000	0001	0009	0036	0116	0271	0781	1341	1535	1625
15	0000	0001	0007	0030	0099	0235	0705	1258	1468	1615
20	0000	0001	0006	0025	0085	0205	0635	1174	1397	1595
25	0000	0001	0005	0021	0072	0178	0570	1092	1322	1564
30	0000	0001	0004	0018	0062	0154	0511	1011	1246	1526
35	0000	0001	0003	0015	0053	0134	0457	0934	1170	1480
40	0000	0000	0003	0012	0045	0116	0408	0859	1094	1428
45	0000	0000	0002	0010	0038	0100	0363	0789	1020	1372
50	0000	0000	0002	0009	0033	0087	0323	0722	0947	1312
55	0000	0000	0002	0007	0028	0075	0287	0659	0877	1249
60	0000	0000	0001	0006	0024	0065	0255	0600	0810	1185
65	0000	0000	0001	0005	0020	0056	0226	0545	0746	1120
70	0000	0000	0001	0004	0017	0048	0200	0495	0685	1055
75	0000	0000	0001	0004	0015	0042	0177	0448	0628	0990
80	0000	0000	0001	0003	0013	0036	0156	0405	0574	0927
85	0000	0000	0001	0003	0011	0031	0138	0365	0524	0866
90	0000	0000	0000	0002	0009	0027	0122	0329	0478	0806
95	0000	0000	0000	0002	0008	0023	0107	0297	0435	0749
4.00	0000	0000	0000	0002	0007	0020	0095	0267	0395	0694
05	0000	0000	0000	0001	0006	0017	0083	0240	0358	0642
10	0000	0000	0000	0001	0005	0015	0073	0215	0325	0593
15	0000	0000	0000	0001	0004	0013	0065	0193	0294	0547
20	0000	0000	0000	0001	0004	0011	0057	0173	0266	0503
25	0000	0000	0000	0001	0003	0010	0050	0155	0240	0462
30	0000	0000	0000	0001	0003	0009	0044	0138	0216	0424
35	0000	0000	0000	0001	0002	0007	0039	0124	0195	0388
40	0000	0000	0000	0000	0002	0006	0034	0111	0176	0355
45	0000	0000	0000	0000	0002	0006	0030	0099	0158	0325
50	0000	0000	0000	0000	0002	0005	0027	0088	0143	0297
55	0000	0000	0000	0000	0001	0004	0023	0079	0128	0271
60	0000	0000	0000	0000	0001	0004	0021	0070	0115	0247
65	0000	0000	0000	0000	0001	0003	0018	0063	0104	0225
70	0000	0000	0000	0000	0001	0003	0016	0056	0093	0204
75	0000	0000	0000	0000	0001	0002	0014	0050	0084	0186
80	0000	0000	0000	0000	0001	0002	0012	0045	0075	0169
85	0000	0000	0000	0000	0001	0002	0011	0040	0067	0153
90	0000	0000	0000	0000	0000	0002	0010	0036	0060	0139
95	0000	0000	0000	0000	0000	0001	0009	0032	0054	0126
5.00	0000	0000	0000	0000	0000	0001	0008	0028	0049	0115
05	0000	0000	0000	0000	0000	0001	0007	0025	0044	0104
10	0000	0000	0000	0000	0000	0001	0006	0023	0039	0094
15	0000	0000	0000	0000	0000	0001	0005	0020	0035	0085
20	0000	0000	0000	0000	0000	0001	0005	0018	0032	0077
25	0000	0000	0000	0000	0000	0001	0004	0016	0028	0070
30	0000	0000	0000	0000	0000	0001	0004	0014	0025	0063
35	0000	0000	0000	0000	0000	0000	0003	0013	0023	0057
40	0000	0000	0000	0000	0000	0000	0003	0011	0020	0052
45	0000	0000	0000	0000	0000	0000	0003	0010	0018	0047

PROBABILITY DENSITY OF t, THE NON-CENTRAL t-STATISTIC, TABLED AS A FUNCTION OF $t/\sqrt{f}$.

f is the number of degrees of freedom; the non-centrality parameter is $\sqrt{f+1}\,K_p$.
K_p is the standardized normal deviate exceeded with probability p.

$t/\sqrt{f}$ \ p	.2500	.1500	.1000	.0650	.0400	.0250	.0100	.0040	.0025	.0010
5.50	0000	0000	0000	0000	0000	0000	0002	0009	0017	0043
55	0000	0000	0000	0000	0000	0000	0002	0008	0015	0039
60	0000	0000	0000	0000	0000	0000	0002	0007	0013	0035
65	0000	0000	0000	0000	0000	0000	0002	0007	0012	0032
70	0000	0000	0000	0000	0000	0000	0001	0006	0011	0029
75	0000	0000	0000	0000	0000	0000	0001	0005	0010	0026
80	0000	0000	0000	0000	0000	0000	0001	0005	0009	0023
85	0000	0000	0000	0000	0000	0000	0001	0004	0008	0021
90	0000	0000	0000	0000	0000	0000	0001	0004	0007	0019
95	0000	0000	0000	0000	0000	0000	0001	0003	0006	0017
6.00	0000	0000	0000	0000	0000	0000	0001	0003	0006	0016
05	0000	0000	0000	0000	0000	0000	0001	0003	0005	0014
10	0000	0000	0000	0000	0000	0000	0001	0002	0005	0013
15	0000	0000	0000	0000	0000	0000	0000	0002	0004	0012
20	0000	0000	0000	0000	0000	0000	0000	0002	0004	0011
25	0000	0000	0000	0000	0000	0000	0000	0002	0003	0010
30	0000	0000	0000	0000	0000	0000	0000	0002	0003	0009
35	0000	0000	0000	0000	0000	0000	0000	0001	0003	0008
40	0000	0000	0000	0000	0000	0000	0000	0001	0002	0007
45	0000	0000	0000	0000	0000	0000	0000	0001	0002	0007
50	0000	0000	0000	0000	0000	0000	0000	0001	0002	0006
55	0000	0000	0000	0000	0000	0000	0000	0001	0002	0005
60	0000	0000	0000	0000	0000	0000	0000	0001	0002	0005
65	0000	0000	0000	0000	0000	0000	0000	0001	0001	0004
70	0000	0000	0000	0000	0000	0000	0000	0001	0001	0004
75	0000	0000	0000	0000	0000	0000	0000	0001	0001	0004
80	0000	0000	0000	0000	0000	0000	0000	0001	0001	0003
85	0000	0000	0000	0000	0000	0000	0000	0001	0001	0003
90	0000	0000	0000	0000	0000	0000	0000	0000	0001	0003
95	0000	0000	0000	0000	0000	0000	0000	0000	0001	0002
7.00	0000	0000	0000	0000	0000	0000	0000	0000	0001	0002
05	0000	0000	0000	0000	0000	0000	0000	0000	0001	0002
10	0000	0000	0000	0000	0000	0000	0000	0000	0001	0002
15	0000	0000	0000	0000	0000	0000	0000	0000	0001	0002
20	0000	0000	0000	0000	0000	0000	0000	0000	0001	0002
25	0000	0000	0000	0000	0000	0000	0000	0000	0000	0001
30	0000	0000	0000	0000	0000	0000	0000	0000	0000	0001
35	0000	0000	0000	0000	0000	0000	0000	0000	0000	0001
40	0000	0000	0000	0000	0000	0000	0000	0000	0000	0001
45	0000	0000	0000	0000	0000	0000	0000	0000	0000	0001
50	0000	0000	0000	0000	0000	0000	0000	0000	0000	0001
55	0000	0000	0000	0000	0000	0000	0000	0000	0000	0001
60	0000	0000	0000	0000	0000	0000	0000	0000	0000	0001
65	0000	0000	0000	0000	0000	0000	0000	0000	0000	0001
70	0000	0000	0000	0000	0000	0000	0000	0000	0000	0001
75	0000	0000	0000	0000	0000	0000	0000	0000	0000	0001
80	0000	0000	0000	0000	0000	0000	0000	0000	0000	0001
85	0000	0000	0000	0000	0000	0000	0000	0000	0000	0000

PROBABILITY DENSITY OF t, THE NON-CENTRAL t-STATISTIC, TABLED AS A FUNCTION OF $t/\sqrt{f}$.

f is the number of degrees of freedom; the non-centrality parameter is $\sqrt{f+1}\,K_p$.
K_p is the standardized normal deviate exceeded with probability p.

DEGREES OF FREEDOM **17**

$t/\sqrt{f}$ \ p	.2500	.1500	.1000	.0650	.0400	.0250	.0100	.0040	.0025	.0010
-0.40	0000	0000	0000	0000	0000	0000	0000	0000	0000	0000
35	0001	0000	0000	0000	0000	0000	0000	0000	0000	0000
30	0001	0000	0000	0000	0000	0000	0000	0000	0000	0000
25	0002	0000	0000	0000	0000	0000	0000	0000	0000	0000
20	0005	0000	0000	0000	0000	0000	0000	0000	0000	0000
15	0009	0000	0000	0000	0000	0000	0000	0000	0000	0000
10	0019	0000	0000	0000	0000	0000	0000	0000	0000	0000
05	0035	0000	0000	0000	0000	0000	0000	0000	0000	0000
0.00	0066	0000	0000	0000	0000	0000	0000	0000	0000	0000
05	0117	0001	0000	0000	0000	0000	0000	0000	0000	0000
10	0201	0001	0000	0000	0000	0000	0000	0000	0000	0000
15	0331	0003	0000	0000	0000	0000	0000	0000	0000	0000
20	0522	0008	0000	0000	0000	0000	0000	0000	0000	0000
25	0783	0017	0000	0000	0000	0000	0000	0000	0000	0000
30	1118	0034	0001	0000	0000	0000	0000	0000	0000	0000
35	1520	0067	0002	0000	0000	0000	0000	0000	0000	0000
40	1966	0122	0005	0000	0000	0000	0000	0000	0000	0000
45	2421	0213	0012	0000	0000	0000	0000	0000	0000	0000
50	2844	0350	0026	0001	0000	0000	0000	0000	0000	0000
55	3193	0543	0052	0002	0000	0000	0000	0000	0000	0000
60	3433	0799	0097	0006	0000	0000	0000	0000	0000	0000
65	3544	1114	0171	0013	0000	0000	0000	0000	0000	0000
70	3523	1475	0283	0027	0001	0000	0000	0000	0000	0000
75	3383	1860	0442	0053	0003	0000	0000	0000	0000	0000
80	3146	2239	0653	0097	0007	0000	0000	0000	0000	0000
85	2843	2583	0914	0168	0015	0001	0000	0000	0000	0000
90	2502	2864	1217	0271	0030	0002	0000	0000	0000	0000
95	2151	3059	1547	0415	0057	0006	0000	0000	0000	0000
1.00	1811	3159	1881	0602	0100	0012	0000	0000	0000	0000
05	1497	3163	2198	0830	0167	0025	0000	0000	0000	0000
10	1217	3080	2474	1092	0263	0046	0001	0000	0000	0000
15	0975	2922	2691	1376	0392	0081	0002	0000	0000	0000
20	0772	2710	2837	1666	0556	0134	0004	0000	0000	0000
25	0604	2462	2907	1944	0753	0210	0009	0000	0000	0000
30	0469	2195	2902	2194	0978	0312	0017	0000	0000	0000
35	0361	1925	2830	2401	1221	0444	0031	0001	0000	0000
40	0276	1663	2702	2554	1470	0605	0053	0003	0000	0000
45	0210	1419	2531	2648	1713	0790	0088	0005	0001	0000
50	0159	1196	2330	2682	1937	0995	0137	0010	0002	0000
55	0120	0998	2111	2659	2130	1212	0204	0019	0005	0000
60	0090	0825	1887	2586	2283	1429	0291	0032	0009	0000
65	0068	0677	1665	2472	2391	1639	0400	0053	0016	0001
70	0051	0551	1453	2325	2452	1830	0529	0084	0028	0002
75	0038	0446	1256	2156	2466	1994	0676	0126	0045	0005
80	0028	0360	1076	1974	2438	2126	0838	0182	0070	0009
85	0021	0289	0914	1786	2371	2221	1009	0253	0106	0015
90	0016	0231	0771	1599	2273	2277	1183	0340	0153	0025
95	0012	0184	0647	1418	2150	2296	1354	0443	0213	0040
2.00	0009	0146	0539	1247	2010	2279	1515	0561	0287	0061
05	0007	0116	0448	1088	1858	2230	1661	0690	0375	0090
10	0005	0092	0370	0943	1701	2155	1786	0829	0477	0128
15	0004	0073	0305	0813	1543	2057	1888	0972	0590	0177
20	0003	0057	0251	0696	1389	1943	1964	1115	0712	0236
25	0002	0045	0205	0594	1240	1817	2012	1255	0841	0307
30	0002	0036	0168	0504	1101	1684	2034	1386	0973	0389
35	0001	0028	0137	0426	0971	1548	2031	1505	1104	0480
40	0001	0022	0111	0360	0852	1413	2004	1609	1230	0581
45	0001	0018	0091	0302	0744	1280	1956	1695	1348	0688

PROBABILITY DENSITY OF t, THE NON-CENTRAL t-STATISTIC, TABLED AS A FUNCTION OF $t/\sqrt{f}$.

f is the number of degrees of freedom; the non-centrality parameter is $\sqrt{f+1}\,K_p$.
K_p is the standardized normal deviate exceeded with probability p.

DEGREES OF FREEDOM **17**

$t/\sqrt{f}$ \ p	.2500	.1500	.1000	.0650	.0400	.0250	.0100	.0040	.0025	.0010
2.50	0001	0014	0074	0253	0646	1153	1891	1762	1455	0800
55	0000	0011	0060	0212	0560	1033	1812	1809	1548	0913
60	0000	0009	0049	0177	0483	0920	1722	1835	1625	1025
65	0000	0007	0039	0148	0415	0816	1623	1842	1685	1133
70	0000	0005	0032	0123	0356	0721	1520	1831	1728	1235
75	0000	0004	0026	0102	0305	0634	1414	1803	1753	1328
80	0000	0003	0021	0085	0260	0556	1308	1761	1761	1411
85	0000	0003	0017	0071	0222	0486	1203	1706	1753	1481
90	0000	0002	0014	0059	0189	0424	1101	1641	1730	1538
95	0000	0002	0011	0049	0160	0369	1003	1568	1694	1581
3.00	0000	0001	0009	0040	0136	0320	0910	1488	1647	1610
05	0000	0001	0008	0033	0115	0277	0822	1405	1590	1625
10	0000	0001	0006	0028	0098	0239	0740	1319	1525	1627
15	0000	0001	0005	0023	0083	0207	0664	1232	1455	1617
20	0000	0001	0004	0019	0070	0178	0595	1146	1380	1595
25	0000	0000	0003	0016	0059	0153	0531	1061	1303	1563
30	0000	0000	0003	0013	0050	0132	0472	0978	1224	1522
35	0000	0000	0002	0011	0042	0113	0420	0899	1144	1474
40	0000	0000	0002	0009	0036	0097	0372	0824	1066	1420
45	0000	0000	0002	0008	0030	0084	0329	0752	0989	1360
50	0000	0000	0001	0006	0026	0072	0291	0685	0915	1297
55	0000	0000	0001	0005	0022	0061	0257	0622	0843	1232
60	0000	0000	0001	0004	0018	0053	0226	0563	0775	1165
65	0000	0000	0001	0004	0015	0045	0199	0509	0710	1098
70	0000	0000	0001	0003	0013	0039	0175	0459	0649	1030
75	0000	0000	0000	0003	0011	0033	0153	0413	0592	0964
80	0000	0000	0000	0002	0009	0028	0135	0371	0539	0899
85	0000	0000	0000	0002	0008	0024	0118	0333	0489	0836
90	0000	0000	0000	0002	0007	0021	0103	0298	0444	0775
95	0000	0000	0000	0001	0006	0018	0091	0267	0401	0717
4.00	0000	0000	0000	0001	0005	0015	0079	0239	0363	0662
05	0000	0000	0000	0001	0004	0013	0069	0213	0327	0609
10	0000	0000	0000	0001	0003	0011	0061	0190	0295	0560
15	0000	0000	0000	0001	0003	0010	0053	0169	0265	0514
20	0000	0000	0000	0001	0003	0008	0046	0151	0238	0470
25	0000	0000	0000	0000	0002	0007	0041	0134	0214	0430
30	0000	0000	0000	0000	0002	0006	0035	0119	0192	0393
35	0000	0000	0000	0000	0002	0005	0031	0106	0172	0358
40	0000	0000	0000	0000	0001	0005	0027	0094	0154	0326
45	0000	0000	0000	0000	0001	0004	0024	0084	0138	0296
50	0000	0000	0000	0000	0001	0003	0021	0074	0123	0269
55	0000	0000	0000	0000	0001	0003	0018	0066	0110	0244
60	0000	0000	0000	0000	0001	0003	0016	0058	0098	0221
65	0000	0000	0000	0000	0001	0002	0014	0052	0088	0200
70	0000	0000	0000	0000	0001	0002	0012	0046	0079	0181
75	0000	0000	0000	0000	0000	0002	0011	0041	0070	0164
80	0000	0000	0000	0000	0000	0001	0009	0036	0063	0148
85	0000	0000	0000	0000	0000	0001	0008	0032	0056	0134
90	0000	0000	0000	0000	0000	0001	0007	0028	0050	0121
95	0000	0000	0000	0000	0000	0001	0006	0025	0044	0109
5.00	0000	0000	0000	0000	0000	0001	0005	0022	0040	0098
05	0000	0000	0000	0000	0000	0001	0005	0020	0035	0089
10	0000	0000	0000	0000	0000	0001	0004	0017	0031	0080
15	0000	0000	0000	0000	0000	0001	0004	0016	0028	0072
20	0000	0000	0000	0000	0000	0000	0003	0014	0025	0065
25	0000	0000	0000	0000	0000	0000	0003	0012	0022	0058
30	0000	0000	0000	0000	0000	0000	0003	0011	0020	0053
35	0000	0000	0000	0000	0000	0000	0002	0010	0018	0047
40	0000	0000	0000	0000	0000	0000	0002	0009	0016	0043
45	0000	0000	0000	0000	0000	0000	0002	0008	0014	0038

PROBABILITY DENSITY OF t, THE NON-CENTRAL t-STATISTIC, TABLED AS A FUNCTION OF $t/\sqrt{f}$.

f is the number of degrees of freedom; the non-centrality parameter is $\sqrt{f+1}\,K_p$.
K_p is the standardized normal deviate exceeded with probability p.

DEGREES OF FREEDOM **17**

$t/\sqrt{f}$ \ p	.2500	.1500	.1000	.0650	.0400	.0250	.0100	.0040	.0025	.0010
5.50	0000	0000	0000	0000	0000	0000	0002	0007	0013	0034
55	0000	0000	0000	0000	0000	0000	0001	0006	0011	0031
60	0000	0000	0000	0000	0000	0000	0001	0005	0010	0028
65	0000	0000	0000	0000	0000	0000	0001	0005	0009	0025
70	0000	0000	0000	0000	0000	0000	0001	0004	0008	0023
75	0000	0000	0000	0000	0000	0000	0001	0004	0007	0020
80	0000	0000	0000	0000	0000	0000	0001	0003	0006	0018
85	0000	0000	0000	0000	0000	0000	0001	0003	0006	0016
90	0000	0000	0000	0000	0000	0000	0001	0003	0005	0015
95	0000	0000	0000	0000	0000	0000	0000	0002	0005	0013
6.00	0000	0000	0000	0000	0000	0000	0000	0002	0004	0012
05	0000	0000	0000	0000	0000	0000	0000	0002	0004	0011
10	0000	0000	0000	0000	0000	0000	0000	0002	0003	0010
15	0000	0000	0000	0000	0000	0000	0000	0001	0003	0009
20	0000	0000	0000	0000	0000	0000	0000	0001	0003	0008
25	0000	0000	0000	0000	0000	0000	0000	0001	0002	0007
30	0000	0000	0000	0000	0000	0000	0000	0001	0002	0006
35	0000	0000	0000	0000	0000	0000	0000	0001	0002	0006
40	0000	0000	0000	0000	0000	0000	0000	0001	0002	0005
45	0000	0000	0000	0000	0000	0000	0000	0001	0002	0005
50	0000	0000	0000	0000	0000	0000	0000	0001	0001	0004
55	0000	0000	0000	0000	0000	0000	0000	0001	0001	0004
60	0000	0000	0000	0000	0000	0000	0000	0001	0001	0003
65	0000	0000	0000	0000	0000	0000	0000	0000	0001	0003
70	0000	0000	0000	0000	0000	0000	0000	0000	0001	0003
75	0000	0000	0000	0000	0000	0000	0000	0000	0001	0003
80	0000	0000	0000	0000	0000	0000	0000	0000	0001	0002
85	0000	0000	0000	0000	0000	0000	0000	0000	0001	0002
90	0000	0000	0000	0000	0000	0000	0000	0000	0001	0002
95	0000	0000	0000	0000	0000	0000	0000	0000	0001	0002
7.00	0000	0000	0000	0000	0000	0000	0000	0000	0000	0002
05	0000	0000	0000	0000	0000	0000	0000	0000	0000	0001
10	0000	0000	0000	0000	0000	0000	0000	0000	0000	0001
15	0000	0000	0000	0000	0000	0000	0000	0000	0000	0001
20	0000	0000	0000	0000	0000	0000	0000	0000	0000	0001
25	0000	0000	0000	0000	0000	0000	0000	0000	0000	0001
30	0000	0000	0000	0000	0000	0000	0000	0000	0000	0001
35	0000	0000	0000	0000	0000	0000	0000	0000	0000	0001
40	0000	0000	0000	0000	0000	0000	0000	0000	0000	0001
45	0000	0000	0000	0000	0000	0000	0000	0000	0000	0001
50	0000	0000	0000	0000	0000	0000	0000	0000	0000	0001
55	0000	0000	0000	0000	0000	0000	0000	0000	0000	0001
60	0000	0000	0000	0000	0000	0000	0000	0000	0000	0000

PROBABILITY DENSITY OF t, THE NON-CENTRAL t-STATISTIC, TABLED AS A FUNCTION OF $t/\sqrt{f}$.

f is the number of degrees of freedom; the non-centrality parameter is $\sqrt{f+1}\,K_p$.
K_p is the standardized normal deviate exceeded with probability p.

DEGREES OF FREEDOM **18**

$t/\sqrt{f}$ \ p	.2500	.1500	.1000	.0650	.0400	.0250	.0100	.0040	.0025	.0010
-0.35	0000	0000	0000	0000	0000	0000	0000	0000	0000	0000
30	0001	0000	0000	0000	0000	0000	0000	0000	0000	0000
25	0002	0000	0000	0000	0000	0000	0000	0000	0000	0000
20	0003	0000	0000	0000	0000	0000	0000	0000	0000	0000
15	0007	0000	0000	0000	0000	0000	0000	0000	0000	0000
10	0014	0000	0000	0000	0000	0000	0000	0000	0000	0000
05	0027	0000	0000	0000	0000	0000	0000	0000	0000	0000
0.00	0052	0000	0000	0000	0000	0000	0000	0000	0000	0000
05	0096	0000	0000	0000	0000	0000	0000	0000	0000	0000
10	0171	0001	0000	0000	0000	0000	0000	0000	0000	0000
15	0289	0002	0000	0000	0000	0000	0000	0000	0000	0000
20	0467	0006	0000	0000	0000	0000	0000	0000	0000	0000
25	0718	0012	0000	0000	0000	0000	0000	0000	0000	0000
30	1046	0026	0001	0000	0000	0000	0000	0000	0000	0000
35	1447	0053	0002	0000	0000	0000	0000	0000	0000	0000
40	1900	0102	0004	0000	0000	0000	0000	0000	0000	0000
45	2369	0182	0009	0000	0000	0000	0000	0000	0000	0000
50	2808	0308	0020	0001	0000	0000	0000	0000	0000	0000
55	3174	0491	0041	0002	0000	0000	0000	0000	0000	0000
60	3427	0738	0080	0004	0000	0000	0000	0000	0000	0000
65	3546	1048	0145	0010	0000	0000	0000	0000	0000	0000
70	3525	1411	0248	0021	0001	0000	0000	0000	0000	0000
75	3378	1803	0397	0042	0002	0000	0000	0000	0000	0000
80	3130	2195	0599	0081	0005	0000	0000	0000	0000	0000
85	2813	2553	0855	0143	0011	0001	0000	0000	0000	0000
90	2459	2847	1157	0238	0023	0002	0000	0000	0000	0000
95	2097	3054	1491	0373	0046	0004	0000	0000	0000	0000
1.00	1749	3161	1834	0552	0084	0009	0000	0000	0000	0000
05	1430	3166	2162	0776	0143	0019	0000	0000	0000	0000
10	1150	3078	2451	1037	0231	0037	0000	0000	0000	0000
15	0910	2913	2679	1324	0353	0067	0001	0000	0000	0000
20	0711	2691	2834	1621	0511	0114	0003	0000	0000	0000
25	0549	2432	2908	1909	0704	0183	0006	0000	0000	0000
30	0420	2155	2904	2169	0927	0279	0013	0000	0000	0000
35	0319	1877	2829	2386	1173	0405	0024	0001	0000	0000
40	0240	1609	2695	2548	1428	0560	0044	0002	0000	0000
45	0180	1360	2515	2648	1678	0744	0073	0004	0001	0000
50	0134	1136	2305	2684	1911	0949	0118	0008	0002	0000
55	0100	0938	2078	2661	2112	1168	0179	0014	0003	0000
60	0074	0768	1846	2584	2274	1392	0261	0026	0007	0000
65	0054	0623	1618	2464	2388	1608	0365	0044	0012	0001
70	0040	0502	1402	2310	2453	1807	0490	0070	0022	0002
75	0030	0402	1202	2134	2469	1979	0635	0108	0037	0003
80	0022	0320	1021	1944	2439	2117	0797	0160	0059	0007
85	0016	0254	0860	1749	2369	2218	0969	0226	0090	0012
90	0012	0200	0719	1557	2266	2278	1147	0310	0133	0020
95	0009	0158	0597	1372	2137	2298	1323	0409	0189	0033
2.00	0006	0124	0493	1198	1990	2280	1490	0525	0259	0051
05	0005	0097	0405	1038	1832	2229	1642	0653	0344	0077
10	0004	0076	0331	0892	1669	2150	1774	0792	0443	0111
15	0003	0059	0270	0763	1506	2048	1881	0938	0555	0156
20	0002	0046	0219	0648	1348	1928	1961	1085	0677	0212
25	0001	0036	0178	0548	1197	1797	2013	1229	0807	0279
30	0001	0028	0144	0461	1055	1658	2036	1365	0942	0359
35	0001	0022	0116	0386	0924	1518	2032	1489	1076	0449
40	0001	0017	0093	0323	0805	1378	2004	1598	1206	0549
45	0000	0013	0075	0269	0698	1242	1954	1688	1329	0656

PROBABILITY DENSITY OF t, THE NON-CENTRAL t-STATISTIC, TABLED AS A FUNCTION OF $t/\sqrt{f}$.

f is the number of degrees of freedom; the non-centrality parameter is $\sqrt{f+1}\,K_p$.
K_p is the standardized normal deviate exceeded with probability p.

DEGREES OF FREEDOM **18**

$t/\sqrt{f}$ \ p	.2500	.1500	.1000	.0650	.0400	.0250	.0100	.0040	.0025	.0010
2.50	0000	0010	0060	0223	0602	1112	1886	1759	1440	0769
55	0000	0008	0048	0185	0517	0990	1803	1808	1538	0884
60	0000	0006	0039	0153	0442	0877	1708	1836	1619	0999
65	0000	0005	0031	0126	0377	0772	1606	1844	1683	1111
70	0000	0004	0025	0104	0321	0678	1498	1832	1728	1217
75	0000	0003	0020	0086	0272	0592	1388	1803	1754	1314
80	0000	0002	0016	0070	0230	0515	1278	1759	1763	1400
85	0000	0002	0013	0058	0195	0447	1171	1701	1754	1474
90	0000	0001	0010	0048	0164	0387	1066	1633	1730	1534
95	0000	0001	0008	0039	0138	0334	0966	1556	1693	1579
3.00	0000	0001	0007	0032	0116	0287	0872	1473	1643	1610
05	0000	0001	0005	0026	0098	0247	0784	1386	1583	1626
10	0000	0001	0004	0022	0082	0212	0701	1297	1516	1629
15	0000	0000	0004	0018	0069	0181	0626	1207	1442	1618
20	0000	0000	0003	0015	0058	0155	0557	1118	1364	1595
25	0000	0000	0002	0012	0048	0132	0494	1031	1283	1561
30	0000	0000	0002	0010	0040	0113	0437	0946	1201	1519
35	0000	0000	0002	0008	0034	0096	0385	0866	1119	1468
40	0000	0000	0001	0007	0028	0082	0339	0789	1039	1411
45	0000	0000	0001	0006	0024	0070	0298	0717	0960	1349
50	0000	0000	0001	0005	0020	0059	0262	0649	0884	1283
55	0000	0000	0001	0004	0017	0050	0229	0586	0811	1215
60	0000	0000	0001	0003	0014	0043	0201	0528	0742	1145
65	0000	0000	0000	0003	0012	0036	0175	0475	0677	1076
70	0000	0000	0000	0002	0010	0031	0153	0426	0616	1006
75	0000	0000	0000	0002	0008	0026	0133	0381	0559	0938
80	0000	0000	0000	0001	0007	0022	0116	0340	0506	0871
85	0000	0000	0000	0001	0006	0019	0101	0304	0457	0807
90	0000	0000	0000	0001	0005	0016	0088	0270	0412	0745
95	0000	0000	0000	0001	0004	0014	0076	0240	0371	0686
4.00	0000	0000	0000	0001	0003	0012	0066	0214	0333	0631
05	0000	0000	0000	0001	0003	0010	0058	0189	0299	0578
10	0000	0000	0000	0000	0002	0008	0050	0168	0267	0529
15	0000	0000	0000	0000	0002	0007	0043	0149	0239	0483
20	0000	0000	0000	0000	0002	0006	0038	0131	0214	0440
25	0000	0000	0000	0000	0001	0005	0033	0116	0191	0400
30	0000	0000	0000	0000	0001	0004	0028	0103	0170	0363
35	0000	0000	0000	0000	0001	0004	0025	0091	0152	0330
40	0000	0000	0000	0000	0001	0003	0021	0080	0135	0299
45	0000	0000	0000	0000	0001	0003	0019	0071	0120	0270
50	0000	0000	0000	0000	0001	0002	0016	0062	0107	0244
55	0000	0000	0000	0000	0001	0002	0014	0055	0095	0220
60	0000	0000	0000	0000	0000	0002	0012	0048	0084	0198
65	0000	0000	0000	0000	0000	0001	0011	0043	0075	0179
70	0000	0000	0000	0000	0000	0001	0009	0037	0066	0161
75	0000	0000	0000	0000	0000	0001	0008	0033	0059	0145
80	0000	0000	0000	0000	0000	0001	0007	0029	0052	0130
85	0000	0000	0000	0000	0000	0001	0006	0026	0046	0117
90	0000	0000	0000	0000	0000	0001	0005	0023	0041	0105
95	0000	0000	0000	0000	0000	0001	0005	0020	0036	0094
5.00	0000	0000	0000	0000	0000	0001	0004	0017	0032	0084
05	0000	0000	0000	0000	0000	0000	0003	0015	0028	0076
10	0000	0000	0000	0000	0000	0000	0003	0014	0025	0068
15	0000	0000	0000	0000	0000	0000	0003	0012	0022	0061
20	0000	0000	0000	0000	0000	0000	0002	0011	0020	0054
25	0000	0000	0000	0000	0000	0000	0002	0009	0018	0049
30	0000	0000	0000	0000	0000	0000	0002	0008	0016	0044
35	0000	0000	0000	0000	0000	0000	0002	0007	0014	0039
40	0000	0000	0000	0000	0000	0000	0001	0006	0012	0035
45	0000	0000	0000	0000	0000	0000	0001	0006	0011	0031

PROBABILITY DENSITY OF t, THE NON-CENTRAL t-STATISTIC, TABLED AS A FUNCTION OF $t/\sqrt{f}$.

f is the number of degrees of freedom; the non-centrality parameter is $\sqrt{f+1}\,K_p$.
K_p is the standardized normal deviate exceeded with probability p.

DEGREES OF FREEDOM **18**

$t/\sqrt{f}$ \ p	.2500	.1500	.1000	.0650	.0400	.0250	.0100	.0040	.0025	.0010
5.50	0000	0000	0000	0000	0000	0000	0001	0005	0010	0028
55	0000	0000	0000	0000	0000	0000	0001	0004	0008	0025
60	0000	0000	0000	0000	0000	0000	0001	0004	0008	0022
65	0000	0000	0000	0000	0000	0000	0001	0003	0007	0020
70	0000	0000	0000	0000	0000	0000	0001	0003	0006	0018
75	0000	0000	0000	0000	0000	0000	0001	0003	0005	0016
80	0000	0000	0000	0000	0000	0000	0000	0002	0005	0014
85	0000	0000	0000	0000	0000	0000	0000	0002	0004	0013
90	0000	0000	0000	0000	0000	0000	0000	0002	0004	0011
95	0000	0000	0000	0000	0000	0000	0000	0002	0003	0010
6.00	0000	0000	0000	0000	0000	0000	0000	0001	0003	0009
05	0000	0000	0000	0000	0000	0000	0000	0001	0003	0008
10	0000	0000	0000	0000	0000	0000	0000	0001	0002	0007
15	0000	0000	0000	0000	0000	0000	0000	0001	0002	0007
20	0000	0000	0000	0000	0000	0000	0000	0001	0002	0006
25	0000	0000	0000	0000	0000	0000	0000	0001	0002	0005
30	0000	0000	0000	0000	0000	0000	0000	0001	0001	0005
35	0000	0000	0000	0000	0000	0000	0000	0001	0001	0004
40	0000	0000	0000	0000	0000	0000	0000	0001	0001	0004
45	0000	0000	0000	0000	0000	0000	0000	0000	0001	0003
50	0000	0000	0000	0000	0000	0000	0000	0000	0001	0003
55	0000	0000	0000	0000	0000	0000	0000	0000	0001	0003
60	0000	0000	0000	0000	0000	0000	0000	0000	0001	0002
65	0000	0000	0000	0000	0000	0000	0000	0000	0001	0002
70	0000	0000	0000	0000	0000	0000	0000	0000	0001	0002
75	0000	0000	0000	0000	0000	0000	0000	0000	0001	0002
80	0000	0000	0000	0000	0000	0000	0000	0000	0000	0002
85	0000	0000	0000	0000	0000	0000	0000	0000	0000	0001
90	0000	0000	0000	0000	0000	0000	0000	0000	0000	0001
95	0000	0000	0000	0000	0000	0000	0000	0000	0000	0001
7.00	0000	0000	0000	0000	0000	0000	0000	0000	0000	0001
05	0000	0000	0000	0000	0000	0000	0000	0000	0000	0001
10	0000	0000	0000	0000	0000	0000	0000	0000	0000	0001
15	0000	0000	0000	0000	0000	0000	0000	0000	0000	0001
20	0000	0000	0000	0000	0000	0000	0000	0000	0000	0001
25	0000	0000	0000	0000	0000	0000	0000	0000	0000	0001
30	0000	0000	0000	0000	0000	0000	0000	0000	0000	0001
35	0000	0000	0000	0000	0000	0000	0000	0000	0000	0001
40	0000	0000	0000	0000	0000	0000	0000	0000	0000	0000

PROBABILITY DENSITY OF t, THE NON-CENTRAL t-STATISTIC, TABLED AS A FUNCTION OF $t/\sqrt{f}$.

f is the number of degrees of freedom; the non-centrality parameter is $\sqrt{f+1}\,K_p$.
K_p is the standardized normal deviate exceeded with probability p.

DEGREES OF FREEDOM **19**

$t/\sqrt{f}$ \ p	.2500	.1500	.1000	.0650	.0400	.0250	.0100	.0040	.0025	.0010
-0.30	0000	0000	0000	0000	0000	0000	0000	0000	0000	0000
25	0001	0000	0000	0000	0000	0000	0000	0000	0000	0000
20	0002	0000	0000	0000	0000	0000	0000	0000	0000	0000
15	0005	0000	0000	0000	0000	0000	0000	0000	0000	0000
10	0010	0000	0000	0000	0000	0000	0000	0000	0000	0000
05	0021	0000	0000	0000	0000	0000	0000	0000	0000	0000
0.00	0042	0000	0000	0000	0000	0000	0000	0000	0000	0000
05	0079	0000	0000	0000	0000	0000	0000	0000	0000	0000
10	0145	0001	0000	0000	0000	0000	0000	0000	0000	0000
15	0253	0002	0000	0000	0000	0000	0000	0000	0000	0000
20	0419	0004	0000	0000	0000	0000	0000	0000	0000	0000
25	0658	0009	0000	0000	0000	0000	0000	0000	0000	0000
30	0979	0020	0000	0000	0000	0000	0000	0000	0000	0000
35	1378	0043	0001	0000	0000	0000	0000	0000	0000	0000
40	1836	0084	0003	0000	0000	0000	0000	0000	0000	0000
45	2317	0156	0007	0000	0000	0000	0000	0000	0000	0000
50	2773	0272	0015	0000	0000	0000	0000	0000	0000	0000
55	3155	0444	0033	0001	0000	0000	0000	0000	0000	0000
60	3422	0682	0066	0003	0000	0000	0000	0000	0000	0000
65	3547	0987	0124	0007	0000	0000	0000	0000	0000	0000
70	3527	1349	0217	0016	0000	0000	0000	0000	0000	0000
75	3373	1748	0356	0034	0001	0000	0000	0000	0000	0000
80	3113	2150	0550	0067	0004	0000	0000	0000	0000	0000
85	2783	2522	0800	0122	0008	0000	0000	0000	0000	0000
90	2416	2830	1100	0209	0018	0001	0000	0000	0000	0000
95	2043	3048	1437	0335	0037	0003	0000	0000	0000	0000
1.00	1688	3161	1788	0507	0070	0007	0000	0000	0000	0000
05	1367	3168	2127	0725	0123	0015	0000	0000	0000	0000
10	1086	3077	2428	0985	0204	0029	0000	0000	0000	0000
15	0850	2904	2667	1274	0318	0055	0001	0000	0000	0000
20	0655	2672	2830	1577	0469	0097	0002	0000	0000	0000
25	0500	2403	2910	1874	0658	0160	0005	0000	0000	0000
30	0377	2116	2906	2144	0880	0249	0010	0000	0000	0000
35	0282	1830	2828	2371	1126	0369	0019	0000	0000	0000
40	0209	1556	2687	2541	1386	0519	0035	0001	0000	0000
45	0154	1304	2500	2647	1644	0700	0062	0003	0000	0000
50	0113	1079	2281	2686	1884	0905	0101	0006	0001	0000
55	0083	0883	2045	2662	2095	1127	0157	0011	0002	0000
60	0060	0715	1806	2582	2264	1355	0234	0021	0005	0000
65	0044	0574	1572	2456	2385	1577	0333	0036	0009	0001
70	0032	0457	1352	2296	2454	1784	0454	0059	0017	0001
75	0023	0362	1150	2112	2471	1964	0597	0093	0030	0002
80	0017	0285	0968	1915	2440	2109	0758	0140	0049	0005
85	0012	0223	0808	1714	2366	2215	0932	0202	0077	0009
90	0009	0174	0669	1516	2259	2278	1113	0282	0116	0016
95	0006	0135	0551	1327	2124	2300	1293	0378	0168	0027
2.00	0005	0105	0450	1151	1971	2282	1466	0491	0234	0043
05	0003	0081	0366	0989	1807	2228	1624	0618	0315	0065
10	0002	0062	0297	0844	1638	2145	1762	0758	0412	0097
15	0002	0048	0239	0716	1471	2038	1874	0905	0522	0138
20	0001	0037	0192	0603	1308	1913	1958	1055	0644	0190
25	0001	0028	0154	0505	1154	1776	2013	1203	0775	0255
30	0001	0022	0123	0421	1011	1633	2038	1344	0911	0331
35	0001	0017	0098	0350	0880	1487	2034	1473	1048	0419
40	0000	0013	0078	0290	0761	1344	2005	1586	1183	0518
45	0000	0010	0062	0239	0654	1205	1952	1681	1310	0625

PROBABILITY DENSITY OF t, THE NON-CENTRAL t-STATISTIC, TABLED AS A FUNCTION OF $t/\sqrt{f}$.

f is the number of degrees of freedom; the non-centrality parameter is $\sqrt{f+1}\,K_p$.
K_p is the standardized normal deviate exceeded with probability p.

DEGREES OF FREEDOM **19**

$t/\sqrt{f}$ \ p	.2500	.1500	.1000	.0650	.0400	.0250	.0100	.0040	.0025	.0010
2.50	0000	0008	0049	0196	0560	1073	1881	1756	1426	0739
55	0000	0006	0039	0161	0477	0949	1794	1808	1528	0856
60	0000	0005	0031	0132	0405	0835	1695	1838	1613	0974
65	0000	0004	0025	0108	0343	0731	1588	1846	1680	1089
70	0000	0003	0020	0088	0289	0637	1476	1834	1727	1199
75	0000	0002	0016	0072	0243	0553	1363	1803	1755	1300
80	0000	0002	0012	0058	0204	0478	1250	1757	1764	1390
85	0000	0001	0010	0048	0171	0411	1139	1696	1756	1467
90	0000	0001	0008	0039	0143	0353	1032	1625	1731	1530
95	0000	0001	0006	0031	0119	0303	0931	1544	1691	1578
3.00	0000	0001	0005	0026	0099	0258	0835	1458	1639	1611
05	0000	0000	0004	0021	0083	0220	0747	1368	1576	1628
10	0000	0000	0003	0017	0069	0187	0665	1275	1506	1630
15	0000	0000	0002	0014	0057	0159	0589	1182	1429	1619
20	0000	0000	0002	0011	0047	0135	0521	1091	1347	1595
25	0000	0000	0002	0009	0039	0114	0459	1001	1264	1560
30	0000	0000	0001	0007	0033	0097	0404	0915	1179	1515
35	0000	0000	0001	0006	0027	0082	0354	0833	1095	1462
40	0000	0000	0001	0005	0022	0069	0310	0756	1012	1402
45	0000	0000	0001	0004	0019	0058	0270	0683	0931	1337
50	0000	0000	0001	0003	0015	0049	0236	0616	0854	1269
55	0000	0000	0000	0003	0013	0041	0205	0553	0780	1198
60	0000	0000	0000	0002	0011	0035	0178	0496	0710	1126
65	0000	0000	0000	0002	0009	0029	0154	0443	0645	1054
70	0000	0000	0000	0001	0007	0025	0134	0395	0584	0982
75	0000	0000	0000	0001	0006	0021	0116	0351	0527	0912
80	0000	0000	0000	0001	0005	0018	0100	0312	0474	0844
85	0000	0000	0000	0001	0004	0015	0086	0277	0426	0779
90	0000	0000	0000	0001	0004	0012	0075	0245	0382	0716
95	0000	0000	0000	0001	0003	0011	0064	0216	0342	0657
4.00	0000	0000	0000	0000	0002	0009	0056	0191	0306	0601
05	0000	0000	0000	0000	0002	0007	0048	0168	0272	0549
10	0000	0000	0000	0000	0002	0006	0041	0148	0243	0499
15	0000	0000	0000	0000	0001	0005	0036	0130	0216	0454
20	0000	0000	0000	0000	0001	0004	0031	0115	0192	0411
25	0000	0000	0000	0000	0001	0004	0026	0101	0170	0372
30	0000	0000	0000	0000	0001	0003	0023	0088	0151	0337
35	0000	0000	0000	0000	0001	0003	0020	0078	0133	0304
40	0000	0000	0000	0000	0001	0002	0017	0068	0118	0274
45	0000	0000	0000	0000	0000	0002	0015	0060	0104	0246
50	0000	0000	0000	0000	0000	0002	0012	0052	0092	0221
55	0000	0000	0000	0000	0000	0001	0011	0046	0081	0199
60	0000	0000	0000	0000	0000	0001	0009	0040	0072	0178
65	0000	0000	0000	0000	0000	0001	0008	0035	0063	0159
70	0000	0000	0000	0000	0000	0001	0007	0031	0056	0143
75	0000	0000	0000	0000	0000	0001	0006	0027	0049	0128
80	0000	0000	0000	0000	0000	0001	0005	0023	0043	0114
85	0000	0000	0000	0000	0000	0001	0004	0020	0038	0102
90	0000	0000	0000	0000	0000	0000	0004	0018	0034	0091
95	0000	0000	0000	0000	0000	0000	0003	0016	0030	0081
5.00	0000	0000	0000	0000	0000	0000	0003	0014	0026	0072
05	0000	0000	0000	0000	0000	0000	0002	0012	0023	0064
10	0000	0000	0000	0000	0000	0000	0002	0010	0020	0057
15	0000	0000	0000	0000	0000	0000	0002	0009	0018	0051
20	0000	0000	0000	0000	0000	0000	0002	0008	0016	0046
25	0000	0000	0000	0000	0000	0000	0001	0007	0014	0041
30	0000	0000	0000	0000	0000	0000	0001	0006	0012	0036
35	0000	0000	0000	0000	0000	0000	0001	0005	0011	0032
40	0000	0000	0000	0000	0000	0000	0001	0005	0009	0029
45	0000	0000	0000	0000	0000	0000	0001	0004	0008	0025

PROBABILITY DENSITY OF t, THE NON-CENTRAL t-STATISTIC, TABLED AS A FUNCTION OF $t/\sqrt{f}$.

f is the number of degrees of freedom; the non-centrality parameter is $\sqrt{f+1}\,K_p$.
K_p is the standardized normal deviate exceeded with probability p.

DEGREES OF FREEDOM **19**

$t/\sqrt{f}$ \ p	.2500	.1500	.1000	.0650	.0400	.0250	.0100	.0040	.0025	.0010
5.50	0000	0000	0000	0000	0000	0000	0001	0004	0007	0023
55	0000	0000	0000	0000	0000	0000	0001	0003	0006	0020
60	0000	0000	0000	0000	0000	0000	0001	0003	0006	0018
65	0000	0000	0000	0000	0000	0000	0000	0002	0005	0016
70	0000	0000	0000	0000	0000	0000	0000	0002	0004	0014
75	0000	0000	0000	0000	0000	0000	0000	0002	0004	0013
80	0000	0000	0000	0000	0000	0000	0000	0002	0003	0011
85	0000	0000	0000	0000	0000	0000	0000	0001	0003	0010
90	0000	0000	0000	0000	0000	0000	0000	0001	0003	0009
95	0000	0000	0000	0000	0000	0000	0000	0001	0002	0008
6.00	0000	0000	0000	0000	0000	0000	0000	0001	0002	0007
05	0000	0000	0000	0000	0000	0000	0000	0001	0002	0006
10	0000	0000	0000	0000	0000	0000	0000	0001	0002	0006
15	0000	0000	0000	0000	0000	0000	0000	0001	0001	0005
20	0000	0000	0000	0000	0000	0000	0000	0001	0001	0004
25	0000	0000	0000	0000	0000	0000	0000	0001	0001	0004
30	0000	0000	0000	0000	0000	0000	0000	0000	0001	0004
35	0000	0000	0000	0000	0000	0000	0000	0000	0001	0003
40	0000	0000	0000	0000	0000	0000	0000	0000	0001	0003
45	0000	0000	0000	0000	0000	0000	0000	0000	0001	0002
50	0000	0000	0000	0000	0000	0000	0000	0000	0001	0002
55	0000	0000	0000	0000	0000	0000	0000	0000	0001	0002
60	0000	0000	0000	0000	0000	0000	0000	0000	0000	0002
65	0000	0000	0000	0000	0000	0000	0000	0000	0000	0002
70	0000	0000	0000	0000	0000	0000	0000	0000	0000	0001
75	0000	0000	0000	0000	0000	0000	0000	0000	0000	0001
80	0000	0000	0000	0000	0000	0000	0000	0000	0000	0001
85	0000	0000	0000	0000	0000	0000	0000	0000	0000	0001
90	0000	0000	0000	0000	0000	0000	0000	0000	0000	0001
95	0000	0000	0000	0000	0000	0000	0000	0000	0000	0001
7.00	0000	0000	0000	0000	0000	0000	0000	0000	0000	0001
05	0000	0000	0000	0000	0000	0000	0000	0000	0000	0001
10	0000	0000	0000	0000	0000	0000	0000	0000	0000	0001
15	0000	0000	0000	0000	0000	0000	0000	0000	0000	0001
20	0000	0000	0000	0000	0000	0000	0000	0000	0000	0000

PROBABILITY DENSITY OF t, THE NON-CENTRAL t-STATISTIC, TABLED AS A FUNCTION OF $t/\sqrt{f}$.

f is the number of degrees of freedom; the non-centrality parameter is $\sqrt{f+1}\,K_p$.
K_p is the standardized normal deviate exceeded with probability p.

$t/\sqrt{f}$ \ p	.2500	.1500	.1000	.0650	.0400	.0250	.0100	.0040	.0025	.0010
-0.30	0000	0000	0000	0000	0000	0000	0000	0000	0000	0000
25	0001	0000	0000	0000	0000	0000	0000	0000	0000	0000
20	0002	0000	0000	0000	0000	0000	0000	0000	0000	0000
15	0003	0000	0000	0000	0000	0000	0000	0000	0000	0000
10	0008	0000	0000	0000	0000	0000	0000	0000	0000	0000
05	0016	0000	0000	0000	0000	0000	0000	0000	0000	0000
0.00	0033	0000	0000	0000	0000	0000	0000	0000	0000	0000
05	0065	0000	0000	0000	0000	0000	0000	0000	0000	0000
10	0123	0000	0000	0000	0000	0000	0000	0000	0000	0000
15	0221	0001	0000	0000	0000	0000	0000	0000	0000	0000
20	0375	0003	0000	0000	0000	0000	0000	0000	0000	0000
25	0604	0007	0000	0000	0000	0000	0000	0000	0000	0000
30	0916	0016	0000	0000	0000	0000	0000	0000	0000	0000
35	1313	0034	0001	0000	0000	0000	0000	0000	0000	0000
40	1774	0070	0002	0000	0000	0000	0000	0000	0000	0000
45	2266	0134	0005	0000	0000	0000	0000	0000	0000	0000
50	2737	0239	0012	0000	0000	0000	0000	0000	0000	0000
55	3136	0401	0026	0001	0000	0000	0000	0000	0000	0000
60	3416	0629	0054	0002	0000	0000	0000	0000	0000	0000
65	3549	0929	0105	0005	0000	0000	0000	0000	0000	0000
70	3528	1291	0190	0012	0000	0000	0000	0000	0000	0000
75	3368	1694	0320	0027	0001	0000	0000	0000	0000	0000
80	3097	2107	0504	0055	0003	0000	0000	0000	0000	0000
85	2753	2492	0748	0104	0006	0000	0000	0000	0000	0000
90	2374	2814	1046	0183	0014	0001	0000	0000	0000	0000
95	1991	3042	1385	0301	0030	0002	0000	0000	0000	0000
1.00	1630	3162	1743	0465	0058	0005	0000	0000	0000	0000
05	1306	3170	2092	0678	0106	0011	0000	0000	0000	0000
10	1026	3075	2405	0935	0179	0024	0000	0000	0000	0000
15	0793	2895	2655	1226	0286	0046	0001	0000	0000	0000
20	0604	2653	2827	1534	0431	0082	0001	0000	0000	0000
25	0454	2373	2911	1839	0614	0139	0003	0000	0000	0000
30	0338	2077	2908	2120	0834	0222	0007	0000	0000	0000
35	0249	1783	2826	2356	1082	0336	0015	0000	0000	0000
40	0182	1505	2680	2535	1345	0481	0029	0001	0000	0000
45	0132	1250	2484	2646	1609	0659	0052	0002	0000	0000
50	0096	1025	2257	2688	1859	0863	0087	0004	0001	0000
55	0069	0830	2013	2663	2078	1086	0138	0009	0002	0000
60	0049	0665	1766	2580	2255	1318	0210	0016	0004	0000
65	0035	0528	1528	2448	2382	1547	0303	0029	0007	0000
70	0025	0416	1304	2281	2454	1761	0421	0050	0014	0001
75	0018	0325	1100	2090	2472	1948	0561	0080	0024	0002
80	0013	0253	0919	1886	2440	2100	0721	0123	0041	0004
85	0009	0196	0760	1679	2364	2211	0895	0181	0066	0007
90	0007	0151	0624	1476	2251	2278	1079	0256	0101	0013
95	0005	0116	0508	1284	2111	2301	1264	0349	0149	0022
2.00	0003	0088	0411	1105	1952	2283	1442	0459	0211	0035
05	0002	0068	0331	0943	1782	2227	1606	0585	0289	0056
10	0002	0051	0265	0799	1608	2140	1749	0724	0382	0084
15	0001	0039	0212	0671	1436	2028	1867	0873	0491	0122
20	0001	0030	0168	0561	1270	1898	1955	1025	0612	0171
25	0001	0023	0133	0466	1114	1756	2013	1177	0743	0232
30	0000	0017	0105	0385	0969	1608	2039	1323	0881	0305
35	0000	0013	0083	0317	0837	1458	2036	1457	1021	0391
40	0000	0010	0065	0260	0719	1310	2005	1575	1159	0489
45	0000	0007	0051	0212	0614	1169	1950	1674	1291	0596

PROBABILITY DENSITY OF t, THE NON-CENTRAL t-STATISTIC, TABLED AS A FUNCTION OF $t/\sqrt{f}$.

f is the number of degrees of freedom; the non-centrality parameter is $\sqrt{f+1}\,K_p$.
K_p is the standardized normal deviate exceeded with probability p.

DEGREES OF FREEDOM **20**

$t/\sqrt{f}$ \ p	.2500	.1500	.1000	.0650	.0400	.0250	.0100	.0040	.0025	.0010
2.50	0000	0006	0040	0173	0521	1035	1875	1752	1411	0710
55	0000	0004	0032	0140	0441	0910	1784	1807	1517	0829
60	0000	0003	0025	0114	0371	0796	1682	1838	1607	0950
65	0000	0003	0020	0092	0311	0692	1571	1847	1677	1068
70	0000	0002	0015	0074	0260	0599	1455	1835	1727	1181
75	0000	0001	0012	0060	0217	0516	1338	1803	1756	1286
80	0000	0001	0009	0048	0181	0443	1221	1754	1766	1380
85	0000	0001	0007	0039	0150	0378	1108	1691	1757	1460
90	0000	0001	0006	0031	0124	0322	0999	1616	1731	1526
95	0000	0001	0005	0025	0103	0274	0897	1533	1689	1577
3.00	0000	0000	0004	0020	0085	0232	0800	1443	1635	1611
05	0000	0000	0003	0016	0070	0196	0711	1349	1569	1629
10	0000	0000	0002	0013	0058	0166	0630	1253	1495	1632
15	0000	0000	0002	0011	0047	0140	0555	1158	1415	1620
20	0000	0000	0001	0009	0039	0117	0488	1064	1331	1595
25	0000	0000	0001	0007	0032	0099	0427	0973	1244	1558
30	0000	0000	0001	0006	0026	0083	0373	0885	1157	1511
35	0000	0000	0001	0004	0022	0069	0325	0802	1070	1456
40	0000	0000	0001	0004	0018	0058	0282	0724	0986	1393
45	0000	0000	0000	0003	0015	0049	0245	0651	0903	1326
50	0000	0000	0000	0002	0012	0041	0212	0584	0825	1255
55	0000	0000	0000	0002	0010	0034	0183	0522	0750	1181
60	0000	0000	0000	0002	0008	0028	0158	0465	0680	1107
65	0000	0000	0000	0001	0007	0024	0136	0413	0614	1033
70	0000	0000	0000	0001	0006	0020	0117	0366	0553	0959
75	0000	0000	0000	0001	0005	0017	0101	0324	0497	0888
80	0000	0000	0000	0001	0004	0014	0086	0286	0445	0818
85	0000	0000	0000	0001	0003	0012	0074	0252	0398	0752
90	0000	0000	0000	0000	0003	0010	0063	0222	0355	0689
95	0000	0000	0000	0000	0002	0008	0054	0195	0316	0629
4.00	0000	0000	0000	0000	0002	0007	0047	0171	0280	0573
05	0000	0000	0000	0000	0001	0006	0040	0150	0249	0520
10	0000	0000	0000	0000	0001	0005	0034	0131	0220	0472
15	0000	0000	0000	0000	0001	0004	0029	0115	0195	0426
20	0000	0000	0000	0000	0001	0003	0025	0100	0172	0385
25	0000	0000	0000	0000	0001	0003	0021	0087	0152	0347
30	0000	0000	0000	0000	0001	0002	0018	0076	0134	0312
35	0000	0000	0000	0000	0000	0002	0016	0066	0118	0280
40	0000	0000	0000	0000	0000	0002	0013	0058	0103	0251
45	0000	0000	0000	0000	0000	0001	0011	0050	0091	0224
50	0000	0000	0000	0000	0000	0001	0010	0044	0080	0201
55	0000	0000	0000	0000	0000	0001	0008	0038	0070	0179
60	0000	0000	0000	0000	0000	0001	0007	0033	0061	0160
65	0000	0000	0000	0000	0000	0001	0006	0029	0054	0142
70	0000	0000	0000	0000	0000	0001	0005	0025	0047	0127
75	0000	0000	0000	0000	0000	0000	0004	0022	0041	0113
80	0000	0000	0000	0000	0000	0000	0004	0019	0036	0100
85	0000	0000	0000	0000	0000	0000	0003	0016	0032	0089
90	0000	0000	0000	0000	0000	0000	0003	0014	0028	0079
95	0000	0000	0000	0000	0000	0000	0002	0012	0024	0070
5.00	0000	0000	0000	0000	0000	0000	0002	0011	0021	0062
05	0000	0000	0000	0000	0000	0000	0002	0009	0019	0055
10	0000	0000	0000	0000	0000	0000	0002	0008	0016	0049
15	0000	0000	0000	0000	0000	0000	0001	0007	0014	0043
20	0000	0000	0000	0000	0000	0000	0001	0006	0012	0038
25	0000	0000	0000	0000	0000	0000	0001	0005	0011	0034
30	0000	0000	0000	0000	0000	0000	0001	0005	0009	0030
35	0000	0000	0000	0000	0000	0000	0001	0004	0008	0026
40	0000	0000	0000	0000	0000	0000	0001	0004	0007	0023
45	0000	0000	0000	0000	0000	0000	0001	0003	0006	0021

PROBABILITY DENSITY OF t, THE NON-CENTRAL t-STATISTIC, TABLED AS A FUNCTION OF $t/\sqrt{f}$.

f is the number of degrees of freedom; the non-centrality parameter is $\sqrt{f+1}\,K_p$.
K_p is the standardized normal deviate exceeded with probability p.

DEGREES OF FREEDOM **20**

$t/\sqrt{f}$ \ p	.2500	.1500	.1000	.0650	.0400	.0250	.0100	.0040	.0025	.0010
5.50	0000	0000	0000	0000	0000	0000	0000	0003	0006	0018
55	0000	0000	0000	0000	0000	0000	0000	0002	0005	0016
60	0000	0000	0000	0000	0000	0000	0000	0002	0004	0014
65	0000	0000	0000	0000	0000	0000	0000	0002	0004	0013
70	0000	0000	0000	0000	0000	0000	0000	0002	0003	0011
75	0000	0000	0000	0000	0000	0000	0000	0001	0003	0010
80	0000	0000	0000	0000	0000	0000	0000	0001	0003	0009
85	0000	0000	0000	0000	0000	0000	0000	0001	0002	0008
90	0000	0000	0000	0000	0000	0000	0000	0001	0002	0007
95	0000	0000	0000	0000	0000	0000	0000	0001	0002	0006
6.00	0000	0000	0000	0000	0000	0000	0000	0001	0001	0005
05	0000	0000	0000	0000	0000	0000	0000	0001	0001	0005
10	0000	0000	0000	0000	0000	0000	0000	0001	0001	0004
15	0000	0000	0000	0000	0000	0000	0000	0000	0001	0004
20	0000	0000	0000	0000	0000	0000	0000	0000	0001	0003
25	0000	0000	0000	0000	0000	0000	0000	0000	0001	0003
30	0000	0000	0000	0000	0000	0000	0000	0000	0001	0003
35	0000	0000	0000	0000	0000	0000	0000	0000	0001	0002
40	0000	0000	0000	0000	0000	0000	0000	0000	0001	0002
45	0000	0000	0000	0000	0000	0000	0000	0000	0000	0002
50	0000	0000	0000	0000	0000	0000	0000	0000	0000	0002
55	0000	0000	0000	0000	0000	0000	0000	0000	0000	0001
60	0000	0000	0000	0000	0000	0000	0000	0000	0000	0001
65	0000	0000	0000	0000	0000	0000	0000	0000	0000	0001
70	0000	0000	0000	0000	0000	0000	0000	0000	0000	0001
75	0000	0000	0000	0000	0000	0000	0000	0000	0000	0001
80	0000	0000	0000	0000	0000	0000	0000	0000	0000	0001
85	0000	0000	0000	0000	0000	0000	0000	0000	0000	0001
90	0000	0000	0000	0000	0000	0000	0000	0000	0000	0001
95	0000	0000	0000	0000	0000	0000	0000	0000	0000	0001
7.00	0000	0000	0000	0000	0000	0000	0000	0000	0000	0000

PROBABILITY DENSITY OF t, THE NON-CENTRAL t-STATISTIC, TABLED AS A FUNCTION OF $t/\sqrt{f}$.

f is the number of degrees of freedom; the non-centrality parameter is $\sqrt{f+1}\,K_p$.
K_p is the standardized normal deviate exceeded with probability p.

DEGREES OF FREEDOM **21**

$t/\sqrt{f}$ \ p	.2500	.1500	.1000	.0650	.0400	.0250	.0100	.0040	.0025	.0010
-0.25	0000	0000	0000	0000	0000	0000	0000	0000	0000	0000
20	0001	0000	0000	0000	0000	0000	0000	0000	0000	0000
15	0002	0000	0000	0000	0000	0000	0000	0000	0000	0000
10	0006	0000	0000	0000	0000	0000	0000	0000	0000	0000
05	0012	0000	0000	0000	0000	0000	0000	0000	0000	0000
0.00	0026	0000	0000	0000	0000	0000	0000	0000	0000	0000
05	0054	0000	0000	0000	0000	0000	0000	0000	0000	0000
10	0105	0000	0000	0000	0000	0000	0000	0000	0000	0000
15	0193	0001	0000	0000	0000	0000	0000	0000	0000	0000
20	0336	0002	0000	0000	0000	0000	0000	0000	0000	0000
25	0553	0005	0000	0000	0000	0000	0000	0000	0000	0000
30	0857	0012	0000	0000	0000	0000	0000	0000	0000	0000
35	1250	0027	0000	0000	0000	0000	0000	0000	0000	0000
40	1715	0058	0001	0000	0000	0000	0000	0000	0000	0000
45	2216	0115	0004	0000	0000	0000	0000	0000	0000	0000
50	2702	0211	0009	0000	0000	0000	0000	0000	0000	0000
55	3117	0362	0021	0000	0000	0000	0000	0000	0000	0000
60	3410	0581	0045	0001	0000	0000	0000	0000	0000	0000
65	3550	0874	0089	0004	0000	0000	0000	0000	0000	0000
70	3529	1234	0166	0010	0000	0000	0000	0000	0000	0000
75	3362	1642	0287	0022	0001	0000	0000	0000	0000	0000
80	3080	2064	0463	0046	0002	0000	0000	0000	0000	0000
85	2724	2462	0700	0089	0005	0000	0000	0000	0000	0000
90	2332	2797	0995	0161	0011	0001	0000	0000	0000	0000
95	1941	3036	1335	0271	0024	0001	0000	0000	0000	0000
1.00	1574	3163	1699	0427	0048	0004	0000	0000	0000	0000
05	1248	3171	2058	0634	0091	0009	0000	0000	0000	0000
10	0969	3073	2382	0888	0158	0019	0000	0000	0000	0000
15	0740	2885	2643	1179	0257	0038	0000	0000	0000	0000
20	0556	2634	2823	1492	0395	0070	0001	0000	0000	0000
1.25	0413	2344	2912	1805	0574	0122	0002	0000	0000	0000
30	0303	2039	2910	2095	0791	0199	0006	0000	0000	0000
35	0220	1738	2825	2342	1039	0306	0012	0000	0000	0000
40	0159	1455	2672	2528	1306	0446	0023	0001	0000	0000
45	0114	1199	2468	2645	1576	0620	0043	0001	0000	0000
50	0081	0973	2232	2689	1833	0823	0074	0003	0001	0000
55	0057	0781	1981	2664	2060	1047	0121	0007	0001	0000
60	0040	0619	1728	2577	2245	1283	0188	0013	0003	0000
65	0028	0486	1484	2440	2378	1518	0277	0024	0006	0000
70	0020	0379	1258	2266	2454	1738	0390	0042	0011	0001
75	0014	0293	1053	2068	2474	1933	0527	0069	0021	0001
80	0010	0225	0871	1857	2441	2092	0685	0108	0034	0003
85	0007	0172	0715	1644	2361	2208	0861	0162	0056	0005
90	0005	0131	0581	1437	2244	2279	1046	0233	0088	0010
95	0003	0099	0469	1242	2098	2303	1235	0322	0132	0018
2.00	0002	0075	0376	1062	1933	2284	1418	0429	0191	0029
05	0002	0056	0299	0900	1757	2226	1588	0554	0265	0047
10	0001	0042	0237	0756	1578	2135	1737	0693	0355	0073
15	0001	0032	0187	0630	1401	2019	1860	0842	0461	0107
20	0001	0024	0147	0522	1232	1883	1952	0997	0582	0153
25	0000	0018	0116	0430	1074	1736	2012	1152	0713	0211
30	0000	0013	0090	0352	0928	1583	2040	1302	0852	0282
35	0000	0010	0070	0287	0797	1429	2037	1441	0995	0366
40	0000	0008	00[illegible]5	0233	0679	1278	2005	1564	1136	0461
45	0000	0006	0043	0189	0576	1133	1948	1667	1272	0568
50	0000	0004	0033	0152	0485	0998	1870	1749	1397	0683
55	0000	0003	0026	0122	0407	0872	1775	1806	1507	0803
60	0000	0002	0020	0098	0340	0758	1668	1839	1600	0926
65	0000	0002	0015	0079	0283	0655	1553	1849	1673	1047
70	0000	0001	0012	0063	0234	0563	1434	1836	1726	1163

PROBABILITY DENSITY OF t, THE NON-CENTRAL t-STATISTIC, TABLED AS A FUNCTION OF $t/\sqrt{f}$.

f is the number of degrees of freedom; the non-centrality parameter is $\sqrt{f+1}\,K_p$.
K_p is the standardized normal deviate exceeded with probability p.

$t/\sqrt{f}$ \ p	.2500	.1500	.1000	.0650	.0400	.0250	.0100	.0040	.0025	.0010
2.75	0000	0001	0009	0050	0194	0481	1313	1802	1757	1272
80	0000	0001	0007	0040	0160	0410	1194	1751	1767	1369
85	0000	0001	0006	0032	0132	0348	1078	1686	1758	1453
90	0000	0000	0004	0025	0108	0294	0968	1608	1730	1522
95	0000	0000	0003	0020	0088	0248	0864	1521	1687	1575
3.00	0000	0000	0003	0016	0072	0209	0767	1428	1630	1611
05	0000	0000	0002	0013	0059	0175	0678	1331	1562	1630
10	0000	0000	0002	0010	0048	0147	0596	1232	1485	1633
15	0000	0000	0001	0008	0039	0122	0523	1134	1402	1621
20	0000	0000	0001	0006	0032	0102	0456	1038	1315	1594
25	0000	0000	0001	0005	0026	0085	0397	0945	1226	1556
30	0000	0000	0001	0004	0021	0071	0345	0856	1136	1507
35	0000	0000	0000	0003	0017	0059	0298	0772	1047	1449
40	0000	0000	0000	0003	0014	0049	0258	0694	0960	1384
45	0000	0000	0000	0002	0012	0041	0222	0621	0876	1314
50	0000	0000	0000	0002	0009	0034	0191	0554	0797	1241
55	0000	0000	0000	0001	0008	0028	0164	0492	0721	1165
60	0000	0000	0000	0001	0006	0023	0140	0436	0651	1088
65	0000	0000	0000	0001	0005	0019	0120	0386	0585	1012
70	0000	0000	0000	0001	0004	0016	0102	0340	0524	0936
75	0000	0000	0000	0001	0003	0013	0087	0299	0468	0863
80	0000	0000	0000	0000	0003	0011	0074	0262	0417	0793
85	0000	0000	0000	0000	0002	0009	0063	0230	0371	0726
90	0000	0000	0000	0000	0002	0007	0054	0201	0329	0662
95	0000	0000	0000	0000	0002	0006	0046	0175	0291	0602
4.00	0000	0000	0000	0000	0001	0005	0039	0153	0257	0546
05	0000	0000	0000	0000	0001	0004	0033	0133	0227	0493
10	0000	0000	0000	0000	0001	0004	0028	0116	0200	0445
15	0000	0000	0000	0000	0001	0003	0024	0101	0176	0401
20	0000	0000	0000	0000	0001	0002	0020	0087	0154	0360
4.25	0000	0000	0000	0000	0000	0002	0017	0076	0135	0322
30	0000	0000	0000	0000	0000	0002	0015	0066	0118	0288
35	0000	0000	0000	0000	0000	0001	0012	0057	0104	0258
40	0000	0000	0000	0000	0000	0001	0010	0049	0090	0230
45	0000	0000	0000	0000	0000	0001	0009	0042	0079	0204
50	0000	0000	0000	0000	0000	0001	0008	0037	0069	0182
55	0000	0000	0000	0000	0000	0001	0006	0032	0060	0161
60	0000	0000	0000	0000	0000	0001	0005	0027	0052	0143
65	0000	0000	0000	0000	0000	0000	0005	0024	0046	0127
70	0000	0000	0000	0000	0000	0000	0004	0020	0040	0112
75	0000	0000	0000	0000	0000	0000	0003	0018	0035	0099
80	0000	0000	0000	0000	0000	0000	0003	0015	0030	0088
85	0000	0000	0000	0000	0000	0000	0002	0013	0026	0077
90	0000	0000	0000	0000	0000	0000	0002	0011	0023	0068
95	0000	0000	0000	0000	0000	0000	0002	0010	0020	0060
5.00	0000	0000	0000	0000	0000	0000	0001	0008	0017	0053
05	0000	0000	0000	0000	0000	0000	0001	0007	0015	0047
10	0000	0000	0000	0000	0000	0000	0001	0006	0013	0041
15	0000	0000	0000	0000	0000	0000	0001	0005	0011	0036
20	0000	0000	0000	0000	0000	0000	0001	0005	0010	0032
25	0000	0000	0000	0000	0000	0000	0001	0004	0009	0028
30	0000	0000	0000	0000	0000	0000	0001	0003	0007	0025
35	0000	0000	0000	0000	0000	0000	0000	0003	0006	0022
40	0000	0000	0000	0000	0000	0000	0000	0003	0006	0019
45	0000	0000	0000	0000	0000	0000	0000	0002	0005	0017
50	0000	0000	0000	0000	0000	0000	0000	0002	0004	0015
55	0000	0000	0000	0000	0000	0000	0000	0002	0004	0013
60	0000	0000	0000	0000	0000	0000	0000	0001	0003	0011
65	0000	0000	0000	0000	0000	0000	0000	0001	0003	0010
70	0000	0000	0000	0000	0000	0000	0000	0001	0002	0009

PROBABILITY DENSITY OF t, THE NON-CENTRAL t-STATISTIC, TABLED AS A FUNCTION OF $t/\sqrt{f}$.

f is the number of degrees of freedom; the non-centrality parameter is $\sqrt{f+1}\,K_p$.
K_p is the standardized normal deviate exceeded with probability p.

DEGREES OF FREEDOM **21**

$t/\sqrt{f}$ \ p	.2500	.1500	.1000	.0650	.0400	.0250	.0100	.0040	.0025	.0010
5.75	0000	0000	0000	0000	0000	0000	0000	0001	0002	0008
80	0000	0000	0000	0000	0000	0000	0000	0001	0002	0007
85	0000	0000	0000	0000	0000	0000	0000	0001	0002	0006
90	0000	0000	0000	0000	0000	0000	0000	0001	0001	0005
95	0000	0000	0000	0000	0000	0000	0000	0001	0001	0005
6.00	0000	0000	0000	0000	0000	0000	0000	0000	0001	0004
05	0000	0000	0000	0000	0000	0000	0000	0000	0001	0004
10	0000	0000	0000	0000	0000	0000	0000	0000	0001	0003
15	0000	0000	0000	0000	0000	0000	0000	0000	0001	0003
20	0000	0000	0000	0000	0000	0000	0000	0000	0001	0002
25	0000	0000	0000	0000	0000	0000	0000	0000	0001	0002
30	0000	0000	0000	0000	0000	0000	0000	0000	0000	0002
35	0000	0000	0000	0000	0000	0000	0000	0000	0000	0002
40	0000	0000	0000	0000	0000	0000	0000	0000	0000	0001
45	0000	0000	0000	0000	0000	0000	0000	0000	0000	0001
50	0000	0000	0000	0000	0000	0000	0000	0000	0000	0001
55	0000	0000	0000	0000	0000	0000	0000	0000	0000	0001
60	0000	0000	0000	0000	0000	0000	0000	0000	0000	0001
65	0000	0000	0000	0000	0000	0000	0000	0000	0000	0001
70	0000	0000	0000	0000	0000	0000	0000	0000	0000	0001
75	0000	0000	0000	0000	0000	0000	0000	0000	0000	0001
80	0000	0000	0000	0000	0000	0000	0000	0000	0000	0001
85	0000	0000	0000	0000	0000	0000	0000	0000	0000	0000

PROBABILITY DENSITY OF t, THE NON-CENTRAL t-STATISTIC, TABLED AS A FUNCTION OF $t/\sqrt{f}$.

f is the number of degrees of freedom; the non-centrality parameter is $\sqrt{f+1}\,K_p$.
K_p is the standardized normal deviate exceeded with probability p.

DEGREES OF FREEDOM **22**

$t/\sqrt{f}$ \ p	.2500	.1500	.1000	.0650	.0400	.0250	.0100	.0040	.0025	.0010
-0.25	0000	0000	0000	0000	0000	0000	0000	0000	0000	0000
20	0001	0000	0000	0000	0000	0000	0000	0000	0000	0000
15	0002	0000	0000	0000	0000	0000	0000	0000	0000	0000
10	0004	0000	0000	0000	0000	0000	0000	0000	0000	0000
05	0010	0000	0000	0000	0000	0000	0000	0000	0000	0000
0.00	0021	0000	0000	0000	0000	0000	0000	0000	0000	0000
05	0044	0000	0000	0000	0000	0000	0000	0000	0000	0000
10	0089	0000	0000	0000	0000	0000	0000	0000	0000	0000
15	0168	0000	0000	0000	0000	0000	0000	0000	0000	0000
20	0301	0001	0000	0000	0000	0000	0000	0000	0000	0000
25	0507	0004	0000	0000	0000	0000	0000	0000	0000	0000
30	0802	0009	0000	0000	0000	0000	0000	0000	0000	0000
35	1190	0022	0000	0000	0000	0000	0000	0000	0000	0000
40	1657	0048	0001	0000	0000	0000	0000	0000	0000	0000
45	2167	0098	0003	0000	0000	0000	0000	0000	0000	0000
50	2667	0186	0007	0000	0000	0000	0000	0000	0000	0000
55	3097	0327	0016	0000	0000	0000	0000	0000	0000	0000
60	3404	0537	0037	0001	0000	0000	0000	0000	0000	0000
65	3551	0823	0076	0003	0000	0000	0000	0000	0000	0000
70	3530	1180	0145	0007	0000	0000	0000	0000	0000	0000
75	3356	1591	0257	0017	0000	0000	0000	0000	0000	0000
80	3063	2022	0424	0038	0001	0000	0000	0000	0000	0000
85	2694	2432	0654	0076	0003	0000	0000	0000	0000	0000
90	2291	2780	0945	0141	0009	0000	0000	0000	0000	0000
95	1891	3030	1287	0243	0019	0001	0000	0000	0000	0000
1.00	1519	3163	1656	0392	0040	0003	0000	0000	0000	0000
05	1192	3173	2024	0592	0078	0007	0000	0000	0000	0000
10	0916	3070	2359	0843	0139	0015	0000	0000	0000	0000
15	0691	2876	2631	1135	0232	0031	0000	0000	0000	0000
20	0513	2615	2819	1451	0363	0060	0001	0000	0000	0000
1.25	0375	2315	2913	1771	0536	0106	0002	0000	0000	0000
30	0272	2002	2911	2071	0750	0177	0004	0000	0000	0000
35	0194	1694	2823	2327	0998	0278	0009	0000	0000	0000
40	0138	1407	2664	2521	1267	0413	0019	0000	0000	0000
45	0097	1149	2452	2644	1543	0583	0036	0001	0000	0000
50	0068	0925	2208	2691	1807	0784	0064	0002	0000	0000
55	0048	0734	1949	2665	2043	1010	0107	0005	0001	0000
60	0033	0576	1690	2574	2235	1249	0168	0010	0002	0000
65	0023	0448	1442	2432	2374	1489	0252	0020	0004	0000
70	0016	0345	121[illegible]	2251	2455	1716	0361	0035	0009	0000
75	0011	0263	1007	2046	2476	1918	0495	0059	0016	0001
80	0008	0200	0827	1829	2441	2083	0651	0095	0029	0002
85	0005	0151	0672	1611	2358	2205	0827	0145	0048	0004
90	0004	0113	0541	1399	2236	2278	1015	0212	0077	0008
95	0002	0085	0432	1201	2084	2304	1207	0297	0118	0014
2.00	0002	0063	0343	1020	1913	2285	1395	0401	0172	0025
05	0001	0047	0271	0858	1732	2224	1570	0524	0243	0040
10	0001	0035	0212	0715	1548	2130	1724	0662	0330	0063
15	0001	0026	0166	0591	1368	2009	1852	0812	0434	0095
20	0000	0019	0129	0486	1196	1868	1949	0969	0553	0137
25	0000	0014	0100	0396	1036	1716	2012	1128	0684	0192
30	0000	0010	0077	0322	0890	1558	2041	1281	0825	0260
35	0000	0008	0060	0260	0758	1400	2038	1425	0969	0341
40	0000	0006	0046	0209	0642	1246	2005	1552	1114	0436
45	0000	0004	0035	0168	0540	1099	1945	1660	1253	0541
50	0000	0003	0027	0134	0451	0962	1864	1745	1382	0656
55	0000	0002	0021	0107	0376	0836	1766	1805	1497	0777
60	0000	0002	0016	0085	0311	0722	1655	1840	1593	0902
65	0000	0001	0012	0067	0257	0619	1536	1850	1670	1026
70	0000	0001	0009	0053	0211	0529	1413	1937	1725	1146

PROBABILITY DENSITY OF t, THE NON-CENTRAL t-STATISTIC, TABLED AS A FUNCTION OF $t/\sqrt{f}$

f is the number of degrees of freedom; the non-centrality parameter is $\sqrt{f+1}\,K_p$.
K_p is the standardized normal deviate exceeded with probability p.

DEGREES OF FREEDOM **22**

$t/\sqrt{f}$ \ p	.2500	.1500	.1000	.0650	.0400	.0250	.0100	.0040	.0025	.0010
2.75	0000	0001	0007	0042	0173	0449	1289	1802	1758	1258
80	0000	0001	0005	0033	0141	0380	1167	1749	1769	1359
85	0000	0000	0004	0026	0115	0320	1049	1680	1759	1446
90	0000	0000	0003	0021	0094	0269	0937	1599	1730	1518
95	0000	0000	0002	0016	0076	0225	0832	1509	1685	1573
3.00	0000	0000	0002	0013	0062	0188	0735	1413	1626	1611
05	0000	0000	0001	0010	0050	0156	0646	1312	1555	1631
10	0000	0000	0001	0008	0041	0130	0565	1211	1475	1634
15	0000	0000	0001	0006	0033	0107	0492	1111	1389	1621
20	0000	0000	0001	0005	0026	0089	0427	1012	1299	1594
25	0000	0000	0001	0004	0021	0073	0369	0918	1207	1554
30	0000	0000	0000	0003	0017	0061	0319	0828	1114	1503
35	0000	0000	0000	0002	0014	0050	0274	0743	1023	1443
40	0000	0000	0000	0002	0011	0041	0235	0665	0935	1375
45	0000	0000	0000	0002	0009	0034	0201	0592	0850	1303
50	0000	0000	0000	0001	0007	0028	0172	0525	0769	1226
55	0000	0000	0000	0001	0006	0023	0146	0464	0693	1148
60	0000	0000	0000	0001	0005	0019	0124	0409	0623	1069
65	0000	0000	0000	0001	0004	0015	0106	0360	0557	0991
70	0000	0000	0000	0000	0003	0013	0090	0315	0497	0914
75	0000	0000	0000	0000	0003	0010	0076	0276	0442	0840
80	0000	0000	0000	0000	0002	0009	0064	0241	0392	0768
85	0000	0000	0000	0000	0002	0007	0054	0209	0346	0700
90	0000	0000	0000	0000	0001	0006	0046	0182	0305	0636
95	0000	0000	0000	0000	0001	0005	0039	0158	0269	0576
4.00	0000	0000	0000	0000	0001	0004	0033	0137	0236	0520
05	0000	0000	0000	0000	0001	0003	0027	0118	0207	0468
10	0000	0000	0000	0000	0001	0003	0023	0102	0181	0420
15	0000	0000	0000	0000	0000	0002	0020	0088	0158	0376
20	0000	0000	0000	0000	0000	0002	0016	0076	0138	0336
4.25	0000	0000	0000	0000	0000	0001	0014	0066	0120	0300
30	0000	0000	0000	0000	0000	0001	0012	0056	0105	0267
35	0000	0000	0000	0000	0000	0001	0010	0049	0091	0237
40	0000	0000	0000	0000	0000	0001	0008	0042	0079	0210
45	0000	0000	0000	0000	0000	0001	0007	0036	0069	0186
50	0000	0000	0000	0000	0000	0001	0006	0031	0060	0165
55	0000	0000	0000	0000	0000	0000	0005	0026	0052	0146
60	0000	0000	0000	0000	0000	0000	0004	0023	0045	0128
65	0000	0000	0000	0000	0000	0000	0004	0019	0039	0113
70	0000	0000	0000	0000	0000	0000	0003	0017	0033	0100
75	0000	0000	0000	0000	0000	0000	0002	0014	0029	0088
80	0000	0000	0000	0000	0000	0000	0002	0012	0025	0077
85	0000	0000	0000	0000	0000	0000	0002	0010	0022	0068
90	0000	0000	0000	0000	0000	0000	0002	0009	0019	0059
95	0000	0000	0000	0000	0000	0000	0001	0008	0016	0052
5.00	0000	0000	0000	0000	0000	0000	0001	0007	0014	0046
05	0000	0000	0000	0000	0000	0000	0001	0006	0012	0040
10	0000	0000	0000	0000	0000	0000	0001	0005	0010	0035
15	0000	0000	0000	0000	0000	0000	0001	0004	0009	0031
20	0000	0000	0000	0000	0000	0000	0001	0004	0008	0027
25	0000	0000	0000	0000	0000	0000	0000	0003	0007	0023
30	0000	0000	0000	0000	0000	0000	0000	0003	0006	0020
35	0000	0000	0000	0000	0000	0000	0000	0002	0005	0018
40	0000	0000	0000	0000	0000	0000	0000	0002	0004	0016
45	0000	0000	0000	0000	0000	0000	0000	0002	0004	0014
50	0000	0000	0000	0000	0000	0000	0000	0001	0003	0012
55	0000	0000	0000	0000	0000	0000	0000	0001	0003	0010
60	0000	0000	0000	0000	0000	0000	0000	0001	0002	0009
65	0000	0000	0000	0000	0000	0000	0000	0001	0002	0008
70	0000	0000	0000	0000	0000	0000	0000	0001	0002	0007

PROBABILITY DENSITY OF t, THE NON-CENTRAL t-STATISTIC, TABLED AS A FUNCTION OF $t/\sqrt{f}$.

f is the number of degrees of freedom; the non-centrality parameter is $\sqrt{f+1}\,K_p$.
K_p is the standardized normal deviate exceeded with probability p.

DEGREES OF FREEDOM **22**

$t/\sqrt{f}$ \ p	.2500	.1500	.1000	.0650	.0400	.0250	.0100	.0040	.0025	.0010
5.75	0000	0000	0000	0000	0000	0000	0000	0001	0002	0006
80	0000	0000	0000	0000	0000	0000	0000	0001	0001	0005
85	0000	0000	0000	0000	0000	0000	0000	0001	0001	0005
90	0000	0000	0000	0000	0000	0000	0000	0000	0001	0004
95	0000	0000	0000	0000	0000	0000	0000	0000	0001	0004
6.00	0000	0000	0000	0000	0000	0000	0000	0000	0001	0003
05	0000	0000	0000	0000	0000	0000	0000	0000	0001	0003
10	0000	0000	0000	0000	0000	0000	0000	0000	0001	0002
15	0000	0000	0000	0000	0000	0000	0000	0000	0000	0002
20	0000	0000	0000	0000	0000	0000	0000	0000	0000	0002
25	0000	0000	0000	0000	0000	0000	0000	0000	0000	0002
30	0000	0000	0000	0000	0000	0000	0000	0000	0000	0001
35	0000	0000	0000	0000	0000	0000	0000	0000	0000	0001
40	0000	0000	0000	0000	0000	0000	0000	0000	0000	0001
45	0000	0000	0000	0000	0000	0000	0000	0000	0000	0001
50	0000	0000	0000	0000	0000	0000	0000	0000	0000	0001
55	0000	0000	0000	0000	0000	0000	0000	0000	0000	0001
60	0000	0000	0000	0000	0000	0000	0000	0000	0000	0001
65	0000	0000	0000	0000	0000	0000	0000	0000	0000	0001
70	0000	0000	0000	0000	0000	0000	0000	0000	0000	0000

PROBABILITY DENSITY OF t, THE NON-CENTRAL t-STATISTIC, TABLED AS A FUNCTION OF $t/\sqrt{f}$.

f is the number of degrees of freedom; the non-centrality parameter is $\sqrt{f+1}\,K_{\hat{p}}$.
K_p is the standardized normal deviate exceeded with probability p.

$t/\sqrt{f}$ \ p	.2500	.1500	.1000	.0650	.0400	.0250	.0100	.0040	.0025	.0010
-0.25	0000	0000	0000	0000	0000	0000	0000	0000	0000	0000
20	0001	0000	0000	0000	0000	0000	0000	0000	0000	0000
15	0001	0000	0000	0000	0000	0000	0000	0000	0000	0000
10	0003	0000	0000	0000	0000	0000	0000	0000	0000	0000
05	0007	0000	0000	0000	0000	0000	0000	0000	0000	0000
0.00	0017	0000	0000	0000	0000	0000	0000	0000	0000	0000
05	0037	0000	0000	0000	0000	0000	0000	0000	0000	0000
10	0075	0000	0000	0000	0000	0000	0000	0000	0000	0000
15	0147	0000	0000	0000	0000	0000	0000	0000	0000	0000
20	0270	0001	0000	0000	0000	0000	0000	0000	0000	0000
25	0465	0003	0000	0000	0000	0000	0000	0000	0000	0000
30	0751	0007	0000	0000	0000	0000	0000	0000	0000	0000
35	1133	0018	0000	0000	0000	0000	0000	0000	0000	0000
40	1601	0040	0001	0000	0000	0000	0000	0000	0000	0000
45	2119	0084	0002	0000	0000	0000	0000	0000	0000	0000
50	2633	0164	0005	0000	0000	0000	0000	0000	0000	0000
55	3078	0295	0013	0000	0000	0000	0000	0000	0000	0000
60	3397	0496	0030	0001	0000	0000	0000	0000	0000	0000
65	3552	0774	0065	0002	0000	0000	0000	0000	0000	0000
70	3531	1129	0127	0006	0000	0000	0000	0000	0000	0000
75	3350	1542	0231	0014	0000	0000	0000	0000	0000	0000
80	3047	1981	0389	0031	0001	0000	0000	0000	0000	0000
85	2665	2403	0612	0065	0003	0000	0000	0000	0000	0000
90	2251	2763	0899	0124	0007	0000	0000	0000	0000	0000
95	1843	3024	1240	0219	0016	0001	0000	0000	0000	0000
1.00	1467	3163	1614	0360	0034	0002	0000	0000	0000	0000
05	1139	3174	1991	0553	0067	0005	0000	0000	0000	0000
10	0865	3068	2336	0800	0122	0012	0000	0000	0000	0000
15	0645	2866	2619	1091	0209	0026	0000	0000	0000	0000
20	0472	2596	2815	1412	0333	0051	0000	0000	0000	0000
1.25	0341	2287	2913	1738	0501	0093	0001	0000	0000	0000
30	0243	1965	2912	2046	0711	0158	0003	0000	0000	0000
35	0172	1651	2821	2312	0958	0253	0007	0000	0000	0000
40	0120	1361	2655	2515	1230	0383	0016	0000	0000	0000
45	0083	1102	2437	2643	1511	0548	0030	0001	0000	0000
50	0058	0878	2184	2692	1782	0747	0055	0002	0000	0000
55	0040	0690	1918	2665	2026	0973	0094	0004	0001	0000
60	0027	0536	1653	2571	2226	1215	0151	0008	0001	0000
65	0018	0412	1401	2424	2371	1460	0230	0016	0003	0000
70	0013	0314	1169	2236	2455	1694	0334	0029	0007	0000
75	0009	0237	0963	2024	2477	1902	0465	0051	0013	0001
80	0006	0178	0784	1801	2441	2074	0619	0083	0024	0002
85	0004	0133	0632	1577	2355	2201	0795	0129	0041	0003
90	0003	0098	0504	1362	2228	2278	0984	0192	0067	0006
95	0002	0073	0399	1162	2071	2306	1179	0274	0104	0012
2.00	0001	0054	0313	0980	1894	2285	1372	0375	0156	0020
05	0001	0039	0245	0818	1708	2223	1552	0496	0223	0034
10	0001	0029	0190	0676	1519	2124	1712	0633	0307	0055
15	0000	0021	0147	0555	1335	1999	1845	0783	0408	0084
20	0000	0015	0113	0452	1161	1854	1946	0942	0525	0123
25	0000	0011	0087	0365	0999	1696	2012	1103	0656	0175
30	0000	0008	0066	0294	0852	1534	2042	1261	0797	0240
35	0000	0006	0051	0235	0721	1372	2039	1409	0944	0319
40	0000	0004	0038	0188	0606	1215	2005	1541	1092	0411
45	0000	0003	0029	0149	0506	1066	1943	1653	1235	0516
50	0000	0002	0022	0118	0420	0928	1858	1741	1368	0630
55	0000	0002	0017	0093	0347	0801	1756	1804	1486	0753
60	0000	0001	0013	0073	0285	0687	1641	1841	1587	0879
65	0000	0001	0010	0057	0233	0586	1519	1851	1667	1005
70	0000	0001	0007	0045	0190	0497	1392	1837	1724	1128

PROBABILITY DENSITY OF t, THE NON-CENTRAL t-STATISTIC, TABLED AS A FUNCTION OF $t/\sqrt{f}$.

f is the number of degrees of freedom; the non-centrality parameter is $\sqrt{f+1}\,K_p$.
K_p is the standardized normal deviate exceeded with probability p.

$t/\sqrt{f}$ \ p	.2500	.1500	.1000	.0650	.0400	.0250	.0100	.0040	.0025	.0010
2.75	0000	0000	0006	0035	0155	0419	1265	1801	1758	1244
80	0000	0000	0004	0028	0125	0352	1140	1746	1770	1348
85	0000	0000	0003	0021	0101	0294	1020	1675	1760	1439
90	0000	0000	0002	0017	0082	0245	0907	1591	1730	1514
95	0000	0000	0002	0013	0066	0204	0801	1497	1683	1571
3.00	0000	0000	0001	0010	0053	0169	0704	1398	1621	1611
05	0000	0000	0001	0008	0042	0139	0615	1294	1548	1632
10	0000	0000	0001	0006	0034	0115	0535	1190	1465	1635
15	0000	0000	0001	0005	0027	0094	0463	1088	1376	1622
20	0000	0000	0000	0004	0022	0077	0400	0987	1283	1594
25	0000	0000	0000	0003	0017	0063	0344	0891	1188	1552
30	0000	0000	0000	0002	0014	0052	0294	0801	1093	1499
35	0000	0000	0000	0002	0011	0042	0251	0715	1000	1436
40	0000	0000	0000	0001	0009	0035	0214	0636	0910	1366
45	0000	0000	0000	0001	0007	0028	0182	0564	0824	1291
50	0000	0000	0000	0001	0006	0023	0154	0498	0743	1212
55	0000	0000	0000	0001	0005	0019	0131	0438	0667	1132
60	0000	0000	0000	0001	0004	0015	0110	0384	0596	1051
65	0000	0000	0000	0000	0003	0012	0093	0335	0531	0971
70	0000	0000	0000	0000	0002	0010	0078	0292	0471	0892
75	0000	0000	0000	0000	0002	0008	0066	0254	0416	0817
80	0000	0000	0000	0000	0001	0007	0055	0220	0367	0744
85	0000	0000	0000	0000	0001	0005	0046	0191	0323	0676
90	0000	0000	0000	0000	0001	0004	0039	0165	0283	0611
95	0000	0000	0000	0000	0001	0004	0033	0142	0248	0551
4.00	0000	0000	0000	0000	0001	0003	0027	0122	0217	0495
05	0000	0000	0000	0000	0000	0002	0023	0105	0189	0444
10	0000	0000	0000	0000	0000	0002	0019	0090	0164	0397
15	0000	0000	0000	0000	0000	0002	0016	0077	0143	0354
20	0000	0000	0000	0000	0000	0001	0013	0066	0124	0314
4.25	0000	0000	0000	0000	0000	0001	0011	0057	0107	0279
30	0000	0000	0000	0000	0000	0001	0009	0049	0093	0247
35	0000	0000	0000	0000	0000	0001	0008	0041	0080	0218
40	0000	0000	0000	0000	0000	0001	0007	0035	0069	0193
45	0000	0000	0000	0000	0000	0000	0005	0030	0060	0170
50	0000	0000	0000	0000	0000	0000	0005	0026	0052	0149
55	0000	0000	0000	0000	0000	0000	0004	0022	0044	0131
60	0000	0000	0000	0000	0000	0000	0003	0019	0038	0115
65	0000	0000	0000	0000	0000	0000	0003	0016	0033	0101
70	0000	0000	0000	0000	0000	0000	0002	0014	0028	0088
75	0000	0000	0000	0000	0000	0000	0002	0012	0024	0077
80	0000	0000	0000	0000	0000	0000	0002	0010	0021	0067
85	0000	0000	0000	0000	0000	0000	0001	0008	0018	0059
90	0000	0000	0000	0000	0000	0000	0001	0007	0015	0051
95	0000	0000	0000	0000	0000	0000	0001	0006	0013	0045
5.00	0000	0000	0000	0000	0000	0000	0001	0005	0011	0039
05	0000	0000	0000	0000	0000	0000	0001	0004	0010	0034
10	0000	0000	0000	0000	0000	0000	0001	0004	0008	0030
15	0000	0000	0000	0000	0000	0000	0000	0003	0007	0026
20	0000	0000	0000	0000	0000	0000	0000	0003	0006	0022
25	0000	0000	0000	0000	0000	0000	0000	0002	0005	0019
30	0000	0000	0000	0000	0000	0000	0000	0002	0005	0017
35	0000	0000	0000	0000	0000	0000	0000	0002	0004	0015
40	0000	0000	0000	0000	0000	0000	0000	0001	0003	0013
45	0000	0000	0000	0000	0000	0000	0000	0001	0003	0011
50	0000	0000	0000	0000	0000	0000	0000	0001	0002	0010
55	0000	0000	0000	0000	0000	0000	0000	0001	0002	0008
60	0000	0000	0000	0000	0000	0000	0000	0001	0002	0007
65	0000	0000	0000	0000	0000	0000	0000	0001	0002	0006
70	0000	0000	0000	0000	0000	0000	0000	0001	0001	0006

PROBABILITY DENSITY OF t, THE NON-CENTRAL t-STATISTIC, TABLED AS A FUNCTION OF $t/\sqrt{f}$.

f is the number of degrees of freedom; the non-centrality parameter is $\sqrt{f+1}\,K_p$.
K_p is the standardized normal deviate exceeded with probability p.

DEGREES OF FREEDOM **23**

$t/\sqrt{f}$ \ p	.2500	.1500	.1000	.0650	.0400	.0250	.0100	.0040	.0025	.0010
5.75	0000	0000	0000	0000	0000	0000	0000	0000	0001	0005
80	0000	0000	0000	0000	0000	0000	0000	0000	0001	0004
85	0000	0000	0000	0000	0000	0000	0000	0000	0001	0004
90	0000	0000	0000	0000	0000	0000	0000	0000	0001	0003
95	0000	0000	0000	0000	0000	0000	0000	0000	0001	0003
6.00	0000	0000	0000	0000	0000	0000	0000	0000	0001	0002
05	0000	0000	0000	0000	0000	0000	0000	0000	0000	0002
10	0000	0000	0000	0000	0000	0000	0000	0000	0000	0002
15	0000	0000	0000	0000	0000	0000	0000	0000	0000	0002
20	0000	0000	0000	0000	0000	0000	0000	0000	0000	0001
25	0000	0000	0000	0000	0000	0000	0000	0000	0000	0001
30	0000	0000	0000	0000	0000	0000	0000	0000	0000	0001
35	0000	0000	0000	0000	0000	0000	0000	0000	0000	0001
40	0000	0000	0000	0000	0000	0000	0000	0000	0000	0001
45	0000	0000	0000	0000	0000	0000	0000	0000	0000	0001
50	0000	0000	0000	0000	0000	0000	0000	0000	0000	0001
55	0000	0000	0000	0000	0000	0000	0000	0000	0000	0001
60	0000	0000	0000	0000	0000	0000	0000	0000	0000	0000

PROBABILITY DENSITY OF t, THE NON-CENTRAL t-STATISTIC, TABLED AS A FUNCTION OF $t/\sqrt{f}$.

f is the number of degrees of freedom; the non-centrality parameter is $\sqrt{f+1}\,K_p$.
K_p is the standardized normal deviate exceeded with probability p.

DEGREES OF FREEDOM **24**

$t/\sqrt{f}$ \ p	.2500	.1500	.1000	.0650	.0400	.0250	.0100	.0040	.0025	.0010
-0.20	0000	0000	0000	0000	0000	0000	0000	0000	0000	0000
15	0001	0000	0000	0000	0000	0000	0000	0000	0000	0000
10	0002	0000	0000	0000	0000	0000	0000	0000	0000	0000
05	0006	0000	0000	0000	0000	0000	0000	0000	0000	0000
0.00	0013	0000	0000	0000	0000	0000	0000	0000	0000	0000
05	0030	0000	0000	0000	0000	0000	0000	0000	0000	0000
10	0064	0000	0000	0000	0000	0000	0000	0000	0000	0000
15	0128	0000	0000	0000	0000	0000	0000	0000	0000	0000
20	0242	0001	0000	0000	0000	0000	0000	0000	0000	0000
25	0427	0002	0000	0000	0000	0000	0000	0000	0000	0000
30	0702	0006	0000	0000	0000	0000	0000	0000	0000	0000
35	1079	0014	0000	0000	0000	0000	0000	0000	0000	0000
40	1547	0033	0000	0000	0000	0000	0000	0000	0000	0000
45	2072	0072	0001	0000	0000	0000	0000	0000	0000	0000
50	2599	0144	0004	0000	0000	0000	0000	0000	0000	0000
55	3059	0267	0010	0000	0000	0000	0000	0000	0000	0000
60	3391	0458	0025	0000	0000	0000	0000	0000	0000	0000
65	3553	0728	0055	0002	0000	0000	0000	0000	0000	0000
70	3531	1079	0111	0004	0000	0000	0000	0000	0000	0000
75	3344	1494	0207	0011	0000	0000	0000	0000	0000	0000
80	3030	1941	0357	0026	0001	0000	0000	0000	0000	0000
85	2636	2374	0572	0055	0002	0000	0000	0000	0000	0000
90	2211	2746	0854	0108	0005	0000	0000	0000	0000	0000
95	1795	3017	1195	0196	0013	0001	0000	0000	0000	0000
1.00	1416	3163	1573	0330	0028	0002	0000	0000	0000	0000
05	1088	3176	1958	0517	0057	0004	0000	0000	0000	0000
10	0817	3065	2313	0759	0107	0010	0000	0000	0000	0000
15	0601	2856	2606	1050	0188	0021	0000	0000	0000	0000
20	0435	2576	2811	1373	0306	0043	0000	0000	0000	0000
1.25	0310	2258	2914	1706	0468	0081	0001	0000	0000	0000
30	0218	1929	2913	2022	0674	0141	0003	0000	0000	0000
35	0152	1609	2818	2297	0920	0231	0006	0000	0000	0000
40	0105	1316	2647	2508	1194	0355	0013	0000	0000	0000
45	0072	1056	2421	2641	1479	0516	0025	0001	0000	0000
50	0049	0834	2161	2693	1757	0712	0047	0001	0000	0000
55	0033	0649	1887	2666	2009	0938	0082	0003	0000	0000
60	0022	0499	1616	2569	2216	1183	0135	0007	0001	0000
65	0015	0379	1361	2415	2367	1432	0210	0013	0003	0000
70	0010	0286	1128	2221	2455	1672	0310	0025	0005	0000
75	0007	0213	0921	2003	2478	1887	0436	0044	0011	0000
80	0004	0158	0744	1773	2441	2065	0589	0073	0020	0001
85	0003	0117	0594	1545	2352	2197	0763	0116	0035	0002
90	0002	0085	0469	1326	2220	2278	0954	0175	0058	0005
95	0001	0062	0368	1124	2058	2307	1152	0253	0093	0009
2.00	0001	0045	0286	0941	1875	2286	1349	0351	0140	0017
05	0001	0033	0221	0779	1683	2221	1534	0469	0204	0029
10	0000	0024	0170	0639	1490	2119	1699	0605	0285	0047
15	0000	0017	0130	0520	1303	1989	1837	0756	0383	0074
20	0000	0012	0099	0420	1126	1839	1942	0916	0499	0111
25	0000	0009	0075	0337	0963	1677	2011	1080	0629	0159
30	0000	0006	0057	0269	0817	1510	2043	1241	0771	0221
35	0000	0005	0043	0213	0687	1344	2040	1393	0920	0298
40	0000	0003	0032	0168	0573	1184	2004	1529	1070	0388
45	0000	0002	0024	0132	0475	1034	1940	1645	1216	0491
50	0000	0002	0018	0104	0391	0895	1852	1737	1353	0606
55	0000	0001	0014	0081	0320	0768	1747	1803	1476	0729
60	0000	0001	0010	0063	0261	0655	1628	1841	1580	0856
65	0000	0001	0008	0049	0212	0555	1501	1852	1663	0985
70	0000	0000	0006	0038	0171	0467	1371	1938	1723	1111

PROBABILITY DENSITY OF t, THE NON-CENTRAL t-STATISTIC, TABLED AS A FUNCTION OF $t/\sqrt{f}$.

f is the number of degrees of freedom; the non-centrality parameter is $\sqrt{f+1}\,K_p$.
K_p is the standardized normal deviate exceeded with probability p.

$t/\sqrt{f}$ \ p	.2500	.1500	.1000	.0650	.0400	.0250	.0100	.0040	.0025	.0010
2.75	0000	0000	0004	0029	0138	0391	1241	1800	1759	1230
80	0000	0000	0003	0023	0111	0326	1114	1743	1771	1338
85	0000	0000	0002	0018	0089	0271	0992	1669	1760	1432
90	0000	0000	0002	0014	0071	0224	0878	1582	1730	1510
95	0000	0000	0001	0011	0057	0184	0771	1485	1681	1570
3.00	0000	0000	0001	0008	0045	0151	0674	1383	1617	1611
05	0000	0000	0001	0006	0036	0124	0586	1277	1540	1633
10	0000	0000	0001	0005	0028	0101	0507	1170	1455	1636
15	0000	0000	0000	0004	0023	0083	0436	1065	1363	1622
20	0000	0000	0000	0003	0018	0067	0374	0963	1267	1593
25	0000	0000	0000	0002	0014	0055	0319	0866	1170	1549
30	0000	0000	0000	0002	0011	0044	0272	0774	1073	1494
35	0000	0000	0000	0001	0009	0036	0231	0689	0978	1429
40	0000	0000	0000	0001	0007	0029	0195	0610	0887	1357
45	0000	0000	0000	0001	0006	0024	0165	0537	0799	1280
50	0000	0000	0000	0001	0004	0019	0139	0472	0717	1199
55	0000	0000	0000	0000	0004	0015	0117	0413	0641	1116
60	0000	0000	0000	0000	0003	0012	0098	0360	0570	1033
65	0000	0000	0000	0000	0002	0010	0082	0313	0505	0951
70	0000	0000	0000	0000	0002	0008	0068	0271	0446	0871
75	0000	0000	0000	0000	0001	0007	0057	0234	0393	0794
80	0000	0000	0000	0000	0001	0005	0048	0202	0344	0721
85	0000	0000	0000	0000	0001	0004	0040	0174	0301	0652
90	0000	0000	0000	0000	0001	0003	0033	0149	0263	0587
95	0000	0000	0000	0000	0001	0003	0027	0128	0229	0527
4.00	0000	0000	0000	0000	0000	0002	0023	0109	0199	0472
05	0000	0000	0000	0000	0000	0002	0019	0094	0172	0421
10	0000	0000	0000	0000	0000	0001	0016	0080	0149	0374
15	0000	0000	0000	0000	0000	0001	0013	0068	0129	0332
20	0000	0000	0000	0000	0000	0001	0011	0058	0111	0294
4.25	0000	0000	0000	0000	0000	0001	0009	0049	0096	0260
30	0000	0000	0000	0000	0000	0001	0007	0042	0082	0229
35	0000	0000	0000	0000	0000	0001	0006	0035	0071	0201
40	0000	0000	0000	0000	0000	0000	0005	0030	0061	0177
45	0000	0000	0000	0000	0000	0000	0004	0026	0052	0155
50	0000	0000	0000	0000	0000	0000	0004	0022	0045	0135
55	0000	0000	0000	0000	0000	0000	0003	0018	0038	0118
60	0000	0000	0000	0000	0000	0000	0002	0015	0033	0103
65	0000	0000	0000	0000	0000	0000	0002	0013	0028	0090
70	0000	0000	0000	0000	0000	0000	0002	0011	0024	0078
75	0000	0000	0000	0000	0000	0000	0001	0009	0020	0068
80	0000	0000	0000	0000	0000	0000	0001	0008	0017	0059
85	0000	0000	0000	0000	0000	0000	0001	0007	0015	0051
90	0000	0000	0000	0000	0000	0000	0001	0006	0013	0045
95	0000	0000	0000	0000	0000	0000	0001	0005	0011	0039
5.00	0000	0000	0000	0000	0000	0000	0001	0004	0009	0033
05	0000	0000	0000	0000	0000	0000	0000	0003	0008	0029
10	0000	0000	0000	0000	0000	0000	0000	0003	0007	0025
15	0000	0000	0000	0000	0000	0000	0000	0002	0006	0022
20	0000	0000	0000	0000	0000	0000	0000	0002	0005	0019
25	0000	0000	0000	0000	0000	0000	0000	0002	0004	0016
30	0000	0000	0000	0000	0000	0000	0000	0001	0004	0014
35	0000	0000	0000	0000	0000	0000	0000	0001	0003	0012
40	0000	0000	0000	0000	0000	0000	0000	0001	0003	0010
45	0000	0000	0000	0000	0000	0000	0000	0001	0002	0009
50	0000	0000	0000	0000	0000	0000	0000	0001	0002	0008
55	0000	0000	0000	0000	0000	0000	0000	0001	0002	0007
60	0000	0000	0000	0000	0000	0000	0000	0001	0001	0006
65	0000	0000	0000	0000	0000	0000	0000	0000	0001	0005
70	0000	0000	0000	0000	0000	0000	0000	0000	0001	0004

PROBABILITY DENSITY OF t, THE NON-CENTRAL t-STATISTIC, TABLED AS A FUNCTION OF $t/\sqrt{f}$.

f is the number of degrees of freedom; the non-centrality parameter is $\sqrt{f+1}\,K_p$.
K_p is the standardized normal deviate exceeded with probability p.

$t/\sqrt{f}$ \ p	.2500	.1500	.1000	.0650	.0400	.0250	.0100	.0040	.0025	.0010
5.75	0000	0000	0000	0000	0000	0000	0000	0000	0001	0004
80	0000	0000	0000	0000	0000	0000	0000	0000	0001	0003
85	0000	0000	0000	0000	0000	0000	0000	0000	0001	0003
90	0000	0000	0000	0000	0000	0000	0000	0000	0001	0002
95	0000	0000	0000	0000	0000	0000	0000	0000	0000	0002
6.00	0000	0000	0000	0000	0000	0000	0000	0000	0000	0002
05	0000	0000	0000	0000	0000	0000	0000	0000	0000	0002
10	0000	0000	0000	0000	0000	0000	0000	0000	0000	0001
15	0000	0000	0000	0000	0000	0000	0000	0000	0000	0001
20	0000	0000	0000	0000	0000	0000	0000	0000	0000	0001
25	0000	0000	0000	0000	0000	0000	0000	0000	0000	0001
30	0000	0000	0000	0000	0000	0000	0000	0000	0000	0001
35	0000	0000	0000	0000	0000	0000	0000	0000	0000	0001
40	0000	0000	0000	0000	0000	0000	0000	0000	0000	0001
45	0000	0000	0000	0000	0000	0000	0000	0000	0000	0000

PROBABILITY DENSITY OF t, THE NON-CENTRAL t-STATISTIC, TABLED AS A FUNCTION OF $t/\sqrt{f}$.

f is the number of degrees of freedom; the non-centrality parameter is $\sqrt{f+1}\,K_p$.
K_p is the standardized normal deviate exceeded with probability p.

DEGREES OF FREEDOM **29**

$t/\sqrt{f}$ \ p	.2500	.1500	.1000	.0650	.0400	.0250	.0100	.0040	.0025	.0010
-0.10	0001	0000	0000	0000	0000	0000	0000	0000	0000	0000
05	0002	0000	0000	0000	0000	0000	0000	0000	0000	0000
0.00	0004	0000	0000	0000	0000	0000	0000	0000	0000	0000
05	0011	0000	0000	0000	0000	0000	0000	0000	0000	0000
10	0028	0000	0000	0000	0000	0000	0000	0000	0000	0000
15	0065	0000	0000	0000	0000	0000	0000	0000	0000	0000
20	0140	0000	0000	0000	0000	0000	0000	0000	0000	0000
25	0276	0000	0000	0000	0000	0000	0000	0000	0000	0000
30	0503	0002	0000	0000	0000	0000	0000	0000	0000	0000
35	0844	0005	0000	0000	0000	0000	0000	0000	0000	0000
40	1302	0013	0000	0000	0000	0000	0000	0000	0000	0000
45	1852	0033	0000	0000	0000	0000	0000	0000	0000	0000
50	2433	0076	0001	0000	0000	0000	0000	0000	0000	0000
55	2963	0160	0003	0000	0000	0000	0000	0000	0000	0000
60	3357	0307	0009	0000	0000	0000	0000	0000	0000	0000
65	3554	0537	0024	0000	0000	0000	0000	0000	0000	0000
70	3533	0863	0057	0001	0000	0000	0000	0000	0000	0000
75	3313	1277	0120	0004	0000	0000	0000	0000	0000	0000
80	2946	1750	0232	0010	0000	0000	0000	0000	0000	0000
85	2495	2231	0409	0025	0000	0000	0000	0000	0000	0000
90	2022	2661	0662	0056	0001	0000	0000	0000	0000	0000
95	1576	2984	0993	0115	0004	0000	0000	0000	0000	0000
1.00	1186	3162	1383	0215	0011	0000	0000	0000	0000	0000
05	0865	3180	1801	0368	0026	0001	0000	0000	0000	0000
10	0614	3052	2202	0585	0057	0003	0000	0000	0000	0000
15	0425	2806	2544	0864	0111	0008	0000	0000	0000	0000
20	0289	2482	2789	1194	0199	0019	0000	0000	0000	0000
1.25	0192	2120	2915	1552	0333	0041	0000	0000	0000	0000
30	0126	1756	2917	1906	0516	0080	0001	0000	0000	0000
35	0082	1414	2806	2222	0750	0144	0002	0000	0000	0000
40	0052	1111	2605	2472	1027	0242	0005	0000	0000	0000
45	0033	0854	2342	2633	1330	0380	0010	0000	0000	0000
50	0021	0644	2045	2698	1637	0560	0022	0000	0000	0000
55	0013	0477	1740	2667	1924	0780	0043	0001	0000	0000
60	0008	0348	1445	2553	2166	1031	0078	0002	0000	0000
65	0005	0251	1176	2373	2346	1299	0132	0005	0001	0000
70	0003	0178	0939	2147	2453	1565	0211	0010	0002	0000
75	0002	0126	0738	1898	2483	1811	0319	0020	0004	0000
80	0001	0088	0571	1641	2440	2020	0457	0038	0008	0000
85	0001	0061	0436	1391	2335	2177	0625	0066	0016	0001
90	0000	0042	0329	1159	2181	2275	0817	0108	0029	0002
95	0000	0029	0246	0950	1991	2311	1025	0169	0051	0003
2.00	0000	0020	0182	0768	1783	2287	1239	0251	0084	0007
05	0000	0013	0133	0613	1567	2211	1447	0355	0132	0013
10	0000	0009	0097	0484	1354	2091	1637	0482	0197	0023
15	0000	0006	0071	0378	1153	1938	1799	0630	0282	0040
20	0000	0004	0051	0292	0968	1765	1924	0794	0386	0064
25	0000	0003	0036	0225	0803	1581	2007	0968	0511	0100
30	0000	0002	0026	0171	0659	1395	2047	1145	0652	0148
35	0000	0001	0019	0130	0535	1214	2044	1316	0806	0211
40	0000	0001	0013	0098	0431	1043	2002	1472	0967	0290
45	0000	0001	0009	0073	0344	0886	1926	1608	1128	0386
50	0000	0000	0007	0055	0273	0745	1822	1717	1283	0496
55	0000	0000	0005	0041	0215	0621	1699	1796	1424	0619
60	0000	0000	0003	0030	0168	0513	1562	1842	1546	0752
65	0000	0000	0002	0022	0131	0420	1418	1856	1645	0890
70	0000	0000	0002	0016	0101	0342	1272	1840	1716	1028

PROBABILITY DENSITY OF t, THE NON-CENTRAL t-STATISTIC, TABLED AS A FUNCTION OF $t/\sqrt{f}$.

f is the number of degrees of freedom; the non-centrality parameter is $\sqrt{f+1}\,K_p$.
K_p is the standardized normal deviate exceeded with probability p.

$t/\sqrt{f}$ \ p	.2500	.1500	.1000	.0650	.0400	.0250	.0100	.0040	.0025	.0010
2.75	0000	0000	0001	0012	0078	0276	1129	1795	1760	1162
80	0000	0000	0001	0009	0060	0222	0992	1727	1775	1286
85	0000	0000	0001	0007	0046	0178	0863	1640	1763	1396
90	0000	0000	0000	0005	0035	0142	0745	1539	1726	1488
95	0000	0000	0000	0004	0027	0112	0639	1427	1668	1559
3.00	0000	0000	0000	0003	0020	0089	0543	1310	1593	1609
05	0000	0000	0000	0002	0016	0070	0459	1190	1503	1635
10	0000	0000	0000	0001	0012	0055	0386	1072	1404	1640
15	0000	0000	0000	0001	0009	0043	0322	0958	1299	1624
20	0000	0000	0000	0001	0007	0034	0268	0849	1190	1589
25	0000	0000	0000	0001	0005	0026	0222	0747	1082	1537
30	0000	0000	0000	0000	0004	0020	0183	0654	0975	1472
35	0000	0000	0000	0000	0003	0016	0150	0568	0873	1396
40	0000	0000	0000	0000	0002	0012	0123	0491	0776	1312
45	0000	0000	0000	0000	0002	0010	0100	0422	0685	1223
50	0000	0000	0000	0000	0001	0007	0082	0361	0602	1131
55	0000	0000	0000	0000	0001	0006	0066	0308	0526	1038
60	0000	0000	0000	0000	0001	0004	0054	0261	0457	0946
65	0000	0000	0000	0000	0001	0003	0043	0221	0395	0857
70	0000	0000	0000	0000	0000	0003	0035	0186	0341	0771
75	0000	0000	0000	0000	0000	0002	0028	0156	0292	0691
80	0000	0000	0000	0000	0000	0002	0023	0131	0250	0615
85	0000	0000	0000	0000	0000	0001	0018	0109	0213	0545
90	0000	0000	0000	0000	0000	0001	0015	0091	0181	0481
95	0000	0000	0000	0000	0000	0001	0012	0076	0153	0423
4.00	0000	0000	0000	0000	0000	0001	0009	0063	0129	0370
05	0000	0000	0000	0000	0000	0000	0008	0052	0109	0323
10	0000	0000	0000	0000	0000	0000	0006	0043	0092	0280
15	0000	0000	0000	0000	0000	0000	0005	0035	0077	0243
20	0000	0000	0000	0000	0000	0000	0004	0029	0064	0210
4.25	0000	0000	0000	0000	0000	0000	0003	0024	0054	0181
30	0000	0000	0000	0000	0000	0000	0002	0020	0045	0155
35	0000	0000	0000	0000	0000	0000	0002	0016	0037	0133
40	0000	0000	0000	0000	0000	0000	0002	0013	0031	0114
45	0000	0000	0000	0000	0000	0000	0001	0011	0026	0097
50	0000	0000	0000	0000	0000	0000	0001	0009	0021	0083
55	0000	0000	0000	0000	0000	0000	0001	0007	0018	0070
60	0000	0000	0000	0000	0000	0000	0001	0006	0015	0060
65	0000	0000	0000	0000	0000	0000	0001	0005	0012	0051
70	0000	0000	0000	0000	0000	0000	0000	0004	0010	0043
75	0000	0000	0000	0000	0000	0000	0000	0003	0008	0036
80	0000	0000	0000	0000	0000	0000	0000	0003	0007	0031
85	0000	0000	0000	0000	0000	0000	0000	0002	0006	0026
90	0000	0000	0000	0000	0000	0000	0000	0002	0005	0022
95	0000	0000	0000	0000	0000	0000	0000	0001	0004	0018
5.00	0000	0000	0000	0000	0000	0000	0000	0001	0003	0015
05	0000	0000	0000	0000	0000	0000	0000	0001	0003	0013
10	0000	0000	0000	0000	0000	0000	0000	0001	0002	0011
15	0000	0000	0000	0000	0000	0000	0000	0001	0002	0009
20	0000	0000	0000	0000	0000	0000	0000	0001	0002	0008
25	0000	0000	0000	0000	0000	0000	0000	0000	0001	0007
30	0000	0000	0000	0000	0000	0000	0000	0000	0001	0005
35	0000	0000	0000	0000	0000	0000	0000	0000	0001	0005
40	0000	0000	0000	0000	0000	0000	0000	0000	0001	0004
45	0000	0000	0000	0000	0000	0000	0000	0000	0001	0003
50	0000	0000	0000	0000	0000	0000	0000	0000	0000	0003
55	0000	0000	0000	0000	0000	0000	0000	0000	0000	0002
60	0000	0000	0000	0000	0000	0000	0000	0000	0000	0002
65	0000	0000	0000	0000	0000	0000	0000	0000	0000	0002
70	0000	0000	0000	0000	0000	0000	0000	0000	0000	0001

PROBABILITY DENSITY OF t, THE NON-CENTRAL t-STATISTIC, TABLED AS A FUNCTION OF $t/\sqrt{f}$.

f is the number of degrees of freedom; the non-centrality parameter is $\sqrt{f+1}\,K_p$.
K_p is the standardized normal deviate exceeded with probability p.

DEGREES OF FREEDOM **29**

$t/\sqrt{f}$ \ p	.2500	.1500	.1000	.0650	.0400	.0250	.0100	.0040	.0025	.0010
5.75	0000	0000	0000	0000	0000	0000	0000	0000	0000	0001
80	0000	0000	0000	0000	0000	0000	0000	0000	0000	0001
85	0000	0000	0000	0000	0000	0000	0000	0000	0000	0001
90	0000	0000	0000	0000	0000	0000	0000	0000	0000	0001
95	0000	0000	0000	0000	0000	0000	0000	0000	0000	0001
6.00	0000	0000	0000	0000	0000	0000	0000	0000	0000	0000

PROBABILITY DENSITY OF t, THE NON-CENTRAL t-STATISTIC, TABLED AS A FUNCTION OF $t/\sqrt{f}$.

f is the number of degrees of freedom; the non-centrality parameter is $\sqrt{f+1}\,K_p$.
K_p is the standardized normal deviate exceeded with probability p.

DEGREES OF FREEDOM **34**

$t/\sqrt{f}$ \ p	.2500	.1500	.1000	.0650	.0400	.0250	.0100	.0040	.0025	.0010
-0.05	0000	0000	0000	0000	0000	0000	0000	0000	0000	0000
0.00	0001	0000	0000	0000	0000	0000	0000	0000	0000	0000
05	0004	0000	0000	0000	0000	0000	0000	0000	0000	0000
10	0012	0000	0000	0000	0000	0000	0000	0000	0000	0000
15	0033	0000	0000	0000	0000	0000	0000	0000	0000	0000
20	0081	0000	0000	0000	0000	0000	0000	0000	0000	0000
25	0179	0000	0000	0000	0000	0000	0000	0000	0000	0000
30	0360	0000	0000	0000	0000	0000	0000	0000	0000	0000
35	0659	0002	0000	0000	0000	0000	0000	0000	0000	0000
40	1095	0005	0000	0000	0000	0000	0000	0000	0000	0000
45	1654	0015	0000	0000	0000	0000	0000	0000	0000	0000
50	2277	0040	0000	0000	0000	0000	0000	0000	0000	0000
55	2869	0096	0000	0000	0000	0000	0000	0000	0000	0000
60	3322	0206	0004	0000	0000	0000	0000	0000	0000	0000
65	3554	0396	0011	0000	0000	0000	0000	0000	0000	0000
70	3533	0689	0029	0000	0000	0000	0000	0000	0000	0000
75	3280	1090	0070	0001	0000	0000	0000	0000	0000	0000
80	2862	1577	0150	0004	0000	0000	0000	0000	0000	0000
85	2360	2097	0292	0011	0000	0000	0000	0000	0000	0000
90	1848	2578	0513	0029	0000	0000	0000	0000	0000	0000
95	1383	2950	0824	0067	0001	0000	0000	0000	0000	0000
1.00	0993	3158	1215	0139	0005	0000	0000	0000	0000	0000
05	0688	3183	1655	0262	0012	0000	0000	0000	0000	0000
10	0461	3036	2096	0451	0030	0001	0000	0000	0000	0000
15	0301	2755	2482	0711	0065	0003	0000	0000	0000	0000
20	0191	2389	2766	1038	0130	0009	0000	0000	0000	0000
1.25	0119	1990	2915	1411	0236	0021	0000	0000	0000	0000
30	0073	1598	2919	1794	0395	0045	0000	0000	0000	0000
35	0044	1242	2793	2149	0612	0090	0001	0000	0000	0000
40	0026	0938	2562	2435	0883	0165	0002	0000	0000	0000
45	0015	0690	2264	2624	1195	0280	0004	0000	0000	0000
50	0009	0497	1934	2701	1524	0440	0010	0000	0000	0000
55	0005	0350	1603	2667	1841	0649	0022	0000	0000	0000
60	0003	0242	1292	2535	2116	0899	0045	0001	0000	0000
65	0002	0165	1016	2329	2325	1177	0083	0002	0000	0000
70	0001	0111	0782	2075	2450	1465	0144	0004	0001	0000
75	0001	0074	0590	1797	2487	1738	0233	0010	0001	0000
80	0000	0049	0438	1517	2438	1975	0355	0020	0003	0000
85	0000	0032	0320	1252	2317	2157	0511	0038	0007	0000
90	0000	0021	0230	1012	2140	2271	0699	0067	0015	0000
95	0000	0013	0164	0803	1926	2314	0912	0113	0028	0001
2.00	0000	0008	0115	0627	1693	2288	1138	0179	0050	0003
05	0000	0005	0080	0482	1457	2199	1365	0269	0085	0006
10	0000	0003	0056	0366	1230	2062	1577	0384	0136	0011
15	0000	0002	0038	0274	1020	1888	1761	0525	0207	0021
20	0000	0001	0026	0204	0832	1693	1905	0688	0299	0037
25	0000	0001	0018	0150	0669	1490	2002	0868	0414	0062
30	0000	0001	0012	0109	0531	1287	2049	1056	0551	0099
35	0000	0000	0008	0079	0417	1095	2046	1242	0706	0150
40	0000	0000	0005	0057	0324	0918	1998	1417	0873	0217
45	0000	0000	0004	0040	0249	0759	1910	1571	1046	0302
50	0000	0000	0002	0029	0190	0620	1792	1697	1215	0406
55	0000	0000	0002	0020	0144	0501	1651	1788	1373	0526
60	0000	0000	0001	0014	0108	0401	1497	1843	1512	0660
65	0000	0000	0001	0010	0081	0318	1338	1859	1625	0803
70	0000	0000	0000	0007	0060	0250	1179	1840	1709	0951

PROBABILITY DENSITY OF t, THE NON-CENTRAL t-STATISTIC, TABLED AS A FUNCTION OF $t/\sqrt{f}$.

f is the number of degrees of freedom; the non-centrality parameter is $\sqrt{f+1}\,K_p$.
K_p is the standardized normal deviate exceeded with probability p.

DEGREES OF FREEDOM **34**

$t/\sqrt{f}$ \ p	.2500	.1500	.1000	.0650	.0400	.0250	.0100	.0040	.0025	.0010
2.75	0000	0000	0000	0005	0044	0195	1026	1789	1760	1097
80	0000	0000	0000	0003	0033	0151	0882	1711	1778	1235
85	0000	0000	0000	0002	0024	0117	0751	1611	1764	1359
90	0000	0000	0000	0002	0017	0090	0633	1495	1722	1465
95	0000	0000	0000	0001	0013	0068	0528	1370	1655	1548
3.00	0000	0000	0000	0001	0009	0052	0438	1240	1568	1605
05	0000	0000	0000	0001	0007	0039	0360	1109	1466	1637
10	0000	0000	0000	0000	0005	0030	0294	0982	1354	1643
15	0000	0000	0000	0000	0004	0022	0238	0861	1237	1624
20	0000	0000	0000	0000	0003	0017	0192	0748	1117	1584
25	0000	0000	0000	0000	0002	0012	0154	0645	0999	1524
30	0000	0000	0000	0000	0001	0009	0123	0552	0886	1449
35	0000	0000	0000	0000	0001	0007	0098	0468	0778	1363
40	0000	0000	0000	0000	0001	0005	0077	0395	0679	1268
45	0000	0000	0000	0000	0000	0004	0061	0331	0587	1168
50	0000	0000	0000	0000	0000	0003	0048	0276	0505	1066
55	0000	0000	0000	0000	0000	0002	0038	0229	0431	0965
60	0000	0000	0000	0000	0000	0002	0029	0189	0366	0866
65	0000	0000	0000	0000	0000	0001	0023	0156	0309	0771
70	0000	0000	0000	0000	0000	0001	0018	0127	0260	0683
75	0000	0000	0000	0000	0000	0001	0014	0104	0217	0600
80	0000	0000	0000	0000	0000	0000	0011	0084	0181	0524
85	0000	0000	0000	0000	0000	0000	0008	0068	0150	0455
90	0000	0000	0000	0000	0000	0000	0006	0055	0124	0394
95	0000	0000	0000	0000	0000	0000	0005	0045	0102	0338
4.00	0000	0000	0000	0000	0000	0000	0004	0036	0084	0290
05	0000	0000	0000	0000	0000	0000	0003	0029	0069	0247
10	0000	0000	0000	0000	0000	0000	0002	0023	0056	0210
15	0000	0000	0000	0000	0000	0000	0002	0018	0046	0177
20	0000	0000	0000	0000	0000	0000	0001	0015	0037	0150
4.25	0000	0000	0000	0000	0000	0000	0001	0012	0030	0126
30	0000	0000	0000	0000	0000	0000	0001	0009	0024	0105
35	0000	0000	0000	0000	0000	0000	0001	0007	0020	0088
40	0000	0000	0000	0000	0000	0000	0000	0006	0016	0073
45	0000	0000	0000	0000	0000	0000	0000	0005	0013	0061
50	0000	0000	0000	0000	0000	0000	0000	0004	0010	0051
55	0000	0000	0000	0000	0000	0000	0000	0003	0008	0042
60	0000	0000	0000	0000	0000	0000	0000	0002	0007	0035
65	0000	0000	0000	0000	0000	0000	0000	0002	0005	0029
70	0000	0000	0000	0000	0000	0000	0000	0001	0004	0024
75	0000	0000	0000	0000	0000	0000	0000	0001	0003	0019
80	0000	0000	0000	0000	0000	0000	0000	0001	0003	0016
85	0000	0000	0000	0000	0000	0000	0000	0001	0002	0013
90	0000	0000	0000	0000	0000	0000	0000	0001	0002	0011
95	0000	0000	0000	0000	0000	0000	0000	0000	0001	0009
5.00	0000	0000	0000	0000	0000	0000	0000	0000	0001	0007
05	0000	0000	0000	0000	0000	0000	0000	0000	0001	0006
10	0000	0000	0000	0000	0000	0000	0000	0000	0001	0005
15	0000	0000	0000	0000	0000	0000	0000	0000	0001	0004
20	0000	0000	0000	0000	0000	0000	0000	0000	0000	0003
25	0000	0000	0000	0000	0000	0000	0000	0000	0000	0003
30	0000	0000	0000	0000	0000	0000	0000	0000	0000	0002
35	0000	0000	0000	0000	0000	0000	0000	0000	0000	0002
40	0000	0000	0000	0000	0000	0000	0000	0000	0000	0001
45	0000	0000	0000	0000	0000	0000	0000	0000	0000	0001
50	0000	0000	0000	0000	0000	0000	0000	0000	0000	0001
55	0000	0000	0000	0000	0000	0000	0000	0000	0000	0001
60	0000	0000	0000	0000	0000	0000	0000	0000	0000	0001
65	0000	0000	0000	0000	0000	0000	0000	0000	0000	0001
70	0000	0000	0000	0000	0000	0000	0000	0000	0000	0000

PROBABILITY DENSITY OF t, THE NON-CENTRAL t-STATISTIC, TABLED AS A FUNCTION OF $t/\sqrt{f}$.

f is the number of degrees of freedom; the non-centrality parameter is $\sqrt{f+1}\,K_p$.
K_p is the standardized normal deviate exceeded with probability p.

DEGREES OF FREEDOM **39**

$t/\sqrt{f}$ \ p	.2500	.1500	.1000	.0650	.0400	.0250	.0100	.0040	.0025	.0010
0.00	0000	0000	0000	0000	0000	0000	0000	0000	0000	0000
05	0002	0000	0000	0000	0000	0000	0000	0000	0000	0000
10	0005	0000	0000	0000	0000	0000	0000	0000	0000	0000
15	0017	0000	0000	0000	0000	0000	0000	0000	0000	0000
20	0046	0000	0000	0000	0000	0000	0000	0000	0000	0000
25	0116	0000	0000	0000	0000	0000	0000	0000	0000	0000
30	0258	0000	0000	0000	0000	0000	0000	0000	0000	0000
35	0515	0001	0000	0000	0000	0000	0000	0000	0000	0000
40	0921	0002	0000	0000	0000	0000	0000	0000	0000	0000
45	1477	0007	0000	0000	0000	0000	0000	0000	0000	0000
50	2130	0021	0000	0000	0000	0000	0000	0000	0000	0000
55	2776	0058	0000	0000	0000	0000	0000	0000	0000	0000
60	3286	0138	0001	0000	0000	0000	0000	0000	0000	0000
65	3553	0292	0005	0000	0000	0000	0000	0000	0000	0000
70	3531	0550	0015	0000	0000	0000	0000	0000	0000	0000
75	3247	0931	0040	0000	0000	0000	0000	0000	0000	0000
80	2780	1421	0097	0002	0000	0000	0000	0000	0000	0000
85	2231	1970	0208	0005	0000	0000	0000	0000	0000	0000
90	1689	2497	0398	0015	0000	0000	0000	0000	0000	0000
95	1213	2915	0684	0039	0001	0000	0000	0000	0000	0000
1.00	0831	3154	1067	0091	0002	0000	0000	0000	0000	0000
05	0546	3185	1521	0187	0006	0000	0000	0000	0000	0000
10	0346	3020	1994	0347	0016	0000	0000	0000	0000	0000
15	0213	2704	2421	0585	0039	0001	0000	0000	0000	0000
20	0127	2300	2742	0902	0085	0004	0000	0000	0000	0000
25	0074	1867	2913	1282	0168	0010	0000	0000	0000	0000
30	0042	1454	2921	1689	0302	0026	0000	0000	0000	0000
35	0024	1091	2778	2077	0499	0056	0000	0000	0000	0000
40	0013	0792	2520	2398	0759	0113	0001	0000	0000	0000
45	0007	0558	2189	2613	1074	0206	0002	0000	0000	0000
1.50	0004	0383	1829	2703	1419	0346	0005	0000	0000	0000
55	0002	0257	1476	2665	1762	0539	0012	0000	0000	0000
60	0001	0169	1154	2517	2067	0783	0026	0000	0000	0000
65	0001	0109	0878	2286	2302	1067	0052	0001	0000	0000
70	0000	0069	0651	2004	2446	1370	0098	0002	0000	0000
75	0000	0044	0472	1701	2489	1667	0170	0004	0000	0000
80	0000	0027	0336	1402	2435	1930	0275	0010	0001	0000
85	0000	0017	0234	1127	2298	2135	0418	0021	0003	0000
90	0000	0010	0161	0884	2100	2267	0598	0042	0007	0000
95	0000	0006	0109	0679	1862	2317	0810	0075	0015	0000
2.00	0000	0004	0073	0512	1608	2287	1045	0128	0030	0001
05	0000	0002	0048	0379	1355	2187	1286	0203	0055	0003
10	0000	0001	0032	0277	1116	2032	1518	0306	0094	0006
15	0000	0001	0021	0199	0901	1839	1722	0438	0151	0011
20	0000	0000	0013	0142	0715	1624	1885	0596	0231	0022
25	0000	0000	0009	0100	0558	1403	1996	0778	0336	0039
30	0000	0000	0006	0069	0428	1188	2050	0973	0465	0066
35	0000	0000	0004	0048	0325	0988	2048	1172	0618	0106
40	0000	0000	0002	0033	0243	0807	1993	1363	0788	0162
45	0000	0000	0001	0022	0180	0650	1894	1534	0969	0237
50	0000	0000	0001	0015	0132	0516	1761	1675	1150	0332
55	0000	0000	0001	0010	0096	0405	1604	1779	1323	0446
60	0000	0000	0000	0007	0069	0314	1435	1842	1478	0579
65	0000	0000	0000	0005	0050	0241	1262	1862	1605	0725
70	0000	0000	0000	0003	0035	0183	1092	1840	1700	0880
75	0000	0000	0000	0002	0025	0138	0932	1782	1759	1036
80	0000	0000	0000	0001	0018	0103	0785	1694	1780	1186
85	0000	0000	0000	0001	0012	0077	0653	1582	1765	1324
90	0000	0000	0000	0001	0009	0057	0537	1453	1717	1442
95	0000	0000	0000	0000	0006	0042	0437	1315	1641	1536

PROBABILITY DENSITY OF t, THE NON-CENTRAL t-STATISTIC, TABLED AS A FUNCTION OF $t/\sqrt{f}$.

f is the number of degrees of freedom; the non-centrality parameter is $\sqrt{f+1}\,K_p$.
K_p is the standardized normal deviate exceeded with probability p.

$t/\sqrt{f}$ \ p	.2500	.1500	.1000	.0650	.0400	.0250	.0100	.0040	.0025	.0010
3.00	0000	0000	0000	0000	0004	0030	0352	1173	1543	1602
05	0000	0000	0000	0000	0003	0022	0282	1033	1430	1638
10	0000	0000	0000	0000	0002	0016	0223	0899	1306	1645
15	0000	0000	0000	0000	0001	0012	0176	0774	1177	1624
20	0000	0000	0000	0000	0001	0008	0138	0659	1048	1578
25	0000	0000	0000	0000	0001	0006	0107	0556	0923	1511
30	0000	0000	0000	0000	0000	0004	0083	0465	0804	1426
35	0000	0000	0000	0000	0000	0003	0064	0386	0694	1329
40	0000	0000	0000	0000	0000	0002	0049	0318	0593	1224
45	0000	0000	0000	0000	0000	0002	0037	0200	0503	1115
50	0000	0000	0000	0000	0000	0001	0028	0211	0423	1004
55	0000	0000	0000	0000	0000	0001	0021	0171	0353	0896
60	0000	0000	0000	0000	0000	0001	0016	0137	0293	0792
65	0000	0000	0000	0000	0000	0000	0012	0110	0242	0694
70	0000	0000	0000	0000	0000	0000	0009	0087	0198	0604
75	0000	0000	0000	0000	0000	0000	0007	0069	0162	0521
80	0000	0000	0000	0000	0000	0000	0005	0055	0131	0447
85	0000	0000	0000	0000	0000	0000	0004	0043	0106	0380
90	0000	0000	0000	0000	0000	0000	0003	0034	0085	0322
95	0000	0000	0000	0000	0000	0000	0002	0026	0068	0271
4.00	0000	0000	0000	0000	0000	0000	0002	0020	0055	0227
05	0000	0000	0000	0000	0000	0000	0001	0016	0043	0189
10	0000	0000	0000	0000	0000	0000	0001	0012	0034	0157
15	0000	0000	0000	0000	0000	0000	0001	0010	0027	0130
20	0000	0000	0000	0000	0000	0000	0000	0007	0022	0107
25	0000	0000	0000	0000	0000	0000	0000	0006	0017	0087
30	0000	0000	0000	0000	0000	0000	0000	0004	0013	0071
35	0000	0000	0000	0000	0000	0000	0000	0003	0010	0058
40	0000	0000	0000	0000	0000	0000	0000	0003	0008	0047
45	0000	0000	0000	0000	0000	0000	0000	0002	0006	0038
4.50	0000	0000	0000	0000	0000	0000	0000	0002	0005	0031
55	0000	0000	0000	0000	0000	0000	0000	0001	0004	0025
60	0000	0000	0000	0000	0000	0000	0000	0001	0003	0020
65	0000	0000	0000	0000	0000	0000	0000	0001	0002	0016
70	0000	0000	0000	0000	0000	0000	0000	0001	0002	0013
75	0000	0000	0000	0000	0000	0000	0000	0000	0001	0010
80	0000	0000	0000	0000	0000	0000	0000	0000	0001	0008
85	0000	0000	0000	0000	0000	0000	0000	0000	0001	0007
90	0000	0000	0000	0000	0000	0000	0000	0000	0001	0005
95	0000	0000	0000	0000	0000	0000	0000	0000	0001	0004
5.00	0000	0000	0000	0000	0000	0000	0000	0000	0000	0003
05	0000	0000	0000	0000	0000	0000	0000	0000	0000	0003
10	0000	0000	0000	0000	0000	0000	0000	0000	0000	0002
15	0000	0000	0000	0000	0000	0000	0000	0000	0000	0002
20	0000	0000	0000	0000	0000	0000	0000	0000	0000	0001
25	0000	0000	0000	0000	0000	0000	0000	0000	0000	0001
30	0000	0000	0000	0000	0000	0000	0000	0000	0000	0001
35	0000	0000	0000	0000	0000	0000	0000	0000	0000	0001
40	0000	0000	0000	0000	0000	0000	0000	0000	0000	0001
45	0000	0000	0000	0000	0000	0000	0000	0000	0000	0000

PROBABILITY DENSITY OF t, THE NON-CENTRAL t-STATISTIC, TABLED AS A FUNCTION OF $t/\sqrt{f}$.

f is the number of degrees of freedom; the non-centrality parameter is $\sqrt{f+1}\,K_p$.
K_p is the standardized normal deviate exceeded with probability p.

DEGREES OF FREEDOM **44**

$t/\sqrt{f}$ \ p	.2500	.1500	.1000	.0650	.0400	.0250	.0100	.0040	.0025	.0010
0.05	0001	0000	0000	0000	0000	0000	0000	0000	0000	0000
10	0002	0000	0000	0000	0000	0000	0000	0000	0000	0000
15	0009	0000	0000	0000	0000	0000	0000	0000	0000	0000
20	0027	0000	0000	0000	0000	0000	0000	0000	0000	0000
25	0075	0000	0000	0000	0000	0000	0000	0000	0000	0000
30	0185	0000	0000	0000	0000	0000	0000	0000	0000	0000
35	0402	0000	0000	0000	0000	0000	0000	0000	0000	0000
40	0774	0001	0000	0000	0000	0000	0000	0000	0000	0000
45	1318	0003	0000	0000	0000	0000	0000	0000	0000	0000
50	1992	0011	0000	0000	0000	0000	0000	0000	0000	0000
55	2686	0035	0000	0000	0000	0000	0000	0000	0000	0000
60	3250	0092	0001	0000	0000	0000	0000	0000	0000	0000
65	3551	0215	0002	0000	0000	0000	0000	0000	0000	0000
70	3529	0439	0008	0000	0000	0000	0000	0000	0000	0000
75	3213	0794	0023	0000	0000	0000	0000	0000	0000	0000
80	2700	1280	0063	0001	0000	0000	0000	0000	0000	0000
85	2110	1850	0149	0002	0000	0000	0000	0000	0000	0000
90	1543	2418	0308	0008	0000	0000	0000	0000	0000	0000
95	1064	2880	0568	0023	0000	0000	0000	0000	0000	0000
1.00	0696	3150	0937	0059	0001	0000	0000	0000	0000	0000
05	0434	3187	1397	0133	0003	0000	0000	0000	0000	0000
10	0260	3003	1896	0267	0008	0000	0000	0000	0000	0000
15	0150	2654	2361	0481	0023	0000	0000	0000	0000	0000
20	0084	2213	2718	0784	0055	0002	0000	0000	0000	0000
25	0046	1751	2911	1165	0119	0005	0000	0000	0000	0000
30	0024	1322	2922	1590	0231	0014	0000	0000	0000	0000
35	0013	0958	2763	2007	0406	0035	0000	0000	0000	0000
40	0007	0668	2477	2361	0653	0077	0000	0000	0000	0000
45	0003	0451	2115	2602	0964	0151	0001	0000	0000	0000
1.50	0002	0295	1729	2704	1320	0272	0002	0000	0000	0000
55	0001	0189	1359	2663	1685	0448	0006	0000	0000	0000
60	0000	0118	1031	2499	2018	0682	0015	0000	0000	0000
65	0000	0072	0758	2243	2280	0966	0033	0000	0000	0000
70	0000	0043	0542	1935	2442	1281	0067	0001	0000	0000
75	0000	0026	0377	1610	2491	1598	0124	0002	0000	0000
80	0000	0015	0257	1296	2431	1885	0213	0005	0001	0000
85	0000	0009	0172	1013	2279	2113	0341	0012	0001	0000
90	0000	0005	0113	0772	2060	2261	0511	0026	0004	0000
95	0000	0003	0073	0574	1800	2319	0720	0050	0009	0000
2.00	0000	0002	0046	0417	1526	2286	0959	0091	0018	0000
05	0000	0001	0029	0298	1259	2175	1212	0154	0035	0001
10	0000	0000	0018	0209	1013	2003	1461	0244	0065	0003
15	0000	0000	0011	0144	0797	1790	1684	0365	0111	0006
20	0000	0000	0007	0098	0614	1557	1865	0517	0179	0013
25	0000	0000	0004	0066	0464	1321	1990	0696	0272	0024
30	0000	0000	0003	0044	0345	1096	2051	0896	0393	0044
35	0000	0000	0002	0029	0253	0890	2049	1105	0541	0075
40	0000	0000	0001	0019	0183	0710	1988	1310	0711	0121
45	0000	0000	0001	0012	0131	0557	1877	1497	0897	0186
50	0000	0000	0000	0008	0092	0430	1730	1654	1089	0271
55	0000	0000	0000	0005	0064	0327	1559	1770	1275	0379
60	0000	0000	0000	0003	0045	0245	1375	1841	1444	0507
65	0000	0000	0000	0002	0031	0182	1190	1863	1585	0654
70	0000	0000	0000	0001	0021	0134	1012	1840	1692	0813
75	0000	0000	0000	0001	0014	0097	0846	1775	1758	0977
80	0000	0000	0000	0001	0010	0070	0698	1676	1782	1139
85	0000	0000	0000	0000	0006	0050	0567	1552	1765	1288
90	0000	0000	0000	0000	0004	0036	0455	1411	1711	1419
95	0000	0000	0000	0000	0003	0025	0361	1261	1627	1523

PROBABILITY DENSITY OF t, THE NON-CENTRAL t-STATISTIC, TABLED AS A FUNCTION OF $t/\sqrt{f}$.

f is the number of degrees of freedom; the non-centrality parameter is $\sqrt{f+1}\,K_p$.
K_p is the standardized normal deviate exceeded with probability p.

DEGREES OF FREEDOM **44**

$t/\sqrt{f}$ \ P	.2500	.1500	.1000	.0650	.0400	.0250	.0100	.0040	.0025	.0010
3.00	0000	0000	0000	0000	0002	0018	0284	1110	1519	1597
05	0000	0000	0000	0000	0001	0012	0221	0962	1393	1638
10	0000	0000	0000	0000	0001	0009	0170	0823	1258	1646
15	0000	0000	0000	0000	0001	0006	0130	0695	1120	1623
20	0000	0000	0000	0000	0000	0004	0099	0581	0983	1572
25	0000	0000	0000	0000	0000	0003	0074	0480	0852	1497
30	0000	0000	0000	0000	0000	0002	0056	0392	0730	1403
35	0000	0000	0000	0000	0000	0001	0041	0318	0618	1296
40	0000	0000	0000	0000	0000	0001	0031	0256	0518	1182
45	0000	0000	0000	0000	0000	0001	0023	0204	0431	1064
50	0000	0000	0000	0000	0000	0000	0017	0161	0355	0946
55	0000	0000	0000	0000	0000	0000	0012	0127	0290	0832
60	0000	0000	0000	0000	0000	0000	0009	0099	0235	0725
65	0000	0000	0000	0000	0000	0000	0006	0077	0189	0625
70	0000	0000	0000	0000	0000	0000	0005	0060	0151	0534
75	0000	0000	0000	0000	0000	0000	0003	0046	0120	0452
80	0000	0000	0000	0000	0000	0000	0002	0035	0095	0380
85	0000	0000	0000	0000	0000	0000	0002	0027	0075	0318
90	0000	0000	0000	0000	0000	0000	0001	0020	0058	0263
95	0000	0000	0000	0000	0000	0000	0001	0015	0046	0217
4.00	0000	0000	0000	0000	0000	0000	0001	0012	0035	0178
05	0000	0000	0000	0000	0000	0000	0000	0009	0027	0145
10	0000	0000	0000	0000	0000	0000	0000	0007	0021	0117
15	0000	0000	0000	0000	0000	0000	0000	0005	0016	0095
20	0000	0000	0000	0000	0000	0000	0000	0004	0012	0076
25	0000	0000	0000	0000	0000	0000	0000	0003	0010	0061
30	0000	0000	0000	0000	0000	0000	0000	0002	0007	0048
35	0000	0000	0000	0000	0000	0000	0000	0002	0006	0038
40	0000	0000	0000	0000	0000	0000	0000	0001	0004	0030
45	0000	0000	0000	0000	0000	0000	0000	0001	0003	0024
4.50	0000	0000	0000	0000	0000	0000	0000	0001	0002	0019
55	0000	0000	0000	0000	0000	0000	0000	0000	0002	0015
60	0000	0000	0000	0000	0000	0000	0000	0000	0001	0012
65	0000	0000	0000	0000	0000	0000	0000	0000	0001	0009
70	0000	0000	0000	0000	0000	0000	0000	0000	0001	0007
75	0000	0000	0000	0000	0000	0000	0000	0000	0001	0005
80	0000	0000	0000	0000	0000	0000	0000	0000	0000	0004
85	0000	0000	0000	0000	0000	0000	0000	0000	0000	0003
90	0000	0000	0000	0000	0000	0000	0000	0000	0000	0003
95	0000	0000	0000	0000	0000	0000	0000	0000	0000	0002
5.00	0000	0000	0000	0000	0000	0000	0000	0000	0000	0002
05	0000	0000	0000	0000	0000	0000	0000	0000	0000	0001
10	0000	0000	0000	0000	0000	0000	0000	0000	0000	0001
15	0000	0000	0000	0000	0000	0000	0000	0000	0000	0001
20	0000	0000	0000	0000	0000	0000	0000	0000	0000	0001
25	0000	0000	0000	0000	0000	0000	0000	0000	0000	0000

PROBABILITY DENSITY OF t, THE NON-CENTRAL t-STATISTIC, TABLED AS A FUNCTION OF $t/\sqrt{f}$.

f is the number of degrees of freedom; the non-centrality parameter is $\sqrt{f+1}\,K_p$.
K_p is the standardized normal deviate exceeded with probability p.

DEGREES OF FREEDOM **49**

$t/\sqrt{f}$ \ p	.2500	.1500	.1000	.0650	.0400	.0250	.0100	.0040	.0025	.0010
0.05	0000	0000	0000	0000	0000	0000	0000	0000	0000	0000
10	0001	0000	0000	0000	0000	0000	0000	0000	0000	0000
15	0004	0000	0000	0000	0000	0000	0000	0000	0000	0000
20	0015	0000	0000	0000	0000	0000	0000	0000	0000	0000
25	0048	0000	0000	0000	0000	0000	0000	0000	0000	0000
30	0132	0000	0000	0000	0000	0000	0000	0000	0000	0000
35	0314	0000	0000	0000	0000	0000	0000	0000	0000	0000
40	0651	0000	0000	0000	0000	0000	0000	0000	0000	0000
45	1176	0001	0000	0000	0000	0000	0000	0000	0000	0000
50	1863	0006	0000	0000	0000	0000	0000	0000	0000	0000
55	2599	0021	0000	0000	0000	0000	0000	0000	0000	0000
60	3214	0062	0000	0000	0000	0000	0000	0000	0000	0000
65	3549	0158	0001	0000	0000	0000	0000	0000	0000	0000
70	3527	0351	0004	0000	0000	0000	0000	0000	0000	0000
75	3179	0678	0014	0000	0000	0000	0000	0000	0000	0000
80	2622	1152	0041	0000	0000	0000	0000	0000	0000	0000
85	1994	1737	0106	0001	0000	0000	0000	0000	0000	0000
90	1409	2341	0239	0004	0000	0000	0000	0000	0000	0000
95	0933	2845	0471	0013	0000	0000	0000	0000	0000	0000
1.00	0582	3145	0823	0038	0000	0000	0000	0000	0000	0000
05	0345	3187	1283	0095	0001	0000	0000	0000	0000	0000
10	0195	2986	1803	0205	0004	0000	0000	0000	0000	0000
15	0106	2604	2302	0396	0013	0000	0000	0000	0000	0000
20	0056	2129	2693	0681	0036	0001	0000	0000	0000	0000
25	0028	1642	2909	1058	0085	0003	0000	0000	0000	0000
30	0014	1203	2922	1496	0177	0008	0000	0000	0000	0000
35	0007	0841	2748	1940	0331	0022	0000	0000	0000	0000
40	0003	0564	2434	2324	0561	0052	0000	0000	0000	0000
45	0002	0364	2044	2591	0866	0111	0000	0000	0000	0000
1.50	0001	0228	1634	2705	1228	0214	0001	0000	0000	0000
55	0000	0138	1251	2661	1612	0372	0003	0000	0000	0000
60	0000	0082	0921	2480	1970	0594	0009	0000	0000	0000
65	0000	0047	0654	2201	2257	0875	0021	0000	0000	0000
70	0000	0027	0451	1868	2437	1198	0045	0000	0000	0000
75	0000	0015	0302	1523	2493	1532	0090	0001	0000	0000
80	0000	0008	0197	1198	2427	1841	0165	0003	0000	0000
85	0000	0005	0126	0911	2260	2092	0279	0007	0001	0000
90	0000	0002	0079	0673	2020	2256	0437	0016	0002	0000
95	0000	0001	0049	0485	1740	2320	0640	0034	0005	0000
2.00	0000	0001	0029	0340	1449	2284	0880	0065	0011	0000
05	0000	0000	0018	0234	1170	2162	1142	0116	0023	0000
10	0000	0000	0010	0158	0919	1973	1405	0194	0045	0001
15	0000	0000	0006	0105	0704	1742	1647	0304	0081	0003
20	0000	0000	0004	0068	0527	1493	1845	0447	0138	0007
25	0000	0000	0002	0044	0387	1244	1983	0624	0220	0015
30	0000	0000	0001	0028	0278	1011	2052	0826	0332	0029
35	0000	0000	0001	0018	0197	0803	2050	1042	0473	0053
40	0000	0000	0000	0011	0137	0624	1982	1259	0642	0091
45	0000	0000	0000	0007	0094	0476	1861	1461	0831	0145
50	0000	0000	0000	0004	0064	0357	1700	1632	1030	0222
55	0000	0000	0000	0003	0043	0264	1514	1761	1228	0321
60	0000	0000	0000	0002	0029	0192	1317	1839	1411	0445
65	0000	0000	0000	0001	0019	0138	1122	1865	1565	0590
70	0000	0000	0000	0001	0012	0098	0937	1839	1683	0751
75	0000	0000	0000	0000	0008	0069	0769	1767	1756	0922
80	0000	0000	0000	0000	0005	0048	0620	1659	1783	1093
85	0000	0000	0000	0000	0003	0033	0493	1523	1765	1254
90	0000	0000	0000	0000	0002	0023	0386	1370	1706	1396
95	0000	0000	0000	0000	0001	0015	0299	1210	1613	1511

PROBABILITY DENSITY OF t, THE NON-CENTRAL t-STATISTIC, TABLED AS A FUNCTION OF $t/\sqrt{f}$.

f is the number of degrees of freedom; the non-centrality parameter is $\sqrt{f+1}\,K_p$.
K_p is the standardized normal deviate exceeded with probability p.

DEGREES OF FREEDOM **49**

$t/\sqrt{f}$ \ p	.2500	.1500	.1000	.0650	.0400	.0250	.0100	.0040	.0025	.0010
3.00	0000	0000	0000	0000	0001	0010	0228	1049	1494	1593
05	0000	0000	0000	0000	0001	0007	0173	0896	1358	1638
10	0000	0000	0000	0000	0000	0005	0129	0753	1213	1648
15	0000	0000	0000	0000	0000	0003	0096	0624	1065	1622
20	0000	0000	0000	0000	0000	0002	0071	0511	0922	1565
25	0000	0000	0000	0000	0000	0001	0052	0414	0786	1482
30	0000	0000	0000	0000	0000	0001	0037	0331	0662	1380
35	0000	0000	0000	0000	0000	0001	0027	0262	0551	1264
40	0000	0000	0000	0000	0000	0000	0019	0206	0453	1141
45	0000	0000	0000	0000	0000	0000	0014	0160	0369	1015
50	0000	0000	0000	0000	0000	0000	0010	0123	0297	0891
55	0000	0000	0000	0000	0000	0000	0007	0095	0237	0773
60	0000	0000	0000	0000	0000	0000	0005	0072	0188	0663
65	0000	0000	0000	0000	0000	0000	0003	0054	0148	0562
70	0000	0000	0000	0000	0000	0000	0002	0041	0115	0472
75	0000	0000	0000	0000	0000	0000	0002	0031	0089	0393
80	0000	0000	0000	0000	0000	0000	0001	0023	0069	0324
85	0000	0000	0000	0000	0000	0000	0001	0017	0053	0265
90	0000	0000	0000	0000	0000	0000	0001	0012	0040	0215
95	0000	0000	0000	0000	0000	0000	0000	0009	0030	0174
4.00	0000	0000	0000	0000	0000	0000	0000	0007	0023	0139
05	0000	0000	0000	0000	0000	0000	0000	0005	0017	0111
10	0000	0000	0000	0000	0000	0000	0000	0004	0013	0088
15	0000	0000	0000	0000	0000	0000	0000	0003	0010	0069
20	0000	0000	0000	0000	0000	0000	0000	0002	0007	0054
25	0000	0000	0000	0000	0000	0000	0000	0001	0005	0042
30	0000	0000	0000	0000	0000	0000	0000	0001	0004	0033
35	0000	0000	0000	0000	0000	0000	0000	0001	0003	0025
40	0000	0000	0000	0000	0000	0000	0000	0001	0002	0020
45	0000	0000	0000	0000	0000	0000	0000	0000	0002	0015
4.50	0000	0000	0000	0000	0000	0000	0000	0000	0001	0011
55	0000	0000	0000	0000	0000	0000	0000	0000	0001	0009
60	0000	0000	0000	0000	0000	0000	0000	0000	0001	0007
65	0000	0000	0000	0000	0000	0000	0000	0000	0000	0005
70	0000	0000	0000	0000	0000	0000	0000	0000	0000	0004
75	0000	0000	0000	0000	0000	0000	0000	0000	0000	0003
80	0000	0000	0000	0000	0000	0000	0000	0000	0000	0002
85	0000	0000	0000	0000	0000	0000	0000	0000	0000	0002
90	0000	0000	0000	0000	0000	0000	0000	0000	0000	0001
95	0000	0000	0000	0000	0000	0000	0000	0000	0000	0001
5.00	0000	0000	0000	0000	0000	0000	0000	0000	0000	0001
05	0000	0000	0000	0000	0000	0000	0000	0000	0000	0001
10	0000	0000	0000	0000	0000	0000	0000	0000	0000	0000

PROBABILITY INTEGRAL OF t, THE NON-CENTRAL t-STATISTIC.

PROBABILITY INTEGRAL OF t, THE NON-CENTRAL t-STATISTIC. THIS TABLE GIVES Pr $[t/\sqrt{f} \leq x]$.

f is the number of degrees of freedom; the non-centrality parameter is $\sqrt{f+1}\,K_p$.
K_p is the standardized normal deviate exceeded with probability p.

DEGREES OF FREEDOM **2**

x \ p	.2500	.1500	.1000	.0650	.0400	.0250	.0100	.0040	.0025	.0010
- 7.50	0.0005									
- 7.45	0.0005									
- 7.40	0.0005									
- 7.35	0.0005									
- 7.30	0.0005									
- 7.25	0.0005									
- 7.20	0.0005									
- 7.15	0.0005									
- 7.10	0.0005									
- 7.05	0.0005									
- 7.00	0.0005									
- 6.95	0.0005									
- 6.90	0.0005									
- 6.85	0.0005									
- 6.80	0.0006									
- 6.75	0.0006									
- 6.70	0.0006									
- 6.65	0.0006									
- 6.60	0.0006									
- 6.55	0.0006									
- 6.50	0.0006									
- 6.45	0.0006									
- 6.40	0.0006									
- 6.35	0.0006									
- 6.30	0.0006									
- 6.25	0.0007									
- 6.20	0.0007									
- 6.15	0.0007									
- 6.10	0.0007									
- 6.05	0.0007									
- 6.00	0.0007									
- 5.95	0.0007									
- 5.90	0.0007									
- 5.85	0.0007									
- 5.80	0.0008									
- 5.75	0.0008									
- 5.70	0.0008									
- 5.65	0.0008									
- 5.60	0.0008									
- 5.55	0.0008									
- 5.50	0.0008									
- 5.45	0.0009									
- 5.40	0.0009									
- 5.35	0.0009									
- 5.30	0.0009									
- 5.25	0.0009									
- 5.20	0.0009									
- 5.15	0.0010									
- 5.10	0.0010									
- 5.05	0.0010									
- 5.00	0.0010									
- 4.95	0.0010									
- 4.90	0.0011									
- 4.85	0.0011									
- 4.80	0.0011									
- 4.75	0.0011									
- 4.70	0.0011									
- 4.65	0.0012									
- 4.60	0.0012									
- 4.55	0.0012									

PROBABILITY INTEGRAL OF t, THE NON-CENTRAL t-STATISTIC. THIS TABLE GIVES Pr $[t/\sqrt{f} \leq x]$.

f is the number of degrees of freedom; the non-centrality parameter is $\sqrt{f+1}\,K_p$.
K_p is the standardized normal deviate exceeded with probability p.

DEGREES OF FREEDOM **2**

x \ p	.2500	.1500	.1000	.0650	.0400	.0250	.0100	.0040	.0025	.0010
-4.50	0.0012									
-4.45	0.0013									
-4.40	0.0013									
-4.35	0.0013									
-4.30	0.0014									
-4.25	0.0014									
-4.20	0.0014									
-4.15	0.0015									
-4.10	0.0015									
-4.05	0.0015									
-4.00	0.0016									
-3.95	0.0016									
-3.90	0.0016									
-3.85	0.0017									
-3.80	0.0017									
-3.75	0.0018									
-3.70	0.0018									
-3.65	0.0019									
-3.60	0.0019									
-3.55	0.0020									
-3.50	0.0020	0.0004								
-3.45	0.0021	0.0004								
-3.40	0.0021	0.0004								
-3.35	0.0022	0.0004								
-3.30	0.0023	0.0005								
-3.25	0.0023	0.0005								
-3.20	0.0024	0.0005								
-3.15	0.0025	0.0005								
-3.10	0.0026	0.0005								
-3.05	0.0026	0.0005								
-3.00	0.0027	0.0006								
-2.95	0.0028	0.0006								
-2.90	0.0029	0.0006								
-2.85	0.0030	0.0006								
-2.80	0.0031	0.0006								
-2.75	0.0032	0.0007								
-2.70	0.0033	0.0007								
-2.65	0.0035	0.0007								
-2.60	0.0036	0.0007								
-2.55	0.0037	0.0008								
-2.50	0.0039	0.0008								
-2.45	0.0040	0.0008								
-2.40	0.0042	0.0009								
-2.35	0.0043	0.0009								
-2.30	0.0045	0.0009								
-2.25	0.0047	0.0010								
-2.20	0.0049	0.0010								
-2.15	0.0051	0.0011								
-2.10	0.0053	0.0011								
-2.05	0.0056	0.0012								
-2.00	0.0058	0.0012								
-1.95	0.0061	0.0013								
-1.90	0.0064	0.0013								
-1.85	0.0067	0.0014								
-1.80	0.0071	0.0015								
-1.75	0.0074	0.0015	0.0004							
-1.70	0.0078	0.0016	0.0005							
-1.65	0.0082	0.0017	0.0005							
-1.60	0.0087	0.0018	0.0005							
-1.55	0.0092	0.0019	0.0006							

PROBABILITY INTEGRAL OF t, THE NON-CENTRAL t-STATISTIC. THIS TABLE GIVES Pr $[t/\sqrt{f} \leq x]$.

f is the number of degrees of freedom; the non-centrality parameter is $\sqrt{f+1}\,K_p$.

K_p is the standardized normal deviate exceeded with probability p.

DEGREES OF FREEDOM **2**

x \ p	.2500	.1500	.1000	.0650	.0400	.0250	.0100	.0040	.0025	.0010
- 1.50	0.0097	0.0020	0.0006							
- 1.45	0.0103	0.0022	0.0006							
- 1.40	0.0109	0.0023	0.0007							
- 1.35	0.0116	0.0025	0.0007							
- 1.30	0.0124	0.0026	0.0008							
- 1.25	0.0132	0.0028	0.0008							
- 1.20	0.0141	0.0030	0.0009							
- 1.15	0.0150	0.0032	0.0009							
- 1.10	0.0161	0.0035	0.0010							
- 1.05	0.0173	0.0037	0.0011							
- 1.00	0.0186	0.0040	0.0012							
- 0.95	0.0201	0.0044	0.0013							
- 0.90	0.0217	0.0048	0.0014							
- 0.85	0.0235	0.0052	0.0016	0.0004						
- 0.80	0.0254	0.0057	0.0017	0.0005						
- 0.75	0.0277	0.0062	0.0019	0.0005						
- 0.70	0.0301	0.0068	0.0021	0.0006						
- 0.65	0.0329	0.0075	0.0023	0.0006						
- 0.60	0.0360	0.0083	0.0026	0.0007						
- 0.55	0.0395	0.0093	0.0029	0.0008						
- 0.50	0.0435	0.0103	0.0032	0.0009						
- 0.45	0.0479	0.0115	0.0036	0.0011						
- 0.40	0.0528	0.0129	0.0041	0.0012						
- 0.35	0.0584	0.0146	0.0047	0.0014						
- 0.30	0.0647	0.0165	0.0054	0.0016						
- 0.25	0.0718	0.0187	0.0062	0.0019	0.0005					
- 0.20	0.0797	0.0212	0.0071	0.0022	0.0006					
- 0.15	0.0885	0.0242	0.0083	0.0026	0.0007					
- 0.10	0.0984	0.0277	0.0096	0.0031	0.0008					
- 0.05	0.1093	0.0317	0.0113	0.0036	0.0010					
0.00	0.1214	0.0363	0.0132	0.0044	0.0012					
0.05	0.1346	0.0417	0.0155	0.0053	0.0015					
0.10	0.1490	0.0478	0.0183	0.0064	0.0019	0.0006				
0.15	0.1646	0.0549	0.0216	0.0077	0.0023	0.0007				
0.20	0.1814	0.0629	0.0255	0.0094	0.0029	0.0009				
0.25	0.1994	0.0719	0.0300	0.0114	0.0037	0.0012				
0.30	0.2184	0.0820	0.0354	0.0139	0.0046	0.0016				
0.35	0.2383	0.0932	0.0415	0.0168	0.0058	0.0020				
0.40	0.2591	0.1055	0.0485	0.0204	0.0074	0.0027				
0.45	0.2805	0.1190	0.0565	0.0246	0.0092	0.0035	0.0005			
0.50	0.3025	0.1335	0.0655	0.0295	0.0115	0.0045	0.0007			
0.55	0.3249	0.1491	0.0756	0.0352	0.0143	0.0058	0.0009			
0.60	0.3475	0.1657	0.0866	0.0418	0.0177	0.0075	0.0013			
0.65	0.3702	0.1831	0.0987	0.0492	0.0217	0.0095	0.0018			
0.70	0.3929	0.2013	0.1118	0.0576	0.0264	0.0120	0.0025	0.0005		
0.75	0.4153	0.2202	0.1259	0.0670	0.0318	0.0150	0.0033	0.0007		
0.80	0.4375	0.2395	0.1408	0.0772	0.0380	0.0186	0.0044	0.0010	0.0005	
0.85	0.4593	0.2593	0.1565	0.0884	0.0450	0.0229	0.0058	0.0014	0.0007	
0.90	0.4805	0.2794	0.1729	0.1005	0.0529	0.0277	0.0075	0.0020	0.0010	
0.95	0.5012	0.2997	0.1900	0.1134	0.0615	0.0333	0.0097	0.0027	0.0014	0.0004
1.00	0.5213	0.3200	0.2075	0.1270	0.0711	0.0397	0.0122	0.0036	0.0019	0.0005
1.05	0.5407	0.3402	0.2255	0.1414	0.0814	0.0467	0.0152	0.0048	0.0026	0.0008
1.10	0.5595	0.3604	0.2437	0.1564	0.0925	0.0546	0.0188	0.0063	0.0035	0.0011
1.15	0.5775	0.3803	0.2622	0.1719	0.1043	0.0631	0.0229	0.0080	0.0047	0.0016
1.20	0.5949	0.3999	0.2808	0.1879	0.1168	0.0724	0.0276	0.0102	0.0061	0.0022
1.25	0.6115	0.4192	0.2994	0.2043	0.1299	0.0824	0.0329	0.0127	0.0078	0.0029
1.30	0.6274	0.4381	0.3180	0.2210	0.1435	0.0931	0.0388	0.0157	0.0098	0.0039
1.35	0.6426	0.4565	0.3364	0.2378	0.1577	0.1043	0.0453	0.0191	0.0122	0.0050
1.40	0.6571	0.4744	0.3547	0.2548	0.1722	0.1162	0.0524	0.0230	0.0150	0.0064
1.45	0.6710	0.4919	0.3727	0.2719	0.1871	0.1285	0.0601	0.0275	0.0182	0.0081

PROBABILITY INTEGRAL OF t, THE NON-CENTRAL t-STATISTIC. THIS TABLE GIVES Pr $[t/\sqrt{f} \leq x]$.

f is the number of degrees of freedom; the non-centrality parameter is $\sqrt{f+1}\,K_p$.
K_p is the standardized normal deviate exceeded with probability p.

DEGREES OF FREEDOM **2**

x \ p	.2500	.1500	.1000	.0650	.0400	.0250	.0100	.0040	.0025	.0010
1.50	0.6842	0.5088	0.3905	0.2890	0.2023	0.1413	0.0684	0.0324	0.0219	0.0101
1.55	0.6968	0.5252	0.4079	0.3060	0.2176	0.1545	0.0773	0.0378	0.0261	0.0125
1.60	0.7088	0.5411	0.4250	0.3229	0.2331	0.1681	0.0867	0.0438	0.0307	0.0152
1.65	0.7202	0.5564	0.4417	0.3397	0.2487	0.1819	0.0966	0.0503	0.0357	0.0182
1.70	0.7311	0.5712	0.4580	0.3562	0.2644	0.1960	0.1069	0.0572	0.0413	0.0217
1.75	0.7414	0.5855	0.4739	0.3725	0.2800	0.2102	0.1177	0.0647	0.0473	0.0255
1.80	0.7513	0.5993	0.4894	0.3886	0.2956	0.2246	0.1289	0.0726	0.0538	0.0298
1.85	0.7606	0.6125	0.5044	0.4044	0.3111	0.2391	0.1403	0.0810	0.0608	0.0345
1.90	0.7696	0.6252	0.5190	0.4198	0.3265	0.2536	0.1521	0.0897	0.0681	0.0396
1.95	0.7781	0.6375	0.5331	0.4349	0.3417	0.2681	0.1641	0.0989	0.0759	0.0451
2.00	0.7862	0.6492	0.5468	0.4497	0.3567	0.2825	0.1764	0.1084	0.0841	0.0510
2.05	0.7939	0.6605	0.5601	0.4642	0.3714	0.2969	0.1888	0 1183	0.0927	0.0573
2.10	0.8012	0.6714	0.5729	0.4782	0.3860	0.3112	0.2013	0.1284	0.1016	0.0639
2.15	0.8082	0.6818	0.5853	0.4920	0.4003	0.3254	0.2140	0.1389	0.1108	0.0710
2.20	0.8149	0.6919	0.5973	0.5053	0.4143	0.3394	0.2267	0.1495	0.1203	0.0783
2.25	0.8212	0.7015	0.6089	0.5183	0.4281	0.3532	0.2395	0.1604	0.1301	0.0860
2.30	0.8273	0.7108	0.6201	0.5309	0.4416	0.3669	0.2523	0.1714	0.1401	0.0941
2.35	0.8331	0.7196	0.6310	0.5432	0.4547	0.3804	0.2651	0.1826	0.1503	0.1024
2.40	0.8386	0.7282	0.6414	0.5551	0.4676	0.3936	0.2778	0.1938	0.1607	0.1109
2.45	0.8439	0.7364	0.6515	0.5667	0.4802	0.4066	0.2905	0.2052	0.1712	0.1197
2.50	0.8489	0.7442	0.6612	0.5779	0.4925	0.4194	0.3030	0.2167	0.1819	0.1287
2.55	0.8538	0.7518	0.6706	0.5887	0.5045	0.4319	0.3155	0.2282	0.1926	0.1380
2.60	0.8584	0.7591	0.6797	0.5993	0.5161	0.4442	0.3279	0.2397	0.2035	0.1474
2.65	0.8628	0.7661	0.6884	0.6095	0.5275	0.4562	0.3401	0.2512	0.2144	0.1569
2.70	0.8670	0.7728	0.6969	0.6194	0.5386	0.4680	0.3522	0.2627	0.2254	0.1666
2.75	0.8710	0.7792	0.7050	0.6290	0.5494	0.4795	0.3642	0.2741	0.2364	0.1764
2.80	0.8749	0.7855	0.7129	0.6383	0.5599	0.4907	0.3759	0.2855	0.2473	0.1863
2.85	0.8786	0.7914	0.7205	0.6473	0.5701	0.5017	0.3875	0.2968	0.2583	0.1963
2.90	0.8822	0.7972	0.7278	0.6560	0.5800	0.5125	0.3990	0.3081	0.2692	0.2063
2.95	0.8856	0.8027	0.7349	0.6645	0.5897	0.5230	0.4102	0.3193	0.2801	0.2164
3.00	0.8888	0.8081	0.7417	0.6727	0.5991	0.5332	0.4213	0.3303	0.2910	0.2265
3.05	0.8920	0.8132	0.7483	0.6806	0.6082	0.5432	0.4321	0.3413	0.3017	0.2366
3.10	0.8950	0.8181	0.7547	0.6883	0.6171	0.5529	0.4428	0.3521	0.3124	0.2467
3.15	0.8979	0.8229	0.7608	0.6957	0.6257	0.5624	0.4533	0.3628	0.3230	0.2568
3.20	0.9007	0.8275	0.7668	0.7029	0.6341	0.5716	0.4636	0.3734	0.3335	0.2669
3.25	0.9033	0.8319	0.7725	0.7099	0.6422	0.5807	0.4736	0.3838	0.3439	0.2769
3.30	0.9059	0.8362	0.7781	0.7167	0.6501	0.5895	0.4835	0.3941	0.3542	0.2869
3.35	0.9084	0.8403	0.7834	0.7233	0.6578	0.5980	0.4932	0.4042	0.3643	0.2968
3.40	0.9108	0.8443	0.7886	0.7296	0.6653	0.6064	0.5027	0.4142	0.3744	0.3067
3.45	0.9131	0.8481	0.7936	0.7358	0.6726	0.6145	0.5120	0.4240	0.3843	0.3165
3.50	0.9153	0.8518	0.7985	0.7418	0.6796	0.6224	0.5211	0.4337	0.3941	0.3262
3.55	0.9174	0.8554	0.8032	0.7476	0.6865	0.6302	0.5300	0.4432	0.4037	0.3358
3.60	0.9195	0.8589	0.8078	0.7532	0.6932	0.6377	0.5387	0.4525	0.4132	0.3453
3.65	0.9215	0.8622	0.8122	0.7586	0.6997	0.6450	0.5472	0.4617	0.4226	0.3548
3.70	0.9234	0.8654	0.8164	0.7639	0.7060	0.6521	0.5555	0.4707	0.4318	0.3641
3.75	0.9252	0.8686	0.8206	0.7690	0.7121	0.6591	0.5637	0.4796	0.4409	0.3733
3.80	0.9270	0.8716	0.8246	0.7740	0.7180	0.6659	0.5716	0.4883	0.4498	0.3824
3.85	0.9287	0.8745	0.8284	0.7788	0.7238	0.6725	0.5794	0.4969	0.4586	0.3914
3.90	0.9304	0.8773	0.8322	0.7835	0.7295	0.6789	0.5871	0.5053	0.4672	0.4003
3.95	0.9320	0.8801	0.8358	0.7881	0.7349	0.6851	0.5945	0.5135	0.4757	0.4091
4.00	0.9335	0.8827	0.8394	0.7925	0.7403	0.6912	0.6018	0.5216	0.4841	0.4177
4.05	0.9350	0.8853	0.8428	0.7968	0.7455	0.6972	0.6089	0.5295	0.4923	0.4263
4.10	0.9365	0.8878	0.8461	0.8010	0.7505	0.7030	0.6159	0.5373	0.5003	0.4347
4.15	0.9379	0.8902	0.8493	0.8050	0.7554	0.7086	0.6227	0.5449	0.5082	0.4430
4.20	0.9393	0.8925	0.8525	0.8089	0.7602	0.7141	0.6293	0.5524	0.5160	0.4511
4.25	0.9406	0.8948	0.8555	0.8128	0.7648	0.7195	0.6358	0.5597	0.5236	0.4591
4.30	0.9419	0.8970	0.8584	0.8165	0.7693	0.7247	0.6422	0.5669	0.5311	0.4671
4.35	0.9431	0.8991	0.8613	0.8201	0.7738	0.7298	0.6484	0.5739	0.5385	0.4749
4.40	0.9443	0.9012	0.8641	0.8236	0.7780	0.7348	0.6545	0.5808	0.5457	0.4825
4.45	0.9455	0.9032	0.8668	0.8270	0.7822	0.7396	0.6604	0.5876	0.5528	0.4901

PROBABILITY INTEGRAL OF t, THE NON-CENTRAL t-STATISTIC. THIS TABLE GIVES Pr $[t/\sqrt{f} \leq x]$.

f is the number of degrees of freedom; the non-centrality parameter is $\sqrt{f+1}\,K_p$.
K_p is the standardized normal deviate exceeded with probability p.

DEGREES OF FREEDOM **2**

x \ p	.2500	.1500	.1000	.0650	.0400	.0250	.0100	.0040	.0025	.0010
4.50	0.9466	0.9052	0.8694	0.8304	0.7863	0.7443	0.6662	0.5942	0.5597	0.4975
4.55	0.9477	0.9070	0.8720	0.8336	0.7902	0.7489	0.6719	0.6007	0.5666	0.5048
4.60	0.9487	0.9089	0.8744	0.8367	0.7941	0.7534	0.6775	0.6070	0.5732	0.5120
4.65	0.9498	0.9107	0.8768	0.8398	0.7979	0.7578	0.6829	0.6133	0.5798	0.5190
4.70	0.9508	0.9124	0.8792	0.8428	0.8015	0.7621	0.6882	0.6194	0.5862	0.5260
4.75	0.9517	0.9141	0.8815	0.8457	0.8051	0.7663	0.6934	0.6254	0.5926	0.5328
4.80	0.9527	0.9157	0.8837	0.8485	0.8086	0.7703	0.6984	0.6312	0.5988	0.5395
4.85	0.9536	0.9173	0.8858	0.8513	0.8120	0.7743	0.7034	0.6370	0.6048	0.5461
4.90	0.9545	0.9189	0.8880	0.8540	0.8153	0.7782	0.7082	0.6426	0.6108	0.5526
4.95	0.9553	0.9204	0.8900	0.8566	0.8185	0.7820	0.7130	0.6481	0.6167	0.5590
5.00	0.9562	0.9219	0.8920	0.8591	0.8217	0.7857	0.7176	0.6535	0.6224	0.5652
5.05	0.9570	0.9233	0.8939	0.8616	0.8247	0.7893	0.7221	0.6588	0.6280	0.5714
5.10	0.9578	0.9247	0.8958	0.8640	0.8277	0.7928	0.7266	0.6640	0.6335	0.5774
5.15	0.9586	0.9260	0.8977	0.8664	0.8307	0.7962	0.7309	0.6691	0.6389	0.5834
5.20	0.9593	0.9274	0.8995	0.8687	0.8335	0.7996	0.7352	0.6741	0.6443	0.5892
5.25	0.9601	0.9286	0.9012	0.8709	0.8363	0.8029	0.7393	0.6790	0.6495	0.5949
5.30	0.9608	0.9299	0.9029	0.8731	0.8390	0.8061	0.7434	0.6837	0.6546	0.6006
5.35	0.9615	0.9311	0.9046	0.8753	0.8417	0.8092	0.7474	0.6884	0.6596	0.6061
5.40	0.9622	0.9323	0.9062	0.8773	0.8443	0.8123	0.7512	0.6930	0.6645	0.6115
5.45	0.9628	0.9335	0.9078	0.8794	0.8468	0.8153	0.7551	0.6975	0.6693	0.6169
5.50	0.9635	0.9346	0.9093	0.8814	0.8493	0.8182	0.7588	0.7020	0.6740	0.6221
5.55	0.9641	0.9357	0.9108	0.8833	0.8517	0.8211	0.7624	0.7063	0.6787	0.6273
5.60	0.9647	0.9368	0.9123	0.8852	0.8541	0.8239	0.7660	0.7105	0.6832	0.6323
5.65	0.9653	0.9378	0.9137	0.8870	0.8564	0.8266	0.7695	0.7147	0.6877	0.6373
5.70	0.9659	0.9388	0.9151	0.8889	0.8586	0.8293	0.7729	0.7188	0.6921	0.6422
5.75	0.9664	0.9398	0.9165	0.8906	0.8608	0.8319	0.7763	0.7228	0.6964	0.6470
5.80	0.9670	0.9408	0.9179	0.8923	0.8630	0.8344	0.7796	0.7267	0.7006	0.6517
5.85	0.9675	0.9418	0.9192	0.8940	0.8651	0.8369	0.7828	0.7306	0.7047	0.6563
5.90	0.9681	0.9427	0.9204	0.8957	0.8672	0.8394	0.7860	0.7343	0.7088	0.6609
5.95	0.9686	0.9436	0.9217	0.8973	0.8692	0.8418	0.7890	0.7380	0.7128	0.6654
6.00	0.9691	0.9445	0.9229	0.8989	0.8711	0.8441	0.7921	0.7417	0.7167	0.6698
6.05	0.9696	0.9454	0.9241	0.9004	0.8731	0.8464	0.7950	0.7452	0.7205	0.6741
6.10	0.9700	0.9462	0.9252	0.9019	0.8750	0.8487	0.7979	0.7487	0.7243	0.6783
6.15	0.9705	0.9470	0.9264	0.9034	0.8768	0.8509	0.8008	0.7522	0.7280	0.6825
6.20	0.9710	0.9479	0.9275	0.9048	0.8786	0.8530	0.8036	0.7555	0.7316	0.6866
6.25	0.9714	0.9486	0.9286	0.9062	0.8804	0.8552	0.8063	0.7588	0.7352	0.6906
6.30	0.9719	0.9494	0.9296	0.9076	0.8821	0.8572	0.8090	0.7621	0.7387	0.6946
6.35	0.9723	0.9502	0.9307	0.9090	0.8838	0.8593	0.8116	0.7652	0.7421	0.6984
6.40	0.9727	0.9509	0.9317	0.9103	0.8855	0.8612	0.8142	0.7684	0.7455	0.7023
6.45	0.9731	0.9516	0.9327	0.9116	0.8871	0.8632	0.8167	0.7714	0.7488	0.7060
6.50	0.9735	0.9523	0.9337	0.9128	0.8887	0.8651	0.8192	0.7744	0.7520	0.7097
6.55	0.9739	0.9530	0.9346	0.9141	0.8903	0.8670	0.8217	0.7774	0.7552	0.7133
6.60	0.9743	0.9537	0.9356	0.9153	0.8918	0.8688	0.8240	0.7803	0.7584	0.7169
6.65	0.9746	0.9544	0.9365	0.9165	0.8933	0.8706	0.8264	0.7831	0.7614	0.7204
6.70	0.9750	0.9550	0.9374	0.9176	0.8948	0.8723	0.8287	0.7859	0.7645	0.7239
6.75	0.9754	0.9557	0.9383	0.9188	0.8962	0.8741	0.8309	0.7886	0.7674	0.7272
6.80	0.9757	0.9563	0.9391	0.9199	0.8976	0.8758	0.8332	0.7913	0.7704	0.7306
6.85	0.9761	0.9569	0.9400	0.9210	0.8990	0.8774	0.8353	0.7940	0.7732	0.7339
6.90	0.9764	0.9575	0.9408	0.9221	0.9004	0.8790	0.8375	0.7966	0.7761	0.7371
6.95	0.9767	0.9581	0.9416	0.9231	0.9017	0.8806	0.8395	0.7991	0.7788	0.7402
7.00	0.9771	0.9587	0.9424	0.9242	0.9030	0.8822	0.8416	0.8016	0.7815	0.7434
7.05	0.9774	0.9592	0.9432	0.9252	0.9043	0.8837	0.8436	0.8041	0.7842	0.7464
7.10	0.9777	0.9598	0.9439	0.9262	0.9055	0.8852	0.8456	0.8065	0.7868	0.7494
7.15	0.9780	0.9603	0.9447	0.9271	0.9068	0.8867	0.8475	0.8089	0.7894	0.7524
7.20	0.9783	0.9608	0.9454	0.9281	0.9080	0.8882	0.8494	0.8112	0.7920	0.7553
7.25	0.9786	0.9614	0.9461	0.9290	0.9091	0.8896	0.8513	0.8135	0.7945	0.7582
7.30	0.9789	0.9619	0.9468	0.9299	0.9103	0.8910	0.8531	0.8158	0.7969	0.7610
7.35	0.9791	0.9624	0.9475	0.9308	0.9114	0.8924	0.8550	0.8180	0.7993	0.7638
7.40	0.9794	0.9629	0.9482	0.9317	0.9126	0.8937	0.8567	0.8202	0.8017	0.7665
7.45	0.9797	0.9633	0.9489	0.9326	0.9137	0.8950	0.8585	0.8223	0.8040	0.7692

PROBABILITY INTEGRAL OF t, THE NON-CENTRAL t-STATISTIC. THIS TABLE GIVES Pr $[t/\sqrt{f} \leq x]$.

f is the number of degrees of freedom; the non-centrality parameter is $\sqrt{f+1}\,K_p$.

K_p is the standardized normal deviate exceeded with probability p.

DEGREES OF FREEDOM 2

x \ p	.2500	.1500	.1000	.0650	.0400	.0250	.0100	.0040	.0025	.0010
7.50	0.9799	0.9638	0.9495	0.9334	0.9147	0.8963	0.8602	0.8244	0.8063	0.7718
7.55	0.9802	0.9643	0.9501	0.9343	0.9158	0.8976	0.8619	0.8265	0.8086	0.7744
7.60	0.9804	0.9647	0.9508	0.9351	0.9168	0.8988	0.8635	0.8285	0.8108	0.7770
7.65	0.9807	0.9652	0.9514	0.9359	0.9179	0.9001	0.8651	0.8305	0.8130	0.7795
7.70	0.9809	0.9656	0.9520	0.9367	0.9189	0.9013	0.8667	0.8325	0.8151	0.7820
7.75	0.9812	0.9660	0.9526	0.9375	0.9199	0.9025	0.8683	0.8344	0.8172	0.7844
7.80	0.9814	0.9665	0.9532	0.9382	0.9208	0.9036	0.8699	0.8363	0.8193	0.7868
7.85	0.9816	0.9669	0.9537	0.9390	0.9218	0.9048	0.8714	0.8382	0.8214	0.7892
7.90	0.9819	0.9673	0.9543	0.9397	0.9227	0.9059	0.8729	0.8400	0.8234	0.7915
7.95	0.9821	0.9677	0.9549	0.9404	0.9236	0.9070	0.8743	0.8418	0.8254	0.7938
8.00	0.9823	0.9681	0.9554	0.9412	0.9245	0.9081	0.8758	0.8436	0.8273	0.7960
8.05	0.9825	0.9684	0.9559	0.9419	0.9254	0.9092	0.8772	0.8454	0.8292	0.7982
8.10	0.9827	0.9688	0.9564	0.9425	0.9263	0.9102	0.8786	0.8471	0.8311	0.8004
8.15	0.9829	0.9692	0.9570	0.9432	0.9271	0.9113	0.8800	0.8488	0.8330	0.8026
8.20	0.9831	0.9696	0.9575	0.9439	0.9280	0.9123	0.8813	0.8505	0.8348	0.8047
8.25	0.9833	0.9699	0.9580	0.9445	0.9288	0.9133	0.8826	0.8521	0.8366	0.8068
8.30	0.9835	0.9703	0.9585	0.9452	0.9296	0.9142	0.8839	0.8537	0.8384	0.8088
8.35	0.9837	0.9706	0.9589	0.9458	0.9304	0.9152	0.8852	0.8553	0.8401	0.8109
8.40	0.9839	0.9710	0.9594	0.9464	0.9312	0.9162	0.8865	0.8569	0.8418	0.8129
8.45	0.9841	0.9713	0.9599	0.9470	0.9320	0.9171	0.8877	0.8584	0.8435	0.8148
8.50	0.9843	0.9716	0.9603	0.9476	0.9327	0.9180	0.8890	0.8599	0.8452	0.8167
8.55	0.9845	0.9719	0.9608	0.9482	0.9335	0.9189	0.8902	0.8614	0.8468	0.8187
8.60	0.9847	0.9723	0.9612	0.9488	0.9342	0.9198	0.8914	0.8629	0.8484	0.8205
8.65	0.9848	0.9726	0.9616	0.9493	0.9349	0.9207	0.8925	0.8643	0.8500	0.8224
8.70	0.9850	0.9729	0.9621	0.9499	0.9357	0.9215	0.8937	0.8658	0.8516	0.8242
8.75	0.9852	0.9732	0.9625	0.9505	0.9364	0.9224	0.8948	0.8672	0.8531	0.8260
8.80	0.9853	0.9735	0.9629	0.9510	0.9370	0.9232	0.8959	0.8686	0.8546	0.8278
8.85	0.9855	0.9738	0.9633	0.9515	0.9377	0.9241	0.8970	0.8699	0.8561	0.8295
8.90	0.9857	0.9740	0.9637	0.9520	0.9384	0.9249	0.8981	0.8713	0.8576	0.8312
8.95	0.9858	0.9743	0.9641	0.9526	0.9390	0.9257	0.8992	0.8726	0.8590	0.8329
9.00	0.9860	0.9746	0.9645	0.9531	0.9397	0.9264	0.9002	0.8739	0.8605	0.8346
9.05	0.9861	0.9749	0.9649	0.9536	0.9403	0.9272	0.9012	0.8752	0.8619	0.8362
9.10	0.9863	0.9752	0.9652	0.9541	0.9410	0.9280	0.9023	0.8764	0.8633	0.8379
9.15	0.9864	0.9754	0.9656	0.9545	0.9416	0.9287	0.9033	0.8777	0.8646	0.8394
9.20	0.9866	0.9757	0.9660	0.9550	0.9422	0.9295	0.9042	0.8789	0.8660	0.8410
9.25	0.9867	0.9759	0.9663	0.9555	0.9428	0.9302	0.9052	0.8801	0.8673	0.8426
9.30	0.9868	0.9762	0.9667	0.9560	0.9434	0.9309	0.9062	0.8813	0.8686	0.8441
9.35	0.9870	0.9764	0.9670	0.9564	0.9440	0.9316	0.9071	0.8825	0.8699	0.8456
9.40	0.9871	0.9767	0.9674	0.9569	0.9445	0.9323	0.9080	0.8837	0.8712	0.8471
9.45	0.9872	0.9769	0.9677	0.9573	0.9451	0.9330	0.9090	0.8848	0.8724	0.8486
9.50	0.9874	0.9772	0.9680	0.9577	0.9456	0.9336	0.9099	0.8859	0.8737	0.8500
9.55	0.9875	0.9774	0.9684	0.9582	0.9462	0.9343	0.9107	0.8870	0.8749	0.8515
9.60	0.9876	0.9776	0.9687	0.9586	0.9467	0.9350	0.9116	0.8881	0.8761	0.8529
9.65	0.9878	0.9778	0.9690	0.9590	0.9473	0.9356	0.9125	0.8892	0.8773	0.8543
9.70	0.9879	0.9781	0.9693	0.9594	0.9478	0.9362	0.9133	0.8903	0.8785	0.8556
9.75	0.9880	0.9783	0.9696	0.9598	0.9483	0.9369	0.9142	0.8913	0.8796	0.8570
9.80	0.9881	0.9785	0.9699	0.9602	0.9488	0.9375	0.9150	0.8923	0.8808	0.8583
9.85	0.9882	0.9787	0.9702	0.9606	0.9493	0.9381	0.9158	0.8934	0.8819	0.8596
9.90	0.9884	0.9789	0.9705	0.9610	0.9498	0.9387	0.9166	0.8944	0.8830	0.8609
9.95	0.9885	0.9791	0.9708	0.9614	0.9503	0.9393	0.9174	0.8954	0.8841	0.8622
10.00	0.9886	0.9793	0.9711	0.9617	0.9508	0.9399	0.9182	0.8963	0.8852	0.8635
10.05	0.9887	0.9795	0.9714	0.9621	0.9512	0.9404	0.9190	0.8973	0.8862	0.8647
10.10	0.9888	0.9797	0.9716	0.9625	0.9517	0.9410	0.9197	0.8983	0.8873	0.8659
10.15	0.9889	0.9799	0.9719	0.9628	0.9522	0.9416	0.9205	0.8992	0.8883	0.8672
10.20	0.9890	0.9801	0.9722	0.9632	0.9526	0.9421	0.9212	0.9001	0.8893	0.8684
10.25	0.9891	0.9803	0.9724	0.9635	0.9531	0.9426	0.9219	0.9010	0.8903	0.8695
10.30	0.9892	0.9805	0.9727	0.9639	0.9535	0.9432	0.9227	0.9019	0.8913	0.8707
10.35	0.9893	0.9807	0.9730	0.9642	0.9539	0.9437	0.9234	0.9028	0.8923	0.8719
10.40	0.9894	0.9809	0.9732	0.9645	0.9544	0.9442	0.9241	0.9037	0.8933	0.8730
10.45	0.9895	0.9811	0.9735	0.9649	0.9548	0.9447	0.9248	0.9046	0.8942	0.8741

PROBABILITY INTEGRAL OF t, THE NON-CENTRAL t-STATISTIC. THIS TABLE GIVES Pr $[t/\sqrt{f} \leq x]$.

f is the number of degrees of freedom; the non-centrality parameter is $\sqrt{f+1}\, K_p$.

K_p is the standardized normal deviate exceeded with probability p.

x \ p	.2500	.1500	.1000	.0650	.0400	.0250	.0100	.0040	.0025	.0010
10.50	0.9896	0.9812	0.9737	0.9652	0.9552	0.9452	0.9254	0.9054	0.8952	0.8752
10.55	0.9897	0.9814	0.9739	0.9655	0.9556	0.9457	0.9261	0.9063	0.8961	0.8763
10.60	0.9898	0.9816	0.9742	0.9658	0.9560	0.9462	0.9268	0.9071	0.8970	0.8774
10.65	0.9899	0.9817	0.9744	0.9661	0.9564	0.9467	0.9274	0.9079	0.8979	0.8785
10.70	0.9900	0.9819	0.9747	0.9665	0.9568	0.9472	0.9281	0.9087	0.8988	0.8795
10.75	0.9901	0.9821	0.9749	0.9668	0.9572	0.9477	0.9287	0.9095	0.8997	0.8806
10.80	0.9902	0.9822	0.9751	0.9671	0.9576	0.9481	0.9293	0.9103	0.9006	0.8816
10.85	0.9903	0.9824	0.9753	0.9674	0.9580	0.9486	0.9300	0.9111	0.9014	0.8826
10.90	0.9904	0.9826	0.9756	0.9676	0.9583	0.9491	0.9306	0.9119	0.9023	0.8836
10.95	0.9905	0.9827	0.9758	0.9679	0.9587	0.9495	0.9312	0.9126	0.9031	0.8846
11.00	0.9905	0.9829	0.9760	0.9682	0.9591	0.9499	0.9318	0.9134	0.9039	0.8856
11.05	0.9906	0.9830	0.9762	0.9685	0.9594	0.9504	0.9324	0.9141	0.9048	0.8865
11.10	0.9907	0.9832	0.9764	0.9688	0.9598	0.9508	0.9330	0.9149	0.9056	0.8875
11.15	0.9908	0.9833	0.9766	0.9690	0.9601	0.9512	0.9335	0.9156	0.9064	0.8884
11.20	0.9909	0.9835	0.9768	0.9693	0.9605	0.9517	0.9341	0.9163	0.9072	0.8894
11.25	0.9910	0.9836	0.9770	0.9696	0.9608	0.9521	0.9347	0.9170	0.9079	0.8903
11.30	0.9910	0.9838	0.9772	0.9698	0.9612	0.9525	0.9352	0.9177	0.9087	0.8912
11.35	0.9911	0.9839	0.9774	0.9701	0.9615	0.9529	0.9358	0.9184	0.9095	0.8921
11.40	0.9912	0.9840	0.9776	0.9704	0.9618	0.9533	0.9363	0.9191	0.9102	0.8930
11.45	0.9913	0.9842	0.9778	0.9706	0.9621	0.9537	0.9368	0.9197	0.9109	0.8938
11.50	0.9913	0.9843	0.9780	0.9709	0.9625	0.9541	0.9374	0.9204	0.9117	0.8947
11.55	0.9914	0.9844	0.9782	0.9711	0.9628	0.9545	0.9379	0.9211	0.9124	0.8956
11.60	0.9915	0.9846	0.9784	0.9714	0.9631	0.9548	0.9384	0.9217	0.9131	0.8964
11.65	0.9916	0.9847	0.9786	0.9716	0.9634	0.9552	0.9389	0.9223	0.9138	0.8972
11.70	0.9916	0.9848	0.9787	0.9718	0.9637	0.9556	0.9394	0.9230	0.9145	0.8981
11.75	0.9917	0.9850	0.9789	0.9721	0.9640	0.9560	0.9399	0.9236	0.9152	0.8989
11.80	0.9918	0.9851	0.9791	0.9723	0.9643	0.9563	0.9404	0.9242	0.9159	0.8997
11.85	0.9918	0.9852	0.9793	0.9725	0.9646	0.9567	0.9409	0.9248	0.9166	0.9005
11.90	0.9919	0.9853	0.9794	0.9727	0.9649	0.9570	0.9413	0.9254	0.9172	0.9013
11.95	0.9920	0.9854	0.9796	0.9730	0.9652	0.9574	0.9418	0.9260	0.9179	0.9020
12.00	0.9920	0.9856	0.9798	0.9732	0.9654	0.9577	0.9423	0.9266	0.9185	0.9028
12.05	0.9921	0.9857	0.9799	0.9734	0.9657	0.9581	0.9427	0.9272	0.9192	0.9036
12.10	0.9922	0.9858	0.9801	0.9736	0.9660	0.9584	0.9432	0.9278	0.9198	0.9043
12.15	0.9922	0.9859	0.9803	0.9738	0.9663	0.9587	0.9437	0.9283	0.9204	0.9051
12.20	0.9923	0.9860	0.9804	0.9740	0.9665	0.9591	0.9441	0.9289	0.9211	0.9058
12.25	0.9924	0.9861	0.9806	0.9743	0.9668	0.9594	0.9445	0.9295	0.9217	0.9065
12.30	0.9924	0.9863	0.9807	0.9745	0.9671	0.9597	0.9450	0.9300	0.9223	0.9072
12.35	0.9925	0.9864	0.9809	0.9747	0.9673	0.9600	0.9454	0.9305	0.9229	0.9079
12.40	0.9925	0.9865	0.9810	0.9749	0.9676	0.9603	0.9458	0.9311	0.9235	0.9086
12.45	0.9926	0.9866	0.9812	0.9751	0.9678	0.9606	0.9462	0.9316	0.9241	0.9093
12.50	0.9927	0.9867	0.9813	0.9753	0.9681	0.9609	0.9467	0.9321	0.9246	0.9100
12.55	0.9927	0.9868	0.9815	0.9754	0.9683	0.9613	0.9471	0.9326	0.9252	0.9107
12.60	0.9928	0.9869	0.9816	0.9756	0.9686	0.9616	0.9475	0.9332	0.9258	0.9114
12.65	0.9928	0.9870	0.9818	0.9758	0.9688	0.9618	0.9479	0.9337	0.9263	0.9120
12.70	0.9929	0.9871	0.9819	0.9760	0.9691	0.9621	0.9483	0.9342	0.9269	0.9127
12.75	0.9929	0.9872	0.9820	0.9762	0.9693	0.9624	0.9487	0.9347	0.9274	0.9133
12.80	0.9930	0.9873	0.9822	0.9764	0.9695	0.9627	0.9491	0.9352	0.9280	0.9140
12.85	0.9931	0.9874	0.9823	0.9766	0.9698	0.9630	0.9494	0.9356	0.9285	0.9146
12.90	0.9931	0.9875	0.9824	0.9767	0.9700	0.9633	0.9498	0.9361	0.9290	0.9152
12.95	0.9932	0.9876	0.9826	0.9769	0.9702	0.9636	0.9502	0.9366	0.9296	0.9159
13.00	0.9932	0.9877	0.9827	0.9771	0.9705	0.9638	0.9506	0.9371	0.9301	0.9165
13.05	0.9933	0.9878	0.9828	0.9773	0.9707	0.9641	0.9509	0.9375	0.9306	0.9171
13.10	0.9933	0.9879	0.9830	0.9774	0.9709	0.9644	0.9513	0.9380	0.9311	0.9177
13.15	0.9934	0.9880	0.9831	0.9776	0.9711	0.9646	0.9516	0.9384	0.9316	0.9183
13.20	0.9934	0.9880	0.9832	0.9778	0.9713	0.9649	0.9520	0.9389	0.9321	0.9189
13.25	0.9935	0.9881	0.9834	0.9779	0.9715	0.9651	0.9524	0.9393	0.9326	0.9195
13.30	0.9935	0.9882	0.9835	0.9781	0.9717	0.9654	0.9527	0.9398	0.9331	0.9200
13.35	0.9936	0.9883	0.9836	0.9783	0.9720	0.9657	0.9530	0.9402	0.9336	0.9206
13.40	0.9936	0.9884	0.9837	0.9784	0.9722	0.9659	0.9534	0.9406	0.9340	0.9212
13.45	0.9937	0.9885	0.9838	0.9786	0.9724	0.9662	0.9537	0.9411	0.9345	0.9217

PROBABILITY INTEGRAL OF t, THE NON-CENTRAL t-STATISTIC. THIS TABLE GIVES Pr $[t/\sqrt{f} \leq x]$.

f is the number of degrees of freedom; the non-centrality parameter is $\sqrt{f+1}\,K_p$.

K_p is the standardized normal deviate exceeded with probability p.

DEGREES OF FREEDOM **2**

x \ p	.2500	.1500	.1000	.0650	.0400	.0250	.0100	.0040	.0025	.0010
13.50	0.9937	0.9886	0.9840	0.9787	0.9726	0.9664	0.9541	0.9415	0.9350	0.9223
13.55	0.9937	0.9886	0.9841	0.9789	0.9728	0.9666	0.9544	0.9419	0.9354	0.9228
13.60	0.9938	0.9887	0.9842	0.9790	0.9730	0.9669	0.9547	0.9423	0.9359	0.9234
13.65	0.9938	0.9888	0.9843	0.9792	0.9731	0.9671	0.9550	0.9427	0.9363	0.9239
13.70	0.9939	0.9889	0.9844	0.9793	0.9733	0.9674	0.9553	0.9431	0.9368	0.9244
13.75	0.9939	0.9890	0.9845	0.9795	0.9735	0.9676	0.9557	0.9435	0.9372	0.9250
13.80	0.9940	0.9891	0.9846	0.9796	0.9737	0.9678	0.9560	0.9439	0.9377	0.9255
13.85	0.9940	0.9891	0.9847	0.9798	0.9739	0.9680	0.9563	0.9443	0.9381	0.9260
13.90	0.9941	0.9892	0.9849	0.9799	0.9741	0.9683	0.9566	0.9447	0.9385	0.9265
13.95	0.9941	0.9893	0.9850	0.9801	0.9743	0.9685	0.9569	0.9451	0.9390	0.9270
14.00	0.9941	0.9894	0.9851	0.9802	0.9745	0.9687	0.9572	0.9454	0.9394	0.9275
14.05	0.9942	0.9894	0.9852	0.9803	0.9746	0.9689	0.9575	0.9458	0.9398	0.9280
14.10	0.9942	0.9895	0.9853	0.9805	0.9748	0.9691	0.9578	0.9462	0.9402	0.9285
14.15	0.9943	0.9896	0.9854	0.9806	0.9750	0.9694	0.9581	0.9466	0.9406	0.9290
14.20	0.9943	0.9897	0.9855	0.9807	0.9752	0.9696	0.9584	0.9469	0.9410	0.9295
14.25	0.9943	0.9897	0.9856	0.9809	0.9753	0.9698	0.9586	0.9473	0.9414	0.9299
14.30	0.9944	0.9898	0.9857	0.9810	0.9755	0.9700	0.9589	0.9476	0.9418	0.9304
14.35	0.9944	0.9899	0.9858	0.9811	0.9757	0.9702	0.9592	0.9480	0.9422	0.9309
14.40	0.9945	0.9899	0.9859	0.9813	0.9758	0.9704	0.9595	0.9483	0.9426	0.9313
14.45	0.9945	0.9900	0.9860	0.9814	0.9760	0.9706	0.9598	0.9487	0.9430	0.9318
14.50	0.9945	0.9901	0.9861	0.9815	0.9762	0.9708	0.9600	0.9490	0.9433	0.9322
14.55	0.9946	0.9901	0.9862	0.9816	0.9763	0.9710	0.9603	0.9494	0.9437	0.9327
14.60	0.9946	0.9902	0.9863	0.9818	0.9765	0.9712	0.9606	0.9497	0.9441	0.9331
14.65	0.9946	0.9903	0.9863	0.9819	0.9766	0.9714	0.9608	0.9500	0.9445	0.9336
14.70	0.9947	0.9903	0.9864	0.9820	0.9768	0.9716	0.9611	0.9504	0.9448	0.9340
14.75	0.9947	0.9904	0.9865	0.9821	0.9769	0.9717	0.9613	0.9507	0.9452	0.9344
14.80	0.9948	0.9905	0.9866	0.9823	0.9771	0.9719	0.9616	0.9510	0.9455	0.9348
14.85	0.9948	0.9905	0.9867	0.9824	0.9772	0.9721	0.9618	0.9513	0.9459	0.9353
14.90	0.9948	0.9906	0.9868	0.9825	0.9774	0.9723	0.9621	0.9517	0.9463	0.9357
14.95	0.9949	0.9907	0.9869	0.9826	0.9775	0.9725	0.9623	0.9520	0.9466	0.9361
15.00	0.9949	0.9907	0.9870	0.9827	0.9777	0.9727	0.9626	0.9523	0.9469	0.9365
15.05	0.9949	0.9908	0.9871	0.9828	0.9778	0.9728	0.9628	0.9526	0.9473	0.9369
15.10	0.9950	0.9908	0.9871	0.9829	0.9780	0.9730	0.9631	0.9529	0.9476	0.9373
15.15	0.9950	0.9909	0.9872	0.9831	0.9781	0.9732	0.9633	0.9532	0.9480	0.9377
15.20	0.9950	0.9910	0.9873	0.9832	0.9783	0.9734	0.9635	0.9535	0.9483	0.9381
15.25	0.9951	0.9910	0.9874	0.9833	0.9784	0.9735	0.9638	0.9538	0.9486	0.9385
15.30	0.9951	0.9911	0.9875	0.9834	0.9785	0.9737	0.9640	0.9541	0.9489	0.9389
15.35	0.9951	0.9911	0.9876	0.9835	0.9787	0.9739	0.9642	0.9544	0.9493	0.9393
15.40	0.9952	0.9912	0.9876	0.9836	0.9788	0.9740	0.9645	0.9547	0.9496	0.9396
15.45	0.9952	0.9912	0.9877	0.9837	0.9790	0.9742	0.9647	0.9549	0.9499	0.9400
15.50	0.9952	0.9913	0.9878	0.9838	0.9791	0.9744	0.9649	0.9552	0.9502	0.9404
15.55	0.9952	0.9914	0.9879	0.9839	0.9792	0.9745	0.9651	0.9555	0.9505	0.9408
15.60	0.9953	0.9914	0.9879	0.9840	0.9794	0.9747	0.9653	0.9558	0.9508	0.9411
15.65	0.9953	0.9915	0.9880	0.9841	0.9795	0.9749	0.9656	0.9561	0.9511	0.9415
15.70	0.9953	0.9915	0.9881	0.9842	0.9796	0.9750	0.9658	0.9563	0.9514	0.9419
15.75	0.9954	0.9916	0.9882	0.9843	0.9797	0.9752	0.9660	0.9566	0.9517	0.9422
15.80	0.9954	0.9916	0.9882	0.9844	0.9799	0.9753	0.9662	0.9569	0.9520	0.9426
15.85	0.9954	0.9917	0.9883	0.9845	0.9800	0.9755	0.9664	0.9571	0.9523	0.9429
15.90	0.9954	0.9917	0.9884	0.9846	0.9801	0.9756	0.9666	0.9574	0.9526	0.9433
15.95	0.9955	0.9918	0.9885	0.9847	0.9802	0.9758	0.9668	0.9577	0.9529	0.9436
16.00	0.9955	0.9918	0.9885	0.9848	0.9804	0.9759	0.9670	0.9579	0.9532	0.9439
16.05	0.9955	0.9919	0.9886	0.9849	0.9805	0.9761	0.9672	0.9582	0.9535	0.9443
16.10	0.9956	0.9919	0.9887	0.9850	0.9806	0.9762	0.9674	0.9584	0.9538	0.9446
16.15	0.9956	0.9920	0.9887	0.9851	0.9807	0.9764	0.9676	0.9587	0.9540	0.9449
16.20	0.9956	0.9920	0.9888	0.9852	0.9808	0.9765	0.9678	0.9589	0.9543	0.9453
16.25	0.9956	0.9921	0.9889	0.9852	0.9809	0.9766	0.9680	0.9592	0.9546	0.9456
16.30	0.9957	0.9921	0.9889	0.9853	0.9811	0.9768	0.9682	0.9594	0.9549	0.9459
16.35	0.9957	0.9922	0.9890	0.9854	0.9812	0.9769	0.9684	0.9597	0.9551	0.9462
16.40	0.9957	0.9922	0.9891	0.9855	0.9813	0.9771	0.9686	0.9599	0.9554	0.9466
16.45	0.9957	0.9923	0.9891	0.9856	0.9814	0.9772	0.9688	0.9601	0.9557	0.9469

PROBABILITY INTEGRAL OF t, THE NON-CENTRAL t-STATISTIC. THIS TABLE GIVES Pr $[t/\sqrt{f} \leq x]$.

f is the number of degrees of freedom; the non-centrality parameter is $\sqrt{f+1}\,K_p$.

K_p is the standardized normal deviate exceeded with probability p.

DEGREES OF FREEDOM **2**

x \ p	.2500	.1500	.1000	.0650	.0400	.0250	.0100	.0040	.0025	.0010
16.50	0.9958	0.9923	0.9892	0.9857	0.9815	0.9773	0.9690	0.9604	0.9559	0.9472
16.55	0.9958	0.9924	0.9893	0.9858	0.9816	0.9775	0.9691	0.9606	0.9562	0.9475
16.60	0.9958	0.9924	0.9893	0.9859	0.9817	0.9776	0.9693	0.9608	0.9564	0.9478
16.65	0.9958	0.9925	0.9894	0.9859	0.9818	0.9777	0.9695	0.9611	0.9567	0.9481
16.70	0.9959	0.9925	0.9895	0.9860	0.9819	0.9779	0.9697	0.9613	0.9569	0.9484
16.75	0.9959	0.9925	0.9895	0.9861	0.9821	0.9780	0.9699	0.9615	0.9572	0.9487
16.80	0.9959	0.9926	0.9896	0.9862	0.9822	0.9781	0.9700	0.9617	0.9574	0.9490
16.85	0.9959	0.9926	0.9896	0.9863	0.9823	0.9783	0.9702	0.9620	0.9577	0.9493
16.90	0.9960	0.9927	0.9897	0.9863	0.9824	0.9784	0.9704	0.9622	0.9579	0.9496
16.95	0.9960	0.9927	0.9898	0.9864	0.9825	0.9785	0.9705	0.9624	0.9582	0.9499
17.00	0.9960	0.9928	0.9898	0.9865	0.9826	0.9786	0.9707	0.9626	0.9584	0.9502
17.05	0.9960	0.9928	0.9899	0.9866	0.9827	0.9788	0.9709	0.9628	0.9586	0.9504
17.10	0.9961	0.9928	0.9899	0.9867	0.9828	0.9789	0.9711	0.9630	0.9589	0.9507
17.15	0.9961	0.9929	0.9900	0.9867	0.9829	0.9790	0.9712	0.9632	0.9591	0.9510
17.20	0.9961	0.9929	0.9901	0.9868	0.9830	0.9791	0.9714	0.9635	0.9593	0.9513
17.25	0.9961	0.9930	0.9901	0.9869	0.9831	0.9792	0.9715	0.9637	0.9596	0.9516
17.30	0.9962	0.9930	0.9902	0.9870	0.9832	0.9794	0.9717	0.9639	0.9598	0.9518
17.35	0.9962	0.9930	0.9902	0.9870	0.9833	0.9795	0.9719	0.9641	0.9600	0.9521
17.40	0.9962	0.9931	0.9903	0.9871	0.9834	0.9796	0.9720	0.9643	0.9603	0.9524
17.45	0.9962	0.9931	0.9903	0.9872	0.9835	0.9797	0.9722	0.9645	0.9605	0.9526
17.50	0.9962	0.9932	0.9904	0.9873	0.9835	0.9798	0.9723	0.9647	0.9607	0.9529
17.55	0.9963	0.9932	0.9905	0.9873	0.9836	0.9799	0.9725	0.9649	0.9609	0.9532
17.60	0.9963	0.9932	0.9905	0.9874	0.9837	0.9800	0.9726	0.9651	0.9611	0.9534
17.65	0.9963	0.9933	0.9906	0.9875	0.9838	0.9802	0.9728	0.9653	0.9613	0.9537
17.70	0.9963	0.9933	0.9906	0.9875	0.9839	0.9803	0.9729	0.9654	0.9616	0.9539
17.75	0.9963	0.9934	0.9907	0.9876	0.9840	0.9804	0.9731	0.9656	0.9618	0.9542
17.80	0.9964	0.9934	0.9907	0.9877	0.9841	0.9805	0.9732	0.9658	0.9620	0.9544
17.85	0.9964	0.9934	0.9908	0.9878	0.9842	0.9806	0.9734	0.9660	0.9622	0.9547
17.90	0.9964	0.9935	0.9908	0.9878	0.9843	0.9807	0.9735	0.9662	0.9624	0.9549
17.95	0.9964	0.9935	0.9909	0.9879	0.9844	0.9808	0.9737	0.9664	0.9626	0.9552
18.00	0.9964	0.9935	0.9909	0.9880	0.9844	0.9809	0.9738	0.9666	0.9628	0.9554
18.05	0.9965	0.9936	0.9910	0.9880	0.9845	0.9810	0.9740	0.9667	0.9630	0.9556
18.10	0.9965	0.9936	0.9910	0.9881	0.9846	0.9811	0.9741	0.9669	0.9632	0.9559
18.15	0.9965	0.9936	0.9911	0.9881	0.9847	0.9812	0.9743	0.9671	0.9634	0.9561
18.20	0.9965	0.9937	0.9911	0.9882	0.9848	0.9813	0.9744	0.9673	0.9636	0.9564
18.25	0.9965	0.9937	0.9912	0.9883	0.9849	0.9814	0.9745	0.9675	0.9638	0.9566
18.30	0.9966	0.9937	0.9912	0.9883	0.9849	0.9815	0.9747	0.9676	0.9640	0.9568
18.35	0.9966	0.9938	0.9913	0.9884	0.9850	0.9816	0.9748	0.9678	0.9642	0.9571
18.40	0.9966	0.9938	0.9913	0.9885	0.9851	0.9817	0.9749	0.9680	0.9644	0.9573
18.45	0.9966	0.9938	0.9914	0.9885	0.9852	0.9818	0.9751	0.9681	0.9646	0.9575
18.50	0.9966	0.9939	0.9914	0.9886	0.9853	0.9819	0.9752	0.9683	0.9647	0.9577
18.55	0.9967	0.9939	0.9914	0.9887	0.9853	0.9820	0.9753	0.9685	0.9649	0.9580
18.60	0.9967	0.9939	0.9915	0.9887	0.9854	0.9821	0.9755	0.9687	0.9651	0.9582
18.65	0.9967	0.9940	0.9915	0.9888	0.9855	0.9822	0.9756	0.9688	0.9653	0.9584
18.70	0.9967	0.9940	0.9916	0.9888	0.9856	0.9823	0.9757	0.9690	0.9655	0.9586
18.75	0.9967	0.9940	0.9916	0.9889	0.9856	0.9824	0.9759	0.9691	0.9657	0.9588
18.80	0.9967	0.9941	0.9917	0.9889	0.9857	0.9825	0.9760	0.9693	0.9658	0.9590
18.85	0.9968	0.9941	0.9917	0.9890	0.9858	0.9826	0.9761	0.9695	0.9660	0.9592
18.90	0.9968	0.9941	0.9918	0.9891	0.9859	0.9827	0.9762	0.9696	0.9662	0.9595
18.95	0.9968	0.9942	0.9918	0.9891	0.9859	0.9828	0.9764	0.9698	0.9664	0.9597
19.00	0.9968	0.9942	0.9918	0.9892	0.9860	0.9828	0.9765	0.9699	0.9665	0.9599
19.05	0.9968	0.9942	0.9919	0.9892	0.9861	0.9829	0.9766	0.9701	0.9667	0.9601
19.10	0.9968	0.9943	0.9919	0.9893	0.9862	0.9830	0.9767	0.9702	0.9669	0.9603
19.15	0.9969	0.9943	0.9920	0.9893	0.9862	0.9831	0.9768	0.9704	0.9671	0.9605
19.20	0.9969	0.9943	0.9920	0.9894	0.9863	0.9832	0.9770	0.9705	0.9672	0.9607
19.25	0.9969	0.9943	0.9921	0.9895	0.9864	0.9833	0.9771	0.9707	0.9674	0.9609
19.30	0.9969	0.9944	0.9921	0.9895	0.9864	0.9834	0.9772	0.9708	0.9676	0.9611
19.35	0.9969	0.9944	0.9921	0.9896	0.9865	0.9835	0.9773	0.9710	0.9677	0.9613
19.40	0.9969	0.9944	0.9922	0.9896	0.9866	0.9835	0.9774	0.9711	0.9679	0.9615
19.45	0.9970	0.9945	0.9922	0.9897	0.9866	0.9836	0.9775	0.9713	0.9680	0.9617

PROBABILITY INTEGRAL OF t, THE NON-CENTRAL t-STATISTIC. THIS TABLE GIVES Pr $[t/\sqrt{f} \leq x]$.

f is the number of degrees of freedom; the non-centrality parameter is $\sqrt{f+1}\,K_p$.

K_p is the standardized normal deviate exceeded with probability p.

DEGREES OF FREEDOM **2**

x \ p	.2500	.1500	.1000	.0650	.0400	.0250	.0100	.0040	.0025	.0010
19.50	0.9970	0.9945	0.9923	0.9897	0.9867	0.9837	0.9776	0.9714	0.9682	0.9619
19.55	0.9970	0.9945	0.9923	0.9898	0.9868	0.9838	0.9778	0.9716	0.9684	0.9621
19.60	0.9970	0.9945	0.9923	0.9898	0.9869	0.9839	0.9779	0.9717	0.9685	0.9622
19.65	0.9970	0.9946	0.9924	0.9899	0.9869	0.9840	0.9780	0.9719	0.9687	0.9624
19.70	0.9970	0.9946	0.9924	0.9899	0.9870	0.9840	0.9781	0.9720	0.9688	0.9626
19.75	0.9970	0.9946	0.9924	0.9900	0.9870	0.9841	0.9782	0.9721	0.9690	0.9628
19.80	0.9971	0.9947	0.9925	0.9900	0.9871	0.9842	0.9783	0.9723	0.9691	0.9630
19.85	0.9971	0.9947	0.9925	0.9901	0.9872	0.9843	0.9784	0.9724	0.9693	0.9632
19.90	0.9971	0.9947	0.9926	0.9901	0.9872	0.9843	0.9785	0.9725	0.9694	0.9633
19.95	0.9971	0.9947	0.9926	0.9902	0.9873	0.9844	0.9786	0.9727	0.9696	0.9635
20.00	0.9971	0.9948	0.9926	0.9902	0.9874	0.9845	0.9787	0.9728	0.9697	0.9637
20.05	0.9971	0.9948	0.9927	0.9903	0.9874	0.9846	0.9788	0.9730	0.9699	0.9639
20.10	0.9971	0.9948	0.9927	0.9903	0.9875	0.9847	0.9789	0.9731	0.9700	0.9641
20.15	0.9972	0.9948	0.9927	0.9904	0.9876	0.9847	0.9790	0.9732	0.9702	0.9642
20.20	0.9972	0.9949	0.9928	0.9904	0.9876	0.9848	0.9792	0.9733	0.9703	0.9644
20.25	0.9972	0.9949	0.9928	0.9905	0.9877	0.9849	0.9793	0.9735	0.9705	0.9646
20.30	0.9972	0.9949	0.9928	0.9905	0.9877	0.9850	0.9794	0.9736	0.9706	0.9647
20.35	0.9972	0.9949	0.9929	0.9906	0.9878	0.9850	0.9795	0.9737	0.9708	0.9649
20.40	0.9972	0.9950	0.9929	0.9906	0.9879	0.9851	0.9796	0.9739	0.9709	0.9651
20.45	0.9972	0.9950	0.9930	0.9906	0.9879	0.9852	0.9797	0.9740	0.9710	0.9653
20.50	0.9973	0.9950	0.9930	0.9907	0.9880	0.9852	0.9797	0.9741	0.9712	0.9654
20.55	0.9973	0.9950	0.9930	0.9907	0.9880	0.9853	0.9798	0.9742	0.9713	0.9656
20.60	0.9973	0.9951	0.9931	0.9908	0.9881	0.9854	0.9799	0.9744	0.9715	0.9657
20.65	0.9973	0.9951	0.9931	0.9908	0.9881	0.9855	0.9800	0.9745	0.9716	0.9659
20.70	0.9973	0.9951	0.9931	0.9909	0.9882	0.9855	0.9801	0.9746	0.9717	0.9661
20.75	0.9973	0.9951	0.9932	0.9909	0.9883	0.9856	0.9802	0.9747	0.9719	0.9662
20.80	0.9973	0.9952	0.9932	0.9910	0.9883	0.9857	0.9803	0.9748	0.9720	0.9664
20.85	0.9973	0.9952	0.9932	0.9910	0.9884	0.9857	0.9804	0.9750	0.9721	0.9665
20.90	0.9974	0.9952	0.9932	0.9910	0.9884	0.9858	0.9805	0.9751	0.9723	0.9667
20.95	0.9974	0.9952	0.9933	0.9701	0.9885	0.9859	0.9806	0.9752	0.9724	0.9669
21.00	0.9974	0.9952	0.9933	0.9911	0.9885	0.9859	0.9807	0.9753	0.9725	0.9670
21.05	0.9974	0.9953	0.9933	0.9912	0.9886	0.9860	0.9808	0.9754	0.9726	0.9672
21.10	0.9974	0.9953	0.9934	0.9912	0.9886	0.9861	0.9809	0.9755	0.9728	0.9673
21.15	0.9974	0.9953	0.9934	0.9913	0.9887	0.9861	0.9810	0.9757	0.9729	0.9675
21.20	0.9974	0.9953	0.9934	0.9913	0.9887	0.9862	0.9810	0.9758	0.9730	0.9676
21.25	0.9974	0.9954	0.9935	0.9913	0.9888	0.9863	0.9811	0.9759	0.9731	0.9678
21.30	0.9975	0.9954	0.9935	0.9914	0.9889	0.9863	0.9812	0.9760	0.9733	0.9679
21.35	0.9975	0.9954	0.9935	0.9914	0.9889	0.9864	0.9813	0.9761	0.9734	0.9681
21.40	0.9975	0.9954	0.9936	0.9915	0.9890	0.9864	0.9814	0.9762	0.9735	0.9682
21.45	0.9975	0.9954	0.9936	0.9915	0.9890	0.9865	0.9815	0.9763	0.9736	0.9684
21.50	0.9975	0.9955	0.9936	0.9915	0.9891	0.9866	0.9816	0.9764	0.9738	0.9685
21.55	0.9975	0.9955	0.9936	0.9916	0.9891	0.9866	0.9817	0.9765	0.9739	0.9686
21.60	0.9975	0.9955	0.9937	0.9916	0.9892	0.9867	0.9817	0.9766	0.9740	0.9688
21.65	0.9975	0.9955	0.9937	0.9916	0.9892	0.9868	0.9818	0.9767	0.9741	0.9689
21.70	0.9976	0.9955	0.9937	0.9917	0.9893	0.9868	0.9819	0.9769	0.9742	0.9691
21.75	0.9976	0.9956	0.9938	0.9917	0.9893	0.9869	0.9820	0.9770	0.9743	0.9692
21.80	0.9976	0.9956	0.9938	0.9918	0.9894	0.9869	0.9821	0.9771	0.9745	0.9694
21.85	0.9976	0.9956	0.9938	0.9918	0.9894	0.9870	0.9821	0.9772	0.9746	0.9695
21.90	0.9976	0.9956	0.9938	0.9918	0.9894	0.9871	0.9822	0.9773	0.9747	0.9696
21.95	0.9976	0.9956	0.9939	0.9919	0.9895	0.9871	0.9823	0.9774	0.9748	0.9698
22.00	0.9976	0.9957	0.9939	0.9919	0.9895	0.9872	0.9824	0.9775	0.9749	0.9699
22.05	0.9976	0.9957	0.9939	0.9919	0.9896	0.9872	0.9825	0.9776	0.9750	0.9700
22.10	0.9976	0.9957	0.9940	0.9920	0.9896	0.9873	0.9825	0.9777	0.9751	0.9702
22.15	0.9976	0.9957	0.9940	0.9920	0.9897	0.9873	0.9826	0.9778	0.9753	0.9703
22.20	0.9977	0.9957	0.9940	0.9921	0.9897	0.9874	0.9827	0.9779	0.9754	0.9704
22.25	0.9977	0.9958	0.9940	0.9921	0.9898	0.9875	0.9828	0.9780	0.9755	0.9706
22.30	0.9977	0.9958	0.9941	0.9921	0.9898	0.9875	0.9829	0.9781	0.9756	0.9707
22.35	0.9977	0.9958	0.9941	0.9922	0.9899	0.9876	0.9829	0.9782	0.9757	0.9708
22.40	0.9977	0.9958	0.9941	0.9922	0.9899	0.9876	0.9830	0.9783	0.9758	0.9709
22.45	0.9977	0.9958	0.9941	0.9922	0.9900	0.9877	0.9831	0.9784	0.9759	0.9711

PROBABILITY INTEGRAL OF t, THE NON-CENTRAL t-STATISTIC. THIS TABLE GIVES $Pr\ [t/\sqrt{f} \leq x]$.

f is the number of degrees of freedom; the non-centrality parameter is $\sqrt{f+1}\, K_p$.

K_p is the standardized normal deviate exceeded with probability p.

DEGREES OF FREEDOM **2**

x \ p	.2500	.1500	.1000	.0650	.0400	.0250	.0100	.0040	.0025	.0010
22.50	0.9977	0.9959	0.9942	0.9923	0.9900	0.9877	0.9832	0.9784	0.9760	0.9712
22.55	0.9977	0.9959	0.9942	0.9923	0.9900	0.9878	0.9832	0.9785	0.9761	0.9713
22.60	0.9977	0.9959	0.9942	0.9923	0.9901	0.9878	0.9833	0.9786	0.9762	0.9714
22.65	0.9978	0.9959	0.9942	0.9924	0.9901	0.9879	0.9834	0.9787	0.9763	0.9716
22.70	0.9978	0.9959	0.9943	0.9924	0.9902	0.9879	0.9834	0.9788	0.9764	0.9717
22.75	0.9978	0.9959	0.9943	0.9924	0.9902	0.9880	0.9835	0.9789	0.9765	0.9718
22.80	0.9978	0.9960	0.9943	0.9925	0.9903	0.9880	0.9836	0.9790	0.9766	0.9719
22.85	0.9978	0.9960	0.9943	0.9925	0.9903	0.9881	0.9837	0.9791	0.9767	0.9721
22.90	0.9978	0.9960	0.9944	0.9925	0.9903	0.9882	0.9837	0.9792	0.9768	0.9722
22.95	0.9978	0.9960	0.9944	0.9926	0.9904	0.9882	0.9838	0.9793	0.9769	0.9723
23.00	0.9978	0.9960	0.9944	0.9926	0.9904	0.9883	0.9839	0.9794	0.9770	0.9724
23.05	0.9978	0.9961	0.9944	0.9926	0.9905	0.9883	0.9839	0.9795	0.9771	0.9725
23.10	0.9978	0.9961	0.9945	0.9927	0.9905	0.9884	0.9840	0.9795	0.9772	0.9727
23.15	0.9978	0.9961	0.9945	0.9927	0.9906	0.9884	0.9841	0.9796	0.9773	0.9728
23.20	0.9979	0.9961	0.9945	0.9927	0.9906	0.9885	0.9841	0.9797	0.9774	0.9729
23.25	0.9979	0.9961	0.9945	0.9928	0.9906	0.9885	0.9842	0.9798	0.9775	0.9730
23.30	0.9979	0.9961	0.9946	0.9928	0.9907	0.9886	0.9843	0.9799	0.9776	0.9731
23.35	0.9979	0.9962	0.9946	0.9928	0.9907	0.9886	0.9843	0.9800	0.9777	0.9732
23.40	0.9979	0.9962	0.9946	0.9928	0.9907	0.9886	0.9844	0.9801	0.9778	0.9733
23.45	0.9979	0.9962	0.9946	0.9929	0.9908	0.9887	0.9845	0.9801	0.9779	0.9734
23.50	0.9979	0.9962	0.9947	0.9929	0.9908	0.9887	0.9845	0.9802	0.9780	0.9736
23.55	0.9979	0.9962	0.9947	0.9929	0.9909	0.9888	0.9846	0.9803	0.9781	0.9737
23.60	0.9979	0.9962	0.9947	0.9930	0.9909	0.9888	0.9847	0.9804	0.9782	0.9738
23.65	0.9979	0.9962	0.9947	0.9930	0.9909	0.9889	0.9847	0.9805	0.9783	0.9739
23.70	0.9979	0.9963	0.9947	0.9930	0.9910	0.9889	0.9848	0.9806	0.9783	0.9740
23.75	0.9980	0.9963	0.9948	0.9931	0.9910	0.9890	0.9849	0.9806	0.9784	0.9741
23.80	0.9980	0.9963	0.9948	0.9931	0.9911	0.9890	0.9849	0.9807	0.9785	0.9742
23.85	0.9980	0.9963	0.9948	0.9931	0.9911	0.9891	0.9850	0.9808	0.9786	0.9743
23.90	0.9980	0.9963	0.9948	0.9931	0.9911	0.9891	0.9851	0.9809	0.9787	0.9744
23.95	0.9980	0.9963	0.9948	0.9932	0.9912	0.9892	0.9851	0.9810	0.9788	0.9745
24.00	0.9980	0.9964	0.9949	0.9932	0.9912	0.9892	0.9852	0.9810	0.9789	0.9746
24.05	0.9980	0.9964	0.9949	0.9932	0.9912	0.9892	0.9852	0.9811	0.9790	0.9747
24.10	0.9980	0.9964	0.9949	0.9933	0.9913	0.9893	0.9853	0.9812	0.9790	0.9748
24.15	0.9980	0.9964	0.9949	0.9933	0.9913	0.9893	0.9854	0.9813	0.9791	0.9749
24.20	0.9980	0.9964	0.9950	0.9933	0.9913	0.9894	0.9854	0.9813	0.9792	0.9750
24.25	0.9980	0.9964	0.9950	0.9933	0.9914	0.9894	0.9855	0.9814	0.9793	0.9751
24.30	0.9980	0.9964	0.9950	0.9934	0.9914	0.9895	0.9855	0.9815	0.9794	0.9752
24.35	0.9981	0.9965	0.9950	0.9934	0.9915	0.9895	0.9856	0.9816	0.9795	0.9753
24.40	0.9981	0.9965	0.9950	0.9934	0.9915	0.9896	0.9857	0.9816	0.9796	0.9754
24.45	0.9981	0.9965	0.9951	0.9934	0.9915	0.9896	0.9857	0.9817	0.9796	0.9755
24.50	0.9981	0.9965	0.9951	0.9935	0.9916	0.9896	0.9858	0.9818	0.9797	0.9756
24.55	0.9981	0.9965	0.9951	0.9935	0.9916	0.9897	0.9858	0.9819	0.9798	0.9757
24.60	0.9981	0.9965	0.9951	0.9935	0.9916	0.9897	0.9859	0.9819	0.9799	0.9758
24.65	0.9981	0.9965	0.9951	0.9935	0.9917	0.9898	0.9859	0.9820	0.9800	0.9759
24.70	0.9981	0.9966	0.9952	0.9936	0.9917	0.9898	0.9860	0.9821	0.9800	0.9760
24.75	0.9981	0.9966	0.9952	0.9936	0.9917	0.9898	0.9861	0.9822	0.9801	0.9761
24.80	0.9981	0.9966	0.9952	0.9936	0.9918	0.9899	0.9861	0.9822	0.9802	0.9762
24.85	0.9981	0.9966	0.9952	0.9937	0.9918	0.9899	0.9862	0.9823	0.9803	0.9763
24.90	0.9981	0.9966	0.9952	0.9937	0.9918	0.9900	0.9862	0.9824	0.9804	0.9764
24.95	0.9981	0.9966	0.9953	0.9937	0.9919	0.9900	0.9863	0.9824	0.9804	0.9765
25.00	0.9982	0.9966	0.9953	0.9937	0.9919	0.9900	0.9863	0.9825	0.9805	0.9766
25.05	0.9982	0.9967	0.9953	0.9938	0.9919	0.9901	0.9864	0.9826	0.9806	0.9767
25.10	0.9982	0.9967	0.9953	0.9938	0.9920	0.9901	0.9864	0.9826	0.9807	0.9768
25.15	0.9982	0.9967	0.9953	0.9938	0.9920	0.9902	0.9865	0.9827	0.9807	0.9769
25.20	0.9982	0.9967	0.9953	0.9938	0.9920	0.9902	0.9865	0.9828	0.9808	0.9770
25.25	0.9982	0.9967	0.9954	0.9939	0.9920	0.9902	0.9866	0.9828	0.9809	0.9771
25.30	0.9982	0.9967	0.9954	0.9939	0.9921	0.9903	0.9866	0.9829	0.9810	0.9771
25.35	0.9982	0.9967	0.9954	0.9939	0.9921	0.9903	0.9867	0.9830	0.9810	0.9772
25.40	0.9982	0.9967	0.9954	0.9939	0.9921	0.9904	0.9868	0.9830	0.9811	0.9773
25.45	0.9982	0.9968	0.9954	0.9939	0.9922	0.9904	0.9868	0.9831	0.9812	0.9774

PROBABILITY INTEGRAL OF t, THE NON-CENTRAL t-STATISTIC. THIS TABLE GIVES Pr $[t/\sqrt{f} \leq x]$.

f is the number of degrees of freedom; the non-centrality parameter is $\sqrt{f+1}\,K_p$.
K_p is the standardized normal deviate exceeded with probability p.

DEGREES OF FREEDOM **2**

x \ p	.2500	.1500	.1000	.0650	.0400	.0250	.0100	.0040	.0025	.0010
25.50	0.9982	0.9968	0.9955	0.9940	0.9922	0.9904	0.9869	0.9832	0.9813	0.9775
25.55	0.9982	0.9968	0.9955	0.9940	0.9922	0.9905	0.9869	0.9832	0.9813	0.9776
25.60	0.9982	0.9968	0.9955	0.9940	0.9923	0.9905	0.9870	0.9833	0.9814	0.9777
25.65	0.9982	0.9968	0.9955	0.9940	0.9923	0.9905	0.9870	0.9834	0.9815	0.9778
25.70	0.9983	0.9968	0.9955	0.9941	0.9923	0.9906	0.9871	0.9834	0.9815	0.9778
25.75	0.9983	0.9968	0.9955	0.9941	0.9924	0.9906	0.9871	0.9835	0.9816	0.9779
25.80	0.9983	0.9968	0.9956	0.9941	0.9924	0.9906	0.9872	0.9836	0.9817	0.9780
25.85	0.9983	0.9969	0.9956	0.9941	0.9924	0.9907	0.9872	0.9836	0.9818	0.9781
25.90	0.9983	0.9969	0.9956	0.9942	0.9924	0.9907	0.9873	0.9837	0.9818	0.9782
25.95	0.9983	0.9969	0.9956	0.9942	0.9925	0.9908	0.9873	0.9837	0.9819	0.9783
26.00	0.9983	0.9969	0.9956	0.9942	0.9925	0.9908	0.9874	0.9838	0.9820	0.9783
26.05	0.9983	0.9969	0.9956	0.9942	0.9925	0.9908	0.9874	0.9839	0.9820	0.9784
26.10	0.9983	0.9969	0.9957	0.9942	0.9926	0.9909	0.9874	0.9839	0.9821	0.9785
26.15	0.9983	0.9969	0.9957	0.9943	0.9926	0.9909	0.9875	0.9840	0.9822	0.9786
26.20	0.9983	0.9969	0.9957	0.9943	0.9926	0.9909	0.9875	0.9841	0.9822	0.9787
26.25	0.9983	0.9970	0.9957	0.9943	0.9926	0.9910	0.9876	0.9841	0.9823	0.9787
26.30	0.9983	0.9970	0.9957	0.9943	0.9927	0.9910	0.9876	0.9842	0.9824	0.9788
26.35	0.9983	0.9970	0.9957	0.9944	0.9927	0.9910	0.9877	0.9842	0.9824	0.9789
26.40	0.9983	0.9970	0.9958	0.9944	0.9927	0.9911	0.9877	0.9843	0.9825	0.9790
26.45	0.9983	0.9970	0.9958	0.9944	0.9928	0.9911	0.9878	0.9844	0.9826	0.9791
26.50	0.9984	0.9970	0.9958	0.9944	0.9928	0.9911	0.9878	0.9844	0.9826	0.9791
26.55	0.9984	0.9970	0.9958	0.9944	0.9928	0.9912	0.9879	0.9845	0.9827	0.9792
26.60	0.9984	0.9970	0.9958	0.9945	0.9928	0.9912	0.9879	0.9845	0.9828	0.9793
26.65	0.9984	0.9970	0.9958	0.9945	0.9929	0.9912	0.9880	0.9846	0.9828	0.9794
26.70	0.9984	0.9971	0.9958	0.9945	0.9929	0.9913	0.9880	0.9846	0.9829	0.9794
26.75	0.9984	0.9971	0.9959	0.9945	0.9929	0.9913	0.9880	0 9847	0.9830	0.9795
26.80	0.9984	0.9971	0.9959	0.9945	0.9929	0.9913	0.9881	0.9848	0.9830	0.9796
26.85	0.9984	0.9971	0.9959	0.9946	0.9930	0.9914	0.9881	0.9848	0.9831	0.9797
26.90	0.9984	0.9971	0.9959	0.9946	0.9930	0.9914	0.9882	0.9849	0.9831	0.9797
26.95	0.9984	0.9971	0.9959	0.9946	0.9930	0.9914	0.9882	0.9849	0.9832	0.9798
27.00	0.9984	0.9971	0.9959	0.9946	0.9930	0.9915	0.9883	0.9850	0.9833	0.9799
27.05	0.9984	0.9971	0.9960	0.9946	0.9931	0.9915	0.9883	0.9850	0.9833	0.9800
27.10	0.9984	0.9971	0.9960	0.9947	0.9931	0.9915	0.9883	0.9851	0.9834	0.9800
27.15	0.9984	0.9972	0.9960	0.9947	0.9931	0.9916	0.9884	0.9851	0.9834	0.9801
27.20	0.9984	0.9972	0.9960	0.9947	0.9931	0.9916	0.9884	0.9852	0.9835	0.9802
27.25	0.9984	0.9972	0.9960	0.9947	0.9932	0.9916	0.9885	0.9852	0.9836	0.9803
27.30	0.9985	0.9972	0.9960	0.9947	0.9932	0.9916	0.9885	0.9853	0.9836	0.9803
27.35	0.9985	0.9972	0.9960	0.9948	0.9932	0.9917	0.9886	0.9854	0.9837	0.9804
27.40	0.9985	0.9972	0.9961	0.9948	0.9932	0.9917	0.9886	0.9854	0.9837	0.9805
27.45	0.9985	0.9972	0.9961	0.9948	0.9933	0.9917	0.9886	0.9855	0.9838	0.9805
27.50	0.9985	0.9972	0.9961	0.9948	0.9933	0.9918	0.9887	0.9855	0.9839	0.9806
27.55	0.9985	0.9972	0.9961	0.9948	0.9933	0.9918	0.9887	0.9856	0.9839	0.9807
27.60	0.9985	0.9972	0.9961	0.9948	0.9933	0.9918	0.9888	0.9856	0.9840	0.9808
27.65	0.9985	0.9973	0.9961	0.9949	0.9934	0.9919	0.9888	0.9857	0.9840	0.9808
27.70	0.9985	0.9973	0.9961	0.9949	0.9934	0.9919	0.9888	0.9857	0.9841	0.9809
27.75	0.9985	0.9973	0.9962	0.9949	0.9934	0.9919	0.9889	0.9858	0.9841	0.9810
27.80	0.9985	0.9973	0.9962	0.9949	0.9934	0.9919	0.9889	0.9858	0.9842	0.9810
27.85	0.9985	0.9973	0.9962	0.9949	0.9935	0.9920	0.9890	0.9859	0.9843	0.9811
27.90	0.9985	0.9973	0.9962	0.9950	0.9935	0.9920	0.9890	0.9859	0.9843	0.9812
27.95	0.9985	0.9973	0.9962	0.9950	0.9935	0.9920	0.9890	0.9860	0.9844	0.9812
28.00	0.9985	0.9973	0.9962	0.9950	0.9935	0.9921	0.9891	0.9860	0.9844	0.9813
28.05	0.9985	0.9973	0.9962	0.9950	0.9936	0.9921	0.9891	0.9861	0.9845	0.9814
28.10	0.9985	0.9973	0.9963	0.9950	0.9936	0.9921	0.9892	0.9861	0.9845	0.9814
28.15	0.9985	0.9973	0.9963	0.9950	0.9936	0.9921	0.9892	0.9862	0.9846	0.9815
28.20	0.9985	0.9974	0.9963	0.9951	0.9936	0.9922	0.9892	0.9862	0.9846	0.9816
28.25	0.9986	0.9974	0.9963	0.9951	0.9936	0.9922	0.9893	0.9863	0.9847	0.9816
28.30	0.9986	0.9974	0.9963	0.9951	0.9937	0.9922	0.9893	0.9863	0.9848	0.9817
28.35	0.9986	0.9974	0.9963	0.9951	0.9937	0.9922	0.9893	0.9864	0.9848	0.9817
28.40	0.9986	0.9974	0.9963	0.9951	0.9937	0.9923	0.9894	0.9864	0.9849	0.9818
28.45	0.9986	0.9974	0.9963	0.9952	0.9937	0.9923	0.9894	0.9865	0.9849	0.9819

PROBABILITY INTEGRAL OF t, THE NON-CENTRAL t-STATISTIC. THIS TABLE GIVES Pr $[t/\sqrt{f} \leq x]$.

f is the number of degrees of freedom; the non-centrality parameter is $\sqrt{f+1}\,K_p$.

K_p is the standardized normal deviate exceeded with probability p.

DEGREES OF FREEDOM **2**

x \ p	.2500	.1500	.1000	.0650	.0400	.0250	.0100	.0040	.0025	.0010
28.50	0.9986	0.9974	0.9964	0.9952	0.9938	0.9923	0.9895	0.9865	0.9850	0.9819
28.55	0.9986	0.9974	0.9964	0.9952	0.9938	0.9924	0.9895	0.9866	0.9850	0.9820
28.60	0.9986	0.9974	0.9964	0.9952	0.9938	0.9924	0.9895	0.9866	0.9851	0.9821
28.65	0.9986	0.9974	0.9964	0.9952	0.9938	0.9924	0.9896	0.9866	0.9851	0.9821
28.70	0.9986	0.9974	0.9964	0.9952	0.9938	0.9924	0.9896	0.9867	0.9852	0.9822
28.75	0.9986	0.9975	0.9964	0.9953	0.9939	0.9925	0.9896	0.9867	0.9852	0.9822
28.80	0.9986	0.9975	0.9964	0.9953	0.9939	0.9925	0.9897	0.9868	0.9853	0.9823
28.85	0.9986	0.9975	0.9964	0.9953	0.9939	0.9925	0.9897	0.9868	0.9853	0.9824
28.90	0.9986	0.9975	0.9965	0.9953	0.9939	0.9925	0.9897	0.9869	0.9854	0.9824
28.95	0.9986	0.9975	0.9965	0.9953	0.9939	0.9926	0.9898	0.9869	0.9854	0.9825
29.00	0.9986	0.9975	0.9965	0.9953	0.9940	0.9926	0.9898	0.9870	0.9855	0.9825
29.05	0.9986	0.9975	0.9965	0.9953	0.9940	0.9926	0.9899	0.9870	0.9855	0.9826
29.10	0.9986	0.9975	0.9965	0.9954	0.9940	0.9926	0.9899	0.9871	0.9856	0.9827
29.15	0.9986	0.9975	0.9965	0.9954	0.9940	0.9927	0.9899	0.9871	0.9856	0.9827
29.20	0.9986	0.9975	0.9965	0.9954	0.9940	0.9927	0.9900	0.9871	0.9857	0.9828
29.25	0.9987	0.9975	0.9965	0.9954	0.9941	0.9927	0.9900	0.9872	0.9857	0.9828
29.30	0.9987	0.9976	0.9965	0.9954	0.9941	0.9927	0.9900	0.9872	0.9858	0.9829
29.35	0.9987	0.9976	0.9966	0.9954	0.9941	0.9928	0.9901	0.9873	0.9858	0.9830
29.40	0.9987	0.9976	0.9966	0.9955	0.9941	0.9928	0.9901	0.9873	0.9859	0.9830
29.45	0.9987	0.9976	0.9966	0.9955	0.9941	0.9928	0.9901	0.9874	0.9859	0.9831
29.50	0.9987	0.9976	0.9966	0.9955	0.9942	0.9928	0.9902	0.9874	0.9860	0.9831
29.55	0.9987	0.9976	0.9966	0.9955	0.9942	0.9929	0.9902	0.9874	0.9860	0.9832
29.60	0.9987	0.9976	0.9966	0.9955	0.9942	0.9929	0.9902	0.9875	0.9861	0.9832
29.65	0.9987	0.9976	0.9966	0.9955	0.9942	0.9929	0.9903	0.9875	0.9861	0.9833
29.70	0.9987	0.9976	0.9966	0.9955	0.9942	0.9929	0.9903	0.9876	0.9861	0.9834
29.75	0.9987	0.9976	0.9967	0.9956	0.9943	0.9930	0.9903	0.9876	0.9862	0.9834
29.80	0.9987	0.9976	0.9967	0.9956	0.9943	0.9930	0.9904	0.9876	0.9862	0.9835
29.85	0.9987	0.9976	0.9967	0.9956	0.9943	0.9930	0.9904	0.9877	0.9863	0.9835
29.90	0.9987	0.9976	0.9967	0.9956	0.9943	0.9930	0.9904	0.9877	0.9863	0.9836
29.95	0.9987	0.9977	0.9967	0.9956	0.9943	0.9931	0.9904	0.9878	0.9864	0.9836
30.00	0.9987	0.9977	0.9967	0.9956	0.9944	0.9931	0.9905	0.9878	0.9864	0.9837
30.05	0.9987	0.9977	0.9967	0.9957	0.9944	0.9931	0.9905	0.9878	0.9865	0.9837
30.10	0.9987	0.9977	0.9967	0.9957	0.9944	0.9931	0.9905	0.9879	0.9865	0.9838
30.15	0.9987	0.9977	0.9967	0.9957	0.9944	0.9931	0.9906	0.9879	0.9866	0.9838
30.20	0.9987	0.9977	0.9968	0.9957	0.9944	0.9932	0.9906	0.9880	0.9866	0.9839
30.25	0.9987	0.9977	0.9968	0.9957	0.9945	0.9932	0.9906	0.9880	0.9866	0.9839
30.30	0.9987	0.9977	0.9968	0.9957	0.9945	0.9932	0.9907	0.9880	0.9867	0.9840
30.35	0.9987	0.9977	0.9968	0.9957	0.9945	0.9932	0.9907	0.9881	0.9867	0.9841
30.40	0.9987	0.9977	0.9968	0.9958	0.9945	0.9933	0.9907	0.9881	0.9868	0.9841
30.45	0.9988	0.9977	0.9968	0.9958	0.9945	0.9933	0.9908	0.9882	0.9868	0.9842
30.50	0.9988	0.9977	0.9968	0.9958	0.9945	0.9933	0.9908	0.9882	0.9869	0.9842
30.55	0.9988	0.9977	0.9968	0.9958	0.9946	0.9933	0.9908	0.9882	0.9869	0.9843
30.60	0.9988	0.9978	0.9968	0.9958	0.9946	0.9933	0.9908	0.9883	0.9869	0.9843
30.65	0.9988	0.9978	0.9968	0.9958	0.9946	0.9934	0.9909	0.9883	0.9870	0.9844
30.70	0.9988	0.9978	0.9969	0.9958	0.9946	0.9934	0.9909	0.9884	0.9870	0.9844
30.75	0.9988	0.9978	0.9969	0.9958	0.9946	0.9934	0.9909	0.9884	0.9871	0.9845
30.80	0.9988	0.9978	0.9969	0.9959	0.9946	0.9934	0.9910	0.9884	0.9871	0.9845
30.85	0.9988	0.9978	0.9969	0.9959	0.9947	0.9934	0.9910	0.9885	0.9872	0.9846
30.90	0.9988	0.9978	0.9969	0.9959	0.9947	0.9935	0.9910	0.9885	0.9872	0.9846
30.95	0.9988	0.9978	0.9969	0.9959	0.9947	0.9935	0.9911	0.9885	0.9872	0.9847
31.00	0.9988	0.9978	0.9969	0.9959	0.9947	0.9935	0.9911	0.9886	0.9873	0.9847
31.05	0.9988	0.9978	0.9969	0.9959	0.9947	0.9935	0.9911	0.9886	0.9873	0.9848
31.10	0.9988	0.9978	0.9969	0.9959	0.9947	0.9936	0.9911	0.9887	0.9874	0.9848
31.15	0.9988	0.9978	0.9969	0.9960	0.9948	0.9936	0.9912	0.9887	0.9874	0.9849
31.20	0.9988	0.9978	0.9970	0.9960	0.9948	0.9936	0.9912	0.9887	0.9874	0.9849
31.25	0.9988	0.9978	0.9970	0.9960	0.9948	0.9936	0.9912	0.9888	0.9875	0.9849
31.30	0.9988	0.9979	0.9970	0.9960	0.9948	0.9936	0.9913	0.9888	0.9875	0.9850
31.35	0.9988	0.9979	0.9970	0.9960	0.9948	0.9937	0.9913	0.9888	0.9876	0.9850
31.40	0.9988	0.9979	0.9970	0.9960	0.9948	0.9937	0.9913	0.9889	0.9876	0.9851
31.45	0.9988	0.9979	0.9970	0.9960	0.9949	0.9937	0.9913	0.9889	0.9876	0.9851

PROBABILITY INTEGRAL OF t, THE NON-CENTRAL t-STATISTIC. THIS TABLE GIVES Pr $[t/\sqrt{f} \leq x]$.

f is the number of degrees of freedom; the non-centrality parameter is $\sqrt{f+1}\,K_p$.

K_p is the standardized normal deviate exceeded with probability p.

DEGREES OF FREEDOM **2**

x \ p	.2500	.1500	.1000	.0650	.0400	.0250	.0100	.0040	.0025	.0010
31.50	0.9988	0.9979	0.9970	0.9960	0.9949	0.9937	0.9914	0.9889	0.9877	0.9852
31.55	0.9988	0.9979	0.9970	0.9961	0.9949	0.9937	0.9914	0.9890	0.9877	0.9852
31.60	0.9988	0.9979	0.9970	0.9961	0.9949	0.9938	0.9914	0.9890	0.9878	0.9853
31.65	0.9988	0.9979	0.9970	0.9961	0.9949	0.9938	0.9914	0.9890	0.9878	0.9853
31.70	0.9989	0.9979	0.9970	0.9961	0.9949	0.9938	0.9915	0.9891	0.9878	0.9854
31.75	0.9989	0.9979	0.9971	0.9961	0.9950	0.9938	0.9915	0.9891	0.9879	0.9854
31.80	0.9989	0.9979	0.9971	0.9961	0.9950	0.9938	0.9915	0.9891	0.9879	0.9855
31.85	0.9989	0.9979	0.9971	0.9961	0.9950	0.9939	0.9915	0.9892	0.9879	0.9855
31.90	0.9989	0.9979	0.9971	0.9961	0.9950	0.9939	0.9916	0.9892	0.9880	0.9856
31.95	0.9989	0.9979	0.9971	0.9962	0.9950	0.9939	0.9916	0.9892	0.9880	0.9856
32.00	0.9989	0.9979	0.9971	0.9962	0.9950	0.9939	0.9916	0.9893	0.9881	0.9856
32.05	0.9989	0.9980	0.9971	0.9962	0.9951	0.9939	0.9917	0.9893	0.9881	0.9857
32.10	0.9989	0.9980	0.9971	0.9962	0.9951	0.9939	0.9917	0.9893	0.9881	0.9857
32.15	0.9989	0.9980	0.9971	0.9962	0.9951	0.9940	0.9917	0.9894	0.9882	0.9858
32.20	0.9989	0.9980	0.9971	0.9962	0.9951	0.9940	0.9917	0.9894	0.9882	0.9858
32.25	0.9989	0.9980	0.9971	0.9962	0.9951	0.9940	0.9918	0.9894	0.9882	0.9859
32.30	0.9989	0.9980	0.9972	0.9962	0.9951	0.9940	0.9918	0.9895	0.9883	0.9859
32.35	0.9989	0.9980	0.9972	0.9962	0.9951	0.9940	0.9918	0.9895	0.9883	0.9859
32.40	0.9989	0.9980	0.9972	0.9963	0.9952	0.9941	0.9918	0.9895	0.9883	0.9860
32.45	0.9989	0.9980	0.9972	0.9963	0.9952	0.9941	0.9919	0.9896	0.9884	0.9860
32.50	0.9989	0.9980	0.9972	0.9963	0.9952	0.9941	0.9919	0.9896	0.9884	0.9861
32.55	0.9989	0.9980	0.9972	0.9963	0.9952	0.9941	0.9919	0.9896	0.9885	0.9861
32.60	0.9989	0.9980	0.9972	0.9963	0.9952	0.9941	0.9919	0.9897	0.9885	0.9862
32.65	0.9989	0.9980	0.9972	0.9963	0.9952	0.9941	0.9920	0.9897	0.9885	0.9862
32.70	0.9989	0.9980	0.9972	0.9963	0.9952	0.9942	0.9920	0.9897	0.9886	0.9862
32.75	0.9989	0.9980	0.9972	0.9963	0.9953	0.9942	0.9920	0.9898	0.9886	0.9863
32.80	0.9989	0.9980	0.9972	0.9963	0.9953	0.9942	0.9920	0.9898	0.9886	0.9863
32.85	0.9989	0.9981	0.9972	0.9964	0.9953	0.9942	0.9921	0.9898	0.9887	0.9864
32.90	0.9989	0.9981	0.9973	0.9964	0.9953	0.9942	0.9921	0.9899	0.9887	0.9864
32.95	0.9989	0.9981	0.9973	0.9964	0.9953	0.9943	0.9921	0.9899	0.9887	0.9865
33.00	0.9989	0.9981	0.9973	0.9964	0.9953	0.9943	0.9921	0.9899	0.9888	0.9865
33.05	0.9989	0.9981	0.9973	0.9964	0.9953	0.9943	0.9921	0.9899	0.9888	0.9865
33.10	0.9989	0.9981	0.9973	0.9964	0.9954	0.9943	0.9922	0.9900	0.9888	0.9866
33.15	0.9989	0.9981	0.9973	0.9964	0.9954	0.9943	0.9922	0.9900	0.9889	0.9866
33.20	0.9990	0.9981	0.9973	0.9964	0.9954	0.9943	0.9922	0.9900	0.9889	0.9867
33.25	0.9990	0.9981	0.9973	0.9964	0.9954	0.9944	0.9922	0.9901	0.9889	0.9867
33.30	0.9990	0.9981	0.9973	0.9965	0.9954	0.9944	0.9923	0.9901	0.9890	0.9867
33.35	0.9990	0.9981	0.9973	0.9965	0.9954	0.9944	0.9923	0.9901	0.9890	0.9868
33.40	0.9990	0.9981	0.9973	0.9965	0.9954	0.9944	0.9923	0.9902	0.9890	0.9868
33.45	0.9990	0.9981	0.9973	0.9965	0.9955	0.9944	0.9923	0.9902	0.9891	0.9868
33.50	0.9990	0.9981	0.9974	0.9965	0.9955	0.9944	0.9924	0.9902	0.9891	0.9869
33.55	0.9990	0.9981	0.9974	0.9965	0.9955	0.9945	0.9924	0.9902	0.9891	0.9869
33.60	0.9990	0.9981	0.9974	0.9965	0.9955	0.9945	0.9924	0.9903	0.9892	0.9870
33.65	0.9990	0.9981	0.9974	0.9965	0.9955	0.9945	0.9924	0.9903	0.9892	0.9870
33.70	0.9990	0.9981	0.9974	0.9965	0.9955	0.9945	0.9924	0.9903	0.9892	0.9870
33.75	0.9990	0.9982	0.9974	0.9966	0.9955	0.9945	0.9925	0.9904	0.9893	0.9871
33.80	0.9990	0.9982	0.9974	0.9966	0.9956	0.9945	0.9925	0.9904	0.9893	0.9871
33.85	0.9990	0.9982	0.9974	0.9966	0.9956	0.9946	0.9925	0.9904	0.9893	0.9872
33.90	0.9990	0.9982	0.9974	0.9966	0.9956	0.9946	0.9925	0.9904	0.9893	0.9872
33.95	0.9990	0.9982	0.9974	0.9966	0.9956	0.9946	0.9926	0.9905	0.9894	0.9872
34.00	0.9990	0.9982	0.9974	0.9966	0.9956	0.9946	0.9926	0.9905	0.9894	0.9873
34.05	0.9990	0.9982	0.9974	0.9966	0.9956	0.9946	0.9926	0.9905	0.9894	0.9873
34.10	0.9990	0.9982	0.9974	0.9966	0.9956	0.9946	0.9926	0.9905	0.9895	0.9873
34.15	0.9990	0.9982	0.9975	0.9966	0.9956	0.9946	0.9926	0.9906	0.9895	0.9874
34.20	0.9990	0.9982	0.9975	0.9966	0.9957	0.9947	0.9927	0.9906	0.9895	0.9874
34.25	0.9990	0.9982	0.9975	0.9967	0.9957	0.9947	0.9927	0.9906	0.9896	0.9875
34.30	0.9990	0.9982	0.9975	0.9967	0.9957	0.9947	0.9927	0.9907	0.9896	0.9875
34.35	0.9990	0.9982	0.9975	0.9967	0.9957	0.9947	0.9927	0.9907	0.9896	0.9875
34.40	0.9990	0.9982	0.9975	0.9967	0.9957	0.9947	0.9927	0.9907	0.9897	0.9876
34.45	0.9990	0.9982	0.9975	0.9967	0.9957	0.9947	0.9928	0.9907	0.9897	0.9876

PROBABILITY INTEGRAL OF t, THE NON-CENTRAL t-STATISTIC. THIS TABLE GIVES Pr $[t/\sqrt{f} \leq x]$.

f is the number of degrees of freedom; the non-centrality parameter is $\sqrt{f+1}\,K_p$.

K_p is the standardized normal deviate exceeded with probability p.

DEGREES OF FREEDOM **2**

x \ p	.2500	.1500	.1000	.0650	.0400	.0250	.0100	.0040	.0025	.0010
34.50	0.9990	0.9982	0.9975	0.9967	0.9957	0.9948	0.9928	0.9908	0.9897	0.9876
34.55	0.9990	0.9982	0.9975	0.9967	0.9957	0.9948	0.9928	0.9908	0.9897	0.9877
34.60	0.9990	0.9982	0.9975	0.9967	0.9958	0.9948	0.9928	0.9908	0.9898	0.9877
34.65	0.9990	0.9982	0.9975	0.9967	0.9958	0.9948	0.9929	0.9908	0.9898	0.9877
34.70	0.9990	0.9983	0.9975	0.9967	0.9958	0.9948	0.9929	0.9909	0.9898	0.9878
34.75	0.9990	0.9983	0.9975	0.9967	0.9958	0.9948	0.9929	0.9909	0.9899	0.9878
34.80	0.9990	0.9983	0.9975	0.9968	0.9958	0.9948	0.9929	0.9909	0.9899	0.9878
34.85	0.9990	0.9983	0.9976	0.9968	0.9958	0.9949	0.9929	0.9909	0.9899	0.9879
34.90	0.9991	0.9983	0.9976	0.9968	0.9958	0.9949	0.9930	0.9910	0.9899	0.9879
34.95	0.9991	0.9983	0.9976	0.9968	0.9958	0.9949	0.9930	0.9910	0.9900	0.9879
35.00	0.9991	0.9983	0.9976	0.9968	0.9959	0.9949	0.9930	0.9910	0.9900	0.9880
35.05	0.9991	0.9983	0.9976	0.9968	0.9959	0.9949	0.9930	0.9911	0.9900	0.9880
35.10	0.9991	0.9983	0.9976	0.9968	0.9959	0.9949	0.9930	0.9911	0.9901	0.9880
35.15	0.9991	0.9983	0.9976	0.9968	0.9959	0.9949	0.9931	0.9911	0.9901	0.9881
35.20	0.9991	0.9983	0.9976	0.9968	0.9959	0.9950	0.9931	0.9911	0.9901	0.9881
35.25	0.9991	0.9983	0.9976	0.9968	0.9959	0.9950	0.9931	0.9912	0.9901	0.9881
35.30	0.9991	0.9983	0.9976	0.9968	0.9959	0.9950	0.9931	0.9912	0.9902	0.9882
35.35	0.9991	0.9983	0.9976	0.9969	0.9959	0.9950	0.9931	0.9912	0.9902	0.9882
35.40	0.9991	0.9983	0.9976	0.9969	0.9959	0.9950	0.9932	0.9912	0.9902	0.9882
35.45	0.9991	0.9983	0.9976	0.9969	0.9960	0.9950	0.9932	0.9913	0.9903	0.9883
35.50	0.9991	0.9983	0.9976	0.9969	0.9960	0.9950	0.9932	0.9913	0.9903	0.9883
35.55	0.9991	0.9983	0.9976	0.9969	0.9960	0.9951	0.9932	0.9913	0.9903	0.9883
35.60	0.9991	0.9983	0.9977	0.9969	0.9960	0.9951	0.9932	0.9913	0.9903	0.9884
35.65	0.9991	0.9983	0.9977	0.9969	0.9960	0.9951	0.9932	0.9913	0.9904	0.9884
35.70	0.9991	0.9983	0.9977	0.9969	0.9960	0.9951	0.9933	0.9914	0.9904	0.9884
35.75	0.9991	0.9984	0.9977	0.9969	0.9960	0.9951	0.9933	0.9914	0.9904	0.9885
35.80	0.9991	0.9984	0.9977	0.9969	0.9960	0.9951	0.9933	0.9914	0.9904	0.9885
35.85	0.9991	0.9984	0.9977	0.9969	0.9960	0.9951	0.9933	0.9914	0.9905	0.9885
35.90	0.9991	0.9984	0.9977	0.9970	0.9961	0.9952	0.9933	0.9915	0.9905	0.9886
35.95	0.9991	0.9984	0.9977	0.9970	0.9961	0.9952	0.9934	0.9915	0.9905	0.9886
36.00	0.9991	0.9984	0.9977	0.9970	0.9961	0.9952	0.9934	0.9915	0.9905	0.9886
36.05	0.9991	0.9984	0.9977	0.9970	0.9961	0.9952	0.9934	0.9915	0.9906	0.9887
36.10	0.9991	0.9984	0.9977	0.9970	0.9961	0.9952	0.9934	0.9916	0.9906	0.9887
36.15	0.9991	0.9984	0.9977	0.9970	0.9961	0.9952	0.9934	0.9916	0.9906	0.9887
36.20	0.9991	0.9984	0.9977	0.9970	0.9961	0.9952	0.9935	0.9916	0.9907	0.9888
36.25	0.9991	0.9984	0.9977	0.9970	0.9961	0.9952	0.9935	0.9916	0.9907	0.9888
36.30	0.9991	0.9984	0.9977	0.9970	0.9961	0.9953	0.9935	0.9917	0.9907	0.9888
36.35	0.9991	0.9984	0.9977	0.9970	0.9962	0.9953	0.9935	0.9917	0.9907	0.9889
36.40	0.9991	0.9984	0.9978	0.9970	0.9962	0.9953	0.9935	0.9917	0.9908	0.9889
36.45	0.9991	0.9984	0.9978	0.9970	0.9962	0.9953	0.9935	0.9917	0.9908	0.9889
36.50	0.9991	0.9984	0.9978	0.9971	0.9962	0.9953	0.9936	0.9917	0.9908	0.9889
36.55	0.9991	0.9984	0.9978	0.9971	0.9962	0.9953	0.9936	0.9918	0.9908	0.9890
36.60	0.9991	0.9984	0.9978	0.9971	0.9962	0.9953	0.9936	0.9918	0.9909	0.9890
36.65	0.9991	0.9984	0.9978	0.9971	0.9962	0.9954	0.9936	0.9918	0.9909	0.9890
36.70	0.9991	0.9984	0.9978	0.9971	0.9962	0.9954	0.9936	0.9918	0.9909	0.9891
36.75	0.9991	0.9984	0.9978	0.9971	0.9962	0.9954	0.9936	0.9919	0.9909	0.9891
36.80	0.9991	0.9984	0.9978	0.9971	0.9962	0.9954	0.9937	0.9919	0.9910	0.9891
36.85	0.9991	0.9985	0.9978	0.9971	0.9963	0.9954	0.9937	0.9919	0.9910	0.9891
36.90	0.9992	0.9985	0.9978	0.9971	0.9963	0.9954	0.9937	0.9919	0.9910	0.9892
36.95	0.9992	0.9985	0.9978	0.9971	0.9963	0.9954	0.9937	0.9919	0.9910	0.9892
37.00	0.9992	0.9985	0.9978	0.9971	0.9963	0.9954	0.9937	0.9920	0.9910	0.9892
37.05	0.9992	0.9985	0.9978	0.9971	0.9963	0.9955	0.9937	0.9920	0.9911	0.9893
37.10	0.9992	0.9985	0.9978	0.9971	0.9963	0.9955	0.9938	0.9920	0.9911	0.9893
37.15	0.9992	0.9985	0.9978	0.9972	0.9963	0.9955	0.9938	0.9920	0.9911	0.9893
37.20	0.9992	0.9985	0.9979	0.9972	0.9963	0.9955	0.9938	0.9921	0.9911	0.9894
37.25	0.9992	0.9985	0.9979	0.9972	0.9963	0.9955	0.9938	0.9921	0.9912	0.9894
37.30	0.9992	0.9985	0.9979	0.9972	0.9963	0.9955	0.9938	0.9921	0.9912	0.9894
37.35	0.9992	0.9985	0.9979	0.9972	0.9964	0.9955	0.9938	0.9921	0.9912	0.9894
37.40	0.9992	0.9985	0.9979	0.9972	0.9964	0.9955	0.9939	0.9921	0.9912	0.9895
37.45	0.9992	0.9985	0.9979	0.9972	0.9964	0.9955	0.9939	0.9922	0.9913	0.9895

PROBABILITY INTEGRAL OF t, THE NON-CENTRAL t-STATISTIC. THIS TABLE GIVES Pr $[t/\sqrt{f} \leq x]$.

f is the number of degrees of freedom; the non-centrality parameter is $\sqrt{f+1}\,K_p$.

K_p is the standardized normal deviate exceeded with probability p.

DEGREES OF FREEDOM **2**

x \ p	.2500	.1500	.1000	.0650	.0400	.0250	.0100	.0040	.0025	.0010
37.50	0.9992	0.9985	0.9979	0.9972	0.9964	0.9956	0.9939	0.9922	0.9913	0.9895
37.55	0.9992	0.9985	0.9979	0.9972	0.9964	0.9956	0.9939	0.9922	0.9913	0.9895
37.60	0.9992	0.9985	0.9979	0.9972	0.9964	0.9956	0.9939	0.9922	0.9913	0.9896
37.65	0.9992	0.9985	0.9979	0.9972	0.9964	0.9956	0.9939	0.9922	0.9914	0.9896
37.70	0.9992	0.9985	0.9979	0.9972	0.9964	0.9956	0.9940	0.9923	0.9914	0.9896
37.75	0.9992	0.9985	0.9979	0.9972	0.9964	0.9956	0.9940	0.9923	0.9914	0.9897
37.80	0.9992	0.9985	0.9979	0.9972	0.9964	0.9956	0.9940	0.9923	0.9914	0.9897
37.85	0.9992	0.9985	0.9979	0.9973	0.9965	0.9956	0.9940	0.9923	0.9914	0.9897
37.90	0.9992	0.9985	0.9979	0.9973	0.9965	0.9957	0.9940	0.9923	0.9915	0.9897
37.95	0.9992	0.9985	0.9979	0.9973	0.9965	0.9957	0.9940	0.9924	0.9915	0.9898
38.00	0.9992	0.9985	0.9979	0.9973	0.9965	0.9957	0.9941	0.9924	0.9915	0.9898
38.05	0.9992	0.9985	0.9979	0.9973	0.9965	0.9957	0.9941	0.9924	0.9915	0.9898
38.10	0.9992	0.9986	0.9979	0.9973	0.9965	0.9957	0.9941	0.9924	0.9916	0.9898
38.15	0.9992	0.9986	0.9980	0.9973	0.9965	0.9957	0.9941	0.9924	0.9916	0.9899
38.20	0.9992	0.9986	0.9980	0.9973	0.9965	0.9957	0.9941	0.9925	0.9916	0.9899
38.25	0.9992	0.9986	0.9980	0.9973	0.9965	0.9957	0.9941	0.9925	0.9916	0.9899
38.30	0.9992	0.9986	0.9980	0.9973	0.9965	0.9957	0.9941	0.9925	0.9916	0.9900
38.35	0.9992	0.9986	0.9980	0.9973	0.9965	0.9958	0.9942	0.9925	0.9917	0.9900
38.40	0.9992	0.9986	0.9980	0.9973	0.9966	0.9958	0.9942	0.9925	0.9917	0.9900
38.45	0.9992	0.9986	0.9980	0.9973	0.9966	0.9958	0.9942	0.9926	0.9917	0.9900
38.50	0.9992	0.9986	0.9980	0.9973	0.9966	0.9958	0.9942	0.9926	0.9917	0.9901
38.55	0.9992	0.9986	0.9980	0.9974	0.9966	0.9958	0.9942	0.9926	0.9917	0.9901
38.60	0.9992	0.9986	0.9980	0.9974	0.9966	0.9958	0.9942	0.9926	0.9918	0.9901
38.65	0.9992	0.9986	0.9980	0.9974	0.9966	0.9958	0.9943	0.9926	0.9918	0.9901
38.70	0.9992	0.9986	0.9980	0.9974	0.9966	0.9958	0.9943	0.9927	0.9918	0.9902
38.75	0.9992	0.9986	0.9980	0.9974	0.9966	0.9958	0.9943	0.9927	0.9918	0.9902
38.80	0.9992	0.9986	0.9980	0.9974	0.9966	0.9959	0.9943	0.9927	0.9919	0.9902
38.85	0.9992	0.9986	0.9980	0.9974	0.9966	0.9959	0.9943	0.9927	0.9919	0.9902
38.90	0.9992	0.9986	0.9980	0.9974	0.9966	0.9959	0.9943	0.9927	0.9919	0.9903
38.95	0.9992	0.9986	0.9980	0.9974	0.9966	0.9959	0.9943	0.9927	0.9919	0.9903
39.00	0.9992	0.9986	0.9980	0.9974	0.9967	0.9959	0.9944	0.9928	0.9919	0.9903
39.05	0.9992	0.9986	0.9980	0.9974	0.9967	0.9959	0.9944	0.9928	0.9920	0.9903
39.10	0.9992	0.9986	0.9981	0.9974	0.9967	0.9959	0.9944	0.9928	0.9920	0.9904
39.15	0.9992	0.9986	0.9981	0.9974	0.9967	0.9959	0.9944	0.9928	0.9920	0.9904
39.20	0.9992	0.9986	0.9981	0.9974	0.9967	0.9959	0.9944	0.9928	0.9920	0.9904
39.25	0.9992	0.9986	0.9981	0.9974	0.9967	0.9959	0.9944	0.9929	0.9920	0.9904
39.30	0.9993	0.9986	0.9981	0.9975	0.9967	0.9960	0.9944	0.9929	0.9921	0.9905
39.35	0.9993	0.9986	0.9981	0.9975	0.9967	0.9960	0.9945	0.9929	0.9921	0.9905
39.40	0.9993	0.9986	0.9981	0.9975	0.9967	0.9960	0.9945	0.9929	0.9921	0.9905
39.45	0.9993	0.9986	0.9981	0.9975	0.9967	0.9960	0.9945	0.9929	0.9921	0.9905
39.50	0.9993	0.9987	0.9981	0.9975	0.9967	0.9960	0.9945	0.9929	0.9921	0.9905
39.55	0.9993	0.9987	0.9981	0.9975	0.9967	0.9960	0.9945	0.9930	0.9922	0.9906
39.60	0.9993	0.9987	0.9981	0.9975	0.9968	0.9960	0.9945	0.9930	0.9922	0.9906
39.65	0.9993	0.9987	0.9981	0.9975	0.9968	0.9960	0.9945	0.9930	0.9922	0.9906
39.70	0.9993	0.9987	0.9981	0.9975	0.9968	0.9960	0.9946	0.9930	0.9922	0.9906
39.75	0.9993	0.9987	0.9981	0.9975	0.9968	0.9960	0.9946	0.9930	0.9922	0.9907
39.80	0.9993	0.9987	0.9981	0.9975	0.9968	0.9961	0.9946	0.9931	0.9923	0.9907
39.85	0.9993	0.9987	0.9981	0.9975	0.9968	0.9961	0.9946	0.9931	0.9923	0.9907
39.90	0.9993	0.9987	0.9981	0.9975	0.9968	0.9961	0.9946	0.9931	0.9923	0.9907
39.95	0.9993	0.9987	0.9981	0.9975	0.9968	0.9961	0.9946	0.9931	0.9923	0.9908
40.00	0.9993	0.9987	0.9981	0.9975	0.9968	0.9961	0.9946	0.9931	0.9923	0.9908
40.05	0.9993	0.9987	0.9981	0.9975	0.9968	0.9961	0.9946	0.9931	0.9924	0.9908
40.10	0.9993	0.9987	0.9981	0.9976	0.9968	0.9961	0.9947	0.9932	0.9924	0.9908
40.15	0.9993	0.9987	0.9982	0.9976	0.9968	0.9961	0.9947	0.9932	0.9924	0.9909
40.20	0.9993	0.9987	0.9982	0.9976	0.9969	0.9961	0.9947	0.9932	0.9924	0.9909
40.25	0.9993	0.9987	0.9982	0.9976	0.9969	0.9961	0.9947	0.9932	0.9924	0.9909
40.30	0.9993	0.9987	0.9982	0.9976	0.9969	0.9962	0.9947	0.9932	0.9924	0.9909
40.35	0.9993	0.9987	0.9982	0.9976	0.9969	0.9962	0.9947	0.9932	0.9925	0.9909
40.40	0.9993	0.9987	0.9982	0.9976	0.9969	0.9962	0.9947	0.9933	0.9925	0.9910
40.45	0.9993	0.9987	0.9982	0.9976	0.9969	0.9962	0.9948	0.9933	0.9925	0.9910

PROBABILITY INTEGRAL OF t, THE NON-CENTRAL t-STATISTIC. THIS TABLE GIVES Pr $[t/\sqrt{f} \leq x]$.

f is the number of degrees of freedom; the non-centrality parameter is $\sqrt{f+1}\,K_p$.
K_p is the standardized normal deviate exceeded with probability p.

DEGREES OF FREEDOM **2**

x \ p	.2500	.1500	.1000	.0650	.0400	.0250	.0100	.0040	.0025	.0010
40.50	0.9993	0.9987	0.9982	0.9976	0.9969	0.9962	0.9948	0.9933	0.9925	0.9910
40.55	0.9993	0.9987	0.9982	0.9976	0.9969	0.9962	0.9948	0.9933	0.9925	0.9910
40.60	0.9993	0.9987	0.9982	0.9976	0.9969	0.9962	0.9948	0.9933	0.9926	0.9911
40.65	0.9993	0.9987	0.9982	0.9976	0.9969	0.9962	0.9948	0.9933	0.9926	0.9911
40.70	0.9993	0.9987	0.9982	0.9976	0.9969	0.9962	0.9948	0.9934	0.9926	0.9911
40.75	0.9993	0.9987	0.9982	0.9976	0.9969	0.9962	0.9948	0.9934	0.9926	0.9911
40.80	0.9993	0.9987	0.9982	0.9976	0.9969	0.9962	0.9948	0.9934	0.9926	0.9911
40.85	0.9993	0.9987	0.9982	0.9976	0.9970	0.9963	0.9949	0.9934	0.9926	0.9912
40.90	0.9993	0.9987	0.9982	0.9976	0.9970	0.9963	0.9949	0.9934	0.9927	0.9912
40.95	0.9993	0.9987	0.9982	0.9977	0.9970	0.9963	0.9949	0.9934	0.9927	0.9912
41.00	0.9993	0.9987	0.9982	0.9977	0.9970	0.9963	0.9949	0.9935	0.9927	0.9912
41.05	0.9993	0.9988	0.9982	0.9977	0.9970	0.9963	0.9949	0.9935	0.9927	0.9912
41.10	0.9993	0.9988	0.9982	0.9977	0.9970	0.9963	0.9949	0.9935	0.9927	0.9913
41.15	0.9993	0.9988	0.9982	0.9977	0.9970	0.9963	0.9949	0.9935	0.9928	0.9913
41.20	0.9993	0.9988	0.9982	0.9977	0.9970	0.9963	0.9949	0.9935	0.9928	0.9913
41.25	0.9993	0.9988	0.9982	0.9977	0.9970	0.9963	0.9950	0.9935	0.9928	0.9913
41.30	0.9993	0.9988	0.9983	0.9977	0.9970	0.9963	0.9950	0.9935	0.9928	0.9914
41.35	0.9993	0.9988	0.9983	0.9977	0.9970	0.9963	0.9950	0.9936	0.9928	0.9914
41.40	0.9993	0.9988	0.9983	0.9977	0.9970	0.9964	0.9950	0.9936	0.9928	0.9914
41.45	0.9993	0.9988	0.9983	0.9977	0.9970	0.9964	0.9950	0.9936	0.9929	0.9914
41.50	0.9993	0.9988	0.9983	0.9977	0.9970	0.9964	0.9950	0.9936	0.9929	0.9914
41.55	0.9993	0.9988	0.9983	0.9977	0.9971	0.9964	0.9950	0.9936	0.9929	0.9915
41.60	0.9993	0.9988	0.9983	0.9977	0.9971	0.9964	0.9950	0.9936	0.9929	0.9915
41.65	0.9993	0.9988	0.9983	0.9977	0.9971	0.9964	0.9950	0.9937	0.9929	0.9915
41.70	0.9993	0.9988	0.9983	0.9977	0.9971	0.9964	0.9951	0.9937	0.9929	0.9915
41.75	0.9993	0.9988	0.9983	0.9977	0.9971	0.9964	0.9951	0.9937	0.9930	0.9915
41.80	0.9993	0.9988	0.9983	0.9977	0.9971	0.9964	0.9951	0.9937	0.9930	0.9916
41.85	0.9993	0.9988	0.9983	0.9978	0.9971	0.9964	0.9951	0.9937	0.9930	0.9916
41.90	0.9993	0.9988	0.9983	0.9978	0.9971	0.9964	0.9951	0.9937	0.9930	0.9916
41.95	0.9993	0.9988	0.9983	0.9978	0.9971	0.9964	0.9951	0.9937	0.9930	0.9916
42.00	0.9993	0.9988	0.9983	0.9978	0.9971	0.9965	0.9951	0.9938	0.9930	0.9916
42.05	0.9993	0.9988	0.9983	0.9978	0.9971	0.9965	0.9951	0.9938	0.9931	0.9917
42.10	0.9993	0.9988	0.9983	0.9978	0.9971	0.9965	0.9952	0.9938	0.9931	0.9917
42.15	0.9993	0.9988	0.9983	0.9978	0.9971	0.9965	0.9952	0.9938	0.9931	0.9917
42.20	0.9994	0.9988	0.9983	0.9978	0.9971	0.9965	0.9952	0.9938	0.9931	0.9917
42.25	0.9994	0.9988	0.9983	0.9978	0.9972	0.9965	0.9952	0.9938	0.9931	0.9917
42.30	0.9994	0.9988	0.9983	0.9978	0.9972	0.9965	0.9952	0.9938	0.9931	0.9918
42.35	0.9994	0.9988	0.9983	0.9978	0.9972	0.9965	0.9952	0.9939	0.9932	0.9918
42.40	0.9994	0.9988	0.9983	0.9978	0.9972	0.9965	0.9952	0.9939	0.9932	0.9918
42.45	0.9994	0.9988	0.9983	0.9978	0.9972	0.9965	0.9952	0.9939	0.9932	0.9918
42.50	0.9994	0.9988	0.9983	0.9978	0.9972	0.9965	0.9952	0.9939	0.9932	0.9918
42.55	0.9994	0.9988	0.9984	0.9978	0.9972	0.9965	0.9953	0.9939	0.9932	0.9918
42.60	0.9994	0.9988	0.9984	0.9978	0.9972	0.9966	0.9953	0.9939	0.9932	0.9919
42.65	0.9994	0.9988	0.9984	0.9978	0.9972	0.9966	0.9953	0.9939	0.9933	0.9919
42.70	0.9994	0.9988	0.9984	0.9978	0.9972	0.9966	0.9953	0.9940	0.9933	0.9919
42.75	0.9994	0.9988	0.9984	0.9978	0.9972	0.9966	0.9953	0.9940	0.9933	0.9919
42.80	0.9994	0.9989	0.9984	0.9979	0.9972	0.9966	0.9953	0.9940	0.9933	0.9919
42.85	0.9994	0.9989	0.9984	0.9979	0.9972	0.9966	0.9953	0.9940	0.9933	0.9920
42.90	0.9994	0.9989	0.9984	0.9979	0.9972	0.9966	0.9953	0.9940	0.9933	0.9920
42.95	0.9994	0.9989	0.9984	0.9979	0.9972	0.9966	0.9953	0.9940	0.9933	0.9920
43.00	0.9994	0.9989	0.9984	0.9979	0.9972	0.9966	0.9954	0.9940	0.9934	0.9920
43.05	0.9994	0.9989	0.9984	0.9979	0.9973	0.9966	0.9954	0.9941	0.9934	0.9920
43.10	0.9994	0.9989	0.9984	0.9979	0.9973	0.9966	0.9954	0.9941	0.9934	0.9921
43.15	0.9994	0.9989	0.9984	0.9979	0.9973	0.9966	0.9954	0.9941	0.9934	0.9921
43.20	0.9994	0.9989	0.9984	0.9979	0.9973	0.9967	0.9954	0.9941	0.9934	0.9921
43.25	0.9994	0.9989	0.9984	0.9979	0.9973	0.9967	0.9954	0.9941	0.9934	0.9921
43.30	0.9994	0.9989	0.9984	0.9979	0.9973	0.9967	0.9954	0.9941	0.9935	0.9921
43.35	0.9994	0.9989	0.9984	0.9979	0.9973	0.9967	0.9954	0.9941	0.9935	0.9921
43.40	0.9994	0.9989	0.9984	0.9979	0.9973	0.9967	0.9954	0.9942	0.9935	0.9922
43.45	0.9994	0.9989	0.9984	0.9979	0.9973	0.9967	0.9954	0.9942	0.9935	0.9922

PROBABILITY INTEGRAL OF t, THE NON-CENTRAL t-STATISTIC. THIS TABLE GIVES Pr $[t/\sqrt{f} \leq x]$.

f is the number of degrees of freedom; the non-centrality parameter is $\sqrt{f+1}\,K_p$.

K_p is the standardized normal deviate exceeded with probability p.

DEGREES OF FREEDOM **2**

x \ p	.2500	.1500	.1000	.0650	.0400	.0250	.0100	.0040	.0025	.0010
43.50	0.9994	0.9989	0.9984	0.9979	0.9973	0.9967	0.9955	0.9942	0.9935	0.9922
43.55	0.9994	0.9989	0.9984	0.9979	0.9973	0.9967	0.9955	0.9942	0.9935	0.9922
43.60	0.9994	0.9989	0.9984	0.9979	0.9973	0.9967	0.9955	0.9942	0.9935	0.9922
43.65	0.9994	0.9989	0.9984	0.9979	0.9973	0.9967	0.9955	0.9942	0.9936	0.9923
43.70	0.9994	0.9989	0.9984	0.9979	0.9973	0.9967	0.9955	0.9942	0.9936	0.9923
43.75	0.9994	0.9989	0.9984	0.9979	0.9973	0.9967	0.9955	0.9942	0.9936	0.9923
43.80	0.9994	0.9989	0.9984	0.9980	0.9973	0.9967	0.9955	0.9943	0.9936	0.9923
43.85	0.9994	0.9989	0.9984	0.9980	0.9974	0.9967	0.9955	0.9943	0.9936	0.9923
43.90	0.9994	0.9989	0.9984	0.9980	0.9974	0.9968	0.9955	0.9943	0.9936	0.9923
43.95	0.9994	0.9989	0.9985	0.9980	0.9974	0.9968	0.9956	0.9943	0.9936	0.9924
44.00	0.9994	0.9989	0.9985	0.9980	0.9974	0.9968	0.9956	0.9943	0.9937	0.9924
44.05	0.9994	0.9989	0.9985	0.9980	0.9974	0.9968	0.9956	0.9943	0.9937	0.9924
44.10	0.9994	0.9989	0.9985	0.9980	0.9974	0.9968	0.9956	0.9943	0.9937	0.9924
44.15	0.9994	0.9989	0.9985	0.9980	0.9974	0.9968	0.9956	0.9943	0.9937	0.9924
44.20	0.9994	0.9989	0.9985	0.9980	0.9974	0.9968	0.9956	0.9944	0.9937	0.9924
44.25	0.9994	0.9989	0.9985	0.9980	0.9974	0.9968	0.9956	0.9944	0.9937	0.9925
44.30	0.9994	0.9989	0.9985	0.9980	0.9974	0.9968	0.9956	0.9944	0.9937	0.9925
44.35	0.9994	0.9989	0.9985	0.9980	0.9974	0.9968	0.9956	0.9944	0.9938	0.9925
44.40	0.9994	0.9989	0.9985	0.9980	0.9974	0.9968	0.9956	0.9944	0.9938	0.9925
44.45	0.9994	0.9989	0.9985	0.9980	0.9974	0.9968	0.9957	0.9944	0.9938	0.9925
44.50	0.9994	0.9989	0.9985	0.9980	0.9974	0.9968	0.9957	0.9944	0.9938	0.9925
44.55	0.9994	0.9989	0.9985	0.9980	0.9974	0.9969	0.9957	0.9945	0.9938	0.9926
44.60	0.9994	0.9989	0.9985	0.9980	0.9974	0.9969	0.9957	0.9945	0.9938	0.9926
44.65	0.9994	0.9989	0.9985	0.9980	0.9974	0.9969	0.9957	0.9945	0.9938	0.9926
44.70	0.9994	0.9989	0.9985	0.9980	0.9975	0.9969	0.9957	0.9945	0.9939	0.9926
44.75	0.9994	0.9989	0.9985	0.9980	0.9975	0.9969	0.9957	0.9945	0.9939	0.9926
44.80	0.9994	0.9990	0.9985	0.9980	0.9975	0.9969	0.9957	0.9945	0.9939	0.9926
44.85	0.9994	0.9990	0.9985	0.9980	0.9975	0.9969	0.9957	0.9945	0.9939	0.9927
44.90	0.9994	0.9990	0.9985	0.9980	0.9975	0.9969	0.9957	0.9945	0.9939	0.9927
44.95	0.9994	0.9990	0.9985	0.9981	0.9975	0.9969	0.9957	0.9945	0.9939	0.9927
45.00	0.9994	0.9990	0.9985	0.9981	0.9975	0.9969	0.9958	0.9946	0.9939	0.9927
45.05	0.9994	0.9990	0.9985	0.9981	0.9975	0.9969	0.9958	0.9946	0.9940	0.9927
45.10	0.9994	0.9990	0.9985	0.9981	0.9975	0.9969	0.9958	0.9946	0.9940	0.9927
45.15	0.9994	0.9990	0.9985	0.9981	0.9975	0.9969	0.9958	0.9946	0.9940	0.9928
45.20	0.9994	0.9990	0.9985	0.9981	0.9975	0.9969	0.9958	0.9946	0.9940	0.9928
45.25	0.9994	0.9990	0.9985	0.9981	0.9975	0.9969	0.9958	0.9946	0.9940	0.9928
45.30	0.9994	0.9990	0.9985	0.9981	0.9975	0.9970	0.9958	0.9946	0.9940	0.9928
45.35	0.9994	0.9990	0.9985	0.9981	0.9975	0.9970	0.9958	0.9946	0.9940	0.9928
45.40	0.9994	0.9990	0.9985	0.9981	0.9975	0.9970	0.9958	0.9947	0.9940	0.9928
45.45	0.9994	0.9990	0.9986	0.9981	0.9975	0.9970	0.9958	0.9947	0.9941	0.9929
45.50	0.9994	0.9990	0.9986	0.9981	0.9975	0.9970	0.9958	0.9947	0.9941	0.9929
45.55	0.9994	0.9990	0.9986	0.9981	0.9975	0.9970	0.9959	0.9947	0.9941	0.9929
45.60	0.9994	0.9990	0.9986	0.9981	0.9976	0.9970	0.9959	0.9947	0.9941	0.9929
45.65	0.9994	0.9990	0.9986	0.9981	0.9976	0.9970	0.9959	0.9947	0.9941	0.9929
45.70	0.9994	0.9990	0.9986	0.9981	0.9976	0.9970	0.9959	0.9947	0.9941	0.9929
45.75	0.9994	0.9990	0.9986	0.9981	0.9976	0.9970	0.9959	0.9947	0.9941	0.9929
45.80	0.9994	0.9990	0.9986	0.9981	0.9976	0.9970	0.9959	0.9947	0.9941	0.9930
45.85	0.9995	0.9990	0.9986	0.9981	0.9976	0.9970	0.9959	0.9948	0.9942	0.9930
45.90	0.9995	0.9990	0.9986	0.9981	0.9976	0.9970	0.9959	0.9948	0.9942	0.9930
45.95	0.9995	0.9990	0.9986	0.9981	0.9976	0.9970	0.9959	0.9948	0.9942	0.9930
46.00	0.9995	0.9990	0.9986	0.9981	0.9976	0.9970	0.9959	0.9948	0.9942	0.9930
46.05	0.9995	0.9990	0.9986	0.9981	0.9976	0.9971	0.9959	0.9948	0.9942	0.9930
46.10	0.9995	0.9990	0.9986	0.9981	0.9976	0.9971	0.9960	0.9948	0.9942	0.9931
46.15	0.9995	0.9990	0.9986	0.9982	0.9976	0.9971	0.9960	0.9948	0.9942	0.9931
46.20	0.9995	0.9990	0.9986	0.9982	0.9976	0.9971	0.9960	0.9948	0.9942	0.9931
46.25	0.9995	0.9990	0.9986	0.9982	0.9976	0.9971	0.9960	0.9948	0.9943	0.9931
46.30	0.9995	0.9990	0.9986	0.9982	0.9976	0.9971	0.9960	0.9949	0.9943	0.9931
46.35	0.9995	0.9990	0.9986	0.9982	0.9976	0.9971	0.9960	0.9949	0.9943	0.9931
46.40	0.9995	0.9990	0.9986	0.9982	0.9976	0.9971	0.9960	0.9949	0.9943	0.9931
46.45	0.9995	0.9990	0.9986	0.9982	0.9976	0.9971	0.9960	0.9949	0.9943	0.9932

PROBABILITY INTEGRAL OF t, THE NON-CENTRAL t-STATISTIC. THIS TABLE GIVES Pr $[t/\sqrt{f} \leq x]$.

f is the number of degrees of freedom; the non-centrality parameter is $\sqrt{f+1}\,K_p$.

K_p is the standardized normal deviate exceeded with probability p.

DEGREES OF FREEDOM **2**

x \ p	.2500	.1500	.1000	.0650	.0400	.0250	.0100	.0040	.0025	.0010
46.50	0.9995	0.9990	0.9986	0.9982	0.9976	0.9971	0.9960	0.9949	0.9943	0.9932
46.55	0.9995	0.9990	0.9986	0.9982	0.9977	0.9971	0.9960	0.9949	0.9943	0.9932
46.60	0.9995	0.9990	0.9986	0.9982	0.9977	0.9971	0.9960	0.9949	0.9943	0.9932
46.65	0.9995	0.9990	0.9986	0.9982	0.9977	0.9971	0.9961	0.9949	0.9944	0.9932
46.70	0.9995	0.9990	0.9986	0.9982	0.9977	0.9971	0.9961	0.9949	0.9944	0.9932
46.75	0.9995	0.9990	0.9986	0.9982	0.9977	0.9971	0.9961	0.9950	0.9944	0.9932
46.80	0.9995	0.9990	0.9986	0.9982	0.9977	0.9971	0.9961	0.9950	0.9944	0.9933
46.85	0.9995	0.9990	0.9986	0.9982	0.9977	0.9972	0.9961	0.9950	0.9944	0.9933
46.90	0.9995	0.9990	0.9986	0.9982	0.9977	0.9972	0.9961	0.9950	0.9944	0.9933
46.95	0.9995	0.9990	0.9986	0.9982	0.9977	0.9972	0.9961	0.9950	0.9944	0.9933
47.00	0.9995	0.9990	0.9986	0.9982	0.9977	0.9972	0.9961	0.9950	0.9944	0.9933
47.05	0.9995	0.9990	0.9986	0.9982	0.9977	0.9972	0.9961	0.9950	0.9945	0.9933
47.10	0.9995	0.9991	0.9986	0.9982	0.9977	0.9972	0.9961	0.9950	0.9945	0.9933
47.15	0.9995	0.9991	0.9987	0.9982	0.9977	0.9972	0.9961	0.9950	0.9945	0.9934
47.20	0.9995	0.9991	0.9987	0.9982	0.9977	0.9972	0.9961	0.9951	0.9945	0.9934
47.25	0.9995	0.9991	0.9987	0.9982	0.9977	0.9972	0.9961	0.9951	0.9945	0.9934
47.30	0.9995	0.9991	0.9987	0.9982	0.9977	0.9972	0.9962	0.9951	0.9945	0.9934
47.35	0.9995	0.9991	0.9987	0.9982	0.9977	0.9972	0.9962	0.9951	0.9945	0.9934
47.40	0.9995	0.9991	0.9987	0.9982	0.9977	0.9972	0.9962	0.9951	0.9945	0.9934
47.45	0.9995	0.9991	0.9987	0.9983	0.9977	0.9972	0.9962	0.9951	0.9945	0.9934
47.50	0.9995	0.9991	0.9987	0.9983	0.9977	0.9972	0.9962	0.9951	0.9946	0.9935
47.55	0.9995	0.9991	0.9987	0.9983	0.9977	0.9972	0.9962	0.9951	0.9946	0.9935
47.60	0.9995	0.9991	0.9987	0.9983	0.9978	0.9972	0.9962	0.9951	0.9946	0.9935
47.65	0.9995	0.9991	0.9987	0.9983	0.9978	0.9972	0.9962	0.9951	0.9946	0.9935
47.70	0.9995	0.9991	0.9987	0.9983	0.9978	0.9973	0.9962	0.9952	0.9946	0.9935
47.75	0.9995	0.9991	0.9987	0.9983	0.9978	0.9973	0.9962	0.9952	0.9946	0.9935
47.80	0.9995	0.9991	0.9987	0.9983	0.9978	0.9973	0.9962	0.9952	0.9946	0.9935
47.85	0.9995	0.9991	0.9987	0.9983	0.9978	0.9973	0.9962	0.9952	0.9946	0.9935
47.90	0.9995	0.9991	0.9987	0.9983	0.9978	0.9973	0.9963	0.9952	0.9946	0.9936
47.95	0.9995	0.9991	0.9987	0.9983	0.9978	0.9973	0.9963	0.9952	0.9947	0.9936
48.00	0.9995	0.9991	0.9987	0.9983	0.9978	0.9973	0.9963	0.9952	0.9947	0.9936
48.05	0.9995	0.9991	0.9987	0.9983	0.9978	0.9973	0.9963	0.9952	0.9947	0.9936
48.10	0.9995	0.9991	0.9987	0.9983	0.9978	0.9973	0.9963	0.9952	0.9947	0.9936
48.15	0.9995	0.9991	0.9987	0.9983	0.9978	0.9973	0.9963	0.9952	0.9947	0.9936
48.20	0.9995	0.9991	0.9987	0.9983	0.9978	0.9973	0.9963	0.9953	0.9947	0.9936
48.25	0.9995	0.9991	0.9987	0.9983	0.9978	0.9973	0.9963	0.9953	0.9947	0.9937
48.30	0.9995	0.9991	0.9987	0.9983	0.9978	0.9973	0.9963	0.9953	0.9947	0.9937
48.35		0.9991	0.9987	0.9983	0.9978	0.9973	0.9963	0.9953	0.9947	0.9937
48.40		0.9991	0.9987	0.9983	0.9978	0.9973	0.9963	0.9953	0.9948	0.9937
48.45		0.9991	0.9987	0.9983	0.9978	0.9973	0.9963	0.9953	0.9948	0.9937
48.50		0.9991	0.9987	0.9983	0.9978	0.9973	0.9963	0.9953	0.9948	0.9937
48.55		0.9991	0.9987	0.9983	0.9978	0.9973	0.9964	0.9953	0.9948	0.9937
48.60		0.9991	0.9987	0.9983	0.9978	0.9974	0.9964	0.9953	0.9948	0.9937
48.65		0.9991	0.9987	0.9983	0.9978	0.9974	0.9964	0.9953	0.9948	0.9938
48.70		0.9991	0.9987	0.9983	0.9979	0.9974	0.9964	0.9954	0.9948	0.9938
48.75		0.9991	0.9987	0.9983	0.9979	0.9974	0.9964	0.9954	0.9948	0.9938
48.80		0.9991	0.9987	0.9983	0.9979	0.9974	0.9964	0.9954	0.9948	0.9938
48.85		0.9991	0.9987	0.9984	0.9979	0.9974	0.9964	0.9954	0.9949	0.9938
48.90		0.9991	0.9987	0.9984	0.9979	0.9974	0.9964	0.9954	0.9949	0.9938
48.95		0.9991	0.9987	0.9984	0.9979	0.9974	0.9964	0.9954	0.9949	0.9938
49.00		0.9991	0.9988	0.9984	0.9979	0.9974	0.9964	0.9954	0.9949	0.9938
49.05		0.9991	0.9988	0.9984	0.9979	0.9974	0.9964	0.9954	0.9949	0.9939
49.10		0.9991	0.9988	0.9984	0.9979	0.9974	0.9964	0.9954	0.9949	0.9939
49.15		0.9991	0.9988	0.9984	0.9979	0.9974	0.9964	0.9954	0.9949	0.9939
49.20		0.9991	0.9988	0.9984	0.9979	0.9974	0.9964	0.9954	0.9949	0.9939
49.25		0.9991	0.9988	0.9984	0.9979	0.9974	0.9965	0.9955	0.9949	0.9939
49.30		0.9991	0.9988	0.9984	0.9979	0.9974	0.9965	0.9955	0.9949	0.9939
49.35		0.9991	0.9988	0.9984	0.9979	0.9974	0.9965	0.9955	0.9950	0.9939
49.40		0.9991	0.9988	0.9984	0.9979	0.9974	0.9965	0.9955	0.9950	0.9939
49.45		0.9991	0.9988	0.9984	0.9979	0.9974	0.9965	0.9955	0.9950	0.9940

PROBABILITY INTEGRAL OF t, THE NON-CENTRAL t-STATISTIC. THIS TABLE GIVES Pr $[t/\sqrt{f} \leq x]$.

f is the number of degrees of freedom; the non-centrality parameter is $\sqrt{f+1}\,K_p$.
K_p is the standardized normal deviate exceeded with probability p.

DEGREES OF FREEDOM, **2**

x \ p	.2500	.1500	.1000	.0650	.0400	.0250	.0100	.0040	.0025	.0010
49.50		0.9991	0.9988	0.9984	0.9979	0.9974	0.9965	0.9955	0.9950	0.9940
49.55		0.9991	0.9988	0.9984	0.9979	0.9975	0.9965	0.9955	0.9950	0.9940
49.60		0.9991	0.9988	0.9984	0.9979	0.9975	0.9965	0.9955	0.9950	0.9940
49.65		0.9991	0.9988	0.9984	0.9979	0.9975	0.9965	0.9955	0.9950	0.9940
49.70		0.9991	0.9988	0.9984	0.9979	0.9975	0.9965	0.9955	0.9950	0.9940
49.75		0.9991	0.9988	0.9984	0.9979	0.9975	0.9965	0.9955	0.9950	0.9940
49.80		0.9992	0.9988	0.9984	0.9979	0.9975	0.9965	0.9956	0.9950	0.9940
49.85		0.9992	0.9988	0.9984	0.9980	0.9975	0.9965	0.9956	0.9951	0.9941
49.90		0.9992	0.9988	0.9984	0.9980	0.9975	0.9965	0.9956	0.9951	0.9941
49.95		0.9992	0.9988	0.9984	0.9980	0.9975	0.9966	0.9956	0.9951	0.9941
50.00		0.9992	0.9988	0.9984	0.9980	0.9975	0.9966	0.9956	0.9951	0.9941
50.05		0.9992	0.9988	0.9984	0.9980	0.9975	0.9966	0.9956	0.9951	0.9941
50.10		0.9992	0.9988	0.9984	0.9980	0.9975	0.9966	0.9956	0.9951	0.9941
50.15		0.9992	0.9988	0.9984	0.9980	0.9975	0.9966	0.9956	0.9951	0.9941
50.20		0.9992	0.9988	0.9984	0.9980	0.9975	0.9966	0.9956	0.9951	0.9941
50.25		0.9992	0.9988	0.9984	0.9980	0.9975	0.9966	0.9956	0.9951	0.9941
50.30		0.9992	0.9988	0.9984	0.9980	0.9975	0.9966	0.9956	0.9951	0.9942
50.35		0.9992	0.9988	0.9984	0.9980	0.9975	0.9966	0.9957	0.9952	0.9942
50.40		0.9992	0.9988	0.9985	0.9980	0.9975	0.9966	0.9957	0.9952	0.9942
50.45		0.9992	0.9988	0.9985	0.9980	0.9975	0.9966	0.9957	0.9952	0.9942
50.50		0.9992	0.9988	0.9985	0.9980	0.9975	0.9966	0.9957	0.9952	0.9942
50.55		0.9992	0.9988	0.9985	0.9980	0.9976	0.9966	0.9957	0.9952	0.9942
50.60		0.9992	0.9988	0.9985	0.9980	0.9976	0.9966	0.9957	0.9952	0.9942
50.65		0.9992	0.9988	0.9985	0.9980	0.9976	0.9966	0.9957	0.9952	0.9942
50.70		0.9992	0.9988	0.9985	0.9980	0.9976	0.9967	0.9957	0.9952	0.9943
50.75		0.9992	0.9988	0.9985	0.9980	0.9976	0.9967	0.9957	0.9952	0.9943
50.80		0.9992	0.9988	0.9985	0.9980	0.9976	0.9967	0.9957	0.9952	0.9943
50.85		0.9992	0.9988	0.9985	0.9980	0.9976	0.9967	0.9957	0.9952	0.9943
50.90		0.9992	0.9988	0.9985	0.9980	0.9976	0.9967	0.9957	0.9953	0.9943
50.95		0.9992	0.9988	0.9985	0.9980	0.9976	0.9967	0.9958	0.9953	0.9943
51.00		0.9992	0.9988	0.9985	0.9980	0.9976	0.9967	0.9958	0.9953	0.9943
51.05		0.9992	0.9988	0.9985	0.9980	0.9976	0.9967	0.9958	0.9953	0.9943
51.10		0.9992	0.9988	0.9985	0.9980	0.9976	0.9967	0.9958	0.9953	0.9943
51.15		0.9992	0.9989	0.9985	0.9981	0.9976	0.9967	0.9958	0.9953	0.9944
51.20		0.9992	0.9989	0.9985	0.9981	0.9976	0.9967	0.9958	0.9953	0.9944
51.25		0.9992	0.9989	0.9985	0.9981	0.9976	0.9967	0.9958	0.9953	0.9944
51.30		0.9992	0.9989	0.9985	0.9981	0.9976	0.9967	0.9958	0.9953	0.9944
51.35		0.9992	0.9989	0.9985	0.9981	0.9976	0.9967	0.9958	0.9953	0.9944
51.40		0.9992	0.9989	0.9985	0.9981	0.9976	0.9967	0.9958	0.9953	0.9944
51.45		0.9992	0.9989	0.9985	0.9981	0.9976	0.9968	0.9958	0.9954	0.9944
51.50		0.9992	0.9989	0.9985	0.9981	0.9976	0.9968	0.9958	0.9954	0.9944
51.55		0.9992	0.9989	0.9985	0.9981	0.9976	0.9968	0.9959	0.9954	0.9944
51.60		0.9992	0.9989	0.9985	0.9981	0.9977	0.9968	0.9959	0.9954	0.9944
51.65		0.9992	0.9989	0.9985	0.9981	0.9977	0.9968	0.9959	0.9954	0.9945
51.70		0.9992	0.9989	0.9985	0.9981	0.9977	0.9968	0.9959	0.9954	0.9945
51.75		0.9992	0.9989	0.9985	0.9981	0.9977	0.9968	0.9959	0.9954	0.9945
51.80		0.9992	0.9989	0.9985	0.9981	0.9977	0.9968	0.9959	0.9954	0.9945
51.85		0.9992	0.9989	0.9985	0.9981	0.9977	0.9968	0.9959	0.9954	0.9945
51.90		0.9992	0.9989	0.9985	0.9981	0.9977	0.9968	0.9959	0.9954	0.9945
51.95		0.9992	0.9989	0.9985	0.9981	0.9977	0.9968	0.9959	0.9954	0.9945
52.00		0.9992	0.9989	0.9985	0.9981	0.9977	0.9968	0.9959	0.9955	0.9945
52.05		0.9992	0.9989	0.9985	0.9981	0.9977	0.9968	0.9959	0.9955	0.9945
52.10		0.9992	0.9989	0.9986	0.9981	0.9977	0.9968	0.9959	0.9955	0.9946
52.15		0.9992	0.9989	0.9986	0.9981	0.9977	0.9968	0.9959	0.9955	0.9946
52.20		0.9992	0.9989	0.9986	0.9981	0.9977	0.9968	0.9960	0.9955	0.9946
52.25		0.9992	0.9989	0.9986	0.9981	0.9977	0.9968	0.9960	0.9955	0.9946
52.30		0.9992	0.9989	0.9986	0.9981	0.9977	0.9969	0.9960	0.9955	0.9946
52.35		0.9992	0.9989	0.9986	0.9981	0.9977	0.9969	0.9960	0.9955	0.9946
52.40		0.9992	0.9989	0.9986	0.9981	0.9977	0.9969	0.9960	0.9955	0.9946
52.45		0.9992	0.9989	0.9986	0.9981	0.9977	0.9969	0.9960	0.9955	0.9946

PROBABILITY INTEGRAL OF t, THE NON-CENTRAL t-STATISTIC. THIS TABLE GIVES Pr $[t/\sqrt{f} \leq x]$.

f is the number of degrees of freedom; the non-centrality parameter is $\sqrt{f+1}\,K_p$.

K_p is the standardized normal deviate exceeded with probability p.

DEGREES OF FREEDOM **2**

x \ p	.2500	.1500	.1000	.0650	.0400	.0250	.0100	.0040	.0025	.0010
52.50		0.9992	0.9989	0.9986	0.9982	0.9977	0.9969	0.9960	0.9955	0.9946
52.55		0.9992	0.9989	0.9986	0.9982	0.9977	0.9969	0.9960	0.9956	0.9946
52.60		0.9992	0.9989	0.9986	0.9982	0.9977	0.9969	0.9960	0.9956	0.9947
52.65		0.9992	0.9989	0.9986	0.9982	0.9977	0.9969	0.9960	0.9956	0.9947
52.70		0.9992	0.9989	0.9986	0.9982	0.9977	0.9969	0.9960	0.9956	0.9947
52.75		0.9992	0.9989	0.9986	0.9982	0.9978	0.9969	0.9960	0.9956	0.9947
52.80		0.9992	0.9989	0.9986	0.9982	0.9978	0.9969	0.9960	0.9956	0.9947
52.85		0.9992	0.9989	0.9986	0.9982	0.9978	0.9969	0.9961	0.9956	0.9947
52.90		0.9992	0.9989	0.9986	0.9982	0.9978	0.9969	0.9961	0.9956	0.9947
52.95		0.9992	0.9989	0.9986	0.9982	0.9978	0.9969	0.9961	0.9956	0.9947
53.00		0.9993	0.9989	0.9986	0.9982	0.9978	0.9969	0.9961	0.9956	0.9947
53.05		0.9993	0.9989	0.9986	0.9982	0.9978	0.9969	0.9961	0.9956	0.9947
53.10		0.9993	0.9989	0.9986	0.9982	0.9978	0.9969	0.9961	0.9956	0.9948
53.15		0.9993	0.9989	0.9986	0.9982	0.9978	0.9970	0.9961	0.9956	0.9948
53.20		0.9993	0.9989	0.9986	0.9982	0.9978	0.9970	0.9961	0.9957	0.9948
53.25		0.9993	0.9989	0.9986	0.9982	0.9978	0.9970	0.9961	0.9957	0.9948
53.30		0.9993	0.9989	0.9986	0.9982	0.9978	0.9970	0.9961	0.9957	0.9948
53.35		0.9993	0.9989	0.9986	0.9982	0.9978	0.9970	0.9961	0.9957	0.9948
53.40		0.9993	0.9989	0.9986	0.9982	0.9978	0.9970	0.9961	0.9957	0.9948
53.45		0.9993	0.9989	0.9986	0.9982	0.9978	0.9970	0.9961	0.9957	0.9948
53.50		0.9993	0.9989	0.9986	0.9982	0.9978	0.9970	0.9961	0.9957	0.9948
53.55		0.9993	0.9989	0.9986	0.9982	0.9978	0.9970	0.9962	0.9957	0.9948
53.60		0.9993	0.9990	0.9986	0.9982	0.9978	0.9970	0.9962	0.9957	0.9949
53.65		0.9993	0.9990	0.9986	0.9982	0.9978	0.9970	0.9962	0.9957	0.9949
53.70		0.9993	0.9990	0.9986	0.9982	0.9978	0.9970	0.9962	0.9957	0.9949
53.75		0.9993	0.9990	0.9986	0.9982	0.9978	0.9970	0.9962	0.9957	0.9949
53.80		0.9993	0.9990	0.9986	0.9982	0.9978	0.9970	0.9962	0.9958	0.9949
53.85		0.9993	0.9990	0.9986	0.9982	0.9978	0.9970	0.9962	0.9958	0.9949
53.90		0.9993	0.9990	0.9986	0.9982	0.9978	0.9970	0.9962	0.9958	0.9949
53.95		0.9993	0.9990	0.9986	0.9983	0.9979	0.9970	0.9962	0.9958	0.9949
54.00		0.9993	0.9990	0.9987	0.9983	0.9979	0.9970	0.9962	0.9958	0.9949
54.05		0.9993	0.9990	0.9987	0.9983	0.9979	0.9971	0.9962	0.9958	0.9949
54.10		0.9993	0.9990	0.9987	0.9983	0.9979	0.9971	0.9962	0.9958	0.9949
54.15		0.9993	0.9990	0.9987	0.9983	0.9979	0.9971	0.9962	0.9958	0.9950
54.20		0.9993	0.9990	0.9987	0.9983	0.9979	0.9971	0.9962	0.9958	0.9950
54.25		0.9993	0.9990	0.9987	0.9983	0.9979	0.9971	0.9963	0.9958	0.9950
54.30		0.9993	0.9990	0.9987	0.9983	0.9979	0.9971	0.9963	0.9958	0.9950
54.35		0.9993	0.9990	0.9987	0.9983	0.9979	0.9971	0.9963	0.9958	0.9950
54.40		0.9993	0.9990	0.9987	0.9983	0.9979	0.9971	0.9963	0.9958	0.9950
54.45		0.9993	0.9990	0.9987	0.9983	0.9979	0.9971	0.9963	0.9959	0.9950
54.50		0.9993	0.9990	0.9987	0.9983	0.9979	0.9971	0.9963	0.9959	0.9950
54.55		0.9993	0.9990	0.9987	0.9983	0.9979	0.9971	0.9963	0.9959	0.9950
54.60		0.9993	0.9990	0.9987	0.9983	0.9979	0.9971	0.9963	0.9959	0.9950
54.65		0.9993	0.9990	0.9987	0.9983	0.9979	0.9971	0.9963	0.9959	0.9950
54.70		0.9993	0.9990	0.9987	0.9983	0.9979	0.9971	0.9963	0.9959	0.9951
54.75		0.9993	0.9990	0.9987	0.9983	0.9979	0.9971	0.9963	0.9959	0.9951
54.80		0.9993	0.9990	0.9987	0.9983	0.9979	0.9971	0.9963	0.9959	0.9951
54.85		0.9993	0.9990	0.9987	0.9983	0.9979	0.9971	0.9963	0.9959	0.9951
54.90		0.9993	0.9990	0.9987	0.9983	0.9979	0.9971	0.9963	0.9959	0.9951
54.95		0.9993	0.9990	0.9987	0.9983	0.9979	0.9972	0.9963	0.9959	0.9951
55.00		0.9993	0.9990	0.9987	0.9983	0.9979	0.9972	0.9964	0.9959	0.9951
55.05		0.9993	0.9990	0.9987	0.9983	0.9979	0.9972	0.9964	0.9959	0.9951
55.10		0.9993	0.9990	0.9987	0.9983	0.9979	0.9972	0.9964	0.9960	0.9951
55.15		0.9993	0.9990	0.9987	0.9983	0.9979	0.9972	0.9964	0.9960	0.9951
55.20		0.9993	0.9990	0.9987	0.9983	0.9979	0.9972	0.9964	0.9960	0.9951
55.25		0.9993	0.9990	0.9987	0.9983	0.9980	0.9972	0.9964	0.9960	0.9952
55.30		0.9993	0.9990	0.9987	0.9983	0.9980	0.9972	0.9964	0.9960	0.9952
55.35		0.9993	0.9990	0.9987	0.9983	0.9980	0.9972	0.9964	0.9960	0.9952
55.40		0.9993	0.9990	0.9987	0.9983	0.9980	0.9972	0.9964	0.9960	0.9952
55.45		0.9993	0.9990	0.9987	0.9983	0.9980	0.9972	0.9964	0.9960	0.9952

PROBABILITY INTEGRAL OF t, THE NON-CENTRAL t-STATISTIC. THIS TABLE GIVES Pr $[t/\sqrt{f} \leq x]$.

f is the number of degrees of freedom; the non-centrality parameter is $\sqrt{f+1}\,K_p$.
K_p is the standardized normal deviate exceeded with probability p.

DEGREES OF FREEDOM **2**

x \ p	.2500	.1500	.1000	.0650	.0400	.0250	.0100	.0040	.0025	.0010
55.50		0.9993	0.9990	0.9987	0.9983	0.9980	0.9972	0.9964	0.9960	0.9952
55.55		0.9993	0.9990	0.9987	0.9984	0.9980	0.9972	0.9964	0.9960	0.9952
55.60		0.9993	0.9990	0.9987	0.9984	0.9980	0.9972	0.9964	0.9960	0.9952
55.65		0.9993	0.9990	0.9987	0.9984	0.9980	0.9972	0.9964	0.9960	0.9952
55.70		0.9993	0.9990	0.9987	0.9984	0.9980	0.9972	0.9964	0.9960	0.9952
55.75		0.9993	0.9990	0.9987	0.9984	0.9980	0.9972	0.9965	0.9960	0.9952
55.80		0.9993	0.9990	0.9987	0.9984	0.9980	0.9972	0.9965	0.9961	0.9953
55.85		0.9993	0.9990	0.9987	0.9984	0.9980	0.9972	0.9965	0.9961	0.9953
55.90		0.9993	0.9990	0.9987	0.9984	0.9980	0.9972	0.9965	0.9961	0.9953
55.95		0.9993	0.9990	0.9987	0.9984	0.9980	0.9973	0.9965	0.9961	0.9953
56.00		0.9993	0.9990	0.9987	0.9984	0.9980	0.9973	0.9965	0.9961	0.9953
56.05		0.9993	0.9990	0.9987	0.9984	0.9980	0.9973	0.9965	0.9961	0.9953
56.10		0.9993	0.9990	0.9987	0.9984	0.9980	0.9973	0.9965	0.9961	0.9953
56.15		0.9993	0.9990	0.9988	0.9984	0.9980	0.9973	0.9965	0.9961	0.9953
56.20		0.9993	0.9990	0.9988	0.9984	0.9980	0.9973	0.9965	0.9961	0.9953
56.25		0.9993	0.9990	0.9988	0.9984	0.9980	0.9973	0.9965	0.9961	0.9953
56.30		0.9993	0.9990	0.9988	0.9984	0.9980	0.9973	0.9965	0.9961	0.9953
56.35		0.9993	0.9990	0.9988	0.9984	0.9980	0.9973	0.9965	0.9961	0.9953
56.40		0.9993	0.9990	0.9988	0.9984	0.9980	0.9973	0.9965	0.9961	0.9954
56.45		0.9993	0.9991	0.9988	0.9984	0.9980	0.9973	0.9965	0.9961	0.9954
56.50		0.9993	0.9991	0.9988	0.9984	0.9980	0.9973	0.9965	0.9961	0.9954
56.55		0.9993	0.9991	0.9988	0.9984	0.9980	0.9973	0.9966	0.9962	0.9954
56.60		0.9993	0.9991	0.9988	0.9984	0.9980	0.9973	0.9966	0.9962	0.9954
56.65		0.9993	0.9991	0.9988	0.9984	0.9981	0.9973	0.9966	0.9962	0.9954
56.70		0.9993	0.9991	0.9988	0.9984	0.9981	0.9973	0.9966	0.9962	0.9954
56.75		0.9993	0.9991	0.9988	0.9984	0.9981	0.9973	0.9966	0.9962	0.9954
56.80		0.9993	0.9991	0.9988	0.9984	0.9981	0.9973	0.9966	0.9962	0.9954
56.85		0.9993	0.9991	0.9988	0.9984	0.9981	0.9973	0.9966	0.9962	0.9954
56.90		0.9993	0.9991	0.9988	0.9984	0.9981	0.9973	0.9966	0.9962	0.9954
56.95		0.9994	0.9991	0.9988	0.9984	0.9981	0.9973	0.9966	0.9962	0.9954
57.00		0.9994	0.9991	0.9988	0.9984	0.9981	0.9974	0.9966	0.9962	0.9954
57.05		0.9994	0.9991	0.9988	0.9984	0.9981	0.9974	0.9966	0.9962	0.9955
57.10		0.9994	0.9991	0.9988	0.9984	0.9981	0.9974	0.9966	0.9962	0.9955
57.15		0.9994	0.9991	0.9988	0.9984	0.9981	0.9974	0.9966	0.9962	0.9955
57.20		0.9994	0.9991	0.9988	0.9984	0.9981	0.9974	0.9966	0.9962	0.9955
57.25		0.9994	0.9991	0.9988	0.9984	0.9981	0.9974	0.9966	0.9962	0.9955
57.30		0.9994	0.9991	0.9988	0.9984	0.9981	0.9974	0.9966	0.9963	0.9955
57.35		0.9994	0.9991	0.9988	0.9985	0.9981	0.9974	0.9966	0.9963	0.9955
57.40		0.9994	0.9991	0.9988	0.9985	0.9981	0.9974	0.9967	0.9963	0.9955
57.45		0.9994	0.9991	0.9988	0.9985	0.9981	0.9974	0.9967	0.9963	0.9955
57.50		0.9994	0.9991	0.9988	0.9985	0.9981	0.9974	0.9967	0.9963	0.9955
57.55		0.9994	0.9991	0.9988	0.9985	0.9981	0.9974	0.9967	0.9963	0.9955
57.60		0.9994	0.9991	0.9988	0.9985	0.9981	0.9974	0.9967	0.9963	0.9955
57.65		0.9994	0.9991	0.9988	0.9985	0.9981	0.9974	0.9967	0.9963	0.9956
57.70		0.9994	0.9991	0.9988	0.9985	0.9981	0.9974	0.9967	0.9963	0.9956
57.75		0.9994	0.9991	0.9988	0.9985	0.9981	0.9974	0.9967	0.9963	0.9956
57.80		0.9994	0.9991	0.9988	0.9985	0.9981	0.9974	0.9967	0.9963	0.9956
57.85		0.9994	0.9991	0.9988	0.9985	0.9981	0.9974	0.9967	0.9963	0.9956
57.90		0.9994	0.9991	0.9988	0.9985	0.9981	0.9974	0.9967	0.9963	0.9956
57.95		0.9994	0.9991	0.9988	0.9985	0.9981	0.9974	0.9967	0.9963	0.9956
58.00		0.9994	0.9991	0.9988	0.9985	0.9981	0.9974	0.9967	0.9963	0.9956
58.05		0.9994	0.9991	0.9988	0.9985	0.9981	0.9974	0.9967	0.9964	0.9956
58.10		0.9994	0.9991	0.9988	0.9985	0.9981	0.9975	0.9967	0.9964	0.9956
58.15		0.9994	0.9991	0.9988	0.9985	0.9982	0.9975	0.9967	0.9964	0.9956
58.20		0.9994	0.9991	0.9988	0.9985	0.9982	0.9975	0.9967	0.9964	0.9956
58.25		0.9994	0.9991	0.9988	0.9985	0.9982	0.9975	0.9967	0.9964	0.9956
58.30		0.9994	0.9991	0.9988	0.9985	0.9982	0.9975	0.9968	0.9964	0.9956
58.35		0.9994	0.9991	0.9988	0.9985	0.9982	0.9975	0.9968	0.9964	0.9957
58.40		0.9994	0.9991	0.9988	0.9985	0.9982	0.9975	0.9968	0.9964	0.9957
58.45		0.9994	0.9991	0.9988	0.9985	0.9982	0.9975	0.9968	0.9964	0.9957

PROBABILITY INTEGRAL OF t, THE NON-CENTRAL t-STATISTIC. THIS TABLE GIVES $Pr\ [t/\sqrt{f} \leq x]$.

f is the number of degrees of freedom; the non-centrality parameter is $\sqrt{f+1}\,K_p$.

K_p is the standardized normal deviate exceeded with probability p.

DEGREES OF FREEDOM **2**

x \ p	.2500	.1500	.1000	.0650	.0400	.0250	.0100	.0040	.0025	.0010
58.50		0.9994	0.9991	0.9989	0.9985	0.9982	0.9975	0.9968	0.9964	0.9957
58.55		0.9994	0.9991	0.9989	0.9985	0.9982	0.9975	0.9968	0.9964	0.9957
58.60		0.9994	0.9991	0.9989	0.9985	0.9982	0.9975	0.9968	0.9964	0.9957
58.65		0.9994	0.9991	0.9989	0.9985	0.9982	0.9975	0.9968	0.9964	0.9957
58.70		0.9994	0.9991	0.9989	0.9985	0.9982	0.9975	0.9968	0.9964	0.9957
58.75		0.9994	0.9991	0.9989	0.9985	0.9982	0.9975	0.9968	0.9964	0.9957
58.80		0.9994	0.9991	0.9989	0.9985	0.9982	0.9975	0.9968	0.9964	0.9957
58.85		0.9994	0.9991	0.9989	0.9985	0.9982	0.9975	0.9968	0.9964	0.9957
58.90		0.9994	0.9991	0.9989	0.9985	0.9982	0.9975	0.9968	0.9965	0.9957
58.95		0.9994	0.9991	0.9989	0.9985	0.9982	0.9975	0.9968	0.9965	0.9957
59.00		0.9994	0.9991	0.9989	0.9985	0.9982	0.9975	0.9968	0.9965	0.9958
59.05		0.9994	0.9991	0.9989	0.9985	0.9982	0.9975	0.9968	0.9965	0.9958
59.10		0.9994	0.9991	0.9989	0.9985	0.9982	0.9975	0.9968	0.9965	0.9958
59.15		0.9994	0.9991	0.9989	0.9985	0.9982	0.9975	0.9968	0.9965	0.9958
59.20		0.9994	0.9991	0.9989	0.9985	0.9982	0.9975	0.9969	0.9965	0.9958
59.25		0.9994	0.9991	0.9989	0.9985	0.9982	0.9975	0.9969	0.9965	0.9958
59.30		0.9994	0.9991	0.9989	0.9986	0.9982	0.9976	0.9969	0.9965	0.9958
59.35		0.9994	0.9991	0.9989	0.9986	0.9982	0.9976	0.9969	0.9965	0.9958
59.40		0.9994	0.9991	0.9989	0.9986	0.9982	0.9976	0.9969	0.9965	0.9958
59.45		0.9994	0.9991	0.9989	0.9986	0.9982	0.9976	0.9969	0.9965	0.9958
59.50		0.9994	0.9991	0.9989	0.9986	0.9982	0.9976	0.9969	0.9965	0.9958
59.55		0.9994	0.9991	0.9989	0.9986	0.9982	0.9976	0.9969	0.9965	0.9958
59.60		0.9994	0.9991	0.9989	0.9986	0.9982	0.9976	0.9969	0.9965	0.9958
59.65		0.9994	0.9991	0.9989	0.9986	0.9982	0.9976	0.9969	0.9965	0.9958
59.70		0.9994	0.9991	0.9989	0.9986	0.9982	0.9976	0.9969	0.9966	0.9959
59.75		0.9994	0.9992	0.9989	0.9986	0.9982	0.9976	0.9969	0.9966	0.9959
59.80		0.9994	0.9992	0.9989	0.9986	0.9983	0.9976	0.9969	0.9966	0.9959
59.85		0.9994	0.9992	0.9989	0.9986	0.9983	0.9976	0.9969	0.9966	0.9959
59.90		0.9994	0.9992	0.9989	0.9986	0.9983	0.9976	0.9969	0.9966	0.9959
59.95		0.9994	0.9992	0.9989	0.9986	0.9983	0.9976	0.9969	0.9966	0.9959
60.00		0.9994	0.9992	0.9989	0.9986	0.9983	0.9976	0.9969	0.9966	0.9959
60.05		0.9994	0.9992	0.9989	0.9986	0.9983	0.9976	0.9969	0.9966	0.9959
60.10		0.9994	0.9992	0.9989	0.9986	0.9983	0.9976	0.9969	0.9966	0.9959
60.15		0.9994	0.9992	0.9989	0.9986	0.9983	0.9976	0.9970	0.9966	0.9959
60.20		0.9994	0.9992	0.9989	0.9986	0.9983	0.9976	0.9970	0.9966	0.9959
60.25		0.9994	0.9992	0.9989	0.9986	0.9983	0.9976	0.9970	0.9966	0.9959
60.30		0.9994	0.9992	0.9989	0.9986	0.9983	0.9976	0.9970	0.9966	0.9959
60.35		0.9994	0.9992	0.9989	0.9986	0.9983	0.9976	0.9970	0.9966	0.9959
60.40		0.9994	0.9992	0.9989	0.9986	0.9983	0.9976	0.9970	0.9966	0.9959
60.45		0.9994	0.9992	0.9989	0.9986	0.9983	0.9976	0.9970	0.9966	0.9960
60.50		0.9994	0.9992	0.9989	0.9986	0.9983	0.9976	0.9970	0.9966	0.9960
60.55		0.9994	0.9992	0.9989	0.9986	0.9983	0.9977	0.9970	0.9966	0.9960
60.60		0.9994	0.9992	0.9989	0.9986	0.9983	0.9977	0.9970	0.9967	0.9960
60.65		0.9994	0.9992	0.9989	0.9986	0.9983	0.9977	0.9970	0.9967	0.9960
60.70		0.9994	0.9992	0.9989	0.9986	0.9983	0.9977	0.9970	0.9967	0.9960
60.75		0.9994	0.9992	0.9989	0.9986	0.9983	0.9977	0.9970	0.9967	0.9960
60.80		0.9994	0.9992	0.9989	0.9986	0.9983	0.9977	0.9970	0.9967	0.9960
60.85		0.9994	0.9992	0.9989	0.9986	0.9983	0.9977	0.9970	0.9967	0.9960
60.90		0.9994	0.9992	0.9989	0.9986	0.9983	0.9977	0.9970	0.9967	0.9960
60.95		0.9994	0.9992	0.9989	0.9986	0.9983	0.9977	0.9970	0.9967	0.9960
61.00		0.9994	0.9992	0.9989	0.9986	0.9983	0.9977	0.9970	0.9967	0.9960
61.05		0.9994	0.9992	0.9989	0.9986	0.9983	0.9977	0.9970	0.9967	0.9960
61.10		0.9994	0.9992	0.9989	0.9986	0.9983	0.9977	0.9970	0.9967	0.9960
61.15		0.9994	0.9992	0.9989	0.9986	0.9983	0.9977	0.9970	0.9967	0.9960
61.20		0.9994	0.9992	0.9989	0.9986	0.9983	0.9977	0.9971	0.9967	0.9961
61.25		0.9994	0.9992	0.9990	0.9986	0.9983	0.9977	0.9971	0.9967	0.9961
61.30		0.9994	0.9992	0.9990	0.9986	0.9983	0.9977	0.9971	0.9967	0.9961
61.35		0.9994	0.9992	0.9990	0.9986	0.9983	0.9977	0.9971	0.9967	0.9961
61.40		0.9994	0.9992	0.9990	0.9986	0.9983	0.9977	0.9971	0.9967	0.9961
61.45		0.9994	0.9992	0.9990	0.9987	0.9983	0.9977	0.9971	0.9967	0.9961

PROBABILITY INTEGRAL OF t, THE NON-CENTRAL t-STATISTIC. THIS TABLE GIVES Pr $[t/\sqrt{f} \leq x]$.

f is the number of degrees of freedom; the non-centrality parameter is $\sqrt{f+1}\,K_p$.

K_p is the standardized normal deviate exceeded with probability p.

DEGREES OF FREEDOM **2**

x \ p	.2500	.1500	.1000	.0650	.0400	.0250	.0100	.0040	.0025	.0010
61.50		0.9994	0.9992	0.9990	0.9987	0.9983	0.9977	0.9971	0.9967	0.9961
61.55		0.9994	0.9992	0.9990	0.9987	0.9983	0.9977	0.9971	0.9968	0.9961
61.60		0.9994	0.9992	0.9990	0.9987	0.9984	0.9977	0.9971	0.9968	0.9961
61.65		0.9994	0.9992	0.9990	0.9987	0.9984	0.9977	0.9971	0.9968	0.9961
61.70		0.9994	0.9992	0.9990	0.9987	0.9984	0.9977	0.9971	0.9968	0.9961
61.75		0.9994	0.9992	0.9990	0.9987	0.9984	0.9977	0.9971	0.9968	0.9961
61.80		0.9994	0.9992	0.9990	0.9987	0.9984	0.9977	0.9971	0.9968	0.9961
61.85		0.9994	0.9992	0.9990	0.9987	0.9984	0.9978	0.9971	0.9968	0.9961
61.90		0.9995	0.9992	0.9990	0.9987	0.9984	0.9978	0.9971	0.9968	0.9961
61.95		0.9995	0.9992	0.9990	0.9987	0.9984	0.9978	0.9971	0.9968	0.9961
62.00		0.9995	0.9992	0.9990	0.9987	0.9984	0.9978	0.9971	0.9968	0.9962
62.05		0.9995	0.9992	0.9990	0.9987	0.9984	0.9978	0.9971	0.9968	0.9962
62.10		0.9995	0.9992	0.9990	0.9987	0.9984	0.9978	0.9971	0.9968	0.9962
62.15		0.9995	0.9992	0.9990	0.9987	0.9984	0.9978	0.9971	0.9968	0.9962
62.20		0.9995	0.9992	0.9990	0.9987	0.9984	0.9978	0.9971	0.9968	0.9962
62.25		0.9995	0.9992	0.9990	0.9987	0.9984	0.9978	0.9972	0.9968	0.9962
62.30		0.9995	0.9992	0.9990	0.9987	0.9984	0.9978	0.9972	0.9968	0.9962
62.35		0.9995	0.9992	0.9990	0.9987	0.9984	0.9978	0.9972	0.9968	0.9962
62.40		0.9995	0.9992	0.9990	0.9987	0.9984	0.9978	0.9972	0.9968	0.9962
62.45		0.9995	0.9992	0.9990	0.9987	0.9984	0.9978	0.9972	0.9968	0.9962
62.50		0.9995	0.9992	0.9990	0.9987	0.9984	0.9978	0.9972	0.9969	0.9962
62.55		0.9995	0.9992	0.9990	0.9987	0.9984	0.9978	0.9972	0.9969	0.9962
62.60		0.9995	0.9992	0.9990	0.9987	0.9984	0.9978	0.9972	0.9969	0.9962
62.65		0.9995	0.9992	0.9990	0.9987	0.9984	0.9978	0.9972	0.9969	0.9962
62.70		0.9995	0.9992	0.9990	0.9987	0.9984	0.9978	0.9972	0.9969	0.9962
62.75		0.9995	0.9992	0.9990	0.9987	0.9984	0.9978	0.9972	0.9969	0.9962
62.80		0.9995	0.9992	0.9990	0.9987	0.9984	0.9978	0.9972	0.9969	0.9962
62.85		0.9995	0.9992	0.9990	0.9987	0.9984	0.9978	0.9972	0.9969	0.9963
62.90		0.9995	0.9992	0.9990	0.9987	0.9984	0.9978	0.9972	0.9969	0.9963
62.95		0.9995	0.9992	0.9990	0.9987	0.9984	0.9978	0.9972	0.9969	0.9963
63.00		0.9995	0.9992	0.9990	0.9987	0.9984	0.9978	0.9972	0.9969	0.9963
63.05		0.9995	0.9992	0.9990	0.9987	0.9984	0.9978	0.9972	0.9969	0.9963
63.10		0.9995	0.9992	0.9990	0.9987	0.9984	0.9978	0.9972	0.9969	0.9963
63.15		0.9995	0.9992	0.9990	0.9987	0.9984	0.9978	0.9972	0.9969	0.9963
63.20		0.9995	0.9992	0.9990	0.9987	0.9984	0.9978	0.9972	0.9969	0.9963
63.25		0.9995	0.9992	0.9990	0.9987	0.9984	0.9978	0.9972	0.9969	0.9963
63.30		0.9995	0.9992	0.9990	0.9987	0.9984	0.9979	0.9972	0.9969	0.9963
63.35		0.9995	0.9992	0.9990	0.9987	0.9984	0.9979	0.9973	0.9969	0.9963
63.40		0.9995	0.9992	0.9990	0.9987	0.9984	0.9979	0.9973	0.9969	0.9963
63.45		0.9995	0.9992	0.9990	0.9987	0.9984	0.9979	0.9973	0.9969	0.9963
63.50		0.9995	0.9992	0.9990	0.9987	0.9984	0.9979	0.9973	0.9970	0.9963
63.55		0.9995	0.9992	0.9990	0.9987	0.9985	0.9979	0.9973	0.9970	0.9963
63.60		0.9995	0.9992	0.9990	0.9987	0.9985	0.9979	0.9973	0.9970	0.9963
63.65		0.9995	0.9992	0.9990	0.9987	0.9985	0.9979	0.9973	0.9970	0.9963
63.70		0.9995	0.9992	0.9990	0.9987	0.9985	0.9979	0.9973	0.9970	0.9964
63.75		0.9995	0.9992	0.9990	0.9987	0.9985	0.9979	0.9973	0.9970	0.9964
63.80		0.9995	0.9993	0.9990	0.9987	0.9985	0.9979	0.9973	0.9970	0.9964
63.85		0.9995	0.9993	0.9990	0.9988	0.9985	0.9979	0.9973	0.9970	0.9964
63.90		0.9995	0.9993	0.9990	0.9988	0.9985	0.9979	0.9973	0.9970	0.9964
63.95		0.9995	0.9993	0.9990	0.9988	0.9985	0.9979	0.9973	0.9970	0.9964
64.00		0.9995	0.9993	0.9990	0.9988	0.9985	0.9979	0.9973	0.9970	0.9964
64.05		0.9995	0.9993	0.9990	0.9988	0.9985	0.9979	0.9973	0.9970	0.9964
64.10		0.9995	0.9993	0.9990	0.9988	0.9985	0.9979	0.9973	0.9970	0.9964
64.15		0.9995	0.9993	0.9990	0.9988	0.9985	0.9979	0.9973	0.9970	0.9964
64.20		0.9995	0.9993	0.9990	0.9988	0.9985	0.9979	0.9973	0.9970	0.9964
64.25		0.9995	0.9993	0.9990	0.9988	0.9985	0.9979	0.9973	0.9970	0.9964
64.30		0.9995	0.9993	0.9990	0.9988	0.9985	0.9979	0.9973	0.9970	0.9964
64.35		0.9995	0.9993	0.9990	0.9988	0.9985	0.9979	0.9973	0.9970	0.9964
64.40		0.9995	0.9993	0.9991	0.9988	0.9985	0.9979	0.9973	0.9970	0.9964
64.45		0.9995	0.9993	0.9991	0.9988	0.9985	0.9979	0.9973	0.9970	0.9964

PROBABILITY INTEGRAL OF t, THE NON-CENTRAL t-STATISTIC. THIS TABLE GIVES Pr $[t/\sqrt{f} \leq x]$.

f is the number of degrees of freedom; the non-centrality parameter is $\sqrt{f+1}\,K_p$.
K_p is the standardized normal deviate exceeded with probability p.

DEGREES OF FREEDOM **2**

x \ P	.2500	.1500	.1000	.0650	.0400	.0250	.0100	.0040	.0025	.0010
64.50		0.9995	0.9993	0.9991	0.9988	0.9985	0.9979	0.9973	0.9970	0.9964
64.55		0.9995	0.9993	0.9991	0.9988	0.9985	0.9979	0.9974	0.9970	0.9964
64.60		0.9995	0.9993	0.9991	0.9988	0.9985	0.9979	0.9974	0.9971	0.9965
64.65		0.9995	0.9993	0.9991	0.9988	0.9985	0.9979	0.9974	0.9971	0.9965
64.70		0.9995	0.9993	0.9991	0.9988	0.9985	0.9979	0.9974	0.9971	0.9965
64.75		0.9995	0.9993	0.9991	0.9988	0.9985	0.9979	0.9974	0.9971	0.9965
64.80		0.9995	0.9993	0.9991	0.9988	0.9985	0.9980	0.9974	0.9971	0.9965
64.85		0.9995	0.9993	0.9991	0.9988	0.9985	0.9980	0.9974	0.9971	0.9965
64.90		0.9995	0.9993	0.9991	0.9988	0.9985	0.9980	0.9974	0.9971	0.9965
64.95		0.9995	0.9993	0.9991	0.9988	0.9985	0.9980	0.9974	0.9971	0.9965
65.00		0.9995	0.9993	0.9991	0.9988	0.9985	0.9980	0.9974	0.9971	0.9965
65.05		0.9995	0.9993	0.9991	0.9988	0.9985	0.9980	0.9974	0.9971	0.9965
65.10		0.9995	0.9993	0.9991	0.9988	0.9985	0.9980	0.9974	0.9971	0.9965
65.15		0.9995	0.9993	0.9991	0.9988	0.9985	0.9980	0.9974	0.9971	0.9965
65.20		0.9995	0.9993	0.9991	0.9988	0.9985	0.9980	0.9974	0.9971	0.9965
65.25		0.9995	0.9993	0.9991	0.9988	0.9985	0.9980	0.9974	0.9971	0.9965
65.30		0.9995	0.9993	0.9991	0.9988	0.9985	0.9980	0.9974	0.9971	0.9965
65.35		0.9995	0.9993	0.9991	0.9988	0.9985	0.9980	0.9974	0.9971	0.9965
65.40		0.9995	0.9993	0.9991	0.9988	0.9985	0.9980	0.9974	0.9971	0.9965
65.45			0.9993	0.9991	0.9988	0.9985	0.9980	0.9974	0.9971	0.9965
65.50			0.9993	0.9991	0.9988	0.9985	0.9980	0.9974	0.9971	0.9966
65.55			0.9993	0.9991	0.9988	0.9985	0.9980	0.9974	0.9971	0.9966
65.60			0.9993	0.9991	0.9988	0.9985	0.9980	0.9974	0.9971	0.9966
65.65			0.9993	0.9991	0.9988	0.9985	0.9980	0.9974	0.9971	0.9966
65.70			0.9993	0.9991	0.9988	0.9986	0.9980	0.9974	0.9972	0.9966
65.75			0.9993	0.9991	0.9988	0.9986	0.9980	0.9974	0.9972	0.9966
65.80			0.9993	0.9991	0.9988	0.9986	0.9980	0.9975	0.9972	0.9966
65.85			0.9993	0.9991	0.9988	0.9986	0.9980	0.9975	0.9972	0.9966
65.90			0.9993	0.9991	0.9988	0.9986	0.9980	0.9975	0.9972	0.9966
65.95			0.9993	0.9991	0.9988	0.9986	0.9980	0.9975	0.9972	0.9966
66.00			0.9993	0.9991	0.9988	0.9986	0.9980	0.9975	0.9972	0.9966
66.05			0.9993	0.9991	0.9988	0.9986	0.9980	0.9975	0.9972	0.9966
66.10			0.9993	0.9991	0.9988	0.9986	0.9980	0.9975	0.9972	0.9966
66.15			0.9993	0.9991	0.9988	0.9986	0.9980	0.9975	0.9972	0.9966
66.20			0.9993	0.9991	0.9988	0.9986	0.9980	0.9975	0.9972	0.9966
66.25			0.9993	0.9991	0.9988	0.9986	0.9980	0.9975	0.9972	0.9966
66.30			0.9993	0.9991	0.9988	0.9986	0.9980	0.9975	0.9972	0.9966
66.35			0.9993	0.9991	0.9988	0.9986	0.9980	0.9975	0.9972	0.9966
66.40			0.9993	0.9991	0.9988	0.9986	0.9980	0.9975	0.9972	0.9966
66.45			0.9993	0.9991	0.9988	0.9986	0.9981	0.9975	0.9972	0.9966
66.50			0.9993	0.9991	0.9988	0.9986	0.9981	0.9975	0.9972	0.9967
66.55			0.9993	0.9991	0.9989	0.9986	0.9981	0.9975	0.9972	0.9967
66.60			0.9993	0.9991	0.9989	0.9986	0.9981	0.9975	0.9972	0.9967
66.65			0.9993	0.9991	0.9989	0.9986	0.9981	0.9975	0.9972	0.9967
66.70			0.9993	0.9991	0.9989	0.9986	0.9981	0.9975	0.9972	0.9967
66.75			0.9993	0.9991	0.9989	0.9986	0.9981	0.9975	0.9972	0.9967
66.80			0.9993	0.9991	0.9989	0.9986	0.9981	0.9975	0.9972	0.9967
66.85			0.9993	0.9991	0.9989	0.9986	0.9981	0.9975	0.9972	0.9967
66.90			0.9993	0.9991	0.9989	0.9986	0.9981	0.9975	0.9973	0.9967
66.95			0.9993	0.9991	0.9989	0.9986	0.9981	0.9975	0.9973	0.9967
67.00			0.9993	0.9991	0.9989	0.9986	0.9981	0.9975	0.9973	0.9967
67.05			0.9993	0.9991	0.9989	0.9986	0.9981	0.9975	0.9973	0.9967
67.10			0.9993	0.9991	0.9989	0.9986	0.9981	0.9975	0.9973	0.9967
67.15			0.9993	0.9991	0.9989	0.9986	0.9981	0.9976	0.9973	0.9967
67.20			0.9993	0.9991	0.9989	0.9986	0.9981	0.9976	0.9973	0.9967
67.25			0.9993	0.9991	0.9989	0.9986	0.9981	0.9976	0.9973	0.9967
67.30			0.9993	0.9991	0.9989	0.9986	0.9981	0.9976	0.9973	0.9967
67.35			0.9993	0.9991	0.9989	0.9986	0.9981	0.9976	0.9973	0.9967
67.40			0.9993	0.9991	0.9989	0.9986	0.9981	0.9976	0.9973	0.9967
67.45			0.9993	0.9991	0.9989	0.9986	0.9981	0.9976	0.9973	0.9967

PROBABILITY INTEGRAL OF t, THE NON-CENTRAL t-STATISTIC. THIS TABLE GIVES Pr $[t/\sqrt{f} \leq x]$.

f is the number of degrees of freedom; the non-centrality parameter is $\sqrt{f+1}\,K_p$.

K_p is the standardized normal deviate exceeded with probability p.

DEGREES OF FREEDOM **2**

x \ p	.2500	.1500	.1000	.0650	.0400	.0250	.0100	.0040	.0025	.0010
67.50			0.9993	0.9991	0.9989	0.9986	0.9981	0.9976	0.9973	0.9968
67.55			0.9993	0.9991	0.9989	0.9986	0.9981	0.9976	0.9973	0.9968
67.60			0.9993	0.9991	0.9989	0.9986	0.9981	0.9976	0.9973	0.9968
67.65			0.9993	0.9991	0.9989	0.9986	0.9981	0.9976	0.9973	0.9968
67.70			0.9993	0.9991	0.9989	0.9986	0.9981	0.9976	0.9973	0.9968
67.75			0.9993	0.9991	0.9989	0.9986	0.9981	0.9976	0.9973	0.9968
67.80			0.9993	0.9991	0.9989	0.9986	0.9981	0.9976	0.9973	0.9968
67.85			0.9993	0.9991	0.9989	0.9986	0.9981	0.9976	0.9973	0.9968
67.90			0.9993	0.9991	0.9989	0.9986	0.9981	0.9976	0.9973	0.9968
67.95			0.9993	0.9991	0.9989	0.9986	0.9981	0.9976	0.9973	0.9968
68.00			0.9993	0.9991	0.9989	0.9986	0.9981	0.9976	0.9973	0.9968
68.05			0.9993	0.9991	0.9989	0.9986	0.9981	0.9976	0.9973	0.9968
68.10			0.9993	0.9992	0.9989	0.9987	0.9981	0.9976	0.9973	0.9968
68.15			0.9993	0.9992	0.9989	0.9987	0.9981	0.9976	0.9974	0.9968
68.20			0.9993	0.9992	0.9989	0.9987	0.9981	0.9976	0.9974	0.9968
68.25			0.9993	0.9992	0.9989	0.9987	0.9982	0.9976	0.9974	0.9968
68.30			0.9993	0.9992	0.9989	0.9987	0.9982	0.9976	0.9974	0.9968
68.35			0.9993	0.9992	0.9989	0.9987	0.9982	0.9976	0.9974	0.9968
68.40			0.9993	0.9992	0.9989	0.9987	0.9982	0.9976	0.9974	0.9968
68.45			0.9993	0.9992	0.9989	0.9987	0.9982	0.9976	0.9974	0.9968
68.50			0.9993	0.9992	0.9989	0.9987	0.9982	0.9976	0.9974	0.9968
68.55			0.9993	0.9992	0.9989	0.9987	0.9982	0.9977	0.9974	0.9969
68.60			0.9993	0.9992	0.9989	0.9987	0.9982	0.9977	0.9974	0.9969
68.65			0.9993	0.9992	0.9989	0.9987	0.9982	0.9977	0.9974	0.9969
68.70			0.9993	0.9992	0.9989	0.9987	0.9982	0.9977	0.9974	0.9969
68.75			0.9993	0.9992	0.9989	0.9987	0.9982	0.9977	0.9974	0.9969
68.80			0.9994	0.9992	0.9989	0.9987	0.9982	0.9977	0.9974	0.9969
68.85			0.9994	0.9992	0.9989	0.9987	0.9982	0.9977	0.9974	0.9969
68.90			0.9994	0.9992	0.9989	0.9987	0.9982	0.9977	0.9974	0.9969
68.95			0.9994	0.9992	0.9989	0.9987	0.9982	0.9977	0.9974	0.9969
69.00			0.9994	0.9992	0.9989	0.9987	0.9982	0.9977	0.9974	0.9969
69.05			0.9994	0.9992	0.9989	0.9987	0.9982	0.9977	0.9974	0.9969
69.10			0.9994	0.9992	0.9989	0.9987	0.9982	0.9977	0.9974	0.9969
69.15			0.9994	0.9992	0.9989	0.9987	0.9982	0.9977	0.9974	0.9969
69.20			0.9994	0.9992	0.9989	0.9987	0.9982	0.9977	0.9974	0.9969
69.25			0.9994	0.9992	0.9989	0.9987	0.9982	0.9977	0.9974	0.9969
69.30			0.9994	0.9992	0.9989	0.9987	0.9982	0.9977	0.9974	0.9969
69.35			0.9994	0.9992	0.9989	0.9987	0.9982	0.9977	0.9974	0.9969
69.40			0.9994	0.9992	0.9989	0.9987	0.9982	0.9977	0.9974	0.9969
69.45			0.9994	0.9992	0.9989	0.9987	0.9982	0.9977	0.9974	0.9969
69.50			0.9994	0.9992	0.9989	0.9987	0.9982	0.9977	0.9975	0.9969
69.55			0.9994	0.9992	0.9989	0.9987	0.9982	0.9977	0.9975	0.9969
69.60			0.9994	0.9992	0.9989	0.9987	0.9982	0.9977	0.9975	0.9969
69.65			0.9994	0.9992	0.9990	0.9987	0.9982	0.9977	0.9975	0.9969
69.70			0.9994	0.9992	0.9990	0.9987	0.9982	0.9977	0.9975	0.9970
69.75			0.9994	0.9992	0.9990	0.9987	0.9982	0.9977	0.9975	0.9970
69.80			0.9994	0.9992	0.9990	0.9987	0.9982	0.9977	0.9975	0.9970
69.85			0.9994	0.9992	0.9990	0.9987	0.9982	0.9977	0.9975	0.9970
69.90			0.9994	0.9992	0.9990	0.9987	0.9982	0.9977	0.9975	0.9970
69.95			0.9994	0.9992	0.9990	0.9987	0.9982	0.9977	0.9975	0.9970
70.00			0.9994	0.9992	0.9990	0.9987	0.9982	0.9977	0.9975	0.9970
70.05			0.9994	0.9992	0.9990	0.9987	0.9982	0.9978	0.9975	0.9970
70.10			0.9994	0.9992	0.9990	0.9987	0.9982	0.9978	0.9975	0.9970
70.15			0.9994	0.9992	0.9990	0.9987	0.9983	0.9978	0.9975	0.9970
70.20			0.9994	0.9992	0.9990	0.9987	0.9983	0.9978	0.9975	0.9970
70.25			0.9994	0.9992	0.9990	0.9987	0.9983	0.9978	0.9975	0.9970
70.30			0.9994	0.9992	0.9990	0.9987	0.9983	0.9978	0.9975	0.9970
70.35			0.9994	0.9992	0.9990	0.9987	0.9983	0.9978	0.9975	0.9970
70.40			0.9994	0.9992	0.9990	0.9987	0.9983	0.9978	0.9975	0.9970
70.45			0.9994	0.9992	0.9990	0.9987	0.9983	0.9978	0.9975	0.9970

PROBABILITY INTEGRAL OF t, THE NON-CENTRAL t-STATISTIC. THIS TABLE GIVES Pr $[t/\sqrt{f} \leq x]$.

f is the number of degrees of freedom; the non-centrality parameter is $\sqrt{f+1}\,K_p$.

K_p is the standardized normal deviate exceeded with probability p.

DEGREES OF FREEDOM **2**

x \ P	.2500	.1500	.1000	.0650	.0400	.0250	.0100	.0040	.0025	.0010
70.50			0.9994	0.9992	0.9990	0.9987	0.9983	0.9978	0.9975	0.9970
70.55			0.9994	0.9992	0.9990	0.9987	0.9983	0.9978	0.9975	0.9970
70.60			0.9994	0.9992	0.9990	0.9987	0.9983	0.9978	0.9975	0.9970
70.65			0.9994	0.9992	0.9990	0.9987	0.9983	0.9978	0.9975	0.9970
70.70			0.9994	0.9992	0.9990	0.9987	0.9983	0.9978	0.9975	0.9970
70.75			0.9994	0.9992	0.9990	0.9987	0.9983	0.9978	0.9975	0.9970
70.80			0.9994	0.9992	0.9990	0.9988	0.9983	0.9978	0.9975	0.9970
70.85			0.9994	0.9992	0.9990	0.9988	0.9983	0.9978	0.9975	0.9971
70.90			0.9994	0.9992	0.9990	0.9988	0.9983	0.9978	0.9976	0.9971
70.95			0.9994	0.9992	0.9990	0.9988	0.9983	0.9978	0.9976	0.9971
71.00			0.9994	0.9992	0.9990	0.9988	0.9983	0.9978	0.9976	0.9971
71.05			0.9994	0.9992	0.9990	0.9988	0.9983	0.9978	0.9976	0.9971
71.10			0.9994	0.9992	0.9990	0.9988	0.9983	0.9978	0.9976	0.9971
71.15			0.9994	0.9992	0.9990	0.9988	0.9983	0.9978	0.9976	0.9971
71.20			0.9994	0.9992	0.9990	0.9988	0.9983	0.9978	0.9976	0.9971
71.25			0.9994	0.9992	0.9990	0.9988	0.9983	0.9978	0.9976	0.9971
71.30			0.9994	0.9992	0.9990	0.9988	0.9983	0.9978	0.9976	0.9971
71.35			0.9994	0.9992	0.9990	0.9988	0.9983	0.9978	0.9976	0.9971
71.40			0.9994	0.9992	0.9990	0.9988	0.9983	0.9978	0.9976	0.9971
71.45			0.9994	0.9992	0.9990	0.9988	0.9983	0.9978	0.9976	0.9971
71.50			0.9994	0.9992	0.9990	0.9988	0.9983	0.9978	0.9976	0.9971
71.55			0.9994	0.9992	0.9990	0.9988	0.9983	0.9978	0.9976	0.9971
71.60			0.9994	0.9992	0.9990	0.9988	0.9983	0.9978	0.9976	0.9971
71.65			0.9994	0.9992	0.9990	0.9988	0.9983	0.9978	0.9976	0.9971
71.70			0.9994	0.9992	0.9990	0.9988	0.9983	0.9979	0.9976	0.9971
71.75			0.9994	0.9992	0.9990	0.9988	0.9983	0.9979	0.9976	0.9971
71.80			0.9994	0.9992	0.9990	0.9988	0.9983	0.9979	0.9976	0.9971
71.85			0.9994	0.9992	0.9990	0.9988	0.9983	0.9979	0.9976	0.9971
71.90			0.9994	0.9992	0.9990	0.9988	0.9983	0.9979	0.9976	0.9971
71.95			0.9994	0.9992	0.9990	0.9988	0.9983	0.9979	0.9976	0.9971
72.00			0.9994	0.9992	0.9990	0.9988	0.9983	0.9979	0.9976	0.9971
72.05			0.9994	0.9992	0.9990	0.9988	0.9983	0.9979	0.9976	0.9971
72.10			0.9994	0.9992	0.9990	0.9988	0.9983	0.9979	0.9976	0.9971
72.15			0.9994	0.9992	0.9990	0.9988	0.9983	0.9979	0.9976	0.9972
72.20			0.9994	0.9992	0.9990	0.9988	0.9983	0.9979	0.9976	0.9972
72.25			0.9994	0.9992	0.9990	0.9988	0.9984	0.9979	0.9976	0.9972
72.30			0.9994	0.9992	0.9990	0.9988	0.9984	0.9979	0.9976	0.9972
72.35			0.9994	0.9992	0.9990	0.9988	0.9984	0.9979	0.9976	0.9972
72.40			0.9994	0.9992	0.9990	0.9988	0.9984	0.9979	0.9977	0.9972
72.45			0.9994	0.9992	0.9990	0.9988	0.9984	0.9979	0.9977	0.9972
72.50			0.9994	0.9993	0.9990	0.9988	0.9984	0.9979	0.9977	0.9972
72.55			0.9994	0.9993	0.9990	0.9988	0.9984	0.9979	0.9977	0.9972
72.60			0.9994	0.9993	0.9990	0.9988	0.9984	0.9979	0.9977	0.9972
72.65			0.9994	0.9993	0.9990	0.9988	0.9984	0.9979	0.9977	0.9972
72.70			0.9994	0.9993	0.9990	0.9988	0.9984	0.9979	0.9977	0.9972
72.75			0.9994	0.9993	0.9990	0.9988	0.9984	0.9979	0.9977	0.9972
72.80			0.9994	0.9993	0.9990	0.9988	0.9984	0.9979	0.9977	0.9972
72.85			0.9994	0.9993	0.9990	0.9988	0.9984	0.9979	0.9977	0.9972
72.90			0.9994	0.9993	0.9990	0.9988	0.9984	0.9979	0.9977	0.9972
72.95			0.9994	0.9993	0.9990	0.9988	0.9984	0.9979	0.9977	0.9972
73.00			0.9994	0.9993	0.9990	0.9988	0.9984	0.9979	0.9977	0.9972
73.05			0.9994	0.9993	0.9990	0.9988	0.9984	0.9979	0.9977	0.9972
73.10			0.9994	0.9993	0.9990	0.9988	0.9984	0.9979	0.9977	0.9972
73.15			0.9994	0.9993	0.9990	0.9988	0.9984	0.9979	0.9977	0.9972
73.20			0.9994	0.9993	0.9990	0.9988	0.9984	0.9979	0.9977	0.9972
73.25			0.9994	0.9993	0.9991	0.9988	0.9984	0.9979	0.9977	0.9972
73.30			0.9994	0.9993	0.9991	0.9988	0.9984	0.9979	0.9977	0.9972
73.35			0.9994	0.9993	0.9991	0.9988	0.9984	0.9979	0.9977	0.9972
73.40			0.9994	0.9993	0.9991	0.9988	0.9984	0.9980	0.9977	0.9973
73.45			0.9994	0.9993	0.9991	0.9988	0.9984	0.9980	0.9977	0.9973

PROBABILITY INTEGRAL OF t, THE NON-CENTRAL t-STATISTIC. THIS TABLE GIVES Pr $[t/\sqrt{f} \leq x]$.

f is the number of degrees of freedom; the non-centrality parameter is $\sqrt{f+1}\,K_p$.

K_p is the standardized normal deviate exceeded with probability p.

DEGREES OF FREEDOM **2**

x \ p	.2500	.1500	.1000	.0650	.0400	.0250	.0100	.0040	.0025	.0010
73.50			0.9994	0.9993	0.9991	0.9988	0.9984	0.9980	0.9977	0.9973
73.55			0.9994	0.9993	0.9991	0.9988	0.9984	0.9980	0.9977	0.9973
73.60			0.9994	0.9993	0.9991	0.9988	0.9984	0.9980	0.9977	0.9973
73.65			0.9994	0.9993	0.9991	0.9988	0.9984	0.9980	0.9977	0.9973
73.70			0.9994	0.9993	0.9991	0.9988	0.9984	0.9980	0.9977	0.9973
73.75			0.9994	0.9993	0.9991	0.9988	0.9984	0.9980	0.9977	0.9973
73.80			0.9994	0.9993	0.9991	0.9989	0.9984	0.9980	0.9977	0.9973
73.85			0.9994	0.9993	0.9991	0.9989	0.9984	0.9980	0.9977	0.9973
73.90			0.9994	0.9993	0.9991	0.9989	0.9984	0.9980	0.9977	0.9973
73.95			0.9994	0.9993	0.9991	0.9989	0.9984	0.9980	0.9977	0.9973
74.00			0.9994	0.9993	0.9991	0.9989	0.9984	0.9980	0.9978	0.9973
74.05			0.9994	0.9993	0.9991	0.9989	0.9984	0.9980	0.9978	0.9973
74.10			0.9994	0.9993	0.9991	0.9989	0.9984	0.9980	0.9978	0.9973
74.15			0.9994	0.9993	0.9991	0.9989	0.9984	0.9980	0.9978	0.9973
74.20			0.9994	0.9993	0.9991	0.9989	0.9984	0.9980	0.9978	0.9973
74.25			0.9994	0.9993	0.9991	0.9989	0.9984	0.9980	0.9978	0.9973
74.30			0.9994	0.9993	0.9991	0.9989	0.9984	0.9980	0.9978	0.9973
74.35			0.9994	0.9993	0.9991	0.9989	0.9984	0.9980	0.9978	0.9973
74.40			0.9994	0.9993	0.9991	0.9989	0.9984	0.9980	0.9978	0.9973
74.45			0.9994	0.9993	0.9991	0.9989	0.9984	0.9980	0.9978	0.9973
74.50			0.9994	0.9993	0.9991	0.9989	0.9984	0.9980	0.9978	0.9973
74.55			0.9994	0.9993	0.9991	0.9989	0.9985	0.9980	0.9978	0.9973
74.60			0.9994	0.9993	0.9991	0.9989	0.9985	0.9980	0.9978	0.9973
74.65			0.9994	0.9993	0.9991	0.9989	0.9985	0.9980	0.9978	0.9973
74.70			0.9994	0.9993	0.9991	0.9989	0.9985	0.9980	0.9978	0.9973
74.75			0.9994	0.9993	0.9991	0.9989	0.9985	0.9980	0.9978	0.9974
74.80			0.9994	0.9993	0.9991	0.9989	0.9985	0.9980	0.9978	0.9974
74.85			0.9994	0.9993	0.9991	0.9989	0.9985	0.9980	0.9978	0.9974
74.90			0.9994	0.9993	0.9991	0.9989	0.9985	0.9980	0.9978	0.9974
74.95			0.9994	0.9993	0.9991	0.9989	0.9985	0.9980	0.9978	0.9974
75.00			0.9994	0.9993	0.9991	0.9989	0.9985	0.9980	0.9978	0.9974
75.05			0.9994	0.9993	0.9991	0.9989	0.9985	0.9980	0.9978	0.9974
75.10			0.9994	0.9993	0.9991	0.9989	0.9985	0.9980	0.9978	0.9974
75.15			0.9994	0.9993	0.9991	0.9989	0.9985	0.9980	0.9978	0.9974
75.20			0.9995	0.9993	0.9991	0.9989	0.9985	0.9980	0.9978	0.9974
75.25			0.9995	0.9993	0.9991	0.9989	0.9985	0.9981	0.9978	0.9974
75.30			0.9995	0.9993	0.9991	0.9989	0.9985	0.9981	0.9978	0.9974
75.35			0.9995	0.9993	0.9991	0.9989	0.9985	0.9981	0.9978	0.9974
75.40			0.9995	0.9993	0.9991	0.9989	0.9985	0.9981	0.9978	0.9974
75.45			0.9995	0.9993	0.9991	0.9989	0.9985	0.9981	0.9978	0.9974
75.50			0.9995	0.9993	0.9991	0.9989	0.9985	0.9981	0.9978	0.9974
75.55			0.9995	0.9993	0.9991	0.9989	0.9985	0.9981	0.9978	0.9974
75.60			0.9995	0.9993	0.9991	0.9989	0.9985	0.9981	0.9978	0.9974
75.65			0.9995	0.9993	0.9991	0.9989	0.9985	0.9981	0.9978	0.9974
75.70			0.9995	0.9993	0.9991	0.9989	0.9985	0.9981	0.9979	0.9974
75.75			0.9995	0.9993	0.9991	0.9989	0.9985	0.9981	0.9979	0.9974
75.80			0.9995	0.9993	0.9991	0.9989	0.9985	0.9981	0.9979	0.9974
75.85			0.9995	0.9993	0.9991	0.9989	0.9985	0.9981	0.9979	0.9974
75.90			0.9995	0.9993	0.9991	0.9989	0.9985	0.9981	0.9979	0.9974
75.95			0.9995	0.9993	0.9991	0.9989	0.9985	0.9981	0.9979	0.9974
76.00			0.9995	0.9993	0.9991	0.9989	0.9985	0.9981	0.9979	0.9974
76.05			0.9995	0.9993	0.9991	0.9989	0.9985	0.9981	0.9979	0.9974
76.10			0.9995	0.9993	0.9991	0.9989	0.9985	0.9981	0.9979	0.9974
76.15			0.9995	0.9993	0.9991	0.9989	0.9985	0.9981	0.9979	0.9974
76.20			0.9995	0.9993	0.9991	0.9989	0.9985	0.9981	0.9979	0.9975
76.25			0.9995	0.9993	0.9991	0.9989	0.9985	0.9981	0.9979	0.9975
76.30			0.9995	0.9993	0.9991	0.9989	0.9985	0.9981	0.9979	0.9975
76.35			0.9995	0.9993	0.9991	0.9989	0.9985	0.9981	0.9979	0.9975
76.40			0.9995	0.9993	0.9991	0.9989	0.9985	0.9981	0.9979	0.9975
76.45			0.9995	0.9993	0.9991	0.9989	0.9985	0.9981	0.9979	0.9975

PROBABILITY INTEGRAL OF t, THE NON-CENTRAL t-STATISTIC. THIS TABLE GIVES Pr $[t/\sqrt{f} \leq x]$.

f is the number of degrees of freedom; the non-centrality parameter is $\sqrt{f+1}\,K_p$.

K_p is the standardized normal deviate exceeded with probability p.

DEGREES OF FREEDOM **2**

x \ p	.2500	.1500	.1000	.0650	.0400	.0250	.0100	.0040	.0025	.0010
76.50			0.9995	0.9993	0.9991	0.9989	0.9985	0.9981	0.9979	0.9975
76.55			0.9995	0.9993	0.9991	0.9989	0.9985	0.9981	0.9979	0.9975
76.60			0.9995	0.9993	0.9991	0.9989	0.9985	0.9981	0.9979	0.9975
76.65			0.9995	0.9993	0.9991	0.9989	0.9985	0.9981	0.9979	0.9975
76.70			0.9995	0.9993	0.9991	0.9989	0.9985	0.9981	0.9979	0.9975
76.75			0.9995	0.9993	0.9991	0.9989	0.9985	0.9981	0.9979	0.9975
76.80			0.9995	0.9993	0.9991	0.9989	0.9985	0.9981	0.9979	0.9975
76.85			0.9995	0.9993	0.9991	0.9989	0.9985	0.9981	0.9979	0.9975
76.90			0.9995	0.9993	0.9991	0.9989	0.9985	0.9981	0.9979	0.9975
76.95			0.9995	0.9993	0.9991	0.9989	0.9985	0.9981	0.9979	0.9975
77.00			0.9995	0.9993	0.9991	0.9989	0.9985	0.9981	0.9979	0.9975
77.05			0.9995	0.9993	0.9991	0.9989	0.9986	0.9981	0.9979	0.9975
77.10			0.9995	0.9993	0.9991	0.9989	0.9986	0.9981	0.9979	0.9975
77.15			0.9995	0.9993	0.9991	0.9989	0.9986	0.9981	0.9979	0.9975
77.20			0.9995	0.9993	0.9991	0.9989	0.9986	0.9981	0.9979	0.9975
77.25			0.9995	0.9993	0.9991	0.9990	0.9986	0.9981	0.9979	0.9975
77.30			0.9995	0.9993	0.9991	0.9990	0.9986	0.9982	0.9979	0.9975
77.35			0.9995	0.9993	0.9991	0.9990	0.9986	0.9982	0.9979	0.9975
77.40			0.9995	0.9993	0.9991	0.9990	0.9986	0.9982	0.9979	0.9975
77.45			0.9995	0.9993	0.9992	0.9990	0.9986	0.9982	0.9979	0.9975
77.50			0.9995	0.9993	0.9992	0.9990	0.9986	0.9982	0.9980	0.9975
77.55			0.9995	0.9993	0.9992	0.9990	0.9986	0.9982	0.9980	0.9975
77.60			0.9995	0.9993	0.9992	0.9990	0.9986	0.9982	0.9980	0.9975
77.65			0.9995	0.9993	0.9992	0.9990	0.9986	0.9982	0.9980	0.9975
77.70			0.9995	0.9993	0.9992	0.9990	0.9986	0.9982	0.9980	0.9975
77.75			0.9995	0.9993	0.9992	0.9990	0.9986	0.9982	0.9980	0.9976
77.80			0.9995	0.9993	0.9992	0.9990	0.9986	0.9982	0.9980	0.9976
77.85			0.9995	0.9994	0.9992	0.9990	0.9986	0.9982	0.9980	0.9976
77.90			0.9995	0.9994	0.9992	0.9990	0.9986	0.9982	0.9980	0.9976
77.95			0.9995	0.9994	0.9992	0.9990	0.9986	0.9982	0.9980	0.9976
78.00			0.9995	0.9994	0.9992	0.9990	0.9986	0.9982	0.9980	0.9976
78.05			0.9995	0.9994	0.9992	0.9990	0.9986	0.9982	0.9980	0.9976
78.10			0.9995	0.9994	0.9992	0.9990	0.9986	0.9982	0.9980	0.9976
78.15			0.9995	0.9994	0.9992	0.9990	0.9986	0.9982	0.9980	0.9976
78.20			0.9995	0.9994	0.9992	0.9990	0.9986	0.9982	0.9980	0.9976
78.25			0.9995	0.9994	0.9992	0.9990	0.9986	0.9982	0.9980	0.9976
78.30			0.9995	0.9994	0.9992	0.9990	0.9986	0.9982	0.9980	0.9976
78.35			0.9995	0.9994	0.9992	0.9990	0.9986	0.9982	0.9980	0.9976
78.40			0.9995	0.9994	0.9992	0.9990	0.9986	0.9982	0.9980	0.9976
78.45			0.9995	0.9994	0.9992	0.9990	0.9986	0.9982	0.9980	0.9976
78.50			0.9995	0.9994	0.9992	0.9990	0.9986	0.9982	0.9980	0.9976
78.55			0.9995	0.9994	0.9992	0.9990	0.9986	0.9982	0.9980	0.9976
78.60			0.9995	0.9994	0.9992	0.9990	0.9986	0.9982	0.9980	0.9976
78.65			0.9995	0.9994	0.9992	0.9990	0.9986	0.9982	0.9980	0.9976
78.70			0.9995	0.9994	0.9992	0.9990	0.9986	0.9982	0.9980	0.9976
78.75			0.9995	0.9994	0.9992	0.9990	0.9986	0.9982	0.9980	0.9976
78.80			0.9995	0.9994	0.9992	0.9990	0.9986	0.9982	0.9980	0.9976
78.85			0.9995	0.9994	0.9992	0.9990	0.9986	0.9982	0.9980	0.9976
78.90			0.9995	0.9994	0.9992	0.9990	0.9986	0.9982	0.9980	0.9976
78.95			0.9995	0.9994	0.9992	0.9990	0.9986	0.9982	0.9980	0.9976
79.00			0.9995	0.9994	0.9992	0.9990	0.9986	0.9982	0.9980	0.9976
79.05			0.9995	0.9994	0.9992	0.9990	0.9986	0.9982	0.9980	0.9976
79.10			0.9995	0.9994	0.9992	0.9990	0.9986	0.9982	0.9980	0.9976
79.15			0.9995	0.9994	0.9992	0.9990	0.9986	0.9982	0.9980	0.9976
79.20			0.9995	0.9994	0.9992	0.9990	0.9986	0.9982	0.9980	0.9976
79.25			0.9995	0.9994	0.9992	0.9990	0.9986	0.9982	0.9980	0.9976
79.30			0.9995	0.9994	0.9992	0.9990	0.9986	0.9982	0.9980	0.9976
79.35			0.9995	0.9994	0.9992	0.9990	0.9986	0.9982	0.9980	0.9976
79.40			0.9995	0.9994	0.9992	0.9990	0.9986	0.9982	0.9980	0.9977
79.45			0.9995	0.9994	0.9992	0.9990	0.9986	0.9983	0.9981	0.9977

PROBABILITY INTEGRAL OF t, THE NON-CENTRAL t-STATISTIC. THIS TABLE GIVES Pr $[t/\sqrt{f} \leq x]$.

f is the number of degrees of freedom; the non-centrality parameter is $\sqrt{f+1}\,K_p$.

K_p is the standardized normal deviate exceeded with probability p.

DEGREES OF FREEDOM **2**

x \ p	.2500	.1500	.1000	.0650	.0400	.0250	.0100	.0040	.0025	.0010
79.50			0.9995	0.9994	0.9992	0.9990	0.9986	0.9983	0.9981	0.9977
79.55			0.9995	0.9994	0.9992	0.9990	0.9986	0.9983	0.9981	0.9977
79.60			0.9995	0.9994	0.9992	0.9990	0.9986	0.9983	0.9981	0.9977
79.65			0.9995	0.9994	0.9992	0.9990	0.9986	0.9983	0.9981	0.9977
79.70			0.9995	0.9994	0.9992	0.9990	0.9986	0.9983	0.9981	0.9977
79.75				0.9994	0.9992	0.9990	0.9986	0.9983	0.9981	0.9977
79.80				0.9994	0.9992	0.9990	0.9986	0.9983	0.9981	0.9977
79.85				0.9994	0.9992	0.9990	0.9986	0.9983	0.9981	0.9977
79.90				0.9994	0.9992	0.9990	0.9987	0.9983	0.9981	0.9977
79.95				0.9994	0.9992	0.9990	0.9987	0.9983	0.9981	0.9977
80.00				0.9994	0.9992	0.9990	0.9987	0.9983	0.9981	0.9977
80.05				0.9994	0.9992	0.9990	0.9987	0.9983	0.9981	0.9977
80.10				0.9994	0.9992	0.9990	0.9987	0.9983	0.9981	0.9977
80.15				0.9994	0.9992	0.9990	0.9987	0.9983	0.9981	0.9977
80.20				0.9994	0.9992	0.9990	0.9987	0.9983	0.9981	0.9977
80.25				0.9994	0.9992	0.9990	0.9987	0.9983	0.9981	0.9977
80.30				0.9994	0.9992	0.9990	0.9987	0.9983	0.9981	0.9977
80.35				0.9994	0.9992	0.9990	0.9987	0.9983	0.9981	0.9977
80.40				0.9994	0.9992	0.9990	0.9987	0.9983	0.9981	0.9977
80.45				0.9994	0.9992	0.9990	0.9987	0.9983	0.9981	0.9977
80.50				0.9994	0.9992	0.9990	0.9987	0.9983	0.9981	0.9977
80.55				0.9994	0.9992	0.9990	0.9987	0.9983	0.9981	0.9977
80.60				0.9994	0.9992	0.9990	0.9987	0.9983	0.9981	0.9977
80.65				0.9994	0.9992	0.9990	0.9987	0.9983	0.9981	0.9977
80.70				0.9994	0.9992	0.9990	0.9987	0.9983	0.9981	0.9977
80.75				0.9994	0.9992	0.9990	0.9987	0.9983	0.9981	0.9977
80.80				0.9994	0.9992	0.9990	0.9987	0.9983	0.9981	0.9977
80.85				0.9994	0.9992	0.9990	0.9987	0.9983	0.9981	0.9977
80.90				0.9994	0.9992	0.9990	0.9987	0.9983	0.9981	0.9977
80.95				0.9994	0.9992	0.9990	0.9987	0.9983	0.9981	0.9977
81.00				0.9994	0.9992	0.9990	0.9987	0.9983	0.9981	0.9977
81.05				0.9994	0.9992	0.9990	0.9987	0.9983	0.9981	0.9977
81.10				0.9994	0.9992	0.9990	0.9987	0.9983	0.9981	0.9977
81.15				0.9994	0.9992	0.9990	0.9987	0.9983	0.9981	0.9978
81.20				0.9994	0.9992	0.9991	0.9987	0.9983	0.9981	0.9978
81.25				0.9994	0.9992	0.9991	0.9987	0.9983	0.9981	0.9978
81.30				0.9994	0.9992	0.9991	0.9987	0.9983	0.9981	0.9978
81.35				0.9994	0.9992	0.9991	0.9987	0.9983	0.9981	0.9978
81.40				0.9994	0.9992	0.9991	0.9987	0.9983	0.9981	0.9978
81.45				0.9994	0.9992	0.9991	0.9987	0.9983	0.9981	0.9978
81.50				0.9994	0.9992	0.9991	0.9987	0.9983	0.9981	0.9978
81.55				0.9994	0.9992	0.9991	0.9987	0.9983	0.9981	0.9978
81.60				0.9994	0.9992	0.9991	0.9987	0.9983	0.9982	0.9978
81.65				0.9994	0.9992	0.9991	0.9987	0.9983	0.9982	0.9978
81.70				0.9994	0.9992	0.9991	0.9987	0.9983	0.9982	0.9978
81.75				0.9994	0.9992	0.9991	0.9987	0.9983	0.9982	0.9978
81.80				0.9994	0.9992	0.9991	0.9987	0.9984	0.9982	0.9978
81.85				0.9994	0.9992	0.9991	0.9987	0.9984	0.9982	0.9978
81.90				0.9994	0.9992	0.9991	0.9987	0.9984	0.9982	0.9978
81.95				0.9994	0.9992	0.9991	0.9987	0.9984	0.9982	0.9978
82.00				0.9994	0.9992	0.9991	0.9987	0.9984	0.9982	0.9978
82.05				0.9994	0.9992	0.9991	0.9987	0.9984	0.9982	0.9978
82.10				0.9994	0.9992	0.9991	0.9987	0.9984	0.9982	0.9978
82.15				0.9994	0.9992	0.9991	0.9987	0.9984	0.9982	0.9978
82.20				0.9994	0.9992	0.9991	0.9987	0.9984	0.9982	0.9978
82.25				0.9994	0.9992	0.9991	0.9987	0.9984	0.9982	0.9978
82.30				0.9994	0.9992	0.9991	0.9987	0.9984	0.9982	0.9978
82.35				0.9994	0.9992	0.9991	0.9987	0.9984	0.9982	0.9978
82.40				0.9994	0.9992	0.9991	0.9987	0.9984	0.9982	0.9978
82.45				0.9994	0.9993	0.9991	0.9987	0.9984	0.9982	0.9978

PROBABILITY INTEGRAL OF t, THE NON-CENTRAL t-STATISTIC. THIS TABLE GIVES Pr $[t/\sqrt{f} \leq x]$.

f is the number of degrees of freedom; the non-centrality parameter is $\sqrt{f+1}\,K_p$.
K_p is the standardized normal deviate exceeded with probability p.

DEGREES OF FREEDOM **2**

x \ p	.2500	.1500	.1000	.0650	.0400	.0250	.0100	.0040	.0025	.0010
82.50				0.9994	0.9993	0.9991	0.9987	0.9984	0.9982	0.9978
82.55				0.9994	0.9993	0.9991	0.9987	0.9984	0.9982	0.9978
82.60				0.9994	0.9993	0.9991	0.9987	0.9984	0.9982	0.9978
82.65				0.9994	0.9993	0.9991	0.9987	0.9984	0.9982	0.9978
82.70				0.9994	0.9993	0.9991	0.9987	0.9984	0.9982	0.9978
82.75				0.9994	0.9993	0.9991	0.9987	0.9984	0.9982	0.9978
82.80				0.9994	0.9993	0.9991	0.9987	0.9984	0.9982	0.9978
82.85				0.9994	0.9993	0.9991	0.9987	0.9984	0.9982	0.9978
82.90				0.9994	0.9993	0.9991	0.9987	0.9984	0.9982	0.9978
82.95				0.9994	0.9993	0.9991	0.9987	0.9984	0.9982	0.9978
83.00				0.9994	0.9993	0.9991	0.9987	0.9984	0.9982	0.9979
83.05				0.9994	0.9993	0.9991	0.9988	0.9984	0.9982	0.9979
83.10				0.9994	0.9993	0.9991	0.9988	0.9984	0.9982	0.9979
83.15				0.9994	0.9993	0.9991	0.9988	0.9984	0.9982	0.9979
83.20				0.9994	0.9993	0.9991	0.9988	0.9984	0.9982	0.9979
83.25				0.9994	0.9993	0.9991	0.9988	0.9984	0.9982	0.9979
83.30				0.9994	0.9993	0.9991	0.9988	0.9984	0.9982	0.9979
83.35				0.9994	0.9993	0.9991	0.9988	0.9984	0.9982	0.9979
83.40				0.9994	0.9993	0.9991	0.9988	0.9984	0.9982	0.9979
83.45				0.9994	0.9993	0.9991	0.9988	0.9984	0.9982	0.9979
83.50				0.9994	0.9993	0.9991	0.9988	0.9984	0.9982	0.9979
83.55				0.9994	0.9993	0.9991	0.9988	0.9984	0.9982	0.9979
83.60				0.9994	0.9993	0.9991	0.9988	0.9984	0.9982	0.9979
83.65				0.9994	0.9993	0.9991	0.9988	0.9984	0.9982	0.9979
83.70				0.9994	0.9993	0.9991	0.9988	0.9984	0.9982	0.9979
83.75				0.9994	0.9993	0.9991	0.9988	0.9984	0.9982	0.9979
83.80				0.9994	0.9993	0.9991	0.9988	0.9984	0.9982	0.9979
83.85				0.9994	0.9993	0.9991	0.9988	0.9984	0.9982	0.9979
83.90				0.9994	0.9993	0.9991	0.9988	0.9984	0.9983	0.9979
83.95				0.9994	0.9993	0.9991	0.9988	0.9984	0.9983	0.9979
84.00				0.9994	0.9993	0.9991	0.9988	0.9984	0.9983	0.9979
84.05				0.9994	0.9993	0.9991	0.9988	0.9984	0.9983	0.9979
84.10				0.9994	0.9993	0.9991	0.9988	0.9984	0.9983	0.9979
84.15				0.9994	0.9993	0.9991	0.9988	0.9984	0.9983	0.9979
84.20				0.9994	0.9993	0.9991	0.9988	0.9984	0.9983	0.9979
84.25				0.9994	0.9993	0.9991	0.9988	0.9984	0.9983	0.9979
84.30				0.9994	0.9993	0.9991	0.9988	0.9984	0.9983	0.9979
84.35				0.9994	0.9993	0.9991	0.9988	0.9984	0.9983	0.9979
84.40				0.9994	0.9993	0.9991	0.9988	0.9985	0.9983	0.9979
84.45				0.9994	0.9993	0.9991	0.9988	0.9985	0.9983	0.9979
84.50				0.9994	0.9993	0.9991	0.9988	0.9985	0.9983	0.9979
84.55				0.9994	0.9993	0.9991	0.9988	0.9985	0.9983	0.9979
84.60				0.9994	0.9993	0.9991	0.9988	0.9985	0.9983	0.9979
84.65				0.9995	0.9993	0.9991	0.9988	0.9985	0.9983	0.9979
84.70				0.9995	0.9993	0.9991	0.9988	0.9985	0.9983	0.9979
84.75				0.9995	0.9993	0.9991	0.9988	0.9985	0.9983	0.9979
84.80				0.9995	0.9993	0.9991	0.9988	0.9985	0.9983	0.9979
84.85				0.9995	0.9993	0.9991	0.9988	0.9985	0.9983	0.9979
84.90				0.9995	0.9993	0.9991	0.9988	0.9985	0.9983	0.9979
84.95				0.9995	0.9993	0.9991	0.9988	0.9985	0.9983	0.9979
85.00				0.9995	0.9993	0.9991	0.9988	0.9985	0.9983	0.9980
85.05				0.9995	0.9993	0.9991	0.9988	0.9985	0.9983	0.9980
85.10				0.9995	0.9993	0.9991	0.9988	0.9985	0.9983	0.9980
85.15				0.9995	0.9993	0.9991	0.9988	0.9985	0.9983	0.9980
85.20				0.9995	0.9993	0.9991	0.9988	0.9985	0.9983	0.9980
85.25				0.9995	0.9993	0.9991	0.9988	0.9985	0.9983	0.9980
85.30				0.9995	0.9993	0.9991	0.9988	0.9985	0.9983	0.9980
85.35				0.9995	0.9993	0.9991	0.9988	0.9985	0.9983	0.9980
85.40				0.9995	0.9993	0.9991	0.9988	0.9985	0.9983	0.9980
85.45				0.9995	0.9993	0.9991	0.9988	0.9985	0.9983	0.9980

PROBABILITY INTEGRAL OF t, THE NON-CENTRAL t-STATISTIC. THIS TABLE GIVES Pr $[t/\sqrt{f} \leq x]$.

f is the number of degrees of freedom; the non-centrality parameter is $\sqrt{f+1}\,K_p$.

K_p is the standardized normal deviate exceeded with probability p.

DEGREES OF FREEDOM **2**

x \ p	.2500	.1500	.1000	.0650	.0400	.0250	.0100	.0040	.0025	.0010
85.50				0.9995	0.9993	0.9991	0.9988	0.9985	0.9983	0.9980
85.55				0.9995	0.9993	0.9991	0.9988	0.9985	0.9983	0.9980
85.60				0.9995	0.9993	0.9991	0.9988	0.9985	0.9983	0.9980
85.65				0.9995	0.9993	0.9991	0.9988	0.9985	0.9983	0.9980
85.70				0.9995	0.9993	0.9991	0.9988	0.9985	0.9983	0.9980
85.75				0.9995	0.9993	0.9991	0.9988	0.9985	0.9983	0.9980
85.80				0.9995	0.9993	0.9991	0.9988	0.9985	0.9983	0.9980
85.85				0.9995	0.9993	0.9992	0.9988	0.9985	0.9983	0.9980
85.90				0.9995	0.9993	0.9992	0.9988	0.9985	0.9983	0.9980
85.95				0.9995	0.9993	0.9992	0.9988	0.9985	0.9983	0.9980
86.00				0.9995	0.9993	0.9992	0.9988	0.9985	0.9983	0.9980
86.05				0.9995	0.9993	0.9992	0.9988	0.9985	0.9983	0.9980
86.10				0.9995	0.9993	0.9992	0.9988	0.9985	0.9983	0.9980
86.15				0.9995	0.9993	0.9992	0.9988	0.9985	0.9983	0.9980
86.20				0.9995	0.9993	0.9992	0.9988	0.9985	0.9983	0.9980
86.25				0.9995	0.9993	0.9992	0.9988	0.9985	0.9983	0.9980
86.30				0.9995	0.9993	0.9992	0.9988	0.9985	0.9983	0.9980
86.35				0.9995	0.9993	0.9992	0.9988	0.9985	0.9983	0.9980
86.40				0.9995	0.9993	0.9992	0.9988	0.9985	0.9984	0.9980
86.45				0.9995	0.9993	0.9992	0.9988	0.9985	0.9984	0.9980
86.50				0.9995	0.9993	0.9992	0.9988	0.9985	0.9984	0.9980
86.55				0.9995	0.9993	0.9992	0.9989	0.9985	0.9984	0.9980
86.60				0.9995	0.9993	0.9992	0.9989	0.9985	0.9984	0.9980
86.65				0.9995	0.9993	0.9992	0.9989	0.9985	0.9984	0.9980
86.70				0.9995	0.9993	0.9992	0.9989	0.9985	0.9984	0.9980
86.75				0.9995	0.9993	0.9992	0.9989	0.9985	0.9984	0.9980
86.80				0.9995	0.9993	0.9992	0.9989	0.9985	0.9984	0.9980
86.85				0.9995	0.9993	0.9992	0.9989	0.9985	0.9984	0.9980
86.90				0.9995	0.9993	0.9992	0.9989	0.9985	0.9984	0.9980
86.95				0.9995	0.9993	0.9992	0.9989	0.9985	0.9984	0.9980
87.00				0.9995	0.9993	0.9992	0.9989	0.9985	0.9984	0.9980
87.05				0.9995	0.9993	0.9992	0.9989	0.9985	0.9984	0.9980
87.10				0.9995	0.9993	0.9992	0.9989	0.9985	0.9984	0.9980
87.15				0.9995	0.9993	0.9992	0.9989	0.9985	0.9984	0.9980
87.20				0.9995	0.9993	0.9992	0.9989	0.9985	0.9984	0.9981
87.25				0.9995	0.9993	0.9992	0.9989	0.9985	0.9984	0.9981
87.30				0.9995	0.9993	0.9992	0.9989	0.9986	0.9984	0.9981
87.35				0.9995	0.9993	0.9992	0.9989	0.9986	0.9984	0.9981
87.40				0.9995	0.9993	0.9992	0.9989	0.9986	0.9984	0.9981
87.45				0.9995	0.9993	0.9992	0.9989	0.9986	0.9984	0.9981
87.50				0.9995	0.9993	0.9992	0.9989	0.9986	0.9984	0.9981
87.55				0.9995	0.9993	0.9992	0.9989	0.9986	0.9984	0.9981
87.60				0.9995	0.9993	0.9992	0.9989	0.9986	0.9984	0.9981
87.65				0.9995	0.9993	0.9992	0.9989	0.9986	0.9984	0.9981
87.70				0.9995	0.9993	0.9992	0.9989	0.9986	0.9984	0.9981
87.75				0.9995	0.9993	0.9992	0.9989	0.9986	0.9984	0.9981
87.80				0.9995	0.9993	0.9992	0.9989	0.9986	0.9984	0.9981
87.85				0.9995	0.9993	0.9992	0.9989	0.9986	0.9984	0.9981
87.90				0.9995	0.9993	0.9992	0.9989	0.9986	0.9984	0.9981
87.95				0.9995	0.9993	0.9992	0.9989	0.9986	0.9984	0.9981
88.00				0.9995	0.9993	0.9992	0.9989	0.9986	0.9984	0.9981
88.05				0.9995	0.9993	0.9992	0.9989	0.9986	0.9984	0.9981
88.10				0.9995	0.9993	0.9992	0.9989	0.9986	0.9984	0.9981
88.15				0.9995	0.9993	0.9992	0.9989	0.9986	0.9984	0.9981
88.20				0.9995	0.9993	0.9992	0.9989	0.9986	0.9984	0.9981
88.25				0.9995	0.9993	0.9992	0.9989	0.9986	0.9984	0.9981
88.30				0.9995	0.9993	0.9992	0.9989	0.9986	0.9984	0.9981
88.35				0.9995	0.9993	0.9992	0.9989	0.9986	0.9984	0.9981
88.40				0.9995	0.9993	0.9992	0.9989	0.9986	0.9984	0.9981
88.45				0.9995	0.9993	0.9992	0.9989	0.9986	0.9984	0.9981

PROBABILITY INTEGRAL OF t, THE NON-CENTRAL t-STATISTIC. THIS TABLE GIVES Pr $[t/\sqrt{f} \leq x]$.

f is the number of degrees of freedom; the non-centrality parameter is $\sqrt{f+1}\,K_p$.

K_p is the standardized normal deviate exceeded with probability p.

DEGREES OF FREEDOM **2**

x \ p	.2500	.1500	.1000	.0650	.0400	.0250	.0100	.0040	.0025	.0010
88.50				0.9995	0.9993	0.9992	0.9989	0.9986	0.9984	0.9981
88.55				0.9995	0.9994	0.9992	0.9989	0.9986	0.9984	0.9981
88.60				0.9995	0.9994	0.9992	0.9989	0.9986	0.9984	0.9981
88.65				0.9995	0.9994	0.9992	0.9989	0.9986	0.9984	0.9981
88.70				0.9995	0.9994	0.9992	0.9989	0.9986	0.9984	0.9981
88.75				0.9995	0.9994	0.9992	0.9989	0.9986	0.9984	0.9981
88.80				0.9995	0.9994	0.9992	0.9989	0.9986	0.9984	0.9981
88.85				0.9995	0.9994	0.9992	0.9989	0.9986	0.9984	0.9981
88.90				0.9995	0.9994	0.9992	0.9989	0.9986	0.9984	0.9981
88.95					0.9994	0.9992	0.9989	0.9986	0.9984	0.9981
89.00					0.9994	0.9992	0.9989	0.9986	0.9984	0.9981
89.05					0.9994	0.9992	0.9989	0.9986	0.9984	0.9981
89.10					0.9994	0.9992	0.9989	0.9986	0.9985	0.9981
89.15					0.9994	0.9992	0.9989	0.9986	0.9985	0.9981
89.20					0.9994	0.9992	0.9989	0.9986	0.9985	0.9981
89.25					0.9994	0.9992	0.9989	0.9986	0.9985	0.9981
89.30					0.9994	0.9992	0.9989	0.9986	0.9985	0.9981
89.35					0.9994	0.9992	0.9989	0.9986	0.9985	0.9981
89.40					0.9994	0.9992	0.9989	0.9986	0.9985	0.9981
89.45					0.9994	0.9992	0.9989	0.9986	0.9985	0.9981
89.50					0.9994	0.9992	0.9989	0.9986	0.9985	0.9982
89.55					0.9994	0.9992	0.9989	0.9986	0.9985	0.9982
89.60					0.9994	0.9992	0.9989	0.9986	0.9985	0.9982
89.65					0.9994	0.9992	0.9989	0.9986	0.9985	0.9982
89.70					0.9994	0.9992	0.9989	0.9986	0.9985	0.9982
89.75					0.9994	0.9992	0.9989	0.9986	0.9985	0.9982
89.80					0.9994	0.9992	0.9989	0.9986	0.9985	0.9982
89.85					0.9994	0.9992	0.9989	0.9986	0.9985	0.9982
89.90					0.9994	0.9992	0.9989	0.9986	0.9985	0.9982
89.95					0.9994	0.9992	0.9989	0.9986	0.9985	0.9982
90.00					0.9994	0.9992	0.9989	0.9986	0.9985	0.9982
90.05					0.9994	0.9992	0.9989	0.9986	0.9985	0.9982
90.10					0.9994	0.9992	0.9989	0.9986	0.9985	0.9982
90.15					0.9994	0.9992	0.9989	0.9986	0.9985	0.9982
90.20					0.9994	0.9992	0.9989	0.9986	0.9985	0.9982
90.25					0.9994	0.9992	0.9989	0.9986	0.9985	0.9982
90.30					0.9994	0.9992	0.9989	0.9986	0.9985	0.9982
90.35					0.9994	0.9992	0.9989	0.9986	0.9985	0.9982
90.40					0.9994	0.9992	0.9989	0.9986	0.9985	0.9982
90.45					0.9994	0.9992	0.9989	0.9986	0.9985	0.9982
90.50					0.9994	0.9992	0.9989	0.9987	0.9985	0.9982
90.55					0.9994	0.9992	0.9990	0.9987	0.9985	0.9982
90.60					0.9994	0.9992	0.9990	0.9987	0.9985	0.9982
90.65					0.9994	0.9992	0.9990	0.9987	0.9985	0.9982
90.70					0.9994	0.9992	0.9990	0.9987	0.9985	0.9982
90.75					0.9994	0.9992	0.9990	0.9987	0.9985	0.9982
90.80					0.9994	0.9992	0.9990	0.9987	0.9985	0.9982
90.85					0.9994	0.9992	0.9990	0.9987	0.9985	0.9982
90.90					0.9994	0.9992	0.9990	0.9987	0.9985	0.9982
90.95					0.9994	0.9992	0.9990	0.9987	0.9985	0.9982
91.00					0.9994	0.9992	0.9990	0.9987	0.9985	0.9982
91.05					0.9994	0.9992	0.9990	0.9987	0.9985	0.9982
91.10					0.9994	0.9992	0.9990	0.9987	0.9985	0.9982
91.15					0.9994	0.9992	0.9990	0.9987	0.9985	0.9982
91.20					0.9994	0.9992	0.9990	0.9987	0.9985	0.9982
91.25					0.9994	0.9992	0.9990	0.9987	0.9985	0.9982
91.30					0.9994	0.9992	0.9990	0.9987	0.9985	0.9982
91.35					0.9994	0.9992	0.9990	0.9987	0.9985	0.9982
91.40					0.9994	0.9993	0.9990	0.9987	0.9985	0.9982
91.45					0.9994	0.9993	0.9990	0.9987	0.9985	0.9982

PROBABILITY INTEGRAL OF t, THE NON-CENTRAL t-STATISTIC. THIS TABLE GIVES Pr $[t/\sqrt{f} \leq x]$.

f is the number of degrees of freedom; the non-centrality parameter is $\sqrt{f+1}\,K_p$.

K_p is the standardized normal deviate exceeded with probability p.

DEGREES OF FREEDOM **2**

x \ p	.2500	.1500	.1000	.0650	.0400	.0250	.0100	.0040	.0025	.0010
91.50					0.9994	0.9993	0.9990	0.9987	0.9985	0.9982
91.55					0.9994	0.9993	0.9990	0.9987	0.9985	0.9982
91.60					0.9994	0.9993	0.9990	0.9987	0.9985	0.9982
91.65					0.9994	0.9993	0.9990	0.9987	0.9985	0.9982
91.70					0.9994	0.9993	0.9990	0.9987	0.9985	0.9982
91.75					0.9994	0.9993	0.9990	0.9987	0.9985	0.9982
91.80					0.9994	0.9993	0.9990	0.9987	0.9985	0.9982
91.85					0.9994	0.9993	0.9990	0.9987	0.9985	0.9982
91.90					0.9994	0.9993	0.9990	0.9987	0.9985	0.9982
91.95					0.9994	0.9993	0.9990	0.9987	0.9985	0.9982
92.00					0.9994	0.9993	0.9990	0.9987	0.9985	0.9983
92.05					0.9994	0.9993	0.9990	0.9987	0.9985	0.9983
92.10					0.9994	0.9993	0.9990	0.9987	0.9985	0.9983
92.15					0.9994	0.9993	0.9990	0.9987	0.9985	0.9983
92.20					0.9994	0.9993	0.9990	0.9987	0.9986	0.9983
92.25					0.9994	0.9993	0.9990	0.9987	0.9986	0.9983
92.30					0.9994	0.9993	0.9990	0.9987	0.9986	0.9983
92.35					0.9994	0.9993	0.9990	0.9987	0.9986	0.9983
92.40					0.9994	0.9993	0.9990	0.9987	0.9986	0.9983
92.45					0.9994	0.9993	0.9990	0.9987	0.9986	0.9983
92.50					0.9994	0.9993	0.9990	0.9987	0.9986	0.9983
92.55					0.9994	0.9993	0.9990	0.9987	0.9986	0.9983
92.60					0.9994	0.9993	0.9990	0.9987	0.9986	0.9983
92.65					0.9994	0.9993	0.9990	0.9987	0.9986	0.9983
92.70					0.9994	0.9993	0.9990	0.9987	0.9986	0.9983
92.75					0.9994	0.9993	0.9990	0.9987	0.9986	0.9983
92.80					0.9994	0.9993	0.9990	0.9987	0.9986	0.9983
92.85					0.9994	0.9993	0.9990	0.9987	0.9986	0.9983
92.90					0.9994	0.9993	0.9990	0.9987	0.9986	0.9983
92.95					0.9994	0.9993	0.9990	0.9987	0.9986	0.9983
93.00					0.9994	0.9993	0.9990	0.9987	0.9986	0.9983
93.05					0.9994	0.9993	0.9990	0.9987	0.9986	0.9983
93.10					0.9994	0.9993	0.9990	0.9987	0.9986	0.9983
93.15					0.9994	0.9993	0.9990	0.9987	0.9986	0.9983
93.20					0.9994	0.9993	0.9990	0.9987	0.9986	0.9983
93.25					0.9994	0.9993	0.9990	0.9987	0.9986	0.9983
93.30					0.9994	0.9993	0.9990	0.9987	0.9986	0.9983
93.35					0.9994	0.9993	0.9990	0.9987	0.9986	0.9983
93.40					0.9994	0.9993	0.9990	0.9987	0.9986	0.9983
93.45					0.9994	0.9993	0.9990	0.9987	0.9986	0.9983
93.50					0.9994	0.9993	0.9990	0.9987	0.9986	0.9983
93.55					0.9994	0.9993	0.9990	0.9987	0.9986	0.9983
93.60					0.9994	0.9993	0.9990	0.9987	0.9986	0.9983
93.65					0.9994	0.9993	0.9990	0.9987	0.9986	0.9983
93.70					0.9994	0.9993	0.9990	0.9987	0.9986	0.9983
93.75					0.9994	0.9993	0.9990	0.9987	0.9986	0.9983
93.80					0.9994	0.9993	0.9990	0.9987	0.9986	0.9983
93.85					0.9994	0.9993	0.9990	0.9987	0.9986	0.9983
93.90					0.9994	0.9993	0.9990	0.9987	0.9986	0.9983
93.95					0.9994	0.9993	0.9990	0.9987	0.9986	0.9983
94.00					0.9994	0.9993	0.9990	0.9988	0.9986	0.9983
94.05					0.9994	0.9993	0.9990	0.9988	0.9986	0.9983
94.10					0.9994	0.9993	0.9990	0.9988	0.9986	0.9983
94.15					0.9994	0.9993	0.9990	0.9988	0.9986	0.9983
94.20					0.9994	0.9993	0.9990	0.9988	0.9986	0.9983
94.25					0.9994	0.9993	0.9990	0.9988	0.9986	0.9983
94.30					0.9994	0.9993	0.9990	0.9988	0.9986	0.9983
94.35					0.9994	0.9993	0.9990	0.9988	0.9986	0.9983
94.40					0.9994	0.9993	0.9990	0.9988	0.9986	0.9983
94.45					0.9994	0.9993	0.9990	0.9988	0.9986	0.9983

PROBABILITY INTEGRAL OF t, THE NON-CENTRAL t-STATISTIC. THIS TABLE GIVES Pr $[t/\sqrt{f} \leq x]$.

f is the number of degrees of freedom; the non-centrality parameter is $\sqrt{f+1}\,K_p$.

K_p is the standardized normal deviate exceeded with probability p.

DEGREES OF FREEDOM **2**

x \ p	.2500	.1500	.1000	.0650	.0400	.0250	.0100	.0040	.0025	.0010
94.50					0.9994	0.9993	0.9990	0.9988	0.9986	0.9983
94.55					0.9994	0.9993	0.9990	0.9988	0.9986	0.9983
94.60					0.9994	0.9993	0.9990	0.9988	0.9986	0.9983
94.65					0.9994	0.9993	0.9990	0.9988	0.9986	0.9983
94.70					0.9994	0.9993	0.9990	0.9988	0.9986	0.9983
94.75					0.9994	0.9993	0.9990	0.9988	0.9986	0.9983
94.80					0.9994	0.9993	0.9990	0.9988	0.9986	0.9984
94.85					0.9994	0.9993	0.9990	0.9988	0.9986	0.9984
94.90					0.9994	0.9993	0.9990	0.9988	0.9986	0.9984
94.95					0.9994	0.9993	0.9990	0.9988	0.9986	0.9984
95.00					0.9994	0.9993	0.9990	0.9988	0.9986	0.9984
95.05					0.9994	0.9993	0.9990	0.9988	0.9986	0.9984
95.10					0.9994	0.9993	0.9990	0.9988	0.9986	0.9984
95.15					0.9994	0.9993	0.9990	0.9988	0.9986	0.9984
95.20					0.9994	0.9993	0.9990	0.9988	0.9986	0.9984
95.25					0.9994	0.9993	0.9991	0.9988	0.9986	0.9984
95.30					0.9994	0.9993	0.9991	0.9988	0.9986	0.9984
95.35					0.9994	0.9993	0.9991	0.9988	0.9986	0.9984
95.40					0.9994	0.9993	0.9991	0.9988	0.9986	0.9984
95.45					0.9994	0.9993	0.9991	0.9988	0.9986	0.9984
95.50					0.9994	0.9993	0.9991	0.9988	0.9987	0.9984
95.55					0.9994	0.9993	0.9991	0.9988	0.9987	0.9984
95.60					0.9994	0.9993	0.9991	0.9988	0.9987	0.9984
95.65					0.9994	0.9993	0.9991	0.9988	0.9987	0.9984
95.70					0.9994	0.9993	0.9991	0.9988	0.9987	0.9984
95.75					0.9994	0.9993	0.9991	0.9988	0.9987	0.9984
95.80					0.9994	0.9993	0.9991	0.9988	0.9987	0.9984
95.85					0.9994	0.9993	0.9991	0.9988	0.9987	0.9984
95.90					0.9994	0.9993	0.9991	0.9988	0.9987	0.9984
95.95					0.9994	0.9993	0.9991	0.9988	0.9987	0.9984
96.00					0.9994	0.9993	0.9991	0.9988	0.9987	0.9984
96.05					0.9994	0.9993	0.9991	0.9988	0.9987	0.9984
96.10					0.9994	0.9993	0.9991	0.9988	0.9987	0.9984
96.15					0.9994	0.9993	0.9991	0.9988	0.9987	0.9984
96.20					0.9994	0.9993	0.9991	0.9988	0.9987	0.9984
96.25					0.9995	0.9993	0.9991	0.9988	0.9987	0.9984
96.30					0.9994	0.9993	0.9991	0.9988	0.9987	0.9984
96.35					0.9995	0.9993	0.9991	0.9988	0.9987	0.9984
96.40					0.9995	0.9993	0.9991	0.9988	0.9987	0.9984
96.45					0.9995	0.9993	0.9991	0.9988	0.9987	0.9984
96.50					0.9995	0.9993	0.9991	0.9988	0.9987	0.9984
96.55					0.9995	0.9993	0.9991	0.9988	0.9987	0.9984
96.60					0.9995	0.9993	0.9991	0.9988	0.9987	0.9984
96.65					0.9995	0.9993	0.9991	0.9988	0.9987	0.9984
96.70					0.9995	0.9993	0.9991	0.9988	0.9987	0.9984
96.75					0.9995	0.9993	0.9991	0.9988	0.9987	0.9984
96.80					0.9995	0.9993	0.9991	0.9988	0.9987	0.9984
96.85					0.9995	0.9993	0.9991	0.9988	0.9987	0.9984
96.90					0.9995	0.9993	0.9991	0.9988	0.9987	0.9984
96.95					0.9995	0.9993	0.9991	0.9988	0.9987	0.9984
97.00					0.9995	0.9993	0.9991	0.9988	0.9987	0.9984
97.05					0.9995	0.9993	0.9991	0.9988	0.9987	0.9984
97.10					0.9995	0.9993	0.9991	0.9988	0.9987	0.9984
97.15					0.9995	0.9993	0.9991	0.9988	0.9987	0.9984
97.20					0.9995	0.9993	0.9991	0.9988	0.9987	0.9984
97.25					0.9995	0.9993	0.9991	0.9988	0.9987	0.9984
97.30					0.9995	0.9993	0.9991	0.9988	0.9987	0.9984
97.35					0.9995	0.9993	0.9991	0.9988	0.9987	0.9984
97.40					0.9995	0.9993	0.9991	0.9988	0.9987	0.9984
97.45					0.9995	0.9993	0.9991	0.9988	0.9987	0.9984

PROBABILITY INTEGRAL OF t, THE NON-CENTRAL t-STATISTIC. THIS TABLE GIVES Pr $[t/\sqrt{f} \leq x]$.

f is the number of degrees of freedom; the non-centrality parameter is $\sqrt{f+1}\,K_p$.

K_p is the standardized normal deviate exceeded with probability p.

DEGREES OF FREEDOM 2

x \ P	.2500	.1500	.1000	.0650	.0400	.0250	.0100	.0040	.0025	.0010
97.50					0.9995	0.9993	0.9991	0.9988	0.9987	0.9984
97.55					0.9995	0.9993	0.9991	0.9988	0.9987	0.9984
97.60					0.9995	0.9993	0.9991	0.9988	0.9987	0.9984
97.65					0.9995	0.9993	0.9991	0.9988	0.9987	0.9984
97.70					0.9995	0.9993	0.9991	0.9988	0.9987	0.9984
97.75					0.9995	0.9993	0.9991	0.9988	0.9987	0.9984
97.80					0.9995	0.9993	0.9991	0.9988	0.9987	0.9985
97.85					0.9995	0.9993	0.9991	0.9988	0.9987	0.9985
97.90					0.9995	0.9993	0.9991	0.9988	0.9987	0.9985
97.95					0.9995	0.9993	0.9991	0.9988	0.9987	0.9985
98.00					0.9995	0.9993	0.9991	0.9989	0.9987	0.9985
98.05					0.9995	0.9993	0.9991	0.9989	0.9987	0.9985
98.10					0.9995	0.9993	0.9991	0.9989	0.9987	0.9985
98.15					0.9995	0.9994	0.9991	0.9989	0.9987	0.9985
98.20					0.9995	0.9994	0.9991	0.9989	0.9987	0.9985
98.25					0.9995	0.9994	0.9991	0.9989	0.9987	0.9985
98.30					0.9995	0.9994	0.9991	0.9989	0.9987	0.9985
98.35					0.9995	0.9994	0.9991	0.9989	0.9987	0.9985
98.40					0.9995	0.9994	0.9991	0.9989	0.9987	0.9985
98.45					0.9995	0.9994	0.9991	0.9989	0.9987	0.9985
98.50					0.9995	0.9994	0.9991	0.9989	0.9987	0.9985
98.55					0.9995	0.9994	0.9991	0.9989	0.9987	0.9985
98.60					0.9995	0.9994	0.9991	0.9989	0.9987	0.9985
98.65					0.9995	0.9994	0.9991	0.9989	0.9987	0.9985
98.70					0.9995	0.9994	0.9991	0.9989	0.9987	0.9985
98.75					0.9995	0.9994	0.9991	0.9989	0.9987	0.9985
98.80					0.9995	0.9994	0.9991	0.9989	0.9987	0.9985
98.85					0.9995	0.9994	0.9991	0.9989	0.9987	0.9985
98.90					0.9995	0.9994	0.9991	0.9989	0.9987	0.9985
98.95					0.9995	0.9994	0.9991	0.9989	0.9987	0.9985
99.00					0.9995	0.9994	0.9991	0.9989	0.9987	0.9985
99.05					0.9995	0.9994	0.9991	0.9989	0.9987	0.9985
99.10					0.9995	0.9994	0.9991	0.9989	0.9987	0.9985
99.15					0.9995	0.9994	0.9991	0.9989	0.9987	0.9985
99.20					0.9995	0.9994	0.9991	0.9989	0.9987	0.9985
99.25					0.9995	0.9994	0.9991	0.9989	0.9987	0.9985
99.30					0.9995	0.9994	0.9991	0.9989	0.9988	0.9985
99.35					0.9995	0.9994	0.9991	0.9989	0.9988	0.9985
99.40					0.9995	0.9994	0.9991	0.9989	0.9988	0.9985
99.45					0.9995	0.9994	0.9991	0.9989	0.9988	0.9985
99.50					0.9995	0.9994	0.9991	0.9989	0.9988	0.9985
99.55					0.9995	0.9994	0.9991	0.9989	0.9988	0.9985
99.60					0.9995	0.9994	0.9991	0.9989	0.9988	0.9985
99.65					0.9995	0.9994	0.9991	0.9989	0.9988	0.9985
99.70					0.9995	0.9994	0.9991	0.9989	0.9988	0.9985
99.75					0.9995	0.9994	0.9991	0.9989	0.9988	0.9985
99.80					0.9995	0.9994	0.9991	0.9989	0.9988	0.9985
99.85					0.9995	0.9994	0.9991	0.9989	0.9988	0.9985
99.90					0.9995	0.9994	0.9991	0.9989	0.9988	0.9985
99.95					0.9995	0.9994	0.9991	0.9989	0.9988	0.9985
100.00					0.9995	0.9994	0.9991	0.9989	0.9988	0.9985
100.05					0.9995	0.9994	0.9991	0.9989	0.9988	0.9985
100.10					0.9995	0.9994	0.9991	0.9989	0.9988	0.9985
100.15					0.9995	0.9994	0.9991	0.9989	0.9988	0.9985
100.20					0.9995	0.9994	0.9991	0.9989	0.9988	0.9985
100.25					0.9995	0.9994	0.9991	0.9989	0.9988	0.9985
100.30					0.9995	0.9994	0.9991	0.9989	0.9988	0.9985
100.35					0.9995	0.9994	0.9991	0.9989	0.9988	0.9985
100.40					0.9995	0.9994	0.9991	0.9989	0.9988	0.9985
100.45					0.9995	0.9994	0.9991	0.9989	0.9988	0.9985

PROBABILITY INTEGRAL OF t, THE NON-CENTRAL t-STATISTIC. THIS TABLE GIVES Pr $[t/\sqrt{f} \leq x]$.

f is the number of degrees of freedom; the non-centrality parameter is $\sqrt{f+1}\, K_p$.

K_p is the standardized normal deviate exceeded with probability p.

DEGREES OF FREEDOM **2**

x \ p	.2500	.1500	.1000	.0650	.0400	.0250	.0100	.0040	.0025	.0010
100.50					0.9995	0.9994	0.9991	0.9989	0.9988	0.9985
100.55					0.9995	0.9994	0.9991	0.9989	0.9988	0.9985
100.60					0.9995	0.9994	0.9991	0.9989	0.9988	0.9985
100.65					0.9995	0.9994	0.9992	0.9989	0.9988	0.9985
100.70					0.9995	0.9994	0.9992	0.9989	0.9988	0.9985
100.75					0.9995	0.9994	0.9992	0.9989	0.9988	0.9985
100.80					0.9995	0.9994	0.9992	0.9989	0.9988	0.9985
100.85					0.9995	0.9994	0.9992	0.9989	0.9988	0.9985
100.90					0.9995	0.9994	0.9992	0.9989	0.9988	0.9985
100.95					0.9995	0.9994	0.9992	0.9989	0.9988	0.9985
101.00					0.9995	0.9994	0.9992	0.9989	0.9988	0.9985
101.05					0.9995	0.9994	0.9992	0.9989	0.9988	0.9985
101.10					0.9995	0.9994	0.9992	0.9989	0.9988	0.9986
101.15					0.9995	0.9994	0.9992	0.9989	0.9988	0.9986
101.20					0.9995	0.9994	0.9992	0.9989	0.9988	0.9986
101.25						0.9994	0.9992	0.9989	0.9988	0.9986
101.30						0.9994	0.9992	0.9989	0.9988	0.9986
101.35						0.9994	0.9992	0.9989	0.9988	0.9986
101.40						0.9994	0.9992	0.9989	0.9988	0.9986
101.45						0.9994	0.9992	0.9989	0.9988	0.9986
101.50						0.9994	0.9992	0.9989	0.9988	0.9986
101.55						0.9994	0.9992	0.9989	0.9988	0.9986
101.60						0.9994	0.9992	0.9989	0.9988	0.9986
101.65						0.9994	0.9992	0.9989	0.9988	0.9986
101.70						0.9994	0.9992	0.9989	0.9988	0.9986
101.75						0.9994	0.9992	0.9989	0.9988	0.9986
101.80						0.9994	0.9992	0.9989	0.9988	0.9986
101.85						0.9994	0.9992	0.9989	0.9988	0.9986
101.90						0.9994	0.9992	0.9989	0.9988	0.9986
101.95						0.9994	0.9992	0.9989	0.9988	0.9986
102.00						0.9994	0.9992	0.9989	0.9988	0.9986
102.05						0.9994	0.9992	0.9989	0.9988	0.9986
102.10						0.9994	0.9992	0.9989	0.9988	0.9986
102.15						0.9994	0.9992	0.9989	0.9988	0.9986
102.20						0.9994	0.9992	0.9989	0.9988	0.9986
102.25						0.9994	0.9992	0.9989	0.9988	0.9986
102.30						0.9994	0.9992	0.9989	0.9988	0.9986
102.35						0.9994	0.9992	0.9989	0.9988	0.9986
102.40						0.9994	0.9992	0.9989	0.9988	0.9986
102.45						0.9994	0.9992	0.9989	0.9988	0.9986
102.50						0.9994	0.9992	0.9989	0.9988	0.9986
102.55						0.9994	0.9992	0.9989	0.9988	0.9986
102.60						0.9994	0.9992	0.9990	0.9988	0.9986
102.65						0.9994	0.9992	0.9990	0.9988	0.9986
102.70						0.9994	0.9992	0.9990	0.9988	0.9986
102.75						0.9994	0.9992	0.9990	0.9988	0.9986
102.80						0.9994	0.9992	0.9990	0.9988	0.9986
102.85						0.9994	0.9992	0.9990	0.9988	0.9986
102.90						0.9994	0.9992	0.9990	0.9988	0.9986
102.95						0.9994	0.9992	0.9990	0.9988	0.9986
103.00						0.9994	0.9992	0.9990	0.9988	0.9986
103.05						0.9994	0.9992	0.9990	0.9988	0.9986
103.10						0.9994	0.9992	0.9990	0.9988	0.9986
103.15						0.9994	0.9992	0.9990	0.9988	0.9986
103.20						0.9994	0.9992	0.9990	0.9988	0.9986
103.25						0.9994	0.9992	0.9990	0.9988	0.9986
103.30						0.9994	0.9992	0.9990	0.9988	0.9986
103.35						0.9994	0.9992	0.9990	0.9988	0.9986
103.40						0.9994	0.9992	0.9990	0.9988	0.9986
103.45						0.9994	0.9992	0.9990	0.9988	0.9986

PROBABILITY INTEGRAL OF t, THE NON-CENTRAL t-STATISTIC. THIS TABLE GIVES Pr $[t/\sqrt{f} \leq x]$

f is the number of degrees of freedom; the non-centrality parameter is $\sqrt{f+1}\,K_p$.
K_p is the standardized normal deviate exceeded with probability p.

DEGREES OF FREEDOM **2**

x \ p	.2500	.1500	.1000	.0650	.0400	.0250	.0100	.0040	.0025	.0010
103.50						0.9994	0.9992	0.9990	0.9989	0.9986
103.55						0.9994	0.9992	0.9990	0.9989	0.9986
103.60						0.9994	0.9992	0.9990	0.9989	0.9986
103.65						0.9994	0.9992	0.9990	0.9989	0.9986
103.70						0.9994	0.9992	0.9990	0.9989	0.9986
103.75						0.9994	0.9992	0.9990	0.9989	0.9986
103.80						0.9994	0.9992	0.9990	0.9989	0.9986
103.85						0.9994	0.9992	0.9990	0.9989	0.9986
103.90						0.9994	0.9992	0.9990	0.9989	0.9986
103.95						0.9994	0.9992	0.9990	0.9989	0.9986
104.00						0.9994	0.9992	0.9990	0.9989	0.9986
104.05						0.9994	0.9992	0.9990	0.9989	0.9986
104.10						0.9994	0.9992	0.9990	0.9989	0.9986
104.15						0.9994	0.9992	0.9990	0.9989	0.9986
104.20						0.9994	0.9992	0.9990	0.9989	0.9986
104.25						0.9994	0.9992	0.9990	0.9989	0.9986
104.30						0.9994	0.9992	0.9990	0.9989	0.9986
104.35						0.9994	0.9992	0.9990	0.9989	0.9986
104.40						0.9994	0.9992	0.9990	0.9989	0.9986
104.45						0.9994	0.9992	0.9990	0.9989	0.9986
104.50						0.9994	0.9992	0.9990	0.9989	0.9986
104.55						0.9994	0.9992	0.9990	0.9989	0.9986
104.60						0.9994	0.9992	0.9990	0.9989	0.9986
104.65						0.9994	0.9992	0.9990	0.9989	0.9986
104.70						0.9994	0.9992	0.9990	0.9989	0.9986
104.75						0.9994	0.9992	0.9990	0.9989	0.9986
104.80						0.9994	0.9992	0.9990	0.9989	0.9987
104.85						0.9994	0.9992	0.9990	0.9989	0.9987
104.90						0.9994	0.9992	0.9990	0.9989	0.9987
104.95						0.9994	0.9992	0.9990	0.9989	0.9987
105.00						0.9994	0.9992	0.9990	0.9989	0.9987
105.05						0.9994	0.9992	0.9990	0.9989	0.9987
105.10						0.9994	0.9992	0.9990	0.9989	0.9987
105.15						0.9994	0.9992	0.9990	0.9989	0.9987
105.20						0.9994	0.9992	0.9990	0.9989	0.9987
105.25						0.9994	0.9992	0.9990	0.9989	0.9987
105.30						0.9994	0.9992	0.9990	0.9989	0.9987
105.35						0.9994	0.9992	0.9990	0.9989	0.9987
105.40						0.9994	0.9992	0.9990	0.9989	0.9987
105.45						0.9994	0.9992	0.9990	0.9989	0.9987
105.50						0.9994	0.9992	0.9990	0.9989	0.9987
105.55						0.9994	0.9992	0.9990	0.9989	0.9987
105.60						0.9994	0.9992	0.9990	0.9989	0.9987
105.65						0.9994	0.9992	0.9990	0.9989	0.9987
105.70						0.9994	0.9992	0.9990	0.9989	0.9987
105.75						0.9994	0.9992	0.9990	0.9989	0.9987
105.80						0.9994	0.9992	0.9990	0.9989	0.9987
105.85						0.9994	0.9992	0.9990	0.9989	0.9987
105.90						0.9994	0.9992	0.9990	0.9989	0.9987
105.95						0.9994	0.9992	0.9990	0.9989	0.9987
106.00						0.9994	0.9992	0.9990	0.9989	0.9987
106.05						0.9994	0.9992	0.9990	0.9989	0.9987
106.10						0.9994	0.9992	0.9990	0.9989	0.9987
106.15						0.9994	0.9992	0.9990	0.9989	0.9987
106.20						0.9994	0.9992	0.9990	0.9989	0.9987
106.25						0.9994	0.9992	0.9990	0.9989	0.9987
106.30						0.9994	0.9992	0.9990	0.9989	0.9987
106.35						0.9994	0.9992	0.9990	0.9989	0.9987
106.40						0.9994	0.9992	0.9990	0.9989	0.9987
106.45						0.9994	0.9992	0.9990	0.9989	0.9987

PROBABILITY INTEGRAL OF t, THE NON-CENTRAL t-STATISTIC. THIS TABLE GIVES Pr $[t/\sqrt{f} \leq x]$.

f is the number of degrees of freedom; the non-centrality parameter is $\sqrt{f+1}\,K_p$.

K_p is the standardized normal deviate exceeded with probability p.

DEGREES OF FREEDOM **2**

x \ p	.2500	.1500	.1000	.0650	.0400	.0250	.0100	.0040	.0025	.0010
106.50						0.9994	0.9992	0.9990	0.9989	0.9987
106.55						0.9994	0.9992	0.9990	0.9989	0.9987
106.60						0.9994	0.9992	0.9990	0.9989	0.9987
106.65						0.9994	0.9992	0.9990	0.9989	0.9987
106.70						0.9995	0.9992	0.9990	0.9989	0.9987
106.75						0.9995	0.9992	0.9990	0.9989	0.9987
106.80						0.9995	0.9992	0.9990	0.9989	0.9987
106.85						0.9995	0.9992	0.9990	0.9989	0.9987
106.90						0.9995	0.9992	0.9990	0.9989	0.9987
106.95						0.9995	0.9992	0.9990	0.9989	0.9987
107.00						0.9995	0.9992	0.9990	0.9989	0.9987
107.05						0.9995	0.9992	0.9990	0.9989	0.9987
107.10						0.9995	0.9992	0.9990	0.9989	0.9987
107.15						0.9995	0.9993	0.9990	0.9989	0.9987
107.20						0.9995	0.9993	0.9990	0.9989	0.9987
107.25						0.9995	0.9993	0.9990	0.9989	0.9987
107.30						0.9995	0.9993	0.9990	0.9989	0.9987
107.35						0.9995	0.9993	0.9990	0.9989	0.9987
107.40						0.9995	0.9993	0.9990	0.9989	0.9987
107.45						0.9995	0.9993	0.9990	0.9989	0.9987
107.50						0.9995	0.9993	0.9990	0.9989	0.9987
107.55						0.9995	0.9993	0.9990	0.9989	0.9987
107.60						0.9995	0.9993	0.9990	0.9989	0.9987
107.65						0.9995	0.9993	0.9990	0.9989	0.9987
107.70						0.9995	0.9993	0.9990	0.9989	0.9987
107.75						0.9995	0.9993	0.9990	0.9989	0.9987
107.80						0.9995	0.9993	0.9990	0.9989	0.9987
107.85						0.9995	0.9993	0.9991	0.9989	0.9987
107.90						0.9995	0.9993	0.9991	0.9989	0.9987
107.95						0.9995	0.9993	0.9991	0.9989	0.9987
108.00						0.9995	0.9993	0.9991	0.9989	0.9987
108.05						0.9995	0.9993	0.9991	0.9989	0.9987
108.10						0.9995	0.9993	0.9991	0.9989	0.9987
108.15						0.9995	0.9993	0.9991	0.9989	0.9987
108.20						0.9995	0.9993	0.9991	0.9989	0.9987
108.25						0.9995	0.9993	0.9991	0.9989	0.9987
108.30						0.9995	0.9993	0.9991	0.9990	0.9987
108.35						0.9995	0.9993	0.9991	0.9990	0.9987
108.40						0.9995	0.9993	0.9991	0.9990	0.9987
108.45						0.9995	0.9993	0.9991	0.9990	0.9987
108.50						0.9995	0.9993	0.9991	0.9990	0.9987
108.55						0.9995	0.9993	0.9991	0.9990	0.9987
108.60						0.9995	0.9993	0.9991	0.9990	0.9987
108.65						0.9995	0.9993	0.9991	0.9990	0.9987
108.70						0.9995	0.9993	0.9991	0.9990	0.9987
108.75						0.9995	0.9993	0.9991	0.9990	0.9987
108.80						0.9995	0.9993	0.9991	0.9990	0.9987
108.85						0.9995	0.9993	0.9991	0.9990	0.9987
108.90						0.9995	0.9993	0.9991	0.9990	0.9988
108.95						0.9995	0.9993	0.9991	0.9990	0.9988
109.00						0.9995	0.9993	0.9991	0.9990	0.9988
109.05						0.9995	0.9993	0.9991	0.9990	0.9988
109.10						0.9995	0.9993	0.9991	0.9990	0.9988
109.15						0.9995	0.9993	0.9991	0.9990	0.9988
109.20						0.9995	0.9993	0.9991	0.9990	0.9988
109.25						0.9995	0.9993	0.9991	0.9990	0.9988
109.30						0.9995	0.9993	0.9991	0.9990	0.9988
109.35						0.9995	0.9993	0.9991	0.9990	0.9988
109.40						0.9995	0.9993	0.9991	0.9990	0.9988
109.45						0.9995	0.9993	0.9991	0.9990	0.9988

PROBABILITY INTEGRAL OF t, THE NON-CENTRAL t-STATISTIC. THIS TABLE GIVES $\Pr\,[t/\sqrt{f} \leq x]$.

f is the number of degrees of freedom; the non-centrality parameter is $\sqrt{f+1}\,K_p$.

K_p is the standardized normal deviate exceeded with probability p.

DEGREES OF FREEDOM **2**

x \ p	.2500	.1500	.1000	.0650	.0400	.0250	.0100	.0040	.0025	.0010
109.50						0.9995	0.9993	0.9991	0.9990	0.9988
109.55						0.9995	0.9993	0.9991	0.9990	0.9988
109.60						0.9995	0.9993	0.9991	0.9990	0.9988
109.65						0.9995	0.9993	0.9991	0.9990	0.9988
109.70						0.9995	0.9993	0.9991	0.9990	0.9988
109.75						0.9995	0.9993	0.9991	0.9990	0.9988
109.80						0.9995	0.9993	0.9991	0.9990	0.9988
109.85						0.9995	0.9993	0.9991	0.9990	0.9988
109.90						0.9995	0.9993	0.9991	0.9990	0.9988
109.95						0.9995	0.9993	0.9991	0.9990	0.9988
110.00						0.9995	0.9993	0.9991	0.9990	0.9988
110.05						0.9995	0.9993	0.9991	0.9990	0.9988
110.10						0.9995	0.9993	0.9991	0.9990	0.9988
110.15						0.9995	0.9993	0.9991	0.9990	0.9988
110.20						0.9995	0.9993	0.9991	0.9990	0.9988
110.25						0.9995	0.9993	0.9991	0.9990	0.9988
110.30						0.9995	0.9993	0.9991	0.9990	0.9988
110.35						0.9995	0.9993	0.9991	0.9990	0.9988
110.40						0.9995	0.9993	0.9991	0.9990	0.9988
110.45						0.9995	0.9993	0.9991	0.9990	0.9988
110.50						0.9995	0.9993	0.9991	0.9990	0.9988
110.55						0.9995	0.9993	0.9991	0.9990	0.9988
110.60						0.9995	0.9993	0.9991	0.9990	0.9988
110.65						0.9995	0.9993	0.9991	0.9990	0.9988
110.70						0.9995	0.9993	0.9991	0.9990	0.9988
110.75						0.9995	0.9993	0.9991	0.9990	0.9988
110.80						0.9995	0.9993	0.9991	0.9990	0.9988
110.85						0.9995	0.9993	0.9991	0.9990	0.9988
110.90						0.9995	0.9993	0.9991	0.9990	0.9988
110.95						0.9995	0.9993	0.9991	0.9990	0.9988
111.00						0.9995	0.9993	0.9991	0.9990	0.9988
111.05						0.9995	0.9993	0.9991	0.9990	0.9988
111.10						0.9995	0.9993	0.9991	0.9990	0.9988
111.15						0.9995	0.9993	0.9991	0.9990	0.9988
111.20						0.9995	0.9993	0.9991	0.9990	0.9988
111.25						0.9995	0.9993	0.9991	0.9990	0.9988
111.30						0.9995	0.9993	0.9991	0.9990	0.9988
111.35						0.9995	0.9993	0.9991	0.9990	0.9988
111.40						0.9995	0.9993	0.9991	0.9990	0.9988
111.45						0.9995	0.9993	0.9991	0.9990	0.9988
111.50						0.9995	0.9993	0.9991	0.9990	0.9988
111.55						0.9995	0.9993	0.9991	0.9990	0.9988
111.60						0.9995	0.9993	0.9991	0.9990	0.9988
111.65						0.9995	0.9993	0.9991	0.9990	0.9988
111.70						0.9995	0.9993	0.9991	0.9990	0.9988
111.75						0.9995	0.9993	0.9991	0.9990	0.9988
111.80						0.9995	0.9993	0.9991	0.9990	0.9988
111.85						0.9995	0.9993	0.9991	0.9990	0.9988
111.90						0.9995	0.9993	0.9991	0.9990	0.9988
111.95						0.9995	0.9993	0.9991	0.9990	0.9988
112.00						0.9995	0.9993	0.9991	0.9990	0.9988
112.05						0.9995	0.9993	0.9991	0.9990	0.9988
112.10						0.9995	0.9993	0.9991	0.9990	0.9988
112.15						0.9995	0.9993	0.9991	0.9990	0.9988
112.20						0.9995	0.9993	0.9991	0.9990	0.9988
112.25							0.9993	0.9991	0.9990	0.9988
112.30							0.9993	0.9991	0.9990	0.9988
112.35							0.9993	0.9991	0.9990	0.9988
112.40							0.9993	0.9991	0.9990	0.9988
112.45							0.9993	0.9991	0.9990	0.9988

PROBABILITY INTEGRAL OF t, THE NON-CENTRAL t-STATISTIC. THIS TABLE GIVES Pr $[t/\sqrt{f} \leq x]$.

f is the number of degrees of freedom; the non-centrality parameter is $\sqrt{f+1}\,K_p$.

K_p is the standardized normal deviate exceeded with probability p.

DEGREES OF FREEDOM **2**

x \ p	.2500	.1500	.1000	.0650	.0400	.0250	.0100	.0040	.0025	.0010
112.50							0.9993	0.9991	0.9990	0.9988
112.55							0.9993	0.9991	0.9990	0.9988
112.60							0.9993	0.9991	0.9990	0.9988
112.65							0.9993	0.9991	0.9990	0.9988
112.70							0.9993	0.9991	0.9990	0.9988
112.75							0.9993	0.9991	0.9990	0.9988
112.80							0.9993	0.9991	0.9990	0.9988
112.85							0.9993	0.9991	0.9990	0.9988
112.90							0.9993	0.9991	0.9990	0.9988
112.95							0.9993	0.9991	0.9990	0.9988
113.00							0.9993	0.9991	0.9990	0.9988
113.05							0.9993	0.9991	0.9990	0.9988
113.10							0.9993	0.9991	0.9990	0.9988
113.15							0.9993	0.9991	0.9990	0.9988
113.20							0.9993	0.9991	0.9990	0.9988
113.25							0.9993	0.9991	0.9990	0.9988
113.30							0.9993	0.9991	0.9990	0.9988
113.35							0.9993	0.9991	0.9990	0.9988
113.40							0.9993	0.9991	0.9990	0.9988
113.45							0.9993	0.9991	0.9990	0.9988
113.50							0.9993	0.9991	0.9990	0.9988
113.55							0.9993	0.9991	0.9990	0.9989
113.60							0.9993	0.9991	0.9990	0.9989
113.65							0.9993	0.9991	0.9990	0.9989
113.70							0.9993	0.9991	0.9990	0.9989
113.75							0.9993	0.9991	0.9990	0.9989
113.80							0.9993	0.9991	0.9990	0.9989
113.85							0.9993	0.9991	0.9990	0.9989
113.90							0.9993	0.9991	0.9991	0.9989
113.95							0.9993	0.9991	0.9991	0.9989
114.00							0.9993	0.9992	0.9991	0.9989
114.05							0.9993	0.9992	0.9991	0.9989
114.10							0.9993	0.9992	0.9991	0.9989
114.15							0.9993	0.9992	0.9991	0.9989
114.20							0.9993	0.9992	0.9991	0.9989
114.25							0.9993	0.9992	0.9991	0.9989
114.30							0.9993	0.9992	0.9991	0.9989
114.35							0.9993	0.9992	0.9991	0.9989
114.40							0.9993	0.9992	0.9991	0.9989
114.45							0.9993	0.9992	0.9991	0.9989
114.50							0.9993	0.9992	0.9991	0.9989
114.55							0.9993	0.9992	0.9991	0.9989
114.60							0.9993	0.9992	0.9991	0.9989
114.65							0.9993	0.9992	0.9991	0.9989
114.70							0.9993	0.9992	0.9991	0.9989
114.75							0.9993	0.9992	0.9991	0.9989
114.80							0.9993	0.9992	0.9991	0.9989
114.85							0.9993	0.9992	0.9991	0.9989
114.90							0.9993	0.9992	0.9991	0.9989
114.95							0.9993	0.9992	0.9991	0.9989
115.00							0.9993	0.9992	0.9991	0.9989
115.05							0.9993	0.9992	0.9991	0.9989
115.10							0.9993	0.9992	0.9991	0.9989
115.15							0.9994	0.9992	0.9991	0.9989
115.20							0.9994	0.9992	0.9991	0.9989
115.25							0.9994	0.9992	0.9991	0.9989
115.30							0.9994	0.9992	0.9991	0.9989
115.35							0.9994	0.9992	0.9991	0.9989
115.40							0.9994	0.9992	0.9991	0.9989
115.45							0.9994	0.9992	0.9991	0.9989

PROBABILITY INTEGRAL OF t, THE NON-CENTRAL t-STATISTIC. THIS TABLE GIVES Pr $[t/\sqrt{f} \leq x]$.

f is the number of degrees of freedom; the non-centrality parameter is $\sqrt{f+1}\,K_p$.
K_p is the standardized normal deviate exceeded with probability p.

DEGREES OF FREEDOM **2**

x \ p	.2500	.1500	.1000	.0650	.0400	.0250	.0100	.0040	.0025	.0010
115.50							0.9994	0.9992	0.9991	0.9989
115.55							0.9994	0.9992	0.9991	0.9989
115.60							0.9994	0.9992	0.9991	0.9989
115.65							0.9994	0.9992	0.9991	0.9989
115.70							0.9994	0.9992	0.9991	0.9989
115.75							0.9994	0.9992	0.9991	0.9989
115.80							0.9994	0.9992	0.9991	0.9989
115.85							0.9994	0.9992	0.9991	0.9989
115.90							0.9994	0.9992	0.9991	0.9989
115.95							0.9994	0.9992	0.9991	0.9989
116.00							0.9994	0.9992	0.9991	0.9989
116.05							0.9994	0.9992	0.9991	0.9989
116.10							0.9994	0.9992	0.9991	0.9989
116.15							0.9994	0.9992	0.9991	0.9989
116.20							0.9994	0.9992	0.9991	0.9989
116.25							0.9994	0.9992	0.9991	0.9989
116.30							0.9994	0.9992	0.9991	0.9989
116.35							0.9994	0.9992	0.9991	0.9989
116.40							0.9994	0.9992	0.9991	0.9989
116.45							0.9994	0.9992	0.9991	0.9989
116.50							0.9994	0.9992	0.9991	0.9989
116.55							0.9994	0.9992	0.9991	0.9989
116.60							0.9994	0.9992	0.9991	0.9989
116.65							0.9994	0.9992	0.9991	0.9989
116.70							0.9994	0.9992	0.9991	0.9989
116.75							0.9994	0.9992	0.9991	0.9989
116.80							0.9994	0.9992	0.9991	0.9989
116.85							0.9994	0.9992	0.9991	0.9989
116.90							0.9994	0.9992	0.9991	0.9989
116.95							0.9994	0.9992	0.9991	0.9989
117.00							0.9994	0.9992	0.9991	0.9989
117.05							0.9994	0.9992	0.9991	0.9989
117.10							0.9994	0.9992	0.9991	0.9989
117.15							0.9994	0.9992	0.9991	0.9989
117.20							0.9994	0.9992	0.9991	0.9989
117.25							0.9994	0.9992	0.9991	0.9989
117.30							0.9994	0.9992	0.9991	0.9989
117.35							0.9994	0.9992	0.9991	0.9989
117.40							0.9994	0.9992	0.9991	0.9989
117.45							0.9994	0.9992	0.9991	0.9989
117.50							0.9994	0.9992	0.9991	0.9989
117.55							0.9994	0.9992	0.9991	0.9989
117.60							0.9994	0.9992	0.9991	0.9989
117.65							0.9994	0.9992	0.9991	0.9989
117.70							0.9994	0.9992	0.9991	0.9989
117.75							0.9994	0.9992	0.9991	0.9989
117.80							0.9994	0.9992	0.9991	0.9989
117.85							0.9994	0.9992	0.9991	0.9989
117.90							0.9994	0.9992	0.9991	0.9989
117.95							0.9994	0.9992	0.9991	0.9989
118.00							0.9994	0.9992	0.9991	0.9989
118.05							0.9994	0.9992	0.9991	0.9989
118.10							0.9994	0.9992	0.9991	0.9989
118.15							0.9994	0.9992	0.9991	0.9989
118.20							0.9994	0.9992	0.9991	0.9989
118.25							0.9994	0.9992	0.9991	0.9989
118.30							0.9994	0.9992	0.9991	0.9989
118.35							0.9994	0.9992	0.9991	0.9989
118.40							0.9994	0.9992	0.9991	0.9989
118.45							0.9994	0.9992	0.9991	0.9989

PROBABILITY INTEGRAL OF t, THE NON-CENTRAL t-STATISTIC. THIS TABLE GIVES Pr $[t/\sqrt{f} \leq x]$.

f is the number of degrees of freedom; the non-centrality parameter is $\sqrt{f+1}\,K_p$.

K_p is the standardized normal deviate exceeded with probability p.

DEGREES OF FREEDOM **2**

x \ p	.2500	.1500	.1000	.0650	.0400	.0250	.0100	.0040	.0025	.0010
118.50							0.9994	0.9992	0.9991	0.9989
118.55							0.9994	0.9992	0.9991	0.9989
118.60							0.9994	0.9992	0.9991	0.9989
118.65							0.9994	0.9992	0.9991	0.9989
118.70							0.9994	0.9992	0.9991	0.9989
118.75							0.9994	0.9992	0.9991	0.9989
118.80							0.9994	0.9992	0.9991	0.9990
118.85							0.9994	0.9992	0.9991	0.9990
118.90							0.9994	0.9992	0.9991	0.9990
118.95							0.9994	0.9992	0.9991	0.9990
119.00							0.9994	0.9992	0.9991	0.9990
119.05							0.9994	0.9992	0.9991	0.9990
119.10							0.9994	0.9992	0.9991	0.9990
119.15							0.9994	0.9992	0.9991	0.9990
119.20							0.9994	0.9992	0.9991	0.9990
119.25							0.9994	0.9992	0.9991	0.9990
119.30							0.9994	0.9992	0.9991	0.9990
119.35							0.9994	0.9992	0.9991	0.9990
119.40							0.9994	0.9992	0.9991	0.9990
119.45							0.9994	0.9992	0.9991	0.9990
119.50							0.9994	0.9992	0.9991	0.9990
119.55							0.9994	0.9992	0.9991	0.9990
119.60							0.9994	0.9992	0.9991	0.9990
119.65							0.9994	0.9992	0.9991	0.9990
119.70							0.9994	0.9992	0.9991	0.9990
119.75							0.9994	0.9992	0.9991	0.9990
119.80							0.9994	0.9992	0.9991	0.9990
119.85							0.9994	0.9992	0.9991	0.9990
119.90							0.9994	0.9992	0.9991	0.9990
119.95							0.9994	0.9992	0.9991	0.9990
120.00							0.9994	0.9992	0.9991	0.9990
120.05							0.9994	0.9992	0.9991	0.9990
120.10							0.9994	0.9992	0.9991	0.9990
120.15							0.9994	0.9992	0.9991	0.9990
120.20							0.9994	0.9992	0.9991	0.9990
120.25							0.9994	0.9992	0.9991	0.9990
120.30							0.9994	0.9992	0.9991	0.9990
120.35							0.9994	0.9992	0.9991	0.9990
120.40							0.9994	0.9992	0.9992	0.9990
120.45							0.9994	0.9992	0.9992	0.9990
120.50							0.9994	0.9992	0.9992	0.9990
120.55							0.9994	0.9992	0.9992	0.9990
120.60							0.9994	0.9992	0.9992	0.9990
120.65							0.9994	0.9992	0.9992	0.9990
120.70							0.9994	0.9992	0.9992	0.9990
120.75							0.9994	0.9992	0.9992	0.9990
120.80							0.9994	0.9992	0.9992	0.9990
120.85							0.9994	0.9992	0.9992	0.9990
120.90							0.9994	0.9992	0.9992	0.9990
120.95							0.9994	0.9992	0.9992	0.9990
121.00							0.9994	0.9992	0.9992	0.9990
121.05							0.9994	0.9992	0.9992	0.9990
121.10							0.9994	0.9992	0.9992	0.9990
121.15							0.9994	0.9992	0.9992	0.9990
121.20							0.9994	0.9992	0.9992	0.9990
121.25							0.9994	0.9992	0.9992	0.9990
121.30							0.9994	0.9992	0.9992	0.9990
121.35							0.9994	0.9992	0.9992	0.9990
121.40							0.9994	0.9993	0.9992	0.9990
121.45							0.9994	0.9993	0.9992	0.9990

PROBABILITY INTEGRAL OF t, THE NON-CENTRAL t-STATISTIC. THIS TABLE GIVES Pr $[t/\sqrt{f} \leq x]$.

f is the number of degrees of freedom; the non-centrality parameter is $\sqrt{f+1}\,K_p$.
K_p is the standardized normal deviate exceeded with probability p.

DEGREES OF FREEDOM **2**

x \ p	.2500	.1500	.1000	.0650	.0400	.0250	.0100	.0040	.0025	.0010
121.50							0.9994	0.9993	0.9992	0.9990
121.55							0.9994	0.9993	0.9992	0.9990
121.60							0.9994	0.9993	0.9992	0.9990
121.65							0.9994	0.9993	0.9992	0.9990
121.70							0.9994	0.9993	0.9992	0.9990
121.75							0.9994	0.9993	0.9992	0.9990
121.80							0.9994	0.9993	0.9992	0.9990
121.85							0.9994	0.9993	0.9992	0.9990
121.90							0.9994	0.9993	0.9992	0.9990
121.95							0.9994	0.9993	0.9992	0.9990
122.00							0.9994	0.9993	0.9992	0.9990
122.05							0.9994	0.9993	0.9992	0.9990
122.10							0.9994	0.9993	0.9992	0.9990
122.15							0.9994	0.9993	0.9992	0.9990
122.20							0.9994	0.9993	0.9992	0.9990
122.25							0.9994	0.9993	0.9992	0.9990
122.30							0.9994	0.9993	0.9992	0.9990
122.35							0.9994	0.9993	0.9992	0.9990
122.40							0.9994	0.9993	0.9992	0.9990
122.45							0.9994	0.9993	0.9992	0.9990
122.50							0.9994	0.9993	0.9992	0.9990
122.55							0.9994	0.9993	0.9992	0.9990
122.60							0.9994	0.9993	0.9992	0.9990
122.65							0.9994	0.9993	0.9992	0.9990
122.70							0.9994	0.9993	0.9992	0.9990
122.75							0.9994	0.9993	0.9992	0.9990
122.80							0.9994	0.9993	0.9992	0.9990
122.85							0.9994	0.9993	0.9992	0.9990
122.90							0.9994	0.9993	0.9992	0.9990
122.95							0.9994	0.9993	0.9992	0.9990
123.00							0.9994	0.9993	0.9992	0.9990
123.05							0.9994	0.9993	0.9992	0.9990
123.10							0.9994	0.9993	0.9992	0.9990
123.15							0.9994	0.9993	0.9992	0.9990
123.20							0.9994	0.9993	0.9992	0.9990
123.25							0.9994	0.9993	0.9992	0.9990
123.30							0.9994	0.9993	0.9992	0.9990
123.35							0.9994	0.9993	0.9992	0.9990
123.40							0.9994	0.9993	0.9992	0.9990
123.45							0.9994	0.9993	0.9992	0.9990
123.50							0.9994	0.9993	0.9992	0.9990
123.55							0.9994	0.9993	0.9992	0.9990
123.60							0.9994	0.9993	0.9992	0.9990
123.65							0.9994	0.9993	0.9992	0.9990
123.70							0.9994	0.9993	0.9992	0.9990
123.75							0.9994	0.9993	0.9992	0.9990
123.80							0.9994	0.9993	0.9992	0.9990
123.85							0.9994	0.9993	0.9992	0.9990
123.90							0.9994	0.9993	0.9992	0.9990
123.95							0.9994	0.9993	0.9992	0.9990
124.00							0.9994	0.9993	0.9992	0.9990
124.05							0.9994	0.9993	0.9992	0.9990
124.10							0.9994	0.9993	0.9992	0.9990
124.15							0.9994	0.9993	0.9992	0.9990
124.20							0.9994	0.9993	0.9992	0.9990
124.25							0.9994	0.9993	0.9992	0.9990
124.30							0.9994	0.9993	0.9992	0.9990
124.35							0.9994	0.9993	0.9992	0.9990
124.40							0.9994	0.9993	0.9992	0.9990
124.45							0.9994	0.9993	0.9992	0.9990

PROBABILITY INTEGRAL OF t, THE NON-CENTRAL t-STATISTIC. THIS TABLE GIVES Pr $[t/\sqrt{f} \leq x]$.

f is the number of degrees of freedom; the non-centrality parameter is $\sqrt{f+1}\,K_p$.

K_p is the standardized normal deviate exceeded with probability p.

DEGREES OF FREEDOM **2**

x \ p	.2500	.1500	.1000	.0650	.0400	.0250	.0100	.0040	.0025	.0010
124.50							0.9994	0.9993	0.9992	0.9990
124.55							0.9994	0.9993	0.9992	0.9990
124.60							0.9994	0.9993	0.9992	0.9990
124.65							0.9994	0.9993	0.9992	0.9990
124.70							0.9994	0.9993	0.9992	0.9990
124.75							0.9994	0.9993	0.9992	0.9990
124.80							0.9994	0.9993	0.9992	0.9990
124.85							0.9994	0.9993	0.9992	0.9990
124.90							0.9994	0.9993	0.9992	0.9991
124.95							0.9994	0.9993	0.9992	0.9991
125.00							0.9994	0.9993	0.9992	0.9991
125.05							0.9994	0.9993	0.9992	0.9991
125.10							0.9994	0.9993	0.9992	0.9991
125.15							0.9994	0.9993	0.9992	0.9991
125.20							0.9995	0.9993	0.9992	0.9991
125.25							0.9995	0.9993	0.9992	0.9991
125.30							0.9995	0.9993	0.9992	0.9991
125.35							0.9995	0.9993	0.9992	0.9991
125.40							0.9995	0.9993	0.9992	0.9991
125.45							0.9995	0.9993	0.9992	0.9991
125.50							0.9995	0.9993	0.9992	0.9991
125.55							0.9995	0.9993	0.9992	0.9991
125.60							0.9995	0.9993	0.9992	0.9991
125.65							0.9995	0.9993	0.9992	0.9991
125.70							0.9995	0.9993	0.9992	0.9991
125.75							0.9995	0.9993	0.9992	0.9991
125.80							0.9995	0.9993	0.9992	0.9991
125.85							0.9995	0.9993	0.9992	0.9991
125.90							0.9995	0.9993	0.9992	0.9991
125.95							0.9995	0.9993	0.9992	0.9991
126.00							0.9995	0.9993	0.9992	0.9991
126.05							0.9995	0.9993	0.9992	0.9991
126.10							0.9995	0.9993	0.9992	0.9991
126.15							0.9995	0.9993	0.9992	0.9991
126.20							0.9995	0.9993	0.9992	0.9991
126.25							0.9995	0.9993	0.9992	0.9991
126.30							0.9995	0.9993	0.9992	0.9991
126.35							0.9995	0.9993	0.9992	0.9991
126.40							0.9995	0.9993	0.9992	0.9991
126.45							0.9995	0.9993	0.9992	0.9991
126.50							0.9995	0.9993	0.9992	0.9991
126.55							0.9995	0.9993	0.9992	0.9991
126.60							0.9995	0.9993	0.9992	0.9991
126.65							0.9995	0.9993	0.9992	0.9991
126.70							0.9995	0.9993	0.9992	0.9991
126.75							0.9995	0.9993	0.9992	0.9991
126.80							0.9995	0.9993	0.9992	0.9991
126.85							0.9995	0.9993	0.9992	0.9991
126.90							0.9995	0.9993	0.9992	0.9991
126.95							0.9995	0.9993	0.9992	0.9991
127.00							0.9995	0.9993	0.9992	0.9991
127.05							0.9995	0.9993	0.9992	0.9991
127.10							0.9995	0.9993	0.9992	0.9991
127.15							0.9995	0.9993	0.9992	0.9991
127.20							0.9995	0.9993	0.9992	0.9991
127.25							0.9995	0.9993	0.9992	0.9991
127.30							0.9995	0.9993	0.9992	0.9991
127.35							0.9995	0.9993	0.9992	0.9991
127.40							0.9995	0.9993	0.9992	0.9991
127.45							0.9995	0.9993	0.9992	0.9991

PROBABILITY INTEGRAL OF t, THE NON-CENTRAL t-STATISTIC. THIS TABLE GIVES Pr $[t/\sqrt{f} \leq x]$.

f is the number of degrees of freedom; the non-centrality parameter is $\sqrt{f+1}\,K_p$.
K_p is the standardized normal deviate exceeded with probability p.

DEGREES OF FREEDOM **2**

x \ p	.2500	.1500	.1000	.0650	.0400	.0250	.0100	.0040	.0025	.0010
127.50							0.9995	0.9993	0.9992	0.9991
127.55							0.9995	0.9993	0.9992	0.9991
127.60							0.9995	0.9993	0.9992	0.9991
127.65							0.9995	0.9993	0.9992	0.9991
127.70							0.9995	0.9993	0.9992	0.9991
127.75							0.9995	0.9993	0.9992	0.9991
127.80							0.9995	0.9993	0.9992	0.9991
127.85							0.9995	0.9993	0.9992	0.9991
127.90							0.9995	0.9993	0.9992	0.9991
127.95							0.9995	0.9993	0.9992	0.9991
128.00							0.9995	0.9993	0.9992	0.9991
128.05							0.9995	0.9993	0.9992	0.9991
128.10							0.9995	0.9993	0.9993	0.9991
128.15							0.9995	0.9993	0.9993	0.9991
128.20							0.9995	0.9993	0.9993	0.9991
128.25							0.9995	0.9993	0.9993	0.9991
128.30							0.9995	0.9993	0.9993	0.9991
128.35							0.9995	0.9993	0.9993	0.9991
128.40							0.9995	0.9993	0.9993	0.9991
128.45							0.9995	0.9993	0.9993	0.9991
128.50							0.9995	0.9993	0.9993	0.9991
128.55							0.9995	0.9993	0.9993	0.9991
128.60							0.9995	0.9993	0.9993	0.9991
128.65							0.9995	0.9993	0.9993	0.9991
128.70							0.9995	0.9993	0.9993	0.9991
128.75							0.9995	0.9993	0.9993	0.9991
128.80							0.9995	0.9993	0.9993	0.9991
128.85							0.9995	0.9993	0.9993	0.9991
128.90							0.9995	0.9993	0.9993	0.9991
128.95							0.9995	0.9993	0.9993	0.9991
129.00							0.9995	0.9993	0.9993	0.9991
129.05							0.9995	0.9993	0.9993	0.9991
129.10							0.9995	0.9993	0.9993	0.9991
129.15							0.9995	0.9993	0.9993	0.9991
129.20							0.9995	0.9993	0.9993	0.9991
129.25							0.9995	0.9993	0.9993	0.9991
129.30							0.9995	0.9993	0.9993	0.9991
129.35							0.9995	0.9993	0.9993	0.9991
129.40							0.9995	0.9993	0.9993	0.9991
129.45							0.9995	0.9993	0.9993	0.9991
129.50							0.9995	0.9993	0.9993	0.9991
129.55							0.9995	0.9993	0.9993	0.9991
129.60							0.9995	0.9993	0.9993	0.9991
129.65							0.9995	0.9993	0.9993	0.9991
129.70							0.9995	0.9993	0.9993	0.9991
129.75							0.9995	0.9993	0.9993	0.9991
129.80							0.9995	0.9993	0.9993	0.9991
129.85							0.9995	0.9993	0.9993	0.9991
129.90							0.9995	0.9993	0.9993	0.9991
129.95							0.9995	0.9993	0.9993	0.9991
130.00							0.9995	0.9993	0.9993	0.9991
130.05							0.9995	0.9993	0.9993	0.9991
130.10							0.9995	0.9993	0.9993	0.9991
130.15							0.9995	0.9993	0.9993	0.9991
130.20							0.9995	0.9993	0.9993	0.9991
130.25							0.9995	0.9993	0.9993	0.9991
130.30							0.9995	0.9993	0.9993	0.9991
130.35							0.9995	0.9993	0.9993	0.9991
130.40							0.9995	0.9994	0.9993	0.9991
130.45							0.9995	0.9994	0.9993	0.9991

PROBABILITY INTEGRAL OF t, THE NON-CENTRAL t-STATISTIC. THIS TABLE GIVES Pr $[t/\sqrt{f} \leq x]$.

f is the number of degrees of freedom; the non-centrality parameter is $\sqrt{f+1}\,K_p$.
K_p is the standardized normal deviate exceeded with probability p.

DEGREES OF FREEDOM **2**

x \ p	.2500	.1500	.1000	.0650	.0400	.0250	.0100	.0040	.0025	.0010
130.50							0.9995	0.9994	0.9993	0.9991
130.55							0.9995	0.9994	0.9993	0.9991
130.60							0.9995	0.9994	0.9993	0.9991
130.65							0.9995	0.9994	0.9993	0.9991
130.70							0.9995	0.9994	0.9993	0.9991
130.75							0.9995	0.9994	0.9993	0.9991
130.80							0.9995	0.9994	0.9993	0.9991
130.85							0.9995	0.9994	0.9993	0.9991
130.90							0.9995	0.9994	0.9993	0.9991
130.95							0.9995	0.9994	0.9993	0.9991
131.00							0.9995	0.9994	0.9993	0.9991
131.05							0.9995	0.9994	0.9993	0.9991
131.10							0.9995	0.9994	0.9993	0.9991
131.15							0.9995	0.9994	0.9993	0.9991
131.20							0.9995	0.9994	0.9993	0.9991
131.25							0.9995	0.9994	0.9993	0.9991
131.30							0.9995	0.9994	0.9993	0.9991
131.35							0.9995	0.9994	0.9993	0.9991
131.40							0.9995	0.9994	0.9993	0.9991
131.45							0.9995	0.9994	0.9993	0.9991
131.50							0.9995	0.9994	0.9993	0.9991
131.55							0.9995	0.9994	0.9993	0.9991
131.60							0.9995	0.9994	0.9993	0.9991
131.65							0.9995	0.9994	0.9993	0.9991
131.70							0.9995	0.9994	0.9993	0.9991
131.75							0.9995	0.9994	0.9993	0.9991
131.80							0.9995	0.9994	0.9993	0.9991
131.85							0.9995	0.9994	0.9993	0.9991
131.90							0.9995	0.9994	0.9993	0.9991
131.95							0.9995	0.9994	0.9993	0.9991
132.00							0.9995	0.9994	0.9993	0.9991
132.05							0.9995	0.9994	0.9993	0.9991
132.10							0.9995	0.9994	0.9993	0.9992
132.15							0.9995	0.9994	0.9993	0.9992
132.20							0.9995	0.9994	0.9993	0.9992
132.25							0.9995	0.9994	0.9993	0.9992
132.30							0.9995	0.9994	0.9993	0.9992
132.35							0.9995	0.9994	0.9993	0.9992
132.40							0.9995	0.9994	0.9993	0.9992
132.45							0.9995	0.9994	0.9993	0.9992
132.50							0.9995	0.9994	0.9993	0.9992
132.55							0.9995	0.9994	0.9993	0.9992
132.60							0.9995	0.9994	0.9993	0.9992
132.65							0.9995	0.9994	0.9993	0.9992
132.70							0.9995	0.9994	0.9993	0.9992
132.75							0.9995	0.9994	0.9993	0.9992
132.80							0.9995	0.9994	0.9993	0.9992
132.85							0.9995	0.9994	0.9993	0.9992
132.90							0.9995	0.9994	0.9993	0.9992
132.95							0.9995	0.9994	0.9993	0.9992
133.00							0.9995	0.9994	0.9993	0.9992
133.05							0.9995	0.9994	0.9993	0.9992
133.10							0.9995	0.9994	0.9993	0.9992
133.15							0.9995	0.9994	0.9993	0.9992
133.20							0.9995	0.9994	0.9993	0.9992
133.25							0.9995	0.9994	0.9993	0.9992
133.30							0.9995	0.9994	0.9993	0.9992
133.35							0.9995	0.9994	0.9993	0.9992
133.40							0.9995	0.9994	0.9993	0.9992
133.45							0.9995	0.9994	0.9993	0.9992

PROBABILITY INTEGRAL OF t, THE NON-CENTRAL t-STATISTIC. THIS TABLE GIVES Pr $[t/\sqrt{f} \leq x]$.

f is the number of degrees of freedom; the non-centrality parameter is $\sqrt{f+1}\,K_p$.

K_p is the standardized normal deviate exceeded with probability p.

DEGREES OF FREEDOM **2**

x \ p	.2500	.1500	.1000	.0650	.0400	.0250	.0100	.0040	.0025	.0010
133.50							0.9995	0.9994	0.9993	0.9992
133.55							0.9995	0.9994	0.9993	0.9992
133.60							0.9995	0.9994	0.9993	0.9992
133.65							0.9995	0.9994	0.9993	0.9992
133.70							0.9995	0.9994	0.9993	0.9992
133.75							0.9995	0.9994	0.9993	0.9992
133.80							0.9995	0.9994	0.9993	0.9992
133.85								0.9994	0.9993	0.9992
133.90								0.9994	0.9993	0.9992
133.95								0.9994	0.9993	0.9992
134.00								0.9994	0.9993	0.9992
134.05								0.9994	0.9993	0.9992
134.10								0.9994	0.9993	0.9992
134.15								0.9994	0.9993	0.9992
134.20								0.9994	0.9993	0.9992
134.25								0.9994	0.9993	0.9992
134.30								0.9994	0.9993	0.9992
134.35								0.9994	0.9993	0.9992
134.40								0.9994	0.9993	0.9992
134.45								0.9994	0.9993	0.9992
134.50								0.9994	0.9993	0.9992
134.55								0.9994	0.9993	0.9992
134.60								0.9994	0.9993	0.9992
134.65								0.9994	0.9993	0.9992
134.70								0.9994	0.9993	0.9992
134.75								0.9994	0.9993	0.9992
134.80								0.9994	0.9993	0.9992
134.85								0.9994	0.9993	0.9992
134.90								0.9994	0.9993	0.9992
134.95								0.9994	0.9993	0.9992
135.00								0.9994	0.9993	0.9992
135.05								0.9994	0.9993	0.9992
135.10								0.9994	0.9993	0.9992
135.15								0.9994	0.9993	0.9992
135.20								0.9994	0.9993	0.9992
135.25								0.9994	0.9993	0.9992
135.30								0.9994	0.9993	0.9992
135.35								0.9994	0.9993	0.9992
135.40								0.9994	0.9993	0.9992
135.45								0.9994	0.9993	0.9992
135.50								0.9994	0.9993	0.9992
135.55								0.9994	0.9993	0.9992
135.60								0.9994	0.9993	0.9992
135.65								0.9994	0.9993	0.9992
135.70								0.9994	0.9993	0.9992
135.75								0.9994	0.9993	0.9992
135.80								0.9994	0.9993	0.9992
135.85								0.9994	0.9993	0.9992
135.90								0.9994	0.9993	0.9992
135.95								0.9994	0.9993	0.9992
136.00								0.9994	0.9993	0.9992
136.05								0.9994	0.9993	0.9992
136.10								0.9994	0.9993	0.9992
136.15								0.9994	0.9993	0.9992
136.20								0.9994	0.9993	0.9992
136.25								0.9994	0.9993	0.9992
136.30								0.9994	0.9993	0.9992
136.35								0.9994	0.9993	0.9992
136.40								0.9994	0.9993	0.9992
136.45								0.9994	0.9993	0.9992

PROBABILITY INTEGRAL OF t, THE NON-CENTRAL t-STATISTIC. THIS TABLE GIVES Pr $[t/\sqrt{f} \leq x]$.

f is the number of degrees of freedom; the non-centrality parameter is $\sqrt{f+1}\,K_p$.

K_p is the standardized normal deviate exceeded with probability p.

DEGREES OF FREEDOM **2**

x \ p	.2500	.1500	.1000	.0650	.0400	.0250	.0100	.0040	.0025	.0010
136.50								0.9994	0.9993	0.9992
136.55								0.9994	0.9993	0.9992
136.60								0.9994	0.9993	0.9992
136.65								0.9994	0.9993	0.9992
136.70								0.9994	0.9993	0.9992
136.75								0.9994	0.9993	0.9992
136.80								0.9994	0.9993	0.9992
136.85								0.9994	0.9993	0.9992
136.90								0.9994	0.9993	0.9992
136.95								0.9994	0.9993	0.9992
137.00								0.9994	0.9993	0.9992
137.05								0.9994	0.9993	0.9992
137.10								0.9994	0.9993	0.9992
137.15								0.9994	0.9993	0.9992
137.20								0.9994	0.9993	0.9992
137.25								0.9994	0.9993	0.9992
137.30								0.9994	0.9993	0.9992
137.35								0.9994	0.9993	0.9992
137.40								0.9994	0.9993	0.9992
137.45								0.9994	0.9993	0.9992
137.50								0.9994	0.9993	0.9992
137.55								0.9994	0.9993	0.9992
137.60								0.9994	0.9994	0.9992
137.65								0.9994	0.9994	0.9992
137.70								0.9994	0.9994	0.9992
137.75								0.9994	0.9994	0.9992
137.80								0.9994	0.9994	0.9992
137.85								0.9994	0.9994	0.9992
137.90								0.9994	0.9994	0.9992
137.95								0.9994	0.9994	0.9992
138.00								0.9994	0.9994	0.9992
138.05								0.9994	0.9994	0.9992
138.10								0.9994	0.9994	0.9992
138.15								0.9994	0.9994	0.9992
138.20								0.9994	0.9994	0.9992
138.25								0.9994	0.9994	0.9992
138.30								0.9994	0.9994	0.9992
138.35								0.9994	0.9994	0.9992
138.40								0.9994	0.9994	0.9992
138.45								0.9994	0.9994	0.9992
138.50								0.9994	0.9994	0.9992
138.55								0.9994	0.9994	0.9992
138.60								0.9994	0.9994	0.9992
138.65								0.9994	0.9994	0.9992
138.70								0.9994	0.9994	0.9992
138.75								0.9994	0.9994	0.9992
138.80								0.9994	0.9994	0.9992
138.85								0.9994	0.9994	0.9992
138.90								0.9994	0.9994	0.9992
138.95								0.9994	0.9994	0.9992
139.00								0.9994	0.9994	0.9992
139.05								0.9994	0.9994	0.9992
139.10								0.9994	0.9994	0.9992
139.15								0.9994	0.9994	0.9992
139.20								0.9994	0.9994	0.9992
139.25								0.9994	0.9994	0.9992
139.30								0.9994	0.9994	0.9992
139.35								0.9994	0.9994	0.9992
139.40								0.9994	0.9994	0.9992
139.45								0.9994	0.9994	0.9992

PROBABILITY INTEGRAL OF t, THE NON-CENTRAL t-STATISTIC. THIS TABLE GIVES Pr $[t/\sqrt{f} \leq x]$.

f is the number of degrees of freedom; the non-centrality parameter is $\sqrt{f+1}\,K_p$.

K_p is the standardized normal deviate exceeded with probability p.

DEGREES OF FREEDOM **2**

x \ p	.2500	.1500	.1000	.0650	.0400	.0250	.0100	.0040	.0025	.0010
139.50								0.9994	0.9994	0.9992
139.55								0.9994	0.9994	0.9992
139.60								0.9994	0.9994	0.9992
139.65								0.9994	0.9994	0.9992
139.70								0.9994	0.9994	0.9992
139.75								0.9994	0.9994	0.9992
139.80								0.9994	0.9994	0.9992
139.85								0.9994	0.9994	0.9992
139.90								0.9994	0.9994	0.9992
139.95								0.9994	0.9994	0.9992
140.00								0.9994	0.9994	0.9992
140.05								0.9994	0.9994	0.9992
140.10								0.9994	0.9994	0.9992
140.15								0.9994	0.9994	0.9992
140.20								0.9994	0.9994	0.9992
140.25								0.9994	0.9994	0.9992
140.30								0.9994	0.9994	0.9992
140.35								0.9994	0.9994	0.9992
140.40								0.9994	0.9994	0.9992
140.45								0.9994	0.9994	0.9992
140.50								0.9994	0.9994	0.9992
140.55								0.9994	0.9994	0.9992
140.60								0.9994	0.9994	0.9993
140.65								0.9994	0.9994	0.9993
140.70								0.9994	0.9994	0.9993
140.75								0.9994	0.9994	0.9993
140.80								0.9994	0.9994	0.9993
140.85								0.9994	0.9994	0.9993
140.90								0.9994	0.9994	0.9993
140.95								0.9994	0.9994	0.9993
141.00								0.9994	0.9994	0.9993
141.05								0.9994	0.9994	0.9993
141.10								0.9994	0.9994	0.9993
141.15								0.9994	0.9994	0.9993
141.20								0.9994	0.9994	0.9993
141.25								0.9994	0.9994	0.9993
141.30								0.9994	0.9994	0.9993
141.35								0.9994	0.9994	0.9993
141.40								0.9994	0.9994	0.9993
141.45								0.9994	0.9994	0.9993
141.50								0.9994	0.9994	0.9993
141.55								0.9994	0.9994	0.9993
141.60								0.9994	0.9994	0.9993
141.65								0.9994	0.9994	0.9993
141.70								0.9995	0.9994	0.9993
141.75								0.9995	0.9994	0.9993
141.80								0.9995	0.9994	0.9993
141.85								0.9995	0.9994	0.9993
141.90								0.9995	0.9994	0.9993
141.95								0.9995	0.9994	0.9993
142.00								0.9995	0.9994	0.9993
142.05								0.9995	0.9994	0.9993
142.10								0.9995	0.9994	0.9993
142.15								0.9995	0.9994	0.9993
142.20								0.9995	0.9994	0.9993
142.25								0.9995	0.9994	0.9993
142.30								0.9995	0.9994	0.9993
142.35								0.9995	0.9994	0.9993
142.40								0.9995	0.9994	0.9993
142.45								0.9995	0.9994	0.9993

PROBABILITY INTEGRAL OF t, THE NON-CENTRAL t-STATISTIC. THIS TABLE GIVES Pr $[t/\sqrt{f} \leq x]$.

f is the number of degrees of freedom; the non-centrality parameter is $\sqrt{f+1}\,K_p$.
K_p is the standardized normal deviate exceeded with probability p.

DEGREES OF FREEDOM **2**

x \ p	.2500	.1500	.1000	.0650	.0400	.0250	.0100	.0040	.0025	.0010
142.50								0.9995	0.9994	0.9993
142.55								0.9995	0.9994	0.9993
142.60								0.9995	0.9994	0.9993
142.65								0.9995	0.9994	0.9993
142.70								0.9995	0.9994	0.9993
142.75								0.9995	0.9994	0.9993
142.80								0.9995	0.9994	0.9993
142.85								0.9995	0.9994	0.9993
142.90								0.9995	0.9994	0.9993
142.95								0.9995	0.9994	0.9993
143.00								0.9995	0.9994	0.9993
143.05								0.9995	0.9994	0.9993
143.10								0.9995	0.9994	0.9993
143.15								0.9995	0.9994	0.9993
143.20								0.9995	0.9994	0.9993
143.25								0.9995	0.9994	0.9993
143.30								0.9995	0.9994	0.9993
143.35								0.9995	0.9994	0.9993
143.40								0.9995	0.9994	0.9993
143.45								0.9995	0.9994	0.9993
143.50								0.9995	0.9994	0.9993
143.55								0.9995	0.9994	0.9993
143.60								0.9995	0.9994	0.9993
143.65								0.9995	0.9994	0.9993
143.70								0.9995	0.9994	0.9993
143.75								0.9995	0.9994	0.9993
143.80								0.9995	0.9994	0.9993
143.85								0.9995	0.9994	0.9993
143.90								0.9995	0.9994	0.9993
143.95								0.9995	0.9994	0.9993
144.00								0.9995	0.9994	0.9993
144.05								0.9995	0.9994	0.9993
144.10								0.9995	0.9994	0.9993
144.15								0.9995	0.9994	0.9993
144.20								0.9995	0.9994	0.9993
144.25								0.9995	0.9994	0.9993
144.30								0.9995	0.9994	0.9993
144.35								0.9995	0.9994	0.9993
144.40								0.9995	0.9994	0.9993
144.45								0.9995	0.9994	0.9993
144.50								0.9995	0.9994	0.9993
144.55								0.9995	0.9994	0.9993
144.60								0.9995	0.9994	0.9993
144.65								0.9995	0.9994	0.9993
144.70								0.9995	0.9994	0.9993
144.75								0.9995	0.9994	0.9993
144.80								0.9995	0.9994	0.9993
144.85								0.9995	0.9994	0.9993
144.90								0.9995	0.9994	0.9993
144.95								0.9995	0.9994	0.9993
145.00								0.9995	0.9994	0.9993
145.05								0.9995	0.9994	0.9993
145.10								0.9995	0.9994	0.9993
145.15								0.9995	0.9994	0.9993
145.20								0.9995	0.9994	0.9993
145.25								0.9995	0.9994	0.9993
145.30								0.9995	0.9994	0.9993
145.35								0.9995	0.9994	0.9993
145.40								0.9995	0.9994	0.9993
145.45								0.9995	0.9994	0.9993

PROBABILITY INTEGRAL OF t, THE NON-CENTRAL t-STATISTIC. THIS TABLE GIVES Pr $[t/\sqrt{f} \leq x]$.

f is the number of degrees of freedom; the non-centrality parameter is $\sqrt{f+1}\,K_p$.

K_p is the standardized normal deviate exceeded with probability p.

DEGREES OF FREEDOM **2**

x \ p	.2500	.1500	.1000	.0650	.0400	.0250	.0100	.0040	.0025	.0010
145.50								0.9995	0.9994	0.9993
145.55								0.9995	0.9994	0.9993
145.60								0.9995	0.9994	0.9993
145.65								0.9995	0.9994	0.9993
145.70								0.9995	0.9994	0.9993
145.75								0.9995	0.9994	0.9993
145.80								0.9995	0.9994	0.9993
145.85								0.9995	0.9994	0.9993
145.90								0.9995	0.9994	0.9993
145.95								0.9995	0.9994	0.9993
146.00								0.9995	0.9994	0.9993
146.05								0.9995	0.9994	0.9993
146.10								0.9995	0.9994	0.9993
146.15								0.9995	0.9994	0.9993
146.20								0.9995	0.9994	0.9993
146.25								0.9995	0.9994	0.9993
146.30								0.9995	0.9994	0.9993
146.35								0.9995	0.9994	0.9993
146.40								0.9995	0.9994	0.9993
146.45								0.9995	0.9994	0.9993
146.50								0.9995	0.9994	0.9993
146.55								0.9995	0.9994	0.9993
146.60								0.9995	0.9994	0.9993
146.65								0.9995	0.9994	0.9993
146.70								0.9995	0.9994	0.9993
146.75								0.9995	0.9994	0.9993
146.80								0.9995	0.9994	0.9993
146.85								0.9995	0.9994	0.9993
146.90								0.9995	0.9994	0.9993
146.95								0.9995	0.9994	0.9993
147.00								0.9995	0.9994	0.9993
147.05								0.9995	0.9994	0.9993
147.10								0.9995	0.9994	0.9993
147.15								0.9995	0.9994	0.9993
147.20								0.9995	0.9994	0.9993
147.25								0.9995	0.9994	0.9993
147.30								0.9995	0.9994	0.9993
147.35								0.9995	0.9994	0.9993
147.40								0.9995	0.9994	0.9993
147.45								0.9995	0.9994	0.9993
147.50								0.9995	0.9994	0.9993
147.55								0.9995	0.9994	0.9993
147.60								0.9995	0.9994	0.9993
147.65								0.9995	0.9994	0.9993
147.70								0.9995	0.9994	0.9993
147.75								0.9995	0.9994	0.9993
147.80								0.9995	0.9994	0.9993
147.85								0.9995	0.9994	0.9993
147.90								0.9995	0.9994	0.9993
147.95								0.9995	0.9994	0.9993
148.00								0.9995	0.9994	0.9993
148.05								0.9995	0.9994	0.9993
148.10								0.9995	0.9994	0.9993
148.15								0.9995	0.9994	0.9993
148.20								0.9995	0.9994	0.9993
148.25								0.9995	0.9994	0.9993
148.30								0.9995	0.9994	0.9993
148.35								0.9995	0.9994	0.9993
148.40								0.9995	0.9994	0.9993
148.45								0.9995	0.9994	0.9993

PROBABILITY INTEGRAL OF t, THE NON-CENTRAL t-STATISTIC. THIS TABLE GIVES Pr $[t/\sqrt{f} \leq x]$.

f is the number of degrees of freedom; the non-centrality parameter is $\sqrt{f+1}\,K_p$.
K_p is the standardized normal deviate exceeded with probability p.

DEGREES OF FREEDOM **2**

x \ p	.2500	.1500	.1000	.0650	.0400	.0250	.0100	.0040	.0025	.0010
148.50								0.9995	0.9994	0.9993
148.55								0.9995	0.9994	0.9993
148.60								0.9995	0.9994	0.9993
148.65								0.9995	0.9994	0.9993
148.70								0.9995	0.9994	0.9993
148.75								0.9995	0.9994	0.9993
148.80									0.9994	0.9993
148.85									0.9994	0.9993
148.90									0.9994	0.9993
148.95									0.9994	0.9993
149.00									0.9994	0.9993
149.05									0.9994	0.9993
149.10									0.9994	0.9993
149.15									0.9994	0.9993
149.20									0.9994	0.9993
149.25									0.9994	0.9993
149.30									0.9994	0.9993
149.35									0.9994	0.9993
149.40									0.9994	0.9993
149.45									0.9994	0.9993
149.50									0.9994	0.9993
149.55									0.9994	0.9993
149.60									0.9995	0.9993
149.65									0.9995	0.9993
149.70									0.9995	0.9993
149.75									0.9995	0.9993
149.80									0.9995	0.9993
149.85									0.9995	0.9993
149.90									0.9995	0.9993
149.95									0.9995	0.9993
150.00									0.9995	0.9993
150.05									0.9995	0.9993
150.10									0.9995	0.9993
150.15									0.9995	0.9993
150.20									0.9995	0.9993
150.25									0.9995	0.9993
150.30									0.9995	0.9993
150.35									0.9995	0.9993
150.40									0.9995	0.9993
150.45									0.9995	0.9993
150.50									0.9995	0.9993
150.55									0.9995	0.9993
150.60									0.9995	0.9993
150.65									0.9995	0.9993
150.70									0.9995	0.9993
150.75									0.9995	0.9993
150.80									0.9995	0.9993
150.85									0.9995	0.9993
150.90									0.9995	0.9993
150.95									0.9995	0.9993
151.00									0.9995	0.9993
151.05									0.9995	0.9994
151.10									0.9995	0.9994
151.15									0.9995	0.9994
151.20									0.9995	0.9994
151.25									0.9995	0.9994
151.30									0.9995	0.9994
151.35									0.9995	0.9994
151.40									0.9995	0.9994
151.45									0.9995	0.9994

PROBABILITY INTEGRAL OF t, THE NON-CENTRAL t-STATISTIC. THIS TABLE GIVES Pr $[t/\sqrt{f} \leq x]$.

f is the number of degrees of freedom; the non-centrality parameter is $\sqrt{f+1}\,K_p$.

K_p is the standardized normal deviate exceeded with probability p.

DEGREES OF FREEDOM **2**

x \ p	.2500	.1500	.1000	.0650	.0400	.0250	.0100	.0040	.0025	.0010
151.50									0.9995	0.9994
151.55									0.9995	0.9994
151.60									0.9995	0.9994
151.65									0.9995	0.9994
151.70									0.9995	0.9994
151.75									0.9995	0.9994
151.80									0.9995	0.9994
151.85									0.9995	0.9994
151.90									0.9995	0.9994
151.95									0.9995	0.9994
152.00									0.9995	0.9994
152.05									0.9995	0.9994
152.10									0.9995	0.9994
152.15									0.9995	0.9994
152.20									0.9995	0.9994
152.25									0.9995	0.9994
152.30									0.9995	0.9994
152.35									0.9995	0.9994
152.40									0.9995	0.9994
152.45									0.9995	0.9994
152.50									0.9995	0.9994
152.55									0.9995	0.9994
152.60									0.9995	0.9994
152.65									0.9995	0.9994
152.70									0.9995	0.9994
152.75									0.9995	0.9994
152.80									0.9995	0.9994
152.85									0.9995	0.9994
152.90									0.9995	0.9994
152.95									0.9995	0.9994
153.00									0.9995	0.9994
153.05									0.9995	0.9994
153.10									0.9995	0.9994
153.15									0.9995	0.9994
153.20									0.9995	0.9994
153.25									0.9995	0.9994
153.30									0.9995	0.9994
153.35									0.9995	0.9994
153.40									0.9995	0.9994
153.45									0.9995	0.9994
153.50									0.9995	0.9994
153.55									0.9995	0.9994
153.60									0.9995	0.9994
153.65									0.9995	0.9994
153.70									0.9995	0.9994
153.75									0.9995	0.9994
153.80									0.9995	0.9994
153.85									0.9995	0.9994
153.90									0.9995	0.9994
153.95									0.9995	0.9994
154.00									0.9995	0.9994
154.05									0.9995	0.9994
154.10									0.9995	0.9994
154.15									0.9995	0.9994
154.20									0.9995	0.9994
154.25									0.9995	0.9994
154.30									0.9995	0.9994
154.35									0.9995	0.9994
154.40									0.9995	0.9994
154.45									0.9995	0.9994

PROBABILITY INTEGRAL OF t, THE NON-CENTRAL t-STATISTIC. THIS TABLE GIVES Pr $[t/\sqrt{f} \leq x]$.

f is the number of degrees of freedom; the non-centrality parameter is $\sqrt{f+1}\, K_p$.

K_p is the standardized normal deviate exceeded with probability p.

DEGREES OF FREEDOM **2**

x \ p	.2500	.1500	.1000	.0650	.0400	.0250	.0100	.0040	.0025	.0010
154.50									0.9995	0.9994
154.55									0.9995	0.9994
154.60									0.9995	0.9994
154.65									0.9995	0.9994
154.70									0.9995	0.9994
154.75									0.9995	0.9994
154.80									0.9995	0.9994
154.85									0.9995	0.9994
154.90									0.9995	0.9994
154.95									0.9995	0.9994
155.00									0.9995	0.9994
155.05									0.9995	0.9994
155.10									0.9995	0.9994
155.15									0.9995	0.9994
155.20									0.9995	0.9994
155.25									0.9995	0.9994
155.30									0.9995	0.9994
155.35									0.9995	0.9994
155.40									0.9995	0.9994
155.45									0.9995	0.9994
155.50									0.9995	0.9994
155.55									0.9995	0.9994
155.60									0.9995	0.9994
155.65									0.9995	0.9994
155.70									0.9995	0.9994
155.75									0.9995	0.9994
155.80									0.9995	0.9994
155.85									0.9995	0.9994
155.90									0.9995	0.9994
155.95									0.9995	0.9994
156.00									0.9995	0.9994
156.05									0.9995	0.9994
156.10									0.9995	0.9994
156.15									0.9995	0.9994
156.20									0.9995	0.9994
156.25									0.9995	0.9994
156.30									0.9995	0.9994
156.35									0.9995	0.9994
156.40									0.9995	0.9994
156.45									0.9995	0.9994
156.50									0.9995	0.9994
156.55									0.9995	0.9994
156.60									0.9995	0.9994
156.65									0.9995	0.9994
156.70									0.9995	0.9994
156.75									0.9995	0.9994
156.80									0.9995	0.9994
156.85									0.9995	0.9994
156.90									0.9995	0.9994
156.95									0.9995	0.9994
157.00									0.9995	0.9994
157.05									0.9995	0.9994
157.10									0.9995	0.9994
157.15									0.9995	0.9994
157.20									0.9995	0.9994
157.25									0.9995	0.9994
157.30									0.9995	0.9994
157.35									0.9995	0.9994
157.40									0.9995	0.9994
157.45									0.9995	0.9994

PROBABILITY INTEGRAL OF t, THE NON-CENTRAL t-STATISTIC. THIS TABLE GIVES Pr $[t/\sqrt{f} \leq x]$.

f is the number of degrees of freedom; the non-centrality parameter is $\sqrt{f+1}\,K_p$.

K_p is the standardized normal deviate exceeded with probability p.

DEGREES OF FREEDOM **2**

x \ p	.2500	.1500	.1000	.0650	.0400	.0250	.0100	.0040	.0025	.0010
157.50									0.9995	0.9994
157.55									0.9995	0.9994
157.60									0.9995	0.9994
157.65									0.9995	0.9994
157.70										0.9994
157.75										0.9994
157.80										0.9994
157.85										0.9994
157.90										0.9994
157.95										0.9994
158.00										0.9994
158.05										0.9994
158.10										0.9994
158.15										0.9994
158.20										0.9994
158.25										0.9994
158.30										0.9994
158.35										0.9994
158.40										0.9994
158.45										0.9994
158.50										0.9994
158.55										0.9994
158.60										0.9994
158.65										0.9994
158.70										0.9994
158.75										0.9994
158.80										0.9994
158.85										0.9994
158.90										0.9994
158.95										0.9994
159.00										0.9994
159.05										0.9994
159.10										0.9994
159.15										0.9994
159.20										0.9994
159.25										0.9994
159.30										0.9994
159.35										0.9994
159.40										0.9994
159.45										0.9994
159.50										0.9994
159.55										0.9994
159.60										0.9994
159.65										0.9994
159.70										0.9994
159.75										0.9994
159.80										0.9994
159.85										0.9994
159.90										0.9994
159.95										0.9994
160.00										0.9994
160.05										0.9994
160.10										0.9994
160.15										0.9994
160.20										0.9994
160.25										0.9994
160.30										0.9994
160.35										0.9994
160.40										0.9994
160.45										0.9994

PROBABILITY INTEGRAL OF t, THE NON-CENTRAL t-STATISTIC. THIS TABLE GIVES Pr $[t/\sqrt{f} \leq x]$.

f is the number of degrees of freedom; the non-centrality parameter is $\sqrt{f+1}\,K_p$.

K_p is the standardized normal deviate exceeded with probability p.

DEGREES OF FREEDOM **2**

x \ p	.2500	.1500	.1000	.0650	.0400	.0250	.0100	.0040	.0025	.0010
160.50										0.9994
160.55										0.9994
160.60										0.9994
160.65										0.9994
160.70										0.9994
160.75										0.9994
160.80										0.9994
160.85										0.9994
160.90										0.9994
160.95										0.9994
161.00										0.9994
161.05										0.9994
161.10										0.9994
161.15										0.9994
161.20										0.9994
161.25										0.9994
161.30										0.9994
161.35										0.9994
161.40										0.9994
161.45										0.9994
161.50										0.9994
161.55										0.9994
161.60										0.9994
161.65										0.9994
161.70										0.9994
161.75										0.9994
161.80										0.9994
161.85										0.9994
161.90										0.9994
161.95										0.9994
162.00										0.9994
162.05										0.9994
162.10										0.9994
162.15										0.9994
162.20										0.9994
162.25										0.9994
162.30										0.9994
162.35										0.9994
162.40										0.9994
162.45										0.9994
162.50										0.9994
162.55										0.9994
162.60										0.9994
162.65										0.9994
162.70										0.9994
162.75										0.9994
162.80										0.9994
162.85										0.9994
162.90										0.9994
162.95										0.9994
163.00										0.9994
163.05										0.9994
163.10										0.9994
163.15										0.9994
163.20										0.9994
163.25										0.9994
163.30										0.9994
163.35										0.9994
163.40										0.9994
163.45										0.9994

PROBABILITY INTEGRAL OF t, THE NON-CENTRAL t-STATISTIC. THIS TABLE GIVES Pr $[t/\sqrt{f} \leq x]$.

f is the number of degrees of freedom; the non-centrality parameter is $\sqrt{f+1}\,K_p$.

K_p is the standardized normal deviate exceeded with probability p.

DEGREES OF FREEDOM **2**

x \ p	.2500	.1500	.1000	.0650	.0400	.0250	.0100	.0040	.0025	.0010
163.50										0.9994
163.55										0.9994
163.60										0.9994
163.65										0.9994
163.70										0.9994
163.75										0.9994
163.80										0.9994
163.85										0.9994
163.90										0.9994
163.95										0.9994
164.00										0.9994
164.05										0.9994
164.10										0.9994
164.15										0.9994
164.20										0.9995
164.25										0.9995
164.30										0.9995
164.35										0.9995
164.40										0.9995
164.45										0.9995
164.50										0.9995
164.55										0.9995
164.60										0.9995
164.65										0.9995
164.70										0.9995
164.75										0.9995
164.80										0.9995
164.85										0.9995
164.90										0.9995
164.95										0.9995
165.00										0.9995
165.05										0.9995
165.10										0.9995
165.15										0.9995
165.20										0.9995
165.25										0.9995
165.30										0.9995
165.35										0.9995
165.40										0.9995
165.45										0.9995
165.50										0.9995
165.55										0.9995
165.60										0.9995
165.65										0.9995
165.70										0.9995
165.75										0.9995
165.80										0.9995
165.85										0.9995
165.90										0.9995
165.95										0.9995
166.00										0.9995
166.05										0.9995
166.10										0.9995
166.15										0.9995
166.20										0.9995
166.25										0.9995
166.30										0.9995
166.35										0.9995
166.40										0.9995
166.45										0.9995

PROBABILITY INTEGRAL OF t, THE NON-CENTRAL t-STATISTIC. THIS TABLE GIVES Pr $[t/\sqrt{f} \leq x]$.

f is the number of degrees of freedom; the non-centrality parameter is $\sqrt{f+1}\,K_p$.

K_p is the standardized normal deviate exceeded with probability p.

DEGREES OF FREEDOM **2**

x \ p	.2500	.1500	.1000	.0650	.0400	.0250	.0100	.0040	.0025	.0010
166.50										0.9995
166.55										0.9995
166.60										0.9995
166.65										0.9995
166.70										0.9995
166.75										0.9995
166.80										0.9995
166.85										0.9995
166.90										0.9995
166.95										0.9995
167.00										0.9995
167.05										0.9995
167.10										0.9995
167.15										0.9995
167.20										0.9995
167.25										0.9995
167.30										0.9995
167.35										0.9995
167.40										0.9995
167.45										0.9995
167.50										0.9995
167.55										0.9995
167.60										0.9995
167.65										0.9995
167.70										0.9995
167.75										0.9995
167.80										0.9995
167.85										0.9995
167.90										0.9995
167.95										0.9995
168.00										0.9995
168.05										0.9995
168.10										0.9995
168.15										0.9995
168.20										0.9995
168.25										0.9995
168.30										0.9995
168.35										0.9995
168.40										0.9995
168.45										0.9995
168.50										0.9995
168.55										0.9995
168.60										0.9995
168.65										0.9995
168.70										0.9995
168.75										0.9995
168.80										0.9995
168.85										0.9995
168.90										0.9995
168.95										0.9995
169.00										0.9995
169.05										0.9995
169.10										0.9995
169.15										0.9995
169.20										0.9995
169.25										0.9995
169.30										0.9995
169.35										0.9995
169.40										0.9995
169.45										0.9995

PROBABILITY INTEGRAL OF t, THE NON-CENTRAL t-STATISTIC. THIS TABLE GIVES Pr $[t/\sqrt{f} \leq x]$.

f is the number of degrees of freedom; the non-centrality parameter is $\sqrt{f+1}\,K_p$.

K_p is the standardized normal deviate exceeded with probability p.

DEGREES OF FREEDOM **2**

x \ p	.2500	.1500	.1000	.0650	.0400	.0250	.0100	.0040	.0025	.0010
169.50										0.9995
169.55										0.9995
169.60										0.9995
169.65										0.9995
169.70										0.9995
169.75										0.9995
169.80										0.9995
169.85										0.9995
169.90										0.9995
169.95										0.9995
170.00										0.9995
170.05										0.9995
170.10										0.9995
170.15										0.9995
170.20										0.9995
170.25										0.9995
170.30										0.9995
170.35										0.9995
170.40										0.9995
170.45										0.9995
170.50										0.9995
170.55										0.9995
170.60										0.9995
170.65										0.9995
170.70										0.9995
170.75										0.9995
170.80										0.9995
170.85										0.9995
170.90										0.9995
170.95										0.9995
171.00										0.9995
171.05										0.9995
171.10										0.9995
171.15										0.9995
171.20										0.9995
171.25										0.9995
171.30										0.9995
171.35										0.9995
171.40										0.9995
171.45										0.9995
171.50										0.9995
171.55										0.9995
171.60										0.9995
171.65										0.9995
171.70										0.9995
171.75										0.9995
171.80										0.9995
171.85										0.9995
171.90										0.9995
171.95										0.9995
172.00										0.9995
172.05										0.9995
172.10										0.9995
172.15										0.9995
172.20										0.9995
172.25										0.9995
172.30										0.9995
172.35										0.9995
172.40										0.9995

PROBABILITY INTEGRAL OF t, THE NON-CENTRAL t-STATISTIC. THIS TABLE GIVES Pr $[t/\sqrt{f} \leq x]$.

f is the number of degrees of freedom; the non-centrality parameter is $\sqrt{f+1}\, K_p$.

K_p is the standardized normal deviate exceeded with probability p.

DEGREES OF FREEDOM 3

x \ p	.2500	.1500	.1000	.0650	.0400	.0250	.0100	.0040	.0025	.0010
-5.60	0.0000	0.0000	0.0000	0.0000	0.0000	0.0000	0.0000	0.0000	0.0000	0.0000
-5.55	0.0001	0.0000	0.0000	0.0000	0.0000	0.0000	0.0000	0.0000	0.0000	0.0000
-5.50	0.0001	0.0000	0.0000	0.0000	0.0000	0.0000	0.0000	0.0000	0.0000	0.0000
-5.45	0.0001	0.0000	0.0000	0.0000	0.0000	0.0000	0.0000	0.0000	0.0000	0.0000
-5.40	0.0001	0.0000	0.0000	0.0000	0.0000	0.0000	0.0000	0.0000	0.0000	0.0000
-5.35	0.0001	0.0000	0.0000	0.0000	0.0000	0.0000	0.0000	0.0000	0.0000	0.0000
-5.30	0.0001	0.0000	0.0000	0.0000	0.0000	0.0000	0.0000	0.0000	0.0000	0.0000
-5.25	0.0001	0.0000	0.0000	0.0000	0.0000	0.0000	0.0000	0.0000	0.0000	0.0000
-5.20	0.0001	0.0000	0.0000	0.0000	0.0000	0.0000	0.0000	0.0000	0.0000	0.0000
-5.15	0.0001	0.0000	0.0000	0.0000	0.0000	0.0000	0.0000	0.0000	0.0000	0.0000
-5.10	0.0001	0.0000	0.0000	0.0000	0.0000	0.0000	0.0000	0.0000	0.0000	0.0000
-5.05	0.0001	0.0000	0.0000	0.0000	0.0000	0.0000	0.0000	0.0000	0.0000	0.0000
-5.00	0.0001	0.0000	0.0000	0.0000	0.0000	0.0000	0.0000	0.0000	0.0000	0.0000
-4.95	0.0001	0.0000	0.0000	0.0000	0.0000	0.0000	0.0000	0.0000	0.0000	0.0000
-4.90	0.0001	0.0000	0.0000	0.0000	0.0000	0.0000	0.0000	0.0000	0.0000	0.0000
-4.85	0.0001	0.0000	0.0000	0.0000	0.0000	0.0000	0.0000	0.0000	0.0000	0.0000
-4.80	0.0001	0.0000	0.0000	0.0000	0.0000	0.0000	0.0000	0.0000	0.0000	0.0000
-4.75	0.0001	0.0000	0.0000	0.0000	0.0000	0.0000	0.0000	0.0000	0.0000	0.0000
-4.70	0.0001	0.0000	0.0000	0.0000	0.0000	0.0000	0.0000	0.0000	0.0000	0.0000
-4.65	0.0001	0.0000	0.0000	0.0000	0.0000	0.0000	0.0000	0.0000	0.0000	0.0000
-4.60	0.0001	0.0000	0.0000	0.0000	0.0000	0.0000	0.0000	0.0000	0.0000	0.0000
-4.55	0.0001	0.0000	0.0000	0.0000	0.0000	0.0000	0.0000	0.0000	0.0000	0.0000
-4.50	0.0001	0.0000	0.0000	0.0000	0.0000	0.0000	0.0000	0.0000	0.0000	0.0000
-4.45	0.0001	0.0000	0.0000	0.0000	0.0000	0.0000	0.0000	0.0000	0.0000	0.0000
-4.40	0.0001	0.0000	0.0000	0.0000	0.0000	0.0000	0.0000	0.0000	0.0000	0.0000
-4.35	0.0001	0.0000	0.0000	0.0000	0.0000	0.0000	0.0000	0.0000	0.0000	0.0000
-4.30	0.0001	0.0000	0.0000	0.0000	0.0000	0.0000	0.0000	0.0000	0.0000	0.0000
-4.25	0.0001	0.0000	0.0000	0.0000	0.0000	0.0000	0.0000	0.0000	0.0000	0.0000
-4.20	0.0001	0.0000	0.0000	0.0000	0.0000	0.0000	0.0000	0.0000	0.0000	0.0000
-4.15	0.0001	0.0000	0.0000	0.0000	0.0000	0.0000	0.0000	0.0000	0.0000	0.0000
-4.10	0.0001	0.0000	0.0000	0.0000	0.0000	0.0000	0.0000	0.0000	0.0000	0.0000
-4.05	0.0001	0.0000	0.0000	0.0000	0.0000	0.0000	0.0000	0.0000	0.0000	0.0000
-4.00	0.0001	0.0000	0.0000	0.0000	0.0000	0.0000	0.0000	0.0000	0.0000	0.0000
-3.95	0.0001	0.0000	0.0000	0.0000	0.0000	0.0000	0.0000	0.0000	0.0000	0.0000
-3.90	0.0002	0.0000	0.0000	0.0000	0.0000	0.0000	0.0000	0.0000	0.0000	0.0000
-3.85	0.0002	0.0000	0.0000	0.0000	0.0000	0.0000	0.0000	0.0000	0.0000	0.0000
-3.80	0.0002	0.0000	0.0000	0.0000	0.0000	0.0000	0.0000	0.0000	0.0000	0.0000
-3.75	0.0002	0.0000	0.0000	0.0000	0.0000	0.0000	0.0000	0.0000	0.0000	0.0000
-3.70	0.0002	0.0000	0.0000	0.0000	0.0000	0.0000	0.0000	0.0000	0.0000	0.0000
-3.65	0.0002	0.0000	0.0000	0.0000	0.0000	0.0000	0.0000	0.0000	0.0000	0.0000
-3.60	0.0002	0.0000	0.0000	0.0000	0.0000	0.0000	0.0000	0.0000	0.0000	0.0000
-3.55	0.0002	0.0000	0.0000	0.0000	0.0000	0.0000	0.0000	0.0000	0.0000	0.0000
-3.50	0.0002	0.0000	0.0000	0.0000	0.0000	0.0000	0.0000	0.0000	0.0000	0.0000
-3.45	0.0002	0.0000	0.0000	0.0000	0.0000	0.0000	0.0000	0.0000	0.0000	0.0000
-3.40	0.0002	0.0000	0.0000	0.0000	0.0000	0.0000	0.0000	0.0000	0.0000	0.0000
-3.35	0.0002	0.0000	0.0000	0.0000	0.0000	0.0000	0.0000	0.0000	0.0000	0.0000
-3.30	0.0003	0.0000	0.0000	0.0000	0.0000	0.0000	0.0000	0.0000	0.0000	0.0000
-3.25	0.0003	0.0000	0.0000	0.0000	0.0000	0.0000	0.0000	0.0000	0.0000	0.0000
-3.20	0.0003	0.0000	0.0000	0.0000	0.0000	0.0000	0.0000	0.0000	0.0000	0.0000
-3.15	0.0003	0.0000	0.0000	0.0000	0.0000	0.0000	0.0000	0.0000	0.0000	0.0000
-3.10	0.0003	0.0000	0.0000	0.0000	0.0000	0.0000	0.0000	0.0000	0.0000	0.0000
-3.05	0.0003	0.0000	0.0000	0.0000	0.0000	0.0000	0.0000	0.0000	0.0000	0.0000
-3.00	0.0003	0.0000	0.0000	0.0000	0.0000	0.0000	0.0000	0.0000	0.0000	0.0000
-2.95	0.0004	0.0000	0.0000	0.0000	0.0000	0.0000	0.0000	0.0000	0.0000	0.0000
-2.90	0.0004	0.0000	0.0000	0.0000	0.0000	0.0000	0.0000	0.0000	0.0000	0.0000
-2.85	0.0004	0.0000	0.0000	0.0000	0.0000	0.0000	0.0000	0.0000	0.0000	0.0000
-2.80	0.0004	0.0001	0.0000	0.0000	0.0000	0.0000	0.0000	0.0000	0.0000	0.0000

PROBABILITY INTEGRAL OF t, THE NON-CENTRAL t-STATISTIC. THIS TABLE GIVES Pr $[t/\sqrt{f} \leq x]$.

f is the number of degrees of freedom; the non-centrality parameter is $\sqrt{f+1}\,K_p$.

K_p is the standardized normal deviate exceeded with probability p.

DEGREES OF FREEDOM **3**

x \ p	.2500	.1500	.1000	.0650	.0400	.0250	.0100	.0040	.0025	.0010
-2.75	0.0004	0.0001	0.0000	0.0000	0.0000	0.0000	0.0000	0.0000	0.0000	0.0000
-2.70	0.0005	0.0001	0.0000	0.0000	0.0000	0.0000	0.0000	0.0000	0.0000	0.0000
-2.65	0.0005	0.0001	0.0000	0.0000	0.0000	0.0000	0.0000	0.0000	0.0000	0.0000
-2.60	0.0005	0.0001	0.0000	0.0000	0.0000	0.0000	0.0000	0.0000	0.0000	0.0000
-2.55	0.0005	0.0001	0.0000	0.0000	0.0000	0.0000	0.0000	0.0000	0.0000	0.0000
-2.50	0.0006	0.0001	0.0000	0.0000	0.0000	0.0000	0.0000	0.0000	0.0000	0.0000
-2.45	0.0006	0.0001	0.0000	0.0000	0.0000	0.0000	0.0000	0.0000	0.0000	0.0000
-2.40	0.0006	0.0001	0.0000	0.0000	0.0000	0.0000	0.0000	0.0000	0.0000	0.0000
-2.35	0.0007	0.0001	0.0000	0.0000	0.0000	0.0000	0.0000	0.0000	0.0000	0.0000
-2.30	0.0007	0.0001	0.0000	0.0000	0.0000	0.0000	0.0000	0.0000	0.0000	0.0000
-2.25	0.0008	0.0001	0.0000	0.0000	0.0000	0.0000	0.0000	0.0000	0.0000	0.0000
-2.20	0.0008	0.0001	0.0000	0.0000	0.0000	0.0000	0.0000	0.0000	0.0000	0.0000
-2.15	0.0009	0.0001	0.0000	0.0000	0.0000	0.0000	0.0000	0.0000	0.0000	0.0000
-2.10	0.0009	0.0001	0.0000	0.0000	0.0000	0.0000	0.0000	0.0000	0.0000	0.0000
-2.05	0.0010	0.0001	0.0000	0.0000	0.0000	0.0000	0.0000	0.0000	0.0000	0.0000
-2.00	0.0011	0.0001	0.0000	0.0000	0.0000	0.0000	0.0000	0.0000	0.0000	0.0000
-1.95	0.0011	0.0001	0.0000	0.0000	0.0000	0.0000	0.0000	0.0000	0.0000	0.0000
-1.90	0.0012	0.0001	0.0000	0.0000	0.0000	0.0000	0.0000	0.0000	0.0000	0.0000
-1.85	0.0013	0.0002	0.0000	0.0000	0.0000	0.0000	0.0000	0.0000	0.0000	0.0000
-1.80	0.0014	0.0002	0.0000	0.0000	0.0000	0.0000	0.0000	0.0000	0.0000	0.0000
-1.75	0.0015	0.0002	0.0000	0.0000	0.0000	0.0000	0.0000	0.0000	0.0000	0.0000
-1.70	0.0016	0.0002	0.0000	0.0000	0.0000	0.0000	0.0000	0.0000	0.0000	0.0000
-1.65	0.0018	0.0002	0.0000	0.0000	0.0000	0.0000	0.0000	0.0000	0.0000	0.0000
-1.60	0.0019	0.0002	0.0000	0.0000	0.0000	0.0000	0.0000	0.0000	0.0000	0.0000
-1.55	0.0021	0.0003	0.0001	0.0000	0.0000	0.0000	0.0000	0.0000	0.0000	0.0000
-1.50	0.0023	0.0003	0.0001	0.0000	0.0000	0.0000	0.0000	0.0000	0.0000	0.0000
-1.45	0.0025	0.0003	0.0001	0.0000	0.0000	0.0000	0.0000	0.0000	0.0000	0.0000
-1.40	0.0027	0.0003	0.0001	0.0000	0.0000	0.0000	0.0000	0.0000	0.0000	0.0000
-1.35	0.0029	0.0004	0.0001	0.0000	0.0000	0.0000	0.0000	0.0000	0.0000	0.0000
-1.30	0.0032	0.0004	0.0001	0.0000	0.0000	0.0000	0.0000	0.0000	0.0000	0.0000
-1.25	0.0035	0.0005	0.0001	0.0000	0.0000	0.0000	0.0000	0.0000	0.0000	0.0000
-1.20	0.0039	0.0005	0.0001	0.0000	0.0000	0.0000	0.0000	0.0000	0.0000	0.0000
-1.15	0.0043	0.0006	0.0001	0.0000	0.0000	0.0000	0.0000	0.0000	0.0000	0.0000
-1.10	0.0048	0.0006	0.0001	0.0000	0.0000	0.0000	0.0000	0.0000	0.0000	0.0000
-1.05	0.0053	0.0007	0.0001	0.0000	0.0000	0.0000	0.0000	0.0000	0.0000	0.0000
-1.00	0.0059	0.0008	0.0002	0.0000	0.0000	0.0000	0.0000	0.0000	0.0000	0.0000
-0.95	0.0066	0.0009	0.0002	0.0000	0.0000	0.0000	0.0000	0.0000	0.0000	0.0000
-0.90	0.0074	0.0010	0.0002	0.0000	0.0000	0.0000	0.0000	0.0000	0.0000	0.0000
-0.85	0.0083	0.0011	0.0002	0.0000	0.0000	0.0000	0.0000	0.0000	0.0000	0.0000
-0.80	0.0094	0.0013	0.0003	0.0001	0.0000	0.0000	0.0000	0.0000	0.0000	0.0000
-0.75	0.0106	0.0015	0.0003	0.0001	0.0000	0.0000	0.0000	0.0000	0.0000	0.0000
-0.70	0.0120	0.0017	0.0004	0.0001	0.0000	0.0000	0.0000	0.0000	0.0000	0.0000
-0.65	0.0137	0.0020	0.0004	0.0001	0.0000	0.0000	0.0000	0.0000	0.0000	0.0000
-0.60	0.0156	0.0023	0.0005	0.0001	0.0000	0.0000	0.0000	0.0000	0.0000	0.0000
-0.55	0.0179	0.0026	0.0006	0.0001	0.0000	0.0000	0.0000	0.0000	0.0000	0.0000
-0.50	0.0205	0.0031	0.0007	0.0001	0.0000	0.0000	0.0000	0.0000	0.0000	0.0000
-0.45	0.0236	0.0036	0.0008	0.0002	0.0000	0.0000	0.0000	0.0000	0.0000	0.0000
-0.40	0.0273	0.0043	0.0010	0.0002	0.0000	0.0000	0.0000	0.0000	0.0000	0.0000
-0.35	0.0315	0.0051	0.0012	0.0002	0.0000	0.0000	0.0000	0.0000	0.0000	0.0000
-0.30	0.0365	0.0061	0.0014	0.0003	0.0001	0.0000	0.0000	0.0000	0.0000	0.0000
-0.25	0.0423	0.0074	0.0017	0.0004	0.0001	0.0000	0.0000	0.0000	0.0000	0.0000
-0.20	0.0491	0.0089	0.0021	0.0005	0.0001	0.0000	0.0000	0.0000	0.0000	0.0000
-0.15	0.0570	0.0107	0.0026	0.0006	0.0001	0.0000	0.0000	0.0000	0.0000	0.0000
-0.10	0.0661	0.0130	0.0033	0.0007	0.0001	0.0000	0.0000	0.0000	0.0000	0.0000
-0.05	0.0766	0.0157	0.0041	0.0010	0.0002	0.0000	0.0000	0.0000	0.0000	0.0000
0.00	0.0887	0.0191	0.0052	0.0012	0.0002	0.0001	0.0000	0.0000	0.0000	0.0000
0.05	0.1023	0.0232	0.0065	0.0016	0.0003	0.0001	0.0000	0.0000	0.0000	0.0000
0.10	0.1177	0.0281	0.0082	0.0021	0.0004	0.0001	0.0000	0.0000	0.0000	0.0000
0.15	0.1348	0.0341	0.0104	0.0028	0.0006	0.0001	0.0000	0.0000	0.0000	0.0000
0.20	0.1537	0.0411	0.0131	0.0036	0.0008	0.0002	0.0000	0.0000	0.0000	0.0000

PROBABILITY INTEGRAL OF t, THE NON-CENTRAL t-STATISTIC. THIS TABLE GIVES Pr $[t/\sqrt{f} \leq x]$.

f is the number of degrees of freedom; the non-centrality parameter is $\sqrt{f+1}\,K_p$.

K_p is the standardized normal deviate exceeded with probability p.

DEGREES OF FREEDOM **3**

x \ p	.2500	.1500	.1000	.0650	.0400	.0250	.0100	.0040	.0025	.0010
0.25	0.1744	0.0495	0.0165	0.0048	0.0011	0.0003	0.0000	0.0000	0.0000	0.0000
0.30	0.1968	0.0592	0.0206	0.0063	0.0016	0.0004	0.0000	0.0000	0.0000	0.0000
0.35	0.2207	0.0704	0.0257	0.0082	0.0021	0.0006	0.0000	0.0000	0.0000	0.0000
0.40	0.2460	0.0832	0.0318	0.0107	0.0029	0.0008	0.0001	0.0000	0.0000	0.0000
0.45	0.2726	0.0976	0.0391	0.0138	0.0040	0.0012	0.0001	0.0000	0.0000	0.0000
0.50	0.3000	0.1137	0.0476	0.0177	0.0054	0.0017	0.0002	0.0000	0.0000	0.0000
0.55	0.3282	0.1314	0.0576	0.0224	0.0073	0.0023	0.0002	0.0000	0.0000	0.0000
0.60	0.3569	0.1506	0.0690	0.0282	0.0097	0.0033	0.0004	0.0000	0.0000	0.0000
0.65	0.3857	0.1712	0.0818	0.0350	0.0127	0.0045	0.0006	0.0001	0.0000	0.0000
0.70	0.4145	0.1932	0.0962	0.0431	0.0164	0.0062	0.0008	0.0001	0.0000	0.0000
0.75	0.4430	0.2162	0.1120	0.0524	0.0210	0.0083	0.0013	0.0002	0.0001	0.0000
0.80	0.4711	0.2402	0.1292	0.0630	0.0265	0.0110	0.0018	0.0003	0.0001	0.0000
0.85	0.4985	0.2650	0.1477	0.0749	0.0331	0.0144	0.0026	0.0004	0.0002	0.0000
0.90	0.5252	0.2903	0.1674	0.0882	0.0407	0.0185	0.0037	0.0007	0.0003	0.0001
0.95	0.5509	0.3159	0.1881	0.1027	0.0495	0.0235	0.0051	0.0010	0.0004	0.0001
1.00	0.5757	0.3417	0.2097	0.1185	0.0594	0.0293	0.0069	0.0015	0.0007	0.0001
1.05	0.5994	0.3674	0.2320	0.1354	0.0705	0.0361	0.0091	0.0022	0.0010	0.0002
1.10	0.6220	0.3930	0.2549	0.1534	0.0828	0.0440	0.0120	0.0031	0.0015	0.0004
1.15	0.6436	0.4183	0.2782	0.1723	0.0961	0.0529	0.0154	0.0042	0.0021	0.0006
1.20	0.6640	0.4431	0.3018	0.1920	0.1106	0.0628	0.0195	0.0057	0.0030	0.0008
1.25	0.6833	0.4673	0.3254	0.2123	0.1260	0.0738	0.0244	0.0076	0.0041	0.0012
1.30	0.7015	0.4909	0.3491	0.2332	0.1424	0.0857	0.0300	0.0099	0.0055	0.0017
1.35	0.7187	0.5138	0.3725	0.2544	0.1595	0.0987	0.0365	0.0127	0.0073	0.0024
1.40	0.7349	0.5360	0.3957	0.2760	0.1774	0.1125	0.0437	0.0161	0.0095	0.0033
1.45	0.7501	0.5573	0.4185	0.2976	0.1959	0.1272	0.0519	0.0201	0.0122	0.0045
1.50	0.7644	0.5779	0.4409	0.3194	0.2148	0.1426	0.0609	0.0247	0.0153	0.0059
1.55	0.7778	0.5976	0.4628	0.3410	0.2342	0.1588	0.0707	0.0300	0.0190	0.0077
1.60	0.7904	0.6164	0.4841	0.3625	0.2538	0.1755	0.0813	0.0360	0.0233	0.0099
1.65	0.8022	0.6345	0.5048	0.3838	0.2736	0.1927	0.0927	0.0426	0.0283	0.0124
1.70	0.8132	0.6517	0.5248	0.4047	0.2935	0.2103	0.1049	0.0500	0.0338	0.0154
1.75	0.8236	0.6681	0.5442	0.4253	0.3134	0.2283	0.1177	0.0581	0.0400	0.0189
1.80	0.8333	0.6837	0.5629	0.4455	0.3333	0.2465	0.1311	0.0669	0.0468	0.0229
1.85	0.8423	0.6986	0.5810	0.4652	0.3530	0.2649	0.1451	0.0764	0.0543	0.0275
1.90	0.8508	0.7127	0.5983	0.4844	0.3725	0.2834	0.1596	0.0865	0.0625	0.0326
1.95	0.8588	0.7261	0.6150	0.5031	0.3918	0.3019	0.1746	0.0972	0.0712	0.0382
2.00	0.8663	0.7388	0.6310	0.5212	0.4107	0.3203	0.1899	0.1085	0.0806	0.0444
2.05	0.8733	0.7509	0.6463	0.5388	0.4293	0.3387	0.2055	0.1204	0.0906	0.0512
2.10	0.8798	0.7624	0.6610	0.5558	0.4476	0.3569	0.2214	0.1327	0.1011	0.0585
2.15	0.8860	0.7732	0.6750	0.5722	0.4654	0.3748	0.2375	0.1455	0.1121	0.0664
2.20	0.8918	0.7835	0.6885	0.5881	0.4828	0.3926	0.2537	0.1587	0.1236	0.0748
2.25	0.8972	0.7933	0.7013	0.6034	0.4997	0.4101	0.2700	0.1722	0.1355	0.0837
2.30	0.9023	0.8026	0.7136	0.6181	0.5162	0.4273	0.2863	0.1861	0.1478	0.0931
2.35	0.9070	0.8114	0.7253	0.6323	0.5323	0.4441	0.3026	0.2002	0.1605	0.1030
2.40	0.9115	0.8197	0.7365	0.6460	0.5478	0.4606	0.3189	0.2145	0.1735	0.1133
2.45	0.9158	0.8276	0.7472	0.6591	0.5629	0.4767	0.3351	0.2289	0.1868	0.1240
2.50	0.9198	0.8350	0.7573	0.6717	0.5775	0.4925	0.3511	0.2435	0.2002	0.1350
2.55	0.9235	0.8421	0.7671	0.6839	0.5916	0.5078	0.3670	0.2582	0.2139	0.1464
2.60	0.9270	0.8489	0.7763	0.6955	0.6053	0.5228	0.3827	0.2729	0.2277	0.1581
2.65	0.9304	0.8553	0.7852	0.7067	0.6185	0.5373	0.3981	0.2876	0.2416	0.1701
2.70	0.9335	0.8613	0.7936	0.7174	0.6313	0.5515	0.4134	0.3023	0.2556	0.1823
2.75	0.9365	0.8671	0.8017	0.7276	0.6436	0.5652	0.4284	0.3170	0.2697	0.1946
2.80	0.9393	0.8725	0.8093	0.7375	0.6555	0.5786	0.4431	0.3315	0.2837	0.2072
2.85	0.9419	0.8777	0.8167	0.7469	0.6670	0.5915	0.4576	0.3460	0.2978	0.2199
2.90	0.9444	0.8827	0.8237	0.7560	0.6780	0.6041	0.4717	0.3603	0.3118	0.2327
2.95	0.9468	0.8874	0.8303	0.7647	0.6887	0.6162	0.4856	0.3745	0.3257	0.2456
3.00	0.9490	0.8918	0.8367	0.7730	0.6989	0.6280	0.4992	0.3885	0.3395	0.2585
3.05	0.9511	0.8961	0.8428	0.7810	0.7088	0.6394	0.5124	0.4024	0.3532	0.2715
3.10	0.9532	0.9001	0.8486	0.7887	0.7183	0.6504	0.5254	0.4160	0.3668	0.2844
3.15	0.9551	0.9039	0.8542	0.7960	0.7275	0.6611	0.5380	0.4294	0.3803	0.2974
3.20	0.9569	0.9076	0.8595	0.8030	0.7364	0.6714	0.5503	0.4426	0.3936	0.3103

PROBABILITY INTEGRAL OF t, THE NON-CENTRAL t-STATISTIC. THIS TABLE GIVES Pr $[t/\sqrt{f} \leq x]$.

f is the number of degrees of freedom; the non-centrality parameter is $\sqrt{f+1}\,K_p$.
K_p is the standardized normal deviate exceeded with probability p.

DEGREES OF FREEDOM **3**

x \ P	.2500	.1500	.1000	.0650	.0400	.0250	.0100	.0040	.0025	.0010
3.25	0.9586	0.9111	0.8645	0.8098	0.7449	0.6813	0.5623	0.4556	0.4067	0.3231
3.30	0.9602	0.9144	0.8694	0.8163	0.7530	0.6910	0.5740	0.4684	0.4196	0.3359
3.35	0.9618	0.9176	0.8740	0.8225	0.7609	0.7003	0.5854	0.4809	0.4324	0.3485
3.40	0.9632	0.9206	0.8784	0.8284	0.7685	0.7093	0.5965	0.4931	0.4449	0.3611
3.45	0.9646	0.9235	0.8827	0.8341	0.7758	0.7180	0.6073	0.5051	0.4572	0.3735
3.50	0.9659	0.9262	0.8867	0.8396	0.7829	0.7264	0.6178	0.5169	0.4694	0.3859
3.55	0.9672	0.9288	0.8906	0.8449	0.7896	0.7345	0.6280	0.5284	0.4812	0.3980
3.60	0.9684	0.9313	0.8943	0.8499	0.7962	0.7423	0.6379	0.5396	0.4929	0.4100
3.65	0.9696	0.9337	0.8979	0.8548	0.8025	0.7499	0.6475	0.5506	0.5043	0.4219
3.70	0.9707	0.9360	0.9013	0.8595	0.8085	0.7572	0.6568	0.5614	0.5156	0.4336
3.75	0.9717	0.9382	0.9045	0.8639	0.8143	0.7643	0.6659	0.5718	0.5265	0.4451
3.80	0.9727	0.9403	0.9077	0.8682	0.8199	0.7711	0.6747	0.5821	0.5373	0.4564
3.85	0.9737	0.9423	0.9107	0.8724	0.8254	0.7777	0.6833	0.5921	0.5478	0.4676
3.90	0.9746	0.9442	0.9136	0.8763	0.8306	0.7840	0.6916	0.6018	0.5581	0.4785
3.95	0.9754	0.9461	0.9163	0.8802	0.8356	0.7902	0.6996	0.6113	0.5681	0.4893
4.00	0.9763	0.9478	0.9190	0.8838	0.8404	0.7961	0.7075	0.6206	0.5780	0.4999
4.05	0.9771	0.9495	0.9215	0.8874	0.8451	0.8018	0.7150	0.6297	0.5876	0.5103
4.10	0.9778	0.9512	0.9240	0.8908	0.8496	0.8074	0.7224	0.6385	0.5970	0.5205
4.15	0.9786	0.9527	0.9263	0.8940	0.8539	0.8127	0.7296	0.6470	0.6061	0.5305
4.20	0.9793	0.9542	0.9286	0.8972	0.8581	0.8179	0.7365	0.6554	0.6151	0.5403
4.25	0.9799	0.9556	0.9307	0.9002	0.8622	0.8229	0.7432	0.6635	0.6238	0.5499
4.30	0.9806	0.9570	0.9328	0.9031	0.8661	0.8277	0.7497	0.6715	0.6323	0.5593
4.35	0.9812	0.9583	0.9348	0.9059	0.8698	0.8324	0.7561	0.6792	0.6406	0.5685
4.40	0.9818	0.9596	0.9368	0.9087	0.8734	0.8369	0.7622	0.6867	0.6488	0.5775
4.45	0.9823	0.9608	0.9386	0.9113	0.8769	0.8413	0.7682	0.6940	0.6567	0.5863
4.50	0.9829	0.9620	0.9404	0.9138	0.8803	0.8455	0.7740	0.7012	0.6644	0.5950
4.55	0.9834	0.9631	0.9421	0.9162	0.8836	0.8496	0.7796	0.7081	0.6719	0.6034
4.60	0.9839	0.9642	0.9438	0.9185	0.8867	0.8536	0.7850	0.7149	0.6792	0.6117
4.65	0.9844	0.9652	0.9454	0.9208	0.8898	0.8574	0.7903	0.7215	0.6864	0.6198
4.70	0.9848	0.9662	0.9469	0.9230	0.8927	0.8611	0.7954	0.7279	0.6934	0.6277
4.75	0.9853	0.9672	0.9484	0.9251	0.8956	0.8647	0.8004	0.7341	0.7002	0.6354
4.80	0.9857	0.9681	0.9498	0.9271	0.8983	0.8681	0.8052	0.7401	0.7068	0.6430
4.85	0.9861	0.9690	0.9512	0.9291	0.9010	0.8715	0.8099	0.7460	0.7132	0.6504
4.90	0.9865	0.9699	0.9526	0.9310	0.9036	0.8747	0.8145	0.7518	0.7195	0.6576
4.95	0.9869	0.9707	0.9538	0.9328	0.9061	0.8779	0.8189	0.7574	0.7257	0.6647
5.00	0.9873	0.9715	0.9551	0.9346	0.9085	0.8809	0.8232	0.7628	0.7317	0.6716
5.05	0.9876	0.9723	0.9563	0.9363	0.9108	0.8839	0.8274	0.7681	0.7375	0.6783
5.10	0.9880	0.9730	0.9574	0.9379	0.9130	0.8868	0.8314	0.7733	0.7432	0.6849
5.15	0.9883	0.9737	0.9585	0.9395	0.9152	0.8895	0.8353	0.7783	0.7487	0.6913
5.20	0.9886	0.9744	0.9596	0.9411	0.9173	0.8922	0.8392	0.7832	0.7541	0.6976
5.25	0.9889	0.9751	0.9607	0.9425	0.9194	0.8948	0.8429	0.7880	0.7593	0.7037
5.30	0.9892	0.9758	0.9617	0.9440	0.9214	0.8974	0.8465	0.7926	0.7645	0.7097
5.35	0.9895	0.9764	0.9626	0.9454	0.9233	0.8998	0.8500	0.7971	0.7695	0.7156
5.40	0.9898	0.9770	0.9636	0.9467	0.9251	0.9022	0.8534	0.8015	0.7743	0.7213
5.45	0.9900	0.9776	0.9645	0.9481	0.9269	0.9045	0.8567	0.8058	0.7791	0.7269
5.50	0.9903	0.9781	0.9654	0.9493	0.9287	0.9067	0.8599	0.8099	0.7837	0.7323
5.55	0.9905	0.9787	0.9662	0.9505	0.9304	0.9089	0.8630	0.8140	0.7882	0.7377
5.60	0.9908	0.9792	0.9671	0.9517	0.9320	0.9110	0.8661	0.8179	0.7926	0.7429
5.65	0.9910	0.9797	0.9679	0.9529	0.9336	0.9130	0.8690	0.8218	0.7969	0.7480
5.70	0.9912	0.9802	0.9686	0.9540	0.9352	0.9150	0.8719	0.8255	0.8011	0.7529
5.75	0.9914	0.9807	0.9694	0.9551	0.9367	0.9169	0.8747	0.8292	0.8051	0.7578
5.80	0.9916	0.9812	0.9701	0.9561	0.9381	0.9188	0.8774	0.8327	0.8091	0.7625
5.85	0.9919	0.9816	0.9708	0.9572	0.9395	0.9206	0.8800	0.8362	0.8130	0.7671
5.90	0.9920	0.9821	0.9715	0.9582	0.9409	0.9224	0.8826	0.8396	0.8168	0.7716
5.95	0.9922	0.9825	0.9722	0.9591	0.9422	0.9241	0.8851	0.8429	0.8205	0.7760
6.00	0.9924	0.9829	0.9728	0.9600	0.9435	0.9258	0.8875	0.8461	0.8240	0.7804
6.05	0.9926	0.9833	0.9734	0.9609	0.9448	0.9274	0.8899	0.8492	0.8275	0.7846
6.10	0.9928	0.9837	0.9740	0.9618	0.9460	0.9290	0.8922	0.8522	0.8310	0.7887
6.15	0.9929	0.9841	0.9746	0.9627	0.9472	0.9305	0.8945	0.8552	0.8343	0.7927
6.20	0.9931	0.9844	0.9752	0.9635	0.9483	0.9320	0.8966	0.8581	0.8376	0.7966

PROBABILITY INTEGRAL OF t, THE NON-CENTRAL t-STATISTIC. THIS TABLE GIVES Pr $[t/\sqrt{f} \leq x]$.

f is the number of degrees of freedom; the non-centrality parameter is $\sqrt{f+1}\,K_p$.

K_p is the standardized normal deviate exceeded with probability p.

DEGREES OF FREEDOM **3**

x \ p	.2500	.1500	.1000	.0650	.0400	.0250	.0100	.0040	.0025	.0010
6.25	0.9933	0.9848	0.9758	0.9643	0.9494	0.9334	0.8988	0.8609	0.8407	0.8004
6.30	0.9934	0.9851	0.9763	0.9651	0.9505	0.9349	0.9009	0.8637	0.8438	0.8042
6.35	0.9936	0.9854	0.9768	0.9658	0.9516	0.9362	0.9029	0.8664	0.8469	0.8079
6.40	0.9937	0.9858	0.9773	0.9666	0.9526	0.9376	0.9048	0.8690	0.8498	0.8114
6.45	0.9939	0.9861	0.9778	0.9673	0.9536	0.9389	0.9068	0.8716	0.8527	0.8149
6.50	0.9940	0.9864	0.9783	0.9680	0.9546	0.9401	0.9086	0.8740	0.8555	0.8183
6.55	0.9941	0.9867	0.9788	0.9687	0.9555	0.9413	0.9104	0.8765	0.8582	0.8217
6.60	0.9942	0.9870	0.9792	0.9693	0.9565	0.9425	0.9122	0.8788	0.8609	0.8249
6.65	0.9944	0.9872	0.9796	0.9700	0.9574	0.9437	0.9139	0.8812	0.8635	0.8281
6.70	0.9945	0.9875	0.9801	0.9706	0.9582	0.9448	0.9156	0.8834	0.8661	0.8313
6.75	0.9946	0.9878	0.9805	0.9712	0.9591	0.9460	0.9173	0.8856	0.8686	0.8343
6.80	0.9947	0.9880	0.9809	0.9718	0.9599	0.9470	0.9189	0.8878	0.8710	0.8373
6.85	0.9948	0.9883	0.9813	0.9724	0.9607	0.9481	0.9204	0.8899	0.8734	0.8402
6.90	0.9949	0.9885	0.9817	0.9729	0.9615	0.9491	0.9220	0.8919	0.8757	0.8430
6.95	0.9950	0.9888	0.9820	0.9735	0.9623	0.9501	0.9235	0.8939	0.8780	0.8458
7.00	0.9951	0.9890	0.9824	0.9740	0.9630	0.9511	0.9249	0.8959	0.8802	0.8486
7.05	0.9952	0.9892	0.9828	0.9745	0.9637	0.9520	0.9263	0.8978	0.8824	0.8512
7.10	0.9953	0.9894	0.9831	0.9750	0.9644	0.9529	0.9277	0.8997	0.8845	0.8538
7.15	0.9954	0.9896	0.9834	0.9755	0.9651	0.9538	0.9291	0.9015	0.8866	0.8564
7.20	0.9955	0.9898	0.9838	0.9760	0.9658	0.9547	0.9304	0.9033	0.8886	0.8589
7.25	0.9956	0.9900	0.9841	0.9764	0.9664	0.9556	0.9317	0.9050	0.8906	0.8613
7.30	0.9957	0.9902	0.9844	0.9769	0.9671	0.9564	0.9329	0.9067	0.8925	0.8637
7.35	0.9958	0.9904	0.9847	0.9773	0.9677	0.9572	0.9341	0.9084	0.8944	0.8660
7.40	0.9959	0.9906	0.9850	0.9778	0.9683	0.9580	0.9353	0.9100	0.8962	0.8683
7.45	0.9960	0.9908	0.9853	0.9782	0.9689	0.9588	0.9365	0.9116	0.8980	0.8706
7.50	0.9960	0.9910	0.9856	0.9786	0.9695	0.9596	0.9376	0.9131	0.8998	0.8727
7.55	0.9961	0.9911	0.9858	0.9790	0.9701	0.9603	0.9388	0.9147	0.9015	0.8749
7.60	0.9962	0.9913	0.9861	0.9794	0.9706	0.9610	0.9398	0.9161	0.9032	0.8770
7.65	0.9963	0.9915	0.9864	0.9798	0.9711	0.9617	0.9409	0.9176	0.9049	0.8790
7.70	0.9963	0.9916	0.9866	0.9801	0.9717	0.9624	0.9419	0.9190	0.9065	0.8810
7.75	0.9964	0.9918	0.9869	0.9805	0.9722	0.9631	0.9430	0.9204	0.9081	0.8830
7.80	0.9965	0.9919	0.9871	0.9809	0.9727	0.9637	0.9440	0.9217	0.9096	0.8849
7.85	0.9965	0.9921	0.9873	0.9812	0.9732	0.9644	0.9449	0.9231	0.9111	0.8868
7.90	0.9966	0.9922	0.9876	0.9815	0.9736	0.9650	0.9459	0.9244	0.9126	0.8887
7.95	0.9967	0.9924	0.9878	0.9819	0.9741	0.9656	0.9468	0.9256	0.9141	0.8905
8.00	0.9967	0.9925	0.9880	0.9822	0.9746	0.9662	0.9477	0.9269	0.9155	0.8922
8.05	0.9968	0.9926	0.9882	0.9825	0.9750	0.9668	0.9486	0.9281	0.9169	0.8940
8.10	0.9968	0.9928	0.9884	0.9828	0.9754	0.9674	0.9495	0.9293	0.9183	0.8957
8.15	0.9969	0.9929	0.9886	0.9831	0.9759	0.9679	0.9503	0.9305	0.9196	0.8973
8.20	0.9969	0.9930	0.9888	0.9834	0.9763	0.9685	0.9511	0.9316	0.9209	0.8990
8.25	0.9970	0.9932	0.9890	0.9837	0.9767	0.9690	0.9520	0.9327	0.9222	0.9006
8.30	0.9971	0.9933	0.9892	0.9840	0.9771	0.9695	0.9528	0.9338	0.9234	0.9021
8.35	0.9971	0.9934	0.9894	0.9843	0.9775	0.9700	0.9535	0.9349	0.9246	0.9037
8.40	0.9972	0.9935	0.9896	0.9845	0.9779	0.9705	0.9543	0.9359	0.9258	0.9052
8.45	0.9972	0.9936	0.9898	0.9848	0.9782	0.9710	0.9550	0.9369	0.9270	0.9067
8.50	0.9972	0.9937	0.9899	0.9850	0.9786	0.9715	0.9558	0.9379	0.9282	0.9081
8.55	0.9973	0.9938	0.9901	0.9853	0.9789	0.9720	0.9565	0.9389	0.9293	0.9095
8.60	0.9973	0.9939	0.9903	0.9855	0.9793	0.9724	0.9572	0.9399	0.9304	0.9109
8.65	0.9974	0.9940	0.9904	0.9858	0.9796	0.9729	0.9579	0.9408	0.9315	0.9123
8.70	0.9974	0.9941	0.9906	0.9860	0.9800	0.9733	0.9585	0.9418	0.9325	0.9136
8.75	0.9975	0.9942	0.9907	0.9862	0.9803	0.9738	0.9592	0.9427	0.9336	0.9149
8.80	0.9975	0.9943	0.9909	0.9865	0.9806	0.9742	0.9598	0.9436	0.9346	0.9162
8.85	0.9976	0.9944	0.9910	0.9867	0.9809	0.9746	0.9605	0.9444	0.9356	0.9174
8.90	0.9976	0.9945	0.9912	0.9869	0.9812	0.9750	0.9611	0.9453	0.9366	0.9187
8.95	0.9976	0.9946	0.9913	0.9871	0.9815	0.9754	0.9617	0.9461	0.9375	0.9199
9.00	0.9977	0.9947	0.9915	0.9873	0.9818	0.9758	0.9623	0.9469	0.9385	0.9211
9.05	0.9977	0.9948	0.9916	0.9875	0.9821	0.9762	0.9629	0.9477	0.9394	0.9222
9.10	0.9977	0.9949	0.9917	0.9877	0.9824	0.9765	0.9634	0.9485	0.9403	0.9234
9.15	0.9978	0.9949	0.9919	0.9879	0.9827	0.9769	0.9640	0.9493	0.9412	0.9245
9.20	0.9978	0.9950	0.9920	0.9881	0.9829	0.9772	0.9645	0.9501	0.9421	0.9256

PROBABILITY INTEGRAL OF t, THE NON-CENTRAL t-STATISTIC. THIS TABLE GIVES Pr $[t/\sqrt{f} \leq x]$.

f is the number of degrees of freedom; the non-centrality parameter is $\sqrt{f+1}\,K_p$.
K_p is the standardized normal deviate exceeded with probability p.

DEGREES OF FREEDOM **3**

x \ p	.2500	.1500	.1000	.0650	.0400	.0250	.0100	.0040	.0025	.0010
9.25	0.9979	0.9951	0.9921	0.9883	0.9832	0.9776	0.9651	0.9508	0.9429	0.9267
9.30	0.9979	0.9952	0.9922	0.9885	0.9834	0.9779	0.9656	0.9515	0.9438	0.9277
9.35	0.9979	0.9953	0.9924	0.9886	0.9837	0.9783	0.9661	0.9522	0.9446	0.9288
9.40	0.9980	0.9953	0.9925	0.9888	0.9839	0.9786	0.9666	0.9529	0.9454	0.9298
9.45	0.9980	0.9954	0.9926	0.9890	0.9842	0.9789	0.9671	0.9536	0.9462	0.9308
9.50	0.9980	0.9955	0.9927	0.9891	0.9844	0.9792	0.9676	0.9543	0.9470	0.9318
9.55	0.9980	0.9955	0.9928	0.9893	0.9847	0.9795	0.9681	0.9550	0.9477	0.9328
9.60	0.9981	0.9956	0.9929	0.9895	0.9849	0.9798	0.9685	0.9556	0.9485	0.9337
9.65	0.9981	0.9957	0.9930	0.9896	0.9851	0.9801	0.9690	0.9563	0.9492	0.9346
9.70	0.9981	0.9957	0.9931	0.9898	0.9853	0.9804	0.9694	0.9569	0.9499	0.9356
9.75	0.9982	0.9958	0.9932	0.9899	0.9856	0.9807	0.9699	0.9575	0.9506	0.9364
9.80	0.9982	0.9959	0.9933	0.9901	0.9858	0.9810	0.9703	0.9581	0.9513	0.9373
9.85	0.9982	0.9959	0.9934	0.9902	0.9860	0.9813	0.9707	0.9587	0.9520	0.9382
9.90	0.9982	0.9960	0.9935	0.9904	0.9862	0.9815	0.9712	0.9593	0.9527	0.9390
9.95	0.9983	0.9960	0.9936	0.9905	0.9864	0.9818	0.9716	0.9598	0.9533	0.9399
10.00	0.9983	0.9961	0.9937	0.9907	0.9866	0.9821	0.9720	0.9604	0.9540	0.9407
10.05	0.9983	0.9962	0.9938	0.9908	0.9868	0.9823	0.9724	0.9609	0.9546	0.9415
10.10	0.9983	0.9962	0.9939	0.9909	0.9870	0.9826	0.9727	0.9615	0.9552	0.9423
10.15	0.9984	0.9963	0.9940	0.9910	0.9871	0.9828	0.9731	0.9620	0.9558	0.9431
10.20	0.9984	0.9963	0.9941	0.9912	0.9873	0.9831	0.9735	0.9625	0.9564	0.9438
10.25	0.9984	0.9964	0.9942	0.9913	0.9875	0.9833	0.9739	0.9630	0.9570	0.9446
10.30	0.9984	0.9964	0.9942	0.9914	0.9877	0.9835	0.9742	0.9635	0.9576	0.9453
10.35	0.9985	0.9965	0.9943	0.9915	0.9878	0.9838	0.9746	0.9640	0.9582	0.9460
10.40	0.9985	0.9965	0.9944	0.9917	0.9880	0.9840	0.9749	0.9645	0.9587	0.9468
10.45	0.9985	0.9966	0.9945	0.9918	0.9882	0.9842	0.9753	0.9650	0.9593	0.9475
10.50	0.9985	0.9966	0.9946	0.9919	0.9883	0.9844	0.9756	0.9655	0.9598	0.9481
10.55	0.9985	0.9967	0.9946	0.9920	0.9885	0.9846	0.9759	0.9659	0.9604	0.9488
10.60	0.9986	0.9967	0.9947	0.9921	0.9887	0.9848	0.9762	0.9664	0.9609	0.9495
10.65	0.9986	0.9968	0.9948	0.9922	0.9888	0.9850	0.9766	0.9668	0.9614	0.9501
10.70	0.9986	0.9968	0.9949	0.9923	0.9890	0.9852	0.9769	0.9673	0.9619	0.9508
10.75	0.9986	0.9968	0.9949	0.9924	0.9891	0.9854	0.9772	0.9677	0.9624	0.9514
10.80	0.9986	0.9969	0.9950	0.9925	0.9893	0.9856	0.9775	0.9681	0.9629	0.9520
10.85	0.9987	0.9969	0.9951	0.9926	0.9894	0.9858	0.9778	0.9685	0.9634	0.9527
10.90	0.9987	0.9970	0.9951	0.9927	0.9895	0.9860	0.9781	0.9689	0.9638	0.9533
10.95	0.9987	0.9970	0.9952	0.9928	0.9897	0.9862	0.9784	0.9693	0.9643	0.9538
11.00	0.9987	0.9971	0.9953	0.9929	0.9898	0.9864	0.9786	0.9697	0.9648	0.9544
11.05	0.9987	0.9971	0.9953	0.9930	0.9899	0.9866	0.9789	0.9701	0.9652	0.9550
11.10	0.9987	0.9971	0.9954	0.9931	0.9901	0.9867	0.9792	0.9705	0.9656	0.9556
11.15	0.9988	0.9972	0.9954	0.9932	0.9902	0.9869	0.9794	0.9709	0.9661	0.9561
11.20	0.9988	0.9972	0.9955	0.9933	0.9903	0.9871	0.9797	0.9712	0.9665	0.9567
11.25	0.9988	0.9972	0.9956	0.9934	0.9905	0.9872	0.9800	0.9716	0.9669	0.9572
11.30	0.9988	0.9973	0.9956	0.9935	0.9906	0.9874	0.9802	0.9719	0.9673	0.9577
11.35	0.9988	0.9973	0.9957	0.9935	0.9907	0.9876	0.9805	0.9723	0.9677	0.9582
11.40	0.9988	0.9973	0.9957	0.9936	0.9908	0.9877	0.9807	0.9726	0.9681	0.9588
11.45	0.9989	0.9974	0.9958	0.9937	0.9909	0.9879	0.9810	0.9730	0.9685	0.9593
11.50	0.9989	0.9974	0.9958	0.9938	0.9911	0.9880	0.9812	0.9733	0.9689	0.9597
11.55	0.9989	0.9974	0.9959	0.9939	0.9912	0.9882	0.9814	0.9736	0.9693	0.9602
11.60	0.9989	0.9975	0.9959	0.9939	0.9913	0.9883	0.9817	0.9740	0.9697	0.9607
11.65	0.9989	0.9975	0.9960	0.9940	0.9914	0.9885	0.9819	0.9743	0.9700	0.9612
11.70	0.9989	0.9975	0.9960	0.9941	0.9915	0.9886	0.9821	0.9746	0.9704	0.9616
11.75	0.9989	0.9976	0.9961	0.9942	0.9916	0.9888	0.9823	0.9749	0.9707	0.9621
11.80	0.9990	0.9976	0.9961	0.9942	0.9917	0.9889	0.9825	0.9752	0.9711	0.9626
11.85	0.9990	0.9976	0.9962	0.9943	0.9918	0.9890	0.9827	0.9755	0.9714	0.9630
11.90	0.9990	0.9977	0.9962	0.9944	0.9919	0.9892	0.9830	0.9758	0.9718	0.9634
11.95	0.9990	0.9977	0.9963	0.9944	0.9920	0.9893	0.9832	0.9761	0.9721	0.9639
12.00	0.9990	0.9977	0.9963	0.9945	0.9921	0.9894	0.9834	0.9764	0.9724	0.9643
12.05	0.9990	0.9978	0.9964	0.9946	0.9922	0.9895	0.9836	0.9766	0.9728	0.9647
12.10	0.9990	0.9978	0.9964	0.9946	0.9923	0.9897	0.9838	0.9769	0.9731	0.9651
12.15	0.9990	0.9978	0.9965	0.9947	0.9924	0.9898	0.9839	0.9772	0.9734	0.9655
12.20	0.9991	0.9978	0.9965	0.9948	0.9925	0.9899	0.9841	0.9775	0.9737	0.9659

PROBABILITY INTEGRAL OF t, THE NON-CENTRAL t-STATISTIC. THIS TABLE GIVES Pr $[t/\sqrt{f} \leq x]$.

f is the number of degrees of freedom; the non-centrality parameter is $\sqrt{f+1}\,K_p$.

K_p is the standardized normal deviate exceeded with probability p.

DEGREES OF FREEDOM **3**

x \ p	.2500	.1500	.1000	.0650	.0400	.0250	.0100	.0040	.0025	.0010
12.25	0.9991	0.9979	0.9965	0.9948	0.9926	0.9900	0.9843	0.9777	0.9740	0.9663
12.30	0.9991	0.9979	0.9966	0.9949	0.9927	0.9902	0.9845	0.9780	0.9743	0.9667
12.35	0.9991	0.9979	0.9966	0.9950	0.9927	0.9903	0.9847	0.9782	0.9746	0.9671
12.40	0.9991	0.9979	0.9967	0.9950	0.9928	0.9904	0.9849	0.9785	0.9749	0.9674
12.45	0.9991	0.9980	0.9967	0.9951	0.9929	0.9905	0.9850	0.9787	0.9752	0.9678
12.50	0.9991	0.9980	0.9967	0.9951	0.9930	0.9906	0.9852	0.9790	0.9755	0.9682
12.55	0.9991	0.9980	0.9968	0.9952	0.9931	0.9907	0.9854	0.9792	0.9758	0.9685
12.60	0.9991	0.9980	0.9968	0.9952	0.9932	0.9908	0.9856	0.9794	0.9760	0.9689
12.65	0.9991	0.9981	0.9969	0.9953	0.9932	0.9909	0.9857	0.9797	0.9763	0.9692
12.70	0.9992	0.9981	0.9969	0.9954	0.9933	0.9910	0.9859	0.9799	0.9766	0.9696
12.75	0.9992	0.9981	0.9969	0.9954	0.9934	0.9911	0.9860	0.9801	0.9768	0.9699
12.80	0.9992	0.9981	0.9970	0.9955	0.9935	0.9912	0.9862	0.9804	0.9771	0.9702
12.85	0.9992	0.9981	0.9970	0.9955	0.9935	0.9913	0.9864	0.9806	0.9773	0.9706
12.90	0.9992	0.9982	0.9970	0.9956	0.9936	0.9914	0.9865	0.9808	0.9776	0.9709
12.95	0.9992	0.9982	0.9971	0.9956	0.9937	0.9915	0.9867	0.9810	0.9778	0.9712
13.00	0.9992	0.9982	0.9971	0.9957	0.9938	0.9916	0.9868	0.9812	0.9781	0.9715
13.05	0.9992	0.9982	0.9971	0.9957	0.9938	0.9917	0.9870	0.9814	0.9783	0.9718
13.10	0.9992	0.9982	0.9972	0.9958	0.9939	0.9918	0.9871	0.9816	0.9786	0.9721
13.15	0.9992	0.9983	0.9972	0.9958	0.9940	0.9919	0.9872	0.9818	0.9788	0.9724
13.20	0.9992	0.9983	0.9972	0.9959	0.9940	0.9920	0.9874	0.9820	0.9790	0.9727
13.25	0.9993	0.9983	0.9973	0.9959	0.9941	0.9921	0.9875	0.9822	0.9792	0.9730
13.30	0.9993	0.9983	0.9973	0.9959	0.9942	0.9922	0.9877	0.9824	0.9795	0.9733
13.35	0.9993	0.9983	0.9973	0.9960	0.9942	0.9923	0.9878	0.9826	0.9797	0.9736
13.40	0.9993	0.9984	0.9974	0.9960	0.9943	0.9923	0.9879	0.9828	0.9799	0.9739
13.45	0.9993	0.9984	0.9974	0.9961	0.9943	0.9924	0.9880	0.9830	0.9801	0.9742
13.50	0.9993	0.9984	0.9974	0.9961	0.9944	0.9925	0.9882	0.9831	0.9803	0.9744
13.55	0.9993	0.9984	0.9974	0.9962	0.9945	0.9926	0.9883	0.9833	0.9805	0.9747
13.60	0.9993	0.9984	0.9975	0.9962	0.9945	0.9927	0.9884	0.9835	0.9807	0.9750
13.65	0.9993	0.9984	0.9975	0.9962	0.9946	0.9927	0.9885	0.9837	0.9809	0.9752
13.70	0.9993	0.9985	0.9975	0.9963	0.9946	0.9928	0.9887	0.9838	0.9811	0.9755
13.75	0.9993	0.9985	0.9975	0.9963	0.9947	0.9929	0.9888	0.9840	0.9813	0.9757
13.80	0.9993	0.9985	0.9976	0.9964	0.9948	0.9930	0.9889	0.9842	0.9815	0.9760
13.85	0.9993	0.9985	0.9976	0.9964	0.9948	0.9930	0.9890	0.9844	0.9817	0.9762
13.90	0.9994	0.9985	0.9976	0.9964	0.9949	0.9931	0.9891	0.9845	0.9819	0.9765
13.95	0.9994	0.9985	0.9976	0.9965	0.9949	0.9932	0.9893	0.9847	0.9821	0.9767
14.00	0.9994	0.9986	0.9977	0.9965	0.9950	0.9933	0.9894	0.9848	0.9823	0.9770
14.05	0.9994	0.9986	0.9977	0.9966	0.9950	0.9933	0.9895	0.9850	0.9825	0.9772
14.10	0.9994	0.9986	0.9977	0.9966	0.9951	0.9934	0.9896	0.9851	0.9826	0.9774
14.15	0.9994	0.9986	0.9977	0.9966	0.9951	0.9935	0.9897	0.9853	0.9828	0.9777
14.20	0.9994	0.9986	0.9978	0.9967	0.9952	0.9935	0.9898	0.9854	0.9830	0.9779
14.25	0.9994	0.9986	0.9978	0.9967	0.9952	0.9936	0.9899	0.9856	0.9832	0.9781
14.30	0.9994	0.9986	0.9978	0.9967	0.9953	0.9937	0.9900	0.9857	0.9833	0.9783
14.35	0.9994	0.9987	0.9978	0.9968	0.9953	0.9937	0.9901	0.9859	0.9835	0.9785
14.40	0.9994	0.9987	0.9979	0.9968	0.9954	0.9938	0.9902	0.9860	0.9837	0.9787
14.45	0.9994	0.9987	0.9979	0.9968	0.9954	0.9939	0.9903	0.9862	0.9838	0.9790
14.50	0.9994	0.9987	0.9979	0.9969	0.9955	0.9939	0.9904	0.9863	0.9840	0.9792
14.55	0.9994	0.9987	0.9979	0.9969	0.9955	0.9940	0.9905	0.9864	0.9842	0.9794
14.60	0.9994	0.9987	0.9979	0.9969	0.9956	0.9940	0.9906	0.9866	0.9843	0.9796
14.65	0.9994	0.9987	0.9980	0.9970	0.9956	0.9941	0.9907	0.9867	0.9845	0.9798
14.70	0.9995	0.9988	0.9980	0.9970	0.9956	0.9942	0.9908	0.9868	0.9846	0.9800
14.75	0.9995	0.9988	0.9980	0.9970	0.9957	0.9942	0.9909	0.9870	0.9848	0.9802
14.80	0.9995	0.9988	0.9980	0.9970	0.9957	0.9943	0.9910	0.9871	0.9849	0.9804
14.85	0.9995	0.9988	0.9980	0.9971	0.9958	0.9943	0.9910	0.9872	0.9851	0.9806
14.90	0.9995	0.9988	0.9981	0.9971	0.9958	0.9944	0.9911	0.9873	0.9852	0.9807
14.95	0.9995	0.9988	0.9981	0.9971	0.9959	0.9944	0.9912	0.9875	0.9854	0.9809
15.00	0.9995	0.9988	0.9981	0.9972	0.9959	0.9945	0.9913	0.9876	0.9855	0.9811
15.05	0.9995	0.9988	0.9981	0.9972	0.9959	0.9946	0.9914	0.9877	0.9856	0.9813
15.10	0.9995	0.9988	0.9981	0.9972	0.9960	0.9946	0.9915	0.9878	0.9858	0.9815
15.15	0.9995	0.9989	0.9982	0.9972	0.9960	0.9947	0.9916	0.9879	0.9859	0.9816
15.20	0.9995	0.9989	0.9982	0.9973	0.9961	0.9947	0.9916	0.9881	0.9860	0.9818

PROBABILITY INTEGRAL OF t, THE NON-CENTRAL t-STATISTIC. THIS TABLE GIVES Pr $[t/\sqrt{f} \leq x]$.

f is the number of degrees of freedom; the non-centrality parameter is $\sqrt{f+1}\,K_p$.

K_p is the standardized normal deviate exceeded with probability p.

DEGREES OF FREEDOM **3**

x \ p	.2500	.1500	.1000	.0650	.0400	.0250	.0100	.0040	.0025	.0010
15.25	0.9995	0.9989	0.9982	0.9973	0.9961	0.9948	0.9917	0.9882	0.9862	0.9820
15.30	0.9995	0.9989	0.9982	0.9973	0.9961	0.9948	0.9918	0.9883	0.9863	0.9822
15.35	0.9995	0.9989	0.9982	0.9973	0.9962	0.9949	0.9919	0.9884	0.9864	0.9823
15.40	0.9995	0.9989	0.9982	0.9974	0.9962	0.9949	0.9920	0.9885	0.9866	0.9825
15.45	0.9995	0.9989	0.9983	0.9974	0.9962	0.9950	0.9920	0.9886	0.9867	0.9827
15.50	0.9995	0.9989	0.9983	0.9974	0.9963	0.9950	0.9921	0.9887	0.9868	0.9828
15.55	0.9995	0.9989	0.9983	0.9974	0.9963	0.9951	0.9922	0.9888	0.9869	0.9830
15.60	0.9995	0.9990	0.9983	0.9975	0.9964	0.9951	0.9923	0.9889	0.9871	0.9831
15.65	0.9995	0.9990	0.9983	0.9975	0.9964	0.9951	0.9923	0.9890	0.9872	0.9833
15.70	0.9995	0.9990	0.9983	0.9975	0.9964	0.9952	0.9924	0.9891	0.9873	0.9834
15.75	0.9996	0.9990	0.9984	0.9975	0.9965	0.9952	0.9925	0.9892	0.9874	0.9836
15.80	0.9996	0.9990	0.9984	0.9976	0.9965	0.9953	0.9925	0.9893	0.9875	0.9837
15.85	0.9996	0.9990	0.9984	0.9976	0.9965	0.9953	0.9926	0.9894	0.9876	0.9839
15.90	0.9996	0.9990	0.9984	0.9976	0.9966	0.9954	0.9927	0.9895	0.9878	0.9840
15.95	0.9996	0.9990	0.9984	0.9976	0.9966	0.9954	0.9927	0.9896	0.9879	0.9842
16.00	0.9996	0.9990	0.9984	0.9977	0.9966	0.9955	0.9928	0.9897	0.9880	0.9843
16.05	0.9996	0.9990	0.9984	0.9977	0.9966	0.9955	0.9929	0.9898	0.9881	0.9845
16.10	0.9996	0.9990	0.9985	0.9977	0.9967	0.9955	0.9929	0.9899	0.9882	0.9846
16.15	0.9996	0.9991	0.9985	0.9977	0.9967	0.9956	0.9930	0.9900	0.9883	0.9847
16.20	0.9996	0.9991	0.9985	0.9977	0.9967	0.9956	0.9931	0.9901	0.9884	0.9849
16.25	0.9996	0.9991	0.9985	0.9978	0.9968	0.9957	0.9931	0.9902	0.9885	0.9850
16.30	0.9996	0.9991	0.9985	0.9978	0.9968	0.9957	0.9932	0.9903	0.9886	0.9851
16.35	0.9996	0.9991	0.9985	0.9978	0.9968	0.9957	0.9932	0.9903	0.9887	0.9853
16.40	0.9996	0.9991	0.9985	0.9978	0.9969	0.9958	0.9933	0.9904	0.9888	0.9854
16.45	0.9996	0.9991	0.9986	0.9978	0.9969	0.9958	0.9934	0.9905	0.9889	0.9855
16.50	0.9996	0.9991	0.9986	0.9979	0.9969	0.9958	0.9934	0.9906	0.9890	0.9857
16.55	0.9996	0.9991	0.9986	0.9979	0.9969	0.9959	0.9935	0.9907	0.9891	0.9858
16.60	0.9996	0.9991	0.9986	0.9979	0.9970	0.9959	0.9935	0.9908	0.9892	0.9859
16.65	0.9996	0.9991	0.9986	0.9979	0.9970	0.9960	0.9936	0.9908	0.9893	0.9860
16.70	0.9996	0.9991	0.9986	0.9979	0.9970	0.9960	0.9937	0.9909	0.9894	0.9862
16.75	0.9996	0.9992	0.9986	0.9980	0.9970	0.9960	0.9937	0.9910	0.9895	0.9863
16.80	0.9996	0.9992	0.9986	0.9980	0.9971	0.9961	0.9938	0.9911	0.9896	0.9864
16.85	0.9996	0.9992	0.9987	0.9980	0.9971	0.9961	0.9938	0.9912	0.9897	0.9865
16.90	0.9996	0.9992	0.9987	0.9980	0.9971	0.9961	0.9939	0.9912	0.9898	0.9866
16.95	0.9996	0.9992	0.9987	0.9980	0.9971	0.9962	0.9939	0.9913	0.9898	0.9867
17.00	0.9996	0.9992	0.9987	0.9980	0.9972	0.9962	0.9940	0.9914	0.9899	0.9869
17.05	0.9996	0.9992	0.9987	0.9981	0.9972	0.9962	0.9940	0.9915	0.9900	0.9870
17.10	0.9996	0.9992	0.9987	0.9981	0.9972	0.9963	0.9941	0.9915	0.9901	0.9871
17.15	0.9997	0.9992	0.9987	0.9981	0.9972	0.9963	0.9941	0.9916	0.9902	0.9872
17.20	0.9997	0.9992	0.9987	0.9981	0.9973	0.9963	0.9942	0.9917	0.9903	0.9873
17.25	0.9997	0.9992	0.9987	0.9981	0.9973	0.9964	0.9942	0.9917	0.9903	0.9874
17.30	0.9997	0.9992	0.9988	0.9981	0.9973	0.9964	0.9943	0.9918	0.9904	0.9875
17.35	0.9997	0.9992	0.9988	0.9982	0.9973	0.9964	0.9943	0.9919	0.9905	0.9876
17.40	0.9997	0.9992	0.9988	0.9982	0.9974	0.9965	0.9944	0.9920	0.9906	0.9877
17.45	0.9997	0.9993	0.9988	0.9982	0.9974	0.9965	0.9944	0.9920	0.9907	0.9878
17.50	0.9997	0.9993	0.9988	0.9982	0.9974	0.9965	0.9945	0.9921	0.9907	0.9879
17.55	0.9997	0.9993	0.9988	0.9982	0.9974	0.9965	0.9945	0.9922	0.9908	0.9880
17.60	0.9997	0.9993	0.9988	0.9982	0.9974	0.9966	0.9946	0.9922	0.9909	0.9881
17.65	0.9997	0.9993	0.9988	0.9983	0.9975	0.9966	0.9946	0.9923	0.9910	0.9882
17.70	0.9997	0.9993	0.9988	0.9983	0.9975	0.9966	0.9947	0.9923	0.9910	0.9883
17.75	0.9997	0.9993	0.9989	0.9983	0.9975	0.9967	0.9947	0.9924	0.9911	0.9884
17.80	0.9997	0.9993	0.9989	0.9983	0.9975	0.9967	0.9947	0.9925	0.9912	0.9885
17.85	0.9997	0.9993	0.9989	0.9983	0.9976	0.9967	0.9948	0.9925	0.9913	0.9886
17.90	0.9997	0.9993	0.9989	0.9983	0.9976	0.9967	0.9948	0.9926	0.9913	0.9887
17.95	0.9997	0.9993	0.9989	0.9983	0.9976	0.9968	0.9949	0.9927	0.9914	0.9888
18.00	0.9997	0.9993	0.9989	0.9983	0.9976	0.9968	0.9949	0.9927	0.9915	0.9889
18.05	0.9997	0.9993	0.9989	0.9984	0.9976	0.9968	0.9950	0.9928	0.9915	0.9890
18.10	0.9997	0.9993	0.9989	0.9984	0.9977	0.9968	0.9950	0.9928	0.9916	0.9890
18.15	0.9997	0.9993	0.9989	0.9984	0.9977	0.9969	0.9950	0.9929	0.9917	0.9891
18.20	0.9997	0.9993	0.9989	0.9984	0.9977	0.9969	0.9951	0.9930	0.9918	0.9892

PROBABILITY INTEGRAL OF t, THE NON-CENTRAL t-STATISTIC. THIS TABLE GIVES Pr $[t/\sqrt{f} \leq x]$.

f is the number of degrees of freedom; the non-centrality parameter is $\sqrt{f+1}\,K_p$.

K_p is the standardized normal deviate exceeded with probability p.

DEGREES OF FREEDOM **3**

x \ p	.2500	.1500	.1000	.0650	.0400	.0250	.0100	.0040	.0025	.0010
18.25	0.9997	0.9993	0.9989	0.9984	0.9977	0.9969	0.9951	0.9930	0.9918	0.9893
18.30	0.9997	0.9994	0.9990	0.9984	0.9977	0.9969	0.9952	0.9931	0.9919	0.9894
18.35	0.9997	0.9994	0.9990	0.9984	0.9977	0.9970	0.9952	0.9931	0.9919	0.9895
18.40	0.9997	0.9994	0.9990	0.9985	0.9978	0.9970	0.9952	0.9932	0.9920	0.9896
18.45	0.9997	0.9994	0.9990	0.9985	0.9978	0.9970	0.9953	0.9932	0.9921	0.9896
18.50	0.9997	0.9994	0.9990	0.9985	0.9978	0.9970	0.9953	0.9933	0.9921	0.9897
18.55	0.9997	0.9994	0.9990	0.9985	0.9978	0.9971	0.9953	0.9933	0.9922	0.9898
18.60	0.9997	0.9994	0.9990	0.9985	0.9978	0.9971	0.9954	0.9934	0.9923	0.9899
18.65	0.9997	0.9994	0.9990	0.9985	0.9979	0.9971	0.9954	0.9934	0.9923	0.9900
18.70	0.9997	0.9994	0.9990	0.9985	0.9979	0.9971	0.9955	0.9935	0.9924	0.9900
18.75	0.9997	0.9994	0.9990	0.9985	0.9979	0.9972	0.9955	0.9935	0.9924	0.9901
18.80	0.9997	0.9994	0.9990	0.9986	0.9979	0.9972	0.9955	0.9936	0.9925	0.9902
18.85	0.9997	0.9994	0.9990	0.9986	0.9979	0.9972	0.9956	0.9936	0.9926	0.9903
18.90	0.9997	0.9994	0.9990	0.9986	0.9979	0.9972	0.9956	0.9937	0.9926	0.9904
18.95	0.9997	0.9994	0.9991	0.9986	0.9980	0.9972	0.9956	0.9937	0.9927	0.9904
19.00	0.9997	0.9994	0.9991	0.9986	0.9980	0.9973	0.9957	0.9938	0.9927	0.9905
19.05	0.9997	0.9994	0.9991	0.9986	0.9980	0.9973	0.9957	0.9938	0.9928	0.9906
19.10	0.9997	0.9994	0.9991	0.9986	0.9980	0.9973	0.9957	0.9939	0.9928	0.9906
19.15	0.9997	0.9994	0.9991	0.9986	0.9980	0.9973	0.9958	0.9939	0.9929	0.9907
19.20	0.9998	0.9994	0.9991	0.9986	0.9980	0.9974	0.9958	0.9940	0.9930	0.9908
19.25	0.9998	0.9994	0.9991	0.9986	0.9980	0.9974	0.9958	0.9940	0.9930	0.9909
19.30	0.9998	0.9994	0.9991	0.9987	0.9981	0.9974	0.9959	0.9941	0.9931	0.9909
19.35	0.9998	0.9994	0.9991	0.9987	0.9981	0.9974	0.9959	0.9941	0.9931	0.9910
19.40	0.9998	0.9995	0.9991	0.9987	0.9981	0.9974	0.9959	0.9942	0.9932	0.9911
19.45	0.9998	0.9995	0.9991	0.9987	0.9981	0.9975	0.9960	0.9942	0.9932	0.9911
19.50	0.9998	0.9995	0.9991	0.9987	0.9981	0.9975	0.9960	0.9942	0.9933	0.9912
19.55	0.9998	0.9995	0.9991	0.9987	0.9981	0.9975	0.9960	0.9943	0.9933	0.9913
19.60	0.9998	0.9995	0.9991	0.9987	0.9981	0.9975	0.9960	0.9943	0.9934	0.9913
19.65	0.9998	0.9995	0.9992	0.9987	0.9982	0.9975	0.9961	0.9944	0.9934	0.9914
19.70	0.9998	0.9995	0.9992	0.9987	0.9982	0.9975	0.9961	0.9944	0.9935	0.9915
19.75	0.9998	0.9995	0.9992	0.9987	0.9982	0.9976	0.9961	0.9945	0.9935	0.9915
19.80	0.9998	0.9995	0.9992	0.9988	0.9982	0.9976	0.9962	0.9945	0.9936	0.9916
19.85	0.9998	0.9995	0.9992	0.9988	0.9982	0.9976	0.9962	0.9945	0.9936	0.9916
19.90	0.9998	0.9995	0.9992	0.9988	0.9982	0.9976	0.9962	0.9946	0.9937	0.9917
19.95	0.9998	0.9995	0.9992	0.9988	0.9982	0.9976	0.9962	0.9946	0.9937	0.9918
20.00	0.9998	0.9995	0.9992	0.9988	0.9983	0.9977	0.9963	0.9947	0.9938	0.9918
20.05	0.9998	0.9995	0.9992	0.9988	0.9983	0.9977	0.9963	0.9947	0.9938	0.9919
20.10	0.9998	0.9995	0.9992	0.9988	0.9983	0.9977	0.9963	0.9947	0.9938	0.9919
20.15	0.9998	0.9995	0.9992	0.9988	0.9983	0.9977	0.9964	0.9948	0.9939	0.9920
20.20	0.9998	0.9995	0.9992	0.9988	0.9983	0.9977	0.9964	0.9948	0.9939	0.9921
20.25	0.9998	0.9995	0.9992	0.9988	0.9983	0.9977	0.9964	0.9949	0.9940	0.9921
20.30	0.9998	0.9995	0.9992	0.9988	0.9983	0.9978	0.9964	0.9949	0.9940	0.9922
20.35	0.9998	0.9995	0.9992	0.9989	0.9983	0.9978	0.9965	0.9949	0.9941	0.9922
20.40	0.9998	0.9995	0.9992	0.9989	0.9984	0.9978	0.9965	0.9950	0.9941	0.9923
20.45	0.9998	0.9995	0.9992	0.9989	0.9984	0.9978	0.9965	0.9950	0.9941	0.9923
20.50	0.9998	0.9995	0.9993	0.9989	0.9984	0.9978	0.9965	0.9950	0.9942	0.9924
20.55	0.9998	0.9995	0.9993	0.9989	0.9984	0.9978	0.9966	0.9951	0.9942	0.9925
20.60	0.9998	0.9995	0.9993	0.9989	0.9984	0.9979	0.9966	0.9951	0.9943	0.9925
20.65	0.9998	0.9995	0.9993	0.9989	0.9984	0.9979	0.9966	0.9951	0.9943	0.9926
20.70	0.9998	0.9996	0.9993	0.9989	0.9984	0.9979	0.9966	0.9952	0.9944	0.9926
20.75	0.9998	0.9996	0.9993	0.9989	0.9984	0.9979	0.9967	0.9952	0.9944	0.9927
20.80	0.9998	0.9996	0.9993	0.9989	0.9984	0.9979	0.9967	0.9952	0.9944	0.9927
20.85	0.9998	0.9996	0.9993	0.9989	0.9985	0.9979	0.9967	0.9953	0.9945	0.9928
20.90	0.9998	0.9996	0.9993	0.9989	0.9985	0.9979	0.9967	0.9953	0.9945	0.9928
20.95	0.9998	0.9996	0.9993	0.9990	0.9985	0.9980	0.9968	0.9953	0.9945	0.9929
21.00	0.9998	0.9996	0.9993	0.9990	0.9985	0.9980	0.9968	0.9954	0.9946	0.9929
21.05	0.9998	0.9996	0.9993	0.9990	0.9985	0.9980	0.9968	0.9954	0.9946	0.9930
21.10	0.9998	0.9996	0.9993	0.9990	0.9985	0.9980	0.9968	0.9954	0.9947	0.9930
21.15	0.9998	0.9996	0.9993	0.9990	0.9985	0.9980	0.9968	0.9955	0.9947	0.9931
21.20	0.9998	0.9996	0.9993	0.9990	0.9985	0.9980	0.9969	0.9955	0.9947	0.9931

PROBABILITY INTEGRAL OF t, THE NON-CENTRAL t-STATISTIC. THIS TABLE GIVES Pr $[t/\sqrt{f} \leq x]$.

f is the number of degrees of freedom; the non-centrality parameter is $\sqrt{f+1}\,K_p$.

K_p is the standardized normal deviate exceeded with probability p.

DEGREES OF FREEDOM **3**

x \ p	.2500	.1500	.1000	.0650	.0400	.0250	.0100	.0040	.0025	.0010
21.25	0.9998	0.9996	0.9993	0.9990	0.9985	0.9980	0.9969	0.9955	0.9948	0.9932
21.30	0.9998	0.9996	0.9993	0.9990	0.9986	0.9981	0.9969	0.9956	0.9948	0.9932
21.35	0.9998	0.9996	0.9993	0.9990	0.9986	0.9981	0.9969	0.9956	0.9948	0.9933
21.40	0.9998	0.9996	0.9993	0.9990	0.9986	0.9981	0.9970	0.9956	0.9949	0.9933
21.45	0.9998	0.9996	0.9993	0.9990	0.9986	0.9981	0.9970	0.9957	0.9949	0.9933
21.50	0.9998	0.9996	0.9994	0.9990	0.9986	0.9981	0.9970	0.9957	0.9950	0.9934
21.55	0.9998	0.9996	0.9994	0.9990	0.9986	0.9981	0.9970	0.9957	0.9950	0.9934
21.60	0.9998	0.9996	0.9994	0.9990	0.9986	0.9981	0.9970	0.9957	0.9950	0.9935
21.65	0.9998	0.9996	0.9994	0.9991	0.9986	0.9981	0.9971	0.9958	0.9951	0.9935
21.70	0.9998	0.9996	0.9994	0.9991	0.9986	0.9982	0.9971	0.9958	0.9951	0.9936
21.75	0.9998	0.9996	0.9994	0.9991	0.9986	0.9982	0.9971	0.9958	0.9951	0.9936
21.80	0.9998	0.9996	0.9994	0.9991	0.9987	0.9982	0.9971	0.9959	0.9952	0.9937
21.85	0.9998	0.9996	0.9994	0.9991	0.9987	0.9982	0.9971	0.9959	0.9952	0.9937
21.90	0.9998	0.9996	0.9994	0.9991	0.9987	0.9982	0.9972	0.9959	0.9952	0.9937
21.95	0.9998	0.9996	0.9994	0.9991	0.9987	0.9982	0.9972	0.9959	0.9953	0.9938
22.00	0.9998	0.9996	0.9994	0.9991	0.9987	0.9982	0.9972	0.9960	0.9953	0.9938
22.05	0.9998	0.9996	0.9994	0.9991	0.9987	0.9982	0.9972	0.9960	0.9953	0.9939
22.10	0.9998	0.9996	0.9994	0.9991	0.9987	0.9983	0.9972	0.9960	0.9953	0.9939
22.15	0.9998	0.9996	0.9994	0.9991	0.9987	0.9983	0.9973	0.9961	0.9954	0.9939
22.20	0.9998	0.9996	0.9994	0.9991	0.9987	0.9983	0.9973	0.9961	0.9954	0.9940
22.25	0.9998	0.9996	0.9994	0.9991	0.9987	0.9983	0.9973	0.9961	0.9954	0.9940
22.30	0.9998	0.9996	0.9994	0.9991	0.9987	0.9983	0.9973	0.9961	0.9955	0.9941
22.35	0.9998	0.9996	0.9994	0.9991	0.9987	0.9983	0.9973	0.9962	0.9955	0.9941
22.40	0.9998	0.9996	0.9994	0.9991	0.9988	0.9983	0.9973	0.9962	0.9955	0.9941
22.45	0.9998	0.9996	0.9994	0.9991	0.9988	0.9983	0.9974	0.9962	0.9956	0.9942
22.50	0.9998	0.9996	0.9994	0.9992	0.9988	0.9983	0.9974	0.9962	0.9956	0.9942
22.55	0.9998	0.9997	0.9994	0.9992	0.9988	0.9984	0.9974	0.9963	0.9956	0.9943
22.60	0.9998	0.9997	0.9994	0.9992	0.9988	0.9984	0.9974	0.9963	0.9956	0.9943
22.65	0.9998	0.9997	0.9994	0.9992	0.9988	0.9984	0.9974	0.9963	0.9957	0.9943
22.70	0.9998	0.9997	0.9995	0.9992	0.9988	0.9984	0.9974	0.9963	0.9957	0.9944
22.75	0.9998	0.9997	0.9995	0.9992	0.9988	0.9984	0.9975	0.9964	0.9957	0.9944
22.80	0.9998	0.9997	0.9995	0.9992	0.9988	0.9984	0.9975	0.9964	0.9958	0.9944
22.85	0.9998	0.9997	0.9995	0.9992	0.9988	0.9984	0.9975	0.9964	0.9958	0.9945
22.90	0.9999	0.9997	0.9995	0.9992	0.9988	0.9984	0.9975	0.9964	0.9958	0.9945
22.95	0.9999	0.9997	0.9995	0.9992	0.9988	0.9984	0.9975	0.9964	0.9958	0.9946
23.00	0.9999	0.9997	0.9995	0.9992	0.9989	0.9985	0.9975	0.9965	0.9959	0.9946
23.05	0.9999	0.9997	0.9995	0.9992	0.9989	0.9985	0.9976	0.9965	0.9959	0.9946
23.10	0.9999	0.9997	0.9995	0.9992	0.9989	0.9985	0.9976	0.9965	0.9959	0.9947
23.15	0.9999	0.9997	0.9995	0.9992	0.9989	0.9985	0.9976	0.9965	0.9959	0.9947
23.20	0.9999	0.9997	0.9995	0.9992	0.9989	0.9985	0.9976	0.9966	0.9960	0.9947
23.25	0.9999	0.9997	0.9995	0.9992	0.9989	0.9985	0.9976	0.9966	0.9960	0.9948
23.30	0.9999	0.9997	0.9995	0.9992	0.9989	0.9985	0.9976	0.9966	0.9960	0.9948
23.35	0.9999	0.9997	0.9995	0.9992	0.9989	0.9985	0.9977	0.9966	0.9960	0.9948
23.40	0.9999	0.9997	0.9995	0.9992	0.9989	0.9985	0.9977	0.9966	0.9961	0.9949
23.45	0.9999	0.9997	0.9995	0.9993	0.9989	0.9985	0.9977	0.9967	0.9961	0.9949
23.50	0.9999	0.9997	0.9995	0.9993	0.9989	0.9986	0.9977	0.9967	0.9961	0.9949
23.55	0.9999	0.9997	0.9995	0.9993	0.9989	0.9986	0.9977	0.9967	0.9961	0.9950
23.60	0.9999	0.9997	0.9995	0.9993	0.9989	0.9986	0.9977	0.9967	0.9962	0.9950
23.65	0.9999	0.9997	0.9995	0.9993	0.9989	0.9986	0.9977	0.9968	0.9962	0.9950
23.70	0.9999	0.9997	0.9995	0.9993	0.9990	0.9986	0.9978	0.9968	0.9962	0.9950
23.75	0.9999	0.9997	0.9995	0.9993	0.9990	0.9986	0.9978	0.9968	0.9962	0.9951
23.80	0.9999	0.9997	0.9995	0.9993	0.9990	0.9986	0.9978	0.9968	0.9963	0.9951
23.85	0.9999	0.9997	0.9995	0.9993	0.9990	0.9986	0.9978	0.9968	0.9963	0.9951
23.90	0.9999	0.9997	0.9995	0.9993	0.9990	0.9986	0.9978	0.9969	0.9963	0.9952
23.95	0.9999	0.9997	0.9995	0.9993	0.9990	0.9986	0.9978	0.9969	0.9963	0.9952
24.00	0.9999	0.9997	0.9995	0.9993	0.9990	0.9986	0.9978	0.9969	0.9964	0.9952
24.05	0.9999	0.9997	0.9995	0.9993	0.9990	0.9986	0.9978	0.9969	0.9964	0.9953
24.10	0.9999	0.9997	0.9995	0.9993	0.9990	0.9987	0.9979	0.9969	0.9964	0.9953
24.15	0.9999	0.9997	0.9995	0.9993	0.9990	0.9987	0.9979	0.9969	0.9964	0.9953
24.20	0.9999	0.9997	0.9995	0.9993	0.9990	0.9987	0.9979	0.9970	0.9964	0.9953

PROBABILITY INTEGRAL OF t, THE NON-CENTRAL t-STATISTIC. THIS TABLE GIVES Pr $[t/\sqrt{f} \leq x]$.

f is the number of degrees of freedom; the non-centrality parameter is $\sqrt{f+1}\,K_p$.
K_p is the standardized normal deviate exceeded with probability p.

DEGREES OF FREEDOM 3

x \ p	.2500	.1500	.1000	.0650	.0400	.0250	.0100	.0040	.0025	.0010
24.25	0.9999	0.9997	0.9995	0.9993	0.9990	0.9987	0.9979	0.9970	0.9965	0.9954
24.30	0.9999	0.9997	0.9996	0.9993	0.9990	0.9987	0.9979	0.9970	0.9965	0.9954
24.35	0.9999	0.9997	0.9996	0.9993	0.9990	0.9987	0.9979	0.9970	0.9965	0.9954
24.40	0.9999	0.9997	0.9996	0.9993	0.9990	0.9987	0.9979	0.9970	0.9965	0.9955
24.45	0.9999	0.9997	0.9996	0.9993	0.9990	0.9987	0.9980	0.9971	0.9966	0.9955
24.50	0.9999	0.9997	0.9996	0.9993	0.9991	0.9987	0.9980	0.9971	0.9966	0.9955
24.55	0.9999	0.9997	0.9996	0.9993	0.9991	0.9987	0.9980	0.9971	0.9966	0.9955
24.60	0.9999	0.9997	0.9996	0.9994	0.9991	0.9987	0.9980	0.9971	0.9966	0.9956
24.65	0.9999	0.9997	0.9996	0.9994	0.9991	0.9987	0.9980	0.9971	0.9966	0.9956
24.70	0.9999	0.9997	0.9996	0.9994	0.9991	0.9988	0.9980	0.9971	0.9967	0.9956
24.75	0.9999	0.9997	0.9996	0.9994	0.9991	0.9988	0.9980	0.9972	0.9967	0.9956
24.80	0.9999	0.9997	0.9996	0.9994	0.9991	0.9988	0.9980	0.9972	0.9967	0.9957
24.85	0.9999	0.9997	0.9996	0.9994	0.9991	0.9988	0.9980	0.9972	0.9967	0.9957
24.90	0.9999	0.9997	0.9996	0.9994	0.9991	0.9988	0.9981	0.9972	0.9967	0.9957
24.95	0.9999	0.9997	0.9996	0.9994	0.9991	0.9988	0.9981	0.9972	0.9968	0.9957
25.00	0.9999	0.9997	0.9996	0.9994	0.9991	0.9988	0.9981	0.9972	0.9968	0.9958
25.05	0.9999	0.9997	0.9996	0.9994	0.9991	0.9988	0.9981	0.9973	0.9968	0.9958
25.10	0.9999	0.9997	0.9996	0.9994	0.9991	0.9988	0.9981	0.9973	0.9968	0.9958
25.15	0.9999	0.9997	0.9996	0.9994	0.9991	0.9988	0.9981	0.9973	0.9968	0.9958
25.20	0.9999	0.9998	0.9996	0.9994	0.9991	0.9988	0.9981	0.9973	0.9968	0.9959
25.25	0.9999	0.9998	0.9996	0.9994	0.9991	0.9988	0.9981	0.9973	0.9969	0.9959
25.30	0.9999	0.9998	0.9996	0.9994	0.9991	0.9988	0.9982	0.9973	0.9969	0.9959
25.35	0.9999	0.9998	0.9996	0.9994	0.9991	0.9988	0.9982	0.9974	0.9969	0.9959
25.40	0.9999	0.9998	0.9996	0.9994	0.9991	0.9989	0.9982	0.9974	0.9969	0.9960
25.45	0.9999	0.9998	0.9996	0.9994	0.9992	0.9989	0.9982	0.9974	0.9969	0.9960
25.50	0.9999	0.9998	0.9996	0.9994	0.9992	0.9989	0.9982	0.9974	0.9970	0.9960
25.55	0.9999	0.9998	0.9996	0.9994	0.9992	0.9989	0.9982	0.9974	0.9970	0.9960
25.60	0.9999	0.9998	0.9996	0.9994	0.9992	0.9989	0.9982	0.9974	0.9970	0.9961
25.65	0.9999	0.9998	0.9996	0.9994	0.9992	0.9989	0.9982	0.9974	0.9970	0.9961
25.70	0.9999	0.9998	0.9996	0.9994	0.9992	0.9989	0.9982	0.9975	0.9970	0.9961
25.75	0.9999	0.9998	0.9996	0.9994	0.9992	0.9989	0.9982	0.9975	0.9970	0.9961
25.80	0.9999	0.9998	0.9996	0.9994	0.9992	0.9989	0.9983	0.9975	0.9971	0.9961
25.85	0.9999	0.9998	0.9996	0.9994	0.9992	0.9989	0.9983	0.9975	0.9971	0.9962
25.90	0.9999	0.9998	0.9996	0.9994	0.9992	0.9989	0.9983	0.9975	0.9971	0.9962
25.95	0.9999	0.9998	0.9996	0.9994	0.9992	0.9989	0.9983	0.9975	0.9971	0.9962
26.00	0.9999	0.9998	0.9996	0.9995	0.9992	0.9989	0.9983	0.9975	0.9971	0.9962
26.05	0.9999	0.9998	0.9996	0.9995	0.9992	0.9989	0.9983	0.9976	0.9971	0.9963
26.10	0.9999	0.9998	0.9996	0.9995	0.9992	0.9989	0.9983	0.9976	0.9972	0.9963
26.15	0.9999	0.9998	0.9996	0.9995	0.9992	0.9989	0.9983	0.9976	0.9972	0.9963
26.20	0.9999	0.9998	0.9996	0.9995	0.9992	0.9990	0.9983	0.9976	0.9972	0.9963
26.25	0.9999	0.9998	0.9996	0.9995	0.9992	C.9990	0.9983	0.9976	0.9972	0.9963
26.30	0.9999	0.9998	0.9996	0.9995	0.9992	0.9990	0.9984	0.9976	0.9972	0.9964
26.35	0.9999	0.9998	0.9996	0.9995	0.9992	0.9990	0.9984	0.9976	0.9972	0.9964
26.40	0.9999	0.9998	0.9997	0.9995	0.9992	0.9990	0.9984	0.9977	0.9973	0.9964
26.45	0.9999	0.9998	0.9997	0.9995	0.9992	0.9990	0.9984	0.9977	0.9973	0.9964
26.50	0.9999	0.9998	0.9997	0.9995	0.9993	0.9990	0.9984	0.9977	0.9973	0.9964
26.55	0.9999	0.9998	0.9997	0.9995	0.9993	0.9990	0.9984	0.9977	0.9973	0.9965
26.60	0.9999	0.9998	0.9997	0.9995	0.9993	0.9990	0.9984	0.9977	0.9973	0.9965
26.65	0.9999	0.9998	0.9997	0.9995	0.9993	0.9990	0.9984	0.9977	0.9973	0.9965
26.70	0.9999	0.9998	0.9997	0.9995	0.9993	0.9990	0.9984	0.9977	0.9973	0.9965
26.75	0.9999	0.9998	0.9997	0.9995	0.9993	0.9990	0.9984	0.9977	0.9974	0.9965
26.80	0.9999	0.9998	0.9997	0.9995	0.9993	0.9990	0.9984	0.9978	0.9974	0.9966
26.85	0.9999	0.9998	0.9997	0.9995	0.9993	0.9990	0.9985	0.9978	0.9974	0.9966
26.90	0.9999	0.9998	0.9997	0.9995	0.9993	0.9990	0.9985	0.9978	0.9974	0.9966
26.95	0.9999	0.9998	0.9997	0.9995	0.9993	0.9990	0.9985	0.9978	0.9974	0.9966
27.00	0.9999	0.9998	0.9997	0.9995	0.9993	0.9990	0.9985	0.9978	0.9974	0.9966
27.05	0.9999	0.9998	0.9997	0.9995	0.9993	0.9991	0.9985	0.9978	0.9974	0.9967
27.10	0.9999	0.9998	0.9997	0.9995	0.9993	0.9991	0.9985	0.9978	0.9975	0.9967
27.15	0.9999	0.9998	0.9997	0.9995	0.9993	0.9991	0.9985	0.9978	0.9975	0.9967
27.20	0.9999	0.9998	0.9997	0.9995	0.9993	0.9991	0.9985	0.9979	0.9975	0.9967

PROBABILITY INTEGRAL OF t, THE NON-CENTRAL t-STATISTIC. THIS TABLE GIVES Pr $[t/\sqrt{f} \leq x]$.

f is the number of degrees of freedom; the non-centrality parameter is $\sqrt{f+1}\,K_p$.

K_p is the standardized normal deviate exceeded with probability p.

DEGREES OF FREEDOM **3**

x \ p	.2500	.1500	.1000	.0650	.0400	.0250	.0100	.0040	.0025	.0010
27.25	0.9999	0.9998	0.9997	0.9995	0.9993	0.9991	0.9985	0.9979	0.9975	0.9967
27.30	0.9999	0.9998	0.9997	0.9995	0.9993	0.9991	0.9985	0.9979	0.9975	0.9967
27.35	0.9999	0.9998	0.9997	0.9995	0.9993	0.9991	0.9985	0.9979	0.9975	0.9968
27.40	0.9999	0.9998	0.9997	0.9995	0.9993	0.9991	0.9985	0.9979	0.9975	0.9968
27.45	0.9999	0.9998	0.9997	0.9995	0.9993	0.9991	0.9986	0.9979	0.9976	0.9968
27.50	0.9999	0.9998	0.9997	0.9995	0.9993	0.9991	0.9986	0.9979	0.9976	0.9968
27.55	0.9999	0.9998	0.9997	0.9995	0.9993	0.9991	0.9986	0.9979	0.9976	0.9968
27.60	0.9999	0.9998	0.9997	0.9995	0.9993	0.9991	0.9986	0.9979	0.9976	0.9968
27.65	0.9999	0.9998	0.9997	0.9995	0.9993	0.9991	0.9986	0.9980	0.9976	0.9969
27.70	0.9999	0.9998	0.9997	0.9995	0.9993	0.9991	0.9986	0.9980	0.9976	0.9969
27.75	0.9999	0.9998	0.9997	0.9996	0.9993	0.9991	0.9986	0.9980	0.9976	0.9969
27.80	0.9999	0.9998	0.9997	0.9996	0.9994	0.9991	0.9986	0.9980	0.9976	0.9969
27.85	0.9999	0.9998	0.9997	0.9996	0.9994	0.9991	0.9986	0.9980	0.9977	0.9969
27.90	0.9999	0.9998	0.9997	0.9996	0.9994	0.9991	0.9986	0.9980	0.9977	0.9969
27.95	0.9999	0.9998	0.9997	0.9996	0.9994	0.9991	0.9986	0.9980	0.9977	0.9970
28.00	0.9999	0.9998	0.9997	0.9996	0.9994	0.9991	0.9986	0.9980	0.9977	0.9970
28.05	0.9999	0.9998	0.9997	0.9996	0.9994	0.9991	0.9986	0.9980	0.9977	0.9970
28.10	0.9999	0.9998	0.9997	0.9996	0.9994	0.9992	0.9987	0.9981	0.9977	0.9970
28.15	0.9999	0.9998	0.9997	0.9996	0.9994	0.9992	0.9987	0.9981	0.9977	0.9970
28.20	0.9999	0.9998	0.9997	0.9996	0.9994	0.9992	0.9987	0.9981	0.9977	0.9970
28.25	0.9999	0.9998	0.9997	0.9996	0.9994	0.9992	0.9987	0.9981	0.9978	0.9971
28.30	0.9999	0.9998	0.9997	0.9996	0.9994	0.9992	0.9987	0.9981	0.9978	0.9971
28.35	0.9999	0.9998	0.9997	0.9996	0.9994	0.9992	0.9987	0.9981	0.9978	0.9971
28.40	0.9999	0.9998	0.9997	0.9996	0.9994	0.9992	0.9987	0.9981	0.9978	0.9971
28.45	0.9999	0.9998	0.9997	0.9996	0.9994	0.9992	0.9987	0.9981	0.9978	0.9971
28.50	0.9999	0.9998	0.9997	0.9996	0.9994	0.9992	0.9987	0.9981	0.9978	0.9971
28.55	0.9999	0.9998	0.9997	0.9996	0.9994	0.9992	0.9987	0.9981	0.9978	0.9971
28.60	0.9999	0.9998	0.9997	0.9996	0.9994	0.9992	0.9987	0.9982	0.9978	0.9972
28.65	0.9999	0.9998	0.9997	0.9996	0.9994	0.9992	0.9987	0.9982	0.9978	0.9972
28.70	0.9999	0.9998	0.9997	0.9996	0.9994	0.9992	0.9987	0.9982	0.9979	0.9972
28.75	0.9999	0.9998	0.9997	0.9996	0.9994	0.9992	0.9987	0.9982	0.9979	0.9972
28.80	0.9999	0.9998	0.9997	0.9996	0.9994	0.9992	0.9987	0.9982	0.9979	0.9972
28.85	0.9999	0.9998	0.9997	0.9996	0.9994	0.9992	0.9988	0.9982	0.9979	0.9972
28.90	0.9999	0.9998	0.9997	0.9996	0.9994	0.9992	0.9988	0.9982	0.9979	0.9973
28.95	0.9999	0.9998	0.9997	0.9996	0.9994	0.9992	0.9988	0.9982	0.9979	0.9973
29.00	0.9999	0.9998	0.9997	0.9996	0.9994	0.9992	0.9988	0.9982	0.9979	0.9973
29.05	0.9999	0.9998	0.9997	0.9996	0.9994	0.9992	0.9988	0.9982	0.9979	0.9973
29.10	0.9999	0.9998	0.9997	0.9996	0.9994	0.9992	0.9988	0.9982	0.9979	0.9973
29.15	0.9999	0.9998	0.9997	0.9996	0.9994	0.9992	0.9988	0.9983	0.9980	0.9973
29.20	0.9999	0.9998	0.9997	0.9996	0.9994	0.9992	0.9988	0.9983	0.9980	0.9973
29.25	0.9999	0.9998	0.9997	0.9996	0.9994	0.9993	0.9988	0.9983	0.9980	0.9973
29.30	0.9999	0.9998	0.9997	0.9996	0.9994	0.9993	0.9988	0.9983	0.9980	0.9974
29.35	0.9999	0.9998	0.9997	0.9996	0.9994	0.9993	0.9988	0.9983	0.9980	0.9974
29.40	0.9999	0.9998	0.9997	0.9996	0.9995	0.9993	0.9988	0.9983	0.9980	0.9974
29.45	0.9999	0.9998	0.9997	0.9996	0.9995	0.9993	0.9988	0.9983	0.9980	0.9974
29.50	0.9999	0.9998	0.9998	0.9996	0.9995	0.9993	0.9988	0.9983	0.9980	0.9974
29.55	0.9999	0.9998	0.9998	0.9996	0.9995	0.9993	0.9988	0.9983	0.9980	0.9974
29.60	0.9999	0.9998	0.9998	0.9996	0.9995	0.9993	0.9988	0.9983	0.9980	0.9974
29.65	0.9999	0.9998	0.9998	0.9996	0.9995	0.9993	0.9989	0.9983	0.9981	0.9975
29.70	0.9999	0.9998	0.9998	0.9996	0.9995	0.9993	0.9989	0.9984	0.9981	0.9975
29.75	0.9999	0.9998	0.9998	0.9996	0.9995	0.9993	0.9989	0.9984	0.9981	0.9975
29.80	0.9999	0.9998	0.9998	0.9996	0.9995	0.9993	0.9989	0.9984	0.9981	0.9975
29.85	0.9999	0.9998	0.9998	0.9996	0.9995	0.9993	0.9989	0.9984	0.9981	0.9975
29.90	0.9999	0.9999	0.9998	0.9996	0.9995	0.9993	0.9989	0.9984	0.9981	0.9975
29.95	0.9999	0.9999	0.9998	0.9996	0.9995	0.9993	0.9989	0.9984	0.9981	0.9975
30.00	0.9999	0.9999	0.9998	0.9996	0.9995	0.9993	0.9989	0.9984	0.9981	0.9975
30.05	0.9999	0.9999	0.9998	0.9996	0.9995	0.9993	0.9939	0.9984	0.9981	0.9976
30.10	0.9999	0.9999	0.9998	0.9996	0.9995	0.9993	0.9989	0.9984	0.9981	0.9976
30.15	0.9999	0.9999	0.9998	0.9997	0.9995	0.9993	0.9989	0.9984	0.9982	0.9976
30.20	0.9999	0.9999	0.9998	0.9997	0.9995	0.9993	0.9989	0.9984	0.9982	0.9976

PROBABILITY INTEGRAL OF t, THE NON-CENTRAL t-STATISTIC. THIS TABLE GIVES Pr $[t/\sqrt{f} \leq x]$.

f is the number of degrees of freedom; the non-centrality parameter is $\sqrt{f+1}\,K_p$.
K_p is the standardized normal deviate exceeded with probability p.

DEGREES OF FREEDOM **3**

x \ p	.2500	.1500	.1000	.0650	.0400	.0250	.0100	.0040	.0025	.0010
30.25	0.9999	0.9999	0.9998	0.9997	0.9995	0.9993	0.9989	0.9984	0.9982	0.9976
30.30	0.9999	0.9999	0.9998	0.9997	0.9995	0.9993	0.9989	0.9984	0.9982	0.9976
30.35	0.9999	0.9999	0.9998	0.9997	0.9995	0.9993	0.9989	0.9985	0.9982	0.9976
30.40	0.9999	0.9999	0.9998	0.9997	0.9995	0.9993	0.9989	0.9985	0.9982	0.9976
30.45	0.9999	0.9999	0.9998	0.9997	0.9995	0.9993	0.9989	0.9985	0.9982	0.9976
30.50	0.9999	0.9999	0.9998	0.9997	0.9995	0.9993	0.9989	0.9985	0.9982	0.9977
30.55	0.9999	0.9999	0.9998	0.9997	0.9995	0.9993	0.9989	0.9985	0.9982	0.9977
30.60	0.9999	0.9999	0.9998	0.9997	0.9995	0.9993	0.9990	0.9985	0.9982	0.9977
30.65	0.9999	0.9999	0.9998	0.9997	0.9995	0.9993	0.9990	0.9985	0.9982	0.9977
30.70	0.9999	0.9999	0.9998	0.9997	0.9995	0.9994	0.9990	0.9985	0.9983	0.9977
30.75	0.9999	0.9999	0.9998	0.9997	0.9995	0.9994	0.9990	0.9985	0.9983	0.9977
30.80	0.9999	0.9999	0.9998	0.9997	0.9995	0.9994	0.9990	0.9985	0.9983	0.9977
30.85	0.9999	0.9999	0.9998	0.9997	0.9995	0.9994	0.9990	0.9985	0.9983	0.9977
30.90	0.9999	0.9999	0.9998	0.9997	0.9995	0.9994	0.9990	0.9985	0.9983	0.9977
30.95	0.9999	0.9999	0.9998	0.9997	0.9995	0.9994	0.9990	0.9985	0.9983	0.9978
31.00	0.9999	0.9999	0.9998	0.9997	0.9995	0.9994	0.9990	0.9986	0.9983	0.9978
31.05	0.9999	0.9999	0.9998	0.9997	0.9995	0.9994	0.9990	0.9986	0.9983	0.9978
31.10	0.9999	0.9999	0.9998	0.9997	0.9995	0.9994	0.9990	0.9986	0.9983	0.9978
31.15	0.9999	0.9999	0.9998	0.9997	0.9995	0.9994	0.9990	0.9986	0.9983	0.9978
31.20	0.9999	0.9999	0.9998	0.9997	0.9995	0.9994	0.9990	0.9986	0.9983	0.9978
31.25	0.9999	0.9999	0.9998	0.9997	0.9995	0.9994	0.9990	0.9986	0.9983	0.9978
31.30	0.9999	0.9999	0.9998	0.9997	0.9995	0.9994	0.9990	0.9986	0.9983	0.9978
31.35	0.9999	0.9999	0.9998	0.9997	0.9995	0.9994	0.9990	0.9986	0.9984	0.9978
31.40	0.9999	0.9999	0.9998	0.9997	0.9996	0.9994	0.9990	0.9986	0.9984	0.9979
31.45	0.9999	0.9999	0.9998	0.9997	0.9996	0.9994	0.9990	0.9986	0.9984	0.9979
31.50	0.9999	0.9999	0.9998	0.9997	0.9996	0.9994	0.9990	0.9986	0.9984	0.9979
31.55	0.9999	0.9999	0.9998	0.9997	0.9996	0.9994	0.9990	0.9986	0.9984	0.9979
31.60	0.9999	0.9999	0.9998	0.9997	0.9996	0.9994	0.9991	0.9986	0.9984	0.9979
31.65	0.9999	0.9999	0.9998	0.9997	0.9996	0.9994	0.9991	0.9986	0.9984	0.9979
31.70	0.9999	0.9999	0.9998	0.9997	0.9996	0.9994	0.9991	0.9986	0.9984	0.9979
31.75	0.9999	0.9999	0.9998	0.9997	0.9996	0.9994	0.9991	0.9987	0.9984	0.9979
31.80	0.9999	0.9999	0.9998	0.9997	0.9996	0.9994	0.9991	0.9987	0.9984	0.9979
31.85	0.9999	0.9999	0.9998	0.9997	0.9996	0.9994	0.9991	0.9987	0.9984	0.9979
31.90	0.9999	0.9999	0.9998	0.9997	0.9996	0.9994	0.9991	0.9987	0.9984	0.9980
31.95	0.9999	0.9999	0.9998	0.9997	0.9996	0.9994	0.9991	0.9987	0.9984	0.9980
32.00	0.9999	0.9999	0.9998	0.9997	0.9996	0.9994	0.9991	0.9987	0.9985	0.9980
32.05	0.9999	0.9999	0.9998	0.9997	0.9996	0.9994	0.9991	0.9987	0.9985	0.9980
32.10	0.9999	0.9999	0.9998	0.9997	0.9996	0.9994	0.9991	0.9987	0.9985	0.9980
32.15	0.9999	0.9999	0.9998	0.9997	0.9996	0.9994	0.9991	0.9987	0.9985	0.9980
32.20	0.9999	0.9999	0.9998	0.9997	0.9996	0.9994	0.9991	0.9987	0.9985	0.9980
32.25	0.9999	0.9999	0.9998	0.9997	0.9996	0.9994	0.9991	0.9987	0.9985	0.9980
32.30	0.9999	0.9999	0.9998	0.9997	0.9996	0.9994	0.9991	0.9987	0.9985	0.9980
32.35	0.9999	0.9999	0.9998	0.9997	0.9996	0.9994	0.9991	0.9987	0.9985	0.9980
32.40	0.9999	0.9999	0.9998	0.9997	0.9996	0.9994	0.9991	0.9987	0.9985	0.9980
32.45	0.9999	0.9999	0.9998	0.9997	0.9996	0.9995	0.9991	0.9987	0.9985	0.9981
32.50	0.9999	0.9999	0.9998	0.9997	0.9996	0.9995	0.9991	0.9987	0.9985	0.9981
32.55	0.9999	0.9999	0.9998	0.9997	0.9996	0.9995	0.9991	0.9987	0.9985	0.9981
32.60	0.9999	0.9999	0.9998	0.9997	0.9996	0.9995	0.9991	0.9988	0.9985	0.9981
32.65	0.9999	0.9999	0.9998	0.9997	0.9996	0.9995	0.9991	0.9988	0.9985	0.9981
32.70	0.9999	0.9999	0.9998	0.9997	0.9996	0.9995	0.9991	0.9988	0.9986	0.9981
32.75	0.9999	0.9999	0.9998	0.9997	0.9996	0.9995	0.9991	0.9988	0.9986	0.9981
32.80	0.9999	0.9999	0.9998	0.9997	0.9996	0.9995	0.9992	0.9988	0.9986	0.9981
32.85	0.9999	0.9999	0.9998	0.9997	0.9996	0.9995	0.9992	0.9988	0.9986	0.9981
32.90	0.9999	0.9999	0.9998	0.9997	0.9996	0.9995	0.9992	0.9988	0.9986	0.9981
32.95	0.9999	0.9999	0.9998	0.9997	0.9996	0.9995	0.9992	0.9988	0.9986	0.9981
33.00	0.9999	0.9999	0.9998	0.9997	0.9996	0.9995	0.9992	0.9988	0.9986	0.9981
33.05	0.9999	0.9999	0.9998	0.9997	0.9996	0.9995	0.9992	0.9988	0.9986	0.9982
33.10	0.9999	0.9999	0.9998	0.9997	0.9996	0.9995	0.9992	0.9988	0.9986	0.9982
33.15	0.9999	0.9999	0.9998	0.9997	0.9996	0.9995	0.9992	0.9988	0.9986	0.9982
33.20	0.9999	0.9999	0.9998	0.9997	0.9996	0.9995	0.9992	0.9988	0.9986	0.9982

PROBABILITY INTEGRAL OF t, THE NON-CENTRAL t-STATISTIC. THIS TABLE GIVES Pr $[t/\sqrt{f} \leq x]$.

f is the number of degrees of freedom; the non-centrality parameter is $\sqrt{f+1}\,K_p$.

K_p is the standardized normal deviate exceeded with probability p.

DEGREES OF FREEDOM **3**

x \ p	.2500	.1500	.1000	.0650	.0400	.0250	.0100	.0040	.0025	.0010
33.25	0.9999	0.9999	0.9998	0.9997	0.9996	0.9995	0.9992	0.9988	0.9986	0.9982
33.30	0.9999	0.9999	0.9998	0.9997	0.9996	0.9995	0.9992	0.9988	0.9986	0.9982
33.35	0.9999	0.9999	0.9998	0.9997	0.9996	0.9995	0.9992	0.9988	0.9986	0.9982
33.40	0.9999	0.9999	0.9998	0.9997	0.9996	0.9995	0.9992	0.9988	0.9986	0.9982
33.45	0.9999	0.9999	0.9998	0.9997	0.9996	0.9995	0.9992	0.9988	0.9986	0.9982
33.50	0.9999	0.9999	0.9998	0.9997	0.9996	0.9995	0.9992	0.9989	0.9987	0.9982
33.55	0.9999	0.9999	0.9998	0.9997	0.9996	0.9995	0.9992	0.9989	0.9987	0.9982
33.60	0.9999	0.9999	0.9998	0.9997	0.9996	0.9995	0.9992	0.9989	0.9987	0.9982
33.65	0.9999	0.9999	0.9998	0.9998	0.9996	0.9995	0.9992	0.9989	0.9987	0.9983
33.70	0.9999	0.9999	0.9998	0.9998	0.9996	0.9995	0.9992	0.9989	0.9987	0.9983
33.75	0.9999	0.9999	0.9998	0.9998	0.9996	0.9995	0.9992	0.9989	0.9987	0.9983
33.80	0.9999	0.9999	0.9998	0.9998	0.9996	0.9995	0.9992	0.9989	0.9987	0.9983
33.85	0.9999	0.9999	0.9998	0.9998	0.9996	0.9995	0.9992	0.9989	0.9987	0.9983
33.90	0.9999	0.9999	0.9998	0.9998	0.9996	0.9995	0.9992	0.9989	0.9987	0.9983
33.95	0.9999	0.9999	0.9998	0.9998	0.9996	0.9995	0.9992	0.9989	0.9987	0.9983
34.00	0.9999	0.9999	0.9998	0.9998	0.9996	0.9995	0.9992	0.9989	0.9987	0.9983
34.05	0.9999	0.9999	0.9998	0.9998	0.9997	0.9995	0.9992	0.9989	0.9987	0.9983
34.10	1.0000	0.9999	0.9998	0.9998	0.9997	0.9995	0.9992	0.9989	0.9987	0.9983
34.15	1.0000	0.9999	0.9998	0.9998	0.9997	0.9995	0.9992	0.9989	0.9987	0.9983
34.20	1.0000	0.9999	0.9998	0.9998	0.9997	0.9995	0.9993	0.9989	0.9987	0.9983
34.25	1.0000	0.9999	0.9998	0.9998	0.9997	0.9995	0.9993	0.9989	0.9987	0.9983
34.30	1.0000	0.9999	0.9998	0.9998	0.9997	0.9995	0.9993	0.9989	0.9987	0.9984
34.35	1.0000	0.9999	0.9998	0.9998	0.9997	0.9995	0.9993	0.9989	0.9988	0.9984
34.40	1.0000	0.9999	0.9998	0.9998	0.9997	0.9995	0.9993	0.9989	0.9988	0.9984
34.45	1.0000	0.9999	0.9998	0.9998	0.9997	0.9995	0.9993	0.9989	0.9988	0.9984
34.50	1.0000	0.9999	0.9998	0.9998	0.9997	0.9995	0.9993	0.9989	0.9988	0.9984
34.55	1.0000	0.9999	0.9998	0.9998	0.9997	0.9995	0.9993	0.9990	0.9988	0.9984
34.60	1.0000	0.9999	0.9998	0.9998	0.9997	0.9995	0.9993	0.9990	0.9988	0.9984
34.65	1.0000	0.9999	0.9998	0.9998	0.9997	0.9996	0.9993	0.9990	0.9988	0.9984
34.70	1.0000	0.9999	0.9998	0.9998	0.9997	0.9996	0.9993	0.9990	0.9988	0.9984
34.75	1.0000	0.9999	0.9998	0.9998	0.9997	0.9996	0.9993	0.9990	0.9988	0.9984
34.80	1.0000	0.9999	0.9998	0.9998	0.9997	0.9996	0.9993	0.9990	0.9988	0.9984
34.85	1.0000	0.9999	0.9999	0.9998	0.9997	0.9996	0.9993	0.9990	0.9988	0.9984
34.90	1.0000	0.9999	0.9999	0.9998	0.9997	0.9996	0.9993	0.9990	0.9988	0.9984
34.95	1.0000	0.9999	0.9999	0.9998	0.9997	0.9996	0.9993	0.9990	0.9988	0.9984
35.00	1.0000	0.9999	0.9999	0.9998	0.9997	0.9996	0.9993	0.9990	0.9988	0.9984
35.05	1.0000	0.9999	0.9999	0.9998	0.9997	0.9996	0.9993	0.9990	0.9988	0.9985
35.10	1.0000	0.9999	0.9999	0.9998	0.9997	0.9996	0.9993	0.9990	0.9988	0.9985
35.15	1.0000	0.9999	0.9999	0.9998	0.9997	0.9996	0.9993	0.9990	0.9988	0.9985
35.20	1.0000	0.9999	0.9999	0.9998	0.9997	0.9996	0.9993	0.9990	0.9988	0.9985
35.25	1.0000	0.9999	0.9999	0.9998	0.9997	0.9996	0.9993	0.9990	0.9988	0.9985
35.30	1.0000	0.9999	0.9999	0.9998	0.9997	0.9996	0.9993	0.9990	0.9988	0.9985
35.35	1.0000	0.9999	0.9999	0.9998	0.9997	0.9996	0.9993	0.9990	0.9989	0.9985
35.40	1.0000	0.9999	0.9999	0.9998	0.9997	0.9996	0.9993	0.9990	0.9989	0.9985
35.45	1.0000	0.9999	0.9999	0.9998	0.9997	0.9996	0.9993	0.9990	0.9989	0.9985
35.50	1.0000	0.9999	0.9999	0.9998	0.9997	0.9996	0.9993	0.9990	0.9989	0.9985
35.55	1.0000	0.9999	0.9999	0.9998	0.9997	0.9996	0.9993	0.9990	0.9989	0.9985
35.60	1.0000	0.9999	0.9999	0.9998	0.9997	0.9996	0.9993	0.9990	0.9989	0.9985
35.65	1.0000	0.9999	0.9999	0.9998	0.9997	0.9996	0.9993	0.9990	0.9989	0.9985
35.70	1.0000	0.9999	0.9999	0.9998	0.9997	0.9996	0.9993	0.9991	0.9989	0.9985
35.75	1.0000	0.9999	0.9999	0.9998	0.9997	0.9996	0.9993	0.9991	0.9989	0.9985
35.80	1.0000	0.9999	0.9999	0.9998	0.9997	0.9996	0.9993	0.9991	0.9989	0.9986
35.85	1.0000	0.9999	0.9999	0.9998	0.9997	0.9996	0.9994	0.9991	0.9989	0.9986
35.90	1.0000	0.9999	0.9999	0.9998	0.9997	0.9996	0.9994	0.9991	0.9989	0.9986
35.95	1.0000	0.9999	0.9999	0.9998	0.9997	0.9996	0.9994	0.9991	0.9989	0.9986
36.00	1.0000	0.9999	0.9999	0.9998	0.9997	0.9996	0.9994	0.9991	0.9989	0.9986
36.05	1.0000	0.9999	0.9999	0.9998	0.9997	0.9996	0.9994	0.9991	0.9989	0.9986
36.10	1.0000	0.9999	0.9999	0.9998	0.9997	0.9996	0.9994	0.9991	0.9989	0.9986
36.15	1.0000	0.9999	0.9999	0.9998	0.9997	0.9996	0.9994	0.9991	0.9989	0.9986
36.20	1.0000	0.9999	0.9999	0.9998	0.9997	0.9996	0.9994	0.9991	0.9989	0.9986

PROBABILITY INTEGRAL OF t, THE NON-CENTRAL t-STATISTIC. THIS TABLE GIVES Pr $[t/\sqrt{f} \leq x]$.

f is the number of degrees of freedom; the non-centrality parameter is $\sqrt{f+1}\, K_p$.

K_p is the standardized normal deviate exceeded with probability p.

DEGREES OF FREEDOM **3**

x \ p	.2500	.1500	.1000	.0650	.0400	.0250	.0100	.0040	.0025	.0010
36.25	1.0000	0.9999	0.9999	0.9998	0.9997	0.9996	0.9994	0.9991	0.9989	0.9986
36.30	1.0000	0.9999	0.9999	0.9998	0.9997	0.9996	0.9994	0.9991	0.9989	0.9986
36.35	1.0000	0.9999	0.9999	0.9998	0.9997	0.9996	0.9994	0.9991	0.9989	0.9986
36.40	1.0000	0.9999	0.9999	0.9998	0.9997	0.9996	0.9994	0.9991	0.9990	0.9986
36.45	1.0000	0.9999	0.9999	0.9998	0.9997	0.9996	0.9994	0.9991	0.9990	0.9986
36.50	1.0000	0.9999	0.9999	0.9998	0.9997	0.9996	0.9994	0.9991	0.9990	0.9986
36.55	1.0000	0.9999	0.9999	0.9998	0.9997	0.9996	0.9994	0.9991	0.9990	0.9986
36.60	1.0000	0.9999	0.9999	0.9998	0.9997	0.9996	0.9994	0.9991	0.9990	0.9986
36.65	1.0000	0.9999	0.9999	0.9998	0.9997	0.9996	0.9994	0.9991	0.9990	0.9986
36.70	1.0000	0.9999	0.9999	0.9998	0.9997	0.9996	0.9994	0.9991	0.9990	0.9987
36.75	1.0000	0.9999	0.9999	0.9998	0.9997	0.9996	0.9994	0.9991	0.9990	0.9987
36.80	1.0000	0.9999	0.9999	0.9998	0.9997	0.9996	0.9994	0.9991	0.9990	0.9987
36.85	1.0000	0.9999	0.9999	0.9998	0.9997	0.9996	0.9994	0.9991	0.9990	0.9987
36.90	1.0000	0.9999	0.9999	0.9998	0.9997	0.9996	0.9994	0.9991	0.9990	0.9987
36.95	1.0000	0.9999	0.9999	0.9998	0.9997	0.9996	0.9994	0.9991	0.9990	0.9987
37.00	1.0000	0.9999	0.9999	0.9998	0.9997	0.9996	0.9994	0.9991	0.9990	0.9987
37.05	1.0000	0.9999	0.9999	0.9998	0.9997	0.9996	0.9994	0.9992	0.9990	0.9987
37.10	1.0000	0.9999	0.9999	0.9998	0.9997	0.9996	0.9994	0.9992	0.9990	0.9987
37.15	1.0000	0.9999	0.9999	0.9998	0.9997	0.9996	0.9994	0.9992	0.9990	0.9987
37.20	1.0000	0.9999	0.9999	0.9998	0.9997	0.9996	0.9994	0.9992	0.9990	0.9987
37.25	1.0000	0.9999	0.9999	0.9998	0.9997	0.9996	0.9994	0.9992	0.9990	0.9987
37.30	1.0000	0.9999	0.9999	0.9998	0.9997	0.9996	0.9994	0.9992	0.9990	0.9987
37.35	1.0000	0.9999	0.9999	0.9998	0.9997	0.9996	0.9994	0.9992	0.9990	0.9987
37.40	1.0000	0.9999	0.9999	0.9998	0.9997	0.9996	0.9994	0.9992	0.9990	0.9987
37.45	1.0000	0.9999	0.9999	0.9998	0.9997	0.9996	0.9994	0.9992	0.9990	0.9987
37.50	1.0000	0.9999	0.9999	0.9998	0.9997	0.9996	0.9994	0.9992	0.9990	0.9987
37.55	1.0000	0.9999	0.9999	0.9998	0.9997	0.9996	0.9994	0.9992	0.9990	0.9987
37.60	1.0000	0.9999	0.9999	0.9998	0.9997	0.9997	0.9994	0.9992	0.9990	0.9987
37.65	1.0000	0.9999	0.9999	0.9998	0.9997	0.9997	0.9994	0.9992	0.9991	0.9988
37.70	1.0000	0.9999	0.9999	0.9998	0.9997	0.9997	0.9994	0.9992	0.9991	0.9988
37.75	1.0000	0.9999	0.9999	0.9998	0.9997	0.9997	0.9994	0.9992	0.9991	0.9988
37.80	1.0000	0.9999	0.9999	0.9998	0.9997	0.9997	0.9994	0.9992	0.9991	0.9988
37.85	1.0000	0.9999	0.9999	0.9998	0.9997	0.9997	0.9995	0.9992	0.9991	0.9988
37.90	1.0000	0.9999	0.9999	0.9998	0.9997	0.9997	0.9995	0.9992	0.9991	0.9988
37.95	1.0000	0.9999	0.9999	0.9998	0.9997	0.9997	0.9995	0.9992	0.9991	0.9988
38.00	1.0000	0.9999	0.9999	0.9998	0.9997	0.9997	0.9995	0.9992	0.9991	0.9988
38.05	1.0000	0.9999	0.9999	0.9998	0.9998	0.9997	0.9995	0.9992	0.9991	0.9988
38.10	1.0000	0.9999	0.9999	0.9998	0.9998	0.9997	0.9995	0.9992	0.9991	0.9988
38.15	1.0000	0.9999	0.9999	0.9998	0.9998	0.9997	0.9995	0.9992	0.9991	0.9988
38.20	1.0000	0.9999	0.9999	0.9998	0.9998	0.9997	0.9995	0.9992	0.9991	0.9988
38.25	1.0000	0.9999	0.9999	0.9998	0.9998	0.9997	0.9995	0.9992	0.9991	0.9988
38.30	1.0000	0.9999	0.9999	0.9998	0.9998	0.9997	0.9995	0.9992	0.9991	0.9988
38.35	1.0000	0.9999	0.9999	0.9998	0.9998	0.9997	0.9995	0.9992	0.9991	0.9988
38.40	1.0000	0.9999	0.9999	0.9998	0.9998	0.9997	0.9995	0.9992	0.9991	0.9988
38.45	1.0000	0.9999	0.9999	0.9998	0.9998	0.9997	0.9995	0.9992	0.9991	0.9988
38.50	1.0000	0.9999	0.9999	0.9998	0.9998	0.9997	0.9995	0.9992	0.9991	0.9988
38.55	1.0000	0.9999	0.9999	0.9998	0.9998	0.9997	0.9995	0.9992	0.9991	0.9988
38.60	1.0000	0.9999	0.9999	0.9998	0.9998	0.9997	0.9995	0.9993	0.9991	0.9988
38.65	1.0000	0.9999	0.9999	0.9998	0.9998	0.9997	0.9995	0.9993	0.9991	0.9988
38.70	1.0000	0.9999	0.9999	0.9998	0.9998	0.9997	0.9995	0.9993	0.9991	0.9989
38.75	1.0000	0.9999	0.9999	0.9998	0.9998	0.9997	0.9995	0.9993	0.9991	0.9989
38.80	1.0000	0.9999	0.9999	0.9998	0.9998	0.9997	0.9995	0.9993	0.9991	0.9989
38.85	1.0000	0.9999	0.9999	0.9998	0.9998	0.9997	0.9995	0.9993	0.9991	0.9989
38.90	1.0000	0.9999	0.9999	0.9998	0.9998	0.9997	0.9995	0.9993	0.9991	0.9989
38.95	1.0000	0.9999	0.9999	0.9998	0.9998	0.9997	0.9995	0.9993	0.9991	0.9989
39.00	1.0000	0.9999	0.9999	0.9998	0.9998	0.9997	0.9995	0.9993	0.9991	0.9989
39.05	1.0000	0.9999	0.9999	0.9998	0.9998	0.9997	0.9995	0.9993	0.9992	0.9989
39.10	1.0000	0.9999	0.9999	0.9998	0.9998	0.9997	0.9995	0.9993	0.9992	0.9989
39.15	1.0000	0.9999	0.9999	0.9998	0.9998	0.9997	0.9995	0.9993	0.9992	0.9989
39.20	1.0000	0.9999	0.9999	0.9998	0.9998	0.9997	0.9995	0.9993	0.9992	0.9989

PROBABILITY INTEGRAL OF t, THE NON-CENTRAL t-STATISTIC. THIS TABLE GIVES Pr $[t/\sqrt{f} \leq x]$.

f is the number of degrees of freedom; the non-centrality parameter is $\sqrt{f+1}\,K_p$.

K_p is the standardized normal deviate exceeded with probability p.

DEGREES OF FREEDOM **3**

x \ p	.2500	.1500	.1000	.0650	.0400	.0250	.0100	.0040	.0025	.0010
39.25	1.0000	0.9999	0.9999	0.9998	0.9998	0.9997	0.9995	0.9993	0.9992	0.9989
39.30	1.0000	0.9999	0.9999	0.9998	0.9998	0.9997	0.9995	0.9993	0.9992	0.9989
39.35	1.0000	0.9999	0.9999	0.9998	0.9998	0.9997	0.9995	0.9993	0.9992	0.9989
39.40	1.0000	0.9999	0.9999	0.9998	0.9998	0.9997	0.9995	0.9993	0.9992	0.9989
39.45			0.9999	0.9998	0.9998	0.9997	0.9995	0.9993	0.9992	0.9989
39.50			0.9999	0.9998	0.9998	0.9997	0.9995	0.9993	0.9992	0.9989
39.55			0.9999	0.9998	0.9998	0.9997	0.9995	0.9993	0.9992	0.9989
39.60			0.9999	0.9998	0.9998	0.9997	0.9995	0.9993	0.9992	0.9989
39.65			0.9999	0.9998	0.9998	0.9997	0.9995	0.9993	0.9992	0.9989
39.70			0.9999	0.9999	0.9998	0.9997	0.9995	0.9993	0.9992	0.9989
39.75			0.9999	0.9999	0.9998	0.9997	0.9995	0.9993	0.9992	0.9989
39.80			0.9999	0.9999	0.9998	0.9997	0.9995	0.9993	0.9992	0.9989
39.85			0.9999	0.9999	0.9998	0.9997	0.9995	0.9993	0.9992	0.9990
39.90			0.9999	0.9999	0.9998	0.9997	0.9995	0.9993	0.9992	0.9990
39.95			0.9999	0.9999	0.9998	0.9997	0.9995	0.9993	0.9992	0.9990
40.00			0.9999	0.9999	0.9998	0.9997	0.9995	0.9993	0.9992	0.9990
40.05			0.9999	0.9999	0.9998	0.9997	0.9995	0.9993	0.9992	0.9990
40.10			0.9999	0.9999	0.9998	0.9997	0.9995	0.9993	0.9992	0.9990
40.15			0.9999	0.9999	0.9998	0.9997	0.9995	0.9993	0.9992	0.9990
40.20			0.9999	0.9999	0.9998	0.9997	0.9995	0.9993	0.9992	0.9990
40.25			0.9999	0.9999	0.9998	0.9997	0.9995	0.9993	0.9992	0.9990
40.30			0.9999	0.9999	0.9998	0.9997	0.9995	0.9993	0.9992	0.9990
40.35			0.9999	0.9999	0.9998	0.9997	0.9995	0.9993	0.9992	0.9990
40.40			0.9999	0.9999	0.9998	0.9997	0.9995	0.9993	0.9992	0.9990
40.45			0.9999	0.9999	0.9998	0.9997	0.9996	0.9993	0.9992	0.9990
40.50			0.9999	0.9999	0.9998	0.9997	0.9996	0.9994	0.9992	0.9990
40.55			0.9999	0.9999	0.9998	0.9997	0.9996	0.9994	0.9992	0.9990
40.60			0.9999	0.9999	0.9998	0.9997	0.9996	0.9994	0.9992	0.9990
40.65			0.9999	0.9999	0.9998	0.9997	0.9996	0.9994	0.9992	0.9990
40.70			0.9999	0.9999	0.9998	0.9997	0.9996	0.9994	0.9993	0.9990
40.75			0.9999	0.9999	0.9998	0.9997	0.9996	0.9994	0.9993	0.9990
40.80			0.9999	0.9999	0.9998	0.9997	0.9996	0.9994	0.9993	0.9990
40.85			0.9999	0.9999	0.9998	0.9997	0.9996	0.9994	0.9993	0.9990
40.90			0.9999	0.9999	0.9998	0.9997	0.9996	0.9994	0.9993	0.9990
40.95			0.9999	0.9999	0.9998	0.9997	0.9996	0.9994	0.9993	0.9990
41.00			0.9999	0.9999	0.9998	0.9997	0.9996	0.9994	0.9993	0.9990
41.05			0.9999	0.9999	0.9998	0.9997	0.9996	0.9994	0.9993	0.9990
41.10			0.9999	0.9999	0.9998	0.9997	0.9996	0.9994	0.9993	0.9990
41.15			0.9999	0.9999	0.9998	0.9997	0.9996	0.9994	0.9993	0.9990
41.20			0.9999	0.9999	0.9998	0.9997	0.9996	0.9994	0.9993	0.9991
41.25			0.9999	0.9999	0.9998	0.9997	0.9996	0.9994	0.9993	0.9991
41.30			0.9999	0.9999	0.9998	0.9997	0.9996	0.9994	0.9993	0.9991
41.35			0.9999	0.9999	0.9998	0.9997	0.9996	0.9994	0.9993	0.9991
41.40			0.9999	0.9999	0.9998	0.9997	0.9996	0.9994	0.9993	0.9991
41.45			0.9999	0.9999	0.9998	0.9997	0.9996	0.9994	0.9993	0.9991
41.50			0.9999	0.9999	0.9998	0.9997	0.9996	0.9994	0.9993	0.9991
41.55			0.9999	0.9999	0.9998	0.9997	0.9996	0.9994	0.9993	0.9991
41.60			0.9999	0.9999	0.9998	0.9997	0.9996	0.9994	0.9993	0.9991
41.65			0.9999	0.9999	0.9998	0.9997	0.9996	0.9994	0.9993	0.9991
41.70			0.9999	0.9999	0.9998	0.9997	0.9996	0.9994	0.9993	0.9991
41.75			0.9999	0.9999	0.9998	0.9997	0.9996	0.9994	0.9993	0.9991
41.80			0.9999	0.9999	0.9998	0.9997	0.9996	0.9994	0.9993	0.9991
41.85			0.9999	0.9999	0.9998	0.9997	0.9996	0.9994	0.9993	0.9991
41.90			0.9999	0.9999	0.9998	0.9997	0.9996	0.9994	0.9993	0.9991
41.95			0.9999	0.9999	0.9998	0.9998	0.9996	0.9994	0.9993	0.9991
42.00			0.9999	0.9999	0.9998	0.9998	0.9996	0.9994	0.9993	0.9991
42.05			0.9999	0.9999	0.9998	0.9998	0.9996	0.9994	0.9993	0.9991
42.10			0.9999	0.9999	0.9998	0.9998	0.9996	0.9994	0.9993	0.9991
42.15			0.9999	0.9999	0.9998	0.9998	0.9996	0.9994	0.9993	0.9991
42.20			0.9999	0.9999	0.9998	0.9998	0.9996	0.9994	0.9993	0.9991

PROBABILITY INTEGRAL OF t, THE NON-CENTRAL t-STATISTIC. THIS TABLE GIVES Pr $[t/\sqrt{f} \leq x]$.

f is the number of degrees of freedom; the non-centrality parameter is $\sqrt{f+1}\,K_p$.
K_p is the standardized normal deviate exceeded with probability p.

DEGREES OF FREEDOM **3**

x \ p	.2500	.1500	.1000	.0650	.0400	.0250	.0100	.0040	.0025	.0010
42.25			0.9999	0.9999	0.9998	0.9998	0.9996	0.9994	0.9993	0.9991
42.30			0.9999	0.9999	0.9998	0.9998	0.9996	0.9994	0.9993	0.9991
42.35			0.9999	0.9999	0.9998	0.9998	0.9996	0.9994	0.9993	0.9991
42.40			0.9999	0.9999	0.9998	0.9998	0.9996	0.9994	0.9993	0.9991
42.45			0.9999	0.9999	0.9998	0.9998	0.9996	0.9994	0.9993	0.9991
42.50			0.9999	0.9999	0.9998	0.9998	0.9996	0.9994	0.9993	0.9991
42.55			0.9999	0.9999	0.9998	0.9998	0.9996	0.9994	0.9993	0.9991
42.60			0.9999	0.9999	0.9998	0.9998	0.9996	0.9994	0.9993	0.9991
42.65			0.9999	0.9999	0.9998	0.9998	0.9996	0.9994	0.9993	0.9991
42.70			0.9999	0.9999	0.9998	0.9998	0.9996	0.9994	0.9994	0.9991
42.75			0.9999	0.9999	0.9998	0.9998	0.9996	0.9995	0.9994	0.9992
42.80			0.9999	0.9999	0.9998	0.9998	0.9996	0.9995	0.9994	0.9992
42.85			0.9999	0.9999	0.9998	0.9998	0.9996	0.9995	0.9994	0.9992
42.90			0.9999	0.9999	0.9998	0.9998	0.9996	0.9995	0.9994	0.9992
42.95			0.9999	0.9999	0.9998	0.9998	0.9996	0.9995	0.9994	0.9992
43.00			0.9999	0.9999	0.9998	0.9998	0.9996	0.9995	0.9994	0.9992
43.05			0.9999	0.9999	0.9998	0.9998	0.9996	0.9995	0.9994	0.9992
43.10			0.9999	0.9999	0.9998	0.9998	0.9996	0.9995	0.9994	0.9992
43.15			0.9999	0.9999	0.9998	0.9998	0.9996	0.9995	0.9994	0.9992
43.20			0.9999	0.9999	0.9998	0.9998	0.9996	0.9995	0.9994	0.9992
43.25			0.9999	0.9999	0.9998	0.9998	0.9996	0.9995	0.9994	0.9992
43.30			0.9999	0.9999	0.9998	0.9998	0.9996	0.9995	0.9994	0.9992
43.35			0.9999	0.9999	0.9998	0.9998	0.9996	0.9995	0.9994	0.9992
43.40			0.9999	0.9999	0.9998	0.9998	0.9996	0.9995	0.9994	0.9992
43.45			0.9999	0.9999	0.9998	0.9998	0.9996	0.9995	0.9994	0.9992
43.50			0.9999	0.9999	0.9998	0.9998	0.9996	0.9995	0.9994	0.9992
43.55			0.9999	0.9999	0.9998	0.9998	0.9996	0.9995	0.9994	0.9992
43.60			0.9999	0.9999	0.9998	0.9998	0.9996	0.9995	0.9994	0.9992
43.65			0.9999	0.9999	0.9998	0.9998	0.9996	0.9995	0.9994	0.9992
43.70			0.9999	0.9999	0.9998	0.9998	0.9996	0.9995	0.9994	0.9992
43.75				0.9999	0.9998	0.9998	0.9996	0.9995	0.9994	0.9992
43.80				0.9999	0.9998	0.9998	0.9996	0.9995	0.9994	0.9992
43.85				0.9999	0.9998	0.9998	0.9996	0.9995	0.9994	0.9992
43.90				0.9999	0.9998	0.9998	0.9997	0.9995	0.9994	0.9992
43.95				0.9999	0.9998	0.9998	0.9997	0.9995	0.9994	0.9992
44.00				0.9999	0.9998	0.9998	0.9997	0.9995	0.9994	0.9992
44.05				0.9999	0.9998	0.9998	0.9997	0.9995	0.9994	0.9992
44.10				0.9999	0.9998	0.9998	0.9997	0.9995	0.9994	0.9992
44.15				0.9999	0.9998	0.9998	0.9997	0.9995	0.9994	0.9992
44.20				0.9999	0.9998	0.9998	0.9997	0.9995	0.9994	0.9992
44.25				0.9999	0.9998	0.9998	0.9997	0.9995	0.9994	0.9992
44.30				0.9999	0.9998	0.9998	0.9997	0.9995	0.9994	0.9992
44.35				0.9999	0.9998	0.9998	0.9997	0.9995	0.9994	0.9992
44.40				0.9999	0.9998	0.9998	0.9997	0.9995	0.9994	0.9992
44.45				0.9999	0.9998	0.9998	0.9997	0.9995	0.9994	0.9992
44.50				0.9999	0.9998	0.9998	0.9997	0.9995	0.9994	0.9992
44.55				0.9999	0.9998	0.9998	0.9997	0.9995	0.9994	0.9993
44.60				0.9999	0.9998	0.9998	0.9997	0.9995	0.9994	0.9993
44.65				0.9999	0.9998	0.9998	0.9997	0.9995	0.9994	0.9993
44.70				0.9999	0.9998	0.9998	0.9997	0.9995	0.9994	0.9993
44.75				0.9999	0.9999	0.9998	0.9997	0.9995	0.9994	0.9993
44.80				0.9999	0.9999	0.9998	0.9997	0.9995	0.9994	0.9993
44.85				0.9999	0.9999	0.9998	0.9997	0.9995	0.9994	0.9993
44.90				0.9999	0.9999	0.9998	0.9997	0.9995	0.9994	0.9993
44.95				0.9999	0.9999	0.9998	0.9997	0.9995	0.9994	0.9993
45.00				0.9999	0.9999	0.9998	0.9997	0.9995	0.9994	0.9993
45.05				0.9999	0.9999	0.9998	0.9997	0.9995	0.9994	0.9993
45.10				0.9999	0.9999	0.9998	0.9997	0.9995	0.9995	0.9993
45.15				0.9999	0.9999	0.9998	0.9997	0.9995	0.9995	0.9993
45.20				0.9999	0.9999	0.9998	0.9997	0.9995	0.9995	0.9993

PROBABILITY INTEGRAL OF t, THE NON-CENTRAL t-STATISTIC. THIS TABLE GIVES Pr $[t/\sqrt{f} \leq x]$.

f is the number of degrees of freedom; the non-centrality parameter is $\sqrt{f+1}\, K_p$.

K_p is the standardized normal deviate exceeded with probability p.

DEGREES OF FREEDOM **3**

x \ p	.2500	.1500	.1000	.0650	.0400	.0250	.0100	.0040	.0025	.0010
45.25				0.9999	0.9999	0.9998	0.9997	0.9995	0.9995	0.9993
45.30				0.9999	0.9999	0.9998	0.9997	0.9995	0.9995	0.9993
45.35				0.9999	0.9999	0.9998	0.9997	0.9995	0.9995	0.9993
45.40				0.9999	0.9999	0.9998	0.9997	0.9995	0.9995	0.9993
45.45				0.9999	0.9999	0.9998	0.9997	0.9995	0.9995	0.9993
45.50				0.9999	0.9999	0.9998	0.9997	0.9995	0.9995	0.9993
45.55				0.9999	0.9999	0.9998	0.9997	0.9995	0.9995	0.9993
45.60				0.9999	0.9999	0.9998	0.9997	0.9995	0.9995	0.9993
45.65				0.9999	0.9999	0.9998	0.9997	0.9995	0.9995	0.9993
45.70				0.9999	0.9999	0.9998	0.9997	0.9996	0.9995	0.9993
45.75				0.9999	0.9999	0.9998	0.9997	0.9996	0.9995	0.9993
45.80				0.9999	0.9999	0.9998	0.9997	0.9996	0.9995	0.9993
45.85				0.9999	0.9999	0.9998	0.9997	0.9996	0.9995	0.9993
45.90				0.9999	0.9999	0.9998	0.9997	0.9996	0.9995	0.9993
45.95				0.9999	0.9999	0.9998	0.9997	0.9996	0.9995	0.9993
46.00				0.9999	0.9999	0.9998	0.9997	0.9996	0.9995	0.9993
46.05				0.9999	0.9999	0.9998	0.9997	0.9996	0.9995	0.9993
46.10				0.9999	0.9999	0.9998	0.9997	0.9996	0.9995	0.9993
46.15				0.9999	0.9999	0.9998	0.9997	0.9996	0.9995	0.9993
46.20				0.9999	0.9999	0.9998	0.9997	0.9996	0.9995	0.9993
46.25				0.9999	0.9999	0.9998	0.9997	0.9996	0.9995	0.9993
46.30				0.9999	0.9999	0.9998	0.9997	0.9996	0.9995	0.9993
46.35				0.9999	0.9999	0.9998	0.9997	0.9996	0.9995	0.9993
46.40				0.9999	0.9999	0.9998	0.9997	0.9996	0.9995	0.9993
46.45				0.9999	0.9999	0.9998	0.9997	0.9996	0.9995	0.9993
46.50				0.9999	0.9999	0.9998	0.9997	0.9996	0.9995	0.9993
46.55				0.9999	0.9999	0.9998	0.9997	0.9996	0.9995	0.9993
46.60				0.9999	0.9999	0.9998	0.9997	0.9996	0.9995	0.9993
46.65				0.9999	0.9999	0.9998	0.9997	0.9996	0.9995	0.9993
46.70				0.9999	0.9999	0.9998	0.9997	0.9996	0.9995	0.9994
46.75					0.9999	0.9998	0.9997	0.9996	0.9995	0.9994
46.80					0.9999	0.9998	0.9997	0.9996	0.9995	0.9994
46.85					0.9999	0.9998	0.9997	0.9996	0.9995	0.9994
46.90					0.9999	0.9998	0.9997	0.9996	0.9995	0.9994
46.95					0.9999	0.9998	0.9997	0.9996	0.9995	0.9994
47.00					0.9999	0.9998	0.9997	0.9996	0.9995	0.9994
47.05					0.9999	0.9998	0.9997	0.9996	0.9995	0.9994
47.10					0.9999	0.9998	0.9997	0.9996	0.9995	0.9994
47.15					0.9999	0.9998	0.9997	0.9996	0.9995	0.9994
47.20					0.9999	0.9998	0.9997	0.9996	0.9995	0.9994
47.25					0.9999	0.9998	0.9997	0.9996	0.9995	0.9994
47.30					0.9999	0.9998	0.9997	0.9996	0.9995	0.9994
47.35					0.9999	0.9998	0.9997	0.9996	0.9995	0.9994
47.40					0.9999	0.9998	0.9997	0.9996	0.9995	0.9994
47.45					0.9999	0.9998	0.9997	0.9996	0.9995	0.9994
47.50					0.9999	0.9998	0.9997	0.9996	0.9995	0.9994
47.55					0.9999	0.9998	0.9997	0.9996	0.9995	0.9994
47.60					0.9999	0.9998	0.9997	0.9996	0.9995	0.9994
47.65					0.9999	0.9998	0.9997	0.9996	0.9995	0.9994
47.70					0.9999	0.9998	0.9997	0.9996	0.9995	0.9994
47.75					0.9999	0.9998	0.9997	0.9996	0.9995	0.9994
47.80					0.9999	0.9998	0.9997	0.9996	0.9995	0.9994
47.85					0.9999	0.9998	0.9997	0.9996	0.9995	0.9994
47.90					0.9999	0.9998	0.9997	0.9996	0.9995	0.9994
47.95					0.9999	0.9998	0.9997	0.9996	0.9995	0.9994
48.00					0.9999	0.9998	0.9997	0.9996	0.9995	0.9994
48.05					0.9999	0.9998	0.9997	0.9996	0.9995	0.9994
48.10					0.9999	0.9998	0.9997	0.9996	0.9995	0.9994
48.15					0.9999	0.9998	0.9997	0.9996	0.9996	0.9994
48.20					0.9999	0.9998	0.9997	0.9996	0.9996	0.9994

PROBABILITY INTEGRAL OF t, THE NON-CENTRAL t-STATISTIC. THIS TABLE GIVES Pr $[t/\sqrt{f} \leq x]$.

f is the number of degrees of freedom; the non-centrality parameter is $\sqrt{f+1}\,K_p$.

K_p is the standardized normal deviate exceeded with probability p.

DEGREES OF FREEDOM **3**

x \ p	.2500	.1500	.1000	.0650	.0400	.0250	.0100	.0040	.0025	.0010
48.25					0.9999	0.9998	0.9997	0.9996	0.9996	0.9994
48.30					0.9999	0.9998	0.9997	0.9996	0.9996	0.9994
48.35					0.9999	0.9998	0.9997	0.9996	0.9996	0.9994
48.40					0.9999	0.9998	0.9997	0.9996	0.9996	0.9994
48.45					0.9999	0.9998	0.9997	0.9996	0.9996	0.9994
48.50					0.9999	0.9998	0.9997	0.9996	0.9996	0.9994
48.55					0.9999	0.9998	0.9997	0.9996	0.9996	0.9994
48.60					0.9999	0.9998	0.9997	0.9996	0.9996	0.9994
48.65					0.9999	0.9998	0.9997	0.9996	0.9996	0.9994
48.70					0.9999	0.9998	0.9997	0.9996	0.9996	0.9994
48.75					0.9999	0.9998	0.9997	0.9996	0.9996	0.9994
48.80					0.9999	0.9998	0.9997	0.9996	0.9996	0.9994
48.85					0.9999	0.9998	0.9997	0.9996	0.9996	0.9994
48.90					0.9999	0.9998	0.9997	0.9996	0.9996	0.9994
48.95					0.9999	0.9998	0.9998	0.9996	0.9996	0.9994
49.00					0.9999	0.9998	0.9998	0.9996	0.9996	0.9994
49.05					0.9999	0.9998	0.9998	0.9996	0.9996	0.9994
49.10					0.9999	0.9998	0.9998	0.9996	0.9996	0.9994
49.15					0.9999	0.9998	0.9998	0.9996	0.9996	0.9994
49.20					0.9999	0.9998	0.9998	0.9996	0.9996	0.9994
49.25					0.9999	0.9998	0.9998	0.9996	0.9996	0.9994
49.30					0.9999	0.9998	0.9998	0.9996	0.9996	0.9994
49.35					0.9999	0.9999	0.9998	0.9996	0.9996	0.9995
49.40					0.9999	0.9999	0.9998	0.9996	0.9996	0.9995
49.45					0.9999	0.9999	0.9998	0.9996	0.9996	0.9995
49.50					0.9999	0.9999	0.9998	0.9996	0.9996	0.9995
49.55					0.9999	0.9999	0.9998	0.9996	0.9996	0.9995
49.60					0.9999	0.9999	0.9998	0.9997	0.9996	0.9995
49.65					0.9999	0.9999	0.9998	0.9997	0.9996	0.9995
49.70					0.9999	0.9999	0.9998	0.9997	0.9996	0.9995
49.75						0.9999	0.9998	0.9997	0.9996	0.9995
49.80						0.9999	0.9998	0.9997	0.9996	0.9995
49.85						0.9999	0.9998	0.9997	0.9996	0.9995
49.90						0.9999	0.9998	0.9997	0.9996	0.9995
49.95						0.9999	0.9998	0.9997	0.9996	0.9995
50.00						0.9999	0.9998	0.9997	0.9996	0.9995
50.05						0.9999	0.9998	0.9997	0.9996	0.9995
50.10						0.9999	0.9998	0.9997	0.9996	0.9995
50.15						0.9999	0.9998	0.9997	0.9996	0.9995
50.20						0.9999	0.9998	0.9997	0.9996	0.9995
50.25						0.9999	0.9998	0.9997	0.9996	0.9995
50.30						0.9999	0.9998	0.9997	0.9996	0.9995
50.35						0.9999	0.9998	0.9997	0.9996	0.9995
50.40						0.9999	0.9998	0.9997	0.9996	0.9995
50.45						0.9999	0.9998	0.9997	0.9996	0.9995
50.50						0.9999	0.9998	0.9997	0.9996	0.9995
50.55						0.9999	0.9998	0.9997	0.9996	0.9995
50.60						0.9999	0.9998	0.9997	0.9996	0.9995
50.65						0.9999	0.9998	0.9997	0.9996	0.9995
50.70						0.9999	0.9998	0.9997	0.9996	0.9995
50.75						0.9999	0.9998	0.9997	0.9996	0.9995
50.80						0.9999	0.9998	0.9997	0.9996	0.9995
50.85						0.9999	0.9998	0.9997	0.9996	0.9995
50.90						0.9999	0.9998	0.9997	0.9996	0.9995
50.95						0.9999	0.9998	0.9997	0.9996	0.9995
51.00						0.9999	0.9998	0.9997	0.9996	0.9995
51.05						0.9999	0.9998	0.9997	0.9996	0.9995
51.10						0.9999	0.9998	0.9997	0.9996	0.9995
51.15						0.9999	0.9998	0.9997	0.9996	0.9995
51.20						0.9999	0.9998	0.9997	0.9996	0.9995

PROBABILITY INTEGRAL OF t, THE NON-CENTRAL t-STATISTIC. THIS TABLE GIVES Pr $[t/\sqrt{f} \leq x]$.

f is the number of degrees of freedom; the non-centrality parameter is $\sqrt{f+1}\,K_p$.
K_p is the standardized normal deviate exceeded with probability p.

DEGREES OF FREEDOM **3**

x \ p	.2500	.1500	.1000	.0650	.0400	.0250	.0100	.0040	.0025	.0010
51.25						0.9999	0.9998	0.9997	0.9996	0.9995
51.30						0.9999	0.9998	0.9997	0.9996	0.9995
51.35						0.9999	0.9998	0.9997	0.9996	0.9995
51.40						0.9999	0.9998	0.9997	0.9996	0.9995
51.45						0.9999	0.9998	0.9997	0.9996	0.9995
51.50						0.9999	0.9998	0.9997	0.9996	0.9995
51.55						0.9999	0.9998	0.9997	0.9996	0.9995
51.60						0.9999	0.9998	0.9997	0.9996	0.9995
51.65						0.9999	0.9998	0.9997	0.9996	0.9995
51.70						0.9999	0.9998	0.9997	0.9996	0.9995
51.75						0.9999	0.9998	0.9997	0.9996	0.9995
51.80						0.9999	0.9998	0.9997	0.9996	0.9995
51.85						0.9999	0.9998	0.9997	0.9996	0.9995
51.90						0.9999	0.9998	0.9997	0.9996	0.9995
51.95						0.9999	0.9998	0.9997	0.9996	0.9995
52.00						0.9999	0.9998	0.9997	0.9996	0.9995
52.05						0.9999	0.9998	0.9997	0.9996	0.9995
52.10						0.9999	0.9998	0.9997	0.9996	0.9995
52.15						0.9999	0.9998	0.9997	0.9996	0.9995
52.20						0.9999	0.9998	0.9997	0.9997	0.9995
52.25						0.9999	0.9998	0.9997	0.9997	0.9995
52.30						0.9999	0.9998	0.9997	0.9997	0.9995
52.35						0.9999	0.9998	0.9997	0.9997	0.9995
52.40						0.9999	0.9998	0.9997	0.9997	0.9995
52.45						0.9999	0.9998	0.9997	0.9997	0.9995
52.50						0.9999	0.9998	0.9997	0.9997	0.9995
52.55						0.9999	0.9998	0.9997	0.9997	0.9995
52.60						0.9999	0.9998	0.9997	0.9997	0.9995
52.65						0.9999	0.9998	0.9997	0.9997	0.9995
52.70						0.9999	0.9998	0.9997	0.9997	0.9996
52.75						0.9999	0.9998	0.9997	0.9997	0.9996
52.80						0.9999	0.9998	0.9997	0.9997	0.9996
52.85						0.9999	0.9998	0.9997	0.9997	0.9996
52.90						0.9999	0.9998	0.9997	0.9997	0.9996
52.95						0.9999	0.9998	0.9997	0.9997	0.9996
53.00						0.9999	0.9998	0.9997	0.9997	0.9996
53.05						0.9999	0.9998	0.9997	0.9997	0.9996
53.10						0.9999	0.9998	0.9997	0.9997	0.9996
53.15						0.9999	0.9998	0.9997	0.9997	0.9996
53.20						0.9999	0.9998	0.9997	0.9997	0.9996
53.25						0.9999	0.9998	0.9997	0.9997	0.9996
53.30						0.9999	0.9998	0.9997	0.9997	0.9996
53.35						0.9999	0.9998	0.9997	0.9997	0.9996
53.40						0.9999	0.9998	0.9997	0.9997	0.9996
53.45						0.9999	0.9998	0.9997	0.9997	0.9996
53.50						0.9999	0.9998	0.9997	0.9997	0.9996
53.55						0.9999	0.9998	0.9997	0.9997	0.9996
53.60						0.9999	0.9998	0.9997	0.9997	0.9996
53.65						0.9999	0.9998	0.9997	0.9997	0.9996
53.70						0.9999	0.9998	0.9997	0.9997	0.9996
53.75						0.9999	0.9998	0.9997	0.9997	0.9996
53.80						0.9999	0.9998	0.9997	0.9997	0.9996
53.85						0.9999	0.9998	0.9997	0.9997	0.9996
53.90						0.9999	0.9998	0.9997	0.9997	0.9996
53.95						0.9999	0.9998	0.9997	0.9997	0.9996
54.00						0.9999	0.9998	0.9997	0.9997	0.9996
54.05						0.9999	0.9998	0.9997	0.9997	0.9996
54.10						0.9999	0.9998	0.9997	0.9997	0.9996
54.15						0.9999	0.9998	0.9997	0.9997	0.9996
54.20						0.9999	0.9998	0.9997	0.9997	0.9996

PROBABILITY INTEGRAL OF t, THE NON-CENTRAL t-STATISTIC. THIS TABLE GIVES Pr $[t/\sqrt{f} \leq x]$.

f is the number of degrees of freedom; the non-centrality parameter is $\sqrt{f+1}\,K_p$.
K_p is the standardized normal deviate exceeded with probability p.

DEGREES OF FREEDOM **3**

x \ p	.2500	.1500	.1000	.0650	.0400	.0250	.0100	.0040	.0025	.0010
54.25						0.9999	0.9998	0.9997	0.9997	0.9996
54.30						0.9999	0.9998	0.9997	0.9997	0.9996
54.35						0.9999	0.9998	0.9997	0.9997	0.9996
54.40						0.9999	0.9998	0.9997	0.9997	0.9996
54.45						0.9999	0.9998	0.9997	0.9997	0.9996
54.50						0.9999	0.9998	0.9997	0.9997	0.9996
54.55						0.9999	0.9998	0.9997	0.9997	0.9996
54.60						0.9999	0.9998	0.9997	0.9997	0.9996
54.65						0.9999	0.9998	0.9997	0.9997	0.9996
54.70						0.9999	0.9998	0.9997	0.9997	0.9996
54.75						0.9999	0.9998	0.9997	0.9997	0.9996
54.80						0.9999	0.9998	0.9997	0.9997	0.9996
54.85						0.9999	0.9998	0.9997	0.9997	0.9996
54.90						0.9999	0.9998	0.9997	0.9997	0.9996
54.95						0.9999	0.9998	0.9997	0.9997	0.9996
55.00						0.9999	0.9998	0.9997	0.9997	0.9996
55.05						0.9999	0.9998	0.9997	0.9997	0.9996
55.10						0.9999	0.9998	0.9997	0.9997	0.9996
55.15						0.9999	0.9998	0.9997	0.9997	0.9996
55.20						0.9999	0.9998	0.9997	0.9997	0.9996
55.25						0.9999	0.9998	0.9997	0.9997	0.9996
55.30						0.9999	0.9998	0.9998	0.9997	0.9996
55.35						0.9999	0.9998	0.9998	0.9997	0.9996
55.40						0.9999	0.9998	0.9998	0.9997	0.9996
55.45						0.9999	0.9998	0.9998	0.9997	0.9996
55.50						0.9999	0.9998	0.9998	0.9997	0.9996
55.55						0.9999	0.9998	0.9998	0.9997	0.9996
55.60						0.9999	0.9998	0.9998	0.9997	0.9996
55.65						0.9999	0.9998	0.9998	0.9997	0.9996
55.70						0.9999	0.9998	0.9998	0.9997	0.9996
55.75						0.9999	0.9998	0.9998	0.9997	0.9996
55.80						0.9999	0.9998	0.9998	0.9997	0.9996
55.85						0.9999	0.9998	0.9998	0.9997	0.9996
55.90						0.9999	0.9998	0.9998	0.9997	0.9996
55.95						0.9999	0.9998	0.9998	0.9997	0.9996
56.00						0.9999	0.9998	0.9998	0.9997	0.9996
56.05						0.9999	0.9998	0.9998	0.9997	0.9996
56.10						0.9999	0.9998	0.9998	0.9997	0.9996
56.15						0.9999	0.9998	0.9998	0.9997	0.9996
56.20						0.9999	0.9998	0.9998	0.9997	0.9996
56.25						0.9999	0.9998	0.9998	0.9997	0.9996
56.30						0.9999	0.9998	0.9998	0.9997	0.9996
56.35						0.9999	0.9998	0.9998	0.9997	0.9996
56.40						0.9999	0.9998	0.9998	0.9997	0.9996
56.45						0.9999	0.9998	0.9998	0.9997	0.9996
56.50						0.9999	0.9998	0.9998	0.9997	0.9996
56.55						0.9999	0.9998	0.9998	0.9997	0.9996
56.60						0.9999	0.9998	0.9998	0.9997	0.9996
56.65						0.9999	0.9998	0.9998	0.9997	0.9996
56.70						0.9999	0.9998	0.9998	0.9997	0.9996
56.75						0.9999	0.9998	0.9998	0.9997	0.9996
56.80						0.9999	0.9998	0.9998	0.9997	0.9996
56.85						0.9999	0.9998	0.9998	0.9997	0.9996
56.90						0.9999	0.9998	0.9998	0.9997	0.9996
56.95						0.9999	0.9998	0.9998	0.9997	0.9996
57.00						0.9999	0.9998	0.9998	0.9997	0.9996
57.05						0.9999	0.9998	0.9998	0.9997	0.9996
57.10						0.9999	0.9998	0.9998	0.9997	0.9996
57.15						0.9999	0.9998	0.9998	0.9997	0.9996
57.20						0.9999	0.9998	0.9998	0.9997	0.9997

PROBABILITY INTEGRAL OF t, THE NON-CENTRAL t-STATISTIC. THIS TABLE GIVES Pr $[t/\sqrt{f} \leq x]$.

f is the number of degrees of freedom; the non-centrality parameter is $\sqrt{f+1}\,K_p$.

K_p is the standardized normal deviate exceeded with probability p.

DEGREES OF FREEDOM **3**

x \ p	.2500	.1500	.1000	.0650	.0400	.0250	.0100	.0040	.0025	.0010
57.25							0.9998	0.9998	0.9997	0.9997
57.30							0.9998	0.9998	0.9997	0.9997
57.35							0.9998	0.9998	0.9997	0.9997
57.40							0.9998	0.9998	0.9997	0.9997
57.45							0.9998	0.9998	0.9997	0.9997
57.50							0.9998	0.9998	0.9997	0.9997
57.55							0.9998	0.9998	0.9997	0.9997
57.60							0.9998	0.9998	0.9997	0.9997
57.65							0.9998	0.9998	0.9997	0.9997
57.70							0.9998	0.9998	0.9997	0.9997
57.75							0.9998	0.9998	0.9997	0.9997
57.80							0.9998	0.9998	0.9997	0.9997
57.85							0.9998	0.9998	0.9997	0.9997
57.90							0.9998	0.9998	0.9997	0.9997
57.95							0.9998	0.9998	0.9997	0.9997
58.00							0.9998	0.9998	0.9997	0.9997
58.05							0.9998	0.9998	0.9997	0.9997
58.10							0.9998	0.9998	0.9997	0.9997
58.15							0.9998	0.9998	0.9997	0.9997
58.20							0.9998	0.9998	0.9998	0.9997
58.25							0.9998	0.9998	0.9997	0.9997
58.30							0.9998	0.9998	0.9998	0.9997
58.35							0.9998	0.9998	0.9998	0.9997
58.40							0.9998	0.9998	0.9998	0.9997
58.45							0.9998	0.9998	0.9998	0.9997
58.50							0.9998	0.9998	0.9998	0.9997
58.55							0.9998	0.9998	0.9998	0.9997
58.60							0.9998	0.9998	0.9998	0.9997
58.65							0.9998	0.9998	0.9998	0.9997
58.70							0.9998	0.9998	0.9998	0.9997
58.75							0.9998	0.9998	0.9998	0.9997
58.80							0.9998	0.9998	0.9998	0.9997
58.85							0.9998	0.9998	0.9998	0.9997
58.90							0.9998	0.9998	0.9998	0.9997
58.95							0.9998	0.9998	0.9998	0.9997
59.00							0.9998	0.9998	0.9998	0.9997
59.05							0.9998	0.9998	0.9998	0.9997
59.10							0.9998	0.9998	0.9998	0.9997
59.15							0.9998	0.9998	0.9998	0.9997
59.20							0.9998	0.9998	0.9998	0.9997
59.25							0.9998	0.9998	0.9998	0.9997
59.30							0.9998	0.9998	0.9998	0.9997
59.35							0.9998	0.9998	0.9998	0.9997
59.40							0.9998	0.9998	0.9998	0.9997
59.45							0.9998	0.9998	0.9998	0.9997
59.50							0.9998	0.9998	0.9998	0.9997
59.55							0.9998	0.9998	0.9998	0.9997
59.60							0.9998	0.9998	0.9998	0.9997
59.65							0.9998	0.9998	0.9998	0.9997
59.70							0.9998	0.9998	0.9998	0.9997
59.75							0.9998	0.9998	0.9998	0.9997
59.80							0.9998	0.9998	0.9998	0.9997
59.85							0.9998	0.9998	0.9998	0.9997
59.90							0.9998	0.9998	0.9998	0.9997
59.95							0.9998	0.9998	0.9998	0.9997
60.00							0.9998	0.9998	0.9998	0.9997
60.05							0.9998	0.9998	0.9998	0.9997
60.10							0.9998	0.9998	0.9998	0.9997
60.15							0.9998	0.9998	0.9998	0.9997
60.20							0.9998	0.9998	0.9998	0.9997

PROBABILITY INTEGRAL OF t, THE NON-CENTRAL t-STATISTIC. THIS TABLE GIVES Pr $[t/\sqrt{f} \leq x]$.

f is the number of degrees of freedom; the non-centrality parameter is $\sqrt{f+1}\,K_p$.

K_p is the standardized normal deviate exceeded with probability p.

DEGREES OF FREEDOM **3**

x \ p	.2500	.1500	.1000	.0650	.0400	.0250	.0100	.0040	.0025	.0010
60.25							0.9998	0.9998	0.9998	0.9997
60.30							0.9998	0.9998	0.9998	0.9997
60.35							0.9998	0.9998	0.9998	0.9997
60.40							0.9998	0.9998	0.9998	0.9997
60.45							0.9998	0.9998	0.9998	0.9997
60.50							0.9998	0.9998	0.9998	0.9997
60.55							0.9998	0.9998	0.9998	0.9997
60.60							0.9998	0.9998	0.9998	0.9997
60.65							0.9998	0.9998	0.9998	0.9997
60.70							0.9998	0.9998	0.9998	0.9997
60.75							0.9998	0.9998	0.9998	0.9997
60.80							0.9998	0.9998	0.9998	0.9997
60.85							0.9998	0.9998	0.9998	0.9997
60.90							0.9998	0.9998	0.9998	0.9997
60.95							0.9998	0.9998	0.9998	0.9997
61.00							0.9998	0.9998	0.9998	0.9997
61.05							0.9998	0.9998	0.9998	0.9997
61.10							0.9999	0.9998	0.9998	0.9997
61.15							0.9998	0.9998	0.9998	0.9997
61.20							0.9999	0.9998	0.9998	0.9997
61.25							0.9999	0.9998	0.9998	0.9997
61.30							0.9999	0.9998	0.9998	0.9997
61.35							0.9999	0.9998	0.9998	0.9997
61.40							0.9999	0.9998	0.9998	0.9997
61.45							0.9999	0.9998	0.9998	0.9997
61.50							0.9999	0.9998	0.9998	0.9997
61.55							0.9999	0.9998	0.9998	0.9997
61.60							0.9999	0.9998	0.9998	0.9997
61.65							0.9999	0.9998	0.9998	0.9997
61.70							0.9999	0.9998	0.9998	0.9997
61.75								0.9998	0.9998	0.9997
61.80								0.9998	0.9998	0.9997
61.85								0.9998	0.9998	0.9997
61.90								0.9998	0.9998	0.9997
61.95								0.9998	0.9998	0.9997
62.00								0.9998	0.9998	0.9997
62.05								0.9998	0.9998	0.9997
62.10								0.9998	0.9998	0.9997
62.15								0.9998	0.9998	0.9997
62.20								0.9998	0.9998	0.9997
62.25								0.9998	0.9998	0.9997
62.30								0.9998	0.9998	0.9997
62.35								0.9998	0.9998	0.9997
62.40								0.9998	0.9998	0.9997
62.45								0.9998	0.9998	0.9997
62.50								0.9998	0.9998	0.9997
62.55								0.9998	0.9998	0.9997
62.60								0.9998	0.9998	0.9997
62.65								0.9998	0.9998	0.9997
62.70								0.9998	0.9998	0.9997
62.75								0.9998	0.9998	0.9997
62.80								0.9998	0.9998	0.9997
62.85								0.9998	0.9998	0.9997
62.90								0.9998	0.9998	0.9997
62.95								0.9998	0.9998	0.9997
63.00								0.9998	0.9998	0.9997
63.05								0.9998	0.9998	0.9997
63.10								0.9998	0.9998	0.9997
63.15								0.9998	0.9998	0.9997
63.20								0.9998	0.9998	0.9997

PROBABILITY INTEGRAL OF t, THE NON-CENTRAL t-STATISTIC. THIS TABLE GIVES Pr $[t/\sqrt{f} \leq x]$.

f is the number of degrees of freedom; the non-centrality parameter is $\sqrt{f+1}\,K_p$.

K_p is the standardized normal deviate exceeded with probability p.

DEGREES OF FREEDOM **3**

x \ p	.2500	.1500	.1000	.0650	.0400	.0250	.0100	.0040	.0025	.0010
63.25								0.9998	0.9998	0.9997
63.30								0.9998	0.9998	0.9997
63.35								0.9998	0.9998	0.9997
63.40								0.9998	0.9998	0.9997
63.45								0.9998	0.9998	0.9997
63.50								0.9998	0.9998	0.9997
63.55								0.9998	0.9998	0.9997
63.60								0.9998	0.9998	0.9997
63.65								0.9998	0.9998	0.9997
63.70								0.9998	0.9998	0.9997
63.75								0.9998	0.9998	0.9998
63.80								0.9998	0.9998	0.9998
63.85								0.9998	0.9998	0.9998
63.90								0.9998	0.9998	0.9998
63.95								0.9998	0.9998	0.9998
64.00								0.9998	0.9998	0.9998
64.05								0.9998	0.9998	0.9998
64.10								0.9998	0.9998	0.9998
64.15								0.9998	0.9998	0.9998
64.20								0.9998	0.9998	0.9998
64.25								0.9998	0.9998	0.9998
64.30								0.9998	0.9998	0.9998
64.35								0.9998	0.9998	0.9998
64.40								0.9998	0.9998	0.9998
64.45								0.9998	0.9998	0.9998
64.50								0.9998	0.9998	0.9998
64.55								0.9998	0.9998	0.9998
64.60								0.9998	0.9998	0.9998
64.65								0.9998	0.9998	0.9998
64.70								0.9998	0.9998	0.9998
64.75								0.9998	0.9998	0.9998
64.80								0.9998	0.9998	0.9998
64.85								0.9998	0.9998	0.9998
64.90								0.9998	0.9998	0.9998
64.95								0.9998	0.9998	0.9998
65.00								0.9998	0.9998	0.9998
65.05								0.9998	0.9998	0.9998
65.10								0.9998	0.9998	0.9998
65.15								0.9998	0.9998	0.9998
65.20								0.9998	0.9998	0.9998
65.25								0.9998	0.9998	0.9998
65.30								0.9998	0.9998	0.9998
65.35								0.9998	0.9998	0.9998
65.40								0.9998	0.9998	0.9998
65.45								0.9998	0.9998	0.9998

PROBABILITY INTEGRAL OF t, THE NON-CENTRAL t-STATISTIC. THIS TABLE GIVES Pr $[t/\sqrt{f} \leq x]$.

f is the number of degrees of freedom; the non-centrality parameter is $\sqrt{f+1}\,K_p$.
K_p is the standardized normal deviate exceeded with probability p.

DEGREES OF FREEDOM **4**

x \ p	.2500	.1500	.1000	.0650	.0400	.0250	.0100	.0040	.0025	.0010
- 2.90	0.0000	0.0000	0.0000	0.0000	0.0000	0.0000	0.0000	0.0000	0.0000	0.0000
- 2.85	0.0001	0.0000	0.0000	0.0000	0.0000	0.0000	0.0000	0.0000	0.0000	0.0000
- 2.80	0.0001	0.0000	0.0000	0.0000	0.0000	0.0000	0.0000	0.0000	0.0000	0.0000
- 2.75	0.0001	0.0000	0.0000	0.0000	0.0000	0.0000	0.0000	0.0000	0.0000	0.0000
- 2.70	0.0001	0.0000	0.0000	0.0000	0.0000	0.0000	0.0000	0.0000	0.0000	0.0000
- 2.65	0.0001	0.0000	0.0000	0.0000	0.0000	0.0000	0.0000	0.0000	0.0000	0.0000
- 2.60	0.0001	0.0000	0.0000	0.0000	0.0000	0.0000	0.0000	0.0000	0.0000	0.0000
- 2.55	0.0001	0.0000	0.0000	0.0000	0.0000	0.0000	0.0000	0.0000	0.0000	0.0000
- 2.50	0.0001	0.0000	0.0000	0.0000	0.0000	0.0000	0.0000	0.0000	0.0000	0.0000
- 2.45	0.0001	0.0000	0.0000	0.0000	0.0000	0.0000	0.0000	0.0000	0.0000	0.0000
- 2.40	0.0001	0.0000	0.0000	0.0000	0.0000	0.0000	0.0000	0.0000	0.0000	0.0000
- 2.35	0.0001	0.0000	0.0000	0.0000	0.0000	0.0000	0.0000	0.0000	0.0000	0.0000
- 2.30	0.0001	0.0000	0.0000	0.0000	0.0000	0.0000	0.0000	0.0000	0.0000	0.0000
- 2.25	0.0001	0.0000	0.0000	0.0000	0.0000	0.0000	0.0000	0.0000	0.0000	0.0000
- 2.20	0.0001	0.0000	0.0000	0.0000	0.0000	0.0000	0.0000	0.0000	0.0000	0.0000
- 2.15	0.0001	0.0000	0.0000	0.0000	0.0000	0.0000	0.0000	0.0000	0.0000	0.0000
- 2.10	0.0002	0.0000	0.0000	0.0000	0.0000	0.0000	0.0000	0.0000	0.0000	0.0000
- 2.05	0.0002	0.0000	0.0000	0.0000	0.0000	0.0000	0.0000	0.0000	0.0000	0.0000
- 2.00	0.0002	0.0000	0.0000	0.0000	0.0000	0.0000	0.0000	0.0000	0.0000	0.0000
- 1.95	0.0002	0.0000	0.0000	0.0000	0.0000	0.0000	0.0000	0.0000	0.0000	0.0000
- 1.90	0.0002	0.0000	0.0000	0.0000	0.0000	0.0000	0.0000	0.0000	0.0000	0.0000
- 1.85	0.0003	0.0000	0.0000	0.0000	0.0000	0.0000	0.0000	0.0000	0.0000	0.0000
- 1.80	0.0003	0.0000	0.0000	0.0000	0.0000	0.0000	0.0000	0.0000	0.0000	0.0000
- 1.75	0.0003	0.0000	0.0000	0.0000	0.0000	0.0000	0.0000	0.0000	0.0000	0.0000
- 1.70	0.0004	0.0000	0.0000	0.0000	0.0000	0.0000	0.0000	0.0000	0.0000	0.0000
- 1.65	0.0004	0.0000	0.0000	0.0000	0.0000	0.0000	0.0000	0.0000	0.0000	0.0000
- 1.60	0.0004	0.0000	0.0000	0.0000	0.0000	0.0000	0.0000	0.0000	0.0000	0.0000
- 1.55	0.0005	0.0000	0.0000	0.0000	0.0000	0.0000	0.0000	0.0000	0.0000	0.0000
- 1.50	0.0005	0.0000	0.0000	0.0000	0.0000	0.0000	0.0000	0.0000	0.0000	0.0000
- 1.45	0.0006	0.0000	0.0000	0.0000	0.0000	0.0000	0.0000	0.0000	0.0000	0.0000
- 1.40	0.0007	0.0001	0.0000	0.0000	0.0000	0.0000	0.0000	0.0000	0.0000	0.0000
- 1.35	0.0008	0.0001	0.0000	0.0000	0.0000	0.0000	0.0000	0.0000	0.0000	0.0000
- 1.30	0.0009	0.0001	0.0000	0.0000	0.0000	0.0000	0.0000	0.0000	0.0000	0.0000
- 1.25	0.0010	0.0001	0.0000	0.0000	0.0000	0.0000	0.0000	0.0000	0.0000	0.0000
- 1.20	0.0011	0.0001	0.0000	0.0000	0.0000	0.0000	0.0000	0.0000	0.0000	0.0000
- 1.15	0.0013	0.0001	0.0000	0.0000	0.0000	0.0000	0.0000	0.0000	0.0000	0.0000
- 1.10	0.0015	0.0001	0.0000	0.0000	0.0000	0.0000	0.0000	0.0000	0.0000	0.0000
- 1.05	0.0017	0.0001	0.0000	0.0000	0.0000	0.0000	0.0000	0.0000	0.0000	0.0000
- 1.00	0.0019	0.0002	0.0000	0.0000	0.0000	0.0000	0.0000	0.0000	0.0000	0.0000
- 0.95	0.0023	0.0002	0.0000	0.0000	0.0000	0.0000	0.0000	0.0000	0.0000	0.0000
- 0.90	0.0026	0.0002	0.0000	0.0000	0.0000	0.0000	0.0000	0.0000	0.0000	0.0000
- 0.85	0.0031	0.0003	0.0000	0.0000	0.0000	0.0000	0.0000	0.0000	0.0000	0.0000
- 0.80	0.0036	0.0003	0.0000	0.0000	0.0000	0.0000	0.0000	0.0000	0.0000	0.0000
- 0.75	0.0042	0.0004	0.0001	0.0000	0.0000	0.0000	0.0000	0.0000	0.0000	0.0000
- 0.70	0.0050	0.0004	0.0001	0.0000	0.0000	0.0000	0.0000	0.0000	0.0000	0.0000
- 0.65	0.0059	0.0005	0.0001	0.0000	0.0000	0.0000	0.0000	0.0000	0.0000	0.0000
- 0.60	0.0070	0.0006	0.0001	0.0000	0.0000	0.0000	0.0000	0.0000	0.0000	0.0000
- 0.55	0.0083	0.0008	0.0001	0.0000	0.0000	0.0000	0.0000	0.0000	0.0000	0.0000
- 0.50	0.0100	0.0010	0.0001	0.0000	0.0000	0.0000	0.0000	0.0000	0.0000	0.0000
- 0.45	0.0120	0.0012	0.0002	0.0000	0.0000	0.0000	0.0000	0.0000	0.0000	0.0000
- 0.40	0.0144	0.0015	0.0002	0.0000	0.0000	0.0000	0.0000	0.0000	0.0000	0.0000
- 0.35	0.0174	0.0019	0.0003	0.0000	0.0000	0.0000	0.0000	0.0000	0.0000	0.0000
- 0.30	0.0211	0.0024	0.0004	0.0001	0.0000	0.0000	0.0000	0.0000	0.0000	0.0000
- 0.25	0.0255	0.0030	0.0005	0.0001	0.0000	0.0000	0.0000	0.0000	0.0000	0.0000
- 0.20	0.0309	0.0038	0.0007	0.0001	0.0000	0.0000	0.0000	0.0000	0.0000	0.0000
- 0.15	0.0374	0.0048	0.0009	0.0001	0.0000	0.0000	0.0000	0.0000	0.0000	0.0000
- 0.10	0.0453	0.0062	0.0012	0.0002	0.0000	0.0000	0.0000	0.0000	0.0000	0.0000
- 0.05	0.0547	0.0080	0.0016	0.0003	0.0000	0.0000	0.0000	0.0000	0.0000	0.0000

PROBABILITY INTEGRAL OF t, THE NON-CENTRAL t-STATISTIC. THIS TABLE GIVES Pr $[t/\sqrt{f} \leq x]$.

f is the number of degrees of freedom; the non-centrality parameter is $\sqrt{f+1}\,K_p$.

K_p is the standardized normal deviate exceeded with probability p.

x \ p	.2500	.1500	.1000	.0650	.0400	.0250	.0100	.0040	.0025	.0010
0.00	0.0657	0.0102	0.0021	0.0004	0.0001	0.0000	0.0000	0.0000	0.0000	0.0000
0.05	0.0788	0.0131	0.0028	0.0005	0.0001	0.0000	0.0000	0.0000	0.0000	0.0000
0.10	0.0939	0.0168	0.0038	0.0007	0.0001	0.0000	0.0000	0.0000	0.0000	0.0000
0.15	0.1113	0.0214	0.0051	0.0010	0.0002	0.0000	0.0000	0.0000	0.0000	0.0000
0.20	0.1310	0.0272	0.0068	0.0014	0.0002	0.0000	0.0000	0.0000	0.0000	0.0000
0.25	0.1531	0.0343	0.0091	0.0020	0.0004	0.0001	0.0000	0.0000	0.0000	0.0000
0.30	0.1775	0.0430	0.0121	0.0029	0.0005	0.0001	0.0000	0.0000	0.0000	0.0000
0.35	0.2040	0.0534	0.0160	0.0041	0.0008	0.0002	0.0000	0.0000	0.0000	0.0000
0.40	0.2327	0.0657	0.0209	0.0056	0.0012	0.0002	0.0000	0.0000	0.0000	0.0000
0.45	0.2630	0.0800	0.0271	0.0078	0.0018	0.0004	0.0000	0.0000	0.0000	0.0000
0.50	0.2948	0.0964	0.0346	0.0106	0.0026	0.0006	0.0000	0.0000	0.0000	0.0000
0.55	0.3277	0.1150	0.0437	0.0142	0.0037	0.0009	0.0001	0.0000	0.0000	0.0000
0.60	0.3612	0.1356	0.0545	0.0189	0.0053	0.0014	0.0001	0.0000	0.0000	0.0000
0.65	0.3952	0.1583	0.0672	0.0247	0.0074	0.0021	0.0002	0.0000	0.0000	0.0000
0.70	0.4291	0.1828	0.0818	0.0319	0.0101	0.0032	0.0003	0.0000	0.0000	0.0000
0.75	0.4626	0.2089	0.0983	0.0405	0.0137	0.0045	0.0005	0.0000	0.0000	0.0000
0.80	0.4955	0.2365	0.1166	0.0506	0.0183	0.0064	0.0007	0.0001	0.0000	0.0000
0.85	0.5275	0.2651	0.1368	0.0624	0.0239	0.0089	0.0012	0.0001	0.0000	0.0000
0.90	0.5584	0.2946	0.1586	0.0760	0.0308	0.0121	0.0018	0.0002	0.0001	0.0000
0.95	0.5881	0.3246	0.1820	0.0912	0.0390	0.0162	0.0026	0.0004	0.0001	0.0000
1.00	0.6163	0.3549	0.2066	0.1080	0.0485	0.0212	0.0038	0.0006	0.0002	0.0000
1.05	0.6431	0.3852	0.2324	0.1265	0.0596	0.0273	0.0054	0.0010	0.0004	0.0001
1.10	0.6684	0.4153	0.2590	0.1464	0.0722	0.0346	0.0074	0.0015	0.0006	0.0001
1.15	0.6922	0.4450	0.2862	0.1676	0.0862	0.0431	0.0101	0.0022	0.0010	0.0002
1.20	0.7145	0.4740	0.3139	0.1901	0.1017	0.0529	0.0134	0.0031	0.0014	0.0003
1.25	0.7353	0.5023	0.3417	0.2135	0.1185	0.0641	0.0176	0.0044	0.0021	0.0005
1.30	0.7547	0.5296	0.3695	0.2377	0.1366	0.0765	0.0225	0.0061	0.0030	0.0008
1.35	0.7727	0.5560	0.3971	0.2625	0.1559	0.0903	0.0284	0.0082	0.0043	0.0011
1.40	0.7894	0.5813	0.4243	0.2877	0.1762	0.1053	0.0353	0.0109	0.0058	0.0017
1.45	0.8049	0.6055	0.4510	0.3132	0.1975	0.1214	0.0433	0.0142	0.0079	0.0024
1.50	0.8192	0.6285	0.4772	0.3388	0.2194	0.1387	0.0523	0.0182	0.0104	0.0034
1.55	0.8324	0.6505	0.5025	0.3643	0.2419	0.1569	0.0623	0.0230	0.0135	0.0046
1.60	0.8446	0.6712	0.5271	0.3896	0.2649	0.1760	0.0734	0.0285	0.0172	0.0062
1.65	0.8559	0.6909	0.5509	0.4146	0.2882	0.1958	0.0856	0.0349	0.0215	0.0082
1.70	0.8662	0.7094	0.5738	0.4391	0.3116	0.2163	0.0988	0.0421	0.0267	0.0106
1.75	0.8758	0.7269	0.5957	0.4631	0.3351	0.2372	0.1129	0.0502	0.0325	0.0135
1.80	0.8846	0.7433	0.6167	0.4866	0.3585	0.2586	0.1279	0.0593	0.0392	0.0170
1.85	0.8928	0.7588	0.6367	0.5094	0.3817	0.2802	0.1437	0.0692	0.0467	0.0211
1.90	0.9003	0.7733	0.6558	0.5315	0.4047	0.3019	0.1602	0.0799	0.0550	0.0258
1.95	0.9072	0.7869	0.6740	0.5528	0.4272	0.3237	0.1774	0.0915	0.0641	0.0311
2.00	0.9135	0.7996	0.6913	0.5734	0.4494	0.3455	0.1951	0.1040	0.0740	0.0371
2.05	0.9194	0.8116	0.7077	0.5932	0.4710	0.3671	0.2133	0.1171	0.0847	0.0439
2.10	0.9248	0.8228	0.7232	0.6122	0.4922	0.3885	0.2318	0.1310	0.0961	0.0513
2.15	0.9299	0.8333	0.7380	0.6305	0.5127	0.4097	0.2507	0.1455	0.1083	0.0595
2.20	0.9345	0.8431	0.7519	0.6479	0.5327	0.4305	0.2698	0.1606	0.1211	0.0683
2.25	0.9388	0.8522	0.7650	0.6646	0.5520	0.4509	0.2890	0.1761	0.1346	0.0778
2.30	0.9427	0.8608	0.7775	0.6806	0.5707	0.4709	0.3083	0.1922	0.1486	0.0880
2.35	0.9464	0.8689	0.7892	0.6958	0.5887	0.4904	0.3275	0.2086	0.1632	0.0989
2.40	0.9498	0.8764	0.8003	0.7103	0.6061	0.5094	0.3467	0.2253	0.1782	0.1103
2.45	0.9529	0.8834	0.8108	0.7241	0.6229	0.5279	0.3658	0.2423	0.1935	0.1223
2.50	0.9558	0.8900	0.8206	0.7372	0.6389	0.5458	0.3847	0.2595	0.2093	0.1348
2.55	0.9585	0.8962	0.8300	0.7497	0.6544	0.5632	0.4033	0.2768	0.2253	0.1478
2.60	0.9610	0.9019	0.8387	0.7616	0.6692	0.5801	0.4217	0.2942	0.2415	0.1612
2.65	0.9634	0.9073	0.8470	0.7729	0.6834	0.5964	0.4398	0.3116	0.2579	0.1750
2.70	0.9655	0.9124	0.8548	0.7836	0.6970	0.6121	0.4576	0.3289	0.2744	0.1892
2.75	0.9675	0.9171	0.8622	0.7938	0.7100	0.6273	0.4750	0.3463	0.2910	0.2036
2.80	0.9694	0.9216	0.8691	0.8035	0.7225	0.6419	0.4920	0.3634	0.3076	0.2183
2.85	0.9712	0.9258	0.8757	0.8126	0.7344	0.6560	0.5086	0.3805	0.3242	0.2332
2.90	0.9728	0.9297	0.8819	0.8213	0.7458	0.6696	0.5249	0.3973	0.3408	0.2483
2.95	0.9743	0.9334	0.8877	0.8296	0.7566	0.6826	0.5407	0.4140	0.3572	0.2635

PROBABILITY INTEGRAL OF t, THE NON-CENTRAL t-STATISTIC. THIS TABLE GIVES Pr $[t/\sqrt{f} \leq x]$.

f is the number of degrees of freedom; the non-centrality parameter is $\sqrt{f+1}\,K_p$.
K_p is the standardized normal deviate exceeded with probability p.

DEGREES OF FREEDOM **4**

x \ p	.2500	.1500	.1000	.0650	.0400	.0250	.0100	.0040	.0025	.0010
3.00	0.9757	0.9368	0.8932	0.8374	0.7670	0.6951	0.5561	0.4304	0.3735	0.2788
3.05	0.9771	0.9401	0.8984	0.8449	0.7769	0.7071	0.5710	0.4465	0.3897	0.2941
3.10	0.9783	0.9431	0.9033	0.8519	0.7864	0.7187	0.5855	0.4624	0.4057	0.3094
3.15	0.9795	0.9460	0.9079	0.8586	0.7954	0.7298	0.5996	0.4779	0.4214	0.3247
3.20	0.9805	0.9487	0.9123	0.8650	0.8040	0.7404	0.6133	0.4932	0.4370	0.3400
3.25	0.9816	0.9512	0.9165	0.8710	0.8122	0.7506	0.6265	0.5081	0.4522	0.3552
3.30	0.9825	0.9536	0.9204	0.8768	0.8201	0.7603	0.6393	0.5227	0.4673	0.3702
3.35	0.9834	0.9559	0.9241	0.8822	0.8275	0.7697	0.6517	0.5369	0.4820	0.3852
3.40	0.9843	0.9580	0.9276	0.8874	0.8347	0.7787	0.6636	0.5508	0.4965	0.3999
3.45	0.9850	0.9600	0.9309	0.8923	0.8415	0.7873	0.6752	0.5644	0.5106	0.4145
3.50	0.9858	0.9619	0.9340	0.8969	0.8480	0.7955	0.6864	0.5776	0.5245	0.4289
3.55	0.9865	0.9637	0.9370	0.9014	0.8541	0.8034	0.6972	0.5904	0.5380	0.4431
3.60	0.9871	0.9653	0.9398	0.9056	0.8600	0.8109	0.7076	0.6029	0.5512	0.4571
3.65	0.9878	0.9669	0.9424	0.9096	0.8657	0.8181	0.7176	0.6151	0.5641	0.4709
3.70	0.9883	0.9684	0.9450	0.9134	0.8711	0.8251	0.7273	0.6269	0.5767	0.4844
3.75	0.9889	0.9699	0.9473	0.9170	0.8762	0.8317	0.7366	0.6384	0.5890	0.4976
3.80	0.9894	0.9712	0.9496	0.9204	0.8811	0.8380	0.7456	0.6495	0.6009	0.5107
3.85	0.9899	0.9725	0.9518	0.9237	0.8858	0.8441	0.7543	0.6603	0.6126	0.5234
3.90	0.9903	0.9737	0.9538	0.9268	0.8902	0.8499	0.7627	0.6707	0.6239	0.5359
3.95	0.9908	0.9748	0.9558	0.9298	0.8945	0.8555	0.7707	0.6809	0.6349	0.5481
4.00	0.9912	0.9759	0.9576	0.9326	0.8986	0.8608	0.7785	0.6907	0.6456	0.5600
4.05	0.9916	0.9769	0.9594	0.9353	0.9024	0.8660	0.7859	0.7002	0.6560	0.5717
4.10	0.9920	0.9779	0.9610	0.9379	0.9062	0.8709	0.7931	0.7094	0.6660	0.5831
4.15	0.9923	0.9788	0.9626	0.9403	0.9097	0.8756	0.8001	0.7184	0.6758	0.5942
4.20	0.9926	0.9797	0.9641	0.9426	0.9131	0.8801	0.8068	0.7270	0.6853	0.6051
4.25	0.9929	0.9805	0.9656	0.9449	0.9163	0.8844	0.8132	0.7354	0.6946	0.6157
4.30	0.9932	0.9813	0.9669	0.9470	0.9194	0.8885	0.8194	0.7435	0.7035	0.6260
4.35	0.9935	0.9821	0.9682	0.9490	0.9224	0.8925	0.8253	0.7513	0.7122	0.6361
4.40	0.9938	0.9828	0.9695	0.9509	0.9253	0.8963	0.8311	0.7589	0.7206	0.6459
4.45	0.9940	0.9835	0.9706	0.9528	0.9280	0.8999	0.8366	0.7662	0.7288	0.6554
4.50	0.9943	0.9841	0.9718	0.9545	0.9306	0.9034	0.8419	0.7733	0.7367	0.6647
4.55	0.9945	0.9848	0.9728	0.9562	0.9331	0.9068	0.8471	0.7801	0.7443	0.6737
4.60	0.9947	0.9854	0.9739	0.9578	0.9354	0.9100	0.8520	0.7868	0.7518	0.6825
4.65	0.9949	0.9859	0.9748	0.9594	0.9377	0.9131	0.8568	0.7932	0.7590	0.6911
4.70	0.9951	0.9865	0.9758	0.9609	0.9399	0.9160	0.8614	0.7994	0.7659	0.6994
4.75	0.9953	0.9870	0.9767	0.9623	0.9420	0.9189	0.8658	0.8054	0.7727	0.7075
4.80	0.9955	0.9875	0.9775	0.9636	0.9440	0.9216	0.8700	0.8112	0.7792	0.7154
4.85	0.9957	0.9879	0.9783	0.9649	0.9460	0.9242	0.8741	0.8168	0.7856	0.7230
4.90	0.9958	0.9884	0.9791	0.9661	0.9478	0.9268	0.8781	0.8222	0.7917	0.7304
4.95	0.9960	0.9888	0.9799	0.9673	0.9496	0.9292	0.8819	0.8274	0.7976	0.7376
5.00	0.9961	0.9892	0.9806	0.9684	0.9513	0.9315	0.8856	0.8325	0.8034	0.7446
5.05	0.9963	0.9896	0.9813	0.9695	0.9529	0.9337	0.8891	0.8374	0.8090	0.7515
5.10	0.9964	0.9900	0.9819	0.9706	0.9545	0.9359	0.8925	0.8421	0.8144	0.7581
5.15	0.9965	0.9903	0.9826	0.9716	0.9560	0.9379	0.8958	0.8467	0.8196	0.7645
5.20	0.9967	0.9906	0.9832	0.9725	0.9574	0.9399	0.8990	0.8511	0.8246	0.7707
5.25	0.9968	0.9910	0.9837	0.9734	0.9588	0.9418	0.9020	0.8554	0.8295	0.7768
5.30	0.9969	0.9913	0.9843	0.9743	0.9602	0.9437	0.9050	0.8595	0.8343	0.7827
5.35	0.9970	0.9916	0.9848	0.9752	0.9614	0.9455	0.9078	0.8635	0.8389	0.7884
5.40	0.9971	0.9919	0.9853	0.9760	0.9627	0.9472	0.9106	0.8674	0.8433	0.7939
5.45	0.9972	0.9921	0.9858	0.9768	0.9639	0.9488	0.9132	0.8711	0.8476	0.7993
5.50	0.9973	0.9924	0.9863	0.9775	0.9650	0.9504	0.9158	0.8748	0.8518	0.8046
5.55	0.9974	0.9927	0.9867	0.9782	0.9661	0.9519	0.9183	0.8783	0.8559	0.8096
5.60	0.9975	0.9929	0.9871	0.9789	0.9671	0.9534	0.9206	0.8816	0.8598	0.8146
5.65	0.9976	0.9931	0.9876	0.9796	0.9682	0.9548	0.9229	0.8849	0.8636	0.8194
5.70	0.9977	0.9934	0.9880	0.9802	0.9691	0.9561	0.9252	0.8881	0.8673	0.8240
5.75	0.9977	0.9936	0.9883	0.9808	0.9701	0.9575	0.9273	0.8912	0.8708	0.8285
5.80	0.9978	0.9938	0.9887	0.9814	0.9710	0.9587	0.9294	0.8942	0.8743	0.8329
5.85	0.9979	0.9940	0.9891	0.9820	0.9719	0.9599	0.9314	0.8970	0.8776	0.8372
5.90	0.9979	0.9942	0.9894	0.9826	0.9727	0.9611	0.9333	0.8998	0.8809	0.8413
5.95	0.9980	0.9943	0.9897	0.9831	0.9735	0.9622	0.9352	0.9025	0.8840	0.8453

PROBABILITY INTEGRAL OF t, THE NON-CENTRAL t-STATISTIC. THIS TABLE GIVES Pr $[t/\sqrt{f} \leq x]$.

f is the number of degrees of freedom; the non-centrality parameter is $\sqrt{f+1}\,K_p$.

K_p is the standardized normal deviate exceeded with probability p.

DEGREES OF FREEDOM **4**

x \ p	.2500	.1500	.1000	.0650	.0400	.0250	.0100	.0040	.0025	.0010
6.00	0.9981	0.9945	0.9900	0.9836	0.9743	0.9633	0.9370	0.9051	0.8871	0.8492
6.05	0.9981	0.9947	0.9903	0.9841	0.9751	0.9644	0.9388	0.9077	0.8900	0.8530
6.10	0.9982	0.9948	0.9906	0.9846	0.9758	0.9654	0.9404	0.9101	0.8929	0.8567
6.15	0.9982	0.9950	0.9909	0.9850	0.9765	0.9664	0.9421	0.9125	0.8957	0.8603
6.20	0.9983	0.9952	0.9912	0.9855	0.9772	0.9674	0.9437	0.9148	0.8984	0.8637
6.25	0.9983	0.9953	0.9914	0.9859	0.9778	0.9683	0.9452	0.9170	0.9010	0.8671
6.30	0.9984	0.9954	0.9917	0.9863	0.9784	0.9692	0.9467	0.9192	0.9035	0.8704
6.35	0.9984	0.9956	0.9919	0.9867	0.9791	0.9700	0.9481	0.9213	0.9060	0.8736
6.40	0.9985	0.9957	0.9922	0.9871	0.9796	0.9708	0.9495	0.9233	0.9084	0.8766
6.45	0.9985	0.9958	0.9924	0.9874	0.9802	0.9716	0.9508	0.9253	0.9107	0.8797
6.50	0.9986	0.9959	0.9926	0.9878	0.9808	0.9724	0.9521	0.9272	0.9129	0.8826
6.55	0.9986	0.9961	0.9928	0.9881	0.9813	0.9732	0.9534	0.9291	0.9151	0.8854
6.60	0.9987	0.9962	0.9930	0.9885	0.9818	0.9739	0.9546	0.9309	0.9172	0.8882
6.65	0.9987	0.9963	0.9932	0.9888	0.9823	0.9746	0.9558	0.9326	0.9193	0.8908
6.70	0.9987	0.9964	0.9934	0.9891	0.9828	0.9753	0.9569	0.9343	0.9212	0.8934
6.75	0.9988	0.9965	0.9936	0.9894	0.9832	0.9759	0.9580	0.9359	0.9232	0.8960
6.80	0.9988	0.9966	0.9938	0.9897	0.9837	0.9766	0.9591	0.9375	0.9251	0.8984
6.85	0.9988	0.9967	0.9939	0.9899	0.9841	0.9772	0.9602	0.9391	0.9269	0.9008
6.90	0.9989	0.9968	0.9941	0.9902	0.9845	0.9778	0.9612	0.9406	0.9286	0.9032
6.95	0.9989	0.9969	0.9943	0.9905	0.9849	0.9783	0.9622	0.9420	0.9304	0.9054
7.00	0.9989	0.9969	0.9944	0.9907	0.9853	0.9789	0.9631	0.9434	0.9320	0.9076
7.05	0.9990	0.9970	0.9946	0.9910	0.9857	0.9794	0.9640	0.9448	0.9337	0.9098
7.10	0.9990	0.9971	0.9947	0.9912	0.9861	0.9800	0.9649	0.9461	0.9352	0.9118
7.15	0.9990	0.9972	0.9948	0.9914	0.9864	0.9805	0.9658	0.9474	0.9368	0.9139
7.20	0.9990	0.9973	0.9950	0.9917	0.9868	0.9810	0.9666	0.9487	0.9382	0.9158
7.25	0.9991	0.9973	0.9951	0.9919	0.9871	0.9814	0.9674	0.9499	0.9397	0.9178
7.30	0.9991	0.9974	0.9952	0.9921	0.9875	0.9819	0.9682	0.9511	0.9411	0.9196
7.35	0.9991	0.9975	0.9954	0.9923	0.9878	0.9824	0.9690	0.9522	0.9425	0.9214
7.40	0.9991	0.9975	0.9955	0.9925	0.9881	0.9828	0.9697	0.9534	0.9438	0.9232
7.45	0.9992	0.9976	0.9956	0.9927	0.9884	0.9832	0.9705	0.9544	0.9451	0.9249
7.50	0.9992	0.9977	0.9957	0.9928	0.9887	0.9836	0.9712	0.9555	0.9463	0.9266
7.55	0.9992	0.9977	0.9958	0.9930	0.9889	0.9840	0.9718	0.9565	0.9476	0.9282
7.60	0.9992	0.9978	0.9959	0.9932	0.9892	0.9844	0.9725	0.9575	0.9488	0.9298
7.65	0.9992	0.9978	0.9960	0.9934	0.9895	0.9848	0.9731	0.9585	0.9499	0.9314
7.70	0.9993	0.9979	0.9961	0.9935	0.9897	0.9851	0.9738	0.9594	0.9510	0.9329
7.75	0.9993	0.9979	0.9962	0.9937	0.9900	0.9855	0.9744	0.9603	0.9521	0.9343
7.80	0.9993	0.9980	0.9963	0.9938	0.9902	0.9858	0.9750	0.9612	0.9532	0.9357
7.85	0.9993	0.9980	0.9964	0.9940	0.9904	0.9862	0.9755	0.9621	0.9542	0.9371
7.90	0.9993	0.9981	0.9965	0.9941	0.9907	0.9865	0.9761	0.9630	0.9552	0.9385
7.95	0.9993	0.9981	0.9966	0.9943	0.9909	0.9868	0.9766	0.9638	0.9562	0.9398
8.00	0.9994	0.9982	0.9966	0.9944	0.9911	0.9871	0.9772	0.9646	0.9572	0.9411
8.05	0.9994	0.9982	0.9967	0.9945	0.9913	0.9874	0.9777	0.9654	0.9581	0.9423
8.10	0.9994	0.9983	0.9968	0.9947	0.9915	0.9877	0.9782	0.9661	0.9590	0.9436
8.15	0.9994	0.9983	0.9969	0.9948	0.9917	0.9880	0.9787	0.9669	0.9599	0.9448
8.20	0.9994	0.9983	0.9969	0.9949	0.9919	0.9882	0.9791	0.9676	0.9608	0.9459
8.25	0.9994	0.9984	0.9970	0.9950	0.9921	0.9885	0.9796	0.9683	0.9616	0.9470
8.30	0.9994	0.9984	0.9971	0.9951	0.9922	0.9888	0.9800	0.9690	0.9624	0.9481
8.35	0.9995	0.9985	0.9972	0.9952	0.9924	0.9890	0.9805	0.9696	0.9632	0.9492
8.40	0.9995	0.9985	0.9972	0.9954	0.9926	0.9893	0.9809	0.9703	0.9640	0.9503
8.45	0.9995	0.9985	0.9973	0.9955	0.9928	0.9895	0.9813	0.9709	0.9647	0.9513
8.50	0.9995	0.9986	0.9973	0.9956	0.9929	0.9897	0.9817	0.9715	0.9655	0.9523
8.55	0.9995	0.9986	0.9974	0.9957	0.9931	0.9899	0.9821	0.9721	0.9662	0.9533
8.60	0.9995	0.9986	0.9975	0.9958	0.9932	0.9902	0.9825	0.9727	0.9669	0.9542
8.65	0.9995	0.9987	0.9975	0.9958	0.9934	0.9904	0.9829	0.9733	0.9676	0.9551
8.70	0.9995	0.9987	0.9976	0.9959	0.9935	0.9906	0.9832	0.9738	0.9682	0.9560
8.75	0.9996	0.9987	0.9976	0.9960	0.9937	0.9908	0.9836	0.9743	0.9689	0.9569
8.80	0.9996	0.9987	0.9977	0.9961	0.9938	0.9910	0.9839	0.9749	0.9695	0.9578
8.85	0.9996	0.9988	0.9977	0.9962	0.9939	0.9912	0.9843	0.9754	0.9701	0.9586
8.90	0.9996	0.9988	0.9978	0.9963	0.9941	0.9914	0.9846	0.9759	0.9707	0.9594
8.95	0.9996	0.9988	0.9978	0.9964	0.9942	0.9915	0.9849	0.9764	0.9713	0.9602

PROBABILITY INTEGRAL OF t, THE NON-CENTRAL t-STATISTIC. THIS TABLE GIVES Pr $[t/\sqrt{f} \leq x]$.

f is the number of degrees of freedom; the non-centrality parameter is $\sqrt{f+1}\,K_p$.
K_p is the standardized normal deviate exceeded with probability p.

DEGREES OF FREEDOM **4**

x \ p	.2500	.1500	.1000	.0650	.0400	.0250	.0100	.0040	.0025	.0010
9.00	0.9996	0.9988	0.9979	0.9964	0.9943	0.9917	0.9852	0.9769	0.9719	0.9610
9.05	0.9996	0.9989	0.9979	0.9965	0.9944	0.9919	0.9855	0.9773	0.9725	0.9618
9.10	0.9996	0.9989	0.9980	0.9966	0.9945	0.9920	0.9858	0.9778	0.9730	0.9625
9.15	0.9996	0.9989	0.9980	0.9967	0.9946	0.9922	0.9861	0.9782	0.9735	0.9632
9.20	0.9996	0.9989	0.9980	0.9967	0.9948	0.9924	0.9864	0.9786	0.9740	0.9639
9.25	0.9996	0.9990	0.9981	0.9968	0.9949	0.9925	0.9866	0.9791	0.9746	0.9646
9.30	0.9996	0.9990	0.9981	0.9969	0.9950	0.9927	0.9869	0.9795	0.9750	0.9653
9.35	0.9997	0.9990	0.9982	0.9969	0.9951	0.9928	0.9872	0.9799	0.9755	0.9660
9.40	0.9997	0.9990	0.9982	0.9970	0.9952	0.9930	0.9874	0.9803	0.9760	0.9666
9.45	0.9997	0.9990	0.9982	0.9970	0.9953	0.9931	0.9877	0.9806	0.9765	0.9673
9.50	0.9997	0.9991	0.9983	0.9971	0.9954	0.9932	0.9879	0.9810	0.9769	0.9679
9.55	0.9997	0.9991	0.9983	0.9972	0.9955	0.9934	0.9881	0.9814	0.9774	0.9685
9.60	0.9997	0.9991	0.9983	0.9972	0.9955	0.9935	0.9884	0.9817	0.9778	0.9691
9.65	0.9997	0.9991	0.9984	0.9973	0.9956	0.9936	0.9886	0.9821	0.9782	0.9696
9.70	0.9997	0.9991	0.9984	0.9973	0.9957	0.9938	0.9888	0.9824	0.9786	0.9702
9.75	0.9997	0.9992	0.9984	0.9974	0.9958	0.9939	0.9890	0.9828	0.9790	0.9707
9.80	0.9997	0.9992	0.9985	0.9974	0.9959	0.9940	0.9892	0.9831	0.9794	0.9713
9.85	0.9997	0.9992	0.9985	0.9975	0.9960	0.9941	0.9894	0.9834	0.9798	0.9718
9.90	0.9997	0.9992	0.9985	0.9975	0.9960	0.9942	0.9896	0.9837	0.9802	0.9723
9.95	0.9997	0.9992	0.9986	0.9976	0.9961	0.9943	0.9898	0.9840	0.9805	0.9728
10.00	0.9997	0.9992	0.9986	0.9976	0.9962	0.9944	0.9900	0.9843	0.9809	0.9733
10.05	0.9997	0.9992	0.9986	0.9977	0.9963	0.9946	0.9902	0.9846	0.9812	0.9738
10.10	0.9997	0.9993	0.9986	0.9977	0.9963	0.9947	0.9904	0.9849	0.9816	0.9743
10.15	0.9997	0.9993	0.9987	0.9978	0.9964	0.9948	0.9906	0.9851	0.9819	0.9747
10.20	0.9998	0.9993	0.9987	0.9978	0.9965	0.9949	0.9907	0.9854	0.9822	0.9752
10.25	0.9998	0.9993	0.9987	0.9978	0.9965	0.9949	0.9909	0.9857	0.9825	0.9756
10.30	0.9998	0.9993	0.9987	0.9979	0.9966	0.9950	0.9911	0.9859	0.9829	0.9760
10.35	0.9998	0.9993	0.9988	0.9979	0.9967	0.9951	0.9912	0.9862	0.9832	0.9765
10.40	0.9998	0.9993	0.9988	0.9980	0.9967	0.9952	0.9914	0.9864	0.9835	0.9769
10.45	0.9998	0.9994	0.9988	0.9980	0.9968	0.9953	0.9916	0.9867	0.9838	0.9773
10.50	0.9998	0.9994	0.9988	0.9980	0.9968	0.9954	0.9917	0.9869	0.9840	0.9777
10.55	0.9998	0.9994	0.9989	0.9981	0.9969	0.9955	0.9919	0.9871	0.9843	0.9781
10.60	0.9998	0.9994	0.9989	0.9981	0.9970	0.9956	0.9920	0.9874	0.9846	0.9784
10.65	0.9998	0.9994	0.9989	0.9981	0.9970	0.9956	0.9921	0.9876	0.9849	0.9788
10.70	0.9998	0.9994	0.9989	0.9982	0.9971	0.9957	0.9923	0.9878	0.9851	0.9792
10.75	0.9998	0.9994	0.9989	0.9982	0.9971	0.9958	0.9924	0.9880	0.9854	0.9795
10.80	0.9998	0.9994	0.9990	0.9982	0.9972	0.9959	0.9925	0.9882	0.9856	0.9799
10.85	0.9998	0.9994	0.9990	0.9983	0.9972	0.9959	0.9927	0.9884	0.9859	0.9802
10.90	0.9998	0.9995	0.9990	0.9983	0.9973	0.9960	0.9928	0.9886	0.9861	0.9805
10.95	0.9998	0.9995	0.9990	0.9983	0.9973	0.9961	0.9929	0.9888	0.9864	0.9809
11.00	0.9998	0.9995	0.9990	0.9984	0.9974	0.9961	0.9931	0.9890	0.9866	0.9812
11.05	0.9998	0.9995	0.9990	0.9984	0.9974	0.9962	0.9932	0.9892	0.9868	0.9815
11.10	0.9998	0.9995	0.9991	0.9984	0.9975	0.9963	0.9933	0.9894	0.9870	0.9818
11.15	0.9998	0.9995	0.9991	0.9984	0.9975	0.9963	0.9934	0.9896	0.9872	0.9821
11.20	0.9998	0.9995	0.9991	0.9985	0.9975	0.9964	0.9935	0.9897	0.9875	0.9824
11.25	0.9998	0.9995	0.9991	0.9985	0.9976	0.9965	0.9936	0.9899	0.9877	0.9827
11.30	0.9998	0.9995	0.9991	0.9985	0.9976	0.9965	0.9937	0.9901	0.9879	0.9830
11.35	0.9998	0.9995	0.9991	0.9985	0.9977	0.9966	0.9938	0.9902	0.9881	0.9833
11.40	0.9998	0.9995	0.9992	0.9986	0.9977	0.9966	0.9939	0.9904	0.9883	0.9835
11.45	0.9998	0.9996	0.9992	0.9986	0.9977	0.9967	0.9940	0.9906	0.9885	0.9838
11.50	0.9998	0.9996	0.9992	0.9986	0.9978	0.9968	0.9941	0.9907	0.9886	0.9840
11.55	0.9998	0.9996	0.9992	0.9986	0.9978	0.9968	0.9942	0.9909	0.9888	0.9843
11.60	0.9998	0.9996	0.9992	0.9987	0.9979	0.9969	0.9943	0.9910	0.9890	0.9846
11.65	0.9999	0.9996	0.9992	0.9987	0.9979	0.9969	0.9944	0.9912	0.9892	0.9848
11.70	0.9999	0.9996	0.9992	0.9987	0.9979	0.9970	0.9945	0.9913	0.9894	0.9850
11.75	0.9999	0.9996	0.9992	0.9987	0.9980	0.9970	0.9946	0.9914	0.9895	0.9853
11.80	0.9999	0.9996	0.9993	0.9988	0.9980	0.9971	0.9947	0.9916	0.9897	0.9855
11.85	0.9999	0.9996	0.9993	0.9988	0.9980	0.9971	0.9948	0.9917	0.9899	0.9857
11.90	0.9999	0.9996	0.9993	0.9988	0.9981	0.9972	0.9949	0.9918	0.9900	0.9860
11.95	0.9999	0.9996	0.9993	0.9988	0.9981	0.9972	0.9949	0.9920	0.9902	0.9862

PROBABILITY INTEGRAL OF t, THE NON-CENTRAL t-STATISTIC. THIS TABLE GIVES Pr $[t/\sqrt{f} \leq x]$.

f is the number of degrees of freedom; the non-centrality parameter is $\sqrt{f+1}\,K_p$.

K_p is the standardized normal deviate exceeded with probability p.

DEGREES OF FREEDOM **4**

x \ P	.2500	.1500	.1000	.0650	.0400	.0250	.0100	.0040	.0025	.0010
12.00	0.9999	0.9996	0.9993	0.9988	0.9981	0.9972	0.9950	0.9921	0.9903	0.9864
12.05	0.9999	0.9996	0.9993	0.9989	0.9982	0.9973	0.9951	0.9922	0.9905	0.9866
12.10	0.9999	0.9996	0.9993	0.9989	0.9982	0.9973	0.9952	0.9923	0.9906	0.9868
12.15	0.9999	0.9996	0.9993	0.9989	0.9982	0.9974	0.9953	0.9925	0.9908	0.9870
12.20	0.9999	0.9996	0.9994	0.9989	0.9982	0.9974	0.9953	0.9926	0.9909	0.9872
12.25	0.9999	0.9997	0.9994	0.9989	0.9983	0.9975	0.9954	0.9927	0.9911	0.9874
12.30	0.9999	0.9997	0.9994	0.9989	0.9983	0.9975	0.9955	0.9928	0.9912	0.9876
12.35	0.9999	0.9997	0.9994	0.9990	0.9983	0.9975	0.9955	0.9929	0.9913	0.9878
12.40	0.9999	0.9997	0.9994	0.9990	0.9983	0.9976	0.9956	0.9930	0.9915	0.9880
12.45	0.9999	0.9997	0.9994	0.9990	0.9984	0.9976	0.9957	0.9931	0.9916	0.9882
12.50	0.9999	0.9997	0.9994	0.9990	0.9984	0.9977	0.9957	0.9932	0.9917	0.9883
12.55	0.9999	0.9997	0.9994	0.9990	0.9984	0.9977	0.9958	0.9933	0.9918	0.9885
12.60	0.9999	0.9997	0.9994	0.9990	0.9984	0.9977	0.9959	0.9934	0.9920	0.9887
12.65	0.9999	0.9997	0.9994	0.9991	0.9985	0.9978	0.9959	0.9935	0.9921	0.9888
12.70	0.9999	0.9997	0.9994	0.9991	0.9985	0.9978	0.9960	0.9936	0.9922	0.9890
12.75	0.9999	0.9997	0.9995	0.9991	0.9985	0.9978	0.9961	0.9937	0.9923	0.9892
12.80	0.9999	0.9997	0.9995	0.9991	0.9985	0.9979	0.9961	0.9938	0.9924	0.9893
12.85	0.9999	0.9997	0.9995	0.9991	0.9986	0.9979	0.9962	0.9939	0.9925	0.9895
12.90	0.9999	0.9997	0.9995	0.9991	0.9986	0.9979	0.9962	0.9940	0.9927	0.9896
12.95	0.9999	0.9997	0.9995	0.9991	0.9986	0.9980	0.9963	0.9941	0.9928	0.9898
13.00	0.9999	0.9997	0.9995	0.9991	0.9986	0.9980	0.9963	0.9942	0.9929	0.9899
13.05	0.9999	0.9997	0.9995	0.9992	0.9986	0.9980	0.9964	0.9943	0.9930	0.9901
13.10	0.9999	0.9997	0.9995	0.9992	0.9987	0.9980	0.9965	0.9943	0.9931	0.9902
13.15	0.9999	0.9997	0.9995	0.9992	0.9987	0.9981	0.9965	0.9944	0.9932	0.9904
13.20	0.9999	0.9997	0.9995	0.9992	0.9987	0.9981	0.9966	0.9945	0.9933	0.9905
13.25	0.9999	0.9997	0.9995	0.9992	0.9987	0.9981	0.9966	0.9946	0.9934	0.9906
13.30	0.9999	0.9998	0.9995	0.9992	0.9987	0.9982	0.9967	0.9947	0.9935	0.9908
13.35	0.9999	0.9998	0.9995	0.9992	0.9988	0.9982	0.9967	0.9947	0.9936	0.9909
13.40	0.9999	0.9998	0.9996	0.9992	0.9988	0.9982	0.9967	0.9948	0.9936	0.9910
13.45	0.9999	0.9998	0.9996	0.9993	0.9988	0.9982	0.9968	0.9949	0.9937	0.9912
13.50	0.9999	0.9998	0.9996	0.9993	0.9988	0.9983	0.9968	0.9950	0.9938	0.9913
13.55	0.9999	0.9998	0.9996	0.9993	0.9988	0.9983	0.9969	0.9950	0.9939	0.9914
13.60	0.9999	0.9998	0.9996	0.9993	0.9988	0.9983	0.9969	0.9951	0.9940	0.9915
13.65	0.9999	0.9998	0.9996	0.9993	0.9989	0.9983	0.9970	0.9952	0.9941	0.9916
13.70	0.9999	0.9998	0.9996	0.9993	0.9989	0.9984	0.9970	0.9952	0.9942	0.9918
13.75	0.9999	0.9998	0.9996	0.9993	0.9989	0.9984	0.9971	0.9953	0.9942	0.9919
13.80	0.9999	0.9998	0.9996	0.9993	0.9989	0.9984	0.9971	0.9954	0.9943	0.9920
13.85	0.9999	0.9998	0.9996	0.9993	0.9989	0.9984	0.9971	0.9954	0.9944	0.9921
13.90	0.9999	0.9998	0.9996	0.9993	0.9989	0.9984	0.9972	0.9955	0.9945	0.9922
13.95	0.9999	0.9998	0.9996	0.9994	0.9990	0.9985	0.9972	0.9956	0.9946	0.9923
14.00	0.9999	0.9998	0.9996	0.9994	0.9990	0.9985	0.9973	0.9956	0.9946	0.9924
14.05	0.9999	0.9998	0.9996	0.9994	0.9990	0.9985	0.9973	0.9957	0.9947	0.9925
14.10	0.9999	0.9998	0.9996	0.9994	0.9990	0.9985	0.9973	0.9957	0.9948	0.9926
14.15	0.9999	0.9998	0.9996	0.9994	0.9990	0.9986	0.9974	0.9958	0.9948	0.9927
14.20	0.9999	0.9998	0.9996	0.9994	0.9990	0.9986	0.9974	0.9959	0.9949	0.9928
14.25	0.9999	0.9998	0.9996	0.9994	0.9990	0.9986	0.9974	0.9959	0.9950	0.9929
14.30	0.9999	0.9998	0.9997	0.9994	0.9991	0.9986	0.9975	0.9960	0.9951	0.9930
14.35	0.9999	0.9998	0.9997	0.9994	0.9991	0.9986	0.9975	0.9960	0.9951	0.9931
14.40	0.9999	0.9998	0.9997	0.9994	0.9991	0.9986	0.9975	0.9961	0.9952	0.9932
14.45	0.9999	0.9998	0.9997	0.9994	0.9991	0.9987	0.9976	0.9961	0.9952	0.9933
14.50	0.9999	0.9998	0.9997	0.9994	0.9991	0.9987	0.9976	0.9962	0.9953	0.9934
14.55	0.9999	0.9998	0.9997	0.9995	0.9991	0.9987	0.9976	0.9962	0.9954	0.9934
14.60	0.9999	0.9998	0.9997	0.9995	0.9991	0.9987	0.9977	0.9963	0.9954	0.9935
14.65	0.9999	0.9998	0.9997	0.9995	0.9991	0.9987	0.9977	0.9963	0.9955	0.9936
14.70	0.9999	0.9998	0.9997	0.9995	0.9992	0.9988	0.9977	0.9964	0.9956	0.9937
14.75	0.9999	0.9998	0.9997	0.9995	0.9992	0.9988	0.9978	0.9964	0.9956	0.9938
14.80	0.9999	0.9998	0.9997	0.9995	0.9992	0.9988	0.9978	0.9965	0.9957	0.9939
14.85	0.9999	0.9998	0.9997	0.9995	0.9992	0.9988	0.9978	0.9965	0.9957	0.9939
14.90	0.9999	0.9998	0.9997	0.9995	0.9992	0.9988	0.9978	0.9966	0.9958	0.9940
14.95	0.9999	0.9998	0.9997	0.9995	0.9992	0.9988	0.9979	0.9966	0.9958	0.9941

PROBABILITY INTEGRAL OF t, THE NON-CENTRAL t-STATISTIC. THIS TABLE GIVES Pr $[t/\sqrt{f} \leq x]$.

f is the number of degrees of freedom; the non-centrality parameter is $\sqrt{f+1}\,K_p$.
K_p is the standardized normal deviate exceeded with probability p.

DEGREES OF FREEDOM **4**

x \ p	.2500	.1500	.1000	.0650	.0400	.0250	.0100	.0040	.0025	.0010
15.00	0.9999	0.9998	0.9997	0.9995	0.9992	0.9988	0.9979	0.9966	0.9959	0.9942
15.05	0.9999	0.9998	0.9997	0.9995	0.9992	0.9989	0.9979	0.9967	0.9959	0.9942
15.10	0.9999	0.9998	0.9997	0.9995	0.9992	0.9989	0.9980	0.9967	0.9960	0.9943
15.15	0.9999	0.9999	0.9997	0.9995	0.9992	0.9989	0.9980	0.9968	0.9960	0.9944
15.20	0.9999	0.9999	0.9997	0.9995	0.9993	0.9989	0.9980	0.9968	0.9961	0.9945
15.25	0.9999	0.9999	0.9997	0.9995	0.9993	0.9989	0.9980	0.9969	0.9961	0.9945
15.30	0.9999	0.9999	0.9997	0.9996	0.9993	0.9989	0.9981	0.9969	0.9962	0.9946
15.35	0.9999	0.9999	0.9997	0.9996	0.9993	0.9989	0.9981	0.9969	0.9962	0.9947
15.40	0.9999	0.9999	0.9997	0.9996	0.9993	0.9990	0.9981	0.9970	0.9963	0.9947
15.45	1.0000	0.9999	0.9997	0.9996	0.9993	0.9990	0.9981	0.9970	0.9963	0.9948
15.50	1.0000	0.9999	0.9997	0.9996	0.9993	0.9990	0.9982	0.9970	0.9964	0.9949
15.55	1.0000	0.9999	0.9998	0.9996	0.9993	0.9990	0.9982	0.9971	0.9964	0.9949
15.60	1.0000	0.9999	0.9998	0.9996	0.9993	0.9990	0.9982	0.9971	0.9965	0.9950
15.65	1.0000	0.9999	0.9998	0.9996	0.9993	0.9990	0.9982	0.9972	0.9965	0.9950
15.70	1.0000	0.9999	0.9998	0.9996	0.9993	0.9990	0.9982	0.9972	0.9966	0.9951
15.75	1.0000	0.9999	0.9998	0.9996	0.9994	0.9991	0.9983	0.9972	0.9966	0.9952
15.80	1.0000	0.9999	0.9998	0.9996	0.9994	0.9991	0.9983	0.9973	0.9966	0.9952
15.85	1.0000	0.9999	0.9998	0.9996	0.9994	0.9991	0.9983	0.9973	0.9967	0.9953
15.90	1.0000	0.9999	0.9998	0.9996	0.9994	0.9991	0.9983	0.9973	0.9967	0.9953
15.95	1.0000	0.9999	0.9998	0.9996	0.9994	0.9991	0.9984	0.9974	0.9968	0.9954
16.00	1.0000	0.9999	0.9998	0.9996	0.9994	0.9991	0.9984	0.9974	0.9968	0.9955
16.05	1.0000	0.9999	0.9998	0.9996	0.9994	0.9991	0.9984	0.9974	0.9968	0.9955
16.10	1.0000	0.9999	0.9998	0.9996	0.9994	0.9991	0.9984	0.9975	0.9969	0.9956
16.15	1.0000	0.9999	0.9998	0.9996	0.9994	0.9991	0.9984	0.9975	0.9969	0.9956
16.20	1.0000	0.9999	0.9998	0.9996	0.9994	0.9992	0.9984	0.9975	0.9969	0.9957
16.25	1.0000	0.9999	0.9998	0.9996	0.9994	0.9992	0.9985	0.9975	0.9970	0.9957
16.30	1.0000	0.9999	0.9998	0.9997	0.9994	0.9992	0.9985	0.9976	0.9970	0.9958
16.35	1.0000	0.9999	0.9998	0.9997	0.9994	0.9992	0.9985	0.9976	0.9971	0.9958
16.40	1.0000	0.9999	0.9998	0.9997	0.9995	0.9992	0.9985	0.9976	0.9971	0.9959
16.45	1.0000	0.9999	0.9998	0.9997	0.9995	0.9992	0.9985	0.9977	0.9971	0.9959
16.50	1.0000	0.9999	0.9998	0.9997	0.9995	0.9992	0.9986	0.9977	0.9972	0.9960
16.55	1.0000	0.9999	0.9998	0.9997	0.9995	0.9992	0.9986	0.9977	0.9972	0.9960
16.60	1.0000	0.9999	0.9998	0.9997	0.9995	0.9992	0.9986	0.9977	0.9972	0.9961
16.65	1.0000	0.9999	0.9998	0.9997	0.9995	0.9992	0.9986	0.9978	0.9973	0.9961
16.70	1.0000	0.9999	0.9998	0.9997	0.9995	0.9992	0.9986	0.9978	0.9973	0.9961
16.75	1.0000	0.9999	0.9998	0.9997	0.9995	0.9993	0.9986	0.9978	0.9973	0.9962
16.80	1.0000	0.9999	0.9998	0.9997	0.9995	0.9993	0.9987	0.9978	0.9973	0.9962
16.85	1.0000	0.9999	0.9998	0.9997	0.9995	0.9993	0.9987	0.9979	0.9974	0.9963
16.90	1.0000	0.9999	0.9998	0.9997	0.9995	0.9993	0.9987	0.9979	0.9974	0.9963
16.95	1.0000	0.9999	0.9998	0.9997	0.9995	0.9993	0.9987	0.9979	0.9974	0.9964
17.00	1.0000	0.9999	0.9998	0.9997	0.9995	0.9993	0.9987	0.9979	0.9975	0.9964
17.05	1.0000	0.9999	0.9998	0.9997	0.9995	0.9993	0.9987	0.9980	0.9975	0.9964
17.10	1.0000	0.9999	0.9998	0.9997	0.9995	0.9993	0.9987	0.9980	0.9975	0.9965
17.15	1.0000	0.9999	0.9998	0.9997	0.9995	0.9993	0.9988	0.9980	0.9976	0.9965
17.20	1.0000	0.9999	0.9998	0.9997	0.9995	0.9993	0.9988	0.9980	0.9976	0.9966
17.25	1.0000	0.9999	0.9998	0.9997	0.9996	0.9993	0.9988	0.9981	0.9976	0.9966
17.30	1.0000	0.9999	0.9998	0.9997	0.9996	0.9993	0.9988	0.9981	0.9976	0.9966
17.35	1.0000	0.9999	0.9998	0.9997	0.9996	0.9994	0.9988	0.9981	0.9977	0.9967
17.40	1.0000	0.9999	0.9998	0.9997	0.9996	0.9994	0.9988	0.9981	0.9977	0.9967
17.45	1.0000	0.9999	0.9998	0.9997	0.9996	0.9994	0.9988	0.9981	0.9977	0.9967
17.50	1.0000	0.9999	0.9998	0.9997	0.9996	0.9994	0.9989	0.9982	0.9977	0.9968
17.55	1.0000	0.9999	0.9998	0.9997	0.9996	0.9994	0.9989	0.9982	0.9978	0.9968
17.60	1.0000	0.9999	0.9998	0.9997	0.9996	0.9994	0.9989	0.9982	0.9978	0.9969
17.65	1.0000	0.9999	0.9999	0.9997	0.9996	0.9994	0.9989	0.9982	0.9978	0.9969
17.70	1.0000	0.9999	0.9999	0.9998	0.9996	0.9994	0.9989	0.9982	0.9978	0.9969
17.75	1.0000	0.9999	0.9999	0.9998	0.9996	0.9994	0.9989	0.9983	0.9979	0.9970
17.80	1.0000	0.9999	0.9999	0.9998	0.9996	0.9994	0.9989	0.9983	0.9979	0.9970
17.85	1.0000	0.9999	0.9999	0.9998	0.9996	0.9994	0.9989	0.9983	0.9979	0.9970
17.90	1.0000	0.9999	0.9999	0.9998	0.9996	0.9994	0.9990	0.9983	0.9979	0.9971
17.95	1.0000	0.9999	0.9999	0.9998	0.9996	0.9994	0.9990	0.9983	0.9980	0.9971

PROBABILITY INTEGRAL OF t, THE NON-CENTRAL t-STATISTIC. THIS TABLE GIVES Pr $[t/\sqrt{f} \leq x]$.

f is the number of degrees of freedom; the non-centrality parameter is $\sqrt{f+1}\,K_p$.

K_p is the standardized normal deviate exceeded with probability p.

DEGREES OF FREEDOM **4**

x \ p	.2500	1500	.1000	.0650	.0400	.0250	.0100	.0040	.0025	.0010
18.00	1.0000	0.9999	0.9999	0.9998	0.9996	0.9994	0.9990	0.9984	0.9980	0.9971
18.05	1.0000	0.9999	0.9999	0.9998	0.9996	0.9994	0.9990	0.9984	0.9980	0.9971
18.10	1.0000	0.9999	0.9999	0.9998	0.9996	0.9995	0.9990	0.9984	0.9980	0.9972
18.15	1.0000	0.9999	0.9999	0.9998	0.9996	0.9995	0.9990	0.9984	0.9980	0.9972
18.20	1.0000	0.9999	0.9999	0.9998	0.9996	0.9995	0.9990	0.9984	0.9981	0.9972
18.25	1.0000	0.9999	0.9999	0.9998	0.9996	0.9995	0.9990	0.9984	0.9981	0.9973
18.30	1.0000	0.9999	0.9999	0.9998	0.9996	0.9995	0.9990	0.9985	0.9981	0.9973
18.35	1.0000	0.9999	0.9999	0.9998	0.9996	0.9995	0.9991	0.9985	0.9981	0.9973
18.40	1.0000	0.9999	0.9999	0.9998	0.9997	0.9995	0.9991	0.9985	0.9981	0.9974
18.45	1.0000	0.9999	0.9999	0.9998	0.9997	0.9995	0.9991	0.9985	0.9982	0.9974
18.50	1.0000	0.9999	0.9999	0.9998	0.9997	0.9995	0.9991	0.9985	0.9982	0.9974
18.55	1.0000	0.9999	0.9999	0.9998	0.9997	0.9995	0.9991	0.9985	0.9982	0.9974
18.60	1.0000	0.9999	0.9999	0.9998	0.9997	0.9995	0.9991	0.9986	0.9982	0.9975
18.65	1.0000	0.9999	0.9999	0.9998	0.9997	0.9995	0.9991	0.9986	0.9982	0.9975
18.70	1.0000	0.9999	0.9999	0.9998	0.9997	0.9995	0.9991	0.9986	0.9983	0.9975
18.75	1.0000	0.9999	0.9999	0.9998	0.9997	0.9995	0.9991	0.9986	0.9983	0.9975
18.80	1.0000	0.9999	0.9999	0.9998	0.9997	0.9995	0.9991	0.9986	0.9983	0.9976
18.85	1.0000	0.9999	0.9999	0.9998	0.9997	0.9995	0.9991	0.9986	0.9983	0.9976
18.90	1.0000	0.9999	0.9999	0.9998	0.9997	0.9995	0.9992	0.9986	0.9983	0.9976
18.95	1.0000	0.9999	0.9999	0.9998	0.9997	0.9995	0.9992	0.9987	0.9983	0.9976
19.00	1.0000	0.9999	0.9999	0.9998	0.9997	0.9995	0.9992	0.9987	0.9984	0.9977
19.05	1.0000	0.9999	0.9999	0.9998	0.9997	0.9996	0.9992	0.9987	0.9984	0.9977
19.10	1.0000	0.9999	0.9999	0.9998	0.9997	0.9996	0.9992	0.9987	0.9984	0.9977
19.15	1.0000	0.9999	0.9999	0.9998	0.9997	0.9996	0.9992	0.9987	0.9984	0.9977
19.20	1.0000	0.9999	0.9999	0.9998	0.9997	0.9996	0.9992	0.9987	0.9984	0.9978
19.25	1.0000	0.9999	0.9999	0.9998	0.9997	0.9996	0.9992	0.9987	0.9984	0.9978
19.30	1.0000	0.9999	0.9999	0.9998	0.9997	0.9996	0.9992	0.9988	0.9985	0.9978
19.35	1.0000	0.9999	0.9999	0.9998	0.9997	0.9996	0.9992	0.9988	0.9985	0.9978
19.40	1.0000	0.9999	0.9999	0.9998	0.9997	0.9996	0.9992	0.9988	0.9985	0.9978
19.45	1.0000	0.9999	0.9999	0.9998	0.9997	0.9996	0.9992	0.9988	0.9985	0.9979
19.50	1.0000	0.9999	0.9999	0.9998	0.9997	0.9996	0.9993	0.9988	0.9985	0.9979
19.55	1.0000	0.9999	0.9999	0.9998	0.9997	0.9996	0.9993	0.9988	0.9985	0.9979
19.60	1.0000	0.9999	0.9999	0.9998	0.9997	0.9996	0.9993	0.9988	0.9986	0.9979
19.65	1.0000	0.9999	0.9999	0.9998	0.9997	0.9996	0.9993	0.9988	0.9986	0.9980
19.70	1.0000	0.9999	0.9999	0.9998	0.9997	0.9996	0.9993	0.9988	0.9986	0.9980
19.75	1.0000	0.9999	0.9999	0.9998	0.9997	0.9996	0.9993	0.9989	0.9986	0.9980
19.80	1.0000	0.9999	0.9999	0.9998	0.9997	0.9996	0.9993	0.9989	0.9986	0.9980
19.85	1.0000	0.9999	0.9999	0.9998	0.9997	0.9996	0.9993	0.9989	0.9986	0.9980
19.90	1.0000	1.0000	0.9999	0.9998	0.9997	0.9996	0.9993	0.9989	0.9986	0.9981
19.95	1.0000	1.0000	0.9999	0.9998	0.9997	0.9996	0.9993	0.9989	0.9986	0.9981
20.00	1.0000	1.0000	0.9999	0.9998	0.9998	0.9996	0.9993	0.9989	0.9987	0.9981
20.05	1.0000	1.0000	0.9999	0.9998	0.9998	0.9996	0.9993	0.9989	0.9987	0.9981
20.10	1.0000	1.0000	0.9999	0.9999	0.9998	0.9996	0.9993	0.9989	0.9987	0.9981
20.15	1.0000	1.0000	0.9999	0.9999	0.9998	0.9996	0.9993	0.9989	0.9987	0.9981
20.20	1.0000	1.0000	0.9999	0.9999	0.9998	0.9996	0.9994	0.9990	0.9987	0.9982
20.25	1.0000	1.0000	0.9999	0.9999	0.9998	0.9996	0.9994	0.9990	0.9987	0.9982
20.30	1.0000	1.0000	0.9999	0.9999	0.9998	0.9997	0.9994	0.9990	0.9987	0.9982
20.35	1.0000	1.0000	0.9999	0.9999	0.9998	0.9997	0.9994	0.9990	0.9987	0.9982
20.40	1.0000	1.0000	0.9999	0.9999	0.9998	0.9997	0.9994	0.9990	0.9988	0.9982
20.45	1.0000	1.0000	0.9999	0.9999	0.9998	0.9997	0.9994	0.9990	0.9988	0.9983
20.50	1.0000	1.0000	0.9999	0.9999	0.9998	0.9997	0.9994	0.9990	0.9988	0.9983
20.55	1.0000	1.0000	0.9999	0.9999	0.9998	0.9997	0.9994	0.9990	0.9988	0.9983
20.60	1.0000	1.0000	0.9999	0.9999	0.9998	0.9997	0.9994	0.9990	0.9988	0.9983
20.65	1.0000	1.0000	0.9999	0.9999	0.9998	0.9997	0.9994	0.9990	0.9988	0.9983
20.70	1.0000	1.0000	0.9999	0.9999	0.9998	0.9997	0.9994	0.9991	0.9988	0.9983
20.75	1.0000	1.0000	0.9999	0.9999	0.9998	0.9997	0.9994	0.9991	0.9988	0.9983
20.80	1.0000	1.0000	0.9999	0.9999	0.9998	0.9997	0.9994	0.9991	0.9989	0.9984
20.85	1.0000	1.0000	0.9999	0.9999	0.9998	0.9997	0.9994	0.9991	0.9989	0.9984
20.90	1.0000	1.0000	0.9999	0.9999	0.9998	0.9997	0.9994	0.9991	0.9989	0.9984
20.95	1.0000	1.0000	0.9999	0.9999	0.9998	0.9997	0.9994	0.9991	0.9989	0.9984

PROBABILITY INTEGRAL OF t, THE NON-CENTRAL t-STATISTIC. THIS TABLE GIVES Pr $[t/\sqrt{f} \leq x]$.

f is the number of degrees of freedom; the non-centrality parameter is $\sqrt{f+1}\, K_p$.

K_p is the standardized normal deviate exceeded with probability p.

x \ p	.2500	.1500	.1000	.0650	.0400	.0250	.0100	.0040	.0025	.0010
21.00	1.0000	1.0000	0.9999	0.9999	0.9998	0.9997	0.9994	0.9991	0.9989	0.9984
21.05	1.0000	1.0000	0.9999	0.9999	0.9998	0.9997	0.9995	0.9991	0.9989	0.9984
21.10	1.0000	1.0000	0.9999	0.9999	0.9998	0.9997	0.9995	0.9991	0.9989	0.9985
21.15	1.0000	1.0000	0.9999	0.9999	0.9998	0.9997	0.9995	0.9991	0.9989	0.9985
21.20	1.0000	1.0000	0.9999	0.9999	0.9998	0.9997	0.9995	0.9991	0.9989	0.9985
21.25	1.0000	1.0000	0.9999	0.9999	0.9998	0.9997	0.9995	0.9991	0.9989	0.9985
21.30	1.0000	1.0000	0.9999	0.9999	0.9998	0.9997	0.9995	0.9992	0.9990	0.9985
21.35	1.0000	1.0000	0.9999	0.9999	0.9998	0.9997	0.9995	0.9992	0.9990	0.9985
21.40	1.0000	1.0000	0.9999	0.9999	0.9998	0.9997	0.9995	0.9992	0.9990	0.9985
21.45	1.0000	1.0000	0.9999	0.9999	0.9998	0.9997	0.9995	0.9992	0.9990	0.9985
21.50	1.0000	1.0000	0.9999	0.9999	0.9998	0.9997	0.9995	0.9992	0.9990	0.9986
21.55	1.0000	1.0000	0.9999	0.9999	0.9998	0.9997	0.9995	0.9992	0.9990	0.9986
21.60	1.0000	1.0000	0.9999	0.9999	0.9998	0.9997	0.9995	0.9992	0.9990	0.9986
21.65	1.0000	1.0000	0.9999	0.9999	0.9998	0.9997	0.9995	0.9992	0.9990	0.9986
21.70	1.0000	1.0000	0.9999	0.9999	0.9998	0.9997	0.9995	0.9992	0.9990	0.9986
21.75	1.0000	1.0000	0.9999	0.9999	0.9998	0.9997	0.9995	0.9992	0.9990	0.9986
21.80	1.0000	1.0000	0.9999	0.9999	0.9998	0.9997	0.9995	0.9992	0.9990	0.9986
21.85	1.0000	1.0000	0.9999	0.9999	0.9998	0.9997	0.9995	0.9992	0.9991	0.9987
21.90	1.0000	1.0000	0.9999	0.9999	0.9998	0.9997	0.9995	0.9992	0.9991	0.9987
21.95	1.0000	1.0000	0.9999	0.9999	0.9998	0.9997	0.9995	0.9992	0.9991	0.9987
22.00	1.0000	1.0000	0.9999	0.9999	0.9998	0.9997	0.9995	0.9993	0.9991	0.9987
22.05	1.0000	1.0000	0.9999	0.9999	0.9998	0.9998	0.9995	0.9993	0.9991	0.9987
22.10	1.0000	1.0000	0.9999	0.9999	0.9998	0.9998	0.9995	0.9993	0.9991	0.9987
22.15	1.0000	1.0000	0.9999	0.9999	0.9998	0.9998	0.9996	0.9993	0.9991	0.9987
22.20	1.0000	1.0000	0.9999	0.9999	0.9998	0.9998	0.9996	0.9993	0.9991	0.9987
22.25	1.0000	1.0000	0.9999	0.9999	0.9998	0.9998	0.9996	0.9993	0.9991	0.9987
22.30	1.0000	1.0000	0.9999	0.9999	0.9998	0.9998	0.9996	0.9993	0.9991	0.9988
22.35	1.0000	1.0000	0.9999	0.9999	0.9998	0.9998	0.9996	0.9993	0.9991	0.9988
22.40	1.0000	1.0000	0.9999	0.9999	0.9998	0.9998	0.9996	0.9993	0.9991	0.9988
22.45	1.0000	1.0000	0.9999	0.9999	0.9998	0.9998	0.9996	0.9993	0.9992	0.9988
22.50	1.0000	1.0000	0.9999	0.9999	0.9998	0.9998	0.9996	0.9993	0.9992	0.9988
22.55	1.0000	1.0000	0.9999	0.9999	0.9998	0.9998	0.9996	0.9993	0.9992	0.9988
22.60	1.0000	1.0000	0.9999	0.9999	0.9998	0.9998	0.9996	0.9993	0.9992	0.9988
22.65	1.0000	1.0000	0.9999	0.9999	0.9999	0.9998	0.9996	0.9993	0.9992	0.9988
22.70	1.0000	1.0000	0.9999	0.9999	0.9999	0.9998	0.9996	0.9993	0.9992	0.9988
22.75	1.0000	1.0000	0.9999	0.9999	0.9999	0.9998	0.9996	0.9993	0.9992	0.9989
22.80	1.0000	1.0000	0.9999	0.9999	0.9999	0.9998	0.9996	0.9994	0.9992	0.9989
22.85	1.0000	1.0000	0.9999	0.9999	0.9999	0.9998	0.9996	0.9994	0.9992	0.9989
22.90	1.0000	1.0000	0.9999	0.9999	0.9999	0.9998	0.9996	0.9994	0.9992	0.9989
22.95	1.0000	1.0000	0.9999	0.9999	0.9999	0.9998	0.9996	0.9994	0.9992	0.9989
23.00	1.0000	1.0000	0.9999	0.9999	0.9999	0.9998	0.9996	0.9994	0.9992	0.9989
23.05	1.0000	1.0000	0.9999	0.9999	0.9999	0.9998	0.9996	0.9994	0.9992	0.9989
23.10	1.0000	1.0000	1.0000	0.9999	0.9999	0.9998	0.9996	0.9994	0.9992	0.9989
23.15	1.0000	1.0000	1.0000	0.9999	0.9999	0.9998	0.9996	0.9994	0.9992	0.9989
23.20	1.0000	1.0000	1.0000	0.9999	0.9999	0.9998	0.9996	0.9994	0.9993	0.9989
23.25	1.0000	1.0000	1.0000	0.9999	0.9999	0.9998	0.9996	0.9994	0.9993	0.9989
23.30	1.0000	1.0000	1.0000	0.9999	0.9999	0.9998	0.9996	0.9994	0.9993	0.9990
23.35	1.0000	1.0000	1.0000	0.9999	0.9999	0.9998	0.9996	0.9994	0.9993	0.9990
23.40	1.0000	1.0000	1.0000	0.9999	0.9999	0.9998	0.9996	0.9994	0.9993	0.9990
23.45	1.0000	1.0000	1.0000	0.9999	0.9999	0.9998	0.9996	0.9994	0.9993	0.9990
23.50	1.0000	1.0000	1.0000	0.9999	0.9999	0.9998	0.9996	0.9994	0.9993	0.9990
23.55	1.0000	1.0000	1.0000	0.9999	0.9999	0.9998	0.9996	0.9994	0.9993	0.9990
23.60	1.0000	1.0000	1.0000	0.9999	0.9999	0.9998	0.9997	0.9994	0.9993	0.9990
23.65	1.0000	1.0000	1.0000	0.9999	0.9999	0.9998	0.9997	0.9994	0.9993	0.9990
23.70	1.0000	1.0000	1.0000	0.9999	0.9999	0.9998	0.9997	0.9994	0.9993	0.9990
23.75	1.0000	1.0000	1.0000	0.9999	0.9999	0.9998	0.9997	0.9995	0.9993	0.9990
23.80	1.0000	1.0000	1.0000	0.9999	0.9999	0.9998	0.9997	0.9995	0.9993	0.9990
23.85	1.0000	1.0000	1.0000	0.9999	0.9999	0.9998	0.9997	0.9995	0.9993	0.9990
23.90	1.0000	1.0000	1.0000	0.9999	0.9999	0.9998	0.9997	0.9995	0.9993	0.9991
23.95	1.0000	1.0000	1.0000	0.9999	0.9999	0.9998	0.9997	0.9995	0.9993	0.9991

PROBABILITY INTEGRAL OF t, THE NON-CENTRAL t-STATISTIC. THIS TABLE GIVES Pr $[t/\sqrt{f} \leq x]$.

f is the number of degrees of freedom; the non-centrality parameter is $\sqrt{f+1}\,K_p$.

K_p is the standardized normal deviate exceeded with probability p.

DEGREES OF FREEDOM **4**

x \ p	.2500	.1500	.1000	.0650	.0400	.0250	.0100	.0040	.0025	.0010
24.00	1.0000	1.0000	1.0000	0.9999	0.9999	0.9998	0.9997	0.9995	0.9994	0.9991
24.05	1.0000	1.0000	1.0000	0.9999	0.9999	0.9998	0.9997	0.9995	0.9994	0.9991
24.10	1.0000	1.0000	1.0000	0.9999	0.9999	0.9998	0.9997	0.9995	0.9994	0.9991
24.15	1.0000	1.0000	1.0000	0.9999	0.9999	0.9998	0.9997	0.9995	0.9994	0.9991
24.20	1.0000	1.0000	1.0000	0.9999	0.9999	0.9998	0.9997	0.9995	0.9994	0.9991
24.25	1.0000	1.0000	1.0000	0.9999	0.9999	0.9998	0.9997	0.9995	0.9994	0.9991
24.30	1.0000	1.0000	1.0000	0.9999	0.9999	0.9998	0.9997	0.9995	0.9994	0.9991
24.35	1.0000	1.0000	1.0000	0.9999	0.9999	0.9998	0.9997	0.9995	0.9994	0.9991
24.40	1.0000	1.0000	1.0000	0.9999	0.9999	0.9998	0.9997	0.9995	0.9994	0.9991
24.45	1.0000	1.0000	1.0000	0.9999	0.9999	0.9998	0.9997	0.9995	0.9994	0.9991
24.50	1.0000	1.0000	1.0000	0.9999	0.9999	0.9998	0.9997	0.9995	0.9994	0.9991
24.55	1.0000	1.0000	1.0000	0.9999	0.9999	0.9998	0.9997	0.9995	0.9994	0.9991
24.60	1.0000	1.0000	1.0000	0.9999	0.9999	0.9998	0.9997	0.9995	0.9994	0.9992
24.65	1.0000	1.0000	1.0000	0.9999	0.9999	0.9998	0.9997	0.9995	0.9994	0.9992
24.70	1.0000	1.0000	1.0000	0.9999	0.9999	0.9998	0.9997	0.9995	0.9994	0.9992
24.75	1.0000	1.0000	1.0000	0.9999	0.9999	0.9998	0.9997	0.9995	0.9994	0.9992
24.80	1.0000	1.0000	1.0000	0.9999	0.9999	0.9998	0.9997	0.9995	0.9994	0.9992
24.85	1.0000	1.0000	1.0000	0.9999	0.9999	0.9998	0.9997	0.9995	0.9994	0.9992
24.90	1.0000	1.0000	1.0000	0.9999	0.9999	0.9998	0.9997	0.9995	0.9994	0.9992
24.95	1.0000	1.0000	1.0000	0.9999	0.9999	0.9998	0.9997	0.9995	0.9994	0.9992
25.00	1.0000	1.0000	1.0000	0.9999	0.9999	0.9999	0.9997	0.9996	0.9994	0.9992
25.05	1.0000	1.0000	1.0000	0.9999	0.9999	0.9999	0.9997	0.9996	0.9995	0.9992
25.10	1.0000	1.0000	1.0000	0.9999	0.9999	0.9999	0.9997	0.9996	0.9995	0.9992
25.15	1.0000	1.0000	1.0000	0.9999	0.9999	0.9999	0.9997	0.9996	0.9995	0.9992
25.20	1.0000	1.0000	1.0000	0.9999	0.9999	0.9999	0.9997	0.9996	0.9995	0.9992
25.25	1.0000	1.0000	1.0000	0.9999	0.9999	0.9999	0.9997	0.9996	0.9995	0.9992
25.30	1.0000	1.0000	1.0000	0.9999	0.9999	0.9999	0.9997	0.9996	0.9995	0.9992
25.35	1.0000	1.0000	1.0000	0.9999	0.9999	0.9999	0.9997	0.9996	0.9995	0.9993
25.40	1.0000	1.0000	1.0000	0.9999	0.9999	0.9999	0.9997	0.9996	0.9995	0.9993
25.45	1.0000	1.0000	1.0000	0.9999	0.9999	0.9999	0.9997	0.9996	0.9995	0.9993
25.50	1.0000	1.0000	1.0000	0.9999	0.9999	0.9999	0.9997	0.9996	0.9995	0.9993
25.55	1.0000	1.0000	1.0000	0.9999	0.9999	0.9999	0.9997	0.9996	0.9995	0.9993
25.60	1.0000	1.0000	1.0000	0.9999	0.9999	0.9999	0.9997	0.9996	0.9995	0.9993
25.65	1.0000	1.0000	1.0000	0.9999	0.9999	0.9999	0.9998	0.9996	0.9995	0.9993
25.70	1.0000	1.0000	1.0000	0.9999	0.9999	0.9999	0.9998	0.9996	0.9995	0.9993
25.75	1.0000	1.0000	1.0000	0.9999	0.9999	0.9999	0.9998	0.9996	0.9995	0.9993
25.80	1.0000	1.0000	1.0000	0.9999	0.9999	0.9999	0.9998	0.9996	0.9995	0.9993
25.85	1.0000	1.0000	1.0000	0.9999	0.9999	0.9999	0.9998	0.9996	0.9995	0.9993
25.90	1.0000	1.0000	1.0000	0.9999	0.9999	0.9999	0.9998	0.9996	0.9995	0.9993
25.95	1.0000	1.0000	1.0000	0.9999	0.9999	0.9999	0.9998	0.9996	0.9995	0.9993
26.00	1.0000	1.0000	1.0000	0.9999	0.9999	0.9999	0.9998	0.9996	0.9995	0.9993
26.05	1.0000	1.0000	1.0000	0.9999	0.9999	0.9999	0.9998	0.9996	0.9995	0.9993
26.10	1.0000	1.0000	1.0000	0.9999	0.9999	0.9999	0.9998	0.9996	0.9995	0.9993
26.15	1.0000	1.0000	1.0000	1.0000	0.9999	0.9999	0.9998	0.9996	0.9995	0.9993
26.20	1.0000	1.0000	1.0000	1.0000	0.9999	0.9999	0.9998	0.9996	0.9995	0.9993
26.25	1.0000	1.0000	1.0000	1.0000	0.9999	0.9999	0.9998	0.9996	0.9995	0.9993
26.30	1.0000	1.0000	1.0000	1.0000	0.9999	0.9999	0.9998	0.9996	0.9995	0.9994
26.35	1.0000	1.0000	1.0000	1.0000	0.9999	0.9999	0.9998	0.9996	0.9996	0.9994
26.40	1.0000	1.0000	1.0000	1.0000	0.9999	0.9999	0.9998	0.9996	0.9996	0.9994
26.45	1.0000	1.0000	1.0000	1.0000	0.9999	0.9999	0.9998	0.9996	0.9996	0.9994
26.50	1.0000	1.0000	1.0000	1.0000	0.9999	0.9999	0.9998	0.9996	0.9996	0.9994
26.55	1.0000	1.0000	1.0000	1.0000	0.9999	0.9999	0.9998	0.9996	0.9996	0.9994
26.60	1.0000	1.0000	1.0000	1.0000	0.9999	0.9999	0.9998	0.9997	0.9996	0.9994
26.65	1.0000	1.0000	1.0000	1.0000	0.9999	0.9999	0.9998	0.9997	0.9996	0.9994
26.70	1.0000	1.0000	1.0000	1.0000	0.9999	0.9999	0.9998	0.9997	0.9996	0.9994
26.75	1.0000	1.0000	1.0000	1.0000	0.9999	0.9999	0.9998	0.9997	0.9996	0.9994
26.80	1.0000	1.0000	1.0000	1.0000	0.9999	0.9999	0.9998	0.9997	0.9996	0.9994
26.85	1.0000	1.0000	1.0000	1.0000	0.9999	0.9999	0.9998	0.9997	0.9996	0.9994
26.90	1.0000	1.0000	1.0000	1.0000	0.9999	0.9999	0.9998	0.9997	0.9996	0.9994
26.95	1.0000	1.0000	1.0000	1.0000	0.9999	0.9999	0.9998	0.9997	0.9996	0.9994

PROBABILITY INTEGRAL OF t, THE NON-CENTRAL t-STATISTIC. THIS TABLE GIVES Pr $[t/\sqrt{f} \leq x]$.

f is the number of degrees of freedom; the non-centrality parameter is $\sqrt{f+1}\,K_p$.
K_p is the standardized normal deviate exceeded with probability p.

DEGREES OF FREEDOM **4**

x \ p	.2500	.1500	.1000	.0650	.0400	.0250	.0100	.0040	.0025	.0010
27.00	1.0000	1.0000	1.0000	1.0000	0.9999	0.9999	0.9998	0.9997	0.9996	0.9994
27.05	1.0000	1.0000	1.0000	1.0000	0.9999	0.9999	0.9998	0.9997	0.9996	0.9994
27.10	1.0000	1.0000	1.0000	1.0000	0.9999	0.9999	0.9998	0.9997	0.9996	0.9994
27.15	1.0000	1.0000	1.0000	1.0000	0.9999	0.9999	0.9998	0.9997	0.9996	0.9994
27.20	1.0000	1.0000	1.0000	1.0000	0.9999	0.9999	0.9998	0.9997	0.9996	0.9994
27.25	1.0000	1.0000	1.0000	1.0000	0.9999	0.9999	0.9998	0.9997	0.9996	0.9994
27.30	1.0000	1.0000	1.0000	1.0000	0.9999	0.9999	0.9998	0.9997	0.9996	0.9994
27.35	1.0000	1.0000	1.0000	1.0000	0.9999	0.9999	0.9998	0.9997	0.9996	0.9994
27.40	1.0000	1.0000	1.0000	1.0000	0.9999	0.9999	0.9998	0.9997	0.9996	0.9995
27.45	1.0000	1.0000	1.0000	1.0000	0.9999	0.9999	0.9998	0.9997	0.9996	0.9995
27.50	1.0000	1.0000	1.0000	1.0000	0.9999	0.9999	0.9998	0.9997	0.9996	0.9995
27.55	1.0000	1.0000	1.0000	1.0000	0.9999	0.9999	0.9998	0.9997	0.9996	0.9995
27.60	1.0000	1.0000	1.0000	1.0000	0.9999	0.9999	0.9998	0.9997	0.9996	0.9995
27.65	1.0000	1.0000	1.0000	1.0000	0.9999	0.9999	0.9998	0.9997	0.9996	0.9995
27.70	1.0000	1.0000	1.0000	1.0000	0.9999	0.9999	0.9998	0.9997	0.9996	0.9995
27.75	1.0000	1.0000	1.0000	1.0000	0.9999	0.9999	0.9998	0.9997	0.9996	0.9995
27.80	1.0000	1.0000	1.0000	1.0000	0.9999	0.9999	0.9998	0.9997	0.9996	0.9995
27.85	1.0000	1.0000	1.0000	1.0000	0.9999	0.9999	0.9998	0.9997	0.9996	0.9995
27.90	1.0000	1.0000	1.0000	1.0000	0.9999	0.9999	0.9998	0.9997	0.9996	0.9995
27.95	1.0000	1.0000	1.0000	1.0000	0.9999	0.9999	0.9998	0.9997	0.9996	0.9995
28.00	1.0000	1.0000	1.0000	1.0000	0.9999	0.9999	0.9998	0.9997	0.9997	0.9995
28.05	1.0000	1.0000	1.0000	1.0000	0.9999	0.9999	0.9998	0.9997	0.9997	0.9995
28.10	1.0000	1.0000	1.0000	1.0000	0.9999	0.9999	0.9998	0.9997	0.9997	0.9995
28.15	1.0000	1.0000	1.0000	1.0000	0.9999	0.9999	0.9998	0.9997	0.9997	0.9995
28.20	1.0000	1.0000	1.0000	1.0000	0.9999	0.9999	0.9998	0.9997	0.9997	0.9995
28.25	1.0000	1.0000	1.0000	1.0000	0.9999	0.9999	0.9998	0.9997	0.9997	0.9995
28.30	1.0000	1.0000	1.0000	1.0000	0.9999	0.9999	0.9998	0.9997	0.9997	0.9995
28.35	1.0000	1.0000	1.0000	1.0000	0.9999	0.9999	0.9998	0.9997	0.9997	0.9995
28.40	1.0000	1.0000	1.0000	1.0000	0.9999	0.9999	0.9998	0.9997	0.9997	0.9995
28.45	1.0000	1.0000	1.0000	1.0000	0.9999	0.9999	0.9998	0.9997	0.9997	0.9995
28.50	1.0000	1.0000	1.0000	1.0000	0.9999	0.9999	0.9998	0.9997	0.9997	0.9995
28.55	1.0000	1.0000	1.0000	1.0000	0.9999	0.9999	0.9998	0.9997	0.9997	0.9995
28.60	1.0000	1.0000	1.0000	1.0000	0.9999	0.9999	0.9998	0.9997	0.9997	0.9995
28.65	1.0000	1.0000	1.0000	1.0000	0.9999	0.9999	0.9998	0.9997	0.9997	0.9995
28.70	1.0000	1.0000	1.0000	1.0000	0.9999	0.9999	0.9998	0.9997	0.9997	0.9995
28.75	1.0000	1.0000	1.0000	1.0000	0.9999	0.9999	0.9998	0.9997	0.9997	0.9995
28.80	1.0000	1.0000	1.0000	1.0000	0.9999	0.9999	0.9998	0.9997	0.9997	0.9995
28.85	1.0000	1.0000	1.0000	1.0000	0.9999	0.9999	0.9998	0.9997	0.9997	0.9996
28.90	1.0000	1.0000	1.0000	1.0000	0.9999	0.9999	0.9998	0.9998	0.9997	0.9996
28.95	1.0000	1.0000	1.0000	1.0000	0.9999	0.9999	0.9998	0.9998	0.9997	0.9996
29.00	1.0000	1.0000	1.0000	1.0000	0.9999	0.9999	0.9998	0.9998	0.9997	0.9996
29.05	1.0000	1.0000	1.0000	1.0000	0.9999	0.9999	0.9999	0.9998	0.9997	0.9996
29.10	1.0000	1.0000	1.0000	1.0000	0.9999	0.9999	0.9999	0.9998	0.9997	0.9996
29.15	1.0000	1.0000	1.0000	1.0000	0.9999	0.9999	0.9999	0.9998	0.9997	0.9996
29.20	1.0000	1.0000	1.0000	1.0000	0.9999	0.9999	0.9999	0.9998	0.9997	0.9996
29.25	1.0000	1.0000	1.0000	1.0000	0.9999	0.9999	0.9999	0.9998	0.9997	0.9996
29.30	1.0000	1.0000	1.0000	1.0000	0.9999	0.9999	0.9999	0.9998	0.9997	0.9996
29.35	1.0000	1.0000	1.0000	1.0000	0.9999	0.9999	0.9999	0.9998	0.9997	0.9996
29.40	1.0000	1.0000	1.0000	1.0000	0.9999	0.9999	0.9999	0.9998	0.9997	0.9996
29.45	1.0000	1.0000	1.0000	1.0000	1.0000	0.9999	0.9999	0.9998	0.9997	0.9996
29.50	1.0000	1.0000	1.0000	1.0000	1.0000	0.9999	0.9999	0.9998	0.9997	0.9996
29.55	1.0000	1.0000	1.0000	1.0000	1.0000	0.9999	0.9999	0.9998	0.9997	0.9996
29.60	1.0000	1.0000	1.0000	1.0000	1.0000	0.9999	0.9999	0.9998	0.9997	0.9996
29.65	1.0000	1.0000	1.0000	1.0000	1.0000	0.9999	0.9999	0.9998	0.9997	0.9996
29.70	1.0000	1.0000	1.0000	1.0000	1.0000	0.9999	0.9999	0.9998	0.9997	0.9996
29.75	1.0000	1.0000	1.0000	1.0000	1.0000	0.9999	0.9999	0.9998	0.9997	0.9996
29.80	1.0000	1.0000	1.0000	1.0000	1.0000	0.9999	0.9999	0.9998	0.9997	0.9996
29.85	1.0000	1.0000	1.0000	1.0000	1.0000	0.9999	0.9999	0.9998	0.9997	0.9996
29.90	1.0000	1.0000	1.0000	1.0000	1.0000	0.9999	0.9999	0.9998	0.9997	0.9996
29.95	1.0000	1.0000	1.0000	1.0000	1.0000	0.9999	0.9999	0.9998	0.9997	0.9996

PROBABILITY INTEGRAL OF t, THE NON-CENTRAL t-STATISTIC. THIS TABLE GIVES Pr $[t/\sqrt{f} \leq x]$.

f is the number of degrees of freedom; the non-centrality parameter is $\sqrt{f+1}\,K_p$.

K_p is the standardized normal deviate exceeded with probability p.

DEGREES OF FREEDOM **4**

x \ P	.2500	.1500	.1000	.0650	.0400	.0250	.0100	.0040	.0025	.0010
30.00	1.0000	1.0000	1.0000	1.0000	1.0000	0.9999	0.9999	0.9998	0.9997	0.9996
30.05	1.0000	1.0000	1.0000	1.0000	1.0000	0.9999	0.9999	0.9998	0.9997	0.9996
30.10	1.0000	1.0000	1.0000	1.0000	1.0000	0.9999	0.9999	0.9998	0.9997	0.9996
30.15	1.0000	1.0000	1.0000	1.0000	1.0000	0.9999	0.9999	0.9998	0.9997	0.9996
30.20	1.0000	1.0000	1.0000	1.0000	1.0000	0.9999	0.9999	0.9998	0.9997	0.9996
30.25	1.0000	1.0000	1.0000	1.0000	1.0000	0.9999	0.9999	0.9998	0.9997	0.9996
30.30	1.0000	1.0000	1.0000	1.0000	1.0000	0.9999	0.9999	0.9998	0.9997	0.9996
30.35	1.0000	1.0000	1.0000	1.0000	1.0000	0.9999	0.9999	0.9998	0.9997	0.9996
30.40	1.0000	1.0000	1.0000	1.0000	1.0000	0.9999	0.9999	0.9998	0.9997	0.9996
30.45	1.0000	1.0000	1.0000	1.0000	1.0000	0.9999	0.9999	0.9998	0.9998	0.9996
30.50	1.0000	1.0000	1.0000	1.0000	1.0000	0.9999	0.9999	0.9998	0.9998	0.9996
30.55	1.0000	1.0000	1.0000	1.0000	1.0000	0.9999	0.9999	0.9998	0.9998	0.9996
30.60	1.0000	1.0000	1.0000	1.0000	1.0000	0.9999	0.9999	0.9998	0.9998	0.9996
30.65	1.0000	1.0000	1.0000	1.0000	1.0000	0.9999	0.9999	0.9998	0.9998	0.9996
30.70	1.0000	1.0000	1.0000	1.0000	1.0000	0.9999	0.9999	0.9998	0.9998	0.9997
30.75	1.0000	1.0000	1.0000	1.0000	1.0000	0.9999	0.9999	0.9998	0.9998	0.9997
30.80	1.0000	1.0000	1.0000	1.0000	1.0000	0.9999	0.9999	0.9998	0.9998	0.9997
30.85	1.0000	1.0000	1.0000	1.0000	1.0000	0.9999	0.9999	0.9998	0.9998	0.9997
30.90	1.0000	1.0000	1.0000	1.0000	1.0000	0.9999	0.9999	0.9998	0.9998	0.9997
30.95	1.0000	1.0000	1.0000	1.0000	1.0000	0.9999	0.9999	0.9998	0.9998	0.9997
31.00	1.0000	1.0000	1.0000	1.0000	1.0000	0.9999	0.9999	0.9998	0.9998	0.9997
31.05	1.0000	1.0000	1.0000	1.0000	1.0000	0.9999	0.9999	0.9998	0.9998	0.9997
31.10	1.0000	1.0000	1.0000	1.0000	1.0000	0.9999	0.9999	0.9998	0.9998	0.9997
31.15	1.0000	1.0000	1.0000	1.0000	1.0000	0.9999	0.9999	0.9998	0.9998	0.9997
31.20	1.0000	1.0000	1.0000	1.0000	1.0000	0.9999	0.9999	0.9998	0.9998	0.9997
31.25	1.0000	1.0000	1.0000	1.0000	1.0000	0.9999	0.9999	0.9998	0.9998	0.9997
31.30	1.0000	1.0000	1.0000	1.0000	1.0000	0.9999	0.9999	0.9998	0.9998	0.9997
31.35	1.0000	1.0000	1.0000	1.0000	1.0000	0.9999	0.9999	0.9998	0.9998	0.9997
31.40	1.0000	1.0000	1.0000	1.0000	1.0000	0.9999	0.9999	0.9998	0.9998	0.9997
31.45	1.0000	1.0000	1.0000	1.0000	1.0000	0.9999	0.9999	0.9998	0.9998	0.9997
31.50	1.0000	1.0000	1.0000	1.0000	1.0000	0.9999	0.9999	0.9998	0.9998	0.9997
31.55	1.0000	1.0000	1.0000	1.0000	1.0000	0.9999	0.9999	0.9998	0.9998	0.9997
31.60	1.0000	1.0000	1.0000	1.0000	1.0000	0.9999	0.9999	0.9998	0.9998	0.9997
31.65	1.0000	1.0000	1.0000	1.0000	1.0000	0.9999	0.9999	0.9998	0.9998	0.9997
31.70	1.0000	1.0000	1.0000	1.0000	1.0000	0.9999	0.9999	0.9998	0.9998	0.9997
31.75	1.0000	1.0000	1.0000	1.0000	1.0000	0.9999	0.9999	0.9998	0.9998	0.9997
31.80	1.0000	1.0000	1.0000	1.0000	1.0000	0.9999	0.9999	0.9998	0.9998	0.9997
31.85	1.0000	1.0000	1.0000	1.0000	1.0000	0.9999	0.9999	0.9998	0.9998	0.9997
31.90	1.0000	1.0000	1.0000	1.0000	1.0000	0.9999	0.9999	0.9998	0.9998	0.9997
31.95	1.0000	1.0000	1.0000	1.0000	1.0000	0.9999	0.9999	0.9998	0.9998	0.9997
32.00	1.0000	1.0000	1.0000	1.0000	1.0000	0.9999	0.9999	0.9998	0.9998	0.9997
32.05	1.0000	1.0000	1.0000	1.0000	1.0000	0.9999	0.9999	0.9998	0.9998	0.9997
32.10	1.0000	1.0000	1.0000	1.0000	1.0000	0.9999	0.9999	0.9998	0.9998	0.9997
32.15	1.0000	1.0000	1.0000	1.0000	1.0000	0.9999	0.9999	0.9998	0.9998	0.9997
32.20	1.0000	1.0000	1.0000	1.0000	1.0000	0.9999	0.9999	0.9998	0.9998	0.9997
32.25	1.0000	1.0000	1.0000	1.0000	1.0000	0.9999	0.9999	0.9998	0.9998	0.9997
32.30	1.0000	1.0000	1.0000	1.0000	1.0000	0.9999	0.9999	0.9998	0.9998	0.9997
32.35	1.0000	1.0000	1.0000	1.0000	1.0000	0.9999	0.9999	0.9998	0.9998	0.9997
32.40	1.0000	1.0000	1.0000	1.0000	1.0000	0.9999	0.9999	0.9998	0.9998	0.9997
32.45	1.0000	1.0000	1.0000	1.0000	1.0000	0.9999	0.9999	0.9998	0.9998	0.9997
32.50	1.0000	1.0000	1.0000	1.0000	1.0000	1.0000	0.9999	0.9998	0.9998	0.9997
32.55	1.0000	1.0000	1.0000	1.0000	1.0000	1.0000	0.9999	0.9998	0.9998	0.9997
32.60	1.0000	1.0000	1.0000	1.0000	1.0000	1.0000	0.9999	0.9998	0.9998	0.9997
32.65	1.0000	1.0000	1.0000	1.0000	1.0000	1.0000	0.9999	0.9998	0.9998	0.9997
32.70	1.0000	1.0000	1.0000	1.0000	1.0000	1.0000	0.9999	0.9999	0.9998	0.9997
32.75	1.0000	1.0000	1.0000	1.0000	1.0000	1.0000	0.9999	0.9999	0.9998	0.9997
32.80	1.0000	1.0000	1.0000	1.0000	1.0000	1.0000	0.9999	0.9999	0.9998	0.9997
32.85	1.0000	1.0000	1.0000	1.0000	1.0000	1.0000	0.9999	0.9999	0.9998	0.9997
32.90	1.0000	1.0000	1.0000	1.0000	1.0000	1.0000	0.9999	0.9999	0.9998	0.9997
32.95	1.0000	1.0000	1.0000	1.0000	1.0000	1.0000	0.9999	0.9999	0.9998	0.9997

PROBABILITY INTEGRAL OF t, THE NON-CENTRAL t-STATISTIC. THIS TABLE GIVES Pr $[t/\sqrt{f} \leq x]$.

f is the number of degrees of freedom; the non-centrality parameter is $\sqrt{f+1}\,K_p$.

K_p is the standardized normal deviate exceeded with probability p.

DEGREES OF FREEDOM **4**

x \ p	.2500	.1500	.1000	.0650	.0400	.0250	.0100	.0040	.0025	.0010
33.00	1.0000	1.0000	1.0000	1.0000	1.0000	1.0000	0.9999	0.9999	0.9998	0.9997
33.05	1.0000	1.0000	1.0000	1.0000	1.0000	1.0000	0.9999	0.9999	0.9998	0.9997
33.10	1.0000	1.0000	1.0000	1.0000	1.0000	1.0000	0.9999	0.9999	0.9998	0.9997
33.15	1.0000	1.0000	1.0000	1.0000	1.0000	1.0000	0.9999	0.9999	0.9998	0.9997
33.20	1.0000	1.0000	1.0000	1.0000	1.0000	1.0000	0.9999	0.9999	0.9998	0.9997
33.25	1.0000	1.0000	1.0000	1.0000	1.0000	1.0000	0.9999	0.9999	0.9998	0.9997
33.30	1.0000	1.0000	1.0000	1.0000	1.0000	1.0000	0.9999	0.9999	0.9998	0.9997
33.35	1.0000	1.0000	1.0000	1.0000	1.0000	1.0000	0.9999	0.9999	0.9998	0.9998
33.40	1.0000	1.0000	1.0000	1.0000	1.0000	1.0000	0.9999	0.9999	0.9998	0.9998
33.45	1.0000	1.0000	1.0000	1.0000	1.0000	1.0000	0.9999	0.9999	0.9998	0.9998
33.50	1.0000	1.0000	1.0000	1.0000	1.0000	1.0000	0.9999	0.9999	0.9998	0.9998
33.55	1.0000	1.0000	1.0000	1.0000	1.0000	1.0000	0.9999	0.9999	0.9998	0.9998
33.60	1.0000	1.0000	1.0000	1.0000	1.0000	1.0000	0.9999	0.9999	0.9998	0.9998
33.65	1.0000	1.0000	1.0000	1.0000	1.0000	1.0000	0.9999	0.9999	0.9998	0.9998
33.70	1.0000	1.0000	1.0000	1.0000	1.0000	1.0000	0.9999	0.9999	0.9998	0.9998
33.75	1.0000	1.0000	1.0000	1.0000	1.0000	1.0000	0.9999	0.9999	0.9998	0.9998
33.80	1.0000	1.0000	1.0000	1.0000	1.0000	1.0000	0.9999	0.9999	0.9998	0.9998
33.85	1.0000	1.0000	1.0000	1.0000	1.0000	1.0000	0.9999	0.9999	0.9998	0.9998
33.90	1.0000	1.0000	1.0000	1.0000	1.0000	1.0000	0.9999	0.9999	0.9998	0.9998
33.95	1.0000	1.0000	1.0000	1.0000	1.0000	1.0000	0.9999	0.9999	0.9998	0.9998
34.00	1.0000	1.0000	1.0000	1.0000	1.0000	1.0000	0.9999	0.9999	0.9998	0.9998
34.05	1.0000	1.0000	1.0000	1.0000	1.0000	1.0000	0.9999	0.9999	0.9998	0.9998
34.10	1.0000	1.0000	1.0000	1.0000	1.0000	1.0000	0.9999	0.9999	0.9998	0.9998
34.15	1.0000	1.0000	1.0000	1.0000	1.0000	1.0000	0.9999	0.9999	0.9998	0.9998
34.20	1.0000	1.0000	1.0000	1.0000	1.0000	1.0000	0.9999	0.9999	0.9998	0.9998
34.25	1.0000	1.0000	1.0000	1.0000	1.0000	1.0000	0.9999	0.9999	0.9998	0.9998
34.30	1.0000	1.0000	1.0000	1.0000	1.0000	1.0000	0.9999	0.9999	0.9998	0.9998
34.35	1.0000	1.0000	1.0000	1.0000	1.0000	1.0000	0.9999	0.9999	0.9998	0.9998
34.40	1.0000	1.0000	1.0000	1.0000	1.0000	1.0000	0.9999	0.9999	0.9998	0.9998
34.45	1.0000	1.0000	1.0000	1.0000	1.0000	1.0000	0.9999	0.9999	0.9998	0.9998
34.50	1.0000	1.0000	1.0000	1.0000	1.0000	1.0000	0.9999	0.9999	0.9999	0.9998
34.55	1.0000	1.0000	1.0000	1.0000	1.0000	1.0000	0.9999	0.9999	0.9999	0.9998
34.60	1.0000	1.0000	1.0000	1.0000	1.0000	1.0000	0.9999	0.9999	0.9999	0.9998
34.65	1.0000	1.0000	1.0000	1.0000	1.0000	1.0000	0.9999	0.9999	0.9999	0.9998
34.70	1.0000	1.0000	1.0000	1.0000	1.0000	1.0000	0.9999	0.9999	0.9999	0.9998
34.75	1.0000	1.0000	1.0000	1.0000	1.0000	1.0000	0.9999	0.9999	0.9999	0.9998
34.80	1.0000	1.0000	1.0000	1.0000	1.0000	1.0000	0.9999	0.9999	0.9999	0.9998
34.85	1.0000	1.0000	1.0000	1.0000	1.0000	1.0000	0.9999	0.9999	0.9999	0.9998
34.90	1.0000	1.0000	1.0000	1.0000	1.0000	1.0000	0.9999	0.9999	0.9999	0.9998
34.95	1.0000	1.0000	1.0000	1.0000	1.0000	1.0000	0.9999	0.9999	0.9999	0.9998
35.00	1.0000	1.0000	1.0000	1.0000	1.0000	1.0000	0.9999	0.9999	0.9999	0.9998
35.05	1.0000	1.0000	1.0000	1.0000	1.0000	1.0000	0.9999	0.9999	0.9999	0.9998
35.10	1.0000	1.0000	1.0000	1.0000	1.0000	1.0000	0.9999	0.9999	0.9999	0.9998
35.15	1.0000	1.0000	1.0000	1.0000	1.0000	1.0000	0.9999	0.9999	0.9999	0.9998
35.20	1.0000	1.0000	1.0000	1.0000	1.0000	1.0000	0.9999	0.9999	0.9999	0.9998
35.25	1.0000	1.0000	1.0000	1.0000	1.0000	1.0000	0.9999	0.9999	0.9999	0.9998
35.30	1.0000	1.0000	1.0000	1.0000	1.0000	1.0000	0.9999	0.9999	0.9999	0.9998
35.35	1.0000	1.0000	1.0000	1.0000	1.0000	1.0000	0.9999	0.9999	0.9999	0.9998
35.40	1.0000	1.0000	1.0000	1.0000	1.0000	1.0000	0.9999	0.9999	0.9999	0.9998
35.45	1.0000	1.0000	1.0000	1.0000	1.0000	1.0000	0.9999	0.9999	0.9999	0.9998
35.50	1.0000	1.0000	1.0000	1.0000	1.0000	1.0000	0.9999	0.9999	0.9999	0.9998
35.55	1.0000	1.0000	1.0000	1.0000	1.0000	1.0000	0.9999	0.9999	0.9999	0.9998
35.60	1.0000	1.0000	1.0000	1.0000	1.0000	1.0000	0.9999	0.9999	0.9999	0.9998
35.65	1.0000	1.0000	1.0000	1.0000	1.0000	1.0000	0.9999	0.9999	0.9999	0.9998
35.70	1.0000	1.0000	1.0000	1.0000	1.0000	1.0000	0.9999	0.9999	0.9999	0.9998
35.75	1.0000	1.0000	1.0000	1.0000	1.0000	1.0000	0.9999	0.9999	0.9999	0.9998
35.80	1.0000	1.0000	1.0000	1.0000	1.0000	1.0000	0.9999	0.9999	0.9999	0.9998
35.85	1.0000	1.0000	1.0000	1.0000	1.0000	1.0000	0.9999	0.9999	0.9999	0.9998
35.90	1.0000	1.0000	1.0000	1.0000	1.0000	1.0000	0.9999	0.9999	0.9999	0.9998
35.95	1.0000	1.0000	1.0000	1.0000	1.0000	1.0000	0.9999	0.9999	0.9999	0.9998

PROBABILITY INTEGRAL OF t, THE NON-CENTRAL t-STATISTIC. THIS TABLE GIVES Pr $[t/\sqrt{f} \leq x]$.

f is the number of degrees of freedom; the non-centrality parameter is $\sqrt{f+1}\,K_p$.

K_p is the standardized normal deviate exceeded with probability p.

DEGREES OF FREEDOM 4

x \ p	.2500	.1500	.1000	.0650	.0400	.0250	.0100	.0040	.0025	.0010
36.00	1.0000	1.0000	1.0000	1.0000	1.0000	1.0000	0.9999	0.9999	0.9999	0.9998
36.05	1.0000	1.0000	1.0000	1.0000	1.0000	1.0000	0.9999	0.9999	0.9999	0.9998
36.10	1.0000	1.0000	1.0000	1.0000	1.0000	1.0000	0.9999	0.9999	0.9999	0.9998
36.15	1.0000	1.0000	1.0000	1.0000	1.0000	1.0000	0.9999	0.9999	0.9999	0.9998
36.20	1.0000	1.0000	1.0000	1.0000	1.0000	1.0000	0.9999	0.9999	0.9999	0.9998
36.25	1.0000	1.0000	1.0000	1.0000	1.0000	1.0000	0.9999	0.9999	0.9999	0.9998
36.30	1.0000	1.0000	1.0000	1.0000	1.0000	1.0000	0.9999	0.9999	0.9999	0.9998
36.35	1.0000	1.0000	1.0000	1.0000	1.0000	1.0000	0.9999	0.9999	0.9999	0.9998
36.40	1.0000	1.0000	1.0000	1.0000	1.0000	1.0000	0.9999	0.9999	0.9999	0.9998
36.45	1.0000	1.0000	1.0000	1.0000	1.0000	1.0000	0.9999	0.9999	0.9999	0.9998
36.50	1.0000	1.0000	1.0000	1.0000	1.0000	1.0000	0.9999	0.9999	0.9999	0.9998
36.55	1.0000	1.0000	1.0000	1.0000	1.0000	1.0000	0.9999	0.9999	0.9999	0.9998
36.60	1.0000	1.0000	1.0000	1.0000	1.0000	1.0000	0.9999	0.9999	0.9999	0.9998
36.65	1.0000	1.0000	1.0000	1.0000	1.0000	1.0000	0.9999	0.9999	0.9999	0.9998
36.70	1.0000	1.0000	1.0000	1.0000	1.0000	1.0000	0.9999	0.9999	0.9999	0.9998
36.75	1.0000	1.0000	1.0000	1.0000	1.0000	1.0000	0.9999	0.9999	0.9999	0.9998
36.80	1.0000	1.0000	1.0000	1.0000	1.0000	1.0000	0.9999	0.9999	0.9999	0.9998
36.85	1.0000	1.0000	1.0000	1.0000	1.0000	1.0000	0.9999	0.9999	0.9999	0.9998
36.90	1.0000	1.0000	1.0000	1.0000	1.0000	1.0000	0.9999	0.9999	0.9999	0.9998
36.95	1.0000	1.0000	1.0000	1.0000	1.0000	1.0000	0.9999	0.9999	0.9999	0.9998
37.00	1.0000	1.0000	1.0000	1.0000	1.0000	1.0000	0.9999	0.9999	0.9999	0.9998
37.05	1.0000	1.0000	1.0000	1.0000	1.0000	1.0000	0.9999	0.9999	0.9999	0.9998
37.10	1.0000	1.0000	1.0000	1.0000	1.0000	1.0000	0.9999	0.9999	0.9999	0.9998
37.15	1.0000	1.0000	1.0000	1.0000	1.0000	1.0000	0.9999	0.9999	0.9999	0.9998
37.20	1.0000	1.0000	1.0000	1.0000	1.0000	1.0000	0.9999	0.9999	0.9999	0.9998
37.25	1.0000	1.0000	1.0000	1.0000	1.0000	1.0000	0.9999	0.9999	0.9999	0.9998
37.30	1.0000	1.0000	1.0000	1.0000	1.0000	1.0000	0.9999	0.9999	0.9999	0.9998
37.35	1.0000	1.0000	1.0000	1.0000	1.0000	1.0000	0.9999	0.9999	0.9999	0.9998
37.40	1.0000	1.0000	1.0000	1.0000	1.0000	1.0000	0.9999	0.9999	0.9999	0.9998
37.45	1.0000	1.0000	1.0000	1.0000	1.0000	1.0000	1.0000	0.9999	0.9999	0.9998
37.50	1.0000	1.0000	1.0000	1.0000	1.0000	1.0000	1.0000	0.9999	0.9999	0.9998
37.55	1.0000	1.0000	1.0000	1.0000	1.0000	1.0000	1.0000	0.9999	0.9999	0.9998
37.60	1.0000	1.0000	1.0000	1.0000	1.0000	1.0000	1.0000	0.9999	0.9999	0.9998
37.65	1.0000	1.0000	1.0000	1.0000	1.0000	1.0000	1.0000	0.9999	0.9999	0.9998
37.70	1.0000	1.0000	1.0000	1.0000	1.0000	1.0000	1.0000	0.9999	0.9999	0.9998
37.75	1.0000	1.0000	1.0000	1.0000	1.0000	1.0000	1.0000	0.9999	0.9999	0.9998
37.80	1.0000	1.0000	1.0000	1.0000	1.0000	1.0000	1.0000	0.9999	0.9999	0.9999
37.85	1.0000	1.0000	1.0000	1.0000	1.0000	1.0000	1.0000	0.9999	0.9999	0.9999
37.90	1.0000	1.0000	1.0000	1.0000	1.0000	1.0000	1.0000	0.9999	0.9999	0.9999
37.95	1.0000	1.0000	1.0000	1.0000	1.0000	1.0000	1.0000	0.9999	0.9999	0.9999
38.00	1.0000	1.0000	1.0000	1.0000	1.0000	1.0000	1.0000	0.9999	0.9999	0.9999
38.05	1.0000	1.0000	1.0000	1.0000	1.0000	1.0000	1.0000	0.9999	0.9999	0.9999
38.10	1.0000	1.0000	1.0000	1.0000	1.0000	1.0000	1.0000	0.9999	0.9999	0.9999
38.15	1.0000	1.0000	1.0000	1.0000	1.0000	1.0000	1.0000	0.9999	0.9999	0.9999
38.20	1.0000	1.0000	1.0000	1.0000	1.0000	1.0000	1.0000	0.9999	0.9999	0.9999
38.25	1.0000	1.0000	1.0000	1.0000	1.0000	1.0000	1.0000	0.9999	0.9999	0.9999
38.30	1.0000	1.0000	1.0000	1.0000	1.0000	1.0000	1.0000	0.9999	0.9999	0.9999
38.35	1.0000	1.0000	1.0000	1.0000	1.0000	1.0000	1.0000	0.9999	0.9999	0.9999
38.40	1.0000	1.0000	1.0000	1.0000	1.0000	1.0000	1.0000	0.9999	0.9999	0.9999
38.45	1.0000	1.0000	1.0000	1.0000	1.0000	1.0000	1.0000	0.9999	0.9999	0.9999
38.50	1.0000	1.0000	1.0000	1.0000	1.0000	1.0000	1.0000	0.9999	0.9999	0.9999
38.55	1.0000	1.0000	1.0000	1.0000	1.0000	1.0000	1.0000	0.9999	0.9999	0.9999
38.60	1.0000	1.0000	1.0000	1.0000	1.0000	1.0000	1.0000	0.9999	0.9999	0.9999
38.65	1.0000	1.0000	1.0000	1.0000	1.0000	1.0000	1.0000	0.9999	0.9999	0.9999
38.70	1.0000	1.0000	1.0000	1.0000	1.0000	1.0000	1.0000	0.9999	0.9999	0.9999
38.75	1.0000	1.0000	1.0000	1.0000	1.0000	1.0000	1.0000	0.9999	0.9999	0.9999
38.80	1.0000	1.0000	1.0000	1.0000	1.0000	1.0000	1.0000	0.9999	0.9999	0.9999
38.85	1.0000	1.0000	1.0000	1.0000	1.0000	1.0000	1.0000	0.9999	0.9999	0.9999
38.90	1.0000	1.0000	1.0000	1.0000	1.0000	1.0000	1.0000	0.9999	0.9999	0.9999
38.95	1.0000	1.0000	1.0000	1.0000	1.0000	1.0000	1.0000	0.9999	0.9999	0.9999

PROBABILITY INTEGRAL OF t, THE NON-CENTRAL t-STATISTIC. THIS TABLE GIVES Pr [$t/\sqrt{f} \leq x$].

f is the number of degrees of freedom; the non-centrality parameter is $\sqrt{f+1}\,K_p$.

K_p is the standardized normal deviate exceeded with probability p.

DEGREES OF FREEDOM **4**

x \ p	.2500	.1500	.1000	.0650	.0400	.0250	.0100	.0040	.0025	.0010
39.00	1.0000	1.0000	1.0000	1.0000	1.0000	1.0000	1.0000	0.9999	0.9999	0.9999
39.05	1.0000	1.0000	1.0000	1.0000	1.0000	1.0000	1.0000	0.9999	0.9999	0.9999
39.10	1.0000	1.0000	1.0000	1.0000	1.0000	1.0000	1.0000	0.9999	0.9999	0.9999
39.15	1.0000	1.0000	1.0000	1.0000	1.0000	1.0000	1.0000	0.9999	0.9999	0.9999
39.20	1.0000	1.0000	1.0000	1.0000	1.0000	1.0000	1.0000	0.9999	0.9999	0.9999
39.25	1.0000	1.0000	1.0000	1.0000	1.0000	1.0000	1.0000	0.9999	0.9999	0.9999
39.30	1.0000	1.0000	1.0000	1.0000	1.0000	1.0000	1.0000	0.9999	0.9999	0.9999
39.35	1.0000	1.0000	1.0000	1.0000	1.0000	1.0000	1.0000	0.9999	0.9999	0.9999
39.40	1.0000	1.0000	1.0000	1.0000	1.0000	1.0000	1.0000	0.9999	0.9999	0.9999
39.45	1.0000	1.0000	1.0000	1.0000	1.0000	1.0000	1.0000	0.9999	0.9999	0.9999
39.50	1.0000	1.0000	1.0000	1.0000	1.0000	1.0000	1.0000	0.9999	0.9999	0.9999
39.55	1.0000	1.0000	1.0000	1.0000	1.0000	1.0000	1.0000	0.9999	0.9999	0.9999
39.60	1.0000	1.0000	1.0000	1.0000	1.0000	1.0000	1.0000	0.9999	0.9999	0.9999
39.65	1.0000	1.0000	1.0000	1.0000	1.0000	1.0000	1.0000	0.9999	0.9999	0.9999
39.70	1.0000	1.0000	1.0000	1.0000	1.0000	1.0000	1.0000	0.9999	0.9999	0.9999
39.75	1.0000	1.0000	1.0000	1.0000	1.0000	1.0000	1.0000	0.9999	0.9999	0.9999
39.80	1.0000	1.0000	1.0000	1.0000	1.0000	1.0000	1.0000	0.9999	0.9999	0.9999
39.85	1.0000	1.0000	1.0000	1.0000	1.0000	1.0000	1.0000	0.9999	0.9999	0.9999
39.90	1.0000	1.0000	1.0000	1.0000	1.0000	1.0000	1.0000	0.9999	0.9999	0.9999
39.95	1.0000	1.0000	1.0000	1.0000	1.0000	1.0000	1.0000	0.9999	0.9999	0.9999
40.00	1.0000	1.0000	1.0000	1.0000	1.0000	1.0000	1.0000	0.9999	0.9999	0.9999
40.05	1.0000	1.0000	1.0000	1.0000	1.0000	1.0000	1.0000	0.9999	0.9999	0.9999
40.10	1.0000	1.0000	1.0000	1.0000	1.0000	1.0000	1.0000	0.9999	0.9999	0.9999
40.15	1.0000	1.0000	1.0000	1.0000	1.0000	1.0000	1.0000	0.9999	0.9999	0.9999
40.20	1.0000	1.0000	1.0000	1.0000	1.0000	1.0000	1.0000	0.9999	0.9999	0.9999
40.25	1.0000	1.0000	1.0000	1.0000	1.0000	1.0000	1.0000	0.9999	0.9999	0.9999
40.30	1.0000	1.0000	1.0000	1.0000	1.0000	1.0000	1.0000	0.9999	0.9999	0.9999
40.35	1.0000	1.0000	1.0000	1.0000	1.0000	1.0000	1.0000	0.9999	0.9999	0.9999
40.40	1.0000	1.0000	1.0000	1.0000	1.0000	1.0000	1.0000	0.9999	0.9999	0.9999
40.45	1.0000	1.0000	1.0000	1.0000	1.0000	1.0000	1.0000	0.9999	0.9999	0.9999
40.50	1.0000	1.0000	1.0000	1.0000	1.0000	1.0000	1.0000	0.9999	0.9999	0.9999
40.55	1.0000	1.0000	1.0000	1.0000	1.0000	1.0000	1.0000	0.9999	0.9999	0.9999
40.60	1.0000	1.0000	1.0000	1.0000	1.0000	1.0000	1.0000	0.9999	0.9999	0.9999
40.65	1.0000	1.0000	1.0000	1.0000	1.0000	1.0000	1.0000	0.9999	0.9999	0.9999
40.70	1.0000	1.0000	1.0000	1.0000	1.0000	1.0000	1.0000	0.9999	0.9999	0.9999
40.75	1.0000	1.0000	1.0000	1.0000	1.0000	1.0000	1.0000	0.9999	0.9999	0.9999
40.80	1.0000	1.0000	1.0000	1.0000	1.0000	1.0000	1.0000	0.9999	0.9999	0.9999
40.85	1.0000	1.0000	1.0000	1.0000	1.0000	1.0000	1.0000	0.9999	0.9999	0.9999
40.90	1.0000	1.0000	1.0000	1.0000	1.0000	1.0000	1.0000	0.9999	0.9999	0.9999
40.95	1.0000	1.0000	1.0000	1.0000	1.0000	1.0000	1.0000	0.9999	0.9999	0.9999
41.00	1.0000	1.0000	1.0000	1.0000	1.0000	1.0000	1.0000	0.9999	0.9999	0.9999
41.05	1.0000	1.0000	1.0000	1.0000	1.0000	1.0000	1.0000	0.9999	0.9999	0.9999
41.10	1.0000	1.0000	1.0000	1.0000	1.0000	1.0000	1.0000	0.9999	0.9999	0.9999
41.15	1.0000	1.0000	1.0000	1.0000	1.0000	1.0000	1.0000	0.9999	0.9999	0.9999
41.20	1.0000	1.0000	1.0000	1.0000	1.0000	1.0000	1.0000	0.9999	0.9999	0.9999
41.25	1.0000	1.0000	1.0000	1.0000	1.0000	1.0000	1.0000	0.9999	0.9999	0.9999
41.30	1.0000	1.0000	1.0000	1.0000	1.0000	1.0000	1.0000	0.9999	0.9999	0.9999
41.35	1.0000	1.0000	1.0000	1.0000	1.0000	1.0000	1.0000	0.9999	0.9999	0.9999
41.40	1.0000	1.0000	1.0000	1.0000	1.0000	1.0000	1.0000	0.9999	0.9999	0.9999
41.45	1.0000	1.0000	1.0000	1.0000	1.0000	1.0000	1.0000	0.9999	0.9999	0.9999
41.50	1.0000	1.0000	1.0000	1.0000	1.0000	1.0000	1.0000	0.9999	0.9999	0.9999
41.55	1.0000	1.0000	1.0000	1.0000	1.0000	1.0000	1.0000	0.9999	0.9999	0.9999
41.60	1.0000	1.0000	1.0000	1.0000	1.0000	1.0000	1.0000	0.9999	0.9999	0.9999
41.65	1.0000	1.0000	1.0000	1.0000	1.0000	1.0000	1.0000	0.9999	0.9999	0.9999
41.70	1.0000	1.0000	1.0000	1.0000	1.0000	1.0000	1.0000	0.9999	0.9999	0.9999
41.75	1.0000	1.0000	1.0000	1.0000	1.0000	1.0000	1.0000	0.9999	0.9999	0.9999
41.80	1.0000	1.0000	1.0000	1.0000	1.0000	1.0000	1.0000	0.9999	0.9999	0.9999
41.85	1.0000	1.0000	1.0000	1.0000	1.0000	1.0000	1.0000	0.9999	0.9999	0.9999
41.90	1.0000	1.0000	1.0000	1.0000	1.0000	1.0000	1.0000	0.9999	0.9999	0.9999
41.95	1.0000	1.0000	1.0000	1.0000	1.0000	1.0000	1.0000	0.9999	0.9999	0.9999

PROBABILITY INTEGRAL OF t, THE NON-CENTRAL t-STATISTIC. THIS TABLE GIVES Pr $[t/\sqrt{f} \leq x]$.

f is the number of degrees of freedom; the non-centrality parameter is $\sqrt{f+1}\,K_p$.

K_p is the standardized normal deviate exceeded with probability p.

DEGREES OF FREEDOM **4**

x \ p	.2500	.1500	.1000	.0650	.0400	.0250	.0100	.0040	.0025	.0010
42.00	1.0000	1.0000	1.0000	1.0000	1.0000	1.0000	1.0000	0.9999	0.9999	0.9999
42.05	1.0000	1.0000	1.0000	1.0000	1.0000	1.0000	1.0000	0.9999	0.9999	0.9999
42.10	1.0000	1.0000	1.0000	1.0000	1.0000	1.0000	1.0000	0.9999	0.9999	0.9999
42.15	1.0000	1.0000	1.0000	1.0000	1.0000	1.0000	1.0000	0.9999	0.9999	0.9999
42.20	1.0000	1.0000	1.0000	1.0000	1.0000	1.0000	1.0000	0.9999	0.9999	0.9999
42.25	1.0000	1.0000	1.0000	1.0000	1.0000	1.0000	1.0000	0.9999	0.9999	0.9999
42.30	1.0000	1.0000	1.0000	1.0000	1.0000	1.0000	1.0000	1.0000	0.9999	0.9999
42.35	1.0000	1.0000	1.0000	1.0000	1.0000	1.0000	1.0000	1.0000	0.9999	0.9999
42.40	1.0000	1.0000	1.0000	1.0000	1.0000	1.0000	1.0000	1.0000	0.9999	0.9999
42.45	1.0000	1.0000	1.0000	1.0000	1.0000	1.0000	1.0000	1.0000	0.9999	0.9999
42.50	1.0000	1.0000	1.0000	1.0000	1.0000	1.0000	1.0000	1.0000	0.9999	0.9999
42.55	1.0000	1.0000	1.0000	1.0000	1.0000	1.0000	1.0000	1.0000	0.9999	0.9999
42.60	1.0000	1.0000	1.0000	1.0000	1.0000	1.0000	1.0000	1.0000	0.9999	0.9999
42.65	1.0000	1.0000	1.0000	1.0000	1.0000	1.0000	1.0000	1.0000	0.9999	0.9999
42.70	1.0000	1.0000	1.0000	1.0000	1.0000	1.0000	1.0000	1.0000	0.9999	0.9999
42.75	1.0000	1.0000	1.0000	1.0000	1.0000	1.0000	1.0000	1.0000	0.9999	0.9999
42.80	1.0000	1.0000	1.0000	1.0000	1.0000	1.0000	1.0000	1.0000	0.9999	0.9999
42.85	1.0000	1.0000	1.0000	1.0000	1.0000	1.0000	1.0000	1.0000	0.9999	0.9999
42.90	1.0000	1.0000	1.0000	1.0000	1.0000	1.0000	1.0000	1.0000	0.9999	0.9999
42.95	1.0000	1.0000	1.0000	1.0000	1.0000	1.0000	1.0000	1.0000	0.9999	0.9999
43.00	1.0000	1.0000	1.0000	1.0000	1.0000	1.0000	1.0000	1.0000	0.9999	0.9999
43.05	1.0000	1.0000	1.0000	1.0000	1.0000	1.0000	1.0000	1.0000	0.9999	0.9999
43.10	1.0000	1.0000	1.0000	1.0000	1.0000	1.0000	1.0000	1.0000	0.9999	0.9999
43.15	1.0000	1.0000	1.0000	1.0000	1.0000	1.0000	1.0000	1.0000	0.9999	0.9999
43.20	1.0000	1.0000	1.0000	1.0000	1.0000	1.0000	1.0000	1.0000	0.9999	0.9999
43.25	1.0000	1.0000	1.0000	1.0000	1.0000	1.0000	1.0000	1.0000	0.9999	0.9999
43.30	1.0000	1.0000	1.0000	1.0000	1.0000	1.0000	1.0000	1.0000	0.9999	0.9999
43.35	1.0000	1.0000	1.0000	1.0000	1.0000	1.0000	1.0000	1.0000	0.9999	0.9999
43.40	1.0000	1.0000	1.0000	1.0000	1.0000	1.0000	1.0000	1.0000	0.9999	0.9999
43.45	1.0000	1.0000	1.0000	1.0000	1.0000	1.0000	1.0000	1.0000	0.9999	0.9999
43.50	1.0000	1.0000	1.0000	1.0000	1.0000	1.0000	1.0000	1.0000	0.9999	0.9999
43.55	1.0000	1.0000	1.0000	1.0000	1.0000	1.0000	1.0000	1.0000	0.9999	0.9999
43.60	1.0000	1.0000	1.0000	1.0000	1.0000	1.0000	1.0000	1.0000	0.9999	0.9999
43.65	1.0000	1.0000	1.0000	1.0000	1.0000	1.0000	1.0000	1.0000	0.9999	0.9999
43.70	1.0000	1.0000	1.0000	1.0000	1.0000	1.0000	1.0000	1.0000	0.9999	0.9999
43.75	1.0000	1.0000	1.0000	1.0000	1.0000	1.0000	1.0000	1.0000	0.9999	0.9999
43.80	1.0000	1.0000	1.0000	1.0000	1.0000	1.0000	1.0000	1.0000	0.9999	0.9999
43.85	1.0000	1.0000	1.0000	1.0000	1.0000	1.0000	1.0000	1.0000	0.9999	0.9999
43.90	1.0000	1.0000	1.0000	1.0000	1.0000	1.0000	1.0000	1.0000	0.9999	0.9999
43.95	1.0000	1.0000	1.0000	1.0000	1.0000	1.0000	1.0000	1.0000	0.9999	0.9999
44.00	1.0000	1.0000	1.0000	1.0000	1.0000	1.0000	1.0000	1.0000	0.9999	0.9999
44.05	1.0000	1.0000	1.0000	1.0000	1.0000	1.0000	1.0000	1.0000	0.9999	0.9999
44.10	1.0000	1.0000	1.0000	1.0000	1.0000	1.0000	1.0000	1.0000	0.9999	0.9999
44.15	1.0000	1.0000	1.0000	1.0000	1.0000	1.0000	1.0000	1.0000	0.9999	0.9999
44.20	1.0000	1.0000	1.0000	1.0000	1.0000	1.0000	1.0000	1.0000	0.9999	0.9999
44.25	1.0000	1.0000	1.0000	1.0000	1.0000	1.0000	1.0000	1.0000	0.9999	0.9999
44.30	1.0000	1.0000	1.0000	1.0000	1.0000	1.0000	1.0000	1.0000	0.9999	0.9999
44.35	1.0000	1.0000	1.0000	1.0000	1.0000	1.0000	1.0000	1.0000	0.9999	0.9999
44.40	1.0000	1.0000	1.0000	1.0000	1.0000	1.0000	1.0000	1.0000	0.9999	0.9999
44.45	1.0000	1.0000	1.0000	1.0000	1.0000	1.0000	1.0000	1.0000	0.9999	0.9999
44.50	1.0000	1.0000	1.0000	1.0000	1.0000	1.0000	1.0000	1.0000	1.0000	0.9999
44.55	1.0000	1.0000	1.0000	1.0000	1.0000	1.0000	1.0000	1.0000	1.0000	0.9999
44.60	1.0000	1.0000	1.0000	1.0000	1.0000	1.0000	1.0000	1.0000	1.0000	0.9999
44.65	1.0000	1.0000	1.0000	1.0000	1.0000	1.0000	1.0000	1.0000	1.0000	0.9999
44.70	1.0000	1.0000	1.0000	1.0000	1.0000	1.0000	1.0000	1.0000	1.0000	0.9999
44.75	1.0000	1.0000	1.0000	1.0000	1.0000	1.0000	1.0000	1.0000	1.0000	0.9999
44.80	1.0000	1.0000	1.0000	1.0000	1.0000	1.0000	1.0000	1.0000	1.0000	0.9999
44.85	1.0000	1.0000	1.0000	1.0000	1.0000	1.0000	1.0000	1.0000	1.0000	0.9999
44.90	1.0000	1.0000	1.0000	1.0000	1.0000	1.0000	1.0000	1.0000	1.0000	0.9999
44.95	1.0000	1.0000	1.0000	1.0000	1.0000	1.0000	1.0000	1.0000	1.0000	0.9999

PROBABILITY INTEGRAL OF t, THE NON-CENTRAL t-STATISTIC. THIS TABLE GIVES Pr $[t/\sqrt{f} \leq x]$.

f is the number of degrees of freedom; the non-centrality parameter is $\sqrt{f+1}\,K_p$.

K_p is the standardized normal deviate exceeded with probability p.

DEGREES OF FREEDOM **4**

x \ p	.2500	.1500	.1000	.0650	.0400	.0250	.0100	.0040	.0025	.0010
45.00	1.0000	1.0000	1.0000	1.0000	1.0000	1.0000	1.0000	1.0000	1.0000	0.9999
45.05	1.0000	1.0000	1.0000	1.0000	1.0000	1.0000	1.0000	1.0000	1.0000	0.9999
45.10	1.0000	1.0000	1.0000	1.0000	1.0000	1.0000	1.0000	1.0000	1.0000	0.9999
45.15	1.0000	1.0000	1.0000	1.0000	1.0000	1.0000	1.0000	1.0000	1.0000	0.9999
45.20	1.0000	1.0000	1.0000	1.0000	1.0000	1.0000	1.0000	1.0000	1.0000	0.9999
45.25	1.0000	1.0000	1.0000	1.0000	1.0000	1.0000	1.0000	1.0000	1.0000	0.9999
45.30	1.0000	1.0000	1.0000	1.0000	1.0000	1.0000	1.0000	1.0000	1.0000	0.9999
45.35	1.0000	1.0000	1.0000	1.0000	1.0000	1.0000	1.0000	1.0000	1.0000	0.9999
45.40	1.0000	1.0000	1.0000	1.0000	1.0000	1.0000	1.0000	1.0000	1.0000	0.9999
45.45	1.0000	1.0000	1.0000	1.0000	1.0000	1.0000	1.0000	1.0000	1.0000	0.9999
45.50	1.0000	1.0000	1.0000	1.0000	1.0000	1.0000	1.0000	1.0000	1.0000	0.9999
45.55	1.0000	1.0000	1.0000	1.0000	1.0000	1.0000	1.0000	1.0000	1.0000	0.9999
45.60	1.0000	1.0000	1.0000	1.0000	1.0000	1.0000	1.0000	1.0000	1.0000	0.9999
45.65	1.0000	1.0000	1.0000	1.0000	1.0000	1.0000	1.0000	1.0000	1.0000	0.9999
45.70	1.0000	1.0000	1.0000	1.0000	1.0000	1.0000	1.0000	1.0000	1.0000	0.9999
45.75	1.0000	1.0000	1.0000	1.0000	1.0000	1.0000	1.0000	1.0000	1.0000	0.9999
45.80	1.0000	1.0000	1.0000	1.0000	1.0000	1.0000	1.0000	1.0000	1.0000	0.9999
45.85	1.0000	1.0000	1.0000	1.0000	1.0000	1.0000	1.0000	1.0000	1.0000	0.9999
45.90	1.0000	1.0000	1.0000	1.0000	1.0000	1.0000	1.0000	1.0000	1.0000	0.9999
45.95	1.0000	1.0000	1.0000	1.0000	1.0000	1.0000	1.0000	1.0000	1.0000	0.9999
46.00	1.0000	1.0000	1.0000	1.0000	1.0000	1.0000	1.0000	1.0000	1.0000	0.9999
46.05	1.0000	1.0000	1.0000	1.0000	1.0000	1.0000	1.0000	1.0000	1.0000	0.9999
46.10	1.0000	1.0000	1.0000	1.0000	1.0000	1.0000	1.0000	1.0000	1.0000	0.9999
46.15	1.0000	1.0000	1.0000	1.0000	1.0000	1.0000	1.0000	1.0000	1.0000	0.9999
46.20	1.0000	1.0000	1.0000	1.0000	1.0000	1.0000	1.0000	1.0000	1.0000	0.9999
46.25	1.0000	1.0000	1.0000	1.0000	1.0000	1.0000	1.0000	1.0000	1.0000	0.9999
46.30	1.0000	1.0000	1.0000	1.0000	1.0000	1.0000	1.0000	1.0000	1.0000	0.9999
46.35	1.0000	1.0000	1.0000	1.0000	1.0000	1.0000	1.0000	1.0000	1.0000	0.9999
46.40	1.0000	1.0000	1.0000	1.0000	1.0000	1.0000	1.0000	1.0000	1.0000	0.9999
46.45	1.0000	1.0000	1.0000	1.0000	1.0000	1.0000	1.0000	1.0000	1.0000	0.9999
46.50	1.0000	1.0000	1.0000	1.0000	1.0000	1.0000	1.0000	1.0000	1.0000	0.9999
46.55	1.0000	1.0000	1.0000	1.0000	1.0000	1.0000	1.0000	1.0000	1.0000	0.9999
46.60	1.0000	1.0000	1.0000	1.0000	1.0000	1.0000	1.0000	1.0000	1.0000	0.9999
46.65	1.0000	1.0000	1.0000	1.0000	1.0000	1.0000	1.0000	1.0000	1.0000	0.9999
46.70	1.0000	1.0000	1.0000	1.0000	1.0000	1.0000	1.0000	1.0000	1.0000	0.9999
46.75	1.0000	1.0000	1.0000	1.0000	1.0000	1.0000	1.0000	1.0000	1.0000	0.9999
46.80	1.0000	1.0000	1.0000	1.0000	1.0000	1.0000	1.0000	1.0000	1.0000	0.9999
46.85	1.0000	1.0000	1.0000	1.0000	1.0000	1.0000	1.0000	1.0000	1.0000	0.9999
46.90	1.0000	1.0000	1.0000	1.0000	1.0000	1.0000	1.0000	1.0000	1.0000	0.9999
46.95	1.0000	1.0000	1.0000	1.0000	1.0000	1.0000	1.0000	1.0000	1.0000	0.9999
47.00	1.0000	1.0000	1.0000	1.0000	1.0000	1.0000	1.0000	1.0000	1.0000	0.9999
47.05	1.0000	1.0000	1.0000	1.0000	1.0000	1.0000	1.0000	1.0000	1.0000	0.9999
47.10	1.0000	1.0000	1.0000	1.0000	1.0000	1.0000	1.0000	1.0000	1.0000	0.9999
47.15	1.0000	1.0000	1.0000	1.0000	1.0000	1.0000	1.0000	1.0000	1.0000	0.9999
47.20	1.0000	1.0000	1.0000	1.0000	1.0000	1.0000	1.0000	1.0000	1.0000	0.9999
47.25	1.0000	1.0000	1.0000	1.0000	1.0000	1.0000	1.0000	1.0000	1.0000	0.9999
47.30	1.0000	1.0000	1.0000	1.0000	1.0000	1.0000	1.0000	1.0000	1.0000	0.9999
47.35	1.0000	1.0000	1.0000	1.0000	1.0000	1.0000	1.0000	1.0000	1.0000	0.9999
47.40	1.0000	1.0000	1.0000	1.0000	1.0000	1.0000	1.0000	1.0000	1.0000	0.9999
47.45	1.0000	1.0000	1.0000	1.0000	1.0000	1.0000	1.0000	1.0000	1.0000	0.9999
47.50	1.0000	1.0000	1.0000	1.0000	1.0000	1.0000	1.0000	1.0000	1.0000	0.9999
47.55	1.0000	1.0000	1.0000	1.0000	1.0000	1.0000	1.0000	1.0000	1.0000	0.9999
47.60	1.0000	1.0000	1.0000	1.0000	1.0000	1.0000	1.0000	1.0000	1.0000	0.9999
47.65	1.0000	1.0000	1.0000	1.0000	1.0000	1.0000	1.0000	1.0000	1.0000	0.9999
47.70	1.0000	1.0000	1.0000	1.0000	1.0000	1.0000	1.0000	1.0000	1.0000	0.9999
47.75	1.0000	1.0000	1.0000	1.0000	1.0000	1.0000	1.0000	1.0000	1.0000	0.9999
47.80	1.0000	1.0000	1.0000	1.0000	1.0000	1.0000	1.0000	1.0000	1.0000	0.9999
47.85	1.0000	1.0000	1.0000	1.0000	1.0000	1.0000	1.0000	1.0000	1.0000	0.9999
47.90	1.0000	1.0000	1.0000	1.0000	1.0000	1.0000	1.0000	1.0000	1.0000	0.9999
47.95	1.0000	1.0000	1.0000	1.0000	1.0000	1.0000	1.0000	1.0000	1.0000	0.9999

PROBABILITY INTEGRAL OF t, THE NON-CENTRAL t-STATISTIC. THIS TABLE GIVES Pr $[t/\sqrt{f} \leq x]$.

f is the number of degrees of freedom; the non-centrality parameter is $\sqrt{f+1}\,K_p$.

K_p is the standardized normal deviate exceeded with probability p.

DEGREES OF FREEDOM **4**

x \ p	.2500	.1500	.1000	.0650	.0400	.0250	.0100	.0040	.0025	.0010
48.00	1.0000	1.0000	1.0000	1.0000	1.0000	1.0000	1.0000	1.0000	1.0000	0.9999
48.05	1.0000	1.0000	1.0000	1.0000	1.0000	1.0000	1.0000	1.0000	1.0000	0.9999
48.10	1.0000	1.0000	1.0000	1.0000	1.0000	1.0000	1.0000	1.0000	1.0000	0.9999
48.15	1.0000	1.0000	1.0000	1.0000	1.0000	1.0000	1.0000	1.0000	1.0000	0.9999
48.20	1.0000	1.0000	1.0000	1.0000	1.0000	1.0000	1.0000	1.0000	1.0000	0.9999
48.25	1.0000	1.0000	1.0000	1.0000	1.0000	1.0000	1.0000	1.0000	1.0000	0.9999
48.30	1.0000	1.0000	1.0000	1.0000	1.0000	1.0000	1.0000	1.0000	1.0000	0.9999
48.35	1.0000	1.0000	1.0000	1.0000	1.0000	1.0000	1.0000	1.0000	1.0000	0.9999
48.40	1.0000	1.0000	1.0000	1.0000	1.0000	1.0000	1.0000	1.0000	1.0000	0.9999
48.45	1.0000	1.0000	1.0000	1.0000	1.0000	1.0000	1.0000	1.0000	1.0000	0.9999
48.50	1.0000	1.0000	1.0000	1.0000	1.0000	1.0000	1.0000	1.0000	1.0000	0.9999
48.55	1.0000	1.0000	1.0000	1.0000	1.0000	1.0000	1.0000	1.0000	1.0000	0.9999
48.60	1.0000	1.0000	1.0000	1.0000	1.0000	1.0000	1.0000	1.0000	1.0000	0.9999
48.65	1.0000	1.0000	1.0000	1.0000	1.0000	1.0000	1.0000	1.0000	1.0000	0.9999
48.70	1.0000	1.0000	1.0000	1.0000	1.0000	1.0000	1.0000	1.0000	1.0000	0.9999
48.75	1.0000	1.0000	1.0000	1.0000	1.0000	1.0000	1.0000	1.0000	1.0000	0.9999
48.80	1.0000	1.0000	1.0000	1.0000	1.0000	1.0000	1.0000	1.0000	1.0000	0.9999
48.85	1.0000	1.0000	1.0000	1.0000	1.0000	1.0000	1.0000	1.0000	1.0000	0.9999
48.90	1.0000	1.0000	1.0000	1.0000	1.0000	1.0000	1.0000	1.0000	1.0000	1.0000

PROBABILITY INTEGRAL OF t, THE NON-CENTRAL t-STATISTIC. THIS TABLE GIVES Pr $[t/\sqrt{f} \leq x]$.

f is the number of degrees of freedom; the non-centrality parameter is $\sqrt{f+1}\,K_p$.

K_p is the standardized normal deviate exceeded with probability p.

DEGREES OF FREEDOM **5**

x \ p	.2500	.1500	.1000	.0650	.0400	.0250	.0100	.0040	.0025	.0010
-1.90	0.0000	0.0000	0.0000	0.0000	0.0000	0.0000	0.0000	0.0000	0.0000	0.0000
-1.85	0.0001	0.0000	0.0000	0.0000	0.0000	0.0000	0.0000	0.0000	0.0000	0.0000
-1.80	0.0001	0.0000	0.0000	0.0000	0.0000	0.0000	0.0000	0.0000	0.0000	0.0000
-1.75	0.0001	0.0000	0.0000	0.0000	0.0000	0.0000	0.0000	0.0000	0.0000	0.0000
-1.70	0.0001	0.0000	0.0000	0.0000	0.0000	0.0000	0.0000	0.0000	0.0000	0.0000
-1.65	0.0001	0.0000	0.0000	0.0000	0.0000	0.0000	0.0000	0.0000	0.0000	0.0000
-1.60	0.0001	0.0000	0.0000	0.0000	0.0000	0.0000	0.0000	0.0000	0.0000	0.0000
-1.55	0.0001	0.0000	0.0000	0.0000	0.0000	0.0000	0.0000	0.0000	0.0000	0.0000
-1.50	0.0001	0.0000	0.0000	0.0000	0.0000	0.0000	0.0000	0.0000	0.0000	0.0000
-1.45	0.0002	0.0000	0.0000	0.0000	0.0000	0.0000	0.0000	0.0000	0.0000	0.0000
-1.40	0.0002	0.0000	0.0000	0.0000	0.0000	0.0000	0.0000	0.0000	0.0000	0.0000
-1.35	0.0002	0.0000	0.0000	0.0000	0.0000	0.0000	0.0000	0.0000	0.0000	0.0000
-1.30	0.0002	0.0000	0.0000	0.0000	0.0000	0.0000	0.0000	0.0000	0.0000	0.0000
-1.25	0.0003	0.0000	0.0000	0.0000	0.0000	0.0000	0.0000	0.0000	0.0000	0.0000
-1.20	0.0003	0.0000	0.0000	0.0000	0.0000	0.0000	0.0000	0.0000	0.0000	0.0000
-1.15	0.0004	0.0000	0.0000	0.0000	0.0000	0.0000	0.0000	0.0000	0.0000	0.0000
-1.10	0.0005	0.0000	0.0000	0.0000	0.0000	0.0000	0.0000	0.0000	0.0000	0.0000
-1.05	0.0006	0.0000	0.0000	0.0000	0.0000	0.0000	0.0000	0.0000	0.0000	0.0000
-1.00	0.0007	0.0000	0.0000	0.0000	0.0000	0.0000	0.0000	0.0000	0.0000	0.0000
-0.95	0.0008	0.0000	0.0000	0.0000	0.0000	0.0000	0.0000	0.0000	0.0000	0.0000
-0.90	0.0009	0.0001	0.0000	0.0000	0.0000	0.0000	0.0000	0.0000	0.0000	0.0000
-0.85	0.0011	0.0001	0.0000	0.0000	0.0000	0.0000	0.0000	0.0000	0.0000	0.0000
-0.80	0.0014	0.0001	0.0000	0.0000	0.0000	0.0000	0.0000	0.0000	0.0000	0.0000
-0.75	0.0017	0.0001	0.0000	0.0000	0.0000	0.0000	0.0000	0.0000	0.0000	0.0000
-0.70	0.0021	0.0001	0.0000	0.0000	0.0000	0.0000	0.0000	0.0000	0.0000	0.0000
-0.65	0.0026	0.0001	0.0000	0.0000	0.0000	0.0000	0.0000	0.0000	0.0000	0.0000
-0.60	0.0032	0.0002	0.0000	0.0000	0.0000	0.0000	0.0000	0.0000	0.0000	0.0000
-0.55	0.0040	0.0002	0.0000	0.0000	0.0000	0.0000	0.0000	0.0000	0.0000	0.0000
-0.50	0.0049	0.0003	0.0000	0.0000	0.0000	0.0000	0.0000	0.0000	0.0000	0.0000
-0.45	0.0062	0.0004	0.0000	0.0000	0.0000	0.0000	0.0000	0.0000	0.0000	0.0000
-0.40	0.0078	0.0005	0.0001	0.0000	0.0000	0.0000	0.0000	0.0000	0.0000	0.0000
-0.35	0.0098	0.0007	0.0001	0.0000	0.0000	0.0000	0.0000	0.0000	0.0000	0.0000
-0.30	0.0124	0.0009	0.0001	0.0000	0.0000	0.0000	0.0000	0.0000	0.0000	0.0000
-0.25	0.0156	0.0012	0.0002	0.0000	0.0000	0.0000	0.0000	0.0000	0.0000	0.0000
-0.20	0.0198	0.0017	0.0002	0.0000	0.0000	0.0000	0.0000	0.0000	0.0000	0.0000
-0.15	0.0249	0.0022	0.0003	0.0000	0.0000	0.0000	0.0000	0.0000	0.0000	0.0000
-0.10	0.0314	0.0030	0.0004	0.0001	0.0000	0.0000	0.0000	0.0000	0.0000	0.0000
-0.05	0.0394	0.0041	0.0006	0.0001	0.0000	0.0000	0.0000	0.0000	0.0000	0.0000
0.00	0.0493	0.0056	0.0009	0.0001	0.0000	0.0000	0.0000	0.0000	0.0000	0.0000
0.05	0.0612	0.0075	0.0012	0.0002	0.0000	0.0000	0.0000	0.0000	0.0000	0.0000
0.10	0.0755	0.0102	0.0018	0.0002	0.0000	0.0000	0.0000	0.0000	0.0000	0.0000
0.15	0.0924	0.0136	0.0025	0.0004	0.0000	0.0000	0.0000	0.0000	0.0000	0.0000
0.20	0.1122	0.0182	0.0036	0.0006	0.0001	0.0000	0.0000	0.0000	0.0000	0.0000
0.25	0.1348	0.0240	0.0051	0.0009	0.0001	0.0000	0.0000	0.0000	0.0000	0.0000
0.30	0.1604	0.0314	0.0072	0.0013	0.0002	0.0000	0.0000	0.0000	0.0000	0.0000
0.35	0.1888	0.0407	0.0101	0.0020	0.0003	0.0000	0.0000	0.0000	0.0000	0.0000
0.40	0.2198	0.0520	0.0138	0.0030	0.0005	0.0001	0.0000	0.0000	0.0000	0.0000
0.45	0.2532	0.0657	0.0188	0.0044	0.0008	0.0001	0.0000	0.0000	0.0000	0.0000
0.50	0.2885	0.0818	0.0252	0.0064	0.0012	0.0002	0.0000	0.0000	0.0000	0.0000
0.55	0.3253	0.1006	0.0332	0.0091	0.0019	0.0004	0.0000	0.0000	0.0000	0.0000
0.60	0.3632	0.1219	0.0431	0.0127	0.0029	0.0006	0.0000	0.0000	0.0000	0.0000
0.65	0.4015	0.1458	0.0551	0.0175	0.0043	0.0010	0.0001	0.0000	0.0000	0.0000
0.70	0.4398	0.1721	0.0693	0.0235	0.0063	0.0016	0.0001	0.0000	0.0000	0.0000
0.75	0.4777	0.2005	0.0859	0.0312	0.0090	0.0025	0.0002	0.0000	0.0000	0.0000
0.80	0.5148	0.2309	0.1047	0.0405	0.0126	0.0037	0.0003	0.0000	0.0000	0.0000
0.85	0.5507	0.2627	0.1258	0.0518	0.0172	0.0055	0.0005	0.0000	0.0000	0.0000
0.90	0.5852	0.2958	0.1491	0.0650	0.0232	0.0079	0.0008	0.0001	0.0000	0.0000
0.95	0.6180	0.3296	0.1744	0.0803	0.0305	0.0111	0.0013	0.0001	0.0000	0.0000

PROBABILITY INTEGRAL OF t, THE NON-CENTRAL t-STATISTIC. THIS TABLE GIVES Pr $[t/\sqrt{f} \leq x]$.

f is the number of degrees of freedom; the non-centrality parameter is $\sqrt{f+1}\,K_p$.

K_p is the standardized normal deviate exceeded with probability p.

DEGREES OF FREEDOM **5**

x \ p	.2500	.1500	.1000	.0650	.0400	.0250	.0100	.0040	.0025	.0010
1.00	0.6490	0.3638	0.2014	0.0976	0.0394	0.0153	0.0021	0.0003	0.0001	0.0000
1.05	0.6781	0.3981	0.2299	0.1169	0.0500	0.0205	0.0031	0.0004	0.0001	0.0000
1.10	0.7052	0.4322	0.2596	0.1382	0.0623	0.0270	0.0046	0.0007	0.0003	0.0000
1.15	0.7305	0.4656	0.2902	0.1611	0.0765	0.0348	0.0066	0.0011	0.0004	0.0001
1.20	0.7539	0.4983	0.3214	0.1857	0.0924	0.0442	0.0092	0.0017	0.0007	0.0001
1.25	0.7754	0.5299	0.3528	0.2115	0.1101	0.0550	0.0125	0.0025	0.0011	0.0002
1.30	0.7952	0.5604	0.3843	0.2385	0.1294	0.0675	0.0168	0.0037	0.0017	0.0003
1.35	0.8133	0.5895	0.4155	0.2663	0.1502	0.0815	0.0220	0.0053	0.0025	0.0005
1.40	0.8299	0.6173	0.4462	0.2947	0.1724	0.0971	0.0282	0.0073	0.0036	0.0008
1.45	0.8450	0.6436	0.4763	0.3235	0.1958	0.1142	0.0356	0.0100	0.0050	0.0013
1.50	0.8588	0.6685	0.5056	0.3525	0.2202	0.1327	0.0443	0.0133	0.0070	0.0019
1.55	0.8713	0.6919	0.5340	0.3813	0.2454	0.1524	0.0542	0.0174	0.0094	0.0027
1.60	0.8827	0.7139	0.5613	0.4100	0.2713	0.1734	0.0654	0.0223	0.0125	0.0039
1.65	0.8930	0.7345	0.5874	0.4382	0.2975	0.1953	0.0778	0.0281	0.0162	0.0053
1.70	0.9024	0.7537	0.6125	0.4658	0.3240	0.2181	0.0915	0.0349	0.0207	0.0072
1.75	0.9109	0.7715	0.6363	0.4928	0.3506	0.2416	0.1064	0.0428	0.0261	0.0095
1.80	0.9186	0.7882	0.6589	0.5189	0.3771	0.2656	0.1225	0.0516	0.0323	0.0124
1.85	0.9256	0.8036	0.6803	0.5442	0.4034	0.2900	0.1396	0.0616	0.0395	0.0159
1.90	0.9320	0.8179	0.7005	0.5686	0.4293	0.3146	0.1577	0.0726	0.0476	0.0201
1.95	0.9377	0.8312	0.7196	0.5921	0.4547	0.3393	0.1766	0.0847	0.0567	0.0250
2.00	0.9430	0.8435	0.7376	0.6145	0.4796	0.3640	0.1963	0.0977	0.0667	0.0306
2.05	0.9477	0.8548	0.7544	0.6360	0.5038	0.3884	0.2166	0.1118	0.0778	0.0370
2.10	0.9520	0.8654	0.7702	0.6564	0.5273	0.4127	0.2374	0.1267	0.0898	0.0443
2.15	0.9560	0.8751	0.7850	0.6759	0.5501	0.4365	0.2587	0.1425	0.1026	0.0523
2.20	0.9595	0.8840	0.7989	0.6943	0.5721	0.4599	0.2802	0.1590	0.1164	0.0613
2.25	0.9628	0.8923	0.8119	0.7118	0.5932	0.4828	0.3020	0.1763	0.1310	0.0710
2.30	0.9657	0.9000	0.8240	0.7284	0.6136	0.5051	0.3238	0.1941	0.1463	0.0816
2.35	0.9684	0.9070	0.8353	0.7441	0.6331	0.5268	0.3457	0.2125	0.1623	0.0930
2.40	0.9709	0.9135	0.8459	0.7589	0.6517	0.5479	0.3674	0.2313	0.1789	0.1052
2.45	0.9731	0.9196	0.8557	0.7728	0.6696	0.5682	0.3890	0.2504	0.1960	0.1181
2.50	0.9752	0.9251	0.8649	0.7860	0.6866	0.5879	0.4103	0.2698	0.2137	0.1317
2.55	0.9770	0.9303	0.8735	0.7983	0.7028	0.6068	0.4314	0.2894	0.2316	0.1459
2.60	0.9787	0.9350	0.8815	0.8100	0.7182	0.6251	0.4521	0.3091	0.2499	0.1607
2.65	0.9803	0.9394	0.8889	0.8210	0.7329	0.6426	0.4724	0.3288	0.2685	0.1760
2.70	0.9817	0.9435	0.8959	0.8313	0.7468	0.6594	0.4923	0.3486	0.2872	0.1918
2.75	0.9831	0.9473	0.9023	0.8410	0.7600	0.6755	0.5117	0.3682	0.3060	0.2079
2.80	0.9843	0.9507	0.9084	0.8501	0.7725	0.6909	0.5306	0.3876	0.3248	0.2245
2.85	0.9854	0.9540	0.9140	0.8586	0.7844	0.7056	0.5489	0.4069	0.3437	0.2413
2.90	0.9864	0.9570	0.9192	0.8667	0.7957	0.7197	0.5668	0.4259	0.3624	0.2583
2.95	0.9873	0.9597	0.9241	0.8742	0.8063	0.7331	0.5841	0.4447	0.3811	0.2755
3.00	0.9882	0.9623	0.9287	0.8813	0.8164	0.7459	0.6008	0.4631	0.3995	0.2928
3.05	0.9890	0.9647	0.9330	0.8880	0.8259	0.7581	0.6170	0.4812	0.4178	0.3102
3.10	0.9897	0.9669	0.9369	0.8942	0.8350	0.7697	0.6326	0.4989	0.4358	0.3276
3.15	0.9904	0.9690	0.9407	0.9001	0.8435	0.7807	0.6477	0.5162	0.4535	0.3450
3.20	0.9910	0.9709	0.9441	0.9056	0.8515	0.7912	0.6622	0.5331	0.4709	0.3623
3.25	0.9916	0.9726	0.9474	0.9108	0.8592	0.8012	0.6762	0.5495	0.4880	0.3795
3.30	0.9922	0.9743	0.9504	0.9157	0.8664	0.8107	0.6896	0.5656	0.5048	0.3965
3.35	0.9926	0.9758	0.9532	0.9202	0.8732	0.8198	0.7026	0.5811	0.5212	0.4134
3.40	0.9931	0.9773	0.9559	0.9245	0.8796	0.8283	0.7150	0.5963	0.5372	0.4301
3.45	0.9935	0.9786	0.9584	0.9286	0.8857	0.8365	0.7269	0.6110	0.5528	0.4465
3.50	0.9939	0.9799	0.9607	0.9324	0.8914	0.8442	0.7383	0.6252	0.5680	0.4627
3.55	0.9943	0.9810	0.9629	0.9359	0.8968	0.8516	0.7493	0.6390	0.5827	0.4786
3.60	0.9946	0.9821	0.9649	0.9393	0.9020	0.8586	0.7598	0.6523	0.5971	0.4942
3.65	0.9950	0.9831	0.9668	0.9425	0.9068	0.8652	0.7698	0.6652	0.6111	0.5095
3.70	0.9953	0.9841	0.9686	0.9455	0.9114	0.8715	0.7795	0.6776	0.6247	0.5245
3.75	0.9955	0.9850	0.9703	0.9483	0.9157	0.8775	0.7887	0.6896	0.6378	0.5391
3.80	0.9958	0.9858	0.9719	0.9509	0.9198	0.8831	0.7975	0.7012	0.6505	0.5535
3.85	0.9960	0.9866	0.9734	0.9534	0.9237	0.8885	0.8059	0.7124	0.6629	0.5675
3.90	0.9963	0.9873	0.9748	0.9558	0.9274	0.8937	0.8140	0.7232	0.6748	0.5811
3.95	0.9965	0.9880	0.9761	0.9580	0.9309	0.8985	0.8217	0.7336	0.6864	0.5944

PROBABILITY INTEGRAL OF t, THE NON-CENTRAL t-STATISTIC. THIS TABLE GIVES Pr $[t/\sqrt{f} \leq x]$.

f is the number of degrees of freedom; the non-centrality parameter is $\sqrt{f+1}\,K_p$.

K_p is the standardized normal deviate exceeded with probability p.

DEGREES OF FREEDOM **5**

x \ p	.2500	.1500	.1000	.0650	.0400	.0250	.0100	.0040	.0025	.0010
4.00	0.9967	0.9886	0.9773	0.9601	0.9342	0.9031	0.8291	0.7436	0.6975	0.6073
4.05	0.9968	0.9892	0.9785	0.9620	0.9373	0.9075	0.8362	0.7532	0.7083	0.6199
4.10	0.9970	0.9898	0.9796	0.9639	0.9402	0.9117	0.8430	0.7625	0.7187	0.6321
4.15	0.9972	0.9903	0.9806	0.9656	0.9430	0.9156	0.8494	0.7714	0.7287	0.6440
4.20	0.9973	0.9908	0.9816	0.9673	0.9456	0.9194	0.8556	0.7799	0.7384	0.6556
4.25	0.9975	0.9913	0.9825	0.9689	0.9481	0.9230	0.8615	0.7882	0.7478	0.6668
4.30	0.9976	0.9917	0.9833	0.9703	0.9505	0.9263	0.8671	0.7961	0.7568	0.6776
4.35	0.9977	0.9921	0.9842	0.9717	0.9528	0.9296	0.8725	0.8037	0.7655	0.6882
4.40	0.9978	0.9925	0.9849	0.9731	0.9549	0.9326	0.8777	0.8110	0.7739	0.6984
4.45	0.9980	0.9929	0.9856	0.9743	0.9569	0.9356	0.8826	0.8181	0.7819	0.7083
4.50	0.9981	0.9932	0.9863	0.9755	0.9589	0.9383	0.8873	0.8248	0.7897	0.7178
4.55	0.9982	0.9936	0.9870	0.9766	0.9607	0.9410	0.8918	0.8313	0.7972	0.7271
4.60	0.9982	0.9939	0.9876	0.9777	0.9624	0.9435	0.8961	0.8376	0.8044	0.7361
4.65	0.9983	0.9942	0.9882	0.9787	0.9641	0.9459	0.9002	0.8436	0.8114	0.7447
4.70	0.9984	0.9944	0.9887	0.9797	0.9656	0.9482	0.9041	0.8493	0.8181	0.7531
4.75	0.9985	0.9947	0.9892	0.9806	0.9671	0.9503	0.9079	0.8548	0.8245	0.7612
4.80	0.9986	0.9950	0.9897	0.9814	0.9685	0.9524	0.9115	0.8601	0.8307	0.7691
4.85	0.9986	0.9952	0.9902	0.9822	0.9699	0.9544	0.9150	0.8652	0.8366	0.7767
4.90	0.9987	0.9954	0.9906	0.9830	0.9711	0.9562	0.9182	0.8701	0.8424	0.7840
4.95	0.9988	0.9956	0.9910	0.9837	0.9723	0.9580	0.9214	0.8748	0.8479	0.7911
5.00	0.9988	0.9958	0.9914	0.9844	0.9735	0.9597	0.9244	0.8794	0.8532	0.7979
5.05	0.9989	0.9960	0.9918	0.9851	0.9746	0.9614	0.9273	0.8837	0.8583	0.8045
5.10	0.9989	0.9962	0.9922	0.9857	0.9756	0.9629	0.9301	0.8879	0.8632	0.8109
5.15	0.9990	0.9963	0.9925	0.9863	0.9766	0.9644	0.9327	0.8919	0.8680	0.8171
5.20	0.9990	0.9965	0.9928	0.9869	0.9776	0.9658	0.9352	0.8957	0.8725	0.8230
5.25	0.9991	0.9967	0.9931	0.9875	0.9785	0.9672	0.9376	0.8994	0.8769	0.8287
5.30	0.9991	0.9968	0.9934	0.9880	0.9794	0.9684	0.9400	0.9029	0.8811	0.8343
5.35	0.9991	0.9969	0.9937	0.9885	0.9802	0.9697	0.9422	0.9064	0.8852	0.8396
5.40	0.9992	0.9971	0.9939	0.9889	0.9810	0.9708	0.9443	0.9096	0.8891	0.8448
5.45	0.9992	0.9972	0.9942	0.9894	0.9817	0.9720	0.9464	0.9128	0.8929	0.8498
5.50	0.9992	0.9973	0.9944	0.9898	0.9824	0.9730	0.9483	0.9158	0.8965	0.8546
5.55	0.9993	0.9974	0.9947	0.9902	0.9831	0.9741	0.9502	0.9187	0.9000	0.8593
5.60	0.9993	0.9975	0.9949	0.9906	0.9838	0.9750	0.9520	0.9215	0.9033	0.8637
5.65	0.9993	0.9976	0.9951	0.9910	0.9844	0.9760	0.9537	0.9242	0.9066	0.8681
5.70	0.9994	0.9977	0.9953	0.9913	0.9850	0.9769	0.9554	0.9268	0.9097	0.8723
5.75	0.9994	0.9978	0.9955	0.9917	0.9856	0.9777	0.9569	0.9293	0.9127	0.8763
5.80	0.9994	0.9979	0.9956	0.9920	0.9861	0.9785	0.9585	0.9317	0.9155	0.8802
5.85	0.9994	0.9980	0.9958	0.9923	0.9866	0.9793	0.9599	0.9340	0.9183	0.8840
5.90	0.9995	0.9981	0.9960	0.9926	0.9871	0.9801	0.9613	0.9362	0.9210	0.8876
5.95	0.9995	0.9981	0.9961	0.9929	0.9876	0.9808	0.9627	0.9383	0.9236	0.8911
6.00	0.9995	0.9982	0.9963	0.9931	0.9880	0.9815	0.9640	0.9403	0.9261	0.8945
6.05	0.9995	0.9983	0.9964	0.9934	0.9885	0.9822	0.9652	0.9423	0.9284	0.8977
6.10	0.9995	0.9983	0.9965	0.9936	0.9889	0.9828	0.9664	0.9442	0.9307	0.9009
6.15	0.9996	0.9984	0.9967	0.9939	0.9893	0.9834	0.9675	0.9460	0.9330	0.9039
6.20	0.9996	0.9985	0.9968	0.9941	0.9897	0.9840	0.9686	0.9478	0.9351	0.9069
6.25	0.9996	0.9985	0.9969	0.9943	0.9900	0.9845	0.9697	0.9495	0.9372	0.9097
6.30	0.9996	0.9986	0.9970	0.9945	0.9904	0.9851	0.9707	0.9511	0.9392	0.9125
6.35	0.9996	0.9986	0.9971	0.9947	0.9907	0.9856	0.9717	0.9527	0.9411	0.9151
6.40	0.9996	0.9987	0.9972	0.9949	0.9911	0.9861	0.9726	0.9542	0.9429	0.9177
6.45	0.9997	0.9987	0.9973	0.9951	0.9914	0.9866	0.9735	0.9557	0.9447	0.9202
6.50	0.9997	0.9988	0.9974	0.9952	0.9917	0.9870	0.9744	0.9571	0.9464	0.9226
6.55	0.9997	0.9988	0.9975	0.9954	0.9920	0.9875	0.9752	0.9584	0.9481	0.9249
6.60	0.9997	0.9989	0.9976	0.9956	0.9922	0.9879	0.9760	0.9597	0.9497	0.9271
6.65	0.9997	0.9989	0.9977	0.9957	0.9925	0.9883	0.9768	0.9610	0.9512	0.9293
6.70	0.9997	0.9989	0.9978	0.9959	0.9927	0.9887	0.9775	0.9622	0.9527	0.9314
6.75	0.9997	0.9990	0.9979	0.9960	0.9930	0.9890	0.9783	0.9634	0.9541	0.9334
6.80	0.9997	0.9990	0.9979	0.9961	0.9932	0.9894	0.9790	0.9645	0.9555	0.9353
6.85	0.9997	0.9991	0.9980	0.9963	0.9934	0.9897	0.9796	0.9656	0.9569	0.9372
6.90	0.9998	0.9991	0.9981	0.9964	0.9937	0.9901	0.9803	0.9666	0.9582	0.9390
6.95	0.9998	0.9991	0.9981	0.9965	0.9939	0.9904	0.9809	0.9676	0.9594	0.9408

PROBABILITY INTEGRAL OF t, THE NON-CENTRAL t-STATISTIC. THIS TABLE GIVES Pr $[t/\sqrt{f} \leq x]$.

f is the number of degrees of freedom; the non-centrality parameter is $\sqrt{f+1}\,K_p$.

K_p is the standardized normal deviate exceeded with probability p.

DEGREES OF FREEDOM **5**

x \ p	.2500	.1500	.1000	.0650	.0400	.0250	.0100	.0040	.0025	.0010
7.00	0.9998	0.9991	0.9982	0.9966	0.9941	0.9907	0.9815	0.9686	0.9606	0.9425
7.05	0.9998	0.9992	0.9983	0.9967	0.9943	0.9910	0.9820	0.9695	0.9618	0.9441
7.10	0.9998	0.9992	0.9983	0.9968	0.9944	0.9913	0.9826	0.9705	0.9629	0.9457
7.15	0.9998	0.9992	0.9984	0.9969	0.9946	0.9916	0.9831	0.9713	0.9640	0.9473
7.20	0.9998	0.9993	0.9984	0.9970	0.9948	0.9918	0.9836	0.9722	0.9650	0.9488
7.25	0.9998	0.9993	0.9985	0.9971	0.9950	0.9921	0.9841	0.9730	0.9661	0.9502
7.30	0.9998	0.9993	0.9985	0.9972	0.9951	0.9923	0.9846	0.9738	0.9670	0.9516
7.35	0.9998	0.9993	0.9986	0.9973	0.9953	0.9926	0.9851	0.9746	0.9680	0.9530
7.40	0.9998	0.9993	0.9986	0.9974	0.9954	0.9928	0.9855	0.9753	0.9689	0.9543
7.45	0.9998	0.9994	0.9987	0.9975	0.9956	0.9930	0.9860	0.9760	0.9698	0.9555
7.50	0.9998	0.9994	0.9987	0.9976	0.9957	0.9932	0.9864	0.9767	0.9707	0.9568
7.55	0.9998	0.9994	0.9987	0.9976	0.9958	0.9934	0.9868	0.9774	0.9715	0.9580
7.60	0.9999	0.9994	0.9988	0.9977	0.9960	0.9936	0.9872	0.9780	0.9723	0.9591
7.65	0.9999	0.9994	0.9988	0.9978	0.9961	0.9938	0.9875	0.9786	0.9731	0.9602
7.70	0.9999	0.9995	0.9989	0.9979	0.9962	0.9940	0.9879	0.9792	0.9738	0.9613
7.75	0.9999	0.9995	0.9989	0.9979	0.9963	0.9942	0.9882	0.9798	0.9745	0.9624
7.80	0.9999	0.9995	0.9989	0.9980	0.9964	0.9943	0.9886	0.9804	0.9752	0.9634
7.85	0.9999	0.9995	0.9990	0.9980	0.9965	0.9945	0.9889	0.9809	0.9759	0.9644
7.90	0.9999	0.9995	0.9990	0.9981	0.9966	0.9947	0.9892	0.9815	0.9766	0.9653
7.95	0.9999	0.9995	0.9990	0.9982	0.9967	0.9948	0.9895	0.9820	0.9772	0.9662
8.00	0.9999	0.9996	0.9990	0.9982	0.9968	0.9950	0.9898	0.9825	0.9778	0.9671
8.05	0.9999	0.9996	0.9991	0.9983	0.9969	0.9951	0.9901	0.9830	0.9784	0.9680
8.10	0.9999	0.9996	0.9991	0.9983	0.9970	0.9953	0.9904	0.9834	0.9790	0.9688
8.15	0.9999	0.9996	0.9991	0.9984	0.9971	0.9954	0.9906	0.9839	0.9796	0.9696
8.20	0.9999	0.9996	0.9992	0.9984	0.9972	0.9955	0.9909	0.9843	0.9801	0.9704
8.25	0.9999	0.9996	0.9992	0.9985	0.9972	0.9956	0.9911	0.9847	0.9806	0.9712
8.30	0.9999	0.9996	0.9992	0.9985	0.9973	0.9958	0.9914	0.9851	0.9812	0.9719
8.35	0.9999	0.9996	0.9992	0.9985	0.9974	0.9959	0.9916	0.9855	0.9817	0.9727
8.40	0.9999	0.9997	0.9993	0.9986	0.9975	0.9960	0.9919	0.9859	0.9821	0.9734
8.45	0.9999	0.9997	0.9993	0.9986	0.9975	0.9961	0.9921	0.9863	0.9826	0.9740
8.50	0.9999	0.9997	0.9993	0.9987	0.9976	0.9962	0.9923	0.9866	0.9830	0.9747
8.55	0.9999	0.9997	0.9993	0.9987	0.9977	0.9963	0.9925	0.9870	0.9835	0.9753
8.60	0.9999	0.9997	0.9993	0.9987	0.9977	0.9964	0.9927	0.9873	0.9839	0.9760
8.65	0.9999	0.9997	0.9994	0.9988	0.9978	0.9965	0.9929	0.9877	0.9843	0.9766
8.70	0.9999	0.9997	0.9994	0.9988	0.9979	0.9966	0.9931	0.9880	0.9847	0.9772
8.75	0.9999	0.9997	0.9994	0.9988	0.9979	0.9967	0.9932	0.9883	0.9851	0.9777
8.80	0.9999	0.9997	0.9994	0.9989	0.9980	0.9968	0.9934	0.9886	0.9855	0.9783
8.85	0.9999	0.9997	0.9994	0.9989	0.9980	0.9969	0.9936	0.9889	0.9859	0.9788
8.90	0.9999	0.9997	0.9994	0.9989	0.9981	0.9970	0.9938	0.9891	0.9862	0.9793
8.95	0.9999	0.9997	0.9995	0.9990	0.9981	0.9970	0.9939	0.9894	0.9865	0.9798
9.00	0.9999	0.9998	0.9995	0.9990	0.9982	0.9971	0.9941	0.9897	0.9869	0.9803
9.05	0.9999	0.9998	0.9995	0.9990	0.9982	0.9972	0.9942	0.9899	0.9872	0.9808
9.10	0.9999	0.9998	0.9995	0.9990	0.9983	0.9973	0.9944	0.9902	0.9875	0.9813
9.15	0.9999	0.9998	0.9995	0.9991	0.9983	0.9973	0.9945	0.9904	0.9878	0.9817
9.20	1.0000	0.9998	0.9995	0.9991	0.9984	0.9974	0.9947	0.9907	0.9881	0.9822
9.25	1.0000	0.9998	0.9995	0.9991	0.9984	0.9975	0.9948	0.9909	0.9884	0.9826
9.30	1.0000	0.9998	0.9995	0.9991	0.9984	0.9975	0.9949	0.9911	0.9887	0.9830
9.35	1.0000	0.9998	0.9996	0.9992	0.9985	0.9976	0.9950	0.9913	0.9890	0.9834
9.40	1.0000	0.9998	0.9996	0.9992	0.9985	0.9976	0.9952	0.9915	0.9892	0.9838
9.45	1.0000	0.9998	0.9996	0.9992	0.9986	0.9977	0.9953	0.9918	0.9895	0.9842
9.50	1.0000	0.9998	0.9996	0.9992	0.9986	0.9978	0.9954	0.9920	0.9897	0.9845
9.55	1.0000	0.9998	0.9996	0.9992	0.9986	0.9978	0.9955	0.9921	0.9900	0.9849
9.60	1.0000	0.9998	0.9996	0.9993	0.9987	0.9979	0.9956	0.9923	0.9902	0.9853
9.65	1.0000	0.9998	0.9996	0.9993	0.9987	0.9979	0.9957	0.9925	0.9904	0.9856
9.70	1.0000	0.9998	0.9996	0.9993	0.9987	0.9980	0.9958	0.9927	0.9907	0.9859
9.75	1.0000	0.9998	0.9996	0.9993	0.9988	0.9980	0.9959	0.9929	0.9909	0.9862
9.80	1.0000	0.9998	0.9997	0.9993	0.9988	0.9981	0.9960	0.9930	0.9911	0.9866
9.85	1.0000	0.9998	0.9997	0.9993	0.9988	0.9981	0.9961	0.9932	0.9913	0.9869
9.90	1.0000	0.9999	0.9997	0.9994	0.9989	0.9982	0.9962	0.9933	0.9915	0.9872
9.95	1.0000	0.9999	0.9997	0.9994	0.9989	0.9982	0.9963	0.9935	0.9917	0.9875

PROBABILITY INTEGRAL OF t, THE NON-CENTRAL t-STATISTIC. THIS TABLE GIVES Pr $[t/\sqrt{f} \leq x]$.

f is the number of degrees of freedom; the non-centrality parameter is $\sqrt{f+1}\,K_p$.

K_p is the standardized normal deviate exceeded with probability p.

DEGREES OF FREEDOM **5**

x \ p	.2500	.1500	.1000	.0650	.0400	.0250	.0100	.0040	.0025	.0010
10.00	1.0000	0.9999	0.9997	0.9994	0.9989	0.9983	0.9964	0.9937	0.9919	0.9877
10.05	1.0000	0.9999	0.9997	0.9994	0.9989	0.9983	0.9965	0.9938	0.9921	0.9880
10.10	1.0000	0.9999	0.9997	0.9994	0.9990	0.9983	0.9966	0.9939	0.9923	0.9883
10.15	1.0000	0.9999	0.9997	0.9994	0.9990	0.9984	0.9966	0.9941	0.9924	0.9885
10.20	1.0000	0.9999	0.9997	0.9995	0.9990	0.9984	0.9967	0.9942	0.9926	0.9888
10.25	1.0000	0.9999	0.9997	0.9995	0.9990	0.9984	0.9968	0.9943	0.9928	0.9890
10.30	1.0000	0.9999	0.9997	0.9995	0.9991	0.9985	0.9969	0.9945	0.9929	0.9893
10.35	1.0000	0.9999	0.9997	0.9995	0.9991	0.9985	0.9969	0.9946	0.9931	0.9895
10.40	1.0000	0.9999	0.9997	0.9995	0.9991	0.9986	0.9970	0.9947	0.9932	0.9897
10.45	1.0000	0.9999	0.9998	0.9995	0.9991	0.9986	0.9971	0.9948	0.9934	0.9900
10.50	1.0000	0.9999	0.9998	0.9995	0.9991	0.9986	0.9971	0.9949	0.9935	0.9902
10.55	1.0000	0.9999	0.9998	0.9995	0.9992	0.9987	0.9972	0.9951	0.9937	0.9904
10.60	1.0000	0.9999	0.9998	0.9995	0.9992	0.9987	0.9973	0.9952	0.9938	0.9906
10.65	1.0000	0.9999	0.9998	0.9996	0.9992	0.9987	0.9973	0.9953	0.9939	0.9908
10.70	1.0000	0.9999	0.9998	0.9996	0.9992	0.9987	0.9974	0.9954	0.9941	0.9910
10.75	1.0000	0.9999	0.9998	0.9996	0.9992	0.9988	0.9974	0.9955	0.9942	0.9912
10.80	1.0000	0.9999	0.9998	0.9996	0.9993	0.9988	0.9975	0.9956	0.9943	0.9914
10.85	1.0000	0.9999	0.9998	0.9996	0.9993	0.9988	0.9975	0.9957	0.9944	0.9916
10.90	1.0000	0.9999	0.9998	0.9996	0.9993	0.9988	0.9976	0.9958	0.9946	0.9917
10.95	1.0000	0.9999	0.9998	0.9996	0.9993	0.9989	0.9977	0.9958	0.9947	0.9919
11.00	1.0000	0.9999	0.9998	0.9996	0.9993	0.9989	0.9977	0.9959	0.9948	0.9921
11.05	1.0000	0.9999	0.9998	0.9996	0.9993	0.9989	0.9978	0.9960	0.9949	0.9922
11.10	1.0000	0.9999	0.9998	0.9996	0.9993	0.9989	0.9978	0.9961	0.9950	0.9924
11.15	1.0000	0.9999	0.9998	0.9997	0.9994	0.9990	0.9978	0.9962	0.9951	0.9926
11.20	1.0000	0.9999	0.9998	0.9997	0.9994	0.9990	0.9979	0.9963	0.9952	0.9927
11.25	1.0000	0.9999	0.9998	0.9997	0.9994	0.9990	0.9979	0.9963	0.9953	0.9929
11.30	1.0000	0.9999	0.9998	0.9997	0.9994	0.9990	0.9980	0.9964	0.9954	0.9930
11.35	1.0000	0.9999	0.9998	0.9997	0.9994	0.9991	0.9980	0.9965	0.9955	0.9931
11.40	1.0000	0.9999	0.9998	0.9997	0.9994	0.9991	0.9981	0.9966	0.9956	0.9933
11.45	1.0000	0.9999	0.9998	0.9997	0.9994	0.9991	0.9981	0.9966	0.9957	0.9934
11.50	1.0000	0.9999	0.9999	0.9997	0.9995	0.9991	0.9981	0.9967	0.9958	0.9936
11.55	1.0000	0.9999	0.9999	0.9997	0.9995	0.9991	0.9982	0.9968	0.9959	0.9937
11.60	1.0000	0.9999	0.9999	0.9997	0.9995	0.9992	0.9982	0.9968	0.9959	0.9938
11.65	1.0000	0.9999	0.9999	0.9997	0.9995	0.9992	0.9983	0.9969	0.9960	0.9939
11.70	1.0000	0.9999	0.9999	0.9997	0.9995	0.9992	0.9983	0.9970	0.9961	0.9941
11.75	1.0000	0.9999	0.9999	0.9997	0.9995	0.9992	0.9983	0.9970	0.9962	0.9942
11.80	1.0000	0.9999	0.9999	0.9997	0.9995	0.9992	0.9984	0.9971	0.9963	0.9943
11.85	1.0000	1.0000	0.9999	0.9997	0.9995	0.9992	0.9984	0.9971	0.9963	0.9944
11.90	1.0000	1.0000	0.9999	0.9998	0.9995	0.9993	0.9984	0.9972	0.9964	0.9945
11.95	1.0000	1.0000	0.9999	0.9998	0.9995	0.9993	0.9985	0.9973	0.9965	0.9946
12.00	1.0000	1.0000	0.9999	0.9998	0.9996	0.9993	0.9985	0.9973	0.9965	0.9947
12.05	1.0000	1.0000	0.9999	0.9998	0.9996	0.9993	0.9985	0.9974	0.9966	0.9948
12.10	1.0000	1.0000	0.9999	0.9998	0.9996	0.9993	0.9986	0.9974	0.9967	0.9949
12.15	1.0000	1.0000	0.9999	0.9998	0.9996	0.9993	0.9986	0.9975	0.9967	0.9950
12.20	1.0000	1.0000	0.9999	0.9998	0.9996	0.9993	0.9986	0.9975	0.9968	0.9951
12.25	1.0000	1.0000	0.9999	0.9998	0.9996	0.9994	0.9986	0.9976	0.9969	0.9952
12.30	1.0000	1.0000	0.9999	0.9998	0.9996	0.9994	0.9987	0.9976	0.9969	0.9953
12.35	1.0000	1.0000	0.9999	0.9998	0.9996	0.9994	0.9987	0.9977	0.9970	0.9954
12.40	1.0000	1.0000	0.9999	0.9998	0.9996	0.9994	0.9987	0.9977	0.9970	0.9955
12.45	1.0000	1.0000	0.9999	0.9998	0.9996	0.9994	0.9987	0.9977	0.9971	0.9956
12.50	1.0000	1.0000	0.9999	0.9998	0.9996	0.9994	0.9988	0.9978	0.9972	0.9956
12.55	1.0000	1.0000	0.9999	0.9998	0.9996	0.9994	0.9988	0.9978	0.9972	0.9957
12.60	1.0000	1.0000	0.9999	0.9998	0.9997	0.9994	0.9988	0.9979	0.9973	0.9958
12.65	1.0000	1.0000	0.9999	0.9998	0.9997	0.9995	0.9988	0.9979	0.9973	0.9959
12.70	1.0000	1.0000	0.9999	0.9998	0.9997	0.9995	0.9989	0.9980	0.9974	0.9960
12.75	1.0000	1.0000	0.9999	0.9998	0.9997	0.9995	0.9989	0.9980	0.9974	0.9960
12.80	1.0000	1.0000	0.9999	0.9998	0.9997	0.9995	0.9989	0.9980	0.9975	0.9961
12.85	1.0000	1.0000	0.9999	0.9998	0.9997	0.9995	0.9989	0.9981	0.9975	0.9962
12.90	1.0000	1.0000	0.9999	0.9998	0.9997	0.9995	0.9989	0.9981	0.9976	0.9962
12.95	1.0000	1.0000	0.9999	0.9998	0.9997	0.9995	0.9990	0.9981	0.9976	0.9963

PROBABILITY INTEGRAL OF t, THE NON-CENTRAL t-STATISTIC. THIS TABLE GIVES Pr $[t/\sqrt{f} \leq x]$.

f is the number of degrees of freedom; the non-centrality parameter is $\sqrt{f+1}\,K_p$.
K_p is the standardized normal deviate exceeded with probability p.

DEGREES OF FREEDOM **5**

x \ p	.2500	.1500	.1000	.0650	.0400	.0250	.0100	.0040	.0025	.0010
13.00	1.0000	1.0000	0.9999	0.9998	0.9997	0.9995	0.9990	0.9982	0.9976	0.9964
13.05	1.0000	1.0000	0.9999	0.9998	0.9997	0.9995	0.9990	0.9982	0.9977	0.9964
13.10	1.0000	1.0000	0.9999	0.9999	0.9997	0.9995	0.9990	0.9982	0.9977	0.9965
13.15	1.0000	1.0000	0.9999	0.9999	0.9997	0.9995	0.9990	0.9983	0.9978	0.9966
13.20	1.0000	1.0000	0.9999	0.9999	0.9997	0.9996	0.9991	0.9983	0.9978	0.9966
13.25	1.0000	1.0000	0.9999	0.9999	0.9997	0.9996	0.9991	0.9983	0.9979	0.9967
13.30	1.0000	1.0000	0.9999	0.9999	0.9997	0.9996	0.9991	0.9984	0.9979	0.9968
13.35	1.0000	1.0000	0.9999	0.9999	0.9997	0.9996	0.9991	0.9984	0.9979	0.9968
13.40	1.0000	1.0000	0.9999	0.9999	0.9998	0.9996	0.9991	0.9984	0.9980	0.9969
13.45	1.0000	1.0000	0.9999	0.9999	0.9998	0.9996	0.9991	0.9984	0.9980	0.9969
13.50	1.0000	1.0000	0.9999	0.9999	0.9998	0.9996	0.9992	0.9985	0.9980	0.9970
13.55	1.0000	1.0000	0.9999	0.9999	0.9998	0.9996	0.9992	0.9985	0.9981	0.9970
13.60	1.0000	1.0000	0.9999	0.9999	0.9998	0.9996	0.9992	0.9985	0.9981	0.9971
13.65	1.0000	1.0000	0.9999	0.9999	0.9998	0.9996	0.9992	0.9986	0.9981	0.9971
13.70	1.0000	1.0000	0.9999	0.9999	0.9998	0.9996	0.9992	0.9986	0.9982	0.9972
13.75	1.0000	1.0000	1.0000	0.9999	0.9998	0.9996	0.9992	0.9986	0.9982	0.9972
13.80	1.0000	1.0000	1.0000	0.9999	0.9998	0.9996	0.9992	0.9986	0.9982	0.9973
13.85	1.0000	1.0000	1.0000	0.9999	0.9998	0.9997	0.9993	0.9987	0.9983	0.9973
13.90	1.0000	1.0000	1.0000	0.9999	0.9998	0.9997	0.9993	0.9987	0.9983	0.9974
13.95	1.0000	1.0000	1.0000	0.9999	0.9998	0.9997	0.9993	0.9987	0.9983	0.9974
14.00	1.0000	1.0000	1.0000	0.9999	0.9998	0.9997	0.9993	0.9987	0.9984	0.9975
14.05	1.0000	1.0000	1.0000	0.9999	0.9998	0.9997	0.9993	0.9987	0.9984	0.9975
14.10	1.0000	1.0000	1.0000	0.9999	0.9998	0.9997	0.9993	0.9988	0.9984	0.9975
14.15	1.0000	1.0000	1.0000	0.9999	0.9998	0.9997	0.9993	0.9988	0.9984	0.9976
14.20	1.0000	1.0000	1.0000	0.9999	0.9998	0.9997	0.9993	0.9988	0.9985	0.9976
14.25	1.0000	1.0000	1.0000	0.9999	0.9998	0.9997	0.9994	0.9988	0.9985	0.9977
14.30	1.0000	1.0000	1.0000	0.9999	0.9998	0.9997	0.9994	0.9989	0.9985	0.9977
14.35	1.0000	1.0000	1.0000	0.9999	0.9998	0.9997	0.9994	0.9989	0.9985	0.9977
14.40	1.0000	1.0000	1.0000	0.9999	0.9998	0.9997	0.9994	0.9989	0.9986	0.9978
14.45	1.0000	1.0000	1.0000	0.9999	0.9998	0.9997	0.9994	0.9989	0.9986	0.9978
14.50	1.0000	1.0000	1.0000	0.9999	0.9998	0.9997	0.9994	0.9989	0.9986	0.9979
14.55	1.0000	1.0000	1.0000	0.9999	0.9998	0.9997	0.9994	0.9989	0.9986	0.9979
14.60	1.0000	1.0000	1.0000	0.9999	0.9998	0.9997	0.9994	0.9990	0.9987	0.9979
14.65	1.0000	1.0000	1.0000	0.9999	0.9998	0.9997	0.9994	0.9990	0.9987	0.9980
14.70	1.0000	1.0000	1.0000	0.9999	0.9999	0.9997	0.9994	0.9990	0.9987	0.9980
14.75	1.0000	1.0000	1.0000	0.9999	0.9999	0.9998	0.9995	0.9990	0.9987	0.9980
14.80	1.0000	1.0000	1.0000	0.9999	0.9999	0.9998	0.9995	0.9990	0.9987	0.9981
14.85	1.0000	1.0000	1.0000	0.9999	0.9999	0.9998	0.9995	0.9990	0.9988	0.9981
14.90	1.0000	1.0000	1.0000	0.9999	0.9999	0.9998	0.9995	0.9991	0.9988	0.9981
14.95	1.0000	1.0000	1.0000	0.9999	0.9999	0.9998	0.9995	0.9991	0.9988	0.9981
15.00	1.0000	1.0000	1.0000	0.9999	0.9999	0.9998	0.9995	0.9991	0.9988	0.9982
15.05	1.0000	1.0000	1.0000	0.9999	0.9999	0.9998	0.9995	0.9991	0.9988	0.9982
15.10	1.0000	1.0000	1.0000	0.9999	0.9999	0.9998	0.9995	0.9991	0.9989	0.9982
15.15	1.0000	1.0000	1.0000	0.9999	0.9999	0.9998	0.9995	0.9991	0.9989	0.9983
15.20	1.0000	1.0000	1.0000	0.9999	0.9999	0.9998	0.9995	0.9992	0.9989	0.9983
15.25	1.0000	1.0000	1.0000	0.9999	0.9999	0.9998	0.9995	0.9992	0.9989	0.9983
15.30	1.0000	1.0000	1.0000	0.9999	0.9999	0.9998	0.9996	0.9992	0.9989	0.9983
15.35	1.0000	1.0000	1.0000	0.9999	0.9999	0.9998	0.9996	0.9992	0.9990	0.9984
15.40	1.0000	1.0000	1.0000	0.9999	0.9999	0.9998	0.9996	0.9992	0.9990	0.9984
15.45	1.0000	1.0000	1.0000	0.9999	0.9999	0.9998	0.9996	0.9992	0.9990	0.9984
15.50	1.0000	1.0000	1.0000	0.9999	0.9999	0.9998	0.9996	0.9992	0.9990	0.9984
15.55	1.0000	1.0000	1.0000	1.0000	0.9999	0.9998	0.9996	0.9992	0.9990	0.9985
15.60	1.0000	1.0000	1.0000	1.0000	0.9999	0.9998	0.9996	0.9993	0.9990	0.9985
15.65	1.0000	1.0000	1.0000	1.0000	0.9999	0.9998	0.9996	0.9993	0.9990	0.9985
15.70	1.0000	1.0000	1.0000	1.0000	0.9999	0.9998	0.9996	0.9993	0.9991	0.9985
15.75	1.0000	1.0000	1.0000	1.0000	0.9999	0.9998	0.9996	0.9993	0.9991	0.9986
15.80	1.0000	1.0000	1.0000	1.0000	0.9999	0.9998	0.9996	0.9993	0.9991	0.9986
15.85	1.0000	1.0000	1.0000	1.0000	0.9999	0.9998	0.9996	0.9993	0.9991	0.9986
15.90	1.0000	1.0000	1.0000	1.0000	0.9999	0.9998	0.9996	0.9993	0.9991	0.9986
15.95	1.0000	1.0000	1.0000	1.0000	0.9999	0.9998	0.9996	0.9993	0.9991	0.9986

PROBABILITY INTEGRAL OF t, THE NON-CENTRAL t-STATISTIC. THIS TABLE GIVES Pr $[t/\sqrt{f} \leq x]$.

f is the number of degrees of freedom; the non-centrality parameter is $\sqrt{f+1}\,K_p$.

K_p is the standardized normal deviate exceeded with probability p.

DEGREES OF FREEDOM **5**

x \ p	.2500	.1500	.1000	.0650	.0400	.0250	.0100	.0040	.0025	.0010
16.00	1.0000	1.0000	1.0000	1.0000	0.9999	0.9998	0.9996	0.9993	0.9991	0.9987
16.05	1.0000	1.0000	1.0000	1.0000	0.9999	0.9998	0.9996	0.9994	0.9992	0.9987
16.10	1.0000	1.0000	1.0000	1.0000	0.9999	0.9998	0.9997	0.9994	0.9992	0.9987
16.15	1.0000	1.0000	1.0000	1.0000	0.9999	0.9999	0.9997	0.9994	0.9992	0.9987
16.20	1.0000	1.0000	1.0000	1.0000	0.9999	0.9999	0.9997	0.9994	0.9992	0.9987
16.25	1.0000	1.0000	1.0000	1.0000	0.9999	0.9999	0.9997	0.9994	0.9992	0.9988
16.30	1.0000	1.0000	1.0000	1.0000	0.9999	0.9999	0.9997	0.9994	0.9992	0.9988
16.35	1.0000	1.0000	1.0000	1.0000	0.9999	0.9999	0.9997	0.9994	0.9992	0.9988
16.40	1.0000	1.0000	1.0000	1.0000	0.9999	0.9999	0.9997	0.9994	0.9992	0.9988
16.45	1.0000	1.0000	1.0000	1.0000	0.9999	0.9999	0.9997	0.9994	0.9993	0.9988
16.50	1.0000	1.0000	1.0000	1.0000	0.9999	0.9999	0.9997	0.9994	0.9993	0.9989
16.55	1.0000	1.0000	1.0000	1.0000	0.9999	0.9999	0.9997	0.9994	0.9993	0.9989
16.60	1.0000	1.0000	1.0000	1.0000	0.9999	0.9999	0.9997	0.9995	0.9993	0.9989
16.65	1.0000	1.0000	1.0000	1.0000	0.9999	0.9999	0.9997	0.9995	0.9993	0.9989
16.70	1.0000	1.0000	1.0000	1.0000	0.9999	0.9999	0.9997	0.9995	0.9993	0.9989
16.75	1.0000	1.0000	1.0000	1.0000	0.9999	0.9999	0.9997	0.9995	0.9993	0.9989
16.80	1.0000	1.0000	1.0000	1.0000	0.9999	0.9999	0.9997	0.9995	0.9993	0.9990
16.85	1.0000	1.0000	1.0000	1.0000	0.9999	0.9999	0.9997	0.9995	0.9993	0.9990
16.90	1.0000	1.0000	1.0000	1.0000	0.9999	0.9999	0.9997	0.9995	0.9994	0.9990
16.95	1.0000	1.0000	1.0000	1.0000	0.9999	0.9999	0.9997	0.9995	0.9994	0.9990
17.00	1.0000	1.0000	1.0000	1.0000	0.9999	0.9999	0.9997	0.9995	0.9994	0.9990
17.05	1.0000	1.0000	1.0000	1.0000	0.9999	0.9999	0.9997	0.9995	0.9994	0.9990
17.10	1.0000	1.0000	1.0000	1.0000	0.9999	0.9999	0.9998	0.9995	0.9994	0.9990
17.15	1.0000	1.0000	1.0000	1.0000	0.9999	0.9999	0.9998	0.9995	0.9994	0.9991
17.20	1.0000	1.0000	1.0000	1.0000	0.9999	0.9999	0.9998	0.9995	0.9994	0.9991
17.25	1.0000	1.0000	1.0000	1.0000	0.9999	0.9999	0.9998	0.9996	0.9994	0.9991
17.30	1.0000	1.0000	1.0000	1.0000	0.9999	0.9999	0.9998	0.9996	0.9994	0.9991
17.35	1.0000	1.0000	1.0000	1.0000	0.9999	0.9999	0.9998	0.9996	0.9994	0.9991
17.40	1.0000	1.0000	1.0000	1.0000	1.0000	0.9999	0.9998	0.9996	0.9994	0.9991
17.45	1.0000	1.0000	1.0000	1.0000	1.0000	0.9999	0.9998	0.9996	0.9994	0.9991
17.50	1.0000	1.0000	1.0000	1.0000	1.0000	0.9999	0.9998	0.9996	0.9995	0.9991
17.55	1.0000	1.0000	1.0000	1.0000	1.0000	0.9999	0.9998	0.9996	0.9995	0.9992
17.60	1.0000	1.0000	1.0000	1.0000	1.0000	0.9999	0.9998	0.9996	0.9995	0.9992
17.65	1.0000	1.0000	1.0000	1.0000	1.0000	0.9999	0.9998	0.9996	0.9995	0.9992
17.70	1.0000	1.0000	1.0000	1.0000	1.0000	0.9999	0.9998	0.9996	0.9995	0.9992
17.75	1.0000	1.0000	1.0000	1.0000	1.0000	0.9999	0.9998	0.9996	0.9995	0.9992
17.80	1.0000	1.0000	1.0000	1.0000	1.0000	0.9999	0.9998	0.9996	0.9995	0.9992
17.85	1.0000	1.0000	1.0000	1.0000	1.0000	0.9999	0.9998	0.9996	0.9995	0.9992
17.90	1.0000	1.0000	1.0000	1.0000	1.0000	0.9999	0.9998	0.9996	0.9995	0.9992
17.95	1.0000	1.0000	1.0000	1.0000	1.0000	0.9999	0.9998	0.9996	0.9995	0.9992
18.00	1.0000	1.0000	1.0000	1.0000	1.0000	0.9999	0.9998	0.9996	0.9995	0.9993
18.05	1.0000	1.0000	1.0000	1.0000	1.0000	0.9999	0.9998	0.9996	0.9995	0.9993
18.10	1.0000	1.0000	1.0000	1.0000	1.0000	0.9999	0.9998	0.9997	0.9995	0.9993
18.15	1.0000	1.0000	1.0000	1.0000	1.0000	0.9999	0.9998	0.9997	0.9996	0.9993
18.20	1.0000	1.0000	1.0000	1.0000	1.0000	0.9999	0.9998	0.9997	0.9996	0.9993
18.25	1.0000	1.0000	1.0000	1.0000	1.0000	0.9999	0.9998	0.9997	0.9996	0.9993
18.30	1.0000	1.0000	1.0000	1.0000	1.0000	0.9999	0.9998	0.9997	0.9996	0.9993
18.35	1.0000	1.0000	1.0000	1.0000	1.0000	0.9999	0.9998	0.9997	0.9996	0.9993
18.40	1.0000	1.0000	1.0000	1.0000	1.0000	0.9999	0.9998	0.9997	0.9996	0.9993
18.45	1.0000	1.0000	1.0000	1.0000	1.0000	0.9999	0.9998	0.9997	0.9996	0.9993
18.50	1.0000	1.0000	1.0000	1.0000	1.0000	0.9999	0.9998	0.9997	0.9996	0.9994
18.55	1.0000	1.0000	1.0000	1.0000	1.0000	0.9999	0.9998	0.9997	0.9996	0.9994
18.60	1.0000	1.0000	1.0000	1.0000	1.0000	0.9999	0.9998	0.9997	0.9996	0.9994
18.65	1.0000	1.0000	1.0000	1.0000	1.0000	0.9999	0.9998	0.9997	0.9996	0.9994
18.70	1.0000	1.0000	1.0000	1.0000	1.0000	0.9999	0.9998	0.9997	0.9996	0.9994
18.75	1.0000	1.0000	1.0000	1.0000	1.0000	0.9999	0.9999	0.9997	0.9996	0.9994
18.80	1.0000	1.0000	1.0000	1.0000	1.0000	0.9999	0.9999	0.9997	0.9996	0.9994
18.85	1.0000	1.0000	1.0000	1.0000	1.0000	0.9999	0.9999	0.9997	0.9996	0.9994
18.90	1.0000	1.0000	1.0000	1.0000	1.0000	0.9999	0.9999	0.9997	0.9996	0.9994
18.95	1.0000	1.0000	1.0000	1.0000	1.0000	0.9999	0.9999	0.9997	0.9996	0.9994

PROBABILITY INTEGRAL OF t, THE NON-CENTRAL t-STATISTIC. THIS TABLE GIVES Pr $[t/\sqrt{f} \leq x]$.

f is the number of degrees of freedom; the non-centrality parameter is $\sqrt{f+1}\,K_p$.

K_p is the standardized normal deviate exceeded with probability p.

DEGREES OF FREEDOM **5**

x \ p	.2500	.1500	.1000	.0650	.0400	.0250	.0100	.0040	.0025	.0010
19.00	1.0000	1.0000	1.0000	1.0000	1.0000	0.9999	0.9999	0.9997	0.9996	0.9994
19.05	1.0000	1.0000	1.0000	1.0000	1.0000	0.9999	0.9999	0.9997	0.9997	0.9994
19.10	1.0000	1.0000	1.0000	1.0000	1.0000	1.0000	0.9999	0.9997	0.9997	0.9995
19.15	1.0000	1.0000	1.0000	1.0000	1.0000	1.0000	0.9999	0.9997	0.9997	0.9995
19.20	1.0000	1.0000	1.0000	1.0000	1.0000	1.0000	0.9999	0.9997	0.9997	0.9995
19.25	1.0000	1.0000	1.0000	1.0000	1.0000	1.0000	0.9999	0.9998	0.9997	0.9995
19.30	1.0000	1.0000	1.0000	1.0000	1.0000	1.0000	0.9999	0.9998	0.9997	0.9995
19.35	1.0000	1.0000	1.0000	1.0000	1.0000	1.0000	0.9999	0.9998	0.9997	0.9995
19.40	1.0000	1.0000	1.0000	1.0000	1.0000	1.0000	0.9999	0.9998	0.9997	0.9995
19.45	1.0000	1.0000	1.0000	1.0000	1.0000	1.0000	0.9999	0.9998	0.9997	0.9995
19.50	1.0000	1.0000	1.0000	1.0000	1.0000	1.0000	0.9999	0.9998	0.9997	0.9995
19.55	1.0000	1.0000	1.0000	1.0000	1.0000	1.0000	0.9999	0.9998	0.9997	0.9995
19.60	1.0000	1.0000	1.0000	1.0000	1.0000	1.0000	0.9999	0.9998	0.9997	0.9995
19.65	1.0000	1.0000	1.0000	1.0000	1.0000	1.0000	0.9999	0.9998	0.9997	0.9995
19.70	1.0000	1.0000	1.0000	1.0000	1.0000	1.0000	0.9999	0.9998	0.9997	0.9995
19.75	1.0000	1.0000	1.0000	1.0000	1.0000	1.0000	0.9999	0.9998	0.9997	0.9995
19.80	1.0000	1.0000	1.0000	1.0000	1.0000	1.0000	0.9999	0.9998	0.9997	0.9995
19.85	1.0000	1.0000	1.0000	1.0000	1.0000	1.0000	0.9999	0.9998	0.9997	0.9996
19.90	1.0000	1.0000	1.0000	1.0000	1.0000	1.0000	0.9999	0.9998	0.9997	0.9996
19.95	1.0000	1.0000	1.0000	1.0000	1.0000	1.0000	0.9999	0.9998	0.9997	0.9996
20.00	1.0000	1.0000	1.0000	1.0000	1.0000	1.0000	0.9999	0.9998	0.9997	0.9996
20.05	1.0000	1.0000	1.0000	1.0000	1.0000	1.0000	0.9999	0.9998	0.9997	0.9996
20.10	1.0000	1.0000	1.0000	1.0000	1.0000	1.0000	0.9999	0.9998	0.9997	0.9996
20.15	1.0000	1.0000	1.0000	1.0000	1.0000	1.0000	0.9999	0.9998	0.9997	0.9996
20.20	1.0000	1.0000	1.0000	1.0000	1.0000	1.0000	0.9999	0.9998	0.9997	0.9996
20.25	1.0000	1.0000	1.0000	1.0000	1.0000	1.0000	0.9999	0.9998	0.9997	0.9996
20.30	1.0000	1.0000	1.0000	1.0000	1.0000	1.0000	0.9999	0.9998	0.9998	0.9996
20.35	1.0000	1.0000	1.0000	1.0000	1.0000	1.0000	0.9999	0.9998	0.9998	0.9996
20.40	1.0000	1.0000	1.0000	1.0000	1.0000	1.0000	0.9999	0.9998	0.9998	0.9996
20.45	1.0000	1.0000	1.0000	1.0000	1.0000	1.0000	0.9999	0.9998	0.9998	0.9996
20.50	1.0000	1.0000	1.0000	1.0000	1.0000	1.0000	0.9999	0.9998	0.9998	0.9996
20.55	1.0000	1.0000	1.0000	1.0000	1.0000	1.0000	0.9999	0.9998	0.9998	0.9996
20.60	1.0000	1.0000	1.0000	1.0000	1.0000	1.0000	0.9999	0.9998	0.9998	0.9996
20.65	1.0000	1.0000	1.0000	1.0000	1.0000	1.0000	0.9999	0.9998	0.9998	0.9996
20.70	1.0000	1.0000	1.0000	1.0000	1.0000	1.0000	0.9999	0.9998	0.9998	0.9996
20.75	1.0000	1.0000	1.0000	1.0000	1.0000	1.0000	0.9999	0.9998	0.9998	0.9996
20.80	1.0000	1.0000	1.0000	1.0000	1.0000	1.0000	0.9999	0.9998	0.9998	0.9996
20.85	1.0000	1.0000	1.0000	1.0000	1.0000	1.0000	0.9999	0.9998	0.9998	0.9997
20.90	1.0000	1.0000	1.0000	1.0000	1.0000	1.0000	0.9999	0.9998	0.9998	0.9997
20.95	1.0000	1.0000	1.0000	1.0000	1.0000	1.0000	0.9999	0.9998	0.9998	0.9997
21.00	1.0000	1.0000	1.0000	1.0000	1.0000	1.0000	0.9999	0.9998	0.9998	0.9997
21.05	1.0000	1.0000	1.0000	1.0000	1.0000	1.0000	0.9999	0.9999	0.9998	0.9997
21.10	1.0000	1.0000	1.0000	1.0000	1.0000	1.0000	0.9999	0.9999	0.9998	0.9997
21.15	1.0000	1.0000	1.0000	1.0000	1.0000	1.0000	0.9999	0.9999	0.9998	0.9997
21.20	1.0000	1.0000	1.0000	1.0000	1.0000	1.0000	0.9999	0.9999	0.9998	0.9997
21.25	1.0000	1.0000	1.0000	1.0000	1.0000	1.0000	0.9999	0.9999	0.9998	0.9997
21.30	1.0000	1.0000	1.0000	1.0000	1.0000	1.0000	0.9999	0.9999	0.9998	0.9997
21.35	1.0000	1.0000	1.0000	1.0000	1.0000	1.0000	0.9999	0.9999	0.9998	0.9997
21.40	1.0000	1.0000	1.0000	1.0000	1.0000	1.0000	0.9999	0.9999	0.9998	0.9997
21.45	1.0000	1.0000	1.0000	1.0000	1.0000	1.0000	0.9999	0.9999	0.9998	0.9997
21.50	1.0000	1.0000	1.0000	1.0000	1.0000	1.0000	0.9999	0.9999	0.9998	0.9997
21.55	1.0000	1.0000	1.0000	1.0000	1.0000	1.0000	0.9999	0.9999	0.9998	0.9997
21.60	1.0000	1.0000	1.0000	1.0000	1.0000	1.0000	0.9999	0.9999	0.9998	0.9997
21.65	1.0000	1.0000	1.0000	1.0000	1.0000	1.0000	0.9999	0.9999	0.9998	0.9997
21.70	1.0000	1.0000	1.0000	1.0000	1.0000	1.0000	0.9999	0.9999	0.9998	0.9997
21.75	1.0000	1.0000	1.0000	1.0000	1.0000	1.0000	0.9999	0.9999	0.9998	0.9997
21.80	1.0000	1.0000	1.0000	1.0000	1.0000	1.0000	0.9999	0.9999	0.9998	0.9997
21.85	1.0000	1.0000	1.0000	1.0000	1.0000	1.0000	0.9999	0.9999	0.9998	0.9997
21.90	1.0000	1.0000	1.0000	1.0000	1.0000	1.0000	0.9999	0.9999	0.9998	0.9997
21.95	1.0000	1.0000	1.0000	1.0000	1.0000	1.0000	0.9999	0.9999	0.9998	0.9997

PROBABILITY INTEGRAL OF t, THE NON-CENTRAL t-STATISTIC. THIS TABLE GIVES Pr $[t/\sqrt{f} \leq x]$.

f is the number of degrees of freedom; the non-centrality parameter is $\sqrt{f+1}\,K_p$.
K_p is the standardized normal deviate exceeded with probability p.

DEGREES OF FREEDOM **5**

x \ p	.2500	.1500	.1000	.0650	.0400	.0250	.0100	.0040	.0025	.0010
22.00	1.0000	1.0000	1.0000	1.0000	1.0000	1.0000	1.0000	0.9999	0.9998	0.9997
22.05	1.0000	1.0000	1.0000	1.0000	1.0000	1.0000	1.0000	0.9999	0.9998	0.9997
22.10	1.0000	1.0000	1.0000	1.0000	1.0000	1.0000	1.0000	0.9999	0.9998	0.9997
22.15	1.0000	1.0000	1.0000	1.0000	1.0000	1.0000	1.0000	0.9999	0.9999	0.9998
22.20	1.0000	1.0000	1.0000	1.0000	1.0000	1.0000	1.0000	0.9999	0.9999	0.9998
22.25	1.0000	1.0000	1.0000	1.0000	1.0000	1.0000	1.0000	0.9999	0.9999	0.9998
22.30	1.0000	1.0000	1.0000	1.0000	1.0000	1.0000	1.0000	0.9999	0.9999	0.9998
22.35	1.0000	1.0000	1.0000	1.0000	1.0000	1.0000	1.0000	0.9999	0.9999	0.9998
22.40	1.0000	1.0000	1.0000	1.0000	1.0000	1.0000	1.0000	0.9999	0.9999	0.9998
22.45	1.0000	1.0000	1.0000	1.0000	1.0000	1.0000	1.0000	0.9999	0.9999	0.9998
22.50	1.0000	1.0000	1.0000	1.0000	1.0000	1.0000	1.0000	0.9999	0.9999	0.9998
22.55	1.0000	1.0000	1.0000	1.0000	1.0000	1.0000	1.0000	0.9999	0.9999	0.9998
22.60	1.0000	1.0000	1.0000	1.0000	1.0000	1.0000	1.0000	0.9999	0.9999	0.9998
22.65	1.0000	1.0000	1.0000	1.0000	1.0000	1.0000	1.0000	0.9999	0.9999	0.9998
22.70	1.0000	1.0000	1.0000	1.0000	1.0000	1.0000	1.0000	0.9999	0.9999	0.9998
22.75	1.0000	1.0000	1.0000	1.0000	1.0000	1.0000	1.0000	0.9999	0.9999	0.9998
22.80	1.0000	1.0000	1.0000	1.0000	1.0000	1.0000	1.0000	0.9999	0.9999	0.9998
22.85	1.0000	1.0000	1.0000	1.0000	1.0000	1.0000	1.0000	0.9999	0.9999	0.9998
22.90	1.0000	1.0000	1.0000	1.0000	1.0000	1.0000	1.0000	0.9999	0.9999	0.9998
22.95	1.0000	1.0000	1.0000	1.0000	1.0000	1.0000	1.0000	0.9999	0.9999	0.9998
23.00	1.0000	1.0000	1.0000	1.0000	1.0000	1.0000	1.0000	0.9999	0.9999	0.9998
23.05	1.0000	1.0000	1.0000	1.0000	1.0000	1.0000	1.0000	0.9999	0.9999	0.9998
23.10	1.0000	1.0000	1.0000	1.0000	1.0000	1.0000	1.0000	0.9999	0.9999	0.9998
23.15	1.0000	1.0000	1.0000	1.0000	1.0000	1.0000	1.0000	0.9999	0.9999	0.9998
23.20	1.0000	1.0000	1.0000	1.0000	1.0000	1.0000	1.0000	0.9999	0.9999	0.9998
23.25	1.0000	1.0000	1.0000	1.0000	1.0000	1.0000	1.0000	0.9999	0.9999	0.9998
23.30	1.0000	1.0000	1.0000	1.0000	1.0000	1.0000	1.0000	0.9999	0.9999	0.9998
23.35	1.0000	1.0000	1.0000	1.0000	1.0000	1.0000	1.0000	0.9999	0.9999	0.9998
23.40	1.0000	1.0000	1.0000	1.0000	1.0000	1.0000	1.0000	0.9999	0.9999	0.9998
23.45	1.0000	1.0000	1.0000	1.0000	1.0000	1.0000	1.0000	0.9999	0.9999	0.9998
23.50	1.0000	1.0000	1.0000	1.0000	1.0000	1.0000	1.0000	0.9999	0.9999	0.9998
23.55	1.0000	1.0000	1.0000	1.0000	1.0000	1.0000	1.0000	0.9999	0.9999	0.9998
23.60	1.0000	1.0000	1.0000	1.0000	1.0000	1.0000	1.0000	0.9999	0.9999	0.9998
23.65	1.0000	1.0000	1.0000	1.0000	1.0000	1.0000	1.0000	0.9999	0.9999	0.9998
23.70	1.0000	1.0000	1.0000	1.0000	1.0000	1.0000	1.0000	0.9999	0.9999	0.9998
23.75	1.0000	1.0000	1.0000	1.0000	1.0000	1.0000	1.0000	0.9999	0.9999	0.9998
23.80	1.0000	1.0000	1.0000	1.0000	1.0000	1.0000	1.0000	0.9999	0.9999	0.9998
23.85	1.0000	1.0000	1.0000	1.0000	1.0000	1.0000	1.0000	0.9999	0.9999	0.9998
23.90	1.0000	1.0000	1.0000	1.0000	1.0000	1.0000	1.0000	0.9999	0.9999	0.9998
23.95	1.0000	1.0000	1.0000	1.0000	1.0000	1.0000	1.0000	0.9999	0.9999	0.9998
24.00	1.0000	1.0000	1.0000	1.0000	1.0000	1.0000	1.0000	0.9999	0.9999	0.9998
24.05	1.0000	1.0000	1.0000	1.0000	1.0000	1.0000	1.0000	0.9999	0.9999	0.9998
24.10	1.0000	1.0000	1.0000	1.0000	1.0000	1.0000	1.0000	0.9999	0.9999	0.9998
24.15	1.0000	1.0000	1.0000	1.0000	1.0000	1.0000	1.0000	0.9999	0.9999	0.9999
24.20	1.0000	1.0000	1.0000	1.0000	1.0000	1.0000	1.0000	0.9999	0.9999	0.9999
24.25	1.0000	1.0000	1.0000	1.0000	1.0000	1.0000	1.0000	0.9999	0.9999	0.9999
24.30	1.0000	1.0000	1.0000	1.0000	1.0000	1.0000	1.0000	0.9999	0.9999	0.9999
24.35	1.0000	1.0000	1.0000	1.0000	1.0000	1.0000	1.0000	0.9999	0.9999	0.9999
24.40	1.0000	1.0000	1.0000	1.0000	1.0000	1.0000	1.0000	0.9999	0.9999	0.9999
24.45	1.0000	1.0000	1.0000	1.0000	1.0000	1.0000	1.0000	0.9999	0.9999	0.9999
24.50	1.0000	1.0000	1.0000	1.0000	1.0000	1.0000	1.0000	0.9999	0.9999	0.9999
24.55	1.0000	1.0000	1.0000	1.0000	1.0000	1.0000	1.0000	0.9999	0.9999	0.9999
24.60	1.0000	1.0000	1.0000	1.0000	1.0000	1.0000	1.0000	1.0000	0.9999	0.9999
24.65	1.0000	1.0000	1.0000	1.0000	1.0000	1.0000	1.0000	1.0000	0.9999	0.9999
24.70	1.0000	1.0000	1.0000	1.0000	1.0000	1.0000	1.0000	1.0000	0.9999	0.9999
24.75	1.0000	1.0000	1.0000	1.0000	1.0000	1.0000	1.0000	1.0000	0.9999	0.9999
24.80	1.0000	1.0000	1.0000	1.0000	1.0000	1.0000	1.0000	1.0000	0.9999	0.9999
24.85	1.0000	1.0000	1.0000	1.0000	1.0000	1.0000	1.0000	1.0000	0.9999	0.9999
24.90	1.0000	1.0000	1.0000	1.0000	1.0000	1.0000	1.0000	1.0000	0.9999	0.9999
24.95	1.0000	1.0000	1.0000	1.0000	1.0000	1.0000	1.0000	1.0000	0.9999	0.9999

PROBABILITY INTEGRAL OF t, THE NON-CENTRAL t-STATISTIC. THIS TABLE GIVES Pr $[t/\sqrt{f} \leq x]$.

f is the number of degrees of freedom; the non-centrality parameter is $\sqrt{f+1}\,K_p$.

K_p is the standardized normal deviate exceeded with probability p.

DEGREES OF FREEDOM **5**

x \ p	.2500	.1500	.1000	.0650	.0400	.0250	.0100	.0040	.0025	.0010
25.00	1.0000	1.0000	1.0000	1.0000	1.0000	1.0000	1.0000	1.0000	0.9999	0.9999
25.05	1.0000	1.0000	1.0000	1.0000	1.0000	1.0000	1.0000	1.0000	0.9999	0.9999
25.10	1.0000	1.0000	1.0000	1.0000	1.0000	1.0000	1.0000	1.0000	0.9999	0.9999
25.15	1.0000	1.0000	1.0000	1.0000	1.0000	1.0000	1.0000	1.0000	0.9999	0.9999
25.20	1.0000	1.0000	1.0000	1.0000	1.0000	1.0000	1.0000	1.0000	0.9999	0.9999
25.25	1.0000	1.0000	1.0000	1.0000	1.0000	1.0000	1.0000	1.0000	0.9999	0.9999
25.30	1.0000	1.0000	1.0000	1.0000	1.0000	1.0000	1.0000	1.0000	0.9999	0.9999
25.35	1.0000	1.0000	1.0000	1.0000	1.0000	1.0000	1.0000	1.0000	0.9999	0.9999
25.40	1.0000	1.0000	1.0000	1.0000	1.0000	1.0000	1.0000	1.0000	0.9999	0.9999
25.45	1.0000	1.0000	1.0000	1.0000	1.0000	1.0000	1.0000	1.0000	0.9999	0.9999
25.50	1.0000	1.0000	1.0000	1.0000	1.0000	1.0000	1.0000	1.0000	0.9999	0.9999
25.55	1.0000	1.0000	1.0000	1.0000	1.0000	1.0000	1.0000	1.0000	0.9999	0.9999
25.60	1.0000	1.0000	1.0000	1.0000	1.0000	1.0000	1.0000	1.0000	0.9999	0.9999
25.65	1.0000	1.0000	1.0000	1.0000	1.0000	1.0000	1.0000	1.0000	0.9999	0.9999
25.70	1.0000	1.0000	1.0000	1.0000	1.0000	1.0000	1.0000	1.0000	0.9999	0.9999
25.75	1.0000	1.0000	1.0000	1.0000	1.0000	1.0000	1.0000	1.0000	0.9999	0.9999
25.80	1.0000	1.0000	1.0000	1.0000	1.0000	1.0000	1.0000	1.0000	0.9999	0.9999
25.85	1.0000	1.0000	1.0000	1.0000	1.0000	1.0000	1.0000	1.0000	1.0000	0.9999
25.90	1.0000	1.0000	1.0000	1.0000	1.0000	1.0000	1.0000	1.0000	1.0000	0.9999
25.95	1.0000	1.0000	1.0000	1.0000	1.0000	1.0000	1.0000	1.0000	1.0000	0.9999
26.00	1.0000	1.0000	1.0000	1.0000	1.0000	1.0000	1.0000	1.0000	1.0000	0.9999
26.05	1.0000	1.0000	1.0000	1.0000	1.0000	1.0000	1.0000	1.0000	1.0000	0.9999
26.10	1.0000	1.0000	1.0000	1.0000	1.0000	1.0000	1.0000	1.0000	1.0000	0.9999
26.15	1.0000	1.0000	1.0000	1.0000	1.0000	1.0000	1.0000	1.0000	1.0000	0.9999
26.20	1.0000	1.0000	1.0000	1.0000	1.0000	1.0000	1.0000	1.0000	1.0000	0.9999
26.25	1.0000	1.0000	1.0000	1.0000	1.0000	1.0000	1.0000	1.0000	1.0000	0.9999
26.30	1.0000	1.0000	1.0000	1.0000	1.0000	1.0000	1.0000	1.0000	1.0000	0.9999
26.35	1.0000	1.0000	1.0000	1.0000	1.0000	1.0000	1.0000	1.0000	1.0000	0.9999
26.40	1.0000	1.0000	1.0000	1.0000	1.0000	1.0000	1.0000	1.0000	1.0000	0.9999
26.45	1.0000	1.0000	1.0000	1.0000	1.0000	1.0000	1.0000	1.0000	1.0000	0.9999
26.50	1.0000	1.0000	1.0000	1.0000	1.0000	1.0000	1.0000	1.0000	1.0000	0.9999
26.55	1.0000	1.0000	1.0000	1.0000	1.0000	1.0000	1.0000	1.0000	1.0000	0.9999
26.60	1.0000	1.0000	1.0000	1.0000	1.0000	1.0000	1.0000	1.0000	1.0000	0.9999
26.65	1.0000	1.0000	1.0000	1.0000	1.0000	1.0000	1.0000	1.0000	1.0000	0.9999
26.70	1.0000	1.0000	1.0000	1.0000	1.0000	1.0000	1.0000	1.0000	1.0000	0.9999
26.75	1.0000	1.0000	1.0000	1.0000	1.0000	1.0000	1.0000	1.0000	1.0000	0.9999
26.80	1.0000	1.0000	1.0000	1.0000	1.0000	1.0000	1.0000	1.0000	1.0000	0.9999
26.85	1.0000	1.0000	1.0000	1.0000	1.0000	1.0000	1.0000	1.0000	1.0000	0.9999
26.90	1.0000	1.0000	1.0000	1.0000	1.0000	1.0000	1.0000	1.0000	1.0000	0.9999
26.95	1.0000	1.0000	1.0000	1.0000	1.0000	1.0000	1.0000	1.0000	1.0000	0.9999
27.00	1.0000	1.0000	1.0000	1.0000	1.0000	1.0000	1.0000	1.0000	1.0000	0.9999
27.05	1.0000	1.0000	1.0000	1.0000	1.0000	1.0000	1.0000	1.0000	1.0000	0.9999
27.10	1.0000	1.0000	1.0000	1.0000	1.0000	1.0000	1.0000	1.0000	1.0000	0.9999
27.15	1.0000	1.0000	1.0000	1.0000	1.0000	1.0000	1.0000	1.0000	1.0000	0.9999
27.20	1.0000	1.0000	1.0000	1.0000	1.0000	1.0000	1.0000	1.0000	1.0000	0.9999
27.25	1.0000	1.0000	1.0000	1.0000	1.0000	1.0000	1.0000	1.0000	1.0000	0.9999
27.30	1.0000	1.0000	1.0000	1.0000	1.0000	1.0000	1.0000	1.0000	1.0000	0.9999
27.35	1.0000	1.0000	1.0000	1.0000	1.0000	1.0000	1.0000	1.0000	1.0000	0.9999
27.40	1.0000	1.0000	1.0000	1.0000	1.0000	1.0000	1.0000	1.0000	1.0000	0.9999
27.45	1.0000	1.0000	1.0000	1.0000	1.0000	1.0000	1.0000	1.0000	1.0000	0.9999
27.50	1.0000	1.0000	1.0000	1.0000	1.0000	1.0000	1.0000	1.0000	1.0000	0.9999
27.55	1.0000	1.0000	1.0000	1.0000	1.0000	1.0000	1.0000	1.0000	1.0000	0.9999
27.60	1.0000	1.0000	1.0000	1.0000	1.0000	1.0000	1.0000	1.0000	1.0000	0.9999
27.65	1.0000	1.0000	1.0000	1.0000	1.0000	1.0000	1.0000	1.0000	1.0000	0.9999
27.70	1.0000	1.0000	1.0000	1.0000	1.0000	1.0000	1.0000	1.0000	1.0000	0.9999
27.75	1.0000	1.0000	1.0000	1.0000	1.0000	1.0000	1.0000	1.0000	1.0000	0.9999
27.80	1.0000	1.0000	1.0000	1.0000	1.0000	1.0000	1.0000	1.0000	1.0000	0.9999
27.85	1.0000	1.0000	1.0000	1.0000	1.0000	1.0000	1.0000	1.0000	1.0000	0.9999
27.90	1.0000	1.0000	1.0000	1.0000	1.0000	1.0000	1.0000	1.0000	1.0000	0.9999
27.95	1.0000	1.0000	1.0000	1.0000	1.0000	1.0000	1.0000	1.0000	1.0000	0.9999
28.00	1.0000	1.0000	1.0000	1.0000	1.0000	1.0000	1.0000	1.0000	1.0000	0.9999
28.05	1.0000	1.0000	1.0000	1.0000	1.0000	1.0000	1.0000	1.0000	1.0000	0.9999
28.10	1.0000	1.0000	1.0000	1.0000	1.0000	1.0000	1.0000	1.0000	1.0000	1.0000

PROBABILITY INTEGRAL OF t, THE NON-CENTRAL t-STATISTIC. THIS TABLE GIVES Pr $[t/\sqrt{f} \leq x]$.

f is the number of degrees of freedom; the non-centrality parameter is $\sqrt{f+1}\,K_p$.

K_p is the standardized normal deviate exceeded with probability p.

DEGREES OF FREEDOM **6**

x \ p	.2500	.1500	.1000	.0650	.0400	.0250	.0100	.0040	.0025	.0010
-1.40	0.0000	0.0000	0.0000	0.0000	0.0000	0.0000	0.0000	0.0000	0.0000	0.0000
-1.35	0.0001	0.0000	0.0000	0.0000	0.0000	0.0000	0.0000	0.0000	0.0000	0.0000
-1.30	0.0001	0.0000	0.0000	0.0000	0.0000	0.0000	0.0000	0.0000	0.0000	0.0000
-1.25	0.0001	0.0000	0.0000	0.0000	0.0000	0.0000	0.0000	0.0000	0.0000	0.0000
-1.20	0.0001	0.0000	0.0000	0.0000	0.0000	0.0000	0.0000	0.0000	0.0000	0.0000
-1.15	0.0001	0.0000	0.0000	0.0000	0.0000	0.0000	0.0000	0.0000	0.0000	0.0000
-1.10	0.0001	0.0000	0.0000	0.0000	0.0000	0.0000	0.0000	0.0000	0.0000	0.0000
-1.05	0.0002	0.0000	0.0000	0.0000	0.0000	0.0000	0.0000	0.0000	0.0000	0.0000
-1.00	0.0002	0.0000	0.0000	0.0000	0.0000	0.0000	0.0000	0.0000	0.0000	0.0000
-0.95	0.0003	0.0000	0.0000	0.0000	0.0000	0.0000	0.0000	0.0000	0.0000	0.0000
-0.90	0.0003	0.0000	0.0000	0.0000	0.0000	0.0000	0.0000	0.0000	0.0000	0.0000
-0.85	0.0004	0.0000	0.0000	0.0000	0.0000	0.0000	0.0000	0.0000	0.0000	0.0000
-0.80	0.0005	0.0000	0.0000	0.0000	0.0000	0.0000	0.0000	0.0000	0.0000	0.0000
-0.75	0.0007	0.0000	0.0000	0.0000	0.0000	0.0000	0.0000	0.0000	0.0000	0.0000
-0.70	0.0009	0.0000	0.0000	0.0000	0.0000	0.0000	0.0000	0.0000	0.0000	0.0000
-0.65	0.0011	0.0000	0.0000	0.0000	0.0000	0.0000	0.0000	0.0000	0.0000	0.0000
-0.60	0.0015	0.0001	0.0000	0.0000	0.0000	0.0000	0.0000	0.0000	0.0000	0.0000
-0.55	0.0019	0.0001	0.0000	0.0000	0.0000	0.0000	0.0000	0.0000	0.0000	0.0000
-0.50	0.0025	0.0001	0.0000	0.0000	0.0000	0.0000	0.0000	0.0000	0.0000	0.0000
-0.45	0.0032	0.0001	0.0000	0.0000	0.0000	0.0000	0.0000	0.0000	0.0000	0.0000
-0.40	0.0042	0.0002	0.0000	0.0000	0.0000	0.0000	0.0000	0.0000	0.0000	0.0000
-0.35	0.0056	0.0003	0.0000	0.0000	0.0000	0.0000	0.0000	0.0000	0.0000	0.0000
-0.30	0.0073	0.0004	0.0000	0.0000	0.0000	0.0000	0.0000	0.0000	0.0000	0.0000
-0.25	0.0097	0.0005	0.0000	0.0000	0.0000	0.0000	0.0000	0.0000	0.0000	0.0000
-0.20	0.0127	0.0007	0.0001	0.0000	0.0000	0.0000	0.0000	0.0000	0.0000	0.0000
-0.15	0.0168	0.0010	0.0001	0.0000	0.0000	0.0000	0.0000	0.0000	0.0000	0.0000
-0.10	0.0220	0.0015	0.0002	0.0000	0.0000	0.0000	0.0000	0.0000	0.0000	0.0000
-0.05	0.0287	0.0021	0.0002	0.0000	0.0000	0.0000	0.0000	0.0000	0.0000	0.0000
0.00	0.0372	0.0031	0.0004	0.0000	0.0000	0.0000	0.0000	0.0000	0.0000	0.0000
0.05	0.0479	0.0044	0.0005	0.0001	0.0000	0.0000	0.0000	0.0000	0.0000	0.0000
0.10	0.0611	0.0062	0.0008	0.0001	0.0000	0.0000	0.0000	0.0000	0.0000	0.0000
0.15	0.0772	0.0087	0.0013	0.0001	0.0000	0.0000	0.0000	0.0000	0.0000	0.0000
0.20	0.0964	0.0122	0.0019	0.0002	0.0000	0.0000	0.0000	0.0000	0.0000	0.0000
0.25	0.1191	0.0169	0.0029	0.0004	0.0000	0.0000	0.0000	0.0000	0.0000	0.0000
0.30	0.1453	0.0231	0.0043	0.0006	0.0001	0.0000	0.0000	0.0000	0.0000	0.0000
0.35	0.1748	0.0312	0.0063	0.0010	0.0001	0.0000	0.0000	0.0000	0.0000	0.0000
0.40	0.2077	0.0414	0.0092	0.0016	0.0002	0.0000	0.0000	0.0000	0.0000	0.0000
0.45	0.2435	0.0541	0.0131	0.0025	0.0003	0.0000	0.0000	0.0000	0.0000	0.0000
0.50	0.2818	0.0696	0.0184	0.0039	0.0006	0.0001	0.0000	0.0000	0.0000	0.0000
0.55	0.3221	0.0880	0.0253	0.0058	0.0010	0.0002	0.0000	0.0000	0.0000	0.0000
0.60	0.3637	0.1095	0.0342	0.0086	0.0016	0.0003	0.0000	0.0000	0.0000	0.0000
0.65	0.4060	0.1341	0.0453	0.0124	0.0025	0.0005	0.0000	0.0000	0.0000	0.0000
0.70	0.4483	0.1616	0.0588	0.0174	0.0039	0.0008	0.0000	0.0000	0.0000	0.0000
0.75	0.4901	0.1919	0.0749	0.0240	0.0059	0.0014	0.0001	0.0000	0.0000	0.0000
0.80	0.5309	0.2245	0.0938	0.0324	0.0086	0.0022	0.0001	0.0000	0.0000	0.0000
0.85	0.5702	0.2591	0.1154	0.0429	0.0124	0.0034	0.0002	0.0000	0.0000	0.0000
0.90	0.6077	0.2952	0.1397	0.0555	0.0174	0.0052	0.0004	0.0000	0.0000	0.0000
0.95	0.6431	0.3324	0.1664	0.0705	0.0238	0.0076	0.0007	0.0001	0.0000	0.0000
1.00	0.6763	0.3702	0.1953	0.0879	0.0319	0.0110	0.0011	0.0001	0.0000	0.0000
1.05	0.7071	0.4081	0.2261	0.1076	0.0418	0.0154	0.0018	0.0002	0.0001	0.0000
1.10	0.7356	0.4456	0.2585	0.1297	0.0536	0.0210	0.0028	0.0003	0.0001	0.0000
1.15	0.7618	0.4825	0.2920	0.1540	0.0675	0.0281	0.0043	0.0006	0.0002	0.0000
1.20	0.7857	0.5183	0.3263	0.1802	0.0835	0.0367	0.0063	0.0009	0.0003	0.0000
1.25	0.8075	0.5529	0.3610	0.2081	0.1016	0.0470	0.0089	0.0015	0.0006	0.0001
1.30	0.8272	0.5860	0.3957	0.2374	0.1217	0.0592	0.0124	0.0023	0.0009	0.0001
1.35	0.8450	0.6175	0.4301	0.2678	0.1436	0.0732	0.0169	0.0034	0.0014	0.0002
1.40	0.8610	0.6472	0.4640	0.2990	0.1673	0.0890	0.0224	0.0049	0.0022	0.0004
1.45	0.8754	0.6752	0.4971	0.3308	0.1925	0.1066	0.0292	0.0070	0.0032	0.0007

PROBABILITY INTEGRAL OF t, THE NON-CENTRAL t-STATISTIC. THIS TABLE GIVES Pr $[t/\sqrt{f} \leq x]$.

f is the number of degrees of freedom; the non-centrality parameter is $\sqrt{f+1}\,K_p$.
K_p is the standardized normal deviate exceeded with probability p.

DEGREES OF FREEDOM **6**

x \ p	.2500	.1500	.1000	.0650	.0400	.0250	.0100	.0040	.0025	.0010
1.50	0.8884	0.7014	0.5291	0.3627	0.2190	0.1259	0.0373	0.0097	0.0046	0.0011
1.55	0.8999	0.7258	0.5600	0.3946	0.2465	0.1469	0.0468	0.0131	0.0066	0.0016
1.60	0.9103	0.7485	0.5896	0.4262	0.2749	0.1693	0.0578	0.0173	0.0090	0.0024
1.65	0.9196	0.7695	0.6177	0.4573	0.3038	0.1930	0.0702	0.0226	0.0122	0.0034
1.70	0.9278	0.7889	0.6445	0.4877	0.3331	0.2177	0.0841	0.0288	0.0160	0.0049
1.75	0.9352	0.8068	0.6697	0.5172	0.3624	0.2434	0.0995	0.0362	0.0208	0.0067
1.80	0.9418	0.8232	0.6935	0.5457	0.3917	0.2698	0.1163	0.0447	0.0265	0.0090
1.85	0.9477	0.8383	0.7158	0.5731	0.4207	0.2967	0.1343	0.0544	0.0331	0.0120
1.90	0.9530	0.8521	0.7367	0.5994	0.4492	0.3238	0.1536	0.0654	0.0409	0.0156
1.95	0.9577	0.8647	0.7563	0.6246	0.4772	0.3511	0.1740	0.0776	0.0497	0.0199
2.00	0.9619	0.8762	0.7744	0.6485	0.5044	0.3783	0.1953	0.0910	0.0597	0.0250
2.05	0.9656	0.8868	0.7914	0.6711	0.5308	0.4054	0.2174	0.1056	0.0708	0.0310
2.10	0.9689	0.8964	0.8070	0.6926	0.5564	0.4321	0.2403	0.1213	0.0830	0.0379
2.15	0.9719	0.9052	0.8216	0.7128	0.5810	0.4583	0.2636	0.1381	0.0964	0.0457
2.20	0.9746	0.9132	0.8351	0.7318	0.6046	0.4840	0.2874	0.1558	0.1107	0.0545
2.25	0.9770	0.9205	0.8475	0.7497	0.6272	0.5091	0.3114	0.1744	0.1261	0.0643
2.30	0.9792	0.9271	0.8591	0.7665	0.6488	0.5334	0.3355	0.1938	0.1424	0.0750
2.35	0.9811	0.9332	0.8697	0.7823	0.6694	0.5569	0.3597	0.2138	0.1596	0.0867
2.40	0.9828	0.9387	0.8795	0.7970	0.6889	0.5797	0.3837	0.2344	0.1775	0.0993
2.45	0.9844	0.9437	0.8886	0.8107	0.7075	0.6015	0.4076	0.2554	0.1962	0.1129
2.50	0.9858	0.9483	0.8969	0.8236	0.7250	0.6226	0.4311	0.2768	0.2154	0.1272
2.55	0.9871	0.9525	0.9046	0.8355	0.7416	0.6427	0.4543	0.2984	0.2351	0.1424
2.60	0.9882	0.9564	0.9117	0.8467	0.7573	0.6619	0.4770	0.3202	0.2552	0.1583
2.65	0.9892	0.9598	0.9182	0.8571	0.7721	0.6803	0.4992	0.3420	0.2757	0.1748
2.70	0.9902	0.9630	0.9243	0.8668	0.7860	0.6978	0.5209	0.3638	0.2963	0.1920
2.75	0.9910	0.9659	0.9298	0.8758	0.7991	0.7145	0.5420	0.3855	0.3171	0.2096
2.80	0.9918	0.9686	0.9349	0.8841	0.8114	0.7303	0.5624	0.4070	0.3380	0.2277
2.85	0.9925	0.9710	0.9397	0.8919	0.8229	0.7453	0.5822	0.4283	0.3588	0.2462
2.90	0.9931	0.9733	0.9440	0.8992	0.8338	0.7595	0.6014	0.4492	0.3796	0.2650
2.95	0.9936	0.9753	0.9480	0.9059	0.8440	0.7730	0.6198	0.4698	0.4002	0.2840
3.00	0.9942	0.9772	0.9517	0.9122	0.8535	0.7858	0.6376	0.4900	0.4205	0.3031
3.05	0.9946	0.9789	0.9551	0.9180	0.8625	0.7978	0.6547	0.5097	0.4406	0.3224
3.10	0.9951	0.9804	0.9583	0.9234	0.8709	0.8092	0.6711	0.5290	0.4604	0.3416
3.15	0.9954	0.9819	0.9612	0.9284	0.8787	0.8199	0.6868	0.5478	0.4799	0.3608
3.20	0.9958	0.9832	0.9639	0.9331	0.8861	0.8301	0.7019	0.5660	0.4989	0.3800
3.25	0.9961	0.9844	0.9664	0.9374	0.8930	0.8397	0.7163	0.5838	0.5176	0.3990
3.30	0.9964	0.9855	0.9687	0.9415	0.8994	0.8487	0.7301	0.6009	0.5358	0.4178
3.35	0.9967	0.9866	0.9708	0.9452	0.9055	0.8572	0.7432	0.6176	0.5535	0.4364
3.40	0.9969	0.9875	0.9728	0.9487	0.9111	0.8652	0.7558	0.6336	0.5708	0.4547
3.45	0.9972	0.9884	0.9746	0.9520	0.9164	0.8727	0.7677	0.6491	0.5875	0.4727
3.50	0.9974	0.9892	0.9763	0.9550	0.9214	0.8799	0.7791	0.6641	0.6038	0.4905
3.55	0.9976	0.9899	0.9778	0.9578	0.9260	0.8865	0.7900	0.6785	0.6196	0.5078
3.60	0.9977	0.9906	0.9793	0.9604	0.9304	0.8928	0.8003	0.6923	0.6348	0.5248
3.65	0.9979	0.9912	0.9806	0.9629	0.9345	0.8988	0.8102	0.7057	0.6496	0.5414
3.70	0.9980	0.9918	0.9819	0.9652	0.9383	0.9044	0.8195	0.7185	0.6638	0.5576
3.75	0.9982	0.9924	0.9830	0.9673	0.9419	0.9096	0.8284	0.7307	0.6776	0.5734
3.80	0.9983	0.9929	0.9841	0.9693	0.9452	0.9146	0.8368	0.7425	0.6908	0.5888
3.85	0.9984	0.9933	0.9851	0.9711	0.9484	0.9192	0.8448	0.7538	0.7036	0.6038
3.90	0.9985	0.9938	0.9860	0.9728	0.9513	0.9236	0.8525	0.7647	0.7158	0.6183
3.95	0.9986	0.9942	0.9869	0.9745	0.9541	0.9277	0.8597	0.7750	0.7277	0.6324
4.00	0.9987	0.9945	0.9877	0.9760	0.9566	0.9316	0.8665	0.7849	0.7390	0.6460
4.05	0.9988	0.9949	0.9884	0.9774	0.9591	0.9353	0.8731	0.7944	0.7499	0.6593
4.10	0.9989	0.9952	0.9891	0.9787	0.9614	0.9387	0.8792	0.8035	0.7604	0.6721
4.15	0.9989	0.9955	0.9898	0.9799	0.9635	0.9420	0.8851	0.8122	0.7704	0.6845
4.20	0.9990	0.9958	0.9904	0.9811	0.9655	0.9451	0.8907	0.8205	0.7801	0.6964
4.25	0.9991	0.9960	0.9909	0.9821	0.9674	0.9479	0.8959	0.8284	0.7893	0.7080
4.30	0.9991	0.9963	0.9915	0.9832	0.9692	0.9507	0.9010	0.8360	0.7982	0.7191
4.35	0.9992	0.9965	0.9920	0.9841	0.9708	0.9532	0.9057	0.8432	0.8067	0.7299
4.40	0.9992	0.9967	0.9924	0.9850	0.9724	0.9556	0.9102	0.8501	0.8148	0.7403
4.45	0.9993	0.9969	0.9929	0.9858	0.9739	0.9579	0.9145	0.8567	0.8226	0.7503

PROBABILITY INTEGRAL OF t, THE NON-CENTRAL t-STATISTIC. THIS TABLE GIVES Pr $[t/\sqrt{f} \leq x]$.

f is the number of degrees of freedom; the non-centrality parameter is $\sqrt{f+1}\, K_p$.
K_p is the standardized normal deviate exceeded with probability p.

DEGREES OF FREEDOM **6**

x \ p	.2500	.1500	.1000	.0650	.0400	.0250	.0100	.0040	.0025	.0010
4.50	0.9993	0.9971	0.9933	0.9866	0.9753	0.9601	0.9186	0.8630	0.8301	0.7599
4.55	0.9994	0.9972	0.9936	0.9873	0.9766	0.9621	0.9224	0.8690	0.8372	0.7691
4.60	0.9994	0.9974	0.9940	0.9880	0.9778	0.9640	0.9261	0.8747	0.8440	0.7781
4.65	0.9994	0.9975	0.9943	0.9887	0.9789	0.9658	0.9295	0.8801	0.8506	0.7866
4.70	0.9995	0.9977	0.9946	0.9893	0.9800	0.9675	0.9328	0.8854	0.8568	0.7949
4.75	0.9995	0.9978	0.9949	0.9898	0.9811	0.9692	0.9360	0.8903	0.8628	0.8028
4.80	0.9995	0.9979	0.9952	0.9904	0.9820	0.9707	0.9389	0.8951	0.8685	0.8105
4.85	0.9996	0.9980	0.9955	0.9909	0.9829	0.9721	0.9417	0.8996	0.8740	0.8178
4.90	0.9996	0.9982	0.9957	0.9913	0.9838	0.9735	0.9444	0.9039	0.8792	0.8249
4.95	0.9996	0.9983	0.9959	0.9918	0.9846	0.9748	0.9470	0.9080	0.8842	0.8316
5.00	0.9996	0.9983	0.9961	0.9922	0.9854	0.9760	0.9494	0.9120	0.8890	0.8381
5.05	0.9996	0.9984	0.9963	0.9926	0.9861	0.9771	0.9517	0.9157	0.8936	0.8444
5.10	0.9997	0.9985	0.9965	0.9930	0.9868	0.9782	0.9538	0.9193	0.8980	0.8504
5.15	0.9997	0.9986	0.9967	0.9933	0.9874	0.9793	0.9559	0.9227	0.9021	0.8561
5.20	0.9997	0.9987	0.9969	0.9937	0.9880	0.9802	0.9579	0.9260	0.9061	0.8616
5.25	0.9997	0.9987	0.9970	0.9940	0.9886	0.9812	0.9597	0.9291	0.9100	0.8670
5.30	0.9997	0.9988	0.9972	0.9943	0.9891	0.9820	0.9615	0.9320	0.9136	0.8720
5.35	0.9997	0.9989	0.9973	0.9945	0.9897	0.9829	0.9632	0.9349	0.9171	0.8769
5.40	0.9998	0.9989	0.9975	0.9948	0.9901	0.9837	0.9648	0.9376	0.9205	0.8816
5.45	0.9998	0.9990	0.9976	0.9951	0.9906	0.9844	0.9664	0.9402	0.9237	0.8861
5.50	0.9998	0.9990	0.9977	0.9953	0.9910	0.9851	0.9678	0.9426	0.9267	0.8904
5.55	0.9998	0.9991	0.9978	0.9955	0.9915	0.9858	0.9692	0.9450	0.9296	0.8946
5.60	0.9998	0.9991	0.9979	0.9957	0.9919	0.9864	0.9705	0.9472	0.9324	0.8986
5.65	0.9998	0.9992	0.9980	0.9959	0.9922	0.9870	0.9718	0.9494	0.9351	0.9024
5.70	0.9998	0.9992	0.9981	0.9961	0.9926	0.9876	0.9730	0.9514	0.9377	0.9061
5.75	0.9998	0.9992	0.9982	0.9963	0.9929	0.9882	0.9741	0.9534	0.9401	0.9096
5.80	0.9998	0.9993	0.9983	0.9965	0.9932	0.9887	0.9752	0.9552	0.9425	0.9130
5.85	0.9998	0.9993	0.9984	0.9966	0.9935	0.9892	0.9762	0.9570	0.9447	0.9162
5.90	0.9999	0.9993	0.9984	0.9968	0.9938	0.9896	0.9772	0.9587	0.9469	0.9193
5.95	0.9999	0.9994	0.9985	0.9969	0.9941	0.9901	0.9782	0.9604	0.9489	0.9223
6.00	0.9999	0.9994	0.9986	0.9971	0.9943	0.9905	0.9791	0.9619	0.9509	0.9252
6.05	0.9999	0.9994	0.9986	0.9972	0.9946	0.9909	0.9799	0.9634	0.9528	0.9279
6.10	0.9999	0.9995	0.9987	0.9973	0.9948	0.9913	0.9807	0.9649	0.9546	0.9305
6.15	0.9999	0.9995	0.9988	0.9974	0.9950	0.9917	0.9815	0.9662	0.9563	0.9331
6.20	0.9999	0.9995	0.9988	0.9975	0.9953	0.9920	0.9823	0.9675	0.9580	0.9355
6.25	0.9999	0.9995	0.9989	0.9977	0.9955	0.9923	0.9830	0.9688	0.9596	0.9379
6.30	0.9999	0.9995	0.9989	0.9978	0.9957	0.9927	0.9837	0.9700	0.9611	0.9401
6.35	0.9999	0.9996	0.9990	0.9978	0.9958	0.9930	0.9843	0.9711	0.9625	0.9423
6.40	0.9999	0.9996	0.9990	0.9979	0.9960	0.9933	0.9849	0.9722	0.9639	0.9443
6.45	0.9999	0.9996	0.9990	0.9980	0.9962	0.9935	0.9855	0.9733	0.9653	0.9463
6.50	0.9999	0.9996	0.9991	0.9981	0.9963	0.9938	0.9861	0.9743	0.9666	0.9482
6.55	0.9999	0.9996	0.9991	0.9982	0.9965	0.9940	0.9866	0.9753	0.9678	0.9501
6.60	0.9999	0.9997	0.9992	0.9983	0.9966	0.9943	0.9871	0.9762	0.9690	0.9518
6.65	0.9999	0.9997	0.9992	0.9983	0.9968	0.9945	0.9876	0.9771	0.9701	0.9535
6.70	0.9999	0.9997	0.9992	0.9984	0.9969	0.9947	0.9881	0.9779	0.9712	0.9552
6.75	0.9999	0.9997	0.9993	0.9985	0.9970	0.9949	0.9886	0.9787	0.9722	0.9567
6.80	0.9999	0.9997	0.9993	0.9985	0.9971	0.9951	0.9890	0.9795	0.9732	0.9582
6.85	0.9999	0.9997	0.9993	0.9986	0.9972	0.9953	0.9894	0.9803	0.9742	0.9597
6.90	0.9999	0.9997	0.9994	0.9986	0.9974	0.9955	0.9898	0.9810	0.9751	0.9611
6.95	0.9999	0.9997	0.9994	0.9987	0.9975	0.9957	0.9902	0.9817	0.9760	0.9624
7.00	0.9999	0.9998	0.9994	0.9987	0.9976	0.9958	0.9906	0.9823	0.9768	0.9637
7.05	0.9999	0.9998	0.9994	0.9988	0.9976	0.9960	0.9909	0.9829	0.9777	0.9649
7.10	0.9999	0.9998	0.9994	0.9988	0.9977	0.9961	0.9912	0.9836	0.9784	0.9661
7.15	1.0000	0.9998	0.9995	0.9989	0.9978	0.9963	0.9916	0.9841	0.9792	0.9673
7.20	1.0000	0.9998	0.9995	0.9989	0.9979	0.9964	0.9919	0.9847	0.9799	0.9684
7.25	1.0000	0.9998	0.9995	0.9990	0.9980	0.9966	0.9922	0.9852	0.9806	0.9694
7.30	1.0000	0.9998	0.9995	0.9990	0.9981	0.9967	0.9924	0.9858	0.9813	0.9705
7.35	1.0000	0.9998	0.9995	0.9990	0.9981	0.9968	0.9927	0.9862	0.9819	0.9714
7.40	1.0000	0.9998	0.9996	0.9991	0.9982	0.9969	0.9930	0.9867	0.9825	0.9724
7.45	1.0000	0.9998	0.9996	0.9991	0.9983	0.9970	0.9932	0.9872	0.9831	0.9733

PROBABILITY INTEGRAL OF t, THE NON-CENTRAL t-STATISTIC. THIS TABLE GIVES Pr $[t/\sqrt{f} \leq x]$.

f is the number of degrees of freedom; the non-centrality parameter is $\sqrt{f+1}\,K_p$.

K_p is the standardized normal deviate exceeded with probability p.

DEGREES OF FREEDOM **6**

x \ p	.2500	.1500	.1000	.0650	.0400	.0250	.0100	.0040	.0025	.0010
7.50	1.0000	0.9998	0.9996	0.9992	0.9983	0.9971	0.9935	0.9876	0.9837	0.9742
7.55	1.0000	0.9998	0.9996	0.9992	0.9984	0.9972	0.9937	0.9880	0.9843	0.9750
7.60	1.0000	0.9998	0.9996	0.9992	0.9985	0.9973	0.9939	0.9884	0.9848	0.9758
7.65	1.0000	0.9999	0.9996	0.9992	0.9985	0.9974	0.9941	0.9888	0.9853	0.9766
7.70	1.0000	0.9999	0.9997	0.9993	0.9986	0.9975	0.9943	0.9892	0.9858	0.9774
7.75	1.0000	0.9999	0.9997	0.9993	0.9986	0.9976	0.9945	0.9896	0.9862	0.9781
7.80	1.0000	0.9999	0.9997	0.9993	0.9987	0.9977	0.9947	0.9899	0.9867	0.9788
7.85	1.0000	0.9999	0.9997	0.9993	0.9987	0.9978	0.9949	0.9903	0.9871	0.9795
7.90	1.0000	0.9999	0.9997	0.9994	0.9988	0.9979	0.9951	0.9906	0.9876	0.9801
7.95	1.0000	0.9999	0.9997	0.9994	0.9988	0.9979	0.9952	0.9909	0.9880	0.9808
8.00	1.0000	0.9999	0.9997	0.9994	0.9988	0.9980	0.9954	0.9912	0.9884	0.9814
8.05	1.0000	0.9999	0.9997	0.9994	0.9989	0.9981	0.9955	0.9915	0.9887	0.9820
8.10	1.0000	0.9999	0.9997	0.9995	0.9989	0.9981	0.9957	0.9918	0.9891	0.9825
8.15	1.0000	0.9999	0.9998	0.9995	0.9990	0.9982	0.9958	0.9920	0.9894	0.9831
8.20	1.0000	0.9999	0.9998	0.9995	0.9990	0.9983	0.9960	0.9923	0.9898	0.9836
8.25	1.0000	0.9999	0.9998	0.9995	0.9990	0.9983	0.9961	0.9925	0.9901	0.9841
8.30	1.0000	0.9999	0.9998	0.9995	0.9991	0.9984	0.9962	0.9928	0.9904	0.9846
8.35	1.0000	0.9999	0.9998	0.9995	0.9991	0.9984	0.9963	0.9930	0.9907	0.9851
8.40	1.0000	0.9999	0.9998	0.9996	0.9991	0.9985	0.9965	0.9932	0.9910	0.9855
8.45	1.0000	0.9999	0.9998	0.9996	0.9991	0.9985	0.9966	0.9934	0.9913	0.9860
8.50	1.0000	0.9999	0.9998	0.9996	0.9992	0.9986	0.9967	0.9936	0.9915	0.9864
8.55	1.0000	0.9999	0.9998	0.9996	0.9992	0.9986	0.9968	0.9938	0.9918	0.9868
8.60	1.0000	0.9999	0.9998	0.9996	0.9992	0.9987	0.9969	0.9940	0.9921	0.9872
8.65	1.0000	0.9999	0.9998	0.9996	0.9993	0.9987	0.9970	0.9942	0.9923	0.9876
8.70	1.0000	0.9999	0.9998	0.9996	0.9993	0.9988	0.9971	0.9944	0.9925	0.9879
8.75	1.0000	0.9999	0.9998	0.9996	0.9993	0.9988	0.9972	0.9946	0.9928	0.9883
8.80	1.0000	0.9999	0.9998	0.9997	0.9993	0.9988	0.9973	0.9947	0.9930	0.9886
8.85	1.0000	0.9999	0.9998	0.9997	0.9993	0.9989	0.9973	0.9949	0.9932	0.9890
8.90	1.0000	0.9999	0.9999	0.9997	0.9994	0.9989	0.9974	0.9950	0.9934	0.9893
8.95	1.0000	0.9999	0.9999	0.9997	0.9994	0.9989	0.9975	0.9952	0.9936	0.9896
9.00	1.0000	0.9999	0.9999	0.9997	0.9994	0.9990	0.9976	0.9953	0.9938	0.9899
9.05	1.0000	0.9999	0.9999	0.9997	0.9994	0.9990	0.9977	0.9955	0.9940	0.9902
9.10	1.0000	0.9999	0.9999	0.9997	0.9994	0.9990	0.9977	0.9956	0.9941	0.9905
9.15	1.0000	0.9999	0.9999	0.9997	0.9995	0.9991	0.9978	0.9957	0.9943	0.9908
9.20	1.0000	0.9999	0.9999	0.9997	0.9995	0.9991	0.9979	0.9959	0.9945	0.9910
9.25	1.0000	1.0000	0.9999	0.9997	0.9995	0.9991	0.9979	0.9960	0.9946	0.9913
9.30	1.0000	1.0000	0.9999	0.9998	0.9995	0.9991	0.9980	0.9961	0.9948	0.9915
9.35	1.0000	1.0000	0.9999	0.9998	0.9995	0.9992	0.9980	0.9962	0.9949	0.9918
9.40	1.0000	1.0000	0.9999	0.9998	0.9995	0.9992	0.9981	0.9963	0.9951	0.9920
9.45	1.0000	1.0000	0.9999	0.9998	0.9996	0.9992	0.9982	0.9964	0.9952	0.9922
9.50	1.0000	1.0000	0.9999	0.9998	0.9996	0.9992	0.9982	0.9965	0.9954	0.9924
9.55	1.0000	1.0000	0.9999	0.9998	0.9996	0.9993	0.9983	0.9966	0.9955	0.9927
9.60	1.0000	1.0000	0.9999	0.9998	0.9996	0.9993	0.9983	0.9967	0.9956	0.9929
9.65	1.0000	1.0000	0.9999	0.9998	0.9996	0.9993	0.9984	0.9968	0.9957	0.9931
9.70	1.0000	1.0000	0.9999	0.9998	0.9996	0.9993	0.9984	0.9969	0.9959	0.9932
9.75	1.0000	1.0000	0.9999	0.9998	0.9996	0.9993	0.9985	0.9970	0.9960	0.9934
9.80	1.0000	1.0000	0.9999	0.9998	0.9996	0.9994	0.9985	0.9971	0.9961	0.9936
9.85	1.0000	1.0000	0.9999	0.9998	0.9996	0.9994	0.9985	0.9972	0.9962	0.9938
9.90	1.0000	1.0000	0.9999	0.9998	0.9997	0.9994	0.9986	0.9972	0.9963	0.9940
9.95	1.0000	1.0000	0.9999	0.9998	0.9997	0.9994	0.9986	0.9973	0.9964	0.9941
10.00	1.0000	1.0000	0.9999	0.9998	0.9997	0.9994	0.9987	0.9974	0.9965	0.9943
10.05	1.0000	1.0000	0.9999	0.9998	0.9997	0.9995	0.9987	0.9975	0.9966	0.9944
10.10	1.0000	1.0000	0.9999	0.9998	0.9997	0.9995	0.9987	0.9975	0.9967	0.9946
10.15	1.0000	1.0000	0.9999	0.9999	0.9997	0.9995	0.9988	0.9976	0.9968	0.9947
10.20	1.0000	1.0000	0.9999	0.9999	0.9997	0.9995	0.9988	0.9977	0.9969	0.9949
10.25	1.0000	1.0000	0.9999	0.9999	0.9997	0.9995	0.9988	0.9977	0.9969	0.9950
10.30	1.0000	1.0000	0.9999	0.9999	0.9997	0.9995	0.9989	0.9978	0.9970	0.9951
10.35	1.0000	1.0000	0.9999	0.9999	0.9997	0.9995	0.9989	0.9978	0.9971	0.9953
10.40	1.0000	1.0000	0.9999	0.9999	0.9997	0.9996	0.9989	0.9979	0.9972	0.9954
10.45	1.0000	1.0000	0.9999	0.9999	0.9997	0.9996	0.9990	0.9980	0.9973	0.9955

PROBABILITY INTEGRAL OF t, THE NON-CENTRAL t-STATISTIC. THIS TABLE GIVES Pr $[t/\sqrt{f} \leq x]$.

f is the number of degrees of freedom; the non-centrality parameter is $\sqrt{f+1}\,K_p$.

K_p is the standardized normal deviate exceeded with probability p.

DEGREES OF FREEDOM **6**

x \ p	.2500	.1500	.1000	.0650	.0400	.0250	.0100	.0040	.0025	.0010
10.50	1.0000	1.0000	0.9999	0.9999	0.9998	0.9996	0.9990	0.9980	0.9973	0.9956
10.55	1.0000	1.0000	0.9999	0.9999	0.9998	0.9996	0.9990	0.9981	0.9974	0.9957
10.60	1.0000	1.0000	0.9999	0.9999	0.9998	0.9996	0.9990	0.9981	0.9975	0.9958
10.65	1.0000	1.0000	0.9999	0.9999	0.9998	0.9996	0.9991	0.9982	0.9975	0.9959
10.70	1.0000	1.0000	1.0000	0.9999	0.9998	0.9996	0.9991	0.9982	0.9976	0.9960
10.75	1.0000	1.0000	1.0000	0.9999	0.9998	0.9996	0.9991	0.9983	0.9977	0.9961
10.80	1.0000	1.0000	1.0000	0.9999	0.9998	0.9996	0.9991	0.9983	0.9977	0.9962
10.85	1.0000	1.0000	1.0000	0.9999	0.9998	0.9996	0.9992	0.9983	0.9978	0.9963
10.90	1.0000	1.0000	1.0000	0.9999	0.9998	0.9997	0.9992	0.9984	0.9978	0.9964
10.95	1.0000	1.0000	1.0000	0.9999	0.9998	0.9997	0.9992	0.9984	0.9979	0.9965
11.00	1.0000	1.0000	1.0000	0.9999	0.9998	0.9997	0.9992	0.9985	0.9979	0.9966
11.05	1.0000	1.0000	1.0000	0.9999	0.9998	0.9997	0.9992	0.9985	0.9980	0.9967
11.10	1.0000	1.0000	1.0000	0.9999	0.9998	0.9997	0.9993	0.9985	0.9980	0.9968
11.15	1.0000	1.0000	1.0000	0.9999	0.9998	0.9997	0.9993	0.9986	0.9981	0.9968
11.20	1.0000	1.0000	1.0000	0.9999	0.9998	0.9997	0.9993	0.9986	0.9981	0.9969
11.25	1.0000	1.0000	1.0000	0.9999	0.9998	0.9997	0.9993	0.9986	0.9982	0.9970
11.30	1.0000	1.0000	1.0000	0.9999	0.9998	0.9997	0.9993	0.9987	0.9982	0.9971
11.35	1.0000	1.0000	1.0000	0.9999	0.9998	0.9997	0.9993	0.9987	0.9983	0.9971
11.40	1.0000	1.0000	1.0000	0.9999	0.9998	0.9997	0.9994	0.9987	0.9983	0.9972
11.45	1.0000	1.0000	1.0000	0.9999	0.9999	0.9997	0.9994	0.9988	0.9984	0.9973
11.50	1.0000	1.0000	1.0000	0.9999	0.9999	0.9997	0.9994	0.9988	0.9984	0.9973
11.55	1.0000	1.0000	1.0000	0.9999	0.9999	0.9998	0.9994	0.9988	0.9984	0.9974
11.60	1.0000	1.0000	1.0000	0.9999	0.9999	0.9998	0.9994	0.9989	0.9985	0.9975
11.65	1.0000	1.0000	1.0000	0.9999	0.9999	0.9998	0.9994	0.9989	0.9985	0.9975
11.70	1.0000	1.0000	1.0000	0.9999	0.9999	0.9998	0.9995	0.9989	0.9985	0.9976
11.75	1.0000	1.0000	1.0000	0.9999	0.9999	0.9998	0.9995	0.9989	0.9986	0.9976
11.80	1.0000	1.0000	1.0000	0.9999	0.9999	0.9998	0.9995	0.9990	0.9986	0.9977
11.85	1.0000	1.0000	1.0000	0.9999	0.9999	0.9998	0.9995	0.9990	0.9986	0.9978
11.90	1.0000	1.0000	1.0000	0.9999	0.9999	0.9998	0.9995	0.9990	0.9987	0.9978
11.95	1.0000	1.0000	1.0000	0.9999	0.9999	0.9998	0.9995	0.9990	0.9987	0.9979
12.00	1.0000	1.0000	1.0000	0.9999	0.9999	0.9998	0.9995	0.9991	0.9987	0.9979
12.05	1.0000	1.0000	1.0000	0.9999	0.9999	0.9998	0.9995	0.9991	0.9988	0.9980
12.10	1.0000	1.0000	1.0000	0.9999	0.9999	0.9998	0.9996	0.9991	0.9988	0.9980
12.15	1.0000	1.0000	1.0000	0.9999	0.9999	0.9998	0.9996	0.9991	0.9988	0.9980
12.20	1.0000	1.0000	1.0000	1.0000	0.9999	0.9998	0.9996	0.9991	0.9989	0.9981
12.25	1.0000	1.0000	1.0000	1.0000	0.9999	0.9998	0.9996	0.9992	0.9989	0.9981
12.30	1.0000	1.0000	1.0000	1.0000	0.9999	0.9998	0.9996	0.9992	0.9989	0.9982
12.35	1.0000	1.0000	1.0000	1.0000	0.9999	0.9998	0.9996	0.9992	0.9989	0.9982
12.40	1.0000	1.0000	1.0000	1.0000	0.9999	0.9998	0.9996	0.9992	0.9990	0.9983
12.45	1.0000	1.0000	1.0000	1.0000	0.9999	0.9998	0.9996	0.9992	0.9990	0.9983
12.50	1.0000	1.0000	1.0000	1.0000	0.9999	0.9998	0.9996	0.9993	0.9990	0.9983
12.55	1.0000	1.0000	1.0000	1.0000	0.9999	0.9999	0.9996	0.9993	0.9990	0.9984
12.60	1.0000	1.0000	1.0000	1.0000	0.9999	0.9999	0.9996	0.9993	0.9990	0.9984
12.65	1.0000	1.0000	1.0000	1.0000	0.9999	0.9999	0.9997	0.9993	0.9991	0.9984
12.70	1.0000	1.0000	1.0000	1.0000	0.9999	0.9999	0.9997	0.9993	0.9991	0.9985
12.75	1.0000	1.0000	1.0000	1.0000	0.9999	0.9999	0.9997	0.9993	0.9991	0.9985
12.80	1.0000	1.0000	1.0000	1.0000	0.9999	0.9999	0.9997	0.9994	0.9991	0.9985
12.85	1.0000	1.0000	1.0000	1.0000	0.9999	0.9999	0.9997	0.9994	0.9991	0.9986
12.90	1.0000	1.0000	1.0000	1.0000	0.9999	0.9999	0.9997	0.9994	0.9992	0.9986
12.95	1.0000	1.0000	1.0000	1.0000	0.9999	0.9999	0.9997	0.9994	0.9992	0.9986
13.00	1.0000	1.0000	1.0000	1.0000	0.9999	0.9999	0.9997	0.9994	0.9992	0.9987
13.05	1.0000	1.0000	1.0000	1.0000	0.9999	0.9999	0.9997	0.9994	0.9992	0.9987
13.10	1.0000	1.0000	1.0000	1.0000	0.9999	0.9999	0.9997	0.9994	0.9992	0.9987
13.15	1.0000	1.0000	1.0000	1.0000	0.9999	0.9999	0.9997	0.9994	0.9993	0.9988
13.20	1.0000	1.0000	1.0000	1.0000	0.9999	0.9999	0.9997	0.9995	0.9993	0.9988
13.25	1.0000	1.0000	1.0000	1.0000	0.9999	0.9999	0.9997	0.9995	0.9993	0.9988
13.30	1.0000	1.0000	1.0000	1.0000	0.9999	0.9999	0.9997	0.9995	0.9993	0.9988
13.35	1.0000	1.0000	1.0000	1.0000	0.9999	0.9999	0.9997	0.9995	0.9993	0.9989
13.40	1.0000	1.0000	1.0000	1.0000	0.9999	0.9999	0.9998	0.9995	0.9993	0.9989
13.45	1.0000	1.0000	1.0000	1.0000	0.9999	0.9999	0.9998	0.9995	0.9993	0.9989

PROBABILITY INTEGRAL OF t, THE NON-CENTRAL t-STATISTIC. THIS TABLE GIVES Pr $[t/\sqrt{f} \leq x]$.

f is the number of degrees of freedom; the non-centrality parameter is $\sqrt{f+1}\,K_p$.

K_p is the standardized normal deviate exceeded with probability p.

DEGREES OF FREEDOM **6**

x \ p	.2500	.1500	.1000	.0650	.0400	.0250	.0100	.0040	.0025	.0010
13.50	1.0000	1.0000	1.0000	1.0000	0.9999	0.9999	0.9998	0.9995	0.9994	0.9989
13.55	1.0000	1.0000	1.0000	1.0000	0.9999	0.9999	0.9998	0.9995	0.9994	0.9989
13.60	1.0000	1.0000	1.0000	1.0000	0.9999	0.9999	0.9998	0.9995	0.9994	0.9990
13.65	1.0000	1.0000	1.0000	1.0000	0.9999	0.9999	0.9998	0.9996	0.9994	0.9990
13.70	1.0000	1.0000	1.0000	1.0000	1.0000	0.9999	0.9998	0.9996	0.9994	0.9990
13.75	1.0000	1.0000	1.0000	1.0000	1.0000	0.9999	0.9998	0.9996	0.9994	0.9990
13.80	1.0000	1.0000	1.0000	1.0000	1.0000	0.9999	0.9998	0.9996	0.9994	0.9991
13.85	1.0000	1.0000	1.0000	1.0000	1.0000	0.9999	0.9998	0.9996	0.9994	0.9991
13.90	1.0000	1.0000	1.0000	1.0000	1.0000	0.9999	0.9998	0.9996	0.9995	0.9991
13.95	1.0000	1.0000	1.0000	1.0000	1.0000	0.9999	0.9998	0.9996	0.9995	0.9991
14.00	1.0000	1.0000	1.0000	1.0000	1.0000	0.9999	0.9998	0.9996	0.9995	0.9991
14.05	1.0000	1.0000	1.0000	1.0000	1.0000	0.9999	0.9998	0.9996	0.9995	0.9991
14.10	1.0000	1.0000	1.0000	1.0000	1.0000	0.9999	0.9998	0.9996	0.9995	0.9992
14.15	1.0000	1.0000	1.0000	1.0000	1.0000	0.9999	0.9998	0.9996	0.9995	0.9992
14.20	1.0000	1.0000	1.0000	1.0000	1.0000	0.9999	0.9998	0.9996	0.9995	0.9992
14.25	1.0000	1.0000	1.0000	1.0000	1.0000	0.9999	0.9998	0.9997	0.9995	0.9992
14.30	1.0000	1.0000	1.0000	1.0000	1.0000	0.9999	0.9998	0.9997	0.9995	0.9992
14.35	1.0000	1.0000	1.0000	1.0000	1.0000	0.9999	0.9998	0.9997	0.9995	0.9992
14.40	1.0000	1.0000	1.0000	1.0000	1.0000	0.9999	0.9998	0.9997	0.9996	0.9993
14.45	1.0000	1.0000	1.0000	1.0000	1.0000	0.9999	0.9998	0.9997	0.9996	0.9993
14.50	1.0000	1.0000	1.0000	1.0000	1.0000	0.9999	0.9998	0.9997	0.9996	0.9993
14.55	1.0000	1.0000	1.0000	1.0000	1.0000	0.9999	0.9998	0.9997	0.9996	0.9993
14.60	1.0000	1.0000	1.0000	1.0000	1.0000	0.9999	0.9999	0.9997	0.9996	0.9993
14.65	1.0000	1.0000	1.0000	1.0000	1.0000	0.9999	0.9999	0.9997	0.9996	0.9993
14.70	1.0000	1.0000	1.0000	1.0000	1.0000	0.9999	0.9999	0.9997	0.9996	0.9993
14.75	1.0000	1.0000	1.0000	1.0000	1.0000	0.9999	0.9999	0.9997	0.9996	0.9994
14.80	1.0000	1.0000	1.0000	1.0000	1.0000	0.9999	0.9999	0.9997	0.9996	0.9994
14.85	1.0000	1.0000	1.0000	1.0000	1.0000	0.9999	0.9999	0.9997	0.9996	0.9994
14.90	1.0000	1.0000	1.0000	1.0000	1.0000	0.9999	0.9999	0.9997	0.9996	0.9994
14.95	1.0000	1.0000	1.0000	1.0000	1.0000	0.9999	0.9999	0.9997	0.9996	0.9994
15.00	1.0000	1.0000	1.0000	1.0000	1.0000	1.0000	0.9999	0.9997	0.9997	0.9994
15.05	1.0000	1.0000	1.0000	1.0000	1.0000	1.0000	0.9999	0.9997	0.9997	0.9994
15.10	1.0000	1.0000	1.0000	1.0000	1.0000	1.0000	0.9999	0.9998	0.9997	0.9994
15.15	1.0000	1.0000	1.0000	1.0000	1.0000	1.0000	0.9999	0.9998	0.9997	0.9994
15.20	1.0000	1.0000	1.0000	1.0000	1.0000	1.0000	0.9999	0.9998	0.9997	0.9995
15.25	1.0000	1.0000	1.0000	1.0000	1.0000	1.0000	0.9999	0.9998	0.9997	0.9995
15.30	1.0000	1.0000	1.0000	1.0000	1.0000	1.0000	0.9999	0.9998	0.9997	0.9995
15.35	1.0000	1.0000	1.0000	1.0000	1.0000	1.0000	0.9999	0.9998	0.9997	0.9995
15.40	1.0000	1.0000	1.0000	1.0000	1.0000	1.0000	0.9999	0.9998	0.9997	0.9995
15.45	1.0000	1.0000	1.0000	1.0000	1.0000	1.0000	0.9999	0.9998	0.9997	0.9995
15.50	1.0000	1.0000	1.0000	1.0000	1.0000	1.0000	0.9999	0.9998	0.9997	0.9995
15.55	1.0000	1.0000	1.0000	1.0000	1.0000	1.0000	0.9999	0.9998	0.9997	0.9995
15.60	1.0000	1.0000	1.0000	1.0000	1.0000	1.0000	0.9999	0.9998	0.9997	0.9995
15.65	1.0000	1.0000	1.0000	1.0000	1.0000	1.0000	0.9999	0.9998	0.9997	0.9995
15.70	1.0000	1.0000	1.0000	1.0000	1.0000	1.0000	0.9999	0.9998	0.9997	0.9995
15.75	1.0000	1.0000	1.0000	1.0000	1.0000	1.0000	0.9999	0.9998	0.9997	0.9996
15.80	1.0000	1.0000	1.0000	1.0000	1.0000	1.0000	0.9999	0.9998	0.9997	0.9996
15.85	1.0000	1.0000	1.0000	1.0000	1.0000	1.0000	0.9999	0.9998	0.9997	0.9996
15.90	1.0000	1.0000	1.0000	1.0000	1.0000	1.0000	0.9999	0.9998	0.9998	0.9996
15.95	1.0000	1.0000	1.0000	1.0000	1.0000	1.0000	0.9999	0.9998	0.9998	0.9996
16.00	1.0000	1.0000	1.0000	1.0000	1.0000	1.0000	0.9999	0.9998	0.9998	0.9996
16.05	1.0000	1.0000	1.0000	1.0000	1.0000	1.0000	0.9999	0.9998	0.9998	0.9996
16.10	1.0000	1.0000	1.0000	1.0000	1.0000	1.0000	0.9999	0.9998	0.9998	0.9996
16.15	1.0000	1.0000	1.0000	1.0000	1.0000	1.0000	0.9999	0.9998	0.9998	0.9996
16.20	1.0000	1.0000	1.0000	1.0000	1.0000	1.0000	0.9999	0.9998	0.9998	0.9996
16.25	1.0000	1.0000	1.0000	1.0000	1.0000	1.0000	0.9999	0.9998	0.9998	0.9996
16.30	1.0000	1.0000	1.0000	1.0000	1.0000	1.0000	0.9999	0.9998	0.9998	0.9996
16.35	1.0000	1.0000	1.0000	1.0000	1.0000	1.0000	0.9999	0.9998	0.9998	0.9996
16.40	1.0000	1.0000	1.0000	1.0000	1.0000	1.0000	0.9999	0.9998	0.9998	0.9996
16.45	1.0000	1.0000	1.0000	1.0000	1.0000	1.0000	0.9999	0.9999	0.9998	0.9997

PROBABILITY INTEGRAL OF t, THE NON-CENTRAL t-STATISTIC. THIS TABLE GIVES Pr $[t/\sqrt{f} \leq x]$.

f is the number of degrees of freedom; the non-centrality parameter is $\sqrt{f+1}\,K_p$.

K_p is the standardized normal deviate exceeded with probability p.

DEGREES OF FREEDOM **6**

x \ p	.2500	.1500	.1000	.0650	.0400	.0250	.0100	.0040	.0025	.0010
16.50	1.0000	1.0000	1.0000	1.0000	1.0000	1.0000	0.9999	0.9999	0.9998	0.9997
16.55	1.0000	1.0000	1.0000	1.0000	1.0000	1.0000	0.9999	0.9999	0.9998	0.9997
16.60	1.0000	1.0000	1.0000	1.0000	1.0000	1.0000	0.9999	0.9999	0.9998	0.9997
16.65	1.0000	1.0000	1.0000	1.0000	1.0000	1.0000	0.9999	0.9999	0.9998	0.9997
16.70	1.0000	1.0000	1.0000	1.0000	1.0000	1.0000	0.9999	0.9999	0.9998	0.9997
16.75	1.0000	1.0000	1.0000	1.0000	1.0000	1.0000	0.9999	0.9999	0.9998	0.9997
16.80	1.0000	1.0000	1.0000	1.0000	1.0000	1.0000	0.9999	0.9999	0.9998	0.9997
16.85	1.0000	1.0000	1.0000	1.0000	1.0000	1.0000	0.9999	0.9999	0.9998	0.9997
16.90	1.0000	1.0000	1.0000	1.0000	1.0000	1.0000	0.9999	0.9999	0.9998	0.9997
16.95	1.0000	1.0000	1.0000	1.0000	1.0000	1.0000	0.9999	0.9999	0.9998	0.9997
17.00	1.0000	1.0000	1.0000	1.0000	1.0000	1.0000	0.9999	0.9999	0.9998	0.9997
17.05	1.0000	1.0000	1.0000	1.0000	1.0000	1.0000	0.9999	0.9999	0.9998	0.9997
17.10	1.0000	1.0000	1.0000	1.0000	1.0000	1.0000	0.9999	0.9999	0.9998	0.9997
17.15	1.0000	1.0000	1.0000	1.0000	1.0000	1.0000	0.9999	0.9999	0.9998	0.9997
17.20	1.0000	1.0000	1.0000	1.0000	1.0000	1.0000	0.9999	0.9999	0.9998	0.9997
17.25	1.0000	1.0000	1.0000	1.0000	1.0000	1.0000	0.9999	0.9999	0.9998	0.9997
17.30	1.0000	1.0000	1.0000	1.0000	1.0000	1.0000	0.9999	0.9999	0.9999	0.9997
17.35	1.0000	1.0000	1.0000	1.0000	1.0000	1.0000	0.9999	0.9999	0.9999	0.9997
17.40	1.0000	1.0000	1.0000	1.0000	1.0000	1.0000	0.9999	0.9999	0.9999	0.9998
17.45	1.0000	1.0000	1.0000	1.0000	1.0000	1.0000	1.0000	0.9999	0.9999	0.9998
17.50	1.0000	1.0000	1.0000	1.0000	1.0000	1.0000	1.0000	0.9999	0.9999	0.9998
17.55	1.0000	1.0000	1.0000	1.0000	1.0000	1.0000	1.0000	0.9999	0.9999	0.9998
17.60	1.0000	1.0000	1.0000	1.0000	1.0000	1.0000	1.0000	0.9999	0.9999	0.9998
17.65	1.0000	1.0000	1.0000	1.0000	1.0000	1.0000	1.0000	0.9999	0.9999	0.9998
17.70	1.0000	1.0000	1.0000	1.0000	1.0000	1.0000	1.0000	0.9999	0.9999	0.9998
17.75	1.0000	1.0000	1.0000	1.0000	1.0000	1.0000	1.0000	0.9999	0.9999	0.9998
17.80	1.0000	1.0000	1.0000	1.0000	1.0000	1.0000	1.0000	0.9999	0.9999	0.9998
17.85	1.0000	1.0000	1.0000	1.0000	1.0000	1.0000	1.0000	0.9999	0.9999	0.9998
17.90	1.0000	1.0000	1.0000	1.0000	1.0000	1.0000	1.0000	0.9999	0.9999	0.9998
17.95	1.0000	1.0000	1.0000	1.0000	1.0000	1.0000	1.0000	0.9999	0.9999	0.9998
18.00	1.0000	1.0000	1.0000	1.0000	1.0000	1.0000	1.0000	0.9999	0.9999	0.9998
18.05	1.0000	1.0000	1.0000	1.0000	1.0000	1.0000	1.0000	0.9999	0.9999	0.9998
18.10	1.0000	1.0000	1.0000	1.0000	1.0000	1.0000	1.0000	0.9999	0.9999	0.9998
18.15	1.0000	1.0000	1.0000	1.0000	1.0000	1.0000	1.0000	0.9999	0.9999	0.9998
18.20	1.0000	1.0000	1.0000	1.0000	1.0000	1.0000	1.0000	0.9999	0.9999	0.9998
18.25	1.0000	1.0000	1.0000	1.0000	1.0000	1.0000	1.0000	0.9999	0.9999	0.9998
18.30	1.0000	1.0000	1.0000	1.0000	1.0000	1.0000	1.0000	0.9999	0.9999	0.9998
18.35	1.0000	1.0000	1.0000	1.0000	1.0000	1.0000	1.0000	0.9999	0.9999	0.9998
18.40	1.0000	1.0000	1.0000	1.0000	1.0000	1.0000	1.0000	0.9999	0.9999	0.9998
18.45	1.0000	1.0000	1.0000	1.0000	1.0000	1.0000	1.0000	0.9999	0.9999	0.9998
18.50	1.0000	1.0000	1.0000	1.0000	1.0000	1.0000	1.0000	0.9999	0.9999	0.9998
18.55	1.0000	1.0000	1.0000	1.0000	1.0000	1.0000	1.0000	0.9999	0.9999	0.9998
18.60	1.0000	1.0000	1.0000	1.0000	1.0000	1.0000	1.0000	0.9999	0.9999	0.9998
18.65	1.0000	1.0000	1.0000	1.0000	1.0000	1.0000	1.0000	0.9999	0.9999	0.9998
18.70	1.0000	1.0000	1.0000	1.0000	1.0000	1.0000	1.0000	0.9999	0.9999	0.9998
18.75	1.0000	1.0000	1.0000	1.0000	1.0000	1.0000	1.0000	0.9999	0.9999	0.9998
18.80	1.0000	1.0000	1.0000	1.0000	1.0000	1.0000	1.0000	0.9999	0.9999	0.9998
18.85	1.0000	1.0000	1.0000	1.0000	1.0000	1.0000	1.0000	0.9999	0.9999	0.9998
18.90	1.0000	1.0000	1.0000	1.0000	1.0000	1.0000	1.0000	0.9999	0.9999	0.9998
18.95	1.0000	1.0000	1.0000	1.0000	1.0000	1.0000	1.0000	0.9999	0.9999	0.9999
19.00	1.0000	1.0000	1.0000	1.0000	1.0000	1.0000	1.0000	0.9999	0.9999	0.9999
19.05	1.0000	1.0000	1.0000	1.0000	1.0000	1.0000	1.0000	0.9999	0.9999	0.9999
19.10	1.0000	1.0000	1.0000	1.0000	1.0000	1.0000	1.0000	0.9999	0.9999	0.9999
19.15	1.0000	1.0000	1.0000	1.0000	1.0000	1.0000	1.0000	0.9999	0.9999	0.9999
19.20	1.0000	1.0000	1.0000	1.0000	1.0000	1.0000	1.0000	0.9999	0.9999	0.9999
19.25	1.0000	1.0000	1.0000	1.0000	1.0000	1.0000	1.0000	0.9999	0.9999	0.9999
19.30	1.0000	1.0000	1.0000	1.0000	1.0000	1.0000	1.0000	0.9999	0.9999	0.9999
19.35	1.0000	1.0000	1.0000	1.0000	1.0000	1.0000	1.0000	0.9999	0.9999	0.9999
19.40	1.0000	1.0000	1.0000	1.0000	1.0000	1.0000	1.0000	0.9999	0.9999	0.9999
19.45	1.0000	1.0000	1.0000	1.0000	1.0000	1.0000	1.0000	0.9999	0.9999	0.9999

PROBABILITY INTEGRAL OF t, THE NON-CENTRAL t-STATISTIC. THIS TABLE GIVES Pr $[t/\sqrt{f} \leq x]$.

f is the number of degrees of freedom; the non-centrality parameter is $\sqrt{f+1}\,K_p$.

K_p is the standardized normal deviate exceeded with probability p.

DEGREES OF FREEDOM **6**

x \ p	.2500	.1500	.1000	.0650	.0400	.0250	.0100	.0040	.0025	.0010
19.50	1.0000	1.0000	1.0000	1.0000	1.0000	1.0000	1.0000	0.9999	0.9999	0.9999
19.55	1.0000	1.0000	1.0000	1.0000	1.0000	1.0000	1.0000	0.9999	0.9999	0.9999
19.60	1.0000	1.0000	1.0000	1.0000	1.0000	1.0000	1.0000	0.9999	0.9999	0.9999
19.65	1.0000	1.0000	1.0000	1.0000	1.0000	1.0000	1.0000	0.9999	0.9999	0.9999
19.70	1.0000	1.0000	1.0000	1.0000	1.0000	1.0000	1.0000	1.0000	0.9999	0.9999
19.75	1.0000	1.0000	1.0000	1.0000	1.0000	1.0000	1.0000	1.0000	0.9999	0.9999
19.80	1.0000	1.0000	1.0000	1.0000	1.0000	1.0000	1.0000	1.0000	0.9999	0.9999
19.85	1.0000	1.0000	1.0000	1.0000	1.0000	1.0000	1.0000	1.0000	0.9999	0.9999
19.90	1.0000	1.0000	1.0000	1.0000	1.0000	1.0000	1.0000	1.0000	0.9999	0.9999
19.95	1.0000	1.0000	1.0000	1.0000	1.0000	1.0000	1.0000	1.0000	0.9999	0.9999
20.00	1.0000	1.0000	1.0000	1.0000	1.0000	1.0000	1.0000	1.0000	0.9999	0.9999
20.05	1.0000	1.0000	1.0000	1.0000	1.0000	1.0000	1.0000	1.0000	0.9999	0.9999
20.10	1.0000	1.0000	1.0000	1.0000	1.0000	1.0000	1.0000	1.0000	0.9999	0.9999
20.15	1.0000	1.0000	1.0000	1.0000	1.0000	1.0000	1.0000	1.0000	0.9999	0.9999
20.20	1.0000	1.0000	1.0000	1.0000	1.0000	1.0000	1.0000	1.0000	0.9999	0.9999
20.25	1.0000	1.0000	1.0000	1.0000	1.0000	1.0000	1.0000	1.0000	0.9999	0.9999
20.30	1.0000	1.0000	1.0000	1.0000	1.0000	1.0000	1.0000	1.0000	0.9999	0.9999
20.35	1.0000	1.0000	1.0000	1.0000	1.0000	1.0000	1.0000	1.0000	0.9999	0.9999
20.40	1.0000	1.0000	1.0000	1.0000	1.0000	1.0000	1.0000	1.0000	0.9999	0.9999
20.45	1.0000	1.0000	1.0000	1.0000	1.0000	1.0000	1.0000	1.0000	0.9999	0.9999
20.50	1.0000	1.0000	1.0000	1.0000	1.0000	1.0000	1.0000	1.0000	0.9999	0.9999
20.55	1.0000	1.0000	1.0000	1.0000	1.0000	1.0000	1.0000	1.0000	0.9999	0.9999
20.60	1.0000	1.0000	1.0000	1.0000	1.0000	1.0000	1.0000	1.0000	0.9999	0.9999
20.65	1.0000	1.0000	1.0000	1.0000	1.0000	1.0000	1.0000	1.0000	1.0000	0.9999
20.70	1.0000	1.0000	1.0000	1.0000	1.0000	1.0000	1.0000	1.0000	1.0000	0.9999
20.75	1.0000	1.0000	1.0000	1.0000	1.0000	1.0000	1.0000	1.0000	1.0000	0.9999
20.80	1.0000	1.0000	1.0000	1.0000	1.0000	1.0000	1.0000	1.0000	1.0000	0.9999
20.85	1.0000	1.0000	1.0000	1.0000	1.0000	1.0000	1.0000	1.0000	1.0000	0.9999
20.90	1.0000	1.0000	1.0000	1.0000	1.0000	1.0000	1.0000	1.0000	1.0000	0.9999
20.95	1.0000	1.0000	1.0000	1.0000	1.0000	1.0000	1.0000	1.0000	1.0000	0.9999
21.00	1.0000	1.0000	1.0000	1.0000	1.0000	1.0000	1.0000	1.0000	1.0000	0.9999
21.05	1.0000	1.0000	1.0000	1.0000	1.0000	1.0000	1.0000	1.0000	1.0000	0.9999
21.10	1.0000	1.0000	1.0000	1.0000	1.0000	1.0000	1.0000	1.0000	1.0000	0.9999
21.15	1.0000	1.0000	1.0000	1.0000	1.0000	1.0000	1.0000	1.0000	1.0000	0.9999
21.20	1.0000	1.0000	1.0000	1.0000	1.0000	1.0000	1.0000	1.0000	1.0000	0.9999
21.25	1.0000	1.0000	1.0000	1.0000	1.0000	1.0000	1.0000	1.0000	1.0000	0.9999
21.30	1.0000	1.0000	1.0000	1.0000	1.0000	1.0000	1.0000	1.0000	1.0000	0.9999
21.35	1.0000	1.0000	1.0000	1.0000	1.0000	1.0000	1.0000	1.0000	1.0000	0.9999
21.40	1.0000	1.0000	1.0000	1.0000	1.0000	1.0000	1.0000	1.0000	1.0000	0.9999
21.45	1.0000	1.0000	1.0000	1.0000	1.0000	1.0000	1.0000	1.0000	1.0000	0.9999
21.50	1.0000	1.0000	1.0000	1.0000	1.0000	1.0000	1.0000	1.0000	1.0000	0.9999
21.55	1.0000	1.0000	1.0000	1.0000	1.0000	1.0000	1.0000	1.0000	1.0000	0.9999
21.60	1.0000	1.0000	1.0000	1.0000	1.0000	1.0000	1.0000	1.0000	1.0000	0.9999
21.65	1.0000	1.0000	1.0000	1.0000	1.0000	1.0000	1.0000	1.0000	1.0000	0.9999
21.70	1.0000	1.0000	1.0000	1.0000	1.0000	1.0000	1.0000	1.0000	1.0000	0.9999
21.75	1.0000	1.0000	1.0000	1.0000	1.0000	1.0000	1.0000	1.0000	1.0000	0.9999
21.80	1.0000	1.0000	1.0000	1.0000	1.0000	1.0000	1.0000	1.0000	1.0000	0.9999
21.85	1.0000	1.0000	1.0000	1.0000	1.0000	1.0000	1.0000	1.0000	1.0000	0.9999
21.90	1.0000	1.0000	1.0000	1.0000	1.0000	1.0000	1.0000	1.0000	1.0000	0.9999
21.95	1.0000	1.0000	1.0000	1.0000	1.0000	1.0000	1.0000	1.0000	1.0000	0.9999
22.00	1.0000	1.0000	1.0000	1.0000	1.0000	1.0000	1.0000	1.0000	1.0000	0.9999
22.05	1.0000	1.0000	1.0000	1.0000	1.0000	1.0000	1.0000	1.0000	1.0000	0.9999
22.10	1.0000	1.0000	1.0000	1.0000	1.0000	1.0000	1.0000	1.0000	1.0000	0.9999
22.15	1.0000	1.0000	1.0000	1.0000	1.0000	1.0000	1.0000	1.0000	1.0000	0.9999
22.20	1.0000	1.0000	1.0000	1.0000	1.0000	1.0000	1.0000	1.0000	1.0000	0.9999
22.25	1.0000	1.0000	1.0000	1.0000	1.0000	1.0000	1.0000	1.0000	1.0000	0.9999
22.30	1.0000	1.0000	1.0000	1.0000	1.0000	1.0000	1.0000	1.0000	1.0000	0.9999
22.35	1.0000	1.0000	1.0000	1.0000	1.0000	1.0000	1.0000	1.0000	1.0000	0.9999
22.40	1.0000	1.0000	1.0000	1.0000	1.0000	1.0000	1.0000	1.0000	1.0000	0.9999
22.45	1.0000	1.0000	1.0000	1.0000	1.0000	1.0000	1.0000	1.0000	1.0000	0.9999

PROBABILITY INTEGRAL OF t, THE NON-CENTRAL t-STATISTIC. THIS TABLE GIVES Pr $[t/\sqrt{f} \leq x]$.

f is the number of degrees of freedom; the non-centrality parameter is $\sqrt{f+1}\,K_p$.

K_p is the standardized normal deviate exceeded with probability p.

DEGREES OF FREEDOM **6**

x \ p	.2500	.1500	.1000	.0650	.0400	.0250	.0100	.0040	.0025	.0010
22.50	1.0000	1.0000	1.0000	1.0000	1.0000	1.0000	1.0000	1.0000	1.0000	0.9999
22.55	1.0000	1.0000	1.0000	1.0000	1.0000	1.0000	1.0000	1.0000	1.0000	0.9999
22.60	1.0000	1.0000	1.0000	1.0000	1.0000	1.0000	1.0000	1.0000	1.0000	0.9999
22.65	1.0000	1.0000	1.0000	1.0000	1.0000	1.0000	1.0000	1.0000	1.0000	0.9999
22.70	1.0000	1.0000	1.0000	1.0000	1.0000	1.0000	1.0000	1.0000	1.0000	1.0000

PROBABILITY INTEGRAL OF t, THE NON-CENTRAL t-STATISTIC. THIS TABLE GIVES Pr $[t/\sqrt{f} \leq x]$.

f is the number of degrees of freedom; the non-centrality parameter is $\sqrt{f+1}\,K_p$.

K_p is the standardized normal deviate exceeded with probability p.

x \ P	.2500	.1500	.1000	.0650	.0400	.0250	.0100	.0040	.0025	.0010
-1.10	0.0000	0.0000	0.0000	0.0000	0.0000	0.0000	0.0000	0.0000	0.0000	0.0000
-1.05	0.0001	0.0000	0.0000	0.0000	0.0000	0.0000	0.0000	0.0000	0.0000	0.0000
-1.00	0.0001	0.0000	0.0000	0.0000	0.0000	0.0000	0.0000	0.0000	0.0000	0.0000
-0.95	0.0001	0.0000	0.0000	0.0000	0.0000	0.0000	0.0000	0.0000	0.0000	0.0000
-0.90	0.0001	0.0000	0.0000	0.0000	0.0000	0.0000	0.0000	0.0000	0.0000	0.0000
-0.85	0.0002	0.0000	0.0000	0.0000	0.0000	0.0000	0.0000	0.0000	0.0000	0.0000
-0.80	0.0002	0.0000	0.0000	0.0000	0.0000	0.0000	0.0000	0.0000	0.0000	0.0000
-0.75	0.0003	0.0000	0.0000	0.0000	0.0000	0.0000	0.0000	0.0000	0.0000	0.0000
-0.70	0.0004	0.0000	0.0000	0.0000	0.0000	0.0000	0.0000	0.0000	0.0000	0.0000
-0.65	0.0005	0.0000	0.0000	0.0000	0.0000	0.0000	0.0000	0.0000	0.0000	0.0000
-0.60	0.0007	0.0000	0.0000	0.0000	0.0000	0.0000	0.0000	0.0000	0.0000	0.0000
-0.55	0.0009	0.0000	0.0000	0.0000	0.0000	0.0000	0.0000	0.0000	0.0000	0.0000
-0.50	0.0013	0.0000	0.0000	0.0000	0.0000	0.0000	0.0000	0.0000	0.0000	0.0000
-0.45	0.0017	0.0000	0.0000	0.0000	0.0000	0.0000	0.0000	0.0000	0.0000	0.0000
-0.40	0.0023	0.0001	0.0000	0.0000	0.0000	0.0000	0.0000	0.0000	0.0000	0.0000
-0.35	0.0032	0.0001	0.0000	0.0000	0.0000	0.0000	0.0000	0.0000	0.0000	0.0000
-0.30	0.0044	0.0001	0.0000	0.0000	0.0000	0.0000	0.0000	0.0000	0.0000	0.0000
-0.25	0.0060	0.0002	0.0000	0.0000	0.0000	0.0000	0.0000	0.0000	0.0000	0.0000
-0.20	0.0083	0.0003	0.0000	0.0000	0.0000	0.0000	0.0000	0.0000	0.0000	0.0000
-0.15	0.0113	0.0005	0.0000	0.0000	0.0000	0.0000	0.0000	0.0000	0.0000	0.0000
-0.10	0.0155	0.0007	0.0001	0.0000	0.0000	0.0000	0.0000	0.0000	0.0000	0.0000
-0.05	0.0210	0.0011	0.0001	0.0000	0.0000	0.0000	0.0000	0.0000	0.0000	0.0000
0.00	0.0282	0.0017	0.0001	0.0000	0.0000	0.0000	0.0000	0.0000	0.0000	0.0000
0.05	0.0376	0.0025	0.0002	0.0000	0.0000	0.0000	0.0000	0.0000	0.0000	0.0000
0.10	0.0496	0.0038	0.0004	0.0000	0.0000	0.0000	0.0000	0.0000	0.0000	0.0000
0.15	0.0647	0.0056	0.0006	0.0001	0.0000	0.0000	0.0000	0.0000	0.0000	0.0000
0.20	0.0832	0.0083	0.0010	0.0001	0.0000	0.0000	0.0000	0.0000	0.0000	0.0000
0.25	0.1055	0.0120	0.0016	0.0002	0.0000	0.0000	0.0000	0.0000	0.0000	0.0000
0.30	0.1318	0.0171	0.0026	0.0003	0.0000	0.0000	0.0000	0.0000	0.0000	0.0000
0.35	0.1622	0.0239	0.0040	0.0005	0.0000	0.0000	0.0000	0.0000	0.0000	0.0000
0.40	0.1964	0.0330	0.0061	0.0009	0.0001	0.0000	0.0000	0.0000	0.0000	0.0000
0.45	0.2342	0.0446	0.0092	0.0014	0.0002	0.0000	0.0000	0.0000	0.0000	0.0000
0.50	0.2751	0.0593	0.0135	0.0023	0.0003	0.0000	0.0000	0.0000	0.0000	0.0000
0.55	0.3184	0.0771	0.0193	0.0037	0.0005	0.0001	0.0000	0.0000	0.0000	0.0000
0.60	0.3634	0.0985	0.0271	0.0058	0.0009	0.0001	0.0000	0.0000	0.0000	0.0000
0.65	0.4093	0.1234	0.0372	0.0088	0.0015	0.0002	0.0000	0.0000	0.0000	0.0000
0.70	0.4553	0.1517	0.0499	0.0129	0.0024	0.0004	0.0000	0.0000	0.0000	0.0000
0.75	0.5006	0.1833	0.0654	0.0186	0.0038	0.0007	0.0000	0.0000	0.0000	0.0000
0.80	0.5447	0.2179	0.0840	0.0260	0.0060	0.0013	0.0001	0.0000	0.0000	0.0000
0.85	0.5871	0.2548	0.1058	0.0355	0.0090	0.0021	0.0001	0.0000	0.0000	0.0000
0.90	0.6272	0.2937	0.1306	0.0474	0.0131	0.0034	0.0002	0.0000	0.0000	0.0000
0.95	0.6648	0.3339	0.1584	0.0618	0.0186	0.0052	0.0004	0.0000	0.0000	0.0000
1.00	0.6998	0.3749	0.1888	0.0790	0.0258	0.0079	0.0006	0.0000	0.0000	0.0000
1.05	0.7320	0.4160	0.2216	0.0989	0.0349	0.0115	0.0011	0.0001	0.0000	0.0000
1.10	0.7614	0.4568	0.2562	0.1215	0.0461	0.0163	0.0018	0.0002	0.0000	0.0000
1.15	0.7880	0.4967	0.2924	0.1466	0.0595	0.0226	0.0028	0.0003	0.0001	0.0000
1.20	0.8121	0.5354	0.3295	0.1742	0.0753	0.0304	0.0043	0.0005	0.0002	0.0000
1.25	0.8337	0.5726	0.3672	0.2038	0.0935	0.0401	0.0064	0.0008	0.0003	0.0000
1.30	0.8530	0.6080	0.4049	0.2351	0.1140	0.0518	0.0092	0.0014	0.0005	0.0001
1.35	0.8703	0.6415	0.4423	0.2679	0.1368	0.0655	0.0130	0.0022	0.0008	0.0001
1.40	0.8855	0.6728	0.4790	0.3017	0.1616	0.0813	0.0178	0.0033	0.0013	0.0002
1.45	0.8991	0.7021	0.5148	0.3361	0.1883	0.0991	0.0239	0.0049	0.0021	0.0004
1.50	0.9110	0.7292	0.5493	0.3708	0.2166	0.1191	0.0313	0.0070	0.0031	0.0006
1.55	0.9215	0.7543	0.5823	0.4054	0.2462	0.1409	0.0403	0.0098	0.0046	0.0009
1.60	0.9308	0.7773	0.6138	0.4397	0.2768	0.1645	0.0509	0.0135	0.0065	0.0015
1.65	0.9390	0.7985	0.6437	0.4734	0.3081	0.1896	0.0631	0.0181	0.0091	0.0022
1.70	0.9462	0.8177	0.6718	0.5062	0.3399	0.2161	0.0770	0.0237	0.0124	0.0033

PROBABILITY INTEGRAL OF t, THE NON-CENTRAL t-STATISTIC. THIS TABLE GIVES Pr $[t/\sqrt{f} \leq x]$.

f is the number of degrees of freedom; the non-centrality parameter is $\sqrt{f+1}\,K_p$.

K_p is the standardized normal deviate exceeded with probability p.

DEGREES OF FREEDOM **7**

x \ p	.2500	.1500	.1000	.0650	.0400	.0250	.0100	.0040	.0025	.0010
1.75	0.9525	0.8353	0.6981	0.5380	0.3718	0.2437	0.0926	0.0305	0.0165	0.0047
1.80	0.9580	0.8512	0.7227	0.5686	0.4036	0.2722	0.1098	0.0385	0.0216	0.0066
1.85	0.9629	0.8657	0.7456	0.5979	0.4351	0.3013	0.1286	0.0479	0.0277	0.0090
1.90	0.9672	0.8788	0.7669	0.6258	0.4660	0.3308	0.1488	0.0587	0.0350	0.0120
1.95	0.9709	0.8906	0.7865	0.6522	0.4962	0.3604	0.1704	0.0708	0.0435	0.0158
2.00	0.9742	0.9013	0.8046	0.6773	0.5256	0.3900	0.1931	0.0844	0.0532	0.0204
2.05	0.9771	0.9109	0.8213	0.7008	0.5539	0.4194	0.2168	0.0993	0.0642	0.0259
2.10	0.9797	0.9196	0.8366	0.7229	0.5812	0.4484	0.2414	0.1156	0.0764	0.0323
2.15	0.9819	0.9274	0.8507	0.7436	0.6074	0.4768	0.2666	0.1331	0.0900	0.0397
2.20	0.9839	0.9344	0.8636	0.7630	0.6324	0.5045	0.2924	0.1517	0.1048	0.0482
2.25	0.9857	0.9407	0.8754	0.7810	0.6562	0.5315	0.3185	0.1714	0.1207	0.0578
2.30	0.9872	0.9464	0.8862	0.7977	0.6787	0.5575	0.3447	0.1921	0.1378	0.0686
2.35	0.9886	0.9515	0.8960	0.8133	0.7000	0.5827	0.3710	0.2136	0.1559	0.0804
2.40	0.9898	0.9561	0.9050	0.8277	0.7202	0.6068	0.3971	0.2357	0.1750	0.0933
2.45	0.9909	0.9603	0.9132	0.8410	0.7391	0.6299	0.4230	0.2585	0.1949	0.1072
2.50	0.9918	0.9640	0.9206	0.8533	0.7569	0.6520	0.4485	0.2817	0.2155	0.1222
2.55	0.9927	0.9674	0.9274	0.8647	0.7737	0.6730	0.4736	0.3051	0.2368	0.1381
2.60	0.9934	0.9704	0.9336	0.8752	0.7893	0.6930	0.4981	0.3288	0.2586	0.1549
2.65	0.9941	0.9731	0.9393	0.8849	0.8039	0.7120	0.5221	0.3526	0.2807	0.1725
2.70	0.9947	0.9756	0.9444	0.8939	0.8176	0.7299	0.5453	0.3763	0.3032	0.1908
2.75	0.9952	0.9778	0.9491	0.9021	0.8304	0.7469	0.5678	0.3999	0.3258	0.2097
2.80	0.9957	0.9798	0.9534	0.9097	0.8422	0.7628	0.5896	0.4232	0.3484	0.2292
2.85	0.9961	0.9816	0.9573	0.9166	0.8533	0.7779	0.6106	0.4463	0.3711	0.2492
2.90	0.9964	0.9832	0.9608	0.9231	0.8636	0.7921	0.6308	0.4689	0.3937	0.2695
2.95	0.9968	0.9847	0.9640	0.9290	0.8732	0.8054	0.6502	0.4912	0.4160	0.2901
3.00	0.9971	0.9860	0.9670	0.9344	0.8821	0.8179	0.6687	0.5129	0.4381	0.3109
3.05	0.9974	0.9872	0.9697	0.9394	0.8904	0.8296	0.6865	0.5341	0.4599	0.3319
3.10	0.9976	0.9883	0.9721	0.9440	0.8980	0.8406	0.7034	0.5547	0.4813	0.3528
3.15	0.9978	0.9893	0.9744	0.9482	0.9052	0.8509	0.7195	0.5748	0.5023	0.3737
3.20	0.9980	0.9902	0.9764	0.9521	0.9118	0.8605	0.7349	0.5942	0.5228	0.3946
3.25	0.9982	0.9910	0.9783	0.9557	0.9179	0.8695	0.7495	0.6129	0.5428	0.4152
3.30	0.9983	0.9918	0.9800	0.9590	0.9236	0.8780	0.7634	0.6310	0.5622	0.4356
3.35	0.9985	0.9925	0.9816	0.9620	0.9289	0.8858	0.7766	0.6484	0.5811	0.4558
3.40	0.9986	0.9931	0.9830	0.9648	0.9338	0.8932	0.7891	0.6652	0.5994	0.4756
3.45	0.9987	0.9936	0.9843	0.9674	0.9383	0.9001	0.8009	0.6813	0.6172	0.4950
3.50	0.9988	0.9941	0.9855	0.9697	0.9426	0.9065	0.8121	0.6968	0.6343	0.5141
3.55	0.9989	0.9946	0.9866	0.9719	0.9465	0.9125	0.8227	0.7116	0.6508	0.5328
3.60	0.9990	0.9950	0.9876	0.9739	0.9501	0.9181	0.8326	0.7257	0.6667	0.5510
3.65	0.9991	0.9954	0.9885	0.9758	0.9535	0.9233	0.8421	0.7393	0.6820	0.5687
3.70	0.9992	0.9958	0.9894	0.9775	0.9566	0.9281	0.8510	0.7522	0.6967	0.5859
3.75	0.9992	0.9961	0.9902	0.9791	0.9595	0.9327	0.8594	0.7646	0.7109	0.6027
3.80	0.9993	0.9964	0.9909	0.9806	0.9622	0.9369	0.8674	0.7764	0.7244	0.6189
3.85	0.9994	0.9967	0.9916	0.9819	0.9647	0.9409	0.8748	0.7876	0.7374	0.6346
3.90	0.9994	0.9969	0.9922	0.9832	0.9670	0.9446	0.8819	0.7983	0.7498	0.6498
3.95	0.9995	0.9971	0.9927	0.9843	0.9692	0.9480	0.8886	0.8085	0.7617	0.6645
4.00	0.9995	0.9973	0.9932	0.9854	0.9712	0.9513	0.8948	0.8181	0.7730	0.6787
4.05	0.9995	0.9975	0.9937	0.9864	0.9730	0.9543	0.9007	0.8273	0.7839	0.6924
4.10	0.9996	0.9977	0.9941	0.9873	0.9748	0.9571	0.9063	0.8361	0.7943	0.7056
4.15	0.9996	0.9979	0.9946	0.9881	0.9764	0.9597	0.9115	0.8444	0.8041	0.7182
4.20	0.9996	0.9980	0.9949	0.9889	0.9779	0.9622	0.9165	0.8523	0.8136	0.7304
4.25	0.9997	0.9982	0.9953	0.9897	0.9793	0.9645	0.9211	0.8598	0.8226	0.7422
4.30	0.9997	0.9983	0.9956	0.9903	0.9806	0.9666	0.9255	0.8669	0.8311	0.7534
4.35	0.9997	0.9984	0.9959	0.9910	0.9818	0.9686	0.9296	0.8736	0.8393	0.7642
4.40	0.9997	0.9985	0.9962	0.9915	0.9829	0.9705	0.9335	0.8800	0.8470	0.7746
4.45	0.9997	0.9986	0.9964	0.9921	0.9840	0.9722	0.9371	0.8861	0.8544	0.7845
4.50	0.9998	0.9987	0.9967	0.9926	0.9850	0.9739	0.9406	0.8918	0.8615	0.7940
4.55	0.9998	0.9988	0.9969	0.9931	0.9859	0.9754	0.9438	0.8973	0.8681	0.8031
4.60	0.9998	0.9989	0.9971	0.9935	0.9867	0.9769	0.9469	0.9024	0.8745	0.8118
4.65	0.9998	0.9990	0.9973	0.9939	0.9875	0.9782	0.9498	0.9073	0.8806	0.8202
4.70	0.9998	0.9990	0.9974	0.9943	0.9883	0.9795	0.9525	0.9120	0.8863	0.8282

PROBABILITY INTEGRAL OF t, THE NON-CENTRAL t-STATISTIC. THIS TABLE GIVES Pr $[t/\sqrt{f} \leq x]$.

f is the number of degrees of freedom; the non-centrality parameter is $\sqrt{f+1}\,K_p$.

K_p is the standardized normal deviate exceeded with probability p.

DEGREES OF FREEDOM **7**

x \ p	.2500	.1500	.1000	.0650	.0400	.0250	.0100	.0040	.0025	.0010
4.75	0.9998	0.9991	0.9976	0.9946	0.9890	0.9806	0.9550	0.9164	0.8918	0.8358
4.80	0.9998	0.9991	0.9977	0.9950	0.9896	0.9817	0.9574	0.9205	0.8970	0.8431
4.85	0.9999	0.9992	0.9979	0.9953	0.9902	0.9828	0.9597	0.9245	0.9019	0.8501
4.90	0.9999	0.9993	0.9980	0.9955	0.9908	0.9838	0.9618	0.9282	0.9066	0.8567
4.95	0.9999	0.9993	0.9981	0.9958	0.9913	0.9847	0.9638	0.9318	0.9111	0.8631
5.00	0.9999	0.9993	0.9983	0.9961	0.9918	0.9855	0.9657	0.9352	0.9153	0.8692
5.05	0.9999	0.9994	0.9984	0.9963	0.9923	0.9863	0.9675	0.9383	0.9193	0.8750
5.10	0.9999	0.9994	0.9985	0.9965	0.9927	0.9871	0.9692	0.9414	0.9231	0.8805
5.15	0.9999	0.9995	0.9985	0.9967	0.9931	0.9878	0.9708	0.9442	0.9268	0.8858
5.20	0.9999	0.9995	0.9986	0.9969	0.9935	0.9885	0.9723	0.9469	0.9302	0.8909
5.25	0.9999	0.9995	0.9987	0.9971	0.9939	0.9891	0.9738	0.9495	0.9335	0.8957
5.30	0.9999	0.9995	0.9988	0.9972	0.9942	0.9897	0.9751	0.9520	0.9366	0.9003
5.35	0.9999	0.9996	0.9989	0.9974	0.9945	0.9902	0.9764	0.9543	0.9396	0.9047
5.40	0.9999	0.9996	0.9989	0.9975	0.9948	0.9907	0.9776	0.9565	0.9424	0.9089
5.45	0.9999	0.9996	0.9990	0.9977	0.9951	0.9912	0.9787	0.9585	0.9451	0.9129
5.50	0.9999	0.9996	0.9990	0.9978	0.9954	0.9917	0.9798	0.9605	0.9476	0.9167
5.55	0.9999	0.9997	0.9991	0.9979	0.9956	0.9921	0.9808	0.9624	0.9500	0.9203
5.60	0.9999	0.9997	0.9991	0.9980	0.9959	0.9925	0.9817	0.9642	0.9523	0.9238
5.65	0.9999	0.9997	0.9992	0.9981	0.9961	0.9929	0.9826	0.9658	0.9545	0.9271
5.70	0.9999	0.9997	0.9992	0.9982	0.9963	0.9933	0.9835	0.9674	0.9566	0.9303
5.75	1.0000	0.9997	0.9993	0.9983	0.9965	0.9936	0.9843	0.9690	0.9586	0.9333
5.80	1.0000	0.9997	0.9993	0.9984	0.9967	0.9940	0.9851	0.9704	0.9604	0.9362
5.85	1.0000	0.9998	0.9994	0.9985	0.9968	0.9943	0.9858	0.9718	0.9622	0.9389
5.90	1.0000	0.9998	0.9994	0.9986	0.9970	0.9946	0.9865	0.9731	0.9639	0.9415
5.95	1.0000	0.9998	0.9994	0.9987	0.9972	0.9948	0.9871	0.9743	0.9655	0.9440
6.00	1.0000	0.9998	0.9995	0.9987	0.9973	0.9951	0.9877	0.9755	0.9671	0.9464
6.05	1.0000	0.9998	0.9995	0.9988	0.9974	0.9953	0.9883	0.9766	0.9685	0.9487
6.10	1.0000	0.9998	0.9995	0.9989	0.9976	0.9956	0.9889	0.9777	0.9699	0.9509
6.15	1.0000	0.9998	0.9995	0.9989	0.9977	0.9958	0.9894	0.9787	0.9713	0.9529
6.20	1.0000	0.9998	0.9996	0.9990	0.9978	0.9960	0.9899	0.9796	0.9725	0.9549
6.25	1.0000	0.9998	0.9996	0.9990	0.9979	0.9962	0.9903	0.9805	0.9737	0.9568
6.30	1.0000	0.9999	0.9996	0.9991	0.9980	0.9964	0.9908	0.9814	0.9749	0.9586
6.35	1.0000	0.9999	0.9996	0.9991	0.9981	0.9965	0.9912	0.9822	0.9760	0.9603
6.40	1.0000	0.9999	0.9996	0.9992	0.9982	0.9967	0.9916	0.9830	0.9770	0.9620
6.45	1.0000	0.9999	0.9997	0.9992	0.9983	0.9969	0.9920	0.9837	0.9780	0.9635
6.50	1.0000	0.9999	0.9997	0.9992	0.9984	0.9970	0.9924	0.9845	0.9789	0.9650
6.55	1.0000	0.9999	0.9997	0.9993	0.9984	0.9971	0.9927	0.9851	0.9798	0.9665
6.60	1.0000	0.9999	0.9997	0.9993	0.9985	0.9973	0.9930	0.9858	0.9807	0.9678
6.65	1.0000	0.9999	0.9997	0.9993	0.9986	0.9974	0.9933	0.9864	0.9815	0.9692
6.70	1.0000	0.9999	0.9997	0.9994	0.9987	0.9975	0.9936	0.9870	0.9823	0.9704
6.75	1.0000	0.9999	0.9997	0.9994	0.9987	0.9976	0.9939	0.9875	0.9830	0.9716
6.80	1.0000	0.9999	0.9998	0.9994	0.9988	0.9977	0.9942	0.9881	0.9837	0.9728
6.85	1.0000	0.9999	0.9998	0.9995	0.9988	0.9978	0.9944	0.9886	0.9844	0.9738
6.90	1.0000	0.9999	0.9998	0.9995	0.9989	0.9979	0.9947	0.9890	0.9850	0.9749
6.95	1.0000	0.9999	0.9998	0.9995	0.9989	0.9980	0.9949	0.9895	0.9857	0.9759
7.00	1.0000	0.9999	0.9998	0.9995	0.9990	0.9981	0.9951	0.9899	0.9863	0.9768
7.05	1.0000	0.9999	0.9998	0.9996	0.9990	0.9982	0.9953	0.9904	0.9868	0.9778
7.10	1.0000	0.9999	0.9998	0.9996	0.9991	0.9983	0.9955	0.9908	0.9873	0.9786
7.15	1.0000	0.9999	0.9998	0.9996	0.9991	0.9984	0.9957	0.9911	0.9879	0.9795
7.20	1.0000	0.9999	0.9998	0.9996	0.9992	0.9984	0.9959	0.9915	0.9883	0.9803
7.25	1.0000	0.9999	0.9998	0.9996	0.9992	0.9985	0.9961	0.9918	0.9888	0.9810
7.30	1.0000	0.9999	0.9998	0.9996	0.9992	0.9986	0.9962	0.9922	0.9893	0.9818
7.35	1.0000	0.9999	0.9999	0.9997	0.9993	0.9986	0.9964	0.9925	0.9897	0.9825
7.40	1.0000	1.0000	0.9999	0.9997	0.9993	0.9987	0.9965	0.9928	0.9901	0.9832
7.45	1.0000	1.0000	0.9999	0.9997	0.9993	0.9987	0.9967	0.9931	0.9905	0.9838
7.50	1.0000	1.0000	0.9999	0.9997	0.9993	0.9988	0.9968	0.9934	0.9909	0.9844
7.55	1.0000	1.0000	0.9999	0.9997	0.9994	0.9988	0.9970	0.9936	0.9912	0.9850
7.60	1.0000	1.0000	0.9999	0.9997	0.9994	0.9989	0.9971	0.9939	0.9916	0.9856
7.65	1.0000	1.0000	0.9999	0.9997	0.9994	0.9989	0.9972	0.9941	0.9919	0.9861
7.70	1.0000	1.0000	0.9999	0.9998	0.9995	0.9990	0.9973	0.9943	0.9922	0.9866

PROBABILITY INTEGRAL OF t, THE NON-CENTRAL t-STATISTIC. THIS TABLE GIVES Pr $[t/\sqrt{f} \leq x]$.

f is the number of degrees of freedom; the non-centrality parameter is $\sqrt{f+1}\,K_p$.

K_p is the standardized normal deviate exceeded with probability p.

DEGREES OF FREEDOM **7**

x \ p	.2500	.1500	.1000	.0650	.0400	.0250	.0100	.0040	.0025	.0010
7.75	1.0000	1.0000	0.9999	0.9998	0.9995	0.9990	0.9974	0.9946	0.9925	0.9871
7.80	1.0000	1.0000	0.9999	0.9998	0.9995	0.9991	0.9975	0.9948	0.9928	0.9876
7.85	1.0000	1.0000	0.9999	0.9998	0.9995	0.9991	0.9976	0.9950	0.9931	0.9881
7.90	1.0000	1.0000	0.9999	0.9998	0.9995	0.9991	0.9977	0.9952	0.9933	0.9885
7.95	1.0000	1.0000	0.9999	0.9998	0.9996	0.9992	0.9978	0.9953	0.9936	0.9889
8.00	1.0000	1.0000	0.9999	0.9998	0.9996	0.9992	0.9979	0.9955	0.9938	0.9893
8.05	1.0000	1.0000	0.9999	0.9998	0.9996	0.9992	0.9980	0.9957	0.9940	0.9897
8.10	1.0000	1.0000	0.9999	0.9998	0.9996	0.9993	0.9980	0.9959	0.9943	0.9901
8.15	1.0000	1.0000	0.9999	0.9998	0.9996	0.9993	0.9981	0.9960	0.9945	0.9905
8.20	1.0000	1.0000	0.9999	0.9998	0.9996	0.9993	0.9982	0.9962	0.9947	0.9908
8.25	1.0000	1.0000	0.9999	0.9998	0.9997	0.9993	0.9983	0.9963	0.9949	0.9911
8.30	1.0000	1.0000	0.9999	0.9998	0.9997	0.9994	0.9983	0.9964	0.9951	0.9915
8.35	1.0000	1.0000	0.9999	0.9999	0.9997	0.9994	0.9984	0.9966	0.9952	0.9918
8.40	1.0000	1.0000	0.9999	0.9999	0.9997	0.9994	0.9984	0.9967	0.9954	0.9920
8.45	1.0000	1.0000	0.9999	0.9999	0.9997	0.9994	0.9985	0.9968	0.9956	0.9923
8.50	1.0000	1.0000	0.9999	0.9999	0.9997	0.9995	0.9986	0.9969	0.9957	0.9926
8.55	1.0000	1.0000	0.9999	0.9999	0.9997	0.9995	0.9986	0.9970	0.9959	0.9929
8.60	1.0000	1.0000	1.0000	0.9999	0.9997	0.9995	0.9987	0.9972	0.9960	0.9931
8.65	1.0000	1.0000	1.0000	0.9999	0.9997	0.9995	0.9987	0.9973	0.9962	0.9933
8.70	1.0000	1.0000	1.0000	0.9999	0.9998	0.9995	0.9988	0.9974	0.9963	0.9936
8.75	1.0000	1.0000	1.0000	0.9999	0.9998	0.9996	0.9988	0.9974	0.9964	0.9938
8.80	1.0000	1.0000	1.0000	0.9999	0.9998	0.9996	0.9989	0.9975	0.9966	0.9940
8.85	1.0000	1.0000	1.0000	0.9999	0.9998	0.9996	0.9989	0.9976	0.9967	0.9942
8.90	1.0000	1.0000	1.0000	0.9999	0.9998	0.9996	0.9989	0.9977	0.9968	0.9944
8.95	1.0000	1.0000	1.0000	0.9999	0.9998	0.9996	0.9990	0.9978	0.9969	0.9946
9.00	1.0000	1.0000	1.0000	0.9999	0.9998	0.9996	0.9990	0.9979	0.9970	0.9948
9.05	1.0000	1.0000	1.0000	0.9999	0.9998	0.9996	0.9990	0.9979	0.9971	0.9950
9.10	1.0000	1.0000	1.0000	0.9999	0.9998	0.9997	0.9991	0.9980	0.9972	0.9951
9.15	1.0000	1.0000	1.0000	0.9999	0.9998	0.9997	0.9991	0.9981	0.9973	0.9953
9.20	1.0000	1.0000	1.0000	0.9999	0.9998	0.9997	0.9991	0.9981	0.9974	0.9954
9.25	1.0000	1.0000	1.0000	0.9999	0.9998	0.9997	0.9992	0.9982	0.9975	0.9956
9.30	1.0000	1.0000	1.0000	0.9999	0.9998	0.9997	0.9992	0.9983	0.9976	0.9957
9.35	1.0000	1.0000	1.0000	0.9999	0.9998	0.9997	0.9992	0.9983	0.9977	0.9959
9.40	1.0000	1.0000	1.0000	0.9999	0.9999	0.9997	0.9993	0.9984	0.9977	0.9960
9.45	1.0000	1.0000	1.0000	0.9999	0.9999	0.9997	0.9993	0.9984	0.9978	0.9961
9.50	1.0000	1.0000	1.0000	0.9999	0.9999	0.9997	0.9993	0.9985	0.9979	0.9963
9.55	1.0000	1.0000	1.0000	0.9999	0.9999	0.9998	0.9993	0.9985	0.9980	0.9964
9.60	1.0000	1.0000	1.0000	0.9999	0.9999	0.9998	0.9993	0.9986	0.9980	0.9965
9.65	1.0000	1.0000	1.0000	0.9999	0.9999	0.9998	0.9994	0.9986	0.9981	0.9966
9.70	1.0000	1.0000	1.0000	0.9999	0.9999	0.9998	0.9994	0.9987	0.9981	0.9967
9.75	1.0000	1.0000	1.0000	1.0000	0.9999	0.9998	0.9994	0.9987	0.9982	0.9968
9.80	1.0000	1.0000	1.0000	1.0000	0.9999	0.9998	0.9994	0.9988	0.9983	0.9969
9.85	1.0000	1.0000	1.0000	1.0000	0.9999	0.9998	0.9994	0.9988	0.9983	0.9970
9.90	1.0000	1.0000	1.0000	1.0000	0.9999	0.9998	0.9995	0.9988	0.9984	0.9971
9.95	1.0000	1.0000	1.0000	1.0000	0.9999	0.9998	0.9995	0.9989	0.9984	0.9972
10.00	1.0000	1.0000	1.0000	1.0000	0.9999	0.9998	0.9995	0.9989	0.9985	0.9973
10.05	1.0000	1.0000	1.0000	1.0000	0.9999	0.9998	0.9995	0.9989	0.9985	0.9974
10.10	1.0000	1.0000	1.0000	1.0000	0.9999	0.9998	0.9995	0.9990	0.9986	0.9975
10.15	1.0000	1.0000	1.0000	1.0000	0.9999	0.9998	0.9995	0.9990	0.9986	0.9975
10.20	1.0000	1.0000	1.0000	1.0000	0.9999	0.9998	0.9996	0.9990	0.9987	0.9976
10.25	1.0000	1.0000	1.0000	1.0000	0.9999	0.9998	0.9996	0.9991	0.9987	0.9977
10.30	1.0000	1.0000	1.0000	1.0000	0.9999	0.9999	0.9996	0.9991	0.9987	0.9978
10.35	1.0000	1.0000	1.0000	1.0000	0.9999	0.9999	0.9996	0.9991	0.9988	0.9978
10.40	1.0000	1.0000	1.0000	1.0000	0.9999	0.9999	0.9996	0.9992	0.9988	0.9979
10.45	1.0000	1.0000	1.0000	1.0000	0.9999	0.9999	0.9996	0.9992	0.9989	0.9980
10.50	1.0000	1.0000	1.0000	1.0000	0.9999	0.9999	0.9996	0.9992	0.9989	0.9980
10.55	1.0000	1.0000	1.0000	1.0000	0.9999	0.9999	0.9997	0.9992	0.9989	0.9981
10.60	1.0000	1.0000	1.0000	1.0000	0.9999	0.9999	0.9997	0.9993	0.9990	0.9981
10.65	1.0000	1.0000	1.0000	1.0000	0.9999	0.9999	0.9997	0.9993	0.9990	0.9982
10.70	1.0000	1.0000	1.0000	1.0000	0.9999	0.9999	0.9997	0.9993	0.9990	0.9982

PROBABILITY INTEGRAL OF t, THE NON-CENTRAL t-STATISTIC. THIS TABLE GIVES Pr $[t/\sqrt{f} \leq x]$.

f is the number of degrees of freedom; the non-centrality parameter is $\sqrt{f+1}\,K_p$.

K_p is the standardized normal deviate exceeded with probability p.

DEGREES OF FREEDOM **7**

x \ p	.2500	.1500	.1000	.0650	.0400	.0250	.0100	.0040	.0025	.0010
10.75	1.0000	1.0000	1.0000	1.0000	0.9999	0.9999	0.9997	0.9993	0.9990	0.9983
10.80	1.0000	1.0000	1.0000	1.0000	0.9999	0.9999	0.9997	0.9993	0.9991	0.9983
10.85	1.0000	1.0000	1.0000	1.0000	0.9999	0.9999	0.9997	0.9994	0.9991	0.9984
10.90	1.0000	1.0000	1.0000	1.0000	0.9999	0.9999	0.9997	0.9994	0.9991	0.9984
10.95	1.0000	1.0000	1.0000	1.0000	0.9999	0.9999	0.9997	0.9994	0.9992	0.9985
11.00	1.0000	1.0000	1.0000	1.0000	1.0000	0.9999	0.9997	0.9994	0.9992	0.9985
11.05	1.0000	1.0000	1.0000	1.0000	1.0000	0.9999	0.9997	0.9994	0.9992	0.9986
11.10	1.0000	1.0000	1.0000	1.0000	1.0000	0.9999	0.9998	0.9995	0.9992	0.9986
11.15	1.0000	1.0000	1.0000	1.0000	1.0000	0.9999	0.9998	0.9995	0.9992	0.9987
11.20	1.0000	1.0000	1.0000	1.0000	1.0000	0.9999	0.9998	0.9995	0.9993	0.9987
11.25	1.0000	1.0000	1.0000	1.0000	1.0000	0.9999	0.9998	0.9995	0.9993	0.9987
11.30	1.0000	1.0000	1.0000	1.0000	1.0000	0.9999	0.9998	0.9995	0.9993	0.9988
11.35	1.0000	1.0000	1.0000	1.0000	1.0000	0.9999	0.9998	0.9995	0.9993	0.9988
11.40	1.0000	1.0000	1.0000	1.0000	1.0000	0.9999	0.9998	0.9995	0.9994	0.9988
11.45	1.0000	1.0000	1.0000	1.0000	1.0000	0.9999	0.9998	0.9996	0.9994	0.9989
11.50	1.0000	1.0000	1.0000	1.0000	1.0000	0.9999	0.9998	0.9996	0.9994	0.9989
11.55	1.0000	1.0000	1.0000	1.0000	1.0000	0.9999	0.9998	0.9996	0.9994	0.9989
11.60	1.0000	1.0000	1.0000	1.0000	1.0000	0.9999	0.9998	0.9996	0.9994	0.9990
11.65	1.0000	1.0000	1.0000	1.0000	1.0000	0.9999	0.9998	0.9996	0.9994	0.9990
11.70	1.0000	1.0000	1.0000	1.0000	1.0000	0.9999	0.9998	0.9996	0.9995	0.9990
11.75	1.0000	1.0000	1.0000	1.0000	1.0000	0.9999	0.9998	0.9996	0.9995	0.9990
11.80	1.0000	1.0000	1.0000	1.0000	1.0000	0.9999	0.9998	0.9996	0.9995	0.9991
11.85	1.0000	1.0000	1.0000	1.0000	1.0000	0.9999	0.9998	0.9996	0.9995	0.9991
11.90	1.0000	1.0000	1.0000	1.0000	1.0000	0.9999	0.9998	0.9997	0.9995	0.9991
11.95	1.0000	1.0000	1.0000	1.0000	1.0000	0.9999	0.9998	0.9997	0.9995	0.9991
12.00	1.0000	1.0000	1.0000	1.0000	1.0000	0.9999	0.9999	0.9997	0.9995	0.9992
12.05	1.0000	1.0000	1.0000	1.0000	1.0000	0.9999	0.9999	0.9997	0.9995	0.9992
12.10	1.0000	1.0000	1.0000	1.0000	1.0000	1.0000	0.9999	0.9997	0.9996	0.9992
12.15	1.0000	1.0000	1.0000	1.0000	1.0000	1.0000	0.9999	0.9997	0.9996	0.9992
12.20	1.0000	1.0000	1.0000	1.0000	1.0000	1.0000	0.9999	0.9997	0.9996	0.9993
12.25	1.0000	1.0000	1.0000	1.0000	1.0000	1.0000	0.9999	0.9997	0.9996	0.9993
12.30	1.0000	1.0000	1.0000	1.0000	1.0000	1.0000	0.9999	0.9997	0.9996	0.9993
12.35	1.0000	1.0000	1.0000	1.0000	1.0000	1.0000	0.9999	0.9997	0.9996	0.9993
12.40	1.0000	1.0000	1.0000	1.0000	1.0000	1.0000	0.9999	0.9997	0.9996	0.9993
12.45	1.0000	1.0000	1.0000	1.0000	1.0000	1.0000	0.9999	0.9997	0.9996	0.9993
12.50	1.0000	1.0000	1.0000	1.0000	1.0000	1.0000	0.9999	0.9998	0.9996	0.9994
12.55	1.0000	1.0000	1.0000	1.0000	1.0000	1.0000	0.9999	0.9998	0.9997	0.9994
12.60	1.0000	1.0000	1.0000	1.0000	1.0000	1.0000	0.9999	0.9998	0.9997	0.9994
12.65	1.0000	1.0000	1.0000	1.0000	1.0000	1.0000	0.9999	0.9998	0.9997	0.9994
12.70	1.0000	1.0000	1.0000	1.0000	1.0000	1.0000	0.9999	0.9998	0.9997	0.9994
12.75	1.0000	1.0000	1.0000	1.0000	1.0000	1.0000	0.9999	0.9998	0.9997	0.9994
12.80	1.0000	1.0000	1.0000	1.0000	1.0000	1.0000	0.9999	0.9998	0.9997	0.9995
12.85	1.0000	1.0000	1.0000	1.0000	1.0000	1.0000	0.9999	0.9998	0.9997	0.9995
12.90	1.0000	1.0000	1.0000	1.0000	1.0000	1.0000	0.9999	0.9998	0.9997	0.9995
12.95	1.0000	1.0000	1.0000	1.0000	1.0000	1.0000	0.9999	0.9998	0.9997	0.9995
13.00	1.0000	1.0000	1.0000	1.0000	1.0000	1.0000	0.9999	0.9998	0.9997	0.9995
13.05	1.0000	1.0000	1.0000	1.0000	1.0000	1.0000	0.9999	0.9998	0.9997	0.9995
13.10	1.0000	1.0000	1.0000	1.0000	1.0000	1.0000	0.9999	0.9998	0.9997	0.9995
13.15	1.0000	1.0000	1.0000	1.0000	1.0000	1.0000	0.9999	0.9998	0.9997	0.9995
13.20	1.0000	1.0000	1.0000	1.0000	1.0000	1.0000	0.9999	0.9998	0.9998	0.9996
13.25	1.0000	1.0000	1.0000	1.0000	1.0000	1.0000	0.9999	0.9998	0.9998	0.9996
13.30	1.0000	1.0000	1.0000	1.0000	1.0000	1.0000	0.9999	0.9998	0.9998	0.9996
13.35	1.0000	1.0000	1.0000	1.0000	1.0000	1.0000	0.9999	0.9998	0.9998	0.9996
13.40	1.0000	1.0000	1.0000	1.0000	1.0000	1.0000	0.9999	0.9998	0.9998	0.9996
13.45	1.0000	1.0000	1.0000	1.0000	1.0000	1.0000	0.9999	0.9998	0.9998	0.9996
13.50	1.0000	1.0000	1.0000	1.0000	1.0000	1.0000	0.9999	0.9999	0.9998	0.9996
13.55	1.0000	1.0000	1.0000	1.0000	1.0000	1.0000	0.9999	0.9999	0.9998	0.9996
13.60	1.0000	1.0000	1.0000	1.0000	1.0000	1.0000	0.9999	0.9999	0.9998	0.9996
13.65	1.0000	1.0000	1.0000	1.0000	1.0000	1.0000	0.9999	0.9999	0.9998	0.9996
13.70	1.0000	1.0000	1.0000	1.0000	1.0000	1.0000	0.9999	0.9999	0.9998	0.9997

PROBABILITY INTEGRAL OF t, THE NON-CENTRAL t-STATISTIC. THIS TABLE GIVES Pr $[t/\sqrt{f} \leq x]$.

f is the number of degrees of freedom; the non-centrality parameter is $\sqrt{f+1}\,K_p$.
K_p is the standardized normal deviate exceeded with probability p.

DEGREES OF FREEDOM **7**

x \ p	.2500	.1500	.1000	.0650	.0400	.0250	.0100	.0040	.0025	.0010
13.75	1.0000	1.0000	1.0000	1.0000	1.0000	1.0000	0.9999	0.9999	0.9998	0.9997
13.80	1.0000	1.0000	1.0000	1.0000	1.0000	1.0000	0.9999	0.9999	0.9998	0.9997
13.85	1.0000	1.0000	1.0000	1.0000	1.0000	1.0000	0.9999	0.9999	0.9998	0.9997
13.90	1.0000	1.0000	1.0000	1.0000	1.0000	1.0000	0.9999	0.9999	0.9998	0.9997
13.95	1.0000	1.0000	1.0000	1.0000	1.0000	1.0000	0.9999	0.9999	0.9998	0.9997
14.00	1.0000	1.0000	1.0000	1.0000	1.0000	1.0000	0.9999	0.9999	0.9998	0.9997
14.05	1.0000	1.0000	1.0000	1.0000	1.0000	1.0000	1.0000	0.9999	0.9998	0.9997
14.10	1.0000	1.0000	1.0000	1.0000	1.0000	1.0000	1.0000	0.9999	0.9998	0.9997
14.15	1.0000	1.0000	1.0000	1.0000	1.0000	1.0000	1.0000	0.9999	0.9998	0.9997
14.20	1.0000	1.0000	1.0000	1.0000	1.0000	1.0000	1.0000	0.9999	0.9999	0.9997
14.25	1.0000	1.0000	1.0000	1.0000	1.0000	1.0000	1.0000	0.9999	0.9999	0.9997
14.30	1.0000	1.0000	1.0000	1.0000	1.0000	1.0000	1.0000	0.9999	0.9999	0.9997
14.35	1.0000	1.0000	1.0000	1.0000	1.0000	1.0000	1.0000	0.9999	0.9999	0.9997
14.40	1.0000	1.0000	1.0000	1.0000	1.0000	1.0000	1.0000	0.9999	0.9999	0.9998
14.45	1.0000	1.0000	1.0000	1.0000	1.0000	1.0000	1.0000	0.9999	0.9999	0.9998
14.50	1.0000	1.0000	1.0000	1.0000	1.0000	1.0000	1.0000	0.9999	0.9999	0.9998
14.55	1.0000	1.0000	1.0000	1.0000	1.0000	1.0000	1.0000	0.9999	0.9999	0.9998
14.60	1.0000	1.0000	1.0000	1.0000	1.0000	1.0000	1.0000	0.9999	0.9999	0.9998
14.65	1.0000	1.0000	1.0000	1.0000	1.0000	1.0000	1.0000	0.9999	0.9999	0.9998
14.70	1.0000	1.0000	1.0000	1.0000	1.0000	1.0000	1.0000	0.9999	0.9999	0.9998
14.75	1.0000	1.0000	1.0000	1.0000	1.0000	1.0000	1.0000	0.9999	0.9999	0.9998
14.80	1.0000	1.0000	1.0000	1.0000	1.0000	1.0000	1.0000	0.9999	0.9999	0.9998
14.85	1.0000	1.0000	1.0000	1.0000	1.0000	1.0000	1.0000	0.9999	0.9999	0.9998
14.90	1.0000	1.0000	1.0000	1.0000	1.0000	1.0000	1.0000	0.9999	0.9999	0.9998
14.95	1.0000	1.0000	1.0000	1.0000	1.0000	1.0000	1.0000	0.9999	0.9999	0.9998
15.00	1.0000	1.0000	1.0000	1.0000	1.0000	1.0000	1.0000	0.9999	0.9999	0.9998
15.05	1.0000	1.0000	1.0000	1.0000	1.0000	1.0000	1.0000	0.9999	0.9999	0.9998
15.10	1.0000	1.0000	1.0000	1.0000	1.0000	1.0000	1.0000	0.9999	0.9999	0.9998
15.15	1.0000	1.0000	1.0000	1.0000	1.0000	1.0000	1.0000	0.9999	0.9999	0.9998
15.20	1.0000	1.0000	1.0000	1.0000	1.0000	1.0000	1.0000	0.9999	0.9999	0.9998
15.25	1.0000	1.0000	1.0000	1.0000	1.0000	1.0000	1.0000	0.9999	0.9999	0.9998
15.30	1.0000	1.0000	1.0000	1.0000	1.0000	1.0000	1.0000	0.9999	0.9999	0.9998
15.35	1.0000	1.0000	1.0000	1.0000	1.0000	1.0000	1.0000	0.9999	0.9999	0.9998
15.40	1.0000	1.0000	1.0000	1.0000	1.0000	1.0000	1.0000	0.9999	0.9999	0.9998
15.45	1.0000	1.0000	1.0000	1.0000	1.0000	1.0000	1.0000	0.9999	0.9999	0.9998
15.50	1.0000	1.0000	1.0000	1.0000	1.0000	1.0000	1.0000	0.9999	0.9999	0.9999
15.55	1.0000	1.0000	1.0000	1.0000	1.0000	1.0000	1.0000	0.9999	0.9999	0.9999
15.60	1.0000	1.0000	1.0000	1.0000	1.0000	1.0000	1.0000	0.9999	0.9999	0.9999
15.65	1.0000	1.0000	1.0000	1.0000	1.0000	1.0000	1.0000	0.9999	0.9999	0.9999
15.70	1.0000	1.0000	1.0000	1.0000	1.0000	1.0000	1.0000	0.9999	0.9999	0.9999
15.75	1.0000	1.0000	1.0000	1.0000	1.0000	1.0000	1.0000	0.9999	0.9999	0.9999
15.80	1.0000	1.0000	1.0000	1.0000	1.0000	1.0000	1.0000	1.0000	0.9999	0.9999
15.85	1.0000	1.0000	1.0000	1.0000	1.0000	1.0000	1.0000	1.0000	0.9999	0.9999
15.90	1.0000	1.0000	1.0000	1.0000	1.0000	1.0000	1.0000	1.0000	0.9999	0.9999
15.95	1.0000	1.0000	1.0000	1.0000	1.0000	1.0000	1.0000	1.0000	0.9999	0.9999
16.00	1.0000	1.0000	1.0000	1.0000	1.0000	1.0000	1.0000	1.0000	0.9999	0.9999
16.05	1.0000	1.0000	1.0000	1.0000	1.0000	1.0000	1.0000	1.0000	0.9999	0.9999
16.10	1.0000	1.0000	1.0000	1.0000	1.0000	1.0000	1.0000	1.0000	0.9999	0.9999
16.15	1.0000	1.0000	1.0000	1.0000	1.0000	1.0000	1.0000	1.0000	0.9999	0.9999
16.20	1.0000	1.0000	1.0000	1.0000	1.0000	1.0000	1.0000	1.0000	0.9999	0.9999
16.25	1.0000	1.0000	1.0000	1.0000	1.0000	1.0000	1.0000	1.0000	0.9999	0.9999
16.30	1.0000	1.0000	1.0000	1.0000	1.0000	1.0000	1.0000	1.0000	0.9999	0.9999
16.35	1.0000	1.0000	1.0000	1.0000	1.0000	1.0000	1.0000	1.0000	0.9999	0.9999
16.40	1.0000	1.0000	1.0000	1.0000	1.0000	1.0000	1.0000	1.0000	0.9999	0.9999
16.45	1.0000	1.0000	1.0000	1.0000	1.0000	1.0000	1.0000	1.0000	0.9999	0.9999
16.50	1.0000	1.0000	1.0000	1.0000	1.0000	1.0000	1.0000	1.0000	0.9999	0.9999
16.55	1.0000	1.0000	1.0000	1.0000	1.0000	1.0000	1.0000	1.0000	0.9999	0.9999
16.60	1.0000	1.0000	1.0000	1.0000	1.0000	1.0000	1.0000	1.0000	1.0000	0.9999
16.65	1.0000	1.0000	1.0000	1.0000	1.0000	1.0000	1.0000	1.0000	1.0000	0.9999
16.70	1.0000	1.0000	1.0000	1.0000	1.0000	1.0000	1.0000	1.0000	1.0000	0.9999

PROBABILITY INTEGRAL OF t, THE NON-CENTRAL t-STATISTIC. THIS TABLE GIVES Pr $[t/\sqrt{f} \leq x]$.

f is the number of degrees of freedom; the non-centrality parameter is $\sqrt{f+1}\,K_p$.
K_p is the standardized normal deviate exceeded with probability p.

DEGREES OF FREEDOM **7**

x \ p	.2500	.1500	.1000	.0650	.0400	.0250	.0100	.0040	.0025	.0010
16.75	1.0000	1.0000	1.0000	1.0000	1.0000	1.0000	1.0000	1.0000	1.0000	0.9999
16.80	1.0000	1.0000	1.0000	1.0000	1.0000	1.0000	1.0000	1.0000	1.0000	0.9999
16.85	1.0000	1.0000	1.0000	1.0000	1.0000	1.0000	1.0000	1.0000	1.0000	0.9999
16.90	1.0000	1.0000	1.0000	1.0000	1.0000	1.0000	1.0000	1.0000	1.0000	0.9999
16.95	1.0000	1.0000	1.0000	1.0000	1.0000	1.0000	1.0000	1.0000	1.0000	0.9999
17.00	1.0000	1.0000	1.0000	1.0000	1.0000	1.0000	1.0000	1.0000	1.0000	0.9999
17.05	1.0000	1.0000	1.0000	1.0000	1.0000	1.0000	1.0000	1.0000	1.0000	0.9999
17.10	1.0000	1.0000	1.0000	1.0000	1.0000	1.0000	1.0000	1.0000	1.0000	0.9999
17.15	1.0000	1.0000	1.0000	1.0000	1.0000	1.0000	1.0000	1.0000	1.0000	0.9999
17.20	1.0000	1.0000	1.0000	1.0000	1.0000	1.0000	1.0000	1.0000	1.0000	0.9999
17.25	1.0000	1.0000	1.0000	1.0000	1.0000	1.0000	1.0000	1.0000	1.0000	0.9999
17.30	1.0000	1.0000	1.0000	1.0000	1.0000	1.0000	1.0000	1.0000	1.0000	0.9999
17.35	1.0000	1.0000	1.0000	1.0000	1.0000	1.0000	1.0000	1.0000	1.0000	0.9999
17.40	1.0000	1.0000	1.0000	1.0000	1.0000	1.0000	1.0000	1.0000	1.0000	0.9999
17.45	1.0000	1.0000	1.0000	1.0000	1.0000	1.0000	1.0000	1.0000	1.0000	0.9999
17.50	1.0000	1.0000	1.0000	1.0000	1.0000	1.0000	1.0000	1.0000	1.0000	0.9999
17.55	1.0000	1.0000	1.0000	1.0000	1.0000	1.0000	1.0000	1.0000	1.0000	0.9999
17.60	1.0000	1.0000	1.0000	1.0000	1.0000	1.0000	1.0000	1.0000	1.0000	0.9999
17.65	1.0000	1.0000	1.0000	1.0000	1.0000	1.0000	1.0000	1.0000	1.0000	0.9999
17.70	1.0000	1.0000	1.0000	1.0000	1.0000	1.0000	1.0000	1.0000	1.0000	0.9999
17.75	1.0000	1.0000	1.0000	1.0000	1.0000	1.0000	1.0000	1.0000	1.0000	0.9999
17.80	1.0000	1.0000	1.0000	1.0000	1.0000	1.0000	1.0000	1.0000	1.0000	0.9999
17.85	1.0000	1.0000	1.0000	1.0000	1.0000	1.0000	1.0000	1.0000	1.0000	0.9999
17.90	1.0000	1.0000	1.0000	1.0000	1.0000	1.0000	1.0000	1.0000	1.0000	0.9999
17.95	1.0000	1.0000	1.0000	1.0000	1.0000	1.0000	1.0000	1.0000	1.0000	0.9999
18.00	1.0000	1.0000	1.0000	1.0000	1.0000	1.0000	1.0000	1.0000	1.0000	0.9999
18.05	1.0000	1.0000	1.0000	1.0000	1.0000	1.0000	1.0000	1.0000	1.0000	0.9999
18.10	1.0000	1.0000	1.0000	1.0000	1.0000	1.0000	1.0000	1.0000	1.0000	0.9999
18.15	1.0000	1.0000	1.0000	1.0000	1.0000	1.0000	1.0000	1.0000	1.0000	0.9999
18.20	1.0000	1.0000	1.0000	1.0000	1.0000	1.0000	1.0000	1.0000	1.0000	1.0000

PROBABILITY INTEGRAL OF t, THE NON-CENTRAL t-STATISTIC. THIS TABLE GIVES Pr $[t/\sqrt{f} \leq x]$.

f is the number of degrees of freedom; the non-centrality parameter is $\sqrt{f+1}\,K_p$.
K_p is the standardized normal deviate exceeded with probability p.

DEGREES OF FREEDOM **8**

x \ p	.2500	.1500	.1000	.0650	.0400	.0250	.0100	.0040	.0025	.0010
-0.90	0.0000	0.0000	0.0000	0.0000	0.0000	0.0000	0.0000	0.0000	0.0000	0.0000
-0.85	0.0001	0.0000	0.0000	0.0000	0.0000	0.0000	0.0000	0.0000	0.0000	0.0000
-0.80	0.0001	0.0000	0.0000	0.0000	0.0000	0.0000	0.0000	0.0000	0.0000	0.0000
-0.75	0.0001	0.0000	0.0000	0.0000	0.0000	0.0000	0.0000	0.0000	0.0000	0.0000
-0.70	0.0002	0.0000	0.0000	0.0000	0.0000	0.0000	0.0000	0.0000	0.0000	0.0000
-0.65	0.0002	0.0000	0.0000	0.0000	0.0000	0.0000	0.0000	0.0000	0.0000	0.0000
-0.60	0.0003	0.0000	0.0000	0.0000	0.0000	0.0000	0.0000	0.0000	0.0000	0.0000
-0.55	0.0005	0.0000	0.0000	0.0000	0.0000	0.0000	0.0000	0.0000	0.0000	0.0000
-0.50	0.0006	0.0000	0.0000	0.0000	0.0000	0.0000	0.0000	0.0000	0.0000	0.0000
-0.45	0.0009	0.0000	0.0000	0.0000	0.0000	0.0000	0.0000	0.0000	0.0000	0.0000
-0.40	0.0013	0.0000	0.0000	0.0000	0.0000	0.0000	0.0000	0.0000	0.0000	0.0000
-0.35	0.0018	0.0000	0.0000	0.0000	0.0000	0.0000	0.0000	0.0000	0.0000	0.0000
-0.30	0.0026	0.0001	0.0000	0.0000	0.0000	0.0000	0.0000	0.0000	0.0000	0.0000
-0.25	0.0038	0.0001	0.0000	0.0000	0.0000	0.0000	0.0000	0.0000	0.0000	0.0000
-0.20	0.0054	0.0001	0.0000	0.0000	0.0000	0.0000	0.0000	0.0000	0.0000	0.0000
-0.15	0.0077	0.0002	0.0000	0.0000	0.0000	0.0000	0.0000	0.0000	0.0000	0.0000
-0.10	0.0109	0.0004	0.0000	0.0000	0.0000	0.0000	0.0000	0.0000	0.0000	0.0000
-0.05	0.0154	0.0006	0.0000	0.0000	0.0000	0.0000	0.0000	0.0000	0.0000	0.0000
0.00	0.0215	0.0009	0.0001	0.0000	0.0000	0.0000	0.0000	0.0000	0.0000	0.0000
0.05	0.0297	0.0015	0.0001	0.0000	0.0000	0.0000	0.0000	0.0000	0.0000	0.0000
0.10	0.0405	0.0023	0.0002	0.0000	0.0000	0.0000	0.0000	0.0000	0.0000	0.0000
0.15	0.0544	0.0036	0.0003	0.0000	0.0000	0.0000	0.0000	0.0000	0.0000	0.0000
0.20	0.0720	0.0056	0.0006	0.0000	0.0000	0.0000	0.0000	0.0000	0.0000	0.0000
0.25	0.0937	0.0085	0.0009	0.0001	0.0000	0.0000	0.0000	0.0000	0.0000	0.0000
0.30	0.1199	0.0126	0.0016	0.0001	0.0000	0.0000	0.0000	0.0000	0.0000	0.0000
0.35	0.1506	0.0185	0.0026	0.0003	0.0000	0.0000	0.0000	0.0000	0.0000	0.0000
0.40	0.1859	0.0264	0.0041	0.0005	0.0000	0.0000	0.0000	0.0000	0.0000	0.0000
0.45	0.2254	0.0369	0.0065	0.0008	0.0001	0.0000	0.0000	0.0000	0.0000	0.0000
0.50	0.2685	0.0506	0.0099	0.0014	0.0001	0.0000	0.0000	0.0000	0.0000	0.0000
0.55	0.3145	0.0677	0.0148	0.0024	0.0003	0.0000	0.0000	0.0000	0.0000	0.0000
0.60	0.3626	0.0886	0.0216	0.0039	0.0005	0.0001	0.0000	0.0000	0.0000	0.0000
0.65	0.4118	0.1135	0.0306	0.0062	0.0009	0.0001	0.0000	0.0000	0.0000	0.0000
0.70	0.4611	0.1424	0.0424	0.0096	0.0015	0.0002	0.0000	0.0000	0.0000	0.0000
0.75	0.5098	0.1751	0.0572	0.0144	0.0025	0.0004	0.0000	0.0000	0.0000	0.0000
0.80	0.5570	0.2112	0.0753	0.0208	0.0041	0.0008	0.0000	0.0000	0.0000	0.0000
0.85	0.6020	0.2502	0.0969	0.0294	0.0065	0.0013	0.0000	0.0000	0.0000	0.0000
0.90	0.6445	0.2916	0.1220	0.0405	0.0099	0.0022	0.0001	0.0000	0.0000	0.0000
0.95	0.6840	0.3346	0.1505	0.0543	0.0146	0.0036	0.0002	0.0000	0.0000	0.0000
1.00	0.7204	0.3785	0.1822	0.0710	0.0209	0.0057	0.0003	0.0000	0.0000	0.0000
1.05	0.7536	0.4226	0.2166	0.0907	0.0292	0.0086	0.0006	0.0000	0.0000	0.0000
1.10	0.7835	0.4664	0.2534	0.1136	0.0396	0.0127	0.0011	0.0001	0.0000	0.0000
1.15	0.8104	0.5092	0.2919	0.1394	0.0525	0.0182	0.0018	0.0001	0.0000	0.0000
1.20	0.8344	0.5505	0.3316	0.1680	0.0679	0.0253	0.0029	0.0003	0.0001	0.0000
1.25	0.8556	0.5900	0.3720	0.1990	0.0860	0.0342	0.0045	0.0005	0.0001	0.0000
1.30	0.8743	0.6274	0.4125	0.2322	0.1067	0.0453	0.0068	0.0008	0.0003	0.0000
1.35	0.8908	0.6625	0.4527	0.2670	0.1300	0.0585	0.0100	0.0014	0.0005	0.0001
1.40	0.9051	0.6952	0.4920	0.3031	0.1557	0.0741	0.0141	0.0022	0.0008	0.0001
1.45	0.9177	0.7254	0.5302	0.3400	0.1837	0.0920	0.0195	0.0034	0.0013	0.0002
1.50	0.9286	0.7532	0.5669	0.3773	0.2135	0.1123	0.0263	0.0051	0.0021	0.0003
1.55	0.9381	0.7786	0.6019	0.4145	0.2449	0.1347	0.0347	0.0074	0.0032	0.0006
1.60	0.9463	0.8018	0.6351	0.4513	0.2776	0.1593	0.0447	0.0105	0.0047	0.0009
1.65	0.9534	0.8228	0.6663	0.4874	0.3111	0.1856	0.0566	0.0144	0.0068	0.0014
1.70	0.9596	0.8417	0.6955	0.5224	0.3452	0.2136	0.0703	0.0195	0.0096	0.0022
1.75	0.9649	0.8587	0.7227	0.5562	0.3795	0.2430	0.0860	0.0257	0.0131	0.0033
1.80	0.9695	0.8740	0.7478	0.5886	0.4136	0.2734	0.1034	0.0332	0.0176	0.0048
1.85	0.9735	0.8878	0.7710	0.6195	0.4474	0.3045	0.1227	0.0421	0.0232	0.0067
1.90	0.9769	0.9000	0.7924	0.6488	0.4805	0.3362	0.1437	0.0525	0.0299	0.0093
1.95	0.9799	0.9110	0.8119	0.6763	0.5128	0.3680	0.1662	0.0645	0.0379	0.0125

PROBABILITY INTEGRAL OF t, THE NON-CENTRAL t-STATISTIC. THIS TABLE GIVES Pr $[t/\sqrt{f} \leq x]$.

f is the number of degrees of freedom; the non-centrality parameter is $\sqrt{f+1}\,K_p$.
K_p is the standardized normal deviate exceeded with probability p.

DEGREES OF FREEDOM **8**

x \ p	.2500	.1500	.1000	.0650	.0400	.0250	.0100	.0040	.0025	.0010
2.00	0.9825	0.9207	0.8298	0.7022	0.5440	0.3998	0.1901	0.0780	0.0473	0.0166
2.05	0.9847	0.9294	0.8460	0.7264	0.5741	0.4314	0.2153	0.0931	0.0580	0.0215
2.10	0.9866	0.9371	0.8608	0.7490	0.6030	0.4624	0.2414	0.1097	0.0702	0.0275
2.15	0.9883	0.9440	0.8743	0.7699	0.6305	0.4928	0.2684	0.1278	0.0838	0.0345
2.20	0.9897	0.9501	0.8865	0.7893	0.6566	0.5224	0.2960	0.1472	0.0988	0.0426
2.25	0.9910	0.9555	0.8975	0.8072	0.6813	0.5510	0.3239	0.1678	0.1152	0.0519
2.30	0.9921	0.9603	0.9074	0.8237	0.7045	0.5787	0.3521	0.1896	0.1328	0.0625
2.35	0.9931	0.9646	0.9164	0.8389	0.7264	0.6052	0.3803	0.2124	0.1517	0.0743
2.40	0.9939	0.9684	0.9245	0.8528	0.7469	0.6305	0.4084	0.2360	0.1718	0.0873
2.45	0.9946	0.9717	0.9318	0.8656	0.7661	0.6546	0.4362	0.2602	0.1928	0.1015
2.50	0.9952	0.9747	0.9384	0.8773	0.7839	0.6776	0.4636	0.2851	0.2147	0.1169
2.55	0.9958	0.9774	0.9444	0.8880	0.8005	0.6993	0.4904	0.3103	0.2373	0.1334
2.60	0.9963	0.9798	0.9497	0.8978	0.8160	0.7198	0.5166	0.3357	0.2605	0.1509
2.65	0.9967	0.9819	0.9546	0.9067	0.8303	0.7391	0.5420	0.3612	0.2843	0.1693
2.70	0.9971	0.9837	0.9589	0.9149	0.8436	0.7573	0.5667	0.3867	0.3083	0.1887
2.75	0.9974	0.9854	0.9628	0.9223	0.8559	0.7743	0.5904	0.4120	0.3326	0.2088
2.80	0.9977	0.9869	0.9663	0.9291	0.8672	0.7903	0.6133	0.4371	0.3570	0.2295
2.85	0.9979	0.9882	0.9695	0.9352	0.8777	0.8052	0.6353	0.4618	0.3814	0.2509
2.90	0.9982	0.9894	0.9724	0.9409	0.8874	0.8191	0.6564	0.4860	0.4056	0.2726
2.95	0.9984	0.9904	0.9749	0.9460	0.8963	0.8321	0.6765	0.5097	0.4296	0.2947
3.00	0.9985	0.9914	0.9773	0.9506	0.9045	0.8442	0.6956	0.5329	0.4533	0.3171
3.05	0.9987	0.9922	0.9793	0.9549	0.9120	0.8555	0.7138	0.5554	0.4766	0.3396
3.10	0.9988	0.9930	0.9812	0.9587	0.9189	0.8660	0.7311	0.5772	0.4995	0.3621
3.15	0.9989	0.9937	0.9829	0.9623	0.9253	0.8757	0.7475	0.5983	0.5219	0.3846
3.20	0.9990	0.9943	0.9845	0.9655	0.9312	0.8847	0.7630	0.6187	0.5436	0.4069
3.25	0.9991	0.9948	0.9859	0.9684	0.9366	0.8931	0.7776	0.6383	0.5648	0.4291
3.30	0.9992	0.9953	0.9871	0.9710	0.9416	0.9009	0.7914	0.6571	0.5854	0.4510
3.35	0.9993	0.9957	0.9883	0.9735	0.9461	0.9081	0.8044	0.6752	0.6052	0.4725
3.40	0.9994	0.9961	0.9893	0.9757	0.9503	0.9148	0.8167	0.6925	0.6244	0.4937
3.45	0.9994	0.9965	0.9902	0.9777	0.9542	0.9210	0.8282	0.7090	0.6429	0.5145
3.50	0.9995	0.9968	0.9911	0.9795	0.9577	0.9267	0.8391	0.7248	0.6607	0.5348
3.55	0.9995	0.9971	0.9918	0.9812	0.9610	0.9320	0.8493	0.7398	0.6778	0.5545
3.60	0.9996	0.9973	0.9925	0.9827	0.9640	0.9369	0.8588	0.7541	0.6942	0.5738
3.65	0.9996	0.9976	0.9932	0.9841	0.9667	0.9414	0.8678	0.7678	0.7099	0.5925
3.70	0.9997	0.9978	0.9937	0.9854	0.9692	0.9456	0.8762	0.7807	0.7249	0.6106
3.75	0.9997	0.9980	0.9943	0.9865	0.9716	0.9495	0.8841	0.7930	0.7393	0.6281
3.80	0.9997	0.9982	0.9947	0.9876	0.9737	0.9531	0.8915	0.8046	0.7529	0.6450
3.85	0.9997	0.9983	0.9952	0.9886	0.9757	0.9564	0.8984	0.8156	0.7660	0.6614
3.90	0.9998	0.9985	0.9956	0.9895	0.9775	0.9595	0.9048	0.8260	0.7784	0.6771
3.95	0.9998	0.9986	0.9959	0.9903	0.9791	0.9624	0.9109	0.8359	0.7902	0.6923
4.00	0.9998	0.9987	0.9963	0.9910	0.9807	0.9650	0.9165	0.8452	0.8015	0.7068
4.05	0.9998	0.9988	0.9966	0.9917	0.9821	0.9675	0.9218	0.8541	0.8121	0.7208
4.10	0.9998	0.9989	0.9968	0.9924	0.9834	0.9697	0.9268	0.8624	0.8223	0.7342
4.15	0.9998	0.9990	0.9971	0.9929	0.9846	0.9718	0.9314	0.8702	0.8319	0.7470
4.20	0.9999	0.9991	0.9973	0.9935	0.9857	0.9738	0.9357	0.8777	0.8410	0.7593
4.25	0.9999	0.9992	0.9975	0.9940	0.9867	0.9756	0.9398	0.8846	0.8496	0.7710
4.30	0.9999	0.9992	0.9977	0.9944	0.9877	0.9772	0.9435	0.8912	0.8578	0.7822
4.35	0.9999	0.9993	0.9979	0.9948	0.9886	0.9788	0.9471	0.8974	0.8655	0.7929
4.40	0.9999	0.9993	0.9980	0.9952	0.9894	0.9802	0.9504	0.9033	0.8728	0.8032
4.45	0.9999	0.9994	0.9982	0.9955	0.9901	0.9816	0.9535	0.9088	0.8798	0.8129
4.50	0.9999	0.9994	0.9983	0.9959	0.9908	0.9828	0.9563	0.9140	0.8863	0.8222
4.55	0.9999	0.9995	0.9984	0.9962	0.9914	0.9839	0.9590	0.9189	0.8925	0.8311
4.60	0.9999	0.9995	0.9986	0.9964	0.9920	0.9850	0.9616	0.9235	0.8983	0.8395
4.65	0.9999	0.9996	0.9987	0.9967	0.9926	0.9860	0.9639	0.9279	0.9039	0.8475
4.70	0.9999	0.9996	0.9988	0.9969	0.9931	0.9869	0.9661	0.9319	0.9091	0.8551
4.75	0.9999	0.9996	0.9988	0.9971	0.9935	0.9878	0.9682	0.9358	0.9140	0.8624
4.80	0.9999	0.9996	0.9989	0.9973	0.9940	0.9886	0.9701	0.9394	0.9187	0.8693
4.85	1.0000	0.9997	0.9990	0.9975	0.9944	0.9893	0.9719	0.9428	0.9231	0.8758
4.90	1.0000	0.9997	0.9991	0.9977	0.9947	0.9900	0.9736	0.9460	0.9273	0.8821
4.95	1.0000	0.9997	0.9991	0.9978	0.9951	0.9906	0.9752	0.9490	0.9312	0.8880

PROBABILITY INTEGRAL OF t, THE NON-CENTRAL t-STATISTIC. THIS TABLE GIVES Pr $[t/\sqrt{f} \leq x]$.

f is the number of degrees of freedom; the non-centrality parameter is $\sqrt{f+1}\,K_p$.

K_p is the standardized normal deviate exceeded with probability p.

DEGREES OF FREEDOM **8**

x \ p	.2500	.1500	.1000	.0650	.0400	.0250	.0100	.0040	.0025	.0010
5.00	1.0000	0.9997	0.9992	0.9980	0.9954	0.9912	0.9767	0.9519	0.9349	0.8936
5.05	1.0000	0.9998	0.9993	0.9981	0.9957	0.9918	0.9780	0.9546	0.9384	0.8989
5.10	1.0000	0.9998	0.9993	0.9982	0.9960	0.9923	0.9793	0.9571	0.9417	0.9040
5.15	1.0000	0.9998	0.9994	0.9984	0.9962	0.9928	0.9806	0.9595	0.9448	0.9088
5.20	1.0000	0.9998	0.9994	0.9985	0.9965	0.9932	0.9817	0.9617	0.9478	0.9134
5.25	1.0000	0.9998	0.9994	0.9986	0.9967	0.9936	0.9828	0.9638	0.9505	0.9177
5.30	1.0000	0.9998	0.9995	0.9987	0.9969	0.9940	0.9838	0.9658	0.9532	0.9218
5.35	1.0000	0.9998	0.9995	0.9987	0.9971	0.9944	0.9847	0.9676	0.9556	0.9257
5.40	1.0000	0.9998	0.9995	0.9988	0.9973	0.9947	0.9856	0.9694	0.9580	0.9294
5.45	1.0000	0.9999	0.9996	0.9989	0.9974	0.9950	0.9864	0.9711	0.9602	0.9329
5.50	1.0000	0.9999	0.9996	0.9990	0.9976	0.9953	0.9872	0.9726	0.9623	0.9362
5.55	1.0000	0.9999	0.9996	0.9990	0.9978	0.9956	0.9879	0.9741	0.9643	0.9393
5.60	1.0000	0.9999	0.9996	0.9991	0.9979	0.9959	0.9886	0.9755	0.9661	0.9423
5.65	1.0000	0.9999	0.9997	0.9992	0.9980	0.9961	0.9892	0.9768	0.9679	0.9452
5.70	1.0000	0.9999	0.9997	0.9992	0.9981	0.9963	0.9898	0.9780	0.9695	0.9479
5.75	1.0000	0.9999	0.9997	0.9993	0.9982	0.9966	0.9904	0.9792	0.9711	0.9504
5.80	1.0000	0.9999	0.9997	0.9993	0.9983	0.9968	0.9909	0.9803	0.9726	0.9528
5.85	1.0000	0.9999	0.9997	0.9993	0.9984	0.9969	0.9914	0.9813	0.9740	0.9551
5.90	1.0000	0.9999	0.9998	0.9994	0.9985	0.9971	0.9919	0.9823	0.9753	0.9573
5.95	1.0000	0.9999	0.9998	0.9994	0.9986	0.9973	0.9923	0.9832	0.9766	0.9594
6.00	1.0000	0.9999	0.9998	0.9994	0.9987	0.9974	0.9927	0.9841	0.9778	0.9613
6.05	1.0000	0.9999	0.9998	0.9995	0.9988	0.9976	0.9931	0.9849	0.9789	0.9632
6.10	1.0000	0.9999	0.9998	0.9995	0.9988	0.9977	0.9935	0.9857	0.9799	0.9650
6.15	1.0000	0.9999	0.9998	0.9995	0.9989	0.9978	0.9938	0.9864	0.9809	0.9666
6.20	1.0000	0.9999	0.9998	0.9996	0.9990	0.9980	0.9942	0.9871	0.9819	0.9682
6.25	1.0000	1.0000	0.9998	0.9996	0.9990	0.9981	0.9945	0.9878	0.9828	0.9697
6.30	1.0000	1.0000	0.9999	0.9996	0.9991	0.9982	0.9948	0.9884	0.9836	0.9712
6.35	1.0000	1.0000	0.9999	0.9996	0.9991	0.9983	0.9950	0.9890	0.9844	0.9725
6.40	1.0000	1.0000	0.9999	0.9997	0.9992	0.9984	0.9953	0.9895	0.9852	0.9738
6.45	1.0000	1.0000	0.9999	0.9997	0.9992	0.9985	0.9955	0.9900	0.9859	0.9751
6.50	1.0000	1.0000	0.9999	0.9997	0.9993	0.9985	0.9958	0.9905	0.9866	0.9762
6.55	1.0000	1.0000	0.9999	0.9997	0.9993	0.9986	0.9960	0.9910	0.9873	0.9773
6.60	1.0000	1.0000	0.9999	0.9997	0.9993	0.9987	0.9962	0.9914	0.9879	0.9784
6.65	1.0000	1.0000	0.9999	0.9997	0.9994	0.9988	0.9964	0.9919	0.9885	0.9794
6.70	1.0000	1.0000	0.9999	0.9998	0.9994	0.9988	0.9966	0.9923	0.9890	0.9803
6.75	1.0000	1.0000	0.9999	0.9998	0.9994	0.9989	0.9967	0.9926	0.9895	0.9812
6.80	1.0000	1.0000	0.9999	0.9998	0.9995	0.9989	0.9969	0.9930	0.9900	0.9821
6.85	1.0000	1.0000	0.9999	0.9998	0.9995	0.9990	0.9971	0.9933	0.9905	0.9829
6.90	1.0000	1.0000	0.9999	0.9998	0.9995	0.9990	0.9972	0.9936	0.9909	0.9837
6.95	1.0000	1.0000	0.9999	0.9998	0.9996	0.9991	0.9973	0.9939	0.9914	0.9844
7.00	1.0000	1.0000	0.9999	0.9998	0.9996	0.9991	0.9975	0.9942	0.9918	0.9851
7.05	1.0000	1.0000	0.9999	0.9998	0.9996	0.9992	0.9976	0.9945	0.9922	0.9858
7.10	1.0000	1.0000	0.9999	0.9998	0.9996	0.9992	0.9977	0.9948	0.9925	0.9864
7.15	1.0000	1.0000	0.9999	0.9999	0.9996	0.9993	0.9978	0.9950	0.9929	0.9870
7.20	1.0000	1.0000	0.9999	0.9999	0.9997	0.9993	0.9979	0.9952	0.9932	0.9876
7.25	1.0000	1.0000	1.0000	0.9999	0.9997	0.9993	0.9980	0.9955	0.9935	0.9882
7.30	1.0000	1.0000	1.0000	0.9999	0.9997	0.9994	0.9981	0.9957	0.9938	0.9887
7.35	1.0000	1.0000	1.0000	0.9999	0.9997	0.9994	0.9982	0.9959	0.9941	0.9892
7.40	1.0000	1.0000	1.0000	0.9999	0.9997	0.9994	0.9983	0.9961	0.9943	0.9896
7.45	1.0000	1.0000	1.0000	0.9999	0.9997	0.9995	0.9984	0.9962	0.9946	0.9901
7.50	1.0000	1.0000	1.0000	0.9999	0.9997	0.9995	0.9984	0.9964	0.9948	0.9905
7.55	1.0000	1.0000	1.0000	0.9999	0.9998	0.9995	0.9985	0.9966	0.9951	0.9909
7.60	1.0000	1.0000	1.0000	0.9999	0.9998	0.9995	0.9986	0.9967	0.9953	0.9913
7.65	1.0000	1.0000	1.0000	0.9999	0.9998	0.9996	0.9987	0.9969	0.9955	0.9917
7.70	1.0000	1.0000	1.0000	0.9999	0.9998	0.9996	0.9987	0.9970	0.9957	0.9920
7.75	1.0000	1.0000	1.0000	0.9999	0.9998	0.9996	0.9988	0.9971	0.9959	0.9924
7.80	1.0000	1.0000	1.0000	0.9999	0.9998	0.9996	0.9988	0.9973	0.9961	0.9927
7.85	1.0000	1.0000	1.0000	0.9999	0.9998	0.9996	0.9989	0.9974	0.9962	0.9930
7.90	1.0000	1.0000	1.0000	0.9999	0.9998	0.9996	0.9989	0.9975	0.9964	0.9933
7.95	1.0000	1.0000	1.0000	0.9999	0.9998	0.9997	0.9990	0.9976	0.9965	0.9936

PROBABILITY INTEGRAL OF t, THE NON-CENTRAL t-STATISTIC. THIS TABLE GIVES Pr $[t/\sqrt{f} \leq x]$.

f is the number of degrees of freedom; the non-centrality parameter is $\sqrt{f+1}\,K_p$.
K_p is the standardized normal deviate exceeded with probability p.

DEGREES OF FREEDOM **8**

x \ p	.2500	.1500	.1000	.0650	.0400	.0250	.0100	.0040	.0025	.0010
8.00	1.0000	1.0000	1.0000	0.9999	0.9998	0.9997	0.9990	0.9977	0.9967	0.9939
8.05	1.0000	1.0000	1.0000	0.9999	0.9999	0.9997	0.9991	0.9978	0.9968	0.9941
8.10	1.0000	1.0000	1.0000	0.9999	0.9999	0.9997	0.9991	0.9979	0.9970	0.9944
8.15	1.0000	1.0000	1.0000	0.9999	0.9999	0.9997	0.9991	0.9980	0.9971	0.9946
8.20	1.0000	1.0000	1.0000	0.9999	0.9999	0.9997	0.9992	0.9981	0.9972	0.9948
8.25	1.0000	1.0000	1.0000	1.0000	0.9999	0.9997	0.9992	0.9982	0.9973	0.9950
8.30	1.0000	1.0000	1.0000	1.0000	0.9999	0.9998	0.9993	0.9982	0.9974	0.9952
8.35	1.0000	1.0000	1.0000	1.0000	0.9999	0.9998	0.9993	0.9983	0.9975	0.9954
8.40	1.0000	1.0000	1.0000	1.0000	0.9999	0.9998	0.9993	0.9984	0.9977	0.9956
8.45	1.0000	1.0000	1.0000	1.0000	0.9999	0.9998	0.9993	0.9984	0.9977	0.9958
8.50	1.0000	1.0000	1.0000	1.0000	0.9999	0.9998	0.9994	0.9985	0.9978	0.9959
8.55	1.0000	1.0000	1.0000	1.0000	0.9999	0.9998	0.9994	0.9986	0.9979	0.9961
8.60	1.0000	1.0000	1.0000	1.0000	0.9999	0.9998	0.9994	0.9986	0.9980	0.9963
8.65	1.0000	1.0000	1.0000	1.0000	0.9999	0.9998	0.9994	0.9987	0.9981	0.9964
8.70	1.0000	1.0000	1.0000	1.0000	0.9999	0.9998	0.9995	0.9987	0.9982	0.9966
8.75	1.0000	1.0000	1.0000	1.0000	0.9999	0.9998	0.9995	0.9988	0.9982	0.9967
8.80	1.0000	1.0000	1.0000	1.0000	0.9999	0.9998	0.9995	0.9988	0.9983	0.9968
8.85	1.0000	1.0000	1.0000	1.0000	0.9999	0.9999	0.9995	0.9989	0.9984	0.9969
8.90	1.0000	1.0000	1.0000	1.0000	0.9999	0.9999	0.9996	0.9989	0.9984	0.9971
8.95	1.0000	1.0000	1.0000	1.0000	0.9999	0.9999	0.9996	0.9990	0.9985	0.9972
9.00	1.0000	1.0000	1.0000	1.0000	0.9999	0.9999	0.9996	0.9990	0.9986	0.9973
9.05	1.0000	1.0000	1.0000	1.0000	0.9999	0.9999	0.9996	0.9991	0.9986	0.9974
9.10	1.0000	1.0000	1.0000	1.0000	0.9999	0.9999	0.9996	0.9991	0.9987	0.9975
9.15	1.0000	1.0000	1.0000	1.0000	0.9999	0.9999	0.9996	0.9991	0.9987	0.9976
9.20	1.0000	1.0000	1.0000	1.0000	0.9999	0.9999	0.9997	0.9992	0.9988	0.9977
9.25	1.0000	1.0000	1.0000	1.0000	0.9999	0.9999	0.9997	0.9992	0.9988	0.9978
9.30	1.0000	1.0000	1.0000	1.0000	1.0000	0.9999	0.9997	0.9992	0.9989	0.9978
9.35	1.0000	1.0000	1.0000	1.0000	1.0000	0.9999	0.9997	0.9993	0.9989	0.9979
9.40	1.0000	1.0000	1.0000	1.0000	1.0000	0.9999	0.9997	0.9993	0.9990	0.9980
9.45	1.0000	1.0000	1.0000	1.0000	1.0000	0.9999	0.9997	0.9993	0.9990	0.9981
9.50	1.0000	1.0000	1.0000	1.0000	1.0000	0.9999	0.9997	0.9993	0.9990	0.9981
9.55	1.0000	1.0000	1.0000	1.0000	1.0000	0.9999	0.9997	0.9994	0.9991	0.9982
9.60	1.0000	1.0000	1.0000	1.0000	1.0000	0.9999	0.9997	0.9994	0.9991	0.9983
9.65	1.0000	1.0000	1.0000	1.0000	1.0000	0.9999	0.9998	0.9994	0.9991	0.9983
9.70	1.0000	1.0000	1.0000	1.0000	1.0000	0.9999	0.9998	0.9994	0.9992	0.9984
9.75	1.0000	1.0000	1.0000	1.0000	1.0000	0.9999	0.9998	0.9995	0.9992	0.9985
9.80	1.0000	1.0000	1.0000	1.0000	1.0000	0.9999	0.9998	0.9995	0.9992	0.9985
9.85	1.0000	1.0000	1.0000	1.0000	1.0000	0.9999	0.9998	0.9995	0.9993	0.9986
9.90	1.0000	1.0000	1.0000	1.0000	1.0000	0.9999	0.9998	0.9995	0.9993	0.9986
9.95	1.0000	1.0000	1.0000	1.0000	1.0000	0.9999	0.9998	0.9995	0.9993	0.9987
10.00	1.0000	1.0000	1.0000	1.0000	1.0000	0.9999	0.9998	0.9995	0.9993	0.9987
10.05	1.0000	1.0000	1.0000	1.0000	1.0000	0.9999	0.9998	0.9996	0.9994	0.9988
10.10	1.0000	1.0000	1.0000	1.0000	1.0000	0.9999	0.9998	0.9996	0.9994	0.9988
10.15	1.0000	1.0000	1.0000	1.0000	1.0000	1.0000	0.9998	0.9996	0.9994	0.9988
10.20	1.0000	1.0000	1.0000	1.0000	1.0000	1.0000	0.9998	0.9996	0.9994	0.9989
10.25	1.0000	1.0000	1.0000	1.0000	1.0000	1.0000	0.9998	0.9996	0.9994	0.9989
10.30	1.0000	1.0000	1.0000	1.0000	1.0000	1.0000	0.9999	0.9996	0.9995	0.9990
10.35	1.0000	1.0000	1.0000	1.0000	1.0000	1.0000	0.9999	0.9997	0.9995	0.9990
10.40	1.0000	1.0000	1.0000	1.0000	1.0000	1.0000	0.9999	0.9997	0.9995	0.9990
10.45	1.0000	1.0000	1.0000	1.0000	1.0000	1.0000	0.9999	0.9997	0.9995	0.9991
10.50	1.0000	1.0000	1.0000	1.0000	1.0000	1.0000	0.9999	0.9997	0.9995	0.9991
10.55	1.0000	1.0000	1.0000	1.0000	1.0000	1.0000	0.9999	0.9997	0.9996	0.9991
10.60	1.0000	1.0000	1.0000	1.0000	1.0000	1.0000	0.9999	0.9997	0.9996	0.9992
10.65	1.0000	1.0000	1.0000	1.0000	1.0000	1.0000	0.9999	0.9997	0.9996	0.9992
10.70	1.0000	1.0000	1.0000	1.0000	1.0000	1.0000	0.9999	0.9997	0.9996	0.9992
10.75	1.0000	1.0000	1.0000	1.0000	1.0000	1.0000	0.9999	0.9997	0.9996	0.9992
10.80	1.0000	1.0000	1.0000	1.0000	1.0000	1.0000	0.9999	0.9997	0.9996	0.9993
10.85	1.0000	1.0000	1.0000	1.0000	1.0000	1.0000	0.9999	0.9998	0.9996	0.9993
10.90	1.0000	1.0000	1.0000	1.0000	1.0000	1.0000	0.9999	0.9998	0.9996	0.9993
10.95	1.0000	1.0000	1.0000	1.0000	1.0000	1.0000	0.9999	0.9998	0.9997	0.9993

PROBABILITY INTEGRAL OF t, THE NON-CENTRAL t-STATISTIC. THIS TABLE GIVES Pr $[t/\sqrt{f} \leq x]$.

f is the number of degrees of freedom; the non-centrality parameter is $\sqrt{f+1}\,K_p$.

K_p is the standardized normal deviate exceeded with probability p.

DEGREES OF FREEDOM **8**

x \ p	.2500	.1500	.1000	.0650	.0400	.0250	.0100	.0040	.0025	.0010
11.00	1.0000	1.0000	1.0000	1.0000	1.0000	1.0000	0.9999	0.9998	0.9997	0.9994
11.05	1.0000	1.0000	1.0000	1.0000	1.0000	1.0000	0.9999	0.9998	0.9997	0.9994
11.10	1.0000	1.0000	1.0000	1.0000	1.0000	1.0000	0.9999	0.9998	0.9997	0.9994
11.15	1.0000	1.0000	1.0000	1.0000	1.0000	1.0000	0.9999	0.9998	0.9997	0.9994
11.20	1.0000	1.0000	1.0000	1.0000	1.0000	1.0000	0.9999	0.9998	0.9997	0.9994
11.25	1.0000	1.0000	1.0000	1.0000	1.0000	1.0000	0.9999	0.9998	0.9997	0.9995
11.30	1.0000	1.0000	1.0000	1.0000	1.0000	1.0000	0.9999	0.9998	0.9997	0.9995
11.35	1.0000	1.0000	1.0000	1.0000	1.0000	1.0000	0.9999	0.9998	0.9997	0.9995
11.40	1.0000	1.0000	1.0000	1.0000	1.0000	1.0000	0.9999	0.9998	0.9997	0.9995
11.45	1.0000	1.0000	1.0000	1.0000	1.0000	1.0000	0.9999	0.9998	0.9998	0.9995
11.50	1.0000	1.0000	1.0000	1.0000	1.0000	1.0000	0.9999	0.9998	0.9998	0.9995
11.55	1.0000	1.0000	1.0000	1.0000	1.0000	1.0000	0.9999	0.9998	0.9998	0.9996
11.60	1.0000	1.0000	1.0000	1.0000	1.0000	1.0000	0.9999	0.9999	0.9998	0.9996
11.65	1.0000	1.0000	1.0000	1.0000	1.0000	1.0000	0.9999	0.9999	0.9998	0.9996
11.70	1.0000	1.0000	1.0000	1.0000	1.0000	1.0000	0.9999	0.9999	0.9998	0.9996
11.75	1.0000	1.0000	1.0000	1.0000	1.0000	1.0000	0.9999	0.9999	0.9998	0.9996
11.80	1.0000	1.0000	1.0000	1.0000	1.0000	1.0000	0.9999	0.9999	0.9998	0.9996
11.85	1.0000	1.0000	1.0000	1.0000	1.0000	1.0000	1.0000	0.9999	0.9998	0.9996
11.90	1.0000	1.0000	1.0000	1.0000	1.0000	1.0000	1.0000	0.9999	0.9998	0.9996
11.95	1.0000	1.0000	1.0000	1.0000	1.0000	1.0000	1.0000	0.9999	0.9998	0.9997
12.00	1.0000	1.0000	1.0000	1.0000	1.0000	1.0000	1.0000	0.9999	0.9998	0.9997
12.05	1.0000	1.0000	1.0000	1.0000	1.0000	1.0000	1.0000	0.9999	0.9998	0.9997
12.10	1.0000	1.0000	1.0000	1.0000	1.0000	1.0000	1.0000	0.9999	0.9998	0.9997
12.15	1.0000	1.0000	1.0000	1.0000	1.0000	1.0000	1.0000	0.9999	0.9998	0.9997
12.20	1.0000	1.0000	1.0000	1.0000	1.0000	1.0000	1.0000	0.9999	0.9999	0.9997
12.25	1.0000	1.0000	1.0000	1.0000	1.0000	1.0000	1.0000	0.9999	0.9999	0.9997
12.30	1.0000	1.0000	1.0000	1.0000	1.0000	1.0000	1.0000	0.9999	0.9999	0.9997
12.35	1.0000	1.0000	1.0000	1.0000	1.0000	1.0000	1.0000	0.9999	0.9999	0.9997
12.40	1.0000	1.0000	1.0000	1.0000	1.0000	1.0000	1.0000	0.9999	0.9999	0.9997
12.45	1.0000	1.0000	1.0000	1.0000	1.0000	1.0000	1.0000	0.9999	0.9999	0.9997
12.50	1.0000	1.0000	1.0000	1.0000	1.0000	1.0000	1.0000	0.9999	0.9999	0.9998
12.55	1.0000	1.0000	1.0000	1.0000	1.0000	1.0000	1.0000	0.9999	0.9999	0.9998
12.60	1.0000	1.0000	1.0000	1.0000	1.0000	1.0000	1.0000	0.9999	0.9999	0.9998
12.65	1.0000	1.0000	1.0000	1.0000	1.0000	1.0000	1.0000	0.9999	0.9999	0.9998
12.70	1.0000	1.0000	1.0000	1.0000	1.0000	1.0000	1.0000	0.9999	0.9999	0.9998
12.75	1.0000	1.0000	1.0000	1.0000	1.0000	1.0000	1.0000	0.9999	0.9999	0.9998
12.80	1.0000	1.0000	1.0000	1.0000	1.0000	1.0000	1.0000	0.9999	0.9999	0.9998
12.85	1.0000	1.0000	1.0000	1.0000	1.0000	1.0000	1.0000	0.9999	0.9999	0.9998
12.90	1.0000	1.0000	1.0000	1.0000	1.0000	1.0000	1.0000	0.9999	0.9999	0.9998
12.95	1.0000	1.0000	1.0000	1.0000	1.0000	1.0000	1.0000	0.9999	0.9999	0.9998
13.00	1.0000	1.0000	1.0000	1.0000	1.0000	1.0000	1.0000	0.9999	0.9999	0.9998
13.05	1.0000	1.0000	1.0000	1.0000	1.0000	1.0000	1.0000	0.9999	0.9999	0.9998
13.10	1.0000	1.0000	1.0000	1.0000	1.0000	1.0000	1.0000	0.9999	0.9999	0.9998
13.15	1.0000	1.0000	1.0000	1.0000	1.0000	1.0000	1.0000	0.9999	0.9999	0.9998
13.20	1.0000	1.0000	1.0000	1.0000	1.0000	1.0000	1.0000	0.9999	0.9999	0.9998
13.25	1.0000	1.0000	1.0000	1.0000	1.0000	1.0000	1.0000	0.9999	0.9999	0.9998
13.30	1.0000	1.0000	1.0000	1.0000	1.0000	1.0000	1.0000	1.0000	0.9999	0.9998
13.35	1.0000	1.0000	1.0000	1.0000	1.0000	1.0000	1.0000	1.0000	0.9999	0.9999
13.40	1.0000	1.0000	1.0000	1.0000	1.0000	1.0000	1.0000	1.0000	0.9999	0.9999
13.45	1.0000	1.0000	1.0000	1.0000	1.0000	1.0000	1.0000	1.0000	0.9999	0.9999
13.50	1.0000	1.0000	1.0000	1.0000	1.0000	1.0000	1.0000	1.0000	0.9999	0.9999
13.55	1.0000	1.0000	1.0000	1.0000	1.0000	1.0000	1.0000	1.0000	0.9999	0.9999
13.60	1.0000	1.0000	1.0000	1.0000	1.0000	1.0000	1.0000	1.0000	0.9999	0.9999
13.65	1.0000	1.0000	1.0000	1.0000	1.0000	1.0000	1.0000	1.0000	0.9999	0.9999
13.70	1.0000	1.0000	1.0000	1.0000	1.0000	1.0000	1.0000	1.0000	0.9999	0.9999
13.75	1.0000	1.0000	1.0000	1.0000	1.0000	1.0000	1.0000	1.0000	0.9999	0.9999
13.80	1.0000	1.0000	1.0000	1.0000	1.0000	1.0000	1.0000	1.0000	0.9999	0.9999
13.85	1.0000	1.0000	1.0000	1.0000	1.0000	1.0000	1.0000	1.0000	0.9999	0.9999
13.90	1.0000	1.0000	1.0000	1.0000	1.0000	1.0000	1.0000	1.0000	0.9999	0.9999
13.95	1.0000	1.0000	1.0000	1.0000	1.0000	1.0000	1.0000	1.0000	0.9999	0.9999

PROBABILITY INTEGRAL OF t, THE NON-CENTRAL t-STATISTIC. THIS TABLE GIVES Pr $[t/\sqrt{f} \leq x]$.

f is the number of degrees of freedom; the non-centrality parameter is $\sqrt{f+1}\,K_p$.

K_p is the standardized normal deviate exceeded with probability p.

DEGREES OF FREEDOM **8**

x \ p	.2500	.1500	.1000	.0650	.0400	.0250	.0100	.0040	.0025	.0010
14.00	1.0000	1.0000	1.0000	1.0000	1.0000	1.0000	1.0000	1.0000	1.0000	0.9999
14.05	1.0000	1.0000	1.0000	1.0000	1.0000	1.0000	1.0000	1.0000	1.0000	0.9999
14.10	1.0000	1.0000	1.0000	1.0000	1.0000	1.0000	1.0000	1.0000	1.0000	0.9999
14.15	1.0000	1.0000	1.0000	1.0000	1.0000	1.0000	1.0000	1.0000	1.0000	0.9999
14.20	1.0000	1.0000	1.0000	1.0000	1.0000	1.0000	1.0000	1.0000	1.0000	0.9999
14.25	1.0000	1.0000	1.0000	1.0000	1.0000	1.0000	1.0000	1.0000	1.0000	0.9999
14.30	1.0000	1.0000	1.0000	1.0000	1.0000	1.0000	1.0000	1.0000	1.0000	0.9999
14.35	1.0000	1.0000	1.0000	1.0000	1.0000	1.0000	1.0000	1.0000	1.0000	0.9999
14.40	1.0000	1.0000	1.0000	1.0000	1.0000	1.0000	1.0000	1.0000	1.0000	0.9999
14.45	1.0000	1.0000	1.0000	1.0000	1.0000	1.0000	1.0000	1.0000	1.0000	0.9999
14.50	1.0000	1.0000	1.0000	1.0000	1.0000	1.0000	1.0000	1.0000	1.0000	0.9999
14.55	1.0000	1.0000	1.0000	1.0000	1.0000	1.0000	1.0000	1.0000	1.0000	0.9999
14.60	1.0000	1.0000	1.0000	1.0000	1.0000	1.0000	1.0000	1.0000	1.0000	0.9999
14.65	1.0000	1.0000	1.0000	1.0000	1.0000	1.0000	1.0000	1.0000	1.0000	0.9999
14.70	1.0000	1.0000	1.0000	1.0000	1.0000	1.0000	1.0000	1.0000	1.0000	0.9999
14.75	1.0000	1.0000	1.0000	1.0000	1.0000	1.0000	1.0000	1.0000	1.0000	0.9999
14.80	1.0000	1.0000	1.0000	1.0000	1.0000	1.0000	1.0000	1.0000	1.0000	0.9999
14.85	1.0000	1.0000	1.0000	1.0000	1.0000	1.0000	1.0000	1.0000	1.0000	0.9999
14.90	1.0000	1.0000	1.0000	1.0000	1.0000	1.0000	1.0000	1.0000	1.0000	0.9999
14.95	1.0000	1.0000	1.0000	1.0000	1.0000	1.0000	1.0000	1.0000	1.0000	0.9999
15.00	1.0000	1.0000	1.0000	1.0000	1.0000	1.0000	1.0000	1.0000	1.0000	0.9999
15.05	1.0000	1.0000	1.0000	1.0000	1.0000	1.0000	1.0000	1.0000	1.0000	0.9999
15.10	1.0000	1.0000	1.0000	1.0000	1.0000	1.0000	1.0000	1.0000	1.0000	0.9999
15.15	1.0000	1.0000	1.0000	1.0000	1.0000	1.0000	1.0000	1.0000	1.0000	0.9999
15.20	1.0000	1.0000	1.0000	1.0000	1.0000	1.0000	1.0000	1.0000	1.0000	0.9999
15.25	1.0000	1.0000	1.0000	1.0000	1.0000	1.0000	1.0000	1.0000	1.0000	0.9999
15.30	1.0000	1.0000	1.0000	1.0000	1.0000	1.0000	1.0000	1.0000	1.0000	1.0000

PROBABILITY INTEGRAL OF t, THE NON-CENTRAL t-STATISTIC. THIS TABLE GIVES Pr $[t/\sqrt{f} \leq x]$.

f is the number of degrees of freedom; the non-centrality parameter is $\sqrt{f+1}\,K_p$.

K_p is the standardized normal deviate exceeded with probability p.

DEGREES OF FREEDOM **9**

x \ p	.2500	.1500	.1000	.0650	.0400	.0250	.0100	.0040	.0025	.0010
- 0.80	0.0000	0.0000	0.0000	0.0000	0.0000	0.0000	0.0000	0.0000	0.0000	0.0000
- 0.75	0.0001	0.0000	0.0000	0.0000	0.0000	0.0000	0.0000	0.0000	0.0000	0.0000
- 0.70	0.0001	0.0000	0.0000	0.0000	0.0000	0.0000	0.0000	0.0000	0.0000	0.0000
- 0.65	0.0001	0.0000	0.0000	0.0000	0.0000	0.0000	0.0000	0.0000	0.0000	0.0000
- 0.60	0.0002	0.0000	0.0000	0.0000	0.0000	0.0000	0.0000	0.0000	0.0000	0.0000
- 0.55	0.0002	0.0000	0.0000	0.0000	0.0000	0.0000	0.0000	0.0000	0.0000	0.0000
- 0.50	0.0003	0.0000	0.0000	0.0000	0.0000	0.0000	0.0000	0.0000	0.0000	0.0000
- 0.45	0.0005	0.0000	0.0000	0.0000	0.0000	0.0000	0.0000	0.0000	0.0000	0.0000
- 0.40	0.0007	0.0000	0.0000	0.0000	0.0000	0.0000	0.0000	0.0000	0.0000	0.0000
- 0.35	0.0011	0.0000	0.0000	0.0000	0.0000	0.0000	0.0000	0.0000	0.0000	0.0000
- 0.30	0.0016	0.0000	0.0000	0.0000	0.0000	0.0000	0.0000	0.0000	0.0000	0.0000
- 0.25	0.0024	0.0000	0.0000	0.0000	0.0000	0.0000	0.0000	0.0000	0.0000	0.0000
- 0.20	0.0036	0.0001	0.0000	0.0000	0.0000	0.0000	0.0000	0.0000	0.0000	0.0000
- 0.15	0.0053	0.0001	0.0000	0.0000	0.0000	0.0000	0.0000	0.0000	0.0000	0.0000
- 0.10	0.0078	0.0002	0.0000	0.0000	0.0000	0.0000	0.0000	0.0000	0.0000	0.0000
- 0.05	0.0114	0.0003	0.0000	0.0000	0.0000	0.0000	0.0000	0.0000	0.0000	0.0000
0.00	0.0165	0.0005	0.0000	0.0000	0.0000	0.0000	0.0000	0.0000	0.0000	0.0000
0.05	0.0235	0.0009	0.0000	0.0000	0.0000	0.0000	0.0000	0.0000	0.0000	0.0000
0.10	0.0331	0.0015	0.0001	0.0000	0.0000	0.0000	0.0000	0.0000	0.0000	0.0000
0.15	0.0459	0.0024	0.0002	0.0000	0.0000	0.0000	0.0000	0.0000	0.0000	0.0000
0.20	0.0625	0.0038	0.0003	0.0000	0.0000	0.0000	0.0000	0.0000	0.0000	0.0000
0.25	0.0834	0.0061	0.0005	0.0000	0.0000	0.0000	0.0000	0.0000	0.0000	0.0000
0.30	0.1092	0.0094	0.0009	0.0001	0.0000	0.0000	0.0000	0.0000	0.0000	0.0000
0.35	0.1401	0.0143	0.0016	0.0001	0.0000	0.0000	0.0000	0.0000	0.0000	0.0000
0.40	0.1761	0.0212	0.0028	0.0003	0.0000	0.0000	0.0000	0.0000	0.0000	0.0000
0.45	0.2169	0.0306	0.0046	0.0005	0.0000	0.0000	0.0000	0.0000	0.0000	0.0000
0.50	0.2620	0.0432	0.0073	0.0009	0.0001	0.0000	0.0000	0.0000	0.0000	0.0000
0.55	0.3105	0.0595	0.0114	0.0016	0.0001	0.0000	0.0000	0.0000	0.0000	0.0000
0.60	0.3614	0.0798	0.0172	0.0027	0.0003	0.0000	0.0000	0.0000	0.0000	0.0000
0.65	0.4137	0.1045	0.0253	0.0044	0.0005	0.0001	0.0000	0.0000	0.0000	0.0000
0.70	0.4662	0.1337	0.0360	0.0071	0.0009	0.0001	0.0000	0.0000	0.0000	0.0000
0.75	0.5180	0.1672	0.0500	0.0111	0.0017	0.0002	0.0000	0.0000	0.0000	0.0000
0.80	0.5680	0.2046	0.0675	0.0167	0.0028	0.0004	0.0000	0.0000	0.0000	0.0000
0.85	0.6155	0.2455	0.0888	0.0244	0.0047	0.0008	0.0000	0.0000	0.0000	0.0000
0.90	0.6601	0.2891	0.1140	0.0346	0.0075	0.0015	0.0000	0.0000	0.0000	0.0000
0.95	0.7012	0.3346	0.1430	0.0476	0.0114	0.0025	0.0001	0.0000	0.0000	0.0000
1.00	0.7387	0.3813	0.1757	0.0638	0.0170	0.0041	0.0002	0.0000	0.0000	0.0000
1.05	0.7726	0.4283	0.2116	0.0833	0.0244	0.0065	0.0004	0.0000	0.0000	0.0000
1.10	0.8029	0.4748	0.2502	0.1062	0.0340	0.0099	0.0007	0.0000	0.0000	0.0000
1.15	0.8297	0.5202	0.2908	0.1324	0.0462	0.0146	0.0012	0.0001	0.0000	0.0000
1.20	0.8534	0.5639	0.3330	0.1618	0.0611	0.0210	0.0020	0.0001	0.0000	0.0000
1.25	0.8741	0.6055	0.3759	0.1941	0.0790	0.0292	0.0032	0.0003	0.0001	0.0000
1.30	0.8920	0.6447	0.4190	0.2289	0.0997	0.0396	0.0051	0.0005	0.0001	0.0000
1.35	0.9076	0.6812	0.4617	0.2656	0.1234	0.0523	0.0076	0.0009	0.0003	0.0000
1.40	0.9210	0.7150	0.5035	0.3038	0.1498	0.0675	0.0112	0.0015	0.0005	0.0001
1.45	0.9325	0.7460	0.5439	0.3430	0.1788	0.0853	0.0159	0.0024	0.0008	0.0001
1.50	0.9424	0.7742	0.5827	0.3827	0.2100	0.1058	0.0221	0.0037	0.0014	0.0002
1.55	0.9509	0.7998	0.6194	0.4223	0.2431	0.1287	0.0298	0.0056	0.0022	0.0003
1.60	0.9581	0.8228	0.6540	0.4614	0.2776	0.1539	0.0393	0.0081	0.0034	0.0006
1.65	0.9642	0.8434	0.6864	0.4997	0.3132	0.1814	0.0507	0.0115	0.0051	0.0009
1.70	0.9695	0.8619	0.7165	0.5368	0.3494	0.2107	0.0642	0.0160	0.0074	0.0015
1.75	0.9739	0.8783	0.7442	0.5725	0.3859	0.2416	0.0797	0.0216	0.0104	0.0023
1.80	0.9777	0.8928	0.7697	0.6065	0.4222	0.2737	0.0973	0.0286	0.0144	0.0035
1.85	0.9809	0.9057	0.7930	0.6388	0.4581	0.3068	0.1169	0.0370	0.0194	0.0051
1.90	0.9837	0.9171	0.8143	0.6692	0.4933	0.3404	0.1384	0.0470	0.0256	0.0072
1.95	0.9860	0.9271	0.8335	0.6976	0.5275	0.3743	0.1618	0.0586	0.0331	0.0100

PROBABILITY INTEGRAL OF t, THE NON-CENTRAL t-STATISTIC. THIS TABLE GIVES Pr $[t/\sqrt{f} \leq x]$.

f is the number of degrees of freedom; the non-centrality parameter is $\sqrt{f+1}\,K_p$.

K_p is the standardized normal deviate exceeded with probability p.

DEGREES OF FREEDOM **9**

x \ p	.2500	.1500	.1000	.0650	.0400	.0250	.0100	.0040	.0025	.0010
2.00	0.9880	0.9360	0.8510	0.7242	0.5605	0.4082	0.1867	0.0720	0.0420	0.0135
2.05	0.9897	0.9437	0.8667	0.7488	0.5921	0.4418	0.2131	0.0871	0.0524	0.0179
2.10	0.9911	0.9506	0.8809	0.7716	0.6223	0.4748	0.2407	0.1039	0.0643	0.0234
2.15	0.9924	0.9565	0.8936	0.7926	0.6509	0.5070	0.2692	0.1224	0.0778	0.0299
2.20	0.9934	0.9618	0.9050	0.8118	0.6780	0.5383	0.2985	0.1424	0.0929	0.0376
2.25	0.9943	0.9664	0.9152	0.8295	0.7034	0.5685	0.3282	0.1638	0.1096	0.0466
2.30	0.9951	0.9704	0.9243	0.8456	0.7272	0.5974	0.3582	0.1866	0.1277	0.0569
2.35	0.9957	0.9740	0.9325	0.8603	0.7494	0.6251	0.3883	0.2105	0.1473	0.0685
2.40	0.9963	0.9771	0.9397	0.8737	0.7701	0.6515	0.4182	0.2354	0.1681	0.0815
2.45	0.9968	0.9798	0.9462	0.8858	0.7893	0.6765	0.4477	0.2611	0.1901	0.0959
2.50	0.9972	0.9822	0.9520	0.8968	0.8070	0.7001	0.4768	0.2874	0.2131	0.1115
2.55	0.9976	0.9842	0.9571	0.9068	0.8234	0.7223	0.5052	0.3142	0.2370	0.1285
2.60	0.9979	0.9861	0.9617	0.9158	0.8386	0.7431	0.5329	0.3413	0.2616	0.1466
2.65	0.9982	0.9877	0.9658	0.9240	0.8525	0.7626	0.5597	0.3684	0.2868	0.1658
2.70	0.9984	0.9891	0.9694	0.9314	0.8652	0.7809	0.5856	0.3956	0.3124	0.1860
2.75	0.9986	0.9903	0.9727	0.9380	0.8770	0.7979	0.6105	0.4226	0.3382	0.2072
2.80	0.9988	0.9914	0.9755	0.9440	0.8877	0.8137	0.6344	0.4492	0.3642	0.2291
2.85	0.9989	0.9924	0.9781	0.9494	0.8975	0.8283	0.6572	0.4754	0.3901	0.2516
2.90	0.9990	0.9932	0.9804	0.9543	0.9065	0.8419	0.6789	0.5011	0.4159	0.2747
2.95	0.9992	0.9940	0.9824	0.9587	0.9147	0.8545	0.6996	0.5262	0.4415	0.2982
3.00	0.9993	0.9947	0.9842	0.9626	0.9222	0.8661	0.7192	0.5506	0.4667	0.3220
3.05	0.9993	0.9952	0.9859	0.9662	0.9290	0.8768	0.7377	0.5743	0.4914	0.3459
3.10	0.9994	0.9958	0.9873	0.9694	0.9352	0.8867	0.7551	0.5972	0.5156	0.3699
3.15	0.9995	0.9962	0.9886	0.9723	0.9409	0.8959	0.7716	0.6192	0.5392	0.3938
3.20	0.9995	0.9966	0.9897	0.9749	0.9461	0.9043	0.7871	0.6404	0.5621	0.4177
3.25	0.9996	0.9970	0.9908	0.9773	0.9508	0.9120	0.8016	0.6607	0.5844	0.4412
3.30	0.9996	0.9973	0.9917	0.9794	0.9550	0.9191	0.8153	0.6801	0.6059	0.4645
3.35	0.9997	0.9976	0.9925	0.9813	0.9590	0.9256	0.8280	0.6987	0.6266	0.4874
3.40	0.9997	0.9978	0.9932	0.9831	0.9625	0.9316	0.8400	0.7164	0.6465	0.5098
3.45	0.9997	0.9980	0.9939	0.9846	0.9657	0.9371	0.8511	0.7332	0.6657	0.5317
3.50	0.9998	0.9982	0.9945	0.9860	0.9687	0.9422	0.8616	0.7491	0.6840	0.5531
3.55	0.9998	0.9984	0.9950	0.9873	0.9714	0.9469	0.8713	0.7643	0.7015	0.5739
3.60	0.9998	0.9986	0.9955	0.9885	0.9738	0.9511	0.8803	0.7786	0.7183	0.5940
3.65	0.9998	0.9987	0.9959	0.9895	0.9760	0.9550	0.8888	0.7922	0.7342	0.6136
3.70	0.9999	0.9988	0.9963	0.9904	0.9781	0.9586	0.8966	0.8050	0.7494	0.6324
3.75	0.9999	0.9990	0.9966	0.9913	0.9799	0.9619	0.9039	0.8171	0.7639	0.6506
3.80	0.9999	0.9991	0.9969	0.9921	0.9816	0.9649	0.9107	0.8284	0.7776	0.6681
3.85	0.9999	0.9991	0.9972	0.9928	0.9831	0.9677	0.9170	0.8392	0.7905	0.6849
3.90	0.9999	0.9992	0.9975	0.9934	0.9845	0.9703	0.9229	0.8493	0.8028	0.7011
3.95	0.9999	0.9993	0.9977	0.9940	0.9858	0.9726	0.9283	0.8587	0.8145	0.7165
4.00	0.9999	0.9994	0.9979	0.9945	0.9870	0.9747	0.9334	0.8677	0.8255	0.7313
4.05	0.9999	0.9994	0.9981	0.9950	0.9880	0.9767	0.9381	0.8760	0.8359	0.7455
4.10	0.9999	0.9995	0.9983	0.9954	0.9890	0.9785	0.9425	0.8839	0.8457	0.7589
4.15	0.9999	0.9995	0.9984	0.9958	0.9899	0.9802	0.9465	0.8912	0.8549	0.7718
4.20	0.9999	0.9996	0.9986	0.9961	0.9907	0.9817	0.9503	0.8981	0.8637	0.7841
4.25	1.0000	0.9996	0.9987	0.9965	0.9915	0.9831	0.9537	0.9046	0.8719	0.7957
4.30	1.0000	0.9996	0.9988	0.9967	0.9921	0.9844	0.9570	0.9107	0.8796	0.8068
4.35	1.0000	0.9997	0.9989	0.9970	0.9928	0.9856	0.9600	0.9163	0.8869	0.8173
4.40	1.0000	0.9997	0.9990	0.9973	0.9933	0.9867	0.9628	0.9217	0.8937	0.8273
4.45	1.0000	0.9997	0.9991	0.9975	0.9938	0.9877	0.9653	0.9266	0.9002	0.8368
4.50	1.0000	0.9997	0.9992	0.9977	0.9943	0.9886	0.9677	0.9313	0.9062	0.8458
4.55	1.0000	0.9998	0.9992	0.9979	0.9948	0.9894	0.9699	0.9356	0.9119	0.8543
4.60	1.0000	0.9998	0.9993	0.9980	0.9952	0.9902	0.9720	0.9397	0.9172	0.8624
4.65	1.0000	0.9998	0.9993	0.9982	0.9955	0.9909	0.9739	0.9435	0.9222	0.8700
4.70	1.0000	0.9998	0.9994	0.9983	0.9959	0.9916	0.9757	0.9471	0.9269	0.8772
4.75	1.0000	0.9998	0.9994	0.9985	0.9962	0.9922	0.9774	0.9504	0.9314	0.8841
4.80	1.0000	0.9999	0.9995	0.9986	0.9965	0.9928	0.9789	0.9535	0.9355	0.8905
4.85	1.0000	0.9999	0.9995	0.9987	0.9967	0.9933	0.9803	0.9564	0.9394	0.8966
4.90	1.0000	0.9999	0.9996	0.9988	0.9970	0.9938	0.9816	0.9592	0.9430	0.9024
4.95	1.0000	0.9999	0.9996	0.9989	0.9972	0.9942	0.9829	0.9617	0.9465	0.9078

PROBABILITY INTEGRAL OF t, THE NON-CENTRAL t-STATISTIC. THIS TABLE GIVES Pr $[t/\sqrt{f} \leq x]$.

f is the number of degrees of freedom; the non-centrality parameter is $\sqrt{f+1}\,K_p$.

K_p is the standardized normal deviate exceeded with probability p.

x \ p	.2500	.1500	.1000	.0650	.0400	.0250	.0100	.0040	.0025	.0010
5.00	1.0000	0.9999	0.9996	0.9990	0.9974	0.9946	0.9840	0.9641	0.9497	0.9130
5.05	1.0000	0.9999	0.9997	0.9990	0.9976	0.9950	0.9851	0.9663	0.9527	0.9178
5.10	1.0000	0.9999	0.9997	0.9991	0.9978	0.9954	0.9860	0.9684	0.9555	0.9224
5.15	1.0000	0.9999	0.9997	0.9992	0.9979	0.9957	0.9870	0.9704	0.9582	0.9267
5.20	1.0000	0.9999	0.9997	0.9992	0.9981	0.9960	0.9878	0.9722	0.9607	0.9308
5.25	1.0000	0.9999	0.9998	0.9993	0.9982	0.9963	0.9886	0.9739	0.9630	0.9347
5.30	1.0000	0.9999	0.9998	0.9993	0.9983	0.9965	0.9893	0.9755	0.9652	0.9383
5.35	1.0000	0.9999	0.9998	0.9994	0.9984	0.9967	0.9900	0.9770	0.9672	0.9417
5.40	1.0000	0.9999	0.9998	0.9994	0.9986	0.9970	0.9907	0.9784	0.9692	0.9449
5.45	1.0000	0.9999	0.9998	0.9995	0.9987	0.9972	0.9913	0.9797	0.9710	0.9480
5.50	1.0000	1.0000	0.9998	0.9995	0.9987	0.9974	0.9918	0.9809	0.9727	0.9509
5.55	1.0000	1.0000	0.9998	0.9995	0.9988	0.9975	0.9923	0.9820	0.9743	0.9536
5.60	1.0000	1.0000	0.9999	0.9996	0.9989	0.9977	0.9928	0.9831	0.9758	0.9561
5.65	1.0000	1.0000	0.9999	0.9996	0.9990	0.9979	0.9933	0.9841	0.9772	0.9585
5.70	1.0000	1.0000	0.9999	0.9996	0.9991	0.9980	0.9937	0.9851	0.9785	0.9608
5.75	1.0000	1.0000	0.9999	0.9997	0.9991	0.9981	0.9941	0.9860	0.9797	0.9629
5.80	1.0000	1.0000	0.9999	0.9997	0.9992	0.9982	0.9944	0.9868	0.9809	0.9650
5.85	1.0000	1.0000	0.9999	0.9997	0.9992	0.9984	0.9948	0.9876	0.9820	0.9669
5.90	1.0000	1.0000	0.9999	0.9997	0.9993	0.9985	0.9951	0.9883	0.9830	0.9687
5.95	1.0000	1.0000	0.9999	0.9997	0.9993	0.9986	0.9954	0.9890	0.9840	0.9704
6.00	1.0000	1.0000	0.9999	0.9998	0.9994	0.9987	0.9957	0.9896	0.9849	0.9720
6.05	1.0000	1.0000	0.9999	0.9998	0.9994	0.9987	0.9959	0.9902	0.9857	0.9735
6.10	1.0000	1.0000	0.9999	0.9998	0.9994	0.9988	0.9962	0.9908	0.9865	0.9749
6.15	1.0000	1.0000	0.9999	0.9998	0.9995	0.9989	0.9964	0.9913	0.9873	0.9762
6.20	1.0000	1.0000	0.9999	0.9998	0.9995	0.9990	0.9966	0.9918	0.9880	0.9775
6.25	1.0000	1.0000	0.9999	0.9998	0.9995	0.9990	0.9968	0.9923	0.9887	0.9787
6.30	1.0000	1.0000	0.9999	0.9998	0.9996	0.9991	0.9970	0.9927	0.9893	0.9798
6.35	1.0000	1.0000	0.9999	0.9999	0.9996	0.9991	0.9972	0.9931	0.9899	0.9809
6.40	1.0000	1.0000	1.0000	0.9999	0.9996	0.9992	0.9974	0.9935	0.9904	0.9819
6.45	1.0000	1.0000	1.0000	0.9999	0.9996	0.9992	0.9975	0.9938	0.9910	0.9828
6.50	1.0000	1.0000	1.0000	0.9999	0.9997	0.9993	0.9976	0.9942	0.9914	0.9837
6.55	1.0000	1.0000	1.0000	0.9999	0.9997	0.9993	0.9978	0.9945	0.9919	0.9846
6.60	1.0000	1.0000	1.0000	0.9999	0.9997	0.9994	0.9979	0.9948	0.9923	0.9854
6.65	1.0000	1.0000	1.0000	0.9999	0.9997	0.9994	0.9980	0.9951	0.9928	0.9861
6.70	1.0000	1.0000	1.0000	0.9999	0.9997	0.9994	0.9981	0.9954	0.9931	0.9868
6.75	1.0000	1.0000	1.0000	0.9999	0.9998	0.9995	0.9982	0.9956	0.9935	0.9875
6.80	1.0000	1.0000	1.0000	0.9999	0.9998	0.9995	0.9983	0.9959	0.9939	0.9882
6.85	1.0000	1.0000	1.0000	0.9999	0.9998	0.9995	0.9984	0.9961	0.9942	0.9888
6.90	1.0000	1.0000	1.0000	0.9999	0.9998	0.9996	0.9985	0.9963	0.9945	0.9893
6.95	1.0000	1.0000	1.0000	0.9999	0.9998	0.9996	0.9986	0.9965	0.9948	0.9899
7.00	1.0000	1.0000	1.0000	0.9999	0.9998	0.9996	0.9987	0.9967	0.9950	0.9904
7.05	1.0000	1.0000	1.0000	0.9999	0.9998	0.9996	0.9988	0.9968	0.9953	0.9909
7.10	1.0000	1.0000	1.0000	0.9999	0.9998	0.9996	0.9988	0.9970	0.9955	0.9913
7.15	1.0000	1.0000	1.0000	0.9999	0.9999	0.9997	0.9989	0.9972	0.9958	0.9918
7.20	1.0000	1.0000	1.0000	0.9999	0.9999	0.9997	0.9989	0.9973	0.9960	0.9922
7.25	1.0000	1.0000	1.0000	1.0000	0.9999	0.9997	0.9990	0.9975	0.9962	0.9926
7.30	1.0000	1.0000	1.0000	1.0000	0.9999	0.9997	0.9991	0.9976	0.9964	0.9929
7.35	1.0000	1.0000	1.0000	1.0000	0.9999	0.9997	0.9991	0.9977	0.9966	0.9933
7.40	1.0000	1.0000	1.0000	1.0000	0.9999	0.9998	0.9992	0.9978	0.9967	0.9936
7.45	1.0000	1.0000	1.0000	1.0000	0.9999	0.9998	0.9992	0.9979	0.9969	0.9939
7.50	1.0000	1.0000	1.0000	1.0000	0.9999	0.9998	0.9992	0.9980	0.9971	0.9942
7.55	1.0000	1.0000	1.0000	1.0000	0.9999	0.9998	0.9993	0.9981	0.9972	0.9945
7.60	1.0000	1.0000	1.0000	1.0000	0.9999	0.9998	0.9993	0.9982	0.9973	0.9947
7.65	1.0000	1.0000	1.0000	1.0000	0.9999	0.9998	0.9994	0.9983	0.9975	0.9950
7.70	1.0000	1.0000	1.0000	1.0000	0.9999	0.9998	0.9994	0.9984	0.9976	0.9952
7.75	1.0000	1.0000	1.0000	1.0000	0.9999	0.9998	0.9994	0.9985	0.9977	0.9955
7.80	1.0000	1.0000	1.0000	1.0000	0.9999	0.9998	0.9994	0.9986	0.9978	0.9957
7.85	1.0000	1.0000	1.0000	1.0000	0.9999	0.9998	0.9995	0.9986	0.9979	0.9959
7.90	1.0000	1.0000	1.0000	1.0000	0.9999	0.9999	0.9995	0.9987	0.9980	0.9961
7.95	1.0000	1.0000	1.0000	1.0000	0.9999	0.9999	0.9995	0.9988	0.9981	0.9963

PROBABILITY INTEGRAL OF t, THE NON-CENTRAL t-STATISTIC. THIS TABLE GIVES $\Pr\,[t/\sqrt{f} \leq x]$.

f is the number of degrees of freedom; the non-centrality parameter is $\sqrt{f+1}\,K_p$.

K_p is the standardized normal deviate exceeded with probability p.

DEGREES OF FREEDOM **9**

x \ p	.2500	.1500	.1000	.0650	.0400	.0250	.0100	.0040	.0025	.0010
8.00	1.0000	1.0000	1.0000	1.0000	0.9999	0.9999	0.9995	0.9988	0.9982	0.9964
8.05	1.0000	1.0000	1.0000	1.0000	0.9999	0.9999	0.9996	0.9989	0.9983	0.9966
8.10	1.0000	1.0000	1.0000	1.0000	0.9999	0.9999	0.9996	0.9989	0.9984	0.9968
8.15	1.0000	1.0000	1.0000	1.0000	1.0000	0.9999	0.9996	0.9990	0.9985	0.9969
8.20	1.0000	1.0000	1.0000	1.0000	1.0000	0.9999	0.9996	0.9990	0.9985	0.9970
8.25	1.0000	1.0000	1.0000	1.0000	1.0000	0.9999	0.9997	0.9991	0.9986	0.9972
8.30	1.0000	1.0000	1.0000	1.0000	1.0000	0.9999	0.9997	0.9991	0.9987	0.9973
8.35	1.0000	1.0000	1.0000	1.0000	1.0000	0.9999	0.9997	0.9992	0.9987	0.9974
8.40	1.0000	1.0000	1.0000	1.0000	1.0000	0.9999	0.9997	0.9992	0.9988	0.9975
8.45	1.0000	1.0000	1.0000	1.0000	1.0000	0.9999	0.9997	0.9992	0.9988	0.9977
8.50	1.0000	1.0000	1.0000	1.0000	1.0000	0.9999	0.9997	0.9993	0.9989	0.9978
8.55	1.0000	1.0000	1.0000	1.0000	1.0000	0.9999	0.9997	0.9993	0.9989	0.9979
8.60	1.0000	1.0000	1.0000	1.0000	1.0000	0.9999	0.9998	0.9993	0.9990	0.9980
8.65	1.0000	1.0000	1.0000	1.0000	1.0000	0.9999	0.9998	0.9994	0.9990	0.9980
8.70	1.0000	1.0000	1.0000	1.0000	1.0000	0.9999	0.9998	0.9994	0.9991	0.9981
8.75	1.0000	1.0000	1.0000	1.0000	1.0000	0.9999	0.9998	0.9994	0.9991	0.9982
8.80	1.0000	1.0000	1.0000	1.0000	1.0000	0.9999	0.9998	0.9995	0.9992	0.9983
8.85	1.0000	1.0000	1.0000	1.0000	1.0000	0.9999	0.9998	0.9995	0.9992	0.9984
8.90	1.0000	1.0000	1.0000	1.0000	1.0000	0.9999	0.9998	0.9995	0.9992	0.9984
8.95	1.0000	1.0000	1.0000	1.0000	1.0000	1.0000	0.9998	0.9995	0.9993	0.9985
9.00	1.0000	1.0000	1.0000	1.0000	1.0000	1.0000	0.9998	0.9995	0.9993	0.9986
9.05	1.0000	1.0000	1.0000	1.0000	1.0000	1.0000	0.9998	0.9996	0.9993	0.9986
9.10	1.0000	1.0000	1.0000	1.0000	1.0000	1.0000	0.9998	0.9996	0.9994	0.9987
9.15	1.0000	1.0000	1.0000	1.0000	1.0000	1.0000	0.9999	0.9996	0.9994	0.9987
9.20	1.0000	1.0000	1.0000	1.0000	1.0000	1.0000	0.9999	0.9996	0.9994	0.9988
9.25	1.0000	1.0000	1.0000	1.0000	1.0000	1.0000	0.9999	0.9996	0.9994	0.9989
9.30	1.0000	1.0000	1.0000	1.0000	1.0000	1.0000	0.9999	0.9997	0.9995	0.9989
9.35	1.0000	1.0000	1.0000	1.0000	1.0000	1.0000	0.9999	0.9997	0.9995	0.9989
9.40	1.0000	1.0000	1.0000	1.0000	1.0000	1.0000	0.9999	0.9997	0.9995	0.9990
9.45	1.0000	1.0000	1.0000	1.0000	1.0000	1.0000	0.9999	0.9997	0.9995	0.9990
9.50	1.0000	1.0000	1.0000	1.0000	1.0000	1.0000	0.9999	0.9997	0.9996	0.9991
9.55	1.0000	1.0000	1.0000	1.0000	1.0000	1.0000	0.9999	0.9997	0.9996	0.9991
9.60	1.0000	1.0000	1.0000	1.0000	1.0000	1.0000	0.9999	0.9997	0.9996	0.9991
9.65	1.0000	1.0000	1.0000	1.0000	1.0000	1.0000	0.9999	0.9997	0.9996	0.9992
9.70	1.0000	1.0000	1.0000	1.0000	1.0000	1.0000	0.9999	0.9998	0.9996	0.9992
9.75	1.0000	1.0000	1.0000	1.0000	1.0000	1.0000	0.9999	0.9998	0.9996	0.9992
9.80	1.0000	1.0000	1.0000	1.0000	1.0000	1.0000	0.9999	0.9998	0.9997	0.9993
9.85	1.0000	1.0000	1.0000	1.0000	1.0000	1.0000	0.9999	0.9998	0.9997	0.9993
9.90	1.0000	1.0000	1.0000	1.0000	1.0000	1.0000	0.9999	0.9998	0.9997	0.9993
9.95	1.0000	1.0000	1.0000	1.0000	1.0000	1.0000	0.9999	0.9998	0.9997	0.9994
10.00	1.0000	1.0000	1.0000	1.0000	1.0000	1.0000	0.9999	0.9998	0.9997	0.9994
10.05	1.0000	1.0000	1.0000	1.0000	1.0000	1.0000	0.9999	0.9998	0.9997	0.9994
10.10	1.0000	1.0000	1.0000	1.0000	1.0000	1.0000	0.9999	0.9998	0.9997	0.9994
10.15	1.0000	1.0000	1.0000	1.0000	1.0000	1.0000	0.9999	0.9998	0.9997	0.9995
10.20	1.0000	1.0000	1.0000	1.0000	1.0000	1.0000	0.9999	0.9998	0.9998	0.9995
10.25	1.0000	1.0000	1.0000	1.0000	1.0000	1.0000	0.9999	0.9998	0.9998	0.9995
10.30	1.0000	1.0000	1.0000	1.0000	1.0000	1.0000	0.9999	0.9999	0.9998	0.9995
10.35	1.0000	1.0000	1.0000	1.0000	1.0000	1.0000	1.0000	0.9999	0.9998	0.9995
10.40	1.0000	1.0000	1.0000	1.0000	1.0000	1.0000	1.0000	0.9999	0.9998	0.9996
10.45	1.0000	1.0000	1.0000	1.0000	1.0000	1.0000	1.0000	0.9999	0.9998	0.9996
10.50	1.0000	1.0000	1.0000	1.0000	1.0000	1.0000	1.0000	0.9999	0.9998	0.9996
10.55	1.0000	1.0000	1.0000	1.0000	1.0000	1.0000	1.0000	0.9999	0.9998	0.9996
10.60	1.0000	1.0000	1.0000	1.0000	1.0000	1.0000	1.0000	0.9999	0.9998	0.9996
10.65	1.0000	1.0000	1.0000	1.0000	1.0000	1.0000	1.0000	0.9999	0.9998	0.9996
10.70	1.0000	1.0000	1.0000	1.0000	1.0000	1.0000	1.0000	0.9999	0.9998	0.9996
10.75	1.0000	1.0000	1.0000	1.0000	1.0000	1.0000	1.0000	0.9999	0.9998	0.9997
10.80	1.0000	1.0000	1.0000	1.0000	1.0000	1.0000	1.0000	0.9999	0.9998	0.9997
10.85	1.0000	1.0000	1.0000	1.0000	1.0000	1.0000	1.0000	0.9999	0.9999	0.9997
10.90	1.0000	1.0000	1.0000	1.0000	1.0000	1.0000	1.0000	0.9999	0.9999	0.9997
10.95	1.0000	1.0000	1.0000	1.0000	1.0000	1.0000	1.0000	0.9999	0.9999	0.9997

PROBABILITY INTEGRAL OF t, THE NON-CENTRAL t-STATISTIC. THIS TABLE GIVES Pr $[t/\sqrt{f} \leq x]$.

f is the number of degrees of freedom; the non-centrality parameter is $\sqrt{f+1}\,K_p$.

K_p is the standardized normal deviate exceeded with probability p.

DEGREES OF FREEDOM **9**

x \ p	.2500	.1500	.1000	.0650	.0400	.0250	.0100	.0040	.0025	.0010
11.00	1.0000	1.0000	1.0000	1.0000	1.0000	1.0000	1.0000	0.9999	0.9999	0.9997
11.05	1.0000	1.0000	1.0000	1.0000	1.0000	1.0000	1.0000	0.9999	0.9999	0.9997
11.10	1.0000	1.0000	1.0000	1.0000	1.0000	1.0000	1.0000	0.9999	0.9999	0.9997
11.15	1.0000	1.0000	1.0000	1.0000	1.0000	1.0000	1.0000	0.9999	0.9999	0.9998
11.20	1.0000	1.0000	1.0000	1.0000	1.0000	1.0000	1.0000	0.9999	0.9999	0.9998
11.25	1.0000	1.0000	1.0000	1.0000	1.0000	1.0000	1.0000	0.9999	0.9999	0.9998
11.30	1.0000	1.0000	1.0000	1.0000	1.0000	1.0000	1.0000	0.9999	0.9999	0.9998
11.35	1.0000	1.0000	1.0000	1.0000	1.0000	1.0000	1.0000	0.9999	0.9999	0.9998
11.40	1.0000	1.0000	1.0000	1.0000	1.0000	1.0000	1.0000	0.9999	0.9999	0.9998
11.45	1.0000	1.0000	1.0000	1.0000	1.0000	1.0000	1.0000	0.9999	0.9999	0.9998
11.50	1.0000	1.0000	1.0000	1.0000	1.0000	1.0000	1.0000	0.9999	0.9999	0.9998
11.55	1.0000	1.0000	1.0000	1.0000	1.0000	1.0000	1.0000	0.9999	0.9999	0.9998
11.60	1.0000	1.0000	1.0000	1.0000	1.0000	1.0000	1.0000	0.9999	0.9999	0.9998
11.65	1.0000	1.0000	1.0000	1.0000	1.0000	1.0000	1.0000	1.0000	0.9999	0.9998
11.70	1.0000	1.0000	1.0000	1.0000	1.0000	1.0000	1.0000	1.0000	0.9999	0.9998
11.75	1.0000	1.0000	1.0000	1.0000	1.0000	1.0000	1.0000	1.0000	0.9999	0.9998
11.80	1.0000	1.0000	1.0000	1.0000	1.0000	1.0000	1.0000	1.0000	0.9999	0.9998
11.85	1.0000	1.0000	1.0000	1.0000	1.0000	1.0000	1.0000	1.0000	0.9999	0.9999
11.90	1.0000	1.0000	1.0000	1.0000	1.0000	1.0000	1.0000	1.0000	0.9999	0.9999
11.95	1.0000	1.0000	1.0000	1.0000	1.0000	1.0000	1.0000	1.0000	0.9999	0.9999
12.00	1.0000	1.0000	1.0000	1.0000	1.0000	1.0000	1.0000	1.0000	0.9999	0.9999
12.05	1.0000	1.0000	1.0000	1.0000	1.0000	1.0000	1.0000	1.0000	0.9999	0.9999
12.10	1.0000	1.0000	1.0000	1.0000	1.0000	1.0000	1.0000	1.0000	0.9999	0.9999
12.15	1.0000	1.0000	1.0000	1.0000	1.0000	1.0000	1.0000	1.0000	0.9999	0.9999
12.20	1.0000	1.0000	1.0000	1.0000	1.0000	1.0000	1.0000	1.0000	0.9999	0.9999
12.25	1.0000	1.0000	1.0000	1.0000	1.0000	1.0000	1.0000	1.0000	0.9999	0.9999
12.30	1.0000	1.0000	1.0000	1.0000	1.0000	1.0000	1.0000	1.0000	1.0000	0.9999
12.35	1.0000	1.0000	1.0000	1.0000	1.0000	1.0000	1.0000	1.0000	1.0000	0.9999
12.40	1.0000	1.0000	1.0000	1.0000	1.0000	1.0000	1.0000	1.0000	1.0000	0.9999
12.45	1.0000	1.0000	1.0000	1.0000	1.0000	1.0000	1.0000	1.0000	1.0000	0.9999
12.50	1.0000	1.0000	1.0000	1.0000	1.0000	1.0000	1.0000	1.0000	1.0000	0.9999
12.55	1.0000	1.0000	1.0000	1.0000	1.0000	1.0000	1.0000	1.0000	1.0000	0.9999
12.60	1.0000	1.0000	1.0000	1.0000	1.0000	1.0000	1.0000	1.0000	1.0000	0.9999
12.65	1.0000	1.0000	1.0000	1.0000	1.0000	1.0000	1.0000	1.0000	1.0000	0.9999
12.70	1.0000	1.0000	1.0000	1.0000	1.0000	1.0000	1.0000	1.0000	1.0000	0.9999
12.75	1.0000	1.0000	1.0000	1.0000	1.0000	1.0000	1.0000	1.0000	1.0000	0.9999
12.80	1.0000	1.0000	1.0000	1.0000	1.0000	1.0000	1.0000	1.0000	1.0000	0.9999
12.85	1.0000	1.0000	1.0000	1.0000	1.0000	1.0000	1.0000	1.0000	1.0000	0.9999
12.90	1.0000	1.0000	1.0000	1.0000	1.0000	1.0000	1.0000	1.0000	1.0000	0.9999
12.95	1.0000	1.0000	1.0000	1.0000	1.0000	1.0000	1.0000	1.0000	1.0000	0.9999
13.00	1.0000	1.0000	1.0000	1.0000	1.0000	1.0000	1.0000	1.0000	1.0000	0.9999
13.05	1.0000	1.0000	1.0000	1.0000	1.0000	1.0000	1.0000	1.0000	1.0000	0.9999
13.10	1.0000	1.0000	1.0000	1.0000	1.0000	1.0000	1.0000	1.0000	1.0000	0.9999
13.15	1.0000	1.0000	1.0000	1.0000	1.0000	1.0000	1.0000	1.0000	1.0000	0.9999
13.20	1.0000	1.0000	1.0000	1.0000	1.0000	1.0000	1.0000	1.0000	1.0000	0.9999
13.25	1.0000	1.0000	1.0000	1.0000	1.0000	1.0000	1.0000	1.0000	1.0000	0.9999
13.30	1.0000	1.0000	1.0000	1.0000	1.0000	1.0000	1.0000	1.0000	1.0000	0.9999
13.35	1.0000	1.0000	1.0000	1.0000	1.0000	1.0000	1.0000	1.0000	1.0000	0.9999
13.40	1.0000	1.0000	1.0000	1.0000	1.0000	1.0000	1.0000	1.0000	1.0000	0.9999
13.45	1.0000	1.0000	1.0000	1.0000	1.0000	1.0000	1.0000	1.0000	1.0000	1.0000

PROBABILITY INTEGRAL OF t, THE NON-CENTRAL t-STATISTIC. THIS TABLE GIVES Pr $[t/\sqrt{f} \leq x]$.

f is the number of degrees of freedom; the non-centrality parameter is $\sqrt{f+1}\,K_p$.

K_p is the standardized normal deviate exceeded with probability p.

DEGREES OF FREEDOM **10**

x \ P	.2500	.1500	.1000	.0650	.0400	.0250	.0100	.0040	.0025	.0010
-0.65	0.0000	0.0000	0.0000	0.0000	0.0000	0.0000	0.0000	0.0000	0.0000	0.0000
-0.60	0.0001	0.0000	0.0000	0.0000	0.0000	0.0000	0.0000	0.0000	0.0000	0.0000
-0.55	0.0001	0.0000	0.0000	0.0000	0.0000	0.0000	0.0000	0.0000	0.0000	0.0000
-0.50	0.0002	0.0000	0.0000	0.0000	0.0000	0.0000	0.0000	0.0000	0.0000	0.0000
-0.45	0.0003	0.0000	0.0000	0.0000	0.0000	0.0000	0.0000	0.0000	0.0000	0.0000
-0.40	0.0004	0.0000	0.0000	0.0000	0.0000	0.0000	0.0000	0.0000	0.0000	0.0000
-0.35	0.0006	0.0000	0.0000	0.0000	0.0000	0.0000	0.0000	0.0000	0.0000	0.0000
-0.30	0.0010	0.0000	0.0000	0.0000	0.0000	0.0000	0.0000	0.0000	0.0000	0.0000
-0.25	0.0015	0.0000	0.0000	0.0000	0.0000	0.0000	0.0000	0.0000	0.0000	0.0000
-0.20	0.0023	0.0000	0.0000	0.0000	0.0000	0.0000	0.0000	0.0000	0.0000	0.0000
-0.15	0.0036	0.0001	0.0000	0.0000	0.0000	0.0000	0.0000	0.0000	0.0000	0.0000
-0.10	0.0056	0.0001	0.0000	0.0000	0.0000	0.0000	0.0000	0.0000	0.0000	0.0000
-0.05	0.0084	0.0002	0.0000	0.0000	0.0000	0.0000	0.0000	0.0000	0.0000	0.0000
0.00	0.0126	0.0003	0.0000	0.0000	0.0000	0.0000	0.0000	0.0000	0.0000	0.0000
0.05	0.0187	0.0005	0.0000	0.0000	0.0000	0.0000	0.0000	0.0000	0.0000	0.0000
0.10	0.0272	0.0009	0.0000	0.0000	0.0000	0.0000	0.0000	0.0000	0.0000	0.0000
0.15	0.0388	0.0015	0.0001	0.0000	0.0000	0.0000	0.0000	0.0000	0.0000	0.0000
0.20	0.0543	0.0026	0.0002	0.0000	0.0000	0.0000	0.0000	0.0000	0.0000	0.0000
0.25	0.0744	0.0043	0.0003	0.0000	0.0000	0.0000	0.0000	0.0000	0.0000	0.0000
0.30	0.0996	0.0070	0.0006	0.0000	0.0000	0.0000	0.0000	0.0000	0.0000	0.0000
0.35	0.1304	0.0111	0.0011	0.0001	0.0000	0.0000	0.0000	0.0000	0.0000	0.0000
0.40	0.1669	0.0170	0.0019	0.0001	0.0000	0.0000	0.0000	0.0000	0.0000	0.0000
0.45	0.2088	0.0254	0.0032	0.0003	0.0000	0.0000	0.0000	0.0000	0.0000	0.0000
0.50	0.2556	0.0370	0.0054	0.0005	0.0000	0.0000	0.0000	0.0000	0.0000	0.0000
0.55	0.3064	0.0523	0.0088	0.0010	0.0001	0.0000	0.0000	0.0000	0.0000	0.0000
0.60	0.3600	0.0720	0.0137	0.0018	0.0001	0.0000	0.0000	0.0000	0.0000	0.0000
0.65	0.4152	0.0963	0.0209	0.0032	0.0003	0.0000	0.0000	0.0000	0.0000	0.0000
0.70	0.4707	0.1256	0.0307	0.0053	0.0006	0.0001	0.0000	0.0000	0.0000	0.0000
0.75	0.5253	0.1596	0.0437	0.0086	0.0011	0.0001	0.0000	0.0000	0.0000	0.0000
0.80	0.5780	0.1982	0.0605	0.0135	0.0020	0.0003	0.0000	0.0000	0.0000	0.0000
0.85	0.6278	0.2406	0.0814	0.0203	0.0034	0.0005	0.0000	0.0000	0.0000	0.0000
0.90	0.6742	0.2863	0.1065	0.0296	0.0056	0.0010	0.0000	0.0000	0.0000	0.0000
0.95	0.7167	0.3342	0.1359	0.0418	0.0090	0.0017	0.0001	0.0000	0.0000	0.0000
1.00	0.7552	0.3835	0.1693	0.0573	0.0138	0.0029	0.0001	0.0000	0.0000	0.0000
1.05	0.7896	0.4331	0.2065	0.0764	0.0204	0.0049	0.0002	0.0000	0.0000	0.0000
1.10	0.8200	0.4823	0.2467	0.0992	0.0293	0.0077	0.0004	0.0000	0.0000	0.0000
1.15	0.8466	0.5302	0.2894	0.1257	0.0407	0.0118	0.0008	0.0000	0.0000	0.0000
1.20	0.8698	0.5761	0.3338	0.1558	0.0551	0.0174	0.0014	0.0001	0.0000	0.0000
1.25	0.8897	0.6196	0.3791	0.1891	0.0725	0.0249	0.0023	0.0002	0.0000	0.0000
1.30	0.9069	0.6604	0.4247	0.2253	0.0932	0.0346	0.0037	0.0003	0.0001	0.0000
1.35	0.9215	0.6981	0.4698	0.2638	0.1171	0.0467	0.0059	0.0006	0.0002	0.0000
1.40	0.9340	0.7328	0.5139	0.3040	0.1440	0.0615	0.0089	0.0010	0.0003	0.0000
1.45	0.9445	0.7643	0.5564	0.3453	0.1739	0.0791	0.0130	0.0017	0.0005	0.0001
1.50	0.9534	0.7927	0.5969	0.3872	0.2063	0.0995	0.0185	0.0027	0.0009	0.0001
1.55	0.9609	0.8182	0.6352	0.4291	0.2408	0.1227	0.0256	0.0042	0.0015	0.0002
1.60	0.9671	0.8410	0.6711	0.4704	0.2771	0.1486	0.0345	0.0063	0.0025	0.0004
1.65	0.9724	0.8612	0.7044	0.5108	0.3146	0.1769	0.0455	0.0092	0.0038	0.0006
1.70	0.9768	0.8790	0.7352	0.5498	0.3528	0.2074	0.0585	0.0131	0.0057	0.0010
1.75	0.9805	0.8947	0.7633	0.5871	0.3914	0.2397	0.0738	0.0182	0.0083	0.0016
1.80	0.9836	0.9085	0.7890	0.6226	0.4298	0.2735	0.0914	0.0246	0.0117	0.0025
1.85	0.9862	0.9205	0.8123	0.6561	0.4677	0.3083	0.1112	0.0325	0.0162	0.0038
1.90	0.9884	0.9310	0.8333	0.6875	0.5048	0.3439	0.1332	0.0420	0.0218	0.0055
1.95	0.9902	0.9402	0.8522	0.7167	0.5407	0.3797	0.1572	0.0533	0.0288	0.0079
2.00	0.9917	0.9481	0.8691	0.7437	0.5753	0.4155	0.1831	0.0664	0.0372	0.0110
2.05	0.9930	0.9550	0.8842	0.7686	0.6083	0.4510	0.2106	0.0814	0.0472	0.0149
2.10	0.9941	0.9610	0.8976	0.7915	0.6397	0.4858	0.2394	0.0983	0.0589	0.0199
2.15	0.9950	0.9662	0.9096	0.8124	0.6693	0.5197	0.2694	0.1171	0.0723	0.0259
2.20	0.9958	0.9706	0.9202	0.8314	0.6971	0.5526	0.3003	0.1375	0.0874	0.0332

PROBABILITY INTEGRAL OF t, THE NON-CENTRAL t-STATISTIC. THIS TABLE GIVES Pr $[t/\sqrt{f} \leq x]$.

f is the number of degrees of freedom; the non-centrality parameter is $\sqrt{f+1}\,K_p$.

K_p is the standardized normal deviate exceeded with probability p.

DEGREES OF FREEDOM 10

x \ p	.2500	.1500	.1000	.0650	.0400	.0250	.0100	.0040	.0025	.0010
2.25	0.9964	0.9745	0.9296	0.8486	0.7231	0.5842	0.3317	0.1597	0.1042	0.0417
2.30	0.9969	0.9779	0.9379	0.8643	0.7473	0.6144	0.3634	0.1833	0.1226	0.0517
2.35	0.9974	0.9808	0.9452	0.8784	0.7697	0.6431	0.3952	0.2082	0.1427	0.0631
2.40	0.9978	0.9833	0.9517	0.8911	0.7904	0.6703	0.4268	0.2343	0.1642	0.0760
2.45	0.9981	0.9855	0.9574	0.9026	0.8095	0.6960	0.4580	0.2613	0.1870	0.0904
2.50	0.9984	0.9873	0.9624	0.9129	0.8271	0.7201	0.4887	0.2890	0.2111	0.1063
2.55	0.9986	0.9890	0.9668	0.9221	0.8431	0.7427	0.5186	0.3173	0.2361	0.1236
2.60	0.9988	0.9904	0.9707	0.9304	0.8578	0.7638	0.5476	0.3459	0.2620	0.1422
2.65	0.9990	0.9916	0.9742	0.9378	0.8712	0.7833	0.5757	0.3746	0.2885	0.1620
2.70	0.9991	0.9927	0.9772	0.9444	0.8834	0.8015	0.6027	0.4033	0.3155	0.1830
2.75	0.9992	0.9936	0.9798	0.9503	0.8945	0.8183	0.6286	0.4319	0.3428	0.2051
2.80	0.9993	0.9944	0.9822	0.9556	0.9046	0.8338	0.6533	0.4600	0.3703	0.2280
2.85	0.9994	0.9951	0.9842	0.9603	0.9137	0.8481	0.6768	0.4877	0.3978	0.2517
2.90	0.9995	0.9957	0.9860	0.9645	0.9220	0.8613	0.6991	0.5147	0.4250	0.2760
2.95	0.9996	0.9962	0.9876	0.9683	0.9295	0.8734	0.7202	0.5411	0.4520	0.3008
3.00	0.9996	0.9967	0.9890	0.9716	0.9363	0.8845	0.7400	0.5666	0.4786	0.3259
3.05	0.9997	0.9971	0.9903	0.9746	0.9425	0.8946	0.7587	0.5913	0.5046	0.3512
3.10	0.9997	0.9974	0.9914	0.9772	0.9480	0.9039	0.7763	0.6151	0.5301	0.3766
3.15	0.9997	0.9977	0.9923	0.9796	0.9530	0.9124	0.7927	0.6380	0.5548	0.4019
3.20	0.9998	0.9980	0.9932	0.9817	0.9575	0.9201	0.8081	0.6599	0.5788	0.4271
3.25	0.9998	0.9982	0.9939	0.9836	0.9616	0.9272	0.8224	0.6808	0.6020	0.4520
3.30	0.9998	0.9984	0.9946	0.9853	0.9653	0.9337	0.8358	0.7007	0.6243	0.4766
3.35	0.9998	0.9986	0.9952	0.9868	0.9686	0.9396	0.8482	0.7196	0.6458	0.5007
3.40	0.9999	0.9988	0.9957	0.9882	0.9716	0.9449	0.8597	0.7375	0.6663	0.5242
3.45	0.9999	0.9989	0.9962	0.9894	0.9743	0.9498	0.8705	0.7545	0.6860	0.5472
3.50	0.9999	0.9990	0.9966	0.9904	0.9767	0.9542	0.8804	0.7706	0.7048	0.5696
3.55	0.9999	0.9991	0.9969	0.9914	0.9789	0.9583	0.8896	0.7857	0.7226	0.5913
3.60	0.9999	0.9992	0.9972	0.9923	0.9809	0.9620	0.8981	0.8000	0.7396	0.6123
3.65	0.9999	0.9993	0.9975	0.9930	0.9827	0.9653	0.9060	0.8134	0.7557	0.6325
3.70	0.9999	0.9994	0.9978	0.9937	0.9843	0.9684	0.9133	0.8259	0.7710	0.6520
3.75	0.9999	0.9995	0.9980	0.9943	0.9857	0.9711	0.9200	0.8377	0.7854	0.6708
3.80	0.9999	0.9995	0.9982	0.9949	0.9871	0.9737	0.9262	0.8488	0.7990	0.6887
3.85	1.0000	0.9996	0.9984	0.9954	0.9883	0.9760	0.9320	0.8592	0.8118	0.7059
3.90	1.0000	0.9996	0.9986	0.9958	0.9893	0.9780	0.9373	0.8689	0.8239	0.7223
3.95	1.0000	0.9996	0.9987	0.9962	0.9903	0.9799	0.9421	0.8779	0.8353	0.7380
4.00	1.0000	0.9997	0.9988	0.9966	0.9912	0.9817	0.9466	0.8864	0.8460	0.7529
4.05	1.0000	0.9997	0.9989	0.9969	0.9920	0.9832	0.9508	0.8943	0.8561	0.7671
4.10	1.0000	0.9997	0.9990	0.9972	0.9927	0.9847	0.9546	0.9016	0.8655	0.7806
4.15	1.0000	0.9998	0.9991	0.9975	0.9933	0.9860	0.9581	0.9085	0.8744	0.7934
4.20	1.0000	0.9998	0.9992	0.9977	0.9939	0.9872	0.9613	0.9148	0.8827	0.8056
4.25	1.0000	0.9998	0.9993	0.9979	0.9945	0.9883	0.9643	0.9208	0.8904	0.8171
4.30	1.0000	0.9998	0.9994	0.9981	0.9950	0.9892	0.9671	0.9263	0.8977	0.8280
4.35	1.0000	0.9998	0.9994	0.9983	0.9954	0.9901	0.9696	0.9315	0.9045	0.8382
4.40	1.0000	0.9999	0.9995	0.9984	0.9958	0.9910	0.9719	0.9363	0.9108	0.8479
4.45	1.0000	0.9999	0.9995	0.9986	0.9962	0.9917	0.9741	0.9407	0.9168	0.8571
4.50	1.0000	0.9999	0.9996	0.9987	0.9965	0.9924	0.9760	0.9448	0.9223	0.8657
4.55	1.0000	0.9999	0.9996	0.9988	0.9968	0.9930	0.9779	0.9487	0.9275	0.8739
4.60	1.0000	0.9999	0.9996	0.9989	0.9971	0.9936	0.9795	0.9523	0.9323	0.8815
4.65	1.0000	0.9999	0.9997	0.9990	0.9973	0.9941	0.9811	0.9556	0.9368	0.8888
4.70	1.0000	0.9999	0.9997	0.9991	0.9975	0.9946	0.9825	0.9587	0.9410	0.8956
4.75	1.0000	0.9999	0.9997	0.9992	0.9977	0.9950	0.9838	0.9615	0.9449	0.9019
4.80	1.0000	0.9999	0.9998	0.9992	0.9979	0.9954	0.9850	0.9642	0.9486	0.9079
4.85	1.0000	0.9999	0.9998	0.9993	0.9981	0.9958	0.9861	0.9667	0.9520	0.9136
4.90	1.0000	0.9999	0.9998	0.9994	0.9982	0.9961	0.9872	0.9690	0.9552	0.9189
4.95	1.0000	1.0000	0.9998	0.9994	0.9984	0.9964	0.9881	0.9711	0.9582	0.9239
5.00	1.0000	1.0000	0.9998	0.9995	0.9985	0.9967	0.9890	0.9731	0.9609	0.9285
5.05	1.0000	1.0000	0.9998	0.9995	0.9986	0.9970	0.9898	0.9749	0.9635	0.9329
5.10	1.0000	1.0000	0.9999	0.9996	0.9987	0.9972	0.9905	0.9766	0.9659	0.9370
5.15	1.0000	1.0000	0.9999	0.9996	0.9988	0.9974	0.9912	0.9782	0.9682	0.9409
5.20	1.0000	1.0000	0.9999	0.9996	0.9989	0.9976	0.9918	0.9797	0.9702	0.9445

PROBABILITY INTEGRAL OF t, THE NON-CENTRAL t-STATISTIC. THIS TABLE GIVES Pr $[t/\sqrt{f} \leq x]$.

f is the number of degrees of freedom; the non-centrality parameter is $\sqrt{f+1}\,K_p$.

K_p is the standardized normal deviate exceeded with probability p.

DEGREES OF FREEDOM **10**

x \ p	.2500	.1500	.1000	.0650	.0400	.0250	.0100	.0040	.0025	.0010
5.25	1.0000	1.0000	0.9999	0.9997	0.9990	0.9978	0.9924	0.9811	0.9722	0.9479
5.30	1.0000	1.0000	0.9999	0.9997	0.9991	0.9980	0.9930	0.9824	0.9740	0.9511
5.35	1.0000	1.0000	0.9999	0.9997	0.9992	0.9981	0.9935	0.9835	0.9757	0.9541
5.40	1.0000	1.0000	0.9999	0.9997	0.9992	0.9983	0.9939	0.9846	0.9773	0.9569
5.45	1.0000	1.0000	0.9999	0.9997	0.9993	0.9984	0.9944	0.9857	0.9788	0.9595
5.50	1.0000	1.0000	0.9999	0.9998	0.9993	0.9985	0.9948	0.9866	0.9801	0.9620
5.55	1.0000	1.0000	0.9999	0.9998	0.9994	0.9986	0.9951	0.9875	0.9814	0.9643
5.60	1.0000	1.0000	0.9999	0.9998	0.9994	0.9987	0.9955	0.9883	0.9826	0.9665
5.65	1.0000	1.0000	0.9999	0.9998	0.9995	0.9988	0.9958	0.9891	0.9837	0.9685
5.70	1.0000	1.0000	0.9999	0.9998	0.9995	0.9989	0.9961	0.9898	0.9847	0.9704
5.75	1.0000	1.0000	1.0000	0.9998	0.9996	0.9990	0.9963	0.9905	0.9857	0.9722
5.80	1.0000	1.0000	1.0000	0.9999	0.9996	0.9990	0.9966	0.9911	0.9866	0.9738
5.85	1.0000	1.0000	1.0000	0.9999	0.9996	0.9991	0.9968	0.9917	0.9875	0.9754
5.90	1.0000	1.0000	1.0000	0.9999	0.9996	0.9992	0.9970	0.9922	0.9883	0.9769
5.95	1.0000	1.0000	1.0000	0.9999	0.9997	0.9992	0.9972	0.9927	0.9890	0.9783
6.00	1.0000	1.0000	1.0000	0.9999	0.9997	0.9993	0.9974	0.9932	0.9897	0.9796
6.05	1.0000	1.0000	1.0000	0.9999	0.9997	0.9993	0.9976	0.9936	0.9903	0.9808
6.10	1.0000	1.0000	1.0000	0.9999	0.9997	0.9994	0.9978	0.9940	0.9909	0.9819
6.15	1.0000	1.0000	1.0000	0.9999	0.9998	0.9994	0.9979	0.9944	0.9915	0.9830
6.20	1.0000	1.0000	1.0000	0.9999	0.9998	0.9995	0.9980	0.9947	0.9920	0.9840
6.25	1.0000	1.0000	1.0000	0.9999	0.9998	0.9995	0.9982	0.9951	0.9925	0.9849
6.30	1.0000	1.0000	1.0000	0.9999	0.9998	0.9995	0.9983	0.9954	0.9930	0.9858
6.35	1.0000	1.0000	1.0000	0.9999	0.9998	0.9996	0.9984	0.9957	0.9934	0.9866
6.40	1.0000	1.0000	1.0000	0.9999	0.9998	0.9996	0.9985	0.9959	0.9938	0.9874
6.45	1.0000	1.0000	1.0000	0.9999	0.9998	0.9996	0.9986	0.9962	0.9942	0.9881
6.50	1.0000	1.0000	1.0000	0.9999	0.9999	0.9996	0.9987	0.9964	0.9945	0.9888
6.55	1.0000	1.0000	1.0000	1.0000	0.9999	0.9997	0.9988	0.9966	0.9948	0.9895
6.60	1.0000	1.0000	1.0000	1.0000	0.9999	0.9997	0.9988	0.9968	0.9951	0.9901
6.65	1.0000	1.0000	1.0000	1.0000	0.9999	0.9997	0.9989	0.9970	0.9954	0.9906
6.70	1.0000	1.0000	1.0000	1.0000	0.9999	0.9997	0.9990	0.9972	0.9957	0.9912
6.75	1.0000	1.0000	1.0000	1.0000	0.9999	0.9997	0.9991	0.9974	0.9960	0.9917
6.80	1.0000	1.0000	1.0000	1.0000	0.9999	0.9998	0.9991	0.9975	0.9962	0.9921
6.85	1.0000	1.0000	1.0000	1.0000	0.9999	0.9998	0.9992	0.9977	0.9964	0.9926
6.90	1.0000	1.0000	1.0000	1.0000	0.9999	0.9998	0.9992	0.9978	0.9966	0.9930
6.95	1.0000	1.0000	1.0000	1.0000	0.9999	0.9998	0.9993	0.9979	0.9968	0.9934
7.00	1.0000	1.0000	1.0000	1.0000	0.9999	0.9998	0.9993	0.9981	0.9970	0.9938
7.05	1.0000	1.0000	1.0000	1.0000	0.9999	0.9998	0.9993	0.9982	0.9972	0.9941
7.10	1.0000	1.0000	1.0000	1.0000	0.9999	0.9998	0.9994	0.9983	0.9973	0.9944
7.15	1.0000	1.0000	1.0000	1.0000	0.9999	0.9999	0.9994	0.9984	0.9975	0.9947
7.20	1.0000	1.0000	1.0000	1.0000	0.9999	0.9999	0.9995	0.9985	0.9976	0.9950
7.25	1.0000	1.0000	1.0000	1.0000	0.9999	0.9999	0.9995	0.9986	0.9978	0.9953
7.30	1.0000	1.0000	1.0000	1.0000	1.0000	0.9999	0.9995	0.9986	0.9979	0.9955
7.35	1.0000	1.0000	1.0000	1.0000	1.0000	0.9999	0.9996	0.9987	0.9980	0.9958
7.40	1.0000	1.0000	1.0000	1.0000	1.0000	0.9999	0.9996	0.9988	0.9981	0.9960
7.45	1.0000	1.0000	1.0000	1.0000	1.0000	0.9999	0.9996	0.9989	0.9982	0.9962
7.50	1.0000	1.0000	1.0000	1.0000	1.0000	0.9999	0.9996	0.9989	0.9983	0.9964
7.55	1.0000	1.0000	1.0000	1.0000	1.0000	0.9999	0.9996	0.9990	0.9984	0.9966
7.60	1.0000	1.0000	1.0000	1.0000	1.0000	0.9999	0.9997	0.9990	0.9985	0.9968
7.65	1.0000	1.0000	1.0000	1.0000	1.0000	0.9999	0.9997	0.9991	0.9986	0.9970
7.70	1.0000	1.0000	1.0000	1.0000	1.0000	0.9999	0.9997	0.9991	0.9987	0.9971
7.75	1.0000	1.0000	1.0000	1.0000	1.0000	0.9999	0.9997	0.9992	0.9987	0.9973
7.80	1.0000	1.0000	1.0000	1.0000	1.0000	0.9999	0.9997	0.9992	0.9988	0.9974
7.85	1.0000	1.0000	1.0000	1.0000	1.0000	0.9999	0.9998	0.9993	0.9989	0.9976
7.90	1.0000	1.0000	1.0000	1.0000	1.0000	0.9999	0.9998	0.9993	0.9989	0.9977
7.95	1.0000	1.0000	1.0000	1.0000	1.0000	0.9999	0.9998	0.9994	0.9990	0.9978
8.00	1.0000	1.0000	1.0000	1.0000	1.0000	0.9999	0.9998	0.9994	0.9990	0.9979
8.05	1.0000	1.0000	1.0000	1.0000	1.0000	1.0000	0.9998	0.9994	0.9991	0.9980
8.10	1.0000	1.0000	1.0000	1.0000	1.0000	1.0000	0.9998	0.9995	0.9991	0.9981
8.15	1.0000	1.0000	1.0000	1.0000	1.0000	1.0000	0.9998	0.9995	0.9992	0.9982
8.20	1.0000	1.0000	1.0000	1.0000	1.0000	1.0000	0.9998	0.9995	0.9992	0.9983

PROBABILITY INTEGRAL OF t, THE NON-CENTRAL t-STATISTIC. THIS TABLE GIVES Pr $[t/\sqrt{f} \leq x]$.

f is the number of degrees of freedom; the non-centrality parameter is $\sqrt{f+1}\,K_p$.

K_p is the standardized normal deviate exceeded with probability p.

DEGREES OF FREEDOM **10**

x \ p	.2500	.1500	.1000	.0650	.0400	.0250	.0100	.0040	.0025	.0010
8.25	1.0000	1.0000	1.0000	1.0000	1.0000	1.0000	0.9998	0.9995	0.9993	0.9984
8.30	1.0000	1.0000	1.0000	1.0000	1.0000	1.0000	0.9999	0.9996	0.9993	0.9985
8.35	1.0000	1.0000	1.0000	1.0000	1.0000	1.0000	0.9999	0.9996	0.9993	0.9986
8.40	1.0000	1.0000	1.0000	1.0000	1.0000	1.0000	0.9999	0.9996	0.9994	0.9986
8.45	1.0000	1.0000	1.0000	1.0000	1.0000	1.0000	0.9999	0.9996	0.9994	0.9987
8.50	1.0000	1.0000	1.0000	1.0000	1.0000	1.0000	0.9999	0.9996	0.9994	0.9988
8.55	1.0000	1.0000	1.0000	1.0000	1.0000	1.0000	0.9999	0.9997	0.9995	0.9988
8.60	1.0000	1.0000	1.0000	1.0000	1.0000	1.0000	0.9999	0.9997	0.9995	0.9989
8.65	1.0000	1.0000	1.0000	1.0000	1.0000	1.0000	0.9999	0.9997	0.9995	0.9989
8.70	1.0000	1.0000	1.0000	1.0000	1.0000	1.0000	0.9999	0.9997	0.9995	0.9990
8.75	1.0000	1.0000	1.0000	1.0000	1.0000	1.0000	0.9999	0.9997	0.9996	0.9990
8.80	1.0000	1.0000	1.0000	1.0000	1.0000	1.0000	0.9999	0.9997	0.9996	0.9991
8.85	1.0000	1.0000	1.0000	1.0000	1.0000	1.0000	0.9999	0.9998	0.9996	0.9991
8.90	1.0000	1.0000	1.0000	1.0000	1.0000	1.0000	0.9999	0.9998	0.9996	0.9992
8.95	1.0000	1.0000	1.0000	1.0000	1.0000	1.0000	0.9999	0.9998	0.9996	0.9992
9.00	1.0000	1.0000	1.0000	1.0000	1.0000	1.0000	0.9999	0.9998	0.9997	0.9992
9.05	1.0000	1.0000	1.0000	1.0000	1.0000	1.0000	0.9999	0.9998	0.9997	0.9993
9.10	1.0000	1.0000	1.0000	1.0000	1.0000	1.0000	0.9999	0.9998	0.9997	0.9993
9.15	1.0000	1.0000	1.0000	1.0000	1.0000	1.0000	0.9999	0.9998	0.9997	0.9994
9.20	1.0000	1.0000	1.0000	1.0000	1.0000	1.0000	0.9999	0.9998	0.9997	0.9994
9.25	1.0000	1.0000	1.0000	1.0000	1.0000	1.0000	0.9999	0.9998	0.9997	0.9994
9.30	1.0000	1.0000	1.0000	1.0000	1.0000	1.0000	0.9999	0.9998	0.9997	0.9994
9.35	1.0000	1.0000	1.0000	1.0000	1.0000	1.0000	1.0000	0.9999	0.9998	0.9995
9.40	1.0000	1.0000	1.0000	1.0000	1.0000	1.0000	1.0000	0.9999	0.9998	0.9995
9.45	1.0000	1.0000	1.0000	1.0000	1.0000	1.0000	1.0000	0.9999	0.9998	0.9995
9.50	1.0000	1.0000	1.0000	1.0000	1.0000	1.0000	1.0000	0.9999	0.9998	0.9995
9.55	1.0000	1.0000	1.0000	1.0000	1.0000	1.0000	1.0000	0.9999	0.9998	0.9996
9.60	1.0000	1.0000	1.0000	1.0000	1.0000	1.0000	1.0000	0.9999	0.9998	0.9996
9.65	1.0000	1.0000	1.0000	1.0000	1.0000	1.0000	1.0000	0.9999	0.9998	0.9996
9.70	1.0000	1.0000	1.0000	1.0000	1.0000	1.0000	1.0000	0.9999	0.9998	0.9996
9.75	1.0000	1.0000	1.0000	1.0000	1.0000	1.0000	1.0000	0.9999	0.9998	0.9996
9.80	1.0000	1.0000	1.0000	1.0000	1.0000	1.0000	1.0000	0.9999	0.9998	0.9996
9.85	1.0000	1.0000	1.0000	1.0000	1.0000	1.0000	1.0000	0.9999	0.9999	0.9997
9.90	1.0000	1.0000	1.0000	1.0000	1.0000	1.0000	1.0000	0.9999	0.9999	0.9997
9.95	1.0000	1.0000	1.0000	1.0000	1.0000	1.0000	1.0000	0.9999	0.9999	0.9997
10.00	1.0000	1.0000	1.0000	1.0000	1.0000	1.0000	1.0000	0.9999	0.9999	0.9997
10.05	1.0000	1.0000	1.0000	1.0000	1.0000	1.0000	1.0000	0.9999	0.9999	0.9997
10.10	1.0000	1.0000	1.0000	1.0000	1.0000	1.0000	1.0000	0.9999	0.9999	0.9997
10.15	1.0000	1.0000	1.0000	1.0000	1.0000	1.0000	1.0000	0.9999	0.9999	0.9997
10.20	1.0000	1.0000	1.0000	1.0000	1.0000	1.0000	1.0000	0.9999	0.9999	0.9998
10.25	1.0000	1.0000	1.0000	1.0000	1.0000	1.0000	1.0000	0.9999	0.9999	0.9998
10.30	1.0000	1.0000	1.0000	1.0000	1.0000	1.0000	1.0000	0.9999	0.9999	0.9998
10.35	1.0000	1.0000	1.0000	1.0000	1.0000	1.0000	1.0000	0.9999	0.9999	0.9998
10.40	1.0000	1.0000	1.0000	1.0000	1.0000	1.0000	1.0000	0.9999	0.9999	0.9998
10.45	1.0000	1.0000	1.0000	1.0000	1.0000	1.0000	1.0000	0.9999	0.9999	0.9998
10.50	1.0000	1.0000	1.0000	1.0000	1.0000	1.0000	1.0000	1.0000	0.9999	0.9998
10.55	1.0000	1.0000	1.0000	1.0000	1.0000	1.0000	1.0000	1.0000	0.9999	0.9998
10.60	1.0000	1.0000	1.0000	1.0000	1.0000	1.0000	1.0000	1.0000	0.9999	0.9998
10.65	1.0000	1.0000	1.0000	1.0000	1.0000	1.0000	1.0000	1.0000	0.9999	0.9998
10.70	1.0000	1.0000	1.0000	1.0000	1.0000	1.0000	1.0000	1.0000	0.9999	0.9998
10.75	1.0000	1.0000	1.0000	1.0000	1.0000	1.0000	1.0000	1.0000	0.9999	0.9999
10.80	1.0000	1.0000	1.0000	1.0000	1.0000	1.0000	1.0000	1.0000	0.9999	0.9999
10.85	1.0000	1.0000	1.0000	1.0000	1.0000	1.0000	1.0000	1.0000	0.9999	0.9999
10.90	1.0000	1.0000	1.0000	1.0000	1.0000	1.0000	1.0000	1.0000	0.9999	0.9999
10.95	1.0000	1.0000	1.0000	1.0000	1.0000	1.0000	1.0000	1.0000	0.9999	0.9999
11.00	1.0000	1.0000	1.0000	1.0000	1.0000	1.0000	1.0000	1.0000	0.9999	0.9999
11.05	1.0000	1.0000	1.0000	1.0000	1.0000	1.0000	1.0000	1.0000	1.0000	0.9999
11.10	1.0000	1.0000	1.0000	1.0000	1.0000	1.0000	1.0000	1.0000	1.0000	0.9999
11.15	1.0000	1.0000	1.0000	1.0000	1.0000	1.0000	1.0000	1.0000	1.0000	0.9999
11.20	1.0000	1.0000	1.0000	1.0000	1.0000	1.0000	1.0000	1.0000	1.0000	0.9999

PROBABILITY INTEGRAL OF t, THE NON-CENTRAL t-STATISTIC. THIS TABLE GIVES Pr $[t/\sqrt{f} \leq x]$.

f is the number of degrees of freedom; the non-centrality parameter is $\sqrt{f+1}\, K_p$.

K_p is the standardized normal deviate exceeded with probability p.

DEGREES OF FREEDOM **10**

x \ p	.2500	.1500	.1000	.0650	.0400	.0250	.0100	.0040	.0025	.0010
11.25	1.0000	1.0000	1.0000	1.0000	1.0000	1.0000	1.0000	1.0000	1.0000	0.9999
11.30	1.0000	1.0000	1.0000	1.0000	1.0000	1.0000	1.0000	1.0000	1.0000	0.9999
11.35	1.0000	1.0000	1.0000	1.0000	1.0000	1.0000	1.0000	1.0000	1.0000	0.9999
11.40	1.0000	1.0000	1.0000	1.0000	1.0000	1.0000	1.0000	1.0000	1.0000	0.9999
11.45	1.0000	1.0000	1.0000	1.0000	1.0000	1.0000	1.0000	1.0000	1.0000	0.9999
11.50	1.0000	1.0000	1.0000	1.0000	1.0000	1.0000	1.0000	1.0000	1.0000	0.9999
11.55	1.0000	1.0000	1.0000	1.0000	1.0000	1.0000	1.0000	1.0000	1.0000	0.9999
11.60	1.0000	1.0000	1.0000	1.0000	1.0000	1.0000	1.0000	1.0000	1.0000	0.9999
11.65	1.0000	1.0000	1.0000	1.0000	1.0000	1.0000	1.0000	1.0000	1.0000	0.9999
11.70	1.0000	1.0000	1.0000	1.0000	1.0000	1.0000	1.0000	1.0000	1.0000	0.9999
11.75	1.0000	1.0000	1.0000	1.0000	1.0000	1.0000	1.0000	1.0000	1.0000	0.9999
11.80	1.0000	1.0000	1.0000	1.0000	1.0000	1.0000	1.0000	1.0000	1.0000	0.9999
11.85	1.0000	1.0000	1.0000	1.0000	1.0000	1.0000	1.0000	1.0000	1.0000	0.9999
11.90	1.0000	1.0000	1.0000	1.0000	1.0000	1.0000	1.0000	1.0000	1.0000	0.9999
11.95	1.0000	1.0000	1.0000	1.0000	1.0000	1.0000	1.0000	1.0000	1.0000	0.9999
12.00	1.0000	1.0000	1.0000	1.0000	1.0000	1.0000	1.0000	1.0000	1.0000	0.9999
12.05	1.0000	1.0000	1.0000	1.0000	1.0000	1.0000	1.0000	1.0000	1.0000	1.0000

PROBABILITY INTEGRAL OF t, THE NON-CENTRAL t-STATISTIC. THIS TABLE GIVES Pr $[t/\sqrt{f} \leq x]$.

f is the number of degrees of freedom; the non-centrality parameter is $\sqrt{f+1}\,K_p$.
K_p is the standardized normal deviate exceeded with probability p.

DEGREES OF FREEDOM **11**

x \ p	.2500	.1500	.1000	.0650	.0400	.0250	.0100	.0040	.0025	.0010
- 0.60	0.0000	0.0000	0.0000	0.0000	0.0000	0.0000	0.0000	0.0000	0.0000	0.0000
- 0.55	0.0001	0.0000	0.0000	0.0000	0.0000	0.0000	0.0000	0.0000	0.0000	0.0000
- 0.50	0.0001	0.0000	0.0000	0.0000	0.0000	0.0000	0.0000	0.0000	0.0000	0.0000
- 0.45	0.0001	0.0000	0.0000	0.0000	0.0000	0.0000	0.0000	0.0000	0.0000	0.0000
- 0.40	0.0002	0.0000	0.0000	0.0000	0.0000	0.0000	0.0000	0.0000	0.0000	0.0000
- 0.35	0.0004	0.0000	0.0000	0.0000	0.0000	0.0000	0.0000	0.0000	0.0000	0.0000
- 0.30	0.0006	0.0000	0.0000	0.0000	0.0000	0.0000	0.0000	0.0000	0.0000	0.0000
- 0.25	0.0010	0.0000	0.0000	0.0000	0.0000	0.0000	0.0000	0.0000	0.0000	0.0000
- 0.20	0.0016	0.0000	0.0000	0.0000	0.0000	0.0000	0.0000	0.0000	0.0000	0.0000
- 0.15	0.0025	0.0000	0.0000	0.0000	0.0000	0.0000	0.0000	0.0000	0.0000	0.0000
- 0.10	0.0040	0.0000	0.0000	0.0000	0.0000	0.0000	0.0000	0.0000	0.0000	0.0000
- 0.05	0.0063	0.0001	0.0000	0.0000	0.0000	0.0000	0.0000	0.0000	0.0000	0.0000
0.00	0.0097	0.0002	0.0000	0.0000	0.0000	0.0000	0.0000	0.0000	0.0000	0.0000
0.05	0.0149	0.0003	0.0000	0.0000	0.0000	0.0000	0.0000	0.0000	0.0000	0.0000
0.10	0.0224	0.0006	0.0000	0.0000	0.0000	0.0000	0.0000	0.0000	0.0000	0.0000
0.15	0.0329	0.0010	0.0000	0.0000	0.0000	0.0000	0.0000	0.0000	0.0000	0.0000
0.20	0.0473	0.0018	0.0001	0.0000	0.0000	0.0000	0.0000	0.0000	0.0000	0.0000
0.25	0.0664	0.0031	0.0002	0.0000	0.0000	0.0000	0.0000	0.0000	0.0000	0.0000
0.30	0.0910	0.0052	0.0004	0.0000	0.0000	0.0000	0.0000	0.0000	0.0000	0.0000
0.35	0.1216	0.0086	0.0007	0.0000	0.0000	0.0000	0.0000	0.0000	0.0000	0.0000
0.40	0.1583	0.0137	0.0013	0.0001	0.0000	0.0000	0.0000	0.0000	0.0000	0.0000
0.45	0.2012	0.0212	0.0023	0.0002	0.0000	0.0000	0.0000	0.0000	0.0000	0.0000
0.50	0.2495	0.0317	0.0040	0.0003	0.0000	0.0000	0.0000	0.0000	0.0000	0.0000
0.55	0.3023	0.0461	0.0067	0.0007	0.0000	0.0000	0.0000	0.0000	0.0000	0.0000
0.60	0.3584	0.0650	0.0110	0.0012	0.0001	0.0000	0.0000	0.0000	0.0000	0.0000
0.65	0.4164	0.0888	0.0173	0.0023	0.0002	0.0000	0.0000	0.0000	0.0000	0.0000
0.70	0.4748	0.1180	0.0262	0.0040	0.0004	0.0000	0.0000	0.0000	0.0000	0.0000
0.75	0.5321	0.1524	0.0383	0.0067	0.0007	0.0001	0.0000	0.0000	0.0000	0.0000
0.80	0.5872	0.1919	0.0543	0.0108	0.0014	0.0002	0.0000	0.0000	0.0000	0.0000
0.85	0.6391	0.2358	0.0746	0.0169	0.0025	0.0003	0.0000	0.0000	0.0000	0.0000
0.90	0.6872	0.2834	0.0995	0.0253	0.0043	0.0006	0.0000	0.0000	0.0000	0.0000
0.95	0.7309	0.3335	0.1291	0.0367	0.0070	0.0012	0.0000	0.0000	0.0000	0.0010
1.00	0.7701	0.3852	0.1632	0.0515	0.0112	0.0021	0.0001	0.0000	0.0000	0.0000
1.05	0.8048	0.4374	0.2014	0.0701	0.0171	0.0037	0.0001	0.0000	0.0000	0.0000
1.10	0.8351	0.4891	0.2431	0.0927	0.0252	0.0060	0.0003	0.0000	0.0000	0.0000
1.15	0.8614	0.5393	0.2876	0.1193	0.0359	0.0095	0.0005	0.0000	0.0000	0.0000
1.20	0.8840	0.5873	0.3342	0.1499	0.0496	0.0145	0.0009	0.0000	0.0000	0.0000
1.25	0.9032	0.6326	0.3818	0.1841	0.0666	0.0213	0.0016	0.0001	0.0000	0.0000
1.30	0.9195	0.6748	0.4297	0.2216	0.0870	0.0302	0.0028	0.0002	0.0000	0.0000
1.35	0.9331	0.7135	0.4771	0.2616	0.1110	0.0417	0.0045	0.0004	0.0001	0.0000
1.40	0.9446	0.7489	0.5233	0.3037	0.1384	0.0560	0.0071	0.0007	0.0002	0.0000
1.45	0.9542	0.7807	0.5677	0.3471	0.1689	0.0733	0.0107	0.0012	0.0003	0.0000
1.50	0.9621	0.8093	0.6100	0.3911	0.2024	0.0936	0.0156	0.0020	0.0006	0.0001
1.55	0.9687	0.8346	0.6497	0.4351	0.2383	0.1170	0.0220	0.0031	0.0011	0.0001
1.60	0.9741	0.8570	0.6867	0.4785	0.2761	0.1434	0.0304	0.0049	0.0018	0.0002
1.65	0.9786	0.8766	0.7208	0.5208	0.3154	0.1724	0.0407	0.0074	0.0028	0.0004
1.70	0.9824	0.8938	0.7521	0.5616	0.3556	0.2040	0.0533	0.0108	0.0044	0.0007
1.75	0.9854	0.9087	0.7805	0.6005	0.3961	0.2376	0.0683	0.0153	0.0066	0.0011
1.80	0.9879	0.9216	0.8062	0.6374	0.4365	0.2728	0.0858	0.0211	0.0096	0.0018
1.85	0.9900	0.9328	0.8293	0.6719	0.4764	0.3093	0.1057	0.0285	0.0135	0.0028
1.90	0.9917	0.9424	0.8500	0.7041	0.5152	0.3466	0.1280	0.0375	0.0187	0.0043
1.95	0.9931	0.9507	0.8684	0.7338	0.5528	0.3843	0.1526	0.0484	0.0251	0.0063
2.00	0.9943	0.9578	0.8847	0.7612	0.5888	0.4220	0.1792	0.0612	0.0330	0.0089
2.05	0.9953	0.9639	0.8991	0.7863	0.6231	0.4592	0.2077	0.0761	0.0426	0.0124
2.10	0.9961	0.9691	0.9118	0.8091	0.6556	0.4958	0.2378	0.0930	0.0539	0.0169
2.15	0.9967	0.9736	0.9230	0.8298	0.6860	0.5313	0.2691	0.1119	0.0671	0.0224
2.20	0.9972	0.9774	0.9328	0.8484	0.7144	0.5656	0.3015	0.1327	0.0821	0.0292

PROBABILITY INTEGRAL OF t, THE NON-CENTRAL t-STATISTIC. THIS TABLE GIVES Pr $[t/\sqrt{f} \leq x]$.

f is the number of degrees of freedom; the non-centrality parameter is $\sqrt{f+1}\,K_p$.

K_p is the standardized normal deviate exceeded with probability p.

DEGREES OF FREEDOM **11**

x \ p	.2500	.1500	.1000	.0650	.0400	.0250	.0100	.0040	.0025	.0010
2.25	0.9977	0.9806	0.9414	0.8652	0.7408	0.5985	0.3345	0.1554	0.0989	0.0374
2.30	0.9981	0.9834	0.9489	0.8803	0.7653	0.6298	0.3678	0.1798	0.1176	0.0470
2.35	0.9984	0.9858	0.9555	0.8938	0.7878	0.6595	0.4012	0.2056	0.1381	0.0582
2.40	0.9986	0.9878	0.9612	0.9059	0.8084	0.6874	0.4344	0.2328	0.1602	0.0709
2.45	0.9989	0.9896	0.9662	0.9166	0.8273	0.7137	0.4672	0.2610	0.1838	0.0852
2.50	0.9990	0.9910	0.9705	0.9262	0.8445	0.7382	0.4994	0.2900	0.2087	0.1012
2.55	0.9992	0.9923	0.9743	0.9347	0.8602	0.7610	0.5307	0.3197	0.2348	0.1187
2.60	0.9993	0.9934	0.9776	0.9422	0.8744	0.7821	0.5610	0.3498	0.2618	0.1377
2.65	0.9994	0.9943	0.9805	0.9489	0.8872	0.8017	0.5903	0.3800	0.2896	0.1581
2.70	0.9995	0.9951	0.9829	0.9548	0.8988	0.8197	0.6183	0.4102	0.3180	0.1798
2.75	0.9996	0.9958	0.9851	0.9600	0.9093	0.8362	0.6450	0.4402	0.3467	0.2027
2.80	0.9996	0.9963	0.9870	0.9647	0.9187	0.8514	0.6705	0.4697	0.3756	0.2265
2.85	0.9997	0.9968	0.9886	0.9687	0.9272	0.8652	0.6945	0.4987	0.4045	0.2513
2.90	0.9997	0.9973	0.9901	0.9723	0.9348	0.8779	0.7173	0.5270	0.4332	0.2768
2.95	0.9998	0.9976	0.9913	0.9755	0.9416	0.8895	0.7386	0.5546	0.4615	0.3028
3.00	0.9998	0.9980	0.9924	0.9783	0.9477	0.9000	0.7587	0.5812	0.4894	0.3291
3.05	0.9998	0.9982	0.9933	0.9808	0.9532	0.9095	0.7775	0.6068	0.5167	0.3558
3.10	0.9999	0.9985	0.9942	0.9830	0.9581	0.9182	0.7950	0.6315	0.5432	0.3825
3.15	0.9999	0.9987	0.9949	0.9849	0.9625	0.9260	0.8113	0.6550	0.5690	0.4091
3.20	0.9999	0.9988	0.9955	0.9866	0.9664	0.9332	0.8265	0.6775	0.5940	0.4356
3.25	0.9999	0.9990	0.9961	0.9881	0.9699	0.9396	0.8405	0.6989	0.6180	0.4617
3.30	0.9999	0.9991	0.9965	0.9895	0.9731	0.9454	0.8535	0.7192	0.6411	0.4875
3.35	0.9999	0.9992	0.9970	0.9907	0.9759	0.9507	0.8656	0.7384	0.6632	0.5127
3.40	0.9999	0.9993	0.9973	0.9917	0.9784	0.9555	0.8767	0.7565	0.6843	0.5374
3.45	0.9999	0.9994	0.9976	0.9926	0.9806	0.9598	0.8869	0.7735	0.7044	0.5614
3.50	1.0000	0.9995	0.9979	0.9934	0.9826	0.9636	0.8963	0.7896	0.7234	0.5846
3.55	1.0000	0.9996	0.9982	0.9941	0.9844	0.9671	0.9050	0.8046	0.7415	0.6071
3.60	1.0000	0.9996	0.9984	0.9948	0.9860	0.9703	0.9130	0.8187	0.7587	0.6289
3.65	1.0000	0.9997	0.9986	0.9953	0.9874	0.9732	0.9203	0.8319	0.7748	0.6497
3.70	1.0000	0.9997	0.9988	0.9958	0.9887	0.9757	0.9270	0.8442	0.7900	0.6698
3.75	1.0000	0.9998	0.9989	0.9963	0.9898	0.9780	0.9332	0.8557	0.8044	0.6890
3.80	1.0000	0.9998	0.9990	0.9967	0.9909	0.9801	0.9389	0.8663	0.8178	0.7073
3.85	1.0000	0.9998	0.9991	0.9970	0.9918	0.9820	0.9440	0.8763	0.8305	0.7248
3.90	1.0000	0.9998	0.9992	0.9974	0.9926	0.9837	0.9488	0.8855	0.8423	0.7414
3.95	1.0000	0.9999	0.9993	0.9976	0.9933	0.9853	0.9531	0.8941	0.8534	0.7572
4.00	1.0000	0.9999	0.9994	0.9979	0.9940	0.9867	0.9571	0.9021	0.8637	0.7722
4.05	1.0000	0.9999	0.9995	0.9981	0.9946	0.9879	0.9607	0.9095	0.8734	0.7864
4.10	1.0000	0.9999	0.9995	0.9983	0.9951	0.9890	0.9640	0.9164	0.8824	0.7998
4.15	1.0000	0.9999	0.9996	0.9985	0.9956	0.9901	0.9671	0.9227	0.8908	0.8125
4.20	1.0000	0.9999	0.9997	0.9986	0.9960	0.9910	0.9698	0.9286	0.8987	0.8244
4.25	1.0000	1.0000	0.9997	0.9988	0.9964	0.9918	0.9724	0.9340	0.9060	0.8357
4.30	1.0000	1.0000	0.9997	0.9989	0.9967	0.9926	0.9747	0.9390	0.9127	0.8463
4.35	1.0000	1.0000	0.9998	0.9990	0.9971	0.9932	0.9768	0.9437	0.9191	0.8563
4.40	1.0000	1.0000	0.9998	0.9991	0.9973	0.9939	0.9787	0.9480	0.9249	0.8657
4.45	1.0000	1.0000	0.9998	0.9992	0.9976	0.9944	0.9805	0.9519	0.9304	0.8745
4.50	1.0000	1.0000	0.9999	0.9993	0.9978	0.9949	0.9821	0.9556	0.9354	0.8827
4.55	1.0000	1.0000	0.9999	0.9993	0.9980	0.9954	0.9836	0.9590	0.9401	0.8905
4.60	1.0000	1.0000	0.9999	0.9994	0.9982	0.9958	0.9850	0.9621	0.9444	0.8977
4.65	1.0000	1.0000	0.9999	0.9995	0.9984	0.9962	0.9862	0.9650	0.9485	0.9045
4.70	1.0000	1.0000	0.9999	0.9995	0.9985	0.9965	0.9873	0.9676	0.9522	0.9108
4.75	1.0000	1.0000	0.9999	0.9996	0.9987	0.9968	0.9884	0.9701	0.9557	0.9168
4.80	1.0000	1.0000	1.0000	0.9996	0.9988	0.9971	0.9893	0.9723	0.9589	0.9223
4.85	1.0000	1.0000	1.0000	0.9996	0.9989	0.9973	0.9902	0.9744	0.9619	0.9275
4.90	1.0000	1.0000	1.0000	0.9997	0.9990	0.9976	0.9910	0.9763	0.9646	0.9324
4.95	1.0000	1.0000	1.0000	0.9997	0.9991	0.9978	0.9917	0.9781	0.9672	0.9369
5.00	1.0000	1.0000	1.0000	0.9997	0.9992	0.9980	0.9924	0.9797	0.9696	0.9411
5.05	1.0000	1.0000	1.0000	0.9997	0.9992	0.9981	0.9930	0.9813	0.9718	0.9450
5.10	1.0000	1.0000	1.0000	0.9998	0.9993	0.9983	0.9935	0.9827	0.9738	0.9487
5.15	1.0000	1.0000	1.0000	0.9998	0.9994	0.9984	0.9941	0.9839	0.9757	0.9522
5.20	1.0000	1.0000	1.0000	0.9998	0.9994	0.9986	0.9945	0.9851	0.9774	0.9554

PROBABILITY INTEGRAL OF t, THE NON-CENTRAL t-STATISTIC. THIS TABLE GIVES Pr $[t/\sqrt{f} \leq x]$.

f is the number of degrees of freedom; the non-centrality parameter is $\sqrt{f+1}\,K_p$.

K_p is the standardized normal deviate exceeded with probability p.

DEGREES OF FREEDOM **11**

x \ p	.2500	.1500	.1000	.0650	.0400	.0250	.0100	.0040	.0025	.0010
5.25	1.0000	1.0000	1.0000	0.9998	0.9995	0.9987	0.9950	0.9862	0.9790	0.9583
5.30	1.0000	1.0000	1.0000	0.9998	0.9995	0.9988	0.9953	0.9872	0.9805	0.9611
5.35	1.0000	1.0000	1.0000	0.9999	0.9996	0.9989	0.9957	0.9882	0.9819	0.9637
5.40	1.0000	1.0000	1.0000	0.9999	0.9996	0.9990	0.9960	0.9890	0.9832	0.9661
5.45	1.0000	1.0000	1.0000	0.9999	0.9996	0.9991	0.9963	0.9898	0.9844	0.9684
5.50	1.0000	1.0000	1.0000	0.9999	0.9997	0.9991	0.9966	0.9906	0.9855	0.9705
5.55	1.0000	1.0000	1.0000	0.9999	0.9997	0.9992	0.9969	0.9913	0.9865	0.9724
5.60	1.0000	1.0000	1.0000	0.9999	0.9997	0.9993	0.9971	0.9919	0.9874	0.9743
5.65	1.0000	1.0000	1.0000	0.9999	0.9997	0.9993	0.9973	0.9925	0.9883	0.9760
5.70	1.0000	1.0000	1.0000	0.9999	0.9998	0.9994	0.9975	0.9930	0.9891	0.9775
5.75	1.0000	1.0000	1.0000	0.9999	0.9998	0.9994	0.9977	0.9935	0.9899	0.9790
5.80	1.0000	1.0000	1.0000	0.9999	0.9998	0.9995	0.9979	0.9940	0.9906	0.9804
5.85	1.0000	1.0000	1.0000	0.9999	0.9998	0.9995	0.9981	0.9944	0.9912	0.9817
5.90	1.0000	1.0000	1.0000	0.9999	0.9998	0.9996	0.9982	0.9948	0.9918	0.9829
5.95	1.0000	1.0000	1.0000	1.0000	0.9998	0.9996	0.9983	0.9952	0.9924	0.9840
6.00	1.0000	1.0000	1.0000	1.0000	0.9999	0.9996	0.9985	0.9955	0.9929	0.9850
6.05	1.0000	1.0000	1.0000	1.0000	0.9999	0.9997	0.9986	0.9958	0.9934	0.9860
6.10	1.0000	1.0000	1.0000	1.0000	0.9999	0.9997	0.9987	0.9961	0.9939	0.9869
6.15	1.0000	1.0000	1.0000	1.0000	0.9999	0.9997	0.9988	0.9964	0.9943	0.9878
6.20	1.0000	1.0000	1.0000	1.0000	0.9999	0.9997	0.9989	0.9966	0.9947	0.9885
6.25	1.0000	1.0000	1.0000	1.0000	0.9999	0.9997	0.9989	0.9969	0.9950	0.9893
6.30	1.0000	1.0000	1.0000	1.0000	0.9999	0.9998	0.9990	0.9971	0.9953	0.9900
6.35	1.0000	1.0000	1.0000	1.0000	0.9999	0.9998	0.9991	0.9973	0.9957	0.9906
6.40	1.0000	1.0000	1.0000	1.0000	0.9999	0.9998	0.9991	0.9975	0.9959	0.9912
6.45	1.0000	1.0000	1.0000	1.0000	0.9999	0.9998	0.9992	0.9976	0.9962	0.9918
6.50	1.0000	1.0000	1.0000	1.0000	0.9999	0.9998	0.9993	0.9978	0.9965	0.9923
6.55	1.0000	1.0000	1.0000	1.0000	0.9999	0.9998	0.9993	0.9979	0.9967	0.9928
6.60	1.0000	1.0000	1.0000	1.0000	0.9999	0.9999	0.9994	0.9981	0.9969	0.9932
6.65	1.0000	1.0000	1.0000	1.0000	0.9999	0.9999	0.9994	0.9982	0.9971	0.9936
6.70	1.0000	1.0000	1.0000	1.0000	1.0000	0.9999	0.9994	0.9983	0.9973	0.9940
6.75	1.0000	1.0000	1.0000	1.0000	1.0000	0.9999	0.9995	0.9984	0.9975	0.9944
6.80	1.0000	1.0000	1.0000	1.0000	1.0000	0.9999	0.9995	0.9985	0.9976	0.9948
6.85	1.0000	1.0000	1.0000	1.0000	1.0000	0.9999	0.9996	0.9986	0.9978	0.9951
6.90	1.0000	1.0000	1.0000	1.0000	1.0000	0.9999	0.9996	0.9987	0.9979	0.9954
6.95	1.0000	1.0000	1.0000	1.0000	1.0000	0.9999	0.9996	0.9988	0.9981	0.9957
7.00	1.0000	1.0000	1.0000	1.0000	1.0000	0.9999	0.9996	0.9989	0.9982	0.9959
7.05	1.0000	1.0000	1.0000	1.0000	1.0000	0.9999	0.9997	0.9989	0.9983	0.9962
7.10	1.0000	1.0000	1.0000	1.0000	1.0000	0.9999	0.9997	0.9990	0.9984	0.9964
7.15	1.0000	1.0000	1.0000	1.0000	1.0000	0.9999	0.9997	0.9991	0.9985	0.9966
7.20	1.0000	1.0000	1.0000	1.0000	1.0000	0.9999	0.9997	0.9991	0.9986	0.9968
7.25	1.0000	1.0000	1.0000	1.0000	1.0000	0.9999	0.9997	0.9992	0.9987	0.9970
7.30	1.0000	1.0000	1.0000	1.0000	1.0000	0.9999	0.9998	0.9992	0.9988	0.9972
7.35	1.0000	1.0000	1.0000	1.0000	1.0000	1.0000	0.9998	0.9993	0.9988	0.9974
7.40	1.0000	1.0000	1.0000	1.0000	1.0000	1.0000	0.9998	0.9993	0.9989	0.9975
7.45	1.0000	1.0000	1.0000	1.0000	1.0000	1.0000	0.9998	0.9994	0.9990	0.9977
7.50	1.0000	1.0000	1.0000	1.0000	1.0000	1.0000	0.9998	0.9994	0.9990	0.9978
7.55	1.0000	1.0000	1.0000	1.0000	1.0000	1.0000	0.9998	0.9994	0.9991	0.9979
7.60	1.0000	1.0000	1.0000	1.0000	1.0000	1.0000	0.9998	0.9995	0.9991	0.9980
7.65	1.0000	1.0000	1.0000	1.0000	1.0000	1.0000	0.9998	0.9995	0.9992	0.9982
7.70	1.0000	1.0000	1.0000	1.0000	1.0000	1.0000	0.9999	0.9995	0.9992	0.9983
7.75	1.0000	1.0000	1.0000	1.0000	1.0000	1.0000	0.9999	0.9996	0.9993	0.9984
7.80	1.0000	1.0000	1.0000	1.0000	1.0000	1.0000	0.9999	0.9996	0.9993	0.9985
7.85	1.0000	1.0000	1.0000	1.0000	1.0000	1.0000	0.9999	0.9996	0.9994	0.9986
7.90	1.0000	1.0000	1.0000	1.0000	1.0000	1.0000	0.9999	0.9996	0.9994	0.9986
7.95	1.0000	1.0000	1.0000	1.0000	1.0000	1.0000	0.9999	0.9997	0.9994	0.9987
8.00	1.0000	1.0000	1.0000	1.0000	1.0000	1.0000	0.9999	0.9997	0.9995	0.9988
8.05	1.0000	1.0000	1.0000	1.0000	1.0000	1.0000	0.9999	0.9997	0.9995	0.9989
8.10	1.0000	1.0000	1.0000	1.0000	1.0000	1.0000	0.9999	0.9997	0.9995	0.9989
8.15	1.0000	1.0000	1.0000	1.0000	1.0000	1.0000	0.9999	0.9997	0.9996	0.9990
8.20	1.0000	1.0000	1.0000	1.0000	1.0000	1.0000	0.9999	0.9998	0.9996	0.9990

PROBABILITY INTEGRAL OF t, THE NON-CENTRAL t-STATISTIC. THIS TABLE GIVES Pr $[t/\sqrt{f} \leq x]$.

f is the number of degrees of freedom; the non-centrality parameter is $\sqrt{f+1}\,K_p$.
K_p is the standardized normal deviate exceeded with probability p.

DEGREES OF FREEDOM **11**

x \ p	.2500	.1500	.1000	.0650	.0400	.0250	.0100	.0040	.0025	.0010
8.25	1.0000	1.0000	1.0000	1.0000	1.0000	1.0000	0.9999	0.9998	0.9996	0.9991
8.30	1.0000	1.0000	1.0000	1.0000	1.0000	1.0000	0.9999	0.9998	0.9996	0.9991
8.35	1.0000	1.0000	1.0000	1.0000	1.0000	1.0000	0.9999	0.9998	0.9997	0.9992
8.40	1.0000	1.0000	1.0000	1.0000	1.0000	1.0000	0.9999	0.9998	0.9997	0.9992
8.45	1.0000	1.0000	1.0000	1.0000	1.0000	1.0000	0.9999	0.9998	0.9997	0.9993
8.50	1.0000	1.0000	1.0000	1.0000	1.0000	1.0000	0.9999	0.9998	0.9997	0.9993
8.55	1.0000	1.0000	1.0000	1.0000	1.0000	1.0000	1.0000	0.9998	0.9997	0.9993
8.60	1.0000	1.0000	1.0000	1.0000	1.0000	1.0000	1.0000	0.9998	0.9997	0.9994
8.65	1.0000	1.0000	1.0000	1.0000	1.0000	1.0000	1.0000	0.9999	0.9998	0.9994
8.70	1.0000	1.0000	1.0000	1.0000	1.0000	1.0000	1.0000	0.9999	0.9998	0.9994
8.75	1.0000	1.0000	1.0000	1.0000	1.0000	1.0000	1.0000	0.9999	0.9998	0.9995
8.80	1.0000	1.0000	1.0000	1.0000	1.0000	1.0000	1.0000	0.9999	0.9998	0.9995
8.85	1.0000	1.0000	1.0000	1.0000	1.0000	1.0000	1.0000	0.9999	0.9998	0.9995
8.90	1.0000	1.0000	1.0000	1.0000	1.0000	1.0000	1.0000	0.9999	0.9998	0.9996
8.95	1.0000	1.0000	1.0000	1.0000	1.0000	1.0000	1.0000	0.9999	0.9998	0.9996
9.00	1.0000	1.0000	1.0000	1.0000	1.0000	1.0000	1.0000	0.9999	0.9998	0.9996
9.05	1.0000	1.0000	1.0000	1.0000	1.0000	1.0000	1.0000	0.9999	0.9998	0.9996
9.10	1.0000	1.0000	1.0000	1.0000	1.0000	1.0000	1.0000	0.9999	0.9999	0.9996
9.15	1.0000	1.0000	1.0000	1.0000	1.0000	1.0000	1.0000	0.9999	0.9999	0.9997
9.20	1.0000	1.0000	1.0000	1.0000	1.0000	1.0000	1.0000	0.9999	0.9999	0.9997
9.25	1.0000	1.0000	1.0000	1.0000	1.0000	1.0000	1.0000	0.9999	0.9999	0.9997
9.30	1.0000	1.0000	1.0000	1.0000	1.0000	1.0000	1.0000	0.9999	0.9999	0.9997
9.35	1.0000	1.0000	1.0000	1.0000	1.0000	1.0000	1.0000	0.9999	0.9999	0.9997
9.40	1.0000	1.0000	1.0000	1.0000	1.0000	1.0000	1.0000	0.9999	0.9999	0.9997
9.45	1.0000	1.0000	1.0000	1.0000	1.0000	1.0000	1.0000	0.9999	0.9999	0.9998
9.50	1.0000	1.0000	1.0000	1.0000	1.0000	1.0000	1.0000	0.9999	0.9999	0.9998
9.55	1.0000	1.0000	1.0000	1.0000	1.0000	1.0000	1.0000	0.9999	0.9999	0.9998
9.60	1.0000	1.0000	1.0000	1.0000	1.0000	1.0000	1.0000	1.0000	0.9999	0.9998
9.65	1.0000	1.0000	1.0000	1.0000	1.0000	1.0000	1.0000	1.0000	0.9999	0.9998
9.70	1.0000	1.0000	1.0000	1.0000	1.0000	1.0000	1.0000	1.0000	0.9999	0.9998
9.75	1.0000	1.0000	1.0000	1.0000	1.0000	1.0000	1.0000	1.0000	0.9999	0.9998
9.80	1.0000	1.0000	1.0000	1.0000	1.0000	1.0000	1.0000	1.0000	0.9999	0.9998
9.85	1.0000	1.0000	1.0000	1.0000	1.0000	1.0000	1.0000	1.0000	0.9999	0.9998
9.90	1.0000	1.0000	1.0000	1.0000	1.0000	1.0000	1.0000	1.0000	0.9999	0.9998
9.95	1.0000	1.0000	1.0000	1.0000	1.0000	1.0000	1.0000	1.0000	0.9999	0.9999
10.00	1.0000	1.0000	1.0000	1.0000	1.0000	1.0000	1.0000	1.0000	0.9999	0.9999
10.05	1.0000	1.0000	1.0000	1.0000	1.0000	1.0000	1.0000	1.0000	0.9999	0.9999
10.10	1.0000	1.0000	1.0000	1.0000	1.0000	1.0000	1.0000	1.0000	0.9999	0.9999
10.15	1.0000	1.0000	1.0000	1.0000	1.0000	1.0000	1.0000	1.0000	1.0000	0.9999
10.20	1.0000	1.0000	1.0000	1.0000	1.0000	1.0000	1.0000	1.0000	1.0000	0.9999
10.25	1.0000	1.0000	1.0000	1.0000	1.0000	1.0000	1.0000	1.0000	1.0000	0.9999
10.30	1.0000	1.0000	1.0000	1.0000	1.0000	1.0000	1.0000	1.0000	1.0000	0.9999
10.35	1.0000	1.0000	1.0000	1.0000	1.0000	1.0000	1.0000	1.0000	1.0000	0.9999
10.40	1.0000	1.0000	1.0000	1.0000	1.0000	1.0000	1.0000	1.0000	1.0000	0.9999
10.45	1.0000	1.0000	1.0000	1.0000	1.0000	1.0000	1.0000	1.0000	1.0000	0.9999
10.50	1.0000	1.0000	1.0000	1.0000	1.0000	1.0000	1.0000	1.0000	1.0000	0.9999
10.55	1.0000	1.0000	1.0000	1.0000	1.0000	1.0000	1.0000	1.0000	1.0000	0.9999
10.60	1.0000	1.0000	1.0000	1.0000	1.0000	1.0000	1.0000	1.0000	1.0000	0.9999
10.65	1.0000	1.0000	1.0000	1.0000	1.0000	1.0000	1.0000	1.0000	1.0000	0.9999
10.70	1.0000	1.0000	1.0000	1.0000	1.0000	1.0000	1.0000	1.0000	1.0000	0.9999
10.75	1.0000	1.0000	1.0000	1.0000	1.0000	1.0000	1.0000	1.0000	1.0000	0.9999
10.80	1.0000	1.0000	1.0000	1.0000	1.0000	1.0000	1.0000	1.0000	1.0000	0.9999
10.85	1.0000	1.0000	1.0000	1.0000	1.0000	1.0000	1.0000	1.0000	1.0000	0.9999
10.90	1.0000	1.0000	1.0000	1.0000	1.0000	1.0000	1.0000	1.0000	1.0000	0.9999
10.95	1.0000	1.0000	1.0000	1.0000	1.0000	1.0000	1.0000	1.0000	1.0000	0.9999
11.00	1.0000	1.0000	1.0000	1.0000	1.0000	1.0000	1.0000	1.0000	1.0000	0.9999
11.05	1.0000	1.0000	1.0000	1.0000	1.0000	1.0000	1.0000	1.0000	1.0000	0.9999
11.10	1.0000	1.0000	1.0000	1.0000	1.0000	1.0000	1.0000	1.0000	1.0000	1.0000

PROBABILITY INTEGRAL OF t, THE NON-CENTRAL t-STATISTIC. THIS TABLE GIVES Pr $[t/\sqrt{f} \leq x]$.

f is the number of degrees of freedom; the non-centrality parameter is $\sqrt{f+1}\,K_p$.
K_p is the standardized normal deviate exceeded with probability p.

DEGREES OF FREEDOM **12**

x \ p	.2500	.1500	.1000	.0650	.0400	.0250	.0100	.0040	.0025	.0010
−0.50	0.0000	0.0000	0.0000	0.0000	0.0000	0.0000	0.0000	0.0000	0.0000	0.0000
−0.45	0.0001	0.0000	0.0000	0.0000	0.0000	0.0000	0.0000	0.0000	0.0000	0.0000
−0.40	0.0001	0.0000	0.0000	0.0000	0.0000	0.0000	0.0000	0.0000	0.0000	0.0000
−0.35	0.0002	0.0000	0.0000	0.0000	0.0000	0.0000	0.0000	0.0000	0.0000	0.0000
−0.30	0.0004	0.0000	0.0000	0.0000	0.0000	0.0000	0.0000	0.0000	0.0000	0.0000
−0.25	0.0006	0.0000	0.0000	0.0000	0.0000	0.0000	0.0000	0.0000	0.0000	0.0000
−0.20	0.0010	0.0000	0.0000	0.0000	0.0000	0.0000	0.0000	0.0000	0.0000	0.0000
−0.15	0.0017	0.0000	0.0000	0.0000	0.0000	0.0000	0.0000	0.0000	0.0000	0.0000
−0.10	0.0029	0.0000	0.0000	0.0000	0.0000	0.0000	0.0000	0.0000	0.0000	0.0000
−0.05	0.0047	0.0000	0.0000	0.0000	0.0000	0.0000	0.0000	0.0000	0.0000	0.0000
0.00	0.0075	0.0001	0.0000	0.0000	0.0000	0.0000	0.0000	0.0000	0.0000	0.0000
0.05	0.0119	0.0002	0.0000	0.0000	0.0000	0.0000	0.0000	0.0000	0.0000	0.0000
0.10	0.0184	0.0004	0.0000	0.0000	0.0000	0.0000	0.0000	0.0000	0.0000	0.0000
0.15	0.0279	0.0007	0.0000	0.0000	0.0000	0.0000	0.0000	0.0000	0.0000	0.0000
0.20	0.0413	0.0012	0.0000	0.0000	0.0000	0.0000	0.0000	0.0000	0.0000	0.0000
0.25	0.0594	0.0022	0.0001	0.0000	0.0000	0.0000	0.0000	0.0000	0.0000	0.0000
0.30	0.0832	0.0039	0.0002	0.0000	0.0000	0.0000	0.0000	0.0000	0.0000	0.0000
0.35	0.1134	0.0067	0.0004	0.0000	0.0000	0.0000	0.0000	0.0000	0.0000	0.0000
0.40	0.1503	0.0110	0.0009	0.0000	0.0000	0.0000	0.0000	0.0000	0.0000	0.0000
0.45	0.1939	0.0177	0.0016	0.0001	0.0000	0.0000	0.0000	0.0000	0.0000	0.0000
0.50	0.2435	0.0273	0.0030	0.0002	0.0000	0.0000	0.0000	0.0000	0.0000	0.0000
0.55	0.2983	0.0407	0.0052	0.0004	0.0000	0.0000	0.0000	0.0000	0.0000	0.0000
0.60	0.3567	0.0587	0.0088	0.0009	0.0000	0.0000	0.0000	0.0000	0.0000	0.0000
0.65	0.4173	0.0820	0.0143	0.0016	0.0001	0.0000	0.0000	0.0000	0.0000	0.0000
0.70	0.4784	0.1109	0.0223	0.0030	0.0002	0.0000	0.0000	0.0000	0.0000	0.0000
0.75	0.5383	0.1456	0.0336	0.0052	0.0005	0.0000	0.0000	0.0000	0.0000	0.0000
0.80	0.5958	0.1858	0.0488	0.0087	0.0010	0.0001	0.0000	0.0000	0.0000	0.0000
0.85	0.6496	0.2310	0.0685	0.0141	0.0018	0.0002	0.0000	0.0000	0.0000	0.0000
0.90	0.6992	0.2803	0.0930	0.0217	0.0032	0.0004	0.0000	0.0000	0.0000	0.0000
0.95	0.7439	0.3326	0.1227	0.0323	0.0055	0.0008	0.0000	0.0000	0.0000	0.0000
1.00	0.7837	0.3866	0.1572	0.0464	0.0091	0.0015	0.0000	0.0000	0.0000	0.0000
1.05	0.8185	0.4412	0.1963	0.0644	0.0143	0.0028	0.0001	0.0000	0.0000	0.0000
1.10	0.8487	0.4952	0.2394	0.0866	0.0217	0.0047	0.0002	0.0000	0.0000	0.0000
1.15	0.8745	0.5476	0.2856	0.1133	0.0317	0.0077	0.0003	0.0000	0.0000	0.0000
1.20	0.8964	0.5976	0.3341	0.1442	0.0447	0.0120	0.0006	0.0000	0.0000	0.0000
1.25	0.9148	0.6445	0.3839	0.1792	0.0612	0.0182	0.0012	0.0001	0.0000	0.0000
1.30	0.9302	0.6879	0.4341	0.2177	0.0813	0.0265	0.0021	0.0001	0.0000	0.0000
1.35	0.9429	0.7276	0.4836	0.2593	0.1052	0.0373	0.0035	0.0002	0.0001	0.0000
1.40	0.9534	0.7635	0.5319	0.3031	0.1329	0.0510	0.0056	0.0004	0.0001	0.0000
1.45	0.9621	0.7956	0.5781	0.3484	0.1641	0.0679	0.0087	0.0008	0.0002	0.0000
1.50	0.9691	0.8240	0.6220	0.3944	0.1984	0.0881	0.0131	0.0014	0.0004	0.0000
1.55	0.9749	0.8491	0.6629	0.4405	0.2356	0.1115	0.0190	0.0024	0.0007	0.0001
1.60	0.9796	0.8709	0.7009	0.4859	0.2749	0.1382	0.0267	0.0038	0.0013	0.0001
1.65	0.9834	0.8900	0.7357	0.5300	0.3159	0.1679	0.0365	0.0059	0.0021	0.0003
1.70	0.9865	0.9064	0.7674	0.5725	0.3579	0.2004	0.0486	0.0089	0.0034	0.0005
1.75	0.9890	0.9205	0.7959	0.6129	0.4003	0.2352	0.0633	0.0129	0.0052	0.0008
1.80	0.9911	0.9326	0.8215	0.6509	0.4426	0.2719	0.0805	0.0182	0.0078	0.0013
1.85	0.9927	0.9429	0.8443	0.6864	0.4842	0.3099	0.1005	0.0250	0.0113	0.0021
1.90	0.9941	0.9517	0.8645	0.7192	0.5248	0.3489	0.1230	0.0335	0.0159	0.0033
1.95	0.9952	0.9592	0.8824	0.7495	0.5639	0.3884	0.1480	0.0440	0.0219	0.0050
2.00	0.9961	0.9655	0.8980	0.7771	0.6013	0.4278	0.1753	0.0564	0.0293	0.0073
2.05	0.9968	0.9709	0.9117	0.8021	0.6367	0.4667	0.2047	0.0711	0.0384	0.0104
2.10	0.9974	0.9754	0.9237	0.8248	0.6701	0.5049	0.2359	0.0879	0.0493	0.0144
2.15	0.9978	0.9792	0.9341	0.8452	0.7013	0.5419	0.2685	0.1069	0.0622	0.0194
2.20	0.9982	0.9825	0.9431	0.8634	0.7302	0.5776	0.3022	0.1280	0.0770	0.0258
2.25	0.9985	0.9852	0.9510	0.8797	0.7569	0.6116	0.3367	0.1511	0.0939	0.0335
2.30	0.9988	0.9875	0.9577	0.8942	0.7815	0.6439	0.3716	0.1761	0.1127	0.0427
2.35	0.9990	0.9894	0.9636	0.9071	0.8040	0.6744	0.4066	0.2028	0.1335	0.0535
2.40	0.9992	0.9910	0.9686	0.9184	0.8245	0.7030	0.4414	0.2310	0.1561	0.0661
2.45	0.9993	0.9924	0.9730	0.9285	0.8431	0.7297	0.4756	0.2603	0.1804	0.0803

PROBABILITY INTEGRAL OF t, THE NON-CENTRAL t-STATISTIC. THIS TABLE GIVES Pr $[t/\sqrt{f} \leq x]$.

f is the number of degrees of freedom; the non-centrality parameter is $\sqrt{f+1}\,K_p$.
K_p is the standardized normal deviate exceeded with probability p.

DEGREES OF FREEDOM **12**

x \ p	.2500	.1500	.1000	.0650	.0400	.0250	.0100	.0040	.0025	.0010
2.50	0.9994	0.9936	0.9767	0.9373	0.8599	0.7545	0.5092	0.2906	0.2061	0.0963
2.55	0.9995	0.9945	0.9799	0.9451	0.8751	0.7775	0.5418	0.3216	0.2332	0.1139
2.60	0.9996	0.9954	0.9827	0.9519	0.8887	0.7986	0.5733	0.3531	0.2613	0.1333
2.65	0.9997	0.9960	0.9851	0.9579	0.9010	0.8180	0.6036	0.3847	0.2903	0.1541
2.70	0.9997	0.9966	0.9871	0.9632	0.9120	0.8358	0.6326	0.4163	0.3200	0.1764
2.75	0.9998	0.9971	0.9889	0.9678	0.9218	0.8520	0.6601	0.4477	0.3500	0.2000
2.80	0.9998	0.9976	0.9904	0.9718	0.9305	0.8667	0.6861	0.4786	0.3803	0.2248
2.85	0.9998	0.9979	0.9917	0.9753	0.9383	0.8801	0.7107	0.5089	0.4105	0.2505
2.90	0.9999	0.9982	0.9928	0.9784	0.9453	0.8922	0.7338	0.5384	0.4405	0.2771
2.95	0.9999	0.9985	0.9938	0.9811	0.9515	0.9032	0.7554	0.5670	0.4701	0.3042
3.00	0.9999	0.9987	0.9946	0.9834	0.9569	0.9131	0.7755	0.5946	0.4992	0.3318
3.05	0.9999	0.9989	0.9953	0.9855	0.9618	0.9221	0.7943	0.6211	0.5277	0.3597
3.10	0.9999	0.9990	0.9959	0.9873	0.9661	0.9301	0.8117	0.6464	0.5553	0.3876
3.15	0.9999	0.9992	0.9965	0.9888	0.9700	0.9374	0.8278	0.6706	0.5821	0.4155
3.20	0.9999	0.9993	0.9969	0.9902	0.9734	0.9439	0.8427	0.6936	0.6079	0.4432
3.25	1.0000	0.9994	0.9973	0.9914	0.9764	0.9497	0.8564	0.7154	0.6327	0.4706
3.30	1.0000	0.9995	0.9977	0.9924	0.9790	0.9550	0.8690	0.7360	0.6564	0.4975
3.35	1.0000	0.9995	0.9980	0.9934	0.9814	0.9597	0.8806	0.7553	0.6791	0.5238
3.40	1.0000	0.9996	0.9982	0.9942	0.9835	0.9639	0.8913	0.7735	0.7006	0.5494
3.45	1.0000	0.9997	0.9985	0.9949	0.9853	0.9677	0.9010	0.7906	0.7210	0.5743
3.50	1.0000	0.9997	0.9987	0.9955	0.9870	0.9710	0.9099	0.8065	0.7404	0.5984
3.55	1.0000	0.9997	0.9988	0.9960	0.9884	0.9740	0.9181	0.8214	0.7586	0.6217
3.60	1.0000	0.9998	0.9990	0.9965	0.9897	0.9767	0.9255	0.8353	0.7758	0.6441
3.65	1.0000	0.9998	0.9991	0.9969	0.9908	0.9792	0.9323	0.8482	0.7919	0.6655
3.70	1.0000	0.9998	0.9992	0.9973	0.9918	0.9813	0.9384	0.8601	0.8071	0.6860
3.75	1.0000	0.9998	0.9993	0.9976	0.9927	0.9832	0.9441	0.8712	0.8212	0.7056
3.80	1.0000	0.9999	0.9994	0.9978	0.9935	0.9850	0.9492	0.8815	0.8345	0.7242
3.85	1.0000	0.9999	0.9995	0.9981	0.9942	0.9865	0.9538	0.8910	0.8469	0.7419
3.90	1.0000	0.9999	0.9995	0.9983	0.9949	0.9879	0.9580	0.8998	0.8584	0.7586
3.95	1.0000	0.9999	0.9996	0.9985	0.9954	0.9891	0.9619	0.9079	0.8691	0.7745
4.00	1.0000	0.9999	0.9996	0.9987	0.9959	0.9902	0.9654	0.9154	0.8791	0.7895
4.05	1.0000	0.9999	0.9997	0.9988	0.9963	0.9912	0.9685	0.9223	0.8883	0.8036
4.10	1.0000	0.9999	0.9997	0.9990	0.9967	0.9921	0.9714	0.9287	0.8969	0.8169
4.15	1.0000	0.9999	0.9997	0.9991	0.9971	0.9929	0.9740	0.9345	0.9049	0.8293
4.20	1.0000	1.0000	0.9998	0.9992	0.9974	0.9936	0.9764	0.9399	0.9122	0.8411
4.25	1.0000	1.0000	0.9998	0.9993	0.9977	0.9943	0.9785	0.9448	0.9191	0.8521
4.30	1.0000	1.0000	0.9998	0.9993	0.9979	0.9948	0.9805	0.9494	0.9254	0.8624
4.35	1.0000	1.0000	0.9998	0.9994	0.9981	0.9953	0.9823	0.9535	0.9312	0.8720
4.40	1.0000	1.0000	0.9999	0.9995	0.9983	0.9958	0.9839	0.9574	0.9366	0.8810
4.45	1.0000	1.0000	0.9999	0.9995	0.9985	0.9962	0.9853	0.9609	0.9416	0.8894
4.50	1.0000	1.0000	0.9999	0.9996	0.9986	0.9966	0.9866	0.9641	0.9461	0.8973
4.55	1.0000	1.0000	0.9999	0.9996	0.9988	0.9969	0.9878	0.9671	0.9504	0.9046
4.60	1.0000	1.0000	0.9999	0.9997	0.9989	0.9972	0.9889	0.9698	0.9543	0.9114
4.65	1.0000	1.0000	0.9999	0.9997	0.9990	0.9975	0.9899	0.9723	0.9579	0.9178
4.70	1.0000	1.0000	0.9999	0.9997	0.9991	0.9977	0.9908	0.9745	0.9612	0.9237
4.75	1.0000	1.0000	0.9999	0.9998	0.9992	0.9979	0.9916	0.9766	0.9642	0.9292
4.80	1.0000	1.0000	0.9999	0.9998	0.9993	0.9981	0.9924	0.9785	0.9670	0.9343
4.85	1.0000	1.0000	1.0000	0.9998	0.9993	0.9983	0.9930	0.9803	0.9696	0.9390
4.90	1.0000	1.0000	1.0000	0.9998	0.9994	0.9985	0.9936	0.9819	0.9720	0.9434
4.95	1.0000	1.0000	1.0000	0.9998	0.9995	0.9986	0.9942	0.9834	0.9742	0.9475
5.00	1.0000	1.0000	1.0000	0.9999	0.9995	0.9987	0.9947	0.9847	0.9762	0.9513
5.05	1.0000	1.0000	1.0000	0.9999	0.9996	0.9989	0.9952	0.9860	0.9781	0.9548
5.10	1.0000	1.0000	1.0000	0.9999	0.9996	0.9990	0.9956	0.9871	0.9798	0.9581
5.15	1.0000	1.0000	1.0000	0.9999	0.9996	0.9991	0.9960	0.9881	0.9813	0.9612
5.20	1.0000	1.0000	1.0000	0.9999	0.9997	0.9991	0.9963	0.9891	0.9828	0.9640
5.25	1.0000	1.0000	1.0000	0.9999	0.9997	0.9992	0.9966	0.9900	0.9841	0.9666
5.30	1.0000	1.0000	1.0000	0.9999	0.9997	0.9993	0.9969	0.9908	0.9854	0.9690
5.35	1.0000	1.0000	1.0000	0.9999	0.9998	0.9994	0.9972	0.9915	0.9865	0.9712
5.40	1.0000	1.0000	1.0000	0.9999	0.9998	0.9994	0.9974	0.9922	0.9875	0.9733
5.45	1.0000	1.0000	1.0000	0.9999	0.9998	0.9995	0.9976	0.9928	0.9885	0.9752

PROBABILITY INTEGRAL OF t, THE NON-CENTRAL t-STATISTIC. THIS TABLE GIVES Pr $[t/\sqrt{f} \leq x]$.

f is the number of degrees of freedom; the non-centrality parameter is $\sqrt{f+1}\,K_p$.

K_p is the standardized normal deviate exceeded with probability p.

x \ p	.2500	.1500	.1000	.0650	.0400	.0250	.0100	.0040	.0025	.0010
5.50	1.0000	1.0000	1.0000	0.9999	0.9998	0.9995	0.9978	0.9934	0.9894	0.9770
5.55	1.0000	1.0000	1.0000	1.0000	0.9998	0.9996	0.9980	0.9939	0.9902	0.9787
5.60	1.0000	1.0000	1.0000	1.0000	0.9998	0.9996	0.9982	0.9944	0.9909	0.9802
5.65	1.0000	1.0000	1.0000	1.0000	0.9999	0.9996	0.9983	0.9948	0.9916	0.9816
5.70	1.0000	1.0000	1.0000	1.0000	0.9999	0.9997	0.9985	0.9952	0.9922	0.9829
5.75	1.0000	1.0000	1.0000	1.0000	0.9999	0.9997	0.9986	0.9956	0.9928	0.9841
5.80	1.0000	1.0000	1.0000	1.0000	0.9999	0.9997	0.9987	0.9959	0.9934	0.9853
5.85	1.0000	1.0000	1.0000	1.0000	0.9999	0.9997	0.9988	0.9962	0.9939	0.9863
5.90	1.0000	1.0000	1.0000	1.0000	0.9999	0.9998	0.9989	0.9965	0.9943	0.9873
5.95	1.0000	1.0000	1.0000	1.0000	0.9999	0.9998	0.9990	0.9968	0.9947	0.9882
6.00	1.0000	1.0000	1.0000	1.0000	0.9999	0.9998	0.9991	0.9970	0.9951	0.9890
6.05	1.0000	1.0000	1.0000	1.0000	0.9999	0.9998	0.9991	0.9973	0.9955	0.9898
6.10	1.0000	1.0000	1.0000	1.0000	0.9999	0.9998	0.9992	0.9975	0.9958	0.9905
6.15	1.0000	1.0000	1.0000	1.0000	0.9999	0.9998	0.9993	0.9976	0.9961	0.9912
6.20	1.0000	1.0000	1.0000	1.0000	0.9999	0.9999	0.9993	0.9978	0.9964	0.9918
6.25	1.0000	1.0000	1.0000	1.0000	1.0000	0.9999	0.9994	0.9980	0.9967	0.9924
6.30	1.0000	1.0000	1.0000	1.0000	1.0000	0.9999	0.9994	0.9981	0.9969	0.9929
6.35	1.0000	1.0000	1.0000	1.0000	1.0000	0.9999	0.9995	0.9983	0.9971	0.9934
6.40	1.0000	1.0000	1.0000	1.0000	1.0000	0.9999	0.9995	0.9984	0.9973	0.9938
6.45	1.0000	1.0000	1.0000	1.0000	1.0000	0.9999	0.9996	0.9985	0.9975	0.9943
6.50	1.0000	1.0000	1.0000	1.0000	1.0000	0.9999	0.9996	0.9986	0.9977	0.9947
6.55	1.0000	1.0000	1.0000	1.0000	1.0000	0.9999	0.9996	0.9987	0.9979	0.9950
6.60	1.0000	1.0000	1.0000	1.0000	1.0000	0.9999	0.9996	0.9988	0.9980	0.9954
6.65	1.0000	1.0000	1.0000	1.0000	1.0000	0.9999	0.9997	0.9989	0.9982	0.9957
6.70	1.0000	1.0000	1.0000	1.0000	1.0000	0.9999	0.9997	0.9990	0.9983	0.9960
6.75	1.0000	1.0000	1.0000	1.0000	1.0000	0.9999	0.9997	0.9991	0.9984	0.9962
6.80	1.0000	1.0000	1.0000	1.0000	1.0000	0.9999	0.9997	0.9991	0.9985	0.9965
6.85	1.0000	1.0000	1.0000	1.0000	1.0000	1.0000	0.9998	0.9992	0.9986	0.9967
6.90	1.0000	1.0000	1.0000	1.0000	1.0000	1.0000	0.9998	0.9992	0.9987	0.9969
6.95	1.0000	1.0000	1.0000	1.0000	1.0000	1.0000	0.9998	0.9993	0.9988	0.9971
7.00	1.0000	1.0000	1.0000	1.0000	1.0000	1.0000	0.9998	0.9993	0.9989	0.9973
7.05	1.0000	1.0000	1.0000	1.0000	1.0000	1.0000	0.9998	0.9994	0.9990	0.9975
7.10	1.0000	1.0000	1.0000	1.0000	1.0000	1.0000	0.9998	0.9994	0.9990	0.9977
7.15	1.0000	1.0000	1.0000	1.0000	1.0000	1.0000	0.9998	0.9995	0.9991	0.9978
7.20	1.0000	1.0000	1.0000	1.0000	1.0000	1.0000	0.9999	0.9995	0.9992	0.9980
7.25	1.0000	1.0000	1.0000	1.0000	1.0000	1.0000	0.9999	0.9995	0.9992	0.9981
7.30	1.0000	1.0000	1.0000	1.0000	1.0000	1.0000	0.9999	0.9996	0.9993	0.9982
7.35	1.0000	1.0000	1.0000	1.0000	1.0000	1.0000	0.9999	0.9996	0.9993	0.9983
7.40	1.0000	1.0000	1.0000	1.0000	1.0000	1.0000	0.9999	0.9996	0.9994	0.9984
7.45	1.0000	1.0000	1.0000	1.0000	1.0000	1.0000	0.9999	0.9997	0.9994	0.9985
7.50	1.0000	1.0000	1.0000	1.0000	1.0000	1.0000	0.9999	0.9997	0.9994	0.9986
7.55	1.0000	1.0000	1.0000	1.0000	1.0000	1.0000	0.9999	0.9997	0.9995	0.9987
7.60	1.0000	1.0000	1.0000	1.0000	1.0000	1.0000	0.9999	0.9997	0.9995	0.9988
7.65	1.0000	1.0000	1.0000	1.0000	1.0000	1.0000	0.9999	0.9997	0.9995	0.9989
7.70	1.0000	1.0000	1.0000	1.0000	1.0000	1.0000	0.9999	0.9998	0.9996	0.9990
7.75	1.0000	1.0000	1.0000	1.0000	1.0000	1.0000	0.9999	0.9998	0.9996	0.9990
7.80	1.0000	1.0000	1.0000	1.0000	1.0000	1.0000	0.9999	0.9998	0.9996	0.9991
7.85	1.0000	1.0000	1.0000	1.0000	1.0000	1.0000	0.9999	0.9998	0.9997	0.9991
7.90	1.0000	1.0000	1.0000	1.0000	1.0000	1.0000	0.9999	0.9998	0.9997	0.9992
7.95	1.0000	1.0000	1.0000	1.0000	1.0000	1.0000	1.0000	0.9998	0.9997	0.9992
8.00	1.0000	1.0000	1.0000	1.0000	1.0000	1.0000	1.0000	0.9998	0.9997	0.9993
8.05	1.0000	1.0000	1.0000	1.0000	1.0000	1.0000	1.0000	0.9998	0.9997	0.9993
8.10	1.0000	1.0000	1.0000	1.0000	1.0000	1.0000	1.0000	0.9999	0.9998	0.9994
8.15	1.0000	1.0000	1.0000	1.0000	1.0000	1.0000	1.0000	0.9999	0.9998	0.9994
8.20	1.0000	1.0000	1.0000	1.0000	1.0000	1.0000	1.0000	0.9999	0.9998	0.9994
8.25	1.0000	1.0000	1.0000	1.0000	1.0000	1.0000	1.0000	0.9999	0.9998	0.9995
8.30	1.0000	1.0000	1.0000	1.0000	1.0000	1.0000	1.0000	0.9999	0.9998	0.9995
8.35	1.0000	1.0000	1.0000	1.0000	1.0000	1.0000	1.0000	0.9999	0.9998	0.9995
8.40	1.0000	1.0000	1.0000	1.0000	1.0000	1.0000	1.0000	0.9999	0.9998	0.9996
8.45	1.0000	1.0000	1.0000	1.0000	1.0000	1.0000	1.0000	0.9999	0.9998	0.9996

PROBABILITY INTEGRAL OF t, THE NON-CENTRAL t-STATISTIC. THIS TABLE GIVES Pr $[t/\sqrt{f} \leq x]$.

f is the number of degrees of freedom; the non-centrality parameter is $\sqrt{f+1}\,K_p$.

K_p is the standardized normal deviate exceeded with probability p.

DEGREES OF FREEDOM **12**

x \ p	.2500	.1500	.1000	.0650	.0400	.0250	.0100	.0040	.0025	.0010
8.50	1.0000	1.0000	1.0000	1.0000	1.0000	1.0000	1.0000	0.9999	0.9999	0.9996
8.55	1.0000	1.0000	1.0000	1.0000	1.0000	1.0000	1.0000	0.9999	0.9999	0.9996
8.60	1.0000	1.0000	1.0000	1.0000	1.0000	1.0000	1.0000	0.9999	0.9999	0.9997
8.65	1.0000	1.0000	1.0000	1.0000	1.0000	1.0000	1.0000	0.9999	0.9999	0.9997
8.70	1.0000	1.0000	1.0000	1.0000	1.0000	1.0000	1.0000	0.9999	0.9999	0.9997
8.75	1.0000	1.0000	1.0000	1.0000	1.0000	1.0000	1.0000	0.9999	0.9999	0.9997
8.80	1.0000	1.0000	1.0000	1.0000	1.0000	1.0000	1.0000	0.9999	0.9999	0.9997
8.85	1.0000	1.0000	1.0000	1.0000	1.0000	1.0000	1.0000	0.9999	0.9999	0.9998
8.90	1.0000	1.0000	1.0000	1.0000	1.0000	1.0000	1.0000	0.9999	0.9999	0.9998
8.95	1.0000	1.0000	1.0000	1.0000	1.0000	1.0000	1.0000	1.0000	0.9999	0.9998
9.00	1.0000	1.0000	1.0000	1.0000	1.0000	1.0000	1.0000	1.0000	0.9999	0.9998
9.05	1.0000	1.0000	1.0000	1.0000	1.0000	1.0000	1.0000	1.0000	0.9999	0.9998
9.10	1.0000	1.0000	1.0000	1.0000	1.0000	1.0000	1.0000	1.0000	0.9999	0.9998
9.15	1.0000	1.0000	1.0000	1.0000	1.0000	1.0000	1.0000	1.0000	0.9999	0.9998
9.20	1.0000	1.0000	1.0000	1.0000	1.0000	1.0000	1.0000	1.0000	0.9999	0.9998
9.25	1.0000	1.0000	1.0000	1.0000	1.0000	1.0000	1.0000	1.0000	0.9999	0.9998
9.30	1.0000	1.0000	1.0000	1.0000	1.0000	1.0000	1.0000	1.0000	0.9999	0.9999
9.35	1.0000	1.0000	1.0000	1.0000	1.0000	1.0000	1.0000	1.0000	0.9999	0.9999
9.40	1.0000	1.0000	1.0000	1.0000	1.0000	1.0000	1.0000	1.0000	1.0000	0.9999
9.45	1.0000	1.0000	1.0000	1.0000	1.0000	1.0000	1.0000	1.0000	1.0000	0.9999
9.50	1.0000	1.0000	1.0000	1.0000	1.0000	1.0000	1.0000	1.0000	1.0000	0.9999
9.55	1.0000	1.0000	1.0000	1.0000	1.0000	1.0000	1.0000	1.0000	1.0000	0.9999
9.60	1.0000	1.0000	1.0000	1.0000	1.0000	1.0000	1.0000	1.0000	1.0000	0.9999
9.65	1.0000	1.0000	1.0000	1.0000	1.0000	1.0000	1.0000	1.0000	1.0000	0.9999
9.70	1.0000	1.0000	1.0000	1.0000	1.0000	1.0000	1.0000	1.0000	1.0000	0.9999
9.75	1.0000	1.0000	1.0000	1.0000	1.0000	1.0000	1.0000	1.0000	1.0000	0.9999
9.80	1.0000	1.0000	1.0000	1.0000	1.0000	1.0000	1.0000	1.0000	1.0000	0.9999
9.85	1.0000	1.0000	1.0000	1.0000	1.0000	1.0000	1.0000	1.0000	1.0000	0.9999
9.90	1.0000	1.0000	1.0000	1.0000	1.0000	1.0000	1.0000	1.0000	1.0000	0.9999
9.95	1.0000	1.0000	1.0000	1.0000	1.0000	1.0000	1.0000	1.0000	1.0000	0.9999
10.00	1.0000	1.0000	1.0000	1.0000	1.0000	1.0000	1.0000	1.0000	1.0000	0.9999
10.05	1.0000	1.0000	1.0000	1.0000	1.0000	1.0000	1.0000	1.0000	1.0000	0.9999
10.10	1.0000	1.0000	1.0000	1.0000	1.0000	1.0000	1.0000	1.0000	1.0000	0.9999
10.15	1.0000	1.0000	1.0000	1.0000	1.0000	1.0000	1.0000	1.0000	1.0000	0.9999
10.20	1.0000	1.0000	1.0000	1.0000	1.0000	1.0000	1.0000	1.0000	1.0000	0.9999
10.25	1.0000	1.0000	1.0000	1.0000	1.0000	1.0000	1.0000	1.0000	1.0000	1.0000

PROBABILITY INTEGRAL OF t, THE NON-CENTRAL t-STATISTIC. THIS TABLE GIVES Pr $[t/\sqrt{f} \leq x]$.

f is the number of degrees of freedom; the non-centrality parameter is $\sqrt{f+1}\,K_p$.

K_p is the standardized normal deviate exceeded with probability p.

DEGREES OF FREEDOM **13**

x \ p	.2500	.1500	.1000	.0650	.0400	.0250	.0100	.0040	.0025	.0010
-0.45	0.0000	0.0000	0.0000	0.0000	0.0000	0.0000	0.0000	0.0000	0.0000	0.0000
-0.40	0.0001	0.0000	0.0000	0.0000	0.0000	0.0000	0.0000	0.0000	0.0000	0.0000
-0.35	0.0001	0.0000	0.0000	0.0000	0.0000	0.0000	0.0000	0.0000	0.0000	0.0000
-0.30	0.0002	0.0000	0.0000	0.0000	0.0000	0.0000	0.0000	0.0000	0.0000	0.0000
-0.25	0.0004	0.0000	0.0000	0.0000	0.0000	0.0000	0.0000	0.0000	0.0000	0.0000
-0.20	0.0007	0.0000	0.0000	0.0000	0.0000	0.0000	0.0000	0.0000	0.0000	0.0000
-0.15	0.0012	0.0000	0.0000	0.0000	0.0000	0.0000	0.0000	0.0000	0.0000	0.0000
-0.10	0.0021	0.0000	0.0000	0.0000	0.0000	0.0000	0.0000	0.0000	0.0000	0.0000
-0.05	0.0035	0.0000	0.0000	0.0000	0.0000	0.0000	0.0000	0.0000	0.0000	0.0000
0.00	0.0058	0.0001	0.0000	0.0000	0.0000	0.0000	0.0000	0.0000	0.0000	0.0000
0.05	0.0095	0.0001	0.0000	0.0000	0.0000	0.0000	0.0000	0.0000	0.0000	0.0000
0.10	0.0152	0.0002	0.0000	0.0000	0.0000	0.0000	0.0000	0.0000	0.0000	0.0000
0.15	0.0237	0.0004	0.0000	0.0000	0.0000	0.0000	0.0000	0.0000	0.0000	0.0000
0.20	0.0360	0.0008	0.0000	0.0000	0.0000	0.0000	0.0000	0.0000	0.0000	0.0000
0.25	0.0532	0.0016	0.0001	0.0000	0.0000	0.0000	0.0000	0.0000	0.0000	0.0000
0.30	0.0762	0.0029	0.0001	0.0000	0.0000	0.0000	0.0000	0.0000	0.0000	0.0000
0.35	0.1059	0.0052	0.0003	0.0000	0.0000	0.0000	0.0000	0.0000	0.0000	0.0000
0.40	0.1428	0.0089	0.0006	0.0000	0.0000	0.0000	0.0000	0.0000	0.0000	0.0000
0.45	0.1869	0.0147	0.0012	0.0001	0.0000	0.0000	0.0000	0.0000	0.0000	0.0000
0.50	0.2378	0.0234	0.0022	0.0001	0.0000	0.0000	0.0000	0.0000	0.0000	0.0000
0.55	0.2943	0.0359	0.0040	0.0003	0.0000	0.0000	0.0000	0.0000	0.0000	0.0000
0.60	0.3550	0.0531	0.0071	0.0006	0.0000	0.0000	0.0000	0.0000	0.0000	0.0000
0.65	0.4181	0.0757	0.0119	0.0012	0.0001	0.0000	0.0000	0.0000	0.0000	0.0000
0.70	0.4817	0.1043	0.0191	0.0022	0.0001	0.0000	0.0000	0.0000	0.0000	0.0000
0.75	0.5441	0.1391	0.0295	0.0041	0.0003	0.0000	0.0000	0.0000	0.0000	0.0000
0.80	0.6038	0.1800	0.0439	0.0071	0.0007	0.0001	0.0000	0.0000	0.0000	0.0000
0.85	0.6595	0.2263	0.0629	0.0117	0.0013	0.0001	0.0000	0.0000	0.0000	0.0000
0.90	0.7103	0.2772	0.0870	0.0186	0.0025	0.0003	0.0000	0.0000	0.0000	0.0000
0.95	0.7560	0.3315	0.1166	0.0284	0.0044	0.0006	0.0000	0.0000	0.0000	0.0000
1.00	0.7962	0.3877	0.1515	0.0418	0.0074	0.0011	0.0000	0.0000	0.0000	0.0000
1.05	0.8310	0.4446	0.1914	0.0591	0.0120	0.0021	0.0000	0.0000	0.0000	0.0000
1.10	0.8609	0.5009	0.2357	0.0810	0.0187	0.0037	0.0001	0.0000	0.0000	0.0000
1.15	0.8862	0.5554	0.2835	0.1075	0.0279	0.0062	0.0002	0.0000	0.0000	0.0000
1.20	0.9073	0.6072	0.3339	0.1387	0.0403	0.0100	0.0004	0.0000	0.0000	0.0000
1.25	0.9249	0.6556	0.3858	0.1743	0.0562	0.0155	0.0008	0.0000	0.0000	0.0000
1.30	0.9393	0.7001	0.4380	0.2139	0.0760	0.0232	0.0015	0.0001	0.0000	0.0000
1.35	0.9511	0.7406	0.4897	0.2568	0.0997	0.0334	0.0027	0.0002	0.0000	0.0000
1.40	0.9607	0.7769	0.5399	0.3022	0.1276	0.0465	0.0045	0.0003	0.0001	0.0000
1.45	0.9685	0.8091	0.5879	0.3494	0.1593	0.0629	0.0071	0.0006	0.0001	0.0000
1.50	0.9748	0.8374	0.6331	0.3974	0.1945	0.0828	0.0110	0.0010	0.0003	0.0000
1.55	0.9798	0.8620	0.6753	0.4454	0.2327	0.1063	0.0163	0.0018	0.0005	0.0000
1.60	0.9839	0.8834	0.7141	0.4927	0.2735	0.1332	0.0235	0.0030	0.0009	0.0001
1.65	0.9871	0.9017	0.7494	0.5386	0.3161	0.1635	0.0327	0.0047	0.0016	0.0002
1.70	0.9897	0.9174	0.7813	0.5826	0.3598	0.1968	0.0443	0.0073	0.0026	0.0003
1.75	0.9917	0.9307	0.8099	0.6243	0.4040	0.2326	0.0586	0.0109	0.0042	0.0006
1.80	0.9934	0.9420	0.8353	0.6634	0.4481	0.2706	0.0756	0.0157	0.0064	0.0010
1.85	0.9947	0.9515	0.8578	0.6997	0.4914	0.3102	0.0954	0.0220	0.0095	0.0016
1.90	0.9958	0.9595	0.8775	0.7332	0.5336	0.3508	0.1181	0.0300	0.0136	0.0026
1.95	0.9966	0.9662	0.8947	0.7637	0.5742	0.3919	0.1435	0.0399	0.0191	0.0040
2.00	0.9973	0.9718	0.9097	0.7915	0.6128	0.4330	0.1714	0.0520	0.0260	0.0059
2.05	0.9978	0.9765	0.9227	0.8165	0.6493	0.4736	0.2016	0.0663	0.0346	0.0086
2.10	0.9982	0.9804	0.9339	0.8389	0.6835	0.5133	0.2338	0.0830	0.0452	0.0122
2.15	0.9986	0.9837	0.9435	0.8589	0.7153	0.5517	0.2676	0.1020	0.0577	0.0169
2.20	0.9988	0.9864	0.9518	0.8767	0.7447	0.5887	0.3026	0.1233	0.0723	0.0227
2.25	0.9991	0.9887	0.9589	0.8924	0.7716	0.6238	0.3385	0.1469	0.0891	0.0300
2.30	0.9992	0.9905	0.9650	0.9063	0.7962	0.6570	0.3749	0.1725	0.1080	0.0388
2.35	0.9994	0.9921	0.9702	0.9184	0.8186	0.6883	0.4114	0.1999	0.1290	0.0493
2.40	0.9995	0.9934	0.9746	0.9291	0.8388	0.7174	0.4476	0.2290	0.1520	0.0615
2.45	0.9996	0.9945	0.9784	0.9385	0.8571	0.7444	0.4834	0.2594	0.1769	0.0756

PROBABILITY INTEGRAL OF t, THE NON-CENTRAL t-STATISTIC. THIS TABLE GIVES Pr $[t/\sqrt{f} \leq x]$.

f is the number of degrees of freedom; the non-centrality parameter is $\sqrt{f+1}\,K_p$.

K_p is the standardized normal deviate exceeded with probability p.

DEGREES OF FREEDOM **13**

x \ p	.2500	.1500	.1000	.0650	.0400	.0250	.0100	.0040	.0025	.0010
2.50	0.9997	0.9954	0.9816	0.9466	0.8735	0.7694	0.5183	0.2909	0.2034	0.0916
2.55	0.9997	0.9961	0.9843	0.9537	0.8881	0.7924	0.5521	0.3232	0.2314	0.1093
2.60	0.9998	0.9968	0.9866	0.9599	0.9012	0.8135	0.5847	0.3559	0.2605	0.1289
2.65	0.9998	0.9973	0.9886	0.9652	0.9129	0.8326	0.6160	0.3889	0.2907	0.1501
2.70	0.9998	0.9977	0.9903	0.9699	0.9232	0.8501	0.6458	0.4219	0.3215	0.1730
2.75	0.9999	0.9981	0.9917	0.9739	0.9324	0.8659	0.6740	0.4546	0.3528	0.1973
2.80	0.9999	0.9984	0.9929	0.9774	0.9405	0.8802	0.7006	0.4867	0.3844	0.2228
2.85	0.9999	0.9986	0.9940	0.9804	0.9477	0.8931	0.7255	0.5182	0.4159	0.2495
2.90	0.9999	0.9988	0.9948	0.9831	0.9540	0.9047	0.7489	0.5488	0.4472	0.2770
2.95	0.9999	0.9990	0.9956	0.9853	0.9596	0.9151	0.7706	0.5784	0.4781	0.3053
3.00	0.9999	0.9992	0.9962	0.9873	0.9645	0.9244	0.7908	0.6069	0.5083	0.3340
3.05	1.0000	0.9993	0.9968	0.9890	0.9688	0.9328	0.8095	0.6342	0.5379	0.3631
3.10	1.0000	0.9994	0.9972	0.9904	0.9726	0.9402	0.8267	0.6603	0.5665	0.3923
3.15	1.0000	0.9995	0.9976	0.9917	0.9759	0.9469	0.8426	0.6850	0.5942	0.4214
3.20	1.0000	0.9996	0.9979	0.9928	0.9788	0.9528	0.8571	0.7084	0.6208	0.4502
3.25	1.0000	0.9996	0.9982	0.9937	0.9814	0.9581	0.8705	0.7305	0.6463	0.4787
3.30	1.0000	0.9997	0.9985	0.9945	0.9836	0.9628	0.8827	0.7513	0.6706	0.5067
3.35	1.0000	0.9997	0.9987	0.9953	0.9856	0.9669	0.8938	0.7708	0.6937	0.5340
3.40	1.0000	0.9998	0.9989	0.9959	0.9874	0.9706	0.9039	0.7890	0.7156	0.5606
3.45	1.0000	0.9998	0.9990	0.9964	0.9889	0.9739	0.9131	0.8059	0.7363	0.5863
3.50	1.0000	0.9998	0.9992	0.9969	0.9902	0.9768	0.9215	0.8217	0.7558	0.6112
3.55	1.0000	0.9999	0.9993	0.9973	0.9914	0.9794	0.9291	0.8364	0.7741	0.6351
3.60	1.0000	0.9999	0.9994	0.9976	0.9924	0.9817	0.9360	0.8500	0.7913	0.6581
3.65	1.0000	0.9999	0.9995	0.9979	0.9933	0.9838	0.9423	0.8626	0.8074	0.6800
3.70	1.0000	0.9999	0.9995	0.9982	0.9941	0.9856	0.9479	0.8742	0.8224	0.7009
3.75	1.0000	0.9999	0.9996	0.9984	0.9948	0.9872	0.9530	0.8849	0.8363	0.7208
3.80	1.0000	0.9999	0.9996	0.9986	0.9954	0.9886	0.9576	0.8947	0.8493	0.7396
3.85	1.0000	0.9999	0.9997	0.9988	0.9959	0.9899	0.9618	0.9038	0.8614	0.7574
3.90	1.0000	0.9999	0.9997	0.9989	0.9964	0.9910	0.9656	0.9121	0.8725	0.7743
3.95	1.0000	1.0000	0.9998	0.9991	0.9968	0.9920	0.9689	0.9198	0.8829	0.7901
4.00	1.0000	1.0000	0.9998	0.9992	0.9972	0.9929	0.9720	0.9268	0.8924	0.8050
4.05	1.0000	1.0000	0.9998	0.9993	0.9975	0.9936	0.9748	0.9332	0.9013	0.8190
4.10	1.0000	1.0000	0.9998	0.9994	0.9978	0.9943	0.9772	0.9390	0.9094	0.8321
4.15	1.0000	1.0000	0.9999	0.9994	0.9981	0.9949	0.9795	0.9444	0.9169	0.8444
4.20	1.0000	1.0000	0.9999	0.9995	0.9983	0.9955	0.9815	0.9493	0.9238	0.8558
4.25	1.0000	1.0000	0.9999	0.9996	0.9985	0.9960	0.9833	0.9538	0.9302	0.8665
4.30	1.0000	1.0000	0.9999	0.9996	0.9986	0.9964	0.9849	0.9579	0.9360	0.8765
4.35	1.0000	1.0000	0.9999	0.9997	0.9988	0.9968	0.9864	0.9616	0.9414	0.8857
4.40	1.0000	1.0000	0.9999	0.9997	0.9989	0.9971	0.9877	0.9650	0.9463	0.8944
4.45	1.0000	1.0000	0.9999	0.9997	0.9990	0.9974	0.9889	0.9681	0.9508	0.9024
4.50	1.0000	1.0000	0.9999	0.9998	0.9991	0.9977	0.9900	0.9709	0.9550	0.9098
4.55	1.0000	1.0000	1.0000	0.9998	0.9992	0.9979	0.9910	0.9735	0.9588	0.9167
4.60	1.0000	1.0000	1.0000	0.9998	0.9993	0.9982	0.9918	0.9758	0.9623	0.9231
4.65	1.0000	1.0000	1.0000	0.9998	0.9994	0.9983	0.9926	0.9780	0.9655	0.9290
4.70	1.0000	1.0000	1.0000	0.9999	0.9995	0.9985	0.9933	0.9799	0.9684	0.9345
4.75	1.0000	1.0000	1.0000	0.9999	0.9995	0.9987	0.9940	0.9817	0.9710	0.9396
4.80	1.0000	1.0000	1.0000	0.9999	0.9996	0.9988	0.9945	0.9833	0.9735	0.9442
4.85	1.0000	1.0000	1.0000	0.9999	0.9996	0.9989	0.9950	0.9848	0.9757	0.9486
4.90	1.0000	1.0000	1.0000	0.9999	0.9997	0.9990	0.9955	0.9861	0.9778	0.9526
4.95	1.0000	1.0000	1.0000	0.9999	0.9997	0.9991	0.9959	0.9873	0.9796	0.9563
5.00	1.0000	1.0000	1.0000	0.9999	0.9997	0.9992	0.9963	0.9884	0.9814	0.9597
5.05	1.0000	1.0000	1.0000	0.9999	0.9998	0.9993	0.9967	0.9894	0.9829	0.9628
5.10	1.0000	1.0000	1.0000	0.9999	0.9998	0.9994	0.9970	0.9904	0.9844	0.9657
5.15	1.0000	1.0000	1.0000	0.9999	0.9998	0.9994	0.9973	0.9912	0.9857	0.9684
5.20	1.0000	1.0000	1.0000	1.0000	0.9998	0.9995	0.9975	0.9920	0.9869	0.9708
5.25	1.0000	1.0000	1.0000	1.0000	0.9998	0.9995	0.9977	0.9926	0.9880	0.9731
5.30	1.0000	1.0000	1.0000	1.0000	0.9999	0.9996	0.9979	0.9933	0.9890	0.9752
5.35	1.0000	1.0000	1.0000	1.0000	0.9999	0.9996	0.9981	0.9939	0.9899	0.9771
5.40	1.0000	1.0000	1.0000	1.0000	0.9999	0.9997	0.9983	0.9944	0.9907	0.9789
5.45	1.0000	1.0000	1.0000	1.0000	0.9999	0.9997	0.9985	0.9949	0.9915	0.9805

PROBABILITY INTEGRAL OF t, THE NON-CENTRAL t-STATISTIC. THIS TABLE GIVES Pr $[t/\sqrt{f} \leq x]$.

f is the number of degrees of freedom; the non-centrality parameter is $\sqrt{f+1}\,K_p$.

K_p is the standardized normal deviate exceeded with probability p.

DEGREES OF FREEDOM **13**

x \ p	.2500	.1500	.1000	.0650	.0400	.0250	.0100	.0040	.0025	.0010
5.50	1.0000	1.0000	1.0000	1.0000	0.9999	0.9997	0.9986	0.9953	0.9922	0.9820
5.55	1.0000	1.0000	1.0000	1.0000	0.9999	0.9997	0.9987	0.9957	0.9928	0.9834
5.60	1.0000	1.0000	1.0000	1.0000	0.9999	0.9998	0.9988	0.9961	0.9934	0.9847
5.65	1.0000	1.0000	1.0000	1.0000	0.9999	0.9998	0.9989	0.9964	0.9940	0.9859
5.70	1.0000	1.0000	1.0000	1.0000	0.9999	0.9998	0.9990	0.9967	0.9944	0.9870
5.75	1.0000	1.0000	1.0000	1.0000	0.9999	0.9998	0.9991	0.9970	0.9949	0.9880
5.80	1.0000	1.0000	1.0000	1.0000	0.9999	0.9998	0.9992	0.9972	0.9953	0.9889
5.85	1.0000	1.0000	1.0000	1.0000	1.0000	0.9999	0.9993	0.9975	0.9957	0.9898
5.90	1.0000	1.0000	1.0000	1.0000	1.0000	0.9999	0.9993	0.9977	0.9960	0.9905
5.95	1.0000	1.0000	1.0000	1.0000	1.0000	0.9999	0.9994	0.9979	0.9964	0.9913
6.00	1.0000	1.0000	1.0000	1.0000	1.0000	0.9999	0.9994	0.9980	0.9966	0.9919
6.05	1.0000	1.0000	1.0000	1.0000	1.0000	0.9999	0.9995	0.9982	0.9969	0.9925
6.10	1.0000	1.0000	1.0000	1.0000	1.0000	0.9999	0.9995	0.9983	0.9972	0.9931
6.15	1.0000	1.0000	1.0000	1.0000	1.0000	0.9999	0.9996	0.9985	0.9974	0.9936
6.20	1.0000	1.0000	1.0000	1.0000	1.0000	0.9999	0.9996	0.9986	0.9976	0.9941
6.25	1.0000	1.0000	1.0000	1.0000	1.0000	0.9999	0.9996	0.9987	0.9978	0.9945
6.30	1.0000	1.0000	1.0000	1.0000	1.0000	0.9999	0.9997	0.9988	0.9980	0.9949
6.35	1.0000	1.0000	1.0000	1.0000	1.0000	0.9999	0.9997	0.9989	0.9981	0.9953
6.40	1.0000	1.0000	1.0000	1.0000	1.0000	1.0000	0.9997	0.9990	0.9983	0.9957
6.45	1.0000	1.0000	1.0000	1.0000	1.0000	1.0000	0.9997	0.9991	0.9984	0.9960
6.50	1.0000	1.0000	1.0000	1.0000	1.0000	1.0000	0.9998	0.9991	0.9985	0.9963
6.55	1.0000	1.0000	1.0000	1.0000	1.0000	1.0000	0.9998	0.9992	0.9986	0.9966
6.60	1.0000	1.0000	1.0000	1.0000	1.0000	1.0000	0.9998	0.9993	0.9987	0.9968
6.65	1.0000	1.0000	1.0000	1.0000	1.0000	1.0000	0.9998	0.9993	0.9988	0.9970
6.70	1.0000	1.0000	1.0000	1.0000	1.0000	1.0000	0.9998	0.9994	0.9989	0.9973
6.75	1.0000	1.0000	1.0000	1.0000	1.0000	1.0000	0.9998	0.9994	0.9990	0.9975
6.80	1.0000	1.0000	1.0000	1.0000	1.0000	1.0000	0.9999	0.9995	0.9991	0.9976
6.85	1.0000	1.0000	1.0000	1.0000	1.0000	1.0000	0.9999	0.9995	0.9991	0.9978
6.90	1.0000	1.0000	1.0000	1.0000	1.0000	1.0000	0.9999	0.9996	0.9992	0.9980
6.95	1.0000	1.0000	1.0000	1.0000	1.0000	1.0000	0.9999	0.9996	0.9993	0.9981
7.00	1.0000	1.0000	1.0000	1.0000	1.0000	1.0000	0.9999	0.9996	0.9993	0.9983
7.05	1.0000	1.0000	1.0000	1.0000	1.0000	1.0000	0.9999	0.9996	0.9994	0.9984
7.10	1.0000	1.0000	1.0000	1.0000	1.0000	1.0000	0.9999	0.9997	0.9994	0.9985
7.15	1.0000	1.0000	1.0000	1.0000	1.0000	1.0000	0.9999	0.9997	0.9995	0.9986
7.20	1.0000	1.0000	1.0000	1.0000	1.0000	1.0000	0.9999	0.9997	0.9995	0.9987
7.25	1.0000	1.0000	1.0000	1.0000	1.0000	1.0000	0.9999	0.9997	0.9995	0.9988
7.30	1.0000	1.0000	1.0000	1.0000	1.0000	1.0000	0.9999	0.9998	0.9996	0.9989
7.35	1.0000	1.0000	1.0000	1.0000	1.0000	1.0000	0.9999	0.9998	0.9996	0.9990
7.40	1.0000	1.0000	1.0000	1.0000	1.0000	1.0000	0.9999	0.9998	0.9996	0.9990
7.45	1.0000	1.0000	1.0000	1.0000	1.0000	1.0000	1.0000	0.9998	0.9997	0.9991
7.50	1.0000	1.0000	1.0000	1.0000	1.0000	1.0000	1.0000	0.9998	0.9997	0.9992
7.55	1.0000	1.0000	1.0000	1.0000	1.0000	1.0000	1.0000	0.9998	0.9997	0.9992
7.60	1.0000	1.0000	1.0000	1.0000	1.0000	1.0000	1.0000	0.9998	0.9997	0.9993
7.65	1.0000	1.0000	1.0000	1.0000	1.0000	1.0000	1.0000	0.9999	0.9997	0.9993
7.70	1.0000	1.0000	1.0000	1.0000	1.0000	1.0000	1.0000	0.9999	0.9998	0.9994
7.75	1.0000	1.0000	1.0000	1.0000	1.0000	1.0000	1.0000	0.9999	0.9998	0.9994
7.80	1.0000	1.0000	1.0000	1.0000	1.0000	1.0000	1.0000	0.9999	0.9998	0.9994
7.85	1.0000	1.0000	1.0000	1.0000	1.0000	1.0000	1.0000	0.9999	0.9998	0.9995
7.90	1.0000	1.0000	1.0000	1.0000	1.0000	1.0000	1.0000	0.9999	0.9998	0.9995
7.95	1.0000	1.0000	1.0000	1.0000	1.0000	1.0000	1.0000	0.9999	0.9998	0.9996
8.00	1.0000	1.0000	1.0000	1.0000	1.0000	1.0000	1.0000	0.9999	0.9998	0.9996
8.05	1.0000	1.0000	1.0000	1.0000	1.0000	1.0000	1.0000	0.9999	0.9999	0.9996
8.10	1.0000	1.0000	1.0000	1.0000	1.0000	1.0000	1.0000	0.9999	0.9999	0.9996
8.15	1.0000	1.0000	1.0000	1.0000	1.0000	1.0000	1.0000	0.9999	0.9999	0.9997
8.20	1.0000	1.0000	1.0000	1.0000	1.0000	1.0000	1.0000	0.9999	0.9999	0.9997
8.25	1.0000	1.0000	1.0000	1.0000	1.0000	1.0000	1.0000	0.9999	0.9999	0.9997
8.30	1.0000	1.0000	1.0000	1.0000	1.0000	1.0000	1.0000	0.9999	0.9999	0.9997
8.35	1.0000	1.0000	1.0000	1.0000	1.0000	1.0000	1.0000	1.0000	0.9999	0.9997
8.40	1.0000	1.0000	1.0000	1.0000	1.0000	1.0000	1.0000	1.0000	0.9999	0.9998
8.45	1.0000	1.0000	1.0000	1.0000	1.0000	1.0000	1.0000	1.0000	0.9999	0.9998

PROBABILITY INTEGRAL OF t, THE NON-CENTRAL t-STATISTIC. THIS TABLE GIVES $Pr\ [t/\sqrt{f} \leq x]$.

f is the number of degrees of freedom; the non-centrality parameter is $\sqrt{f+1}\ K_p$.

K_p is the standardized normal deviate exceeded with probability p.

DEGREES OF FREEDOM **13**

x \ p	.2500	.1500	.1000	.0650	.0400	.0250	.0100	.0040	.0025	.0010
8.50	1.0000	1.0000	1.0000	1.0000	1.0000	1.0000	1.0000	1.0000	0.9999	0.9998
8.55	1.0000	1.0000	1.0000	1.0000	1.0000	1.0000	1.0000	1.0000	0.9999	0.9998
8.60	1.0000	1.0000	1.0000	1.0000	1.0000	1.0000	1.0000	1.0000	0.9999	0.9998
8.65	1.0000	1.0000	1.0000	1.0000	1.0000	1.0000	1.0000	1.0000	0.9999	0.9998
8.70	1.0000	1.0000	1.0000	1.0000	1.0000	1.0000	1.0000	1.0000	0.9999	0.9998
8.75	1.0000	1.0000	1.0000	1.0000	1.0000	1.0000	1.0000	1.0000	0.9999	0.9998
8.80	1.0000	1.0000	1.0000	1.0000	1.0000	1.0000	1.0000	1.0000	1.0000	0.9999
8.85	1.0000	1.0000	1.0000	1.0000	1.0000	1.0000	1.0000	1.0000	1.0000	0.9999
8.90	1.0000	1.0000	1.0000	1.0000	1.0000	1.0000	1.0000	1.0000	1.0000	0.9999
8.95	1.0000	1.0000	1.0000	1.0000	1.0000	1.0000	1.0000	1.0000	1.0000	0.9999
9.00	1.0000	1.0000	1.0000	1.0000	1.0000	1.0000	1.0000	1.0000	1.0000	0.9999
9.05	1.0000	1.0000	1.0000	1.0000	1.0000	1.0000	1.0000	1.0000	1.0000	0.9999
9.10	1.0000	1.0000	1.0000	1.0000	1.0000	1.0000	1.0000	1.0000	1.0000	0.9999
9.15	1.0000	1.0000	1.0000	1.0000	1.0000	1.0000	1.0000	1.0000	1.0000	0.9999
9.20	1.0000	1.0000	1.0000	1.0000	1.0000	1.0000	1.0000	1.0000	1.0000	0.9999
9.25	1.0000	1.0000	1.0000	1.0000	1.0000	1.0000	1.0000	1.0000	1.0000	0.9999
9.30	1.0000	1.0000	1.0000	1.0000	1.0000	1.0000	1.0000	1.0000	1.0000	0.9999
9.35	1.0000	1.0000	1.0000	1.0000	1.0000	1.0000	1.0000	1.0000	1.0000	0.9999
9.40	1.0000	1.0000	1.0000	1.0000	1.0000	1.0000	1.0000	1.0000	1.0000	0.9999
9.45	1.0000	1.0000	1.0000	1.0000	1.0000	1.0000	1.0000	1.0000	1.0000	0.9999
9.50	1.0000	1.0000	1.0000	1.0000	1.0000	1.0000	1.0000	1.0000	1.0000	0.9999
9.55	1.0000	1.0000	1.0000	1.0000	1.0000	1.0000	1.0000	1.0000	1.0000	0.9999
9.60	1.0000	1.0000	1.0000	1.0000	1.0000	1.0000	1.0000	1.0000	1.0000	0.9999
9.65	1.0000	1.0000	1.0000	1.0000	1.0000	1.0000	1.0000	1.0000	1.0000	1.0000

PROBABILITY INTEGRAL OF t, THE NON-CENTRAL t-STATISTIC. THIS TABLE GIVES Pr $[t/\sqrt{f} \leq x]$.

f is the number of degrees of freedom; the non-centrality parameter is $\sqrt{f+1}\,K_p$.

K_p is the standardized normal deviate exceeded with probability p.

DEGREES OF FREEDOM **14**

x \ p	.2500	.1500	.1000	.0650	.0400	.0250	.0100	.0040	.0025	.0010
-0.40	0.0000	0.0000	0.0000	0.0000	0.0000	0.0000	0.0000	0.0000	0.0000	0.0000
-0.35	0.0001	0.0000	0.0000	0.0000	0.0000	0.0000	0.0000	0.0000	0.0000	0.0000
-0.30	0.0001	0.0000	0.0000	0.0000	0.0000	0.0000	0.0000	0.0000	0.0000	0.0000
-0.25	0.0003	0.0000	0.0000	0.0000	0.0000	0.0000	0.0000	0.0000	0.0000	0.0000
-0.20	0.0005	0.0000	0.0000	0.0000	0.0000	0.0000	0.0000	0.0000	0.0000	0.0000
-0.15	0.0008	0.0000	0.0000	0.0000	0.0000	0.0000	0.0000	0.0000	0.0000	0.0000
-0.10	0.0015	0.0000	0.0000	0.0000	0.0000	0.0000	0.0000	0.0000	0.0000	0.0000
-0.05	0.0026	0.0000	0.0000	0.0000	0.0000	0.0000	0.0000	0.0000	0.0000	0.0000
0.00	0.0045	0.0000	0.0000	0.0000	0.0000	0.0000	0.0000	0.0000	0.0000	0.0000
0.05	0.0076	0.0001	0.0000	0.0000	0.0000	0.0000	0.0000	0.0000	0.0000	0.0000
0.10	0.0126	0.0001	0.0000	0.0000	0.0000	0.0000	0.0000	0.0000	0.0000	0.0000
0.15	0.0202	0.0003	0.0000	0.0000	0.0000	0.0000	0.0000	0.0000	0.0000	0.0000
0.20	0.0315	0.0006	0.0000	0.0000	0.0000	0.0000	0.0000	0.0000	0.0000	0.0000
0.25	0.0477	0.0012	0.0000	0.0000	0.0000	0.0000	0.0000	0.0000	0.0000	0.0000
0.30	0.0698	0.0022	0.0001	0.0000	0.0000	0.0000	0.0000	0.0000	0.0000	0.0000
0.35	0.0990	0.0041	0.0002	0.0000	0.0000	0.0000	0.0000	0.0000	0.0000	0.0000
0.40	0.1358	0.0072	0.0004	0.0000	0.0000	0.0000	0.0000	0.0000	0.0000	0.0000
0.45	0.1803	0.0123	0.0008	0.0000	0.0000	0.0000	0.0000	0.0000	0.0000	0.0000
0.50	0.2322	0.0202	0.0016	0.0001	0.0000	0.0000	0.0000	0.0000	0.0000	0.0000
0.55	0.2904	0.0318	0.0031	0.0002	0.0000	0.0000	0.0000	0.0000	0.0000	0.0000
0.60	0.3531	0.0480	0.0057	0.0004	0.0000	0.0000	0.0000	0.0000	0.0000	0.0000
0.65	0.4186	0.0700	0.0098	0.0008	0.0000	0.0000	0.0000	0.0000	0.0000	0.0000
0.70	0.4848	0.0982	0.0163	0.0017	0.0001	0.0000	0.0000	0.0000	0.0000	0.0000
0.75	0.5495	0.1330	0.0259	0.0032	0.0002	0.0000	0.0000	0.0000	0.0000	0.0000
0.80	0.6113	0.1744	0.0395	0.0057	0.0005	0.0000	0.0000	0.0000	0.0000	0.0000
0.85	0.6687	0.2217	0.0578	0.0098	0.0010	0.0001	0.0000	0.0000	0.0000	0.0000
0.90	0.7208	0.2741	0.0814	0.0160	0.0019	0.0002	0.0000	0.0000	0.0000	0.0000
0.95	0.7671	0.3302	0.1108	0.0250	0.0034	0.0004	0.0000	0.0000	0.0000	0.0000
1.00	0.8076	0.3886	0.1459	0.0376	0.0060	0.0008	0.0000	0.0000	0.0000	0.0000
1.05	0.8424	0.4477	0.1865	0.0543	0.0101	0.0016	0.0000	0.0000	0.0000	0.0000
1.10	0.8719	0.5061	0.2320	0.0757	0.0161	0.0029	0.0001	0.0000	0.0000	0.0000
1.15	0.8965	0.5626	0.2813	0.1021	0.0247	0.0050	0.0001	0.0000	0.0000	0.0000
1.20	0.9169	0.6161	0.3335	0.1334	0.0364	0.0083	0.0003	0.0000	0.0000	0.0000
1.25	0.9336	0.6659	0.3873	0.1696	0.0517	0.0133	0.0006	0.0000	0.0000	0.0000
1.30	0.9471	0.7115	0.4416	0.2101	0.0710	0.0203	0.0012	0.0000	0.0000	0.0000
1.35	0.9581	0.7526	0.4953	0.2542	0.0946	0.0298	0.0021	0.0001	0.0000	0.0000
1.40	0.9668	0.7892	0.5473	0.3012	0.1225	0.0424	0.0036	0.0002	0.0000	0.0000
1.45	0.9738	0.8214	0.5969	0.3501	0.1546	0.0583	0.0059	0.0004	0.0001	0.0000
1.50	0.9794	0.8494	0.6435	0.3999	0.1905	0.0779	0.0093	0.0008	0.0002	0.0000
1.55	0.9837	0.8737	0.6867	0.4498	0.2298	0.1013	0.0141	0.0014	0.0004	0.0000
1.60	0.9872	0.8944	0.7263	0.4989	0.2719	0.1284	0.0206	0.0023	0.0007	0.0001
1.65	0.9899	0.9120	0.7621	0.5465	0.3160	0.1591	0.0293	0.0038	0.0012	0.0001
1.70	0.9921	0.9269	0.7941	0.5920	0.3614	0.1931	0.0404	0.0060	0.0020	0.0002
1.75	0.9938	0.9394	0.8227	0.6350	0.4073	0.2300	0.0542	0.0091	0.0033	0.0004
1.80	0.9951	0.9499	0.8478	0.6751	0.4531	0.2692	0.0709	0.0135	0.0052	0.0007
1.85	0.9961	0.9587	0.8698	0.7122	0.4981	0.3102	0.0906	0.0193	0.0079	0.0012
1.90	0.9969	0.9659	0.8890	0.7461	0.5418	0.3524	0.1134	0.0268	0.0117	0.0020
1.95	0.9976	0.9719	0.9056	0.7769	0.5838	0.3951	0.1391	0.0363	0.0166	0.0032
2.00	0.9981	0.9769	0.9199	0.8046	0.6236	0.4378	0.1675	0.0479	0.0231	0.0048
2.05	0.9985	0.9810	0.9321	0.8295	0.6611	0.4799	0.1984	0.0619	0.0313	0.0072
2.10	0.9988	0.9844	0.9426	0.8516	0.6960	0.5211	0.2316	0.0784	0.0413	0.0104
2.15	0.9990	0.9871	0.9515	0.8712	0.7283	0.5609	0.2665	0.0974	0.0535	0.0146
2.20	0.9992	0.9894	0.9591	0.8884	0.7580	0.5990	0.3028	0.1188	0.0679	0.0200
2.25	0.9994	0.9913	0.9655	0.9036	0.7851	0.6352	0.3400	0.1427	0.0845	0.0269
2.30	0.9995	0.9928	0.9709	0.9168	0.8096	0.6692	0.3778	0.1688	0.1035	0.0353
2.35	0.9996	0.9941	0.9755	0.9283	0.8318	0.7011	0.4158	0.1969	0.1246	0.0454
2.40	0.9997	0.9951	0.9794	0.9383	0.8518	0.7307	0.4534	0.2268	0.1480	0.0573
2.45	0.9997	0.9960	0.9827	0.9470	0.8696	0.7580	0.4905	0.2582	0.1734	0.0712

PROBABILITY INTEGRAL OF t, THE NON-CENTRAL t-STATISTIC. THIS TABLE GIVES Pr $[t/\sqrt{f} \leq x]$.

f is the number of degrees of freedom; the non-centrality parameter is $\sqrt{f+1}\,K_p$.
K_p is the standardized normal deviate exceeded with probability p.

DEGREES OF FREEDOM **14**

x \ p	.2500	.1500	.1000	.0650	.0400	.0250	.0100	.0040	.0025	.0010
2.50	0.9998	0.9967	0.9854	0.9544	0.8855	0.7831	0.5267	0.2908	0.2006	0.0871
2.55	0.9998	0.9973	0.9877	0.9609	0.8996	0.8060	0.5617	0.3243	0.2294	0.1049
2.60	0.9999	0.9977	0.9897	0.9665	0.9121	0.8269	0.5954	0.3584	0.2595	0.1246
2.65	0.9999	0.9981	0.9913	0.9712	0.9232	0.8458	0.6275	0.3927	0.2907	0.1462
2.70	0.9999	0.9984	0.9927	0.9753	0.9329	0.8629	0.6581	0.4269	0.3227	0.1695
2.75	0.9999	0.9987	0.9938	0.9789	0.9414	0.8783	0.6869	0.4609	0.3553	0.1944
2.80	0.9999	0.9989	0.9948	0.9819	0.9489	0.8921	0.7139	0.4943	0.3881	0.2207
2.85	1.0000	0.9991	0.9956	0.9845	0.9555	0.9045	0.7392	0.5269	0.4208	0.2482
2.90	1.0000	0.9993	0.9963	0.9867	0.9612	0.9155	0.7627	0.5586	0.4533	0.2768
2.95	1.0000	0.9994	0.9968	0.9886	0.9662	0.9254	0.7845	0.5891	0.4854	0.3061
3.00	1.0000	0.9995	0.9973	0.9902	0.9706	0.9341	0.8047	0.6185	0.5168	0.3359
3.05	1.0000	0.9996	0.9977	0.9916	0.9744	0.9418	0.8232	0.6465	0.5473	0.3661
3.10	1.0000	0.9996	0.9981	0.9928	0.9777	0.9487	0.8402	0.6731	0.5769	0.3964
3.15	1.0000	0.9997	0.9984	0.9938	0.9806	0.9548	0.8558	0.6983	0.6055	0.4267
3.20	1.0000	0.9997	0.9986	0.9947	0.9831	0.9602	0.8700	0.7221	0.6328	0.4566
3.25	1.0000	0.9998	0.9988	0.9954	0.9853	0.9649	0.8829	0.7444	0.6589	0.4862
3.30	1.0000	0.9998	0.9990	0.9961	0.9872	0.9691	0.8946	0.7654	0.6837	0.5152
3.35	1.0000	0.9998	0.9991	0.9966	0.9889	0.9728	0.9053	0.7849	0.7073	0.5435
3.40	1.0000	0.9999	0.9993	0.9971	0.9903	0.9761	0.9149	0.8030	0.7295	0.5709
3.45	1.0000	0.9999	0.9994	0.9975	0.9915	0.9789	0.9236	0.8199	0.7504	0.5975
3.50	1.0000	0.9999	0.9995	0.9978	0.9926	0.9815	0.9315	0.8355	0.7700	0.6231
3.55	1.0000	0.9999	0.9995	0.9981	0.9936	0.9837	0.9386	0.8499	0.7883	0.6476
3.60	1.0000	0.9999	0.9996	0.9984	0.9944	0.9856	0.9450	0.8632	0.8054	0.6711
3.65	1.0000	0.9999	0.9997	0.9986	0.9951	0.9873	0.9507	0.8754	0.8213	0.6935
3.70	1.0000	1.0000	0.9997	0.9988	0.9957	0.9888	0.9558	0.8866	0.8361	0.7147
3.75	1.0000	1.0000	0.9998	0.9989	0.9963	0.9902	0.9605	0.8969	0.8499	0.7348
3.80	1.0000	1.0000	0.9998	0.9991	0.9967	0.9913	0.9646	0.9063	0.8625	0.7538
3.85	1.0000	1.0000	0.9998	0.9992	0.9971	0.9924	0.9683	0.9149	0.8743	0.7717
3.90	1.0000	1.0000	0.9998	0.9993	0.9975	0.9933	0.9717	0.9228	0.8851	0.7885
3.95	1.0000	1.0000	0.9999	0.9994	0.9978	0.9941	0.9746	0.9299	0.8950	0.8043
4.00	1.0000	1.0000	0.9999	0.9995	0.9981	0.9948	0.9773	0.9365	0.9041	0.8191
4.05	1.0000	1.0000	0.9999	0.9995	0.9983	0.9954	0.9797	0.9424	0.9125	0.8329
4.10	1.0000	1.0000	0.9999	0.9996	0.9985	0.9959	0.9818	0.9478	0.9202	0.8458
4.15	1.0000	1.0000	0.9999	0.9997	0.9987	0.9964	0.9837	0.9527	0.9273	0.8578
4.20	1.0000	1.0000	0.9999	0.9997	0.9989	0.9968	0.9854	0.9572	0.9338	0.8690
4.25	1.0000	1.0000	0.9999	0.9997	0.9990	0.9972	0.9870	0.9612	0.9397	0.8793
4.30	1.0000	1.0000	1.0000	0.9998	0.9991	0.9975	0.9883	0.9649	0.9450	0.8889
4.35	1.0000	1.0000	1.0000	0.9998	0.9992	0.9978	0.9895	0.9682	0.9500	0.8978
4.40	1.0000	1.0000	1.0000	0.9998	0.9993	0.9980	0.9906	0.9712	0.9545	0.9060
4.45	1.0000	1.0000	1.0000	0.9998	0.9994	0.9983	0.9916	0.9739	0.9586	0.9136
4.50	1.0000	1.0000	1.0000	0.9999	0.9995	0.9984	0.9925	0.9764	0.9623	0.9207
4.55	1.0000	1.0000	1.0000	0.9999	0.9995	0.9986	0.9933	0.9786	0.9657	0.9271
4.60	1.0000	1.0000	1.0000	0.9999	0.9996	0.9988	0.9939	0.9807	0.9688	0.9331
4.65	1.0000	1.0000	1.0000	0.9999	0.9996	0.9989	0.9946	0.9825	0.9716	0.9386
4.70	1.0000	1.0000	1.0000	0.9999	0.9997	0.9990	0.9951	0.9841	0.9742	0.9437
4.75	1.0000	1.0000	1.0000	0.9999	0.9997	0.9991	0.9956	0.9856	0.9765	0.9483
4.80	1.0000	1.0000	1.0000	0.9999	0.9997	0.9992	0.9961	0.9870	0.9786	0.9526
4.85	1.0000	1.0000	1.0000	0.9999	0.9998	0.9993	0.9965	0.9882	0.9806	0.9565
4.90	1.0000	1.0000	1.0000	1.0000	0.9998	0.9994	0.9968	0.9893	0.9823	0.9601
4.95	1.0000	1.0000	1.0000	1.0000	0.9998	0.9995	0.9971	0.9903	0.9839	0.9635
5.00	1.0000	1.0000	1.0000	1.0000	0.9998	0.9995	0.9974	0.9912	0.9854	0.9665
5.05	1.0000	1.0000	1.0000	1.0000	0.9999	0.9996	0.9977	0.9920	0.9867	0.9693
5.10	1.0000	1.0000	1.0000	1.0000	0.9999	0.9996	0.9979	0.9928	0.9879	0.9719
5.15	1.0000	1.0000	1.0000	1.0000	0.9999	0.9997	0.9981	0.9935	0.9890	0.9742
5.20	1.0000	1.0000	1.0000	1.0000	0.9999	0.9997	0.9983	0.9941	0.9899	0.9763
5.25	1.0000	1.0000	1.0000	1.0000	0.9999	0.9997	0.9985	0.9946	0.9908	0.9783
5.30	1.0000	1.0000	1.0000	1.0000	0.9999	0.9998	0.9986	0.9951	0.9917	0.9801
5.35	1.0000	1.0000	1.0000	1.0000	0.9999	0.9998	0.9988	0.9956	0.9924	0.9818
5.40	1.0000	1.0000	1.0000	1.0000	0.9999	0.9998	0.9989	0.9960	0.9931	0.9833
5.45	1.0000	1.0000	1.0000	1.0000	0.9999	0.9998	0.9990	0.9963	0.9937	0.9847

PROBABILITY INTEGRAL OF t, THE NON-CENTRAL t-STATISTIC. THIS TABLE GIVES Pr $[t/\sqrt{f} \leq x]$.

f is the number of degrees of freedom; the non-centrality parameter is $\sqrt{f+1}\,K_p$.

K_p is the standardized normal deviate exceeded with probability p.

DEGREES OF FREEDOM **14**

x \ p	.2500	.1500	.1000	.0650	.0400	.0250	.0100	.0040	.0025	.0010
5.50	1.0000	1.0000	1.0000	1.0000	1.0000	0.9998	0.9991	0.9967	0.9942	0.9859
5.55	1.0000	1.0000	1.0000	1.0000	1.0000	0.9999	0.9992	0.9970	0.9948	0.9871
5.60	1.0000	1.0000	1.0000	1.0000	1.0000	0.9999	0.9993	0.9972	0.9952	0.9882
5.65	1.0000	1.0000	1.0000	1.0000	1.0000	0.9999	0.9993	0.9975	0.9956	0.9892
5.70	1.0000	1.0000	1.0000	1.0000	1.0000	0.9999	0.9994	0.9977	0.9960	0.9900
5.75	1.0000	1.0000	1.0000	1.0000	1.0000	0.9999	0.9994	0.9979	0.9964	0.9909
5.80	1.0000	1.0000	1.0000	1.0000	1.0000	0.9999	0.9995	0.9981	0.9967	0.9916
5.85	1.0000	1.0000	1.0000	1.0000	1.0000	0.9999	0.9995	0.9983	0.9970	0.9923
5.90	1.0000	1.0000	1.0000	1.0000	1.0000	0.9999	0.9996	0.9984	0.9972	0.9929
5.95	1.0000	1.0000	1.0000	1.0000	1.0000	0.9999	0.9996	0.9986	0.9975	0.9935
6.00	1.0000	1.0000	1.0000	1.0000	1.0000	0.9999	0.9997	0.9987	0.9977	0.9940
6.05	1.0000	1.0000	1.0000	1.0000	1.0000	1.0000	0.9997	0.9988	0.9979	0.9945
6.10	1.0000	1.0000	1.0000	1.0000	1.0000	1.0000	0.9997	0.9989	0.9981	0.9950
6.15	1.0000	1.0000	1.0000	1.0000	1.0000	1.0000	0.9997	0.9990	0.9982	0.9954
6.20	1.0000	1.0000	1.0000	1.0000	1.0000	1.0000	0.9998	0.9991	0.9984	0.9957
6.25	1.0000	1.0000	1.0000	1.0000	1.0000	1.0000	0.9998	0.9992	0.9985	0.9961
6.30	1.0000	1.0000	1.0000	1.0000	1.0000	1.0000	0.9998	0.9992	0.9986	0.9964
6.35	1.0000	1.0000	1.0000	1.0000	1.0000	1.0000	0.9998	0.9993	0.9988	0.9967
6.40	1.0000	1.0000	1.0000	1.0000	1.0000	1.0000	0.9998	0.9994	0.9989	0.9970
6.45	1.0000	1.0000	1.0000	1.0000	1.0000	1.0000	0.9999	0.9994	0.9989	0.9972
6.50	1.0000	1.0000	1.0000	1.0000	1.0000	1.0000	0.9999	0.9995	0.9990	0.9974
6.55	1.0000	1.0000	1.0000	1.0000	1.0000	1.0000	0.9999	0.9995	0.9991	0.9976
6.60	1.0000	1.0000	1.0000	1.0000	1.0000	1.0000	0.9999	0.9996	0.9992	0.9978
6.65	1.0000	1.0000	1.0000	1.0000	1.0000	1.0000	0.9999	0.9996	0.9993	0.9980
6.70	1.0000	1.0000	1.0000	1.0000	1.0000	1.0000	0.9999	0.9996	0.9993	0.9981
6.75	1.0000	1.0000	1.0000	1.0000	1.0000	1.0000	0.9999	0.9997	0.9994	0.9983
6.80	1.0000	1.0000	1.0000	1.0000	1.0000	1.0000	0.9999	0.9997	0.9994	0.9984
6.85	1.0000	1.0000	1.0000	1.0000	1.0000	1.0000	0.9999	0.9997	0.9995	0.9985
6.90	1.0000	1.0000	1.0000	1.0000	1.0000	1.0000	0.9999	0.9997	0.9995	0.9987
6.95	1.0000	1.0000	1.0000	1.0000	1.0000	1.0000	0.9999	0.9998	0.9996	0.9988
7.00	1.0000	1.0000	1.0000	1.0000	1.0000	1.0000	0.9999	0.9998	0.9996	0.9989
7.05	1.0000	1.0000	1.0000	1.0000	1.0000	1.0000	1.0000	0.9998	0.9996	0.9989
7.10	1.0000	1.0000	1.0000	1.0000	1.0000	1.0000	1.0000	0.9998	0.9997	0.9990
7.15	1.0000	1.0000	1.0000	1.0000	1.0000	1.0000	1.0000	0.9998	0.9997	0.9991
7.20	1.0000	1.0000	1.0000	1.0000	1.0000	1.0000	1.0000	0.9998	0.9997	0.9992
7.25	1.0000	1.0000	1.0000	1.0000	1.0000	1.0000	1.0000	0.9999	0.9997	0.9992
7.30	1.0000	1.0000	1.0000	1.0000	1.0000	1.0000	1.0000	0.9999	0.9997	0.9993
7.35	1.0000	1.0000	1.0000	1.0000	1.0000	1.0000	1.0000	0.9999	0.9998	0.9993
7.40	1.0000	1.0000	1.0000	1.0000	1.0000	1.0000	1.0000	0.9999	0.9998	0.9994
7.45	1.0000	1.0000	1.0000	1.0000	1.0000	1.0000	1.0000	0.9999	0.9998	0.9994
7.50	1.0000	1.0000	1.0000	1.0000	1.0000	1.0000	1.0000	0.9999	0.9998	0.9995
7.55	1.0000	1.0000	1.0000	1.0000	1.0000	1.0000	1.0000	0.9999	0.9998	0.9995
7.60	1.0000	1.0000	1.0000	1.0000	1.0000	1.0000	1.0000	0.9999	0.9998	0.9996
7.65	1.0000	1.0000	1.0000	1.0000	1.0000	1.0000	1.0000	0.9999	0.9999	0.9996
7.70	1.0000	1.0000	1.0000	1.0000	1.0000	1.0000	1.0000	0.9999	0.9999	0.9996
7.75	1.0000	1.0000	1.0000	1.0000	1.0000	1.0000	1.0000	0.9999	0.9999	0.9996
7.80	1.0000	1.0000	1.0000	1.0000	1.0000	1.0000	1.0000	0.9999	0.9999	0.9997
7.85	1.0000	1.0000	1.0000	1.0000	1.0000	1.0000	1.0000	0.9999	0.9999	0.9997
7.90	1.0000	1.0000	1.0000	1.0000	1.0000	1.0000	1.0000	1.0000	0.9999	0.9997
7.95	1.0000	1.0000	1.0000	1.0000	1.0000	1.0000	1.0000	1.0000	0.9999	0.9997
8.00	1.0000	1.0000	1.0000	1.0000	1.0000	1.0000	1.0000	1.0000	0.9999	0.9998
8.05	1.0000	1.0000	1.0000	1.0000	1.0000	1.0000	1.0000	1.0000	0.9999	0.9998
8.10	1.0000	1.0000	1.0000	1.0000	1.0000	1.0000	1.0000	1.0000	0.9999	0.9998
8.15	1.0000	1.0000	1.0000	1.0000	1.0000	1.0000	1.0000	1.0000	0.9999	0.9998
8.20	1.0000	1.0000	1.0000	1.0000	1.0000	1.0000	1.0000	1.0000	0.9999	0.9998
8.25	1.0000	1.0000	1.0000	1.0000	1.0000	1.0000	1.0000	1.0000	0.9999	0.9998
8.30	1.0000	1.0000	1.0000	1.0000	1.0000	1.0000	1.0000	1.0000	0.9999	0.9998
8.35	1.0000	1.0000	1.0000	1.0000	1.0000	1.0000	1.0000	1.0000	1.0000	0.9999
8.40	1.0000	1.0000	1.0000	1.0000	1.0000	1.0000	1.0000	1.0000	1.0000	0.9999
8.45	1.0000	1.0000	1.0000	1.0000	1.0000	1.0000	1.0000	1.0000	1.0000	0.9999

PROBABILITY INTEGRAL OF t, THE NON-CENTRAL t-STATISTIC. THIS TABLE GIVES Pr $[t/\sqrt{f} \leq x]$.

f is the number of degrees of freedom; the non-centrality parameter is $\sqrt{f+1}\,K_p$.

K_p is the standardized normal deviate exceeded with probability p.

DEGREES OF FREEDOM **14**

x \ p	.2500	.1500	.1000	.0650	.0400	.0250	.0100	.0040	.0025	.0010
8.50	1.0000	1.0000	1.0000	1.0000	1.0000	1.0000	1.0000	1.0000	1.0000	0.9999
8.55	1.0000	1.0000	1.0000	1.0000	1.0000	1.0000	1.0000	1.0000	1.0000	0.9999
8.60	1.0000	1.0000	1.0000	1.0000	1.0000	1.0000	1.0000	1.0000	1.0000	0.9999
8.65	1.0000	1.0000	1.0000	1.0000	1.0000	1.0000	1.0000	1.0000	1.0000	0.9999
8.70	1.0000	1.0000	1.0000	1.0000	1.0000	1.0000	1.0000	1.0000	1.0000	0.9999
8.75	1.0000	1.0000	1.0000	1.0000	1.0000	1.0000	1.0000	1.0000	1.0000	0.9999
8.80	1.0000	1.0000	1.0000	1.0000	1.0000	1.0000	1.0000	1.0000	1.0000	0.9999
8.85	1.0000	1.0000	1.0000	1.0000	1.0000	1.0000	1.0000	1.0000	1.0000	0.9999
8.90	1.0000	1.0000	1.0000	1.0000	1.0000	1.0000	1.0000	1.0000	1.0000	0.9999
8.95	1.0000	1.0000	1.0000	1.0000	1.0000	1.0000	1.0000	1.0000	1.0000	0.9999
9.00	1.0000	1.0000	1.0000	1.0000	1.0000	1.0000	1.0000	1.0000	1.0000	0.9999
9.05	1.0000	1.0000	1.0000	1.0000	1.0000	1.0000	1.0000	1.0000	1.0000	0.9999
9.10	1.0000	1.0000	1.0000	1.0000	1.0000	1.0000	1.0000	1.0000	1.0000	1.0000

PROBABILITY INTEGRAL OF t, THE NON-CENTRAL t-STATISTIC. THIS TABLE GIVES Pr $[t/\sqrt{f} \leq x]$.

f is the number of degrees of freedom; the non-centrality parameter is $\sqrt{f+1}\, K_p$.

K_p is the standardized normal deviate exceeded with probability p.

DEGREES OF FREEDOM 15

x \ p	.2500	.1500	.1000	.0650	.0400	.0250	.0100	.0040	.0025	.0010
- 0.35	0.0000	0.0000	0.0000	0.0000	0.0000	0.0000	0.0000	0.0000	0.0000	0.0000
- 0.30	0.0001	0.0000	0.0000	0.0000	0.0000	0.0000	0.0000	0.0000	0.0000	0.0000
- 0.25	0.0002	0.0000	0.0000	0.0000	0.0000	0.0000	0.0000	0.0000	0.0000	0.0000
- 0.20	0.0003	0.0000	0.0000	0.0000	0.0000	0.0000	0.0000	0.0000	0.0000	0.0000
- 0.15	0.0006	0.0000	0.0000	0.0000	0.0000	0.0000	0.0000	0.0000	0.0000	0.0000
- 0.10	0.0011	0.0000	0.0000	0.0000	0.0000	0.0000	0.0000	0.0000	0.0000	0.0000
- 0.05	0.0019	0.0000	0.0000	0.0000	0.0000	0.0000	0.0000	0.0000	0.0000	0.0000
0.00	0.0035	0.0000	0.0000	0.0000	0.0000	0.0000	0.0000	0.0000	0.0000	0.0000
0.05	0.0061	0.0000	0.0000	0.0000	0.0000	0.0000	0.0000	0.0000	0.0000	0.0000
0.10	0.0104	0.0001	0.0000	0.0000	0.0000	0.0000	0.0000	0.0000	0.0000	0.0000
0.15	0.0172	0.0002	0.0000	0.0000	0.0000	0.0000	0.0000	0.0000	0.0000	0.0000
0.20	0.0276	0.0004	0.0000	0.0000	0.0000	0.0000	0.0000	0.0000	0.0000	0.0000
0.25	0.0428	0.0008	0.0000	0.0000	0.0000	0.0000	0.0000	0.0000	0.0000	0.0000
0.30	0.0641	0.0017	0.0001	0.0000	0.0000	0.0000	0.0000	0.0000	0.0000	0.0000
0.35	0.0926	0.0032	0.0001	0.0000	0.0000	0.0000	0.0000	0.0000	0.0000	0.0000
0.40	0.1291	0.0058	0.0003	0.0000	0.0000	0.0000	0.0000	0.0000	0.0000	0.0000
0.45	0.1740	0.0103	0.0006	0.0000	0.0000	0.0000	0.0000	0.0000	0.0000	0.0000
0.50	0.2268	0.0174	0.0012	0.0000	0.0000	0.0000	0.0000	0.0000	0.0000	0.0000
0.55	0.2865	0.0281	0.0024	0.0001	0.0000	0.0000	0.0000	0.0000	0.0000	0.0000
0.60	0.3513	0.0435	0.0046	0.0003	0.0000	0.0000	0.0000	0.0000	0.0000	0.0000
0.65	0.4191	0.0647	0.0082	0.0006	0.0000	0.0000	0.0000	0.0000	0.0000	0.0000
0.70	0.4876	0.0924	0.0140	0.0013	0.0001	0.0000	0.0000	0.0000	0.0000	0.0000
0.75	0.5546	0.1272	0.0228	0.0025	0.0001	0.0000	0.0000	0.0000	0.0000	0.0000
0.80	0.6184	0.1689	0.0355	0.0046	0.0003	0.0000	0.0000	0.0000	0.0000	0.0000
0.85	0.6773	0.2172	0.0531	0.0082	0.0007	0.0001	0.0000	0.0000	0.0000	0.0000
0.90	0.7306	0.2710	0.0762	0.0137	0.0014	0.0001	0.0000	0.0000	0.0000	0.0000
0.95	0.7775	0.3289	0.1053	0.0221	0.0027	0.0003	0.0000	0.0000	0.0000	0.0000
1.00	0.8182	0.3893	0.1406	0.0339	0.0049	0.0006	0.0000	0.0000	0.0000	0.0000
1.05	0.8529	0.4505	0.1818	0.0500	0.0085	0.0012	0.0000	0.0000	0.0000	0.0000
1.10	0.8819	0.5111	0.2283	0.0708	0.0139	0.0023	0.0000	0.0000	0.0000	0.0000
1.15	0.9058	0.5695	0.2790	0.0969	0.0218	0.0041	0.0001	0.0000	0.0000	0.0000
1.20	0.9254	0.6246	0.3329	0.1283	0.0328	0.0069	0.0002	0.0000	0.0000	0.0000
1.25	0.9412	0.6757	0.3887	0.1649	0.0475	0.0114	0.0004	0.0000	0.0000	0.0000
1.30	0.9539	0.7221	0.4449	0.2063	0.0663	0.0178	0.0009	0.0000	0.0000	0.0000
1.35	0.9640	0.7638	0.5005	0.2516	0.0897	0.0267	0.0016	0.0001	0.0000	0.0000
1.40	0.9719	0.8006	0.5543	0.3000	0.1176	0.0387	0.0028	0.0001	0.0000	0.0000
1.45	0.9782	0.8327	0.6054	0.3506	0.1500	0.0541	0.0048	0.0003	0.0001	0.0000
1.50	0.9831	0.8604	0.6533	0.4022	0.1866	0.0733	0.0078	0.0006	0.0001	0.0000
1.55	0.9869	0.8841	0.6975	0.4539	0.2269	0.0965	0.0121	0.0010	0.0003	0.0000
1.60	0.9898	0.9042	0.7376	0.5048	0.2702	0.1237	0.0182	0.0018	0.0005	0.0000
1.65	0.9921	0.9211	0.7738	0.5540	0.3157	0.1548	0.0263	0.0030	0.0009	0.0001
1.70	0.9939	0.9353	0.8060	0.6009	0.3627	0.1895	0.0369	0.0049	0.0016	0.0001
1.75	0.9953	0.9470	0.8343	0.6450	0.4103	0.2273	0.0502	0.0077	0.0027	0.0003
1.80	0.9963	0.9567	0.8591	0.6860	0.4578	0.2677	0.0666	0.0116	0.0043	0.0005
1.85	0.9972	0.9647	0.8807	0.7237	0.5044	0.3100	0.0861	0.0169	0.0066	0.0009
1.90	0.9978	0.9713	0.8993	0.7580	0.5495	0.3536	0.1088	0.0240	0.0100	0.0015
1.95	0.9983	0.9766	0.9152	0.7890	0.5928	0.3979	0.1347	0.0330	0.0145	0.0025
2.00	0.9987	0.9810	0.9288	0.8167	0.6337	0.4422	0.1636	0.0442	0.0205	0.0039
2.05	0.9990	0.9846	0.9403	0.8413	0.6720	0.4858	0.1952	0.0578	0.0282	0.0060
2.10	0.9992	0.9875	0.9501	0.8631	0.7076	0.5284	0.2292	0.0741	0.0378	0.0088
2.15	0.9994	0.9898	0.9583	0.8822	0.7404	0.5695	0.2652	0.0929	0.0496	0.0127
2.20	0.9995	0.9918	0.9652	0.8989	0.7703	0.6087	0.3027	0.1145	0.0637	0.0177
2.25	0.9996	0.9933	0.9710	0.9134	0.7974	0.6458	0.3413	0.1386	0.0802	0.0241
2.30	0.9997	0.9946	0.9758	0.9260	0.8219	0.6806	0.3804	0.1651	0.0991	0.0321
2.35	0.9998	0.9956	0.9799	0.9368	0.8439	0.7130	0.4197	0.1938	0.1204	0.0418
2.40	0.9998	0.9964	0.9833	0.9462	0.8635	0.7430	0.4588	0.2245	0.1440	0.0534
2.45	0.9998	0.9971	0.9861	0.9542	0.8808	0.7705	0.4971	0.2569	0.1699	0.0671

PROBABILITY INTEGRAL OF t, THE NON-CENTRAL t-STATISTIC. THIS TABLE GIVES Pr $[t/\sqrt{f} \leq x]$.

f is the number of degrees of freedom; the non-centrality parameter is $\sqrt{f+1}\,K_p$.

K_p is the standardized normal deviate exceeded with probability p.

DEGREES OF FREEDOM **15**

x \ p	.2500	.1500	.1000	.0650	.0400	.0250	.0100	.0040	.0025	.0010
2.50	0.9999	0.9976	0.9884	0.9610	0.8962	0.7956	0.5345	0.2906	0.1977	0.0828
2.55	0.9999	0.9981	0.9904	0.9669	0.9098	0.8185	0.5707	0.3252	0.2273	0.1006
2.60	0.9999	0.9984	0.9920	0.9719	0.9217	0.8392	0.6053	0.3605	0.2583	0.1204
2.65	0.9999	0.9987	0.9933	0.9761	0.9321	0.8578	0.6383	0.3961	0.2906	0.1423
2.70	0.9999	0.9989	0.9944	0.9798	0.9412	0.8745	0.6695	0.4316	0.3237	0.1660
2.75	1.0000	0.9991	0.9954	0.9828	0.9492	0.8894	0.6989	0.4667	0.3574	0.1915
2.80	1.0000	0.9993	0.9961	0.9854	0.9561	0.9027	0.7263	0.5013	0.3914	0.2185
2.85	1.0000	0.9994	0.9968	0.9876	0.9620	0.9145	0.7519	0.5350	0.4253	0.2468
2.90	1.0000	0.9995	0.9973	0.9895	0.9672	0.9250	0.7755	0.5677	0.4590	0.2763
2.95	1.0000	0.9996	0.9977	0.9911	0.9717	0.9343	0.7973	0.5991	0.4922	0.3065
3.00	1.0000	0.9997	0.9981	0.9925	0.9756	0.9424	0.8174	0.6292	0.5247	0.3375
3.05	1.0000	0.9997	0.9984	0.9936	0.9790	0.9496	0.8357	0.6579	0.5562	0.3687
3.10	1.0000	0.9998	0.9987	0.9946	0.9819	0.9559	0.8524	0.6851	0.5867	0.4002
3.15	1.0000	0.9998	0.9989	0.9954	0.9844	0.9615	0.8677	0.7107	0.6160	0.4315
3.20	1.0000	0.9998	0.9991	0.9961	0.9865	0.9664	0.8815	0.7348	0.6440	0.4626
3.25	1.0000	0.9999	0.9992	0.9967	0.9884	0.9706	0.8940	0.7573	0.6707	0.4932
3.30	1.0000	0.9999	0.9993	0.9971	0.9900	0.9744	0.9053	0.7783	0.6960	0.5232
3.35	1.0000	0.9999	0.9994	0.9976	0.9914	0.9776	0.9154	0.7978	0.7199	0.5524
3.40	1.0000	0.9999	0.9995	0.9979	0.9925	0.9805	0.9246	0.8159	0.7423	0.5807
3.45	1.0000	0.9999	0.9996	0.9982	0.9936	0.9829	0.9328	0.8326	0.7633	0.6080
3.50	1.0000	0.9999	0.9997	0.9985	0.9944	0.9851	0.9401	0.8479	0.7830	0.6342
3.55	1.0000	1.0000	0.9997	0.9987	0.9952	0.9870	0.9467	0.8620	0.8013	0.6593
3.60	1.0000	1.0000	0.9998	0.9989	0.9958	0.9887	0.9526	0.8750	0.8183	0.6832
3.65	1.0000	1.0000	0.9998	0.9991	0.9964	0.9901	0.9578	0.8868	0.8341	0.7060
3.70	1.0000	1.0000	0.9998	0.9992	0.9969	0.9914	0.9625	0.8976	0.8486	0.7275
3.75	1.0000	1.0000	0.9999	0.9993	0.9973	0.9924	0.9667	0.9075	0.8621	0.7478
3.80	1.0000	1.0000	0.9999	0.9994	0.9977	0.9934	0.9704	0.9165	0.8744	0.7669
3.85	1.0000	1.0000	0.9999	0.9995	0.9980	0.9942	0.9737	0.9246	0.8858	0.7848
3.90	1.0000	1.0000	0.9999	0.9996	0.9982	0.9950	0.9766	0.9320	0.8962	0.8016
3.95	1.0000	1.0000	0.9999	0.9996	0.9985	0.9956	0.9792	0.9387	0.9057	0.8173
4.00	1.0000	1.0000	0.9999	0.9997	0.9987	0.9961	0.9816	0.9448	0.9144	0.8319
4.05	1.0000	1.0000	0.9999	0.9997	0.9989	0.9966	0.9836	0.9503	0.9224	0.8455
4.10	1.0000	1.0000	1.0000	0.9998	0.9990	0.9970	0.9855	0.9552	0.9296	0.8582
4.15	1.0000	1.0000	1.0000	0.9998	0.9991	0.9974	0.9871	0.9597	0.9363	0.8699
4.20	1.0000	1.0000	1.0000	0.9998	0.9992	0.9977	0.9885	0.9637	0.9423	0.8807
4.25	1.0000	1.0000	1.0000	0.9998	0.9993	0.9980	0.9898	0.9674	0.9477	0.8907
4.30	1.0000	1.0000	1.0000	0.9999	0.9994	0.9982	0.9909	0.9706	0.9527	0.8999
4.35	1.0000	1.0000	1.0000	0.9999	0.9995	0.9985	0.9919	0.9736	0.9572	0.9084
4.40	1.0000	1.0000	1.0000	0.9999	0.9996	0.9986	0.9928	0.9763	0.9613	0.9163
4.45	1.0000	1.0000	1.0000	0.9999	0.9996	0.9988	0.9936	0.9786	0.9650	0.9235
4.50	1.0000	1.0000	1.0000	0.9999	0.9997	0.9989	0.9943	0.9808	0.9684	0.9301
4.55	1.0000	1.0000	1.0000	0.9999	0.9997	0.9991	0.9950	0.9827	0.9714	0.9362
4.60	1.0000	1.0000	1.0000	0.9999	0.9997	0.9992	0.9955	0.9845	0.9741	0.9417
4.65	1.0000	1.0000	1.0000	0.9999	0.9998	0.9993	0.9960	0.9860	0.9766	0.9468
4.70	1.0000	1.0000	1.0000	1.0000	0.9998	0.9994	0.9964	0.9874	0.9789	0.9515
4.75	1.0000	1.0000	1.0000	1.0000	0.9998	0.9994	0.9968	0.9887	0.9809	0.9557
4.80	1.0000	1.0000	1.0000	1.0000	0.9998	0.9995	0.9972	0.9898	0.9827	0.9596
4.85	1.0000	1.0000	1.0000	1.0000	0.9999	0.9996	0.9975	0.9909	0.9844	0.9632
4.90	1.0000	1.0000	1.0000	1.0000	0.9999	0.9996	0.9977	0.9918	0.9859	0.9664
4.95	1.0000	1.0000	1.0000	1.0000	0.9999	0.9997	0.9980	0.9926	0.9873	0.9694
5.00	1.0000	1.0000	1.0000	1.0000	0.9999	0.9997	0.9982	0.9933	0.9885	0.9721
5.05	1.0000	1.0000	1.0000	1.0000	0.9999	0.9997	0.9984	0.9940	0.9896	0.9746
5.10	1.0000	1.0000	1.0000	1.0000	0.9999	0.9998	0.9986	0.9946	0.9906	0.9768
5.15	1.0000	1.0000	1.0000	1.0000	0.9999	0.9998	0.9987	0.9951	0.9915	0.9789
5.20	1.0000	1.0000	1.0000	1.0000	0.9999	0.9998	0.9989	0.9956	0.9923	0.9808
5.25	1.0000	1.0000	1.0000	1.0000	1.0000	0.9998	0.9990	0.9960	0.9930	0.9825
5.30	1.0000	1.0000	1.0000	1.0000	1.0000	0.9999	0.9991	0.9964	0.9937	0.9840
5.35	1.0000	1.0000	1.0000	1.0000	1.0000	0.9999	0.9992	0.9968	0.9943	0.9854
5.40	1.0000	1.0000	1.0000	1.0000	1.0000	0.9999	0.9993	0.9971	0.9948	0.9867
5.45	1.0000	1.0000	1.0000	1.0000	1.0000	0.9999	0.9993	0.9974	0.9953	0.9879

PROBABILITY INTEGRAL OF t, THE NON-CENTRAL t-STATISTIC. THIS TABLE GIVES Pr $[t/\sqrt{f} \leq x]$.

f is the number of degrees of freedom; the non-centrality parameter is $\sqrt{f+1}\,K_p$.

K_p is the standardized normal deviate exceeded with probability p.

DEGREES OF FREEDOM **15**

x \ p	.2500	.1500	.1000	.0650	.0400	.0250	.0100	.0040	.0025	.0010
5.50	1.0000	1.0000	1.0000	1.0000	1.0000	0.9999	0.9994	0.9976	0.9958	0.9890
5.55	1.0000	1.0000	1.0000	1.0000	1.0000	0.9999	0.9995	0.9979	0.9961	0.9900
5.60	1.0000	1.0000	1.0000	1.0000	1.0000	0.9999	0.9995	0.9981	0.9965	0.9908
5.65	1.0000	1.0000	1.0000	1.0000	1.0000	0.9999	0.9996	0.9982	0.9968	0.9916
5.70	1.0000	1.0000	1.0000	1.0000	1.0000	0.9999	0.9996	0.9984	0.9971	0.9924
5.75	1.0000	1.0000	1.0000	1.0000	1.0000	0.9999	0.9997	0.9986	0.9974	0.9930
5.80	1.0000	1.0000	1.0000	1.0000	1.0000	1.0000	0.9997	0.9987	0.9976	0.9937
5.85	1.0000	1.0000	1.0000	1.0000	1.0000	1.0000	0.9997	0.9988	0.9979	0.9942
5.90	1.0000	1.0000	1.0000	1.0000	1.0000	1.0000	0.9997	0.9989	0.9981	0.9947
5.95	1.0000	1.0000	1.0000	1.0000	1.0000	1.0000	0.9998	0.9990	0.9982	0.9952
6.00	1.0000	1.0000	1.0000	1.0000	1.0000	1.0000	0.9998	0.9991	0.9984	0.9956
6.05	1.0000	1.0000	1.0000	1.0000	1.0000	1.0000	0.9998	0.9992	0.9985	0.9960
6.10	1.0000	1.0000	1.0000	1.0000	1.0000	1.0000	0.9998	0.9993	0.9987	0.9963
6.15	1.0000	1.0000	1.0000	1.0000	1.0000	1.0000	0.9999	0.9993	0.9988	0.9966
6.20	1.0000	1.0000	1.0000	1.0000	1.0000	1.0000	0.9999	0.9994	0.9989	0.9969
6.25	1.0000	1.0000	1.0000	1.0000	1.0000	1.0000	0.9999	0.9995	0.9990	0.9972
6.30	1.0000	1.0000	1.0000	1.0000	1.0000	1.0000	0.9999	0.9995	0.9991	0.9974
6.35	1.0000	1.0000	1.0000	1.0000	1.0000	1.0000	0.9999	0.9996	0.9992	0.9976
6.40	1.0000	1.0000	1.0000	1.0000	1.0000	1.0000	0.9999	0.9996	0.9992	0.9978
6.45	1.0000	1.0000	1.0000	1.0000	1.0000	1.0000	0.9999	0.9996	0.9993	0.9980
6.50	1.0000	1.0000	1.0000	1.0000	1.0000	1.0000	0.9999	0.9997	0.9994	0.9982
6.55	1.0000	1.0000	1.0000	1.0000	1.0000	1.0000	0.9999	0.9997	0.9994	0.9983
6.60	1.0000	1.0000	1.0000	1.0000	1.0000	1.0000	0.9999	0.9997	0.9995	0.9985
6.65	1.0000	1.0000	1.0000	1.0000	1.0000	1.0000	0.9999	0.9997	0.9995	0.9986
6.70	1.0000	1.0000	1.0000	1.0000	1.0000	1.0000	1.0000	0.9998	0.9996	0.9987
6.75	1.0000	1.0000	1.0000	1.0000	1.0000	1.0000	1.0000	0.9998	0.9996	0.9988
6.80	1.0000	1.0000	1.0000	1.0000	1.0000	1.0000	1.0000	0.9998	0.9996	0.9989
6.85	1.0000	1.0000	1.0000	1.0000	1.0000	1.0000	1.0000	0.9998	0.9997	0.9990
6.90	1.0000	1.0000	1.0000	1.0000	1.0000	1.0000	1.0000	0.9998	0.9997	0.9991
6.95	1.0000	1.0000	1.0000	1.0000	1.0000	1.0000	1.0000	0.9999	0.9997	0.9992
7.00	1.0000	1.0000	1.0000	1.0000	1.0000	1.0000	1.0000	0.9999	0.9997	0.9992
7.05	1.0000	1.0000	1.0000	1.0000	1.0000	1.0000	1.0000	0.9999	0.9998	0.9993
7.10	1.0000	1.0000	1.0000	1.0000	1.0000	1.0000	1.0000	0.9999	0.9998	0.9994
7.15	1.0000	1.0000	1.0000	1.0000	1.0000	1.0000	1.0000	0.9999	0.9998	0.9994
7.20	1.0000	1.0000	1.0000	1.0000	1.0000	1.0000	1.0000	0.9999	0.9998	0.9995
7.25	1.0000	1.0000	1.0000	1.0000	1.0000	1.0000	1.0000	0.9999	0.9998	0.9995
7.30	1.0000	1.0000	1.0000	1.0000	1.0000	1.0000	1.0000	0.9999	0.9999	0.9995
7.35	1.0000	1.0000	1.0000	1.0000	1.0000	1.0000	1.0000	0.9999	0.9999	0.9996
7.40	1.0000	1.0000	1.0000	1.0000	1.0000	1.0000	1.0000	0.9999	0.9999	0.9996
7.45	1.0000	1.0000	1.0000	1.0000	1.0000	1.0000	1.0000	0.9999	0.9999	0.9996
7.50	1.0000	1.0000	1.0000	1.0000	1.0000	1.0000	1.0000	0.9999	0.9999	0.9997
7.55	1.0000	1.0000	1.0000	1.0000	1.0000	1.0000	1.0000	1.0000	0.9999	0.9997
7.60	1.0000	1.0000	1.0000	1.0000	1.0000	1.0000	1.0000	1.0000	0.9999	0.9997
7.65	1.0000	1.0000	1.0000	1.0000	1.0000	1.0000	1.0000	1.0000	0.9999	0.9997
7.70	1.0000	1.0000	1.0000	1.0000	1.0000	1.0000	1.0000	1.0000	0.9999	0.9998
7.75	1.0000	1.0000	1.0000	1.0000	1.0000	1.0000	1.0000	1.0000	0.9999	0.9998
7.80	1.0000	1.0000	1.0000	1.0000	1.0000	1.0000	1.0000	1.0000	0.9999	0.9998
7.85	1.0000	1.0000	1.0000	1.0000	1.0000	1.0000	1.0000	1.0000	0.9999	0.9998
7.90	1.0000	1.0000	1.0000	1.0000	1.0000	1.0000	1.0000	1.0000	0.9999	0.9998
7.95	1.0000	1.0000	1.0000	1.0000	1.0000	1.0000	1.0000	1.0000	1.0000	0.9998
8.00	1.0000	1.0000	1.0000	1.0000	1.0000	1.0000	1.0000	1.0000	1.0000	0.9999
8.05	1.0000	1.0000	1.0000	1.0000	1.0000	1.0000	1.0000	1.0000	1.0000	0.9999
8.10	1.0000	1.0000	1.0000	1.0000	1.0000	1.0000	1.0000	1.0000	1.0000	0.9999
8.15	1.0000	1.0000	1.0000	1.0000	1.0000	1.0000	1.0000	1.0000	1.0000	0.9999
8.20	1.0000	1.0000	1.0000	1.0000	1.0000	1.0000	1.0000	1.0000	1.0000	0.9999
8.25	1.0000	1.0000	1.0000	1.0000	1.0000	1.0000	1.0000	1.0000	1.0000	0.9999
8.30	1.0000	1.0000	1.0000	1.0000	1.0000	1.0000	1.0000	1.0000	1.0000	0.9999
8.35	1.0000	1.0000	1.0000	1.0000	1.0000	1.0000	1.0000	1.0000	1.0000	0.9999
8.40	1.0000	1.0000	1.0000	1.0000	1.0000	1.0000	1.0000	1.0000	1.0000	0.9999
8.45	1.0000	1.0000	1.0000	1.0000	1.0000	1.0000	1.0000	1.0000	1.0000	0.9999

PROBABILITY INTEGRAL OF t, THE NON-CENTRAL t-STATISTIC. THIS TABLE GIVES Pr $[t/\sqrt{f} \leq x]$.

f is the number of degrees of freedom; the non-centrality parameter is $\sqrt{f+1}\,K_p$.

K_p is the standardized normal deviate exceeded with probability p.

DEGREES OF FREEDOM **15**

x \ p	.2500	.1500	.1000	.0650	.0400	.0250	.0100	.0040	.0025	.0010
8.50	1.0000	1.0000	1.0000	1.0000	1.0000	1.0000	1.0000	1.0000	1.0000	0.9999
8.55	1.0000	1.0000	1.0000	1.0000	1.0000	1.0000	1.0000	1.0000	1.0000	0.9999
8.60	1.0000	1.0000	1.0000	1.0000	1.0000	1.0000	1.0000	1.0000	1.0000	0.9999
8.65	1.0000	1.0000	1.0000	1.0000	1.0000	1.0000	1.0000	1.0000	1.0000	0.9999
8.70	1.0000	1.0000	1.0000	1.0000	1.0000	1.0000	1.0000	1.0000	1.0000	1.0000

PROBABILITY INTEGRAL OF t, THE NON-CENTRAL t-STATISTIC. THIS TABLE GIVES Pr $[t/\sqrt{f} \leq x]$.

f is the number of degrees of freedom; the non-centrality parameter is $\sqrt{f+1}\, K_p$.
K_p is the standardized normal deviate exceeded with probability p.

DEGREES OF FREEDOM **16**

x \ P	.2500	.1500	.1000	.0650	.0400	.0250	.0100	.0040	.0025	.0010
− 0.35	0.0000	0.0000	0.0000	0.0000	0.0000	0.0000	0.0000	0.0000	0.0000	0.0000
− 0.30	0.0001	0.0000	0.0000	0.0000	0.0000	0.0000	0.0000	0.0000	0.0000	0.0000
− 0.25	0.0001	0.0000	0.0000	0.0000	0.0000	0.0000	0.0000	0.0000	0.0000	0.0000
− 0.20	0.0002	0.0000	0.0000	0.0000	0.0000	0.0000	0.0000	0.0000	0.0000	0.0000
− 0.15	0.0004	0.0000	0.0000	0.0000	0.0000	0.0000	0.0000	0.0000	0.0000	0.0000
− 0.10	0.0008	0.0000	0.0000	0.0000	0.0000	0.0000	0.0000	0.0000	0.0000	0.0000
− 0.05	0.0015	0.0000	0.0000	0.0000	0.0000	0.0000	0.0000	0.0000	0.0000	0.0000
0.00	0.0027	0.0000	0.0000	0.0000	0.0000	0.0000	0.0000	0.0000	0.0000	0.0000
0.05	0.0049	0.0000	0.0000	0.0000	0.0000	0.0000	0.0000	0.0000	0.0000	0.0000
0.10	0.0086	0.0001	0.0000	0.0000	0.0000	0.0000	0.0000	0.0000	0.0000	0.0000
0.15	0.0147	0.0001	0.0000	0.0000	0.0000	0.0000	0.0000	0.0000	0.0000	0.0000
0.20	0.0242	0.0003	0.0000	0.0000	0.0000	0.0000	0.0000	0.0000	0.0000	0.0000
0.25	0.0384	0.0006	0.0000	0.0000	0.0000	0.0000	0.0000	0.0000	0.0000	0.0000
0.30	0.0588	0.0013	0.0000	0.0000	0.0000	0.0000	0.0000	0.0000	0.0000	0.0000
0.35	0.0866	0.0025	0.0001	0.0000	0.0000	0.0000	0.0000	0.0000	0.0000	0.0000
0.40	0.1229	0.0047	0.0002	0.0000	0.0000	0.0000	0.0000	0.0000	0.0000	0.0000
0.45	0.1680	0.0086	0.0004	0.0000	0.0000	0.0000	0.0000	0.0000	0.0000	0.0000
0.50	0.2216	0.0150	0.0009	0.0000	0.0000	0.0000	0.0000	0.0000	0.0000	0.0000
0.55	0.2827	0.0249	0.0019	0.0001	0.0000	0.0000	0.0000	0.0000	0.0000	0.0000
0.60	0.3494	0.0394	0.0037	0.0002	0.0000	0.0000	0.0000	0.0000	0.0000	0.0000
0.65	0.4194	0.0598	0.0068	0.0004	0.0000	0.0000	0.0000	0.0000	0.0000	0.0000
0.70	0.4902	0.0870	0.0120	0.0009	0.0000	0.0000	0.0000	0.0000	0.0000	0.0000
0.75	0.5595	0.1216	0.0201	0.0019	0.0001	0.0000	0.0000	0.0000	0.0000	0.0000
0.80	0.6251	0.1637	0.0320	0.0037	0.0002	0.0000	0.0000	0.0000	0.0000	0.0000
0.85	0.6855	0.2128	0.0488	0.0068	0.0005	0.0000	0.0000	0.0000	0.0000	0.0000
0.90	0.7398	0.2678	0.0714	0.0118	0.0011	0.0001	0.0000	0.0000	0.0000	0.0000
0.95	0.7873	0.3274	0.1002	0.0195	0.0021	0.0002	0.0000	0.0000	0.0000	0.0000
1.00	0.8281	0.3898	0.1355	0.0306	0.0040	0.0004	0.0000	0.0000	0.0000	0.0000
1.05	0.8624	0.4531	0.1772	0.0459	0.0071	0.0009	0.0000	0.0000	0.0000	0.0000
1.10	0.8909	0.5157	0.2246	0.0663	0.0120	0.0018	0.0000	0.0000	0.0000	0.0000
1.15	0.9142	0.5759	0.2767	0.0920	0.0193	0.0033	0.0001	0.0000	0.0000	0.0000
1.20	0.9329	0.6326	0.3322	0.1235	0.0296	0.0058	0.0001	0.0000	0.0000	0.0000
1.25	0.9479	0.6848	0.3898	0.1604	0.0437	0.0097	0.0003	0.0000	0.0000	0.0000
1.30	0.9597	0.7321	0.4480	0.2025	0.0620	0.0156	0.0006	0.0000	0.0000	0.0000
1.35	0.9690	0.7742	0.5054	0.2489	0.0850	0.0239	0.0012	0.0000	0.0000	0.0000
1.40	0.9762	0.8111	0.5609	0.2988	0.1129	0.0353	0.0023	0.0001	0.0000	0.0000
1.45	0.9818	0.8431	0.6135	0.3509	0.1455	0.0501	0.0039	0.0002	0.0000	0.0000
1.50	0.9861	0.8704	0.6625	0.4043	0.1827	0.0690	0.0066	0.0004	0.0001	0.0000
1.55	0.9894	0.8936	0.7075	0.4578	0.2239	0.0919	0.0104	0.0008	0.0002	0.0000
1.60	0.9919	0.9130	0.7483	0.5103	0.2683	0.1192	0.0160	0.0014	0.0004	0.0000
1.65	0.9938	0.9292	0.7847	0.5610	0.3153	0.1506	0.0236	0.0024	0.0007	0.0000
1.70	0.9953	0.9426	0.8169	0.6092	0.3638	0.1859	0.0336	0.0041	0.0012	0.0001
1.75	0.9964	0.9535	0.8450	0.6544	0.4130	0.2245	0.0465	0.0065	0.0021	0.0002
1.80	0.9973	0.9625	0.8695	0.6963	0.4621	0.2660	0.0625	0.0100	0.0035	0.0004
1.85	0.9979	0.9698	0.8905	0.7345	0.5102	0.3097	0.0818	0.0149	0.0056	0.0007
1.90	0.9984	0.9757	0.9084	0.7692	0.5568	0.3547	0.1045	0.0214	0.0085	0.0012
1.95	0.9988	0.9805	0.9237	0.8002	0.6012	0.4005	0.1305	0.0300	0.0127	0.0020
2.00	0.9991	0.9844	0.9366	0.8278	0.6432	0.4462	0.1598	0.0407	0.0182	0.0032
2.05	0.9993	0.9875	0.9474	0.8522	0.6823	0.4913	0.1920	0.0540	0.0254	0.0050
2.10	0.9994	0.9900	0.9565	0.8735	0.7185	0.5353	0.2268	0.0700	0.0346	0.0075
2.15	0.9996	0.9920	0.9641	0.8921	0.7516	0.5776	0.2638	0.0887	0.0460	0.0110
2.20	0.9997	0.9936	0.9703	0.9083	0.7817	0.6178	0.3024	0.1102	0.0598	0.0156
2.25	0.9997	0.9948	0.9756	0.9222	0.8089	0.6558	0.3423	0.1345	0.0760	0.0216
2.30	0.9998	0.9959	0.9799	0.9341	0.8332	0.6913	0.3827	0.1614	0.0949	0.0291
2.35	0.9998	0.9967	0.9834	0.9443	0.8549	0.7242	0.4234	0.1908	0.1162	0.0385
2.40	0.9999	0.9973	0.9864	0.9530	0.8740	0.7544	0.4637	0.2222	0.1401	0.0498
2.45	0.9999	0.9979	0.9888	0.9604	0.8910	0.7821	0.5034	0.2555	0.1664	0.0631

PROBABILITY INTEGRAL OF t, THE NON-CENTRAL t-STATISTIC. THIS TABLE GIVES Pr $[t/\sqrt{f} \leq x]$.

f is the number of degrees of freedom; the non-centrality parameter is $\sqrt{f+1}\,K_p$.

K_p is the standardized normal deviate exceeded with probability p.

DEGREES OF FREEDOM **16**

x \ p	.2500	.1500	.1000	.0650	.0400	.0250	.0100	.0040	.0025	.0010
2.50	0.9999	0.9983	0.9908	0.9666	0.9058	0.8072	0.5419	0.2901	0.1948	0.0787
2.55	0.9999	0.9986	0.9924	0.9719	0.9188	0.8299	0.5791	0.3259	0.2251	0.0964
2.60	1.0000	0.9989	0.9938	0.9764	0.9301	0.8504	0.6147	0.3624	0.2570	0.1164
2.65	1.0000	0.9991	0.9949	0.9802	0.9399	0.8686	0.6485	0.3991	0.2902	0.1384
2.70	1.0000	0.9993	0.9958	0.9834	0.9485	0.8849	0.6803	0.4359	0.3244	0.1625
2.75	1.0000	0.9994	0.9965	0.9860	0.9558	0.8993	0.7101	0.4722	0.3592	0.1885
2.80	1.0000	0.9995	0.9971	0.9883	0.9621	0.9121	0.7379	0.5079	0.3944	0.2162
2.85	1.0000	0.9996	0.9976	0.9902	0.9676	0.9234	0.7636	0.5426	0.4295	0.2453
2.90	1.0000	0.9997	0.9980	0.9917	0.9723	0.9333	0.7874	0.5763	0.4643	0.2756
2.95	1.0000	0.9997	0.9984	0.9931	0.9763	0.9420	0.8091	0.6085	0.4986	0.3068
3.00	1.0000	0.9998	0.9987	0.9942	0.9798	0.9496	0.8290	0.6394	0.5320	0.3387
3.05	1.0000	0.9998	0.9989	0.9951	0.9827	0.9563	0.8471	0.6687	0.5645	0.3711
3.10	1.0000	0.9999	0.9991	0.9959	0.9852	0.9621	0.8636	0.6963	0.5959	0.4036
3.15	1.0000	0.9999	0.9992	0.9965	0.9874	0.9671	0.8784	0.7223	0.6259	0.4360
3.20	1.0000	0.9999	0.9994	0.9971	0.9892	0.9715	0.8918	0.7466	0.6546	0.4681
3.25	1.0000	0.9999	0.9995	0.9975	0.9908	0.9753	0.9039	0.7693	0.6817	0.4997
3.30	1.0000	0.9999	0.9996	0.9979	0.9921	0.9787	0.9147	0.7903	0.7074	0.5306
3.35	1.0000	0.9999	0.9996	0.9983	0.9933	0.9815	0.9243	0.8097	0.7316	0.5607
3.40	1.0000	1.0000	0.9997	0.9985	0.9943	0.9840	0.9330	0.8277	0.7542	0.5898
3.45	1.0000	1.0000	0.9997	0.9988	0.9951	0.9862	0.9407	0.8441	0.7754	0.6178
3.50	1.0000	1.0000	0.9998	0.9989	0.9958	0.9880	0.9475	0.8592	0.7950	0.6446
3.55	1.0000	1.0000	0.9998	0.9991	0.9964	0.9896	0.9536	0.8730	0.8133	0.6703
3.60	1.0000	1.0000	0.9998	0.9992	0.9969	0.9910	0.9590	0.8856	0.8301	0.6946
3.65	1.0000	1.0000	0.9999	0.9994	0.9974	0.9922	0.9638	0.8971	0.8457	0.7176
3.70	1.0000	1.0000	0.9999	0.9995	0.9977	0.9933	0.9681	0.9074	0.8600	0.7394
3.75	1.0000	1.0000	0.9999	0.9995	0.9981	0.9942	0.9718	0.9169	0.8731	0.7598
3.80	1.0000	1.0000	0.9999	0.9996	0.9983	0.9950	0.9752	0.9254	0.8851	0.7790
3.85	1.0000	1.0000	0.9999	0.9997	0.9986	0.9956	0.9781	0.9331	0.8961	0.7969
3.90	1.0000	1.0000	0.9999	0.9997	0.9988	0.9962	0.9807	0.9400	0.9061	0.8137
3.95	1.0000	1.0000	1.0000	0.9998	0.9989	0.9967	0.9830	0.9463	0.9152	0.8292
4.00	1.0000	1.0000	1.0000	0.9998	0.9991	0.9971	0.9850	0.9519	0.9235	0.8436
4.05	1.0000	1.0000	1.0000	0.9998	0.9992	0.9975	0.9868	0.9570	0.9310	0.8570
4.10	1.0000	1.0000	1.0000	0.9998	0.9993	0.9978	0.9883	0.9615	0.9378	0.8693
4.15	1.0000	1.0000	1.0000	0.9999	0.9994	0.9981	0.9897	0.9656	0.9440	0.8807
4.20	1.0000	1.0000	1.0000	0.9999	0.9995	0.9984	0.9909	0.9692	0.9496	0.8912
4.25	1.0000	1.0000	1.0000	0.9999	0.9996	0.9986	0.9920	0.9725	0.9547	0.9009
4.30	1.0000	1.0000	1.0000	0.9999	0.9996	0.9988	0.9929	0.9754	0.9592	0.9097
4.35	1.0000	1.0000	1.0000	0.9999	0.9997	0.9989	0.9938	0.9780	0.9633	0.9178
4.40	1.0000	1.0000	1.0000	0.9999	0.9997	0.9991	0.9945	0.9804	0.9670	0.9253
4.45	1.0000	1.0000	1.0000	0.9999	0.9998	0.9992	0.9952	0.9825	0.9704	0.9321
4.50	1.0000	1.0000	1.0000	1.0000	0.9998	0.9993	0.9957	0.9843	0.9734	0.9383
4.55	1.0000	1.0000	1.0000	1.0000	0.9998	0.9994	0.9962	0.9860	0.9761	0.9440
4.60	1.0000	1.0000	1.0000	1.0000	0.9998	0.9995	0.9967	0.9875	0.9785	0.9491
4.65	1.0000	1.0000	1.0000	1.0000	0.9999	0.9995	0.9970	0.9888	0.9807	0.9538
4.70	1.0000	1.0000	1.0000	1.0000	0.9999	0.9996	0.9974	0.9900	0.9827	0.9581
4.75	1.0000	1.0000	1.0000	1.0000	0.9999	0.9996	0.9977	0.9911	0.9844	0.9620
4.80	1.0000	1.0000	1.0000	1.0000	0.9999	0.9997	0.9980	0.9920	0.9860	0.9656
4.85	1.0000	1.0000	1.0000	1.0000	0.9999	0.9997	0.9982	0.9929	0.9875	0.9688
4.90	1.0000	1.0000	1.0000	1.0000	0.9999	0.9998	0.9984	0.9936	0.9887	0.9717
4.95	1.0000	1.0000	1.0000	1.0000	0.9999	0.9998	0.9986	0.9943	0.9899	0.9744
5.00	1.0000	1.0000	1.0000	1.0000	0.9999	0.9998	0.9987	0.9949	0.9909	0.9768
5.05	1.0000	1.0000	1.0000	1.0000	0.9999	0.9998	0.9989	0.9954	0.9918	0.9789
5.10	1.0000	1.0000	1.0000	1.0000	1.0000	0.9998	0.9990	0.9959	0.9927	0.9809
5.15	1.0000	1.0000	1.0000	1.0000	1.0000	0.9999	0.9991	0.9964	0.9934	0.9827
5.20	1.0000	1.0000	1.0000	1.0000	1.0000	0.9999	0.9992	0.9967	0.9941	0.9843
5.25	1.0000	1.0000	1.0000	1.0000	1.0000	0.9999	0.9993	0.9971	0.9947	0.9858
5.30	1.0000	1.0000	1.0000	1.0000	1.0000	0.9999	0.9994	0.9974	0.9952	0.9871
5.35	1.0000	1.0000	1.0000	1.0000	1.0000	0.9999	0.9994	0.9977	0.9957	0.9884
5.40	1.0000	1.0000	1.0000	1.0000	1.0000	0.9999	0.9995	0.9979	0.9961	0.9894
5.45	1.0000	1.0000	1.0000	1.0000	1.0000	0.9999	0.9996	0.9981	0.9965	0.9904

PROBABILITY INTEGRAL OF t, THE NON-CENTRAL t-STATISTIC. THIS TABLE GIVES Pr $[t/\sqrt{f} \leq x]$.

f is the number of degrees of freedom; the non-centrality parameter is $\sqrt{f+1}\,K_p$.

K_p is the standardized normal deviate exceeded with probability p.

DEGREES OF FREEDOM **16**

x \ p	.2500	.1500	.1000	.0650	.0400	.0250	.0100	.0040	.0025	.0010
5.50	1.0000	1.0000	1.0000	1.0000	1.0000	0.9999	0.9996	0.9983	0.9969	0.9913
5.55	1.0000	1.0000	1.0000	1.0000	1.0000	0.9999	0.9997	0.9985	0.9972	0.9921
5.60	1.0000	1.0000	1.0000	1.0000	1.0000	1.0000	0.9997	0.9986	0.9974	0.9929
5.65	1.0000	1.0000	1.0000	1.0000	1.0000	1.0000	0.9997	0.9988	0.9977	0.9935
5.70	1.0000	1.0000	1.0000	1.0000	1.0000	1.0000	0.9998	0.9989	0.9979	0.9941
5.75	1.0000	1.0000	1.0000	1.0000	1.0000	1.0000	0.9998	0.9990	0.9981	0.9947
5.80	1.0000	1.0000	1.0000	1.0000	1.0000	1.0000	0.9998	0.9991	0.9983	0.9952
5.85	1.0000	1.0000	1.0000	1.0000	1.0000	1.0000	0.9998	0.9992	0.9985	0.9956
5.90	1.0000	1.0000	1.0000	1.0000	1.0000	1.0000	0.9998	0.9993	0.9986	0.9960
5.95	1.0000	1.0000	1.0000	1.0000	1.0000	1.0000	0.9999	0.9993	0.9988	0.9964
6.00	1.0000	1.0000	1.0000	1.0000	1.0000	1.0000	0.9999	0.9994	0.9989	0.9967
6.05	1.0000	1.0000	1.0000	1.0000	1.0000	1.0000	0.9999	0.9995	0.9990	0.9970
6.10	1.0000	1.0000	1.0000	1.0000	1.0000	1.0000	0.9999	0.9995	0.9991	0.9973
6.15	1.0000	1.0000	1.0000	1.0000	1.0000	1.0000	0.9999	0.9996	0.9992	0.9975
6.20	1.0000	1.0000	1.0000	1.0000	1.0000	1.0000	0.9999	0.9996	0.9992	0.9978
6.25	1.0000	1.0000	1.0000	1.0000	1.0000	1.0000	0.9999	0.9996	0.9993	0.9980
6.30	1.0000	1.0000	1.0000	1.0000	1.0000	1.0000	0.9999	0.9997	0.9994	0.9982
6.35	1.0000	1.0000	1.0000	1.0000	1.0000	1.0000	0.9999	0.9997	0.9994	0.9983
6.40	1.0000	1.0000	1.0000	1.0000	1.0000	1.0000	0.9999	0.9997	0.9995	0.9985
6.45	1.0000	1.0000	1.0000	1.0000	1.0000	1.0000	0.9999	0.9998	0.9995	0.9986
6.50	1.0000	1.0000	1.0000	1.0000	1.0000	1.0000	1.0000	0.9998	0.9996	0.9987
6.55	1.0000	1.0000	1.0000	1.0000	1.0000	1.0000	1.0000	0.9998	0.9996	0.9988
6.60	1.0000	1.0000	1.0000	1.0000	1.0000	1.0000	1.0000	0.9998	0.9997	0.9989
6.65	1.0000	1.0000	1.0000	1.0000	1.0000	1.0000	1.0000	0.9998	0.9997	0.9990
6.70	1.0000	1.0000	1.0000	1.0000	1.0000	1.0000	1.0000	0.9999	0.9997	0.9991
6.75	1.0000	1.0000	1.0000	1.0000	1.0000	1.0000	1.0000	0.9999	0.9997	0.9992
6.80	1.0000	1.0000	1.0000	1.0000	1.0000	1.0000	1.0000	0.9999	0.9998	0.9993
6.85	1.0000	1.0000	1.0000	1.0000	1.0000	1.0000	1.0000	0.9999	0.9998	0.9993
6.90	1.0000	1.0000	1.0000	1.0000	1.0000	1.0000	1.0000	0.9999	0.9998	0.9994
6.95	1.0000	1.0000	1.0000	1.0000	1.0000	1.0000	1.0000	0.9999	0.9998	0.9994
7.00	1.0000	1.0000	1.0000	1.0000	1.0000	1.0000	1.0000	0.9999	0.9998	0.9995
7.05	1.0000	1.0000	1.0000	1.0000	1.0000	1.0000	1.0000	0.9999	0.9999	0.9995
7.10	1.0000	1.0000	1.0000	1.0000	1.0000	1.0000	1.0000	0.9999	0.9999	0.9996
7.15	1.0000	1.0000	1.0000	1.0000	1.0000	1.0000	1.0000	0.9999	0.9999	0.9996
7.20	1.0000	1.0000	1.0000	1.0000	1.0000	1.0000	1.0000	0.9999	0.9999	0.9996
7.25	1.0000	1.0000	1.0000	1.0000	1.0000	1.0000	1.0000	0.9999	0.9999	0.9997
7.30	1.0000	1.0000	1.0000	1.0000	1.0000	1.0000	1.0000	0.9999	0.9999	0.9997
7.35	1.0000	1.0000	1.0000	1.0000	1.0000	1.0000	1.0000	1.0000	0.9999	0.9997
7.40	1.0000	1.0000	1.0000	1.0000	1.0000	1.0000	1.0000	1.0000	0.9999	0.9997
7.45	1.0000	1.0000	1.0000	1.0000	1.0000	1.0000	1.0000	1.0000	0.9999	0.9998
7.50	1.0000	1.0000	1.0000	1.0000	1.0000	1.0000	1.0000	1.0000	0.9999	0.9998
7.55	1.0000	1.0000	1.0000	1.0000	1.0000	1.0000	1.0000	1.0000	0.9999	0.9998
7.60	1.0000	1.0000	1.0000	1.0000	1.0000	1.0000	1.0000	1.0000	0.9999	0.9998
7.65	1.0000	1.0000	1.0000	1.0000	1.0000	1.0000	1.0000	1.0000	0.9999	0.9998
7.70	1.0000	1.0000	1.0000	1.0000	1.0000	1.0000	1.0000	1.0000	1.0000	0.9998
7.75	1.0000	1.0000	1.0000	1.0000	1.0000	1.0000	1.0000	1.0000	1.0000	0.9999
7.80	1.0000	1.0000	1.0000	1.0000	1.0000	1.0000	1.0000	1.0000	1.0000	0.9999
7.85	1.0000	1.0000	1.0000	1.0000	1.0000	1.0000	1.0000	1.0000	1.0000	0.9999
7.90	1.0000	1.0000	1.0000	1.0000	1.0000	1.0000	1.0000	1.0000	1.0000	0.9999
7.95	1.0000	1.0000	1.0000	1.0000	1.0000	1.0000	1.0000	1.0000	1.0000	0.9999
8.00	1.0000	1.0000	1.0000	1.0000	1.0000	1.0000	1.0000	1.0000	1.0000	0.9999
8.05	1.0000	1.0000	1.0000	1.0000	1.0000	1.0000	1.0000	1.0000	1.0000	0.9999
8.10	1.0000	1.0000	1.0000	1.0000	1.0000	1.0000	1.0000	1.0000	1.0000	0.9999
8.15	1.0000	1.0000	1.0000	1.0000	1.0000	1.0000	1.0000	1.0000	1.0000	0.9999
8.20	1.0000	1.0000	1.0000	1.0000	1.0000	1.0000	1.0000	1.0000	1.0000	0.9999
8.25	1.0000	1.0000	1.0000	1.0000	1.0000	1.0000	1.0000	1.0000	1.0000	0.9999
8.30	1.0000	1.0000	1.0000	1.0000	1.0000	1.0000	1.0000	1.0000	1.0000	0.9999
8.35	1.0000	1.0000	1.0000	1.0000	1.0000	1.0000	1.0000	1.0000	1.0000	0.9999
8.40	1.0000	1.0000	1.0000	1.0000	1.0000	1.0000	1.0000	1.0000	1.0000	0.9999
8.45	1.0000	1.0000	1.0000	1.0000	1.0000	1.0000	1.0000	1.0000	1.0000	1.0000

PROBABILITY INTEGRAL OF t, THE NON-CENTRAL t-STATISTIC. THIS TABLE GIVES Pr $[t/\sqrt{f} \leq x]$.

f is the number of degrees of freedom; the non-centrality parameter is $\sqrt{f+1}\,K_p$.
K_p is the standardized normal deviate exceeded with probability p.

DEGREES OF FREEDOM **17**

x \ p	.2500	.1500	.1000	.0650	.0400	.0250	.0100	.0040	.0025	.0010
-0.30	0.0000	0.0000	0.0000	0.0000	0.0000	0.0000	0.0000	0.0000	0.0000	0.0000
-0.25	0.0001	0.0000	0.0000	0.0000	0.0000	0.0000	0.0000	0.0000	0.0000	0.0000
-0.20	0.0001	0.0000	0.0000	0.0000	0.0000	0.0000	0.0000	0.0000	0.0000	0.0000
-0.15	0.0003	0.0000	0.0000	0.0000	0.0000	0.0000	0.0000	0.0000	0.0000	0.0000
-0.10	0.0006	0.0000	0.0000	0.0000	0.0000	0.0000	0.0000	0.0000	0.0000	0.0000
-0.05	0.0011	0.0000	0.0000	0.0000	0.0000	0.0000	0.0000	0.0000	0.0000	0.0000
0.00	0.0021	0.0000	0.0000	0.0000	0.0000	0.0000	0.0000	0.0000	0.0000	0.0000
0.05	0.0039	0.0000	0.0000	0.0000	0.0000	0.0000	0.0000	0.0000	0.0000	0.0000
0.10	0.0072	0.0000	0.0000	0.0000	0.0000	0.0000	0.0000	0.0000	0.0000	0.0000
0.15	0.0126	0.0001	0.0000	0.0000	0.0000	0.0000	0.0000	0.0000	0.0000	0.0000
0.20	0.0212	0.0002	0.0000	0.0000	0.0000	0.0000	0.0000	0.0000	0.0000	0.0000
0.25	0.0346	0.0004	0.0000	0.0000	0.0000	0.0000	0.0000	0.0000	0.0000	0.0000
0.30	0.0540	0.0009	0.0000	0.0000	0.0000	0.0000	0.0000	0.0000	0.0000	0.0000
0.35	0.0811	0.0019	0.0000	0.0000	0.0000	0.0000	0.0000	0.0000	0.0000	0.0000
0.40	0.1170	0.0038	0.0001	0.0000	0.0000	0.0000	0.0000	0.0000	0.0000	0.0000
0.45	0.1622	0.0072	0.0003	0.0000	0.0000	0.0000	0.0000	0.0000	0.0000	0.0000
0.50	0.2166	0.0129	0.0007	0.0000	0.0000	0.0000	0.0000	0.0000	0.0000	0.0000
0.55	0.2790	0.0220	0.0015	0.0001	0.0000	0.0000	0.0000	0.0000	0.0000	0.0000
0.60	0.3475	0.0358	0.0030	0.0001	0.0000	0.0000	0.0000	0.0000	0.0000	0.0000
0.65	0.4196	0.0554	0.0057	0.0003	0.0000	0.0000	0.0000	0.0000	0.0000	0.0000
0.70	0.4927	0.0820	0.0103	0.0007	0.0000	0.0000	0.0000	0.0000	0.0000	0.0000
0.75	0.5641	0.1164	0.0177	0.0015	0.0001	0.0000	0.0000	0.0000	0.0000	0.0000
0.80	0.6315	0.1586	0.0289	0.0030	0.0002	0.0000	0.0000	0.0000	0.0000	0.0000
0.85	0.6933	0.2084	0.0449	0.0057	0.0004	0.0000	0.0000	0.0000	0.0000	0.0000
0.90	0.7485	0.2647	0.0668	0.0102	0.0008	0.0001	0.0000	0.0000	0.0000	0.0000
0.95	0.7964	0.3259	0.0953	0.0172	0.0017	0.0001	0.0000	0.0000	0.0000	0.0000
1.00	0.8372	0.3902	0.1306	0.0276	0.0033	0.0003	0.0000	0.0000	0.0000	0.0000
1.05	0.8713	0.4555	0.1727	0.0423	0.0060	0.0007	0.0000	0.0000	0.0000	0.0000
1.10	0.8992	0.5200	0.2210	0.0620	0.0104	0.0014	0.0000	0.0000	0.0000	0.0000
1.15	0.9217	0.5820	0.2743	0.0874	0.0171	0.0027	0.0000	0.0000	0.0000	0.0000
1.20	0.9396	0.6402	0.3314	0.1188	0.0268	0.0048	0.0001	0.0000	0.0000	0.0000
1.25	0.9538	0.6935	0.3908	0.1561	0.0402	0.0083	0.0002	0.0000	0.0000	0.0000
1.30	0.9648	0.7415	0.4508	0.1988	0.0580	0.0137	0.0005	0.0000	0.0000	0.0000
1.35	0.9733	0.7840	0.5100	0.2462	0.0806	0.0214	0.0010	0.0000	0.0000	0.0000
1.40	0.9798	0.8210	0.5671	0.2974	0.1084	0.0322	0.0018	0.0001	0.0000	0.0000
1.45	0.9848	0.8527	0.6211	0.3511	0.1412	0.0465	0.0032	0.0001	0.0000	0.0000
1.50	0.9886	0.8796	0.6712	0.4062	0.1789	0.0649	0.0055	0.0003	0.0001	0.0000
1.55	0.9914	0.9022	0.7170	0.4613	0.2209	0.0876	0.0090	0.0006	0.0001	0.0000
1.60	0.9936	0.9209	0.7583	0.5155	0.2664	0.1149	0.0141	0.0011	0.0003	0.0000
1.65	0.9952	0.9363	0.7949	0.5676	0.3147	0.1465	0.0212	0.0020	0.0005	0.0000
1.70	0.9964	0.9490	0.8270	0.6171	0.3647	0.1823	0.0307	0.0034	0.0009	0.0001
1.75	0.9973	0.9592	0.8549	0.6634	0.4155	0.2218	0.0431	0.0055	0.0017	0.0001
1.80	0.9980	0.9675	0.8789	0.7059	0.4661	0.2643	0.0587	0.0086	0.0029	0.0003
1.85	0.9985	0.9742	0.8994	0.7447	0.5157	0.3092	0.0777	0.0131	0.0047	0.0005
1.90	0.9988	0.9795	0.9167	0.7796	0.5636	0.3556	0.1003	0.0192	0.0073	0.0009
1.95	0.9991	0.9837	0.9313	0.8106	0.6093	0.4028	0.1264	0.0272	0.0110	0.0016
2.00	0.9993	0.9871	0.9435	0.8381	0.6522	0.4500	0.1560	0.0376	0.0162	0.0026
2.05	0.9995	0.9898	0.9536	0.8621	0.6920	0.4965	0.1888	0.0504	0.0230	0.0042
2.10	0.9996	0.9919	0.9620	0.8830	0.7287	0.5418	0.2244	0.0661	0.0317	0.0064
2.15	0.9997	0.9936	0.9690	0.9011	0.7622	0.5852	0.2623	0.0847	0.0427	0.0095
2.20	0.9998	0.9950	0.9747	0.9166	0.7924	0.6265	0.3020	0.1062	0.0561	0.0138
2.25	0.9998	0.9960	0.9794	0.9299	0.8195	0.6652	0.3431	0.1306	0.0721	0.0194
2.30	0.9999	0.9968	0.9832	0.9412	0.8436	0.7013	0.3848	0.1579	0.0908	0.0265
2.35	0.9999	0.9975	0.9863	0.9508	0.8649	0.7346	0.4268	0.1877	0.1122	0.0354
2.40	0.9999	0.9980	0.9889	0.9589	0.8837	0.7652	0.4684	0.2198	0.1363	0.0464
2.45	0.9999	0.9984	0.9910	0.9657	0.9001	0.7929	0.5092	0.2539	0.1629	0.0594

PROBABILITY INTEGRAL OF t, THE NON-CENTRAL t-STATISTIC. THIS TABLE GIVES Pr $[t/\sqrt{f} \leq x]$.

f is the number of degrees of freedom; the non-centrality parameter is $\sqrt{f+1}\,K_p$.

K_p is the standardized normal deviate exceeded with probability p.

x \ p	.2500	.1500	.1000	.0650	.0400	.0250	.0100	.0040	.0025	.0010
2.50	1.0000	0.9988	0.9927	0.9714	0.9144	0.8180	0.5489	0.2896	0.1918	0.0748
2.55	1.0000	0.9990	0.9940	0.9762	0.9268	0.8405	0.5871	0.3264	0.2228	0.0924
2.60	1.0000	0.9992	0.9951	0.9802	0.9375	0.8606	0.6235	0.3640	0.2555	0.1124
2.65	1.0000	0.9994	0.9961	0.9835	0.9468	0.8785	0.6580	0.4019	0.2897	0.1347
2.70	1.0000	0.9995	0.9968	0.9863	0.9547	0.8943	0.6904	0.4398	0.3249	0.1591
2.75	1.0000	0.9996	0.9974	0.9886	0.9615	0.9083	0.7207	0.4773	0.3608	0.1855
2.80	1.0000	0.9997	0.9979	0.9905	0.9673	0.9205	0.7487	0.5141	0.3971	0.2138
2.85	1.0000	0.9997	0.9983	0.9921	0.9723	0.9313	0.7746	0.5498	0.4333	0.2436
2.90	1.0000	0.9998	0.9986	0.9935	0.9765	0.9406	0.7984	0.5844	0.4693	0.2748
2.95	1.0000	0.9998	0.9988	0.9946	0.9801	0.9488	0.8200	0.6174	0.5046	0.3069
3.00	1.0000	0.9999	0.9990	0.9955	0.9832	0.9559	0.8397	0.6490	0.5390	0.3398
3.05	1.0000	0.9999	0.9992	0.9962	0.9858	0.9620	0.8576	0.6788	0.5724	0.3732
3.10	1.0000	0.9999	0.9994	0.9969	0.9879	0.9673	0.8737	0.7069	0.6045	0.4067
3.15	1.0000	0.9999	0.9995	0.9974	0.9898	0.9719	0.8882	0.7332	0.6353	0.4402
3.20	1.0000	0.9999	0.9996	0.9978	0.9914	0.9759	0.9011	0.7577	0.6645	0.4733
3.25	1.0000	1.0000	0.9996	0.9982	0.9927	0.9793	0.9127	0.7804	0.6921	0.5059
3.30	1.0000	1.0000	0.9997	0.9985	0.9938	0.9822	0.9230	0.8014	0.7182	0.5377
3.35	1.0000	1.0000	0.9998	0.9987	0.9948	0.9847	0.9322	0.8208	0.7426	0.5686
3.40	1.0000	1.0000	0.9998	0.9989	0.9956	0.9869	0.9404	0.8385	0.7654	0.5984
3.45	1.0000	1.0000	0.9998	0.9991	0.9962	0.9888	0.9476	0.8547	0.7866	0.6271
3.50	1.0000	1.0000	0.9999	0.9993	0.9968	0.9904	0.9540	0.8695	0.8062	0.6545
3.55	1.0000	1.0000	0.9999	0.9994	0.9973	0.9917	0.9596	0.8830	0.8243	0.6806
3.60	1.0000	1.0000	0.9999	0.9995	0.9977	0.9929	0.9646	0.8952	0.8410	0.7053
3.65	1.0000	1.0000	0.9999	0.9996	0.9981	0.9939	0.9690	0.9062	0.8563	0.7286
3.70	1.0000	1.0000	0.9999	0.9996	0.9983	0.9948	0.9728	0.9162	0.8703	0.7505
3.75	1.0000	1.0000	0.9999	0.9997	0.9986	0.9955	0.9762	0.9252	0.8831	0.7711
3.80	1.0000	1.0000	1.0000	0.9997	0.9988	0.9961	0.9792	0.9333	0.8947	0.7903
3.85	1.0000	1.0000	1.0000	0.9998	0.9990	0.9967	0.9818	0.9405	0.9053	0.8082
3.90	1.0000	1.0000	1.0000	0.9998	0.9991	0.9972	0.9840	0.9470	0.9149	0.8248
3.95	1.0000	1.0000	1.0000	0.9998	0.9993	0.9976	0.9860	0.9529	0.9236	0.8401
4.00	1.0000	1.0000	1.0000	0.9999	0.9994	0.9979	0.9878	0.9581	0.9315	0.8543
4.05	1.0000	1.0000	1.0000	0.9999	0.9995	0.9982	0.9893	0.9627	0.9386	0.8674
4.10	1.0000	1.0000	1.0000	0.9999	0.9995	0.9984	0.9906	0.9669	0.9450	0.8795
4.15	1.0000	1.0000	1.0000	0.9999	0.9996	0.9987	0.9918	0.9706	0.9508	0.8906
4.20	1.0000	1.0000	1.0000	0.9999	0.9997	0.9988	0.9928	0.9739	0.9560	0.9007
4.25	1.0000	1.0000	1.0000	0.9999	0.9997	0.9990	0.9937	0.9768	0.9606	0.9100
4.30	1.0000	1.0000	1.0000	0.9999	0.9998	0.9991	0.9945	0.9794	0.9648	0.9185
4.35	1.0000	1.0000	1.0000	1.0000	0.9998	0.9993	0.9952	0.9817	0.9685	0.9262
4.40	1.0000	1.0000	1.0000	1.0000	0.9998	0.9994	0.9958	0.9838	0.9719	0.9332
4.45	1.0000	1.0000	1.0000	1.0000	0.9998	0.9994	0.9963	0.9856	0.9749	0.9396
4.50	1.0000	1.0000	1.0000	1.0000	0.9999	0.9995	0.9968	0.9872	0.9776	0.9455
4.55	1.0000	1.0000	1.0000	1.0000	0.9999	0.9996	0.9972	0.9887	0.9800	0.9507
4.60	1.0000	1.0000	1.0000	1.0000	0.9999	0.9996	0.9975	0.9899	0.9822	0.9555
4.65	1.0000	1.0000	1.0000	1.0000	0.9999	0.9997	0.9978	0.9911	0.9841	0.9599
4.70	1.0000	1.0000	1.0000	1.0000	0.9999	0.9997	0.9981	0.9921	0.9858	0.9638
4.75	1.0000	1.0000	1.0000	1.0000	0.9999	0.9998	0.9983	0.9930	0.9873	0.9674
4.80	1.0000	1.0000	1.0000	1.0000	0.9999	0.9998	0.9985	0.9938	0.9887	0.9706
4.85	1.0000	1.0000	1.0000	1.0000	0.9999	0.9998	0.9987	0.9945	0.9899	0.9735
4.90	1.0000	1.0000	1.0000	1.0000	1.0000	0.9998	0.9989	0.9951	0.9910	0.9761
4.95	1.0000	1.0000	1.0000	1.0000	1.0000	0.9999	0.9990	0.9956	0.9920	0.9785
5.00	1.0000	1.0000	1.0000	1.0000	1.0000	0.9999	0.9991	0.9961	0.9928	0.9806
5.05	1.0000	1.0000	1.0000	1.0000	1.0000	0.9999	0.9992	0.9965	0.9936	0.9825
5.10	1.0000	1.0000	1.0000	1.0000	1.0000	0.9999	0.9993	0.9969	0.9943	0.9843
5.15	1.0000	1.0000	1.0000	1.0000	1.0000	0.9999	0.9994	0.9973	0.9949	0.9858
5.20	1.0000	1.0000	1.0000	1.0000	1.0000	0.9999	0.9995	0.9976	0.9954	0.9872
5.25	1.0000	1.0000	1.0000	1.0000	1.0000	0.9999	0.9995	0.9978	0.9959	0.9885
5.30	1.0000	1.0000	1.0000	1.0000	1.0000	0.9999	0.9996	0.9981	0.9964	0.9896
5.35	1.0000	1.0000	1.0000	1.0000	1.0000	0.9999	0.9996	0.9983	0.9967	0.9907
5.40	1.0000	1.0000	1.0000	1.0000	1.0000	1.0000	0.9997	0.9985	0.9971	0.9916
5.45	1.0000	1.0000	1.0000	1.0000	1.0000	1.0000	0.9997	0.9986	0.9974	0.9924

PROBABILITY INTEGRAL OF t, THE NON-CENTRAL t-STATISTIC. THIS TABLE GIVES Pr $[t/\sqrt{f} \leq x]$.

f is the number of degrees of freedom; the non-centrality parameter is $\sqrt{f+1}\,K_p$.
K_p is the standardized normal deviate exceeded with probability p.

DEGREES OF FREEDOM **17**

x \ p	.2500	.1500	.1000	.0650	.0400	.0250	.0100	.0040	.0025	.0010
5.50	1.0000	1.0000	1.0000	1.0000	1.0000	1.0000	0.9997	0.9988	0.9977	0.9932
5.55	1.0000	1.0000	1.0000	1.0000	1.0000	1.0000	0.9998	0.9989	0.9979	0.9939
5.60	1.0000	1.0000	1.0000	1.0000	1.0000	1.0000	0.9998	0.9990	0.9981	0.9945
5.65	1.0000	1.0000	1.0000	1.0000	1.0000	1.0000	0.9998	0.9991	0.9983	0.9950
5.70	1.0000	1.0000	1.0000	1.0000	1.0000	1.0000	0.9998	0.9992	0.9985	0.9955
5.75	1.0000	1.0000	1.0000	1.0000	1.0000	1.0000	0.9999	0.9993	0.9987	0.9959
5.80	1.0000	1.0000	1.0000	1.0000	1.0000	1.0000	0.9999	0.9994	0.9988	0.9963
5.85	1.0000	1.0000	1.0000	1.0000	1.0000	1.0000	0.9999	0.9995	0.9989	0.9967
5.90	1.0000	1.0000	1.0000	1.0000	1.0000	1.0000	0.9999	0.9995	0.9990	0.9970
5.95	1.0000	1.0000	1.0000	1.0000	1.0000	1.0000	0.9999	0.9996	0.9991	0.9973
6.00	1.0000	1.0000	1.0000	1.0000	1.0000	1.0000	0.9999	0.9996	0.9992	0.9976
6.05	1.0000	1.0000	1.0000	1.0000	1.0000	1.0000	0.9999	0.9996	0.9993	0.9978
6.10	1.0000	1.0000	1.0000	1.0000	1.0000	1.0000	0.9999	0.9997	0.9994	0.9980
6.15	1.0000	1.0000	1.0000	1.0000	1.0000	1.0000	0.9999	0.9997	0.9994	0.9982
6.20	1.0000	1.0000	1.0000	1.0000	1.0000	1.0000	0.9999	0.9997	0.9995	0.9984
6.25	1.0000	1.0000	1.0000	1.0000	1.0000	1.0000	1.0000	0.9998	0.9995	0.9985
6.30	1.0000	1.0000	1.0000	1.0000	1.0000	1.0000	1.0000	0.9998	0.9996	0.9987
6.35	1.0000	1.0000	1.0000	1.0000	1.0000	1.0000	1.0000	0.9998	0.9996	0.9988
6.40	1.0000	1.0000	1.0000	1.0000	1.0000	1.0000	1.0000	0.9998	0.9997	0.9989
6.45	1.0000	1.0000	1.0000	1.0000	1.0000	1.0000	1.0000	0.9998	0.9997	0.9990
6.50	1.0000	1.0000	1.0000	1.0000	1.0000	1.0000	1.0000	0.9999	0.9997	0.9991
6.55	1.0000	1.0000	1.0000	1.0000	1.0000	1.0000	1.0000	0.9999	0.9998	0.9992
6.60	1.0000	1.0000	1.0000	1.0000	1.0000	1.0000	1.0000	0.9999	0.9998	0.9993
6.65	1.0000	1.0000	1.0000	1.0000	1.0000	1.0000	1.0000	0.9999	0.9998	0.9993
6.70	1.0000	1.0000	1.0000	1.0000	1.0000	1.0000	1.0000	0.9999	0.9998	0.9994
6.75	1.0000	1.0000	1.0000	1.0000	1.0000	1.0000	1.0000	0.9999	0.9998	0.9995
6.80	1.0000	1.0000	1.0000	1.0000	1.0000	1.0000	1.0000	0.9999	0.9998	0.9995
6.85	1.0000	1.0000	1.0000	1.0000	1.0000	1.0000	1.0000	0.9999	0.9999	0.9996
6.90	1.0000	1.0000	1.0000	1.0000	1.0000	1.0000	1.0000	0.9999	0.9999	0.9996
6.95	1.0000	1.0000	1.0000	1.0000	1.0000	1.0000	1.0000	0.9999	0.9999	0.9996
7.00	1.0000	1.0000	1.0000	1.0000	1.0000	1.0000	1.0000	0.9999	0.9999	0.9997
7.05	1.0000	1.0000	1.0000	1.0000	1.0000	1.0000	1.0000	1.0000	0.9999	0.9997
7.10	1.0000	1.0000	1.0000	1.0000	1.0000	1.0000	1.0000	1.0000	0.9999	0.9997
7.15	1.0000	1.0000	1.0000	1.0000	1.0000	1.0000	1.0000	1.0000	0.9999	0.9997
7.20	1.0000	1.0000	1.0000	1.0000	1.0000	1.0000	1.0000	1.0000	0.9999	0.9998
7.25	1.0000	1.0000	1.0000	1.0000	1.0000	1.0000	1.0000	1.0000	0.9999	0.9998
7.30	1.0000	1.0000	1.0000	1.0000	1.0000	1.0000	1.0000	1.0000	0.9999	0.9998
7.35	1.0000	1.0000	1.0000	1.0000	1.0000	1.0000	1.0000	1.0000	0.9999	0.9998
7.40	1.0000	1.0000	1.0000	1.0000	1.0000	1.0000	1.0000	1.0000	1.0000	0.9998
7.45	1.0000	1.0000	1.0000	1.0000	1.0000	1.0000	1.0000	1.0000	1.0000	0.9999
7.50	1.0000	1.0000	1.0000	1.0000	1.0000	1.0000	1.0000	1.0000	1.0000	0.9999
7.55	1.0000	1.0000	1.0000	1.0000	1.0000	1.0000	1.0000	1.0000	1.0000	0.9999
7.60	1.0000	1.0000	1.0000	1.0000	1.0000	1.0000	1.0000	1.0000	1.0000	0.9999
7.65	1.0000	1.0000	1.0000	1.0000	1.0000	1.0000	1.0000	1.0000	1.0000	0.9999
7.70	1.0000	1.0000	1.0000	1.0000	1.0000	1.0000	1.0000	1.0000	1.0000	0.9999
7.75	1.0000	1.0000	1.0000	1.0000	1.0000	1.0000	1.0000	1.0000	1.0000	0.9999
7.80	1.0000	1.0000	1.0000	1.0000	1.0000	1.0000	1.0000	1.0000	1.0000	0.9999
7.85	1.0000	1.0000	1.0000	1.0000	1.0000	1.0000	1.0000	1.0000	1.0000	0.9999
7.90	1.0000	1.0000	1.0000	1.0000	1.0000	1.0000	1.0000	1.0000	1.0000	0.9999
7.95	1.0000	1.0000	1.0000	1.0000	1.0000	1.0000	1.0000	1.0000	1.0000	0.9999
8.00	1.0000	1.0000	1.0000	1.0000	1.0000	1.0000	1.0000	1.0000	1.0000	0.9999
8.05	1.0000	1.0000	1.0000	1.0000	1.0000	1.0000	1.0000	1.0000	1.0000	0.9999
8.10	1.0000	1.0000	1.0000	1.0000	1.0000	1.0000	1.0000	1.0000	1.0000	0.9999
8.15	1.0000	1.0000	1.0000	1.0000	1.0000	1.0000	1.0000	1.0000	1.0000	1.0000

PROBABILITY INTEGRAL OF t, THE NON-CENTRAL t-STATISTIC. THIS TABLE GIVES Pr $[t/\sqrt{f} \leq x]$.

f is the number of degrees of freedom; the non-centrality parameter is $\sqrt{f+1}\,K_p$.
K_p is the standardized normal deviate exceeded with probability p.

DEGREES OF FREEDOM **18**

x \ p	.2500	.1500	.1000	.0650	.0400	.0250	.0100	.0040	.0025	.0010
-0.25	0.0000	0.0000	0.0000	0.0000	0.0000	0.0000	0.0000	0.0000	0.0000	0.0000
-0.20	0.0001	0.0000	0.0000	0.0000	0.0000	0.0000	0.0000	0.0000	0.0000	0.0000
-0.15	0.0002	0.0000	0.0000	0.0000	0.0000	0.0000	0.0000	0.0000	0.0000	0.0000
-0.10	0.0004	0.0000	0.0000	0.0000	0.0000	0.0000	0.0000	0.0000	0.0000	0.0000
-0.05	0.0008	0.0000	0.0000	0.0000	0.0000	0.0000	0.0000	0.0000	0.0000	0.0000
0.00	0.0016	0.0000	0.0000	0.0000	0.0000	0.0000	0.0000	0.0000	0.0000	0.0000
0.05	0.0032	0.0000	0.0000	0.0000	0.0000	0.0000	0.0000	0.0000	0.0000	0.0000
0.10	0.0059	0.0000	0.0000	0.0000	0.0000	0.0000	0.0000	0.0000	0.0000	0.0000
0.15	0.0107	0.0001	0.0000	0.0000	0.0000	0.0000	0.0000	0.0000	0.0000	0.0000
0.20	0.0186	0.0001	0.0000	0.0000	0.0000	0.0000	0.0000	0.0000	0.0000	0.0000
0.25	0.0311	0.0003	0.0000	0.0000	0.0000	0.0000	0.0000	0.0000	0.0000	0.0000
0.30	0.0497	0.0007	0.0000	0.0000	0.0000	0.0000	0.0000	0.0000	0.0000	0.0000
0.35	0.0760	0.0015	0.0000	0.0000	0.0000	0.0000	0.0000	0.0000	0.0000	0.0000
0.40	0.1114	0.0031	0.0001	0.0000	0.0000	0.0000	0.0000	0.0000	0.0000	0.0000
0.45	0.1567	0.0061	0.0002	0.0000	0.0000	0.0000	0.0000	0.0000	0.0000	0.0000
0.50	0.2117	0.0112	0.0005	0.0000	0.0000	0.0000	0.0000	0.0000	0.0000	0.0000
0.55	0.2753	0.0195	0.0011	0.0000	0.0000	0.0000	0.0000	0.0000	0.0000	0.0000
0.60	0.3456	0.0325	0.0024	0.0001	0.0000	0.0000	0.0000	0.0000	0.0000	0.0000
0.65	0.4198	0.0513	0.0047	0.0002	0.0000	0.0000	0.0000	0.0000	0.0000	0.0000
0.70	0.4950	0.0773	0.0088	0.0005	0.0000	0.0000	0.0000	0.0000	0.0000	0.0000
0.75	0.5685	0.1114	0.0156	0.0012	0.0000	0.0000	0.0000	0.0000	0.0000	0.0000
0.80	0.6376	0.1538	0.0260	0.0025	0.0001	0.0000	0.0000	0.0000	0.0000	0.0000
0.85	0.7008	0.2042	0.0414	0.0048	0.0003	0.0000	0.0000	0.0000	0.0000	0.0000
0.90	0.7567	0.2616	0.0626	0.0088	0.0006	0.0000	0.0000	0.0000	0.0000	0.0000
0.95	0.8050	0.3244	0.0907	0.0152	0.0013	0.0001	0.0000	0.0000	0.0000	0.0000
1.00	0.8458	0.3905	0.1260	0.0249	0.0027	0.0002	0.0000	0.0000	0.0000	0.0000
1.05	0.8794	0.4578	0.1684	0.0389	0.0050	0.0005	0.0000	0.0000	0.0000	0.0000
1.10	0.9067	0.5242	0.2174	0.0581	0.0090	0.0011	0.0000	0.0000	0.0000	0.0000
1.15	0.9285	0.5878	0.2719	0.0831	0.0151	0.0022	0.0000	0.0000	0.0000	0.0000
1.20	0.9456	0.6474	0.3306	0.1143	0.0242	0.0040	0.0001	0.0000	0.0000	0.0000
1.25	0.9589	0.7017	0.3916	0.1518	0.0370	0.0071	0.0002	0.0000	0.0000	0.0000
1.30	0.9691	0.7504	0.4534	0.1951	0.0543	0.0120	0.0004	0.0000	0.0000	0.0000
1.35	0.9769	0.7932	0.5143	0.2435	0.0765	0.0192	0.0007	0.0000	0.0000	0.0000
1.40	0.9828	0.8301	0.5730	0.2960	0.1041	0.0294	0.0014	0.0000	0.0000	0.0000
1.45	0.9873	0.8616	0.6283	0.3512	0.1370	0.0432	0.0027	0.0001	0.0000	0.0000
1.50	0.9906	0.8880	0.6795	0.4078	0.1752	0.0611	0.0047	0.0002	0.0000	0.0000
1.55	0.9930	0.9100	0.7260	0.4646	0.2179	0.0835	0.0078	0.0004	0.0001	0.0000
1.60	0.9949	0.9280	0.7676	0.5203	0.2645	0.1107	0.0124	0.0009	0.0002	0.0000
1.65	0.9962	0.9427	0.8044	0.5740	0.3140	0.1425	0.0190	0.0016	0.0004	0.0000
1.70	0.9972	0.9546	0.8364	0.6246	0.3655	0.1788	0.0280	0.0028	0.0007	0.0000
1.75	0.9979	0.9642	0.8640	0.6718	0.4177	0.2190	0.0399	0.0046	0.0013	0.0001
1.80	0.9985	0.9718	0.8875	0.7151	0.4699	0.2625	0.0551	0.0075	0.0023	0.0002
1.85	0.9989	0.9778	0.9074	0.7543	0.5209	0.3085	0.0738	0.0115	0.0039	0.0004
1.90	0.9992	0.9826	0.9241	0.7893	0.5701	0.3563	0.0963	0.0172	0.0063	0.0007
1.95	0.9994	0.9864	0.9380	0.8204	0.6169	0.4049	0.1225	0.0248	0.0097	0.0013
2.00	0.9995	0.9894	0.9496	0.8476	0.6607	0.4535	0.1523	0.0347	0.0144	0.0021
2.05	0.9997	0.9917	0.9591	0.8713	0.7012	0.5014	0.1856	0.0471	0.0207	0.0035
2.10	0.9997	0.9935	0.9668	0.8917	0.7384	0.5479	0.2219	0.0624	0.0291	0.0055
2.15	0.9998	0.9950	0.9732	0.9092	0.7721	0.5925	0.2607	0.0808	0.0396	0.0083
2.20	0.9999	0.9961	0.9784	0.9242	0.8023	0.6346	0.3015	0.1022	0.0527	0.0122
2.25	0.9999	0.9969	0.9826	0.9368	0.8293	0.6742	0.3437	0.1268	0.0684	0.0174
2.30	0.9999	0.9976	0.9860	0.9475	0.8532	0.7108	0.3867	0.1543	0.0870	0.0241
2.35	0.9999	0.9981	0.9887	0.9565	0.8741	0.7445	0.4299	0.1846	0.1084	0.0327
2.40	1.0000	0.9985	0.9909	0.9640	0.8925	0.7752	0.4727	0.2174	0.1326	0.0432
2.45	1.0000	0.9989	0.9927	0.9702	0.9084	0.8030	0.5148	0.2523	0.1595	0.0560
2.50	1.0000	0.9991	0.9941	0.9754	0.9221	0.8280	0.5555	0.2889	0.1889	0.0711
2.55	1.0000	0.9993	0.9953	0.9798	0.9340	0.8502	0.5947	0.3267	0.2205	0.0886
2.60	1.0000	0.9994	0.9962	0.9833	0.9441	0.8700	0.6319	0.3654	0.2540	0.1086
2.65	1.0000	0.9996	0.9970	0.9863	0.9528	0.8875	0.6671	0.4045	0.2891	0.1310
2.70	1.0000	0.9997	0.9976	0.9887	0.9602	0.9029	0.7000	0.4435	0.3253	0.1557

PROBABILITY INTEGRAL OF t, THE NON-CENTRAL t-STATISTIC. THIS TABLE GIVES Pr $[t/\sqrt{f} \leq x]$.

f is the number of degrees of freedom; the non-centrality parameter is $\sqrt{f+1}\,K_p$.
K_p is the standardized normal deviate exceeded with probability p.

DEGREES OF FREEDOM **18**

x \ p	.2500	.1500	.1000	.0650	.0400	.0250	.0100	.0040	.0025	.0010
2.75	1.0000	0.9997	0.9980	0.9907	0.9665	0.9163	0.7306	0.4821	0.3622	0.1826
2.80	1.0000	0.9998	0.9984	0.9924	0.9718	0.9280	0.7589	0.5199	0.3996	0.2114
2.85	1.0000	0.9998	0.9987	0.9937	0.9763	0.9382	0.7849	0.5567	0.4369	0.2419
2.90	1.0000	0.9999	0.9990	0.9948	0.9801	0.9471	0.8086	0.5920	0.4739	0.2738
2.95	1.0000	0.9999	0.9992	0.9958	0.9833	0.9547	0.8302	0.6259	0.5102	0.3069
3.00	1.0000	0.9999	0.9993	0.9965	0.9860	0.9613	0.8496	0.6580	0.5456	0.3407
3.05	1.0000	0.9999	0.9995	0.9971	0.9882	0.9669	0.8672	0.6883	0.5799	0.3751
3.10	1.0000	0.9999	0.9996	0.9976	0.9901	0.9718	0.8829	0.7168	0.6127	0.4096
3.15	1.0000	1.0000	0.9996	0.9980	0.9917	0.9760	0.8970	0.7434	0.6441	0.4441
3.20	1.0000	1.0000	0.9997	0.9984	0.9931	0.9795	0.9095	0.7680	0.6739	0.4782
3.25	1.0000	1.0000	0.9998	0.9987	0.9942	0.9826	0.9207	0.7908	0.7020	0.5117
3.30	1.0000	1.0000	0.9998	0.9989	0.9951	0.9852	0.9305	0.8118	0.7283	0.5444
3.35	1.0000	1.0000	0.9998	0.9991	0.9959	0.9874	0.9392	0.8310	0.7529	0.5761
3.40	1.0000	1.0000	0.9999	0.9993	0.9966	0.9893	0.9469	0.8485	0.7758	0.6066
3.45	1.0000	1.0000	0.9999	0.9994	0.9971	0.9909	0.9537	0.8645	0.7970	0.6359
3.50	1.0000	1.0000	0.9999	0.9995	0.9976	0.9922	0.9596	0.8790	0.8166	0.6638
3.55	1.0000	1.0000	0.9999	0.9996	0.9980	0.9934	0.9648	0.8921	0.8345	0.6903
3.60	1.0000	1.0000	0.9999	0.9996	0.9983	0.9944	0.9693	0.9039	0.8510	0.7154
3.65	1.0000	1.0000	0.9999	0.9997	0.9986	0.9952	0.9733	0.9145	0.8660	0.7389
3.70	1.0000	1.0000	1.0000	0.9998	0.9988	0.9959	0.9768	0.9241	0.8797	0.7610
3.75	1.0000	1.0000	1.0000	0.9998	0.9990	0.9965	0.9798	0.9326	0.8922	0.7816
3.80	1.0000	1.0000	1.0000	0.9998	0.9991	0.9970	0.9825	0.9403	0.9035	0.8008
3.85	1.0000	1.0000	1.0000	0.9999	0.9993	0.9975	0.9848	0.9471	0.9137	0.8186
3.90	1.0000	1.0000	1.0000	0.9999	0.9994	0.9979	0.9868	0.9532	0.9229	0.8350
3.95	1.0000	1.0000	1.0000	0.9999	0.9995	0.9982	0.9885	0.9586	0.9312	0.8502
4.00	1.0000	1.0000	1.0000	0.9999	0.9996	0.9984	0.9900	0.9634	0.9386	0.8642
4.05	1.0000	1.0000	1.0000	0.9999	0.9996	0.9987	0.9913	0.9677	0.9453	0.8770
4.10	1.0000	1.0000	1.0000	0.9999	0.9997	0.9989	0.9925	0.9714	0.9513	0.8887
4.15	1.0000	1.0000	1.0000	0.9999	0.9997	0.9990	0.9935	0.9748	0.9567	0.8995
4.20	1.0000	1.0000	1.0000	1.0000	0.9998	0.9992	0.9943	0.9778	0.9615	0.9093
4.25	1.0000	1.0000	1.0000	1.0000	0.9998	0.9993	0.9951	0.9804	0.9658	0.9182
4.30	1.0000	1.0000	1.0000	1.0000	0.9998	0.9994	0.9957	0.9827	0.9696	0.9263
4.35	1.0000	1.0000	1.0000	1.0000	0.9999	0.9995	0.9963	0.9848	0.9730	0.9336
4.40	1.0000	1.0000	1.0000	1.0000	0.9999	0.9996	0.9968	0.9866	0.9760	0.9403
4.45	1.0000	1.0000	1.0000	1.0000	0.9999	0.9996	0.9972	0.9882	0.9787	0.9463
4.50	1.0000	1.0000	1.0000	1.0000	0.9999	0.9997	0.9976	0.9896	0.9811	0.9517
4.55	1.0000	1.0000	1.0000	1.0000	0.9999	0.9997	0.9979	0.9908	0.9832	0.9566
4.60	1.0000	1.0000	1.0000	1.0000	0.9999	0.9998	0.9981	0.9919	0.9851	0.9611
4.65	1.0000	1.0000	1.0000	1.0000	0.9999	0.9998	0.9984	0.9929	0.9868	0.9651
4.70	1.0000	1.0000	1.0000	1.0000	1.0000	0.9998	0.9986	0.9937	0.9883	0.9687
4.75	1.0000	1.0000	1.0000	1.0000	1.0000	0.9998	0.9988	0.9944	0.9896	0.9719
4.80	1.0000	1.0000	1.0000	1.0000	1.0000	0.9999	0.9989	0.9951	0.9908	0.9748
4.85	1.0000	1.0000	1.0000	1.0000	1.0000	0.9999	0.9991	0.9957	0.9919	0.9774
4.90	1.0000	1.0000	1.0000	1.0000	1.0000	0.9999	0.9992	0.9962	0.9928	0.9798
4.95	1.0000	1.0000	1.0000	1.0000	1.0000	0.9999	0.9993	0.9966	0.9936	0.9819
5.00	1.0000	1.0000	1.0000	1.0000	1.0000	0.9999	0.9994	0.9970	0.9943	0.9838
5.05	1.0000	1.0000	1.0000	1.0000	1.0000	0.9999	0.9995	0.9974	0.9950	0.9855
5.10	1.0000	1.0000	1.0000	1.0000	1.0000	0.9999	0.9995	0.9977	0.9955	0.9870
5.15	1.0000	1.0000	1.0000	1.0000	1.0000	0.9999	0.9996	0.9980	0.9960	0.9884
5.20	1.0000	1.0000	1.0000	1.0000	1.0000	1.0000	0.9996	0.9982	0.9965	0.9896
5.25	1.0000	1.0000	1.0000	1.0000	1.0000	1.0000	0.9997	0.9984	0.9969	0.9907
5.30	1.0000	1.0000	1.0000	1.0000	1.0000	1.0000	0.9997	0.9986	0.9972	0.9916
5.35	1.0000	1.0000	1.0000	1.0000	1.0000	1.0000	0.9998	0.9988	0.9975	0.9925
5.40	1.0000	1.0000	1.0000	1.0000	1.0000	1.0000	0.9998	0.9989	0.9978	0.9933
5.45	1.0000	1.0000	1.0000	1.0000	1.0000	1.0000	0.9998	0.9990	0.9981	0.9940
5.50	1.0000	1.0000	1.0000	1.0000	1.0000	1.0000	0.9998	0.9991	0.9983	0.9946
5.55	1.0000	1.0000	1.0000	1.0000	1.0000	1.0000	0.9999	0.9992	0.9985	0.9952
5.60	1.0000	1.0000	1.0000	1.0000	1.0000	1.0000	0.9999	0.9993	0.9986	0.9957
5.65	1.0000	1.0000	1.0000	1.0000	1.0000	1.0000	0.9999	0.9994	0.9988	0.9961
5.70	1.0000	1.0000	1.0000	1.0000	1.0000	1.0000	0.9999	0.9995	0.9989	0.9965

PROBABILITY INTEGRAL OF t, THE NON-CENTRAL t-STATISTIC. THIS TABLE GIVES Pr $[t/\sqrt{f} \leq x]$.

f is the number of degrees of freedom; the non-centrality parameter is $\sqrt{f+1}\,K_p$.

K_p is the standardized normal deviate exceeded with probability p.

DEGREES OF FREEDOM **18**

x \ p	.2500	.1500	.1000	.0650	.0400	.0250	.0100	.0040	.0025	.0010
5.75	1.0000	1.0000	1.0000	1.0000	1.0000	1.0000	0.9999	0.9995	0.9990	0.9969
5.80	1.0000	1.0000	1.0000	1.0000	1.0000	1.0000	0.9999	0.9996	0.9991	0.9972
5.85	1.0000	1.0000	1.0000	1.0000	1.0000	1.0000	0.9999	0.9996	0.9992	0.9975
5.90	1.0000	1.0000	1.0000	1.0000	1.0000	1.0000	0.9999	0.9997	0.9993	0.9978
5.95	1.0000	1.0000	1.0000	1.0000	1.0000	1.0000	0.9999	0.9997	0.9994	0.9980
6.00	1.0000	1.0000	1.0000	1.0000	1.0000	1.0000	0.9999	0.9997	0.9995	0.9982
6.05	1.0000	1.0000	1.0000	1.0000	1.0000	1.0000	1.0000	0.9998	0.9995	0.9984
6.10	1.0000	1.0000	1.0000	1.0000	1.0000	1.0000	1.0000	0.9998	0.9996	0.9985
6.15	1.0000	1.0000	1.0000	1.0000	1.0000	1.0000	1.0000	0.9998	0.9996	0.9987
6.20	1.0000	1.0000	1.0000	1.0000	1.0000	1.0000	1.0000	0.9998	0.9997	0.9988
6.25	1.0000	1.0000	1.0000	1.0000	1.0000	1.0000	1.0000	0.9998	0.9997	0.9989
6.30	1.0000	1.0000	1.0000	1.0000	1.0000	1.0000	1.0000	0.9999	0.9997	0.9991
6.35	1.0000	1.0000	1.0000	1.0000	1.0000	1.0000	1.0000	0.9999	0.9997	0.9991
6.40	1.0000	1.0000	1.0000	1.0000	1.0000	1.0000	1.0000	0.9999	0.9998	0.9992
6.45	1.0000	1.0000	1.0000	1.0000	1.0000	1.0000	1.0000	0.9999	0.9998	0.9993
6.50	1.0000	1.0000	1.0000	1.0000	1.0000	1.0000	1.0000	0.9999	0.9998	0.9994
6.55	1.0000	1.0000	1.0000	1.0000	1.0000	1.0000	1.0000	0.9999	0.9998	0.9994
6.60	1.0000	1.0000	1.0000	1.0000	1.0000	1.0000	1.0000	0.9999	0.9999	0.9995
6.65	1.0000	1.0000	1.0000	1.0000	1.0000	1.0000	1.0000	0.9999	0.9999	0.9995
6.70	1.0000	1.0000	1.0000	1.0000	1.0000	1.0000	1.0000	0.9999	0.9999	0.9996
6.75	1.0000	1.0000	1.0000	1.0000	1.0000	1.0000	1.0000	0.9999	0.9999	0.9996
6.80	1.0000	1.0000	1.0000	1.0000	1.0000	1.0000	1.0000	1.0000	0.9999	0.9997
6.85	1.0000	1.0000	1.0000	1.0000	1.0000	1.0000	1.0000	1.0000	0.9999	0.9997
6.90	1.0000	1.0000	1.0000	1.0000	1.0000	1.0000	1.0000	1.0000	0.9999	0.9997
6.95	1.0000	1.0000	1.0000	1.0000	1.0000	1.0000	1.0000	1.0000	0.9999	0.9998
7.00	1.0000	1.0000	1.0000	1.0000	1.0000	1.0000	1.0000	1.0000	0.9999	0.9998
7.05	1.0000	1.0000	1.0000	1.0000	1.0000	1.0000	1.0000	1.0000	0.9999	0.9998
7.10	1.0000	1.0000	1.0000	1.0000	1.0000	1.0000	1.0000	1.0000	0.9999	0.9998
7.15	1.0000	1.0000	1.0000	1.0000	1.0000	1.0000	1.0000	1.0000	1.0000	0.9998
7.20	1.0000	1.0000	1.0000	1.0000	1.0000	1.0000	1.0000	1.0000	1.0000	0.9998
7.25	1.0000	1.0000	1.0000	1.0000	1.0000	1.0000	1.0000	1.0000	1.0000	0.9999
7.30	1.0000	1.0000	1.0000	1.0000	1.0000	1.0000	1.0000	1.0000	1.0000	0.9999
7.35	1.0000	1.0000	1.0000	1.0000	1.0000	1.0000	1.0000	1.0000	1.0000	0.9999
7.40	1.0000	1.0000	1.0000	1.0000	1.0000	1.0000	1.0000	1.0000	1.0000	0.9999
7.45	1.0000	1.0000	1.0000	1.0000	1.0000	1.0000	1.0000	1.0000	1.0000	0.9999
7.50	1.0000	1.0000	1.0000	1.0000	1.0000	1.0000	1.0000	1.0000	1.0000	0.9999
7.55	1.0000	1.0000	1.0000	1.0000	1.0000	1.0000	1.0000	1.0000	1.0000	0.9999
7.60	1.0000	1.0000	1.0000	1.0000	1.0000	1.0000	1.0000	1.0000	1.0000	0.9999
7.65	1.0000	1.0000	1.0000	1.0000	1.0000	1.0000	1.0000	1.0000	1.0000	0.9999
7.70	1.0000	1.0000	1.0000	1.0000	1.0000	1.0000	1.0000	1.0000	1.0000	0.9999
7.75	1.0000	1.0000	1.0000	1.0000	1.0000	1.0000	1.0000	1.0000	1.0000	0.9999
7.80	1.0000	1.0000	1.0000	1.0000	1.0000	1.0000	1.0000	1.0000	1.0000	0.9999
7.85	1.0000	1.0000	1.0000	1.0000	1.0000	1.0000	1.0000	1.0000	1.0000	1.0000

PROBABILITY INTEGRAL OF t, THE NON-CENTRAL t-STATISTIC. THIS TABLE GIVES Pr $[t/\sqrt{f} \leq x]$.

f is the number of degrees of freedom; the non-centrality parameter is $\sqrt{f+1}\, K_p$.

K_p is the standardized normal deviate exceeded with probability p.

DEGREES OF FREEDOM **19**

x \ p	.2500	.1500	.1000	.0650	.0400	.0250	.0100	.0040	.0025	.0010
- 0.25	0.0000	0.0000	0.0000	0.0000	0.0000	0.0000	0.0000	0.0000	0.0000	0.0000
- 0.20	0.0001	0.0000	0.0000	0.0000	0.0000	0.0000	0.0000	0.0000	0.0000	0.0000
- 0.15	0.0001	0.0000	0.0000	0.0000	0.0000	0.0000	0.0000	0.0000	0.0000	0.0000
- 0.10	0.0003	0.0000	0.0000	0.0000	0.0000	0.0000	0.0000	0.0000	0.0000	0.0000
- 0.05	0.0006	0.0000	0.0000	0.0000	0.0000	0.0000	0.0000	0.0000	0.0000	0.0000
0.00	0.0013	0.0000	0.0000	0.0000	0.0000	0.0000	0.0000	0.0000	0.0000	0.0000
0.05	0.0026	0.0000	0.0000	0.0000	0.0000	0.0000	0.0000	0.0000	0.0000	0.0000
0.10	0.0049	0.0000	0.0000	0.0000	0.0000	0.0000	0.0000	0.0000	0.0000	0.0000
0.15	0.0092	0.0000	0.0000	0.0000	0.0000	0.0000	0.0000	0.0000	0.0000	0.0000
0.20	0.0164	0.0001	0.0000	0.0000	0.0000	0.0000	0.0000	0.0000	0.0000	0.0000
0.25	0.0280	0.0002	0.0000	0.0000	0.0000	0.0000	0.0000	0.0000	0.0000	0.0000
0.30	0.0457	0.0005	0.0000	0.0000	0.0000	0.0000	0.0000	0.0000	0.0000	0.0000
0.35	0.0712	0.0012	0.0000	0.0000	0.0000	0.0000	0.0000	0.0000	0.0000	0.0000
0.40	0.1062	0.0025	0.0001	0.0000	0.0000	0.0000	0.0000	0.0000	0.0000	0.0000
0.45	0.1514	0.0051	0.0002	0.0000	0.0000	0.0000	0.0000	0.0000	0.0000	0.0000
0.50	0.2070	0.0097	0.0004	0.0000	0.0000	0.0000	0.0000	0.0000	0.0000	0.0000
0.55	0.2718	0.0173	0.0009	0.0000	0.0000	0.0000	0.0000	0.0000	0.0000	0.0000
0.60	0.3437	0.0295	0.0019	0.0001	0.0000	0.0000	0.0000	0.0000	0.0000	0.0000
0.65	0.4199	0.0475	0.0039	0.0002	0.0000	0.0000	0.0000	0.0000	0.0000	0.0000
0.70	0.4972	0.0729	0.0076	0.0004	0.0000	0.0000	0.0000	0.0000	0.0000	0.0000
0.75	0.5726	0.1066	0.0137	0.0009	0.0000	0.0000	0.0000	0.0000	0.0000	0.0000
0.80	0.6435	0.1491	0.0235	0.0020	0.0001	0.0000	0.0000	0.0000	0.0000	0.0000
0.85	0.7078	0.2001	0.0381	0.0040	0.0002	0.0000	0.0000	0.0000	0.0000	0.0000
0.90	0.7645	0.2586	0.0587	0.0075	0.0005	0.0000	0.0000	0.0000	0.0000	0.0000
0.95	0.8131	0.3228	0.0863	0.0134	0.0011	0.0001	0.0000	0.0000	0.0000	0.0000
1.00	0.8537	0.3907	0.1215	0.0225	0.0022	0.0002	0.0000	0.0000	0.0000	0.0000
1.05	0.8869	0.4599	0.1642	0.0358	0.0042	0.0004	0.0000	0.0000	0.0000	0.0000
1.10	0.9136	0.5281	0.2139	0.0544	0.0077	0.0009	0.0000	0.0000	0.0000	0.0000
1.15	0.9346	0.5934	0.2696	0.0790	0.0134	0.0018	0.0000	0.0000	0.0000	0.0000
1.20	0.9509	0.6542	0.3296	0.1100	0.0219	0.0034	0.0000	0.0000	0.0000	0.0000
1.25	0.9635	0.7096	0.3923	0.1477	0.0341	0.0061	0.0001	0.0000	0.0000	0.0000
1.30	0.9729	0.7588	0.4558	0.1915	0.0508	0.0105	0.0003	0.0000	0.0000	0.0000
1.35	0.9801	0.8018	0.5185	0.2408	0.0726	0.0172	0.0006	0.0000	0.0000	0.0000
1.40	0.9854	0.8387	0.5787	0.2945	0.1000	0.0268	0.0012	0.0000	0.0000	0.0000
1.45	0.9893	0.8698	0.6353	0.3511	0.1330	0.0401	0.0022	0.0001	0.0000	0.0000
1.50	0.9922	0.8957	0.6874	0.4094	0.1715	0.0575	0.0039	0.0002	0.0000	0.0000
1.55	0.9943	0.9170	0.7345	0.4677	0.2149	0.0796	0.0067	0.0003	0.0001	0.0000
1.60	0.9959	0.9344	0.7765	0.5250	0.2625	0.1067	0.0109	0.0007	0.0001	0.0000
1.65	0.9970	0.9484	0.8133	0.5800	0.3133	0.1386	0.0171	0.0013	0.0003	0.0000
1.70	0.9978	0.9596	0.8451	0.6318	0.3661	0.1753	0.0256	0.0023	0.0006	0.0000
1.75	0.9984	0.9685	0.8724	0.6799	0.4198	0.2162	0.0370	0.0039	0.0011	0.0001
1.80	0.9989	0.9755	0.8954	0.7237	0.4734	0.2606	0.0517	0.0064	0.0019	0.0001
1.85	0.9992	0.9810	0.9147	0.7633	0.5259	0.3078	0.0701	0.0101	0.0033	0.0003
1.90	0.9994	0.9853	0.9308	0.7985	0.5763	0.3569	0.0924	0.0154	0.0054	0.0006
1.95	0.9996	0.9886	0.9441	0.8294	0.6241	0.4068	0.1186	0.0225	0.0084	0.0010
2.00	0.9997	0.9912	0.9549	0.8564	0.6688	0.4568	0.1487	0.0320	0.0128	0.0018
2.05	0.9998	0.9932	0.9638	0.8797	0.7100	0.5060	0.1824	0.0440	0.0187	0.0029
2.10	0.9998	0.9948	0.9710	0.8997	0.7475	0.5537	0.2194	0.0590	0.0266	0.0047
2.15	0.9999	0.9960	0.9768	0.9166	0.7814	0.5994	0.2590	0.0771	0.0368	0.0072
2.20	0.9999	0.9969	0.9815	0.9310	0.8116	0.6424	0.3009	0.0985	0.0495	0.0108
2.25	0.9999	0.9976	0.9853	0.9430	0.8385	0.6827	0.3442	0.1231	0.0649	0.0156
2.30	0.9999	0.9982	0.9883	0.9531	0.8620	0.7198	0.3884	0.1508	0.0833	0.0219
2.35	1.0000	0.9986	0.9907	0.9615	0.8826	0.7538	0.4328	0.1816	0.1046	0.0301
2.40	1.0000	0.9989	0.9926	0.9684	0.9005	0.7847	0.4768	0.2149	0.1289	0.0403
2.45	1.0000	0.9992	0.9941	0.9742	0.9159	0.8124	0.5200	0.2506	0.1561	0.0527
2.50	1.0000	0.9993	0.9953	0.9789	0.9291	0.8372	0.5618	0.2881	0.1860	0.0676
2.55	1.0000	0.9995	0.9963	0.9828	0.9404	0.8593	0.6019	0.3269	0.2182	0.0850
2.60	1.0000	0.9996	0.9970	0.9859	0.9500	0.8787	0.6399	0.3667	0.2524	0.1049
2.65	1.0000	0.9997	0.9977	0.9886	0.9581	0.8957	0.6757	0.4069	0.2883	0.1274
2.70	1.0000	0.9998	0.9981	0.9907	0.9650	0.9106	0.7091	0.4470	0.3255	0.1524

PROBABILITY INTEGRAL OF t, THE NON-CENTRAL t-STATISTIC. THIS TABLE GIVES Pr $[t/\sqrt{f} \leq x]$.

f is the number of degrees of freedom; the non-centrality parameter is $\sqrt{f+1}\,K_p$.

K_p is the standardized normal deviate exceeded with probability p.

DEGREES OF FREEDOM **19**

x \ p	.2500	.1500	.1000	.0650	.0400	.0250	.0100	.0040	.0025	.0010
2.75	1.0000	0.9998	0.9985	0.9924	0.9707	0.9236	0.7400	0.4867	0.3635	0.1796
2.80	1.0000	0.9999	0.9988	0.9938	0.9756	0.9348	0.7685	0.5255	0.4019	0.2089
2.85	1.0000	0.9999	0.9991	0.9950	0.9797	0.9444	0.7945	0.5631	0.4403	0.2401
2.90	1.0000	0.9999	0.9992	0.9959	0.9831	0.9528	0.8182	0.5993	0.4783	0.2728
2.95	1.0000	0.9999	0.9994	0.9967	0.9859	0.9599	0.8396	0.6339	0.5156	0.3067
3.00	1.0000	0.9999	0.9995	0.9973	0.9883	0.9660	0.8588	0.6666	0.5519	0.3415
3.05	1.0000	1.0000	0.9996	0.9978	0.9903	0.9712	0.8760	0.6974	0.5870	0.3768
3.10	1.0000	1.0000	0.9997	0.9982	0.9919	0.9756	0.8914	0.7262	0.6206	0.4123
3.15	1.0000	1.0000	0.9998	0.9985	0.9933	0.9794	0.9050	0.7530	0.6525	0.4478
3.20	1.0000	1.0000	0.9998	0.9988	0.9944	0.9826	0.9171	0.7778	0.6828	0.4828
3.25	1.0000	1.0000	0.9998	0.9990	0.9954	0.9853	0.9278	0.8005	0.7113	0.5172
3.30	1.0000	1.0000	0.9999	0.9992	0.9962	0.9876	0.9372	0.8214	0.7379	0.5507
3.35	1.0000	1.0000	0.9999	0.9993	0.9968	0.9895	0.9454	0.8405	0.7626	0.5832
3.40	1.0000	1.0000	0.9999	0.9995	0.9973	0.9912	0.9527	0.8578	0.7856	0.6144
3.45	1.0000	1.0000	0.9999	0.9996	0.9978	0.9926	0.9590	0.8735	0.8068	0.6443
3.50	1.0000	1.0000	0.9999	0.9996	0.9982	0.9937	0.9645	0.8876	0.8262	0.6727
3.55	1.0000	1.0000	1.0000	0.9997	0.9985	0.9947	0.9693	0.9003	0.8440	0.6995
3.60	1.0000	1.0000	1.0000	0.9998	0.9987	0.9955	0.9734	0.9118	0.8602	0.7249
3.65	1.0000	1.0000	1.0000	0.9998	0.9989	0.9962	0.9771	0.9220	0.8750	0.7486
3.70	1.0000	1.0000	1.0000	0.9998	0.9991	0.9968	0.9802	0.9311	0.8884	0.7708
3.75	1.0000	1.0000	1.0000	0.9999	0.9993	0.9973	0.9829	0.9392	0.9005	0.7915
3.80	1.0000	1.0000	1.0000	0.9999	0.9994	0.9977	0.9852	0.9465	0.9114	0.8106
3.85	1.0000	1.0000	1.0000	0.9999	0.9995	0.9981	0.9873	0.9529	0.9212	0.8283
3.90	1.0000	1.0000	1.0000	0.9999	0.9996	0.9984	0.9890	0.9585	0.9300	0.8446
3.95	1.0000	1.0000	1.0000	0.9999	0.9996	0.9986	0.9905	0.9636	0.9379	0.8595
4.00	1.0000	1.0000	1.0000	0.9999	0.9997	0.9988	0.9918	0.9680	0.9449	0.8732
4.05	1.0000	1.0000	1.0000	1.0000	0.9997	0.9990	0.9930	0.9719	0.9512	0.8858
4.10	1.0000	1.0000	1.0000	1.0000	0.9998	0.9992	0.9939	0.9754	0.9568	0.8972
4.15	1.0000	1.0000	1.0000	1.0000	0.9998	0.9993	0.9948	0.9784	0.9618	0.9076
4.20	1.0000	1.0000	1.0000	1.0000	0.9998	0.9994	0.9955	0.9811	0.9662	0.9170
4.25	1.0000	1.0000	1.0000	1.0000	0.9999	0.9995	0.9961	0.9834	0.9702	0.9255
4.30	1.0000	1.0000	1.0000	1.0000	0.9999	0.9996	0.9967	0.9855	0.9737	0.9332
4.35	1.0000	1.0000	1.0000	1.0000	0.9999	0.9996	0.9971	0.9873	0.9768	0.9402
4.40	1.0000	1.0000	1.0000	1.0000	0.9999	0.9997	0.9975	0.9889	0.9795	0.9465
4.45	1.0000	1.0000	1.0000	1.0000	0.9999	0.9997	0.9978	0.9902	0.9819	0.9521
4.50	1.0000	1.0000	1.0000	1.0000	0.9999	0.9998	0.9981	0.9915	0.9841	0.9572
4.55	1.0000	1.0000	1.0000	1.0000	1.0000	0.9998	0.9984	0.9925	0.9860	0.9618
4.60	1.0000	1.0000	1.0000	1.0000	1.0000	0.9998	0.9986	0.9935	0.9876	0.9659
4.65	1.0000	1.0000	1.0000	1.0000	1.0000	0.9999	0.9988	0.9943	0.9891	0.9696
4.70	1.0000	1.0000	1.0000	1.0000	1.0000	0.9999	0.9990	0.9950	0.9904	0.9729
4.75	1.0000	1.0000	1.0000	1.0000	1.0000	0.9999	0.9991	0.9956	0.9915	0.9758
4.80	1.0000	1.0000	1.0000	1.0000	1.0000	0.9999	0.9992	0.9962	0.9925	0.9784
4.85	1.0000	1.0000	1.0000	1.0000	1.0000	0.9999	0.9993	0.9966	0.9934	0.9808
4.90	1.0000	1.0000	1.0000	1.0000	1.0000	0.9999	0.9994	0.9970	0.9942	0.9829
4.95	1.0000	1.0000	1.0000	1.0000	1.0000	0.9999	0.9995	0.9974	0.9949	0.9848
5.00	1.0000	1.0000	1.0000	1.0000	1.0000	0.9999	0.9996	0.9977	0.9955	0.9864
5.05	1.0000	1.0000	1.0000	1.0000	1.0000	1.0000	0.9996	0.9980	0.9960	0.9879
5.10	1.0000	1.0000	1.0000	1.0000	1.0000	1.0000	0.9997	0.9983	0.9965	0.9892
5.15	1.0000	1.0000	1.0000	1.0000	1.0000	1.0000	0.9997	0.9985	0.9969	0.9904
5.20	1.0000	1.0000	1.0000	1.0000	1.0000	1.0000	0.9997	0.9987	0.9973	0.9915
5.25	1.0000	1.0000	1.0000	1.0000	1.0000	1.0000	0.9998	0.9988	0.9976	0.9924
5.30	1.0000	1.0000	1.0000	1.0000	1.0000	1.0000	0.9998	0.9990	0.9979	0.9932
5.35	1.0000	1.0000	1.0000	1.0000	1.0000	1.0000	0.9998	0.9991	0.9981	0.9940
5.40	1.0000	1.0000	1.0000	1.0000	1.0000	1.0000	0.9999	0.9992	0.9984	0.9946
5.45	1.0000	1.0000	1.0000	1.0000	1.0000	1.0000	0.9999	0.9993	0.9985	0.9952
5.50	1.0000	1.0000	1.0000	1.0000	1.0000	1.0000	0.9999	0.9994	0.9987	0.9958
5.55	1.0000	1.0000	1.0000	1.0000	1.0000	1.0000	0.9999	0.9995	0.9989	0.9962
5.60	1.0000	1.0000	1.0000	1.0000	1.0000	1.0000	0.9999	0.9995	0.9990	0.9966
5.65	1.0000	1.0000	1.0000	1.0000	1.0000	1.0000	0.9999	0.9996	0.9991	0.9970
5.70	1.0000	1.0000	1.0000	1.0000	1.0000	1.0000	0.9999	0.9996	0.9992	0.9973

PROBABILITY INTEGRAL OF t, THE NON-CENTRAL t-STATISTIC. THIS TABLE GIVES Pr $[t/\sqrt{f} \leq x]$.

f is the number of degrees of freedom; the non-centrality parameter is $\sqrt{f+1}\,K_p$.
K_p is the standardized normal deviate exceeded with probability p.

DEGREES OF FREEDOM **19**

x \ p	.2500	.1500	.1000	.0650	.0400	.0250	.0100	.0040	.0025	.0010
5.75	1.0000	1.0000	1.0000	1.0000	1.0000	1.0000	0.9999	0.9997	0.9993	0.9976
5.80	1.0000	1.0000	1.0000	1.0000	1.0000	1.0000	0.9999	0.9997	0.9994	0.9979
5.85	1.0000	1.0000	1.0000	1.0000	1.0000	1.0000	1.0000	0.9997	0.9995	0.9981
5.90	1.0000	1.0000	1.0000	1.0000	1.0000	1.0000	1.0000	0.9998	0.9995	0.9983
5.95	1.0000	1.0000	1.0000	1.0000	1.0000	1.0000	1.0000	0.9998	0.9996	0.9985
6.00	1.0000	1.0000	1.0000	1.0000	1.0000	1.0000	1.0000	0.9998	0.9996	0.9987
6.05	1.0000	1.0000	1.0000	1.0000	1.0000	1.0000	1.0000	0.9998	0.9997	0.9988
6.10	1.0000	1.0000	1.0000	1.0000	1.0000	1.0000	1.0000	0.9999	0.9997	0.9989
6.15	1.0000	1.0000	1.0000	1.0000	1.0000	1.0000	1.0000	0.9999	0.9997	0.9990
6.20	1.0000	1.0000	1.0000	1.0000	1.0000	1.0000	1.0000	0.9999	0.9998	0.9991
6.25	1.0000	1.0000	1.0000	1.0000	1.0000	1.0000	1.0000	0.9999	0.9998	0.9992
6.30	1.0000	1.0000	1.0000	1.0000	1.0000	1.0000	1.0000	0.9999	0.9998	0.9993
6.35	1.0000	1.0000	1.0000	1.0000	1.0000	1.0000	1.0000	0.9999	0.9998	0.9994
6.40	1.0000	1.0000	1.0000	1.0000	1.0000	1.0000	1.0000	0.9999	0.9998	0.9995
6.45	1.0000	1.0000	1.0000	1.0000	1.0000	1.0000	1.0000	0.9999	0.9999	0.9995
6.50	1.0000	1.0000	1.0000	1.0000	1.0000	1.0000	1.0000	0.9999	0.9999	0.9996
6.55	1.0000	1.0000	1.0000	1.0000	1.0000	1.0000	1.0000	0.9999	0.9999	0.9996
6.60	1.0000	1.0000	1.0000	1.0000	1.0000	1.0000	1.0000	1.0000	0.9999	0.9996
6.65	1.0000	1.0000	1.0000	1.0000	1.0000	1.0000	1.0000	1.0000	0.9999	0.9997
6.70	1.0000	1.0000	1.0000	1.0000	1.0000	1.0000	1.0000	1.0000	0.9999	0.9997
6.75	1.0000	1.0000	1.0000	1.0000	1.0000	1.0000	1.0000	1.0000	0.9999	0.9997
6.80	1.0000	1.0000	1.0000	1.0000	1.0000	1.0000	1.0000	1.0000	0.9999	0.9998
6.85	1.0000	1.0000	1.0000	1.0000	1.0000	1.0000	1.0000	1.0000	0.9999	0.9998
6.90	1.0000	1.0000	1.0000	1.0000	1.0000	1.0000	1.0000	1.0000	0.9999	0.9998
6.95	1.0000	1.0000	1.0000	1.0000	1.0000	1.0000	1.0000	1.0000	1.0000	0.9998
7.00	1.0000	1.0000	1.0000	1.0000	1.0000	1.0000	1.0000	1.0000	1.0000	0.9998
7.05	1.0000	1.0000	1.0000	1.0000	1.0000	1.0000	1.0000	1.0000	1.0000	0.9999
7.10	1.0000	1.0000	1.0000	1.0000	1.0000	1.0000	1.0000	1.0000	1.0000	0.9999
7.15	1.0000	1.0000	1.0000	1.0000	1.0000	1.0000	1.0000	1.0000	1.0000	0.9999
7.20	1.0000	1.0000	1.0000	1.0000	1.0000	1.0000	1.0000	1.0000	1.0000	0.9999
7.25	1.0000	1.0000	1.0000	1.0000	1.0000	1.0000	1.0000	1.0000	1.0000	0.9999
7.30	1.0000	1.0000	1.0000	1.0000	1.0000	1.0000	1.0000	1.0000	1.0000	0.9999
7.35	1.0000	1.0000	1.0000	1.0000	1.0000	1.0000	1.0000	1.0000	1.0000	0.9999
7.40	1.0000	1.0000	1.0000	1.0000	1.0000	1.0000	1.0000	1.0000	1.0000	0.9999
7.45	1.0000	1.0000	1.0000	1.0000	1.0000	1.0000	1.0000	1.0000	1.0000	0.9999
7.50	1.0000	1.0000	1.0000	1.0000	1.0000	1.0000	1.0000	1.0000	1.0000	0.9999
7.55	1.0000	1.0000	1.0000	1.0000	1.0000	1.0000	1.0000	1.0000	1.0000	0.9999
7.60	1.0000	1.0000	1.0000	1.0000	1.0000	1.0000	1.0000	1.0000	1.0000	1.0000

PROBABILITY INTEGRAL OF t, THE NON-CENTRAL t-STATISTIC. THIS TABLE GIVES Pr $[t/\sqrt{f} \leq x]$.

f is the number of degrees of freedom; the non-centrality parameter is $\sqrt{f+1}\,K_p$.

K_p is the standardized normal deviate exceeded with probability p.

DEGREES OF FREEDOM **20**

x \ p	.2500	.1500	.1000	.0650	.0400	.0250	.0100	.0040	.0025	.0010
-0.20	0.0000	0.0000	0.0000	0.0000	0.0000	0.0000	0.0000	0.0000	0.0000	0.0000
-0.15	0.0001	0.0000	0.0000	0.0000	0.0000	0.0000	0.0000	0.0000	0.0000	0.0000
-0.10	0.0002	0.0000	0.0000	0.0000	0.0000	0.0000	0.0000	0.0000	0.0000	0.0000
-0.05	0.0005	0.0000	0.0000	0.0000	0.0000	0.0000	0.0000	0.0000	0.0000	0.0000
0.00	0.0010	0.0000	0.0000	0.0000	0.0000	0.0000	0.0000	0.0000	0.0000	0.0000
0.05	0.0021	0.0000	0.0000	0.0000	0.0000	0.0000	0.0000	0.0000	0.0000	0.0000
0.10	0.0041	0.0000	0.0000	0.0000	0.0000	0.0000	0.0000	0.0000	0.0000	0.0000
0.15	0.0079	0.0000	0.0000	0.0000	0.0000	0.0000	0.0000	0.0000	0.0000	0.0000
0.20	0.0144	0.0001	0.0000	0.0000	0.0000	0.0000	0.0000	0.0000	0.0000	0.0000
0.25	0.0252	0.0002	0.0000	0.0000	0.0000	0.0000	0.0000	0.0000	0.0000	0.0000
0.30	0.0420	0.0004	0.0000	0.0000	0.0000	0.0000	0.0000	0.0000	0.0000	0.0000
0.35	0.0668	0.0009	0.0000	0.0000	0.0000	0.0000	0.0000	0.0000	0.0000	0.0000
0.40	0.1012	0.0021	0.0000	0.0000	0.0000	0.0000	0.0000	0.0000	0.0000	0.0000
0.45	0.1464	0.0043	0.0001	0.0000	0.0000	0.0000	0.0000	0.0000	0.0000	0.0000
0.50	0.2024	0.0084	0.0003	0.0000	0.0000	0.0000	0.0000	0.0000	0.0000	0.0000
0.55	0.2683	0.0154	0.0007	0.0000	0.0000	0.0000	0.0000	0.0000	0.0000	0.0000
0.60	0.3418	0.0268	0.0016	0.0000	0.0000	0.0000	0.0000	0.0000	0.0000	0.0000
0.65	0.4199	0.0441	0.0033	0.0001	0.0000	0.0000	0.0000	0.0000	0.0000	0.0000
0.70	0.4993	0.0688	0.0065	0.0003	0.0000	0.0000	0.0000	0.0000	0.0000	0.0000
0.75	0.5767	0.1021	0.0121	0.0007	0.0000	0.0000	0.0000	0.0000	0.0000	0.0000
0.80	0.6491	0.1446	0.0212	0.0016	0.0001	0.0000	0.0000	0.0000	0.0000	0.0000
0.85	0.7146	0.1961	0.0351	0.0034	0.0001	0.0000	0.0000	0.0000	0.0000	0.0000
0.90	0.7720	0.2556	0.0551	0.0065	0.0004	0.0000	0.0000	0.0000	0.0000	0.0000
0.95	0.8207	0.3212	0.0822	0.0118	0.0008	0.0000	0.0000	0.0000	0.0000	0.0000
1.00	0.8612	0.3908	0.1172	0.0203	0.0018	0.0001	0.0000	0.0000	0.0000	0.0000
1.05	0.8939	0.4618	0.1601	0.0330	0.0036	0.0003	0.0000	0.0000	0.0000	0.0000
1.10	0.9199	0.5318	0.2105	0.0510	0.0067	0.0007	0.0000	0.0000	0.0000	0.0000
1.15	0.9402	0.5987	0.2672	0.0751	0.0118	0.0014	0.0000	0.0000	0.0000	0.0000
1.20	0.9557	0.6608	0.3286	0.1059	0.0198	0.0028	0.0000	0.0000	0.0000	0.0000
1.25	0.9675	0.7171	0.3929	0.1437	0.0314	0.0053	0.0001	0.0000	0.0000	0.0000
1.30	0.9763	0.7668	0.4581	0.1880	0.0475	0.0093	0.0002	0.0000	0.0000	0.0000
1.35	0.9828	0.8100	0.5224	0.2381	0.0689	0.0154	0.0004	0.0000	0.0000	0.0000
1.40	0.9876	0.8467	0.5840	0.2929	0.0960	0.0245	0.0009	0.0000	0.0000	0.0000
1.45	0.9911	0.8774	0.6419	0.3510	0.1291	0.0372	0.0018	0.0000	0.0000	0.0000
1.50	0.9936	0.9028	0.6949	0.4107	0.1679	0.0542	0.0033	0.0001	0.0000	0.0000
1.55	0.9954	0.9235	0.7426	0.4707	0.2120	0.0759	0.0058	0.0003	0.0000	0.0000
1.60	0.9967	0.9402	0.7849	0.5294	0.2605	0.1028	0.0096	0.0005	0.0001	0.0000
1.65	0.9977	0.9535	0.8217	0.5857	0.3124	0.1349	0.0153	0.0010	0.0002	0.0000
1.70	0.9983	0.9640	0.8533	0.6386	0.3666	0.1719	0.0234	0.0019	0.0004	0.0000
1.75	0.9988	0.9722	0.8802	0.6875	0.4218	0.2134	0.0343	0.0033	0.0009	0.0000
1.80	0.9991	0.9787	0.9027	0.7320	0.4768	0.2587	0.0486	0.0056	0.0016	0.0001
1.85	0.9994	0.9837	0.9214	0.7718	0.5306	0.3070	0.0667	0.0089	0.0028	0.0002
1.90	0.9996	0.9875	0.9368	0.8071	0.5822	0.3573	0.0887	0.0138	0.0046	0.0004
1.95	0.9997	0.9905	0.9495	0.8379	0.6310	0.4086	0.1149	0.0205	0.0074	0.0008
2.00	0.9998	0.9927	0.9597	0.8646	0.6765	0.4599	0.1452	0.0295	0.0114	0.0014
2.05	0.9998	0.9945	0.9680	0.8875	0.7182	0.5104	0.1793	0.0412	0.0169	0.0024
2.10	0.9999	0.9958	0.9746	0.9069	0.7561	0.5593	0.2169	0.0558	0.0244	0.0040
2.15	0.9999	0.9968	0.9799	0.9233	0.7902	0.6059	0.2573	0.0736	0.0342	0.0063
2.20	0.9999	0.9976	0.9842	0.9371	0.8204	0.6499	0.3001	0.0948	0.0465	0.0095
2.25	1.0000	0.9982	0.9875	0.9485	0.8470	0.6907	0.3446	0.1195	0.0616	0.0140
2.30	1.0000	0.9986	0.9902	0.9580	0.8703	0.7284	0.3899	0.1474	0.0797	0.0200
2.35	1.0000	0.9989	0.9923	0.9658	0.8904	0.7626	0.4355	0.1785	0.1010	0.0277
2.40	1.0000	0.9992	0.9939	0.9723	0.9078	0.7936	0.4807	0.2125	0.1254	0.0376
2.45	1.0000	0.9994	0.9952	0.9775	0.9227	0.8213	0.5250	0.2488	0.1528	0.0497
2.50	1.0000	0.9995	0.9963	0.9818	0.9354	0.8459	0.5678	0.2872	0.1830	0.0643
2.55	1.0000	0.9996	0.9971	0.9853	0.9461	0.8676	0.6087	0.3270	0.2158	0.0815
2.60	1.0000	0.9997	0.9977	0.9882	0.9551	0.8867	0.6475	0.3678	0.2508	0.1014
2.65	1.0000	0.9998	0.9982	0.9904	0.9628	0.9033	0.6839	0.4091	0.2875	0.1239
2.70	1.0000	0.9998	0.9986	0.9923	0.9691	0.9177	0.7177	0.4503	0.3256	0.1491

PROBABILITY INTEGRAL OF t, THE NON-CENTRAL t-STATISTIC. THIS TABLE GIVES Pr $[t/\sqrt{f} \leq x]$.

f is the number of degrees of freedom; the non-centrality parameter is $\sqrt{f+1}\, K_p$.
K_p is the standardized normal deviate exceeded with probability p.

DEGREES OF FREEDOM **20**

x \ p	.2500	.1500	.1000	.0650	.0400	.0250	.0100	.0040	.0025	.0010
2.75	1.0000	0.9999	0.9989	0.9938	0.9745	0.9301	0.7489	0.4910	0.3646	0.1767
2.80	1.0000	0.9999	0.9991	0.9950	0.9789	0.9408	0.7775	0.5308	0.4040	0.2065
2.85	1.0000	0.9999	0.9993	0.9960	0.9826	0.9500	0.8036	0.5693	0.4434	0.2383
2.90	1.0000	0.9999	0.9995	0.9968	0.9856	0.9578	0.8271	0.6063	0.4824	0.2717
2.95	1.0000	1.0000	0.9996	0.9974	0.9882	0.9645	0.8483	0.6415	0.5207	0.3064
3.00	1.0000	1.0000	0.9997	0.9979	0.9902	0.9701	0.8673	0.6748	0.5579	0.3421
3.05	1.0000	1.0000	0.9997	0.9983	0.9920	0.9749	0.8842	0.7060	0.5937	0.3783
3.10	1.0000	1.0000	0.9998	0.9986	0.9934	0.9789	0.8992	0.7351	0.6280	0.4148
3.15	1.0000	1.0000	0.9998	0.9989	0.9946	0.9823	0.9124	0.7621	0.6606	0.4512
3.20	1.0000	1.0000	0.9999	0.9991	0.9955	0.9852	0.9240	0.7869	0.6913	0.4872
3.25	1.0000	1.0000	0.9999	0.9993	0.9963	0.9876	0.9342	0.8097	0.7201	0.5224
3.30	1.0000	1.0000	0.9999	0.9994	0.9970	0.9896	0.9432	0.8305	0.7469	0.5568
3.35	1.0000	1.0000	0.9999	0.9995	0.9975	0.9913	0.9510	0.8493	0.7718	0.5899
3.40	1.0000	1.0000	0.9999	0.9996	0.9979	0.9927	0.9578	0.8664	0.7948	0.6218
3.45	1.0000	1.0000	1.0000	0.9997	0.9983	0.9939	0.9636	0.8817	0.8159	0.6522
3.50	1.0000	1.0000	1.0000	0.9997	0.9986	0.9949	0.9687	0.8955	0.8352	0.6811
3.55	1.0000	1.0000	1.0000	0.9998	0.9989	0.9958	0.9731	0.9079	0.8528	0.7083
3.60	1.0000	1.0000	1.0000	0.9998	0.9991	0.9965	0.9770	0.9189	0.8688	0.7339
3.65	1.0000	1.0000	1.0000	0.9999	0.9992	0.9970	0.9802	0.9287	0.8833	0.7578
3.70	1.0000	1.0000	1.0000	0.9999	0.9994	0.9975	0.9831	0.9374	0.8963	0.7801
3.75	1.0000	1.0000	1.0000	0.9999	0.9995	0.9979	0.9855	0.9451	0.9080	0.8007
3.80	1.0000	1.0000	1.0000	0.9999	0.9996	0.9983	0.9876	0.9520	0.9186	0.8198
3.85	1.0000	1.0000	1.0000	0.9999	0.9996	0.9985	0.9894	0.9580	0.9280	0.8373
3.90	1.0000	1.0000	1.0000	0.9999	0.9997	0.9988	0.9909	0.9633	0.9364	0.8534
3.95	1.0000	1.0000	1.0000	1.0000	0.9997	0.9990	0.9922	0.9679	0.9439	0.8682
4.00	1.0000	1.0000	1.0000	1.0000	0.9998	0.9991	0.9933	0.9720	0.9505	0.8816
4.05	1.0000	1.0000	1.0000	1.0000	0.9998	0.9993	0.9943	0.9756	0.9564	0.8938
4.10	1.0000	1.0000	1.0000	1.0000	0.9999	0.9994	0.9951	0.9787	0.9617	0.9049
4.15	1.0000	1.0000	1.0000	1.0000	0.9999	0.9995	0.9958	0.9815	0.9663	0.9149
4.20	1.0000	1.0000	1.0000	1.0000	0.9999	0.9996	0.9964	0.9839	0.9704	0.9240
4.25	1.0000	1.0000	1.0000	1.0000	0.9999	0.9996	0.9969	0.9859	0.9740	0.9321
4.30	1.0000	1.0000	1.0000	1.0000	0.9999	0.9997	0.9974	0.9878	0.9772	0.9395
4.35	1.0000	1.0000	1.0000	1.0000	0.9999	0.9997	0.9978	0.9894	0.9800	0.9461
4.40	1.0000	1.0000	1.0000	1.0000	0.9999	0.9998	0.9981	0.9907	0.9825	0.9520
4.45	1.0000	1.0000	1.0000	1.0000	1.0000	0.9998	0.9984	0.9920	0.9846	0.9573
4.50	1.0000	1.0000	1.0000	1.0000	1.0000	0.9998	0.9986	0.9930	0.9865	0.9621
4.55	1.0000	1.0000	1.0000	1.0000	1.0000	0.9999	0.9988	0.9939	0.9882	0.9663
4.60	1.0000	1.0000	1.0000	1.0000	1.0000	0.9999	0.9990	0.9947	0.9897	0.9701
4.65	1.0000	1.0000	1.0000	1.0000	1.0000	0.9999	0.9991	0.9954	0.9910	0.9735
4.70	1.0000	1.0000	1.0000	1.0000	1.0000	0.9999	0.9992	0.9960	0.9921	0.9765
4.75	1.0000	1.0000	1.0000	1.0000	1.0000	0.9999	0.9993	0.9965	0.9931	0.9791
4.80	1.0000	1.0000	1.0000	1.0000	1.0000	0.9999	0.9994	0.9970	0.9939	0.9815
4.85	1.0000	1.0000	1.0000	1.0000	1.0000	0.9999	0.9995	0.9974	0.9947	0.9836
4.90	1.0000	1.0000	1.0000	1.0000	1.0000	1.0000	0.9996	0.9977	0.9953	0.9855
4.95	1.0000	1.0000	1.0000	1.0000	1.0000	1.0000	0.9996	0.9980	0.9959	0.9872
5.00	1.0000	1.0000	1.0000	1.0000	1.0000	1.0000	0.9997	0.9983	0.9964	0.9886
5.05	1.0000	1.0000	1.0000	1.0000	1.0000	1.0000	0.9997	0.9985	0.9969	0.9899
5.10	1.0000	1.0000	1.0000	1.0000	1.0000	1.0000	0.9998	0.9987	0.9973	0.9911
5.15	1.0000	1.0000	1.0000	1.0000	1.0000	1.0000	0.9998	0.9988	0.9976	0.9921
5.20	1.0000	1.0000	1.0000	1.0000	1.0000	1.0000	0.9998	0.9990	0.9979	0.9930
5.25	1.0000	1.0000	1.0000	1.0000	1.0000	1.0000	0.9999	0.9991	0.9982	0.9938
5.30	1.0000	1.0000	1.0000	1.0000	1.0000	1.0000	0.9999	0.9992	0.9984	0.9945
5.35	1.0000	1.0000	1.0000	1.0000	1.0000	1.0000	0.9999	0.9993	0.9986	0.9952
5.40	1.0000	1.0000	1.0000	1.0000	1.0000	1.0000	0.9999	0.9994	0.9988	0.9957
5.45	1.0000	1.0000	1.0000	1.0000	1.0000	1.0000	0.9999	0.9995	0.9989	0.9962
5.50	1.0000	1.0000	1.0000	1.0000	1.0000	1.0000	0.9999	0.9996	0.9990	0.9966
5.55	1.0000	1.0000	1.0000	1.0000	1.0000	1.0000	0.9999	0.9996	0.9992	0.9970
5.60	1.0000	1.0000	1.0000	1.0000	1.0000	1.0000	0.9999	0.9997	0.9993	0.9974
5.65	1.0000	1.0000	1.0000	1.0000	1.0000	1.0000	1.0000	0.9997	0.9993	0.9977
5.70	1.0000	1.0000	1.0000	1.0000	1.0000	1.0000	1.0000	0.9997	0.9994	0.9979

PROBABILITY INTEGRAL OF t, THE NON-CENTRAL t-STATISTIC. THIS TABLE GIVES Pr $[t/\sqrt{f} \leq x]$.

f is the number of degrees of freedom; the non-centrality parameter is $\sqrt{f+1}\,K_p$.

K_p is the standardized normal deviate exceeded with probability p.

DEGREES OF FREEDOM **20**

x \ p	.2500	.1500	.1000	.0650	.0400	.0250	.0100	.0040	.0025	.0010
5.75	1.0000	1.0000	1.0000	1.0000	1.0000	1.0000	1.0000	0.9998	0.9995	0.9982
5.80	1.0000	1.0000	1.0000	1.0000	1.0000	1.0000	1.0000	0.9998	0.9996	0.9984
5.85	1.0000	1.0000	1.0000	1.0000	1.0000	1.0000	1.0000	0.9998	0.9996	0.9986
5.90	1.0000	1.0000	1.0000	1.0000	1.0000	1.0000	1.0000	0.9998	0.9997	0.9987
5.95	1.0000	1.0000	1.0000	1.0000	1.0000	1.0000	1.0000	0.9999	0.9997	0.9989
6.00	1.0000	1.0000	1.0000	1.0000	1.0000	1.0000	1.0000	0.9999	0.9997	0.9990
6.05	1.0000	1.0000	1.0000	1.0000	1.0000	1.0000	1.0000	0.9999	0.9998	0.9991
6.10	1.0000	1.0000	1.0000	1.0000	1.0000	1.0000	1.0000	0.9999	0.9998	0.9992
6.15	1.0000	1.0000	1.0000	1.0000	1.0000	1.0000	1.0000	0.9999	0.9998	0.9993
6.20	1.0000	1.0000	1.0000	1.0000	1.0000	1.0000	1.0000	0.9999	0.9998	0.9994
6.25	1.0000	1.0000	1.0000	1.0000	1.0000	1.0000	1.0000	0.9999	0.9999	0.9994
6.30	1.0000	1.0000	1.0000	1.0000	1.0000	1.0000	1.0000	0.9999	0.9999	0.9995
6.35	1.0000	1.0000	1.0000	1.0000	1.0000	1.0000	1.0000	0.9999	0.9999	0.9996
6.40	1.0000	1.0000	1.0000	1.0000	1.0000	1.0000	1.0000	1.0000	0.9999	0.9996
6.45	1.0000	1.0000	1.0000	1.0000	1.0000	1.0000	1.0000	1.0000	0.9999	0.9997
6.50	1.0000	1.0000	1.0000	1.0000	1.0000	1.0000	1.0000	1.0000	0.9999	0.9997
6.55	1.0000	1.0000	1.0000	1.0000	1.0000	1.0000	1.0000	1.0000	0.9999	0.9997
6.60	1.0000	1.0000	1.0000	1.0000	1.0000	1.0000	1.0000	1.0000	0.9999	0.9998
6.65	1.0000	1.0000	1.0000	1.0000	1.0000	1.0000	1.0000	1.0000	0.9999	0.9998
6.70	1.0000	1.0000	1.0000	1.0000	1.0000	1.0000	1.0000	1.0000	0.9999	0.9998
6.75	1.0000	1.0000	1.0000	1.0000	1.0000	1.0000	1.0000	1.0000	1.0000	0.9998
6.80	1.0000	1.0000	1.0000	1.0000	1.0000	1.0000	1.0000	1.0000	1.0000	0.9998
6.85	1.0000	1.0000	1.0000	1.0000	1.0000	1.0000	1.0000	1.0000	1.0000	0.9999
6.90	1.0000	1.0000	1.0000	1.0000	1.0000	1.0000	1.0000	1.0000	1.0000	0.9999
6.95	1.0000	1.0000	1.0000	1.0000	1.0000	1.0000	1.0000	1.0000	1.0000	0.9999
7.00	1.0000	1.0000	1.0000	1.0000	1.0000	1.0000	1.0000	1.0000	1.0000	0.9999
7.05	1.0000	1.0000	1.0000	1.0000	1.0000	1.0000	1.0000	1.0000	1.0000	0.9999
7.10	1.0000	1.0000	1.0000	1.0000	1.0000	1.0000	1.0000	1.0000	1.0000	0.9999
7.15	1.0000	1.0000	1.0000	1.0000	1.0000	1.0000	1.0000	1.0000	1.0000	0.9999
7.20	1.0000	1.0000	1.0000	1.0000	1.0000	1.0000	1.0000	1.0000	1.0000	0.9999
7.25	1.0000	1.0000	1.0000	1.0000	1.0000	1.0000	1.0000	1.0000	1.0000	0.9999
7.30	1.0000	1.0000	1.0000	1.0000	1.0000	1.0000	1.0000	1.0000	1.0000	0.9999
7.35	1.0000	1.0000	1.0000	1.0000	1.0000	1.0000	1.0000	1.0000	1.0000	1.0000

PROBABILITY INTEGRAL OF t, THE NON-CENTRAL t-STATISTIC. THIS TABLE GIVES Pr $[t/\sqrt{f} \leq x]$.

f is the number of degrees of freedom; the non-centrality parameter is $\sqrt{f+1}\,K_p$.

K_p is the standardized normal deviate exceeded with probability p.

DEGREES OF FREEDOM **21**

x \ p	.2500	.1500	.1000	.0650	.0400	.0250	.0100	.0040	.0025	.0010
-0.20	0.0000	0.0000	0.0000	0.0000	0.0000	0.0000	0.0000	0.0000	0.0000	0.0000
-0.15	0.0001	0.0000	0.0000	0.0000	0.0000	0.0000	0.0000	0.0000	0.0000	0.0000
-0.10	0.0002	0.0000	0.0000	0.0000	0.0000	0.0000	0.0000	0.0000	0.0000	0.0000
-0.05	0.0004	0.0000	0.0000	0.0000	0.0000	0.0000	0.0000	0.0000	0.0000	0.0000
0.00	0.0008	0.0000	0.0000	0.0000	0.0000	0.0000	0.0000	0.0000	0.0000	0.0000
0.05	0.0017	0.0000	0.0000	0.0000	0.0000	0.0000	0.0000	0.0000	0.0000	0.0000
0.10	0.0034	0.0000	0.0000	0.0000	0.0000	0.0000	0.0000	0.0000	0.0000	0.0000
0.15	0.0067	0.0000	0.0000	0.0000	0.0000	0.0000	0.0000	0.0000	0.0000	0.0000
0.20	0.0127	0.0000	0.0000	0.0000	0.0000	0.0000	0.0000	0.0000	0.0000	0.0000
0.25	0.0227	0.0001	0.0000	0.0000	0.0000	0.0000	0.0000	0.0000	0.0000	0.0000
0.30	0.0387	0.0003	0.0000	0.0000	0.0000	0.0000	0.0000	0.0000	0.0000	0.0000
0.35	0.0627	0.0007	0.0000	0.0000	0.0000	0.0000	0.0000	0.0000	0.0000	0.0000
0.40	0.0965	0.0017	0.0000	0.0000	0.0000	0.0000	0.0000	0.0000	0.0000	0.0000
0.45	0.1415	0.0036	0.0001	0.0000	0.0000	0.0000	0.0000	0.0000	0.0000	0.0000
0.50	0.1980	0.0072	0.0002	0.0000	0.0000	0.0000	0.0000	0.0000	0.0000	0.0000
0.55	0.2648	0.0137	0.0005	0.0000	0.0000	0.0000	0.0000	0.0000	0.0000	0.0000
0.60	0.3399	0.0243	0.0013	0.0000	0.0000	0.0000	0.0000	0.0000	0.0000	0.0000
0.65	0.4199	0.0409	0.0027	0.0001	0.0000	0.0000	0.0000	0.0000	0.0000	0.0000
0.70	0.5013	0.0649	0.0056	0.0002	0.0000	0.0000	0.0000	0.0000	0.0000	0.0000
0.75	0.5805	0.0978	0.0107	0.0006	0.0000	0.0000	0.0000	0.0000	0.0000	0.0000
0.80	0.6545	0.1403	0.0192	0.0013	0.0000	0.0000	0.0000	0.0000	0.0000	0.0000
0.85	0.7211	0.1922	0.0323	0.0028	0.0001	0.0000	0.0000	0.0000	0.0000	0.0000
0.90	0.7791	0.2526	0.0517	0.0056	0.0003	0.0000	0.0000	0.0000	0.0000	0.0000
0.95	0.8280	0.3196	0.0783	0.0105	0.0007	0.0000	0.0000	0.0000	0.0000	0.0000
1.00	0.8682	0.3909	0.1130	0.0184	0.0015	0.0001	0.0000	0.0000	0.0000	0.0000
1.05	0.9004	0.4637	0.1561	0.0304	0.0030	0.0002	0.0000	0.0000	0.0000	0.0000
1.10	0.9257	0.5354	0.2071	0.0478	0.0058	0.0005	0.0000	0.0000	0.0000	0.0000
1.15	0.9452	0.6038	0.2648	0.0714	0.0105	0.0012	0.0000	0.0000	0.0000	0.0000
1.20	0.9600	0.6671	0.3276	0.1020	0.0179	0.0024	0.0000	0.0000	0.0000	0.0000
1.25	0.9710	0.7242	0.3934	0.1398	0.0289	0.0045	0.0001	0.0000	0.0000	0.0000
1.30	0.9792	0.7744	0.4603	0.1845	0.0445	0.0081	0.0001	0.0000	0.0000	0.0000
1.35	0.9851	0.8177	0.5261	0.2354	0.0654	0.0138	0.0003	0.0000	0.0000	0.0000
1.40	0.9894	0.8542	0.5892	0.2914	0.0922	0.0224	0.0007	0.0000	0.0000	0.0000
1.45	0.9925	0.8846	0.6482	0.3508	0.1253	0.0345	0.0015	0.0000	0.0000	0.0000
1.50	0.9947	0.9094	0.7021	0.4120	0.1644	0.0510	0.0028	0.0001	0.0000	0.0000
1.55	0.9963	0.9294	0.7504	0.4735	0.2090	0.0724	0.0050	0.0002	0.0000	0.0000
1.60	0.9974	0.9454	0.7929	0.5336	0.2585	0.0991	0.0085	0.0004	0.0001	0.0000
1.65	0.9982	0.9580	0.8296	0.5912	0.3115	0.1312	0.0138	0.0008	0.0002	0.0000
1.70	0.9987	0.9679	0.8610	0.6452	0.3670	0.1685	0.0214	0.0016	0.0003	0.0000
1.75	0.9991	0.9756	0.8874	0.6949	0.4236	0.2107	0.0318	0.0028	0.0007	0.0000
1.80	0.9994	0.9815	0.9094	0.7398	0.4800	0.2568	0.0457	0.0048	0.0013	0.0001
1.85	0.9996	0.9860	0.9275	0.7799	0.5351	0.3062	0.0634	0.0079	0.0023	0.0002
1.90	0.9997	0.9894	0.9423	0.8152	0.5879	0.3577	0.0852	0.0124	0.0039	0.0003
1.95	0.9998	0.9920	0.9543	0.8459	0.6377	0.4102	0.1113	0.0187	0.0064	0.0006
2.00	0.9999	0.9940	0.9640	0.8723	0.6839	0.4629	0.1417	0.0272	0.0101	0.0012
2.05	0.9999	0.9955	0.9717	0.8947	0.7261	0.5146	0.1762	0.0385	0.0153	0.0020
2.10	0.9999	0.9966	0.9778	0.9136	0.7643	0.5646	0.2143	0.0527	0.0224	0.0034
2.15	1.0000	0.9975	0.9826	0.9295	0.7985	0.6122	0.2556	0.0703	0.0317	0.0054
2.20	1.0000	0.9981	0.9865	0.9426	0.8286	0.6570	0.2993	0.0913	0.0436	0.0084
2.25	1.0000	0.9986	0.9895	0.9535	0.8550	0.6985	0.3448	0.1160	0.0585	0.0126
2.30	1.0000	0.9990	0.9918	0.9624	0.8779	0.7365	0.3913	0.1441	0.0764	0.0182
2.35	1.0000	0.9992	0.9936	0.9697	0.8977	0.7710	0.4381	0.1755	0.0975	0.0256
2.40	1.0000	0.9994	0.9951	0.9757	0.9146	0.8020	0.4844	0.2100	0.1220	0.0350
2.45	1.0000	0.9996	0.9962	0.9805	0.9289	0.8296	0.5298	0.2471	0.1496	0.0468
2.50	1.0000	0.9997	0.9970	0.9844	0.9410	0.8540	0.5735	0.2862	0.1802	0.0611
2.55	1.0000	0.9998	0.9977	0.9875	0.9512	0.8754	0.6153	0.3270	0.2135	0.0781
2.60	1.0000	0.9998	0.9982	0.9900	0.9598	0.8941	0.6548	0.3688	0.2491	0.0979
2.65	1.0000	0.9999	0.9986	0.9920	0.9669	0.9102	0.6917	0.4111	0.2866	0.1205
2.70	1.0000	0.9999	0.9989	0.9936	0.9728	0.9241	0.7259	0.4534	0.3256	0.1459

PROBABILITY INTEGRAL OF t, THE NON-CENTRAL t-STATISTIC. THIS TABLE GIVES Pr $[t/\sqrt{f} \leq x]$.

f is the number of degrees of freedom; the non-centrality parameter is $\sqrt{f+1}\,K_p$.

K_p is the standardized normal deviate exceeded with probability p.

DEGREES OF FREEDOM **21**

x \ p	.2500	.1500	.1000	.0650	.0400	.0250	.0100	.0040	.0025	.0010
2.75	1.0000	0.9999	0.9992	0.9949	0.9777	0.9361	0.7574	0.4951	0.3656	0.1738
2.80	1.0000	1.0000	0.9994	0.9960	0.9817	0.9463	0.7861	0.5358	0.4060	0.2041
2.85	1.0000	1.0000	0.9995	0.9968	0.9851	0.9549	0.8121	0.5752	0.4464	0.2364
2.90	1.0000	1.0000	0.9996	0.9974	0.9878	0.9623	0.8355	0.6130	0.4864	0.2705
2.95	1.0000	1.0000	0.9997	0.9980	0.9900	0.9685	0.8565	0.6488	0.5256	0.3060
3.00	1.0000	1.0000	0.9998	0.9984	0.9919	0.9737	0.8752	0.6826	0.5636	0.3426
3.05	1.0000	1.0000	0.9998	0.9987	0.9934	0.9781	0.8917	0.7142	0.6002	0.3797
3.10	1.0000	1.0000	0.9999	0.9990	0.9946	0.9818	0.9063	0.7436	0.6351	0.4172
3.15	1.0000	1.0000	0.9999	0.9992	0.9956	0.9848	0.9191	0.7707	0.6682	0.4545
3.20	1.0000	1.0000	0.9999	0.9994	0.9964	0.9874	0.9303	0.7956	0.6994	0.4913
3.25	1.0000	1.0000	1.0000	0.9995	0.9971	0.9895	0.9401	0.8183	0.7285	0.5274
3.30	1.0000	1.0000	1.0000	0.9996	0.9976	0.9913	0.9486	0.8389	0.7555	0.5625
3.35	1.0000	1.0000	1.0000	0.9997	0.9981	0.9928	0.9559	0.8576	0.7805	0.5964
3.40	1.0000	1.0000	1.0000	0.9997	0.9984	0.9940	0.9623	0.8743	0.8035	0.6289
3.45	1.0000	1.0000	1.0000	0.9998	0.9987	0.9951	0.9678	0.8894	0.8245	0.6598
3.50	1.0000	1.0000	1.0000	0.9998	0.9990	0.9959	0.9725	0.9028	0.8437	0.6891
3.55	1.0000	1.0000	1.0000	0.9999	0.9991	0.9966	0.9765	0.9148	0.8611	0.7167
3.60	1.0000	1.0000	1.0000	0.9999	0.9993	0.9972	0.9800	0.9254	0.8768	0.7425
3.65	1.0000	1.0000	1.0000	0.9999	0.9994	0.9977	0.9830	0.9348	0.8909	0.7665
3.70	1.0000	1.0000	1.0000	0.9999	0.9995	0.9981	0.9855	0.9431	0.9036	0.7888
3.75	1.0000	1.0000	1.0000	1.0000	0.9996	0.9984	0.9877	0.9504	0.9150	0.8094
3.80	1.0000	1.0000	1.0000	1.0000	0.9997	0.9987	0.9895	0.9569	0.9251	0.8284
3.85	1.0000	1.0000	1.0000	1.0000	0.9998	0.9989	0.9911	0.9625	0.9341	0.8458
3.90	1.0000	1.0000	1.0000	1.0000	0.9998	0.9991	0.9924	0.9674	0.9421	0.8617
3.95	1.0000	1.0000	1.0000	1.0000	0.9998	0.9992	0.9936	0.9717	0.9492	0.8762
4.00	1.0000	1.0000	1.0000	1.0000	0.9999	0.9994	0.9945	0.9755	0.9555	0.8893
4.05	1.0000	1.0000	1.0000	1.0000	0.9999	0.9995	0.9954	0.9788	0.9611	0.9012
4.10	1.0000	1.0000	1.0000	1.0000	0.9999	0.9996	0.9961	0.9816	0.9659	0.9119
4.15	1.0000	1.0000	1.0000	1.0000	0.9999	0.9996	0.9967	0.9841	0.9702	0.9216
4.20	1.0000	1.0000	1.0000	1.0000	0.9999	0.9997	0.9972	0.9862	0.9740	0.9303
4.25	1.0000	1.0000	1.0000	1.0000	1.0000	0.9998	0.9976	0.9881	0.9773	0.9381
4.30	1.0000	1.0000	1.0000	1.0000	1.0000	0.9998	0.9980	0.9897	0.9802	0.9451
4.35	1.0000	1.0000	1.0000	1.0000	1.0000	0.9998	0.9983	0.9911	0.9828	0.9514
4.40	1.0000	1.0000	1.0000	1.0000	1.0000	0.9999	0.9985	0.9923	0.9850	0.9570
4.45	1.0000	1.0000	1.0000	1.0000	1.0000	0.9999	0.9987	0.9934	0.9869	0.9619
4.50	1.0000	1.0000	1.0000	1.0000	1.0000	0.9999	0.9989	0.9943	0.9886	0.9663
4.55	1.0000	1.0000	1.0000	1.0000	1.0000	0.9999	0.9991	0.9950	0.9901	0.9703
4.60	1.0000	1.0000	1.0000	1.0000	1.0000	0.9999	0.9992	0.9957	0.9914	0.9738
4.65	1.0000	1.0000	1.0000	1.0000	1.0000	0.9999	0.9993	0.9963	0.9925	0.9768
4.70	1.0000	1.0000	1.0000	1.0000	1.0000	1.0000	0.9994	0.9968	0.9935	0.9796
4.75	1.0000	1.0000	1.0000	1.0000	1.0000	1.0000	0.9995	0.9972	0.9943	0.9820
4.80	1.0000	1.0000	1.0000	1.0000	1.0000	1.0000	0.9996	0.9976	0.9951	0.9841
4.85	1.0000	1.0000	1.0000	1.0000	1.0000	1.0000	0.9997	0.9979	0.9957	0.9860
4.90	1.0000	1.0000	1.0000	1.0000	1.0000	1.0000	0.9997	0.9982	0.9963	0.9877
4.95	1.0000	1.0000	1.0000	1.0000	1.0000	1.0000	0.9997	0.9985	0.9967	0.9892
5.00	1.0000	1.0000	1.0000	1.0000	1.0000	1.0000	0.9998	0.9987	0.9972	0.9905
5.05	1.0000	1.0000	1.0000	1.0000	1.0000	1.0000	0.9998	0.9988	0.9975	0.9916
5.10	1.0000	1.0000	1.0000	1.0000	1.0000	1.0000	0.9998	0.9990	0.9979	0.9926
5.15	1.0000	1.0000	1.0000	1.0000	1.0000	1.0000	0.9999	0.9991	0.9981	0.9935
5.20	1.0000	1.0000	1.0000	1.0000	1.0000	1.0000	0.9999	0.9993	0.9984	0.9943
5.25	1.0000	1.0000	1.0000	1.0000	1.0000	1.0000	0.9999	0.9994	0.9986	0.9950
5.30	1.0000	1.0000	1.0000	1.0000	1.0000	1.0000	0.9999	0.9994	0.9988	0.9956
5.35	1.0000	1.0000	1.0000	1.0000	1.0000	1.0000	0.9999	0.9995	0.9989	0.9961
5.40	1.0000	1.0000	1.0000	1.0000	1.0000	1.0000	0.9999	0.9996	0.9991	0.9966
5.45	1.0000	1.0000	1.0000	1.0000	1.0000	1.0000	0.9999	0.9996	0.9992	0.9970
5.50	1.0000	1.0000	1.0000	1.0000	1.0000	1.0000	1.0000	0.9997	0.9993	0.9974
5.55	1.0000	1.0000	1.0000	1.0000	1.0000	1.0000	1.0000	0.9997	0.9994	0.9977
5.60	1.0000	1.0000	1.0000	1.0000	1.0000	1.0000	1.0000	0.9998	0.9995	0.9979
5.65	1.0000	1.0000	1.0000	1.0000	1.0000	1.0000	1.0000	0.9998	0.9995	0.9982
5.70	1.0000	1.0000	1.0000	1.0000	1.0000	1.0000	1.0000	0.9998	0.9996	0.9984

PROBABILITY INTEGRAL OF t, THE NON-CENTRAL t-STATISTIC. THIS TABLE GIVES Pr $[t/\sqrt{f} \leq x]$.

f is the number of degrees of freedom; the non-centrality parameter is $\sqrt{f+1}\,K_p$.

K_p is the standardized normal deviate exceeded with probability p.

DEGREES OF FREEDOM **21**

x \ p	.2500	.1500	.1000	.0650	.0400	.0250	.0100	.0040	.0025	.0010
5.75	1.0000	1.0000	1.0000	1.0000	1.0000	1.0000	1.0000	0.9998	0.9996	0.9986
5.80	1.0000	1.0000	1.0000	1.0000	1.0000	1.0000	1.0000	0.9999	0.9997	0.9988
5.85	1.0000	1.0000	1.0000	1.0000	1.0000	1.0000	1.0000	0.9999	0.9997	0.9989
5.90	1.0000	1.0000	1.0000	1.0000	1.0000	1.0000	1.0000	0.9999	0.9996	0.9990
5.95	1.0000	1.0000	1.0000	1.0000	1.0000	1.0000	1.0000	0.9999	0.9998	0.9992
6.00	1.0000	1.0000	1.0000	1.0000	1.0000	1.0000	1.0000	0.9999	0.9998	0.9993
6.05	1.0000	1.0000	1.0000	1.0000	1.0000	1.0000	1.0000	0.9999	0.9998	0.9993
6.10	1.0000	1.0000	1.0000	1.0000	1.0000	1.0000	1.0000	0.9999	0.9999	0.9994
6.15	1.0000	1.0000	1.0000	1.0000	1.0000	1.0000	1.0000	0.9999	0.9999	0.9995
6.20	1.0000	1.0000	1.0000	1.0000	1.0000	1.0000	1.0000	1.0000	0.9999	0.9995
6.25	1.0000	1.0000	1.0000	1.0000	1.0000	1.0000	1.0000	1.0000	0.9999	0.9996
6.30	1.0000	1.0000	1.0000	1.0000	1.0000	1.0000	1.0000	1.0000	0.9999	0.9996
6.35	1.0000	1.0000	1.0000	1.0000	1.0000	1.0000	1.0000	1.0000	0.9999	0.9997
6.40	1.0000	1.0000	1.0000	1.0000	1.0000	1.0000	1.0000	1.0000	0.9999	0.9997
6.45	1.0000	1.0000	1.0000	1.0000	1.0000	1.0000	1.0000	1.0000	0.9999	0.9998
6.50	1.0000	1.0000	1.0000	1.0000	1.0000	1.0000	1.0000	1.0000	0.9999	0.9998
6.55	1.0000	1.0000	1.0000	1.0000	1.0000	1.0000	1.0000	1.0000	1.0000	0.9998
6.60	1.0000	1.0000	1.0000	1.0000	1.0000	1.0000	1.0000	1.0000	1.0000	0.9998
6.65	1.0000	1.0000	1.0000	1.0000	1.0000	1.0000	1.0000	1.0000	1.0000	0.9998
6.70	1.0000	1.0000	1.0000	1.0000	1.0000	1.0000	1.0000	1.0000	1.0000	0.9999
6.75	1.0000	1.0000	1.0000	1.0000	1.0000	1.0000	1.0000	1.0000	1.0000	0.9999
6.80	1.0000	1.0000	1.0000	1.0000	1.0000	1.0000	1.0000	1.0000	1.0000	0.9999
6.85	1.0000	1.0000	1.0000	1.0000	1.0000	1.0000	1.0000	1.0000	1.0000	0.9999
6.90	1.0000	1.0000	1.0000	1.0000	1.0000	1.0000	1.0000	1.0000	1.0000	0.9999
6.95	1.0000	1.0000	1.0000	1.0000	1.0000	1.0000	1.0000	1.0000	1.0000	0.9999
7.00	1.0000	1.0000	1.0000	1.0000	1.0000	1.0000	1.0000	1.0000	1.0000	0.9999
7.05	1.0000	1.0000	1.0000	1.0000	1.0000	1.0000	1.0000	1.0000	1.0000	0.9999
7.10	1.0000	1.0000	1.0000	1.0000	1.0000	1.0000	1.0000	1.0000	1.0000	0.9999
7.15	1.0000	1.0000	1.0000	1.0000	1.0000	1.0000	1.0000	1.0000	1.0000	1.0000

PROBABILITY INTEGRAL OF t, THE NON-CENTRAL t-STATISTIC. THIS TABLE GIVES Pr $[t/\sqrt{f} \leq x]$.

f is the number of degrees of freedom; the non-centrality parameter is $\sqrt{f+1}\,K_p$.
K_p is the standardized normal deviate exceeded with probability p.

DEGREES OF FREEDOM **22**

x \ p	.2500	.1500	.1000	.0650	.0400	.0250	.0100	.0040	.0025	.0010
-0.15	0.0000	0.0000	0.0000	0.0000	0.0000	0.0000	0.0000	0.0000	0.0000	0.0000
-0.10	0.0001	0.0000	0.0000	0.0000	0.0000	0.0000	0.0000	0.0000	0.0000	0.0000
-0.05	0.0003	0.0000	0.0000	0.0000	0.0000	0.0000	0.0000	0.0000	0.0000	0.0000
0.00	0.0006	0.0000	0.0000	0.0000	0.0000	0.0000	0.0000	0.0000	0.0000	0.0000
0.05	0.0013	0.0000	0.0000	0.0000	0.0000	0.0000	0.0000	0.0000	0.0000	0.0000
0.10	0.0029	0.0000	0.0000	0.0000	0.0000	0.0000	0.0000	0.0000	0.0000	0.0000
0.15	0.0058	0.0000	0.0000	0.0000	0.0000	0.0000	0.0000	0.0000	0.0000	0.0000
0.20	0.0112	0.0000	0.0000	0.0000	0.0000	0.0000	0.0000	0.0000	0.0000	0.0000
0.25	0.0205	0.0001	0.0000	0.0000	0.0000	0.0000	0.0000	0.0000	0.0000	0.0000
0.30	0.0357	0.0002	0.0000	0.0000	0.0000	0.0000	0.0000	0.0000	0.0000	0.0000
0.35	0.0589	0.0006	0.0000	0.0000	0.0000	0.0000	0.0000	0.0000	0.0000	0.0000
0.40	0.0921	0.0014	0.0000	0.0000	0.0000	0.0000	0.0000	0.0000	0.0000	0.0000
0.45	0.1369	0.0030	0.0001	0.0000	0.0000	0.0000	0.0000	0.0000	0.0000	0.0000
0.50	0.1937	0.0063	0.0002	0.0000	0.0000	0.0000	0.0000	0.0000	0.0000	0.0000
0.55	0.2615	0.0122	0.0004	0.0000	0.0000	0.0000	0.0000	0.0000	0.0000	0.0000
0.60	0.3380	0.0221	0.0010	0.0000	0.0000	0.0000	0.0000	0.0000	0.0000	0.0000
0.65	0.4199	0.0379	0.0023	0.0001	0.0000	0.0000	0.0000	0.0000	0.0000	0.0000
0.70	0.5032	0.0613	0.0048	0.0002	0.0000	0.0000	0.0000	0.0000	0.0000	0.0000
0.75	0.5842	0.0937	0.0094	0.0004	0.0000	0.0000	0.0000	0.0000	0.0000	0.0000
0.80	0.6597	0.1361	0.0173	0.0011	0.0000	0.0000	0.0000	0.0000	0.0000	0.0000
0.85	0.7273	0.1884	0.0298	0.0024	0.0001	0.0000	0.0000	0.0000	0.0000	0.0000
0.90	0.7858	0.2497	0.0485	0.0048	0.0002	0.0000	0.0000	0.0000	0.0000	0.0000
0.95	0.8348	0.3180	0.0746	0.0093	0.0005	0.0000	0.0000	0.0000	0.0000	0.0000
1.00	0.8747	0.3909	0.1090	0.0166	0.0012	0.0001	0.0000	0.0000	0.0000	0.0000
1.05	0.9064	0.4654	0.1522	0.0280	0.0025	0.0002	0.0000	0.0000	0.0000	0.0000
1.10	0.9311	0.5388	0.2037	0.0448	0.0050	0.0004	0.0000	0.0000	0.0000	0.0000
1.15	0.9498	0.6087	0.2624	0.0679	0.0093	0.0009	0.0000	0.0000	0.0000	0.0000
1.20	0.9638	0.6732	0.3265	0.0982	0.0162	0.0020	0.0000	0.0000	0.0000	0.0000
1.25	0.9742	0.7310	0.3939	0.1360	0.0267	0.0039	0.0000	0.0000	0.0000	0.0000
1.30	0.9817	0.7817	0.4623	0.1811	0.0417	0.0071	0.0001	0.0000	0.0000	0.0000
1.35	0.9871	0.8250	0.5297	0.2328	0.0621	0.0124	0.0003	0.0000	0.0000	0.0000
1.40	0.9910	0.8613	0.5942	0.2898	0.0886	0.0205	0.0006	0.0000	0.0000	0.0000
1.45	0.9937	0.8912	0.6543	0.3505	0.1216	0.0321	0.0012	0.0000	0.0000	0.0000
1.50	0.9956	0.9155	0.7090	0.4132	0.1609	0.0481	0.0024	0.0001	0.0000	0.0000
1.55	0.9970	0.9348	0.7577	0.4761	0.2061	0.0691	0.0043	0.0001	0.0000	0.0000
1.60	0.9979	0.9501	0.8004	0.5377	0.2564	0.0955	0.0075	0.0003	0.0001	0.0000
1.65	0.9986	0.9621	0.8371	0.5965	0.3106	0.1276	0.0124	0.0007	0.0001	0.0000
1.70	0.9990	0.9713	0.8682	0.6514	0.3673	0.1652	0.0195	0.0013	0.0003	0.0000
1.75	0.9993	0.9784	0.8941	0.7019	0.4253	0.2079	0.0295	0.0024	0.0006	0.0000
1.80	0.9995	0.9838	0.9156	0.7473	0.4830	0.2549	0.0429	0.0042	0.0011	0.0001
1.85	0.9997	0.9879	0.9331	0.7876	0.5394	0.3053	0.0602	0.0069	0.0019	0.0001
1.90	0.9998	0.9910	0.9473	0.8229	0.5933	0.3579	0.0818	0.0111	0.0034	0.0003
1.95	0.9999	0.9933	0.9587	0.8534	0.6440	0.4118	0.1079	0.0170	0.0056	0.0005
2.00	0.9999	0.9951	0.9677	0.8794	0.6909	0.4657	0.1384	0.0252	0.0090	0.0010
2.05	0.9999	0.9963	0.9749	0.9014	0.7337	0.5186	0.1732	0.0360	0.0138	0.0017
2.10	1.0000	0.9973	0.9805	0.9198	0.7721	0.5697	0.2118	0.0498	0.0205	0.0029
2.15	1.0000	0.9980	0.9849	0.9350	0.8063	0.6183	0.2538	0.0671	0.0295	0.0047
2.20	1.0000	0.9985	0.9884	0.9476	0.8364	0.6638	0.2985	0.0880	0.0410	0.0074
2.25	1.0000	0.9989	0.9911	0.9579	0.8625	0.7058	0.3450	0.1126	0.0555	0.0113
2.30	1.0000	0.9992	0.9931	0.9663	0.8851	0.7442	0.3926	0.1408	0.0732	0.0166
2.35	1.0000	0.9994	0.9947	0.9731	0.9043	0.7789	0.4405	0.1726	0.0942	0.0236
2.40	1.0000	0.9996	0.9959	0.9786	0.9207	0.8099	0.4880	0.2075	0.1186	0.0327
2.45	1.0000	0.9997	0.9969	0.9830	0.9346	0.8374	0.5343	0.2452	0.1464	0.0441
2.50	1.0000	0.9998	0.9976	0.9865	0.9462	0.8616	0.5790	0.2852	0.1773	0.0581
2.55	1.0000	0.9998	0.9982	0.9893	0.9558	0.8826	0.6216	0.3269	0.2111	0.0749
2.60	1.0000	0.9999	0.9986	0.9916	0.9639	0.9009	0.6617	0.3697	0.2474	0.0946
2.65	1.0000	0.9999	0.9989	0.9933	0.9705	0.9166	0.6992	0.4130	0.2857	0.1172
2.70	1.0000	0.9999	0.9992	0.9947	0.9760	0.9300	0.7337	0.4563	0.3255	0.1427

PROBABILITY INTEGRAL OF t, THE NON-CENTRAL t-STATISTIC. THIS TABLE GIVES Pr $[t/\sqrt{f} \leq x]$.

f is the number of degrees of freedom; the non-centrality parameter is $\sqrt{f+1}\,K_p$.

K_p is the standardized normal deviate exceeded with probability p.

DEGREES OF FREEDOM **22**

x \ p	.2500	.1500	.1000	.0650	.0400	.0250	.0100	.0040	.0025	.0010
2.75	1.0000	1.0000	0.9994	0.9959	0.9805	0.9415	0.7654	0.4990	0.3664	0.1709
2.80	1.0000	1.0000	0.9995	0.9967	0.9842	0.9512	0.7942	0.5407	0.4078	0.2016
2.85	1.0000	1.0000	0.9996	0.9974	0.9872	0.9594	0.8202	0.5809	0.4492	0.2345
2.90	1.0000	1.0000	0.9997	0.9980	0.9896	0.9662	0.8434	0.6194	0.4902	0.2693
2.95	1.0000	1.0000	0.9998	0.9984	0.9916	0.9720	0.8642	0.6558	0.5302	0.3056
3.00	1.0000	1.0000	0.9998	0.9987	0.9932	0.9768	0.8825	0.6901	0.5691	0.3430
3.05	1.0000	1.0000	0.9999	0.9990	0.9945	0.9809	0.8987	0.7220	0.6064	0.3810
3.10	1.0000	1.0000	0.9999	0.9992	0.9956	0.9842	0.9129	0.7516	0.6419	0.4193
3.15	1.0000	1.0000	0.9999	0.9994	0.9964	0.9870	0.9252	0.7789	0.6755	0.4575
3.20	1.0000	1.0000	1.0000	0.9995	0.9971	0.9893	0.9360	0.8038	0.7071	0.4953
3.25	1.0000	1.0000	1.0000	0.9996	0.9977	0.9912	0.9453	0.8264	0.7364	0.5322
3.30	1.0000	1.0000	1.0000	0.9997	0.9981	0.9927	0.9534	0.8468	0.7637	0.5681
3.35	1.0000	1.0000	1.0000	0.9998	0.9985	0.9940	0.9603	0.8653	0.7887	0.6026
3.40	1.0000	1.0000	1.0000	0.9998	0.9988	0.9951	0.9663	0.8817	0.8117	0.6357
3.45	1.0000	1.0000	1.0000	0.9999	0.9990	0.9960	0.9714	0.8965	0.8326	0.6671
3.50	1.0000	1.0000	1.0000	0.9999	0.9992	0.9967	0.9757	0.9095	0.8516	0.6968
3.55	1.0000	1.0000	1.0000	0.9999	0.9994	0.9973	0.9794	0.9211	0.8687	0.7246
3.60	1.0000	1.0000	1.0000	0.9999	0.9995	0.9978	0.9826	0.9314	0.8841	0.7506
3.65	1.0000	1.0000	1.0000	1.0000	0.9996	0.9982	0.9853	0.9404	0.8980	0.7748
3.70	1.0000	1.0000	1.0000	1.0000	0.9997	0.9985	0.9876	0.9483	0.9103	0.7971
3.75	1.0000	1.0000	1.0000	1.0000	0.9997	0.9988	0.9895	0.9552	0.9213	0.8177
3.80	1.0000	1.0000	1.0000	1.0000	0.9998	0.9990	0.9912	0.9612	0.9311	0.8365
3.85	1.0000	1.0000	1.0000	1.0000	0.9998	0.9992	0.9925	0.9665	0.9397	0.8537
3.90	1.0000	1.0000	1.0000	1.0000	0.9999	0.9993	0.9937	0.9711	0.9473	0.8694
3.95	1.0000	1.0000	1.0000	1.0000	0.9999	0.9994	0.9947	0.9751	0.9541	0.8836
4.00	1.0000	1.0000	1.0000	1.0000	0.9999	0.9995	0.9955	0.9785	0.9600	0.8964
4.05	1.0000	1.0000	1.0000	1.0000	0.9999	0.9996	0.9962	0.9815	0.9652	0.9080
4.10	1.0000	1.0000	1.0000	1.0000	0.9999	0.9997	0.9968	0.9841	0.9697	0.9184
4.15	1.0000	1.0000	1.0000	1.0000	1.0000	0.9997	0.9973	0.9863	0.9737	0.9277
4.20	1.0000	1.0000	1.0000	1.0000	1.0000	0.9998	0.9977	0.9882	0.9772	0.9361
4.25	1.0000	1.0000	1.0000	1.0000	1.0000	0.9998	0.9981	0.9899	0.9802	0.9435
4.30	1.0000	1.0000	1.0000	1.0000	1.0000	0.9999	0.9984	0.9913	0.9828	0.9502
4.35	1.0000	1.0000	1.0000	1.0000	1.0000	0.9999	0.9986	0.9926	0.9851	0.9561
4.40	1.0000	1.0000	1.0000	1.0000	1.0000	0.9999	0.9989	0.9936	0.9871	0.9613
4.45	1.0000	1.0000	1.0000	1.0000	1.0000	0.9999	0.9990	0.9945	0.9889	0.9660
4.50	1.0000	1.0000	1.0000	1.0000	1.0000	0.9999	0.9992	0.9953	0.9904	0.9701
4.55	1.0000	1.0000	1.0000	1.0000	1.0000	0.9999	0.9993	0.9960	0.9917	0.9737
4.60	1.0000	1.0000	1.0000	1.0000	1.0000	1.0000	0.9994	0.9965	0.9928	0.9769
4.65	1.0000	1.0000	1.0000	1.0000	1.0000	1.0000	0.9995	0.9970	0.9938	0.9798
4.70	1.0000	1.0000	1.0000	1.0000	1.0000	1.0000	0.9996	0.9974	0.9946	0.9823
4.75	1.0000	1.0000	1.0000	1.0000	1.0000	1.0000	0.9997	0.9978	0.9953	0.9844
4.80	1.0000	1.0000	1.0000	1.0000	1.0000	1.0000	0.9997	0.9981	0.9960	0.9864
4.85	1.0000	1.0000	1.0000	1.0000	1.0000	1.0000	0.9997	0.9984	0.9965	0.9881
4.90	1.0000	1.0000	1.0000	1.0000	1.0000	1.0000	0.9998	0.9986	0.9970	0.9896
4.95	1.0000	1.0000	1.0000	1.0000	1.0000	1.0000	0.9998	0.9988	0.9974	0.9909
5.00	1.0000	1.0000	1.0000	1.0000	1.0000	1.0000	0.9998	0.9990	0.9978	0.9920
5.05	1.0000	1.0000	1.0000	1.0000	1.0000	1.0000	0.9999	0.9991	0.9981	0.9930
5.10	1.0000	1.0000	1.0000	1.0000	1.0000	1.0000	0.9999	0.9992	0.9983	0.9939
5.15	1.0000	1.0000	1.0000	1.0000	1.0000	1.0000	0.9999	0.9994	0.9985	0.9946
5.20	1.0000	1.0000	1.0000	1.0000	1.0000	1.0000	0.9999	0.9994	0.9987	0.9953
5.25	1.0000	1.0000	1.0000	1.0000	1.0000	1.0000	0.9999	0.9995	0.9989	0.9959
5.30	1.0000	1.0000	1.0000	1.0000	1.0000	1.0000	0.9999	0.9996	0.9991	0.9964
5.35	1.0000	1.0000	1.0000	1.0000	1.0000	1.0000	1.0000	0.9996	0.9992	0.9969
5.40	1.0000	1.0000	1.0000	1.0000	1.0000	1.0000	1.0000	0.9997	0.9993	0.9973
5.45	1.0000	1.0000	1.0000	1.0000	1.0000	1.0000	1.0000	0.9997	0.9994	0.9976
5.50	1.0000	1.0000	1.0000	1.0000	1.0000	1.0000	1.0000	0.9998	0.9995	0.9979
5.55	1.0000	1.0000	1.0000	1.0000	1.0000	1.0000	1.0000	0.9998	0.9995	0.9982
5.60	1.0000	1.0000	1.0000	1.0000	1.0000	1.0000	1.0000	0.9998	0.9996	0.9984
5.65	1.0000	1.0000	1.0000	1.0000	1.0000	1.0000	1.0000	0.9999	0.9997	0.9986
5.70	1.0000	1.0000	1.0000	1.0000	1.0000	1.0000	1.0000	0.9999	0.9997	0.9988

PROBABILITY INTEGRAL OF t, THE NON-CENTRAL t-STATISTIC. THIS TABLE GIVES Pr $[t/\sqrt{f} \leq x]$.

f is the number of degrees of freedom; the non-centrality parameter is $\sqrt{f+1}\,K_p$.
K_p is the standardized normal deviate exceeded with probability p.

DEGREES OF FREEDOM **22**

x \ P	.2500	.1500	.1000	.0650	.0400	.0250	.0100	.0040	.0025	.0010
5.75	1.0000	1.0000	1.0000	1.0000	1.0000	1.0000	1.0000	0.9999	0.9997	0.9989
5.80	1.0000	1.0000	1.0000	1.0000	1.0000	1.0000	1.0000	0.9999	0.9998	0.9991
5.85	1.0000	1.0000	1.0000	1.0000	1.0000	1.0000	1.0000	0.9999	0.9998	0.9992
5.90	1.0000	1.0000	1.0000	1.0000	1.0000	1.0000	1.0000	0.9999	0.9998	0.9993
5.95	1.0000	1.0000	1.0000	1.0000	1.0000	1.0000	1.0000	0.9999	0.9999	0.9994
6.00	1.0000	1.0000	1.0000	1.0000	1.0000	1.0000	1.0000	0.9999	0.9999	0.9994
6.05	1.0000	1.0000	1.0000	1.0000	1.0000	1.0000	1.0000	1.0000	0.9999	0.9995
6.10	1.0000	1.0000	1.0000	1.0000	1.0000	1.0000	1.0000	1.0000	0.9999	0.9996
6.15	1.0000	1.0000	1.0000	1.0000	1.0000	1.0000	1.0000	1.0000	0.9999	0.9996
6.20	1.0000	1.0000	1.0000	1.0000	1.0000	1.0000	1.0000	1.0000	0.9999	0.9997
6.25	1.0000	1.0000	1.0000	1.0000	1.0000	1.0000	1.0000	1.0000	0.9999	0.9997
6.30	1.0000	1.0000	1.0000	1.0000	1.0000	1.0000	1.0000	1.0000	0.9999	0.9997
6.35	1.0000	1.0000	1.0000	1.0000	1.0000	1.0000	1.0000	1.0000	0.9999	0.9998
6.40	1.0000	1.0000	1.0000	1.0000	1.0000	1.0000	1.0000	1.0000	1.0000	0.9998
6.45	1.0000	1.0000	1.0000	1.0000	1.0000	1.0000	1.0000	1.0000	1.0000	0.9998
6.50	1.0000	1.0000	1.0000	1.0000	1.0000	1.0000	1.0000	1.0000	1.0000	0.9998
6.55	1.0000	1.0000	1.0000	1.0000	1.0000	1.0000	1.0000	1.0000	1.0000	0.9999
6.60	1.0000	1.0000	1.0000	1.0000	1.0000	1.0000	1.0000	1.0000	1.0000	0.9999
6.65	1.0000	1.0000	1.0000	1.0000	1.0000	1.0000	1.0000	1.0000	1.0000	0.9999
6.70	1.0000	1.0000	1.0000	1.0000	1.0000	1.0000	1.0000	1.0000	1.0000	0.9999
6.75	1.0000	1.0000	1.0000	1.0000	1.0000	1.0000	1.0000	1.0000	1.0000	0.9999
6.80	1.0000	1.0000	1.0000	1.0000	1.0000	1.0000	1.0000	1.0000	1.0000	0.9999
6.85	1.0000	1.0000	1.0000	1.0000	1.0000	1.0000	1.0000	1.0000	1.0000	0.9999
6.90	1.0000	1.0000	1.0000	1.0000	1.0000	1.0000	1.0000	1.0000	1.0000	0.9999
6.95	1.0000	1.0000	1.0000	1.0000	1.0000	1.0000	1.0000	1.0000	1.0000	0.9999
7.00	1.0000	1.0000	1.0000	1.0000	1.0000	1.0000	1.0000	1.0000	1.0000	1.0000

PROBABILITY INTEGRAL OF t, THE NON-CENTRAL t-STATISTIC. THIS TABLE GIVES Pr $[t/\sqrt{f} \leq x]$.

f is the number of degrees of freedom; the non-centrality parameter is $\sqrt{f+1}\,K_p$.
K_p is the standardized normal deviate exceeded with probability p.

x \ p	.2500	.1500	.1000	.0650	.0400	.0250	.0100	.0040	.0025	.0010
-0.15	0.0000	0.0000	0.0000	0.0000	0.0000	0.0000	0.0000	0.0000	0.0000	0.0000
-0.10	0.0001	0.0000	0.0000	0.0000	0.0000	0.0000	0.0000	0.0000	0.0000	0.0000
-0.05	0.0002	0.0000	0.0000	0.0000	0.0000	0.0000	0.0000	0.0000	0.0000	0.0000
0.00	0.0005	0.0000	0.0000	0.0000	0.0000	0.0000	0.0000	0.0000	0.0000	0.0000
0.05	0.0011	0.0000	0.0000	0.0000	0.0000	0.0000	0.0000	0.0000	0.0000	0.0000
0.10	0.0024	0.0000	0.0000	0.0000	0.0000	0.0000	0.0000	0.0000	0.0000	0.0000
0.15	0.0050	0.0000	0.0000	0.0000	0.0000	0.0000	0.0000	0.0000	0.0000	0.0000
0.20	0.0098	0.0000	0.0000	0.0000	0.0000	0.0000	0.0000	0.0000	0.0000	0.0000
0.25	0.0185	0.0001	0.0000	0.0000	0.0000	0.0000	0.0000	0.0000	0.0000	0.0000
0.30	0.0329	0.0002	0.0000	0.0000	0.0000	0.0000	0.0000	0.0000	0.0000	0.0000
0.35	0.0553	0.0005	0.0000	0.0000	0.0000	0.0000	0.0000	0.0000	0.0000	0.0000
0.40	0.0879	0.0011	0.0000	0.0000	0.0000	0.0000	0.0000	0.0000	0.0000	0.0000
0.45	0.1324	0.0026	0.0000	0.0000	0.0000	0.0000	0.0000	0.0000	0.0000	0.0000
0.50	0.1895	0.0054	0.0001	0.0000	0.0000	0.0000	0.0000	0.0000	0.0000	0.0000
0.55	0.2582	0.0108	0.0003	0.0000	0.0000	0.0000	0.0000	0.0000	0.0000	0.0000
0.60	0.3361	0.0201	0.0008	0.0000	0.0000	0.0000	0.0000	0.0000	0.0000	0.0000
0.65	0.4198	0.0352	0.0019	0.0000	0.0000	0.0000	0.0000	0.0000	0.0000	0.0000
0.70	0.5050	0.0579	0.0041	0.0001	0.0000	0.0000	0.0000	0.0000	0.0000	0.0000
0.75	0.5878	0.0898	0.0083	0.0004	0.0000	0.0000	0.0000	0.0000	0.0000	0.0000
0.80	0.6647	0.1320	0.0156	0.0009	0.0000	0.0000	0.0000	0.0000	0.0000	0.0000
0.85	0.7333	0.1847	0.0275	0.0020	0.0001	0.0000	0.0000	0.0000	0.0000	0.0000
0.90	0.7923	0.2468	0.0455	0.0042	0.0002	0.0000	0.0000	0.0000	0.0000	0.0000
0.95	0.8413	0.3164	0.0710	0.0082	0.0004	0.0000	0.0000	0.0000	0.0000	0.0000
1.00	0.8809	0.3908	0.1052	0.0150	0.0010	0.0000	0.0000	0.0000	0.0000	0.0000
1.05	0.9121	0.4671	0.1485	0.0259	0.0021	0.0001	0.0000	0.0000	0.0000	0.0000
1.10	0.9360	0.5421	0.2005	0.0420	0.0044	0.0003	0.0000	0.0000	0.0000	0.0000
1.15	0.9540	0.6134	0.2600	0.0646	0.0083	0.0008	0.0000	0.0000	0.0000	0.0000
1.20	0.9673	0.6790	0.3254	0.0946	0.0147	0.0017	0.0000	0.0000	0.0000	0.0000
1.25	0.9769	0.7376	0.3942	0.1323	0.0246	0.0033	0.0000	0.0000	0.0000	0.0000
1.30	0.9839	0.7886	0.4643	0.1778	0.0390	0.0063	0.0001	0.0000	0.0000	0.0000
1.35	0.9888	0.8319	0.5332	0.2301	0.0590	0.0112	0.0002	0.0000	0.0000	0.0000
1.40	0.9923	0.8679	0.5990	0.2881	0.0852	0.0187	0.0005	0.0000	0.0000	0.0000
1.45	0.9947	0.8974	0.6601	0.3501	0.1180	0.0298	0.0010	0.0000	0.0000	0.0000
1.50	0.9964	0.9211	0.7155	0.4143	0.1576	0.0453	0.0020	0.0000	0.0000	0.0000
1.55	0.9975	0.9398	0.7647	0.4786	0.2033	0.0659	0.0037	0.0001	0.0000	0.0000
1.60	0.9983	0.9544	0.8075	0.5415	0.2544	0.0921	0.0066	0.0003	0.0000	0.0000
1.65	0.9989	0.9657	0.8441	0.6015	0.3096	0.1242	0.0112	0.0005	0.0001	0.0000
1.70	0.9992	0.9744	0.8749	0.6575	0.3676	0.1620	0.0179	0.0011	0.0002	0.0000
1.75	0.9995	0.9810	0.9004	0.7086	0.4268	0.2052	0.0274	0.0020	0.0004	0.0000
1.80	0.9997	0.9859	0.9213	0.7545	0.4859	0.2530	0.0403	0.0036	0.0009	0.0000
1.85	0.9998	0.9896	0.9382	0.7950	0.5435	0.3043	0.0573	0.0061	0.0016	0.0001
1.90	0.9998	0.9924	0.9518	0.8302	0.5985	0.3581	0.0786	0.0099	0.0029	0.0002
1.95	0.9999	0.9944	0.9626	0.8604	0.6501	0.4132	0.1045	0.0155	0.0049	0.0004
2.00	0.9999	0.9959	0.9711	0.8860	0.6977	0.4683	0.1351	0.0232	0.0080	0.0008
2.05	1.0000	0.9970	0.9777	0.9075	0.7409	0.5224	0.1702	0.0336	0.0125	0.0014
2.10	1.0000	0.9978	0.9829	0.9254	0.7796	0.5746	0.2094	0.0471	0.0188	0.0025
2.15	1.0000	0.9984	0.9869	0.9401	0.8138	0.6241	0.2521	0.0641	0.0274	0.0041
2.20	1.0000	0.9988	0.9900	0.9522	0.8437	0.6703	0.2976	0.0848	0.0385	0.0066
2.25	1.0000	0.9992	0.9924	0.9619	0.8695	0.7129	0.3451	0.1093	0.0527	0.0101
2.30	1.0000	0.9994	0.9942	0.9698	0.8917	0.7516	0.3938	0.1377	0.0701	0.0151
2.35	1.0000	0.9996	0.9956	0.9761	0.9105	0.7865	0.4428	0.1697	0.0910	0.0218
2.40	1.0000	0.9997	0.9967	0.9812	0.9264	0.8175	0.4913	0.2051	0.1154	0.0305
2.45	1.0000	0.9998	0.9975	0.9852	0.9397	0.8448	0.5387	0.2434	0.1433	0.0416
2.50	1.0000	0.9998	0.9981	0.9884	0.9508	0.8687	0.5843	0.2842	0.1745	0.0553
2.55	1.0000	0.9999	0.9986	0.9909	0.9600	0.8894	0.6277	0.3267	0.2087	0.0719
2.60	1.0000	0.9999	0.9989	0.9929	0.9675	0.9072	0.6684	0.3705	0.2456	0.0914
2.65	1.0000	0.9999	0.9992	0.9944	0.9737	0.9224	0.7063	0.4148	0.2847	0.1140
2.70	1.0000	1.0000	0.9994	0.9956	0.9788	0.9354	0.7412	0.4591	0.3254	0.1396

PROBABILITY INTEGRAL OF t, THE NON-CENTRAL t-STATISTIC. THIS TABLE GIVES Pr $[t/\sqrt{f} \leq x]$.

f is the number of degrees of freedom; the non-centrality parameter is $\sqrt{f+1}\,K_p$.
K_p is the standardized normal deviate exceeded with probability p.

DEGREES OF FREEDOM **23**

x \ p	.2500	.1500	.1000	.0650	.0400	.0250	.0100	.0040	.0025	.0010
2.75	1.0000	1.0000	0.9995	0.9966	0.9829	0.9464	0.7731	0.5027	0.3672	0.1681
2.80	1.0000	1.0000	0.9997	0.9974	0.9862	0.9556	0.8019	0.5453	0.4095	0.1992
2.85	1.0000	1.0000	0.9997	0.9979	0.9890	0.9633	0.8278	0.5863	0.4519	0.2326
2.90	1.0000	1.0000	0.9998	0.9984	0.9911	0.9698	0.8509	0.6255	0.4937	0.2681
2.95	1.0000	1.0000	0.9999	0.9987	0.9929	0.9751	0.8713	0.6625	0.5347	0.3051
3.00	1.0000	1.0000	0.9999	0.9990	0.9943	0.9796	0.8893	0.6972	0.5743	0.3433
3.05	1.0000	1.0000	0.9999	0.9992	0.9954	0.9833	0.9051	0.7295	0.6123	0.3822
3.10	1.0000	1.0000	0.9999	0.9994	0.9964	0.9863	0.9189	0.7593	0.6485	0.4214
3.15	1.0000	1.0000	1.0000	0.9995	0.9971	0.9888	0.9309	0.7866	0.6826	0.4605
3.20	1.0000	1.0000	1.0000	0.9996	0.9977	0.9909	0.9412	0.8115	0.7144	0.4991
3.25	1.0000	1.0000	1.0000	0.9997	0.9981	0.9925	0.9501	0.8340	0.7441	0.5368
3.30	1.0000	1.0000	1.0000	0.9998	0.9985	0.9939	0.9577	0.8543	0.7714	0.5734
3.35	1.0000	1.0000	1.0000	0.9998	0.9988	0.9950	0.9643	0.8725	0.7965	0.6086
3.40	1.0000	1.0000	1.0000	0.9999	0.9991	0.9960	0.9698	0.8886	0.8194	0.6422
3.45	1.0000	1.0000	1.0000	0.9999	0.9992	0.9967	0.9746	0.9030	0.8402	0.6741
3.50	1.0000	1.0000	1.0000	0.9999	0.9994	0.9973	0.9786	0.9157	0.8590	0.7041
3.55	1.0000	1.0000	1.0000	0.9999	0.9995	0.9978	0.9820	0.9269	0.8759	0.7322
3.60	1.0000	1.0000	1.0000	1.0000	0.9996	0.9982	0.9849	0.9368	0.8910	0.7584
3.65	1.0000	1.0000	1.0000	1.0000	0.9997	0.9986	0.9873	0.9454	0.9045	0.7826
3.70	1.0000	1.0000	1.0000	1.0000	0.9998	0.9988	0.9894	0.9529	0.9165	0.8049
3.75	1.0000	1.0000	1.0000	1.0000	0.9998	0.9990	0.9911	0.9595	0.9271	0.8254
3.80	1.0000	1.0000	1.0000	1.0000	0.9998	0.9992	0.9925	0.9651	0.9365	0.8441
3.85	1.0000	1.0000	1.0000	1.0000	0.9999	0.9994	0.9937	0.9701	0.9448	0.8612
3.90	1.0000	1.0000	1.0000	1.0000	0.9999	0.9995	0.9948	0.9743	0.9520	0.8766
3.95	1.0000	1.0000	1.0000	1.0000	0.9999	0.9996	0.9956	0.9780	0.9584	0.8905
4.00	1.0000	1.0000	1.0000	1.0000	0.9999	0.9997	0.9963	0.9812	0.9640	0.9030
4.05	1.0000	1.0000	1.0000	1.0000	1.0000	0.9997	0.9969	0.9839	0.9688	0.9143
4.10	1.0000	1.0000	1.0000	1.0000	1.0000	0.9998	0.9974	0.9862	0.9731	0.9244
4.15	1.0000	1.0000	1.0000	1.0000	1.0000	0.9998	0.9979	0.9882	0.9767	0.9333
4.20	1.0000	1.0000	1.0000	1.0000	1.0000	0.9998	0.9982	0.9899	0.9799	0.9413
4.25	1.0000	1.0000	1.0000	1.0000	1.0000	0.9999	0.9985	0.9914	0.9827	0.9485
4.30	1.0000	1.0000	1.0000	1.0000	1.0000	0.9999	0.9987	0.9927	0.9851	0.9548
4.35	1.0000	1.0000	1.0000	1.0000	1.0000	0.9999	0.9989	0.9938	0.9872	0.9603
4.40	1.0000	1.0000	1.0000	1.0000	1.0000	0.9999	0.9991	0.9947	0.9890	0.9653
4.45	1.0000	1.0000	1.0000	1.0000	1.0000	0.9999	0.9993	0.9955	0.9905	0.9696
4.50	1.0000	1.0000	1.0000	1.0000	1.0000	1.0000	0.9994	0.9961	0.9918	0.9734
4.55	1.0000	1.0000	1.0000	1.0000	1.0000	1.0000	0.9995	0.9967	0.9930	0.9768
4.60	1.0000	1.0000	1.0000	1.0000	1.0000	1.0000	0.9996	0.9972	0.9940	0.9797
4.65	1.0000	1.0000	1.0000	1.0000	1.0000	1.0000	0.9996	0.9976	0.9948	0.9823
4.70	1.0000	1.0000	1.0000	1.0000	1.0000	1.0000	0.9997	0.9980	0.9955	0.9846
4.75	1.0000	1.0000	1.0000	1.0000	1.0000	1.0000	0.9997	0.9983	0.9962	0.9866
4.80	1.0000	1.0000	1.0000	1.0000	1.0000	1.0000	0.9998	0.9985	0.9967	0.9883
4.85	1.0000	1.0000	1.0000	1.0000	1.0000	1.0000	0.9998	0.9987	0.9972	0.9898
4.90	1.0000	1.0000	1.0000	1.0000	1.0000	1.0000	0.9998	0.9989	0.9976	0.9911
4.95	1.0000	1.0000	1.0000	1.0000	1.0000	1.0000	0.9999	0.9991	0.9979	0.9923
5.00	1.0000	1.0000	1.0000	1.0000	1.0000	1.0000	0.9999	0.9992	0.9982	0.9933
5.05	1.0000	1.0000	1.0000	1.0000	1.0000	1.0000	0.9999	0.9993	0.9985	0.9942
5.10	1.0000	1.0000	1.0000	1.0000	1.0000	1.0000	0.9999	0.9994	0.9987	0.9949
5.15	1.0000	1.0000	1.0000	1.0000	1.0000	1.0000	0.9999	0.9995	0.9989	0.9956
5.20	1.0000	1.0000	1.0000	1.0000	1.0000	1.0000	0.9999	0.9996	0.9990	0.9962
5.25	1.0000	1.0000	1.0000	1.0000	1.0000	1.0000	1.0000	0.9996	0.9992	0.9967
5.30	1.0000	1.0000	1.0000	1.0000	1.0000	1.0000	1.0000	0.9997	0.9993	0.9971
5.35	1.0000	1.0000	1.0000	1.0000	1.0000	1.0000	1.0000	0.9997	0.9994	0.9975
5.40	1.0000	1.0000	1.0000	1.0000	1.0000	1.0000	1.0000	0.9998	0.9995	0.9978
5.45	1.0000	1.0000	1.0000	1.0000	1.0000	1.0000	1.0000	0.9998	0.9995	0.9981
5.50	1.0000	1.0000	1.0000	1.0000	1.0000	1.0000	1.0000	0.9998	0.9996	0.9983
5.55	1.0000	1.0000	1.0000	1.0000	1.0000	1.0000	1.0000	0.9999	0.9997	0.9986
5.60	1.0000	1.0000	1.0000	1.0000	1.0000	1.0000	1.0000	0.9999	0.9997	0.9987
5.65	1.0000	1.0000	1.0000	1.0000	1.0000	1.0000	1.0000	0.9999	0.9997	0.9989
5.70	1.0000	1.0000	1.0000	1.0000	1.0000	1.0000	1.0000	0.9999	0.9998	0.9990

PROBABILITY INTEGRAL OF t, THE NON-CENTRAL t-STATISTIC. THIS TABLE GIVES Pr $[t/\sqrt{f} \leq x]$.

f is the number of degrees of freedom; the non-centrality parameter is $\sqrt{f+1}\,K_p$.

K_p is the standardized normal deviate exceeded with probability p.

DEGREES OF FREEDOM **23**

x \ P	.2500	.1500	.1000	.0650	.0400	.0250	.0100	.0040	.0025	.0010
5.75	1.0000	1.0000	1.0000	1.0000	1.0000	1.0000	1.0000	0.9999	0.9998	0.9992
5.80	1.0000	1.0000	1.0000	1.0000	1.0000	1.0000	1.0000	0.9999	0.9998	0.9993
5.85	1.0000	1.0000	1.0000	1.0000	1.0000	1.0000	1.0000	0.9999	0.9999	0.9994
5.90	1.0000	1.0000	1.0000	1.0000	1.0000	1.0000	1.0000	1.0000	0.9999	0.9995
5.95	1.0000	1.0000	1.0000	1.0000	1.0000	1.0000	1.0000	1.0000	0.9999	0.9995
6.00	1.0000	1.0000	1.0000	1.0000	1.0000	1.0000	1.0000	1.0000	0.9999	0.9996
6.05	1.0000	1.0000	1.0000	1.0000	1.0000	1.0000	1.0000	1.0000	0.9999	0.9996
6.10	1.0000	1.0000	1.0000	1.0000	1.0000	1.0000	1.0000	1.0000	0.9999	0.9997
6.15	1.0000	1.0000	1.0000	1.0000	1.0000	1.0000	1.0000	1.0000	0.9999	0.9997
6.20	1.0000	1.0000	1.0000	1.0000	1.0000	1.0000	1.0000	1.0000	0.9999	0.9998
6.25	1.0000	1.0000	1.0000	1.0000	1.0000	1.0000	1.0000	1.0000	1.0000	0.9998
6.30	1.0000	1.0000	1.0000	1.0000	1.0000	1.0000	1.0000	1.0000	1.0000	0.9998
6.35	1.0000	1.0000	1.0000	1.0000	1.0000	1.0000	1.0000	1.0000	1.0000	0.9998
6.40	1.0000	1.0000	1.0000	1.0000	1.0000	1.0000	1.0000	1.0000	1.0000	0.9999
6.45	1.0000	1.0000	1.0000	1.0000	1.0000	1.0000	1.0000	1.0000	1.0000	0.9999
6.50	1.0000	1.0000	1.0000	1.0000	1.0000	1.0000	1.0000	1.0000	1.0000	0.9999
6.55	1.0000	1.0000	1.0000	1.0000	1.0000	1.0000	1.0000	1.0000	1.0000	0.9999
6.60	1.0000	1.0000	1.0000	1.0000	1.0000	1.0000	1.0000	1.0000	1.0000	0.9999
6.65	1.0000	1.0000	1.0000	1.0000	1.0000	1.0000	1.0000	1.0000	1.0000	0.9999
6.70	1.0000	1.0000	1.0000	1.0000	1.0000	1.0000	1.0000	1.0000	1.0000	0.9999
6.75	1.0000	1.0000	1.0000	1.0000	1.0000	1.0000	1.0000	1.0000	1.0000	0.9999
6.80	1.0000	1.0000	1.0000	1.0000	1.0000	1.0000	1.0000	1.0000	1.0000	0.9999
6.85	1.0000	1.0000	1.0000	1.0000	1.0000	1.0000	1.0000	1.0000	1.0000	1.0000

PROBABILITY INTEGRAL OF t, THE NON-CENTRAL t-STATISTIC. THIS TABLE GIVES Pr $[t/\sqrt{f} \leq x]$.

f is the number of degrees of freedom; the non-centrality parameter is $\sqrt{f+1}\,K_p$.
K_p is the standardized normal deviate exceeded with probability p.

DEGREES OF FREEDOM **24**

x \ p	.2500	.1500	.1000	.0650	.0400	.0250	.0100	.0040	.0025	.0010
-0.15	0.0000	0.0000	0.0000	0.0000	0.0000	0.0000	0.0000	0.0000	0.0000	0.0000
-0.10	0.0001	0.0000	0.0000	0.0000	0.0000	0.0000	0.0000	0.0000	0.0000	0.0000
-0.05	0.0002	0.0000	0.0000	0.0000	0.0000	0.0000	0.0000	0.0000	0.0000	0.0000
0.00	0.0004	0.0000	0.0000	0.0000	0.0000	0.0000	0.0000	0.0000	0.0000	0.0000
0.05	0.0009	0.0000	0.0000	0.0000	0.0000	0.0000	0.0000	0.0000	0.0000	0.0000
0.10	0.0020	0.0000	0.0000	0.0000	0.0000	0.0000	0.0000	0.0000	0.0000	0.0000
0.15	0.0043	0.0000	0.0000	0.0000	0.0000	0.0000	0.0000	0.0000	0.0000	0.0000
0.20	0.0087	0.0000	0.0000	0.0000	0.0000	0.0000	0.0000	0.0000	0.0000	0.0000
0.25	0.0167	0.0000	0.0000	0.0000	0.0000	0.0000	0.0000	0.0000	0.0000	0.0000
0.30	0.0303	0.0001	0.0000	0.0000	0.0000	0.0000	0.0000	0.0000	0.0000	0.0000
0.35	0.0519	0.0004	0.0000	0.0000	0.0000	0.0000	0.0000	0.0000	0.0000	0.0000
0.40	0.0839	0.0009	0.0000	0.0000	0.0000	0.0000	0.0000	0.0000	0.0000	0.0000
0.45	0.1282	0.0021	0.0000	0.0000	0.0000	0.0000	0.0000	0.0000	0.0000	0.0000
0.50	0.1854	0.0047	0.0001	0.0000	0.0000	0.0000	0.0000	0.0000	0.0000	0.0000
0.55	0.2549	0.0096	0.0003	0.0000	0.0000	0.0000	0.0000	0.0000	0.0000	0.0000
0.60	0.3342	0.0183	0.0007	0.0000	0.0000	0.0000	0.0000	0.0000	0.0000	0.0000
0.65	0.4197	0.0327	0.0016	0.0000	0.0000	0.0000	0.0000	0.0000	0.0000	0.0000
0.70	0.5068	0.0547	0.0036	0.0001	0.0000	0.0000	0.0000	0.0000	0.0000	0.0000
0.75	0.5913	0.0861	0.0074	0.0003	0.0000	0.0000	0.0000	0.0000	0.0000	0.0000
0.80	0.6696	0.1281	0.0141	0.0007	0.0000	0.0000	0.0000	0.0000	0.0000	0.0000
0.85	0.7391	0.1811	0.0254	0.0017	0.0000	0.0000	0.0000	0.0000	0.0000	0.0000
0.90	0.7985	0.2439	0.0427	0.0036	0.0001	0.0000	0.0000	0.0000	0.0000	0.0000
0.95	0.8475	0.3148	0.0677	0.0073	0.0003	0.0000	0.0000	0.0000	0.0000	0.0000
1.00	0.8867	0.3907	0.1016	0.0136	0.0008	0.0000	0.0000	0.0000	0.0000	0.0000
1.05	0.9173	0.4686	0.1448	0.0239	0.0018	0.0001	0.0000	0.0000	0.0000	0.0000
1.10	0.9405	0.5453	0.1973	0.0394	0.0038	0.0003	0.0000	0.0000	0.0000	0.0000
1.15	0.9578	0.6180	0.2577	0.0615	0.0073	0.0006	0.0000	0.0000	0.0000	0.0000
1.20	0.9704	0.6846	0.3242	0.0911	0.0133	0.0014	0.0000	0.0000	0.0000	0.0000
1.25	0.9794	0.7439	0.3945	0.1288	0.0227	0.0029	0.0000	0.0000	0.0000	0.0000
1.30	0.9858	0.7952	0.4661	0.1745	0.0366	0.0055	0.0001	0.0000	0.0000	0.0000
1.35	0.9903	0.8384	0.5365	0.2275	0.0560	0.0100	0.0002	0.0000	0.0000	0.0000
1.40	0.9934	0.8742	0.6036	0.2865	0.0819	0.0171	0.0004	0.0000	0.0000	0.0000
1.45	0.9955	0.9032	0.6657	0.3497	0.1146	0.0277	0.0008	0.0000	0.0000	0.0000
1.50	0.9970	0.9262	0.7219	0.4153	0.1543	0.0427	0.0017	0.0000	0.0000	0.0000
1.55	0.9980	0.9443	0.7715	0.4810	0.2005	0.0629	0.0032	0.0001	0.0000	0.0000
1.60	0.9987	0.9583	0.8144	0.5453	0.2523	0.0888	0.0059	0.0002	0.0000	0.0000
1.65	0.9991	0.9690	0.8508	0.6064	0.3086	0.1208	0.0100	0.0004	0.0001	0.0000
1.70	0.9994	0.9771	0.8812	0.6633	0.3677	0.1589	0.0163	0.0009	0.0002	0.0000
1.75	0.9996	0.9832	0.9062	0.7150	0.4283	0.2025	0.0254	0.0017	0.0004	0.0000
1.80	0.9997	0.9877	0.9266	0.7613	0.4887	0.2510	0.0379	0.0031	0.0007	0.0000
1.85	0.9998	0.9910	0.9429	0.8019	0.5475	0.3033	0.0545	0.0054	0.0014	0.0001
1.90	0.9999	0.9935	0.9559	0.8371	0.6035	0.3582	0.0755	0.0089	0.0025	0.0002
1.95	0.9999	0.9953	0.9661	0.8670	0.6560	0.4145	0.1012	0.0141	0.0043	0.0003
2.00	1.0000	0.9966	0.9740	0.8923	0.7042	0.4708	0.1319	0.0215	0.0071	0.0006
2.05	1.0000	0.9975	0.9802	0.9133	0.7478	0.5261	0.1672	0.0315	0.0113	0.0012
2.10	1.0000	0.9982	0.9850	0.9306	0.7866	0.5793	0.2069	0.0446	0.0173	0.0021
2.15	1.0000	0.9987	0.9886	0.9448	0.8208	0.6297	0.2503	0.0612	0.0254	0.0036
2.20	1.0000	0.9991	0.9914	0.9563	0.8506	0.6766	0.2966	0.0817	0.0362	0.0058
2.25	1.0000	0.9993	0.9935	0.9655	0.8761	0.7197	0.3451	0.1061	0.0500	0.0091
2.30	1.0000	0.9995	0.9952	0.9729	0.8979	0.7587	0.3948	0.1345	0.0671	0.0137
2.35	1.0000	0.9997	0.9964	0.9788	0.9163	0.7936	0.4449	0.1668	0.0878	0.0201
2.40	1.0000	0.9998	0.9973	0.9834	0.9316	0.8246	0.4945	0.2027	0.1122	0.0284
2.45	1.0000	0.9998	0.9980	0.9871	0.9444	0.8517	0.5429	0.2416	0.1402	0.0392
2.50	1.0000	0.9999	0.9985	0.9900	0.9550	0.8753	0.5894	0.2831	0.1717	0.0526
2.55	1.0000	0.9999	0.9989	0.9922	0.9637	0.8957	0.6335	0.3265	0.2064	0.0689
2.60	1.0000	0.9999	0.9992	0.9940	0.9708	0.9131	0.6748	0.3712	0.2439	0.0883
2.65	1.0000	1.0000	0.9994	0.9953	0.9766	0.9279	0.7132	0.4165	0.2836	0.1109
2.70	1.0000	1.0000	0.9995	0.9964	0.9812	0.9403	0.7484	0.4617	0.3251	0.1366

PROBABILITY INTEGRAL OF t, THE NON-CENTRAL t-STATISTIC. THIS TABLE GIVES Pr $[t/\sqrt{f} \leq x]$.

f is the number of degrees of freedom; the non-centrality parameter is $\sqrt{f+1}\,K_p$.
K_p is the standardized normal deviate exceeded with probability p.

DEGREES OF FREEDOM **24**

x \ p	.2500	.1500	.1000	.0650	.0400	.0250	.0100	.0040	.0025	.0010
2.75	1.0000	1.0000	0.9997	0.9972	0.9850	0.9508	0.7803	0.5063	0.3678	0.1653
2.80	1.0000	1.0000	0.9997	0.9978	0.9881	0.9596	0.8092	0.5497	0.4111	0.1967
2.85	1.0000	1.0000	0.9998	0.9983	0.9905	0.9669	0.8350	0.5916	0.4544	0.2307
2.90	1.0000	1.0000	0.9999	0.9987	0.9924	0.9729	0.8578	0.6314	0.4972	0.2668
2.95	1.0000	1.0000	0.9999	0.9990	0.9940	0.9779	0.8780	0.6690	0.5390	0.3045
3.00	1.0000	1.0000	0.9999	0.9992	0.9952	0.9820	0.8957	0.7041	0.5794	0.3435
3.05	1.0000	1.0000	0.9999	0.9994	0.9962	0.9854	0.9111	0.7367	0.6181	0.3833
3.10	1.0000	1.0000	1.0000	0.9996	0.9970	0.9881	0.9245	0.7666	0.6548	0.4233
3.15	1.0000	1.0000	1.0000	0.9997	0.9976	0.9904	0.9360	0.7940	0.6893	0.4633
3.20	1.0000	1.0000	1.0000	0.9997	0.9981	0.9922	0.9459	0.8188	0.7215	0.5027
3.25	1.0000	1.0000	1.0000	0.9998	0.9985	0.9937	0.9544	0.8412	0.7514	0.5412
3.30	1.0000	1.0000	1.0000	0.9998	0.9988	0.9949	0.9616	0.8613	0.7788	0.5785
3.35	1.0000	1.0000	1.0000	0.9999	0.9991	0.9959	0.9678	0.8792	0.8039	0.6143
3.40	1.0000	1.0000	1.0000	0.9999	0.9993	0.9967	0.9730	0.8951	0.8268	0.6485
3.45	1.0000	1.0000	1.0000	0.9999	0.9994	0.9973	0.9774	0.9091	0.8474	0.6808
3.50	1.0000	1.0000	1.0000	1.0000	0.9995	0.9978	0.9811	0.9215	0.8660	0.7111
3.55	1.0000	1.0000	1.0000	1.0000	0.9996	0.9982	0.9842	0.9323	0.8826	0.7395
3.60	1.0000	1.0000	1.0000	1.0000	0.9997	0.9986	0.9868	0.9417	0.8974	0.7658
3.65	1.0000	1.0000	1.0000	1.0000	0.9998	0.9989	0.9890	0.9500	0.9106	0.7901
3.70	1.0000	1.0000	1.0000	1.0000	0.9998	0.9991	0.9909	0.9571	0.9222	0.8124
3.75	1.0000	1.0000	1.0000	1.0000	0.9999	0.9993	0.9924	0.9633	0.9325	0.8328
3.80	1.0000	1.0000	1.0000	1.0000	0.9999	0.9994	0.9937	0.9686	0.9415	0.8513
3.85	1.0000	1.0000	1.0000	1.0000	0.9999	0.9995	0.9947	0.9732	0.9494	0.8681
3.90	1.0000	1.0000	1.0000	1.0000	0.9999	0.9996	0.9956	0.9772	0.9563	0.8833
3.95	1.0000	1.0000	1.0000	1.0000	0.9999	0.9997	0.9964	0.9806	0.9623	0.8969
4.00	1.0000	1.0000	1.0000	1.0000	1.0000	0.9997	0.9970	0.9835	0.9675	0.9092
4.05	1.0000	1.0000	1.0000	1.0000	1.0000	0.9998	0.9975	0.9859	0.9721	0.9201
4.10	1.0000	1.0000	1.0000	1.0000	1.0000	0.9998	0.9979	0.9881	0.9760	0.9298
4.15	1.0000	1.0000	1.0000	1.0000	1.0000	0.9999	0.9983	0.9899	0.9794	0.9385
4.20	1.0000	1.0000	1.0000	1.0000	1.0000	0.9999	0.9986	0.9914	0.9823	0.9461
4.25	1.0000	1.0000	1.0000	1.0000	1.0000	0.9999	0.9988	0.9927	0.9849	0.9529
4.30	1.0000	1.0000	1.0000	1.0000	1.0000	0.9999	0.9990	0.9938	0.9870	0.9589
4.35	1.0000	1.0000	1.0000	1.0000	1.0000	0.9999	0.9992	0.9948	0.9889	0.9641
4.40	1.0000	1.0000	1.0000	1.0000	1.0000	1.0000	0.9993	0.9956	0.9905	0.9687
4.45	1.0000	1.0000	1.0000	1.0000	1.0000	1.0000	0.9994	0.9962	0.9919	0.9728
4.50	1.0000	1.0000	1.0000	1.0000	1.0000	1.0000	0.9995	0.9968	0.9931	0.9763
4.55	1.0000	1.0000	1.0000	1.0000	1.0000	1.0000	0.9996	0.9973	0.9941	0.9794
4.60	1.0000	1.0000	1.0000	1.0000	1.0000	1.0000	0.9997	0.9977	0.9949	0.9822
4.65	1.0000	1.0000	1.0000	1.0000	1.0000	1.0000	0.9997	0.9981	0.9957	0.9845
4.70	1.0000	1.0000	1.0000	1.0000	1.0000	1.0000	0.9998	0.9984	0.9963	0.9866
4.75	1.0000	1.0000	1.0000	1.0000	1.0000	1.0000	0.9998	0.9986	0.9969	0.9884
4.80	1.0000	1.0000	1.0000	1.0000	1.0000	1.0000	0.9998	0.9988	0.9973	0.9899
4.85	1.0000	1.0000	1.0000	1.0000	1.0000	1.0000	0.9999	0.9990	0.9977	0.9913
4.90	1.0000	1.0000	1.0000	1.0000	1.0000	1.0000	0.9999	0.9992	0.9980	0.9924
4.95	1.0000	1.0000	1.0000	1.0000	1.0000	1.0000	0.9999	0.9993	0.9983	0.9935
5.00	1.0000	1.0000	1.0000	1.0000	1.0000	1.0000	0.9999	0.9994	0.9986	0.9943
5.05	1.0000	1.0000	1.0000	1.0000	1.0000	1.0000	0.9999	0.9995	0.9988	0.9951
5.10	1.0000	1.0000	1.0000	1.0000	1.0000	1.0000	0.9999	0.9996	0.9990	0.9958
5.15	1.0000	1.0000	1.0000	1.0000	1.0000	1.0000	1.0000	0.9996	0.9991	0.9963
5.20	1.0000	1.0000	1.0000	1.0000	1.0000	1.0000	1.0000	0.9997	0.9992	0.9968
5.25	1.0000	1.0000	1.0000	1.0000	1.0000	1.0000	1.0000	0.9997	0.9994	0.9973
5.30	1.0000	1.0000	1.0000	1.0000	1.0000	1.0000	1.0000	0.9998	0.9994	0.9976
5.35	1.0000	1.0000	1.0000	1.0000	1.0000	1.0000	1.0000	0.9998	0.9995	0.9980
5.40	1.0000	1.0000	1.0000	1.0000	1.0000	1.0000	1.0000	0.9998	0.9996	0.9982
5.45	1.0000	1.0000	1.0000	1.0000	1.0000	1.0000	1.0000	0.9999	0.9997	0.9985
5.50	1.0000	1.0000	1.0000	1.0000	1.0000	1.0000	1.0000	0.9999	0.9997	0.9987
5.55	1.0000	1.0000	1.0000	1.0000	1.0000	1.0000	1.0000	0.9999	0.9997	0.9989
5.60	1.0000	1.0000	1.0000	1.0000	1.0000	1.0000	1.0000	0.9999	0.9998	0.9990
5.65	1.0000	1.0000	1.0000	1.0000	1.0000	1.0000	1.0000	0.9999	0.9998	0.9991
5.70	1.0000	1.0000	1.0000	1.0000	1.0000	1.0000	1.0000	0.9999	0.9998	0.9993

PROBABILITY INTEGRAL OF t, THE NON-CENTRAL t-STATISTIC. THIS TABLE GIVES Pr $[t/\sqrt{f} \leq x]$.

f is the number of degrees of freedom; the non-centrality parameter is $\sqrt{f+1}\,K_p$.

K_p is the standardized normal deviate exceeded with probability p.

DEGREES OF FREEDOM **24**

x \ p	.2500	.1500	.1000	.0650	.0400	.0250	.0100	.0040	.0025	.0010
5.75	1.0000	1.0000	1.0000	1.0000	1.0000	1.0000	1.0000	0.9999	0.9999	0.9994
5.80	1.0000	1.0000	1.0000	1.0000	1.0000	1.0000	1.0000	1.0000	0.9999	0.9994
5.85	1.0000	1.0000	1.0000	1.0000	1.0000	1.0000	1.0000	1.0000	0.9999	0.9995
5.90	1.0000	1.0000	1.0000	1.0000	1.0000	1.0000	1.0000	1.0000	0.9999	0.9996
5.95	1.0000	1.0000	1.0000	1.0000	1.0000	1.0000	1.0000	1.0000	0.9999	0.9996
6.00	1.0000	1.0000	1.0000	1.0000	1.0000	1.0000	1.0000	1.0000	0.9999	0.9997
6.05	1.0000	1.0000	1.0000	1.0000	1.0000	1.0000	1.0000	1.0000	0.9999	0.9997
6.10	1.0000	1.0000	1.0000	1.0000	1.0000	1.0000	1.0000	1.0000	1.0000	0.9998
6.15	1.0000	1.0000	1.0000	1.0000	1.0000	1.0000	1.0000	1.0000	1.0000	0.9998
6.20	1.0000	1.0000	1.0000	1.0000	1.0000	1.0000	1.0000	1.0000	1.0000	0.9998
6.25	1.0000	1.0000	1.0000	1.0000	1.0000	1.0000	1.0000	1.0000	1.0000	0.9998
6.30	1.0000	1.0000	1.0000	1.0000	1.0000	1.0000	1.0000	1.0000	1.0000	0.9999
6.35	1.0000	1.0000	1.0000	1.0000	1.0000	1.0000	1.0000	1.0000	1.0000	0.9999
6.40	1.0000	1.0000	1.0000	1.0000	1.0000	1.0000	1.0000	1.0000	1.0000	0.9999
6.45	1.0000	1.0000	1.0000	1.0000	1.0000	1.0000	1.0000	1.0000	1.0000	0.9999
6.50	1.0000	1.0000	1.0000	1.0000	1.0000	1.0000	1.0000	1.0000	1.0000	0.9999
6.55	1.0000	1.0000	1.0000	1.0000	1.0000	1.0000	1.0000	1.0000	1.0000	0.9999
6.60	1.0000	1.0000	1.0000	1.0000	1.0000	1.0000	1.0000	1.0000	1.0000	0.9999
6.65	1.0000	1.0000	1.0000	1.0000	1.0000	1.0000	1.0000	1.0000	1.0000	0.9999
6.70	1.0000	1.0000	1.0000	1.0000	1.0000	1.0000	1.0000	1.0000	1.0000	1.0000

PROBABILITY INTEGRAL OF t, THE NON-CENTRAL t-STATISTIC. THIS TABLE GIVES Pr $[t/\sqrt{f} \leq x]$.

f is the number of degrees of freedom; the non-centrality parameter is $\sqrt{f+1}\,K_p$.

K_p is the standardized normal deviate exceeded with probability p.

DEGREES OF FREEDOM **29**

x \ p	.2500	.1500	.1000	.0650	.0400	.0250	.0100	.0040	.0025	.0010
-0.05	0.0000	0.0000	0.0000	0.0000	0.0000	0.0000	0.0000	0.0000	0.0000	0.0000
0.00	0.0001	0.0000	0.0000	0.0000	0.0000	0.0000	0.0000	0.0000	0.0000	0.0000
0.05	0.0003	0.0000	0.0000	0.0000	0.0000	0.0000	0.0000	0.0000	0.0000	0.0000
0.10	0.0008	0.0000	0.0000	0.0000	0.0000	0.0000	0.0000	0.0000	0.0000	0.0000
0.15	0.0020	0.0000	0.0000	0.0000	0.0000	0.0000	0.0000	0.0000	0.0000	0.0000
0.20	0.0047	0.0000	0.0000	0.0000	0.0000	0.0000	0.0000	0.0000	0.0000	0.0000
0.25	0.0101	0.0000	0.0000	0.0000	0.0000	0.0000	0.0000	0.0000	0.0000	0.0000
0.30	0.0203	0.0000	0.0000	0.0000	0.0000	0.0000	0.0000	0.0000	0.0000	0.0000
0.35	0.0382	0.0001	0.0000	0.0000	0.0000	0.0000	0.0000	0.0000	0.0000	0.0000
0.40	0.0668	0.0003	0.0000	0.0000	0.0000	0.0000	0.0000	0.0000	0.0000	0.0000
0.45	0.1091	0.0009	0.0000	0.0000	0.0000	0.0000	0.0000	0.0000	0.0000	0.0000
0.50	0.1668	0.0023	0.0000	0.0000	0.0000	0.0000	0.0000	0.0000	0.0000	0.0000
0.55	0.2397	0.0054	0.0001	0.0000	0.0000	0.0000	0.0000	0.0000	0.0000	0.0000
0.60	0.3252	0.0115	0.0002	0.0000	0.0000	0.0000	0.0000	0.0000	0.0000	0.0000
0.65	0.4187	0.0227	0.0007	0.0000	0.0000	0.0000	0.0000	0.0000	0.0000	0.0000
0.70	0.5146	0.0413	0.0017	0.0000	0.0000	0.0000	0.0000	0.0000	0.0000	0.0000
0.75	0.6072	0.0699	0.0040	0.0001	0.0000	0.0000	0.0000	0.0000	0.0000	0.0000
0.80	0.6917	0.1106	0.0086	0.0003	0.0000	0.0000	0.0000	0.0000	0.0000	0.0000
0.85	0.7650	0.1642	0.0171	0.0007	0.0000	0.0000	0.0000	0.0000	0.0000	0.0000
0.90	0.8258	0.2303	0.0313	0.0017	0.0000	0.0000	0.0000	0.0000	0.0000	0.0000
0.95	0.8742	0.3066	0.0534	0.0040	0.0001	0.0000	0.0000	0.0000	0.0000	0.0000
1.00	0.9112	0.3897	0.0853	0.0083	0.0003	0.0000	0.0000	0.0000	0.0000	0.0000
1.05	0.9387	0.4754	0.1281	0.0160	0.0008	0.0000	0.0000	0.0000	0.0000	0.0000
1.10	0.9584	0.5596	0.1821	0.0287	0.0019	0.0001	0.0000	0.0000	0.0000	0.0000
1.15	0.9723	0.6387	0.2462	0.0481	0.0040	0.0002	0.0000	0.0000	0.0000	0.0000
1.20	0.9818	0.7100	0.3183	0.0757	0.0081	0.0006	0.0000	0.0000	0.0000	0.0000
1.25	0.9882	0.7720	0.3953	0.1127	0.0152	0.0014	0.0000	0.0000	0.0000	0.0000
1.30	0.9924	0.8242	0.4741	0.1592	0.0265	0.0029	0.0000	0.0000	0.0000	0.0000
1.35	0.9952	0.8668	0.5514	0.2149	0.0434	0.0059	0.0000	0.0000	0.0000	0.0000
1.40	0.9970	0.9007	0.6244	0.2783	0.0673	0.0110	0.0001	0.0000	0.0000	0.0000
1.45	0.9981	0.9270	0.6911	0.3472	0.0990	0.0193	0.0003	0.0000	0.0000	0.0000
1.50	0.9988	0.9471	0.7502	0.4192	0.1390	0.0319	0.0007	0.0000	0.0000	0.0000
1.55	0.9993	0.9621	0.8012	0.4917	0.1870	0.0498	0.0016	0.0000	0.0000	0.0000
1.60	0.9996	0.9731	0.8440	0.5621	0.2422	0.0742	0.0032	0.0001	0.0000	0.0000
1.65	0.9997	0.9811	0.8793	0.6285	0.3031	0.1055	0.0059	0.0001	0.0000	0.0000
1.70	0.9998	0.9868	0.9077	0.6894	0.3679	0.1441	0.0105	0.0003	0.0000	0.0000
1.75	0.9999	0.9909	0.9302	0.7439	0.4345	0.1896	0.0176	0.0007	0.0001	0.0000
1.80	0.9999	0.9937	0.9477	0.7916	0.5009	0.2413	0.0279	0.0015	0.0003	0.0000
1.85	1.0000	0.9957	0.9612	0.8324	0.5653	0.2979	0.0425	0.0029	0.0006	0.0000
1.90	1.0000	0.9971	0.9714	0.8666	0.6262	0.3580	0.0618	0.0052	0.0012	0.0000
1.95	1.0000	0.9980	0.9791	0.8950	0.6825	0.4199	0.0866	0.0089	0.0022	0.0001
2.00	1.0000	0.9987	0.9848	0.9181	0.7333	0.4820	0.1171	0.0145	0.0040	0.0002
2.05	1.0000	0.9991	0.9890	0.9366	0.7784	0.5426	0.1533	0.0226	0.0069	0.0005
2.10	1.0000	0.9994	0.9921	0.9513	0.8177	0.6006	0.1949	0.0338	0.0113	0.0010
2.15	1.0000	0.9996	0.9943	0.9629	0.8514	0.6549	0.2412	0.0487	0.0177	0.0018
2.20	1.0000	0.9997	0.9960	0.9719	0.8799	0.7048	0.2914	0.0679	0.0266	0.0032
2.25	1.0000	0.9998	0.9971	0.9788	0.9037	0.7499	0.3444	0.0916	0.0387	0.0054
2.30	1.0000	0.9999	0.9980	0.9841	0.9234	0.7899	0.3991	0.1201	0.0543	0.0087
2.35	1.0000	0.9999	0.9986	0.9881	0.9394	0.8250	0.4543	0.1532	0.0739	0.0135
2.40	1.0000	1.0000	0.9990	0.9911	0.9524	0.8554	0.5088	0.1908	0.0978	0.0202
2.45	1.0000	1.0000	0.9993	0.9934	0.9627	0.8813	0.5618	0.2323	0.1260	0.0292
2.50	1.0000	1.0000	0.9995	0.9951	0.9710	0.9032	0.6123	0.2772	0.1584	0.0411
2.55	1.0000	1.0000	0.9996	0.9964	0.9775	0.9216	0.6597	0.3245	0.1949	0.0561
2.60	1.0000	1.0000	0.9998	0.9974	0.9827	0.9368	0.7036	0.3736	0.2349	0.0745
2.65	1.0000	1.0000	0.9998	0.9981	0.9867	0.9493	0.7438	0.4235	0.2780	0.0966
2.70	1.0000	1.0000	0.9999	0.9986	0.9898	0.9596	0.7800	0.4733	0.3233	0.1224

PROBABILITY INTEGRAL OF t, THE NON-CENTRAL t-STATISTIC. THIS TABLE GIVES Pr $[t/\sqrt{f} \leq x]$.

f is the number of degrees of freedom; the non-centrality parameter is $\sqrt{f+1}\,K_p$.

K_p is the standardized normal deviate exceeded with probability p.

x \ p	.2500	.1500	.1000	.0650	.0400	.0250	.0100	.0040	.0025	.0010
2.75	1.0000	1.0000	0.9999	0.9990	0.9922	0.9679	0.8123	0.5223	0.3701	0.1519
2.80	1.0000	1.0000	0.9999	0.9992	0.9940	0.9746	0.8408	0.5698	0.4178	0.1849
2.85	1.0000	1.0000	1.0000	0.9994	0.9955	0.9799	0.8658	0.6151	0.4655	0.2211
2.90	1.0000	1.0000	1.0000	0.9996	0.9966	0.9842	0.8874	0.6580	0.5125	0.2599
2.95	1.0000	1.0000	1.0000	0.9997	0.9974	0.9876	0.9060	0.6979	0.5582	0.3010
3.00	1.0000	1.0000	1.0000	0.9998	0.9980	0.9903	0.9219	0.7348	0.6022	0.3437
3.05	1.0000	1.0000	1.0000	0.9998	0.9985	0.9924	0.9354	0.7684	0.6439	0.3874
3.10	1.0000	1.0000	1.0000	0.9999	0.9989	0.9941	0.9467	0.7989	0.6831	0.4316
3.15	1.0000	1.0000	1.0000	0.9999	0.9991	0.9954	0.9562	0.8262	0.7194	0.4755
3.20	1.0000	1.0000	1.0000	0.9999	0.9994	0.9964	0.9642	0.8505	0.7530	0.5188
3.25	1.0000	1.0000	1.0000	1.0000	0.9995	0.9972	0.9707	0.8720	0.7835	0.5610
3.30	1.0000	1.0000	1.0000	1.0000	0.9996	0.9979	0.9762	0.8908	0.8112	0.6015
3.35	1.0000	1.0000	1.0000	1.0000	0.9997	0.9983	0.9807	0.9073	0.8361	0.6401
3.40	1.0000	1.0000	1.0000	1.0000	0.9998	0.9987	0.9843	0.9215	0.8583	0.6766
3.45	1.0000	1.0000	1.0000	1.0000	0.9998	0.9990	0.9873	0.9338	0.8779	0.7108
3.50	1.0000	1.0000	1.0000	1.0000	0.9999	0.9992	0.9898	0.9443	0.8952	0.7424
3.55	1.0000	1.0000	1.0000	1.0000	0.9999	0.9994	0.9917	0.9533	0.9104	0.7716
3.60	1.0000	1.0000	1.0000	1.0000	0.9999	0.9995	0.9934	0.9609	0.9236	0.7983
3.65	1.0000	1.0000	1.0000	1.0000	0.9999	0.9996	0.9947	0.9674	0.9351	0.8226
3.70	1.0000	1.0000	1.0000	1.0000	1.0000	0.9997	0.9957	0.9729	0.9450	0.8445
3.75	1.0000	1.0000	1.0000	1.0000	1.0000	0.9998	0.9966	0.9775	0.9535	0.8642
3.80	1.0000	1.0000	1.0000	1.0000	1.0000	0.9998	0.9972	0.9813	0.9608	0.8817
3.85	1.0000	1.0000	1.0000	1.0000	1.0000	0.9999	0.9978	0.9845	0.9670	0.8974
3.90	1.0000	1.0000	1.0000	1.0000	1.0000	0.9999	0.9982	0.9872	0.9723	0.9112
3.95	1.0000	1.0000	1.0000	1.0000	1.0000	0.9999	0.9986	0.9894	0.9768	0.9233
4.00	1.0000	1.0000	1.0000	1.0000	1.0000	0.9999	0.9989	0.9913	0.9806	0.9340
4.05	1.0000	1.0000	1.0000	1.0000	1.0000	1.0000	0.9991	0.9928	0.9838	0.9433
4.10	1.0000	1.0000	1.0000	1.0000	1.0000	1.0000	0.9993	0.9941	0.9864	0.9514
4.15	1.0000	1.0000	1.0000	1.0000	1.0000	1.0000	0.9994	0.9952	0.9887	0.9584
4.20	1.0000	1.0000	1.0000	1.0000	1.0000	1.0000	0.9995	0.9960	0.9906	0.9645
4.25	1.0000	1.0000	1.0000	1.0000	1.0000	1.0000	0.9996	0.9967	0.9922	0.9697
4.30	1.0000	1.0000	1.0000	1.0000	1.0000	1.0000	0.9997	0.9973	0.9935	0.9743
4.35	1.0000	1.0000	1.0000	1.0000	1.0000	1.0000	0.9998	0.9978	0.9946	0.9781
4.40	1.0000	1.0000	1.0000	1.0000	1.0000	1.0000	0.9998	0.9982	0.9955	0.9814
4.45	1.0000	1.0000	1.0000	1.0000	1.0000	1.0000	0.9998	0.9985	0.9963	0.9843
4.50	1.0000	1.0000	1.0000	1.0000	1.0000	1.0000	0.9999	0.9988	0.9969	0.9867
4.55	1.0000	1.0000	1.0000	1.0000	1.0000	1.0000	0.9999	0.9990	0.9975	0.9888
4.60	1.0000	1.0000	1.0000	1.0000	1.0000	1.0000	0.9999	0.9992	0.9979	0.9905
4.65	1.0000	1.0000	1.0000	1.0000	1.0000	1.0000	0.9999	0.9993	0.9983	0.9920
4.70	1.0000	1.0000	1.0000	1.0000	1.0000	1.0000	0.9999	0.9995	0.9986	0.9932
4.75	1.0000	1.0000	1.0000	1.0000	1.0000	1.0000	1.0000	0.9996	0.9988	0.9943
4.80	1.0000	1.0000	1.0000	1.0000	1.0000	1.0000	1.0000	0.9996	0.9990	0.9952
4.85	1.0000	1.0000	1.0000	1.0000	1.0000	1.0000	1.0000	0.9997	0.9992	0.9960
4.90	1.0000	1.0000	1.0000	1.0000	1.0000	1.0000	1.0000	0.9998	0.9993	0.9966
4.95	1.0000	1.0000	1.0000	1.0000	1.0000	1.0000	1.0000	0.9998	0.9994	0.9971
5.00	1.0000	1.0000	1.0000	1.0000	1.0000	1.0000	1.0000	0.9998	0.9995	0.9976
5.05	1.0000	1.0000	1.0000	1.0000	1.0000	1.0000	1.0000	0.9999	0.9996	0.9980
5.10	1.0000	1.0000	1.0000	1.0000	1.0000	1.0000	1.0000	0.9999	0.9997	0.9983
5.15	1.0000	1.0000	1.0000	1.0000	1.0000	1.0000	1.0000	0.9999	0.9997	0.9986
5.20	1.0000	1.0000	1.0000	1.0000	1.0000	1.0000	1.0000	0.9999	0.9998	0.9988
5.25	1.0000	1.0000	1.0000	1.0000	1.0000	1.0000	1.0000	0.9999	0.9998	0.9990
5.30	1.0000	1.0000	1.0000	1.0000	1.0000	1.0000	1.0000	0.9999	0.9999	0.9992
5.35	1.0000	1.0000	1.0000	1.0000	1.0000	1.0000	1.0000	1.0000	0.9999	0.9993
5.40	1.0000	1.0000	1.0000	1.0000	1.0000	1.0000	1.0000	1.0000	0.9999	0.9994
5.45	1.0000	1.0000	1.0000	1.0000	1.0000	1.0000	1.0000	1.0000	0.9999	0.9995
5.50	1.0000	1.0000	1.0000	1.0000	1.0000	1.0000	1.0000	1.0000	0.9999	0.9996
5.55	1.0000	1.0000	1.0000	1.0000	1.0000	1.0000	1.0000	1.0000	0.9999	9.9996
5.60	1.0000	1.0000	1.0000	1.0000	1.0000	1.0000	1.0000	1.0000	0.9999	0.9997
5.65	1.0000	1.0000	1.0000	1.0000	1.0000	1.0000	1.0000	1.0000	1.0000	0.9998
5.70	1.0000	1.0000	1.0000	1.0000	1.0000	1.0000	1.0000	1.0000	1.0000	0.9998

PROBABILITY INTEGRAL OF t, THE NON-CENTRAL t-STATISTIC. THIS TABLE GIVES Pr $[t/\sqrt{f} \leq x]$.

f is the number of degrees of freedom; the non-centrality parameter is $\sqrt{f+1}\, K_p$.
K_p is the standardized normal deviate exceeded with probability p.

DEGREES OF FREEDOM **29**

x \ p	.2500	.1500	.1000	.0650	.0400	.0250	.0100	.0040	.0025	.0010
5.75	1.0000	1.0000	1.0000	1.0000	1.0000	1.0000	1.0000	1.0000	1.0000	0.9998
5.80	1.0000	1.0000	1.0000	1.0000	1.0000	1.0000	1.0000	1.0000	1.0000	0.9999
5.85	1.0000	1.0000	1.0000	1.0000	1.0000	1.0000	1.0000	1.0000	1.0000	0.9999
5.90	1.0000	1.0000	1.0000	1.0000	1.0000	1.0000	1.0000	1.0000	1.0000	0.9999
5.95	1.0000	1.0000	1.0000	1.0000	1.0000	1.0000	1.0000	1.0000	1.0000	0.9999
6.00	1.0000	1.0000	1.0000	1.0000	1.0000	1.0000	1.0000	1.0000	1.0000	0.9999
6.05	1.0000	1.0000	1.0000	1.0000	1.0000	1.0000	1.0000	1.0000	1.0000	0.9999
6.10	1.0000	1.0000	1.0000	1.0000	1.0000	1.0000	1.0000	1.0000	1.0000	0.9999
6.15	1.0000	1.0000	1.0000	1.0000	1.0000	1.0000	1.0000	1.0000	1.0000	1.0000

PROBABILITY INTEGRAL OF t, THE NON-CENTRAL t-STATISTIC. THIS TABLE GIVES Pr $[t/\sqrt{f} \leq x]$.

f is the number of degrees of freedom; the non-centrality parameter is $\sqrt{f+1}\,K_p$.

K_p is the standardized normal deviate exceeded with probability p.

DEGREES OF FREEDOM **34**

x \ p	.2500	.1500	.1000	.0650	.0400	.0250	.0100	.0040	.0025	.0010
0.00	0.0000	0.0000	0.0000	0.0000	0.0000	0.0000	0.0000	0.0000	0.0000	0.0000
0.05	0.0001	0.0000	0.0000	0.0000	0.0000	0.0000	0.0000	0.0000	0.0000	0.0000
0.10	0.0003	0.0000	0.0000	0.0000	0.0000	0.0000	0.0000	0.0000	0.0000	0.0000
0.15	0.0010	0.0000	0.0000	0.0000	0.0000	0.0000	0.0000	0.0000	0.0000	0.0000
0.20	0.0025	0.0000	0.0000	0.0000	0.0000	0.0000	0.0000	0.0000	0.0000	0.0000
0.25	0.0061	0.0000	0.0000	0.0000	0.0000	0.0000	0.0000	0.0000	0.0000	0.0000
0.30	0.0137	0.0000	0.0000	0.0000	0.0000	0.0000	0.0000	0.0000	0.0000	0.0000
0.35	0.0283	0.0000	0.0000	0.0000	0.0000	0.0000	0.0000	0.0000	0.0000	0.0000
0.40	0.0535	0.0001	0.0000	0.0000	0.0000	0.0000	0.0000	0.0000	0.0000	0.0000
0.45	0.0933	0.0004	0.0000	0.0000	0.0000	0.0000	0.0000	0.0000	0.0000	0.0000
0.50	0.1506	0.0012	0.0000	0.0000	0.0000	0.0000	0.0000	0.0000	0.0000	0.0000
0.55	0.2258	0.0030	0.0000	0.0000	0.0000	0.0000	0.0000	0.0000	0.0000	0.0000
0.60	0.3165	0.0073	0.0001	0.0000	0.0000	0.0000	0.0000	0.0000	0.0000	0.0000
0.65	0.4173	0.0158	0.0003	0.0000	0.0000	0.0000	0.0000	0.0000	0.0000	0.0000
0.70	0.5213	0.0314	0.0008	0.0000	0.0000	0.0000	0.0000	0.0000	0.0000	0.0000
0.75	0.6211	0.0571	0.0022	0.0000	0.0000	0.0000	0.0000	0.0000	0.0000	0.0000
0.80	0.7109	0.0958	0.0053	0.0001	0.0000	0.0000	0.0000	0.0000	0.0000	0.0000
0.85	0.7872	0.1493	0.0115	0.0003	0.0000	0.0000	0.0000	0.0000	0.0000	0.0000
0.90	0.8484	0.2177	0.0231	0.0009	0.0000	0.0000	0.0000	0.0000	0.0000	0.0000
0.95	0.8954	0.2986	0.0423	0.0022	0.0000	0.0000	0.0000	0.0000	0.0000	0.0000
1.00	0.9298	0.3881	0.0719	0.0051	0.0001	0.0000	0.0000	0.0000	0.0000	0.0000
1.05	0.9541	0.4810	0.1137	0.0108	0.0003	0.0000	0.0000	0.0000	0.0000	0.0000
1.10	0.9707	0.5720	0.1684	0.0210	0.0009	0.0000	0.0000	0.0000	0.0000	0.0000
1.15	0.9817	0.6567	0.2353	0.0378	0.0023	0.0001	0.0000	0.0000	0.0000	0.0000
1.20	0.9887	0.7318	0.3122	0.0632	0.0050	0.0002	0.0000	0.0000	0.0000	0.0000
1.25	0.9932	0.7957	0.3953	0.0988	0.0102	0.0006	0.0000	0.0000	0.0000	0.0000
1.30	0.9959	0.8479	0.4807	0.1455	0.0193	0.0016	0.0000	0.0000	0.0000	0.0000
1.35	0.9976	0.8892	0.5643	0.2031	0.0338	0.0035	0.0000	0.0000	0.0000	0.0000
1.40	0.9986	0.9209	0.6425	0.2701	0.0555	0.0071	0.0000	0.0000	0.0000	0.0000
1.45	0.9992	0.9445	0.7130	0.3441	0.0857	0.0135	0.0001	0.0000	0.0000	0.0000
1.50	0.9995	0.9617	0.7743	0.4220	0.1254	0.0239	0.0003	0.0000	0.0000	0.0000
1.55	0.9997	0.9739	0.8258	0.5005	0.1745	0.0396	0.0008	0.0000	0.0000	0.0000
1.60	0.9998	0.9825	0.8679	0.5766	0.2323	0.0621	0.0017	0.0000	0.0000	0.0000
1.65	0.9999	0.9883	0.9015	0.6476	0.2973	0.0923	0.0035	0.0001	0.0000	0.0000
1.70	1.0000	0.9923	0.9276	0.7119	0.3671	0.1309	0.0068	0.0001	0.0000	0.0000
1.75	1.0000	0.9950	0.9475	0.7684	0.4393	0.1776	0.0122	0.0003	0.0000	0.0000
1.80	1.0000	0.9968	0.9624	0.8167	0.5112	0.2318	0.0207	0.0007	0.0001	0.0000
1.85	1.0000	0.9979	0.9733	0.8570	0.5807	0.2922	0.0332	0.0016	0.0002	0.0000
1.90	1.0000	0.9987	0.9813	0.8899	0.6458	0.3569	0.0508	0.0031	0.0006	0.0000
1.95	1.0000	0.9991	0.9870	0.9163	0.7051	0.4239	0.0742	0.0056	0.0012	0.0000
2.00	1.0000	0.9995	0.9910	0.9371	0.7579	0.4912	0.1041	0.0098	0.0023	0.0001
2.05	1.0000	0.9997	0.9938	0.9532	0.8038	0.5567	0.1406	0.0163	0.0042	0.0002
2.10	1.0000	0.9998	0.9958	0.9655	0.8430	0.6190	0.1835	0.0257	0.0074	0.0004
2.15	1.0000	0.9999	0.9971	0.9748	0.8757	0.6766	0.2323	0.0389	0.0123	0.0009
2.20	1.0000	0.9999	0.9981	0.9817	0.9026	0.7289	0.2858	0.0566	0.0197	0.0017
2.25	1.0000	1.0000	0.9987	0.9868	0.9245	0.7753	0.3429	0.0793	0.0300	0.0032
2.30	1.0000	1.0000	0.9991	0.9905	0.9419	0.8157	0.4020	0.1073	0.0440	0.0055
2.35	1.0000	1.0000	0.9994	0.9933	0.9557	0.8504	0.4619	0.1408	0.0623	0.0091
2.40	1.0000	1.0000	0.9996	0.9952	0.9664	0.8797	0.5209	0.1796	0.0853	0.0144
2.45	1.0000	1.0000	0.9997	0.9966	0.9748	0.9041	0.5780	0.2232	0.1133	0.0219
2.50	1.0000	1.0000	0.9998	0.9976	0.9811	0.9242	0.6320	0.2709	0.1462	0.0322
2.55	1.0000	1.0000	0.9999	0.9983	0.9860	0.9405	0.6822	0.3218	0.1840	0.0457
2.60	1.0000	1.0000	0.9999	0.9988	0.9896	0.9536	0.7282	0.3748	0.2261	0.0630
2.65	1.0000	1.0000	1.0000	0.9992	0.9923	0.9641	0.7695	0.4289	0.2719	0.0843
2.70	1.0000	1.0000	1.0000	0.9994	0.9944	0.9723	0.8062	0.4829	0.3206	0.1099
2.75	1.0000	1.0000	1.0000	0.9996	0.9959	0.9788	0.8383	0.5359	0.3712	0.1397
2.80	1.0000	1.0000	1.0000	0.9997	0.9970	0.9838	0.8661	0.5870	0.4229	0.1738
2.85	1.0000	1.0000	1.0000	0.9998	0.9978	0.9877	0.8898	0.6354	0.4746	0.2116
2.90	1.0000	1.0000	1.0000	0.9999	0.9984	0.9907	0.9100	0.6807	0.5255	0.2529
2.95	1.0000	1.0000	1.0000	0.9999	0.9988	0.9930	0.9269	0.7225	0.5747	0.2968

PROBABILITY INTEGRAL OF t, THE NON-CENTRAL t-STATISTIC. THIS TABLE GIVES Pr $[t/\sqrt{f} \leq x]$.

f is the number of degrees of freedom; the non-centrality parameter is $\sqrt{f+1}\,K_p$.
K_p is the standardized normal deviate exceeded with probability p.

DEGREES OF FREEDOM **34**

x \ p	.2500	.1500	.1000	.0650	.0400	.0250	.0100	.0040	.0025	.0010
3.00	1.0000	1.0000	1.0000	0.9999	0.9992	0.9947	0.9409	0.7606	0.6218	0.3429
3.05	1.0000	1.0000	1.0000	1.0000	0.9994	0.9960	0.9525	0.7948	0.6660	0.3902
3.10	1.0000	1.0000	1.0000	1.0000	0.9996	0.9970	0.9620	0.8253	0.7072	0.4381
3.15	1.0000	1.0000	1.0000	1.0000	0.9997	0.9978	0.9697	0.8521	0.7449	0.4857
3.20	1.0000	1.0000	1.0000	1.0000	0.9998	0.9983	0.9760	0.8756	0.7793	0.5326
3.25	1.0000	1.0000	1.0000	1.0000	0.9998	0.9988	0.9810	0.8959	0.8101	0.5779
3.30	1.0000	1.0000	1.0000	1.0000	0.9999	0.9991	0.9850	0.9133	0.8376	0.6213
3.35	1.0000	1.0000	1.0000	1.0000	0.9999	0.9993	0.9883	0.9281	0.8618	0.6623
3.40	1.0000	1.0000	1.0000	1.0000	0.9999	0.9995	0.9908	0.9407	0.8830	0.7007
3.45	1.0000	1.0000	1.0000	1.0000	1.0000	0.9996	0.9928	0.9513	0.9014	0.7362
3.50	1.0000	1.0000	1.0000	1.0000	1.0000	0.9997	0.9944	0.9601	0.9173	0.7687
3.55	1.0000	1.0000	1.0000	1.0000	1.0000	0.9998	0.9956	0.9674	0.9310	0.7983
3.60	1.0000	1.0000	1.0000	1.0000	1.0000	0.9998	0.9966	0.9735	0.9426	0.8250
3.65	1.0000	1.0000	1.0000	1.0000	1.0000	0.9999	0.9974	0.9785	0.9524	0.8488
3.70	1.0000	1.0000	1.0000	1.0000	1.0000	0.9999	0.9980	0.9826	0.9607	0.8700
3.75	1.0000	1.0000	1.0000	1.0000	1.0000	0.9999	0.9984	0.9860	0.9676	0.8887
3.80	1.0000	1.0000	1.0000	1.0000	1.0000	1.0000	0.9988	0.9887	0.9734	0.9051
3.85	1.0000	1.0000	1.0000	1.0000	1.0000	1.0000	0.9991	0.9910	0.9782	0.9193
3.90	1.0000	1.0000	1.0000	1.0000	1.0000	1.0000	0.9993	0.9928	0.9822	0.9317
3.95	1.0000	1.0000	1.0000	1.0000	1.0000	1.0000	0.9994	0.9942	0.9855	0.9424
4.00	1.0000	1.0000	1.0000	1.0000	1.0000	1.0000	0.9996	0.9954	0.9882	0.9515
4.05	1.0000	1.0000	1.0000	1.0000	1.0000	1.0000	0.9997	0.9963	0.9904	0.9593
4.10	1.0000	1.0000	1.0000	1.0000	1.0000	1.0000	0.9997	0.9971	0.9923	0.9660
4.15	1.0000	1.0000	1.0000	1.0000	1.0000	1.0000	0.9998	0.9977	0.9937	0.9716
4.20	1.0000	1.0000	1.0000	1.0000	1.0000	1.0000	0.9998	0.9981	0.9949	0.9763
4.25	1.0000	1.0000	1.0000	1.0000	1.0000	1.0000	0.9999	0.9985	0.9959	0.9803
4.30	1.0000	1.0000	1.0000	1.0000	1.0000	1.0000	0.9999	0.9988	0.9967	0.9837
4.35	1.0000	1.0000	1.0000	1.0000	1.0000	1.0000	0.9999	0.9991	0.9974	0.9865
4.40	1.0000	1.0000	1.0000	1.0000	1.0000	1.0000	0.9999	0.9993	0.9979	0.9889
4.45	1.0000	1.0000	1.0000	1.0000	1.0000	1.0000	1.0000	0.9994	0.9983	0.9908
4.50	1.0000	1.0000	1.0000	1.0000	1.0000	1.0000	1.0000	0.9995	0.9986	0.9924
4.55	1.0000	1.0000	1.0000	1.0000	1.0000	1.0000	1.0000	0.9996	0.9989	0.9938
4.60	1.0000	1.0000	1.0000	1.0000	1.0000	1.0000	1.0000	0.9997	0.9991	0.9949
4.65	1.0000	1.0000	1.0000	1.0000	1.0000	1.0000	1.0000	0.9998	0.9993	0.9958
4.70	1.0000	1.0000	1.0000	1.0000	1.0000	1.0000	1.0000	0.9998	0.9994	0.9966
4.75	1.0000	1.0000	1.0000	1.0000	1.0000	1.0000	1.0000	0.9999	0.9995	0.9972
4.80	1.0000	1.0000	1.0000	1.0000	1.0000	1.0000	1.0000	0.9999	0.9996	0.9977
4.85	1.0000	1.0000	1.0000	1.0000	1.0000	1.0000	1.0000	0.9999	0.9997	0.9981
4.90	1.0000	1.0000	1.0000	1.0000	1.0000	1.0000	1.0000	0.9999	0.9998	0.9985
4.95	1.0000	1.0000	1.0000	1.0000	1.0000	1.0000	1.0000	0.9999	0.9998	0.9987
5.00	1.0000	1.0000	1.0000	1.0000	1.0000	1.0000	1.0000	1.0000	0.9998	0.9990
5.05	1.0000	1.0000	1.0000	1.0000	1.0000	1.0000	1.0000	1.0000	0.9999	0.9992
5.10	1.0000	1.0000	1.0000	1.0000	1.0000	1.0000	1.0000	1.0000	0.9999	0.9993
5.15	1.0000	1.0000	1.0000	1.0000	1.0000	1.0000	1.0000	1.0000	0.9999	0.9994
5.20	1.0000	1.0000	1.0000	1.0000	1.0000	1.0000	1.0000	1.0000	0.9999	0.9995
5.25	1.0000	1.0000	1.0000	1.0000	1.0000	1.0000	1.0000	1.0000	0.9999	0.9996
5.30	1.0000	1.0000	1.0000	1.0000	1.0000	1.0000	1.0000	1.0000	1.0000	0.9997
5.35	1.0000	1.0000	1.0000	1.0000	1.0000	1.0000	1.0000	1.0000	1.0000	0.9997
5.40	1.0000	1.0000	1.0000	1.0000	1.0000	1.0000	1.0000	1.0000	1.0000	0.9998
5.45	1.0000	1.0000	1.0000	1.0000	1.0000	1.0000	1.0000	1.0000	1.0000	0.9998
5.50	1.0000	1.0000	1.0000	1.0000	1.0000	1.0000	1.0000	1.0000	1.0000	0.9999
5.55	1.0000	1.0000	1.0000	1.0000	1.0000	1.0000	1.0000	1.0000	1.0000	0.9999
5.60	1.0000	1.0000	1.0000	1.0000	1.0000	1.0000	1.0000	1.0000	1.0000	0.9999
5.65	1.0000	1.0000	1.0000	1.0000	1.0000	1.0000	1.0000	1.0000	1.0000	0.9999
5.70	1.0000	1.0000	1.0000	1.0000	1.0000	1.0000	1.0000	1.0000	1.0000	0.9999
5.75	1.0000	1.0000	1.0000	1.0000	1.0000	1.0000	1.0000	1.0000	1.0000	0.9999
5.80	1.0000	1.0000	1.0000	1.0000	1.0000	1.0000	1.0000	1.0000	1.0000	1.0000

PROBABILITY INTEGRAL OF t, THE NON-CENTRAL t-STATISTIC. THIS TABLE GIVES Pr $[t/\sqrt{f} \leq x]$.

f is the number of degrees of freedom; the non-centrality parameter is $\sqrt{f+1}\,K_p$.

K_p is the standardized normal deviate exceeded with probability p.

DEGREES OF FREEDOM **39**

x \ p	.2500	.1500	.1000	.0650	.0400	.0250	.0100	.0040	.0025	.0010
0.05	0.0000	0.0000	0.0000	0.0000	0.0000	0.0000	0.0000	0.0000	0.0000	0.0000
0.10	0.0001	0.0000	0.0000	0.0000	0.0000	0.0000	0.0000	0.0000	0.0000	0.0000
0.15	0.0005	0.0000	0.0000	0.0000	0.0000	0.0000	0.0000	0.0000	0.0000	0.0000
0.20	0.0014	0.0000	0.0000	0.0000	0.0000	0.0000	0.0000	0.0000	0.0000	0.0000
0.25	0.0038	0.0000	0.0000	0.0000	0.0000	0.0000	0.0000	0.0000	0.0000	0.0000
0.30	0.0093	0.0000	0.0000	0.0000	0.0000	0.0000	0.0000	0.0000	0.0000	0.0000
0.35	0.0211	0.0000	0.0000	0.0000	0.0000	0.0000	0.0000	0.0000	0.0000	0.0000
0.40	0.0431	0.0000	0.0000	0.0000	0.0000	0.0000	0.0000	0.0000	0.0000	0.0000
0.45	0.0801	0.0002	0.0000	0.0000	0.0000	0.0000	0.0000	0.0000	0.0000	0.0000
0.50	0.1363	0.0006	0.0000	0.0000	0.0000	0.0000	0.0000	0.0000	0.0000	0.0000
0.55	0.2131	0.0017	0.0000	0.0000	0.0000	0.0000	0.0000	0.0000	0.0000	0.0000
0.60	0.3083	0.0046	0.0000	0.0000	0.0000	0.0000	0.0000	0.0000	0.0000	0.0000
0.65	0.4158	0.0111	0.0001	0.0000	0.0000	0.0000	0.0000	0.0000	0.0000	0.0000
0.70	0.5271	0.0239	0.0004	0.0000	0.0000	0.0000	0.0000	0.0000	0.0000	0.0000
0.75	0.6336	0.0467	0.0012	0.0000	0.0000	0.0000	0.0000	0.0000	0.0000	0.0000
0.80	0.7280	0.0832	0.0032	0.0000	0.0000	0.0000	0.0000	0.0000	0.0000	0.0000
0.85	0.8063	0.1361	0.0078	0.0001	0.0000	0.0000	0.0000	0.0000	0.0000	0.0000
0.90	0.8674	0.2060	0.0171	0.0004	0.0000	0.0000	0.0000	0.0000	0.0000	0.0000
0.95	0.9125	0.2909	0.0337	0.0012	0.0000	0.0000	0.0000	0.0000	0.0000	0.0000
1.00	0.9442	0.3862	0.0608	0.0031	0.0000	0.0000	0.0000	0.0000	0.0000	0.0000
1.05	0.9654	0.4857	0.1011	0.0073	0.0001	0.0000	0.0000	0.0000	0.0000	0.0000
1.10	0.9792	0.5830	0.1560	0.0155	0.0005	0.0000	0.0000	0.0000	0.0000	0.0000
1.15	0.9878	0.6727	0.2251	0.0298	0.0013	0.0000	0.0000	0.0000	0.0000	0.0000
1.20	0.9930	0.7510	0.3061	0.0528	0.0031	0.0001	0.0000	0.0000	0.0000	0.0000
1.25	0.9960	0.8160	0.3948	0.0868	0.0069	0.0003	0.0000	0.0000	0.0000	0.0000
1.30	0.9978	0.8678	0.4863	0.1332	0.0141	0.0008	0.0000	0.0000	0.0000	0.0000
1.35	0.9988	0.9074	0.5756	0.1921	0.0265	0.0021	0.0000	0.0000	0.0000	0.0000
1.40	0.9993	0.9366	0.6586	0.2622	0.0459	0.0046	0.0000	0.0000	0.0000	0.0000
1.45	0.9996	0.9575	0.7322	0.3407	0.0744	0.0095	0.0000	0.0000	0.0000	0.0000
1.50	0.9998	0.9720	0.7950	0.4241	0.1133	0.0180	0.0001	0.0000	0.0000	0.0000
1.55	0.9999	0.9819	0.8465	0.5082	0.1630	0.0317	0.0004	0.0000	0.0000	0.0000
1.60	0.9999	0.9885	0.8875	0.5894	0.2229	0.0522	0.0009	0.0000	0.0000	0.0000
1.65	1.0000	0.9927	0.9191	0.6645	0.2914	0.0810	0.0021	0.0000	0.0000	0.0000
1.70	1.0000	0.9955	0.9428	0.7316	0.3658	0.1190	0.0044	0.0001	0.0000	0.0000
1.75	1.0000	0.9972	0.9602	0.7895	0.4431	0.1665	0.0085	0.0001	0.0000	0.0000
1.80	1.0000	0.9983	0.9727	0.8379	0.5202	0.2227	0.0154	0.0004	0.0000	0.0000
1.85	1.0000	0.9990	0.9815	0.8773	0.5943	0.2864	0.0261	0.0008	0.0001	0.0000
1.90	1.0000	0.9994	0.9876	0.9086	0.6631	0.3553	0.0418	0.0018	0.0003	0.0000
1.95	1.0000	0.9996	0.9918	0.9329	0.7250	0.4271	0.0638	0.0036	0.0006	0.0000
2.00	1.0000	0.9998	0.9946	0.9514	0.7792	0.4991	0.0927	0.0067	0.0013	0.0000
2.05	1.0000	0.9999	0.9965	0.9652	0.8254	0.5692	0.1291	0.0118	0.0026	0.0001
2.10	1.0000	0.9999	0.9977	0.9754	0.8640	0.6352	0.1729	0.0197	0.0049	0.0002
2.15	1.0000	1.0000	0.9985	0.9827	0.8954	0.6957	0.2236	0.0312	0.0087	0.0005
2.20	1.0000	1.0000	0.9991	0.9880	0.9206	0.7498	0.2800	0.0473	0.0146	0.0010
2.25	1.0000	1.0000	0.9994	0.9917	0.9403	0.7970	0.3408	0.0687	0.0234	0.0019
2.30	1.0000	1.0000	0.9996	0.9943	0.9557	0.8375	0.4041	0.0960	0.0358	0.0035
2.35	1.0000	1.0000	0.9998	0.9961	0.9674	0.8714	0.4682	0.1295	0.0527	0.0061
2.40	1.0000	1.0000	0.9999	0.9974	0.9762	0.8993	0.5314	0.1691	0.0746	0.0103
2.45	1.0000	1.0000	0.9999	0.9982	0.9828	0.9220	0.5922	0.2144	0.1020	0.0165
2.50	1.0000	1.0000	0.9999	0.9988	0.9876	0.9402	0.6494	0.2646	0.1351	0.0253
2.55	1.0000	1.0000	1.0000	0.9992	0.9911	0.9545	0.7020	0.3186	0.1737	0.0374
2.60	1.0000	1.0000	1.0000	0.9995	0.9937	0.9657	0.7494	0.3753	0.2175	0.0533
2.65	1.0000	1.0000	1.0000	0.9997	0.9956	0.9743	0.7915	0.4332	0.2657	0.0737
2.70	1.0000	1.0000	1.0000	0.9998	0.9969	0.9809	0.8283	0.4911	0.3174	0.0987
2.75	1.0000	1.0000	1.0000	0.9998	0.9978	0.9859	0.8599	0.5477	0.3715	0.1286
2.80	1.0000	1.0000	1.0000	0.9999	0.9985	0.9896	0.8866	0.6021	0.4269	0.1633
2.85	1.0000	1.0000	1.0000	0.9999	0.9989	0.9924	0.9090	0.6533	0.4823	0.2026
2.90	1.0000	1.0000	1.0000	1.0000	0.9993	0.9945	0.9275	0.7007	0.5367	0.2458
2.95	1.0000	1.0000	1.0000	1.0000	0.9995	0.9960	0.9427	0.7439	0.5892	0.2923

PROBABILITY INTEGRAL OF t, THE NON-CENTRAL t-STATISTIC. THIS TABLE GIVES $Pr\ [t/\sqrt{f} \leq x]$.

f is the number of degrees of freedom; the non-centrality parameter is $\sqrt{f+1}\ K_p$.

K_p is the standardized normal deviate exceeded with probability p.

DEGREES OF FREEDOM **39**

x \ p	.2500	.1500	.1000	.0650	.0400	.0250	.0100	.0040	.0025	.0010
3.00	1.0000	1.0000	1.0000	1.0000	0.9996	0.9971	0.9550	0.7827	0.6390	0.3414
3.05	1.0000	1.0000	1.0000	1.0000	0.9998	0.9979	0.9648	0.8172	0.6855	0.3920
3.10	1.0000	1.0000	1.0000	1.0000	0.9998	0.9985	0.9727	0.8473	0.7282	0.4434
3.15	1.0000	1.0000	1.0000	1.0000	0.9999	0.9989	0.9789	0.8734	0.7670	0.4945
3.20	1.0000	1.0000	1.0000	1.0000	0.9999	0.9992	0.9838	0.8958	0.8017	0.5445
3.25	1.0000	1.0000	1.0000	1.0000	0.9999	0.9994	0.9876	0.9147	0.8324	0.5928
3.30	1.0000	1.0000	1.0000	1.0000	1.0000	0.9996	0.9905	0.9306	0.8594	0.6387
3.35	1.0000	1.0000	1.0000	1.0000	1.0000	0.9997	0.9928	0.9439	0.8827	0.6817
3.40	1.0000	1.0000	1.0000	1.0000	1.0000	0.9998	0.9946	0.9549	0.9028	0.7216
3.45	1.0000	1.0000	1.0000	1.0000	1.0000	0.9999	0.9959	0.9638	0.9199	0.7581
3.50	1.0000	1.0000	1.0000	1.0000	1.0000	0.9999	0.9969	0.9712	0.9343	0.7912
3.55	1.0000	1.0000	1.0000	1.0000	1.0000	0.9999	0.9977	0.9771	0.9464	0.8209
3.60	1.0000	1.0000	1.0000	1.0000	1.0000	0.9999	0.9983	0.9819	0.9565	0.8472
3.65	1.0000	1.0000	1.0000	1.0000	1.0000	1.0000	0.9987	0.9857	0.9648	0.8704
3.70	1.0000	1.0000	1.0000	1.0000	1.0000	1.0000	0.9990	0.9888	0.9717	0.8906
3.75	1.0000	1.0000	1.0000	1.0000	1.0000	1.0000	0.9993	0.9912	0.9773	0.9082
3.80	1.0000	1.0000	1.0000	1.0000	1.0000	1.0000	0.9995	0.9932	0.9818	0.9233
3.85	1.0000	1.0000	1.0000	1.0000	1.0000	1.0000	0.9996	0.9947	0.9855	0.9362
3.90	1.0000	1.0000	1.0000	1.0000	1.0000	1.0000	0.9997	0.9959	0.9885	0.9471
3.95	1.0000	1.0000	1.0000	1.0000	1.0000	1.0000	0.9998	0.9968	0.9909	0.9564
4.00	1.0000	1.0000	1.0000	1.0000	1.0000	1.0000	0.9998	0.9975	0.9928	0.9641
4.05	1.0000	1.0000	1.0000	1.0000	1.0000	1.0000	0.9999	0.9981	0.9943	0.9706
4.10	1.0000	1.0000	1.0000	1.0000	1.0000	1.0000	0.9999	0.9985	0.9955	0.9760
4.15	1.0000	1.0000	1.0000	1.0000	1.0000	1.0000	0.9999	0.9989	0.9965	0.9804
4.20	1.0000	1.0000	1.0000	1.0000	1.0000	1.0000	0.9999	0.9991	0.9972	0.9841
4.25	1.0000	1.0000	1.0000	1.0000	1.0000	1.0000	1.0000	0.9993	0.9978	0.9871
4.30	1.0000	1.0000	1.0000	1.0000	1.0000	1.0000	1.0000	0.9995	0.9983	0.9896
4.35	1.0000	1.0000	1.0000	1.0000	1.0000	1.0000	1.0000	0.9996	0.9987	0.9916
4.40	1.0000	1.0000	1.0000	1.0000	1.0000	1.0000	1.0000	0.9997	0.9990	0.9932
4.45	1.0000	1.0000	1.0000	1.0000	1.0000	1.0000	1.0000	0.9998	0.9992	0.9946
4.50	1.0000	1.0000	1.0000	1.0000	1.0000	1.0000	1.0000	0.9998	0.9994	0.9957
4.55	1.0000	1.0000	1.0000	1.0000	1.0000	1.0000	1.0000	0.9999	0.9995	0.9965
4.60	1.0000	1.0000	1.0000	1.0000	1.0000	1.0000	1.0000	0.9999	0.9996	0.9972
4.65	1.0000	1.0000	1.0000	1.0000	1.0000	1.0000	1.0000	0.9999	0.9997	0.9978
4.70	1.0000	1.0000	1.0000	1.0000	1.0000	1.0000	1.0000	0.9999	0.9998	0.9982
4.75	1.0000	1.0000	1.0000	1.0000	1.0000	1.0000	1.0000	0.9999	0.9998	0.9986
4.80	1.0000	1.0000	1.0000	1.0000	1.0000	1.0000	1.0000	1.0000	0.9999	0.9989
4.85	1.0000	1.0000	1.0000	1.0000	1.0000	1.0000	1.0000	1.0000	0.9999	0.9991
4.90	1.0000	1.0000	1.0000	1.0000	1.0000	1.0000	1.0000	1.0000	0.9999	0.9993
4.95	1.0000	1.0000	1.0000	1.0000	1.0000	1.0000	1.0000	1.0000	0.9999	0.9994
5.00	1.0000	1.0000	1.0000	1.0000	1.0000	1.0000	1.0000	1.0000	0.9999	0.9995
5.05	1.0000	1.0000	1.0000	1.0000	1.0000	1.0000	1.0000	1.0000	1.0000	0.9996
5.10	1.0000	1.0000	1.0000	1.0000	1.0000	1.0000	1.0000	1.0000	1.0000	0.9997
5.15	1.0000	1.0000	1.0000	1.0000	1.0000	1.0000	1.0000	1.0000	1.0000	0.9998
5.20	1.0000	1.0000	1.0000	1.0000	1.0000	1.0000	1.0000	1.0000	1.0000	0.9998
5.25	1.0000	1.0000	1.0000	1.0000	1.0000	1.0000	1.0000	1.0000	1.0000	0.9999
5.30	1.0000	1.0000	1.0000	1.0000	1.0000	1.0000	1.0000	1.0000	1.0000	0.9999
5.35	1.0000	1.0000	1.0000	1.0000	1.0000	1.0000	1.0000	1.0000	1.0000	0.9999
5.40	1.0000	1.0000	1.0000	1.0000	1.0000	1.0000	1.0000	1.0000	1.0000	0.9999
5.45	1.0000	1.0000	1.0000	1.0000	1.0000	1.0000	1.0000	1.0000	1.0000	0.9999
5.50	1.0000	1.0000	1.0000	1.0000	1.0000	1.0000	1.0000	1.0000	1.0000	1.0000

PROBABILITY INTEGRAL OF t, THE NON-CENTRAL t-STATISTIC. THIS TABLE GIVES Pr $[t/\sqrt{f} \leq x]$.

f is the number of degrees of freedom; the non-centrality parameter is $\sqrt{f+1}\,K_p$.
K_p is the standardized normal deviate exceeded with probability p.

DEGREES OF FREEDOM **44**

x \ p	.2500	.1500	.1000	.0650	.0400	.0250	.0100	.0040	.0025	.0010
0.10	0.0001	0.0000	0.0000	0.0000	0.0000	0.0000	0.0000	0.0000	0.0000	0.0000
0.15	0.0002	0.0000	0.0000	0.0000	0.0000	0.0000	0.0000	0.0000	0.0000	0.0000
0.20	0.0008	0.0000	0.0000	0.0000	0.0000	0.0000	0.0000	0.0000	0.0000	0.0000
0.25	0.0023	0.0000	0.0000	0.0000	0.0000	0.0000	0.0000	0.0000	0.0000	0.0000
0.30	0.0064	0.0000	0.0000	0.0000	0.0000	0.0000	0.0000	0.0000	0.0000	0.0000
0.35	0.0158	0.0000	0.0000	0.0000	0.0000	0.0000	0.0000	0.0000	0.0000	0.0000
0.40	0.0348	0.0000	0.0000	0.0000	0.0000	0.0000	0.0000	0.0000	0.0000	0.0000
0.45	0.0690	0.0001	0.0000	0.0000	0.0000	0.0000	0.0000	0.0000	0.0000	0.0000
0.50	0.1237	0.0003	0.0000	0.0000	0.0000	0.0000	0.0000	0.0000	0.0000	0.0000
0.55	0.2014	0.0010	0.0000	0.0000	0.0000	0.0000	0.0000	0.0000	0.0000	0.0000
0.60	0.3004	0.0030	0.0000	0.0000	0.0000	0.0000	0.0000	0.0000	0.0000	0.0000
0.65	0.4141	0.0078	0.0000	0.0000	0.0000	0.0000	0.0000	0.0000	0.0000	0.0000
0.70	0.5324	0.0183	0.0002	0.0000	0.0000	0.0000	0.0000	0.0000	0.0000	0.0000
0.75	0.6449	0.0384	0.0007	0.0000	0.0000	0.0000	0.0000	0.0000	0.0000	0.0000
0.80	0.7433	0.0725	0.0020	0.0000	0.0000	0.0000	0.0000	0.0000	0.0000	0.0000
0.85	0.8232	0.1243	0.0054	0.0001	0.0000	0.0000	0.0000	0.0000	0.0000	0.0000
0.90	0.8836	0.1952	0.0127	0.0002	0.0000	0.0000	0.0000	0.0000	0.0000	0.0000
0.95	0.9265	0.2835	0.0269	0.0007	0.0000	0.0000	0.0000	0.0000	0.0000	0.0000
1.00	0.9554	0.3841	0.0516	0.0020	0.0000	0.0000	0.0000	0.0000	0.0000	0.0000
1.05	0.9738	0.4898	0.0901	0.0050	0.0001	0.0000	0.0000	0.0000	0.0000	0.0000
1.10	0.9851	0.5930	0.1447	0.0114	0.0002	0.0000	0.0000	0.0000	0.0000	0.0000
1.15	0.9918	0.6871	0.2155	0.0236	0.0007	0.0000	0.0000	0.0000	0.0000	0.0000
1.20	0.9956	0.7680	0.3001	0.0443	0.0019	0.0000	0.0000	0.0000	0.0000	0.0000
1.25	0.9977	0.8337	0.3939	0.0765	0.0047	0.0001	0.0000	0.0000	0.0000	0.0000
1.30	0.9988	0.8845	0.4912	0.1221	0.0104	0.0005	0.0000	0.0000	0.0000	0.0000
1.35	0.9994	0.9221	0.5858	0.1819	0.0207	0.0012	0.0000	0.0000	0.0000	0.0000
1.40	0.9997	0.9489	0.6730	0.2545	0.0381	0.0030	0.0000	0.0000	0.0000	0.0000
1.45	0.9998	0.9672	0.7493	0.3372	0.0648	0.0067	0.0000	0.0000	0.0000	0.0000
1.50	0.9999	0.9795	0.8131	0.4256	0.1026	0.0136	0.0001	0.0000	0.0000	0.0000
1.55	1.0000	0.9874	0.8642	0.5150	0.1524	0.0253	0.0002	0.0000	0.0000	0.0000
1.60	1.0000	0.9924	0.9037	0.6009	0.2140	0.0439	0.0005	0.0000	0.0000	0.0000
1.65	1.0000	0.9955	0.9332	0.6797	0.2855	0.0711	0.0013	0.0000	0.0000	0.0000
1.70	1.0000	0.9973	0.9546	0.7491	0.3641	0.1084	0.0029	0.0000	0.0000	0.0000
1.75	1.0000	0.9985	0.9697	0.8079	0.4462	0.1561	0.0060	0.0001	0.0000	0.0000
1.80	1.0000	0.9991	0.9801	0.8560	0.5282	0.2140	0.0115	0.0002	0.0000	0.0000
1.85	1.0000	0.9995	0.9871	0.8942	0.6065	0.2805	0.0205	0.0005	0.0000	0.0000
1.90	1.0000	0.9997	0.9918	0.9237	0.6786	0.3533	0.0346	0.0011	0.0001	0.0000
1.95	1.0000	0.9998	0.9948	0.9459	0.7427	0.4295	0.0549	0.0023	0.0003	0.0000
2.00	1.0000	0.9999	0.9968	0.9622	0.7979	0.5061	0.0827	0.0046	0.0007	0.0000
2.05	1.0000	1.0000	0.9980	0.9740	0.8440	0.5803	0.1187	0.0086	0.0016	0.0000
2.10	1.0000	1.0000	0.9988	0.9823	0.8816	0.6497	0.1630	0.0151	0.0032	0.0001
2.15	1.0000	1.0000	0.9993	0.9881	0.9115	0.7127	0.2153	0.0251	0.0061	0.0002
2.20	1.0000	1.0000	0.9995	0.9921	0.9348	0.7682	0.2743	0.0396	0.0108	0.0005
2.25	1.0000	1.0000	0.9997	0.9948	0.9526	0.8160	0.3384	0.0597	0.0182	0.0011
2.30	1.0000	1.0000	0.9998	0.9966	0.9660	0.8560	0.4056	0.0860	0.0292	0.0022
2.35	1.0000	1.0000	0.9999	0.9978	0.9758	0.8889	0.4737	0.1192	0.0446	0.0042
2.40	1.0000	1.0000	0.9999	0.9986	0.9830	0.9153	0.5408	0.1593	0.0653	0.0074
2.45	1.0000	1.0000	1.0000	0.9991	0.9881	0.9363	0.6050	0.2059	0.0920	0.0124
2.50	1.0000	1.0000	1.0000	0.9994	0.9918	0.9526	0.6649	0.2583	0.1249	0.0199
2.55	1.0000	1.0000	1.0000	0.9996	0.9944	0.9650	0.7195	0.3152	0.1641	0.0306
2.60	1.0000	1.0000	1.0000	0.9998	0.9962	0.9745	0.7682	0.3752	0.2093	0.0453
2.65	1.0000	1.0000	1.0000	0.9999	0.9974	0.9815	0.8107	0.4367	0.2596	0.0645
2.70	1.0000	1.0000	1.0000	0.9999	0.9982	0.9867	0.8472	0.4983	0.3140	0.0888
2.75	1.0000	1.0000	1.0000	0.9999	0.9988	0.9905	0.8780	0.5583	0.3713	0.1185
2.80	1.0000	1.0000	1.0000	1.0000	0.9992	0.9933	0.9035	0.6157	0.4302	0.1536
2.85	1.0000	1.0000	1.0000	1.0000	0.9995	0.9953	0.9245	0.6693	0.4891	0.1939
2.90	1.0000	1.0000	1.0000	1.0000	0.9997	0.9967	0.9414	0.7184	0.5468	0.2388
2.95	1.0000	1.0000	1.0000	1.0000	0.9998	0.9977	0.9549	0.7628	0.6022	0.2877

PROBABILITY INTEGRAL OF t, THE NON-CENTRAL t-STATISTIC. THIS TABLE GIVES Pr $[t/\sqrt{f} \leq x]$.

f is the number of degrees of freedom; the non-centrality parameter is $\sqrt{f+1}\,K_p$.

K_p is the standardized normal deviate exceeded with probability p.

DEGREES OF FREEDOM **44**

x \ p	.2500	.1500	.1000	.0650	.0400	.0250	.0100	.0040	.0025	.0010
3.00	1.0000	1.0000	1.0000	1.0000	0.9998	0.9984	0.9655	0.8021	0.6545	0.3395
3.05	1.0000	1.0000	1.0000	1.0000	0.9999	0.9989	0.9738	0.8364	0.7028	0.3933
3.10	1.0000	1.0000	1.0000	1.0000	0.9999	0.9992	0.9803	0.8660	0.7468	0.4478
3.15	1.0000	1.0000	1.0000	1.0000	1.0000	0.9995	0.9852	0.8911	0.7862	0.5021
3.20	1.0000	1.0000	1.0000	1.0000	1.0000	0.9996	0.9890	0.9122	0.8211	0.5552
3.25	1.0000	1.0000	1.0000	1.0000	1.0000	0.9997	0.9918	0.9298	0.8515	0.6061
3.30	1.0000	1.0000	1.0000	1.0000	1.0000	0.9998	0.9940	0.9442	0.8777	0.6543
3.35	1.0000	1.0000	1.0000	1.0000	1.0000	0.9999	0.9956	0.9560	0.9000	0.6991
3.40	1.0000	1.0000	1.0000	1.0000	1.0000	0.9999	0.9968	0.9654	0.9188	0.7402
3.45	1.0000	1.0000	1.0000	1.0000	1.0000	0.9999	0.9976	0.9730	0.9345	0.7774
3.50	1.0000	1.0000	1.0000	1.0000	1.0000	1.0000	0.9983	0.9791	0.9475	0.8107
3.55	1.0000	1.0000	1.0000	1.0000	1.0000	1.0000	0.9987	0.9838	0.9582	0.8402
3.60	1.0000	1.0000	1.0000	1.0000	1.0000	1.0000	0.9991	0.9876	0.9669	0.8660
3.65	1.0000	1.0000	1.0000	1.0000	1.0000	1.0000	0.9993	0.9905	0.9739	0.8884
3.70	1.0000	1.0000	1.0000	1.0000	1.0000	1.0000	0.9995	0.9927	0.9795	0.9075
3.75	1.0000	1.0000	1.0000	1.0000	1.0000	1.0000	0.9997	0.9945	0.9840	0.9239
3.80	1.0000	1.0000	1.0000	1.0000	1.0000	1.0000	0.9998	0.9958	0.9875	0.9377
3.85	1.0000	1.0000	1.0000	1.0000	1.0000	1.0000	0.9998	0.9968	0.9903	0.9492
3.90	1.0000	1.0000	1.0000	1.0000	1.0000	1.0000	0.9999	0.9976	0.9925	0.9588
3.95	1.0000	1.0000	1.0000	1.0000	1.0000	1.0000	0.9999	0.9982	0.9942	0.9668
4.00	1.0000	1.0000	1.0000	1.0000	1.0000	1.0000	0.9999	0.9987	0.9956	0.9733
4.05	1.0000	1.0000	1.0000	1.0000	1.0000	1.0000	1.0000	0.9990	0.9966	0.9786
4.10	1.0000	1.0000	1.0000	1.0000	1.0000	1.0000	1.0000	0.9992	0.9974	0.9829
4.15	1.0000	1.0000	1.0000	1.0000	1.0000	1.0000	1.0000	0.9994	0.9980	0.9864
4.20	1.0000	1.0000	1.0000	1.0000	1.0000	1.0000	1.0000	0.9996	0.9985	0.9893
4.25	1.0000	1.0000	1.0000	1.0000	1.0000	1.0000	1.0000	0.9997	0.9989	0.9915
4.30	1.0000	1.0000	1.0000	1.0000	1.0000	1.0000	1.0000	0.9998	0.9991	0.9933
4.35	1.0000	1.0000	1.0000	1.0000	1.0000	1.0000	1.0000	0.9998	0.9993	0.9947
4.40	1.0000	1.0000	1.0000	1.0000	1.0000	1.0000	1.0000	0.9999	0.9995	0.9959
4.45	1.0000	1.0000	1.0000	1.0000	1.0000	1.0000	1.0000	0.9999	0.9996	0.9968
4.50	1.0000	1.0000	1.0000	1.0000	1.0000	1.0000	1.0000	0.9999	0.9997	0.9975
4.55	1.0000	1.0000	1.0000	1.0000	1.0000	1.0000	1.0000	0.9999	0.9998	0.9980
4.60	1.0000	1.0000	1.0000	1.0000	1.0000	1.0000	1.0000	1.0000	0.9998	0.9985
4.65	1.0000	1.0000	1.0000	1.0000	1.0000	1.0000	1.0000	1.0000	0.9999	0.9988
4.70	1.0000	1.0000	1.0000	1.0000	1.0000	1.0000	1.0000	1.0000	0.9999	0.9991
4.75	1.0000	1.0000	1.0000	1.0000	1.0000	1.0000	1.0000	1.0000	0.9999	0.9993
4.80	1.0000	1.0000	1.0000	1.0000	1.0000	1.0000	1.0000	1.0000	0.9999	0.9994
4.85	1.0000	1.0000	1.0000	1.0000	1.0000	1.0000	1.0000	1.0000	1.0000	0.9996
4.90	1.0000	1.0000	1.0000	1.0000	1.0000	1.0000	1.0000	1.0000	1.0000	0.9997
4.95	1.0000	1.0000	1.0000	1.0000	1.0000	1.0000	1.0000	1.0000	1.0000	0.9997
5.00	1.0000	1.0000	1.0000	1.0000	1.0000	1.0000	1.0000	1.0000	1.0000	0.9998
5.05	1.0000	1.0000	1.0000	1.0000	1.0000	1.0000	1.0000	1.0000	1.0000	0.9998
5.10	1.0000	1.0000	1.0000	1.0000	1.0000	1.0000	1.0000	1.0000	1.0000	0.9999
5.15	1.0000	1.0000	1.0000	1.0000	1.0000	1.0000	1.0000	1.0000	1.0000	0.9999
5.20	1.0000	1.0000	1.0000	1.0000	1.0000	1.0000	1.0000	1.0000	1.0000	0.9999
5.25	1.0000	1.0000	1.0000	1.0000	1.0000	1.0000	1.0000	1.0000	1.0000	0.9999
5.30	1.0000	1.0000	1.0000	1.0000	1.0000	1.0000	1.0000	1.0000	1.0000	1.0000

PROBABILITY INTEGRAL OF t, THE NON-CENTRAL t-STATISTIC. THIS TABLE GIVES Pr $[t/\sqrt{f} \leq x]$.

f is the number of degrees of freedom; the non-centrality parameter is $\sqrt{f+1}\, K_p$.
K_p is the standardized normal deviate exceeded with probability p.

DEGREES OF FREEDOM **49**

x \ p	.2500	.1500	.1000	.0650	.0400	.0250	.0100	.0040	.0025	.0010
0.10	0.0000	0.0000	0.0000	0.0000	0.0000	0.0000	0.0000	0.0000	0.0000	0.0000
0.15	0.0001	0.0000	0.0000	0.0000	0.0000	0.0000	0.0000	0.0000	0.0000	0.0000
0.20	0.0004	0.0000	0.0000	0.0000	0.0000	0.0000	0.0000	0.0000	0.0000	0.0000
0.25	0.0014	0.0000	0.0000	0.0000	0.0000	0.0000	0.0000	0.0000	0.0000	0.0000
0.30	0.0044	0.0000	0.0000	0.0000	0.0000	0.0000	0.0000	0.0000	0.0000	0.0000
0.35	0.0118	0.0000	0.0000	0.0000	0.0000	0.0000	0.0000	0.0000	0.0000	0.0000
0.40	0.0282	0.0000	0.0000	0.0000	0.0000	0.0000	0.0000	0.0000	0.0000	0.0000
0.45	0.0596	0.0000	0.0000	0.0000	0.0000	0.0000	0.0000	0.0000	0.0000	0.0000
0.50	0.1124	0.0001	0.0000	0.0000	0.0000	0.0000	0.0000	0.0000	0.0000	0.0000
0.55	0.1906	0.0006	0.0000	0.0000	0.0000	0.0000	0.0000	0.0000	0.0000	0.0000
0.60	0.2930	0.0019	0.0000	0.0000	0.0000	0.0000	0.0000	0.0000	0.0000	0.0000
0.65	0.4123	0.0055	0.0000	0.0000	0.0000	0.0000	0.0000	0.0000	0.0000	0.0000
0.70	0.5372	0.0141	0.0001	0.0000	0.0000	0.0000	0.0000	0.0000	0.0000	0.0000
0.75	0.6553	0.0317	0.0004	0.0000	0.0000	0.0000	0.0000	0.0000	0.0000	0.0000
0.80	0.7573	0.0633	0.0012	0.0000	0.0000	0.0000	0.0000	0.0000	0.0000	0.0000
0.85	0.8381	0.1136	0.0037	0.0000	0.0000	0.0000	0.0000	0.0000	0.0000	0.0000
0.90	0.8974	0.1851	0.0095	0.0001	0.0000	0.0000	0.0000	0.0000	0.0000	0.0000
0.95	0.9380	0.2763	0.0215	0.0004	0.0000	0.0000	0.0000	0.0000	0.0000	0.0000
1.00	0.9642	0.3818	0.0438	0.0012	0.0000	0.0000	0.0000	0.0000	0.0000	0.0000
1.05	0.9801	0.4934	0.0804	0.0034	0.0000	0.0000	0.0000	0.0000	0.0000	0.0000
1.10	0.9893	0.6021	0.1343	0.0085	0.0001	0.0000	0.0000	0.0000	0.0000	0.0000
1.15	0.9945	0.7003	0.2064	0.0187	0.0004	0.0000	0.0000	0.0000	0.0000	0.0000
1.20	0.9972	0.7833	0.2942	0.0373	0.0012	0.0000	0.0000	0.0000	0.0000	0.0000
1.25	0.9986	0.8492	0.3928	0.0675	0.0032	0.0001	0.0000	0.0000	0.0000	0.0000
1.30	0.9993	0.8988	0.4955	0.1121	0.0076	0.0002	0.0000	0.0000	0.0000	0.0000
1.35	0.9997	0.9343	0.5952	0.1723	0.0163	0.0007	0.0000	0.0000	0.0000	0.0000
1.40	0.9999	0.9586	0.6862	0.2472	0.0317	0.0020	0.0000	0.0000	0.0000	0.0000
1.45	0.9999	0.9747	0.7647	0.3336	0.0565	0.0047	0.0000	0.0000	0.0000	0.0000
1.50	1.0000	0.9849	0.8290	0.4267	0.0930	0.0103	0.0000	0.0000	0.0000	0.0000
1.55	1.0000	0.9912	0.8794	0.5211	0.1427	0.0203	0.0001	0.0000	0.0000	0.0000
1.60	1.0000	0.9949	0.9172	0.6114	0.2055	0.0371	0.0003	0.0000	0.0000	0.0000
1.65	1.0000	0.9971	0.9446	0.6935	0.2798	0.0626	0.0008	0.0000	0.0000	0.0000
1.70	1.0000	0.9984	0.9638	0.7648	0.3623	0.0988	0.0019	0.0000	0.0000	0.0000
1.75	1.0000	0.9991	0.9768	0.8242	0.4489	0.1466	0.0042	0.0000	0.0000	0.0000
1.80	1.0000	0.9995	0.9854	0.8717	0.5353	0.2058	0.0086	0.0001	0.0000	0.0000
1.85	1.0000	0.9997	0.9910	0.9084	0.6176	0.2748	0.0162	0.0003	0.0000	0.0000
1.90	1.0000	0.9999	0.9945	0.9360	0.6927	0.3512	0.0286	0.0006	0.0001	0.0000
1.95	1.0000	0.9999	0.9967	0.9562	0.7586	0.4315	0.0473	0.0015	0.0002	0.0000
2.00	1.0000	1.0000	0.9980	0.9705	0.8144	0.5124	0.0738	0.0031	0.0004	0.0000
2.05	1.0000	1.0000	0.9989	0.9804	0.8601	0.5904	0.1092	0.0062	0.0010	0.0000
2.10	1.0000	1.0000	0.9993	0.9872	0.8966	0.6630	0.1538	0.0116	0.0021	0.0000
2.15	1.0000	1.0000	0.9996	0.9918	0.9249	0.7281	0.2073	0.0202	0.0043	0.0001
2.20	1.0000	1.0000	0.9998	0.9947	0.9463	0.7847	0.2686	0.0332	0.0081	0.0003
2.25	1.0000	1.0000	0.9999	0.9967	0.9622	0.8326	0.3358	0.0519	0.0143	0.0007
2.30	1.0000	1.0000	0.9999	0.9979	0.9738	0.8720	0.4066	0.0772	0.0238	0.0014
2.35	1.0000	1.0000	1.0000	0.9987	0.9820	0.9036	0.4786	0.1099	0.0379	0.0028
2.40	1.0000	1.0000	1.0000	0.9992	0.9878	0.9285	0.5493	0.1502	0.0573	0.0053
2.45	1.0000	1.0000	1.0000	0.9995	0.9918	0.9477	0.6167	0.1978	0.0830	0.0094
2.50	1.0000	1.0000	1.0000	0.9997	0.9946	0.9622	0.6791	0.2521	0.1156	0.0157
2.55	1.0000	1.0000	1.0000	0.9998	0.9964	0.9730	0.7354	0.3116	0.1551	0.0252
2.60	1.0000	1.0000	1.0000	0.9999	0.9977	0.9809	0.7849	0.3748	0.2014	0.0385
2.65	1.0000	1.0000	1.0000	0.9999	0.9985	0.9866	0.8276	0.4397	0.2535	0.0566
2.70	1.0000	1.0000	1.0000	1.0000	0.9990	0.9907	0.8636	0.5047	0.3105	0.0800
2.75	1.0000	1.0000	1.0000	1.0000	0.9994	0.9936	0.8934	0.5679	0.3708	0.1093
2.80	1.0000	1.0000	1.0000	1.0000	0.9996	0.9956	0.9176	0.6280	0.4329	0.1445
2.85	1.0000	1.0000	1.0000	1.0000	0.9997	0.9970	0.9370	0.6837	0.4951	0.1856
2.90	1.0000	1.0000	1.0000	1.0000	0.9998	0.9980	0.9523	0.7344	0.5559	0.2321
2.95	1.0000	1.0000	1.0000	1.0000	0.9999	0.9987	0.9643	0.7796	0.6141	0.2830

PROBABILITY INTEGRAL OF t, THE NON-CENTRAL t-STATISTIC. THIS TABLE GIVES Pr $[t/\sqrt{f} \leq x]$.

f is the number of degrees of freedom; the non-centrality parameter is $\sqrt{f+1}\,K_p$.

K_p is the standardized normal deviate exceeded with probability p.

DEGREES OF FREEDOM **49**

x \ p	.2500	.1500	.1000	.0650	.0400	.0250	.0100	.0040	.0025	.0010
3.00	1.0000	1.0000	1.0000	1.0000	0.9999	0.9991	0.9734	0.8191	0.6685	0.3374
3.05	1.0000	1.0000	1.0000	1.0000	1.0000	0.9994	0.9804	0.8531	0.7184	0.3941
3.10	1.0000	1.0000	1.0000	1.0000	1.0000	0.9996	0.9857	0.8819	0.7634	0.4517
3.15	1.0000	1.0000	1.0000	1.0000	1.0000	0.9997	0.9896	0.9060	0.8033	0.5090
3.20	1.0000	1.0000	1.0000	1.0000	1.0000	0.9998	0.9925	0.9258	0.8381	0.5649
3.25	1.0000	1.0000	1.0000	1.0000	1.0000	0.9999	0.9946	0.9419	0.8679	0.6182
3.30	1.0000	1.0000	1.0000	1.0000	1.0000	0.9999	0.9961	0.9549	0.8932	0.6684
3.35	1.0000	1.0000	1.0000	1.0000	1.0000	0.9999	0.9972	0.9653	0.9144	0.7147
3.40	1.0000	1.0000	1.0000	1.0000	1.0000	1.0000	0.9981	0.9734	0.9319	0.7568
3.45	1.0000	1.0000	1.0000	1.0000	1.0000	1.0000	0.9986	0.9798	0.9463	0.7945
3.50	1.0000	1.0000	1.0000	1.0000	1.0000	1.0000	0.9990	0.9847	0.9579	0.8279
3.55	1.0000	1.0000	1.0000	1.0000	1.0000	1.0000	0.9993	0.9885	0.9672	0.8570
3.60	1.0000	1.0000	1.0000	1.0000	1.0000	1.0000	0.9995	0.9914	0.9746	0.8820
3.65	1.0000	1.0000	1.0000	1.0000	1.0000	1.0000	0.9997	0.9936	0.9805	0.9034
3.70	1.0000	1.0000	1.0000	1.0000	1.0000	1.0000	0.9998	0.9953	0.9851	0.9215
3.75	1.0000	1.0000	1.0000	1.0000	1.0000	1.0000	0.9998	0.9965	0.9886	0.9366
3.80	1.0000	1.0000	1.0000	1.0000	1.0000	1.0000	0.9999	0.9974	0.9914	0.9491
3.85	1.0000	1.0000	1.0000	1.0000	1.0000	1.0000	0.9999	0.9981	0.9935	0.9594
3.90	1.0000	1.0000	1.0000	1.0000	1.0000	1.0000	0.9999	0.9986	0.9951	0.9678
3.95	1.0000	1.0000	1.0000	1.0000	1.0000	1.0000	1.0000	0.9990	0.9963	0.9746
4.00	1.0000	1.0000	1.0000	1.0000	1.0000	1.0000	1.0000	0.9993	0.9973	0.9800
4.05	1.0000	1.0000	1.0000	1.0000	1.0000	1.0000	1.0000	0.9995	0.9980	0.9844
4.10	1.0000	1.0000	1.0000	1.0000	1.0000	1.0000	1.0000	0.9996	0.9985	0.9878
4.15	1.0000	1.0000	1.0000	1.0000	1.0000	1.0000	1.0000	0.9997	0.9989	0.9906
4.20	1.0000	1.0000	1.0000	1.0000	1.0000	1.0000	1.0000	0.9998	0.9992	0.9927
4.25	1.0000	1.0000	1.0000	1.0000	1.0000	1.0000	1.0000	0.9999	0.9994	0.9944
4.30	1.0000	1.0000	1.0000	1.0000	1.0000	1.0000	1.0000	0.9999	0.9995	0.9957
4.35	1.0000	1.0000	1.0000	1.0000	1.0000	1.0000	1.0000	0.9999	0.9997	0.9967
4.40	1.0000	1.0000	1.0000	1.0000	1.0000	1.0000	1.0000	0.9999	0.9998	0.9975
4.45	1.0000	1.0000	1.0000	1.0000	1.0000	1.0000	1.0000	1.0000	0.9998	0.9981
4.50	1.0000	1.0000	1.0000	1.0000	1.0000	1.0000	1.0000	1.0000	0.9999	0.9985
4.55	1.0000	1.0000	1.0000	1.0000	1.0000	1.0000	1.0000	1.0000	0.9999	0.9989
4.60	1.0000	1.0000	1.0000	1.0000	1.0000	1.0000	1.0000	1.0000	0.9999	0.9992
4.65	1.0000	1.0000	1.0000	1.0000	1.0000	1.0000	1.0000	1.0000	0.9999	0.9994
4.70	1.0000	1.0000	1.0000	1.0000	1.0000	1.0000	1.0000	1.0000	1.0000	0.9995
4.75	1.0000	1.0000	1.0000	1.0000	1.0000	1.0000	1.0000	1.0000	1.0000	0.9996
4.80	1.0000	1.0000	1.0000	1.0000	1.0000	1.0000	1.0000	1.0000	1.0000	0.9997
4.85	1.0000	1.0000	1.0000	1.0000	1.0000	1.0000	1.0000	1.0000	1.0000	0.9998
4.90	1.0000	1.0000	1.0000	1.0000	1.0000	1.0000	1.0000	1.0000	1.0000	0.9998
4.95	1.0000	1.0000	1.0000	1.0000	1.0000	1.0000	1.0000	1.0000	1.0000	0.9999
5.00	1.0000	1.0000	1.0000	1.0000	1.0000	1.0000	1.0000	1.0000	1.0000	0.9999
5.05	1.0000	1.0000	1.0000	1.0000	1.0000	1.0000	1.0000	1.0000	1.0000	0.9999
5.10	1.0000	1.0000	1.0000	1.0000	1.0000	1.0000	1.0000	1.0000	1.0000	0.9999
5.15	1.0000	1.0000	1.0000	1.0000	1.0000	1.0000	1.0000	1.0000	1.0000	1.0000

PERCENTAGE POINTS OF t, THE NON-CENTRAL t-STATISTIC

PERCENTAGE POINTS OF t, THE NON-CENTRAL t-STATISTIC. THE ENTRIES IN THE TABLE GIVE THE VALUES OF x SUCH THAT $Pr\ [t/\sqrt{f} > x] = \epsilon$.

f is the number of degrees of freedom; the non-centrality parameter is $\sqrt{f+1}\ K_p$.
K_p is the standardized normal deviate exceeded with probability p.

DEGREES OF FREEDOM **2**

p \ ε	.995	.99	.95	.90	.75	.50	.25	.10	.05
.2500	-2.175	-1.474	-.428	-.092	.378	.947	1.793	3.187	4.659
.1500	-.874	-.515	.116	.378	.827	1.474	2.538	4.371	6.337
.1000	-.328	-.087	.410	.655	1.117	1.835	3.063	5.215	7.542
.0650	.033	.216	.655	.898	1.386	2.180	3.571	6.037	8.708
.0400	.319	.469	.882	1.132	1.654	2.531	4.095	6.885	9.921
.0250	.521	.661	1.072	1.331	1.888	2.842	4.562	7.646	11.006
.0100	.823	.956	1.384	1.667	2.291	3.386	5.384	8.990	12.924
.0040	1.057	1.196	1.648	1.956	2.645	3.868	6.118	10.193	14.644
.0025	1.161	1.305	1.771	2.091	2.812	4.098	6.469	10.768	15.466
.0010	1.350	1.498	1.992	2.336	3.116	4.517	7.110	11.820	16.971

DEGREES OF FREEDOM **3**

p \ ε	.995	.99	.95	.90	.75	.50	.25	.10	.05
.2500	-1.080	-.774	-.194	.042	.408	.853	1.450	2.277	3.024
.1500	-.356	-.168	.253	.458	.820	1.320	2.046	3.099	4.064
.1000	-.008	.142	.513	.712	1.089	1.638	2.464	3.680	4.807
.0650	.257	.387	.738	.941	1.340	1.942	2.867	4.247	5.529
.0400	.487	.606	.953	1.164	1.590	2.251	3.281	4.831	6.277
.0250	.667	.783	1.134	1.355	1.810	2.524	3.651	5.354	6.945
.0100	.947	1.067	1.439	1.680	2.189	3.003	4.302	6.278	8.131
.0040	1.179	1.302	1.700	1.963	2.522	3.429	4.884	7.108	9.194
.0025	1.284	1.410	1.822	2.095	2.680	3.631	5.162	7.506	9.707
.0010	1.469	1.602	2.041	2.335	2.967	4.000	5.670	8.231	10.641

DEGREES OF FREEDOM **4**

p \ ε	.995	.99	.95	.90	.75	.50	.25	.10	.05
.2500	-.700	-.500	-.074	.118	.429	.807	1.288	1.898	2.403
.1500	-.142	-.004	.335	.510	.824	1.246	1.821	2.583	3.226
.1000	.146	.266	.580	.755	1.083	1.545	2.193	3.066	3.809
.0650	.381	.490	.797	.977	1.325	1.829	2.551	3.534	4.376
.0400	.592	.698	1.007	1.195	1.568	2.119	2.919	4.018	4.961
.0250	.766	.869	1.186	1.383	1.780	2.375	3.247	4.451	5.487
.0100	1.038	1.148	1.488	1.704	2.148	2.824	3.825	5.217	6.419
.0040	1.269	1.385	1.749	1.984	2.472	3.223	4.342	5.904	7.254
.0025	1.374	1.492	1.870	2.116	2.626	3.412	4.588	6.231	7.655
.0010	1.564	1.688	2.092	2.355	2.906	3.759	5.039	6.833	8.386

DEGREES OF FREEDOM **5**

p \ ε	.995	.99	.95	.90	.75	.50	.25	.10	.05	.01	.005
.2500	-.495	-.346	.003	.170	.445	.780	1.191	1.687	2.076	3.121	3.650
.1500	-.018	.097	.392	.549	.830	1.203	1.690	2.300	2.790	4.119	4.800
.1000	.247	.348	.630	.788	1.084	1.490	2.037	2.732	3.293	4.829	5.625
.0650	.467	.564	.843	1.007	1.321	1.764	2.370	3.149	3.782	5.525	6.418
.0400	.669	.765	1.050	1.222	1.559	2.042	2.712	3.580	4.289	6.250	7.250
.0250	.838	.934	1.228	1.409	1.768	2.288	3.017	3.966	4.743	6.887	8.000
.0100	1.111	1.213	1.530	1.729	2.130	2.720	3.553	4.647	5.545	8.034	9.350
.0040	1.341	1.450	1.791	2.008	2.449	3.103	4.033	5.258	6.265	9.067	10.525
.0025	1.450	1.561	1.914	2.140	2.600	3.286	4.262	5.550	6.610	9.550	11.100
.0010	1.640	1.760	2.136	2.379	2.876	3.619	4.681	6.086	7.243	10.450	12.150

PERCENTAGE POINTS OF t, THE NON-CENTRAL t-STATISTIC. THE ENTRIES IN THE TABLE GIVE THE VALUES OF x SUCH THAT $Pr\ [t/\sqrt{f}>x]=\epsilon$.

f is the number of degrees of freedom; the non-centrality parameter is $\sqrt{f+1}\,K_p$.
K_p is the standardized normal deviate exceeded with probability p.

DEGREES OF FREEDOM 6

p \ ε	.995	.99	.95	.90	.75	.50	.25	.10	.05	.01	.005
.2500	-.370	-.245	.059	.208	.459	.762	1.127	1.550	1.871	2.689	3.087
.1500	.069	.170	.435	.579	.837	1.174	1.603	2.120	2.520	3.557	4.066
.1000	.320	.411	.668	.815	1.087	1.454	1.934	2.520	2.976	4.165	4.767
.0650	.530	.620	.879	1.031	1.321	1.721	2.251	2.906	3.419	4.767	5.430
.0400	.729	.820	1.086	1.246	1.556	1.992	2.576	3.305	3.877	5.388	6.150
.0250	.895	.987	1.263	1.432	1.763	2.232	2.866	3.660	4.287	5.940	6.775
.0100	1.169	1.267	1.565	1.752	2.121	2.652	3.377	4.290	5.013	6.925	7.868
.0040	1.403	1.505	1.828	2.031	2.437	3.025	3.833	4.855	5.665	7.813	8.900
.0025	1.511	1.617	1.952	2.163	2.587	3.203	4.050	5.124	5.977	8.233	9.375
.0010	1.703	1.818	2.175	2.403	2.860	3.527	4.448	5.618	6.547	9.017	10.250

DEGREES OF FREEDOM 7

p \ ε	.995	.99	.95	.90	.75	.50	.25	.10	.05	.01	.005
.2500	-.280	-.170	.101	.238	.470	.749	1.080	1.454	1.729	2.408	2.730
.1500	.134	.224	.470	.603	.844	1.154	1.541	1.994	2.335	3.188	3.600
.1000	.376	.460	.700	.837	1.091	1.429	1.860	2.372	2.760	3.737	4.213
.0650	.583	.666	.910	1.053	1.323	1.690	2.166	2.737	3.173	4.275	4.800
.0400	.779	.864	1.115	1.266	1.556	1.956	2.480	3.114	3.599	4.834	5.434
.0250	.945	1.030	1.293	1.452	1.761	2.192	2.760	3.449	3.980	5.330	5.983
.0100	1.218	1.312	1.596	1.772	2.117	2.604	3.252	4.044	4.654	6.212	6.975
.0040	1.453	1.553	1.860	2.052	2.432	2.970	3.691	4.576	5.260	7.011	7.850
.0025	1.561	1.665	1.984	2.184	2.580	3.144	3.901	4.830	5.550	7.388	8.268
.0010	1.759	1.868	2.210	2.425	2.852	3.463	4.285	5.297	6.079	8.088	9.050

DEGREES OF FREEDOM 8

p \ ε	.995	.99	.95	.90	.75	.50	.25	.10	.05	.01	.005
.2500	-.212	-.113	.135	.263	.479	.740	1.044	1.381	1.625	2.211	2.483
.1500	.186	.270	.498	.624	.850	1.139	1.494	1.900	2.199	2.930	3.269
.1000	.421	.501	.727	.857	1.095	1.410	1.805	2.262	2.603	3.439	3.830
.0650	.626	.705	.935	1.071	1.326	1.668	2.102	2.612	2.993	3.930	4.373
.0400	.820	.901	1.141	1.284	1.558	1.930	2.408	2.972	3.396	4.443	4.937
.0250	.985	1.069	1.319	1.470	1.762	2.162	2.680	3.294	3.757	4.900	5.450
.0100	1.262	1.350	1.623	1.791	2.116	2.568	3.158	3.862	4.394	5.716	6.350
.0040	1.497	1.593	1.889	2.071	2.429	2.929	3.585	4.372	4.967	6.450	7.150
.0025	1.608	1.706	2.013	2.204	2.578	3.101	3.789	4.615	5.241	6.800	7.533
.0010	1.806	1.912	2.240	2.445	2.848	3.415	4.162	5.061	5.742	7.440	8.250

DEGREES OF FREEDOM 9

p \ ε	.995	.99	.95	.90	.75	.50	.25	.10	.05	.01	.005
.2500	-.158	-.068	.164	.283	.487	.732	1.016	1.325	1.544	2.060	2.293
.1500	.229	.307	.522	.642	.855	1.128	1.457	1.827	2.095	2.737	3.030
.1000	.459	.535	.750	.873	1.100	1.396	1.761	2.177	2.482	3.213	3.550
.0650	.663	.737	.958	1.087	1.329	1.650	2.053	2.515	2.856	3.678	4.050
.0400	.856	.934	1.164	1.301	1.560	1.910	2.351	2.863	3.241	4.156	4.572
.0250	1.020	1.101	1.342	1.486	1.763	2.139	2.617	3.174	3.586	4.587	5.050
.0100	1.298	1.385	1.647	1.807	2.116	2.541	3.085	3.723	4.196	5.350	5.883
.0040	1.536	1.630	1.914	2.089	2.429	2.898	3.503	4.214	4.744	6.034	6.634
.0025	1.647	1.744	2.039	2.222	2.577	3.068	3.702	4.448	5.005	6.360	7.000
.0010	1.847	1.950	2.267	2.464	2.846	3.378	4.067	4.879	5.484	6.960	7.650

PERCENTAGE POINTS OF t, THE NON-CENTRAL t-STATISTIC. THE ENTRIES IN THE TABLE GIVE THE VALUES OF x SUCH THAT Pr $[t/\sqrt{f} > x] = \epsilon$.

f is the number of degrees of freedom; the non-centrality parameter is $\sqrt{f+1}\,K_p$.
K_p is the standardized normal deviate exceeded with probability p

DEGREES OF FREEDOM **10**

p \ ε	.995	.99	.95	.90	.75	.50	.25	.10	.05	.01	.005
.2500	-.114	-.028	.187	.301	.494	.727	.993	1.279	1.480	1.944	2.150
.1500	.265	.338	.543	.657	.861	1.118	1.427	1.768	2.013	2.585	2.842
.1000	.492	.564	.770	.888	1.104	1.384	1.726	2.110	2.386	3.038	3.332
.0650	.693	.766	.978	1.102	1.332	1.636	2.012	2.438	2.747	3.480	3.809
.0400	.888	.962	1.183	1.315	1.563	1.893	2.306	2.777	3.119	3.935	4.300
.0250	1.052	1.130	1.362	1.501	1.765	2.121	2.567	3.079	3.452	4.345	4.750
.0100	1.332	1.415	1.668	1.822	2.118	2.519	3.026	3.612	4.040	5.064	5.534
.0040	1.571	1.661	1.936	2.105	2.429	2.873	3.437	4.089	4.568	5.714	6.237
.0025	1.684	1.777	2.063	2.238	2.577	3.041	3.632	4.316	4.820	6.025	6.583
.0010	1.886	1.985	2.292	2.481	2.846	3.349	3.990	4.735	5.282	6.591	7.200

DEGREES OF FREEDOM **11**

p \ ε	.995	.99	.95	.90	.75	.50	.25	.10	.05	.01	.005
.2500	-.077	.004	.208	.316	.500	.722	.974	1.241	1.427	1.850	2.034
.1500	.296	.365	.561	.670	.865	1.111	1.402	1.720	1.945	2.464	2.694
.1000	.520	.590	.788	.901	1.108	1.375	1.696	2.053	2.308	2.896	3.158
.0650	.721	.791	.995	1.114	1.336	1.625	1.979	2.375	2.659	3.320	3.619
.0400	.914	.988	1.201	1.328	1.566	1.880	2.268	2.705	3.020	3.759	4.090
.0250	1.079	1.156	1.380	1.514	1.768	2.106	2.525	3.000	3.343	4.145	4.509
.0100	1.361	1.441	1.688	1.836	2.120	2.501	2.978	3.521	3.914	4.839	5.250
.0040	1.602	1.689	1.957	2.119	2.431	2.852	3.382	3.986	4.425	5.462	5.923
.0025	1.715	1.806	2.084	2.253	2.578	3.019	3.574	4.209	4.670	5.757	6.250
.0010	1.919	2.017	2.314	2.496	2.847	3.325	3.927	4.617	5.118	6.300	6.833

DEGREES OF FREEDOM **12**

p \ ε	.995	.99	.95	.90	.75	.50	.25	.10	.05	.01	.005
.2500	-.044	.031	.226	.329	.506	.718	.957	1.209	1.383	1.772	1.940
.1500	.322	.389	.577	.682	.870	1.105	1.381	1.680	1.890	2.368	2.575
.1000	.545	.612	.803	.913	1.112	1.367	1.672	2.007	2.243	2.786	3.027
.0650	.745	.814	1.011	1.126	1.339	1.616	1.951	2.322	2.585	3.192	3.458
.0400	.940	1.010	1.217	1.340	1.569	1.869	2.237	2.646	2.937	3.613	3.910
.0250	1.106	1.178	1.397	1.526	1.770	2.094	2.491	2.935	3.253	3.991	4.320
.0100	1.387	1.466	1.705	1.849	2.122	2.486	2.937	3.445	3.808	4.655	5.029
.0040	1.631	1.715	1.975	2.132	2.433	2.835	3.336	3.901	4.307	5.250	5.675
.0025	1.745	1.833	2.103	2.267	2.580	3.001	3.526	4.119	4.545	5.537	5.988
.0010	1.950	2.044	2.335	2.511	2.849	3.305	3.874	4.518	4.982	6.064	6.550

DEGREES OF FREEDOM **13**

p \ ε	.995	.99	.95	.90	.75	.50	.25	.10	.05	.01	.005
.2500	-.015	.055	.242	.341	.511	.715	.943	1.182	1.345	1.707	1.862
.1500	.346	.411	.592	.693	.874	1.099	1.362	1.645	1.842	2.285	2.476
.1000	.568	.631	.817	.923	1.115	1.360	1.651	1.967	2.188	2.690	2.911
.0650	.766	.833	1.025	1.137	1.342	1.608	1.927	2.276	2.523	3.085	3.330
.0400	.962	1.030	1.232	1.351	1.572	1.860	2.210	2.595	2.868	3.492	3.766
.0250	1.128	1.200	1.412	1.537	1.773	2.083	2.461	2.879	3.176	3.854	4.158
.0100	1.411	1.489	1.721	1.861	2.124	2.474	2.902	3.380	3.720	4.500	4.850
.0040	1.657	1.738	1.992	2.145	2.435	2.821	3.297	3.829	4.207	5.079	5.462
.0025	1.769	1.857	2.120	2.279	2.582	2.986	3.485	4.042	4.441	5.356	5.762
.0010	1.978	2.071	2.353	2.524	2.851	3.288	3.829	4.435	4.867	5.864	6.313

PERCENTAGE POINTS OF t, THE NON-CENTRAL t-STATISTIC. THE ENTRIES IN THE TABLE GIVE THE VALUES OF x SUCH THAT Pr $[t/\sqrt{f} > x] = \epsilon$.

f is the number of degrees of freedom; the non-centrality parameter is $\sqrt{f+1}\,K_p$.
K_p is the standardized normal deviate exceeded with probability p.

DEGREES OF FREEDOM **14**

p \ ε	.995	.99	.95	.90	.75	.50	.25	.10	.05	.01	.005
.2500	.009	.076	.256	.352	.516	.712	.931	1.158	1.312	1.652	1.796
.1500	.365	.430	.605	.703	.878	1.095	1.347	1.615	1.801	2.214	2.395
.1000	.587	.652	.830	.933	1.119	1.354	1.632	1.932	2.141	2.609	2.812
.0650	.788	.852	1.038	1.146	1.345	1.601	1.906	2.238	2.469	2.993	3.221
.0400	.983	1.049	1.245	1.360	1.574	1.852	2.186	2.552	2.808	3.389	3.642
.0250	1.150	1.219	1.425	1.547	1.776	2.074	2.435	2.831	3.110	3.742	4.016
.0100	1.432	1.508	1.736	1.871	2.127	2.463	2.873	3.325	3.644	4.372	4.690
.0040	1.678	1.762	2.008	2.156	2.437	2.809	3.263	3.766	4.122	4.934	5.290
.0025	1.795	1.879	2.137	2.291	2.584	2.973	3.449	3.977	4.350	5.206	5.574
.0010	2.005	2.094	2.370	2.537	2.853	3.274	3.790	4.363	4.769	5.700	6.100

DEGREES OF FREEDOM **15**

p \ ε	.995	.99	.95	.90	.75	.50	.25	.10	.05	.01	.005
.2500	.031	.096	.269	.361	.520	.709	.920	1.137	1.283	1.604	1.737
.1500	.388	.447	.617	.712	.881	1.091	1.333	1.589	1.764	2.154	2.318
.1000	.606	.668	.842	.942	1.122	1.350	1.617	1.902	2.099	2.539	2.730
.0650	.807	.868	1.050	1.155	1.348	1.595	1.888	2.204	2.423	2.915	3.124
.0400	1.002	1.065	1.257	1.369	1.577	1.845	2.166	2.513	2.755	3.300	3.536
.0250	1.168	1.236	1.438	1.557	1.778	2.067	2.412	2.789	3.053	3.646	3.900
.0100	1.454	1.527	1.749	1.881	2.129	2.454	2.846	3.276	3.577	4.259	4.550
.0040	1.702	1.781	2.022	2.167	2.440	2.798	3.233	3.712	4.047	4.809	5.140
.0025	1.817	1.900	2.152	2.302	2.587	2.962	3.418	3.919	4.272	5.069	5.421
.0010	2.029	2.117	2.386	2.548	2.856	3.261	3.756	4.301	4.683	5.550	5.929

DEGREES OF FREEDOM **16**

p \ ε	.995	.99	.95	.90	.75	.50	.25	.10	.05	.01	.005
.2500	.052	.113	.280	.370	.524	.707	.910	1.118	1.258	1.561	1.689
.1500	.405	.462	.628	.720	.884	1.087	1.321	1.566	1.733	2.100	2.258
.1000	.622	.682	.853	.950	1.125	1.345	1.602	1.876	2.063	2.479	2.655
.0650	.823	.883	1.061	1.164	1.351	1.590	1.872	2.173	2.382	2.844	3.044
.0400	1.017	1.081	1.269	1.378	1.580	1.839	2.147	2.480	2.710	3.224	3.443
.0250	1.186	1.253	1.450	1.566	1.781	2.060	2.392	2.753	3.003	3.563	3.800
.0100	1.474	1.545	1.762	1.891	2.132	2.446	2.823	3.233	3.520	4.162	4.434
.0040	1.720	1.800	2.036	2.177	2.442	2.789	3.207	3.664	3.982	4.700	5.010
.0025	1.837	1.920	2.165	2.312	2.589	2.952	3.390	3.869	4.204	4.955	5.280
.0010	2.050	2.137	2.401	2.559	2.858	3.250	3.726	4.245	4.609	5.430	5.778

DEGREES OF FREEDOM **17**

p \ ε	.995	.99	.95	.90	.75	.50	.25	.10	.05	.01	.005
.2500	.069	.127	.291	.378	.527	.705	.901	1.102	1.235	1.524	1.643
.1500	.419	.476	.637	.728	.887	1.084	1.310	1.545	1.704	2.054	2.200
.1000	.637	.697	.863	.957	1.128	1.341	1.589	1.852	2.031	2.425	2.595
.0650	.837	.898	1.071	1.171	1.354	1.586	1.857	2.147	2.345	2.786	2.970
.0400	1.033	1.096	1.279	1.386	1.582	1.834	2.131	2.450	2.669	3.156	3.361
.0250	1.204	1.267	1.460	1.574	1.784	2.054	2.375	2.720	2.958	3.486	3.713
.0100	1.490	1.561	1.773	1.899	2.134	2.439	2.802	3.196	3.468	4.076	4.335
.0040	1.739	1.817	2.049	2.186	2.444	2.781	3.184	3.621	3.925	4.604	4.891
.0025	1.857	1.938	2.178	2.322	2.592	2.943	3.366	3.824	4.143	4.854	5.160
.0010	2.069	2.157	2.415	2.570	2.860	3.241	3.699	4.196	4.543	5.317	5.650

PERCENTAGE POINTS OF t, THE NON-CENTRAL t-STATISTIC. THE ENTRIES IN THE TABLE GIVE THE VALUES OF x SUCH THAT Pr $[t/\sqrt{f} > x] = \epsilon$.

f is the number of degrees of freedom; the non-centrality parameter is $\sqrt{f+1}\,K_p$.
K_p is the standardized normal deviate exceeded with probability p.

DEGREES OF FREEDOM **18**

p \ ε	.995	.99	.95	.90	.75	.50	.25	.10	.05	.01	.005
.2500	.083	.144	.301	.385	.531	.703	.894	1.087	1.215	1.490	1.603
.1500	.431	.488	.647	.734	.890	1.082	1.300	1.526	1.679	2.012	2.150
.1000	.655	.710	.872	.964	1.130	1.338	1.578	1.830	2.002	2.378	2.536
.0650	.853	.911	1.080	1.178	1.356	1.582	1.844	2.123	2.313	2.732	2.909
.0400	1.050	1.109	1.289	1.393	1.585	1.829	2.117	2.423	2.633	3.097	3.294
.0250	1.219	1.281	1.470	1.581	1.786	2.049	2.359	2.690	2.918	3.420	3.637
.0100	1.506	1.576	1.784	1.907	2.136	2.432	2.784	3.162	3.422	4.000	4.243
.0040	1.758	1.832	2.060	2.195	2.447	2.774	3.163	3.583	3.873	4.516	4.793
.0025	1.875	1.954	2.190	2.331	2.594	2.936	3.344	3.784	4.089	4.766	5.050
.0010	2.088	2.174	2.428	2.579	2.863	3.232	3.675	4.152	4.484	5.217	5.533

DEGREES OF FREEDOM **19**

p \ ε	.995	.99	.95	.90	.75	.50	.25	.10	.05	.01	.005
.2500	.102	.156	.310	.392	.534	.702	.887	1.073	1.197	1.461	1.570
.1500	.448	.502	.656	.741	.893	1.079	1.291	1.509	1.657	1.975	2.107
.1000	.665	.721	.880	.971	1.133	1.335	1.568	1.811	1.976	2.334	2.487
.0650	.866	.923	1.089	1.185	1.359	1.578	1.833	2.101	2.284	2.682	2.850
.0400	1.063	1.122	1.298	1.400	1.587	1.825	2.104	2.398	2.600	3.042	3.229
.0250	1.231	1.295	1.480	1.588	1.788	2.044	2.344	2.664	2.882	3.364	3.568
.0100	1.522	1.590	1.795	1.915	2.139	2.427	2.767	3.131	3.381	3.932	4.163
.0040	1.774	1.849	2.071	2.203	2.449	2.767	3.144	3.549	3.826	4.442	.4.700
.0025	1.891	1.970	2.202	2.340	2.597	2.929	3.324	3.748	4.040	4.684	4.958
.0010	2.107	2.190	2.440	2.588	2.865	3.225	3.653	4.113	4.431	5.133	5.434

DEGREES OF FREEDOM **20**

p \ ε	.995	.99	.95	.90	.75	.50	.25	.10	.05	.01	.005
.2500	.114	.167	.318	.398	.537	.700	.880	1.061	1.180	1.432	1.537
.1500	.460	.513	.663	.747	.896	1.077	1.282	1.494	1.636	1.940	2.068
.1000	.677	.732	.888	.977	1.135	1.332	1.558	1.793	1.952	2.296	2.442
.0650	.875	.935	1.098	1.191	1.361	1.575	1.822	2.081	2.257	2.640	2.800
.0400	1.074	1.134	1.307	1.407	1.589	1.821	2.092	2.377	2.571	2.995	3.171
.0250	1.244	1.307	1.489	1.595	1.791	2.040	2.331	2.639	2.850	3.311	3.505
.0100	1.536	1.604	1.804	1.923	2.141	2.422	2.752	3.103	3.343	3.869	4.093
.0040	1.789	1.863	2.081	2.211	2.452	2.761	3.127	3.517	3.785	4.373	4.621
.0025	1.909	1.984	2.212	2.348	2.599	2.923	3.306	3.715	3.996	4.611	4.875
.0010	2.123	2.206	2.451	2.597	2.868	3.218	3.633	4.077	4.382	5.054	5.334

DEGREES OF FREEDOM **21**

p \ ε	.995	.99	.95	.90	.75	.50	.25	.10	.05	.01	.005
.2500	.125	.178	.325	.404	.539	.699	.874	1.049	1.165	1.408	1.508
.1500	.471	.523	.671	.753	.898	1.075	1.275	1.480	1.617	1.911	2.031
.1000	.691	.742	.896	.983	1.138	1.330	1.550	1.777	1.931	2.260	2.396
.0650	.888	.945	1.105	1.197	1.363	1.572	1.812	2.063	2.233	2.600	2.754
.0400	1.086	1.145	1.314	1.413	1.592	1.818	2.081	2.356	2.544	2.950	3.119
.0250	1.258	1.318	1.497	1.602	1.793	2.036	2.319	2.617	2.821	3.263	3.445
.0100	1.550	1.616	1.813	1.929	2.143	2.417	2.738	3.078	3.309	3.815	4.027
.0040	1.804	1.875	2.091	2.218	2.454	2.756	3.111	3.489	3.747	4.310	4.550
.0025	1.924	1.999	2.223	2.355	2.601	2.917	3.290	3.685	3.956	4.546	4.793
.0010	2.140	2.221	2.462	2.605	2.870	3.212	3.615	4.045	4.338	4.980	5.250

PERCENTAGE POINTS OF t, THE NON-CENTRAL t-STATISTIC. THE ENTRIES IN THE TABLE GIVE THE VALUES OF x SUCH THAT $Pr\ [t/\sqrt{f} > x] = \epsilon$.

f is the number of degrees of freedom; the non-centrality parameter is $\sqrt{f+1}\,K_p$. K_p is the standardized normal deviate exceeded with probability p.

DEGREES OF FREEDOM **22**

p \ ε	.995	.99	.95	.90	.75	.50	.25	.10	.05	.01	.005
.2500	.133	.190	.332	.410	.542	.698	.868	1.039	1.151	1.385	1.483
.1500	.480	.530	.677	.758	.900	1.073	1.268	1.467	1.600	1.882	1.996
.1000	.703	.754	.903	.988	1.140	1.328	1.542	1.763	1.911	2.228	2.362
.0650	.903	.955	1.112	1.203	1.366	1.569	1.803	2.047	2.211	2.564	2.710
.0400	1.100	1.156	1.322	1.418	1.594	1.815	2.070	2.338	2.519	2.909	3.071
.0250	1.269	1.329	1.505	1.608	1.795	2.032	2.308	2.597	2.793	3.217	3.395
.0100	1.563	1.627	1.822	1.936	2.146	2.413	2.725	3.054	3.278	3.763	3.968
.0040	1.816	1.889	2.101	2.225	2.456	2.751	3.097	3.463	3.712	4.253	4.481
.0025	1.936	2.012	2.232	2.363	2.603	2.912	3.274	3.658	3.919	4.486	4.729
.0010	2.157	2.235	2.472	2.613	2.873	3.206	3.599	4.015	4.298	4.914	5.178

DEGREES OF FREEDOM **23**

p \ ε	.995	.99	.95	.90	.75	.50	.25	.10	.05	.01	.005
.2500	.150	.201	.339	.415	.544	.697	.863	1.029	1.138	1.365	1.457
.1500	.489	.543	.684	.763	.902	1.072	1.261	1.455	1.584	1.856	1.968
.1000	.712	.763	.910	.993	1.142	1.326	1.534	1.749	1.893	2.200	2.327
.0650	.911	.965	1.119	1.208	1.368	1.567	1.795	2.032	2.190	2.531	2.674
.0400	1.108	1.165	1.329	1.424	1.596	1.812	2.061	2.321	2.496	2.873	3.032
.0250	1.280	1.338	1.513	1.613	1.797	2.029	2.298	2.579	2.769	3.177	3.350
.0100	1.575	1.638	1.830	1.942	2.148	2.409	2.713	3.033	3.249	3.716	3.912
.0040	1.829	1.901	2.109	2.232	2.458	2.747	3.084	3.439	3.680	4.203	4.417
.0025	1.952	2.024	2.241	2.369	2.606	2.908	3.260	3.633	3.885	4.432	4.664
.0010	2.171	2.249	2.482	2.620	2.875	3.201	3.584	3.988	4.261	4.857	5.107

DEGREES OF FREEDOM **24**

p \ ε	.995	.99	.95	.90	.75	.50	.25	.10	.05	.01	.005
.2500	.158	.209	.346	.420	.547	.696	.859	1.020	1.126	1.346	1.437
.1500	.503	.553	.690	.768	.905	1.070	1.256	1.444	1.569	1.833	1.941
.1000	.716	.770	.916	.998	1.144	1.324	1.527	1.737	1.876	2.173	2.293
.0650	.920	.972	1.126	1.213	1.370	1.565	1.787	2.017	2.171	2.500	2.638
.0400	1.117	1.174	1.336	1.429	1.598	1.809	2.053	2.305	2.475	2.838	2.991
.0250	1.291	1.350	1.520	1.619	1.799	2.026	2.288	2.562	2.746	3.140	3.305
.0100	1.586	1.650	1.838	1.948	2.150	2.406	2.702	3.013	3.223	3.675	3.866
.0040	1.842	1.912	2.117	2.238	2.460	2.743	3.072	3.417	3.650	4.153	4.361
.0025	1.964	2.036	2.250	2.376	2.608	2.903	3.248	3.609	3.854	4.384	4.606
.0010	2.183	2.261	2.491	2.627	2.877	3.197	3.569	3.962	4.228	4.803	5.044

DEGREES OF FREEDOM **29**

p \ ε	.995	.99	.95	.90	.75	.50	.25	.10	.05	.01	.005
.2500	.203	.249	.372	.440	.556	.692	.839	.983	1.077	1.270	1.346
.1500	.540	.585	.717	.788	.913	1.064	1.231	1.399	1.509	1.737	1.831
.1000	.759	.809	.943	1.019	1.153	1.317	1.500	1.685	1.808	2.064	2.167
.0650	.960	1.012	1.154	1.234	1.378	1.556	1.756	1.960	2.095	2.380	2.497
.0400	1.164	1.214	1.365	1.451	1.607	1.799	2.018	2.242	2.390	2.704	2.832
.0250	1.336	1.390	1.550	1.642	1.808	2.015	2.250	2.492	2.653	2.994	3.133
.0100	1.634	1.695	1.871	1.973	2.159	2.392	2.658	2.933	3.116	3.505	3.665
.0040	1.896	1.961	2.154	2.266	2.470	2.727	3.022	3.327	3.531	3.965	4.139
.0025	2.018	2.087	2.287	2.404	2.618	2.887	3.195	3.515	3.729	4.183	4.372
.0010	2.241	2.315	2.531	2.657	2.888	3.178	3.513	3.859	4.091	4.585	4.788

PERCENTAGE POINTS OF t, THE NON-CENTRAL t-STATISTIC. THE ENTRIES IN THE TABLE GIVE THE VALUES OF x SUCH THAT Pr $[t/\sqrt{f}>x]=\epsilon$.

f is the number of degrees of freedom; the non-centrality parameter is $\sqrt{f+1}\,K_p$.
K_p is the standardized normal deviate exceeded with probability p.

DEGREES OF FREEDOM **34**

p \ ε	.995	.99	.95	.90	.75	.50	.25	.10	.05	.01	.005
.2500	.230	.269	.393	.457	.564	.690	.824	.956	1.040	1.212	1.282
.1500	.568	.612	.737	.804	.921	1.060	1.213	1.368	1.464	1.669	1.750
.1000	.789	.835	.965	1.035	1.160	1.311	1.479	1.647	1.757	1.986	2.077
.0650	.997	1.041	1.176	1.251	1.386	1.550	1.733	1.918	2.039	2.293	2.394
.0400	1.200	1.248	1.389	1.469	1.614	1.792	1.992	2.195	2.328	2.607	2.718
.0250	1.372	1.424	1.575	1.661	1.816	2.007	2.222	2.441	2.585	2.887	3.010
.0100	1.673	1.730	1.898	1.994	2.167	2.382	2.626	2.874	3.038	3.383	3.524
.0040	1.939	2.002	2.183	2.288	2.479	2.716	2.986	3.261	3.443	3.827	3.982
.0025	2.064	2.128	2.318	2.427	2.627	2.875	3.157	3.446	3.637	4.040	4.205
.0010	2.289	2.360	2.564	2.682	2.897	3.165	3.471	3.784	3.991	4.428	4.605

DEGREES OF FREEDOM **39**

p \ ε	.995	.99	.95	.90	.75	.50	.25	.10	.05	.01	.005
.2500	.265	.303	.410	.469	.570	.688	.813	.935	1.012	1.169	1.232
.1500	.604	.645	.755	.817	.927	1.057	1.199	1.340	1.430	1.617	1.689
.1000	.824	.865	.983	1.049	1.166	1.308	1.463	1.618	1.719	1.926	2.009
.0650	1.027	1.067	1.194	1.265	1.392	1.545	1.715	1.885	1.996	2.225	2.318
.0400	1.229	1.275	1.408	1.484	1.620	1.787	1.972	2.158	2.280	2.533	2.632
.0250	1.405	1.453	1.595	1.676	1.822	2.001	2.200	2.401	2.533	2.806	2.915
.0100	1.708	1.762	1.920	2.011	2.174	2.375	2.601	2.829	2.978	3.291	3.414
.0040	1.974	2.034	2.207	2.306	2.486	2.708	2.958	3.210	3.376	3.724	3.861
.0025	2.102	2.162	2.343	2.447	2.634	2.866	3.127	3.393	3.567	3.930	4.078
.0010	2.330	2.397	2.591	2.702	2.905	3.155	3.438	3.726	3.915	4.309	4.466

DEGREES OF FREEDOM **44**

p \ ε	.995	.99	.95	.90	.75	.50	.25	.10	.05	.01	.005
.2500	.287	.324	.425	.481	.576	.686	.804	.918	.989	1.135	1.191
.1500	.626	.664	.770	.828	.932	1.055	1.188	1.319	1.403	1.574	1.640
.1000	.846	.885	.997	1.060	1.171	1.305	1.451	1.595	1.688	1.878	1.954
.0650	1.050	1.092	1.210	1.277	1.397	1.542	1.701	1.859	1.961	2.172	2.255
.0400	1.253	1.298	1.424	1.497	1.626	1.783	1.956	2.129	2.242	2.474	2.565
.0250	1.431	1.477	1.612	1.690	1.828	1.996	2.183	2.370	2.491	2.743	2.841
.0100	1.733	1.786	1.939	2.025	2.180	2.369	2.581	2.793	2.930	3.217	3.330
.0040	2.006	2.062	2.228	2.322	2.492	2.701	2.935	3.170	3.323	3.640	3.768
.0025	2.133	2.192	2.364	2.463	2.641	2.859	3.104	3.350	3.511	3.844	3.976
.0010	2.364	2.428	2.613	2.720	2.912	3.148	3.413	3.679	3.854	4.215	4.361

DEGREES OF FREEDOM **49**

p \ ε	.995	.99	.95	.90	.75	.50	.25	.10	.05	.01	.005
.2500	.306	.341	.438	.490	.580	.685	.796	.903	.971	1.107	1.160
.1500	.645	.681	.782	.837	.936	1.053	1.179	1.301	1.380	1.539	1.602
.1000	.865	.902	1.009	1.070	1.176	1.302	1.440	1.576	1.663	1.839	1.909
.0650	1.071	1.110	1.224	1.287	1.402	1.539	1.689	1.837	1.933	2.128	2.206
.0400	1.275	1.317	1.438	1.508	1.631	1.779	1.943	2.105	2.210	2.426	2.510
.0250	1.453	1.498	1.627	1.701	1.833	1.992	2.168	2.344	2.457	2.690	2.783
.0100	1.762	1.810	1.956	2.038	2.185	2.365	2.564	2.763	2.892	3.156	3.262
.0040	2.033	2.085	2.245	2.336	2.498	2.696	2.917	3.137	3.280	3.574	3.690
.0025	2.160	2.217	2.383	2.477	2.647	2.854	3.085	3.315	3.465	3.773	3.896
.0010	2.394	2.456	2.633	2.735	2.918	3.142	3.392	3.641	3.804	4.138	4.271